ENVIRONMENTAL ENGINEERING (Vol. II)

SEWAGE DISPOSAL AND AIR POLLUTION ENGINEERING

Other Publications by the Same Author

(1) "*Water Resources Engineering (Vol. I)—Hydrology, Flood Control and Ground Water Engineering*" (26th Revised 2019 Edition)

(2) "*Water Resources Engineering (Vol. II)—Irrigation Engineering and Hydraulic Structures*" (36th Revised 2020 Edition)

(3) "*Environmental Engineering (Vol. I)—Water Supply Engineering*" (37th Revised 2024 Edition)

(4) "*Physical and Engineering Geology*" (7th Revised Edition—5th Reprint 2018)

(5) "*Geotech Engineering—Soil Mechanics and Foundation Engineering*" (13th Revised 2020 Edition)

(6) "*Ecology and Environmental Studies*" (4th Revised 2015 Edition) (Reprinted in 2022)

(7) "*Environmental Studies and Green Technologies*" (3rd Reprint 2015)

(8) ***"Environmental Studies" for undergraduate courses, as per UGC syllabus*** (Newly Rewritten Revised 2020 Edition)

(9) ***"Elementary Irrigation and Water Resources Engineering"*** (Newly Rewritten 4th Edition, 2021)

Notes :

(*i*) Orders for procuring all the above listed books may be sent to M/s. Khanna Publishers at 4575/15, Onkar House, Opp. Happy School, Ground Floor, Daryaganj, New Delhi-110002, *Phone*: +91-11-45033819 • *Mob.* +91-9811541460

(*ii*) The books in paper format as well as in e-format can also be purchased on line through the website of the publishers *i.e.*, ***www.contactus@khannapublishers.in***

SEWAGE DISPOSAL AND AIR POLLUTION ENGINEERING

[For Civil, Environment, and Agriculture Engineering Degree & Post Graduate Courses ; MSc (Environmental Science) Courses ; Engg. Services and Civil Services Competitive Examinations of UPSC ; GATE examinations for admissions to M. Tech. Courses ; AMIE (Section B) Examinations ; and for Professional Field Engineers.]

By

SANTOSH KUMAR GARG

B.Sc. Engg. (Civil)
Topper (1969 Batch), Delhi University
Life Member—Indian Water Resources Society

Formerly

Superintending Engineer
Flood Control and Irrigation Department,
Govt. of NCT of Delhi

(Revised 43rd Edition 2024)

KHANNA PUBLISHERS®
Investing in Learning®
Operational Office:
4575/15, Onkar House, Ground Floor,
Opp. Happy School, Daryaganj, New Delhi-110 002
Phones : 91-11-45033819 • *Mob.* 91-9811541460, 9650669626
email : contactus@khannapublishers.in

Published by :

Romesh Chander Khanna & Vineet Khanna
for **KHANNA PUBLISHERS**
2-B, Nath Market, Nai Sarak
Delhi- 110 006 (India)

Visit us at : www.khannapublishers.in

Hologram & Description

To all readers of our books: In order to prevent yourself from being defrauded by pirated books, please make sure that there exists a Hologram on the top cover of the book with the specifications indicated below. If you find any book without Hologram of this description, please do not buy that pirated copy, and inform us through e-mail at *contactus@khannapublishers.in*

Thanking you

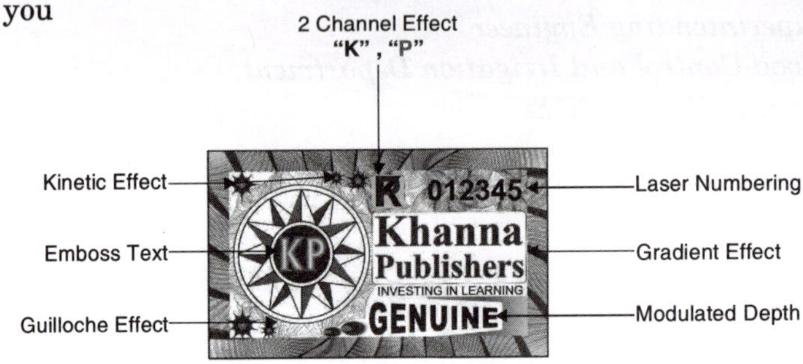

ISBN No. : 978-81-7409-230-4

First Edition : **March, 1979**
Forty Third Edition : **May, 2024 *(Revised)***
First Reprint : 2026

Price : ₹ 849.00

Dedication

Dedicated to those,
who are continuing making efforts,
inspite of several odds,
for the upliftment of our beloved
and beautiful India.

PREFACE TO THE REVISED 43rd EDITION

The first edition of this highly acclaimed book was published in March 1979, and contained hardly 350 pages. During the intervening period exceeding 42 years, more than 3 lakh copies have been sold through its Forty one editions and their reprints. Additional 3300 copies are being printed for this revised 42nd edition.

Due to the continuous and progressive revision and updation of the book, edition after edition, the size of the book has virtually tripled, since its first edition. In order to reduce the number of pages and to control the continuously increasing thickness of the book, we had, in its 24th edition, got the entire book recomposed into a **new size of 24 cm × 16 cm**, as against the previous size of 20 cm × 13 cm. This has helped in reducing the thickness of the book by about 20%, thereby making the book easier to handle and carry. All the drawings were also redrawn on computer to make them of better quality.

The book has been further revised continuously edition after edition. In the last 41st edition, large scale changes in the entire text were made by deleting certain olds obsolete matter and by updating several data, particularly relating to ***Activated Sludge Process (ASP)*** in Chapter 9 and ***Noise Pollution*** (Ch. 20).

The question banks of Engineering Services and Civil Services Exams have also been updated in this revised 43rd edition up to the year 2024.

The author is delighted to find the growing popularity of the book, not only among the average students but also among the toppers of the various engineering colleges of the country and abroad. ***It is also a matter of great pleasure to the author as well as to the students to find questions being set from this book year after year, in the prestigious "Civil Services" and "Engineering Services" Exams. The author expresses his gratitudes to all such learned Professors, Examiners and Paper-setters, who have valued this book worthy of such a high acclaim.***

Acknowledgements. In going through the forty one editions of this book, the author has accumulated enormous amount of debt from students and teachers over the years, who had written about their satisfaction with the book, and extended suggestions for the improvement of its various editions. ***If the book today is so strikingly popular, they all deserve much of the credit. The author certainly wants them to know of his gratitudes and openness to further suggestions, and is hence enlisting their names below :***

Sh. Satish Kumar Aggarwal—a student of M.R. Engineering College Jaipur, Rajasthan ; *Sh. Rajesh Gupta* and *Sh. Ajay Gupta*, students of Jabalpur Engineering College ; *Sh. Ashfaq Ahmed* of Shri Ram Rayons Kota, ***Mr. Biswaroop Ghosh***—an ex student of Bengal Engineering College, Calcutta (who handsomely contributed to the objective section) ; ***Sh. H.S. Bhatia***—Professor Delhi College of Engineering ; *Sh. A.M. Malek*—the then Lecturer S.S. College of Engineering Sidsar, Bhavnagar, Gujarat ; *Sh. Pramod Kumar Tomar* of Maharaja Surajmal Institute, New Delhi ; *Sh. A.P. Thankhavel*—lecturer A.K. College of Engg. Krishnakovil,

Tamilnadu ; *Sh. Shri Ram*—lecturer Madan Mohan Malviya Engg. College Gorakhpur, U.P. ; ***Sh. Pawan Goyal***—a student of MBM Engineering College, Jodhpur, Rajasthan ; *Sh. Krishan Mohan Kansal*—a student of Delhi College of Engineering ; *Mr. Ganesh Sawleshwarkar* of victoria Jubilee Technical Insutitute Bombay ; *Mr. Pradeep Sharma*—a student of I.I.T. Delhi ; *Mr. Azmal Kamal*—a student of Jamia Millia Islamia, New Delhi ; *Mr. Ashok Kumar*—a student of BIT Sindri Dhanbad ; *Mr. R.D. Punia*—Dy Director Technical Education at Regional Engineering College, Kurukshetra, Haryana. ***Dr. V.S. Prasad***—Professor at Regional Engg. College, Calicut ; ***Mr. A.M. Malek***—Asstt. Professor at L.D. College of Engg. Ahmedabad, *Mr. B.J. Bhaskar Reddy* from Hyderabad ; *Mr. Santhanam Krishnan*—an M. Tech. student of IIT Delhi, *Mr. Bibek Bastola* of Nepal ; etc., etc.

The author also wants to thank and acknowledge the receipt of numerous sms's, showering praises on the books of Env. Engg. Vol. I and II both, from several students and teachers, like Ms Rumpa, Mr. Dilip G. Patil, Mr. Sandip Kr. Das (Lecturer), Mr. Manish Kr., Mr. Noopur, etc. etc.

The author further desires to record his special gratitudes and thanks to ***Prof. Kapil Gupta*** of IIT, Bombay, due to whose help, he was able to include photographs of *aerated grit chambers* and *aerated lagoons* as existing in Mumbai, along with giving details of the disposal of sewage into the Mumbai sea.

Thanks are also due to Prof. S.C. Prasad presently Principal & HOD Civil at MIET Meerut (previously HOD at NIT Allahabad) for suggestions, based on which changes have been affected in the chapter of BOD.

The author also records his thanks to the Publishers for promptly printing this revised edition. The author also wants to record his unforgettable sincere gratitudes towards one of his old teacher, ***Prof. R. Naryanan*** (the then HOD Civil Engg. Deptt, Delhi College of Engg. Delhi), who infact provided the inspiration to the author to start writing books, when he was highly depressed in his life at its prime stage, due to having been rejected on medical grounds for a Class I job through the Engg Services Competition, for having lost his Left Lung in early childhood.

Since the book has continuously been revised edition after edition, some typographical errors might still be existing in the book.

The author will certainly be grateful, if such misprints and ommissions if any, are brought to his notice, for making amends in the future editions of the book.

May 1, 2024

Santosh Kr. Garg
Res. : 3A/2 WEA, Sat Nagar
Karol Bagh, New Delhi-110005

Contents

CHAPTERS PAGES

List of Colour Photoes and Chart Figs. Pasted in this book

Conversion of f.p.s., c.g.s., and S.I. Units into M.K.S. Units

Conversion of f.p.s. Units into M.K.S.	*Conversion of c.g.s. Units into M.K.S.*	*Conversion of S.I. Units into M.K.S.*
	For Length	
1 ft = 0.3048 m 1 mile = 1.609 km = 1609 m 1 inch = 0.0254 m	1 cm = 0.01 m 1 micron (μ) = 10^{-6} m	S.I. unit is same *i.e.* *m*
	For Area	
1 ft^2 = 0.093 m^2 1 $mile^2$ = 2.59 km^2 = 259 ha† 1 acre = 0.4347 ha	1 cm^2 = 10^{-4} m^2	S.I. unit is same *i.e.* m^2.
	For Volume and Discharge	
1 ft^3 = 0.0283 m^3 1 acre ft = 1233.6 m^3 ≈ 0.123 ha. m = $\frac{1}{2}$ cusec. day 1 gallon (*i.e.* imp. gallon) = 4.546 litres 1 U.S. gallon = 3.785 litres 1 ft^3/sec (cusec) = 0.0283 m^3/sec (cumec) 1 cusec = 2.447 MLD (Million litres per day) 1 cusec = 0.538 MGD (Million gallon per day)	1 cm^3 = 10^{-6} m^3 1 cm^3/sec = 10^{-6} m^3/sec	S.I. unit of volume is also same *i.e.* m^3. S.I. unit of discharge is also same *i.e.* m^3/s (cumecs)
	For Force	
1 lb* = 0.454 kg* 1 poundal = 0.031 lb* = 0.0141 kg*	1 dyne = $\frac{1}{981}$ gm* = 1.02 × 10^{-6} kg*	1 Newton (N) = $\frac{1}{9.81}$ kg* = 0.102 kg* = 10^5 dynes
	For Pressure	
1 lb/in^2 (psi) = 703 kg*/m^2 1 lb/ft^2 = 4.882 kg*/m^2	1 kg*/cm^2 = 10^4 kg*/m^2 = 10 m of water head	1 Pascal (P_a) = 1 N/m^2 = 0.102 kg*/m^2 1 k.Pa = 1000 Pa = 102 kg*/m^2 = 0.102 m of water head 100 k.Pa = 10.2 m of water head

† ha stands for hectares ; 1 ha = 10^4 m^2

*lb here is lb wt., gm here is gm wt., and kg here is kg wt.

Conversion of f.p.s. Units into M.K.S.	*Conversion of c.g.s. Units into M.K.S.*	*Conversion of S.I. Units into M.K.S.*
	For Density (ρ)	
1 lb/cft = 16.02 kg/m^3	1 gm/cm^3 = 1000 kg/m^3	S.I. unit of density is same *i.e.* kg/m^3.
	For unit wt. or specific wt. (γ or w = $\rho.g$)	
1 lb*/cft = 16.02 kg*/m^3	1 gm*/cm^3 = 1000 kg*/m^3	1 N/m^3 = 0.102 kg*/m^3 1 kN/m^3 = 102 kg*/m^3
	For Energy	
1 ft lb* = 1.356 kg.m^2/sec^2 = 1.356 Joule (J)	1 Erg or 1 g-cm^2/sec^2 10^7 Ergs = 1 kg m^2/sec^2 = 1 Joule	S.I. unit is Joule where 1 Joule = 1 N.m = 1 kg-m^2/sec^2
	For Heat Energy	
1 Btu or BTU = 0.252 k.cal = 1.055 kJ	1 cal = 10^{-3} k.cal	1 kJ = 0.239 k.cal
	For Power	
1 ft lb*/sec = 0.138 kg* m/sec 1 H.P. (fps) = 550 ft* lb/sec = 75.9 kg* m/sec = 746 watts	1 gm* cm/sec = 10^{-5} kg* m/sec 1 H.P. (metric) = 75 kg m/sec = 735 watts	S.I. unit is watt, where 1 Watt (*W*) = 1 J/s = 1 Nm/s = 0.102* kg* m/sec = $\frac{1}{735}$ H.P. (metric)
	For Dynamic Viscosity (η or μ)	
f.p.s unit is lb sec / ft^2, where* 1 lb* sec/ft^2 = 4.887 kg* sec/m^2 (mks)	(*c.g.s. unit is dyne. sec / cm^2* or **Poise** 1 Poise = 0.0102 kg sec/m^2 1 centi poise (cP) = 10^{-2} P	*S.I. unit is N.s / m^2* or Pa.s 1 N.s/m^2 = 1 Pa.s = 10 **Poise** = 0.102 kg* sec/m^2
	For Kinematic Viscosity $\left(\nu = \frac{\mu}{\rho} \text{ or } \frac{\eta}{\rho}\right)$	
1 ft^2/sec = 0.093 m^2/sec	1 cm^2/sec **(Stoke)** = 10^{-4} m^2/sec 1 centi stock = 10^{-2} Stock	*S.I. unit is same as in M.K.S. : i.e. m^2 / s* 1 m^2/s = 10^4 **Stoke** (St)

[**Note.** *w* for water = 1000 kg/m^3 = 9.81 kN/m^3 = 9810 N/m^3.]

*kg here is kg wt., lb here is lb wt., and gm here is gm wt.

—1

Introduction

1.1. Systems of Sanitation

The waste products of a society including the human excreta had been collected, carried and disposed of manually to a safe point of disposal, by the sweepers, since time immemorial. This primitive method of collecting and disposing of the society's wastes, has now been modernised and replaced by a system, in which these wastes are mixed with sufficient quantity of water and carried through closed conduits under the conditions of gravity flow. This mixture of water and waste products, popularly called **sewage,** thus automatically flows up to a place, from where it is disposed of, after giving it suitable treatments ; thus avoiding the carriage of wastes on heads or carts. The treated sewage effluents may be disposed of either in a running body of water, such as a stream, or may be used for irrigating crops.

This modern **water-carried sewerage* system** has completely replaced the **old conservancy system of sanitation** in the developed countries like U.S.A. However, India being a developing country, still uses the old conservancy system at various places, particularly in her villages and smaller towns. The metropolitan cities and a few bigger towns of our country, no doubt, have generally been equipped with the facilities of this modern water carriage sewerage system ; and attempts are being made to equip the remaining cities and towns with this system, as soon as funds become available.

The *modern water-carried sewerage system* is preferred to the old conservancy system, because of its following *advantages* :

(*i*) The water carriage system is more hygienic, because in this system, the society's wastes have not to be collected and carried in buckets or carts, as is required to be done in the conservancy system.

The free carriage of night soil in carts or as head load, which is required in the conservancy system, may pose health hazards to sweepers and other residents, because of the possibilities of flies and insects transmitting disease germs from these accessible carts to the resident's foods and eatables ; whereas, in modern sewerage system, no such danger exists, because the polluted sewage is carried in closed conduits, as soon as it is produced.

(*ii*) In the conservancy system, the waste products are generally buried underground, which may sometimes pollute the city's water supplies, if the water supply pipes happen to pass through such areas or the wells happen to draw water through such areas.

---

*The term sewerage is applied to the art of collecting, treating and finally disposing of the sewage.

(*iii*) In the conservancy system of sanitation, the entire day's human feces are collected and then disposed of in the morning, once a day. Thus, from this type of latrines, pungent smells may continue to pollute the surroundings for the entire day. But since in the water carried system, the human excreta is washed away as soon as it is produced, no such bad smells are produced. Moreover, in the conservancy system of sanitation, the waste waters from bath rooms, wash basins, kitchen sinks, etc. ; is carried through open road side drains, as this is supposed to be not so foul, since it does not contain human excreta. But these road side drains are generally abused by children or adults for passing their stools, particularly at night hours, thus creating foul and more unhygienic conditions. No such problems exist in the water carriage system.

(*iv*) In water carriage system, the sewage is carried through underground pipes (popularly called **sewers**) which owing to their being underground, do not occupy floor area on road sides or impair the beauty of the surroundings. The road side drains carrying foul liquid in the conservancy system, will no doubt pose such problems.

(*v*) The water-carried system may allow the construction of latrines and bath-rooms together [popularly called water-closets (W.C.)], thus occupying lesser space with their compact designs. This system is also very helpful for multistoreyed buildings, where the toilets, one above the other, can be easily constructed, and connected to a single vertical pipe.

Inspite of these advantages of the modern water-carried system, it has not been possible to completely replace the old conservancy system, mainly because huge capital funds (of the order of ₹ 10,000 to 15,000 per person) are required for the construction of such a system. Besides the huge initial investments, the RMO expenses are also high, which make it difficult to replace the simpler and cheaper conservancy system. Moreover, for the functioning of sewerage system, ample amount of water must be made available to the people, and hence, reliable and assured water supply system must, first, be installed, before installing the sewerage system.

1.2. Types of Sewage, and Types of Sewerage Systems

This modern water carriage sewerage system not only helps in removing the domestic* and industrial** wastewaters, but also helps in removing storm water drainage***. The run off resulting from the storms is also sometimes

---

***Domestic sewage** consists of liquid wastes originating from urinals, latrines, bath-rooms, kitchen sinks, wash basins, etc. of the residential, commercial or institutional buildings. This sewage is generally extremely foul, because of the presence of human excreta in it.

****Industrial sewage** consists of liquid wastes originating from the industrial processes of various industries, such as Dyeing, Paper making, Brewing, etc. The quality of the industrial sewage depends largely upon the type of industry and the chemicals used in their process waters. Sometimes, they may be very foul and may require extensive treatment before being disposed of in public sewers.

The sum total of domestic and industrial sewage, may be termed as **sanitary sewage** or simply **sewage**.

***The run-off resulting from the rain storms was used to be called **storm sewage,** but the modern approach is to call it **storm drainage** or simply **drainage,** so as to differentiate it from *sewage*, which is much more foul as compared to the *drainage*, and requires treatment before disposal.

carried through the sewers of the sewerage system, or more generally is carried through separate set of drains (open or closed) directly discharging their drainage waters into a body of water, such as a lake or a river. Since the rain run-off is not so foul as the sewage is, no treatment is generally required to be given to the drainage discharge. When the drainage is taken alongwith sewage, it is called a **combined system** ; and when the drainage and sewage are taken independently of each other through two different sets of conduits, it is called a **separate system.** Sometimes, a part of drainage water, especially that originating from the roofs or paved courtyards of buildings, is allowed to be admitted into the sewers ; and similarly sometimes, the domestic sewage coming out from the residences or institutions, etc., is allowed to be admitted into the drains, the resulting system is called a **partially separate system.**

Strictly speaking, it is generally advantageous and economical to construct a 'separate system' at least in the bigger cities and towns. But in practice, it is generally not possible to attain a 'truly separate system' because some rain water may always find its way into the sewers either through wrong house sewer connections or through open manhole covers. Similarly, wherever the authorities find insufficient sewer capacities, they divert part of the sewage into the storm water drains, thus making most of our existing systems as 'partially separate' only.

In the modern days, a 'separate system' is generally preferred to a 'combined system', although each individual case should be decided separately on merits, keeping the following points into consideration :

(*i*) A separate system will require laying two sets of conduits, whereas, a combined system requires laying only one set of bigger sized conduits, thus making the former system costlier. Moreover, the separate conduits cannot be laid in congested streets and localities, making it physically unfeasible.

(*ii*) The sewer pipes in the combined system are liable to frequent silting during the non-monsoon season (when the flows in them are quite less*) unless they are laid at sufficiently steeper slopes, which, in turn, will make them deeper, requiring more excavation and pumping, thereby making them costlier.

(*iii*) In a combined system, the less-foul drainage water gets mixed with the highly foul sewage water, thus necessitating the treatment of the entire flow, needing more capacity for the treatment plant, thereby making it costlier. Whereas, in the separate system, only the sewage discharge is treated and the drainage discharge is disposed of without any treatment.

(*iv*) In case, flooding and backing up of sewers or drains occur due to excessive rains, more foul and insanitary conditions will prevail in the case of combined sewage than in the case of storm drainage alone.

(*v*) Since the sewer lines are generally laid deep and at steeper slopes, as compared to the storm water surface drains, pumping of sewage and often no

*Since heavy rain storms concentrated for a period of 3 months or so do occur, and there are poor water supplies here in India, the ratio of the 'drainage' to 'sewage' works out to be as high as 20 to 30. Thus, during non-monsoon periods, only 1/20th to 1/30th of the designed discharge will be passing through the sewers, if the combined system has been adopted.

pumping of drainage is required in a separate system. Whereas, the entire discharge will have to be pumped, if the sewage and drainage discharges are mixed together ; thereby making the combined system more costly.

(*vi*) The economy of the two systems must be worked out for each individual project, and the economical system be adopted, if physically feasible.

1.3. Components of a Sewerage System

A sewerage system consists of a network of sewer pipes laid in order to carry the sewage from individual homes to the sewage treatment plant. This network of sewers may consist of *house sewers* (or individual house connections) ; *lateral sewers* ; *branch sewers* (or submains) ; *main sewers* (generally called *trunk sewers*) ; *outfall sewer* (*i.e.* the sewer which transports sewage to the point of treatment) ; etc. *Manholes* are provided in every sewer pipe at suitable intervals, so as to facilitate their cleaning and inspection. In the sewers, which carry the drainage discharge either solely or in combination with sewage, *inlets* called *catch basins* are provided to permit entrance of storm water from street gutters.

In order to avoid the large scale pollution of the water sources and to keep them usable for the downstream people, the original contaminated sewage is not allowed to be discharged directly into the water sources. A complete treatment including *screening, sedimentation, biological filtration* (or *activated sludge treatment*), *sludge digestion,* etc. is therefore, given to this extremely foul sewage, so as to bring down its BOD and concentrations of other constituents to safer values, before discharging it into a national river resource. However, a recent use of sewage is being made for irrigating crops. For this use also, the sewage has to be treated, so as to bring down its constituents to permissible values, as per the requirements of I.S. 3307—1965.

All these aspects are explained in details in subsequent chapters.

1.4. Design and Planning of a Sewerage System

The sewerage system must be properly and skilfully planned and designed, so as to remove the entire sewage effectively and efficiently from the houses, and up to the point of disposal. The sewers must be of adequate size, so as to avoid their overflow and subsequent damages to properties, and health hazards. In order to provide economically adequate sized sewers, it is necessary that the likely sewage discharge be estimated as correctly as possible. The sewer pipes should then be designed to be laid on a slope that will permit reasonable velocity of flow. The flow velocity should neither be so large, as to require heavy excavation and high lift pumping ; nor should it be so small, as to cause the deposition of solids in the sewer, bottoms.

The sewers are generally designed to carry the water from the basements, and should therefore, be atleast 2 to 3 m deep. As far as possible, they should be designed to flow under gravity with $\frac{1}{2}$ or $\frac{3}{4}$ th full. Owing to the requirements of seeking gravity flow, the sewage treatment plant should generally be located in a low lying area. The design of the treatment units also requires good engineering skill. In order to provide adequate and economical treatment, it is necessary to thoroughly study the constituents of the sewage produced in the particular project, and also the quality and other characteristics of the body of water that will receive the sewage. The permissible standards for effluents, and the possible uses of water downstream, should also be studied. The legal bindings, if any, will also have to be taken into consideration, while deciding upon the quantum of treatment required to be given. No fixed standards can be laid for fixing this required treatment, as everything depends upon the exigencies of a particular project.

Since the treatment plant will have to be located at low level ; the flood protection devices both during construction and thereafter, should also have to be taken care of, by the design engineers.

All these design requirements shall be thoroughly discussed in this volume.

1.5. Financing the Sewerage Projects

The full fledged development and functioning of a sewerage system requires enormous capital investments and large amounts of recurring RMO expenses.

The initial money has, of course, to be provided by the Government, but the interest on the capital investments and the depreciation charges as well as the RMO (*i.e.* Running, Maintenance and Operation) expenses must be recovered from the users of these facilities by properly taxing them. However, there cannot be a very rational system of taxing the users of sewerage facilities, because it is not feasible to estimate the quantity of sewage removed from an individual house, and hence cannot be as rationally taxed as can be done for water supplies, which can be measured and taxed according to the volume of water consumed. Consequently, the sewerage services may be financed by adding a certain surcharge, usually a percentage to the water bill. This is somewhat rational in the sense that sewage produced by the house will be directly proportional to the water consumed. Sometimes, charges for sewerage facilities are added, as certain percentage of the annual valuation of the house property, to the property/house tax bills. Nevertheless, the planning engineer must also develop the suitable and logical financing programme, indicating the sources of income and balancing them against depreciation, interest and RMO charges.

1.6. Liability for Damages Caused by Insufficient or Inefficient Sewerage Works

In case of certain damages, the process of claiming compensations through law courts is largely adopted in developed countries like U.S.A. However, Indian people, striving hard for their bread and butter, find it very difficult and almost impossible to go to the long process of redressal through law courts. Moreover, generally, when certain authorities or officials are sued in the court for damages, normally a certain percentage of the claimed amount has to be deposited, making it almost impossible for an ordinary Indian to go to the courts for claiming damages. Nevertheless, we will state here, a few possible chances, when compensations can be successfully claimed, in case certain damages have been caused by insufficient or inefficient sewerage works:

No city governments or public municipalities can be held responsible for not providing sewerage facilities ; but however, when once such facilities have been provided, the officials and municipal bodies will become responsible to certain extent for any damages caused to the health or properties of the citizens, due to its insufficient or inefficient provisions or operations.

If the sewerage system has been constructed, as designed by a competent engineer, normally nothing can be claimed by the sufferers due to inadequate size of sewers. Still, however, damages can always be claimed, if there is some established negligence on the part of the authorities in the maintenance or operation of sewerage works. For example, if the blockage of a sewer is reported to the authorities, who do not take prompt action, then the consequent damages caused by the backing up and overflowing of the sewers, can always be claimed. Similarly, if the leaking sewers happen to pollute water supplies, damages can be claimed for affecting public health, but only if it can be established that the sewerage authorities had a preknowledge of the situation, and they failed to remedy the same. Similarly, if certain manhole cover is left open due to poor maintenance, and if a child happens to fall in it, damages can always be claimed. But again negligence, will have to be established. Even without any negligence, the damages can always be claimed by the property owners for depreciation of their properties, if sewage treatment plant has been established near their properties, which gives off obnoxious smells. Similarly, throwing away of untreated or partially treated sewage in a river resource, may pollute the river water, and pose health hazards to the people utilising such waters, downstream. In such a case also, the affected people can claim compensations.

PROBLEMS

1. Describe conservancy and water-carriage systems. What are the relative advantages and disadvantages of the two system ?
2. Discuss briefly the necessity of replacing the conservancy system by the water carriage system of sanitation.
3. Discuss the relative merits of the separate and the combined systems of sewage, and give the conditions favourable for the adoption of each one of them.
4. Differentiate between :
   (*i*) Domestic sewage, industrial sewage and sanitary sewage ;
   (*ii*) Combined and separate systems of sewage ; and
   (*iii*) Sewage and drainage.
5. Write short notes on :
   (*i*) Financing the sewage projects ;
   (*ii*) Liability for damages for insufficient or inefficient sewerage facilities ;
   (*iii*) Types of sewages ; and
   (*iv*) Systems of sewerage.

—2

Estimating the Design Sewage Discharge

2.1. Estimating Sewage Discharge

The sewage* discharge which has to pass through a sewer must be estimated as correctly as possible ; otherwise the sewers may either prove to be inadequate, resulting in their overflow, or may prove to be of too much of size, resulting in unnecessary wasteful investments. Theoretically speaking, the quantity of sewage (*i.e.*, domestic sewage + industrial sewage) that is likely to enter the municipal sewers under design, should be equal to the quantity of water supplied to the contributing area, from the water-works. But in actual practice, this is not the precise quantity which appears as sewage, but certain additions and subtractions do take place from it, as explained below :

(*i*) **Additions due to unaccounted private water supplies.** The accounted water supplied to the public through the public distribution system (the records of which are easily available from the water-works office), is not necessarily the only water consumed by the public. Some private wells and tubewells may sometimes be used by the public for their domestic needs ; and similarly, certain industries may utilise their own sources of water. This extra quantity of water used by the town is generally small, unless there are large industrial private water uses. This quantity can, however, be estimated by actual field observations.

(*ii*) **Additions due to infiltration.** Whenever, the sewer pipes are laid below the ground water-table, certain amount of ground water generally seeps into them, through their faulty leaky joints or cracks formed in the pipes due to bad materials or poor construction. The quantity of the ground water entering these sewer pipes depends mainly upon the height of the water-table above the sewer invert level, and the nature and extent of faults and fissures present in the sewer pipes. However, if the ground water-table is well below the sewer, the infiltration can occur only after rain, when water is moving down through the soil. In that case, the infiltration quantity will depend upon the permeability of the ground soil. Since these factors cannot be precisely computed, the exact quantity of ground water infiltrating into the sewer pipes cannot be estimated precisely. Only certain nominal allowance, based upon some experimental results, may be made on account of this factor. In U.S.A., an allowance varying from 11,000 to 2,25,000 (average value = 1,14,000) litres per day per kilometre length of sewer pipe, is generally made in high water-table areas.

*It means sanitary sewage alone.

No allowance for infiltration should, however, be made when sewers are provided with under-drains which have free outlets.

Sometimes, the storm water drainage may also infiltrate into the sewers. This inflow cannot be computed easily and generally left unaccounted without making any extra provision for it. This additional water, if happens to enter the sewers, can be accommodated in the extra empty space left at the top in the sewers, which are generally designed as running 3/4th full at maximum designed discharge.

(*iii*) **Subtractions due to water losses.** The water lost, due to leakage in the distribution system and house connections of the water supply scheme, does not reach the consumers, and hence, never appears as sewage.

(*iv*) **Subtractions due to water not entering the sewerage system.** Certain amount of water may be used by the public and industries for such uses which may not produce any sewage at all. For example, the water used in boilers for steam generation ; the water sprinkled over the roads, streets, lawns and gardens ; the water used for automobile washings ; the water consumed in industrial products, such as beverages, etc., the water used in air cooling etc., does not normally produce any sewage.

Net Quantity of Sewage Produced. The net quantity of sewage produced will be equal to the accounted quantity of water supplied from the water-*works plus the additions* due to factors (*i*) and (*ii*) *minus the subtractions* due to factors (*iii*) and (*iv*), described above. *This net value may vary between 70 to 130 per cent of the accounted water supplied from the water-works. In India, however, this value is generally taken as equal to 75 to 80% of the accounted water supplied from the water works.*

DESIGN PERIODS AND FUTURE FORECASTS

2.2. Design Periods for Different Components of a Sewerage Scheme

A sewerage scheme involves the laying of underground sewer pipes and construction of costly treatment units, which cannot be replaced or increased in their capacities easily or conveniently at a later date. For example, addition of sewer pipes at a future date cannot be accomplished without digging the roads and disrupting the traffic. In order to avoid such future complications, and to take care of the future expansions of the city and consequent increase in the quantity of sewage produced, it is necessary to design the various components of the scheme larger than their present day requirements and of such sizes, as to serve the community, satisfactorily, for a reasonable number of years to come. *This future period for which the provision is made in designing the capacities of the various components of the sewerage scheme is known as the design period.*

The design period should neither be too long nor it should be too short, and moreover, it should not exceed the useful life of the component structures. The design period is generally guided by the following considerations :

(*i*) Useful life of component structures, and the chances of their becoming old and obsolete. Design period should not exceed those values.

(*ii*) Ease and difficulty, that is likely to be faced in expansion, if undertaken at future dates. For example, more difficult expansions mean choosing a higher value of the design period.

(*iii*) Amount and availability of additional investment, likely to be incurred for additional provisions. For example if funds are not easily available, then one has to keep a smaller design period.

(*iv*) The rate of interest on the borrowings and the additional money invested. For example, if the interest rate is small, a higher value of the design period may be economically justified, and therefore, adopted.

(*v*) Anticipated rate of population growth, including possible shifts in communities, industries and commercial investments. For example, if the rate of increase of population is less, a higher figure for the design period may be chosen.

The following design periods are often used in designing the different components of a sewerage scheme.

Table 2.1. Design Periods for Different Components of a Sewerage Scheme

S. No. (1)	*Type and name of the component structure* (2)	*Special characteristics and reasons for the selected design period* (3)	*Design period in years* (4)
1.	Lateral sewers, less than 15 cm in dia.	Requirements may change faster in limited area.	Full development
2.	Branches, mains and Trunk sewers	Difficult and costly to enlarge.	40—50
3.	Treatment units	Growth and interest rates being high to moderate.	15—20
4.	Pumping plant	The additional pumps can be installed in future, very easily and also within a short notice.	5—10

2.3. Future Forecasts and Estimating Design Sewage Discharge

The quantity of sewage that is likely to pass through a sewer (Q') at the end of the design period, can be easily computed by multiplying the *per capita production of sewage* (q') *by the expected population at the end of the design period*.

The **per capita sewage** which is produced (q') in a community can be easily estimated by assuming it as 75 to 80 per cent of the per capita water supplied to the public (q). However, it should also be kept in mind that the future increase in population may also increase the per capita water demand, and consequently increasing the per capita production of sewage. The increase in per capita water supply or sewage production with the increase in population, obviously occurs due to improved economical conditions in the city, implying higher standards of living and greater consumption of water. In U.S.A., this increase in per capita water demand and sewage production is generally assumed to be 5% of the percentage increase in population. However, for normal Indian conditions, the following norms may be adopted :

Table 2.2. Variations in Per Capita Water Demand and Sewage Production with Population, in India

S. No. (1)	*Population* (2)	*Per capita water demand in litres/day/person* (*q*) (3)	*Per capita sewage production in litres/day/person* $q' \simeq 80\%$ *of col.* (3) (4)
1.	Less than 20,000	110	90
2.	20,000—50,000	110—150	90—120
3.	50,000—2 lakhs	150—180	120—150
4.	2 lakhs—5 lakhs	180—210	150—170
5.	5 lakhs—10 lakhs	210—240	170—190
6.	Over 10 lakhs	240—270	190—200

The *expected population* at the end of the design period can be estimated by collecting the data of the past populations of several decades from the Census Department, and then by extrapolating the future population by using any one of the different methods, such as :

(*i*) Arithmetical increase method ;
(*ii*) Geometrical increase method ;
(*iii*) Incremental increase method ;
(*iv*) Decreasing rate method ;
(*v*) Simple graphical method ;
(*vi*) Comparative graphical method ;
(*vii*) Master plan method ;
(*viii*) The apportionment method ; and
(*ix*) The logistic curve method.

These methods for forecasting future populations have been described in details in article 2.10 in Vol. I—"Water Supply Engineering", and may again be referred to.

After determining the expected population as well as the per capita sewage contribution of the town, by the end of the design period, the average* quantity of sewage produced in litres/day (then converted to cumecs) can be easily determined by multiplying both these figures.

2.4. Variations in Sewage Flow and their Effects on the Design of Various Components of a Sewerage Scheme

The per capita demand of water (*q*) and the corresponding per capita sewage production (*q'*) so far discussed, are based upon annual flows, and are, therefore, defined as *annual average daily flows per person.* This average value is not sufficient, although very useful, for the design of various

*This quantity will of course be the annual average value, as discussed in article 2.3 in Vol. I—"Water Supply Engineering".

components of a sewerage scheme ; because there are wide variations in the actual flows that take place through the sewers at a given time.

The flows in these sanitary sewers, though fluctuate seasonally, monthly, daily, as well as hourly, with the water consumption*, yet they are sometimes delayed and less pronounced (Fig. 2.1) as they are damped because of the storage space in the sewers and because of the time required for the sewage to reach the point of gauging. In other words, the peaks are flattened, because it requires considerable sewage to fill the sewers to the high flow point, and the high flows from various sections will reach the gauging point after various times of flow. Thus, the time as to when the peak flow occurs will depend upon the flow time in sewers and the type of district served. Hence, if the sewage is gauged near its origin, the peak flow will be quite pronounced ; whereas, if the sewage must travel a long distance before being gauged, the peak will be deferred. *It, therefore, follows that the peak flows (expressed as number of times of their average values) will be much greater for smaller lateral sewers, as compared to these for larger trunk sewers.*

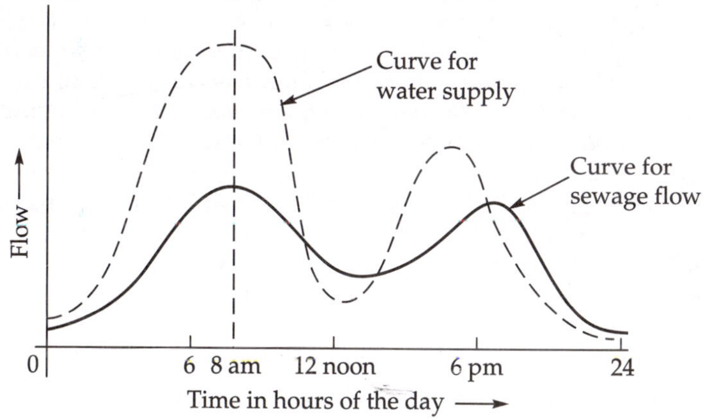

Fig. 2.1. Hourly variation of sewage flow compared to that of water supply.

For areas of moderate sizes, such as involved for branch sewers, the maximum daily or hourly sewage flows, can be expressed as :

Maximum daily flow = 2 times the average daily flow ...(2.1)

Maximum hourly flow = 1.5 times the maximum daily ...(2.2)

= *3 times the average daily.*

However, as pointed out earlier, the peak hourly flows will decrease, as the tributary area increases. Therefore, the peak flow at the outfall of a city sewer system will be much less, usually, about 1.5 times the average. The estimation of maximum hourly flows for different types of sewers, within the city's sewerage system, are given below in table 2.3.

*Please refer article 2.6 on "Variations in Demand", in Vol. I "Water Supply Engineering".

Table 2.3. Hourly Variations in Sewage Flow

S. No.	*Type of Sewer*	*Ratio of maximum flow to average flow*
1.	Trunk mains above 1.25 m in dia.	1.5
2.	Mains upto 1 m in dia.	2.0
3.	Branches upto 0.5 m in dia.	3.0
4.	Laterals and small sewers upto 0.25 m in dia.	4.0

The sizes of the sewers can then be easily designed for carrying the computed maximum hourly flows, with sewers running 3/4th full.*

This peak sewage flow has been connected with the population by certain investigators by the formula

$$Q_{max} = \frac{18 + \sqrt{P}}{4 + \sqrt{P}} Q_{av} \quad ...(2.3)$$

where P = Population in thousands.

The **minimum flow** passing through a sewer is also an important factor in the design of the particular sewer ; because at low flow, the velocity will be reduced considerably, which may cause silting. *Hence, the slope at which the sewer is to be laid has to be decided in accordance with the requirement of minimum permissible velocity at the minimum flow.*

The minimum flows occurring through the sewers during night hours will affect the laterals to a maximum extent, and will affect the mains to a lesser extent. Thus, the minimum flows through laterals, may be even lesser than 25 per cent of the average ; while in the mains, they can be 50 to 70 per cent of the average. For moderate areas, such as involved for branch sewers, the following minimum flows may be assumed :

Minimum daily flow $= \frac{2}{3} \times$ Average daily ...(2.4)

Minimum hourly flow $= \frac{1}{2}$ Minimum daily flow

$= \frac{1}{3}$ *Average daily* ...(2.5)

The sewers must, therefore, be checked for minimum velocities at their minimum hourly flows (i.e. $\frac{1}{3}$ Average daily).

PROBLEMS

1. Briefly discuss the method you will adopt for working out the design capacities of various laterals, branch sewers, and main sewers of a sewerage system.

2. Indicate the ratios of max. flow to average flow, to be adopted in the design of laterals, branches, mains and trunk sewers. How will these calculated maximum flows affect their designs ? Also discuss as to how the minimum flow in the sewers, affects their designs.

3. Write short notes on :

(*i*) Estimating the design sewage discharge ;

(*ii*) Design periods for different components of a sewerage scheme ;

(*iii*) Per-capita sewage ; and

(*iv*) Time variations in sewage flow, and their effects on the design of sewer capacities.

4. What do you mean by variation in flow of sewage ? Explain average flow, dry weather flow, and maximum flow.

---

*Sewer pipes are made only in standard sizes, and where there is a doubt about the insufficient capacity of one size, the next higher size is commonly used.

—3

Estimating the Peak Drainage Discharge

3.1. Introduction

The sewers and the drains of a separate sewerage system should be designed to carry the maximum sewage discharge and the maximum rain run off, respectively. Whereas, the sewers of a combined sewerage system should be designed to carry the sewage discharge plus the rain runoff. The sewers of a combined system should, therefore, in addition to passing this combined maximum flow, should also be capable of passing the low sewage discharge during non-monsoon periods, as **dry weather flow***, with minimum permissible velocities. The partially separate sewers may be designed for carrying the sewage discharge plus part of the storm drainage, particularly that coming from the roofs and courtyards.

In order to design the sewers and the drains properly, it is absolutely necessary to estimate the sewage discharge and the urban storm drainage discharge that are likely to enter the sewers or drains. The methods of estimating the maximum sewage discharge were discussed in the previous chapter ; and here we will discuss the methods of estimating the *maximum rate of storm run-off*, popularly called *peak drainage discharge*.

3.2. The Run-off Process and Peak Run-off Rate

When a rain, falls on a certain area, a part of it is intercepted by the soil, a part of it is evaporated, and the remaining water flows overland towards the valleys, as storm run-off. Since the storm runoff has to be removed through drains or through combined sewers, the drainage engineer must evaluate the peak rate of run off, which can be produced from a certain catchment by the given rain, at any moment. Further, the more intense is the rain, the more will be the peak run-off rate. Hence, a proper and economical value of rain frequency (or recurrence interval**) must be chosen, for which the drains must be designed. The frequency of rainfall to be adopted in design should neither be so large, as to cause too heavy investments, nor should it be so small, as to cause very frequent overflowing of the drains. For Delhi, the experts have recommended a 2 years rain frequency for designing smaller link drains, and 5 years frequency for designing all the major drains.

---

*The rain run-off, *i.e.* drainage discharge, which is produced during monsoon season is generally very high, say 20 to 25 times that of the sewage discharge, called **dry weather flow (D.W.F.)**.

**Please refer "Hydrology and Water Resources Engg." by the same author.

3.3. Estimating Peak Run-off

The peak rate of run-off that is produced from a particular catchment depends upon numerous factors ; such as, the type of precipitation, the intensity and duration of rainfall, the rainfall distribution, the soil moisture deficiency, the direction of the prevailing storm, the climatic conditions, the shape, size and type of catchment basin, etc. etc. Due to these 15 to 20 variables* involved in evaluating the run-off, it is not possible to precisely determine it, even with the help of the most complicated mathematics, as all these variables are interdependent, and run off cannot be easily expressed by an exact equation. Hence, uptil about 40 years ago, the peak run-off rate was used to be estimated by empirical formulas only, even in the developed countries like U.S.A. Different empirical formulas were, therefore, developed for different regions, by the investigators, depending upon their actual experimental works. In recent years, however, a *rational method* has been evolved to estimate the peak drainage discharge. *This method, though called rational, is not rational in the sense that the results given by this formula for larger areas (more than 400 hectares or so) are generally erroneous and misleading.* This method can be applied most precisely to smaller areas (preferably less than 50 hectares or so). For large areas, empirical formulas, are, however, continued to be used, although the most modern method for computing urban storm drainage is by *digital computer simulations.* This is an advanced topic dealt under the subject of "Water Resources Systems Planning" and is beyond the scope of the Undergraduate Courses. The rational formula and other empirical formulas for determining peak drainage discharge are discussed here :

3.3.1. Computing the Peak Drainage Discharge by the Use of Rational Formula. If a rainfall is applied to an impervious surface at a constant rate, the resultant runoff from the surface would finally reach a rate equal to the rainfall. In the beginning, only a certain amount of water will reach the outlet, but after some time, the water will start reaching the outlet from the entire area ; and in this case, the run off rate would become equal to the rate of rainfall. The period after which the entire area will start contributing to the runoff is called the **time of concentration****. The runoff resulting from a rain having a duration lesser than the time of concentration will not be maximum, as the entire area will not contribute to the run-off in this case. Further, it has been established that the runoff is not maximum, even when the duration of the rain is more than the time of concentration, because in such a case, the intensity of rain reduces with the increase in its duration. *In other words, it has been established that the maximum runoff will be obtained from the rain having a duration equal to the time of concentration, and this is called* the **critical rainfall duration.**

---

*Please refer chapter 8 in "Hydrology and Water Resources Engineering" by the same author.

**Time of concentration of a drainage basin may be defined as the time required by the water to reach the outlet from the most remote point of the drainage area.

Based upon these basic principles, the rational formula was evolved, due to the efforts of Fruhling of Germany, Kuichling of America, and later Lloyd Davis of England. This formula states that

$$Q_p = \left(\frac{1}{36}\right) K \cdot p_c \cdot A \qquad \text{...(3.1)}$$

where Q_p = Peak rate of runoff in cumecs.

K = Coefficient of runoff.

A = The catchment area contributing to runoff at the considered point, in hectares.

p_c = **Critical rainfall intensity** of the design frequency *i.e.* the rainfall intensity during the critical rainfall duration equal to the time of concentration, in cm/hr.

Coefficient of Runoff

The coefficient of runoff (K), is in fact, the impervious factor of runoff, representing, the ratio of runoff to precipitation *i.e.*

$$K = \frac{\text{Runoff}}{\text{Precipitation}}. \qquad \text{...(3.1}a\text{)}$$

The value of K increases as the imperviousness of the area increases*, thus tending to make $K = 1$ for perfectly impervious areas. It is generally taken as equal to 0.9 for paved areas and 0.15 for lawns and gardens. The values of K can also be worked out for different localities having different population densities. Various values of K which can be of use in designing storm water drains, are given in Tables 3.1 and 3.2.

Table 3.1. Values of Run-off Coefficient (K) for Various Surfaces

S. No.	*Type of surface*	*Value of K*
1.	Water-tight roof surface	0.70—0.95
2.	Asphalt pavement in good order	0.85—0.90
3.	Stone, brick, wood-block pavement with cemented joints	0.75—0.85
4.	Same as above with uncemented joints	0.50—0.70
5.	Water bond macadam roads	0.25—0.60
6.	Gravel roads and walks	0.15—0.30
7.	Unpaved streets and vacant lands	0.10—0.30
8.	Parks, lawns, gardens, meadows, etc.	0.05—0.25
9.	Wooden lands	0.01—0.20

*Greater is the imperviousness of an area, lesser will be the infiltration, and hence greater will be the runoff.

Table 3.2. Values of Run-off Coefficient (K) for Different Types of Localities

S. No. (1)	*Type of locality* (2)	*Average approximate population density in persons / hectare* (3)	*Value of K* (4)
1.	Extreme suburban areas with 20 to 40 per cent parking and widely detached houses.	75—125	0.35
2.	Suburban areas with widely detached houses.	125—150	0.45—0.55
3.	Areas with 50 per cent attached houses and 50 per cent detached houses.	375—500	0.65
4.	Areas closely built up.	500—625	0.75
5.	Business areas.	more than 625	0.85

Intensity of Rainfall

A rainfall at a place can be completely described if its intensity, duration and frequency is known. The intensity of a rain is the rate at which it is falling, the duration is the time for which it falls with that given intensity, and frequency is the number of times it falls.

The **intensity of a rain** is expressed in cm/hr ; but this rate at which the rain falls, changes continuously throughout the storm period. It may rain 5 cm in a particular one hour giving an average rainfall rate of 5 cm/hr during that particular hour. However, during this particular hour, sometimes the rainfall rate will exceed 5 cm/hr, and sometimes it may be much less than 5 cm/hr.

The intensity of rainfall can be determined with the help of **automatic rain gauges***, such as shown in Fig. 3.1 (*a*) and (*b*). Such gauges automatically

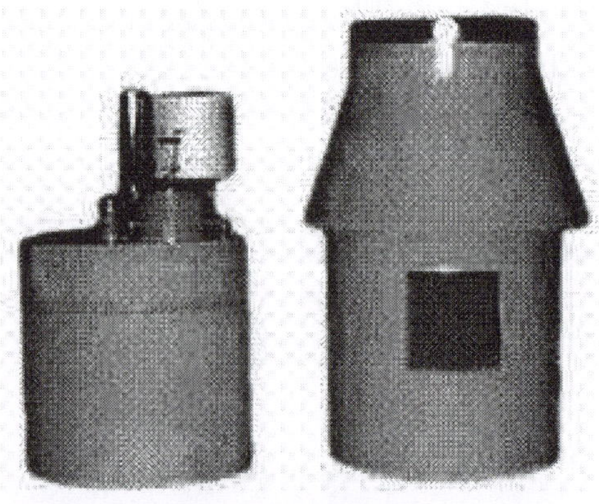

Fig. 3.1. (*a*) Photographic view of an Automatic Rain Gauge (Floating Type), popularly used in India.

*The other type of gauges *i.e.*, **non-recording gauges** collect the falling rain to manually record the mm of rain fallen during the last 24 hours. They are not of much use in calculating the amount of storm runoff from an area as they give only average intensity and not the maximum intensity which lasts for only 10 to 15 minutes during a storm.

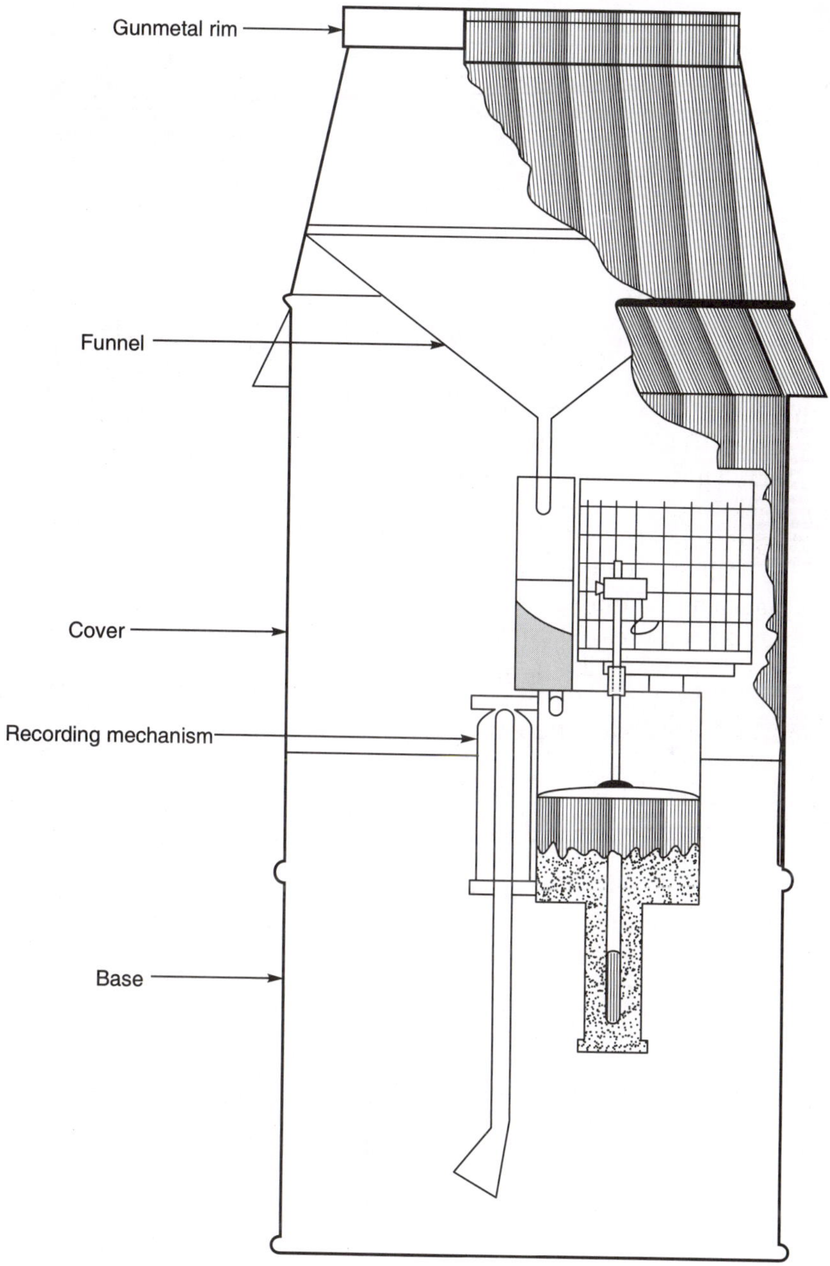

Fig. 3.1. (*b*) Section Through an Automatic Rain Gauge (Floating Type).

record the cumulative amount of water with time on a graph paper. An original chart obtained from such a gauge is shown in Fig. 3.2.

Note. It is evident from this graph, that after the gauge has recorded 10 mm of rain, the bottle gets empty and the line representing cummulative rainfall vertically falls down. Hence, while estimating the amount of rain fallen during any time interval, this fact must be kept in view.

It is found from this graph that between 5 hour of 5.8.75 and 6 hour of 5.8.75, the rainfall during the four intervals of 15 minutes each are about 6.25 + 1.25 = 7.5 mm, 8.75 + 1 = 9.75 mm, 9 + 0.75 = 9.75 mm, and 9.25 mm, respectively ; thus giving a total of 36.25 mm of rainfall in this particular hour.

The average rainfall during this hour will be 3.625 cm/hr and maximum, will be

$$= \frac{9.75}{15/60} = 37.1 \text{ mm/hr or } 3.71 \text{ cm/hr.}$$

Such records of automatic rain gauges are collected over a period of 15 to 20 years. They are then analysed so as to pick up the storm, giving intensities greater than 1 cm/hr or so. The rains fallen during different time intervals such as 5 minutes, 10 minutes, 15 minutes, 20 minutes, 60 minutes, 120 minutes, etc. are worked out for each of the chosen storms. They are then analysed and converted into intensity duration curves, such as shown in Fig. 3.3. The method of analysing and drawing such curves, has been thoroughly explained in "Hydrology and Water Resources Engineering" by the same author. The students may please refer to that book to fully understand the hydrological concepts.

This job of data collection and analysation is being done in India by Indian Meteorological Department (I.M.D.) and readymade Standard Depth-Duration Curves (Fig. 3.4)* of a particular rain gauge station, representing different frequencies, are provided by them to us. These curves can be used to work out the intensity of rainfall having a duration equal to the time of concentration, for the given design frequency.

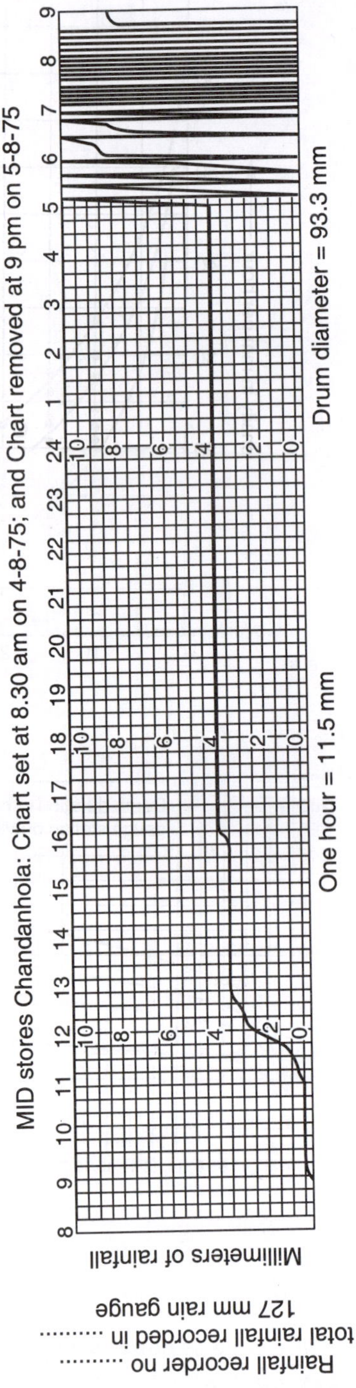

Fig. 3.2

*The curves of Figs. 3.3 and 3.4 differ only in the sense that in Fig. 3.3, y-axis represents intensity in mm/hr, whereas in Fig. 3.4, the y-axis represents the rainfall in mm and should be divided by the duration for obtaining the intensity.

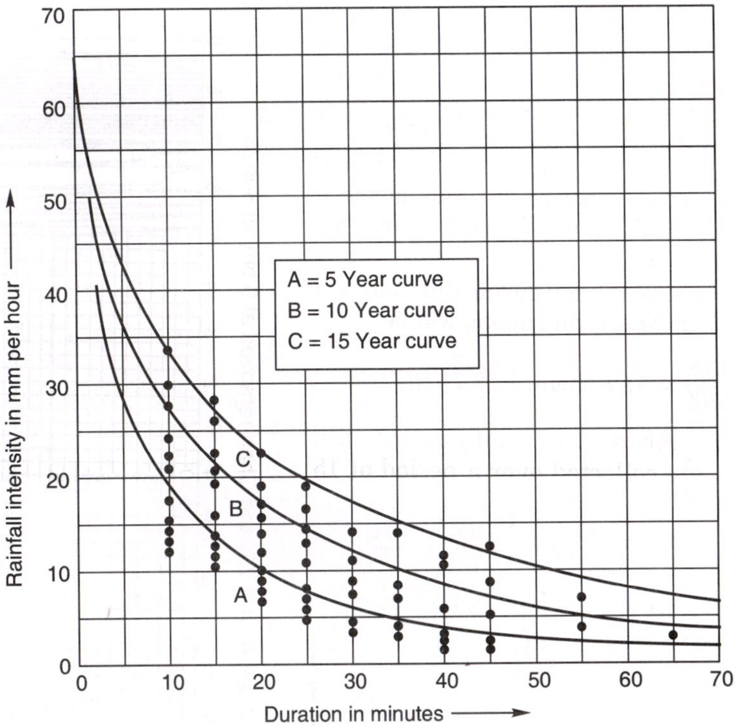

Fig. 3.3. Typical Intensity Duration Curves.

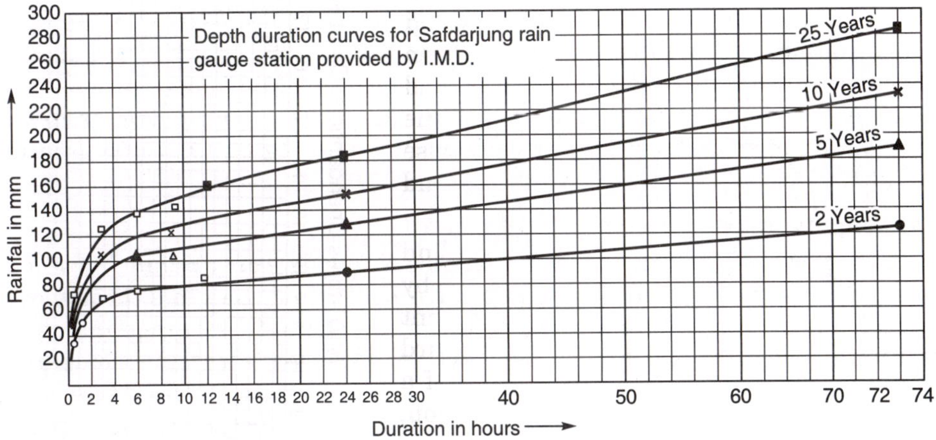

Fig. 3.4. Typical Standard Depth Duration Curves.

The time of concentration

The time of concentration for a given storm water drain generally consists of two parts ; *viz.*,

(*i*) *The inlet time or overland flow time or time of equilibrium* (T_i), *i.e.* the time taken by the water to flow overland from the critical point upto the point where it enters the drain mouth. It may be estimated by using Eq. (3.2) given below :

$$T_i = \left(0.885 \frac{L^3}{H}\right)^{0.385} \qquad \text{...(3.2)}$$

where T_i = Inlet time in hours.

L = Length of overland flow in kilo-metres from the critical point to the mouth of the drain.

H = Total fall of level from the critical point to the mouth of the drain in metres.

L and H can be found from the survey plan of the area, and T_i can be easily calculated.

(*ii*) *The channel flow time or gutter flow time* (T_f), *i.e.* the time taken by the water to flow in the drain channel from the mouth to the considered point. This may be obtained by dividing the length of the drain by the flow velocity in the drain.

$$\therefore \qquad T_f = \frac{\text{Length of the drain}}{\text{Velocity in the drain}} \qquad \text{...(3.3)}$$

The total time of concentration at a given point in the drain, for working out the discharge at that point, can be easily obtained as

$$T_c = T_i + T_f \qquad \text{...(3.4)}$$

The intensity of rainfall during this much of time (for the given design frequency, of course) can be easily obtained from the standard intensity duration curves or DAD curves.

The value of intensity so obtained is still the rainfall intensity at the rain gauge station, and is called the **point rainfall intensity.** In order to make it effective over the entire catchment area (in which this rain gauge station lies), it is necessary to multiply it by a factor called *dispersion factor* or *areal distribution factor*. The resultant value will be nothing but p_c, to be used in Eq. (3.1).

The areal distribution factor

It is a well established fact that the intensity of rainfall recorded at a particular rain gauge station in a catchment is not the same throughout the catchment. As the size of the catchment increases, the average intensity of rainfall over the catchment as a whole goes on decreasing compared to the point intensity recorded at a particular station. Therefore, the **areal distribution factor**, also called, the **dispersion factor**, is always applied to the point rainfall for working out the design rainfall intensity. In case of Delhi, it is seen that the intensity of rainfall varies considerably from one part to another, and as such, the *dispersion factor* reduces considerably with the increase in the catchment area, as shown in Table 3.3.

Table 3.3. Values of Dispersion Factor for Delhi

Area in hectares	*Dispersion or Areal Distribution factor*
0	1.0
200	0.997
400	0.994
800	0.989
1,200	0.984
1,600	0.979
2,000	0.974
2,400	0.972
2,800	0.967
3,200	0.962
3,600	0.956
4,000	0.951
6,000	0.926
8,000	0.903
12,100	0.855
16,200	0.807

In the absence of standard intensity-duration curves, the value of p_c can also be determined in the following two ways :

(*i*) The value of "**one hour rainfall**" of a given frequency at a given place can be found from the ***charts, shown in Fig. 3.5 (a), (b) and (c).***

This value of "one hour rainfall" is multiplied by the areal distribution factor, so as to obtain p_o. The value of p_o is further multiplied by factor $\frac{2}{1+T_c}$ so as obtain p_c ; *viz.*,

$$p_c = p_o \left(\frac{2}{1+T_c}\right)^* \qquad ...(3.5)$$

where T_c = Time of concentration in hours.

(*ii*) Since the intensity of a rain is inversely proportional to the duration of the rain, an intensity duration curve can be represented by a generalised equation of the form

$$p = \frac{a}{T+b} \qquad ...(3.6)$$

where p = Rain intensity in cm/hr.

T = Time in minutes

a and b = constants.

The values of a and b have been found out by the Health Ministry of Britain as 75 and 10 respectively for T varying between 5 to 20 minutes ; and as 100

*The derivation of this formula is given in article 7.17.1.2 of "Water Resources Engg. (Vol I)—Hydrology and Water Resources Engineering" by the same author.

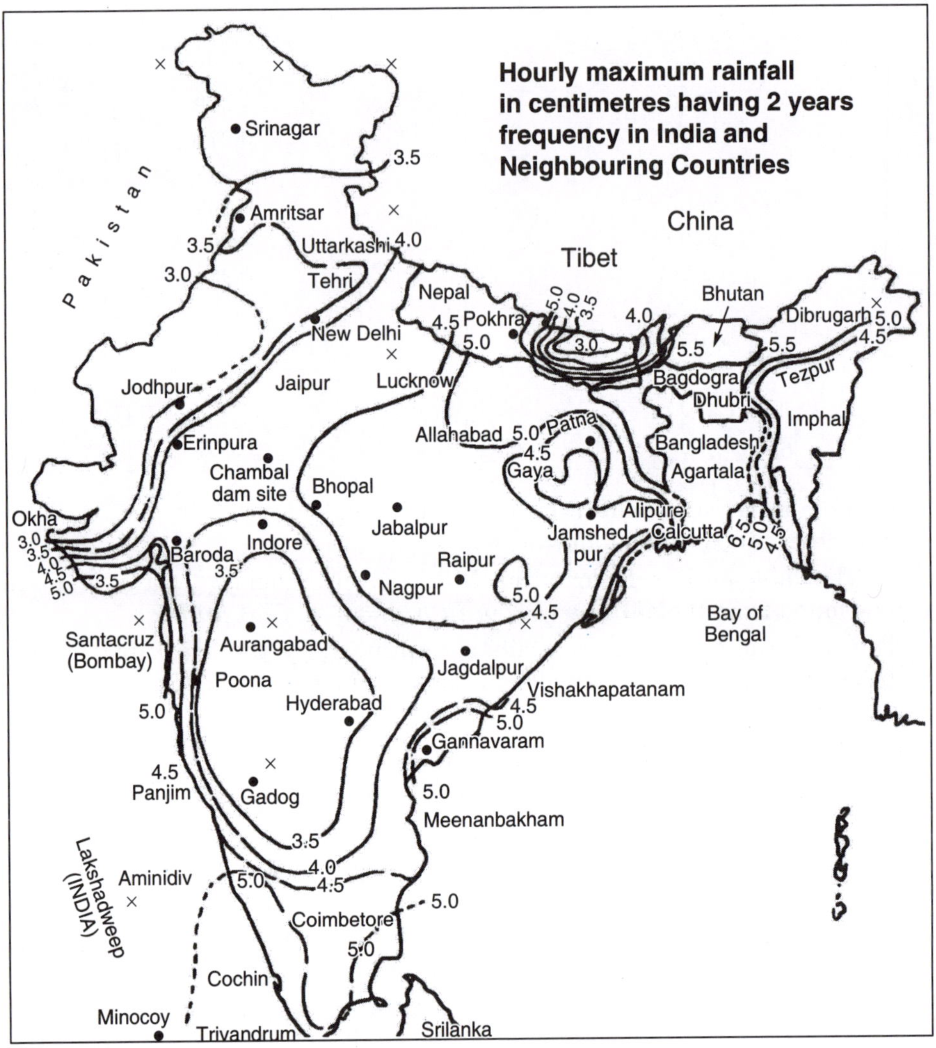
Hourly maximum rainfall in centimetres having 2 years frequency in India and Neighbouring Countries
Srinagar
Pakistan
Amritsar
Uttarkashi
Tehri
New Delhi
Nepal
Pokhra
Tibet
China
Bhutan
Dibrugarh
Tezpur
Imphal
Bagdogra
Dhubri
Bangladesh
Agartala
Jodhpur
Jaipur
Lucknow
Allahabad
Patna
Gaya
Erinpura
Chambal dam site
Bhopal
Jabalpur
Jamshed pur
Alipure
Calcutta
Okha
Baroda
Indore
Raipur
Nagpur
Santacruz (Bombay)
Aurangabad
Poona
Jagdalpur
Vishakhapatanam
Bay of Bengal
Hyderabad
Gannavaram
Panjim
Gadog
Meenanbakham
Lakshadweep (INDIA)
Aminidiv
Coimbetore
Cochin
Minocoy
Trivandrum
Srilanka

Fig. 3.5(*a*)

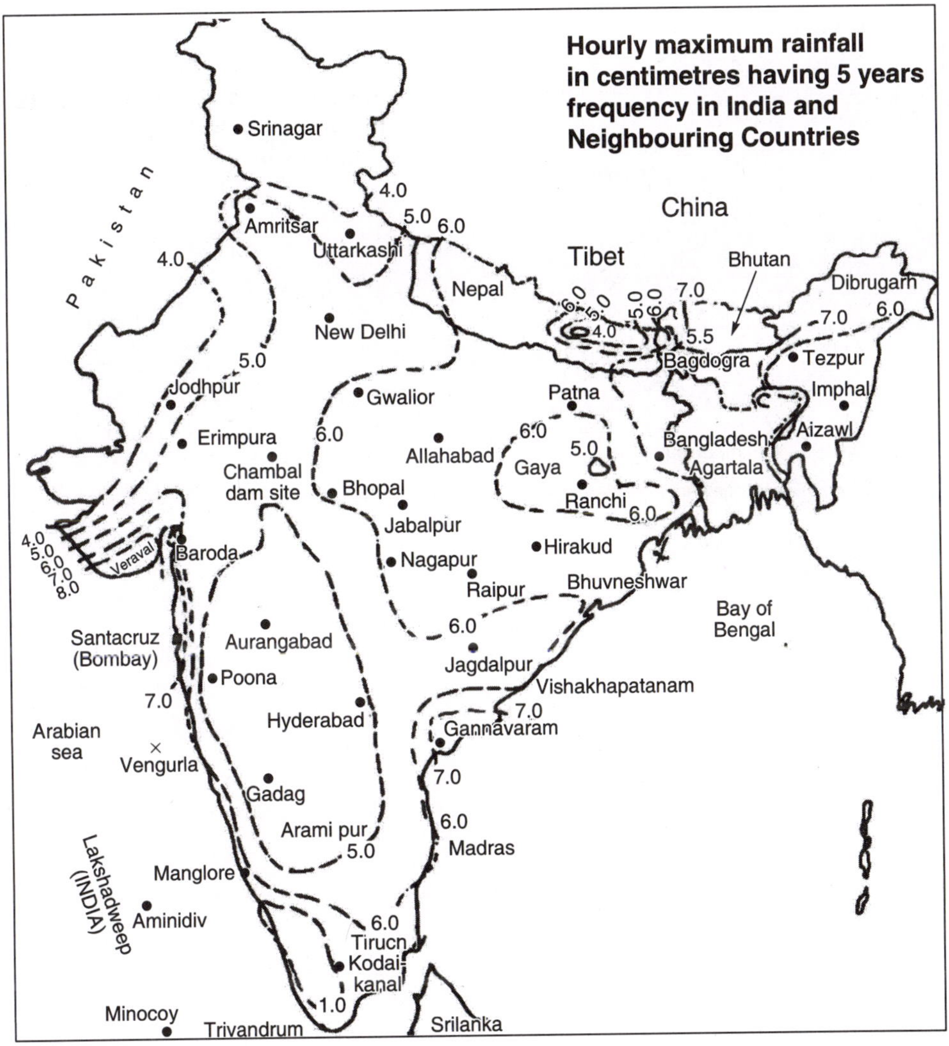
Hourly maximum rainfall in centimetres having 5 years frequency in India and Neighbouring Countries
Srinagar
Pakistan
China
Tibet
Bhutan
Dibrugarh
Nepal
Amritsar
Uttarkashi
New Delhi
Jodhpur
Erimpura
Chambal dam site
Gwalior
Allahabad
Patna
Gaya
Ranchi
Bagdogra
Tezpur
Imphal
Aizawl
Bangladesh
Agartala
Bhopal
Jabalpur
Nagapur
Raipur
Hirakud
Bhuvneshwar
Baroda
Veraval
Santacruz (Bombay)
Aurangabad
Poona
Jagdalpur
Vishakhapatanam
Bay of Bengal
Arabian sea
Vengurla
Hyderabad
Gannavaram
Gadag
Arami pur
Madras
Manglore
Lakshadweep (INDIA)
Aminidiv
Tirucn
Kodai kanal
Minocoy
Trivandrum
Srilanka
4.0
5.0
6.0
7.0
8.0
5.5
1.0

Fig. 3.5(*b*)

Hourly maximum rainfall in centimetres having 50 years frequency in India and Neighbouring Countries
China
Tibet
Pakistan
Srinagar
Amritsar
Nepal
New Delhi
Khathmandu
Bhutan
Dibrugarh
Jodhpur
Jaipur
Lucknow
Bagdogra
Guwahati
Dhubri
Imphal
Gwalior
Allahabad
Patna
Erimpura
Bangladesh
Aizawl
Asansol
Bhopal
Bangawara
Jabalpur
Ranchi
Baroda
Surat
Hirakud
Nagapur
Aurangabad
Santacruz (Bombay)
Jagdalpur
Bay of Bengal
Sholapur
Hyderabad
Gannavaram
Arabian sea
Anantpur
Manglore
Bangalore
Meenabakkam (Madras)
Lakshadweep (INDIA)
Coimbatore
Aminidiv
Cochin
Srilanka
Minicoy
Trivandrum

Fig. 3.5(*c*)

and 20 respectively for T varying between 20 to 100 minutes respectively. The formulas given by them, and generally applicable in England, are, therefore, given as below :

$$p = \frac{75}{T+10} \qquad ...(3.7)$$

(for T varying between 5 to 20 minutes)

and
$$p = \frac{100}{T+20} \qquad ...(3.8)$$

(for T varying between 20 to 100 minutes)

Using T_c in minutes in place of T in Eqs. (3.7) and (3.8), the values of p, *i.e.* p_c can be evaluated.

Besides the above generalised equations, certain other empirical equations have been suggested for determining rainfall intensity, as given below :

(*a*) *For localities where rainfall is frequent.*

$$p = \frac{343}{T+18} \qquad ...(3.9)$$

This formula is applicable for places where heavy and frequent rains occur, and gives an intensity for 5 years frequency.

(*b*) *For rains having frequency of 10 years,* the equation suggested is

$$p = \frac{38}{\sqrt{T}} \qquad ...(3.10)$$

(*c*) *For rains having frequency of 1 year,* the equation suggested is

$$p = \frac{15}{T^{0.620}} \qquad ...(3.11)$$

(*d*) *Kuichling's formula*

$$p = \frac{267}{T+20} \qquad ...(3.12)$$

(for storms having 10 years frequency)

$$p = \frac{305}{T+20} \qquad ...(3.13)$$

(for storms having 15 years frequency)

The equations from (3.7) to (3.13) can, though be used for finding the value of p_c, yet they are very very empirical equations, and are not very reliable. They are, therefore, generally avoided in designing storm water drains in modern days. They may, however, be used when absolutely no rainfall records are available.

3.3.2. Computing the Peak Drainage Discharge by the Use of Empirical Formulas. The rational formula described above is also quite empirical in the sense that the value of K considerably depends upon the judgement of the designer. *Moreover, this method gives reliable results only for smaller areas, and hence used only for the design of drains having catchments less than 400 hectares or so.* For the design of drains having larger catchments (say above 400 hectares or so), it is generally advisable to use the suggested empirical formula for the given region.

Various empirical formulas for calculating storm water run-off have been suggested by various investigators. Some of these formulas are based on local conditions only, and can be adopted only when certain specific requirements are specified. The other formulas are based on experimental studies and results obtained over wide areas, and they can, therefore, be adopted for many localities. Some of the leading empirical formulas are given below :

(*i*) **Burkli-Ziegler formula.** This formula is the oldest empirical formula used for determining the peak run off rate. It was devised by a Swiss engineer for local conditions, but was soon followed in the entire U.S.A. In M.K.S. units, it states that

$$Q_p = \left(\frac{1}{455}\right) K' \cdot p \cdot A \cdot \sqrt{\frac{S_0}{A}} \qquad \text{...(3.14)}$$

where Q_p = The peak runoff in cumecs.

K' = Run off coefficient depending upon the permeability of the surface, and having an average value of 0.7.

p = The maximum rainfall intensity over the entire area, and is usually taken as 2.5 to 7.5 cm/hr.

A = The drainage area in hectares.

S_o = The slope of the ground surface in metres per thousand metres.

(*ii*) **Dicken's formula.** This formula is generally useful for Indian catchments, (and particularly for *North India*) and states that

$$Q_p = CM^{3/4} \qquad \text{...(3.15)}$$

where Q_p = Peak discharge in cumecs

M* = Catchment area in sq. km

C = a constant depending upon all those fifteen to twenty factors which affect the run off.

The value of C must be ascertained for each catchment, and depends upon the nature of the catchment and the intensity of rainfall. An average value of C equal to 11.5 is generally used, and it should be increased for hilly catchments and *vice versa*. Secondly, for the same type of catchment, greater is the rainfall, greater will be the value of C and *vice versa*. The value of C for different zones of India are shown in Fig. 3.6.

(*iii*) **Ryve's formula.** This formula is almost similar to that of Dicken's ; the only difference is in the values of constants. It is generally applicable to *South Indian* catchments, and states that

$$Q_p = C_1 M^{2/3} \qquad \text{...(3.16)}$$

where Q_p, M and C_1 have the same meaning as in Dicken's formula.

---

*The symbol for area has been used as A when expressed in hectares and M when expressed in sq. km.

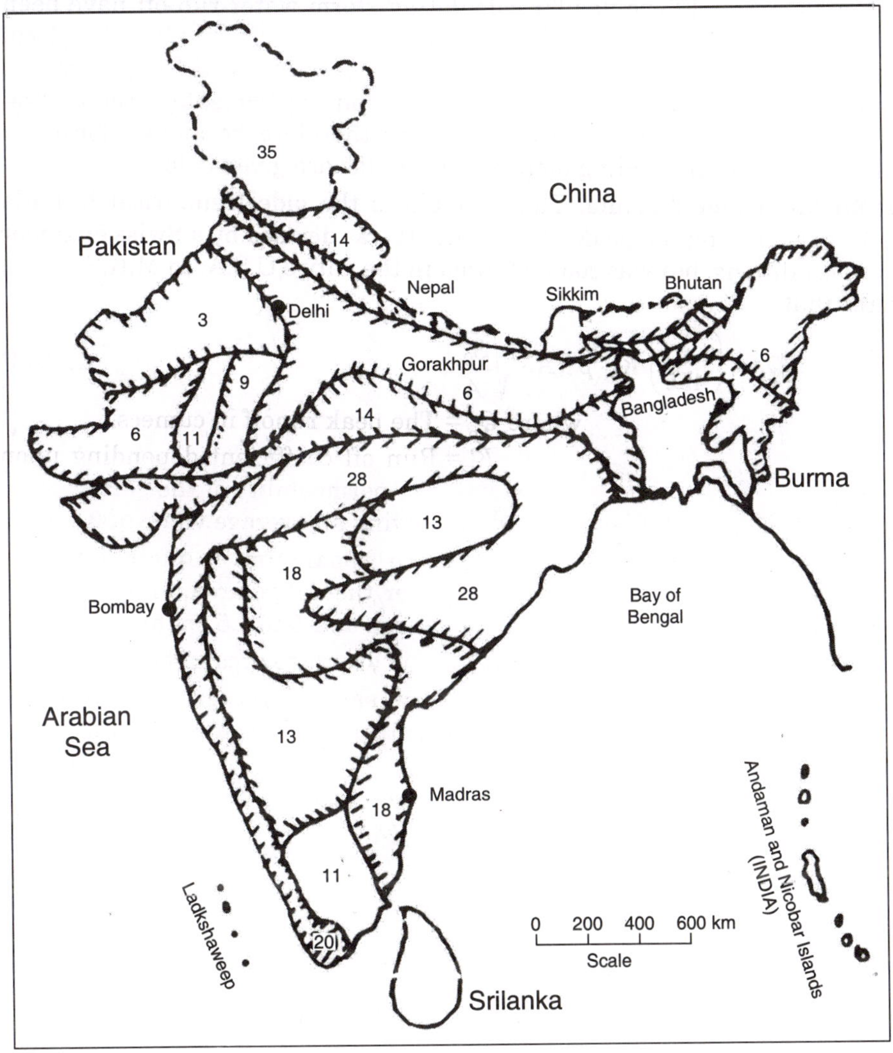

Fig. 3.6. Chart showing values of C in Dicken's formula (Eq. 3.15) for various places in India.

The average value of C_1 to be used is 6.8, with less value for flat catchments and more value for hilly catchments. Different values of C_1 must be ascertained for different catchments, as suggested in Table 3.4.

Table 3.4. Value of C_1 in Ryve's formula

Location of Catchment	*Value of C_1 in Ryve's formula*
Areas within 24 km from the coast	6.8
Areas within 24—16 km from the coast	8.8
Limited areas near hills	10.1

(*iv*) **Inglis formula.** This formula is applicable to the fan shaped catchments in Old Bombay State of India. It states that

$$Q_p = \frac{123\,M}{\sqrt{M + 10.4}} \approx 123\sqrt{M} \qquad ...(3.17)$$

where Q_p and M have the same meaning as given in Dicken's formula.

(*v*) **Nawab Jung Bahadur formula.** This formula has been derived for Hyderabad Deccan catchments. It states that

$$Q_p = C_2 M'^{\left[0.93 - \left(\frac{1}{14}\right)\log M'\right]} \qquad ...(3.18)$$

The value of constant C_2 varies between 48 to 60. M' is the catchment area in acres.

(*vi*) **Dredge or Burge's formula.** This formula is based upon Indian records and states that

$$Q_p = 19.6\,\frac{M}{L^{2/3}} \qquad ...(3.19)$$

where Q_p and M have the same meaning as given earlier.
L is the length of the drainage basin in kilometres.

Example 3.1. *Assuming that the surface on which the rain falls in a district is classified as follows :*

20% of the area consists of roof for which the runoff ratio is 0.9, 20% of the area consists of pavements for which the runoff ratio is 0.85, 5% of the area consists of paved yards of houses for which run off ratio is 0.80, 15% of area consists of macadam roads for which run off ratio is 0.40, 35% of the area consists of lawns, gardens and vegetable plants for which the runoff ratio is 0.10, and the remaining 5% of the area is wooded for which the runoff ratio is 0.05 ; determine the coefficient of runoff for the area.

If the total area of the district is 36 hectares and the maximum rain intensity is taken as 5 cm/hr ; what is the total runoff for the district ?

Solution. The total area A of the district can be considered to be made up of smaller areas A_1, A_2, A_3, ... A_n, having runoff ratios as K_1, K_2, K_3, ... K_n, respectively. Then the runoff ratio (*i.e.*, coefficient of runoff) K for the entire area may be computed by using

$$K = \frac{K_1A_1 + K_2A_2 + K_3A_3 + ... + K_nA_n}{A_1 + A_2 + A_3 + ... + A_n} \qquad ...(3.20)$$

$$= \frac{\Sigma KA}{\Sigma A} = \frac{\Sigma KA}{A}$$

In this example, we have

$$K_1A_1 = \frac{20}{100}\,A(0.90) = 0.18\,A$$

$$K_2A_2 = \frac{20}{100}\,A(0.85) = 0.17\,A$$

$$K_3A_3 = \frac{5}{100}\,A(0.80) = 0.04\,A$$

$$K_4A_4 = \frac{15}{100} A(0.4) = 0.06\,A$$

$$K_5A_5 = \frac{35}{100} A(0.1) = 0.035\,A$$

$$K_6A_6 = \frac{5}{100} A(0.05) = 0.0025\,A.$$

$$\therefore \quad K = \frac{K_1A_1 + K_2A_2 + K_3A_3 + K_4A_4 + K_5A_5 + K_6A_6}{A}$$

$$= \frac{0.18A + 0.17A + 0.04A + 0.06A + 0.035A + 0.0025A}{A}$$

$$= 0.4875.$$

Hence, the runoff factor for the entire area

= 0.4875. **Ans.**

The peak discharge from the area may be computed by using the rational formula given by Eq. (3.1) as

$$Q_p = \frac{1}{36} \,.\, K.p_c.A$$

Here, we have

$K = 0.4875$

p_c = Critical rain intensity = 5.0 cm/hr (given)

A = 36 hectares

$$\therefore \quad Q_p = \frac{1}{36}(0.4875)(5.0)(36.0) = 2.4375 \text{ cumecs.}$$

Say 2.44 cumecs. **Ans.**

Example 3.2. *If in the above example, the density of population is 250 per hectare, and the quota of water supply per day is 225 litres ; calculate the quantity of*

(*a*) *Sewage for which the sewers of a separate system, should be designed.*

(*b*) *Storm water for which the sewers of a partially separate system should be designed.*

Solution. Area of district

= 36 hectares.

Population density = 250 persons per hectare

∴ Population = 36 × 250 = 9000.

Average water supply per day

= 225 litres/person

∴ Average quantity of water supplied to the district per day

= 225 × 9000 litres = 20,25,000 litres = 2,025 cu. m.

∴ Rate of water supplied

$$= \frac{2025}{24 \times 60 \times 60} \text{ cumecs} = 0.0234 \text{ cumecs.}$$

Assuming the sewage discharge as 0.8 times the water supplied, we have

Average rate of sewage produced

$$= 0.8 \times 0.0234 = 0.0187 \text{ cumecs.}$$

Now assuming the peak rate of sewage as three times the average,we have

The peak rate of sewage flow

$$= 3 \times 0.0187 = 0.056 \text{ cumecs.} \quad \textbf{Ans.}$$

Case. (*b*) In case of partially separate system, the storm water from roofs and paved yards of houses will be allowed to enter the sewers. Now, from the previous example, we have

Area of roofs $= \frac{20}{100} \times 36$ hectares = 7.2 hectares.

Coefficient of runoff for roofs = 0.90

Area of pavements $= \frac{5}{100} \times 36 = 1.8$ hectares

Coefficient of runoff for pavements = 0.80.

∴ The discharge from roofs and pavements, as given by rational formula, using p_c = 5 cm/hr. is

$$= \left[\frac{1}{36} \times 0.90 \times 5.0 \times 7.2 + \frac{1}{36} \times 0.80 \times 5.0 \times 1.8\right] \text{ cumecs}$$

$$= 1.1 \text{ cumecs.}$$

Hence, the storm water which must pass through the sewers of a partially separate system = 1.1 cumecs. **Ans.**

Note. This is about 20 times the peak rate of sewage produced (*i.e.* 0.056 cumecs). Moreover, strictly speaking, the sewers of the partially separate system should be designed for carrying this storm water plus the sewage, *i.e.* for a discharge = 1.1. + 0.056 = 1.156 cumecs.

Example 3.3. *The drainage area of one sector of a town is 12 hectares. The classification of the surface of this area is as follows :*

Percent of total surface area	*Type of surface*	*Coefficient of runoff*
20%	*Hard pavement*	*0.85*
20%	*Roof surface*	*0.80*
15%	*Unpaved street*	*0.20*
30%	*Garden and Lawn*	*0.20*
15%	*Wooded area*	*0.15*

If the time of concentration for the area is 30 minutes, find the maximum runoff.

Use the formula

$$R = \frac{900}{t + 60}.$$

Solution. If A represents the total area, then we have

$$K_1A_1 = 0.85 \times 0.20\, A = 0.17\, A$$

$$K_2A_2 = 0.80 \times 0.20\,A = 0.16\,A$$
$$K_3A_3 = 0.20 \times 0.15\,A = 0.03\,A$$
$$K_4A_4 = 0.20 \times 0.30\,A = 0.06\,A$$
$$K_5A_5 = 0.15 \times 0.15\,A = 0.0225\,A$$
$$\Sigma = 0.4425\,A$$

$$K = \frac{\Sigma KA}{A} = \frac{0.4425\,A}{A}$$

or $K = \mathbf{0.4425}$

Now, in the formula of the type

$$R = \frac{900}{t + 60}$$

R is the rainfall intensity, generally expressed in mm/hr and t is the concentration time in minutes.

∴ R = rainfall intensity in mm/hr

$$= \frac{900}{30 + 60} = 10 \text{ mm/hr.} = 1 \text{ cm/hr.}$$

∴ P_c to be used in our Rational formula (Eq. 3.1)

= 1 cm/hr.

Using Rational formula, we have

$$Q_p = \frac{1}{36} K.p_c.A$$

$$= \frac{1}{36} \times 0.4425 \times 1 \times 12 \text{ cumecs.}$$

= 0.1475 cumecs ; say 0.148 cumecs.

∴ Maximum rate of runoff expected from the area

= 0.148 cumecs. **Ans.**

Example 3.4. *A population of 30,000 is residing in a town having an area of 60 hectares. If the average coefficient of runoff for this area is 0.60, and the time of concentration of the design rain is 30 minutes, calculate the discharge for which the sewers of a proposed combined system will be designed for the town in question. Make suitable assumptions where needed.*

Solution. Let us first assume that the town is provided with a planned water supply from the water-works at an average per capita rate equal to 120 litres/day/person*. Also assume that 80% of this water supply will be reaching the sewers as sanitary sewage.

∴ Quantity of sanitary sewage produced per day

$$= \left(\frac{80}{100}\right) 120 \times 30{,}000 \text{ litres}$$

$$= 0.8 \times 120 \times 30 \text{ cu. m} = 2880 \text{ cu. m}$$

*Please see Table 2.2.

Quantity of sanitary sewage produced per second

$$= \frac{2880}{24 \times 60 \times 60} \text{ cumecs} = 0.033 \text{ cumecs}$$

$\therefore$ Average sewage discharge = 0.033 cumecs.

Assuming the maximum sewage discharge to be three times the average, we have

Max. sewage discharge

$= 3 \times 0.033 =$ **0.1 cumecs.**

The storm water discharge can be computed by using Rational formula ; *i.e.*

$$Q_p = \frac{1}{36} K \cdot p_c \cdot A$$

Using $p_c = \dfrac{100}{T + 20}$ (*i.e.* Eq. 3.8)

We have $p_c = \dfrac{100}{30 + 20} = 2$ cm/hr.

$\therefore$ $Q_p = \dfrac{1}{36} \times 0.60 \times 2 \times 60$ cumecs = **2 cumecs.**

Hence, the total peak discharge for which the sewers of the combined system should be designed

= Max. sewage discharge + Max. storm runoff

= 0.1 + 2.0 = **2.1 cumecs. Ans.**

Example 3.5. *A separate sewerage system has to be designed for a suburb near Delhi for a rainfall frequency of 2 years. This town is already provided with adequate water supply from water-works at per capita rate of 200 l/day/person. Compute the maximum storm drainage discharge for which the S.W. drain of a pocket draining an area of 20 hectares will be designed. Also compute the peak design discharge for which the sewers of this pocket will be designed. Make use of hourly rainfall charts (Fig. 3.5) and assume the concentration time (or inlet time) as 20 minutes. The population of the pocket discharging sewage is 9000. Make any other assumption if needed and not given.*

Solution. From Fig. 3.5 (*a*), the maximum one hour rainfall having 2 years frequency near Delhi, = 4 cm ; and from table 3.3, the dispersion factor for an area of 20 hectares ≈ 1.0.

$\therefore$ $p_o = 4.0 \times 1.0 = 4$ cm/hr.

Using Eq. (3.5), we have

$$p_c = p_o \cdot \left(\frac{2}{1 + T_c}\right) \text{; where } T_c \text{ is in hours.}$$

$$= 4 \cdot \left(\frac{2}{1 + \frac{20}{60}}\right) = 4 \cdot \left(\frac{2}{1 + 0.33}\right)$$

$$= \frac{8}{1.33} = 6 \text{ cm/hr.}$$

Further, assuming the coefficient of discharge for this residential pocket as equal to 0.55, we have

The peak run off rate

$$= Q_p = \frac{1}{36} \cdot K \cdot p_c \cdot A$$

$$= \frac{1}{36} \times 0.55 \times 6 \times 20 \text{ cumecs} = 1.83 \text{ cumecs.}$$

Hence, the drainage discharge for which the storm water drain of the pocket shall be designed

= 1.83 cumecs. **Ans.**

Now, the town has been provided with a water supply from the water-works at a per capita rate of 200 litres/day/person (given).

∴ Water supplied per day

= 200 × 9000 litres/day = 1800 cu.m/day

Assuming that 80% of the water supplied appears as sewage, we have

The sewage produced per day

= 0.8 × 1800 cu.m/day = 1440 cu.m/day

∴ Average sewage discharge per second

$$= \frac{1440}{24 \times 60 \times 60} \text{ cumecs} = 0.0167 \text{ cumecs}$$

Assuming the peak sewage discharge as three times the average, we have

Maximum sewage discharge

= 3 × 0.0167 = **0.05 cumecs**

Hence, the sewer of the locality will be designed for a discharge of 0.05 cumecs. **Ans.**

Example 3.6. *The surface water from airport road is drained to the longitudinal side drain from across one half of a bituminous pavement surface of total width 7.0 m, shoulder and adjoining land of width 8.0 m on one side of the drain. On the other side of the drain, water flows across from reserve land with average turf and 2% cross slope towards the side drain, the width of this strip of land being 25 m. The inlet time may be assumed to be 10 min for these conditions. The runoff coefficients of the pavement, shoulder and reserve land with turf are 0.8, 0.25 and 0.35, respectively. The length of the stretch of land parallel to the road from where the water is expected to flow to the side drain is 400 m. Estimate the quantity of runoff flowing in the drain assuming 10 year frequency. The side drain will pass through clayey soil with allowable velocity of flow as 1.33 m/s. Intensity-duration chart for 10 year frequency is*

Duration (mins)	*5*	*10*	*15*	*20*	*30*
Intensity (mm/hour)	*160*	*150*	*125*	*110*	*95*

(Engineering Services, 1997)

Solution. The given conditions are shown in the figure 3.6.

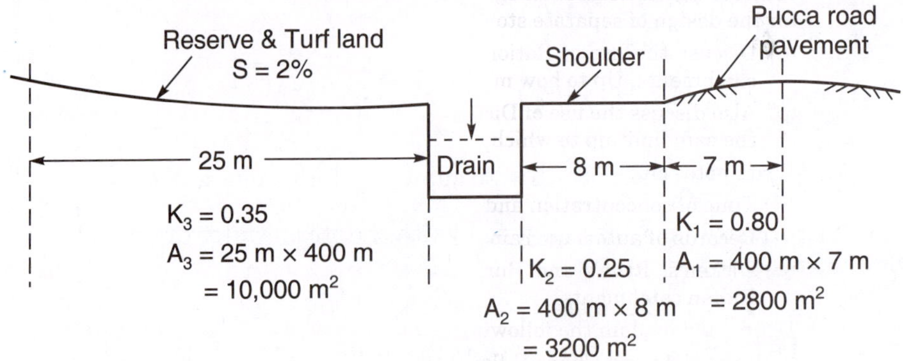

Fig. 3.6

Runoff factor K for the entire area contributing discharge in drain

$$= K = \frac{K_1A_1 + K_2A_2 + K_3A_3}{A_1 + A_2 + A_3}$$

$$= \frac{2800 \times 0.8 + 3200 \times 0.25 + 10000 \times 0.35}{2800 + 3200 + 10000 = 16000}$$

$$= 0.40875$$

Now, Time of conc.

$$T_c = T_i + T_f$$

where, T_i = 10 min (given)

T_f = channel flow time in drain in 400 m

$$\text{length} = \frac{400 \text{ m}}{1.33 \text{ m/s}}$$

= 300 secs = 5 min.

$\therefore \quad T_c$ = 15 min

Corresponding rain intensity (for 15 min duration rain)

= 125 mm/hr

= 12.5 cm/hr

$$Q = \frac{1}{36} \,.\, K \,.\, p_c \,.\, A$$

where A = Area in ha

p_c = is in cm/hr

$$= \frac{1}{36} \times 0.40875 \times 12.5 \times \left(\frac{16000}{10^4}\right)$$

= **0.227 m³/s. Ans.**

PROBLEMS

1. Differentiate between 'Sewage' and 'Drainage'. Discuss the Rational formula for calculating the peak drainage discharge from a given catchment and reaching a particular storm water drain up to a particular point.
2. Differentiate between Sanitary Sewage and Storm Sewage. Suggest other better names, being used in modern days to represent these two terms. Discuss and explain

the Rational formula and its limitations in calculating the quantities of storm sewage. Indicate the off-hand drainage coefficients (per unit of catchment) adopted in Delhi* for the design of separate storm water sewers.

3. (*a*) Discuss the use of Rational formula for the design of drains draining Indian catchments. Up to how much hectares of catchments, it can be safely used ?

(*b*) Also discuss the use of Dicken's formula for the Indian catchments, larger than the safe limit up to which Rational formula can be used.

4. Write notes on :

(*i*) Time of concentration and its significance ;

(*ii*) Records of automatic rain gauges ;

(*iii*) Dicken's, Ryve's and Burge's formulas for estimating drainage discharge of Indian catchments.

5. Define and explain the following terms :

(*i*) Inlet time and Channel flow time ;

(*ii*) One hour rainfall values ;

(*iii*) Areal distribution factor ;

(*iv*) Critical rainfall duration and Critical rainfall intensity.

*1/2 cusecs (ft^3/sec) per acre for urban drains, and 10 cusecs/sq mile for rural drains.

—4

Hydraulic Designs of Sewers and S.W. Drain Sections

4.1. General Introduction

In a separate sewerage system, which is mostly adopted in modern days, the circular sewer pipes are laid below the ground level, sloping continuously at sufficiently steeper gradients towards the outfall point ; and the storm water drains (S.W. drains) are the separately constructed rectangular or trapezoidal surface drains constructed at suitable gradients, and may be covered or kept open. The sewer pipes are designed to carry the maximum quantity of sanitary sewage likely to be produced from the area contributing to the particular sewer ; and the S.W. drains are designed to carry the maximum drainage discharge (*i.e.* the storm run-off that is likely to be produced by the contributing catchment from a rain of design frequency and of a duration equal to the time of concentration, as explained in the previous chapter). The combined sewers are designed to carry the sewage as well as the drainage.

4.2. Difference in the Design of Water Supply Pipes and Sewer Pipes

The hydraulic design of sewers and drains, which means finding out their sections and gradients, is generally carried out on the same lines as that of the water supply pipes. However, there are two major differences between the characteristics of flows in sewers and water supply pipes. These differences are:

(*i*) The water supply pipes carry pure water without containing any kind of solid particles, either organic or inorganic in nature. The sewage, on the other hand, does contain such particles in suspension ; and the heavier of these particles may settle down at the bottom of the sewers, as and when the flow velocity reduces, thus ultimately resulting in the clogging of the sewers. *In order to avoid such clogging or silting of sewers, it is necessary that the sewer pipes be of such a size and laid at such a gradient, as to generate self-cleansing velocities at different possible discharges.* The sewer materials must also be capable of resisting the wear and tear caused due to abrasion of the solid particles present in the sewage, with the interior of the pipe.

(*ii*) The water supply pipes carry water under pressure, and hence, within certain limits, they may be carried up and down the hills and the valleys ; whereas, the sewer pipes carry sewage as gravity conduits (or open channels)*, and they must, therefore, be laid at a continuous gradient in the

*If sewage has to be carried under pressure, it will require elaborate arrangement at each house which so ever is connected to the sewerage system, making it an impractical and impossible proposition.

downward direction up to the outfall point, from where the sewage will be lifted up, treated and disposed of.

4.3. Provision of Freeboard in Sewers and S.W. Drains

The *sanitary sewers*, as pointed out earlier, are designed large enough to carry the maximum sewage discharge while flowing half or three-fourth or two-third full. *Generally, the sewer pipes of sizes less than 0.4 m dia are designed as running half full at maximum discharge, and the sewer pipes greater than 0.4 m in dia are designed as running* $\frac{2}{3}$*rd or* $\frac{3}{4}$*th full at maximum discharge.*

The extra space, thus left, above the designed full supply line, will leave an ample margin, as to act as a factor of safety to counter-act against the factors, such as :

(*i*) low estimates of the average and maximum flows, made due to some wrong data obtained regarding the private water consumption by industries, or public, or about the quantity to be supplied from the water-works at the end of the design period.

(*ii*) large scale infiltration of storm water due to wrong or illegal connections, and that of underground water through cracks or open joints in the sewers.

(*iii*) unforeseen increase in population or water consumption and the consequent increase in sewage production.

The *storm water drains*, on the other hand, are generally not provided with so much margins above their FSLs, because the over-flowing of drains is not so much harmful, as is the overflowing of sewers, mainly because sewage is highly polluted as compared to the storm water. The storm water drains are, therefore, provided with nominal provisions of freeboard* above their designed full supply lines, as given in Table 4.1.

Table 4.1. Values of Freeboard to be adopted for the design of S.W. Drains

Peak discharge in the drain for which designed, in cumecs	*Freeboard to be left in metres*
Below 0.3	0.3
0.3—1.0	0.4
1—5	0.5
5—10	0.6
10—30	0.75
30—150	0.90
more than 150	1.0

*In case of covered drains, the roof slab should be laid at a level above the drain's FSL by an amount equal to the freeboard ; and in case of open drains, the embankments constructed on both sides of the drain must have a level greater than the full supply level by an amount equal to the freeboard.

4.4. Hydraulic Formulas for determining Flow Velocities in Sewers and Drains

The sewers and drains are generally designed as open channels except when it is especially required to design them as flowing under pressure, as is the case of inverted siphons* ; and discharge lines from sewage pumping stations, which always flow under pressure. Various empirical formulas, which have been suggested for determining the gradients necessary to obtain design velocities of flow in sewers or drains are given below :

4.4.1. Chezy's Formula. This formula was evolved by Chezy in 1775, and states that

$$V = c\sqrt{rs} \quad ...(4.1)$$

where V = velocity of flow in the channel in m/sec.

r = hydraulic mean radius of channel, *i.e.* hydraulic mean depth of channel

$$= \frac{a}{p}$$

where a is the area of channel and p is the wetted perimeter of the channel.

For a circular sewer running full, r is given by

$$r = \frac{\frac{\pi D^2}{4}}{\pi D} = \frac{D}{4}$$

where D is the diameter of the sewer.

s = hydraulic gradient, equal to the ground slope for uniform flows, *i.e.* the head drop between the two points divided by the length.

c = a constant, called Chezy's constant.

The Chezy's constant c depends upon various factors, such as the size and the shape of the channel, roughness of the channel surface, the hydraulic characteristics of the channel, etc. The value of c can be obtained by using either the Kutter's formula or the Bazin's formula, as given below. The channel section may then be designed by using

$$Q = A.V$$

where A is the flow area of cross-section of the channel, and V is the flow velocity in the channel.

(*a*) **Kutter's formula.** According to this formula, the value of c to be used in eq. (4.1) is given by

---

*Described in chapter 5.

$$c = \frac{\left(23 + \dfrac{0.00155}{s}\right) + \dfrac{1}{n}}{1 + \left(23 + \dfrac{0.00155}{s}\right) \cdot \dfrac{n}{\sqrt{r}}} \qquad \text{...(4.2)}$$

where n = Rugosity coefficient depending upon the type of the channel surface. Various values of n are given for different surfaces in Table 4.2.

s = Bed slope of the sewer (for uniform flows, of course)

r = Hydraulic mean depth = $\dfrac{a}{p}$.

Table 4.2. Mannings or Kutter's Rugosity Coefficients (n)

S. No.	*Pipe Material*	*Values of n at full depth for*	
		Good interior surface condition	*Fair interior surface condition**
(1)	*(2)*	*(3)*	*(4)*
1.	Salt glazed stoneware pipes	0.012	0.014
2.	Cement concrete pipes	0.013	0.015
3.	Cast iron pipes	0.012	0.013
4.	Brick, unglazed sewers/drains	0.013	0.015
5.	Asbestos cement	0.011	0.012
6.	Plastic (smooth) pipes	0.011	0.011

This formula may be simplified by omitting the term $\dfrac{0.00155}{s}$ from the numerator as well as the denominator, since this will affect the results much less than an increase of 0.001 in the value of n.

(*b*) **Bazin's formula.** According to this formula, the value of c to be used in Chezy's formula is given by

$$c = \frac{157.6}{1.81 + \dfrac{K}{\sqrt{r}}} \qquad \text{...(4.3)}$$

where r = Hydraulic mean depth of channel

$= \dfrac{a}{p}$

K = Bazin's constant as given by Table 4.3.

---

*Usually, the values corresponding to the fair condition of the interior surface are used in the designs.

Table 4.3. Bazin's Constant (K)

S. No. (1)	*Type of the inside surface of the sewer or drain* (2)	*Value of K* (3)
1.	Very smooth surfaces.	0.11
2.	Smooth brick and concrete surfaces.	0.29
3.	Rough brick and concrete surfaces.	0.50
4.	Smooth rubble and masonry surfaces.	0.83
5.	Good earthen channels.	1.54
6.	Rough earthen channels.	3.17

4.4.2. Manning's Formula. This formula was evolved by Manning in 1890, and is the most popular formula used for designing sewers and drains in U.S.A. as well as in India.

This formula has fully satisfied the experimental results and states that

$$V = \frac{1}{n} \cdot r^{2/3} \cdot s^{1/2} \qquad ...(4.4)$$

where n, r and $s$have the same meaning as given earlier. The values of rugosity coefficient (n) are the same as were given by Kutter, and represented in Table 4.2.

4.4.3. Crimp and Burge's Formula. This formula is very simple and is commonly used in England. It states that

$$V = 83.5 \cdot r^{2/3} \cdot s^{1/2} \qquad ...(4.5)$$

This formula is comparable to Manning's formula, having

$$\frac{1}{n} = 83.5; \qquad \text{or } n = 0.012$$

Hence, this formula is identical to Manning's formula for all those sewer materials for which n is taken as 0.012.

4.4.4. William-Hazen's Formula. This formula is generally used for flows under pressure for designing water supply pipes, and is seldom used for designing sewers. It states that

$$V = 0.85\, C_H \cdot r^{0.63} \cdot s^{0.54} \qquad ...(4.6)$$

The values of coefficient C_H may be taken as shown in Table 4.4.

Table 4.4. Values of C_H for William Hazen's formula

S. No. (1)	*Type of pipe material* (2)	*Value of C_H for*	
		New pipes (3)	*Design purposes* (4)
1.	Concrete and R.C.C. pipes	140	110
2.	Cast iron pipes	130	100
3.	Galvanised iron pipes	120	100
4.	Steel pipes with welded joints	140	100

(Contd.)

(1)	(2)	(3)	(4)
5.	Steel pipes with rivetted joints	110	95
6.	Steel pipes with welded joints lined with cement for bituminous enamel	140	110
7.	Asbestos cement pipes	150	120
8.	Plastic pipes	150	120

4.5. Maximum and Minimum Velocities to be generated in Sewers

It was pointed out in the earlier paragraphs that the flow velocities in the sewers should be such that neither the suspended materials in sewage get *silted up* nor the sewage pipe material gets *scoured out*. The first limitation, limits the minimum velocity ; and the second limitation, limits the maximum velocity.

4.5.1. Minimum Velocities. The silting of sewers can be avoided by generating such high velocities that would not permit the solids to settle down ; *i.e.*, the velocity should be such as to cause automatic self-cleansing effect. Such a **self-cleansing velocity,** *i.e.* the velocity which will even scour the deposited particles of a given size, must be developed in the sewers, at least once a day, so as not to allow any deposition in the sewers. The generation of such a *minimum self-cleansing velocity in the sewer, atleast once a day*, is important, because if certain deposition takes place and is not removed, it will obstruct free flow, causing further deposition and finally leading to the complete blocking of the sewer.

Shield's expression for self-cleansing velocity. Self cleansing velocity, which is necessary to cause scouring and suspension of solid particles (heavier than water), can be determined as follows :

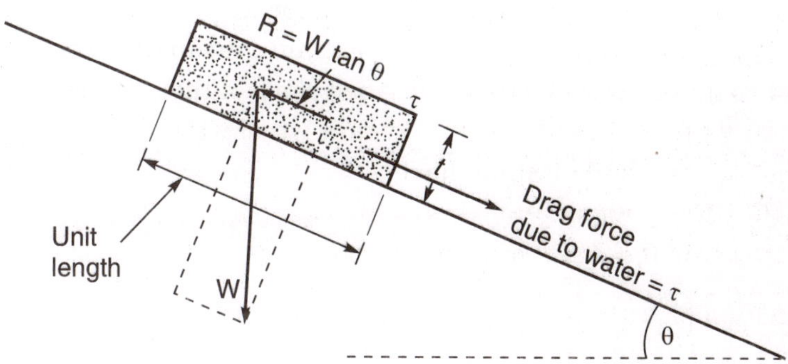

Fig. 4.1

Consider a layer of sediment of unit width and unit length and of thickness t deposited at the invert of a sewer of gradient θ. Let γ_{sub} is the submerged unit weight of the sediment.

Then, the weight of the sediment considered

$$= W = \gamma_{sub}\,(1)(1)t$$

But $$\gamma_{sub} = \gamma_w \left(\frac{G-1}{1+e}\right)^*$$

where γ_w = unit weight of the water.
G = specific gravity of the sediment.
e = void ratio of sediment.

But the porosity of sediment

$$n = \left(\frac{e}{1+e}\right)^*, \text{ where } n = \text{porosity of sediment}$$

$$\therefore \quad 1 - n = \frac{1}{1+e}$$

$$\therefore \quad \gamma_{sub} = \gamma_w (G-1)(1-n)$$

$$\therefore \quad W = \gamma_w (G-1)(1-n)t \qquad ...(i)$$

Now, in order to scour the deposited sediment, and for just causing it to slide down the inclined plane, it is necessary that the drag force (τ) exerted by the flowing water on the surface of the channel equals the frictional resistance R, *i.e.* $\tau = R$;

But $R = W \tan \theta$

and for smaller values of θ, $\tan \theta = \sin \theta$

$$\therefore \quad R = W \sin \theta$$

or $$\tau = R = W \sin \theta \qquad ...(ii)$$

or $$\tau = \gamma_w (G-1)(1-n)t \sin \theta \qquad ...(4.7)$$

But we know that the drag force or the intensity of tractive force (τ) which is exerted by the flowing water on a channel of hydraulic mean depth r is given by

$$\tau = \gamma_w\, r.s^{**} \qquad ...(4.8)$$

where τ = drag force
r = hydraulic mean depth of the channel
s = bed slope of the channel.
γ_w = unit wt. of water.

Equating Eq. (4.7) and (4.8), we have

$$\gamma_w (G-1)(1-n)t \sin \theta = \gamma_w\, r.s.$$

Using $(1-n) \sin \theta$, $= k'$ (a constant), we get

$$(G-1)\, k'.t. = r.s.$$

or $$s = \frac{k'}{r}(G-1)t.$$

For single grains, the volume per unit area (*i.e.* t) becomes a function of the diameter of the grain d' as an inverse measure of the surface area of the individual grains exposed to drag or friction.

---

*Please see equation (4.17) and (4.6) in "*Soil Mechanics and Foundation Engineering*" by the same author.

**Pl. refer eq. (4.1) in "*Irrigation Engg. and Hydraulic Structures*", by the same author.

∴ $$s \propto \frac{k'}{r}(G-1)\,d' \quad \text{(for self cleansing)}$$

or $$s = \frac{k}{r}(G-1)\,d'$$

Hence, the self **cleansing invert slope(s)** is given as:

$$\boxed{s = \frac{k}{r}\,(G-1)\,d'} \qquad \text{...(4.9)}$$

where k is a dimensional constant, indicating an important characteristic of sediment (solids) present in sewage. Its value usually varies from 0.04 (minimum)- applicable to start of scouring of clean grit, to about 0.8 – applicable for full removal of sticky grit. For relatively clean inorganic and organic matters present in sewage, its values are taken at 0.04 and 0.06, respectively. The actual value of *k*, should, however be determined only by experiments for removal of different materials.

Hence, the invert slope at which the sewer will be self-cleansing is given by the Eq. (4.9).

Now, from Chezy's formula, the velocity

$$V = c\sqrt{rs}$$

∴ **Self-cleansing velocity** *(V_s)* is, hence, given as :

$$V_s = c.\,\sqrt{r}\sqrt{\frac{k}{r}(G-1)\,d'} \quad \text{or} \quad V_s = c.\,\sqrt{kd'(G-1)} \qquad \text{...(4.10)}$$

The Chezy's constant (*c*) in the above equation can be equated to $\sqrt{\frac{8g}{f'}}$ by comparing Chezy's formula and Darcy-Weisbach formula*. Therefore, Eq. (4.9) becomes,

---

*From Darcy-Weisbach formula,

$$H_L = \frac{f'LV^2}{2gD} \qquad \text{(where } D \text{ is the pipe dia)}$$

or $$\frac{H_L}{L} = s = \frac{f'V^2}{2gD} \quad \text{or} \quad V = \sqrt{\frac{2gD.s}{f'}}; \qquad \text{...}(i)$$

By Chezy's formula,

$$V = c\sqrt{rs} \qquad \text{...}(ii)$$

Comparing (*i*) and (*ii*), we get

or $$c\sqrt{r} = \sqrt{\frac{2gD}{f'}}$$

But $r = \frac{D}{4}$ (for circular pipes running full)

∴ $$c.\sqrt{\frac{D}{4}} = \sqrt{\frac{2gD}{f'}} \quad \text{or} \quad c = \sqrt{\frac{8g}{f'}}$$

Self cleansing velocity

$$\boxed{V_s = \sqrt{\frac{8g}{f'} kd'(G-1)}} \quad ...(4.11)$$

The usual value of f' for sewer pipes is 0.03.

Similarly, by equating Chezy's formula with Manning's formula $\left(i.e., V = \frac{1}{n} . r^{2/3} \sqrt{s}\right)$, we can get $c = \frac{1}{n} . r^{1/6}$. The Eq. (4.10) then becomes

$$\boxed{V_s = \frac{1}{n} . r^{1/6} \sqrt{k.d'(G-1)}} \quad ...(4.12)$$

(The usual value of n for sewer pipes is 0.013.)

From the above equations, it can be computed that the inorganic sand particles of diameter 1 mm and specific gravity 2.65, can be removed with a minimum velocity of about 0.45 m/sec, and similarly for removing organic particles of 5 mm diameter and sp. gravity 1.2, a minimum velocity of about 0.45 m/sec is required to be generated. *Hence for removing the impurities mostly present in sewage (i.e. sand upto 1 mm diameter and organic particles upto 5 mm dia), it is necessary that a minimum velocity of about 0.45 m/sec and an average velocity of about 0.9 m/sec is developed in sewers.* With this provision, the inorganic solids larger than 1 mm in size and organic solids larger than 5 mm in size, will of course settle down on the invert of the sewer pipe, and will have to be removed by the maintenance crew during regular desilting operations, to be carried out at suitable intervals.

Hence, while finalising the sizes and gradients of the sewers, they must be checked for the minimum velocity that would be generated at minimum hourly discharge (*i.e.* about $\frac{1}{3}$ rd of the average daily discharge).

Such a self-cleansing velocity (0.45 m/sec) at minimum discharge can also be ensured by keeping the minimum design velocity to a value as high as 0.8 m/sec.

Hence, while designing sewers, the flow velocity at full design depth, is generally kept at about 0.8 m/sec or so. And since sewers are generally designed for $\frac{1}{2}$ — $\frac{3}{4}$th full, the velocity at 'running full condition' will even be more than that ; say about 1 m/sec, or so.

The minimum velocity generated in the sewers will not only ensure the adequate transportation of the suspended solids, but will also help in

(*i*) keeping the sewer size under control ;

(*ii*) preventing the sewage from getting stale and decomposed by moving it faster, thereby preventing evolution of foul gases.

Since the velocity of flow developed in a sewer of a particular material depends only upon the hydraulic mean depth of the sewer and the slope on which the sewer has been laid*, values of slopes which are required for generating self-cleansing velocities (0.8 to 1 m/s) in different sized pipes can be easily worked out by using any of the equations from 4.1 to 4.5.

The values of gradients required to generate velocities of 0.75, 0.90 and 1.05 m/sec. for different dia. pipes running full, have been worked out and shown in Table 4.5.

Table 4.5. Sewer Gradient required to generate Self-cleansing Velocities in different sized Pipes (running full)

Dia. of sewer mm	*Gradients 1 in......for developing velocity of*		
	0.75 m/sec	*0.90 m/sec*	*1.05 m/sec*
100	90	75	60
150	150	105	78
225	265	180	135
300	385	270	195
375	520	355	265
450	660	460	340
525	820	570	415
600	970	680	500
675	1,100	790	580
750	1,300	910	670
900	1,700	1,200	850
1,050	2,100	1,450	1,050
1,200	2,500	1,700	1,250

National Building Organisation (N.B.O.) of India has suggested the following gradients as being sufficient for preventing interference of sewage solids with the flow, for small sized sewers :

Table 4.6. N.B.O. Recommendations for Small Sewers

Dia. of the sewer in mm (1)	*Gradient required to generate self cleansing velocity* (2)	*Velocity generated in the sewer when running half full, for which depth, small sewers are usually designed* (3)
100	1 in 60	0.58 m/sec**
150	1 in 100	0.61 m/sec**
225	1 in 120	0.79 m/sec**

4.5.2. Maximum Velocities. The smooth interior surface of a sewer pipe gets scoured due to the continuous abrasion caused by the suspended solids present in sewage. This scouring and wear and tear of the pipe interior is

*Provided the increase in the value of n (Manning's rugosity coefficient) at reduced depths, are ignored.

**These values have been worked out by considering the variations in value of n at lower depths.

much more pronounced at velocities higher than what can be tolerated by the pipe materials. *This wear and tear of the sewer pipes will not only reduce their life spans but will also reduce their carrying capacities.* In order to avoid these complications, it is, therefore, necessary to limit the maximum velocity that will be produced in the sewer pipe at any time.

This limiting or non-scouring velocity will mainly depend upon the material of the sewer, and its values are given in Table 4.7 for different commonly used sewer materials.

Table 4.7. Non-scouring Limiting Velocities in Sewers and Drains

S. No.	*Sewer Material*	*Limiting velocity in m/sec*
1.	Vitrified tiles and glazed bricks*	4.5—5.5
2.	Cast iron sewers	3.5—4.5
3.	Stone ware sewers	3.0—4.0
4.	Cement concrete sewers	2.5—3.0
5.	Ordinary brick-lined sewers	1.5—2.5
6.	Earthen channels	0.6—1.2

The problem of controlling the high velocities generated in the sewers, mainly arises in hilly areas, where the available ground slopes may be very steep. In such places, it may, therefore, be required to limit the design gradient and to adjust the balance available head by constructing drops through manholes, at suitable places along the length of the sewer.

4.6. Effects of Flow Variation on Velocity in a Sewer

The sewage discharge flowing through a sewer does not remain constant all the time, but varies considerably from time to time. Due to this variation in discharge, the depth of flow varies, and hence the hydraulic mean depth (H.M.D. *i.e.* r) varies. Due to the change in hydraulic mean depth, the flow velocity [which directly depends on $(\text{H.M.D.})^{2/3}$] gets affected from time to time.

Since the velocity developed in a sewer of a given section and grade will be less as and when the flow reduces (and the sewer becomes less than half full**), it is necessary to check the sewer for maintaining a minimum velocity of about 0.45 m/sec at the time of minimum (hourly) flow (assumed at about $\frac{1}{3}$rd of average (daily) flow). The designer should also ensure that a velocity of about 0.9 m/sec is developed at least at the time of the maximum flow (assumed as given by Table 2.3) and preferably during the average flow periods also. Moreover, care should be taken to see that at the time of maximum flow, the velocity generated does not exceed the scouring value

**Vitrified tiles and glazed bricks are much more resistant to wear as compared to ordinary bricks or concrete. Therefore, the ordinary brick or concrete sewers are sometimes coated with vitrified tile liners or glazed bricks at their bottoms (where the abrasion is maximum, because sand, grit and gravel are heavy, and travel along the bottom of the sewers).

**Explained in the next article.

given by Table 4.7. However, in flat countries, like indo-Gangetic planes of North India, sewers are designed to develop self cleansing velocities of about 0.9 m/sec at maximum discharge only, and the condition of developing such a velocity at average flow is waived off. This permits flatter gradients, avoiding deep excavations.

On the other hand, if the available ground slopes are neither too steep nor too flat, the condition of developing velocities of about 0.9 m/sec. at average flow may be practically possible and economical, and hence may be insisted upon. In hilly areas permitting too high slopes, the sewers may be designed to develop such velocities of 0.9 m/sec even at minimum discharge, and they may be checked for limiting non-scouring velocities at maximum discharge.

4.7. Hydraulic Characteristics of Circular Sewer Sections Running Full or Partially Full

The circular section is most widely adopted for sewer pipes. They may, however, sometimes be of 'egg shape' or 'horse shoe shape' or 'rectangular shape'. The circular sewers may sometimes run full or may run partially full. When they run full, their hydraulic properties will be as given below :

Area of cross-section

$$(a) = A = \frac{\pi}{4} D^2$$

where D is the dia. of the pipe.

Wetted perimeter

$$(p) = P = \pi D$$

$\therefore$ Hydraulic mean depth

$$r = R = \frac{A}{P} = \frac{\frac{\pi}{4} D^2}{\pi D} = \frac{D}{4}.$$

When the sewers run partially full, at a depth, say d, as shown in Fig. 4.2, the hydraulic elements can be worked out as given below :

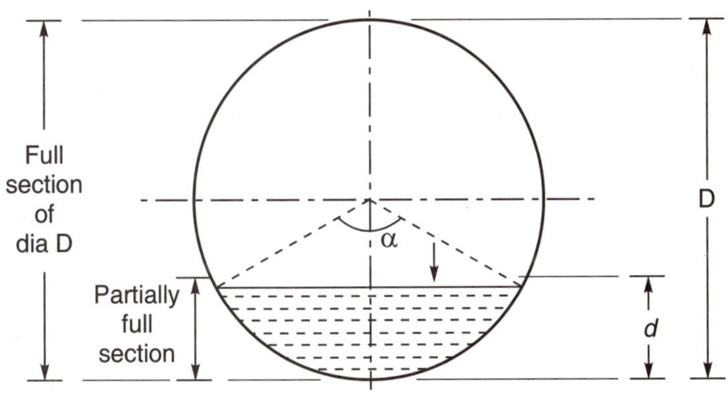

Fig. 4.2. Partially full circular sewer section.

The depth at partial flow

$$= d = \left[\frac{D}{2} - \frac{D}{2} \cos \frac{\alpha}{2}\right]$$

where α is the central angle in degrees, as shown in Fig. 4.2.

∴ Proportionate depth

$$= \frac{d}{D}$$

$$= \frac{1}{2}\left(1 - \cos\frac{\alpha}{2}\right) \quad ...(4.13)$$

Area of cross-section while running partially full

$$= a = \frac{\pi D^2}{4} \cdot \frac{\alpha}{360°} - \frac{D}{2}\cos\frac{\alpha}{2} \cdot \frac{D}{2}\sin\frac{\alpha}{2}$$

$$= \frac{\pi D^2}{4}\left[\frac{\alpha}{360°} - \frac{\sin\alpha}{2\pi}\right] \quad ...(4.14)$$

$$\left[\because \quad \sin\alpha = 2\sin\frac{\alpha}{2} \cdot \cos\frac{\alpha}{2}\right]$$

∴ Proportionate area

$$= \frac{a}{A} = \left[\frac{\alpha}{360°} - \frac{\sin\alpha}{2\pi}\right] \quad ...(4.15)$$

Wetted perimeter, while running partially full

$$= p = \pi D \cdot \frac{\alpha}{360°} \quad ...(4.16)$$

∴ Proportionate perimeter

$$= \frac{p}{P} = \frac{\alpha}{360°} \quad ...(4.17)$$

Hydraulic mean depth (H.M.D.), while running partially full

$$= r = \frac{a}{p}$$

$$= \frac{D}{4}\left[1 - \frac{360° \sin\alpha}{2\pi\alpha}\right] \quad ...(4.18)$$

∴ Proportionate H.M.D.

$$= \frac{r}{R} = \left[1 - \frac{360° \sin\alpha}{2\pi\alpha}\right] \quad ...(4.19)$$

Velocity of flow is given by Manning's formula, as

v = velocity at partial flow

$$= \frac{1}{n} r^{2/3} \sqrt{S_0} \qquad [\because \quad s = S_0, \text{ i.e., bed slope}]$$

∴ V = velocity, when running full

$$= \frac{1}{N} \cdot R^{2/3} \sqrt{S_0}$$

(Bed slope $s = S_0$ remaining constant whether pipe runs full or partially full)

∴ Proportionate velocity

$$= \frac{v}{V} = \frac{N}{n} \cdot \frac{r^{2/3}}{R^{2/3}} \quad ...(4.20)$$

Assuming that roughness coefficient n does not vary with depth, we have

$$n = N.$$

∴ Proportionate velocity

$$= \frac{v}{V} = \frac{r^{2/3}}{R^{2/3}}$$

$$= \left[1 - \frac{360° \sin\alpha}{2\pi\alpha}\right]^{2/3} \quad ...(4.21)$$

Since, discharge is given by $a.v$, therefore,

Discharge when pipe is running partially full

$$= q = av \quad ...(4.22)$$

Discharge when pipe is running full

$$= Q = A.V \quad ...(4.23)$$

∴ Proportionate discharge

$$= \frac{q}{Q} = \frac{av}{A.V} = \frac{a}{A}\cdot\frac{v}{V}$$

$$= \left[\frac{\alpha}{360°} - \frac{\sin\alpha}{2\pi}\right]\left[1 - \frac{360° \sin\alpha}{2\pi\alpha}\right] \quad ...(4.24)$$

In all the above derived equations, except α, every thing is constant, and hence by giving different values to α, all the six proportionate elements can be easily calculated.

By taking proportionate depth (d/D) as reference, values of other elements can be found out from the ready made computed values, shown in Table 4.8.

Table 4.8. Proportionate Values of Hydraulic Elements for Circular Sewers when Flowing Partially Full (without being corrected for variations of roughness with depth)

Proportionate Depth d/D (1)	*Proportionate area* a/A (2)	*Proportionate Wetted perimeter* p/P (3)	*Proportionate H.M.D.* r/R (4)	*Proportionate Velocity* v/V (5)	*Proportionate Discharge* q/Q (6)
1.00	1.00	1.00	1.000	1.000	1.000
0.90	0.949	0.857	1.192	1.124	1.066
0.80	0.858	0.705	1.217	1.140	0.988
0.70	0.748	0.631	1.185	1.120	0.838
0.60	0.626	0.564	1.110	1.072	0.671
0.50	0.500	0.500	1.000	1.000	0.500
0.40	0.373	0.444	0.857	0.902	0.337
0.30	0.252	0.369	0.684	0.776	0.196
0.20	0.143	0.296	0.482	0.615	0.088
0.10	0.052	0.205	0.254	0.401	0.021
0.00	0.000	0.000	0.000	0.000	0.000

From the data given in Table 4.8, it can be seen that the velocities in partially filled circular sewer sections, equal or exceed those in full sections, so long as sewers flow more than half full ; and the maximum velocity is obtained not when the sewer is running full but when the depth of flow is 0.81 times the full depth, and is about 14% greater that when running full.

Similarly, the maximum discharge is obtained not when the sewer is running full, but when the depth is about 0.95 times the full depth, and is about 7% greater than that when running full. But, as the depth of flow goes below half the full depth, the velocities and discharges, both decline, and become lesser than those at full flow. However, the decline in proportionate velocities is not so sharp, as is the decline in discharges, because the area (on which depends discharge) reduces much faster as compared to the hydraulic mean depth (on which depends velocity).

The statements given above are precisely correct only, so long as the roughness (n) is supposed to be independent of depth. However, Sundin has demonstrated theoretically as well as experimentally that the value of rugosity coefficient (n) is not constant, but varies as much as 20% or more with depth, as shown in col. (2) of Table 4.9.

Table 4.9. Hydraulic Particulars of Circular Sewers, accounting Variations of n with depth

Proportionate depth d/D (1)	*Proportionate roughness* n/N (2)	*Proportionate velocity* v/V (3)	*Proportionate discharge* q/Q (4)
1.0	1.00	1.000	1.000
0.9	1.07	1.056	1.020
0.8	1.14	1.003	0.890
0.7	1.18	0.952	0.712
0.6	1.21	0.890	0.557
0.5	1.24	0.810	0.405
0.4	1.27	0.713	0.266
0.3	1.28	0.605	0.153
0.2	1.27	0.486	0.070
0.1	1.22	0.329	0.017

The effect of the variation of n is to reduce the proportionate velocities and discharges at lower depths of flow, *because roughness (n) increases with lower depths*. If these variations of n are also considered, more precise values of proportionate velocities and discharges can be computed out, as shown in Col. (3) and (4) of table 4.9.

These precise proportionate velocities and discharges (Table 4.9), have been plotted, as shown by firm lines in Fig. 4.3, to obtain a standard chart, which is very useful in obtaining different elements, by knowing any one of them. Values of table 4.8 are also plotted by dotted curves (Fig. 4.3), as to obtain proportionate elements, when variations of n are ignored.

By a perusal of Fig. 4.3, *it is found that when roughness is also considered, the velocities equal to or more than those produced in sections flowing full, will be produced so long as the circular sewer sections flow above 80% of depth, as against half the depth (i.e. 50%) when roughness is not considered.*

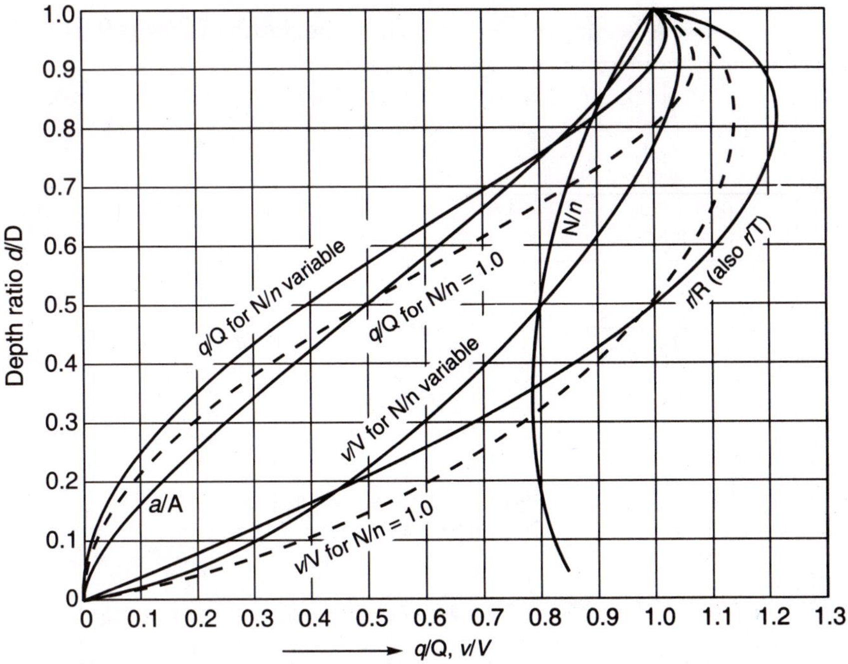

Fig. 4.3. Standard chart for proportionate hydraulic elements for circular sewers.

Nevertheless, sewers flowing with depths between 50% and 80% full, need not be placed on steeper gradients to be as self-cleansing as sewers flowing full. The reason is that velocity and discharge are functions of tractive force intensity, which depends upon the friction coefficient as well as flow velocity. Needed ratios of v_s/V, q_s/Q and s_s/S, where the subscript s denotes self-cleansing equal to that obtained in the full section, can be computed with the help of equation (4.8) on the assumption that equality of tractive force intensity implies equality of cleaning or $\tau = T$.

$$\therefore \quad \gamma_w \, r.s. = \gamma_w \, R.S \, ;$$

where $s = s_s$

or $$s_s = \left(\frac{R}{r}\right) S \qquad \text{...(4.25)}$$

and $$\frac{v_s}{V} = \frac{N}{n}\left(\frac{r}{R}\right)^{2/3} \sqrt{\frac{s_s}{S}}$$

or $$\frac{v_s}{V} = \frac{N}{n} \cdot \left(\frac{r}{R}\right)^{1/6} \qquad \text{...(4.26)}$$

and $$\frac{q_s}{Q} = \frac{N}{n} \cdot \left(\frac{a}{A}\right)\left(\frac{r}{R}\right)^{1/6} \qquad \text{...(4.27)}$$

Various curves for values of $\frac{q_s}{Q}$, $\frac{v_s}{V}$ and $\frac{s_s}{S}$ for different values of $\frac{d}{D}$ are plotted, as shown in Fig. 4.4. *It is confirmed from these figures that the minimum gradients are enough, so long as circular sewers flow more than half full. However, when flow depth reduces to less than 0.3 times the full depth, the*

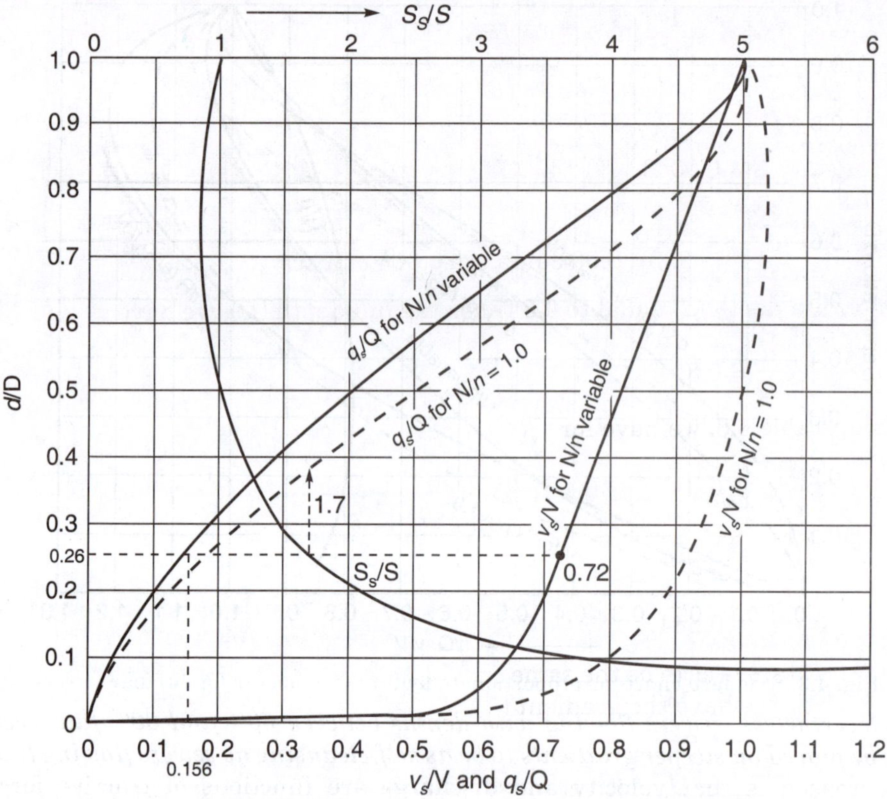

Fig. 4.4. Standard chart for proportionate elements, to ensure self-cleansing equivalent to full flow, in circular sewers.

grades must be increased. So much so, that at a depth of 0.2 times the full depth, the gradient must be doubled, and at a depth of 0.1 times the full depth, the gradient must be trebled. The use of these equations will become more clear when we solve the following examples.

Example 4.1. *A 300 mm dia sewer is to flow at 0.3 depth on a grade ensuring a degree of self-cleansing equivalent to that obtained at full depth at a velocity of 0.90 m/sec. Find the required grade and associated velocity and rate of discharge at this depth. Assume Manning's rugosity coefficient n as 0.013. The variations of n with depth may be neglected.*

Solution. From Manning's formula, we have

$$v = \frac{1}{n} \cdot r^{2/3}\sqrt{s}$$

At full depth, using capital letters, we have

$$V = \frac{1}{N} \cdot R^{2/3}\sqrt{S}$$

Using $V = 0.90$ m/sec.

$N = 0.013$

$$R = \frac{D}{4} = \frac{300}{4} = 75 \text{ mm} = 0.075 \text{ m}$$

We have $0.90 = \frac{1}{0.013}(0.075)^{2/3}\sqrt{S}$

or $$\sqrt{S} = \frac{0.90 \times 0.013}{0.178} = 0.0657$$

or $$S = 0.0043 \quad (i.e.\ 4.3‰)^*$$

and $$Q = A.V.$$

$$= \frac{\pi}{4}(0.3)^2\ 0.90 \text{ cumecs} = \mathbf{0.064} \text{ cumecs}$$

Now, at a depth (d) equal to 0.3 times the full depth (D), we have

$$\frac{d}{D} = 0.3,$$

Using Table 4.8, we have for

$$\frac{d}{D} = 0.3$$ *(Variations of n to be neglected, as given)*

$$\frac{a}{A} = 0.252$$

$$\frac{r}{R} = 0.684$$

Now for the sewer to be the same self-cleansing at 0.3 depth (d), as it will be at full depth, we have the gradient (s_s) required from Eq. (4.25), as ;

$$s_s = \left(\frac{R}{r}\right)S$$

$$\therefore \quad s_s = \frac{1}{0.684}S = \frac{1}{0.684} \times 0.0043$$

$$= 0.0063 \quad (i.e.\ 6.3‰).$$ **Ans.**

Now, the velocity v_s generated at this gradient is given by Eq. (4.26), as

$$v_s = \frac{N}{n} \cdot \left(\frac{r}{R}\right)^{1/6} . V$$

$$= 1 \times (0.684)^{1/6} \times 0.9 \text{ m/sec} = 0.939 \times 0.9$$

$$= 0.846 \text{ m/sec.}$$ **Ans.**

The discharge q_s is given by Eq. (4.27) as

$$q_s = \frac{N}{n}\left(\frac{a}{A}\right)\left(\frac{r}{R}\right)^{1/6} . Q$$

$$\therefore \quad q_s = (1)(0.252)(0.684)^{1/6}(0.064) \text{ cumecs}$$

$$= 0.015 \text{ cumecs.}$$ **Ans.**

Example 4.2. *A 225 mm dia sewer is to discharge 0.005 cumecs at a velocity as self-cleansing as a sewer flowing full at 0.80 m/sec. Find the depth, velocity generated, and the required gradient. Use Manning's rugosity coefficient as 0.013. Also assume that the rugosity coefficient varies with depth. Standard chart of Fig. 4.4 can be used to compute the values of proportionate elements of self cleaning equivalent of full flow.*

---

*‰ is per thousand, just as % is per hundred.

Solution. Using Manning's formula, when sewer runs full at a velocity of 0.8 m/sec, we have

$$V = \frac{1}{N} \cdot R^{2/3} \sqrt{S}$$

$$\therefore \quad 0.8 = \frac{1}{0.013} \cdot \left(\frac{0.225}{4}\right)^{2/3} \sqrt{S}$$

$$\text{or} \quad \sqrt{S} = \frac{0.8 \times 0.013}{0.146} = 0.0712$$

$$\therefore \quad S = 0.0051 \quad (i.e.\ 5.1‰).$$

Now $Q = A.V.$

$$= \frac{\pi}{4} \cdot (0.225)^2 \times 0.8 \text{ cumecs.}$$

$$= 0.0397 \times 0.8$$

$$= 0.032 \text{ cumecs}$$

But $q_s = 0.005$ cumecs (given)

$$\therefore \quad \frac{q_s}{Q} = \frac{0.005}{0.032} = 0.156.$$

Now, from Fig. 4.4 for $\frac{N}{n}$ variable and $\frac{q_s}{Q} = 0.156$, we have

$$\frac{d_s}{D} = 0.26$$

$$\frac{v_s}{V} = 0.72$$

and

$$\frac{s_s}{S} = 1.7.$$

Hence, d_s = depth at equivalent self-cleansing

= 0.26 times full depth

= 0.26 × 0.225 m. = 0.058 m = 58 mm. **Ans.**

v_s = velocity at equivalent self-cleansing

= 0.72 V = 0.72 × 0.8 m/sec.

= 0.58 m/sec. **Ans.**

and v_s = slope at equivalent self-cleansing

= 1.7 × 5.1‰ = 8.6‰. **Ans.**

Example 4.3. *A 350 mm dia sewer is to flow at 0.35 depth on a grade ensuring a degree of self-cleansing equivalent to that obtained at full depth at a velocity of 0.8 m / sec. Find :*

(i) the required grade

(ii) associated velocity

(iii) the rate of discharge at this depth.

Given :

(i) Manning's rugosity coefficient = 0.014

(ii) Proportionate area = 0.315

(iii) Proportionate wetted perimeter = 0.472

(iv) Proportionate HMD (r/R) = 0.7705. *(Civil Services, 1984)*

Solution. At full depth, $V = 0.8$ m/sec, $D = 350$ mm $= 0.35$ m, $N = 0.014$.

At 0.35 depth, $\frac{d}{D} = 0.35$, $\frac{a}{A} = 0.315$, $\frac{p}{P} = 0.472$, $\frac{r}{R} = 0.7705$.

$\therefore$ *At full depth*

$$V = \frac{1}{N} \cdot R^{2/3} \cdot \sqrt{S}$$

or
$$0.8 = \frac{1}{0.014} \cdot \left(\frac{0.35}{4}\right)^{2/3} \cdot \sqrt{S}$$

or
$$\sqrt{S} = 0.0568.$$

or
$$S = 3.234 \times 10^{-3}$$

Now, for a sewer to be the same self-cleansing at 0.35 depth as it will be at full depth, we have the gradient (s_s) required from equation (4.25) as

$$s_s = \left(\frac{R}{r}\right) S$$

$$= \frac{1}{0.7705} \times 3.234 \times 10^{-3} \qquad \left[\because \frac{r}{R} = 0.7705 \text{ given}\right]$$

$$= 4.2 \times 10^{-3}$$

i.e.
$$= 4.2‰.$$

(*i*) Hence, the reqd. grade = **4.2‰. Ans.**

(*ii*) The velocity generated at this gradient at 0.35 depth, is given by equation (4.26), as

$$v_s = \frac{N}{n} \cdot \left(\frac{r}{R}\right)^{1/6} \cdot V$$

$$= 1 \times (0.7705)^{1/6} \cdot 0.8 = \mathbf{0.765 \text{ m/sec.}} \quad \textbf{Ans.}$$

(*iii*) The discharge q_s, is then given by

$$q_s = a.v_s$$

$$= 0.315 \cdot \frac{\pi}{4} \times (0.35)^2 \times 0.765$$

$$= \mathbf{0.023 \text{ cumecs.}} \quad \textbf{Ans.}$$

Example 4.4. (*a*) *A main combined sewer was designed to serve an area of 60 sq. km with an average population of 185 persons/hectare. The average rate of sewage flow is 350 litres/capita/day. The maximum flow is 50% in excess of the average together with the rainfall equivalent of 12 mm in 24 hours, all of which are run off. What should be the capacity of the sewer in cu. m/sec. ?*

(*b*) *Find the minimum velocity and gradient required to transport coarse sand through a sewer of 40 cm dia with sand particles of 1 mm dia and specific gravity 2.65. Assume k for sand = 0.04. The Manning's roughness coefficient(n) for the sewer material may be assumed as 0.012.*

Solution. Total population of the area

= Population density × area

= 185 p/ha × (60×10^2) ha

= 1110×10^3 persons

= 11.1×10^5 persons

Average sewage flow = 350 litres/capita/day

= $350 \times 11.1 \times 10^5$ litres/day

= 388.5×10^6 litres/day

$$= \frac{388.5 \times 10^6}{10^3} \times \frac{1}{24 \times 60 \times 60} \text{ cu.m/sec}$$

= 4.5 cumecs.

$$\text{Storm water flow} = 60 \times 10^6 \times \left(\frac{12}{1000}\right) \frac{1}{24 \times 60 \times 60} \text{ cumecs.}$$

$$\left[i.e. \frac{\text{area in m}^2 \times \text{depth of S.R.O. in m}}{\text{Time of 24 hr. in sec.}} \right]$$

= **8.33 cumecs.**

Maximum sewage flow

= 1.5 × average sewage flow

= 1.5 × 4.5 cumecs = **6.75 cumecs.**

∴ Total maximum flow of the combined sewer

= Max. sewage flow + storm flow

= 6.75 + 8.33 = 15.08 cumecs.

Hence, the capacity of the sewer

= **15.08 cumec. Ans.**

(*b*) Now, the minimum velocity (*i.e.* self cleansing) is given by equation (4.11) as :

$$V_s = \frac{1}{n} . r^{\frac{1}{6}} \sqrt{kd'(G-1)}$$

where, $n = 0.012$, $k = 0.04$, $G = 2.65$

d' = dia of grain = 1 mm

$$r = R = \frac{D}{4} = \frac{0.4 \text{ m}}{4} = 0.1 \text{ m}$$

$$\therefore \quad V_s = \frac{1}{0.012} \times (0.1)^{1/6} \sqrt{0.04 \times \frac{1}{10 \times 100} \times (2.65 - 1)}$$

= $56.77 \times (8.12 \times 10^{-3})$ m/s = 0.46 m/s

∴ Min. velocity required for the sewer to be self cleansing

= **0.46 m/sec. Ans.**

The velocity in sewer, however, is calculated by Manning's formula as :

$$v = \frac{1}{n} r^{2/3} . s^{1/2} \qquad (i.e., \quad \text{Eq. 4.4})$$

Now, assuming the sewer is designed to be running full, we have

$$n = N = 0.012$$

$$r = R = \frac{D}{4} = 0.1 \text{ m}$$

$$\therefore \quad 0.46 = \frac{1}{0.012} (0.1)^{2/3} . S^{1/2}$$

or

$$S = \frac{1}{1523}$$

Hence, the required gradient for the sewer to be self cleansing is 1 in 1523. **Ans.**

4.8. Use of Tables and Nomograms for Hydraulic Computations for the Design of Sewers

The various formulas described earlier for computing the sizes and gradients of sewers required for developing requisite velocities at different discharges, involve a lot of mathematical calculations.

This mathematical work becomes enormous while designing a full fledged sewerage scheme, where such calculations are required to be made for thousands of sewers. To reduce the calculation work, ready- made charts and tables, based upon their original formulas, are generally prepared, kept in the design office, and used.

Fig. 4.5 shows a nomogram based on Manning's formula, for sewers running full, using a value of N as 0.013. This nomogram can be easily used by placing a straight edge across any two known variables (such as velocity and discharge), as to read out the remaining two variables (such as dia and gradient). If the value of N, to be used, is different from 0.013 (say for example it is 0.017) ; then the velocity or discharge obtained from this chart for a value of $N = 0.013$, should be multiplied by a ratio $\frac{0.013^*}{0.017}\left(i.e., \frac{0.013}{N}\right)$ to obtain the real velocity or discharge produced at $N = 0.017$. Similarly, if the discharge or velocity is given, the required parameters can be found by first correcting the given discharge or velocity by multiplying them by $\frac{0.017}{0.013}\left(i.e., \frac{N}{0.013}\right)$ and then placing the straight edge across the two known values.

Santo Crimps tables are the readymade tables based upon the Crimp and Burge's formula (given by Eq. 4.5), and are quite often used in India. A sample page of such tables is shown in table 4.10.

---

*Because at a higher value of N, the velocity and discharge will be less, as the velocity and discharge are inversely proportional to N.

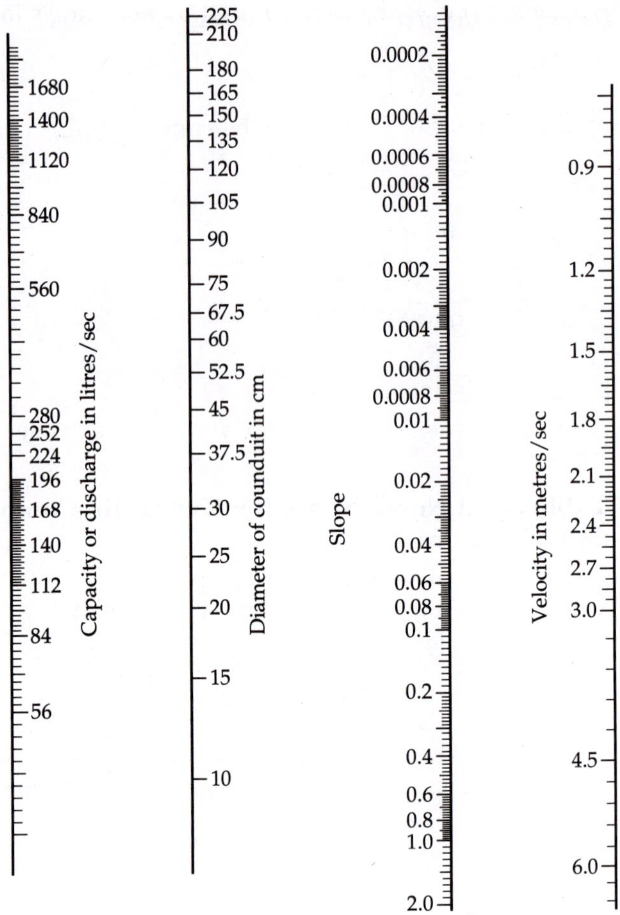

Fig. 4.5. Nomogram based on Manning's formula for $N = 0.013$ for sewers running full.

Table 4.10. Sample Page of Santo-Crimps Tables (for Sewers Running Full)

Sewer Dia in mm	*Grade = 1 in 225*		*Grade = 1 in 100*	
	Velocity in m/sec.	*Discharge in litres/sec.*	*Velocity in m/sec.*	*Discharge in litres/sec.*
100	0.479	3.76	0.560	4.40
150	0.628	11.12	0.942	16.67
200	0.760	23.90	1.140	35.83
250	0.833	43.34	1.324	65.01
300	0.996	70.52	1.494	105.80
375	1.156	127.70	1.735	191.70
450	1.306	208.00	1.960	312.30
525	1.447	313.30	2.170	470.10
600	1.582	448.10	2.373	672.00
675	1.711	611.20	2.566	918.50
750	1.837	812.50	2.754	1219.00
900	2.073	1320.00	3.200	2038.00

Example 4.5. *Determine the size of a circular sewer for a discharge of 600 lps running half-full. Assume i = 0.0001 and n = 0.015.*

Solution. $d = 0.5\ D$

$$\frac{d}{D} = 0.5$$

$$q = 600 \text{ litres/sec.}$$
$$= 0.6 \text{ cumecs}$$
$$S = i = 0.0001$$
$$n = 0.015$$

From table 4.8, at $\frac{d}{D} = 0.5$

we have $\frac{q}{Q} = 0.5$

$$\therefore \quad Q = \frac{q}{0.5} = \frac{0.6}{0.5} = 1.2 \text{ cumecs.}$$

Assuming n not to vary with depth, we have by using Manning's formula,

$$Q = \frac{1}{N} \,.\, AR^{2/3} \,.\, \sqrt{S}$$

or $$1.2 = \frac{1}{0.015}\left(\frac{\pi}{4} . D^2\right)\left(\frac{D}{4}\right)^{2/3} . \sqrt{0.0001}$$

or $$\frac{1.2 \times 0.015 \times 4 \times 2.52 \times 100}{\pi} = D^{8/3}$$

or $$D^{8/3} = 5.78$$

or $$D = (5.78)^{3/8} = (5.78)^{0.375} = 1.93 \text{ m}$$

Hence, the diameter of the sewer required

$= 1.93$ m. **Ans.**

Alternatively when table 4.8 is not available, we may solve the problem as follows :

We know that $$\frac{d}{D} = 0.5$$
$$= \frac{1}{2}\left(1 - \cos\frac{\alpha}{2}\right) \qquad \text{... Refer Eq. (4.13)}$$

or $$\cos\frac{\alpha}{2} = 0$$

or $$\frac{\alpha}{2} = 90°$$

or $$\alpha = 180°.$$

Now, using Eq. (4.14), we have

$$a = \frac{D^2}{4}\left[\frac{\pi\alpha}{360°} - \frac{\sin\alpha}{2}\right]$$

$$= \frac{D^2}{4}\left[\frac{\pi \,.\, 180^\circ}{360^\circ} - \frac{\sin 180^\circ}{2}\right]$$

$$= \frac{D^2}{4}\left[\frac{\pi}{2} - 0\right] = \frac{\pi D^2}{8}.$$

Now, using Eq. (4.16), we have

$$p = \pi D \,\frac{\alpha}{360^\circ}$$

$$= \pi D \,\frac{180^\circ}{360^\circ} = \frac{\pi D}{2}.$$

$$\therefore \qquad r = \frac{a}{p} = \frac{\pi D^2}{8}\left(\frac{1}{\frac{\pi D}{2}}\right) = \frac{D}{4}$$

Using Manning's formula, we have

$$q = \frac{1}{n} \,.\, a \,.\, r^{2/3} \sqrt{S}$$

$$\therefore \qquad 0.6 = \frac{1}{0.015} \cdot \frac{\pi D^2}{8} \cdot \left(\frac{D}{4}\right)^{2/3} \sqrt{0.0001}$$

or $$D^{8/3} = \frac{0.6 \times 0.015 \times 8 \times 2.52 \times 100}{\pi}$$

or $D = 1.93$ m. **Ans.**

Example 4.6. *Design a sewer running 0.7 times full at maximum discharge for a town provided with the separate system, serving a population of 80,000 persons. The water supplied from the water- works to the town is at a rate of 190 litres / person / day. The sewer is made up of brick work plastered smooth with cement mortar (n = 0.013) and the permissible slope is 1 in 600. The variations of n with depth may be neglected. Assume any other data not given and needed.*

Solution.

Population = 80,000

Rate of water supplied

= 190 litres/person/day

Average rate of daily water supplied to the town

= 80,000 × 190 litres/day

Average water supplied (in cumecs)

$$= \frac{80{,}000 \times 190}{24 \times 60 \times 60 \times 1000} \text{ cumecs} = 0.176 \text{ cumecs}$$

Assuming that 80% of water supplied to the town appears as sewage, we have

The average discharge of sewage produced

$$= 0.176 \times 0.8 \text{ cumecs} = \mathbf{0.14 \text{ cumecs.}}$$

Assuming the maximum flow to be three times the average, we have

The maximum sewage discharge

$$= 3 \times 0.14 = 0.42 \text{ cumecs.}$$

Now since the sewer is to be designed as running 0.7 times the full depth at maximum discharge, we have from Table 4.8, for a value of $\frac{d}{D} = 0.7$,

$$\frac{q}{Q} = 0.838$$

Here $q = 0.42$ cumecs

$$\therefore \quad Q = \frac{0.42}{0.838} = 0.5 \text{ cumecs.}$$

Now using Manning's nomogram at full flow and $n = 0.013$ (Fig. 4.5), for known values of

$$Q = 0.5 \text{ cumecs (500 litres/sec.)}$$

and $S = \frac{1}{600}$;

we read the other unknown factors, as

$$D = 0.78 \text{ m}$$

and $V = 1.04$ m/sec.

Then from Table 4.8, for $\frac{d}{D} = 0.7$;

we have $\frac{v}{V} = 1.12$

$$\therefore \quad v = 1.12\ V$$

$$= 1.12 \times 1.04 \text{ m/sec.} = \mathbf{1.17 \text{ m/sec.}}$$

Which is more than the self-cleansing velocity, and hence satisfactory.

Check for minimum flow

Assuming the minimum flow in the sewer to be $\frac{1}{3}$ time the average flow, we have, the minimum flow as

$$q_{min} = \frac{0.14}{3} = 0.047 \text{ cumecs.}$$

From curves of Fig. 4.3, for a discharge ratio of $\frac{q_{min}}{Q}$ as $= \frac{0.42}{0.047} = 0.11$, and $n = N$ as given, we have

$$\text{Depth ratio} = \frac{d_{min}}{D} = 0.23$$

and velocity ratio for this depth ratio is given by

$$\text{Velocity ratio} = \frac{v_{min}}{V} = 0.65$$

∴ The velocity at minimum flow

$$= 0.65 \times 1.04 \text{ m/sec.}$$

$$= 0.68 \text{ m/sec.}$$

which is more than the minimum required of 0.45 m/sec, and hence, satisfactory.

Note. When the velocity at minimum flow (*i.e.* D.W.F.) is not satisfactory, we have either to increase the slope or try with increased dia of the sewer.

Example 4.7. *Calculate the velocity and discharge through a rectangular concrete lined smooth channel 2.4 m wide and 1.2 m deep built to a slope of 1 in 200, when running completely full. Use Bazin's coefficient in Chezy's formula as :*

$$C = \frac{157.6}{1.81 + \frac{K}{\sqrt{r}}}$$

where K = 0.3 for smooth concrete lined surface.

(*Engg. Services, 1968*)

Solution. Area of channel

$$= A = 2.4 \text{ m} \times 1.2 \text{ m} = 2.88 \text{ sq. m.}$$

Wetted perimeter = $P = 2.4 + 1.2 + 1.2 + 2.4^* = 7.2$ m

$$\therefore \quad R = \frac{A}{P} = \frac{2.88}{7.2} = 0.40 \text{ m.}$$

Now, $$C = \frac{157.6}{1.81 + \frac{0.3}{\sqrt{0.40}}} = \frac{157.6}{1.81 + 0.474} = \frac{157.6}{2.284} = 68.99$$

Using, Chezy's formula, we have

$$V = C \cdot \sqrt{RS}$$

$$= 68.99 \sqrt{0.40 \times \frac{1}{200}} = 68.99 \times 0.0447 = 3.08 \text{ m/sec.} \quad \textbf{Ans.}$$

Discharge $Q = A.V$

$$= 3.08 \times 2.88$$

$$= 8.88 \text{ cumecs.} \quad \textbf{Ans.}$$

Example 4.8. *Calculate the diameter and discharge of a circular sewer laid at a slope of 1 in 400 when it is running half full, and with a velocity of 1.9 m / sec. (n in Manning's formula = 0.012).* (*Engg. Services, 1969*)

Solution. When pipe is running half full,

The area of section

$$a = \frac{\pi D^2}{8}$$

---

*Assuming the sewer to be running completely full, and as such, making the roof as a part of wetted perimeter.

The wetted perimeter

$$p = \frac{\pi D}{2}$$

The H.M.D. $r = \frac{D}{4}$.

Using Manning's formula, we have

$$V = \frac{1}{n} \cdot r^{2/3} \cdot \sqrt{S}$$

$$\therefore \qquad 1.9 = \frac{1}{0.012}\left(\frac{D}{4}\right)^{2/3} \cdot \frac{1}{\sqrt{400}}$$

or $\qquad D^{2/3} = 1.9 \times 0.012 \times 2.52 \times 20 = 1.15$

or $\qquad D = (1.15)^{\frac{3}{2} = 1.5} = 1.23$ m.

Hence, use 1.23 m dia sewer. **Ans.**

Discharge $Q = a.v$

$$= \frac{\pi(1.23)^2}{8} \times 1.9 \text{ m}^3/\text{sec}, = 1.13 \text{ cumecs.}$$ **Ans.**

Example 4.9. *Design a sewer to serve a population of 36,000 ; the daily per capita water supply allowance being 135 litres, of which 80 per cent finds its way into the sewer. The slope available for the sewer to be laid is 1 in 625 and the sewer should be designed to carry four times the dry weather flow when running full. What would be the velocity of flow in the sewer when running full ?*

Assume n = 0.012 in Manning's formula. (*Engg. Services, 1970*)

Solution. Population

= 36,000

Per capita water supply

= 135 litres/person/day

∴ Average water supplied daily

= 36,000 × 135 litres/day

Average water supplied in cumecs

$$= \frac{36{,}000 \times 135}{1000 \times 24 \times 60 \times 60} \text{ cumecs}$$

= 0.0562 cumecs.

Average sewage discharge

= 80% of water supplied

= 0.8 × 0.0562 cumecs = 0.045 cumecs.

∴ D.W.F. = 0.045 cumecs.

Maximum discharge for which sewer should be designed running full

= 4 × 0.045 cumec = 0.18 cumecs.

Now, using Manning's formula (and assuming that its Nomogram is not available) we have

$$Q = \frac{1}{N} \cdot A.R^{2/3}. \sqrt{S}$$

(Capital letters being used for running full)

$$\therefore \quad 0.18 = \frac{1}{0.012}\left(\frac{\pi}{4}D^2\right)\left(\frac{D}{4}\right)^{2/3} \frac{1}{\sqrt{625}}$$

or $$\frac{0.18 \times 0.012 \times 4 \times 2.52 \times 25}{\pi} = D^{8/3}$$

or $$D^{8/3} = 0.173$$

or $$D = (0.173)^{\frac{3}{8} = 0.375} = 0.52 \text{ m}$$

Hence, use 0.52 m dia. sewer pipe. **Ans.**

Velocity of flow when running full

$$= V = \frac{Q}{A} = \frac{0.18}{\frac{\pi}{4}(0.52)^2} = 0.85 \text{ m/sec.} \quad \textbf{Ans.}$$

Example 4.10. *Design an outfall circular sewer of the separate system for a town with a population of 1,00,000 persons with a water supply at 180 litres per head per day. The sewer can be laid at a slope of 10 in 10,000 with n = 0.012. A self-cleansing velocity of 0.75 m/sec is to be developed. The dry weather flow may be taken as 1/3 of the maximum discharge. Given the following table :*

Proportionate Depth	*Proportionate Velocity*	*Proportionate Discharge*
0.31	*0.7901*	*0.2086*
0.35	*0.8430*	*0.2629*
0.37	*0.8675*	*0.2981*
0.39	*0.8909*	*0.3217*
0.40	*0.9022*	*0.3370*
0.42	*0.9299*	*0.3682*

(*Engg. Services 1975*)

Solution. Population

= 1,00,000

Average rate of water supply

= 180 litres/person/day

∴ Average rate of water supplied per day

= 1,00,000 × 180 litres

∴ Average rate of water supplied in cumec

$$= \frac{1,00,000 \times 180}{1,000 \times 24 \times 60 \times 60} \text{ cumec} = 0.208 \text{ cumec.}$$

∴ D.W.F. = 0.208 cumec

$\therefore$ Maximum discharge

$$= 3 \times 0.208 \text{ cumecs.} = 0.624 \text{ cumecs}$$

Let us design the sewer as running full at maximum discharge.

Using Manning's formula, we have

$$Q = \frac{1}{N} AR^{2/3} . \sqrt{S}$$

Assuming that the sewer is laid at the available slope of 10 in 10,000 *i.e.* 1 in 1,000, we have

$$S = \frac{1}{1000}$$

Putting the values in Manning's equation, we have

$$0.624 = \frac{1}{0.012}\left(\frac{\pi}{4}D^2\right)\left(\frac{D}{4}\right)^{2/3}\frac{1}{\sqrt{1000}}$$

or $$D^{8/3} = \frac{0.624 \times 0.012 \times 4 \times 2.52 \times 3.16}{\pi}$$

or $$D^{8/3} = 0.758$$

or $$D = (0.758)^{3/8 = 0.375} = 0.915 \text{ m}$$

Now, velocity of flow at full flow

$$V = \frac{Q}{A} = \frac{0.625}{\frac{\pi}{4}(0.915)^2} = 0.95 \text{ m/sec}$$

This is more than 0.75 m/sec, and hence satisfactory.

Let us check for the velocity at D.W.F.

At D.W.F. $$\frac{q}{Q} = \frac{1}{3} = 0.333$$

From the given table, corresponding to this discharge ratio, we find

Depth ratio $$\frac{d}{D} \approx 0.40$$

Velocity ratio $$= \frac{v}{V} \approx 0.9022 \text{ (given)}$$

Hence, the velocity developed at D.W.F.

$$= 0.9022 \times 0.95 \text{ m/sec} = 0.855 \text{ m/sec.}$$

This is more than 0.75 m/sec and, therefore, satisfactory.

Hence, use 0.915 m dia pipe as worked out. **Ans.**

Note. If the velocity at D.W.F. works out to be less than 0.75 m/sec, then increase the design slope or try with increased dia of sewer.

Example 4.11. *A circular sanitary sewer is designed to carry the maximum flow of sewage while flowing 70% full at a velocity of 0.9 m/sec. If the ratio of* $\frac{\textit{Maximum}}{\textit{Average}}$ *and* $\frac{\textit{Average}}{\textit{Minimum}}$ *flow are 3 and 2.5 respectively, find out :*

(*i*) *the proportionate depth of flow, and*

(*ii*) *the velocities of flow generated at the time of :*

(a) average flow, and

(b) minimum flow.

Variations in the value of N with depths may be neglected.

Solution. *At maximum flow,* the $\frac{d}{D}$ ratio is 0.7.

$$\therefore \quad \frac{d_{max}}{D} = 0.7$$

and $\quad v_{max}$ = velocity at the time of maximum flow
= 0.9 m/sec.

From Fig. 4.3, for $\frac{d}{D} = 0.7$, and at $\frac{N}{n} = 1$

we have $\quad \frac{v}{V} = 1.12$

and $\quad \frac{q}{Q} = 0.838.$

Here $\quad v = v_{max} = 0.9$ m/sec.

$$\therefore \quad V = \frac{0.9}{1.12} = 0.8 \text{ m/sec.}$$

and $\quad q_{max}$ = discharge at maximum flow = 0.838 Q.

At average flow,

$$q_{av} = \frac{1}{3} \cdot q_{max} \text{ (given)}$$

$$= \frac{1}{3}(0.838\,Q) = 0.279\,Q.$$

From Fig. 4.3, for $\frac{q}{Q} = 0.279$ and $\frac{N}{n} = 1$ (of course) ;

we have $\quad \frac{d}{D} = \mathbf{0.37}$

and $\quad \frac{v}{V} = 0.87$

$\therefore \quad v_{av}$ = velocity at average flow
= 0.87 V = 0.87 × 0.8 m/sec. = **0.7** m/sec

Hence, while carrying the average discharge, the sewer runs 37% full, and the velocity then generated is 0.7 m/sec. **Ans.**

(b) At minimum flow

$$\frac{q_{av}}{q_{min}} = 2.5$$

$$\therefore \quad q_{min} = \frac{q_{av}}{2.5} = \frac{0.279\,Q}{2.5} = 0.111\,Q$$

Now from Fig. 4.3 for $\frac{q}{Q} = 0.111$, and $\left(\frac{N}{n} = 1 \text{ of course}\right)$, we have

$$\frac{d}{D} = \mathbf{0.23}$$

and $\frac{v}{V} = 0.64$

or v_{min} = Velocity at minimum flow
$= 0.64\ V = 0.64 \times 0.8 = \mathbf{0.51}$ m/sec.

Hence, while carrying the minimum discharge, the sewer runs 23% full, and the velocity then generated is 0.51 m/sec. **Ans.**

Note. From the above example, it follows that in a circular sewer flowing 70% full at the time of maximum discharge, the velocity is reduced by $\left(\frac{0.9-0.7}{0.9}\right)$ = 21% only, when the discharge is reduced by 66.7% ; and the velocity is reduced by $\frac{0.9-0.51}{0.9}$ = 43.3% only, when the discharge reduces by 86.67%. A circular sewer can thus keep a fairly uniform velocity of flow although the discharge may fluctuate considerably. Such sewers are, therefore, quite suitable for carrying sanitary sewage whose flows vary considerably from hour to hour.

Example 4.12. *A combined sewer of a circular section is to be laid to serve a particular area. Calculate the size of this sewer from the following data :*

Area to be served	*= 120 hectares*
Population	*= 1,00,000*
Maximum permissible flow velocity	*= 3 m/sec.*
Time of entry for storm water	*= 10 minutes*
Time of flow in channel	*= 20 minutes*
Per capita water supply	*= 250 litres/day/person*
Coefficient of run-off for the area	*= 0.45*
Hourly, Maximum rainfall for the area at the design frequency	*= 5 cm*

Assume any other data not given, and if needed.

Solution. *Sewage Discharge (i.e.* D.W.F.*) Computations*

Average water supplied

$= 250 \times 1{,}00{,}000$ litres/day

$= \frac{250 \times 1{,}00{,}000}{1000 \times 24 \times 60 \times 60}$ cumecs = 0.289 cumecs

Assuming that 80% of the water supplied appears as sewage, we have

Average sewage discharge

$= 0.8 \times 0.289 = 0.23$ cumecs

A ssuming the maximum sewage discharge to be 3 times the average discharge, we have

Maximum sewage discharge

$= 3 \times 0.23 =$ **0.69 cumecs.**

Storm water discharge computations

Time of concentration

T_c = Time of entry + Time of flow
= (10 + 20) minutes = 30 minutes.

Now, maximum hourly rainfall for the area

$= p_o = 5$ cm/hr.

∴ Using Eq. (3.3), we have

$$p_c = p_o\left(\frac{2}{1+T_c}\right)$$

where T_c is the concentration time in hours

$$= \frac{30}{60} = 0.5 \text{ hour}$$

$$\therefore \quad p_c = 5\left(\frac{2}{1+0.5}\right) = \frac{10}{1.5} = 6.67 \text{ cm/hr.}$$

Now, using rational formula, Eq. (3.1), we have

$$\text{Max. storm run off } = Q_p = \frac{1}{36} K.p_c.A$$

$$= \frac{1}{36} \times 0.45 \times 6.67 \times 120 \text{ cumecs.}$$

$$= 10 \text{ cumecs.}$$

∴ The combined maximum discharge

= Storm run-off + Sewage discharge

= 10 + 0.69 = **10.69 cumecs.**

Now assuming the sewer to be running full at the maximum velocity of 3 m/sec at the time of maximum flow, we have

$$\text{Area required} = \frac{Q}{V} = \frac{10.69}{3} \text{ m}^2 = 3.56 \text{ m}^2$$

∴ Dia of sewer pipe required

$$= \sqrt{\frac{4}{\pi}.3.56} = \sqrt{4.53} = 2.13 \text{ m.}$$

Hence, use a sewer pipe of 2.13 m dia. **Ans.**

Example 4.13. *Using Rational method, rainfall intensity duration curves (Fig. 4.6) and the data given on Fig. 4.7, compute the diameter of the outfall sewer. The length of lines, drainage areas, and inlet times are marked on Fig. 4.7 itself.* *(Engineering Services, 1981)*

Assume :

(*i*) *Runoff coefficient for the entire area = 0.30.*

(*ii*) *Velocity of flow in sewers flowing full = 0.75 m/sec.*

(*iii*) *5 years average frequency curve may be used (Fig. 4.7).*

(*iv*) *Table 1 gives the hydraulic elements for circular pipes flowing full.*

Table 1. Hydraulic Elements for a Circular Pipe Flowing Full

Quantity of flow *l/m*	*Dia of pipe* *mm*	*Slope of pipe* *m/m*	*Velocity* *m/sec*
400	450	0.025	2.7
600	525	0.020	2.8
690	1050	0.00055	0.75
1500	1350	0.001	1.55
2000	1450	0.001	1.20

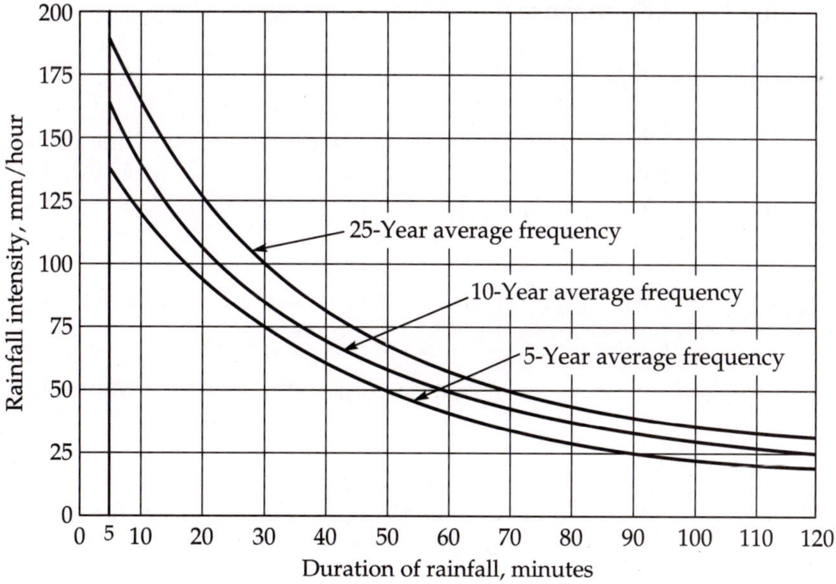

Fig. 4.6 (given)

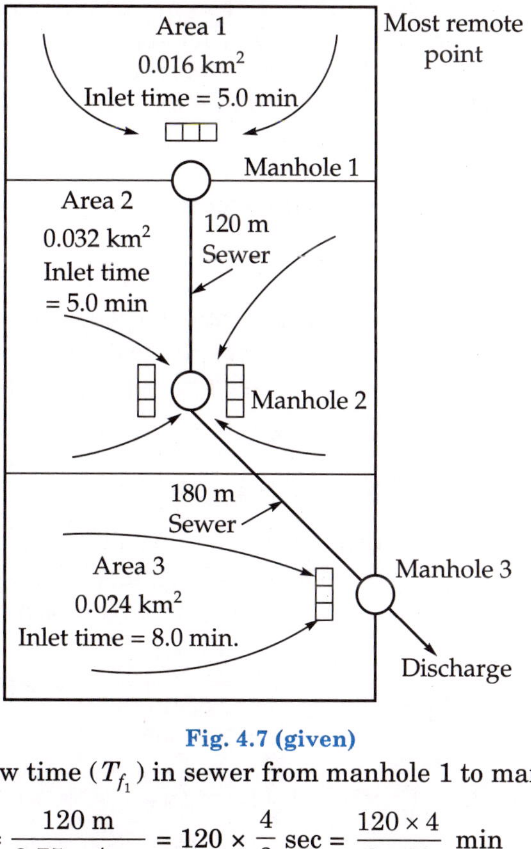

Fig. 4.7 (given)

Solution. Flow time (T_{f_1}) in sewer from manhole 1 to manhole 2

$$= \frac{120 \text{ m}}{0.75 \text{ m/sec}} = 120 \times \frac{4}{3} \text{ sec} = \frac{120 \times 4}{3 \times 60} \text{ min}$$

$$\mathbf{= 2.7 \text{ min.}}$$

Flow time (T_{f_2}) in sewer from manhole 2 to manhole 3

$$= \frac{180 \text{ m}}{0.75 \text{ m/sec}} = 180 \times \frac{4}{3} \text{ sec} = \frac{180 \times 4}{60 \times 3} \text{ min}$$

= 4 min.

(*i*) Time of concentration for area 1 (to contribute wholly at Manhole 3)

$$T_{i_1} + T_{f_1} + T_{f_2} = [5 + 2.7 + 4] \text{ min}$$

$$= \mathbf{11.7\ min} \quad ...(i)$$

(*ii*) Time of concentration for area 2 (to contribute wholly at Manhole 3)

$$T_{i_2} + T_{f_2} = 5 \text{ min} + 4 \text{ min} = \mathbf{9\ min.} \quad ...(ii)$$

(*iii*) Time of concentration for area 3 (to contribute wholly at Manhole 3)

$$T_{i_3} = \mathbf{8\ min.} \quad ...(iii)$$

That means that after 11.7 min, all the three areas 1, 2 and 3 will start contributing fully at Manhole 3. The time of concentration for the entire catchment (area 1 + area 2 + area 3) will, thus, be equal to 11.7 min.

Since the critical rainfall duration is equal to the time of concentration, we can consider, that rain as a critical rain, which will be having a duration of 11.7 min ; and since 5 year frequency is to be considered in designing the sewer, let us workout the rainfall intensity for such a rain having a frequency of 5 years and a duration of 11.7 minutes. Such an intensity is read out from the given intensity duration curves (Fig. 4.6), using the 5 year frequency curve, as **115 mm/hr,** *i.e.* 11.5 cm/hr.

With this critical rain, having intensity of 11.5 cm/hr, we can apply Rational formula, to work out the discharge at the outfall, as :

$$Q = \frac{1}{36} \cdot K \cdot p_c \cdot A$$

where K = run-off coeff. = 0.30 (given)

p_c = critical rain intensity in cm/hr

= 11.5 cm/hr (worked out above)

A = total catchment area in hectares

$$\therefore \quad Q = \frac{1}{36} \times K \cdot [p_c][A_1 + A_2 + A_3]$$

because K is constant throughout the area A_1, A_2 and A_3.

where $A = A_1 + A_2 + A_3$

$= (0.016 + 0.032 + 0.024) \text{ km}^2$

$= \frac{0.072 \times 10^6}{10^4}$ hectares

= 7.2 hectares.

$$\therefore \quad Q = \frac{1}{36} \times 0.30 \times 11.5 \times 7.2$$

= 0.69 cumecs = 690 *l*/sec.

For passing a discharge of 690 *l*/sec, with a sewer flow velocity of 0.75 m/sec, the required sewer from Table 1 should be of **1050 mm dia** laid at a **slope of 0.00055 m/m** *i.e.* 0.55 mm/m of sewer length. **Ans.**

Example 4.14. *A 25 cm diameter sewer with an invert slope of 1 in 400 is running full. Calculate the velocity and rate of flow in the sewer. Is it self cleansing ? Take n = 0.015.* (*Civil Services, 1993*)

Solution. D = dia of sewer = 25 cm = 0.25 m

Area of sewer when running full

$$(A) = \frac{\pi}{4} . D^2$$

$$= \frac{\pi}{4} \times (0.25)^2 \text{ m}^2 = 0.049 \text{ m}^2.$$

$$R = \frac{A}{P} = \frac{\frac{\pi}{4} . D^2}{\pi D} = \frac{D}{4} = \frac{0.25}{4} = 0.0625 \text{ m}$$

$$S = \frac{1}{400}$$

$$N = 0.015$$

Using Manning's equation, we have

$$V = \frac{1}{N} . R^{2/3} . S^{1/2}$$

$$= \frac{1}{0.015} \times (0.0625)^{2/3} . \frac{1}{\sqrt{400}} = \mathbf{0.525 \text{ m/s.}} \quad \textbf{Ans.}$$

$$Q = V.A = 0.525 \times 0.049 \text{ m}^3\text{/s} = \mathbf{0.0257 \text{ m}^3\text{/s.}} \quad \textbf{Ans.}$$

The velocity in the above sewer is 0.525 m/s, which is slightly more than 0.45 m/s, which is the numerical theoretical value of the self-cleansing velocity. Hence, the sewer can be termed as self-cleansing at full flow, although such self-cleansing should be obtained at partial flow also, such as at $\frac{1}{2}$ full. In this case, the velocity at partial flow is likely to be lesser than 0.45 m/s or so, and hence the sewer will no longer remain self-cleansing at partial flow. **Ans.**

4.9. Limitations on Depth of Flow due to Ventilation Considerations

From considerations of ventilation in wastewater flows, sewers should not be designed to run full.

Sewers up to 400 mm dia, may be designed to run at $\frac{1}{2}$ depth ; sewers between 400 to 900 mm dia may be designed for $\frac{2}{3}$ depth ; and larger sewers may be designed for $\frac{3}{4}$ depth, at ultimate peak designed flows.

4.10. Egg Shaped Sewers and Hydraulically Equivalent Sections

Circular sewer sections are most widely used these days in all the cities, which are preferably provided with separate sewerage system. However, other forms of sewer sections, such as, egg shape, horse shoe shape, *U* shape, parabolic shape, rectangular shape, semielliptical shape, etc. may also sometimes be used. The circular sections are generally preferred to all other shapes, because of their following advantages :

(*i*) They can be manufactured most easily and conveniently.

(*ii*) A circular sewer provides the maximum area for a given perimeter, and thus providing the maximum hydraulic mean depth when running full or half full ; and is, therefore, the most efficient section at these flow conditions.

(*iii*) Circular section utilizes the minimum quantities of materials, and is, therefore, the cheapest and most economical.

(*iv*) A circular section, being of uniform curvature all round, offers less opportunities for deposits.

But all these advantages of circular sections are obtained only when the section runs at least half full. When the depth in a circular section goes less than half full, its merits are lost, as the velocity and discharge reduce considerably with the reduction in depth. Lesser the discharge, the poorer the performance.

The circular sections, will therefore, be the best when discharge does not vary too much, and the chances of sewers running with very low depths (less than half) are less.

In combined sewers, however, the variations in discharge are enormous, because the storm run-off is generally 20 to 25 times that of the sewage discharge. Hence, combined sewers will have to run at low discharges of about $\frac{1}{20}$ to $\frac{1}{25}$ times the maximum, during non-monsoon periods.

Egg shaped sewers, such as shown in Figs. 4.8 (*a*) and (*b*), which for low discharges maintain hydraulic depth nearly uniform and give 2 to 15% higher velocities than provided by **hydraulically equivalent*** circular sections

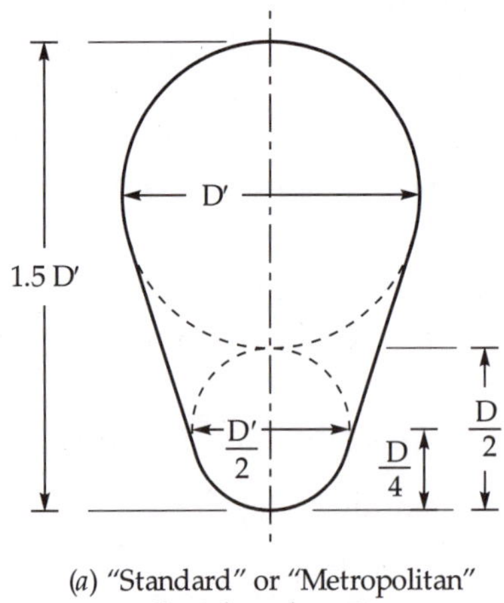

(*a*) "Standard" or "Metropolitan" Egg shaped section

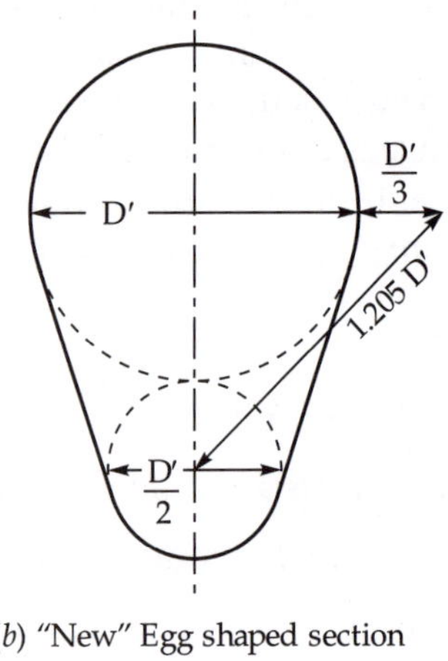

(*b*) "New" Egg shaped section

Fig. 4.8

---

*Two sewers of different shapes are said to be hydraulically equivalent when they discharge at the same rate, while flowing full, on the same grade.

carrying the same low discharges, are, therefore, preferred for combined sewers. However, the increase in velocities is quite small compared to their other disadvantages and, therefore, such sewers, which were quite often used in olden days, are becoming obsolete these days. Their disadvantages over circular sewers are :

(*i*) They are more difficult to construct.

(*ii*) Since the smaller base has to support the weight of the upper broader section, they are less stable.

(*iii*) They require more material and are, therefore, more costly.

Various forms of egg shaped sewers (sometimes called *Ovoid sewers*) had been in use, and the two most common forms are shown in Figs. 4.8 (*a*) and (*b*).

The computations of areas and wetted perimeters of such egg shaped sewers involve complicated mathematical calculations. Therefore, while designing such sewers, the usual practice is to calculate the approximate dia of a hydraulically equivalent circular sewer first, which would give the same discharge when running full at the same gradient ; and then to convert it into dimensions of an egg shaped section having an equal area.

For computing the egg shaped sewer of an equivalent section, the dia of the circular section (D) is multiplied by a constant factor so as to get the top horizontal dia width (D') of the egg shaped section. Thus

$$D' = 0.84\,D \qquad \text{...(4.28)}$$

where D' = width of egg shape section

D = dia of circular sewer of the same cross-sectional area, obtained for passing the requisite discharge.

Knowing D, D' can be easily worked out, and the dimensions of the egg shaped sewer are thus established.

The hydraulic mean depth of egg shape sewers of equivalent section is the same as that of the circular sewers when running full, but it is higher for smaller depths of flow ; and hence the velocity generated in them at smaller depths is also higher than that generated in the equivalent circular sewers. The proportionate velocities at smaller depths in circular as well as in egg shape sewers are given below in Table 4.11.

Table 4.11. Comparison of Circular and Egg Shaped (Standard form) Sewers

Proportionate Depth	*Proportionate Velocity*	
	in Circular sewers	*in Egg shaped sewers**
0.25	0.701	0.698
0.20	0.615	0.627
0.15	0.517	0.544
0.10	0.401	0.440
0.05	0.257	0.295

*The proportionate velocity in egg-shaped sewers becomes higher than that in equivalent circular sewers, as soon as the depth in sewer goes below 0.25 times the full depth.

Example 4.15. *(a) Design a circular sewer so as to cater to a residential colony in town, having the following data :*

Area of the colony = *36 hectares*
Population = *8,000*
Per capita water consumption = *170 lphd*
Critical design rainfall intensity = *4 cm / hr.*
General available ground slope = *1 in 900*

Assume any other data, not given, and if needed.

(b) What will be the dimensions of an equivalent egg shaped sewer if adopted in this case.

Solution. *Sewage Discharge computations*

Average quantity of water consumed per day

$$= 170 \times 8000 \text{ litres/day.}$$

Average quantity of water consumed in cumecs

$$= \frac{170 \times 8000}{1000 \times 24 \times 60 \times 60} \text{ cumecs} = 0.0157 \text{ cumecs.}$$

Assuming that 80% of water consumed appears as sewage, we have

Average quantity of sewage discharge

$$= 0.8 \times 0.0157 \text{ cumecs} = 0.0126 \text{ cumecs.}$$

Assuming the peak sewage discharge to be three times the average discharge, we have

Maximum rate of sewage produced

$$= 3 \times 0.0126 \text{ cumecs} = \mathbf{0.038 \text{ cumecs.}}$$

Storm run-off computations

Assuming the coefficient of run-off (K) for the area as 0.55, we have, by using Rational formula,

Peak storm run-off

$$Q_p = \frac{1}{36} Kp_c.A$$

$$= \frac{1}{36} \times 0.55 \times 4 \times 36 \text{ cumecs} = 2.2 \text{ cumecs.}$$

Combined Maximum discharge

$$= 2.2 + 0.038 = \mathbf{2.238 \text{ cumecs}}$$

Now, assuming that the sewer while carrying this combined peak discharge possesses 10% extra capacity, we have

The design discharge which the sewer should carry while flowing full

$$= \frac{2.238}{0.9} \text{ cumecs} = \mathbf{2.49 \text{ cumecs}}$$

Now, using Manning's formula, we have

$$Q = \frac{1}{N} . A.R^{2/3} \sqrt{S}$$

Using the same gradient as is available, *i.e.* $\frac{1}{900}$ as the first proposition, and Manning's $N = 0.013$ for smooth concrete or vitrified clay sewer, we have

$$2.49 = \frac{1}{0.013}\left(\frac{\pi D^2}{4}\right)\left(\frac{D}{4}\right)^{2/3} \cdot \frac{1}{\sqrt{900}}$$

where D is the dia of the equivalent circular section.

$$\therefore \quad D^{8/3} = \frac{2.49 \times 0.013 \times 4 \times 2.52 \times 30}{\pi}$$

or $$D^{8/3} = 3.12$$

or $$D = (3.12)^{\frac{3}{8} = 0.375} = 1.533 \text{ m ; say } 1.54 \text{ m.}$$

Now, velocity generated

$$= \frac{Q}{A} = \frac{2.49}{\frac{\pi}{4} \cdot (1.54)^2} = 1.33 \text{ m/sec.}$$

This is satisfactory.

Note. The velocity can be increased further by steepening the slope and changing the size of the sewer accordingly. This will no doubt increase the ground excavations but will make the sewer more efficient at low flows, as the presently designed sewer may give very low velocities at low flows during non-monsoon seasons.

Check for lone sewage discharge

When maximum sewage is passing (once a day) in non-monsoon periods, the $\frac{q}{Q}$ will be equal to $\frac{0.038}{2.49} = 0.0152$. For this ratio of $\frac{q}{Q} = 0.0152$, from Fig. 4.3, we have

$$\frac{v}{V} = 0.3.$$

or $$v = 0.3 \times 1.35$$

$$= 0.4 \text{ m/sec.}$$ (which is just sufficient for non-silting)

Hence, in this sewer, deposition will take place during average and minimum lone sewage flow. The efficiency can be further increased by providing a steeper gradient, or by providing egg shaped section, which provide comparatively larger proportionate velocities at low depths.

(*b*) *Equivalent egg-shaped sewer*

Now $D = 1.54$ m

If D' is the width of the standard equivalent egg shaped sewer, then by Eq. (4.28), we have

$$D' = 0.84\, D$$

or $$D' = 0.84 \times 1.54 = 1.295 \text{ m}$$

Say 1.3 m

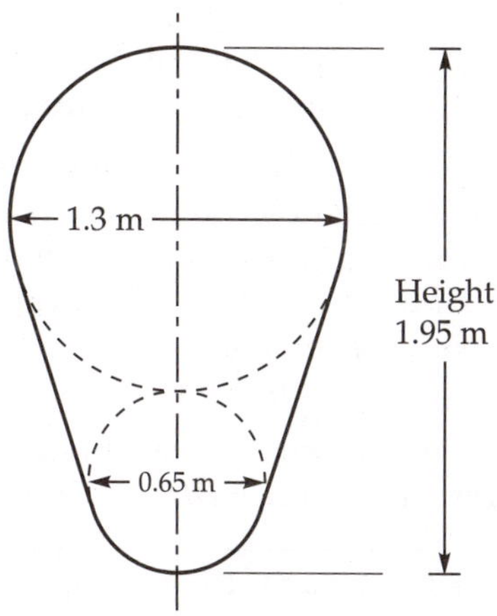

Fig. 4.9

Thus, the top width of the egg shape section

$$= 1.3 \text{ m}$$

and the height or vertical diameter of the egg shape section

$$= 1.5D'$$
$$= 1.5 \times 1.3$$
$$= 1.95 \text{ m.}$$

Hence use a standard egg-shaped section 1.3 m × 1.95 m, as shown in Fig. 4.9.

Example 4.16. *Find the relation between the side of a square section of one sewer and the diameter of a circular section of another sewer when both are hydraulically equivalent.*

Solution. Let D be the diameter of the circular sewer and b be the side of the equivalent square sewer.

The discharging capacity of circular sewer while running full* at a gradient of 1 in S

$$= \frac{1}{N}\left(\frac{\pi}{4}D^2\right)\left(\frac{D}{4}\right)^{2/3} \sqrt{S} \quad ...(i)$$

The discharging capacity of the rectangular section while running full* at a gradient of 1 in S

$$= \frac{1}{N} \cdot b^2 \cdot \left(\frac{b^2}{4b}\right)^{2/3} \sqrt{S} \quad ...(ii)$$

∴ Equating (*i*) and (*ii*), we have

$$\therefore \quad \frac{1}{N}\left(\frac{\pi}{4}D^2\right)\left(\frac{D}{4}\right)^{2/3} \sqrt{S} = \frac{1}{N} b^2 \left(\frac{b}{4}\right)^{2/3} \sqrt{S}$$

or $$\frac{\pi}{4} \cdot \frac{D^{8/3}}{2.52} = \frac{b^{8/3}}{2.52}$$

or $$D^{8/3} = 1.272\, b^{8/3}$$

or $$D = 1.094b.$$

This is the required relation, where b is the side of a square and D is the dia of the circular section. **Ans.**

Example 4.17. *A rectangular sewer with width 1.5 times its depth is hydraulically equivalent to a circular one. Find the relation between the width of the rectangular sewer and the diameter of the circular sewer.*

Solution. Let B and D_1 represents the width and depth of the rectangular sewer, respectively.

$$\therefore \quad B = 1.5\, D_1$$

Now, when this rectangular sewer is running completely full, the area of cross-section $A = BD_1 = 1.5\, D_1^2$.

---

*Assuming that the sewer runs completely full and the roof is also a part of the wetted perimeter : thus giving $A = b^2$; $P = 4b$; and $R = \frac{b^2}{4b} = \frac{b}{4}$. On the other hand, if sewer runs only nearly full, P will be equal to $3b$.

the wetted perimeter P (assuming the roof as part of the wetted perimeter)

$$= 2\,(B + D_1) = 5D_1.$$

$$R = \frac{A}{P} = \frac{1.5D_1^2}{5D_1}\ \frac{A}{P} = \frac{1.5D_1^2}{5D_1} = 0.3D_1$$

$\therefore$ Discharging capacity of rectangular sewer

$$= \frac{1}{N}(1.5D_1^2)(0.3D_1)^{2/3}\sqrt{S} \qquad \text{...}(i)$$

If D is the dia of the circular sewer, then its capacity at full depth

$$= \frac{1}{N}\left(\frac{\pi D^2}{4}\right)\left(\frac{D}{4}\right)^{2/3}\sqrt{S} \qquad \text{...}(ii)$$

For hydraulically equivalent sections, equating (i) and (ii), we get

$$\frac{1}{N}\cdot(1.5D_1)^2\,(0.3D_1)^{2/3}\sqrt{S} = \frac{1}{N}\left(\frac{\pi D^2}{4}\right)\left(\frac{D}{4}\right)^{2/3}\sqrt{S}$$

or $$2.25\,D_1^{8/3}\,(0.448) = \frac{\pi}{4}\cdot\frac{1}{2.52}\cdot D^{8/3}$$

or $$D_1^{8/3} = \frac{\pi}{4\times 2.52\times 2.25\times 0.448}\,D^{8/3}$$

or $$D^{8/3} = 3.24\,D_1^{8/3}$$

or $$D = (3.24)^{\frac{3}{8}=0.375}\,D_1$$

or $$D = 1.565\,D_1$$

But $$B = 1.5\,D_1$$

or $$D_1 = \frac{B}{1.5}$$

$\therefore$ $$D = 1.565\,\frac{B}{1.5}$$

$$D = \mathbf{1.043\ B.}$$

This is the required relation, where D is the dia of the circular sewer and B is the width of the rectangular sewer. **Ans.**

4.11. Design of Storm Water Drains

In a modern separate sewerage system, as pointed out earlier, the storm water is not mixed with the sewage discharge, but is carried separately through storm water drains and disposed of into a stream, lake or ocean.

For accomplishing this process efficiently, storm water is collected in the streets and admitted into the link drains (often kept covered) through inlets, which in-turn discharge into the main drains (often kept open); the main drains finally discharging into some large body of water, such as a river or a stream. Gravity discharge is always preferred, if the FSL of the river or the main drain into which the considered drain is being discharged, permits. However, this is not always possible, because if the FSL of the drain (which has to be less than the N.S.L. of the catchment area drained by it) is lower than the FSL or the HFL of the source of disposal, the gravity flow cannot take place. In such circumstances, the drain discharge has to be disposed of by pumping.

For designing the city's storm water drainage system, first of all, the contour maps of the area are collected, and the positions of various link drains, major drains and sources of disposal are properly planned with due regard to the possible gravity flow. If the gravity flow can be permitted with flatter drain gradient, it is always economical and preferred to pumping. After deciding the alignments of different drains, the catchment areas to be drained by each drain are marked. The peak discharge expected in the drain (reach by reach) is then worked out for each drain, by suitable method (various methods for computing peak discharge are given in chapter 3).

An L-section for the drain is now drawn, which fixes the full supply line of the drain. The full supply line is fixed keeping in view the natural surface level (NSL) of the area to be drained and the permissible outfall full supply level. The FSL line at no place along the length of the drain should go above the NSL line, as otherwise, water-logging may take place in that low lying area. However, sometimes, some encroachment above the NSL may be allowed, and in that case, the water of those few low lying pockets, will be removed by pumping into the gravity drain carrying the major discharge of the catchment.

After fixing the FSL line, and thus the slope of the water surface, the bed line (*i.e.* depth of the drain) has to be fixed. The depth is decided from various considerations, such as :

(*i*) The bed of the considered drain should not at the outfall point go below the bed of the source into which it is discharged.

(*ii*) The depth in uncovered drain, should preferably be kept less than the man's height, as to avoid frequent dangers of drowning.

(*iii*) The depth will sometimes be guided by the availability of land for the width of the drain, *i.e.* when available land width is less, naturally the width of the drain has to be kept less, and thus necessitating the construction of deeper drains.

(*iv*) The drain section should be economical and the velocities generated should be non-silting and non-scouring in nature. Certain empirical formulas relating the width and the depth of the drain have been suggested based on experimental results. They are given below :

(*a*) *For drains up to 15 cumecs*

$$y = 0.5\sqrt{B} \quad \text{...(4.29)}$$

(*b*) *For drains above 15 cumecs*, depths of the following order may be used.

Table 4.12

Discharge Q_p in the drain (cumecs)	*Depth (y) in metres*
15	1.7
30	1.8
75	2.3
150	2.6
300	3.0

(*c*) C.W.C. has recommended a graphical relation for unlined drains representing $\frac{\text{Width}}{\text{Depth}}\left(i.e., =\frac{B}{y}\right)$ ratios for different discharges, as shown in Fig. 4.10.

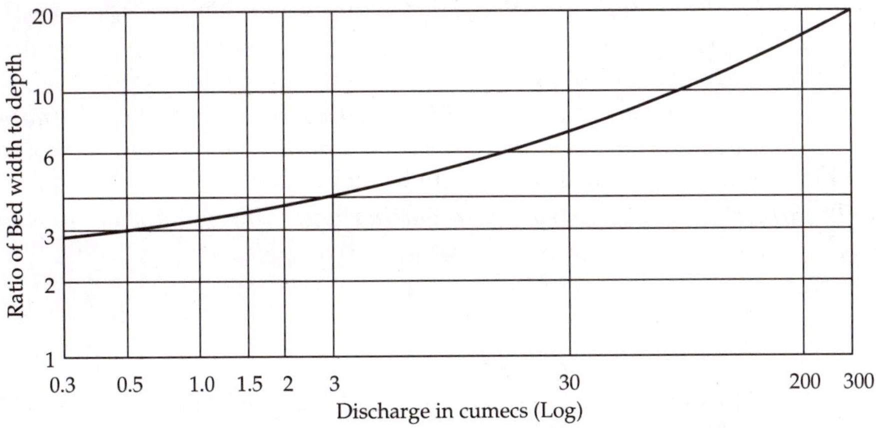

Fig. 4.10

The permissible velocities in unlined and lined drains are given in Table 4.13.

Table 4.13. Permissible Velocities in Drains

S. No.	*Type of soil*	*Max. permissible velocity in m/sec*
	Unlined drains	
1.	Rock and gravel	1.5
2.	Murum, hard soil, etc.	1.0 to 1.1
3.	Sandy loam, black cotton soil, etc.	0.6 to 0.9
4.	Very light loose sand to average sandy soil	0.3 to 0.6
5.	Ordinary soils	0.6 to 0.9
	Lined drains	
1.	Stone pitched	1.5
2.	Burnt clay tile lined	1.8
3.	Cement concrete lined	2 to 2.5

The drain section can finally be designed by using Manning's formula, either as a trapezoidal section (if open and lined or unlined) or as a rectangular section (if covered and lined). A freeboard as suggested in Table 4.1, is then provided above the FSL of the drain, which fixes the bank levels on both sides of the drain.

An outfall structure is also required to be constructed for energy dissipation at the place where any drain outfalls into another major drain. A regulator is also sometimes required at the outfall point, so as to avoid back flow into the drain, when the disposal source (river or stream) is in high floods, and the drain cannot function.

Example 4.18. *Design an unlined trapezoidal section for the outfall reach of an open urban storm water drain, draining a catchment area of 120 hectares. Given the following additional data :*

(*i*) *Inlet time* = *18 minutes.*

(*ii*) *Flow time in the upper reaches of the drain* = *30 minutes.*

(*iii*) *Coefficient of run-off for the area* = *0.6.*

(*iv*) *Design water surface slope* = *1 in 3000.*

(*v*) *The drain has to be designed for a 5 years rain frequency, and is situated near a place for which depth duration curves are available,* from which the rainfall for 48 min duration is read out to be 52 mm for 5 year frequency.

(*vi*) *The drain is to be constructed in cutting with a maximum permissible flow velocity as 0.9 m / sec.*

Solution. Time of concentration,

$$T_c = \text{Inlet time + Flow time}$$

$$= (18 + 30) \text{ minutes} = 48 \text{ minutes.}$$

The critical rainfall corresponding to a duration of 48 minutes for 5 years frequency curve is given to be 52 mm.

$$\therefore \quad p_c = \frac{52 \text{ mm}}{48 \text{ min}} = \frac{5.2}{\frac{48}{60}\text{h}} = 6.5 \text{ cm/hr.}$$

From Rational formula, the peak storm run-off

$$= Q_p = \frac{1}{36} \, . \, Kp_c . A$$

$$= \frac{1}{36} (0.6)(6.5)(120) \text{ cumecs.} = \textbf{13 cumecs.}$$

Assuming a depth (y) of 1.5 m (from Table 4.12) and using Eq. (4.29), let us have the width of the drain (B) as

$$y = 0.5\sqrt{B}$$

or $$1.5 = 0.5\sqrt{B}$$

or $$B = 9 \text{ m}$$

Let us now assume the side slopes of the drain as 1 H : IV, and longitudinal slope as $\frac{1}{3000}$ (given).

Using N = 0.025 in Manning's formula, the discharge through this drain (Fig. 4.11) is given as :

$$Q = \frac{1}{N} \, . \, AR^{2/3} \sqrt{S}$$

where $$A = (B + y)y$$

$$= (9.0 + 1.5)\, 1.5 = 15.75 \text{ sq. m.}$$

$$P = B + 2\sqrt{2}\, . \, y$$

$$= 9.0 + 2\sqrt{2} \times 1.5 = 9.0 + 4.24 = 13.24 \text{ m.}$$

$$R = \frac{A}{P} = \frac{15.75}{13.24} = \textbf{1.19 m.}$$

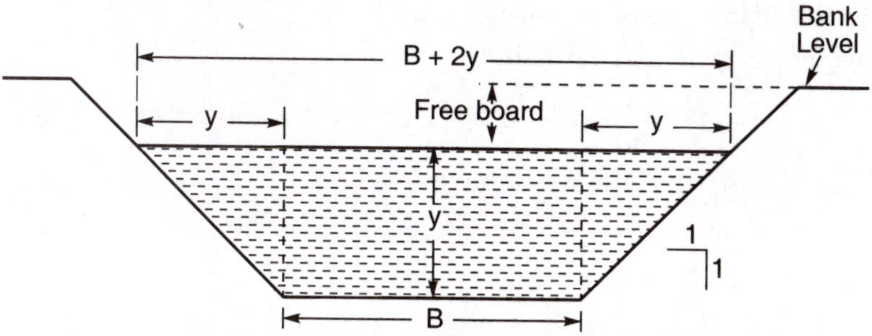

Fig. 4.11

$$\therefore \quad Q = \frac{1}{0.025} \times 15.75\,(1.19)^{2/3}\,\frac{1}{\sqrt{3000}} = 11.5 \times 1.123$$

= 12.93 cumecs, which is slightly less than the required value of 13 cumecs.

Hence, increase the bed width slightly, say use $B = 9.1$ m.

$$\because \quad A = (9.1 + 1.5)\,1.5 = 10.6 \times 1.5 = 15.9 \text{ sq. m.}$$

$$P = 9.1 + 2\sqrt{2} \times 1.5 = 9.1 + 4.24 = 13.34 \text{ m}$$

$$R = \frac{15.9}{13.34} = 1.192 \text{ m}$$

$$\therefore \quad Q = \frac{1}{0.025} \times 15.9(1.192)^{2/3}\,\frac{1}{\sqrt{3000}}$$

= **13.05 cumecs,** which is O.K.

Check for maximum velocity

Velocity generated in the drain at the peak discharge

$$= \frac{Q_p}{A} = \frac{13}{15.9} = 0.81 \text{ m/sec.}$$

(which is less than the maximum permissible velocity of 0.9 m/sec.)

Hence, use a trapezoidal drain section with bed width of 9.1 m and water depth of 1.5 m with 1 : 1 side slopes, laid at a longitudinal slope of 1 in 3000. **Ans.**

Example 4.19. *Design a suitable stone pitched section for a drain reach, required to pass the expected urban drainage discharge from a catchment area of 300 hectares. The maximum hourly design rainfall is 4 cm/hr and the time of concentration for the drain is 1 hour. The FSL line for the proposed drain has been fixed, as to give a longitudinal slope of 1 in 2300.*

Solution. $p_0 = 4$ cm/hr.

$T_c = 1$ hr.

$$p_c = p_0\left(\frac{2}{1+T_c}\right) = 4\left(\frac{2}{1+1}\right) = 4 \text{ cm/hr.}$$

Using $K = 0.55$ in Rational formula, we have,

The expected peak discharge in the drain

$$= Q_p = \frac{1}{36} Kp_c.A = \frac{1}{36} \times 0.55 \times 4 \times 300 = \textbf{18.3 cumecs.}$$

Assuming a depth of 1.5 m, we have

$$y = 0.5\sqrt{B}$$

$$\therefore \quad 1.5 = 0.5\sqrt{B}$$

or $\quad B \approx 9$ m.

Use a trapezoidal drain section with 1 : 1 side slopes a shown in Fig. 4.12.

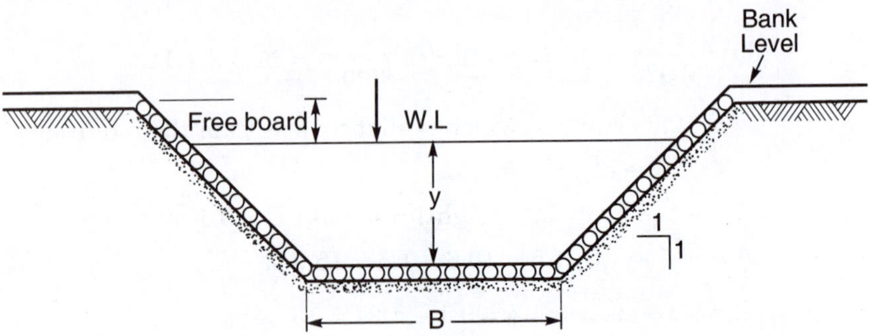

Fig. 4.12

Now $\quad A = (B + y)y = (9.0 + 1.5)1.5 = 15.75$ sq. m.

$$P = B + 2\sqrt{2}\ .y$$

$$= 9.0 + 2\sqrt{2} \times 1.5 = 9.0 + 4.24 = 13.24 \text{ m}$$

$$R = \frac{A}{P} = \frac{15.75}{13.24} = 1.19 \text{ m.}$$

Using Manning's formula and value of n as 0.020, for ordinary dry stone pitching, we have

$$Q = \frac{1}{0.020} \times 15.75 \times (1.19)^{2/3} \frac{1}{\sqrt{2300}}$$

$$= \textbf{18.4 cumecs.}$$

$\approx$ Required (18.3 cumecs) ; $\quad \therefore$ Safe.

Velocity generated $= \dfrac{Q}{A} = \dfrac{18.3}{15.75} = 1.16$ m/sec.

< permissible velocity of 1.5 m/sec.

$\therefore$ Velocity is safe.

Hence, use a trapezoidal drain section with base width of 9 m and water depth as 1.5 m, side slopes 1 : 1, laid at a longitudinal bed slope (parallel to water surface) of 1 in 2300. **Ans.**

PROBLEMS

1. (*a*) Explain briefly the method of finding out the sizes of sanitary sewers when the discharges to be carried through them are known.

(*b*) A circular sanitary sewer is designed to carry the maximum flow of sewage while flowing 70% (*i.e.* at 0.7 depth) full at a velocity of 0.9 m/sec. If the ratio of $\frac{\text{Maximum}}{\text{Average}}$ and $\frac{\text{Average}}{\text{Minimum}}$ flows are 2.5 and 2.0 respectively, find out :

(*i*) the proportionate depth of flow,

(*ii*) the velocities of flow generated at the time of average flow, and at the time of minimum flow. Neglect variations in the value of '*n*', the coefficient of roughness of sewer.

[**Hint.** Follow Example 4.9]

2. (*a*) State three important design criteria for a sanitary sewer (under Indian conditions) in a town with 30,000 population.

(*b*) A 30 cm dia. sewer having an invert slope of 1 in 150 was flowing full. What would be the velocity of flow and discharge ? (n = 0.013). Is the velocity self cleansing ?

What would be the velocity and the discharge when the same is flowing at 0.20 and 0.8 of the full depth.

3. (*a*) Draw two suitable surface drain sections.

(*b*) A sanitary sewer is to serve a uniformly distributed population of 10,000 along a 1,000 m road. The average ground slope for first 500 m is 1 in 400, and for the remaining as 1 in 900. Design the sewer. Give expected peak, average and minimum velocities. Make suitable assumptions, and state them clearly.

4. (*a*) What are the different hydraulic elements and the relation that exists between them, which govern the discharge through a sewer ?

(*b*) Find the diameter of a circular sanitary sewer for the following data :

(*i*) population	= 1,50,000
(*ii*) average sewage flow	= 190 litres/head/day
(*iii*) slope of sewer	= 1 in 1000
(*iv*) value of n in Manning's formula	= 0.013

5. (*a*) Mention the various aspects you would keep in view when designing a sewer.

(*b*) A 30 cm dia. sewer with a invert slope of 1 in 400 is flowing 1/3rd of the full depth. Calculate the velocity and the rate of flow in the sewer. Is it self-cleansing velocity ? Use n = 0.015.

6. (*a*) Discuss the various factors that affect the hydraulics of sewer lines.

(*b*) Mention the conditions when two conduits are hydraulically equivalent.

(*c*) Draw a neat sketch of a standard egg-shaped sewer. Under what circumstances, it is more advantageous compared to a circular one ?

7. (*a*) How does the variation of sewage flow affects its velocity in a circular sewer.

(*b*) Design a sanitary sewer with the following data :

(*i*) Population served	= 25,000.
(*ii*) Expected sewage flow	= 135 1/c/d (average)
(*iii*) Average slope of the ground	= 1 in 500

8. (*a*) Why a circular section is more commonly used in the construction of sewers ? What are the advantages of the egg-shaped section, and under what conditions of flow, does it become more useful ? Will you recommend its use for sanitary sewers, and if not, why ?

9. (*a*) What points should be kept in mind while designing sewers, and how are they designed ?

(*b*) Calculate from Manning's formula the diameter of a circular vitrified clay sewer (n = 0.013), which will just carry 0.05 cumecs, when flowing full at a slope of 1 in 1000.

(*c*) For the sewer in (*b*) above ; if the flow was at 0.6 depth, what would be the discharge in the sewer, and what will be the velocity of flow in partially full sewer, given the following data :

d/D	q/Q	v/V
0.1	0.02	0.30
0.5	0.39	0.80
0.6	0.54	0.88
0.8	0.85	1.01

Where d, q and v are depth of flow, discharge, and velocity respectively for partial flow condition, and D, Q and V are full flow conditions.

10. (*a*) Explain the importance of the following in the design of sewers :

(*i*) Self-cleansing velocity ; and

(*ii*) Non-scouring velocity.

(*b*) A sewer line is laid to serve a community of 150 persons/ha in a mohalla of 30 ha. The average water supply is 225 l/c/d. The available ground slope is 1 in 600. Using Manning's formula with $n = 0.015$, select a suitable diameter of sewer to carry the peak discharge, flowing half-full in the section. Check the velocity for self-cleansing section.

11. (*a*) Name the two factors used as criteria for selection of pipe diameter and slope in design of sewer.

(*b*) Calculate the ratio of discharge of a sewer when flowing at full depth to that when flowing at 3/4 depth.

12. Write detailed notes on :

(*i*) Design of storm water drains ;

(*ii*) Egg-shaped sewers and hydraulically equivalent sections ;

(*iii*) Characteristics of partially full sewers ; and

(*iv*) Maximum and Minimum permissible velocities in sewers.

13. Define and explain the following terms, connected to sewer designs :

(*i*) Self cleansing velocity ;

(*ii*) Non-scouring velocity ;

(*iii*) Hydraulically equivalent section ; and

(*iv*) Circular and Egg-shaped sewer sections.

—5

Sewers, their Construction, Maintenance, and Required Appurtenances

5.1. Introduction

Sewer pipes, as pointed out earlier, are generally circular pipes laid below the ground level, and generally sloping continuously towards the outfall. They are designed to flow under gravity, except for the outfall sewer which carries the treated and pumped sewage into the discharging source, and hence flows full under pressure*.

5.2. Shapes of Sewer Pipes

The sewer pipes are normally circular in section, although some other sections such as basket handle shape, egg shape, horse shoe shape, parabolic shape, semicircular shape, semi elliptical shape, rectangular shape, etc. may also be used under special necessities of a particular project. Out of these remaining shapes also, egg shape sections may be preferred for combined sewers, and rectangular shaped sewers are preferably constructed at site and normally used as independent covered storm water surface drains, and not as sewers. All other forms of sewers are almost out-dated and rarely used these days. The various forms of sewers, which can possibly be constructed and the circumstances in which each one is preferred, are shown in Fig. 5.1 (*a*), (*b*), (*c*), (*d*), (*e*), (*f*), (*g*), (*h*), (*i*) and (*j*).

5.3. Forces Acting on Sewer Pipes

The structural design of the sewer pipes should be such as to enable them to withstand the various forces likely to come on them. The following forces generally come into play in the sewer pipes :

(1) Internal pressure of sewage ;
(2) Pressure due to external loads ;
(3) Temperature stresses ;
(4) Flexural stresses.

These forces are discussed below :

5.3.1. Internal Pressure of Sewage. The pressure exerted by the sewage from inside the pipe when running full is called the internal pressure. Such a pressure may be exerted due to either chancy surcharge or due to overflow of sewers flowing under gravity, or it may be exerted in outfall sewers which have to flow full under pressure. Since most of the sewers of a sewerage

*The requirements of pressure pipes including their designs, materials, construction, etc. have been thoroughly described in Chapter 6 of Vol. I—"Water Supply Engineering".

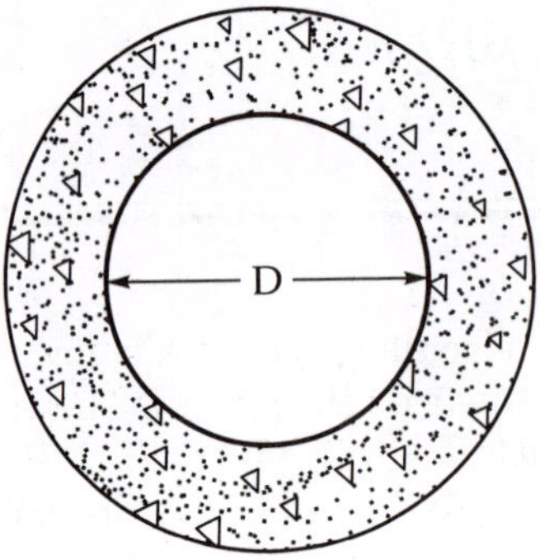

(*a*)
Circular shaped sewer
(most widely used for
all types of sewers)

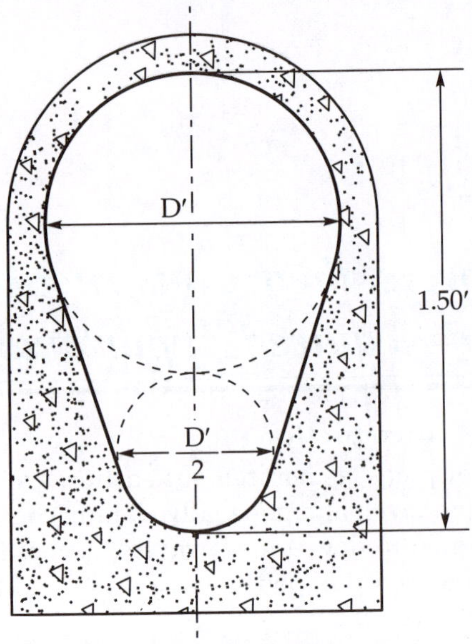

(*b*)
Standard Egg-shaped sewer
(may be preferred for
combined sewers)

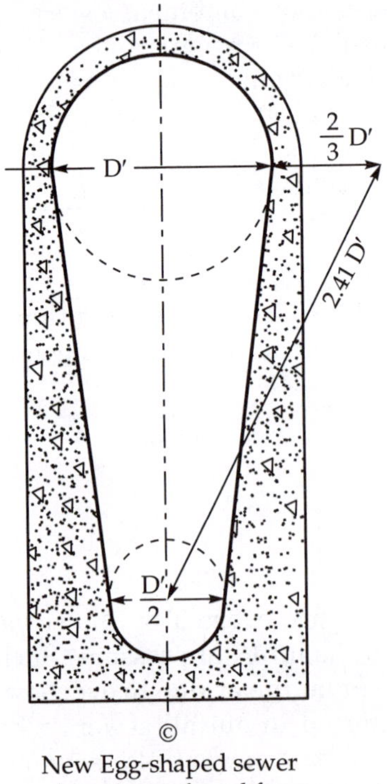

©
New Egg-shaped sewer
(may be preferred for
combined sewers)

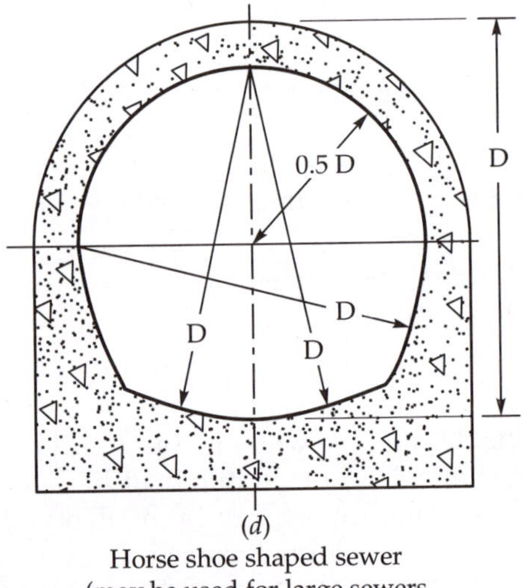

(*d*)
Horse shoe shaped sewer
(may be used for large sewers
with heavy discharges, such as for
Trunk and Outfall sewers)

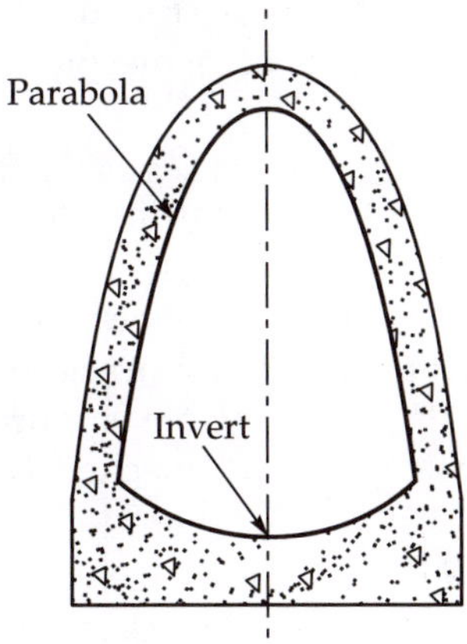

(*e*)
Parabolic shaped Sewer
(may be used for carrying comparatively smaller quantities of sewage.

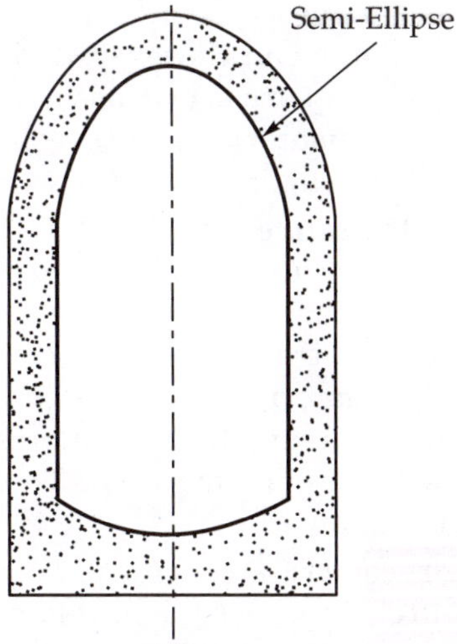

(*f*)
Semi-Elliptical Section
(may be used for soft soils as it is more stable, but useful only for carrying large amounts of sewage, in sewers greater than about 1.8 m in a diameter.

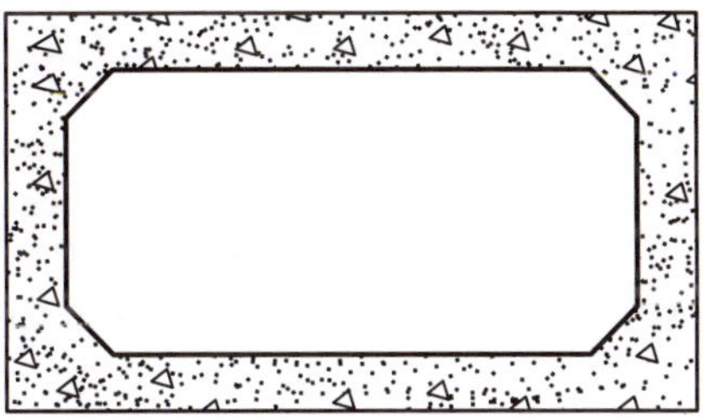

(*g*)
Rectangular shaped section
(generally used for covered storm water drains)

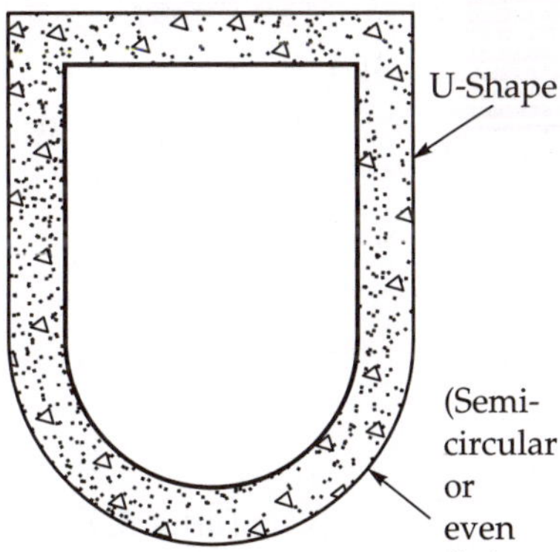

(*h*)
U-shaped section
(may be used for larger sewers and especially in open cuts)

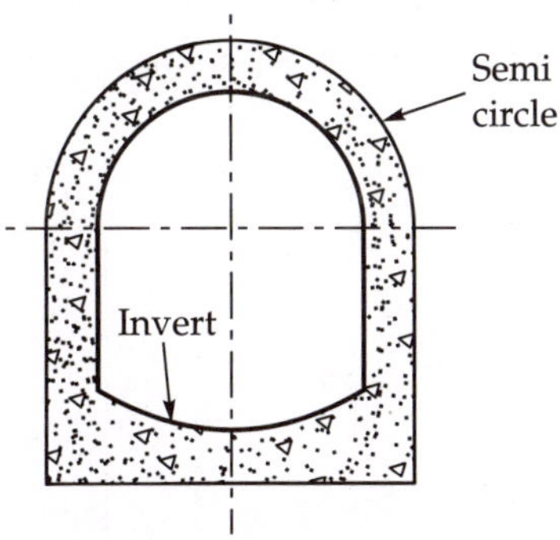

(*I*)
Semi-circular shaped sewer
(out-dated)

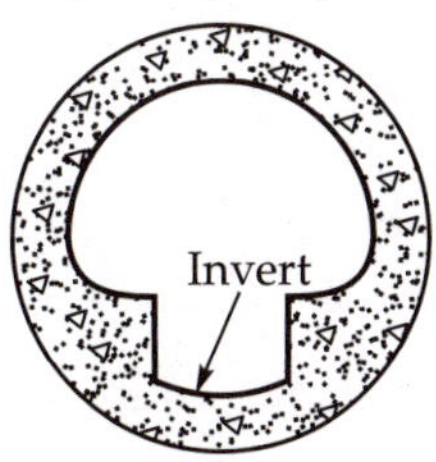

(*j*)
Basket handle shaped sewer
(out-dated)

Fig. 5.1. Different Shapes of Sewer Sections.

scheme are designed as gravity conduits, the internal pressure of water is not of much problem. However, it may be mentioned here that the tendency of the internal pressure is to cause bursting of the pipe and to induce tensile stresses in the pipe material. Hence, when pipes are to be used as pressure pipes, they must be strong in tension.

5.3.2. Pressures due to External Loads. Sewer pipes are mostly buried under the ground and placed in trenches, which are back filled. The weight of the pipe, the weight of the backfill and the superimposed traffic loads if any, will then be transferred to the pipe. This will produce compressive stresses in the pipe material (when the pipe is flowing under no internal pressure) and the material may fail in compression, if it exceeds the allowable compressive stress of the pipe material. The pipes should, therefore, be checked for this possible failure.

This compressive force is the most predominant force for the sewers, since they are sometimes taken very deep, and hence it must be properly evaluated.

The stresses produced due to external loadings can be evaluated by using certain empirical formulas, as given below. These formulas are based on the experiments carried out by Marston, etc. at Iowa State College in U.S.A.

(*a*) For pipes resting on or projecting above the undisturbed ground in cohesion less soils and covered with fills, such as in a highway culvert (Fig. 5.2), the external load likely to come per unit length of pipe (W) is given by

$$W = C_p . \gamma . D^2 \qquad ...(5.1)$$

where C_p = a coefficient whose value depends upon the type of pipe and character of foundation backfill. Typical values of C_p are given in table 5.1.

γ = Sp. wt. of the fill material

D = The external diameter of the pipe = (Internal diameter + 2 × thickness)

Table 5.1. Values of C_p in Eqn. (5.1)

Conditions	*Cover depth / External pipe dia* $= \frac{H}{D}$	*Values of C_p*	
		For rigid pipes	*For flexible pipes*
(1)	(2)	(3)	(4)
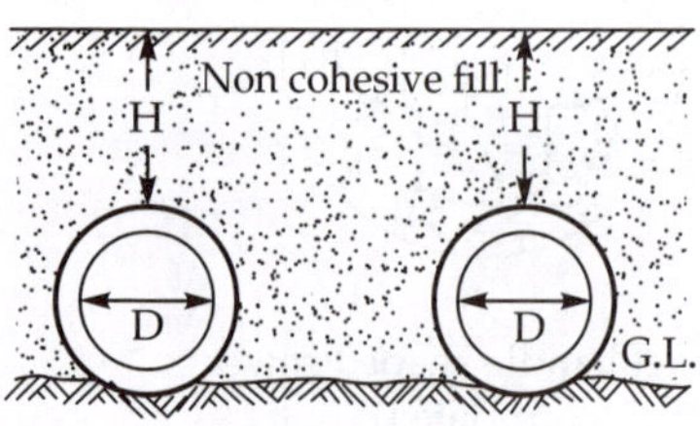 Fig. 5.2	1.0	1.2	1.1
	2.0	2.8	2.6
	3.0	4.7	4.0
	4.0	6.7	5.4
	6.0	11.0	8.2
	8.0	16.0	11.0

(*b*) For *flexible* pipes (such as *steel* pipes) buried in narrow trenches and with thoroughly compacted side fills, such as shown in Fig. 5.3, the external load per unit length of the pipe is given by

$$W = C.\gamma.B.D \qquad ...(5.2)$$

where C = a coefficient, characterising the fill material and the ratio H/B. Typical values are given in table 5.2.

(*c*) For *rigid* pipes (such as concrete, cast iron, vitrified clay, etc.) buried in narrow trenches and thoroughly compacted with cohesionless fills, the external load per unit length of the pipe is given by

$$W = C.\gamma.B^2 \qquad ...(5.3)$$

where C, γ and B have the same meaning as given above.

Table 5.2. Values of C in Eqns. (5.2) and (5.3)

Conditions	*Fill Material*	*Dry Sand and Gravel*	*Saturated top soil*	*Clay*	*Saturated clay*
	Specific wt. in kg/m³	*1600*	*1600*	*1920*	*2080*
	values of $\frac{H}{B}$	C	C	C	C
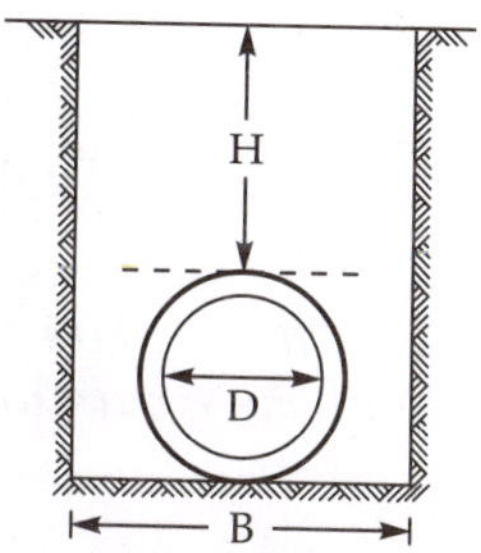	1.0	0.84	0.86	0.88	0.90
	2.0	1.45	1.50	1.55	1.62
	3.0	1.90	2.00	2.10	2.26
	4.0	2.22	2.33	2.49	2.65
	5.0	2.45	2.60	2.80	3.03
	6.0	2.60	2.78	3.04	3.33
	7.0	2.75	2.95	3.23	3.57
	8.0	2.80	3.03	3.37	3.76
Narrow trench such	9.0	2.88	3.11	3.48	3.92
that $B \leq 1.5\,D$	10.0	2.92	3.17	3.56	4.04
Fig. 5.3.	12.0	2.97	3.24	3.68	4.22
	14.0	3.00	3.28	3.75	4.34

(*d*) The amount of superimposed loads (such as traffic load, etc.) which is transmitted to the pipe, can be evaluated by using **Boussinesq's equation**. Assuming the fill surface to be horizontal, the equation is

$$p_t = \frac{3.H^3.P}{2\pi\, Z^5} \qquad ...(5.4)$$

where p_t = unit pressure developed at any point in the fill at a depth H below the surface due to traffic load.

P = Load superimposed.
Z = The slant height of the considered point from the load P.
H = Distance of the top of pipe below the surface of the fill.

The *total traffic load developed on a unit length of conduit* (W') can be found by integrating the Eqn. (5.4) over the projected area of the pipe. This integration can be done by subdividing the projected area of the pipe into small squares, and then computing the load on each sub-division, and finally adding these values to obtain a total.

In case of unsurfaced roads, the impact from moving loads may nearly double the computed loads. Paving considerably reduces the effect of impact. Moreover, the effect of superimposed load decreases rapidly as the depth of cover increases (because Z^5 increases much more than H^3).

The total load per unit length of the pipe from the backfill and from traffic can be found by adding W with W'. The compressive stress produced, which should be checked (when pipe is empty) will then be given by

$$\frac{W + W'}{t} \text{ kg/m}^2 \qquad \text{...(5.5)}$$

(where t is the thickness of the pipe in metres).

Note. There is sometimes the possibility of creation of vacuum in the pipe due to water hammer, which will increase the compressive stresses. The pipe thickness should be checked for that possibility also, so as to avoid collapse in such conditions.

Example 5.1. *A pipe 1 metre in diameter, buried in a trench 1.4 m wide, is backfilled with dry sand. The top of the pipe is 2.5. m below the surface of the fill. The pipe passes at right angles under a one lane road which carries a vehicle whose loading (including impact) consists of two concentrated 800 kg loads located at 1.8 m apart transverse to the roadway. Find the maximum vertical force exerted on a unit length of the pipe, if*

(*a*) *the pipe is made of steel.*

(*b*) *the pipe is a cast iron pipe.*

The thickness of the pipe may be neglected for calculating the external dia of the pipe.

(*c*) *what is the stress produced in steel pipe, if its thickness is 1 cm ?*

Solution. We will, first of all, evaluate the forces caused by *backfill*.

In this question, we have

$D \approx d = 1$ m (neglecting thickness, as given)

$B = 1.4$ m

γ for dry sand = 1600 kg/m^3 (assumed)

$H = 2.5$ m.

Case (*a*) The pipe is of steel—which is a flexible pipe

∴ Using Eqn. (5.2), we have

$$W = C.\gamma.B.D$$

The value of C from table 5.2, for $\frac{H}{B} = \frac{2.5}{1.4}$ = **1.8,** and dry sand is given by

$$= 0.84 + \frac{1.45 - 0.84}{1} \times 0.8 = 0.84 + 0.61 \times 0.8 = 0.84 + 0.49$$

$$= \mathbf{1.33.}$$

Substituting, we get

$$W = 1.33 \times 1600 \times 1.4 \times 1 \text{ kg/m} = \mathbf{2980 \text{ kg/m.}} \quad ...(i)$$

Case (*b*) The pipe is of cast iron, which is a rigid pipe. Using Eqn. (5.3), we have

$$W = C.\,\gamma.\,B^2 = 1.33 \times 1600 \times (1.4)^2 = \mathbf{4170 \text{ kg/m.}} \quad ...(ii)$$

Now, we will evaluate the forces caused by *traffic load.*

Using Eqn. (5.4), we have

p_t = pressure developed due to traffic load

$$= \frac{3H^3 \,.\, P}{2\pi Z^5}$$

Here in this question, we have

Z = The slant height from one of the wheel load to a point on the pipe midway between the load

$$= \sqrt{\left(\frac{\text{space between the wheel}}{2}\right)^2 + H^2}$$

$$= \sqrt{\left(\frac{1.8}{2}\right)^2 + (2.5)^2} = \sqrt{0.81 + 6.25} = \sqrt{7.06}$$

or Z = **2.66 m.**

But $p_t = \frac{3.\,H^3.\,P}{2\pi Z^5}$

Hence, pressure developed due to single wheel load

$$= \frac{3 \times (2.5)^3 \times 800}{2 \times 3.14 \times (2.66)^5} = \frac{3 \times 15.6 \times 800}{2 \times 3.14 \times 134} = \mathbf{44.4 \text{ kg/m}^2.}$$

Total force due to traffic load per metre length of pipe

$W' = p_t \times$ No. of wheels × projected area of the pipe

= (44.4) × 2(1)

[∵ Projected area* ≈ Dia × 1 metre length = 1 m × 1 m = 1 m^2]

W' = 88.8 kg/m length.

or Total external force inclusive of backfill and traffic per metre length of pipe in case (*a*)

$W + W'$ = 2980 + 88.8 = 3068.8 kg/m

Say 3069 kg/m. **Ans.**

---

*The more accurate estimate of superimposed load could have been made by analysing smaller unit areas and then adding the load on each subdivision (Z will then be different in all considered squares). But this labour seems hardly justified in this case.

Total external force inclusive of backfill and traffic per metre length of pipe in case (*b*)

$$= 4170 + 88.8 = 4258.8 \text{ kg/m}$$

Say 4259 kg/m. **Ans.**

Using Eqn. (5.5),

$$\text{Stress produced in steel pipe} = \frac{W + W'}{t} = \frac{4259}{0.01}$$

$$= 4{,}25{,}900 \text{ kg/m}^2 = 42.59 \text{ kg/cm}^2. \quad \textbf{Ans.}$$

5.3.3. Temperature stresses. When pipes are laid above the ground, they are exposed to the atmosphere ; and are, therefore, subjected to temperature changes. They expand during day time and contract at night. If this expansion or contraction is prevented due to fixation or friction over the supports, *longitudinal stresses* are produced in the pipe materials. The amount of these stresses may be calculated as below :

$$\text{Elongation} = L \,.\, \alpha \,.\, T$$

But, Strain = Elongation per unit length

$$= \frac{L \,.\, \alpha \,.\, T}{L}$$

$$= \alpha \,.\, T$$

$$\text{Stress} = E.\ \text{strain} = E_p \,.\, \alpha\, T$$

or

$$f = E_p\, \alpha \,.\, T \qquad \text{...(5.6)}$$

where E_p = Modulus of elasticity of the pipe material

α = Co-efficient of expansion of the material

T = Change in temperature in °C.

Special types of expansion joints at suitable intervals (say 20 m to 30 m or so) must be provided to counteract these stresses. But however, since the sewer pipes are mostly buried underground, these temperature stresses do not come into picture.

5.3.4. Flexural stresses. Sometimes, the sewer pipes may have to be carried supported between trestles, or piers, like beams. Similarly, sometimes the rain water, etc. may wash off the ground from below the pipes resting on the ground, thus exposing them like beams supported between two supports. Under all such circumstances, bending stresses get produced in the pipe, since the pipe acts like a beam with loads resulting from the weight of the pipe, weight of the sewage in the pipe, and any other superimposed loads. The stresses caused by this beam action may be determined by usual methods of structural analysis applied to beams. However, in normal circumstances, the stresses produced are small for smaller spans, and hence often neglected except for long spans or where there are huge superimposed loads.

5.4. Sewer Materials

Vitrified clay (or stone ware), cement concrete, asbestos cement and cast iron are the most common materials used for constructing sewer pipes. While selecting a particular material for constructing sewer pipes, the important **factors which must be considered** are :

(*i*) *Resistance to corrosion*. The sewer pipes are likely to be acted upon by sewer gases, and thus get corroded, due to the presence of acids and other impurities in sewage. The sewer material should therefore, be such as to be resistant to corrosion, and thus to last for a longer life.

(*ii*) *Resistance to abrasion*. When the sewage contains a lot of grit and sand particles, moving at a high velocity at the sewer invert, a lot of wear and tear of the sewer material may be caused due to abrasion. To avoid this erosion or wear and tear of the sewer pipe, the sewer material must be strong enough, so as to withstand such possible abrasions.

(*iii*) *Strength and durability*. The sewer pipes should be strong enough to withstand all the forces that are likely to come on them. Since they are laid well below the ground level, they are subjected to considerable external loads. However, they are generally not subjected to the internal pressure of water (as the water pipes are subjected to). In order to counteract the various forces* to which these pipes are likely to be subjected, sufficient thickness of the pipe material, or suitable reinforcement in the pipes, must be provided. Moreover, the pipe materials must be durable as not to give way quickly due to normal wear and tear, and thus to provide a longer life span and to avoid their frequent replacement.

(*iv*) *Light weight*. The material used for sewers should be light, so that the sewers can be easily handled and transported.

(*v*) *Imperviousness*. The sewer material should be impervious as not to allow any seepage of the sewage from the sewer.

(*vi*) *The economy and cost*. The sewer material must be cheaper and less costly as to cause overall economy in their construction.

(*vii*) *Hydraulically efficient*. The sewer material should be such as to provide a smooth interior surface (with Manning's N as low as possible) so as to provide an hydraulically efficient surface.

Besides cement concrete, asbestos cement and vitrified clay which are the commonly used materials, other materials which may also be used for sewer constructions are bricks, cast iron and plastics. The sewers of different possible materials and their comparative utilities are described below.

(1) **Asbestos Cement Sewers.** Asbestos cement pipes are manufactured from a mixture of asbestos fibre, silica and cement, converted under pressure to a dense homogenous material, possessing considerable strength, called asbestos cement. The asbestos fibre which is thoroughly mixed with cement serves as reinforcement, and provides a strong material. These pipe are normally available in sizes say from 10 to 90 cm in diameter and 4 metres in length.

Jointing. These pipes can be easily assembled without skilled labour, with the help of a special coupling (called Ring Tie coupling or Simplex joint) as shown in Fig. 5.4 (*a*) and (*b*). The assembly consists of a pipe sleeve and two rubber rings, which are compressed between the pipe and the interior of the sleeve. The joint is as resistant to corrosion as the pipe itself, and is flexible enough as to permit as much as 12° deflection, while laying the pipes around curves (in plan).

---

*Described in the previous pages.

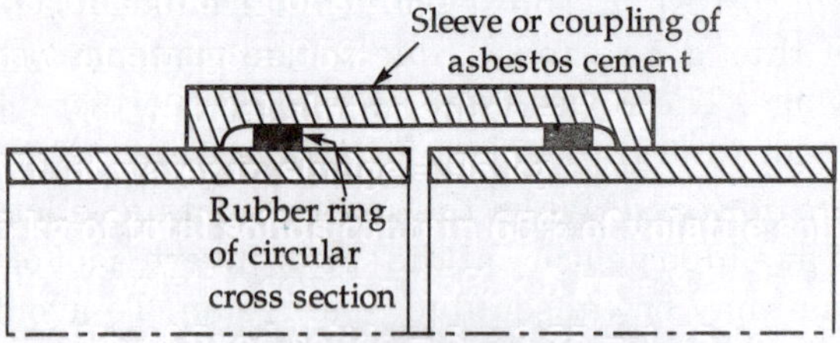

(a)

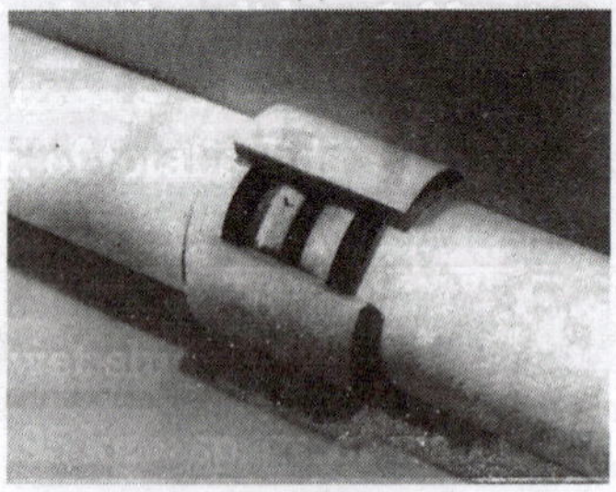

(b)

Fig. 5.4. Simplex joint for A.C. pipes. Rubber O-rings seal the joint and provide a degree of flexibility.

The *advantages* of A.C. pipes are :

(*i*) They are light in weight and hence easy to transport.

(*ii*) They can be easily cut and assembled without skilled labour.

(*iii*) Their interior surface is exceptionally smooth (with Manning's N = 0.011), thus providing an excellent hydraulically efficient sewer.

The *disadvantages* of A.C. pipes are :

(*i*) They are structurally not strong enough to bear the huge compressive stresses induced by the heavy external loads to which the deeply buried sewers may be subjected to.

(*ii*) They are susceptible to corrosion by sulphuric acid from hydrogen sulphide gas generated in sanitary waste water or by some industrial chemicals. The *sulphide corrosion* of asbestos cement as well as cement concrete pipes is a big problem in areas where the sewage is strong, stale and very warm, because under such conditions the bacterial activity responsible for producing hydrogen sulphide gas gets accelerated. Hence in all such cases, *vitrified clay (popularly called stone ware) pipes* should be used for sewers of less than 1 m in diameter, and cement concrete pipes with cast insitu plastic linings may be used for larger diameter sewers.

In view of their disadvantages, asbestos cement pipes are best suited to be used as verticals* for bringing down either the rain water from the roofs, or the comparatively less foul sullage from kitchens and bath rooms situated at the upper floors of the buildings**. The use of A.C. pipes for these purposes in place of cast iron pipes can lead to considerable economy.

(2) Plain Cement Concrete and Reinforced Cement Concrete Sewers. *Plain cement concrete pipes* are manufactured in small sizes (*i.e.*, up to say

---

*Not subjected to any earth pressure.

**Rain water and sullage are less foul, and thus avoiding chances of sulphide corrosion of A.C. pipes.

0.45 m diameter),* while they are reinforced with steel reinforcement for larger diameter pipes. RCC pipes are easily available in sizes up to diameters say 1.8 metres, and may be got manufactured for larger diameters say upto about 4.5 metres, on special orders. These pipes may either be prepared at site by transporting ingredients (*i.e.*, cement, steel, aggregates, water, etc.) or can be manufactured in factories, and then transported to site. They are known as *cast in situ* pipes in the former case, and *precast pipes* in the latter case. Cast insitu pipes are useful when the site conditions are difficult and where it may be difficult to carry the pipes. But since such pipes are cast at site, lesser supervision and check is possible as compared to the case of precast pipes which are cast in the factories and thus subjected to greater quality control and supervision.

The usual mix is $1 : 1\frac{1}{2} : 3$ (1 cement : $1\frac{1}{2}$ fine aggregate : 3 coarse aggregate) with maximum size of aggregate limited to 6 mm. The water cement ratio usually varies between 0.5 to 0.7 depending upon the thickness of the pipe section, and the equipment used for placing and compacting the concrete.

Reinforced cement concrete (R.C.C.) pipes are those concrete pipes which are provided with circumferential reinforcement to carry internal or external stresses, and a nominal longitudinal reinforcement equal to 0.25% of the cross-sectional area of concrete. The circumferential reinforcement is generally provided in three different ways, as shown in Figs. 5.5 (*a*), (*b*) and (*c*).

In a sewer of Fig. 5.5 (*a*), the main circular circumferential reinforcement has been provided near the inner surface of the pipe, so as to withstand the internal forces causing Hoop's tension. Such reinforcements are, therefore, provided in smaller sized pipes (less than 0.8 m in diameter) which are subjected to internal pressure only. In case of larger sized pipes (greater than 0.8 m in diameter) subjected to internal as well as external pressures, two sets of circular circumferential reinforcements (one near the inner face, and one near the outer face) as shown in Fig. 5.5 (*b*), may be provided. However, in case of large diameter sewers subjected to external pressures alone (as is the normal case for sewers), an elliptical cage reinforcement, such as shown in Fig. 5.5 (*c*), may be provided.

The non-pressure R.C.C. pipes are classified according to IS : 458—1988 into the following three categories :

(*i*) *NP2 pipes.* They are light duty R.C.C. non-pressure pipes, normally used for drainage and irrigation use, for culverts carrying light traffic. The thickness of NP2 pipes vary from 25 mm for 8 cm dia pipe to 110 mm for 2.2 m dia pipe.

(*ii*) *NP3 pipes.* They are medium duty non-pressure pipes, normally used for drainage and irrigation use, for culverts carrying medium traffic. The thickness of NP3 pipes vary from 25 mm for 8 cm dia pipe to 215 mm for 2.6 m dia pipe.

(*iii*) *NP4 pipes.* They are heavy duty non-pressure pipes, normally used for drainage and irrigation use, for culverts carrying heavy traffic, such as railway loadings.

* Plain C.C. pipe are non-pressure pipes, and classified as NP1 pipes according to IS 458—1988 ; their thicknesses varying from 2.5 to 3.5 cm. for dia varying from 8 cm to 45 cm.

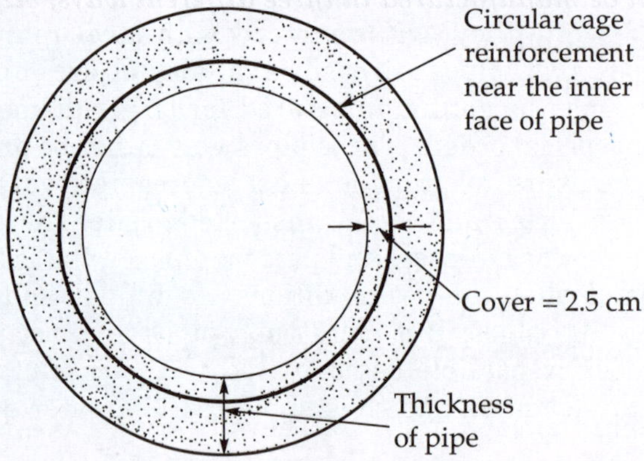

(*a*) Single cage reinforced pipes (only main circumferential reinforcement shown)

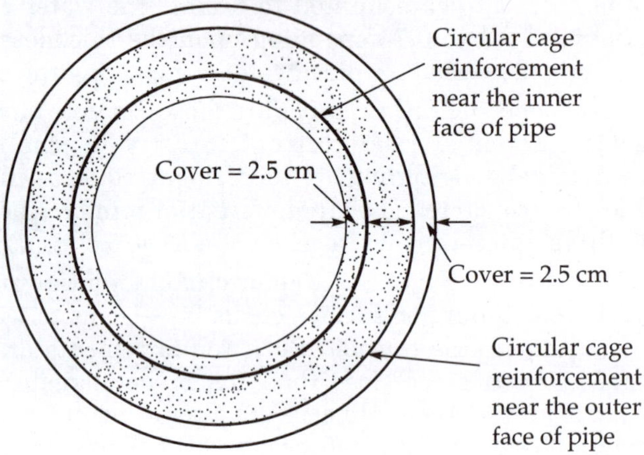

(*b*) Double cage reinforced pipes (only main circumferential reinforcement shown)

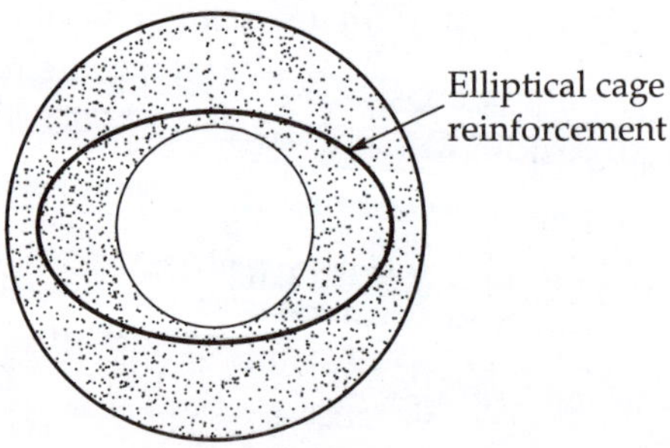

(*c*) Elliptical cage reinforced pipes (only main circumferential reinforcement shown)

Fig. 5.5 Reinforced cement concrete pipes.

The unreinforced as well as reinforced concrete pipes shall be capable of withstanding a test pressure of 7 m head of water.

R.C.C. pressure pipes, classified as P1, P2 and P3 pipes, are generally used for carrying water supplies under pressure and are usually not used as sewers which are designed as gravity conduits.

R.C.C. pipes can be manufactured in three different ways, viz. :

(*a*) Pipes having bar and mesh reinforcement, and concrete poured by usual ordinary methods of concrete pouring and tamped.

(*b*) Pipes made by rotating the mould or the form, rapidly about the pipe axis. The mould contains concrete and fabricated reinforcement. The centrifugal force throws of the concrete, which then spreads in a uniform layer over the internal surface of the mould and embed the reinforcement ; thus providing a high density water-tight concrete surface. This type is known as *centrifugal type*.

(*c*) The third type of pipes are made by lining thin cylindrical steel shells, both internally and externally, with rich cement concrete. These are stronger and more water-tight than the first two. They are known as *cylinder type*.

Hume steel pipes are also the R.C.C. pipes patented under this name, and consist of thin steel shells coated from inside with *cement mortar* by centrifugal process. The thickness of the inside coating varies from 12 mm to 30 mm depending upon the size of the pipe. They are also coated from outside, so as to protect the steel shell from external weather or soil action. The thickness of external coating is 25 mm for pipes up to 1 metre in diameter, and is 37.5 mm for pipes of larger diameters. The thickness of steel shell depends upon the size of the pipe, and also upon the pressures to be borne by the pipe. They are available in sizes ranging between 10 cm to 2.4 m in diameter, and in lengths varying from 0.9 to 2.4 m.

Advantages of concrete pipes are given below :

(*i*) All these different forms of cement concrete pipes are quite strong in tension (for withstanding internal pressures) as well as in compression (for withstanding external loads).

(*ii*) They are quite resistant to erosion and abrasion.

(*iii*) They can be easily moulded and manufactured either at site or in the factory.

(*iv*) They can be made of any desired strength by proper design and proportioning of concrete mixes.

(*v*) Their *precast* forms may be easily used at places where, owing to ground water or running sand conditions, brick sewers or cast at site concrete sewers can not be used.

(*vi*) They prove economical in medium and large sizes, and hence, widely adopted for branch and main sewers.

The *biggest drawback* of the concrete sewers, however, is the fact that they easily get corroded and pitted by the action of sulphuric acid produced from hydrogen sulphide gas (evolved from the stale sewage) or from such other chemicals present in sewage. This not only reduces the life span of the sewers but also reduces their carrying capacities with time. Besides corrosion, they are also susceptible to erosion by sewage containing too much silt and grit.

The concrete sewers can be protected from such actions by lining their interiors with *vitrified clay linings,* as shown in Fig. 5.6. The blocks of vitrified clay, for this purpose, are provided with projections, which project and enter the cement concrete as shown. The joints between adjacent blocks are filled either with rich cement mortar or with bituminous compounds.

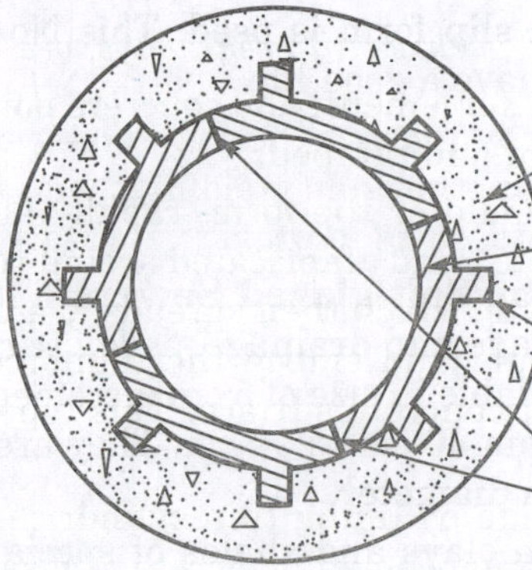

Fig. 5.6. Cement concrete pipe, lined inside with vitrified clay lining.

Other methods of protecting concrete sewers from hydrogen sulphide corrosion are :

(*i*) Prohibiting the entry of wastes containing sulphides.
(*ii*) Reducing the sulphate contents by pre-treating the sewage.
(*iii*) Aerating and chlorinating the sewage.
(*iv*) By adequately ventillating the sewers.
(*v*) By making the sewers to run full.
(*vi*) By adding such chemicals to sewage as may neutralise the already present sulphur compounds.

In view of its merits and drawbacks, the unlined cement concrete sewers are widely used for carrying storm water, which is comparatively free from organic impurities responsible for generating hydrogen sulphide gas. They may, however, be used for branch sewers bringing sewage free from industrial wastes. With protective linings, they are used for almost all the branch and main sewers.

Jointing. The concrete pipe lengths, flowing under gravity, can be easily joined with a *mortar caulked bell and spigot joint*, such as shown in Fig. 5.7.

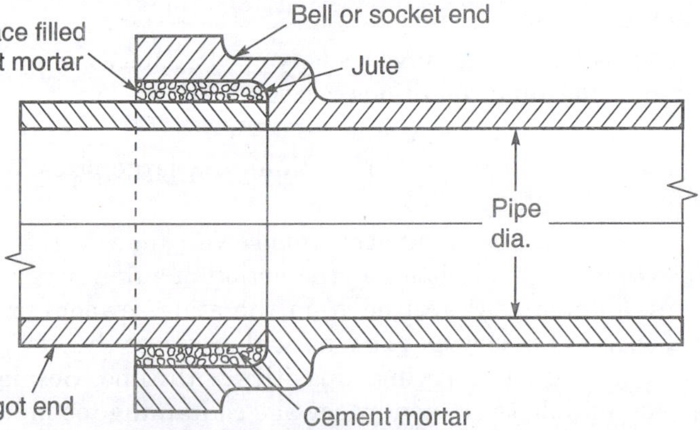

Fig. 5.7. Bell and Spigot Joint.

For gravity flows, even a No-joint concrete pipe has been developed in California. This pipe has been developed in sizes of 0.6 m to 1.8 m in diameter.

A special pipe laying machine with a slip form is used. This No-joint pipe, though not reinforced, is yet found to have a good life.

The R.C.C. pipe lengths are joined by placing the protruding end bars of different lengths butting against one another and welding them, and finally filling the gap with rich cement concrete, so as to provide a water-tight joint.

(3) Vitrified clay or Stoneware or Salt-glazed Sewers. Vitrified clay pipes are widely used for carrying sewage and drainage, as *house connections* as well as *lateral sewers*. They are available in size of 5 cm increments from 10 to 30 cm, and in 7.5 cm increments from 30 cm to 90 cm. They are, however, rarely made in sizes bigger than 90 cm diameter.

These pipes are manufactured from clays and shales of special qualities, which are, first of all, pulverised and mixed thoroughly with water. This mixture is then used for casting standard pipe sections in a pipe press at a pressure of about 8.5 kg/cm^2. These moulded pipe sections are dried in warm air, and then burnt in hot kilns under controlled temperatures. The temperature of the kiln is maintained at about 150°C, in the beginning for several hours, and then raised to about 700°C, and finally to about 1200°C, when fusion or *vitrification* of clay takes place. This makes it very dense and hard. Near the end of the burning period, sodium chloride (*i.e.* common salt) is placed in the kiln. The intensive heat causes the salt to vaporise, which reacts with the clay, forming a thin smooth, hard, and a waterproof glazed layer on the pipe surfaces. These pipes are joined by a bell and a spigot flexible compression joint, in which the precision mated surfaces are in tight contact with one another. These pipe lengths are, therefore, cast with having bell and spigot ends, in lengths of about 0.9 to 1.2 m. The interior surface of the socket end and the exterior surface of the spigot end, are, however, not glazed so as to make a water-tight joint.

The *advantages* of these pipes are :

(*i*) The stone-ware pipes offer the maximum advantage of being highly resistant to sulphide corrosion, and therefore, preferred for carrying polluted sewage and industrial wastes.

(*ii*) Their interiors are very smooth and they are hydraulically very efficient.

(*iii*) They are highly impervious and do not allow any sewage to seep out of them.

(*iv*) They are, though weak in tension, yet quite strong in compression, and hence they are quite suitable for withstanding compressive stresses caused by traffic and back-fills. They are also quite strong to withstand beam action under superimposed loads. So much so, that they can withstand loads of about 4.5 m soil cover, if a pipe length remains hanging between joints due to the removal of soil from below.

They, however, can withstand only very small tensile stresses caused by internal pressures. Hence, they can, though withstand slight tensile stresses caused by some chancy surcharge of gravity sewers, yet cannot be used as sewers flowing under pressure.

(*v*) These pipes are quite cheap, durable, easily available, and can be easily laid and jointed.

(*vi*) They are made non absorbant, so as not to absorb water more than 5% of their own weight, after kept immersed in water for 24 hours.

The *disadvantages* of these pipes are :

(*i*) They are heavy, bulky, and brittle, and, therefore, difficult to transport. Due to this reason, they are cast only in smaller sizes and smaller lengths. Due to their shorter lengths, numerous joints are required in laying such pipes, and due to smaller sizes, they cannot be utilised as branch or main sewers.

(*ii*) They cannot be used as pressure pipes, because they are weak in tension.

(4) Brick sewers. Bricks had been used as sewer material since ancient days. They, however, have now-a-days been almost replaced by cement concrete sewers. However, they may still be used at places where the sewers are required to the constructed at the site and ingradients required for cement concreting may not be easily available. They may also be preferred for constructing large sized combined sewers, or particularly for storm water drains.

Brick sewers are generally plastered on their outer surfaces so as to prevent the entry of tree roots and ground water through the brick joints ; and are lined inside with stone ware or ceramic block so as to render them smooth and hydraulically efficient. The stoneware or ceramic coating also helps in resisting sulphide corrosion which is not possible with the ordinary cement plaster as the same is easily attacked by sewer gases like hydrogen sulphide.

(5) Cast iron sewers. Cast iron pipes are structurally stronger and capable of withstanding greater tensile, compressive, as well as bending stresses, but are costlier, compared to cement concrete or stone ware pipes. They are, therefore, used as sewers, only under special circumstances, such as :

(*i*) For outfall sewers, for rising mains of pumping stations, or for inverted siphons, all running under pressure.

(*ii*) For sewers to be laid below heavy traffic loads, such as those laid below highways or railways.

(*iii*) For sewers carried over piers or trestles while crossing low lying areas.

(*iv*) For sewers which are to be 100% leak proof, so as to avoid possible contamination of under ground water supplies.

Cast iron pipes are though structurally quite stronger and durable, yet can not withstand the corrosive action of gases and other acids present in sewage ; and hence generally lined from inside with cement concrete, or painted with coal tar, etc. Although the sewer pipes are not subjected to high pressures, but still they are made as heavy or even heavier than the water pipes, so as the resist the corrosive action of sewage.

Cast iron pipes can be manufactured by two methods. One is the ordinary *sand moulding method*, and the other is the *centrifugal process*. Pipes, cast in horizontal position by ordinary sand moulding are called *Mcwane pipes*, while those cast vertically are called *pit cast pipes*. Horizontal cast pipes are 100 percent strong in tension and 50 percent stronger in rupture than vertically cast iron pipes. Centrifugally cast iron pipes are made either in sand or metallic moulds. The former are called *Sand spun pipes* and the latter are called *Delavaud pipes*. The spun iron pipes are denser and tougher than the pipes moulded by ordinary methods.

The Indian specifications on cast iron pipes are given in IS : 1536—1967, and its important provisions are given in Vol. I—"Water Supply Engineering". They may be referred to, if needed. The cast iron pipes are generally manufactured in lengths of about 3.5 metres but may also be manufactured up to 6 m or so, on special orders. The different pipe lengths are jointed together by means of *a bell and spigot joint* (such as shown in Fig. 5.7) in which the annular space may be filled either by cement mortar or by lead. Lead may be used in place of cement mortar at places where absolute water tightness is required or where internal pressures are likely to be produced.

(6) Plastic Pipes. During the last decade, plastic pipes have become quite popular for being used as *non-pressure sewer pipes*, in addition to their use as *pressure pipes* for carrying potable as well as raw waters, in water supply systems. They can be easily jointed with ***solvent cement.***

The various types of plastic pipes which have been developed over the last about 10 years, have been described in author's sister publication "Environmental Engineering Vol. I—Water Supply Engineering". As described in that volume, there are four types of plastic pipes, *viz.*

(*i*) *PVC (Poly Venyl Chloride) pipes,*

(*ii*) *CPVC (Chlorinated Poly Venyl Chloride) pipes,*

(*iii*) *UPVC (Unplasticized Poly Venyl Chloride) pipes, and*

(*iv*) *HDPE (High Density Poly-Ethylene) pipes.*

Out of these four types of plastic pipes, PVC and CPVC pipes are usually soft and bendable, *while UPVC pipes are hard and rigid, and hence most widely used for sewers*, as these pipes are as strong as C.I. pipes, though much easy to cut and joint besides being economical, and available in longer lengths of 6 m or so.

Although use of *UPVC pipes* has become very common even in India, for domestic sewers up to 150 mm (6 inch) dia, yet they are being rarely used in larger sizes for municipal sewers in our country.

UPVC pipes have high chemical resistance across its operating temperatures. Due to their long term strength characteristics, high stiffness, and cost effectiveness, UPVC piping systems account for a large proportion of plastic piping installations, including their use in industries like petro-chemicals, fertilizers, etc.

CPVC pipes, on the other hand, are generally being used for house water supply plumbing systems, where hot water applications exist, since such pipes have proven value across high heat and pressure applications that require corrosion resistance. There is almost NIL use of CPVC pipes for sewerage or drainage.

Plain PVC pipes do contain plasticizers*, and are useful for cold water applications, since they cannot resist high temperatures, above 60°C or so. PVC is not only used for making PVC pipes, but is used for making other items, like cables, rain coats, bottles, flooring, credit cards, etc.

HDPE pipes are made up of petroleum, and used for piping systems which require complete protection against corrosion. Such plastics are very elastic

---

**Plasticizer* is a substance, typically a solvent which is added to a synthetic resin, to produce or promote plasticity and flexibility and also to reduce brittleness.

and malleable, and can be moulded or cast into different shapes. That is why HDPE materials are widely used for making daily use items, like chairs, buckets, plastic jugs, bottles, lunch boxes for children, cook and serve bowls for microwave ovens, etc.

HDPE offers the following technical and economical advantages in making HDPE pipes and ducts, as compared to other conventional pipes:

(*i*) *No corrosion and abrasion,* no scale formation. Inert to chemicals and hence can carry acidic and alkaline fluids.

(*ii*) *Leak free joints,* no repairs needed.

(*iii*) *Longer lengths,* minimum joints and fittings. Size up to 50 mm outer dia supplied in coil length up to 100 m.

(*iv*) *Flexible,* no additional joints needed at bends. Easy to coil and transport.

(*v*) *Light weight,* no heavy equipment needed for installation, easy to handle and transport.

(*vi*) *Food grade suitability,* can be used to transport milk, edible oils, etc.

(*vii*) *Lower surge pressure.* As compared to other conventional pipes, HDPE piping system components are subjected to much lower surge pressure.

(*viii*) *Lowest frictional losses,* and hence, lower pumping costs.

(*ix*) *Zero maintenance*

(*x*) *Re-usable,* HDPE pipes can be easily dismantled from a particular location and application, and can be easily re-used at different locations.

HDPE pipes, due to their such advantages are widely used in irrigation piping systems such as for :

(*a*) *Flood irrigation.* (suction and delivery pipes in pump sets).

(*b*) *Sprinkler irrigation* (crops, lawns, golf courses, gardens etc.)

(*c*) *Drip irrigation* (plantations, overheads, nurseries, etc.)

In water supply systems these pipes can be used for potable water supplies in water mains, distribution mains, as well as service pipes. Similarly for sanitary systems, they can be used in domestic sewage systems, besides being used in industrial piping systems, such as for petro chemical industry, fertilizer industry, etc.

HDPE ducts can be used for air-conditioning and refrigeration; extraction of fumes and telecommunication conduits.

HDPE pipes are also used as conduits for cables for electrical installations. They are also used as *drainage pipes* for

(*a*) surface and rain water,

(*b*) waste and water mains

(*c*) sub-soil water.

(7) Lead Sewers. The lead pipes are smooth, soft, and can be easily bent to take odd shapes. They are also not affected by acid or alkaline sewage discharges, and can resist sulphide corrosion. They are, however, very costly.

The lead pipes are occasionally used in smaller sizes (3 to 4 cm diameter) and in smaller lengths in the toilets. They may be used as a down take pipes of flushing cisterns, or as waste pipes from stall urinals and wash basins, or for geyser connections.

5.5. Laying and Testing of Sewer Pipes

During planning a sewerage system for a city or a town, the alignment of various laterals, branches and main sewers are, first of all, decided on a contour map of the area. The populations or catchment areas, to be drained by each pipe, are then marked. The required sizes of the pipe sections and their gradients are then computed, so as to enable each pipe to pass the maximum sewage that may enter into it. Separate L sections are then drawn for each sewer line, showing the invert levels at different R.Ds (Reduced distances) and the typical cross-sections between them. The positions of manholes at suitable intervals along the route of the sewer line are also decided and marked. The positions of other appurtenances* are also decided and marked on the maps.

After finalising all the details of the different sewer lines on the maps, the work is got approved, and finally executed. During execution, the sewer pipes will have to be laid at their designed levels ; the different pipe lengths to be suitably jointed ; and finally tested for their corrected alignments and water tightness. In this article, we are describing the procedures adopted for laying and testing these sewer pipes.

5.5.1. Laying of the Sewer Pipes. All the sewer pipes are generally laid starting from their outfall ends, towards their starting ends. The advantage gained in starting from the tail end, (*i.e.* outfall end) is the utilisation of the tail length even during the initial periods of its construction, thus ensuring that the functioning of the sewerage scheme has not to wait till the completion of the entire scheme. The laying of the sewer pipes is, therefore, started from the outfall end, and proceeded upward by locating the different points along the proposed alignment on the ground. It is a common practice, to first locate the points where manholes are required to be constructed (as per the drawing *i.e.* L-section of sewer), and then laying the sewer pipe between the two manholes.

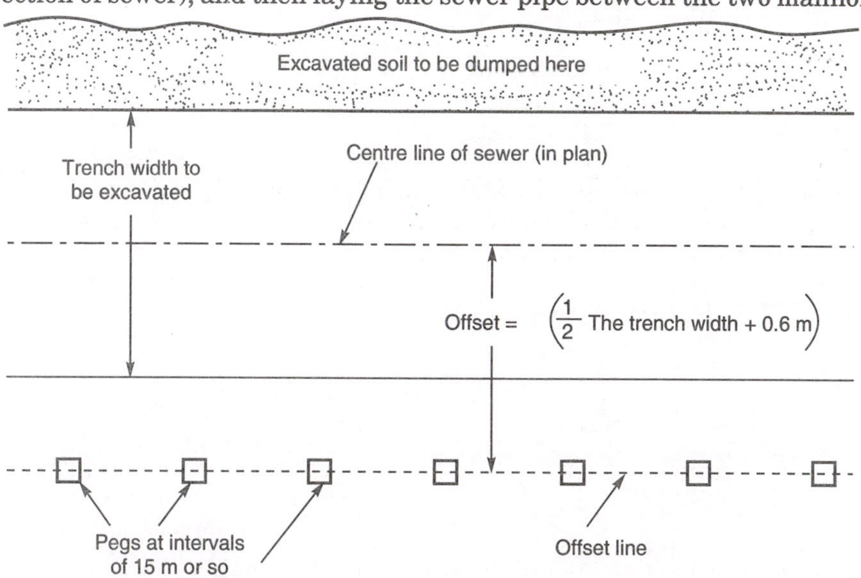

Fig. 5.8

The centre line of the sewer may be best marked by drawing an *offset line* on the ground, parallel to the *centre line* at a suitable fixed horizontal distance of

*Described in next article.

2 to 3 m away from it. This *offset line*, usually offsetted from the centre line of the sewer by a suitable distance of about half the trench width + 0.6 m or so, is therefore, marked on the ground by fixing pegs at intervals of 15 m or so, as shown in Fig. 5.8. This offset line should be drawn on that side of the trench which is not likely to be disturbed by the piling of the excavated earth, and can always be used for finding out the centre line of the sewer (in plan) simply by offsetting the fixed distance from this offset line.

Temporary bench marks along this offset line, at intervals of 200 to 300 m, should also be established by carrying the levels from a G.T.S. bench mark.

Excavating trenches. The excavation work is now started. The road pavements may have to be, first of all, removed. Softer pavements can be removed with pick axes using manual labour ; whereas, the hard concrete pavements may be removed by using mechanical appliances like pneumatic drills or spades. The earth or rock from below the pavements is then excavated so as to dug the required trench. This excavation may be carried out either by pick axes using manual labour, or by using machines like power shovels, boom and bucket excavators, track excavators, continuous bucket excavators, etc. The width of excavation at any level will depend upon the width of the trench at the bottom, and the additions due to side slopes and due to timbering etc. The trench is excavated between two manholes, and the sewer is laid between them. After completing the laying of the sewer between two manholes, further excavations are carried out for laying the pipes between the next consecutive manholes. The process is continued from the outfall end of the sewer towards the uphill, till the entire sewer is laid out.

The excavated material is deposited on one side of the trench ; and the other side being used for the offset line, and also for lowering the sewer pipes and other construction materials into the trench.

The width of the trench at the bottom is generally kept 15 cm more than the dia of the sewer pipe, subjected to a minimum value of about 60 to 75 cm for smaller dia pipes, so as to facilitate laying and jointing of pipe lengths. If sewers are to be cast at site, no extra space is required. If the sewer pipes are not to be embedded in concrete (which is possible for smaller pipes laid in very firm grounds), then the bottom 1/2 portion of the trench is excavated in such a way that it confirms to the shape of the pipe itself. But generally, as pipes are laid in ordinary or softer grounds, they are embedded in concrete, as shown in Fig. 5.9 (*a*) and (*b*). The trench, in all such cases, should also, therefore, go upto the level of the bottom of concrete, as shown in Fig. 5.10.

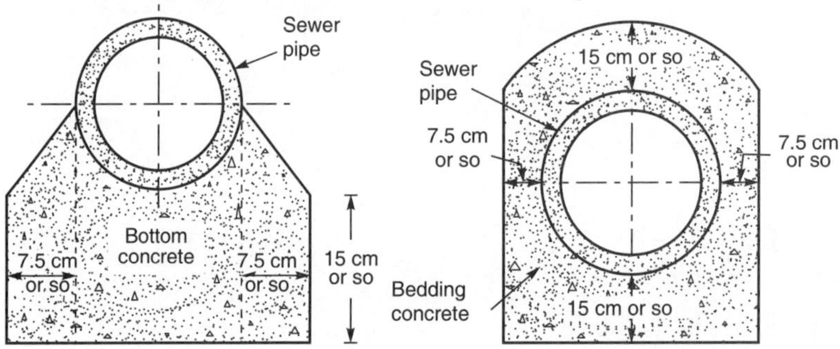

(*a*) Bedding of sewers in ordinary soils

(*b*) Bedding of sewers in softer soils

Fig. 5.9

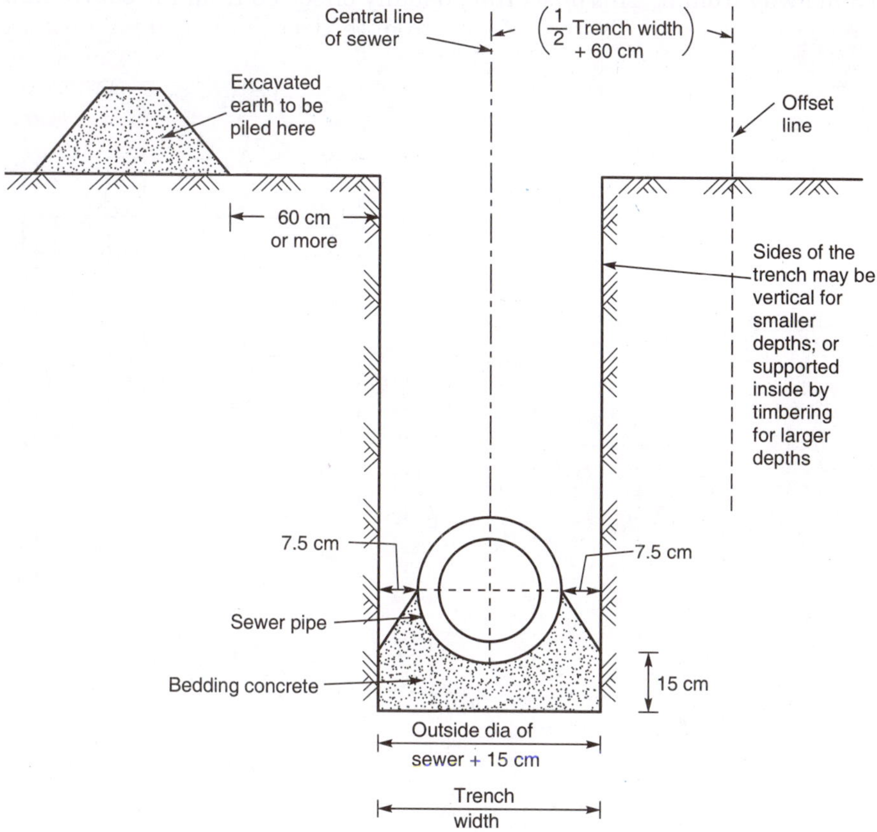

Fig. 5.10. Excavation of Trench for laying Sewers.

In order to accommodate the socket end or the collar of the pipe, and also to facilitate the caulking at the bottom of the pipe with hands, a recess, sufficient for the purpose, is also excavated in the bottom of the trench, at required intervals (equal to each pipe length).

Timbering or Sheeting Trenches. The trenches are excavated with vertical sides or with sloping sides, keeping in view the fact that the soil may cave in, posing dangers to the labourers working in those trenches. When trenches with vertical sides are excavated in depths more than 2 m or so in ordinary soils, it is often necessary to support the soil by sheeting and bracing the trenches. The bracing will absorb the soil pressure, and prevent it from collapsing. The ***sheeting*** or the sheeting boards are the wooden planks which are placed in actual contact with the trench sides, either horizontally or vertically ; whereas the ***braces*** are the cross wooden pieces extending from one side of the trench to the other side, and may also be called ***struts***. The *Ranger* or *Waless* are the timber planks which transfer the load from the sheeting boards to the cross braces by jointing the sheeting boards together, as shown in Fig. 5.11. This entire assembly of sheeting boards, braces and rangers is called *timbering* or *shoring* of the trench.

The extent of timbering required depends upon the type of the soil excavated, and also upon the depth of the excavated trench. For softer soils

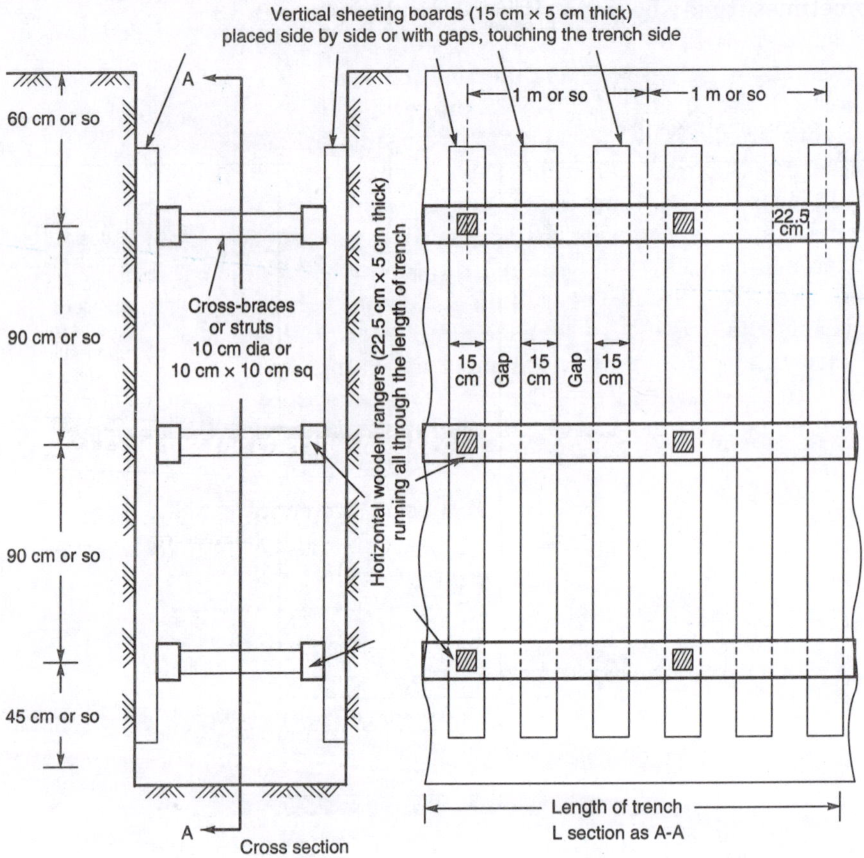

Fig. 5.11. Extent of Timbering required under Ordinary conditions.

and deeper trenches, more elaborate and closer timbering than that of Fig. 5.11, and consisting of a number of closely spaced sheeting boards, rangers and struts, may be required. However, for hard soils and for lesser depths of excavations, lesser timbering consisting of only sheeting boards at good enough intervals and jointed by cross braces without any rangers, such as shown in Fig. 5.12, may prove to be sufficient. In case of extremely soft and water-logged soils, or in places where timber is not easily available, steel sheeting boards may be used in place of wooden sheeting boards. Steel sheeting compared to wooden sheeting is more watertight, durable and stronger, but is more costly. The timbering or sheeting is generally withdrawn after the sewer has been laid, and before the trench is refilled, although

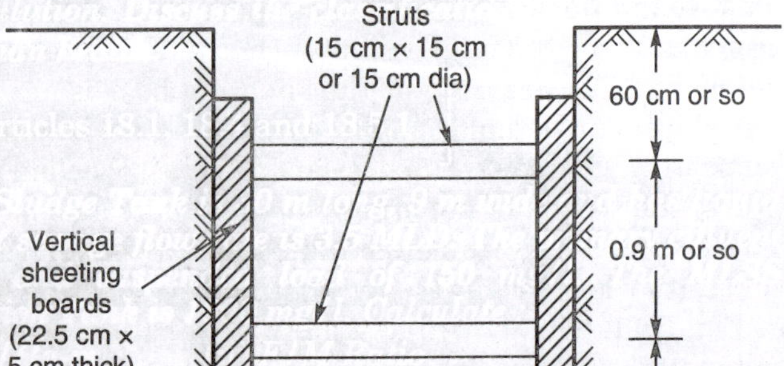

Fig. 5.12. Timbering without Rangers.

sometimes it may be left as it is, particularly in wet soils, where the removal of sheeting may permit the movement of the soil, endangering the sewer, or where, the arch of the sewer is supported against the sheeted sides of the trench. While removing the timbering, the braces and then the rangers are, first of all, loosened and removed. The process requires a lot of care and experience, as otherwise, the sides of the trench may cave-in, endangering the labourers. The sheeting planks are finally removed. If a hoisting engine is available, a loop of rope or a cable is thrown around one or more sheeting planks and they are pulled out. Steel piles have a hole in their upper ends, in which a hook can be placed, and thus pulled out through the cable attached to the hook. The timber sheeting planks can also be removed by hands in the absence of hoisting engine, with the help of a bar and a clamp, as shown in Fig. 5.13.

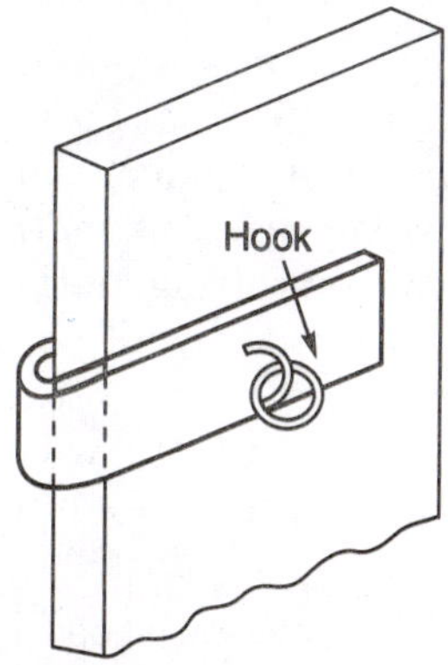

Fig. 5.13. Steel clamp for removing timber sheeting boards.

Dewatering of Trenches. While excavating a trench, the ground water may appear, if the water-table happens to be high or if the sewer happens to be laid very deep. This ground water will create problems in further excavations, if needed, and also in laying and jointing the pipes. Moreover, this ground water is likely to infiltrate into the sewers, thus considerably increasing the load on them. To avoid these troubles, ground water must be removed. The ground water may be removed through an open jointed drain constructed below the sewer trench (Refer Fig. 5.14), which discharges into an independent water course either by gravity or pumping.

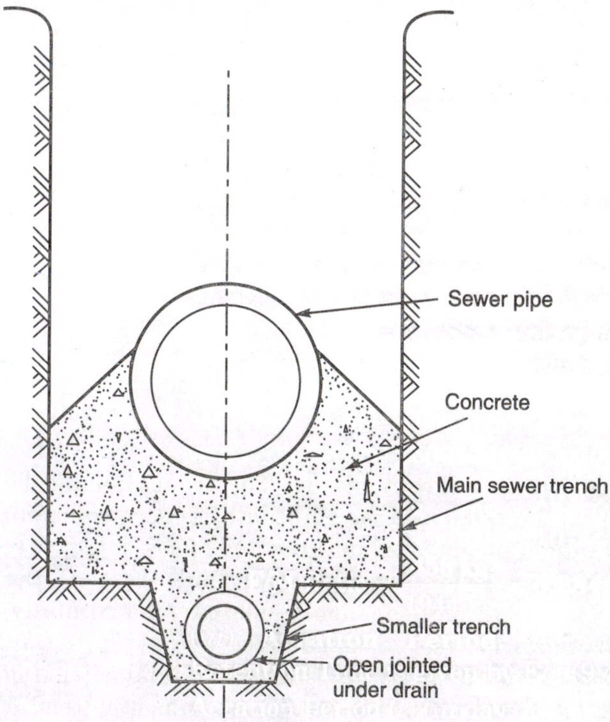

Fig. 5.14

Sometimes, the ground water is removed by pumping out the water from the sump constructed in the trench. The pump in such a case will be operated throughout the 24 hours of the day, as the water keeps on flowing into the trench. This method can be used on smaller jobs and in non sandy soils.

The problem of dewatering in sandy soils becomes enormous, because if the water flow is large and the subsoil is sandy, a *quicksand** is caused, which tends to flow out into the trench, thus undermining the sides of the trench. For larger jobs in such sandy soils, the dewatering is done by means of *well point drainage*. In this method of dewatering, a number of well points are installed by water jetting method, alongside the trench, at suitable intervals of 1.5 m or so. All these well points are connected to a common *header pipe*, which, in turn, is connected to a pump, as shown in Fig. 5.15. The water is removed by pumping, till the water level goes below the bottom of the trench. Each well point should be equipped with a valve, so that it may be cut off from the header, and, thus, to have greater control on the outflow.

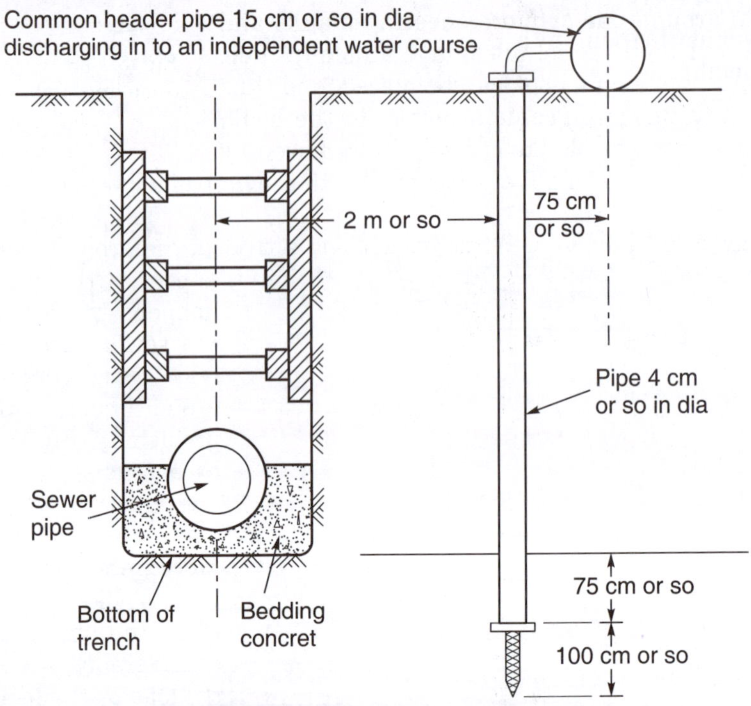

Fig. 5.15. Well point discharge for dewatering trenches.

Shaping the trench bottom to correct levels. The trench should be excavated up to a level equal to the bottom of the embedding concrete or upto the invert level of the sewer pipe if no embedding concrete is provided.

Hence, when the trench excavation reaches the final stages, the bottommost portion of the trench (30 cm or so) should be excavated carefully, so as not to dug more than the required. The designed invert levels at different RDs and the designed slope between them, as per the longitudinal section of the sewer, should, therefore, be precisely transferred to the trench bottom. A simple and a practical method normally adopted for this purpose is described below with reference to Fig. 5.16.

Two vertical posts are first of all driven into the ground at the sides of the trench. A horizontal board spanning the entire width of the trench is now fixed

*A quicksand may be defined as a loose, granular material through which water is moving with a sufficient velocity, so as to separate the grains. and keep them in partial suspension.

on these posts by means of nails and is called **sight rail.** A vertical pencil line is marked on this wooden board with the help of a theodolite, so that this line falls on exact alignment of sewer. A small horizon-tally projecting needle is now fixed on this pencil line at a known reduced level, so that it lies above the invert level of the sewer at that point by a known height say h, and above the bottom of conerete by a height say h'.

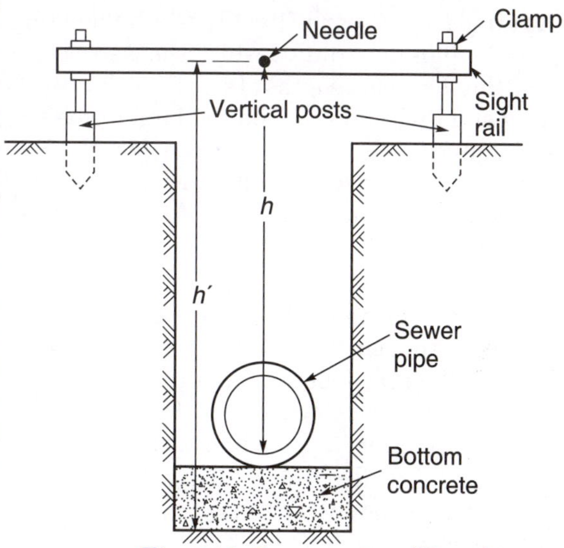

Fig. 5.16. Cross-section of Trench.

Sight rails having such nails fixed on them are thus erected all along the trench at intervals of 10 m or so, and at all junctions and *change points**. A string is now stretched between the needles of consecutive slight rails. This will establish a line in space which will be exactly parallel to the sewer line, and will be at a known height h above the proposed sewer invert. This line may be conveniently used for measuring from it, the exact depth of excavation, the level of top of foundation concrete, and the line and level of the sewer pipe. Usually, a rod of adjustable length, called **boning rod** or a *Traveller* (adjusted to a length of h or h') is used to measure from the string line, so as to reach precisely upto the required invert of sewer, or the bottom of concrete.

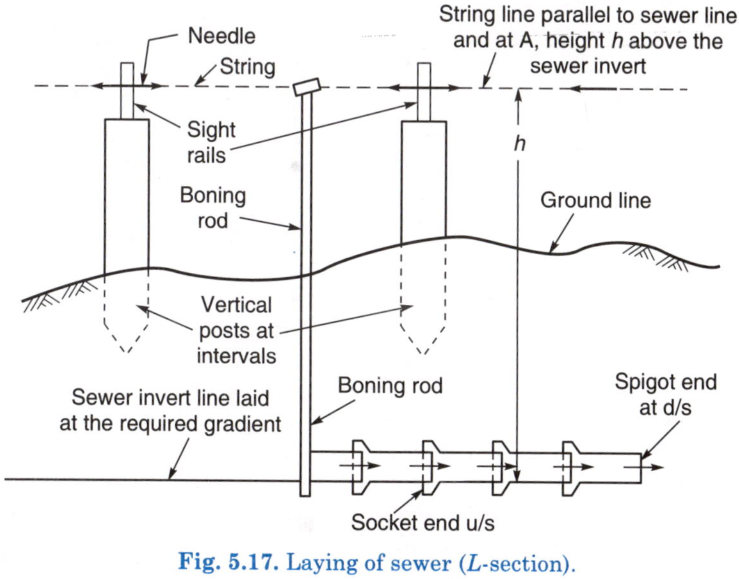

Fig. 5.17. Laying of sewer (*L*-section).

*The points where either the slope or alignment changes.

Laying the sewer pipes. After the bedding concrete has been laid in the required alignment and levels, the sewer pipes are lowered down into the trench. Smaller pipe lengths of sizes less than 40 cm dia can be directly handled by masons and their helpers, and the larger pipe lengths can be lowered by passing ropes around each end of the pipe length. For very large pipes, sometimes, machines may have to be used for lowering down the pipes into the trench.

The sewer pipe lengths are usually laid from the lowest point *with their socket ends facing upstream*, as shown in Fig. 5.17. In this way, the spigot of each pipe can be easily inserted in the socket of the already laid pipe. After fitting the socket and spigot into each other, the proper jointing is done with lead caulking or cement mortar as required. During jointing, care must be taken to see that the pipe lengths remain in true level and alignment, and do not get disturbed till the jointing material hardens. After the sewer has been laid in this way, it is tested for leakage and correct-alignment, as described below :

5.5.2. Testing of the Sewer Pipes.

The sewers after being laid and jointed, are tested for *watertight joints*, and also for *correct straight alignment*, as described below :

(1) **Test for leakage, called Water test.** The sewers are tested, so as to ensure '*no leakage*' through their joints after giving a sufficient time to these joints to set in. For this purpose, the sewer pipe sections are tested between manhole to manhole under a test pressure of about 1.5 m of water head.

In order to carry-out this test on a sewer line between two manholes, the lower end (*i.e.* downstream end) of the sewer is, first of all, plugged, as shown in Fig. 5.18. The water is now filled in the manhole at the upper end, and is allowed to flow through the sewer line. The depth of water in the manhole is maintained to the testing head of about 1.5 m. The sewer line is watched by moving along the trench, and the joints which leak or sweat, are repaired. The leaking pipes, if any, will also be replaced.

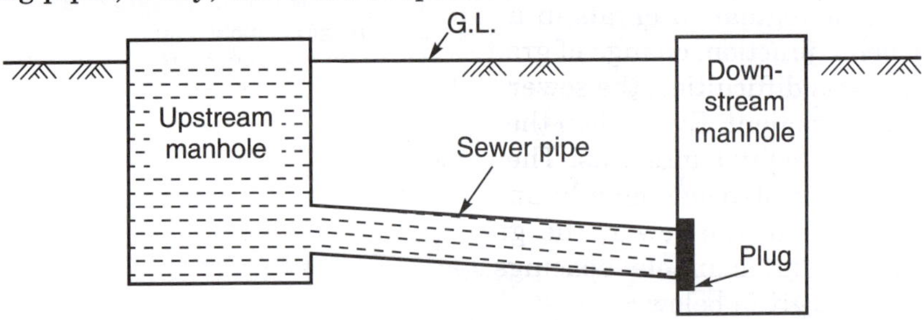

Fig. 5.18

(2) **Test for straightness of alignment and obstruction.** The straightness of the sewer pipe can be tested by placing a mirror at one end of the sewer line and a lamp at the other end. If the pipe line is straight, the full circle of light will be observed. However, if the pipe line is not straight, this would be apparent, and the mirror will also indicate any obstruction in the pipe barrel.

Any obstruction present in the pipe can also be tested by inserting at the upper end of the sewer, a smooth ball of diameter 13 mm less than the internal dia of the sewer pipe. In the absence of any obstruction, such as, yarn or mortar projecting through the joints, etc. ; the ball shall roll down the invert of the sewer pipe and emerge at the lower end.

5.5.3. Back-filling of the Trenches. After the sewer line has been laid and tested, the trenches are back-filled. While back-filling the excavated earth in the trench, the earth should be laid equally on either side of the sewer, and the earth should be filled up in layers of about 15 cm thickness. Each layer should be properly watered, temped and rammed. However, the earth filling above the crown of the sewer pipe should be carefully carried out by hand shoveling in layers and using selected soils (*i.e.* soft earth without any stone rubbish or old lumber). After a few months of exposure, when this top layer gets fully settled, the road pavements may be constructed. This will prevent their subsidence and cracking.

SEWER APPURTENANCES

Sewer appurtenances are those structures which are constructed at suitable intervals along a sewerage system, and help in its efficient operation and maintenance. These devices include :

1. Manholes
2. Drop manholes
3. Lampholes
4. Clean-outs
5. Street inlets called Gullies
6. Catch basins
7. Flushing tanks
8. Grease and Oil traps
9. Inverted siphons
10. Storm regulators.

These devices are discussed in the following pages :

5.6. Manholes

5.6.1. Definition of Monholes. Manholes are masonary or R.C.C. chambers, constructed at suitable intervals along the sewer lines, for providing access into them.

The manholes, thus, help in joining sewer lengths, and also help in their inspection, cleaning and maintenance. If the manhole covers are perforated, they may also assist in ventilating* the sewers, but will cause evolution of pungent gases.

5.6.2. Location and Spacing of Manholes. The manholes are generally provided at regular intervals in a streight sewer line, and also at points of every bend, junction, change of gradient, or change of sewer dia. Unless there are practical difficulties, the sewer line between two manholes is laid straight with even gradient. Even when the sewer line runs straight, the manholes are provided at regular intervals. The spacing between the manholes, in such a case, however, depends mainly upon the size of the sewer line. The larger is the diameter of the sewer, the greater** will be the spacing between the manholes. The manhole spacings, generally adopted, on straight sewer reaches, are given below :

Table 5.3. Manhole spacings as per IS 1742—1960

Size of the Sewer	*Recommended Spacing of Manholes on straight reaches of Sewer lines as per IS 1742—1960*
Dia up to 0.3 m	45 m
Dia upto 0.6 m	75 m
Dia upto 0.9 m	90 m
Dia upto 1.2 m	120 m
Dia upto 1.5 m	250 m
Dia greater than 1.5 m	300 m

*Explained under 'sewer Maintenance'.

**The spacing of manholes is more on large sized sewers, because they can be entered more easily by men for inspection.

The minimum internal dimensions of manhole chambers, as per IS 1742—1960 are also given in Table 5.4.

Table 5.4. Minimum Internal Dimensions for Manhole Chambers as per IS 1742—1960

S. No.	*Depth*	*Min. Size specified*
1.	0.8 m or less	0.75 m × 0.75 m
2.	0.8 m and 2.1 m	1.2 m × 0.9 m
3.	> 2.1 m	Circular chambers with min. dia of 1.4 m ; or rectangular chambers with min. dimensions of 1.2 m × 0.9 m
	Min. wall thickness upto	
	(*a*) 1.5 m depth	20 cm
	(*b*) > 1.5 m depth	30 cm

5.6.3. Classification of Manholes. Depending upon their depth, the manholes may be classified as :

(*i*) Shallow manholes

(*ii*) Normal manholes

(*iii*) Deep manholes.

A shallow manhole is about 0.7 to 0.9 m in depth, and is constructed at the start of a branch sewer or at places, which are not subjected to heavy traffic. Such a manhole is provided with a light cover at its top, and is called an *inspection chamber.*

A normal or medium manhole is about 1.5 m in depth and is constructed either square (1 m × 1 m) or rectangular (1.2 m × 1 m) in cross-section. Its section is not changed with depth, as is done in a deep manhole. Such a manhole is provided with a heavy cover at its top.

A deep manhole is having depth more than 1.5 m. The section of such a manhole is generally not kept the same. The size in the upper portion is reduced by providing an offset, as shown in Fig. 5.19. Steps, etc. are provided in such a manhole for facilitating descending into the manhole, and to enable the workers to go upto its bottom. Such a manhole is provided with a heavy cover at its top.

5.6.4. Component Parts of a Manhole. The typical details of a deep manhole are shown in Fig. 5.19. The various component parts of such a manhole are briefly discussed below :

(*i*) *Access Shaft.* The upper portion of a deep manhole is called access shaft. Its minimum size for a rectangular manhole is about 0.75 × 0.6 m ; and for a circular manhole, the minimum diameter is about 0.6 to 0.75 m. Its depth depends upon the depth of the manhole and the height required for the working chamber.

This upper portion (*i.e.* access shaft) is expanded in the lower portion (*i.e.*, working chamber) by providing an offset by providing R.C.C. slab, or by corbelling or by arching, etc.

(*ii*) *Working chamber.* As explained above, the lower portion of the manhole is known as the working chamber, as it provides a working space for inspecting and cleaning-operations. Its minimum size for a rectangular manhole is about 1.2 m × 0.9 m ; and for a circular manhole, the minimum diameter is about 1.2 m. The height of this chamber should generally be not less than 1.8 m or so.

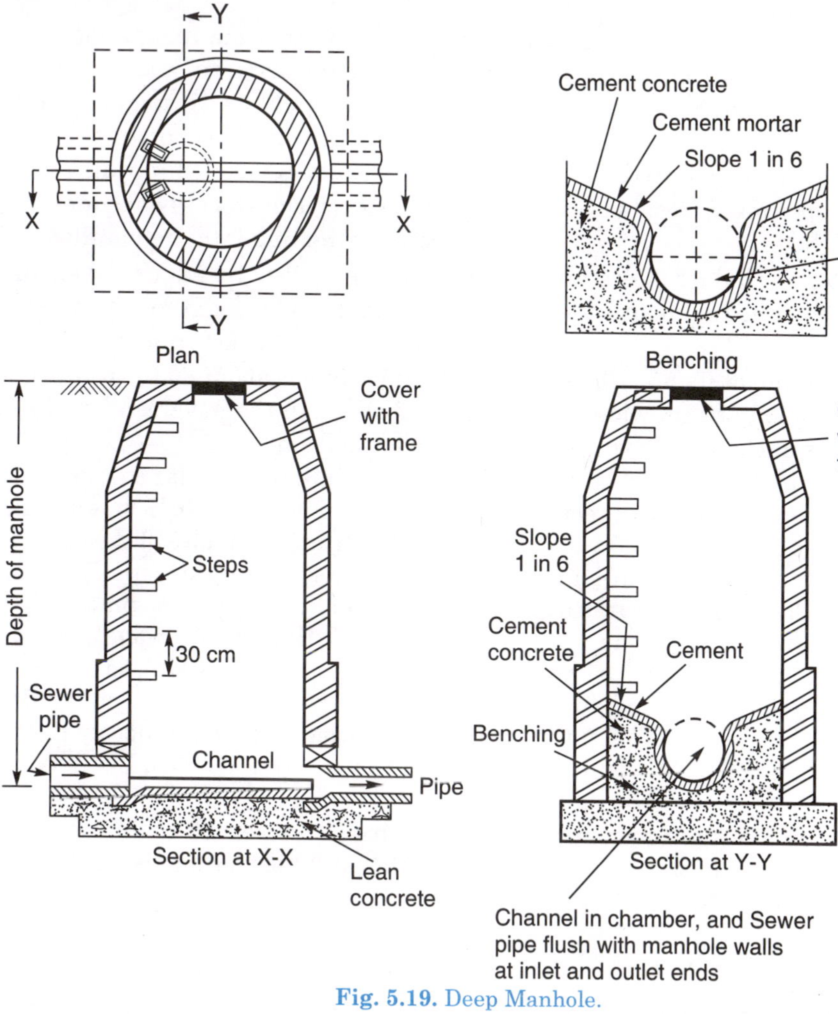

Fig. 5.19. Deep Manhole.

(*iii*) *The Benching i.e. the Bottom or invert portion of manhole*. The bottom portion of the manhole is constructed in cement concrete. A semicircular or a U-shaped channel is generally constructed, and the sides are made to slope towards it, as shown in section Y-Y of Fig. 5.19. The concreting is known as *benching*, and facilitates the entry of sewage into the main sewer.

(*iv*) *The side walls*. The side walls of the manhole are made of brick or stone masonry or R.C.C. The brick masonry walls are simple to construct, and are commonly adopted. The minimum thickness of the brick walls should be 22.5 cm (9″). The approximate thickness may, however, be computed by using the empirical thumb-rule :

$$t = 10 + 4d \text{ (For brick walls)} \quad ...(5.7)$$

where t = Thickness of wall in cm.

d = Depth of excavation in metres.

The thickness of R.C.C. walls will, however, be much less as compared to that of brick walls, and can be designed by the usual structural methods of analysis. The R.C.C. walls, however, prove costlier and require skilled labour, and as such, adopted only under special circumstances.

(*v*) *Steps or Ladders*. As pointed out earlier, steps are generally provided for descending into the manhole. The steps are made of cast iron, and are placed staggered at a horizontal distance of about 20 cm and at a vertical centre to centre distance of about 30 cm. If the steps are made of double width, staggering is not required. The steps should be fully embedded in wall.

For deeper manholes, ladders are provided in place of steps. The ladder will give a high sense of security to the labourers descending into the manhole.

The steps or ladder may start from about 40 cm below the ground level, and may be continued upto about 30 cm above the bottom level of the manhole.

(*vi*) *Cover and Frame*. The manhole is provided with a cast iron cover and a cast iron frame at its top. The thickness of the frame is about 20 to 25 cm, and its base is about 10 cm wide. It is firmly embedded in the pavement, and the cover rests in the groove which is kept inside the frame.

The manhole cover may be rectangular or circular ; the circular covers being very common. The size of a rectangular cover is about 0.6 m × 0.45 m, and that of a circular cover being 0.5 to 0.6 m in diameter. The top surface of the cover is roughened, so as to avoid slipping of the persons walking over it. The top surface of the cover also carries an arrow mark, and the cover is so placed as to place the direction of the arrow in the direction of the flow of sewage. The top level of the cover should flush with the road or the pavement level, so as not to cause any inconvenience to the traffic. Packing or a gasket may also sometimes be placed between the cover and the frame, so as to properly fit the cover in the frame. This is necessary when the cover is loose and is making noise.

The weight of the cover and the frame varies between 90 to 270 kg. The lighter cover is adopted to carry lighter traffic, and heavier one is adopted to carry heavier traffic.

5.7. Drop Manholes

When a branch sewer enters a manhole by more than 0.5 to 0.6 m above the main sewer, the sewage is generally not allowed to fall directly into the manhole, but is brought into it through a down pipe taken from the branch sewer to the bottom of the manhole. If the drop is only a few metres, the down pipe can be kept sloping (at 45° to the ground), and if the drop is more, a vertical pipe is found to be economical.

The manhole, in which a vertical pipe, such as shown in Fig. 5.20 is used, is called a *drop manhole* ; whereas, the one using an inclined pipe is called a *ramp*.

The construction of a drop manhole in place of an ordinary manhole in case a high levelled branch sewer enters a low levelled main sewer, will thus serve the following purposes :

(*i*) the steep gradients which otherwise would have to be given to the branch sewer (so as to bring it low) will be avoided, thus avoiding a lot of earth work excavation ;

(*ii*) the sewage trickling into the manhole from the directly placed branch sewer is likely to fall on persons working in the manhole. This is avoided in drop manholes.

The typical details of a drop manhole are shown in Fig. 5.20. The branch sewer is joined to the manhole through a vertical pipe (*P*). The sewage coming through the branch sewer dips in through the vertical pipe, and trickles over the main sewer channel, just above it, as shown.

A plug is provided at the point where branch sewer, if taken straight, intersects the wall of the manhole. The length of the branch sewer between the vertical pipe and the plug is known as *inspection arm* ; and can be used for inspecting and cleaning the branch sewer after opening the plug.

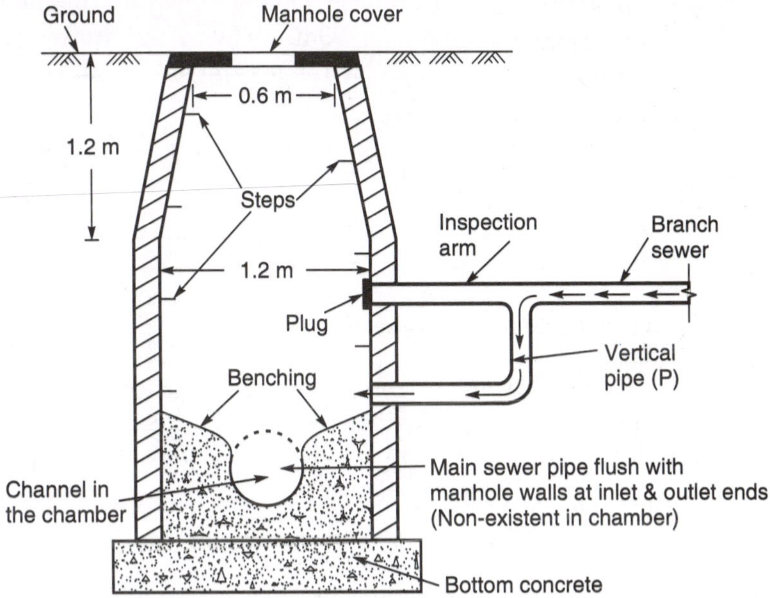

Fig. 5.20. A typical section of circular drop manhole.

5.8. Lamp Holes

Lamp holes are the small openings on sewers to permit the insertion of a lamp into the sewer. The lamp light is then viewed from the upstream as well as the downstream manholes. The obstructed light confirms the obstructions in the sewers.

The typical details of a lamphole are shown in Fig. 5.21. It consists of a vertical cast iron or stone ware pipe (20 to 30 cm in dia) extending from the ground and connected to the sewer line through a T-junction, as shown. This vertical

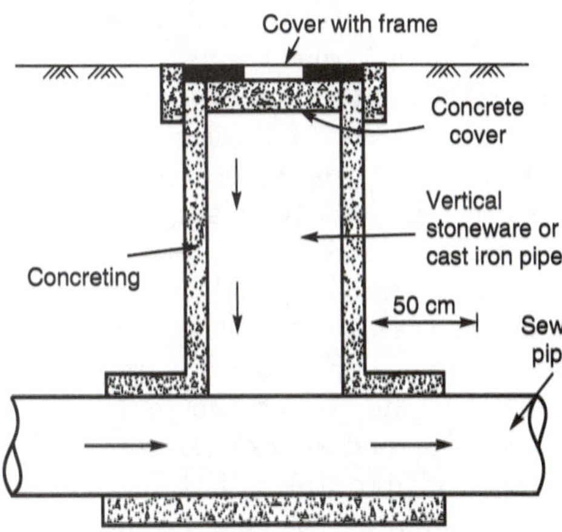

Fig. 5.21. Typical cross-section of a lamphole.

pipe is surrounded by concrete so as to make it stable. A manhole cover is provided at the top of such a pipe, which should be capable of withstanding the traffic load.

The lampholes are especially adoptable when : (*i*) a bend in a sewer is necessary, (*ii*) construction of manholes is difficult, and (*iii*) the spacing of manholes is more than the usual. But however, lampholes are rarely used these days and have become obsolete.

Besides its principal use as an inspection device (by inserting a lamp in it), a lamphole can under some circumstances be used as a flushing device. And also, if its cover is kept perforated, it can be used for ventilation of sewers ; in which case, a lamphole is known as a **fresh air inlet.**

5.9. Clean-Outs

A clean-out is an inclined pipe extending from the ground and connected to the under-ground sewer, as shown in Fig. 5.22. A cleanout is used for cleaning sewer pipes.

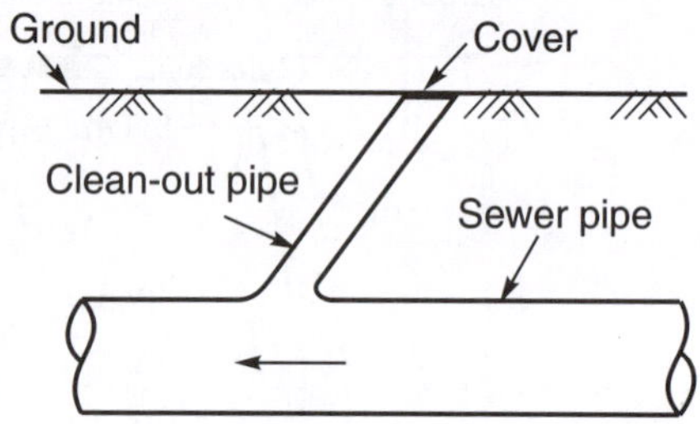

Fig. 5.22. Clean out.

A clean-out is generally provided at the upper ends of lateral sewers in place of manholes.

The functioning of a clean-out is very simple, and consists in removing the top cover and forcing water through the clean-out pipe to lateral sewers to remove obstacles in the sewer line. If obstructions are large enough, a flexible rod may be inserted through the clean-out pipe and pushed forward and backward to remove such obstacles.

5.10. Street Inlets, called Gullies

Inlets are gullies or openings on the road surface at the lowest point for draining rain water from roads, and admitting it into the underground storm water sewers (drains) or combined sewers.

These inlets are, therefore, located along road sides on straight roads at an interval of 30 m to 60 m or so. At intersection points, they are usually located, as shown in Fig. 5.23. In this figure, the slope of the streets is shown by the arrows, and the inlets are placed in such a way that the cross walks will not be flooded. Placing of inlets at the corners not only requires the pedestrians to step across flooded gutters, but also subjects the inlets to considerable traffic wear and damage.

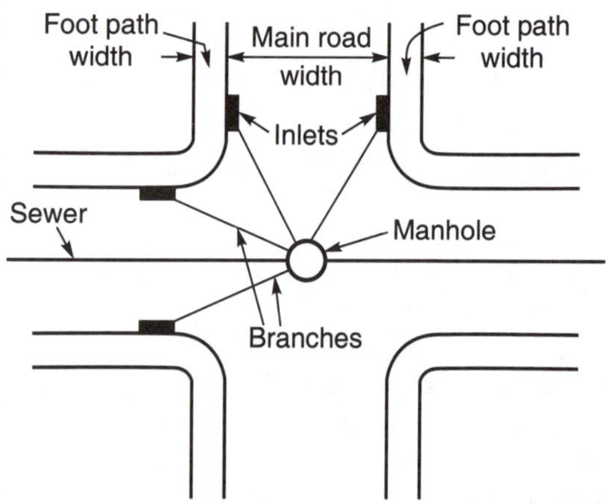

Fig. 5.23. Showing the locations of storm sewer inlets at street intersections. [*Note.* Branch lines enter at manholes. The rrows show the direction of storm runoff in the streets.]

The inlets are connected to the nearby manholes by pipe lines (**branches**) as shown in Fig. 5.23.

A street inlet is a simple concrete box having gratings or openings in vertical or horizontal direction. The inlet having vertical openings is known as *the vertical inlet or the curb inlet*, and the inlet having horizontal openings is known as the *horizontal inlet*. The typical details of both these types of inlets are shown in Figs. 5.24 and 5.25.

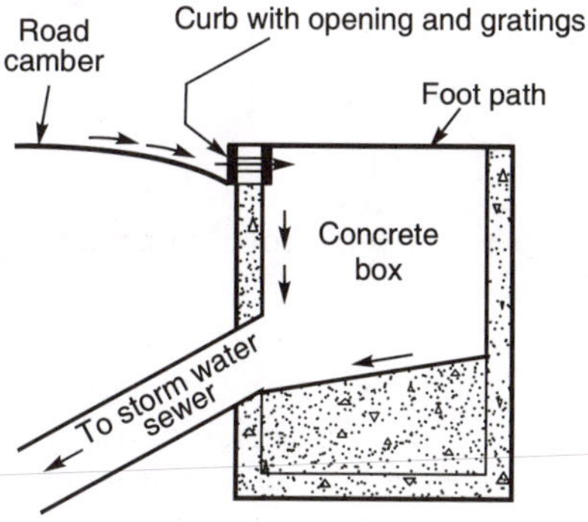

Fig. 5.24. Vertical inlet or curb inlet.

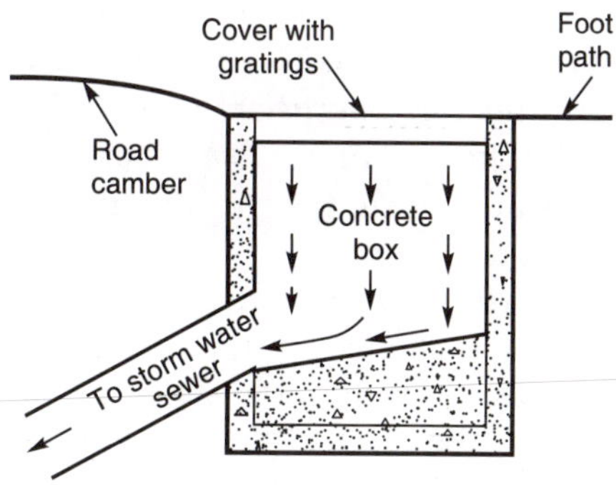

Fig. 5.25. Horizontal inlet.

5.11. Catch Basins or Catch Pits

Catch basins are nothing but street inlets provided with additional small settling basins, as shown in Fig. 5.26. Grit, sand, debris, etc., do settle in these basins, and their entry into the sewer is thus pre-vented. In addition to this, a hood, as shown, is also provided, which prevents the escape of foul gases, which may find its way through the sewer line.

Catch basins need periodical cleaning, as otherwise, the settled organic matter may decompose, producing foul odours, and may also become a breeding place for mosquitoes.

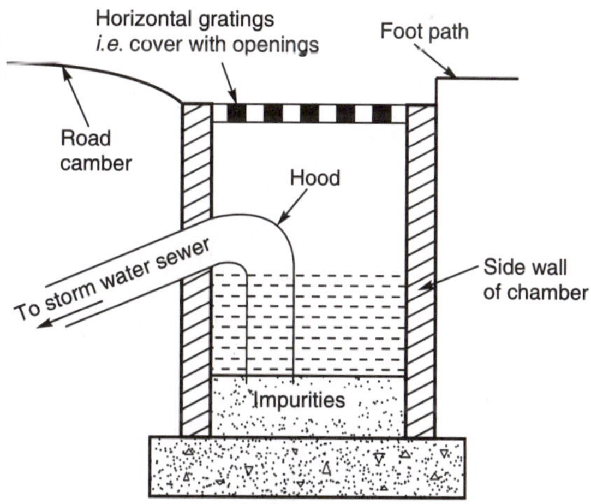

Fig. 5.26. Catch basins or catch pits.

Catch basins were considered necessary in old combined sewerage systems, but, however, in modern days, they are *not* considered as very essential, because the modern well paved streets offer very less grit and debris with storm run off, and the same can be conveyed easily in storm water sewers (drains) laid at suitable gradients to provide self cleansing velocities. Moreover, the problem of eruption of foul gases from S.W. sewers (drains) is very less ; and as such, there is not much necessity of providing such basins in the modern separate sewerage systems.

5.12. Flushing Tanks

Wherever, there are any chances of blockage of sewer pipes, such as in the case of sewers laid on flat gradients not producing self cleansing velocities, or near the dead end points of sewers, flushing devices are installed. These devices store water temporarily, and throw it into the sewer for the purpose of flushing and cleaning the sewer. Such devices are called flushing tanks.

Flushing tanks should have a capacity to store enough water, which may prove to be sufficient for cleaning the sewer line. This capacity is generally kept equal to about one-tenth of the cubical contents of the sewer line served by it.

Two types of *flushing operations* are normally used ; *viz.*

(*i*) *Flushing operation using automatic flushing tank.*

(*ii*) *Hand operated flushing operation.*

The automatic type of flushing tanks are being used more commonly in modern days.

(*i*) **Automatic Flushing Tanks.** In this type of a flushing tank, the flushing operation is carried out automatically at regular intervals. The entry of water is so regulated as to fill the tank upto the discharge point in a period equal to the flushing interval. An overflow pipe is also provided to drain away water in case the tank fails to discharge, and thereby overflows. The typical details of such an automatic tank are shown in Fig. 5.27. With reference to this figure, the working operation of an automatic flushing tank is explained below:

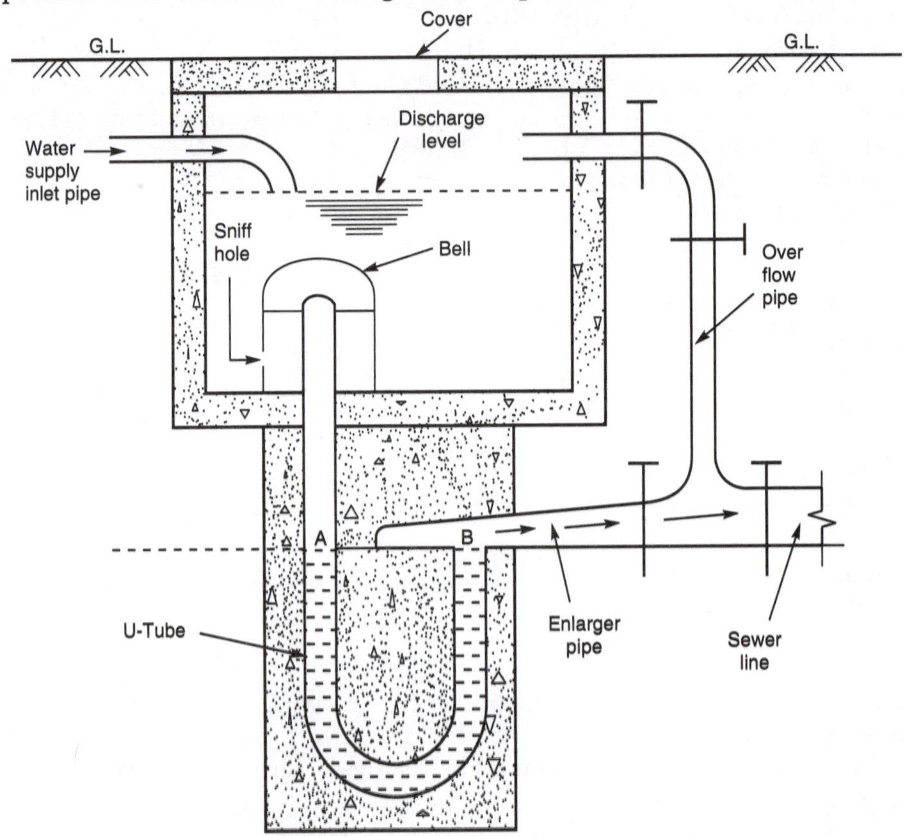

Fig. 5.27. Automatic flushing tank.

Initially, when the tank is empty, the water level stands at *A—B* in the U-tube. As the water enters the tank through the inlet pipe, the water level in the tank slowly goes on rising. The water level in the U-tube remains to this level, *i.e. A—B* till the water level in the tank remains below the level of *sniff hole*. But, however, as the water in the tank goes above the level of the sniff hole, the air is caught and compressed in the bell portion. This compressed air exerts pressure on the surface *A*, and hence the water level gets depressed in this long arm of U-tube. The water level goes on depressing more and more as the tank goes on filling more and more. Ultimately, a stage is reached when the water level in the longer arm of U-tube reaches the bend portion. When this happens, some compressed air gets released through the shorter arm of U-tube and a corresponding quantity of water enters the bell. It is so adjusted that the discharge line is just reached at this stage, and the head of water above the bell becomes greater than that in the shorter arm of U-tube. The compressed air is suddenly removed from the longer arm of U-tube, and a siphoning action starts, which releases the water from the tank into the sewer through the enlarger pipe. The siphonic action continues till the water in the tank falls up to the sniff hole. The air then enters the bell portion through the sniff hole and it breaks the siphonic action. The water level in the two arms of the U-tube again assumes the position *A—B*. The cycle goes on repeating, thus releasing water in the sewer at regular intervals.

(*ii*) **Hand Operated Flushing Operations.** The flushing and cleaning of sewers can be carried out at suitable intervals by means of manual labour. It may be carried out in the following ways :

(*a*) In one method, the outlet end of the manhole is closed by a sluice valve, etc. The sewage entering the manhole from the inlet end will then start collecting in the manbole. When sufficient quantity of sewage gets accumulated, the outlet end of the manhole is suddenly opened, and the sewage is thus allowed to enter the sewer, causing flushing operation.

(*b*) In another method, the inlet end as well as the outlet end of the manhole are closed by sluice valves, etc. The water from outside is now made to enter into the manhole. The flushing of sewer can then be carried out by opening the outlet and the inlet ends.

(*c*) In another method, one end of a hose pipe is connected to a nearby fire hydrant, and the other end may be placed in the manhole to obtain the flushing operation.

5.13. Grease and Oil Traps

Grease and oil traps are those trap chambers which are constructed in a sewerage system to remove oil and grease from the sewage before it enters into the sewer line. Such traps are located near the sources contributing grease and oil to the sewage. They are, therefore, generally located at places, such as, automobile repair work-shops, garages, kitchens of hotels, oil and grease industries, etc.

The removal of oil and grease from the sewage before it enters into the sewer pipe is considered necessary because of the following reasons :

(*i*) The grease and oil, if allowed to enter the sewer, will stick to the sewer sides, and thus reducing the sewer capacity.

(*ii*) The suspended matter which would have, otherwise, flown along with the sewage, also sticks to the sides of the sewer, due to sticky nature of oil and grease ; thus further reducing sewer capacity.

(*iii*) The presence of oil and grease in the sewage adds to the possibilities of explosions in the sewers.

(*iv*) The presence of oil and grease in sewage makes the sewage treatment difficult, as their presence adversely affects the biochemical reactions.

The principle on which oil and grease traps work is simple : *the grease and oil being lighter in weight, float on the top surface of the sewage*. Hence, if an outlet draws the sewage from lower level, grease and oil will get excluded. Based on this principle, the grease and oil trap chambers are designed in such a way that the outlet level is located near the bottom of the chamber, and the inlet level is kept near the top of the chamber, as shown in Fig. 5.28.

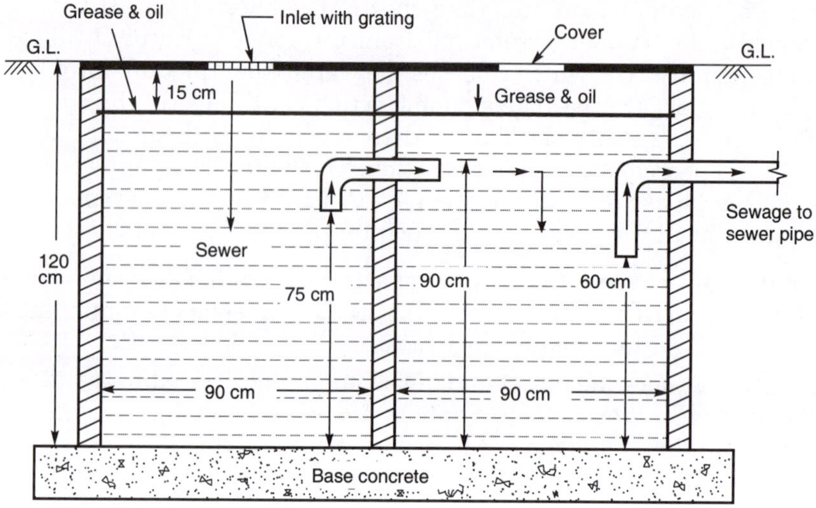

Fig. 5.28. Grease and Oil trap.

If sand is also desired to be excluded from the sewage, dead space should also be kept at the bottom of the chamber for sand to be deposited, as shown in Fig. 5.29.

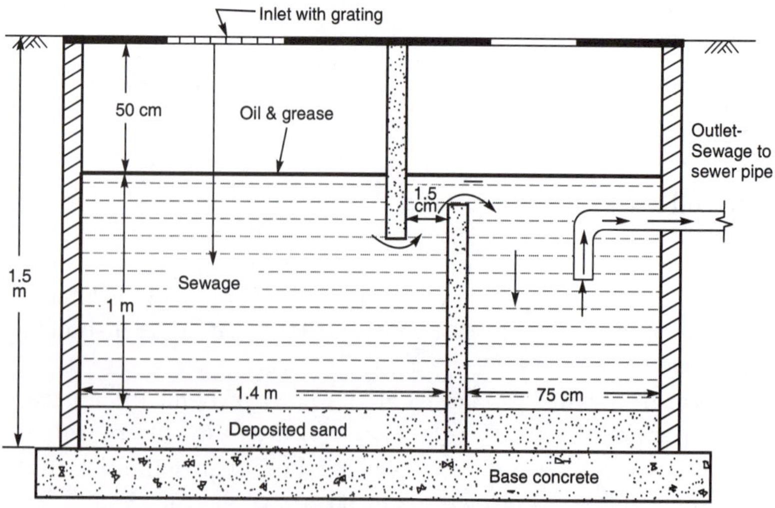

Fig. 5.29. Combined Sand, Grease and Oil trap.

These traps should be cleaned periodically, as otherwise, they would not function properly, and sewage will not flow freely.

5.14. Inverted Siphons

Whenever a sewer pipe has to be dropped below the hydraulic gradient line for passing it beneath a valley, a road, a railway, a stream, a tidal estuary or any other depression in the earth's surface or where it passes beneath some other

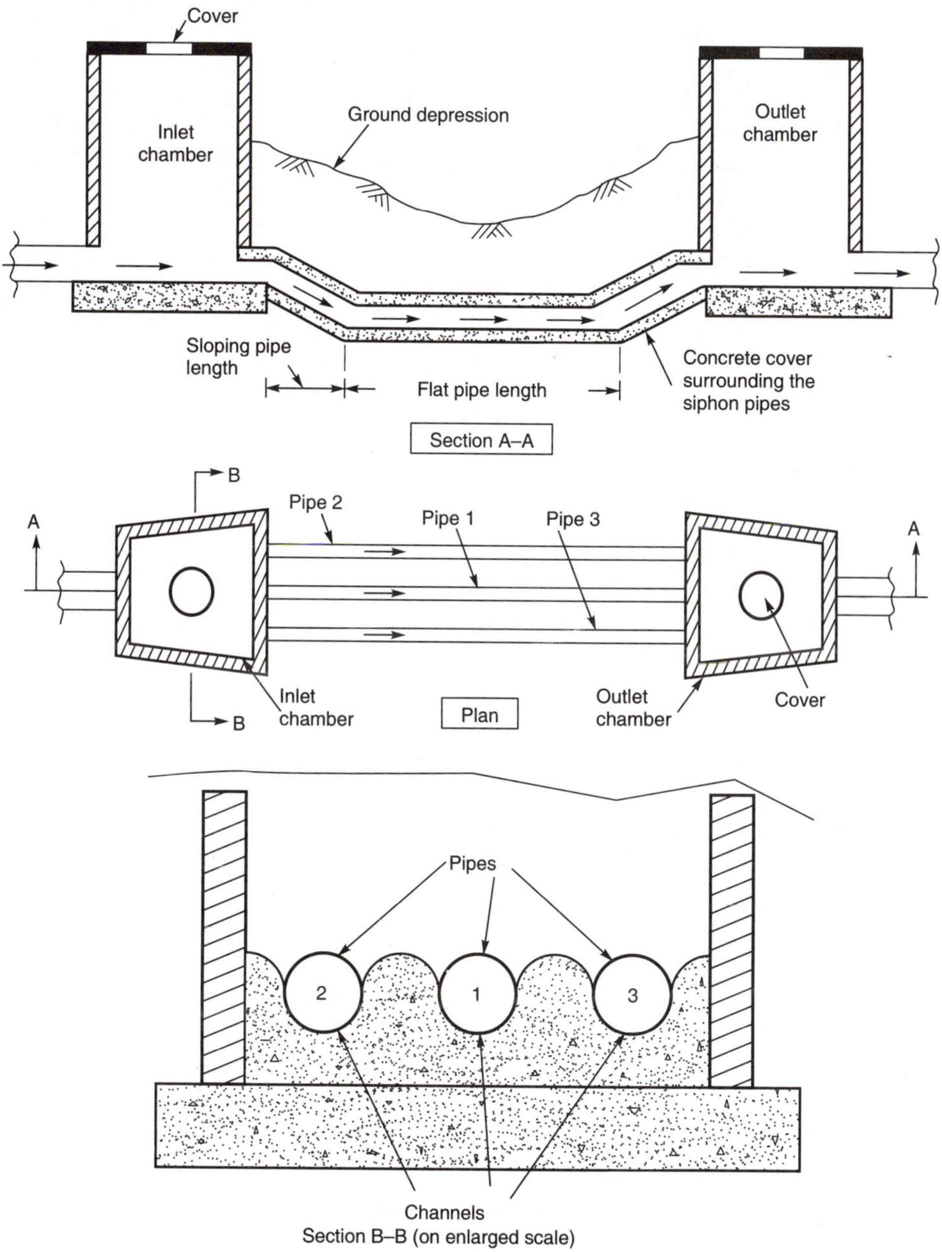

Fig. 5.30. Inverted siphon.

obstructions in its path, it will be known as an *inverted siphon** or a *depressed sewer* or a *sag pipe*. The sewage through such a pipe line will not flow under gravity, but will be flowing under pressure (as in the case of water pipe lines).

An inverted siphon is thus a sewer section constructed lower than the adjacent sewer sections, and it runs full under gravity with pressure greater than the atmosphere.

An inverted siphon is usually made of siphon tubes or pipes made of cast iron or concrete. These pipes are laid between the inlet and the outlet chambers (usually at the same elevation). The pipes between these ends are depressed for passing below obstructions like roads, railways, rivers or any other ground depressions. The inverted siphon laid between the inlet and the outlet chambers consists of two sloping pipe lengths joined by a flat pipe length, as shown in Fig. 5.30.

5.14.1. Design of an Inverted Siphon. The proper design of siphons is very important ; as otherwise, they are likely to be clogged and become inefficient. The siphon should, therefore, be designed to develop a self cleansing velocity of about 0.9 m/sec even during minimum discharge. For this purpose, the inverted siphon is usually made of three pipe sections laid side by side in plan, as shown in Fig. 5.30. One of these three channels is meant for carrying minimum sewage flow, the other one for average sewage flow, and the third one for carrying maximum sewege flow.

The inlet chamber contains three channels one for each pipe section. When channel No. 1 overflows, the sewage enters channel No. 2 and pipe No. 2 starts functioning. Similarly, when channel No. 2 overflows, the sewage enters channel No. 3 and pipe No. 3 starts functioning.

The three barrels as provided above are usually provided for combined sewers. However, for sanitary sewers only two pipe sections may be provided–one for minimum dry weather flow and the other for maximum dry weather flow.

Besides designing the siphons in the manner explained above, the following additional points must be kept in mind while designing these inverted siphons.

(*i*) If the length of the siphon is more, hatch boxes at intervals of about 100 m should be provided for facility of rodding. There should be a vent pipe in the hatch box to prevent the formation of air-locks in the siphon.

(*ii*) The changes of direction of inverted siphons should be easy and gradual.

(*iii*) The design of siphons should be made on the basis of pipes running full under pressure**. It is, therefore, necessary to know the maximum available head (*i.e.* head difference between the water levels of the inlet and the outlet chambers). The greater is this available head, the better self cleansing of the siphon will occur. Also the losses including the loss of head due of friction, losses due to bends, and losses at the entry should be properly calculated, and accordingly adjusted in the design of siphon.

---

*The name inverted siphon has no other justification except that it being the mirrored image of a true siphon.

**There should be no vacuum anywhere in the siphon, *i.e.* the pipes should run full, although due to wide fluctuations in sewage flow, the same is very difficult to achieve.

(*iv*) The inlet chamber should be provided with screens, so as to remove the coarser silt, debris, grit, etc. from the sewage before it enters the siphon pipes.

(*v*) The minimum diameter of the siphon pipe is taken as 15 to 20 cm.

(*vi*) Manholes should be provided at each end of the siphon to enable barrels to be cleaned.

(*vii*) It is advisable to provide a diversion for the siphon. Hence, when the siphon either gets choked, or overflows due to surcharge, the flow of sewage can be diverted.

5.14.2. Demerits of Siphons. The inverted siphons should be avoided as far as possible, because of its following demerits :

(*i*) It is most likely to get silted, as the down gradient is not continuous. The proper design with self cleansing velocity at different discharges is, therefore, of utmost importance. Moreover, when once it gets silted up, it becomes very difficult to clean it up. Sometimes, a chain extending from the inlet chamber to the outlet chamber is provided to stir and keep in suspension the deposited silt in the siphon.

(*ii*) If the inlet chamber is not properly designed, the floating matter present in the sewage will separate out, and will accumulate in this chamber, and thus seriously affecting the proper working of this chamber.

(*iii*) It is not possible to give side connections to the inverted siphons.

An example on the design of an inverted siphon is solved below :

Example 5.2. *Design a three barrel siphon for carrying sewage across a river stream. The total length of the siphon measured along the centre line including slopes is about 80 m. The invert levels at the inlet and the outlet ends of the sewer are 202.38 m and 201.80 m respectively. The average flow of the sewage is 180 litres per second, and the maximum and the minimum flows are 250% and 40% of the average respectively. Assume the minor losses to be about 6 cm.*

Solution. The average flow to be carried by the siphon

= 180 l/s = **0.18 cumecs.**

The maximum flow to be carried by the siphon

= 250% of 0.18 cumecs. = 2.5 × 0.18 = **0.45 cumecs.**

The minimum flow to be carried by the siphon

= 40% of 0.18 cumecs = 0.4 × 0.18 = **0.072 cumecs.**

Max. available head = Difference in the invert levels at the inlet and the outlet

= 202.38 – 201.80 = **0.58 m.**

Flow at minimum discharge. Assuming a flow velocity of 0.9 m/sec at the minimum discharge, we have the dia of the pipe required.

$$Q = A\,.V = \frac{\pi}{4}\,d^2.\,V$$

or $$0.072 = \frac{\pi}{4}\,d^2 \times 0.9$$

or $$d = \sqrt{\frac{0.072 \times 4}{\pi \times 0.9}}$$ = **0.319 m ;** Use 30 cm dia pipe.

Actual velocity generated

$$= \frac{0.072}{\frac{\pi}{4} \times (0.30)^2} \text{ m/sec.} = 1.019 \text{ m/sec} > 0.9 \text{ m/sec.}$$

Hence, the generated velocity is safe and self cleansing.

The head loss in the siphon pipe can be calculated as follows :

Using Manning's formula

$$V = \frac{1}{n} \,.\, R^{2/3} .\, S^{1/2}$$

where S = The hydraulic gradient or loss per metre length of pipe.

$$R = \frac{d}{4} = \frac{0.3}{4} = 0.075 \text{ m}$$

$V = 1.019$ m/sec

$n = 0.013$ (for old cast iron pipe).

$$\therefore \quad 1.019 = \frac{1}{0.013} (0.075)^{2/3} \,.\, S^{1/2}$$

or $$\frac{1.019 \times 0.013}{(0.075)^{2/3}} = S^{1/2}$$

or $$S^{1/2} = \frac{1.019 \times 0.013}{0.178} = 0.074$$

or $$S = 0.0055.$$

Head loss $H_f = S \times L = 0.0055 \times 80 = 0.443$ m.

Total losses $= 0.443 + 0.06^* = 0.506 < 0.58$ m

Hence, we can easily adopt 30 cm dia pipe.

Flow at average discharge. At average discharge, the excess discharge passing through the second pipe

$$= 0.18 - 0.072 \text{ cumecs} = 0.108 \text{ cumecs.}$$

For 0.9 m/sec velocity, the diameter required

$$= \sqrt{\frac{0.108 \times 4}{\pi \times 0.9}} = 0.391 \text{ m. Let us use 38 cm dia pipe.}$$

Actual velocity through the second barrel

$$= \frac{0.108}{\frac{\pi}{4} \times (0.38)^2} \text{ m/sec} = 0.952 \text{ m/sec.}$$

Using Manning's formula, we have

$$0.952 = \frac{1}{0.013} \left(\frac{0.38}{4}\right)^{2/3} .\, S^{1/2}$$

or $$S^{1/2} = \frac{0.952 \times 0.013}{(0.095)^{2/3}} = \frac{0.952 \times 0.013}{0.208} = 0.059$$

---

*Minor losses.

or $S = 0.0035.$

$\therefore \quad H_f = 0.0035 \times 80 = 0.283.$

Total losses $= 0.283 + 0.06 = 0.343 \text{ m} < 0.58 \text{ m}.$

So 38 cm dia pipe is all right.

Flow at maximum discharge. At maximum discharge, the excess discharge passing through the third pipe

$$= 0.45 - 0.18 = 0.27 \text{ cumecs.}$$

Assuming the velocity as 1.5 m/sec in this case, the required diameter

$$= \sqrt{\frac{0.27}{\pi \times 1.5}} = 0.48 \text{ m ; use 48 cm dia pipe.}$$

Actual velocity through this third barrel

$$= \frac{0.27}{\frac{\pi}{4} \times (0.48)^2} = \mathbf{1.49 \ m/sec.}$$

Use Manning's formula

$$1.49 = \frac{1}{0.013}\left(\frac{0.48}{4}\right)^{2/3} . S^{1/2}$$

or $$1.49 = \frac{1}{0.013} \times (0.12)^{2/3} . S^{1/2}$$

or $$S^{1/2} = \frac{1.49 \times 0.013}{0.243} = 0.079$$

or $S = 0.0063.$

$\therefore \quad H_f = 0.0063 \times 80 = 0.508.$

Total losses $= 0.508 + 0.06 = 0.568 \text{ m} < 0.58 \text{ m}.$

$\therefore$ 48 cm dia pipe is O.K.

Hence use pipes of the following sizes :

$$\left.\begin{array}{l}\text{Pipe No. 1} = 30 \text{ cm dia}\\ \text{Pipe No. 2} = 38 \text{ cm dia}\\ \text{Pipe No. 3} = 48 \text{ cm dia}\end{array}\right\}. \quad \textbf{Ans.}$$

5.15. Storm Water Regulators or Storm Relief Works

Storm water regulators are constructed in the combined sewerage systems, *and permit the diversion of excess storm water into a nearby stream*. Say for example, if the combined sewer has been designed for carrying a maximum combined discharge of three times the dry weather flow, the excess quantity of sewage after this limit is crossed, is diverted to some natural stream or river. This excess sewage will be mainly composed of storm water, and, therefore, it will not be foul in character.

Storm regulators may be of the following three types :

(*i*) *Leaping weir :*

(*ii*) *Overflow weir ;* and

(*iii*) *Siphon spillway.*

These three types of storm relief works are briefly discussed below :

5.15.1. Leaping Weir. The leaping weir arrangement consists of an opening in the invert of the storm drain (or combined sewer) through which the normal storm flow is diverted into the intercepting sewer, and the excess flow leaps over the combined sewer to flow into the nearby stream. This arrangement is shown in Fig. 5.31.

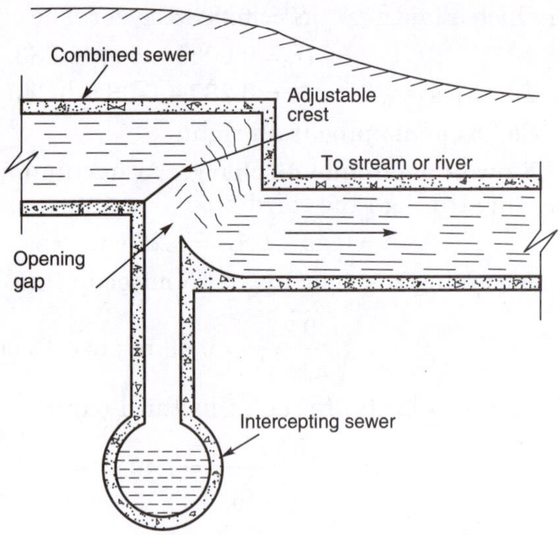

Fig. 5.31. Leaping weir.

When the sewage discharge is small, the sewage will fall directly into the intercepting sewer through the opening. But, however, when the discharge exceeds a certain limit, the excess sewage leaps or jumps across the weir, and it is carried to the natural stream, as shown.

The leaping weir is a good regulator, but in heavy storms, most of the flow may leap over the combined sewer, and only small quantity may be left in the sewer, which may result in low velocity and thus creating silting problems.

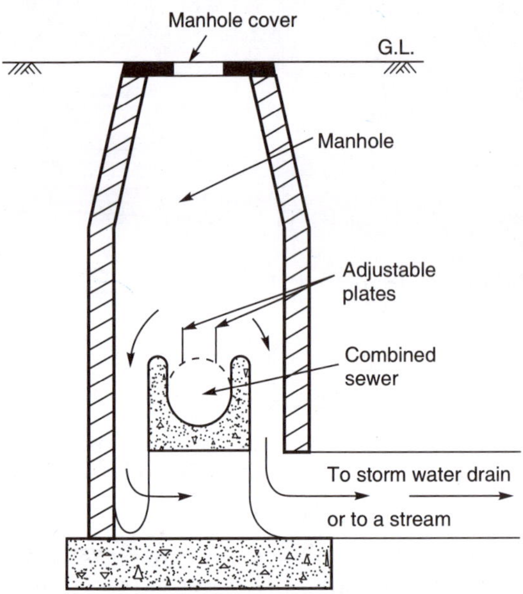

Fig. 5.32. Overflow weir.

5.15.2. Overflow Weir. In this type of arrangement, the excess sewage is allowed to overflow the combined sewer in the manhole, from where it enters into a channel carrying it into a storm water drain or directly into a stream, as shown in Fig. 5.32.

In order to prevent the escape of the floating matter from the combined sewer, adjustable plates, as shown in Fig. 5.32, may be used.

In another type of arrangement, openings at suitable height above the invert of the sewer are provided at suitable intervals along the length of the combined sewer, as shown Fig. 5.33. These openings are then joined to a storm water

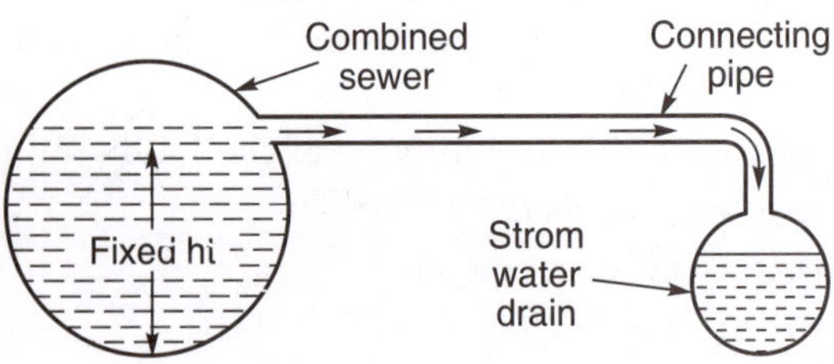

Fig. 5.33. Overflow weir (another type of arrangement).

drain, which is laid near the combined sewer. The excess sewage above this fixed height is thus diverted and conveyed to the natural stream or the river, through the storm water drain.

5.15.3. Siphon Spillway. The siphon spillway arrangement used for diverting excess sewage discharge from the combined sewer, is shown in Fig. 5.34. This method provides the most effective type of a storm relief work. It is an automatic process, and works on the principle of shiphonic action. The shiphonic action starts when the sewage in the combined sewer rises above a fixed level (*i.e.*, the crest level of the siphon) and stops as soon as the sewage falls below this level.

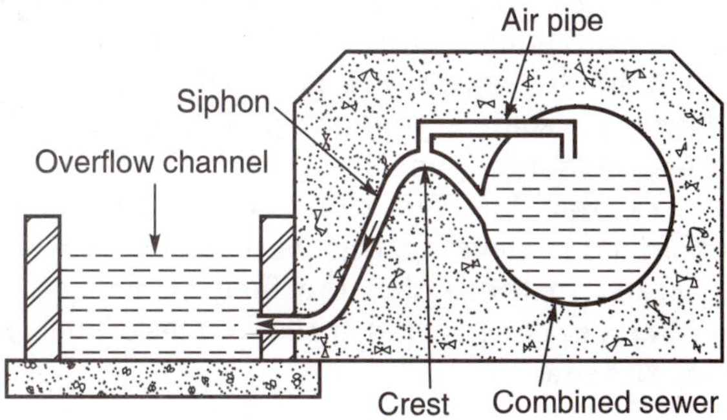

Fig. 5.34. Simphon spillway type of storm regulator

The level of the crest of the siphon is generally kept at the level reached by the flow in the combined sewer during the period of maximum dry weather flow. The siphonic action does not start so long as the level of the sewage in the combined sewer remains lower than this crest level of the siphon. However, where the sewage level in the combined sewer goes beyond the crest level, the mouth of the air pipe gets sealed, and the air contained in the siphon is suddenly removed by the flow. The suction thus developed, starts the siphonic action, and the full flow through the siphon pipe into the storm water drain, immediately gets established. The siphonic action continues till the mouth of the air pipe remains submerged in the sewage flowing in the combined sewer. When the excess sewage is discharged, the level of sewage in the combined sewer falls down, and the mouth of the air pipe gets exposed. The air now enters the siphon pipe, thus breaking the siphonic action and the consequent overflow through the siphon pipe. The process goes on repeating.

As pointed out earlier, this method is automatic and is the most effective out of all other methods. In this arrangement, it is also possible to adopt greater heads which may increase the capacity of the arrangement. Moreover, this arrangement has no moving parts and thus requires the least maintenance. However, it is liable to be clogged due to narrow passages.

MAINTENANCE, CLEANING AND VENTILATION OF SEWERS

5.16. Maintenace of Sewers

Sewer maintenance generally involves their cleaning to keep them free from any clogging and to carry out the repairs to the damaged portions if any, so as to prolong their life and to ensure efficient functioning.

The proper maintenance of sewers is, therefore, absolutely necessary in order to make the sewerage system function efficiently. The sewer maintenance generally includes : their *frequent inspection* and supervision, *measuring the rate of flow, cleaning and flushing, rapairing* the leaking joints or any other damaged portions if any, protecting them against their misuse, preventing explosions, etc.

The most general complaint received in respect of their inefficient functioning is their *clogging*. The sewers get blocked or clogged due to silting and deposition of debris, grit and other floating matter present in the sewage. Although, the sewers are generally designed to flow with a self cleansing velocity, yet some silting always takes place. This silting is more pronounced at low discharges, and in sewers laid on flatter gradients. Greasy and oily matters from kitchens of hotels and restaurants or from such industries, further aid in clogging, because such matters stick to the interior of the sewer pipes and further catch the floating matter. Oil and grease traps should, therefore, be constructed at the sources of such oil and greasy sewage, and also along the sewer line.

Many a times, the general public also throw away their domestic wastes, and garbage, etc. into the manholes, which may find its way into the sewers, thus increasing the chances of their deposition and consequent blockage of the sewer pipes. To avoid this, public opinion should be raised against such habits, and *garbage dumping homes* be made at frequent places, which should regularly be cleaned by municipal sweepers.

Inspite of all precautions taken to avoid their silting, sewers do silt, and, therefore, require frequent cleaning.

5.17. Cleaning of Sewers

Sewers should be periodically cleaned so as to avoid their complete clogging. A full record of the desilting operations should be kept. Proper watch and ward to avoid fake desilting operations is very very important in actual practical life.

For cleaning small sewers which can not be entered into by manual labour, flushing operations are essential. A flexible fire hose with about 2.5 cm or less sized nozzle can be inserted down, and the water discharged under pressure. The water pressure stiffens the rubber hose pipe, and together with the support from the sides of the sewer, make it possible to push the hose in and out in the sewer by 30 m or more, thus helping in flushing and cleaning of the sewer. Flushing can also be carried out by tipping larger volumes of water into the manholes, as explained earlier in article 5.12 under "Flushing Tanks".

Medium sized sewers are generally cleaned by using scraping instruments, which are dragged through the sewer. A commonly used device consists of a steel cylinder of 20 to 25 cm diameter and 60 to 75 cm long, and provided with a sharp cutting teeth on the open forward end. This type of a sewer cleaning rod called *cane rodding* is dragged in the sewer to and fro from the manholes. Sections of the roddings are screwed together in the manhole and pushed down the sewer until the obstruction is reached and removed.

Pills have been successfully used in U.S.A. for cleaning badly clogged sewers. *Pills* are nothing but a series of light floating wooden or rubber hollow balls. When such a ball is rolled into the sewer, it gets struck up at the place where heavy deposition has taken place. It, thus, constricts the passage thereby raising the velocity of flow in its neighbouring area, and thus scouring out the deposited silt. The ball is picked in a net at the next downstream manhole. Progressively, larger and larger sized pills are inserted from one manhole and taken out at the next manhole. The process is repeated till the last pill passed is just 2 to 3 cm less in dia than the dia of the sewer. In order to avoid the entangling and the loss of the pill, a string is generally tied to the pill by which it can be pulled out, if it gets stuck up in the blockage.

Large sized sewers which can be entered into by men are generally cleaned by manual labour. The sweepers do actually enter into the sewer pipes, and remove the deposited materials by hand shovels, etc., collect it up in buckets, and remove it outside the manhole. However, sufficient precautions and care must be taken while entering the sewers, as otherwise, sometimes, foul, poisonous and explosive gases may cause serious hazards to the workers entering the sewer pipes.

Precautions to be taken while entering sewers. Various poisonous and explosive gases which are generally found present in sewers are : hydrogen sulphide (H_2S), carbon dioxide (CO_2), and methane (CH_4) along with petrol vapours. These gases are largely produced when the sewage gets stale and septic, which more commonly happens in hotter climates. In addition to the production of these hazardous gases, oxygen gets consumed by the organic matter in their putrefaction, and thus resulting in the shortage of oxygen inside the sewer, and may consequently cause difficulty in breathing.

In order to avoid the large scale presence of these poisonous and hazardous gases inside the sewers, *ventilation of sewers* is generally adopted. Various methods are adopted for ventilating the sewers, and are discussed in the next article. In addition to the ventilation of sewers, the following precautions should also be taken while allowing the workers to enter the sewers :

(*i*) The very first precaution which should be taken before entering a sewer through a manhole is : to open the cover of the manhole in question as well as the covers of two manholes, one on the upstream and the other on the downstream ; at least half an hour in advance. This will help in obtaining some ventilation and exposure of the sewer to the atmospheric oxygen.

(*ii*) Now, tests should be carried out to detect the presence of any hazardous gases inside the sewer. The following tests may be carried out to detect their presence :

(*a*) H_2S gas may be detected by exposing a sheet of paper moistened with lead acetate for 5 minutes near the sewer entry. If the paper turns black, the presence of H_2S gas is indicated.

(*b*) Presence of CO_2 gas may be detected by lowering a minors safety lamp *near the level of sewage* in the manhole. If the flame extinguishes within 5 minutes, the presence of CO_2 gas is indicated.

(*c*) The presence of methane (CH_4) gas may be detected by lowering the minor's safety lamp in the *upper layers of the sewer* ; methane, being lighter than air, is generally present in the upper layers of sewer. When the gas is present, it forms an explosive mixture with air, and the gauge cylinder of the safety lamp gets filled with the flame. The lamp should be withdrawn immediately. The formation of flames will indicate the presence of methane gas inside the sewer.

When any of these gases are found to be present, special breathing apparatus fitted with hose masks should be used by the workers entering the sewers. Moreover, only permissible safety lamps, rubber boots, and non-sparkling tools should be used.

(*iii*) If the above hazardous gases are absent, a lighted lantern may be lowered down the manhole, so as to test the presence of oxygen. If it burns brilliantly, the sewer can be safe to enter.

(*iv*) In any case, the workers going down for inspection must be tied with ropes to their waists, and they may be held fast by the persons on top, so that they may be immediately pulled up in case of any risks. A responsible officer should preferably be present to direct the operations.

(*v*) *Any smoking or carrying of naked lights inside the sewers are strictly prohibited.*

(*vi*) Necessary warning signals should be erected.

5.18. Ventilation of Sewers

The sewers must be properly ventilated for **the following reasons** :

(*i*) The decomposition and putrefaction of sewage inside the sewers may result in the production of various *sewer gases*, such as, carbon dioxide, carbon monoxide, methane, hydrogen sulphide, ammonia, nitrogen, etc. These gases are disposed of into the atmosphere by exposing the sewage to the outside atmosphere by suitable methods of ventilation. These gases, if not removed, may cause serious problems, and prove hazardous to the workers entering the sewers. Methane gas being highly explosive, if not removed, may even blow off the manhole covers. Moreover, these gases have a tendency to interfere with the flow of sewage.

(*ii*) Another reason for ventilating sewers is to ensure a continuous flow of sewage inside the sewer. This is achieved by ventilation by keeping the surface of sewage in contact with free air and thus preventing the formation of air-locks in the sewage.

5.18.1. Methods of Ventilation. Following methods are adopted for ventilating the sewers :

(*i*) *Use of Ventilating Columns.* In order to achieve proper ventilation, ventilating columns or shafts are generally placed at intervals of 150 to 300 m along the sewer lines. They are also provided at the upper end of every branch sewer, and also at every change in the size of the sewers.

A typical vertical column used for ventilating the sewers is shown in Fig. 5.35.

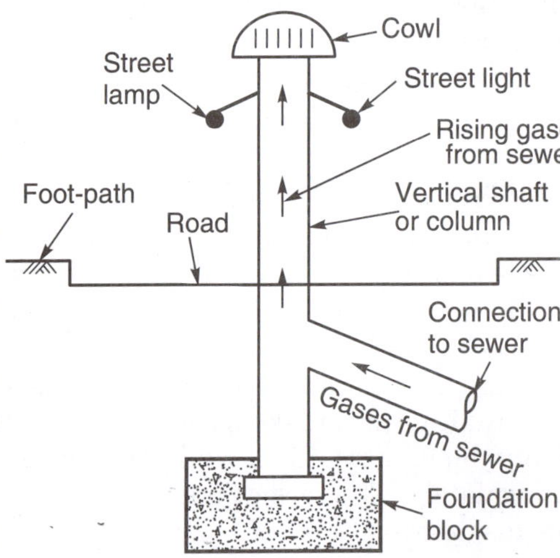

Fig. 5.35. Ventilating column.

It consists of a vertical shaft made by joining cast iron or steel pipe lengths. A foundation block is provided at the bottom end of the shaft in order to keep it steady in a vertical position. A connection is provided in the lower under-ground portion to join it to the sewer as shown. A cowl is provided at the top end of the shaft in order to allow the escape of sewer gases into the atmosphere.

The ventilating columns are generally designed to serve two purposes together: one for ventilating the sewer, and the other for working as a support for street lamps, sign boards, etc. ; in such cases, the presence of ventilating columns will not be detected at a glance by the general public, and hence will not look unaesthetic to the eyes.

The diameter of the ventilating column is preferably kept equal to one third of the dia of the sewer served by it. The ventilating columns should be carried higher than the height of the nearby structures. Moreover, the joints between the pipe lengths used in building the higher vertical column must be made airtight ; as otherwise, sewer gases may leak at low heights, thus causing bad and unpleasant smells.

The top of ventilating columns should be covered with wire mesh or cowl, so as to prevent the birds from building their nests at the top, and to prevent any thing directly falling into the pipe.

The ventilating columns should be located in open places, so that they are exposed to sun-shine for the most part of the day. The heat of sunshine will help in causing proper circulation of air.

(ii) Use of Ventilating Manhole Covers. The manhole covers are sometimes provided with perforations, through which the sewer gets exposed to the atmosphere. This will no doubt help in achieving some ventilation, but will cause more nuisance, as the bad smells continue to erupt from them. Moreover, the openings of the manhole cover will permit admitting large quantities of storm water and other road dust, etc. This method is, therefore, of no practical utility, except that it may be adopted in very isolated places.

(iii) Proper Design of Sewers. The sewers should be properly designed as running half or two third full, thus reserving the top space for the sewer gases. Moreover, the velocity in the sewer should be self-cleansing so that sewage does not stay at one point for longer periods. The proper design of sewers ensures enough ventilation.

(iv) Use of Mechanical Devices. Forced draught is sometimes provided by exhaust fans to expel out the foul gases from the sewers. This method requires mechanical supervision and skill.

(v) Unobstructed Outlets. In the case of storm water drains or sewers, they can also act as partial ventilators.

(vi) House vent and Soil pipes. They may directly help in ventilating house drains and public sewers, particularly where interceptors are not provided on the sewers connecting houses and buildings.

(vii) Sometimes, immediate *artificial ventilation* is resorted to at site before entering a sewer, by blowing fresh air supply into the sewer through mechanical means such as *air blowers* having suitably protected and encased motors and fitted with canvas air pipe suspended into the sewer. The blower, should preferably be situated a little away from the opening, so that it may not act as a source of ignition for the inflammable gases in the sewers.

PROBLEMS

1. Write a short note on the construction and maintenance of sewers, bringing out the salient features like materials used, shapes, and gradients adopted generally, testing, etc.

2. A trench 5 ft. wide and 10 ft. deep has to be dug in loose soil for laying a sewer. Show the best method of shoring to be adopted, giving sketches with dimensions.

3. What are the various types of storm water regulators used in a sewerage system. Explain briefly the working of each of these with sketches.

4. Describe in order the various stages followed in the construction of sewers.

5. (*a*) Draw a neat sketch of a drop man-hole and indicate where it is used.

(*b*) Describe the laying of a sewer line in a trench.

6. Specify the qualities of a good material for constructing sewers. Judging from these requirements, discuss the suitability of :

(*a*) Bricks, (*b*) Cement concrete, and (*c*) Stone ware for sewers.

7. State the types of sewers used in sewerage system, giving their sketches. Comment on their hydraulic properties.

8. Write in detail about (*a*) laying, jointing and testing of sewers ; (*b*) considerations involved in determining the size of sewers.

9. (*a*) What should be the characteristics of materials to be used for sewers.

(*b*) Draw a neat sketch of a sewage ventilator, and explain the necessity of sewage ventilation.

10. Enumerate the various forces that act on a sewer pipe laid underground. Discuss in details, any two of these forces.

11. Write short notes on any four of the following :

(*i*) Manholes, ; (*ii*) Drop manholes ; (*iii*) Eccentric manholes ; (*iv*) Lampholes ; (*v*) Street inlets and gullies ; (*vi*) Catch pits ; (*vii*) Flushing tanks ; (*viii*) Inverted syphons ; (*ix*) Storm water over-flows ; (*x*) Ventilation.

12. Draw a junction manhole with a 22.5 cm diameter sewer at a depth of a 4.5 m below ground level and 10 cm dia. sewer at a depth of 1.5 m below ground level. The outgoing outfall sewer is 60 cm in diameter. Draw plan and section through the manhole.

13. (*a*) What are the common jointing materials used in making joints in sewers, and what should be the necessary qualities with them.

(*b*) Describe the method of making joints in the following types of sewers. Draw sketches to illustrate your answer : (*i*) Stoneware ; (*ii*) Concrete ; and (*iii*) C.I. Rising mains.

(*c*) What are 'poured joints' and 'filled joints' ? State the conditions under which one is suitable to the other.

14. (*a*) Describe the method of construction of large size brick or R.C.C. sewers laid in deep trenches in city areas.

(*b*) What safety precautions would you employ to protect workmen and traffic during such constructions ?

(*c*) How will you go ahead with the construction, if the ground water table is high, or the trenches are to be excavated in running sand ?

15. (*a*) Discuss the points you would keep in mind while deciding upon the programme for construction of sewers in a town ?

(*b*) Describe briefly the laying of pipe sewers by constructing trenches. How will you maintain correctly their gradient and alignment ?

(*c*) Explain how will you test the newly laid sewer lines before bringing them into commission.

16. State the different materials and shapes of which the sewers are made, and explain their merits and demerits. What points are considered in selecting a particular shape and material, and what materials and shape would you use in the following cases :

(*i*) Small size (upto 25 cm diameter) sanitary sewers ;

(*ii*) Medium size (25 to 60 cm diameter) sanitary sewers ;

(*iii*) Large size (> 75 cm diameter) sanitary sewers ; and

(*iv*) Storm sewers : (*a*) small size (*b*) large size.

Give reasons for your choice.

17. (*a*) Discuss and describe the problems commonly encountered in the maintenance of sewers. Also describe the methods employed to tackle these problems.

(*b*) What precautions would you take in the design and construction of sewers so that their maintenance would be simple and easy.

18. What are the different kinds of pipes used in sewerage works ? Compare their advantages and disadvantages. Find out the velocity and discharge of a stoneware pipe flowing full of 0.5 m diameter laid in a gradient of 1 : 300.

19. Write a short note on sewer materials, bringing out the criteria for selection of sewer material.

20. Write short notes on :

(*a*) Testing of new sewers ; and

(*b*) Types of sewers.

21. (*a*) Clearly differentiate between manhole and drop manhole.

(*b*) Design an inverted siphon to cross a 50 m wide canal. The available head loss is 1.0 m and the average sewage flow is 0.5 m^3/s. The maximum and minimum rates of flow may be taken as 300% and 50% of average flow, respectively.

22. Describe briefly and explain the functions of the following :

(*a*) Street inlet ; (*b*) Leaping weir ; (*c*) Drop manhole ; and (*d*) Flushing manhole.

23. Write short notes on any four of the following with neat sketches :

(*i*) Grease and oil trap ; (*ii*) Drop manhole ;

(*iii*) Inverted syphon ; (*iv*) Street inlet ;

(*v*) Lamp-hole ; (*vi*) Flushing tank ;

(*vii*) Catch pit ; and (*viii*) Clean out.

—6

Pumps for Lifting Sewage

6.1. Necessity of Pumping Sewage

Sewage is required to be lifted up from a lower level to a higher level at various places in a sewerage system. Sewage may have to be lifted by pumps under the following circumstances :

(*i*) The sewage from localised low lying pockets in a city has to be pumped, so as to throw it up into the city's sewer pipes flowing under gravity and running at higher elevations.

(*ii*) When the area is flat, the laying of sewers at their designed gradients may involve deeper and deeper excavations in the forward direction of flow. In such circumstances, it may be advisable to lift the sewage at suitable intervals, and then to lay sewers at reasonable depth below the surface, as shown in Fig. 6.1 (*b*).

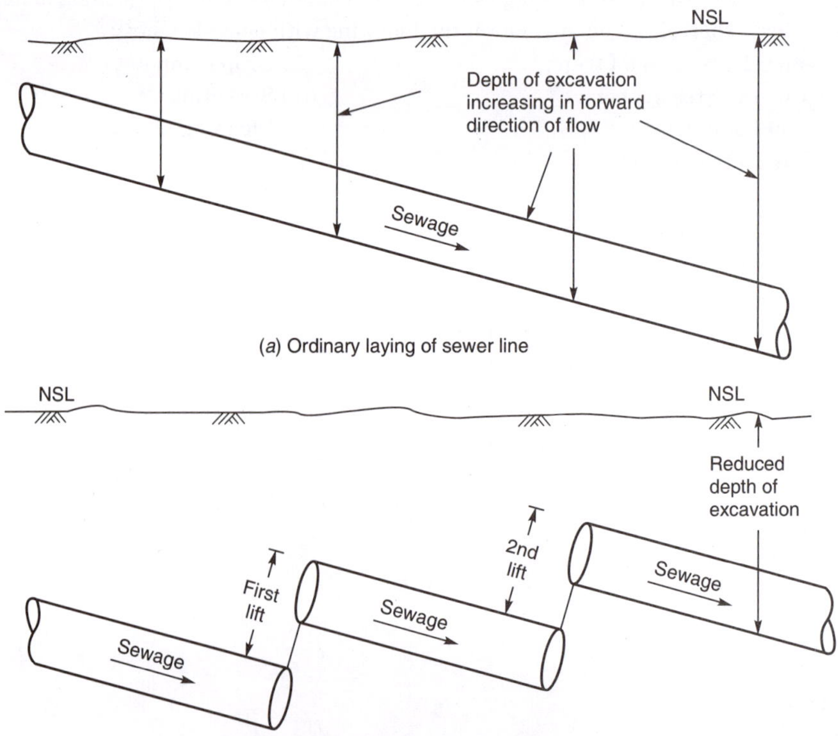

(*a*) Ordinary laying of sewer line

(*b*) Enroute lifting of sewage to reduce the depth of excavation for laying the sewer line

Fig. 6.1

(*iii*) For disposing of the sewage of the basements of large commercial buildings, sewage may have to be pumped, as the street sewer may be higher than the level of the basement floor.

(*iv*) When the outfall sewer is lower than the level of the treatment plant, the sewage will have to be pumped up. Similarly, when the level of the treated sewage coming out from the treatment plant is lower than the level of the source of its disposal, it will have to be pumped before it can be disposed of.

(*v*) In case, a sewer has to go across a high ridge, then instead of driving a tunnel through the ridge, it may prove more economical to pump the sewage into sewers laid across the slope of the ridge at reasonable depths.

6.2. Types of Pumps

Various types of pumps that are commonly used for *lifting water* have been thoroughly discussed in volume I—"Water Supply Engineering", and may be referred to again. Before we discuss here the types of pumps used for *lifting sewage*, it may be mentioned that there are two essential differences between the requirements of a pumping station in a water supply system and a station in a sewerage system :

(1) Sewage cannot be stored except for short periods of time, and the flow is highly variable ; hence in a sewage pumping station provision must be made for pump capacities capable of handling peak flows as they arrive at the station. In order to ensure this, two or more pumps, and in some cases two sources of power, are needed. As against this, water can be stored easily for future use, and as such, water supply pumps may be sized on the basis of average daily or longer use ; the capacity depending on the amount of storage provided.

(2) Sewage contains coarse solids, both floating and suspended, and hence requires special pumps which should not get easily clogged. As against this, water does not contain significant amount of suspended matter, and as such, no specially designed pumps are required for pumping water.

In view of this, it can be easily stated that the **most important characteristic of a sewage pump** is its ability to operate without clogging. As an added precaution, the interior of the pump should be easily accessible for cleaning or removal of obstructions. On the other hand, since the head against which a sewage pump has to work is generally small, efficiency is of minor importance for most installations.

Various types of pumps that are commonly employed for pumping sewage are :

1. *Centrifugal pumps* :
2. *Reciprocating pumps* ; and
3. *Pneumatic ejectors* or *Air pressure pumps*.

These three types of pumps are discussed below :

6.2.1. Centrifugal Pumps. Centrifugal pumps are most widely used for lifting sewage, as they can be easily installed in pits and sumps, and can easily transport the suspended matter present in sewage without getting clogged so often.

The horizontal axis flow type of centrifugal pumps are generally used. They are provided with either open or closed three vane type impellers* (Refer Fig. 6.2). The clearance between the vanes is kept large enough so as to allow any solid matters entering the pump to pass out with the liquid sewage. This helps in preventing the clogging of pumps and damage to the rotor. Such pumps are, therefore, called ***non-clog pumps***.

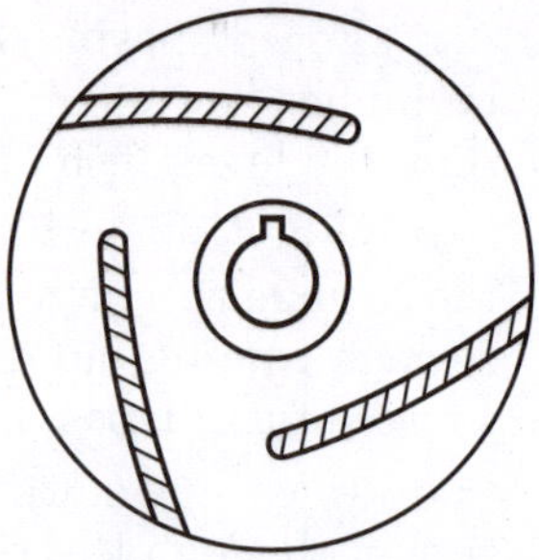

Fig. 6.2. Three vane type impeller of a non-clog centrifugal pump.

In another type of centrifugal pumps, the solid matters present in sewage are broken up (*i.e.* disintegrated) as they pass through the pump impeller. Such pumps are called ***Disintegrating pumps***. They are used when sewage is to be directly disposed of into the sea through the outfall sewer.

Both the above types of centrifugal pumps, no doubt, work at low efficiency (of the order of 30 to 60 per cent), but their use avoids the necessity of installing pre-screens for grit removal before pumping. However, where sewage has, already been screened, ordinary centrifugal pumps of high efficiency may be used.

Other types of centrifugal pumps, such as, *volute, turbine* type, etc. may also sometimes be used depending upon the circumstances.

The centrifugal pumps may be installed either *submerged* beneath the sewage in the *wet well* or a *sump*, or the pumps may be installed in the *dry well* or pit, below the sewage level in the wet well. In the first case, the pump is set with a vertical shaft reaching to the motor placed above the sewage level in a dry space, the pump being always submerged, and therefore, called ***Submergible pump***.

In the latter case, the pump is set in a dry well, and sucks its inlet supplies through the suction pipe from an adjacent wet well containing sewage, as shown in Fig. 6.3. *The pump is installed in such a way that its suction pipe remains below the lowest waste water level in the adjacent well. This will eliminate the need for priming the pump.* The exterior of the pump, in this case, is always dry, and as such, its inspection and maintenance is easy, and the chances of corrosion of the pump casing, shaft and bearings are much less as compared to that in the case of a submergible pump. This type of setting is, therefore, generally preferred.

6.2.2. Reciprocating Pumps. Reciprocating pumps are generally not used these days, because of their high initial cost, difficulty in maintenance, and greater wear and tear of their valves. However, they may sometimes be used for pumping difficult sludges and where large quantity is required to be pumped against low heads. They should, however, be amply protected by screens of 2 cm spacing.

*Open impellers are preferred for pumping sewage containing more solids.

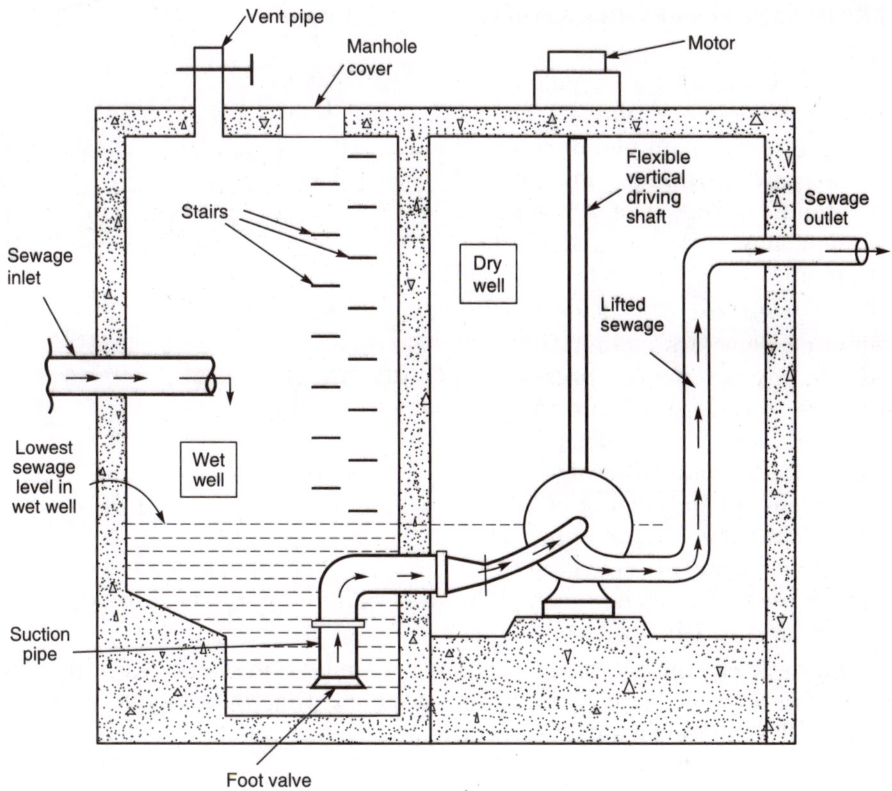

Fig. 6.3. Typical centrifugal pump installation for sewage pumping.

Reciprocating pumps used for pumping sewage are generally of two types :

(*i*) *Ram type* ; and

(*ii*) *Propeller type*.

In the *ram type* of a reciprocating pump, a piston or a plunger moves inside a closed cylinder. On the intake stroke, the liquid enters the cylinder through the intake suction valve. The delivery valve remaining close during intake stroke. On the discharge stroke, the suction valve closes, and the liquid is forced into the delivery pipe through the delivery valve, which opens during the discharge stroke. A *diaphragm pump* is an example of this type.

In the *propeller type* of a reciprocating pump, a multiple blade screw rotor or a propeller moves vertically inside a pump casing, causing the sewage to lift up. The action is somewhat similar to that of a ship's propeller, as it draws water through the inlet guide vanes, and discharges through the outlet guide vanes. The *axial-flow screw pump* is an example of this type.

Diaphragm pump. A diaphragm pump is a ram type of reciprocating pump ; and is shown in Fig. 6.4. It is the most commonly used type of a reciprocating pump. In this pump, a piston or a plunger is attached to the centre of a circular rubber diaphragm, the outer edge of which is bolted to a flange on the pump. The up and down motion of the plunger is permitted by the flexibility of the diaphragm, thus increasing or decreasing the capacity of the pump casing. When the piston moves upward, the liquid flows into the pump through the suction valve ; the delivery valve remaining closed. But, however, when the piston moves downward, the suction valve closes, and the liquid is forced through the delivery pipe by opening the delivery valve. The diaphragm pump is simple, durable and needs no priming. However, after some use, the rubber diaphragm may wear out and need replacement.

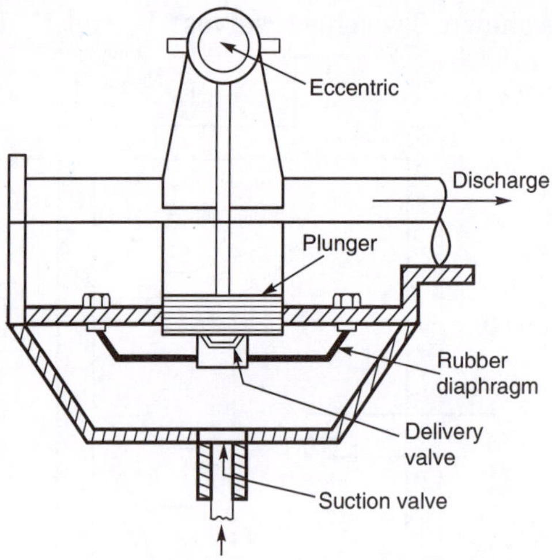

Fig. 6.4. Diaphragm pump a ram type of reciprocating pump.

6.2.3. Pneumatic Ejectors.

Pneumatic ejectors sometimes called *Air ejectors* or *Air pressure pumps* are complete lift units in themselves, and are used for pumping smaller quantities of waste waters, such as for raising waste waters from the basements of buildings and thus discharging it into street sewers. Such a complete unit can be purchased readymade, fabricated for direct installation.

A pneumatic ejector, in general, consists of an air-tight tank into which the waste water flows by gravity, and out of which, the waste water is forced automatically whenever sufficient waste water has accumulated to raise a float, and thus opening the compressed air inlet valve.

A typical air-ejector called ***Shone's air ejector*** is shown in Fig. 6.5. It consists of a cast iron chamber with a spindle having an upper and a lower

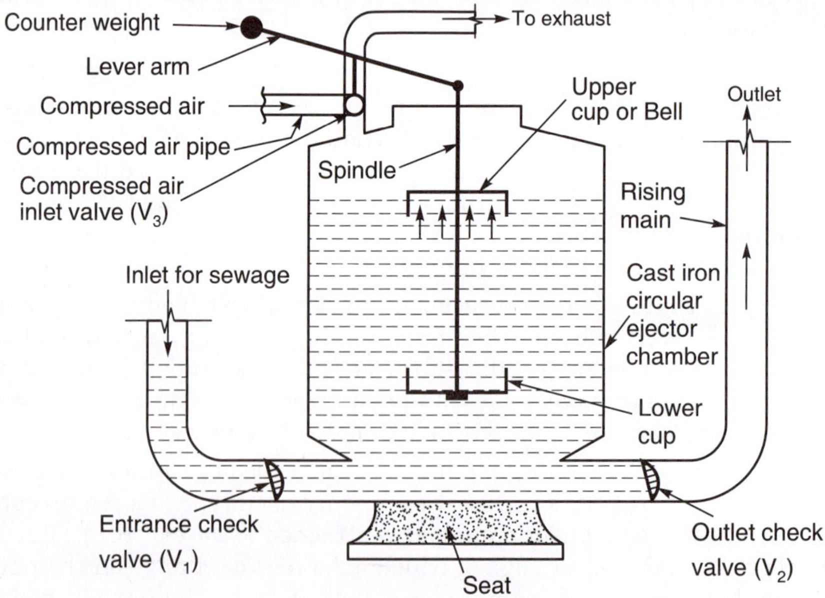

Fig. 6.5. Shone's Air-Ejector.

cup, as shown. Two check valves* V_1 and V_2 are provided at the entrance and exit points, respectively. The ejector chamber rests on a seat, as shown. A lever arrangement with a counter-weight is provided, so as to open the compressed air inlet valve (V_3). The compressed air is supplied through this valve at a pressure of about 1.5 kg/cm^2.

The waste water enters the ejector chamber through the entrance valve (V_1), and slowly rises in the chamber. At this stage, the exit valve (V_2) and compressed air inlet valve (V_3) remain closed. As the waste water level in the chamber rises, the air from the chamber escapes through the exhaust. But when the waste water level in the chamber reaches the *bottom of the upper cup*, the air inside gets entrapped. The further rise of wastewater in the chamber makes this entrapped air to exert vertical pressure on the spindle. This causes the spindle to lift up, and through the lever arrangement, it results in the closing of the exhaust, and opening of the compressed air inlet valve (V_3).

The air under pressure entering the chamber from valve (V_3) forces the wastewater from inside the chamber to rise up in the outlet pipe by opening the exit valve (V_2). At this stage, the valves V_2 and V_3 remain open, but valve V_1 is closed. The wastewater is thus lifted up and discharged from the outlet, till the level of the wastewater falls below the bottom of the upper cup. At this stage, the entrapped air from the upper cup escapes, and the lever arrangement opens the exhaust and closes the compressed air inlet valve (V_3). The exit valve (V_2) also closes, and the entrance valve (V_1) opens to again admit the wastewater. The process goes on repeating.

Pneumatic ejectors are often used in pairs, so that when one is empty, the other is filled.

The *advantages* of air ejectors are :

(*i*) There is no necessity of providing primary treatment arrangement for sewage, when air ejectors are employed.

(*ii*) They are automatic complete units, and require least supervision.

(*iii*) They are useful in conditions, where there are chances of small capacity centrifugal pumps getting clogged, if used.

(*iv*) Such ejectors are not likely to be clogged any where.

(*v*) Compared to the construction of a pumping station, they are found to be economical at places, where smaller quantities of wastewaters are required to be lifted.

The only *disadvantage* of air-ejectors is that they have very low efficiency, of the order of 15 per cent or so.

An example on the design of such an ejector is solved below :

Example 6.1. *A low lying residential colony is having a population of 3000 persons. The colony gets a per capita supply of water at the rate of 140 litres per day. A separate sewerage system for the colony is installed. It is further required to lift the entire sewage of the colony by installing an air-ejector. Assuming the velocity in the main sewer as 0.9 m/sec, and velocity of compressed air as 5 m/sec, design the ejector.*

---

*A check valve, also called reflux valve, opens in one direction only.

Solution. Assuming that the entire water supplied appears as sewage, we have

The average sewage flow

$$= \frac{3000 \times 140}{1000 \times 24 \times 60 \times 60} \text{ cumecs.} = 0.00486 \text{ cumecs.}$$

Assuming the peak flow to be three times the average flow, we have

The peak sewage discharge to be lifted

$= 3 \times 0.00486 = 0.0146$ cumecs.

Design of the Ejector Chamber

Assuming that the ejector fills after every ten minutes, we have

The required capacity of the ejector

$= 0.0146 \times 10 \times 60$ cu-m. = 8.75 cu-m.

Assuming the height of the ejector as 2 m, we have

The cross-sectional area of ejector chamber

$$= \frac{8.75}{2} \text{ m}^2 = 4.375 \text{ m}^2.$$

Diameter of the circular ejector chamber

$$= \sqrt{\frac{4.375 \times 4}{\pi}} = 2.36 \text{ m ; say } 2.5 \text{ m.}$$

Hence use 2.5 m diameter ejector chamber with 2 m height. **Ans.**

Design of the Main-Sewer

Since the peak flow of 0.0146 cumecs flows with a velocity of 0.9 m/sec, we have

The area of x-section of the main

$$= \frac{0.0146}{0.9} = 0.016 \text{ m}^2.$$

$\therefore$ Diameter of the main

$$= \sqrt{\frac{0.016 \times 4}{\pi}} = 0.143 \text{ m.}$$

= 14.3 cm ; Say 15 cm dia.

Hence use 15 cm dia main. **Ans.**

Design of the Compressed Air-pipe

Velocity of the compressed air

= 5 m/sec.

$\therefore$ Area of the pipe required

$$= \frac{0.0146}{5} = 0.0029 \text{ m}^2.$$

$\therefore$ Diameter of air pipe

$$= \sqrt{\frac{0.0029 \times 4}{\pi}} = 0.06 \text{ m} = \mathbf{6 \text{ cm.}} \quad \textbf{Ans.}$$

6.3. Pumping Stations

The building where pumps and other accessories are installed for lifting sewage is called a sewage pumping station. Such a building should be located at a carefully chosen site.

6.3.1. Proper Location. These pumping stations should preferably be located near a natural disposal unit, such as, a stream, a lake, or a river, etc., so that in case of break-down of pumping, the accumulated sewage can overflow into this natural source of disposal. But at the same time, the site should be such that it is not liable to get flooded either due to seepage from the adjoining river stream or due to high floods in that river stream.

6.3.2. Component Parts. A typical sewage pumping station is shown in Fig. 6.6. Such an installation has the following major components :

(1) Grit channel ;

(2) Coarse and fine screens ;

(3) Well, called Sump well ;

(4) Dry well or Pump room ;

(5) Motor room ;

(6) Rising mains ;

(7) Emergency Exit Pipe (Not shown in Fig. 6.6) ; and

(8) Other accessories like starters, valves, flow recorders, etc.

The construction and functioning of these different components is briefly discussed below :

(1) **Grit channel.** The sewage entering the pumping station generally contains a lot of indestructible solid matter, such as, grit, gravel, sand, etc., in addition to the solids in suspension like feces, papers, rags, etc. It is necessary to remove as much of this material as possible before pumping, so as to minimise the wear and tear of the pump impeller and of the rising main.

The solid matter is, therefore, first of all, removed before the sewage enters the pump.

In order to achieve this purpose, the sewage is first passed through grit channel (or grit chamber), where its velocity is considerably reduced to the order of 0.15 to 0.3 m/sec. Grit channel is a long basin with an enlarged cross-section above the invert line of the inlet sewer, as shown, which results in reducing the flow velocity. Due to the reduced velocity, the grit settles down, and is removed by an endless chain to which perforated buckets are fixed. The chain is operated by power, and the grit is dredged into a closed container, till it is removed and disposed of.

The grit channel should have a minimum capacity of one per cent of the daily dry weather flow. Moreover, there should be two similar units, each of which can be used, allowing the other to be cleaned. In small installations, grit is removed once a week ; whereas in larger installations, the removal may be a continuous daily process.

(2) **Screens.** After the sewage has been excluded from the solid grit matter by settling it down in the grit channel, it is passed through the screens ; which help in extracting other matter, such as, excess of rags, sticks, papers, etc. Such matter, if not removed, may also cause the choking of pump impellers.

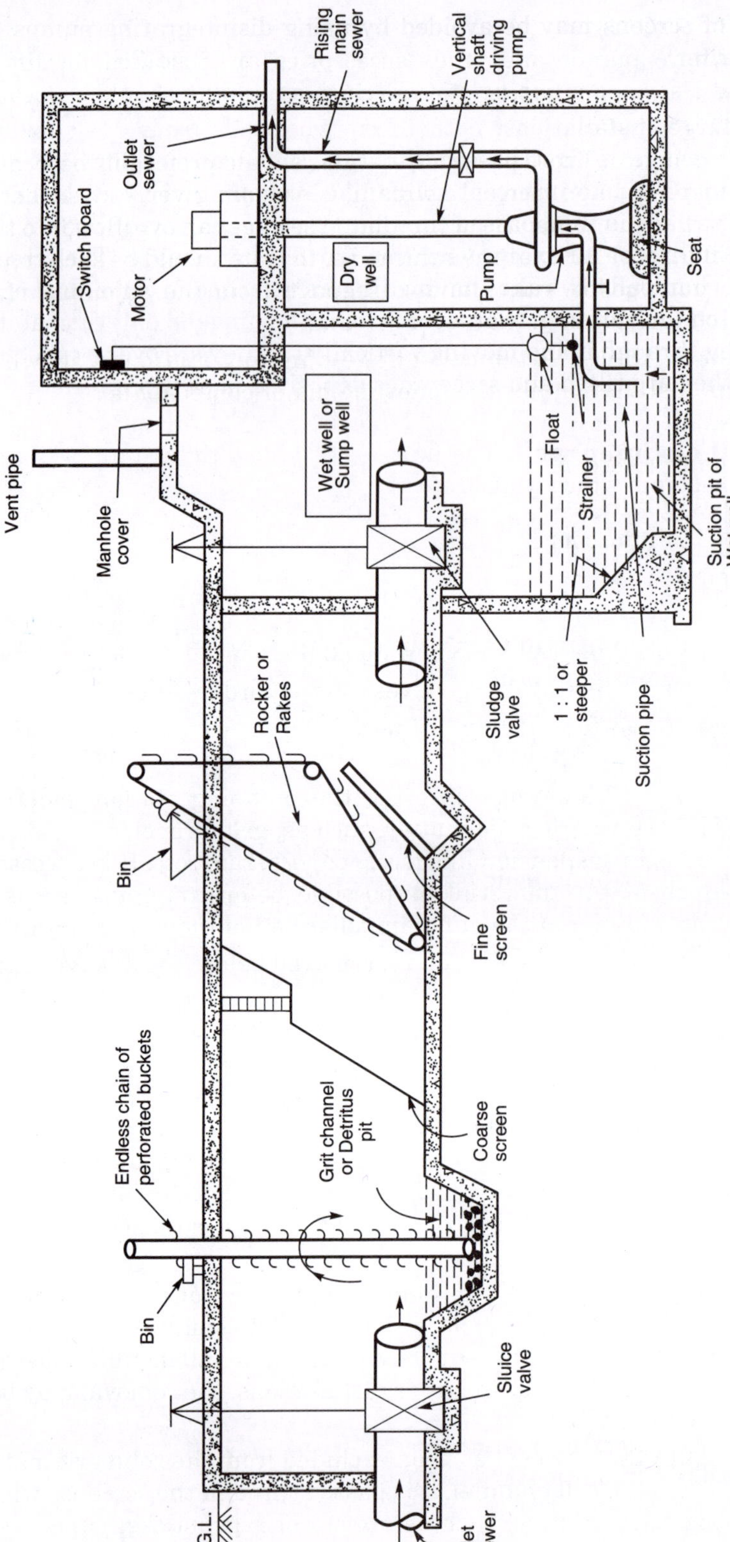

Fig. 6.6. A Typical Sewage Pumping Station.

However, use of screens may be avoided by using disintegrating pumps (as pointed out earlier).

Two types of screens are commonly used—*coarse screens* and *fine screens*. Normally, in large installations, both these types of screens are provided together. The sewage is first passed through coarse screens having clear openings of 5 to 10 cm, to intercept solids like scrubbing brushes, blocks of wood, etc. The sewage is then passed through fine screens having openings of 2.5 to 5 cm, to intercept all except very fine particles of sewage. The screens are cleaned mechanically by rakes having fingers attached to either end of an endless roller chain running over sprockets ; the fingers entering at the bottom, drawing through it and moving vertically till they are over a collecting trough where they are tilted and screenings deposited in sealed bins for quick disposal.

(3) **Wet well or Sump well.** The purpose of providing a sump well is to form a suction pit from which the pump may draw sewage through the suction pipe. It also acts as an equalising basin to minimise the load fluctuations on the pump. A float connected to a switch is provided in this sump well in such a way that when the sewage rises above the float level, the switch gets pressed and pump automatically starts functioning, thus pumping sewage. This designed level of the sewage in the sump well is kept above the pump level, which avoids the necessity of priming of the pump.

The capacity of the sump well should be about 15 to 30 minutes of the peak flow. If its capacity is less, the operation of the power unit shall have to be done at frequent intervals, with the result that the operation becomes expensive (in case of electrically operated plant, it means increased cost of current, as the starting current is more than the full load current).

The floor of the bottom should slope 1 : 1 or steeper. This will help in pushing the sludge into the low point, where the inlet to the suction pipe is located. Sometimes, the capacity of the sump also includes the emptying of the rising main back to the sump at the time of the cleaning of the wells, unless a separate wash-out is provided. When two or more pumps are employed, the pump storage may be suitably split into different interconnected compartments, so that any one of them can be shut off for cleaning and repairs.

(4) **Dry well or Pump well.** In this room, pumps are installed, and the end of their suction pipe is placed in the suction pit or the wet well. The general layout of a pump room should be carefully made, and it may include a repair workshop, office and store room, etc. Enough provision for its possible future extension may also be made.

The number and sizes of pumps required should be carefully worked out, and depends upon the quantity of sewage likely to enter into the station.

The provision for stand-by units should also be made.

(5) **Motor room.** This room is situated above the pump room, and accommodates the electric motor which drives the pumps. The other appurtenances like automatic starters, flow recorders, etc. are also installed in this room.

(6) **Rising mains.** The sewage, after being pumped, is taken to the high levelled gravity sewer through the rising mains. The rising main may be of cast iron or asbestos cement pressure pipes. To avoid siltation, the velocity of flow in the rising main should not be less than 0.75 m/sec at any time of flow ; and to avoid excessive head loss, the maximum velocity of flow in the rising main is usually limited to 1.8 m/sec and never allowed to exceed 3 m/sec. The preferred value, however, is about 1.2 m/sec.

(7) **Emergency exit pipe.** An emergency exit pipe, connecting the sump well with a natural stream or river, is preferably provided at all major pumping stations. Hence, when the sump well overflows due to any reason, the excess sewage can be easily directed through this exit pipe.

(8) **Other accessories.** Besides the above major parts, various other accessories, such as, **automatic starters, check valves, flow recorders,** etc., are needed, and installed suitably in the pumping station.

An example on the design of various components of a pumping station has been solved below :

Example 6.2. *Draw a typical sketch of a sewage pumping station for a town having a population of 50,000 supplied with a per capita water supply of 130 litres per day. The sewage from this town enters the pumping station through a low levelled sewer at R.L. 195.0 m and leaves through the high levelled sewer at R.L. 205.0 m. The town has been provided with a separate sewerage system, and there are no chances of storm water getting infiltrated into sewers. Calculate (a) the size of the rising main, (b) size of the sump well, and (c) B.H.P. of the pump motor required. Assume the length of the rising main as 100 m. Make other suitable assumptions, where needed.*

Solution. For typical sketch of a sewage pumping station, please refer Fig. 6.6.

Now, we will calculate the required size of the rising main and the size of the pump, etc. as below :

Assuming that 80% of the water supplied appears as sewage, we have

The average quantity of sewage produced

$$= 0.8\left[\frac{50{,}000 \times 130}{1000 \times 24 \times 60 \times 60}\right] = 0.06 \text{ cumecs.}$$

Assuming the maximum flow to be three times the average, we have

The peak sewage discharge

$$= 3 \times 0.06 = 0.18 \text{ cumecs.}$$

(*a*) **Design of the rising main.** Assuming the velocity of flow in the rising-main as 1.2 m/sec, we have

The required cross-sectional area of the rising-main

$$= \frac{Q}{V} = \frac{0.18}{1.2} \text{ m}^2 = 0.15 \text{ m}^2$$

Required dia. of the rising main

$$= \sqrt{\frac{0.15 \times 4}{\pi}} = 0.437 \text{ m} = \text{Say } 0.45 \text{ m.}$$

Hence, use 0.45 m dia, rising main.

∴ Velocity in the rising main

$$= \frac{0.18}{\frac{\pi}{4}(0.45)^2} = \frac{0.18}{0.159} = 1.13 \text{ m/sec.} \quad \textbf{Ans.}$$

(*b*) **Design of the sump well.** The sump well can be designed for the peak flow of 15 minutes* (*i.e.*, 15 minutes detention period with peak flow). The capacity of the rising main may also be added to the above quantity, in case the rising main has to be emptied into the sump at the time of cleaning the sump.

Peak flow rate = 0.18 m^3/sec.

∴ Peak flow for 15 minutes

$$= 0.18 \times (15 \times 60) = 162 \text{ cu-m.}$$

Capacity or contents of the rising main

$$= \text{Area} \times \text{Length}$$

$$= \frac{\pi}{4} \times (0.45)^2 \times 100 = 15.9 \text{ cu-m ; } \textbf{Say 16 cu-m.}$$

∴ Total capacity of the sump well

$$= 162 + 16 = \textbf{178 cu-m.}$$

Let us now provide 3 number of sumps, any two taking the maximum flow and the third as stand-by unit, we have the capacity of each sump well

$$= \frac{178}{2} = 89 \text{ cu-m.}$$

Assuming the depth of the sump well as 3 m, we have the reqd. cross-sectional area of the sump well

$$= \frac{\text{Capacity}}{\text{Depth}} = \frac{89}{3} \text{ m}^2 = 29.66 \text{ m}^2.$$

∴ The reqd. dia. of the sump well

$$= \sqrt{\frac{29.66 \times 4}{\pi}} = \textbf{6.15 m.} \quad \textbf{Ans.}$$

(*c*) **Design of pump.** Each pump has to lift 89 cu-m of waste water in 15 minutes.

∴ Capacity of each pump

$$= \frac{89}{15 \times 60} = 0.0988 \text{ cumecs.}$$

The lift of pump reqd.

= R.L. of high levelled sewer

– R.L. of low levelled sewer

+ Depth of sump well

= 205.0 – 195.0 + 3 = **13 m.**

The pump has to work against this left as well as the losses. Let us now evaluate the losses :

*This corresponds to minimum time of 15 minutes for the pumps to run continuously.

Frictional loss of head in the rising main

$$= \frac{4flV^2}{2gd}$$

$$= \frac{4 \times 0.01 \times 100 \times (1.13)^2}{2 \times 9.81 \times 0.45} = 0.58 \text{ m.} \qquad [\text{using } f = 0.01]$$

Assuming the loss in bends, etc. = 0.3 m, we have

Total losses = 0.58 + 0.3 = 0.88 m.

∴ Total lift against which pumps have to work

= 13 + 0.88 = 13.88 m.

∴ H.P. of the pump

$$= \frac{w \,.\, QH}{75} = \frac{1000 \times 0.0988 \times 13.88}{75} = 18.3 \text{ ; say 19 H.P.}$$

Assuming the pump efficiency to be 60%, we have

The Brake Horse Power of the motor reqd.

$$= \frac{19}{0.6} = 31.67 \text{ ; } \textbf{say 32. Ans.}$$

PROBLEMS

1. When is pumping of sewage required in a sewerage system ? How is it more difficult than the pumping of water in water supply schemes ? Draw a sketch of a typical sewage pumping station, showing all the necessary arrangements to pump sewage of a town of one lakh population at the outfall sewer.

2. Propose suitable pumping units to pump (1.0 M.L./day average flow) sewage from 2 km away and 10 m above the level of the outfall sewer level. Suggest suitable size of a rising main. Show on a neat sketch the details of the sewage pumping station.

3. Under what circumstances is pumping of sewage necessary ? What types of pumps would you recommend for pumping sewage and why ?

4. (*a*) Write short note on Shone's ejector.

(*b*) An ejector has to lift 2250 L/min, if the velocity of air in the air pipe is 6 m/sec and that of sewage in the sealed main is 0.75 m/sec, find the diameters of pipes and the capacity of the ejector, supposing, it is filled and emptied 10 times an hour. [**Ans.** Air main = 10 cm ; sewage main = 30 cm ; Ejector = 13.5 m^3]

5. Comment on the following with respect to a sewage pumping station :

(*i*) Capacity of the wet well ;

(*ii*) Selection of the type and capacities of the pump ;

(*iii*) Number of pumps and stand-by capacity of the pumping units ;

(*iv*) Factors governing the size of the rising main ; and

(*v*) Overflow arrangement for the wet well.

6. (*a*) Under what circumstances would you advise pumping of sewage ? Name the various types of lifting devices used in discharging low level sewage to high level sewers, and describe their merits and demerits.

(*b*) Design and draw a dimensioned sketch of sump well and pump house for a low level sewage pumping station dealing with sewage from 15,000 people with a sewage rate of 110 lphd. Invert of the low level sewer is 6 m below the ground.

7. Differentiate between 'ejectors' and 'pumps'. Also describe with a sketch the working of a typical 'ejector'.

8. (*a*) Draw a sketch of a typical sewage pumping station.

(*b*) Mention which type of pump is suitable for sewage pumping. Give reasons.

—7

Quality and Characteristics of Sewage

7.1. Importance of Study

When untreated sewage is discharged into some river stream, floating solids present in the discharged sewage may be washed up on to the shore, near the point of disposal, where they decompose and create foul smells and bad odours. The large amount of organic matter present in the discharged sewage will also consume the dissolved oxygen from the river stream in getting oxidised, and may thus seriously decrease the dissolved oxygen of the river stream, causing fish kills and other undesirable effects. In addition to these effects, the discharged sewage will contaminate the river water with pathogenic* bacteria. *Hence, even though municipal sewage is 99.9 per cent water, it requires treatment, if nuisance is to be avoided.*

The extent and type of treatment required, however, depends upon the character and quality of both sewage and source of disposal. For example, a small community at the seaside might discharge its unaltered sewage directly into the ocean without any ill effects, but if this town/colony was located inland on a small stream, a high degree of treatment might be needed.

In the olden times, the waste waters from a community were not so much contaminated as they are today. The urbanisation, industrial growth, and the improved standards of living, have however, increased the strength and quantity of municipal sewage in recent years to a point where *dilution* alone can no longer be relied upon to prevent the undesirable effects of pollution. In many cases, more advanced treatment of wastewater is essential to prevent undue pollution. This is much more so, when the disposed sewage is likely to contain industrial wastewaters.

Hence, it is absolutely necessary to study the characteristics and behaviour of sewage, to ensure its safe disposal. This study will help us in determining the degree and type of treatment required to be given to a given sewage, and thus avoid the pollution of the source of its disposal.

Before we discuss the various *physical, chemical* and *biological (bacteriological)* characteristics of sewage, let us first discuss as to *how the sewage gets decayed*, and what happens to it with the passage of time.

7.2. Decay or Decomposition of Sewage

Most of the organic matter present in sewage is unstable and decomposes readily through *chemical* as well as *biological* actions. Till the advent of

*Pathogenic bacteria or pathogens are those harmful micro-organisms which cause serious water borne diseases like cholera, typhoid, infectious hepatitis, dysentry, etc. For details, please refer to chapter 8 of Vol. I—"Water Supply Engineering" by the same author.

biological methods, chemical treatment of sewage was, therefore, being adopted to remove the organic matter from the sewage. However, after the advent of biological degradation methods, this type of treatment has become our foremost choice.

The organic matter, which is decomposed by bacteria, under biological action, is called the **biodegradable organic matter.** Most of the organic matter present in sewage is biodegradable, and hence undergoes **biological decomposition,** which can be divided into two types, *i.e.*

(*i*) *aerobic decomposition*, called *aerobic oxidation* ; and

(*ii*) *anaerobic decomposition*, called *putrefaction.*

7.2.1. Aerobic Decomposition. If air or oxygen is available freely to the wastewater in dissolved form, then the biodegradable organic matter will undergo *aerobic decomposition*, caused by *aerobic bacteria** as well as by *facultative bacteria—operating aerobically.*** These bacteria will then utilise the free oxygen as electron acceptor, thereby oxidising the organic matter to stable and unobjectionable end products. The stable end products like **nitrates,** *carbon dioxide*, sulphates are formed, respectively for the three forms of matter, *i.e.* **nitrogenous**, *carbonaceous* and sulphurous matter. Water, heat, and additional bacteria will also be produced in this biological oxidation, which can be represented by the following equations :

$$(i)\ \text{Nitrogenous organic matter} \xrightarrow[\textbf{Aerobics}]{\textit{Oxidation by}} NO_3^- + \underset{\text{(sometimes)}}{NH_3 \uparrow} + \text{Energy} \quad ...(7.1)$$

$$(ii)\ \text{Carbonaceous organic matter} \xrightarrow[\textbf{Aerobics}]{\textit{Oxidation by}} CO_2 \uparrow + H_2O + \text{Energy} \quad ...(7.2)$$

$$(iii)\ \text{Sulphurous organic matter} \xrightarrow[\textbf{Aerobics}]{\textit{Oxidation by}} SO_4^{--} + \text{Energy} \quad ...(7.3)$$

It may also be noted that during the decomposition of nitrogenous organic matter, the ammonia formed in the initial stages, may linger on till the end, depending upon the available oxygen, retention time, temperature, biological activity, etc., because *the facultative bacteria are incapable to break ammonia to nitrates.*

The intermediate products formed in the aerobic oxidation of the three types of organic matter are shown in their respective cycles, in Fig. 7.1 (*a*), (*b*) and (*c*). These cycles are known as **nitrogen cycle, carbon cycle,** and **sulphur cycle**, for *nitrogenous, carbonaceous*, and *sulphurous* organic matter, respectively.

***Aerabic bacteria** are those which flourish in the presence of free dissolved oxygen in wastewater, and consume organic matter for their food, and thereby oxidising it to stable end products.

Anearobic bacteria flourish in the absence of free dissolved oxygen, since they survive by utilising the bounded molecular oxygen in compounds like nitrates (NO_3) and sulphates (SO_4), etc., thereby reducing them to stable end products along with evolution of foul smelling gases like H_2S, CH_4, etc.

****Facultative bacteria** can operate either as *aerobically* or as *anaerobically*. Hence, they can survive and cause decomposition of organic matter, either in the presence or in the absence of free dissolved oxygen in wastewater.

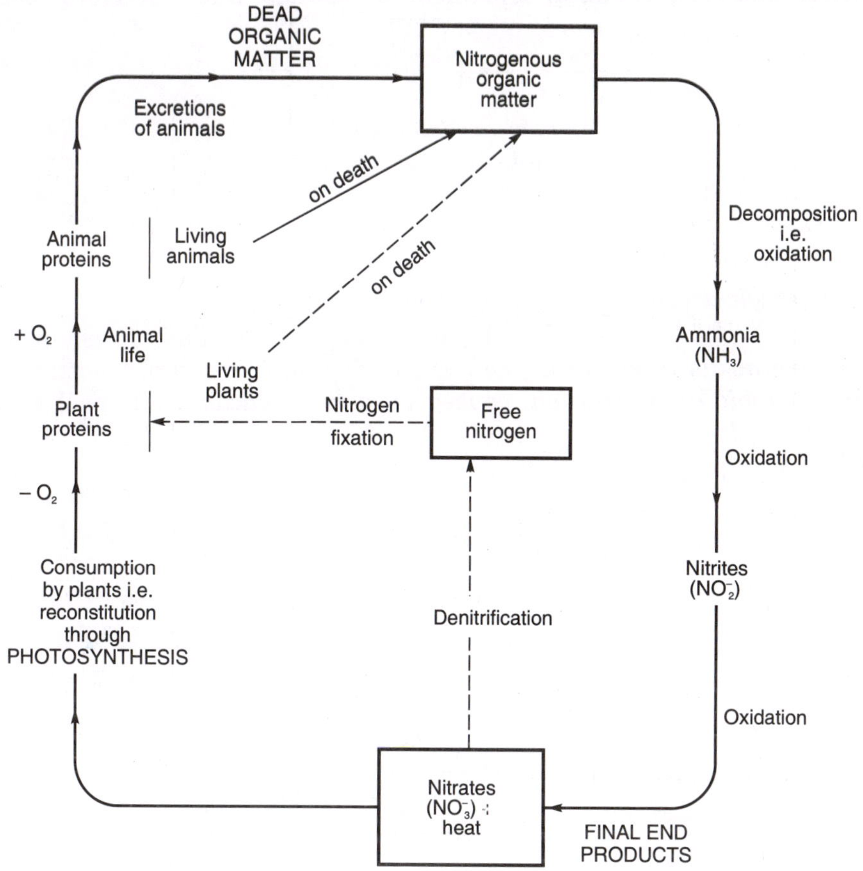

Fig. 7.1. (*a*) Nitrogen cycle under aerobic oxidation.

Steps in the Nitrogen cycle. Nitrogenous organic matter get oxidised to ammonia, then to nitrites, and finally to *nitrates*, which when consumed by plants, through photosynthesis, form plant proteins (plant life). The plant proteins, when consumed by animals, form animal proteins. The wastes produced by animals and their dead bodies, will again form nitrogenous organic matter, thus completing the nitrogen cycle.

There may be some short circuits of the cycle, as shown by dotted lines ; say for example, the dead plants may also, on death, lead to formation of organic matter directly, without changing into animal proteins. Similarly, nitrates on *denitrification* may get converted into free nitrogen (and sometimes to ammonia), which may be converted into plant proteins, as it may be used by certain bacteria residing in the plant roots. This is called ***nitrogen fixation***. **Blue-green algae** also causes nitrogen fixation.

7.2.2. Anaerobic Decomposition. If free dissolved oxygen is not available to the sewage, then anaerobic decomposition, called ***putrefaction***, will occur. **Anaerobic bacteria** as well as **facultative bacteria**—*operating anaerobically*, will then flourish and convert the complex organic matter into simpler organic compounds of nitrogen, carbon, and sulphur. These anaerobic bacteria, infact, survive by extracting and consuming the *bounded molecular oxygen* present in compounds like nitrates (NO_3) and sulphates (SO_4). **Gases** like *ammonia, nitrogen, hydrogen sulphide, methane*, etc. are also evolved in this decomposition, *producing obnoxious odours*. The various stages, at which

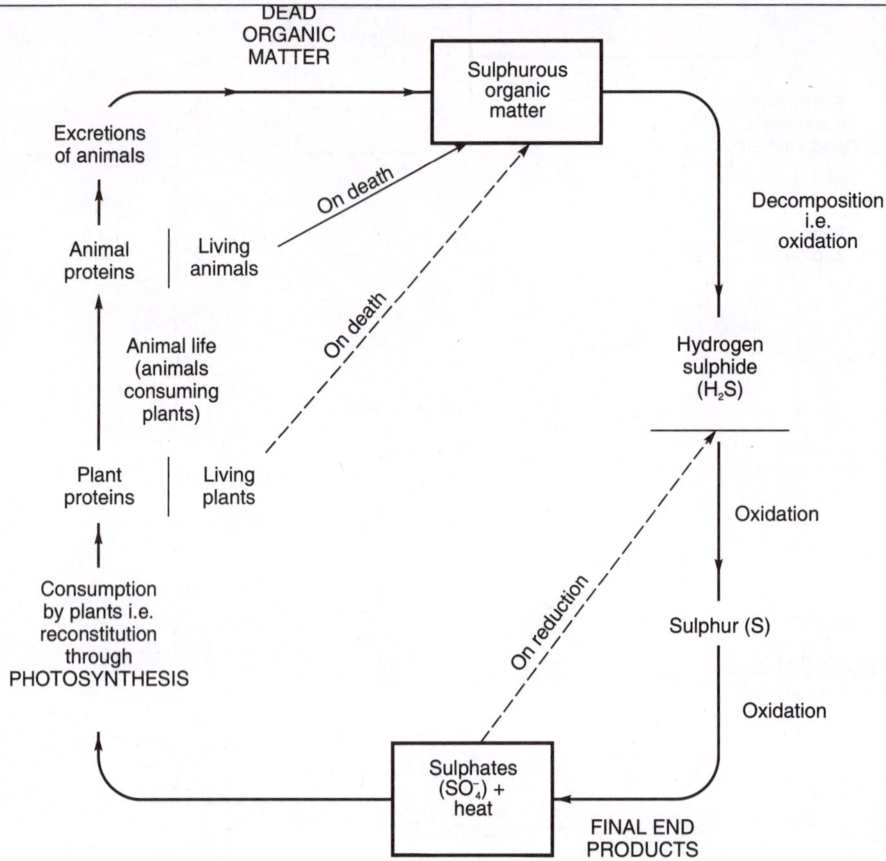

Fig. 7.1. (*b*) Sulphur cycle under aerobic oxidation.

Steps in the sulphur cycle. This cycle is similar to nitrogen cycle. The sulphurous organic matter, on oxidation, produces H_2S gas, which on further oxidation, changes to sulphur, and then finally to sulphates (SO_4^{--}). Sulphates, when consumed by plants through photosynthesis, change into plant proteins ; which when eaten by animals, change into animal proteins. The wastes produced by animals and their dead bodies will again form sulphurous organic matter, thus completing the sulphur cycle.

There may be short circuits in the cycle, as shown by the dotted lines : organic sulphurous matter may be directly formed by the death of the plants, without the formation of animal proteins.

Similarly, sulphates in the absence of oxygen, will be converted into H_2S, by the process of *reduction*.

these gases are evolved are shown in Fig. 7.2, which represents nitrogen, carbon and sulphur cycles together, for the above anaerobic decomposition. The final equations, representing this decomposition, are given below :

$$(i)\ \text{Nitrogenous organic matter} \xrightarrow[\textbf{Anaerobics}]{\textit{Reduction by}} N_2 \uparrow + NH_3 \uparrow + \text{Organic acids} + \text{Heat energy} \quad ...(7.4)$$

$$(ii)\ \text{Carbonaceous organic matter} \xrightarrow[\textbf{Anaerobics}]{\textit{Reduction by}} C_2O \uparrow + \text{Heat (energy)} \quad ...(7.5)$$

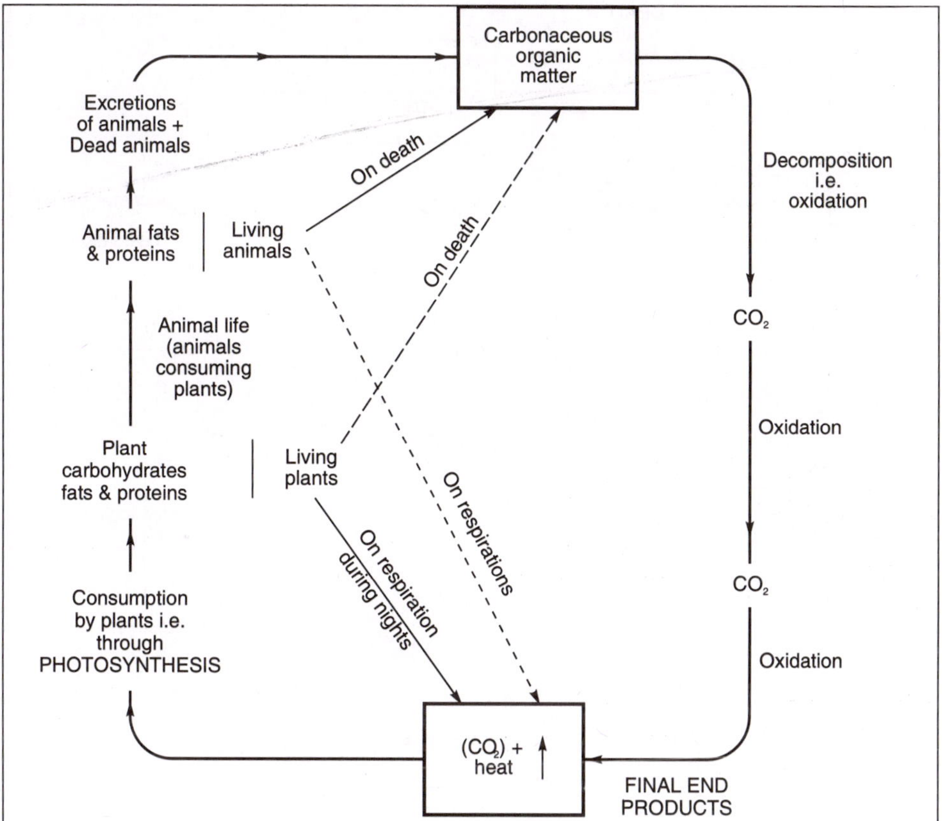

Fig. 7.1. (*c*) Carbon cycle under aerobic oxidation.

Steps in the carbon cycle. The carbonaceous organic matter, on oxidation, releases carbon dioxide, which is its final end product. This CO_2, when used by plants through photosynthesis, gets converted into plant carbohydrates, fats and proteins (sugars) ; which when eaten by animals, change into animal fats and proteins. The wastes produced by animals and their dead bodies will again form carbonaceous organic matter, thus completing the carbon cycle.

There may be short circuits in the cycle, as shown by the dotted lines. Organic carbonaceous matter may be directly formed by the death of plants.

Similarly, the plant life gives off CO_2 at night, and the animal life gives off CO_2 during respiration. Both these *respiration* processes are shown by the dotted lines in the above figure.

(*iii*) $$\text{Sulphurous organic matter} \xrightarrow[\textbf{Anaerobics}]{\textit{Reduction by}} H_2S\uparrow + \text{(Heat Energy)} \qquad ...(7.6)$$

The **organic acids** including alcohols produced in Eq. 7.4, are further converted into **methane gas** ($CH_4\uparrow$), carbon dioxide gas ($CO_2\uparrow$), etc., if **methane-forming bacteria** are also especially present in the sewage. This conversion is represented by the equation :

(*iv*) $$\text{Organic acids} \xrightarrow[\text{anaerobic bacteria}]{\text{methane forming}} CH_4\uparrow + CO_2\uparrow + \underset{\text{(energy)}}{\text{Heat}} \qquad ...(7.7)$$

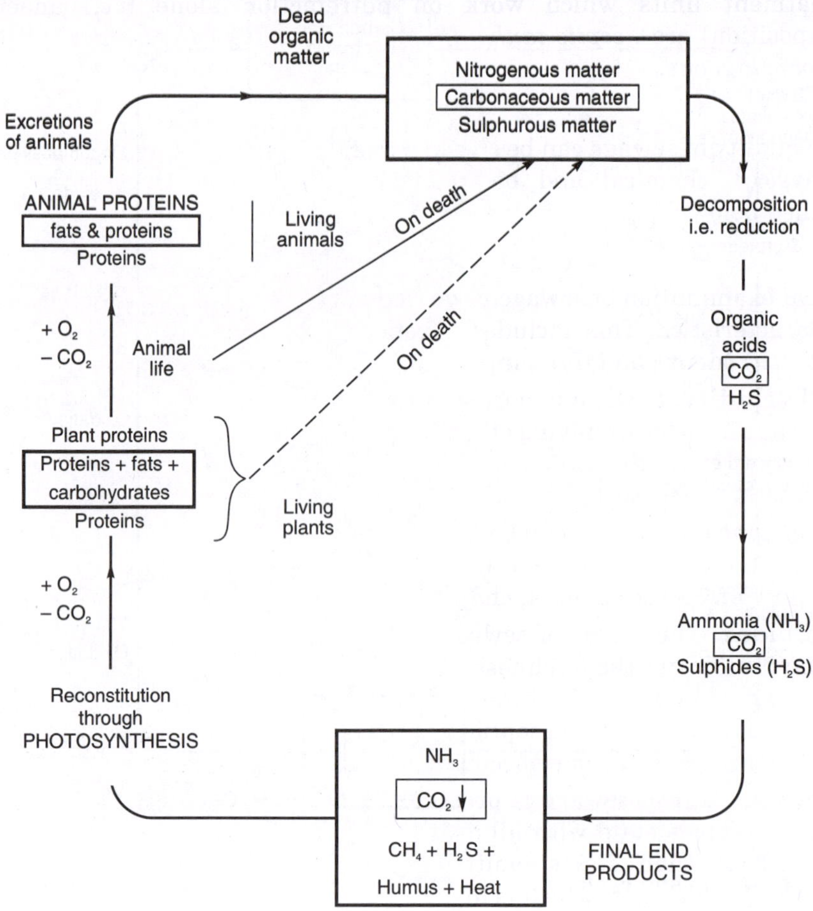

Fig. 7.2. Nitrogen, carbon and sulphur cycles, under anaerobic oxidation.

An understanding of these cycles will help us in determining the **stage of decomposition of sewage** by testing for the products of decay. For example, *a well oxidised sewage will contain nitrates and sulphates, but very little ammonia and hydrogen sulphide. On the other hand, lesser oxidised sewage will contain nitrites and sulphur instead of nitrates and sulphates.*

7.2.3. Basis of Biological Treatment. Numerous bacteria are found in waste waters of a community. So much so, that about 5 to 50 billion bacteria are generally present per litre of sewage. Out of these bacteria, only a small number are harmful to man, and are called **pathogens.** But a large number of bacteria present in sewage, called **non-pathogens**, are, however, important aids, in the process of decomposition, and are thus useful to us in sewage treatment. The fundamental basis of our treatment given to sewage, therefore, is *to provide an environment favourable to the action of the aerobic and anaerobic bacteria*, that stabilise the organic matter present in sewage either through aerobic or anaerobic decomposition.

Treatment units which work on oxidation alone (*i.e.* aerobic decomposition) are : *aeration tanks, contact beds, intermittent sand filters, trickling filters,* and *oxidation ponds.*

Treatment units which work on putrefaction alone (*i.e.*, anaerobic decomposition) are : *septic tanks*, *Imhoff tanks, sludge digestion tanks,* and *anaerobic lagoons*.

CHARACTERISTICS OF SEWAGE

The quality of sewage can be checked and analysed by studying and testing its physical, chemical and bacteriological (biological) characteristics, as explained below :

7.3. Physical Characteristics of Sewage and Their Testing

Physical examination of sewage is carried out in order to determine its physical characteristics. This includes : tests for determining (*i*) turbidity ; (*ii*) colour ; (*iii*) odour ; and (*iv*) temperature. These tests are summarised below :

(*i*) **Turbidity.** Sewage is normally turbid, resembling dirty dish water or wastewater from baths having other floating matter like fecel matter, pieces of paper, cigarette-ends, match-sticks, greases, vegetable debris, fruit skins, soaps, etc., etc. The turbidity increases as sewage becomes stronger.

The degree of turbidity can be measured and tested by **turbidity rods** or by **turbidimeters**, as is done for testing raw water supplies. This has been thoroughly discussed in vol. I, chapter 8, of "Water Supply Engineering".

(*ii*) **Colour.** The colour of sewage can normally be detected by the naked eye, and it indicates the freshness of sewage. If its colour is yellowish, grey, or light brown, it indicates fresh sewage. However, if the colour is black or dark brown, it indicates stale and septic* sewage. Other colours, may also be formed due to the presence of some specific industrial wastes.

(*iii*) **Odour.** Fresh sewage is practically odourless. But, however, in 3 to 4 hours, it becomes stale with all oxygen present in sewage being practically exhausted. It then starts omitting offensive odours, especially that of hydrogen sulphide gas, which is formed due to decomposition of sewage.

The odour of water or wastewater can be measured by a term called the **Threshold odour number (TON)**, which represents the extent of dilution required to just make the sample free of odour.

The minimum odour of the sample that can be detected after successive dilutions with odourless medium, is, thus, known as the *threshold odour*.

The Threshold odour number (TON) can be calculated by the equation :

$$\text{TON} = \frac{V_s + V_D}{V_s} \qquad ...(7.8)$$

where TON = Threshold Odour Number

V_s = Vol. of the sewage

V_D = Vol. of distilled or odourless water added to just make the sewage sample loss its odour.

(*iv*) **Temperature.** The temperature has an effect on the biological activity of bacteria present in sewage, and it also affects the solubility of gases in sewage. In addition, temperature also affects the viscosity of sewage, which, in turn, affects the sedimentation process in its treatment.

*When all the oxygen has disappeared from sewage, it becomes septic.

The normal temperature of sewage is generally slightly higher than the temperature of water, because of additional heat added during the utilisation of water. The average temperature of sewage in India is 20°C, which is near about the ideal temperature for the biological activities. However, when the temperature is more, the dissolved oxygen content (D.O.) of sewage gets reduced.

7.4. Chemical Characteristics of Sewage and Their Testing

Tests conducted for determining the chemical characteristics of sewage help in indicating : the stage of sewage decomposition, its strength, and extent and type of treatment required for making it safe to the point of disposal.

Chemical analysis is, therefore, carried out on sewage in order to determine its chemical characteristics. It includes *tests for determining* :

(*i*) total solids, suspended solids, and settleable solids ;
(*ii*) pH value ;
(*iii*) chloride content ;
(*iv*) nitrogen content ;
(*v*) presence of fats, greases, and oils ;
(*vi*) sulphides, sulphates and H_2S gas ;
(*vii*) dissolved oxygen ;
(*viii*) chemical oxygen demand (C.O.D.) ;
(*ix*) bio-chemical oxygen demand (B.O.D.).

These tests are briefly discussed below :

7.4.1. Total Solids, Suspended Solids and Settleable Solids*. Sewage normally contains very small amount of solids in relation to the huge quantity of water (99.9%). It only contains about 0.05 to 0.1 per cent (*i.e.* 500 to 1000 mg/l) of total solids. Solids present in sewage may be in any of the four forms : suspended solids, dissolved solids, colloidal solids, and settleable solids.

Suspended solids are those solids which remain floating in sewage. *Dissolved solids* are those which remain dissolved in sewage just as salt in water. *Colloidal solids* are finely divided solids remaining either in solution or in suspension. *Settleable solids* are that portion of solid matter which settles out, if sewage is allowed to remain undisturbed for a period of 2 hours. The proportion of these different types of solids is generally found to be as given below :

It has been estimated that about 1000 kg of sewage contains about 0.45 kg of total solids, out of which 0.225 kg is in solution, 0.112 kg is in suspension, and 0.112 kg is settleable.

Further, the solids in sewage comprise of both : the organic as well as inorganic solids. The organic matter works out to be about 45 per cent of the total solids, and the remaining about 55 per cent is the inorganic matter.

---

*Particles more than 1 μm (micron) in size are classified as *suspended solids*, particles less than 1 μm and up to 10^{-3} μm are classified as ***colloidals*** ; and particles smaller than 10^{-3} μm are considered as ***dissolved solids***.

Particles more than 10 μm in size settle without coagulation, and hence are called *settleable solids*. Particles smaller than 10 μm are removed by coagulation.

Inorganic matter consists of minerals and salts, like : sand, gravel, debris, dissolved salts, chlorides, sulphates, etc.

Organic matter consists of : (*i*) carbohydrates such as cellulose, cotton, fibre, starch, sugar, etc. (*ii*) fats and oils received from kitchens, laundries, garages, shops, etc. (*iii*) nitrogenous compounds like proteins and their decomposed products, including wastes from animals, urea, fatty acids, hydrocarbons, etc.

As a general rule, the presence of inorganic solids in sewage is not harmful. They require only mechanical appliances for their removal in the treatment plant. On the other hand, suspended and dissolved organic solids are responsible for creating nuisance, if disposed of, untreated. The amounts of various kinds of solids present in sewage can be determined as follows :

(*a*) The *total amount of solids (S_1 in mg / l)* present in a given sewage can be determined by evaporating a known volume of sewage sample, and weighing the dry residue left. The mass of the residue divided by the volume of the sample evaporated, will represent the total solids in *mg/l*, say S_1.

(*b*) The *suspended solids (S_2)* are those solids which are retained by a filter of 1 μm pores ; and they are, therefore, also called as **non-filterable solids.** Their quantity can be determined by passing a known volume of sewage sample through a *glass-fiber filter apparatus*, and weighing the dry residue left. The mass of the residue divided by the volume of sample filtered, will represent the suspended solids, say (S_2), in *mg/l*.

(*c*) The difference between the total solids (S_1) and the suspended solids (S_2) will represent nothing but **dissolved solids plus colloidals**, and represents the **filterable solids** ; say S_3 where $S_3 = S_1 - S_2$.

(*d*) *Now, the total suspended solids (S_2) may either be volatile or fixed.* In order to determine their proportion, the non-filtered dry residue of step (*b*) above, is burnt and ignited at about 550°C in an electric muffle furnace for about 15 to 20 minutes. Loss of weight due to ignition will represent the **volatile solids** in the sample volume filtered through the filter. Let the volatile suspended solids concentration be S_4 (in *mg/l*).

(*e*) The difference $S_2 - S_4 = S_5$ (say) will evidently represent the **fixed solids**.

(*f*) The quantity of **settleable solids (S_6)** can be determined easily with the help of a specially designed conical glass vessel called **Imhoff cone** (Refer Fig. 7.3). The capacity of the cone is 1 litre, and it is graduated up to about 50 ml.

Sewage is allowed to stand in this Imhoff cone for a period of two hours, and the quantity of solids settled in the bottom of the cone can then be directly read out. However, in order to obtain precise amount of settleable solids, the liquid from the cone should be decanted off, and the settleable solids collected at the bottom of the cone should be dried and weighed.

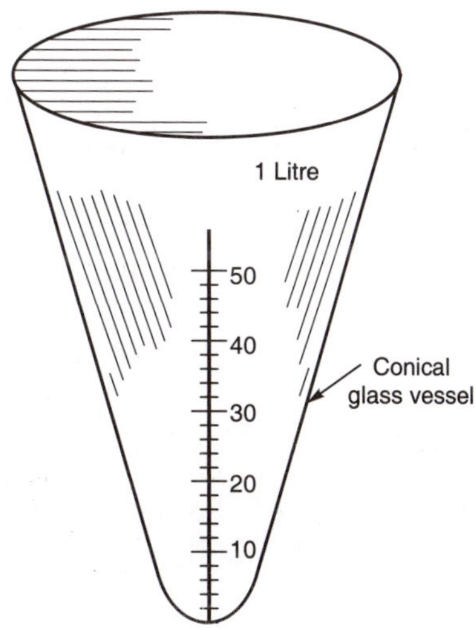

Fig. 7.3. Imhoff cone.

The quantities of different types of solids when determined experimentally as above, will help in detecting the strength of sewage, as given in Table 7.1.

Table 7.1. Typical quantities of Solids present in Domestic Sewage in mg/l

Type of solid	*Strength of Sewage*		
	Weak	*Medium*	*Strong*
Total Solids	400	800	1200
Suspended Solids			
Total	100	200	350
Volatile	75	135	210
Settleable solids	2.5	5	7
Ether soluble matter such as fats, oils and grease.	6	14	20

7.4.2. pH Value. The pH value of sewage indicates the negative log of hydrogen ion concentration present in sewage.

i.e., $pH = -\log H^+ ; \text{ or } H^+ = (10)^{-pH}$ ...(7.9)

It is, thus, an indicator of the alkalinity of sewage. *If the pH value is less than 7, the sewage is acidic, and if the pH value is more than 7, the sewage is alkaline.* The lesser is the pH value, the lesser is the alkalinity.

The determination of pH value of sewage is important, because of the fact that efficiency of certain treatment methods depends upon the availability of a suitable pH value. The pH value can be measured quickly and automatically with the help of a potentiometer, which measures the electrical potential exerted by the hydrogen ions, and thus, indicating their concentration ; as explained in determining the pH value of water in chapter 8, volume I—"Water Supply Engineering".

It may also be mentioned here that the fresh sewage is generally alkaline in nature (with pH more than 7) ; but as time passes, its pH tends to fall due to production of acids by bacterial action in anaerobic or nitrification processes. The pH, however, rises upon treatment of sewage.

7.4.3. Chloride Contents. Chlorides are generally found present in municipal sewage, and are derived from the kitchen wastes, human feces, and urinary discharges, etc. The normal chloride content of domestic sewage is 120 mg/l, whereas, the permissible chloride content for water supplies is 250 mg/l. However, large amounts of chlorides may enter from industries like ice cream plants, meat salting, etc., thus, increasing the chloride contents of sewage. Hence, when the chloride content of a given sewage is found to be high, it indicates the presence of industrial wastes or infiltration of sea water, thereby indicating the strength of sewage.

The chloride content can be measured by titrating the waste water (*i.e.* sewage) with standard silver nitrate solution, using potassium chromate as indicator, as is done for testing water supplies.

7.4.4. Nitrogen Contents. The presence of nitrogen in sewage indicates the presence of organic matter, and may occur in one or more of the following forms :

(*a*) *Free ammonia*, called Ammonia nitrogen*
(*b*) *Albuminoid nitrogen, called Organic nitrogen* } *Kjeldahl nitrogen*
(*c*) *Nitrites* ; and
(*d*) *Nitrates*.

Free Ammonia indicates the very first stage of decomposition of organic matter, thus indicating recently staled sewage.

Albuminoid introgen indicates the quantity of ***organic nitrogen***** present in sewage before the decomposition of organic matter is started. If the organic nitrogen decomposes anaerobically, then it decomposes to ammonia, and if this reacts aerobically, then it converts to ***nitrites*** and finally to ***nitrates***.

The presence of nitrites, thus, indicate partly decomposed sewage (not fully oxidised), whereas, the nitrates indicate the presence of fully oxidised organic matter.

The nitrites thus indicate the intermediate stage of conversion of organic matter of sewage into stable forms, thus indicating the progress of treatment. Their presence will show that treatment given to sewage is still incomplete, and the sewage is stale. Whereas, the presence of nitrates indicate the most stable form of nitrogenous matter contained in sewage, thus indicating the well oxidised and treated sewage.

Even though 'nitrates' have been defined as the non-objectionable final end product in aerobic treatment of sewage, yet its concentrations in potable waters are controlled, because larger concentrations (above 45 ppm ; say), may cause **nitrate poisoning** *in infants*. This happens because there exists lower acidity in the intestines of the infants, which permits growth of nitrate-reducing bacteria, which convert nitrates to nitrites. The nitrites, when absorbed in blood, prove very harmful, because they have greater affinity for hemoglobin than does oxygen, and thus, nitrite replaces oxygen in the blood complex. The body is, therefore, denied oxygen, causing suffocation in extreme cases. Since suffocation or lack of oxygen, causes the body to turn bluish, it may lead the child to turn blue. Hence, this disease, popularly called **blue baby disease**, or technically as ***methaemoglobinemia***, is caused in children by the presence of excess nitrates in water. Fortunately, however, when the child grows, say above 6 months or so, the chances of conversion of nitrate into nitrite gets meagre due to development of the child's intestinal tract, and hence such a disease does not normally occur afterwards.

---

*Ammonia remains a gas, at the temperatures and pressures, normally found in sewage and water systems. This gas (NH_3) exists in equilibrium with the aqueous ionic form, called ammonium (NH_4^+), as

$$NH_3 \uparrow + H_2O \rightleftharpoons NH_4^+ + OH^- \quad ...(7.10a)$$

Its further oxidation occurs as

$$NH_4^+ + 2O_2 \longrightarrow NO_3^- + 2H^+ + H_2O \quad ...(7.10b)$$

*****Organic nitrogen*** evidently represents the nitrogen atoms contained in organic compounds. This includes amino acids, proteins, nucleotides, etc. along with nitrogen bound residues of decomposing plant and animals matter and humus. ***The nitrogen in proteins and urea is thus, organic nitrogen. Organic nitrogen can enter septic systems as bodily wastes and discarded food items, or as components of cleaning agents.***

Fortunately also, the natural oxidation of nitrites into nitrates in the water bodies occurs quickly, and hence *dangerous-nitrite* contents are generally not found in natural water bodies.

All these different forms of nitrogen, present in sewage, can be tested and measured easily as explained below :

The amount of free ammonia present in sewage can be easily measured by simply boiling the sewage, and measuring the ammonia gas which is consequently liberated. The amount of albuminoid nitrogen can be measured by adding strong alkaline solution of potassium permanganate ($KMnO_4$) to the *already boiled** sewage sample and again boiling the same, when ammonia gas is liberated, which is measured, so as to indicate the amount of Albuminoid nitrogen (Organic nitrogen) present in sewage. If however an unboiled sample is used to add $KMnO_4$ before boiling, the evolved ammonia gas will measure the sum total of Ammonia nitrogen as well as Organic nitrogen ; and is known as **Kjeldahl nitrogen**.

The amount of nitrites or nitrates present in sewage sample can be measured by colour matching methods. For nitrites, the colour is developed by adding sulphonilic acid and naphthamine ; whereas for nitrates, the colour is developed by adding phenol-di-sulphonic acid and potassium hydroxide. The colour developed in waste water is finally compared with the standard colours of known concentrations.

7.4.5. Presence of Fats, Oils and Greases. Greases, fats and oils are derived in sewage from the discharges of animals and vegetable matter, or from the industries like garages, kitchens of hotels and restaurants, etc.

Such matter form scum on the top of the sedimentation tanks and clog the voids of the filtering media. They thus interfere with the normal treatment methods, and hence need proper detection and removal.

The amount of fats and greases in a sewage sample is determined by making use of the fact that oils and greases are soluble in ether, and when the ether is evaporated, it leaves behind ether-soluble-matter, which represents the quantity of fats and oils. Hence, in order to estimate their amount, a sample of sewage is, first of all, evaporated. The residual solids left are then mixed with ether (hexane). The solution is then poured off and evaporated, leaving behind the fats and greases as a residue, which can be easily weighed.

7.4.6. Sulphides, Sulphates and Hydrogen Sulphide Gas. The determination of sulphides and sulphates in sewage is rarely called far, although their presence reflects aerobic, and/or anaerobic decomposition.

Sulphides and sulphates are formed due to the decomposition of various sulphur containing substances present in sewage. This decomposition also leads to evolution of hydrogen sulphide gas, causing bad smells and odours, besides causing corrosion of concrete sewer pipes.

In aerobic digestion of sewage, the aerobic and facultative bacteria, oxidise the sulphur and its compounds present in sewage to initially form sulphides, which ultimately break down to form sulphate ions (SO_4^-), which is a stable and an unobjectionable end product. The initial decomposition is associated with formation of H_2S gas, which also ultimately gets oxidised to form sulphate ions.

*Thus removing free ammonia.

In anaerobic digestion of sewage, however, the anaerobic and facultative bacteria reduce the sulphurous matter into sulphides, with evolution of H_2S gas along with methane and carbon dioxide, thus causing very obnoxious smells and odours. If, however, the quantity of H_2S in raw sewage is below 1 ppm, obnoxious odours are not felt.

7.4.7. Dissolved Oxygen (D.O.). The determination of dissolved oxygen present in sewage is very very important, because : while discharging the treated sewage into some river stream, it is necessary to ensure at least 4 ppm of D.O. in it ; as otherwise, fish are likely to be killed, creating nuisance near the vicinity of disposal. To ensure this, D.O. tests are performed during sewage disposal treatment processes.

The D.O. test performed on sewage before treatment, helps in indicating the condition of sewage. It is well known by now, that only very fresh sewage contains some dissolved oxygen, which is soon depleted by aerobic decomposition. Also, the dissolved oxygen in fresh sewage depends upon temperature. *If the temperature of sewage is more, the D.O. content will be less.* The solubility of oxygen in sewage is 95% of that in distilled water.

The D.O. content of sewage is generally determined by the **Winkler's method** which is an oxidation-reduction process carried out chemically to liberate iodine in amount equivalent to the quantity of dissolved oxygen originally present.

7.4.8. Chemical Oxygen Demand (COD). The organic matter present in waste water can be measured in a number of ways ; *volatile solids* determination being a crude measure of organic matter.

Organic matter is most often assessed in terms of oxygen required to completely oxidise the organic matter to CO_2, H_2O and other oxidised species.

The oxygen required to oxidise the organic matter present in a given waste water can be theoretically computed, if the organics present in waste water are known. Thus, if the chemical formulas and the concentrations of the chemical compounds present in water are known to us, we can easily calculate the **theoretical oxygen demand** of each of these compounds by writing the balanced reaction for the compound with oxygen to produce CO_2, H_2O and oxidised inorganic components.

Hence, if the organic compounds and their concentrations are known, the theoretical oxygen demand of the waste water can be accurately calculated, but it is virtually impossible to know the details of the organic compounds present in any natural raw water or a waste water. The chemical oxygen demand (COD) of a raw water or a waste water is, therefore, determined by performing a laboratory test on the given water with a strong oxidant like dichromate solution ; and the theoretical computations of COD are only performed on water solutions prepared with the known amounts of specific organic compounds in laboratory situations to compare the theoretical and test results, and to establish the limitations of the test procedures.

The laboratory determination of COD, as said above, lies in using a strong oxidant like potassium dichromate ($K_2Cr_2O_7$) or potassium permanganate ($KMnO_4$) solution to stabilize the organic matter to determine the molecular oxygen used from the oxidant solution in oxidising the organic matter present in the given waste water.

In order to perform this test, a known quantity of waste water is mixed with a known quantity of standard solution of potassium dichromate, and the mixture is heated. The organic matter is oxidised by $K_2Cr_2O_7$ (in the presence of H_2SO_4*). The resulting solution of $K_2Cr_2O_7$ is titrated, and the oxygen used in oxidising the waste water is determined. This is called the *chemical oxygen demand (C.O.D.)* and is a measure of total organic matter (biodegradable as well as non-biodegradable) present in sewage.

7.4.9. Total Organic Carbon (TOC). Another important method of expressing organic matter is in terms of its carbon content. Carbon is the primary constituent of organic matter, and hence the chemical formula of every organic compound will reflect the extent of carbon present in that compound. Known concentrations of such chemical compounds in a given waste water will thus enable us to theoretically calculate the carbon present in that waste water per litre of solution, as computed in example 7.1.

Example 7.1. *Compute (a) the theoretical oxygen demand ; and (b) the organic carbon concentration of a water that contains the following chemical compounds :*

(*i*) *glucose* $(C_6H_{12}O_6) = 200$ *mg/l ; and*

(*ii*) *benzene* $(C_6H_6) = 25$ *mg/l*

(*c*) *What is the formula weight of the organic matter in this solution ?*

Solution. The equations for the total oxidation of the two constituents are first written as :

$$(i)\quad \underset{180}{C_6H_{12}O_6} + \underset{192}{6O_2} \longrightarrow \underset{264}{6CO_2} + \underset{108}{6H_2O}$$

$$(ii)\quad \underset{78}{C_6H_6} + \underset{240}{7.5O_2} \longrightarrow \underset{78}{6CO_2} + \underset{64}{3H_2O}$$

From the above equations : The theoretical oxygen demand of glucose is found to be $\frac{192}{180}$ = 1.07 mg of O_2/mg of glucose ; and the theoretical oxygen demand of benzene is found to be $\frac{240}{78}$ = 3.08 mg of O_2/mg of benzene.

Hence, the total theoretical oxygen demand of the solution containing 200 mg/l of glucose and 25 mg/e of benzene

$$= [1.07 \times 200 + 3.08 \times 25] \text{ mg/l} = \mathbf{291 \text{ mg/l.}}$$

Hence, COD = 291 mg/l. **Ans.**

(*b*) From the molecular formulas of glucose and benzene, it is found that :

The carbon content of glucose is 72 mgC/180 mg of glucose ; and the carbon content of benzene is 72 mgC/78 mg of benzene. Hence, the total organic carbon content of the solution containing 200 mg/l of glucose and 25 mg/l of benzene

$$= \left[\frac{72}{180} \times 200 + \frac{72}{78} \times 25\right] \text{ mgC/l}$$

$$= (0.4 \times 200 + 0.92 \times 25) = \mathbf{103.1 \text{ mgC/l.}} \quad \textbf{Ans.}$$

(*c*) The concentrations of each atom (*i.e.* C, H and O) in the water containing two organic compounds of given concentrations are now worked out as below :

*Such a strong acid helps to digest (break down) the complex molecules.

$$C = \left(\frac{72}{180} \times 200 + \frac{72}{78} \times 25\right) \text{ mg/l} = 103.1 \text{ mg/l}$$

$$= 103.1 \times 10^{-3} \text{ gm/l} \equiv 103.1 \times 10^{-3} \times \frac{1}{12} \text{ moles of C}$$

$$= \mathbf{8.59 \times 10^{-3} \text{ moles of C.}}$$

$$H = \left(\frac{12}{180} \times 200 + \frac{6}{78} \times 25\right) \text{ mg/l} = 15.26 \text{ mg/l}$$

$$= 15.26 \times 10^{-3} \text{ g/l} \equiv 15.26 \times 10^{-3} \times \frac{1}{1} \text{ moles of H}$$

$$= \mathbf{15.26 \times 10^{-3} \text{ moles of H.}}$$

$$O = \left[\frac{96}{180} \times 200 + 0\right] \text{ mg/l} = 106.67 \text{ mg/l}$$

$$= 106.67 \times 10^{-3} \text{ g/L} \equiv \frac{106.67 \times 10^{-3}}{16} \text{ moles of O}$$

$$= \mathbf{6.67 \times 10^{-3} \text{ moles of O.}}$$

The chemical formula for the organic matter present in water thus becomes : $C_{8.59} \cdot H_{15.26} \cdot O_{6.67}$ at a concentration of 1 milli-moles. **Ans.**

Obviously, this formula does not represent the structure of the organic matter but only the relative amounts of the elements contained in it.

7.4.10. Bio-Chemical Oxygen Demand (B.O.D.). The organic matter, infact, is of two types ; *i.e.* that which is biologically oxidised (*i.e.* oxidised by bacteria) and is called **biologically active** or *biologically degradable* ; and (*ii*) that which cannot be oxidised biologically, and is called **biologically inactive**. Truly speaking, while testing a waste water, we are mainly interested in finding out the amount of biologically active organic matter present in it ; whereas, *the above COD test gives us the total of biologically active as well as biologically inactive organic matter*. Hence, further testing is carried out to determine the **biochemical oxygen demand (B.O.D.)** *of sewage, which directly gives us the amount of* **biologically active organic matter** *present in sewage.*

BOD test constitutes the most important test for waste waters, and is discussed below :

If sufficient oxygen is available in waste water, the useful aerobic bacteria will flourish and cause the aerobic biological decomposition of wastewater, which will continue until oxidation is completed. The amount of oxygen consumed in this process is the BOD. The **BOD** (Biochemical Oxygen Demand) is, hence, **defined** as the oxygen requirement of bacteria to decompose the organic matter (present in a given sample of sewage) under aerobic conditions.

Polluted waters will continue to absorb oxygen for many months, and it is not practically feasible to determine this ultimate oxygen demand in the laboratory. Hence, *the BOD of water during 5 days at 20°C is generally taken as the standard demand, and is about 68% of the total demand*. A 10 day BOD is about 90% of the total.

This standard 5 day BOD, written as BOD_5, or simply as BOD is determined in the laboratory by mixing or diluting a known volume of a sample of wastewater with a known volume of aerated pure water*, and then

*It, however, contains a small amount of sodium bicarbonate and other nutrient salts and having a pH value of 7 to 7.6.

calculating the D.O. of this diluted sample. The diluted sample is then incubated for 5 days at 20°C*. The D.O. of the diluted sample, after this period of incubation, is again calculated. The difference between the initial D.O. value and the final D.O. value will indicate the oxygen consumed by the sewage sample in 5 days (used in causing aerobic decomposition of wastes). The BOD in ppm is then calculated by using the equation

$$\text{BOD or BOD}_5 = \text{D.O. consumed in the test by the diluted sample} \times \left[\frac{\text{Vol. of the diluted sample}}{\text{Vol. of the undiluted sewage sample}}\right] \quad ...(7.11)$$

The factor on the right hand side in the above equation is nothing but the **dilution factor.**

Say for example, 1% diluted sample means : 1 ml of sewage is diluted to make 100 ml of test sample ; and hence dilution factor would be 100, as 1 ml has been diluted by 100 times to make 100 ml. Hence, the multiplying factor of the right hand side would be 100.

In actual experimentation in the laboratory, we generally use 300 ml sized **BOD test bottles**. The given vol. of sample, **say 4 ml** is then placed in the bottle, and mixed with pure aerated water to make 300 ml diluted sample. This sample is incubated at 20°C for 5 days. *Light must be excluded from the incubator to prevent algal growth that may produce oxygen in the bottle.* The DO content before the incubation and after the incubation are thus determined. The BOD_5 of sewage would then be calculated easily as,

$$BOD_5 = \text{DO consumed by the diluted sample} \times \left(\frac{300}{4}\right) \quad ...(7.12a)$$

$$\left(\because \text{ Dilution factor being } \frac{300}{4}\right)$$

The above equation, *i.e.*, Eq. (7.12*a*) can be derived, based on the following basic equation:

$$\left[\text{BOD of diluted sample} \times \text{Vol. of diluted sample}\right] = \left[\text{BOD of sewage sample} \times \text{Vol. of sewage sample} + \text{BOD of dilution water} \times \text{Vol. of dilution water}\right] \quad ...(7.12)$$

But BOD of diluted sample = DO consumed by diluted sample in 5 days incubation.

When distilled water is used for dilution, then the above basic equation becomes :

***Note.** The standard 5 day BOD is measured at 20°C, because the average temperature of water bodies (from where actual decomposition of waste water takes place) is about 20°C in the western world, from where the concept of BOD has originated. Accordingly, the same standard measurement continues to be adopted even in our country, where average temperature of water bodies is about 27°C. It has also been established that since 5 day BOD at 20°C = 3 day BOD at 27°C, we can perform the BOD test only for 3 days at 27°C rather than performing it for 5 days at 20°C. This change offers an advantage of early completion of the test. Accordingly, many research workers have started using the said recommendation of NEERI to perform the standard BOD test for 3 days at 27°C to determine $BOD_{5(20°)}$.

$$\left[\text{DO consumed by diluted sample} \times \text{Vol. of diluted sample (300 ml)}\right]$$

$$= \left[\text{BOD of sewage sample (BOD}_5) \times 4 \text{ ml} + \text{zero} \times 296\right]$$

$$\left[\because \quad \text{BOD of distilled water} = \text{zero} = 0\right]$$

or DO consumed by diluted sample $= \dfrac{4}{300}\ [\text{BOD}_5]$

or BOD_5 of sewage $= \dfrac{300}{4}$ [DO consumed by diluted sample] ...(Eq. 7.12*a*)

This is the same equation as Eqn. (7.12*a*).

Note: *If in place if distilled water, ordinary polluted water having some BOD is used, then the above basic equation (i.e., Eq 7.12) can be used to compute the BOD_5 of sewage; but in that case DO consumed by the feed water (Dilution water) will also have to be determined separately by performing the 5 day DO consumption test on the dilution water (feed water). A numerical example on this type of computation is given under Q. 4(c) (ii) of Engineering Services Questions of year 2016.*

As pointed out earlier, this test is conducted to determine the oxygen demand for the first five days on a number of samples, and their average value is taken as the BOD_5 at the test temperature.

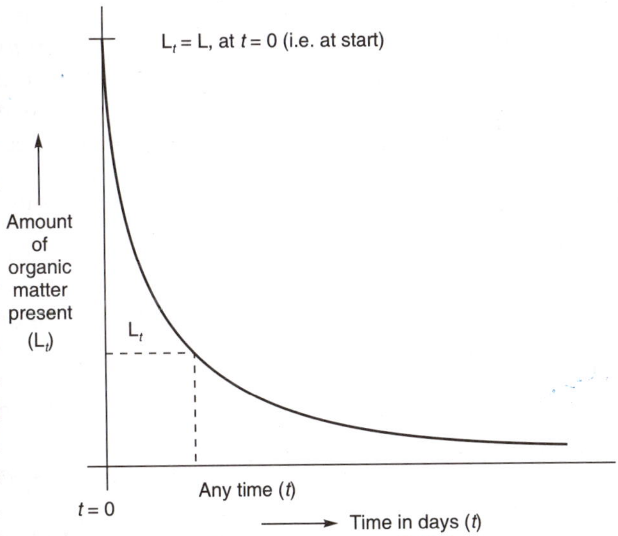

Fig. 7.4. 1st stage BOD curve.

In fact, if the oxygen supply is made available for periods more than 5 days, it is found that the oxygen is consumed rapidly for 6 or 7 days, and then slows down until the end of about 20 days or so. Thereafter, it may again accelerate for sometime, and again slow down to a very low rate for an indefinite period. The first demand during the first 20 days or so, in fact, occurs due to the oxidation of organic matter, and is called **carbonaceous demand** or **first stage demand** or **initial demand.** The latter demand occurs due to biological oxidation of ammonia, and is called *Nitrogenous demand* or *Second-stage demand*.

In fact, a sanitary engineer is more concerned with the first stage demand, since the oxygen consumed in its satisfaction is not recoverable. *Hence, the term BOD is usually used to mean the first stage BOD, i.e.* the demand due to the presence of carbonaceous matter alone.

This carbonaceous demand is represented by a curve, called **1st stage BOD curve**, shown in Fig. 7.4.

The rate at which BOD is satisfied at any time, (*i.e.* the rate of *deoxygenation*) depends on the temperature, and also on the amount and nature of organic matter present in sewage at that time.

Thus, at a certain temperature, the rate of deoxygenation is assumed to be directly proportional to the amount of organic matter present in sewage at that time ; *i.e.*

$$\frac{dL_t}{dt} = -KL_t^{*} \qquad \text{...(7.13)}$$

where L_t = oxygen equivalent of carbonaceous oxidisable organic matter present in sewage after t days from the start of oxidation, in mg/l.

t = time in days.

K = rate constant signifying the rate of oxidation of organic matter, and it depends upon the nature of organic matter and temperature. Its unit is per day.

Integrating Eq. (7.12), we get

$$\int \frac{dL_t}{L_t} = \int -K \,.\, dt$$

or

$$\log_e L_t = -K \,.\, t + C \qquad \text{...(7.13}a\text{)}$$

where C is a constant of integration, and can be evaluated from the boundary conditions at the start *i.e.*

when t = zero (0),

i.e. at start $L_t = L$ (say).**

Substituting in Eq. (7.13), we have

$$\therefore \quad \log_e L = K(0) + C$$

or

$$C = \log_e L$$

Substituting this value of C in Eq. (7.13), we get

$$\log_e L_t = -K \,.\, t + \log_e L \quad \text{or} \quad \log_e L_t - \log_e L = -K \,.\, t$$

or

$$\log_e \frac{L_t}{L} = -K \,.\, t \quad \text{or} \quad 2.3 \log_{10} \frac{L_t}{L} = -K \,.\, t$$

or

$$\log_{10} \frac{L_t}{L} = -\frac{K \,.\, t}{2.3} = -0.434\, K \,.\, t$$

Using

$$0.434\, K = K_D \qquad \text{...(7.14)}$$

where K_D*** is the **De-oxygenation constant** or more strictly, the **BOD rate constant** (on base 10) at the given temperature = 0.434 K

---

*Minus sign indicates that with the passage of time (*i.e.*, increase in t), the value of L_t decreases.

**L is the oxygen equivalent of organic matter present in sewage (at start). It is a fixed quantity. Some authors use the symbol L_0 for this.

***Certain authors use the symbol K_1 for this constant K_D, which is a constant on base 10, as against K which is on base e $\left(i.e.,\ K_D = \frac{K}{2.3}\right)$.

We have $\log_{10} \frac{L_t}{L} = -K_D \cdot t$

or $$\frac{L_t}{L} = (10)^{-K_D \cdot t} \quad ...(7.15)$$

Now, L is the organic matter present at the start of BOD reaction, (expressed as oxygen equivalent) and L_t is the organic matter left after t days ; which means that during t days, the quantity of organic matter oxidised $= L - L_t$.

If Y_t represents the total amount of organic matter oxidised in t days (*i.e.* the BOD of t days), then we have

$$Y_t = L - L_t$$

Taking L out of bracket on R.H.S.

we have $Y_t = L\left[1 - \frac{L_t}{L}\right]$

or $\frac{Y_t}{L} = 1 - \frac{L_t}{L}$ or $\frac{L_t}{L} = 1 - \frac{Y_t}{L}$

Substituting this value of $\frac{L_t}{L}$ in equation (7.15), we get

$1 - \frac{Y_t}{L} = (10)^{-K_D \cdot t}$ or $\frac{L_t}{L} = 1 - (10)^{-K_D \cdot t}$

or $$\boxed{Y_t = L\left[1 - (10)^{-K_D \cdot t}\right]} \quad ...(7.16)$$

This is an important equation. Y_t is the oxygen absorbed in t days, *i.e.* BOD of t days.

The ultimate first stage BOD (Y_u) would be obtained from the above equation, when we substitute $t = \infty$ days in it.

Hence $Y_u = L\left[1 - (10)^{-K_D \cdot \infty}\right]$

or $Y_u = L\left[1 - \frac{1}{(10)^{\infty}}\right] = L[1 - 0]$

or $$Y_u = L \quad ...(7.17)$$

Hence, the ultimate first state *BOD* (Y_u) of a given sewage is equal to the initial oxygen equivalent of the organic matter present in this sewage (L). *This is a fixed quantity, and does not depend upon the temperature of oxidation.*

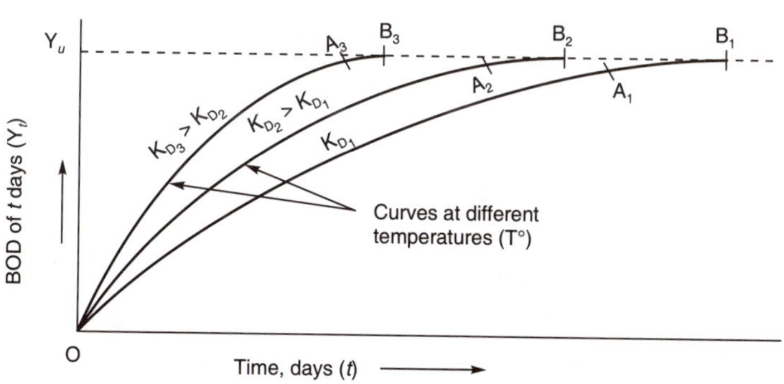

Fig. 7.5. BOD exertion as a function of K_D.

The value of K_D however, determines the speed of the BOD reaction, without influencing the ultimate BOD, as shown in Fig. 7.5. It is found to vary with temperature of sewage, and this relationship is approximately given by the equation

$$\boxed{K_{D(T^\circ)} = K_{D(20^\circ)}[1.047]^{T-20^\circ}} \quad ...(7.18)$$

where $K_{D(20^\circ)}$ = Deoxygenation constant at 20°C. Its numerical value varies between 0.05 to 0.2 per day, depending upon the nature of the organic matter present in sewage. Simple compounds such as sugars and starches are easily utilised by the micro-organisms, and have a high K_D rate, while complex molecules such as phenols are difficult to assimilate and hence have low K_D values. Some typical K_D values are given in Table 7.2.

$K_{D(T^\circ)}$ = Deoxygenation constant at tempera-ture T°C.

Table 7.2. Typical Values of K_D at 20°C for Various Types of Waters and Waste Waters

Water Type	K_D *value per day*
Tap waters	< 0.05
Surface waters	0.05—0.1
Municipal waste waters	0.1—0.15
Treated sewage effluents	0.05—0.1

Equation (7.18) shows that K_D will be higher at higher temperatures, which means that the speed at which BOD is consumed in the oxidation of the organic matter, is higher at higher temperatures. This means that the entire carbonaceous organic matter will get oxidised quickly and in lesser time at higher temperatures.

Equation (7.16) is called the ***first stage equation of BOD reaction***, and is represented graphically by the curve *OAB* of Figs. 7.5 and 7.6.

Fig. 7.5 shows that the coefficient of deoxygenation is different at different temperatures, but finally, Y_u is constant.

By comparing Figs. 7.4 and 7.6, it can be seen that the curve of Fig. 7.6 is nothing but the reciprocal of curve in Fig. 7.4. This is because the oxygen used in satisfying BOD is in direct ratio of the amount of organic matter oxidised, which means reciprocal of the curve of Fig. 7.4.

The portion *AC* of the curve of Fig. 7.6 represents the nitrification stage, which follows the carbonaceous stage, so that the BOD curve for the complete oxidation is represented by *OAC*.

Note. The equations given above are only for the important **first stage BOD** (called ***carbonaceous BOD*** or simply as BOD), and have nothing to do with the **second stage BOD** (called ***Nitrogenous BOD*** or *Nitrogenous oxidation demand* (NOD), with which we are not concerned here.

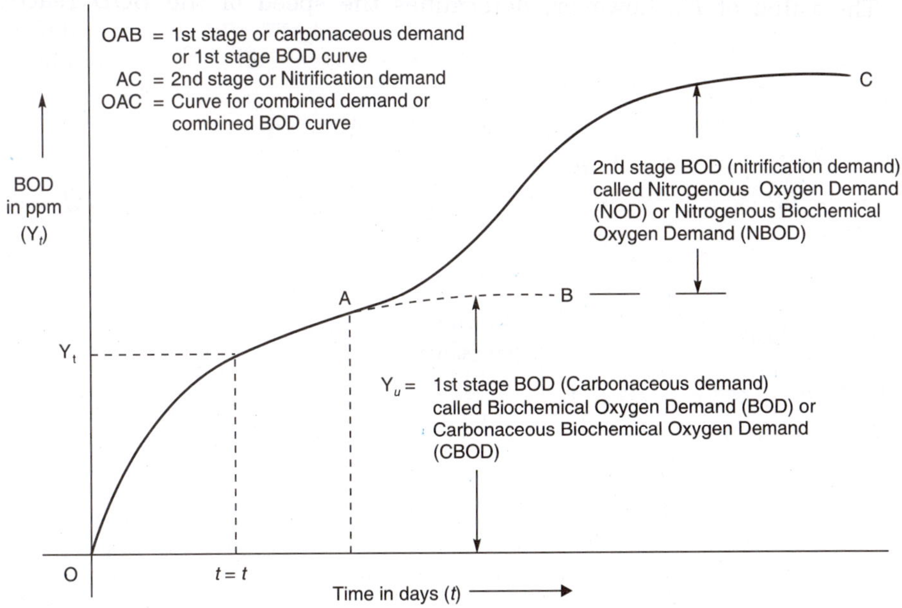

Fig. 7.6

The Eqs. (7.16), (7.17) and (7.18) make it possible to convert the test results from different time periods and temperatures to the standard 5-day 20°C test or *vice versa*, as shown in examples solved below :

Values of 5-days 20°C BOD of municipal wastewaters generally vary between 100 and 500 mg/l, as shown in Table 7.3.

Table 7.3. Typical characteristics of Domestic Sewage in mg/l

S. No.	*Parameter/ characteristic*	*Strength of Sewage*		
		Weak	*Medium*	*Strong*
1.	Total suspended solids (*SS*)	100	200	350
2.	Volatile suspended solids	75	135	210
3.	BOD	100	200	400
4.	COD	175	300	600
5.	TOC*	100	200	400
6.	Ammonia-N	5	10	20
7.	Organic-N	8	20	40
8.	PO_4-P	7	10	20

Example 7.2. *If 2.5 ml of raw sewage has been diluted to 250 ml and the D.O. concentration of the diluted sample at the beginning of the BOD test was 8 mg/l, and 5 mg/l after 5-day incubation at 20°C ; find the BOD of raw sewage.*

Solution. Volume of sample of sewage = 2.5 ml.

Volume of diluted sample = 250 ml.

$$\therefore \quad \text{Dilution ratio} = \frac{250}{2.5} = 100.$$

*Explained in article 7.4.9.

Loss of dissolved oxygen during the test

= D.O. before testing – D.O. after testing = 8 – 5 = 3 mg/l.

Using equations (7.11), we have

∴ BOD of sewage = Loss of oxygen × Dilution factor

= 3 mg/l × 100 = 300 mg/l. **Ans.**

Example 7.3. *A 2% solution of a sewage sample is incubated for 5 days at 20°C. The depletion of oxygen was found to be 4 ppm. Determine the BOD of the sewage.*

Solution. Dilution factor $= \dfrac{100}{\text{Per cent of solution}} = \dfrac{100}{2} = 50$

Depletion of oxygen = 4 ppm.

Using equation (7.11), we have

BOD = Depletion of oxygen × Dilution factor

= 4 ppm × 50 = 200 ppm **Ans.**

7.4.10.1. Laboratory estimation of K_D and L values by Thomas's graphical method. The BOD rate constant K_D used in eqn. (7.16), can be computed from BOD values measured at various times. The sewage samples are tested for BOD (Y_t) at different times (t), such as after 0.5, 1.0, 1.5, 2.0, 3.0, 5.0 days. A graph is now plotted between the values of time t in days on X-axis, and the values of function $\sqrt[3]{\dfrac{t\text{ (in days)}}{Y_t\ i.e.,\text{ BOD (in mg/l)}}}$ on Y-axis, as shown in Fig. 7.6. The best fit line drawn through these points is used to calculate the K_D rate by the following relationship :

$$K_D = 2.61\,\frac{m}{C} \qquad \text{...(7.19)}$$

where K_D = rate constant per day

m = slope of the line.

C = intercept of the line on Y-axis.

Also,
$$L = Y_u = \frac{1}{2.3K_D C^3} \qquad \text{...(7.20)}$$

Example 7.4. *BOD tests performed on 4 ml samples of a waste water through 300 ml standard BOD bottles at different times, have yielded the following results. Compute (a) the value of BOD rate constant ; (b) the value of BOD_5 ; (c) the value of ultimate BOD.*

Bottle No.	*Initial DO in mg/l*	*Incubation period in days*	*Final DO in mg/l*
1	8.4	0.5	7.45
2	8.4	1.0	6.7
3	8.4	1.5	6.1
4	8.4	2.0	5.5
5	8.4	3.0	4.6
6	8.4	5.0	3.6

Solution. The dilution factor in each case = $\frac{300 \text{ ml}}{4 \text{ ml}}$ = **75.**

The given values of initial and final D.O. are used to compute the D.O. drop values, which are multiplied by the dilution factor of 75 to compute BOD values for all the samples, as shown in col. (6) of table 7.4. The values of function $\sqrt[3]{\frac{t}{\text{BOD}}}$ are then computed in col. (7) of this table.

Table 7.4

S. No. / *Bottle No.*	*Initial DO mg/l*	*Incubation time (t) in days*	*Final DO in mg/l*	*DO drop in mg/l col. (2) – col (4)*	*BOD (Y_t) mg/l col. (5) × 75*	$\sqrt[3]{\frac{t}{BOD}}$ $\left(\frac{col.\ 3}{col.\ 6}\right)^{1/3}$
(1)	(2)	(3)	(4)	(5)	(6)	(7)
1	8.4	0.5	7.45	0.95	71.25	0.192
2	8.4	1.0	6.7	1.7	127.5	0.199
3	8.4	1.5	6.1	2.3	172.5	0.206
4	8.4	2.0	5.5	2.9	217.5	0.210
5	8.4	3.0	4.6	3.8	285	0.219
6	8.4	5.0	3.6	4.8	360	0.240

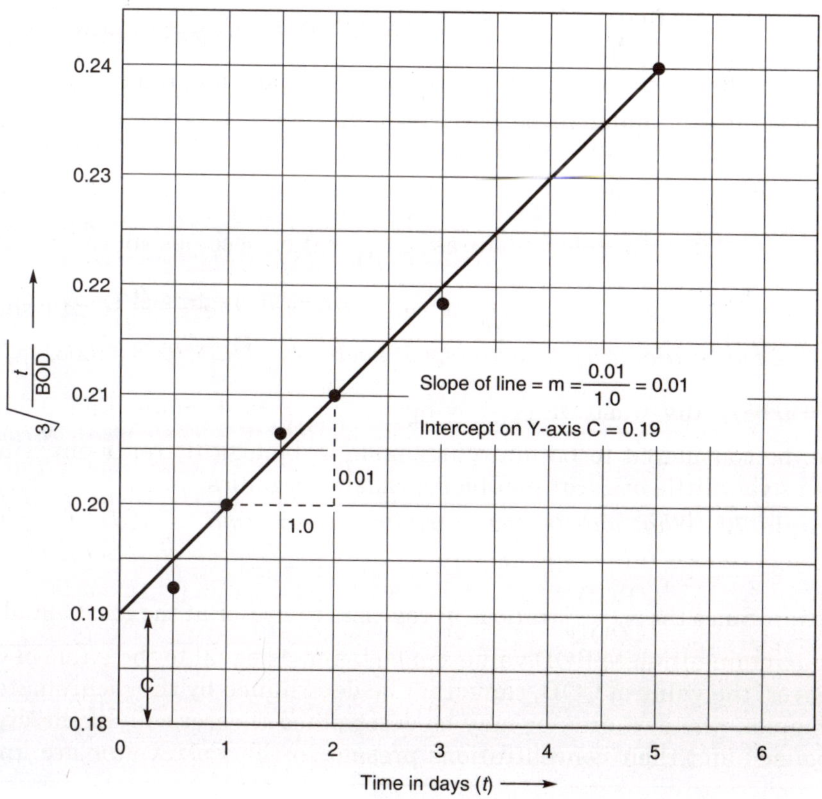

Fig. 7.7

A graph is now plotted between the values of t (col. 3) and the values of function $\sqrt[3]{\frac{t}{\text{BOD}}}$ (col. 7), as shown in Fig. 7.7, to pass a straight line of best fit.

The value of K_D is then computed by using eqn. (7.19) as

$$K_D = 2.61\left(\frac{m}{C}\right)$$

where m = slope of line = 0.01

C = intercept on Y-axis = 0.19

$$\therefore \quad K_D = 2.61 \times \frac{0.01}{0.19} = \mathbf{0.137.}\ \textbf{Ans.}$$

(*b*) The 5 day value of BOD as computed in col. (6) is 360 mg/l. **Ans.**

(*c*) From Eq. (7.20), we have

$$Y_u = L = \frac{1}{2.3\ K_D\ .\ C^3} = \frac{1}{2.3 \times 0.137 \times (0.19)^3} = \mathbf{462.7\ mg/l.}\ \textbf{Ans.}$$

7.4.11. BOD/COD Ratio. Ultimate BOD (BOD_u), as we know, is the oxygen required for oxidising the Biodegradable Organics of a given waste water. COD, on the other hand, is the oxygen required to oxidise the Biodegradable Organics (BO's) as well as the Non-Biodegradable Organics (NBO's) both. COD of a waste water will, therefore, always be more than its BOD_u, since their difference will represent the quantum of NBO's present in the given waste water. The $\frac{\text{BOD}_u}{\text{COD}}$ ratio will, therefore, always be less than 1.0 ; but this value shall approach towards 1.0 with the decreasing amount of NBO's.

If this ratio is found to be between 0.92 to 1.0, the wastewater can be considered to be virtually fully biodegradable.

Since, BOD_u is generally not measured, and only BOD_5 is measured, the ratio $\frac{\text{BOD}_5}{\text{COD}}$, usually referred to as $\frac{\text{BOD}}{\text{COD}}$ ratio, becomes more important. Since BOD_5 is about 68% of BOD_u, *we can easily state that* $\frac{BOD_5}{COD}$ *ratio should, for fully-biodegradable waste waters, vary between (0.92 × 0.68 = 0.63) to (1.0 × 0.68 = 0.68).* Any wastewater, having its $\frac{\text{BOD}}{\text{COD}}$ ratio more than 0.63, can hence, be considered to be quite amenable to biological treatment, since it contains very little or zero* non-biodegradable organics.

Example 7.5. *What will be the maximum upper limit of BOD of a glucose solution of concentration 300 mg/l.* (*Engineering Services, 1998*)

Solution. Since the $\frac{\text{BOD}_u}{\text{COD}}$ ratio is always less than 1 or at the most equal to 1, the maximum ultimate BOD value can be taken as equal to the value of COD. Moreover, the value of COD, can either be determined by the dichromate test for complex wastewaters ; or may be determined theoretically if the organic compounds and their concentrations present in the wastewater are known.

* Non-biodegradable organics will be zero when $\frac{\text{BOD}}{\text{COD}}$ ratio is 0.68.

Such a *theoretical oxygen demand* of an organic compound can be calculated by writing the balanced reaction for the compound with oxygen to produce CO_2, H_2O and oxidised organic components.

In the present case, the water contains only glucose, which is oxidised under the following equation :

$$\underset{\substack{180\\\text{Glucose}}}{C_6H_{12}O_6} + \underset{192}{6O_2} \longrightarrow \underset{264}{6CO_2} + \underset{108}{6H_2O}$$

From this balanced equation of oxidation of glucose, it can be stated that the theoretical oxygen demand for glucose is $\frac{192}{180} = 1.07$ mg of O_2/mg of glucose.

∴ Total theoretical oxygen demand of the 300 mg/l glucose solution

$$= 1.07 \times 300 \text{ mg/l} = 321 \text{ mg/l}$$

This COD can be taken to be equal to the maximum ultimate BOD.

Hence, the maximum ultimate BOD = **321 mg/l. Ans.**

7.4.12. COD/TOC ratio. The total carbonaceous organics present in a given waste-water can be ascertained by computing TOC of the waste-water, by converting the carbonaceous organics to CO_2, which is measured by infra-red analysis, and converted instrumentally to the original organic carbon content. $\frac{\textbf{COD}}{\textbf{TOC}}$ **ratio** is considered to be an important factor in monitoring wastewater treatments.

TOC is related to COD through a carbon-oxygen balance, such as in the oxidation of Glucose,

$$C_6H_{12}O_6 + 6O_2 \longrightarrow 6CO_2 + 6H_2O,$$

we will have, $$\frac{COD}{TOC} = \frac{6 \text{ mol of } O_2}{6 \text{ mol of C}} = \frac{6 \times 32}{6 \times 12} = \mathbf{2.66.}$$

Depending on the organic in question, the $\frac{COD}{TOC}$ ratio may vary from zero for an organic material resistant to dichromate oxidation, to 5.33 for methane. Since the organic content undergoes change during biological oxidation, it can be expected that $\frac{COD}{TOC}$ ratio will also change.

The same rationale will also apply to $\frac{BOD}{TOC}$ ratio. The TOC test is rapid and accurate, but has not become popular in our country, because it needs a costly instrument and sufficient skill in using the same. Moreover, it does not give an idea of Biologically Degradable Organics, and hence cannot be of much use for monitoring the Biological Treatment of wastewaters, which is the most common method adopted in our country for treating wastewaters.

7.5. Population Equivalent

Industrial wastewaters are generally compared with per capita normal domestic wastewaters, so as to rationally charge the industries for the pollution caused by them. The strength of the industrial sewage is, thus, worked out as below :

$$\begin{bmatrix}\text{Standard BOD (BOD}_5)\\ \text{of industrial sewage}\end{bmatrix} = \begin{bmatrix}\text{Standard BOD (BOD}_5)\\ \text{of domestic sewage}\\ \text{per person per day}\end{bmatrix} \times \text{[Population equivalent]} \quad ...(7.21)$$

The average standard BOD_5 of domestic sewage is worked out to be about 0.08 kg/day/person. Hence, if the BOD_5 of sewage coming from an industries is worked out as 300 kg/day, then

The population equivalent

$$= \frac{BOD_5 \text{ of the industry in kg/day}}{0.08 \text{ kg/day/person}} \quad ...(7.22)$$

$$= \frac{300}{0.08} = \mathbf{3750.}$$

The population equivalent, thus, indicates the strength of the industrial wastewaters for estimating the treatment required at the municipal sewage treatment plant, and also helps in assessing realistic charges for this treatment to be charged from the industries instead of charging them simply by the volume of sewage.

7.6. Relative Stability (S_R)

The term *Relative stability* is defined as the ratio of the available oxygen in a given sewage to the required oxygen satisfying first stage BOD. The available oxygen will include dissolved oxygen as well as oxygen present in nitrite and nitrate. It is generally expressed as percentage of the total oxygen required.

The test for relative stability is carried out in the following steps:

(*i*) The waste water is filled in a glass-stoppered bottle and a small quantity of methylene blue is added to it.

(*ii*) The mixture is then incubated at 20°C or 37°C, as per convenience.

(*iii*) During incubation period, the anaerobic bacteria start their function; the available DO is consumed and H_2S gas is produced, which decolourises the mixture. The time t_{37} or t_{20} reqd. for bleaching (discolourising) the blue colour is noted. The relative stability S_R is then calculated using eq. (7.23) or (7.24) as:

$$\text{Relative stability} = S_R = 100\left[1 - (0.794)^{t_{20}}\right] \quad ...(7.23)$$

or

$$S_R = 100\left[1 - (0.605)^{t_{(37)}}\right] \quad ...(7.24)$$

where S_R = The relative stability, $t_{(20)}$ and $t_{(37)}$ represent the time in days for a sewage sample (mixed with a small qty. of methylene blue solution) to *decolourise*, when incubated at 20°C or 37°C*, respectively.

The decolourisation caused by the enzymes produced by anaerobic bacteria, infact, is an indication of the available oxygen in oxidising the unstable organic matter.

The sooner the decolourisation takes place, the earlier the anaerobic conditions develop, which means lesser availability of oxygen. Hence, if the decolourisation takes place sooner (less than 4 days or so), the effluent sample

*In a hot country like India, incubation can be more easily done at 37°C, because here it becomes very costly to maintain the equipments at 20°C.

may be taken as relatively unstable (as relativity is less than about 0.6). But samples which do not decolourise in four days can be taken as relatively stable, and, thus, discharged into the stream without any troubles.

The above test for determining relative stability is suitable only in case of polluted stream waters or sewage effluents (*i.e.* treated sewage). In case of raw sewage, however, the colour is precipitated out due to the presence of colloidal matter in the sewage, and therefore, the test is of little use in that case. This test is, however, very simple, and quickly gives a rough indication of the character of the treated sewage.

7.7. Collection of Sewage Samples for Physical and Chemical Testing

The physical and chemical characteristics of sewage vary from top to bottom of sewage depth, as well as with time as from morning to evening. It therefore becomes difficult to obtain a truly representative sample. A fair compromise is, therefore, made in practice. Samples are taken at a point beneath the surface where the turbulence is thoroughly mixing up the sewage particles. This is called a **grab sample.** Such grab samples are collected at regular intervals during a day. These different samples are now mixed together, and the amount utilised from each specimen is proportional to the rate of flow at the time the specimen was collected. This composite sample is taken for testing, as it represents more nearly, the average true strength of the sewage.

In order to keep a sample cool, so as to prevent the biological action from taking place in the sewage sample, preservatives, such as chloroform, formaldehyde and sulphuric acid are added in the collecting bottle. The characteristics of preservatives should, however, be such as not to affect the result of the analysis to be made. For example, *no preservative should be used on a sample, on which* BOD *determination is to be made.*

7.8. Bacteriological Characteristics and Testing

The bacterial characteristics of sewage are due to the presence of micro-organisms, which include bacteria and other living micro-organisms, such as algae, fungi, protozoa, etc. The bacteria are, however, more active and exist in abundance (5—15 billion per litre of sewage).

Microorganisms are, infact, present in significant quantities (numbers) in sewage, which primarily originates from human feces (including urine). *They are the most diverse group of living organisms on Earth and occupy important niches in the ecosystem.* The micro-organisms are able to survive and proliferate in extreme environments, ranging for example, from low to high pH or temperatures. Their simplicity and minimal survival requirements allow them to exist in diverse situations.

There are various types of micro-organisms. The most fundamental and the simplest wholly contained life systems are **bacteria** *or* **prokaryotes.**

Bacteria *are the most diverse group of micro-organisms*, which are the minute single cell micro-organisms, possessing no defined nucleus and having no chlorophyll to help them manufacture their own food. *They are reproduced by binary fission* and may be of various shapes, such as spherical shape (called *cocci*), straight rod like shape (called *bacillus*), spiral shape (called *spirillum*), and curved shape (called *vibrio*). They are so small in length (of the size of 1 to 4 micron) that they cannot be seen with a naked eye and have to be examined under a microscope.

Viruses are other tiny nano sized, ***non-living infectious particles which reproduce inside the cells of the living hosts.*** When infected, the host cell is forced to rapidly produce thousands of identical copies of the original virus. Since they contain genes, they also mutate to form new viruses, to assemble in the infected host cells.

The viruses are thus, parasitic, as they reproduce only within a host and not outside the host, as bacterias can do, which can reproduce anywhere. However, viruses can survive on surfaces.

They can infect all species of animals including humans, and can even infect bacterias. All viruses consists of a core of either RNA (Ribonucleic Acid) or DNA (Deoxyribo Nucleic Acid) Molecule, surrounded by a fatty coat. RNA viruses cause diseases like ***Influenza, Ebola, Hepatitis A, C, D & E, Measles, SARS, HIV, and COVID-2019.*** They mutate more easily than DNA based viruses, which cause diseases like ***Herps, Smallpox, Chickenpox, and Hapatitis B,*** etc.

It can also be stated that, theoretically even a single virus particle is enough to infect a host and cause disease, but usually more numbers of virus particles forming ***virus load*** *is responsible for causing diseases.* It, infact, depends upon the potency of the virus. ***More potent is the virus, the lesser number of virus particles are needed to infect a host body.***

It is ofcourse true that most of the viruses have a pathogenic relationship with their hosts-meaning they cause diseases ranging from a mild cold to serious conditions like Severe Acute Respiratory Syndroms (SARs). They work by invading the host cell, taking over its cellular machinery and releasing new viral particles that go on to infect more cells and cause illness.

Viruses are not included in the category of micro-organisms. They are much smaller than the bacteria being of the size of 20-400 nano metre*, and cannot be seen singly under an ordinary microscope.

In modern days, however, viruses can be studied under a transmission type of an Electron microscope

Bacteria Types. Most of the bacteria types are harmless, and under certain conditions, beneficial to humans, animals and crops. Such bacteria are called **non-pathogenic bacteria** or ***non-pathogens.*** However, certain other bacteria are the deadly foes of man and animals and may enter their tissues, causing serious diseases. Such bacteria are known as **pathogenic bacteria**** or ***pathogens.***

Most of the bacteria require oxygen for their survival. These bacteria consume dissolved oxygen from the water and decompose the refuge and organic matter present in water or sewage, thus doing their cleaning up job. Besides these bacteria, which survive on oxygen, certain other types of bacteria which can survive with or without oxygen, are also sometimes present. Depending upon the oxygen requirement of the bacteria, they may be classified as follows:

(*i*) Aerobic bacteria — *i.e.*, those which require oxygen for their survival.

(*ii*) Anaerobic bacteria — *i.e.*, those which flourish and thrive in the absence of free oxygen.

(*iii*) Facultative bacteria — *i.e.*, those which can survive with or without free oxygen.

---

*1 nano metre = 10^{-9} metre

**First of all discovered by Louis Pasteur (a Frenchman) and Robert Koca (a German) during 18th century.

Besides the bacteria (***Prokaryotes***), certain other parasitic micro-organisms of higher stage of living do exist, which together are known as **Eukaryotes.** Under this category of ***Eukaryotes***, we do have *protozoa* and *fungi* as the important groups. ***Protozoa*** are the single-celled organisms, and are the lowest and simplest form of animal life. They are bacteria eaters and destroy pathogenic bacteria (pathogens). They can be directly counted under a microscope. ***Fungi*** are generally filamentous and have a true cell wall. They are plants which grow without sunlight and live in other plants or animals, dead or alive. Some fungi are microscopic, while some others may be fairly large.

Rotifers and ***Worms*** are the other important groups of micro-organisms, which may be present in water or sewage. ***Rotifers*** are simple multi-cellular organisms at the first stage above the single-cell bacteria. They have cilia used for movement and food currents, located around their mouth. These aerobic micro-organisms are naturally found in marine and fresh waters in relatively high numbers. ***Worms*** have an elongated body and move with undulating motion. These simple animals are sometimes found in sewage, originating from the diseased man and animals. There are many worms classified into a number of different phyla. *Roundworms* or *nematodes* are estimated to be the second most numerous group of organisms after insects. Flatworms are distinguished by a flat body. The most common free-living flatworms are the *planaria*, found in quiescent waters throughout the world.

Numerous micro-organisms of different types may be found present in wastewater, which primarily originates from human feces (including urine). Diseased humans will evidently contribute pathogens into the resulting sewage. Worms and other micro-organisms may also be introduced into the wastewater from soil or animal wastes.

Bacterial Flora of Human Feces. The chemical and water composition of feces is variable and complex. Different diets and climates also influence the composition of feces and thus, the flora present in it. Likewise, dietary practices and other factors affect the volume and weight of feces. The average per capita wet weight of feces produced on a daily basis around the world from individuals ranges from 91 to 489 g. As water content increases, the volume of fecal material increases. Few organisms are found in urine because of high ammonia content and low pH.

Bacteria dominate amongst the various micro-organisms present in feces. The total bacterial population in 1 g of feces has been estimated to be equivalent to 1.8×10^{18} E-coli (Escherichia Coli) cells; E-coli being one of the indicator organism bacteria. Except for pathogens, fecal bacteria are necessary for digestion. As one would expect, more bacteria in feces are facultative and anaerobic micro-organisms. The environment in the intestines is rich in organic matter, which would require high amounts of oxygen to maintain aerobic bacteria. Maintaining an aerobic environment in the intestines would be counter productive for the host, because the aerobic microbes would oxidise the organic matter and reduce its value for the host.

Table 7.5 lists the average concentrations of major groups of bacteria found in human feces of different countries. It is observed that the bacterial composition of feces is relatively uniform around the world. However, the presence of pathogens is highly variable depending on the occurrence of diseases and carriers of pathogens as well as other variables.

It may also be mentioned here that most of the vast number of bacteria present in sewage are harmless non-pathogenic bacteria. They are useful and helpful in bringing oxidation and decomposition of sewage during sewage

Table 7.5. Fecal Microflora of Different Human Populations*

Diet	*Country*	*Mean $\log_{10}$ number of bacteria per gram of feces*						
		*Entero-bact-eria***	*Strepto-cocci*	*Lacto-bacilli*	*Clost-ridia*	*Bacter-oides*	*Bifido-bacteria*	*Euba-cteria*
Largely carbohydrate	India	7.9	7.3	7.6	5.7	9.2	9.6	9.5
	Japan	9.4	8.1	7.4	5.6	9.4	9.7	9.6
	Uganda	8.0	7.0	7.2	5.1	8.2	9.4	9.3
	Hong Kong	7.0	5.8	6.1	4.7	9.8	9.1	8.5
Mixed Western	England	7.9	5.8	6.5	5.7	9.8	9.9	9.3
	Scotland	7.6	5.3	7.7	5.6	9.8	9.9	9.3
	U.S.A.	7.4	5.9	6.5	5.4	9.7	9.9	9.3
	Denmark	7.0	6.8	6.4	6.3	9.8	9.9	9.3
	Finland	7.0	7.8	8.0	6.2	9.7	9.7	9.5

*From Feachem et. al. (1980).

**Fecal coliform bacteria are in the family of *Enterobacteriaceae*.

treatment processes. A little number of bacteria, however, are disease producing pathogens, and it is they who constitute the real danger to the health of the public.

Since animal wastes also do enter into waste waters, the flora of their feces also become important particularly because animals are a reservoir for some diseases that afflict humans. The microbial flora of the animal feces, as compiled by Geldreich (1978), is given in table 7.6. Table 7.7 also lists the various pathogenic micro-organisms commonly found in raw wastewaters.

Table 7.6. Microbial Flora of Animal Feces*

Animal group	*Average density × 10^{-3} per gram*				
	Fecal coliforms	*Fecal strepto-cocci*	*Clostridium perfringens*	*Bacter-oides*	*Lacto-bacilli*
Farm animals					
Cow	230	1300	0.2	**	0.25
Pig	3300	84000	3.98	500	251000
Sheep	16000	38000	199	**	79
Horse	12.6	6300	**	**	10000
Duck	33000	54000	—	—	—
Chicken	1300	3400	0.25	**	316000
Turkey	290	2800	—	—	—
Pet animals					
Cat	7900	27000	25100	795000	63000
Dog	23000	980000	251000	500000	39.6
Wild animals					
Mice	330	7700	**	795000	1260000
Rabbits	0.020	47	**	396000	**
Chipmunk	148	6000	—	—	—
Humans	13000	3000	1.58	5000000	630000

*From Geldreich (1978).

**Present in negligible amount.

Table 7.7. Pathogenic Micro-organisms Commonly found in Raw Waste Waters

Organism	*Disease*	*Remarks*
(1)	(2)	(3)
Ascaris spp., *Enterobius* spp.	Nematode worms	Danger to man from waste water effluents and dried sludge used as fertilizer
Bacillus anthracis	Anthrax	Found in waste water. Spores are resistant to treatment
Brucella spp.	Brucellosis. Malta fever in man. Contagious abortion in sheep, goats, and cattle	Normally transmitted by infected milk or by contact. Waste water is also suspected
Entamoeba histolytica	Dysentery	Spread by contaminated waters and sludge used as fertilizer. Common in hot climates
Leptospira iceterohae-morrhagiae	Leptospirosis (Weil's disease)	Carried by sewer rats
Mycobacterium tuberculosis	Tuberculosis	Isolated from waste water and polluted streams. Waste water is a possible mode of transmission. Care must be taken with waste water and sludge from sanatoriums
Salmonella paratyphi	Paratyphoid fever	Common in waste water and effluents in times of epidemics
Salmonella typhi	Typhoid fever	Common in waste water and effluents in times of epidemics
Salmonella spp.	Food poisoning	Common in waste water and effluents
Schistosoma spp.	Schistosomiasis	Probably killed by efficient waste water treatment
Shigella spp.	Bacillary dysentery	Polluted waters are main source of infection
Taenia spp.	Tapeworms	Eggs very resistant, present in waste water sludge and waste water effluents. Danger to cattle on land irrigated or land manured with sludge
Vibrio cholerae	Cholera	Transmitted by waste water and polluted waters
Virus	Poliomyelitis, hepatitis	Exact mode of transmission not yet known. Found in effluents from biological waste water-treatment plants

As far as testing of sewage is concerned, it may be clarified that, *the routine bacteriological tests, as performed on water supply samples, are generally not performed on sewage, because of the high concentration of bacteria present in it.* But at the time of epidemiological investigation, certain tests may be useful for separating the pathogenic bacteria. The bacteriological counts may also be useful where the treatment processes are likely to be affected adversely by bactericidal industrial wastewaters. These tests have already been discussed in "Vol-I Water Supply Engg." by the same author.

Example 7.6. *The average sewage flow from a city is 80 × 10⁶ l/d. If the average 5 day BOD is 285 mg/l, compute the total daily 5 day oxygen demand in kg, and the population equivalent of sewage. Assume per capita BOD of sewage per day = 75 g.*

Solution. Quantity of sewage flowing per day = 80×10^6 litres.

Average 5 day BOD = 285 mg/l.

∴ Total daily 5 day oxygen demand

$$= 285 \times (80 \times 10^6) \text{ mg} = 22{,}800 \text{ kg.} \quad \textbf{Ans.}$$

$$\text{Population equivalent} = \left[\frac{20{,}800}{\frac{75}{1000}}\right] = 3{,}04{,}000. \quad \textbf{Ans.}$$

Example 7.7. *If the period of incubation is 10 days at 20°C in the relative conductivity test on sewage, calculate the percentage of relative stability.*

Solution. Using eqn. (7.23), we have

Relative stability $= S$

$$= 100\left[1 - (0.794)^{t_{(20)}}\right]$$

where $t_{(20)}$ = Time in days to decolourise a standard vol. of methylene blue at 20°C = 10 days (given)

$$\therefore \quad S = 100[1 - (0.794)^{10}] = 100\,[1 - 0.0995875] = \mathbf{90.04\%}. \quad \textbf{Ans.}$$

Example 7.8. *Calculate the population equivalent of a city given (i) the average sewage from the city is 95 × 10⁶ l/day, and (ii) the average 5 day BOD is 300 mg/l.*

Solution. Average 5 day BOD = 300 mg/l.

Average sewage flow = 95×10^6 l/day

∴ Total BOD in sewage

$$= 300 \times 95 \times 10^6 \text{ mg/day} = 300 \times 95 \text{ kg/day} = 28500 \text{ kg/day}$$

Population equivalent

$$= \frac{\text{Total 5 day BOD in kg/day}}{0.08}$$

[assuming the domestic sewage quantity to be 0.08 kg/person/day]

$$= \frac{28500}{0.08} = 3{,}56{,}250. \text{ Ans.}$$

Example 7.9. *The BOD_5 of a waste water is 150 mg/l at 20°C. The k value is known to be 0.23 per day. What would BOD_8 be, if the test was run at 15° ?*

Solution. Here, we will first clarify, that the given value of constant k, as equal to 0.23 per day, cannot be the value of deoxygenation constant K_D (on base 10), but it must be the *rate constant (on base e)* which was symbolised by us as 'K', where, $K_D = 0.434\,K$. This is because of the reason that the highest K_D value at 20°C for strong municipal sewage is of the order of 0.1 to 0.15, or slightly less. Hence, the given k value must be the value of K and not of K_D.

Hence, at 20°C, we use

$$K = 0.23 \text{ (given)}$$

$$\therefore \quad K_D = 0.434\,K = 0.434 \times 0.23 = 0.0998 \cong 0.1.$$

Also BOD of 5 days = BOD_5 = 150 mg/l (at 20°C)

Using equation (7.16), we have

$$Y_t = L\left[1-(10)^{-K_D \cdot t}\right]$$

$\therefore$ $$Y_5 = L\left[1-(10)^{-K_D \cdot 5}\right],$$ where Y_5 = BOD of 5 days

or $$150 = L\left[1-(10)^{-0.1\times 5}\right]$$

$$= L\left[1-(10)^{-0.5}\right] = L\left[1-\frac{1}{(10)^{0.5}}\right]$$

$$= L\left[1-\frac{1}{3.16}\right] = L\,[1-0.316] = 0.684\,L$$

$\therefore$ $$L = \frac{150}{0.684} = 219.4$$

or L = **219.4 mg/l.** ...(*i*)

Now, let us find K_D value at 15°C

Using equation (7.18), we have

$$K_{D(T^\circ)} = K_{D(20^\circ)}\,[1.047]^{T-20^\circ}$$

$\therefore$ $$K_{D(15^\circ)} = 0.1[1.047]^{15-20^\circ} = 0.1\,[1.047]^{-5}$$

$$= 0.1\left[\frac{1}{(1.047)^5}\right] = 0.1\left[\frac{1}{1.258}\right] = 0.079$$

Now, again using

$$Y_t = L\left[1-(10)^{-K_D \cdot t}\right],$$ where Y_t is BOD of t days

we have $$Y_8 = 219.4\left[1-(10)^{-0.079\times 8}\right]$$

$$= 219.4\left[1-\frac{1}{(10)^{0.632}}\right] = 219.4\left[1-\frac{1}{4.285}\right]$$

$$= 219.4\,[1-0.233] = 219.4 \times 0.766 = 168.2 \text{ mg/l}$$

Hence $BOD_8 = Y_8$ = 168.2 mg/l. **Ans.**

Example 7.10. *The 5 day 30°C BOD of sewage sample is 110 mg/l. Calculate its 5 days 20°C BOD. Assume the deoxygenation constant at 20°C, K_{20} as 0.1.*

Solution. $K_{D(20^\circ)} = 0.1$

Now, using equation (7.18)

$$K_{D(T^\circ)} = K_{D(20^\circ)}\,[1.047]^{T-20^\circ},$$ we get

$$K_{D(30^\circ)} = 0.1\,[1.047]^{30^\circ-20^\circ} = 0.1\,[1.047]^{10} = 0.158$$ ...(*i*)

Now, using $$Y_t = L\left[1-(10)^{-K_D \cdot t}\right],$$ we get

$$Y_5 = L\left[1-(10)^{-K_D \cdot 5}\right]$$

$\therefore$ $$Y_{5 \text{ at } 30^\circ} = L\left[1-(10)^{-K_D(30^\circ)\times 5}\right]$$

or $$110 = L\left[1-(10)^{-0.158\times 5}\right] = L\left[1-(10)^{-0.79}\right]$$

$$= L\left[1-\frac{1}{(10)^{0.79}}\right] = L\,[1-0.162]$$

or $$110 = L\,(0.838) \quad \text{or } L = \frac{110}{0.838} = 131.3$$

or L = **131.3 mg/l.** ...(*ii*)

Now $$Y_{5\text{ at }20^\circ\text{C}} = L\left[1-(10)^{-K_D(20^\circ)\times 5}\right]$$

$$= 131.3\left[1-(10)^{-0.1\times 5}\right] = 131.3\left[1-\frac{1}{(10)^{0.5}}\right]$$

$$= 131.3 \times (1-0.316) = 131.3 \times 0.684$$

$$= \mathbf{89.8\ mg/l.}\quad \textbf{Ans.}$$

Example 7.11. *Calculate 1 day 37°C BOD of sewage sample whose 5 day 20°C BOD is 100 mg/l. Assume K_D at 20°C as 0.1.*

Solution. 5 day 20°C BOD = 110 mg/l (given)

Now using Eq. (7.16), we have

The BOD at 20°C, say after t = 5 days, is given by

$$Y_t = L\left[1-(10)^{-K_D(20^\circ)\,.\,t}\right]$$

Using Y_t = 100 mg/l (given)

$$K_{D(20^\circ)} = 0.1$$

we have $$100 = L\left[1-(10)^{-0.1\times 5}\right]$$

or $$100 = L\left[1-(10)^{-0.5}\right] = L\left[1-\frac{1}{3.16}\right]$$

$$= L\,[1-0.316] = L\,[0.684]$$

or $$L = \frac{100}{0.684} = \mathbf{146.2\ mg/l.}$$

Now let us work out K_D at 37°C, by using Eq. (7.18) as :

$$K_{D(T^\circ)} = K_{D(20^\circ)}\,[1.047]^{T-20^\circ}$$

or $$K_{D(37^\circ)} = 0.1\,[1.047]^{37-20} = 0.1\,[1.047]^{17}$$

$$= 0.1 \times 2.18 = \mathbf{0.218.}$$

Now, we have to work out Y_t for one day *i.e.* Y_1 at 37°C, using

$$Y_t = L\left[1-(10)^{-K_D\,.\,t}\right]$$

∴ $$Y_1 = L\left[1-(10)^{-K_D\,.\,1}\right]$$

or $$Y_1\text{ (at 37°C)} = 146.2\left[1-(10)^{-K_{D(\text{at }37^\circ\text{C})}\times 1}\right]$$

$$= 146.2\left[1-(10)^{-0.218\times 1}\right] = 146.2\left[1-\frac{1}{(10)^{0.218}}\right]$$

$$= 146.2\left[1-\frac{1}{1.652}\right] = 146.2\,[1-0.605] = 57.75.$$

Hence, Y_1 at 37°C = **57.75 mg/l. Ans.**

Example 7.12. *The BOD of a sewage incubated for one day at 30°C has been found to be 110 mg/l. What will be the 5-day 20°C BOD ? Assume K_1 = 0.1 at 20°C.*

Solution. $Y_{1\,(\text{at }30^\circ)}$ = 110 mg/l. ; $Y_{5\,(\text{at }20^\circ)}$ = ? ; $K_{D\,(20^\circ)}$ = 0.1.

First of all, let us calculate K_D at 30°C, by using Eq. (7.18) *i.e.*

$$K_{D(T)} = K_{D\,(20)}\,[1.047]^{T-20°}$$

or $$K_{D\,(30°)} = 0.1\,[1.047]^{30-20°} = 0.1\,[1.047]^{10}$$
$$= 0.1 \times 1.583 = \mathbf{0.158.}$$

Now using Eq. (7.16), we have

$$Y_t = L\left[1-(10)^{-K_D\,.\,t}\right]$$

At 30°C and for one day, we have

$$Y_{1\,(30°)} = \left[1-(10)^{-K_{D.(30°)}\times 1}\right]L$$

or $$110 = L\left[1-(10)^{-0.158\times 1}\right] = L\left[1-\frac{1}{1.438}\right]$$
$$= L\,[1-0.696] = L\,[0.304]$$

or $$L = \frac{110}{0.304} = \mathbf{361.8\ mg/l.}$$

Now again using $Y_t = L\left[1-(10)^{-K_D\,.\,t}\right]$, we have

$$Y_{5(20°)} = L\left[1-(10)^{-K_{D(20)°}\times 5}\right]$$
$$= L\left[1-(10)^{-0.1\times 5}\right] = L\left[1-\frac{1}{(10)^{0.5}}\right]$$
$$= L\,[1-0.316]$$
$$= 361.8\,[1-0.316] = \mathbf{247.4\ mg/l.}\quad \textbf{Ans.}$$

Example 7.13. *The BOD_5 of a waste has been measured as 600 mg/l. If k_1 = 0.23/day (base e), what is the ultimate BOD_u of the waste. What proportion of the BOD_u would remain unoxidised after 20 days.*

Solution. Use eqn. (7.16) as :

$$Y_t = L\left[1-(10)^{-K_D\,.\,t}\right]$$

Here $K = k_1$ = 0.23/day (given)

$\therefore$ $K_D = 0.434\,K = 0.434 \times 0.23 = 0.1.$

Using t = 5 days, we have

$$Y_5 = \text{BOD of 5 days}$$
$$= 600\text{ mg/l} = L\left[1-(10)^{-0.1\times 5}\right]$$

or $$600\text{ mg/l} = L\left[1-(10)^{-0.5}\right] = L\left[1-\frac{1}{(10)^{0.5}}\right]$$
$$= L\left[1-\frac{1}{3.16}\right] = L\,[1-0.316] = 0.684\,L$$

$\therefore$ $0.684\,L = 600$ mg/l

$\therefore$ $$L = \frac{600}{0.684}\text{ mg/l} = 877.5\text{ mg/l.}$$

Hence, the ultimate BOD = 877.5 mg/l. **Ans.**

Now $$Y_{20} = L\left[1-(10)^{-0.1\times 20}\right] = Y_u\left[1-\frac{1}{(10)^2}\right]$$
$$= Y_u\,[1-0.01] = Y_u\,[0.99]$$

$\therefore \quad Y_{20} = 0.99\, Y_u$

It means that 99% of BOD_u is utilised in 20 days, and hence only 1% of ultimate BOD would be left unoxidised after 20 days. **Ans.**

Example 7.14. *The following observations were made on a 3% dilution of waste water :*

Dissolved oxygen (D.O.) of aerated water used for dilution

= 3.0 mg/l

Dissolved oxygen (D.O.) of diluted sample after 5 days incubation

= 0.8 mg/l

Dissolved oxygen (D.O.) of original sample

= 0.6 mg/l.

Calculate the B.O.D. of 5 days and ultimate BOD of the sample assuming that the deoxygenation coefficient at test temp. is 0.1.

Solution. The 100% contents of the diluted sample consists of 3% wastewater and 97% of aerated water used for dilution.

Hence its D.O. = D.O. of waste water × its content

+ D.O. of dilution water × its content

$= 0.6 \times 0.03 + 3.0 \times 0.97$

$= 0.018 + 2.91 =$ **2.928 mg/l.**

D.O. of the incubated sample after 5 days

= 0.8 mg/l.

Thus, D.O. consumed in oxidising organic matter

$= 2.928 - 0.8 = 2.128$ mg/l.

$\therefore$ B.O.D. of 5 days = D.O. consumed × Dilution factor

$$= 2.128 \times \frac{100}{3} = \mathbf{70.93\ mg/l.}$$

Ultimate B.O.D. is given by L.

Using Eq. (7.16), we have

$$Y_t = L\left[1-(10)^{-K_D\,.\,t}\right]$$

or
$$Y_5 = L\left[1-(10)^{-K_D \times 5}\right] \qquad ...(i)$$

The value of K_D at test temp. is given as 0.1. Substituting the known values in Eq. (*i*) above, we have

$$70.93 = L\left[1-(10)^{-0.1\times 5}\right] = L\,[1-(10)^{-0.5}]$$

$$= L\left[1-\frac{1}{(10)^{0.5}}\right] = L\left[1-\frac{1}{3.16}\right]$$

$$= L\,[1-0.316] = L \times 0.684$$

or
$$L = \frac{70.93}{0.684} = 103.7 \text{ mg/l.} \quad \textbf{Ans.}$$

Example 7.15. *If the per capita contribution of suspended solids and B.O.D. is 90 gm and 55 gm, find the population equivalents of*

(i) A combined system serving 1000 persons and having 75 gm per capita daily of B.O.D. ; and

(ii) 40,000 litres daily of industrial waste water containing 1800 mg/l of suspended solids.

Solution. (*i*) Population served by the combined system

= 1000.

Daily per capita B.O.D. of the combined system

= 75 gm.

Total daily B.O.D. produced by the system

= 1000 × 75 gm = 75,000 gm.

But standard per capita B.O.D. (given)

= 55 gm.

∴ Population equivalent

$$= \frac{75{,}000}{55} = 1364. \quad \textbf{Ans.}$$

(*ii*) 1 litre of industrial wastewater contains 1800 mg of suspended solids.

∴ 40,000 litres of industrial wastewater (daily) will produce

$$= \frac{1800 \times 40{,}000}{10^3} \text{ gm of suspended solids}$$

= 72,000 gm of suspended solids

But standard daily per capita solid contribution is 90 gm.

∴ Population equivalent

$$= \frac{72{,}000}{90} = 800. \ \textbf{Ans.}$$

Example 7.16. *In a test conducted for determining the relative conductivity at 20°C, the period of incubation was found to be 12 days. Calculate the per cent of relative stability.*

Solution. The relative stability (S) at 20°C is given by equation (7.23) as :

$$S = 100\left[1 - (0.794)^{t_{20}}\right]$$

Using $t_{(20)}$ = No. of days at 20°C test temp.

= 12 days.

We get $S = 100\left[1 - (0.794)^{12}\right]$

= 100 [1 – 0.063] = 100 × 0.937 = 93.7%. **Ans.**

Example 7.17. *Change in concentration of organic matter, L, with time, t, is given by* $\frac{dL}{dt} = -K.L.$

Calculate the organic matter remaining after 3 days if the initial concentration was 200 mg/l, and K = 0.4 per day.

Solution. $\frac{dL}{dt} = -KL$ or $\frac{dL}{L} = -K.dt$

Integrating, we have

$$\log_e L = -Kt + C$$

or $$2.3 \log_{10} L = -Kt + C.$$

When $t = 0$ (at start), $L = 200$ mg/l.

$$\therefore \quad 2.3 \log_{10} 200 = 0 + C$$

or $$C = 2.3 \times 2.301 = 5.28.$$

Now, the value of L after 3 days is given by

$$2.3 \log_{10} L = -0.4 \times 3 + C$$

or $$2.3 \log_{10} L = -1.2 + 5.28 = 4.08$$

or $$\log_{10} L = \frac{4.08}{2.3} = 1.773$$

or $$L = 59.3 \text{ mg/l}.$$

Hence the organic matter left after 3 days = 59.3 mg/l. **Ans.**

Example 7.18. *Data from an unseeded domestic waste water BOD test are : 5 ml of waste in 300 ml bottle, initial D.O. of 7.8 mg/l, and 5 days DO equal to 4.3 mg/l. Compute*

(a) the BOD ; and (b) the ultimate BOD, assuming a k-rate of 0.10 per day.

Solution. Initial D.O. = 7.8 mg/l

D.O. after 5 days of incubation = 4.3 mg/l

∴ D.O. consumed in 5 days = 7.8 – 4.3 = 3.5 mg/l

Now, using equation (7.10), we have

BOD_5 of wastewater

$$= \text{D.O. consumed by diluted sample} \times \left[\frac{\text{Vol. of diluted sample}}{\text{Vol. of undiluted sewage used}}\right]$$

$$= 3.5 \text{ mg/l} \times \left[\frac{300 \text{ ml}}{5 \text{ ml}}\right] = \mathbf{210 \text{ mg/l}.} \text{ **Ans.**}$$

Now, using equation $Y_t = L\left[1-(10)^{-K_D \cdot t}\right]$, we have

$$Y_5 = L\left[1-(10)^{-K_D \cdot 5}\right]$$

where $K_D = 0.1$ per day and $Y_5 = 210$ mg/l

$$\therefore \quad 210 = L\left[1-(10)^{-0.1\times 5}\right] = L\left[1-\frac{1}{(10)^{0.5}}\right]$$

$$= L\left[1-0.316\right] = 0.684L$$

or $$L = \frac{210}{0.684} \text{ mg/l} = 307.1 \text{ mg/l}$$

Hence, ultimate BOD (Y_u) = L = **307.1 mg/l. Ans.**

Example 7.19. *A dairy processing about 1,33,000 kg of milk daily produced an average of 246 cubic metre per day of waste water with a BOD of 1400 mg/l. The principal operations are bottling of milk, and making ice cream, with limited production of cheese. Compute the waste water flow and BOD per 1000 kg of milk received, and the equivalent population of the daily waste discharge.*

Solution. Daily milk processed = 1,33,000 kg

Daily wastewater produced = 246 m^3

BOD of wastewater = 1400 mg/l

Evidently,

Wastewater produced per 1000 kg of milk

$$= \frac{246}{1,33,000} \times 1000 \text{ m}^3 = \mathbf{1.85\ m^3.}\quad \textbf{Ans.}$$

BOD of the wastewater = 1400 mg/l

$$= 1400 \times 10^3 \text{ mg/m}^3$$

$$= \frac{1400 \times 10^3}{1000} \text{ gm/m}^3 = \frac{1400 \times 10^3}{10^3 \times 10^3} \text{ kg/m}^3$$

$$= 1.4 \text{ kg/m}^3$$

But vol of waste water produced in processing 1000 kg of milk = 1.85 m^3 (calculated above)

Hence, BOD produced per 1000 kg of milk processed

$$= 1.4 \frac{\text{kg}}{\text{m}^3} \times 1.85 \text{ m}^3 = \mathbf{2.59\ kg.\ Ans.}$$

Daily BOD produced by 246 m^3 of wastewater

$$= 1.4 \text{ kg/m}^3 \times 246 \text{ m}^3 = 344.4 \text{ kg/day.}$$

From equation (7.21), we have

Population equivalent

$$= \frac{\text{BOD of industry in kg/day}}{0.08} = \frac{344.4}{0.08} = \mathbf{4305.\ Ans.}$$

[**Note.** The dairy is producing as much pollution as is likely to be produced by 4305 persons.]

Example 7.20. *The 3 days 15°C BOD of a sample of sewage is 150 mg/l. Draw a graph of 5 day BOD as a function of temperature in the range 10°C to 30°C in steps of 5°C. Assume K_D at 20°C = 0.1.*

Solution. K_D at 20°C = 0.1.

K_D at 15°C is given as :

$$K_{D(T°)} = K_{D(20°)} [1.047]^{T - 20°}$$

or $$K_{D(15°)} = 0.1 [1.047]^{15 - 20°}$$

$$= 0.1 (1.047)^{-5} = \frac{0.1}{(1.047)^5} = \mathbf{0.079}$$

Now, using $Y_{t \text{ at } T°} = L\left[1 - (10)^{-K_D \cdot t}\right]$, we have

$$Y_{3 \text{ at } 15°} = 150 \text{ mg/l}$$

$$\therefore \quad 150 \text{ mg/l} = L\left[1 - (10)^{-0.079 \times 3}\right]$$

$$= L\left[1 - \frac{1}{(10)^{0.237}}\right] = 0.422L$$

or $$L = \mathbf{355.53\ mg/l.}$$

$$K_{D(10°)} = 0.1 [1.047]^{10 - 20} = 0.063$$

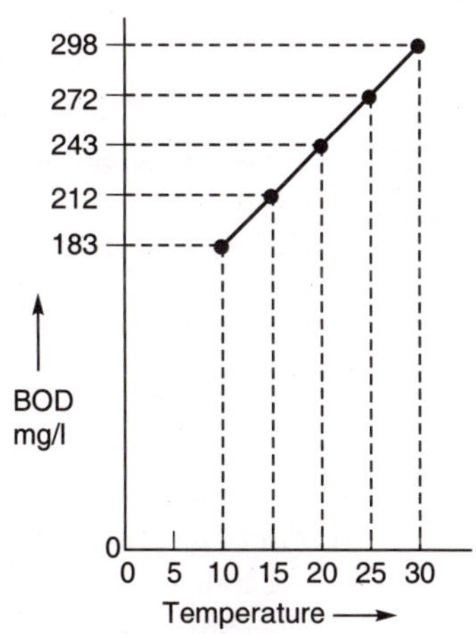

Fig. 7.8

$$K_{D(25°)} = 0.1\ [1.047]^{25-20} = 0.1258$$

$$K_{D(30°)} = 0.1\ [1.047]^{30-20} = 0.1583$$

(*i*) $Y_{5 \text{ at } 10°} = 355.53 \left[1 - (10)^{-0.063 \times 5}\right] = 183$ mg/l

(*ii*) $Y_{5 \text{ at } 15°C} = 355.53 \left[1 - (10)^{-0.079 \times 5}\right] = 212$ mg/l

(*iii*) $Y_{5 \text{ at } 20°C} = 355.53 \left[1 - (10)^{-0.1 \times 5}\right] = 243$ mg/l

(*iv*) $Y_{5 \text{ at } 25°C} = 355.53 \left[1 - (10)^{-0.1258 \times 5}\right] = 272$ mg/l

(*v*) $Y_{5 \text{ at } 30°C} = 355.53 \left[1 - (10)^{-0.1583 \times 5}\right] = 298$ mg/l

These calculated five number BOD values w.r.t. temperature are plotted in Fig. 7.8, so as to obtain the requisite graph, which is almost a straight line. **Ans.**

PROBLEMS

1. (*a*) State the routine tests carried out in the laboratories at sewage treatment and disposal works, and explain the significance of each of them.

(*b*) Mention the common types of organisms found in domestic sewages, and explain why routine biological examination of sewage is usually not carried out, as is done for water supplies.

2. (*a*) Explain the importance of determination of solids in sewage. How do you determine the suspended solids in a given sample of waste water ?

(*b*) What is the difference between BOD and COD ? A 2% solution of a sewage sample is incubated for 5 days at 20°C. The depletion of oxygen was found to be 4 ppm. Determine the BOD of the sewage. [**Ans.** 200 mg/l or 200 ppm]

3. (*a*) Write short note on BOD.

(*b*) Calculate 1 day 37°C BOD of sewage sample whose 5-days 20°C BOD is 100 mg/l.

4. (*a*) Explain population equivalent.

(*b*) Given a waste water containing 300 mg/l of a ketone of chemical formula $CH_3COC_2H_5$.

(*i*) Calculate the COD.

(*ii*) Assume a "*k*" value (base 10) of 0.1 per day.

Calculate the ultimate BOD and the 5 day BOD of the waste.

Note : H = 1, C = 12, O = 16.

5. (*a*) Give in a tabular form, the average composition of domestic sewage of a town.

(*b*) The average sewage flow from a city is 80×10^6 l/d. If the average 5-days BOD is 285 mg/l, compute the total daily 5-day oxygen demand in kg, and the population equivalent of sewage, $k = 0.1$. Assume per capita BOD of the sewage per day = 75 gm.

6. Give a brief account of general composition of sewage. What is the purpose and principle involved in its treatment and disposal ?

7. Explain the importance of the following operations in a laboratory determination of BOD of waste water sample :

(*i*) pH adjustment ;

(*ii*) dilution with aerated water ;

(*iii*) seeding and nutrient addition to dilution water ; and

(*iv*) incubation at controlled temperature.

8. (*a*) State and describe four important tests that may be carried out to know the characteristics of sanitary sewage.

(*b*) Distinguish between aerobic, anaerobic and facultative micro-organisms and their role in the decomposition of sewage.

9. (*a*) Explain the importance of determination of solids in sewage. How do you determine the suspended solids in a given sample of waste water ?

(*b*) A town has an average domestic sewage flow of 31,710 m^3/day with a BOD concentration of 250 ppm. A neighbouring industrial estate adds about 11,325 m^3/day of sewage having 9080 kg of BOD to it. Find out :

(*i*) the concentrations of BOD in industrial and the combined sewage ;

(*ii*) the probable population and per capita flow of sewage ; and

(*iii*) the population equivalents based on the (*i*) BOD load and (*ii*) the hydraulic load.

10. (*a*) What is BOD ? Deduce an expression for BOD with time. What are the factors on which the de-oxygenation constant (*k*) depends ?

(*b*) The 3 day 37°C BOD of a sample of sewage is 300 ppm. What will be its 10 days 20°C BOD, and 5 day 30°C BOD ?

11. Explain the composition of sewage and also the possible methods of measuring the strength of sewage. Write out the laboratory procedure of the experiment to determine the strength of sewage.

12. State the common laboratory tests conducted on sewage and their importance in the treatment and disposal of sewage.

13. (*a*) Name two most important parameters used to characterise sewage and describe their significance.

(*b*) The 7 days 20°C BOD of a sample of sewage is 300 ppm. and its 3 days 37°C BOD is 500 ppm. Find out the value of deoxygenation constant *k*, and then estimate its 5 days 30°C BOD.

14. (*a*) What is BOD ? Deduce an expression for the first stage BOD.

(*b*) If 3.0 ml of raw sewage was diluted to 300 ml and the D.O. concentration of the diluted sample at the beginning was 8.6 mg/l, and 4.6 mg/l at the end of 5 day incubation at 20°C, find the B.O.D. of raw sewage.

(*c*) If the contribution of suspended solids and B.O.D. is 90 gm and 75 gm per capita per day, estimate the population equivalents of

(*i*) a combined system serving 1200 persons and having 125 gm per capita daily of BOD ; and

(*ii*) 50,000 litres daily contribution of industrial waste-water containing 1800 mg/l of suspended solids.

15. Write notes on :

(*i*) Cycles of decay of waste organic substances ;

(*ii*) B.O.D. ;

(*iii*) C.O.D. and B.O.D. ;

(*iv*) Concentration of solids in sewage ;

(*v*) Population Equivalent ;

(*vi*) Compositiion of Municipal Sewage ;

(*vii*) BOD/COD ratio ; and

(*viii*) TOC.

8

Disposing of the Sewage Effluents

Before we discuss, in our next chapter, the various treatments that may be given to raw sewage before disposing it of, it shall be worthwhile to first discuss the various methods and sources of disposal of sewage. The study of the sources of disposal is important, because the amount of treatment required to be given to a sewage depends very much upon the source of disposal, its quality and capacity to tolerate the impurities present in the sewage effluents, without itself getting potentially polluted or becoming less useful.

There are two general methods of disposing of the sewage effluents :

(*a*) *Dilution i.e.* disposal in water ; and

(*b*) *Effluent Irrigation or Broad Irrigation or Sewage Farming, i.e.* disposal on land.

Disposal by dilution is more common of these two methods, which are described below :

DISPOSAL BY DILUTION

Disposal by dilution is the process whereby the treated sewage or the effluent from the sewage treatment plant is discharged into a river stream, or a large body of water, such as a lake or sea. The discharged sewage, in due course of time, is purified by what is known as *self purification process of natural waters*. The degree and amount of treatment given to raw sewage before disposing it of into the river-stream in question, will definitely depend not only upon the quality of raw sewage *but also upon the self purification capacity of the river stream and the intended use of its water*.

8.1. Conditions Favouring Disposal By Dilution

The dilution method for disposing of the sewage can favourably be adopted under the following conditions :

(*i*) When sewage is comparatively fresh (4 to 5 hr old), and free from floating and settleable solids, (or are easily removed by primary treatment).

(*ii*) When the diluting water (*i.e.* the source of disposal) has a high dissolved oxygen (D.O.) content.

(*iii*) Where diluting waters are not used for the purpose of navigation or water supply *for atleast some reasonable distance* on the downstream from the point of sewage disposal.

(*iv*) Where the flow currents of the diluting waters are favourable, causing no deposition, nuisance or destruction of aquatic life. It means that swift forward currents are helpful, as they easily carry away the sewage to the points of unlimited dilution. On the other hand, slow back currents tend to cause sedimentation, resulting in large sludge deposits.

(*v*) When the outfall sewer of the city or the treatment plant is situated near some natural waters having large volumes.

8.2. Standards of Dilution for Discharge of Wastewaters into Rivers

The ratio of the quantity of the diluting water to that of the sewage is known as the *dilution factor* ; and depending upon this factor, 'the Royal Commission Report on Sewage Disposal' has laid down the following standards and degrees of treatment required to be given to a particular sewage.

Table 8.1. Standards of Dilution Based on Royal Commission Report

Dilution factor	*Standards of purification required*
Above 500	No treatment is required. Raw sewage can be directly discharged into the volume of dilution water.
Between 300 to 500	Primary treatment such as *plain sedimentation* should be given to sewage, and the effluents should not contain suspended solids more than 150 ppm.
Between 150 to 300	Treatments such as *sedimentation, screening* and essentially *chemical precipitation* are required. The sewage effluent should not contain suspended solids more than 60 ppm.
Less than 150	Complete thorough treatment should be given to sewage. The sewage effluent should not contain suspended solids more than 30 ppm, and its 5 days B.O.D. at 18.3°C should not exceed 20 ppm.

The above standards have been operative in England since 1912, and had also been followed in India without much variance. However, with the increasing pollution of surface streams due to indiscriminate discharge of domestic and industrial wastewaters, without bothering to look into the available dilution ratios, it has become imperative to limit the concentrations of various pollutants being discharged in to the surface water sources along with the sewage and industrial effluents. The tolerance limits for such constituent pollutants have therefore been prescribed by various countries, including India. These limits are based upon the desirability of giving full fledged treatment to sewage and industrial liquid wastes, up to minimum level of 'secondary treatment'.

The Bureau of Indian Standards (BIS), previously known as Indian Standard Institution (ISI), has therefore laid down its *guiding standards* for sewage effluents, vide IS 4764-1973, and for industrial effluents vide IS : 2490-1974, as shown in table 8.2. These tolerance limits are supposed to be the *national guidelines* for guiding the various *State Pollution Control Boards** for prescribing their legally enforceable standards, depending upon the water

*Indian States like Karnataka, Kerala, Maharashtra, Rajasthan, U.P. and West Bengal have already published their legally enforceable effluent standards for various industries, and other States are in the process of doing so. The standards adopted by these States are almost the same as suggested by BIS(ISI), except for some locally needed changes. GOI has also notified these standards under the Environment (Protection) Rules, 1986, as shown in Table 8.3.

Table 8.2. BIS (ISI) Standards for Discharge of Sewage and Industrial Effluents in Surface Water Sources* and Public Sewers

S. No.	Characteristic of the Effluent	Tolerance limit for Sewage Effluent discharged into Surface Water Sources, as per IS 4764-1973	Tolerance Limit for Industrial effluent discharged into	
			Inland surface waters, as per IS 2490-1974	Public sewers as per IS 3306-1974
(1)	(2)	(3)	(4)	(5)
1.	BOD_5	20 mg/l	30 mg/l	500** mg/l
2.	COD	—	250 mg/l	—
3.	pH value	—	5.5 to 9.0	5.5 to 9.0
4.	Total Suspended Solids (TSS)	30 mg/l	100 mg/l	600 mg/l
5.	Temperature	—	40°C	45°C
6.	Oil and grease	—	10 mg/l	100 mg/l
7.	Phenolic compounds (as Phenol)	—	1 mg/l	5 mg/l
8.	Cyanides (as CN)	—	0.2 mg/l	2 mg/l
9.	Sulphides (as S)	—	2 mg/l	—
10.	Fluorides (as F)	—	2 mg/l	—
11.	Total residual chlorine	—	1 mg/l	—
12.	Insecticides	—	Zero	—
13.	Arsenic (as As)	—	0.2 mg/l	—
14.	Cadmium (as Cd)	—	2 mg/l	—
15.	Chromium, hexavalent (as Cr)	—	0.1 mg/l	2 mg/l
16.	Copper	—	3 mg/l	3 mg/l
17.	Lead	—	0.1 mg/l	1 mg/l
18.	Mercury	—	0.01 mg/l	—
19.	Nickel	—	3 mg/l	2 mg/l
20.	Selenium	—	0.05 mg/l	—
21.	Zinc	—	5 mg/l	15 mg/l
22.	Chlorides (as Cl)	—	—	600 mg/l
23.	% Sodium	—	—	60%
24.	Ammoniacal nitrogen (as N)	—	50 mg/l	50 mg/l
25.	Radioactive materials	—		
	(*i*) α-emitters		10^{-7} μC/ml (micro curie/ml)	—
	(*ii*) β-emitters		10^{-6} μC/ml	—

*Includes Rivers, Estuaries, Streams, Lakes and Reservoirs.

**subject to relaxation or tightening by the local authorities.

quality and dilution available in their respective surface water sources, and the type of effluents produced by the different industries. These standards have also been notified by GOI under Environment (Protection) Rules, 1986, and are shown in Table 8.3.

When the industrial wastewaters are *disposed of in to public sewers*, then also, their quality has to be checked, by following the standards prescribed by IS 3306-1974 (Refer col. (5) of table 8.2) or the standards notified by GOI under Environment (Protection) Rules, 1986, as shown in Table 8.3.

Table 8.3. General Standards for Discharge of Environment Pollutants from Effluents into Surface Water Sources, Public Sewers, and Marine Coasts Under Environment (Protection) Rules, 1986

S. No.	*Characteristic of the effluent i.e. Name of pollutant in the effluent*	*Standard Prescribed under Environment (Protection) Rules, 1986 of GOI for*		
		Inland surface waters	*Public sewers*	*Marine coasts i.e. seas and oceans*
(1)	(2)	(3)	(4)	(5)
1.	*Colour and odour*	All efforts should be made to remove colours and unpleasant odours, as for as possible	All efforts should be made to remove colours and unpleasant odours as for as possible.	All efforts should be made to remove colours and unpleasant odours as far as possible.
2.	Total suspended solids (TSS)	100 mg/l	600 mg/l	(*i*) 100 mg/l for process waste water (*ii*) For cooling water effluent, 10% above total suspended matter of influent
3.	Particle size of suspended solids	Shall pass 850 micron sieve	—	(*a*) Floatable solids : max 3 mm (*b*) Settleable solids : max 850 micron
4.	BOD_5 at 20°C	30 mg/l	350 mg/l	100 mg/l
5.	COD	250 mg/l	—	250 mg/l
6.	pH value	5.5—9.0	5.5—9.0	5.5—9.00
7.	Temperature	Shall not exceed 5°C above the temp. of receiving water	—	Shall not exceed 5°C above the temp. of receiving water.
8.	Oil and grease	10 mg/l	20 mg/l	20 mg/l
9.	Total residual chlorine	1.0 mg/l	—	1.0 mg/l
10.	Ammoniacal Nitrogen (as N)	50 mg/l	50 mg/l	50 mg/l
11.	Total Kjeldahl nitrogen (as N)	100 mg/l	—	100 mg/l

(1)	(2)	(3)	(4)	(5)
12.	Free Ammonia (as NH_3)	5.0 mg/l	—	5.0 mg/l
13.	Phenolic compounds (as C_6H_5OH)	1.0 mg/l	5.0 mg/l	5.0 mg/l
14.	Sulphide (as S)	2.0 mg/l	—	5.0 mg/l
15.	Arsenic (as As)	0.2 mg/l	0.2 mg/l	0.2 mg/l
16.	Mercury (as Hg)	0.01 mg/l	0.01 mg/l	0.01 mg/l
17.	Lead (as Pb)	0.1 mg/l	1.0 mg/l	2.0 mg/l
18.	Cadmium (as Cd)	2.0 mg/l	1.0 mg/l	2.0 mg/l
19.	Hexavalent Chromium (as Cr + 6)	0.1 mg/l	2.0 mg/l	1.0 mg/l
20.	Total Chromium (as Cr)	2.0 mg/l	2.0 mg/l	2.0 mg/l
21.	Copper (as Cu)	3.0 mg/l	3.0 mg/l	3.0 mg/l
22.	Zinc (as Zn)	5.0 mg/l	15 mg/l	15 mg/l
23.	Selenium (as Se)	0.05 mg/l	0.05 mg/l	0.05 mg/l
24.	Nickel (as Ni)	3.0 mg/l	3.0 mg/l	5.0 mg/l
25.	Cyanide (as CN)	0.2 mg/l	2.0 mg/l	0.2 mg/l
26.	Fluoride (as F)	2.0 mg/l	15 mg/l	15 mg/l
27.	Dissolved phosphates (as P)	5.0 mg/l	—	—
28.	Radioactive materials (*a*) Alpha emitters (*b*) Beta emitters	 10^{-7} microcurie/ml 10^{-6} microcurie/ml	 10^{-7} microcurie/ml 10^{-6} microcurie/ml	 10^{-7} microcurie/ml 10^{-6} microcurie/ml
29.	Bio-assay test	90% survival of fish after 96 hrs in 100% effluent	90% survival of fish after 96 hrs in 100% effluent	90% survival of fish after 96 hrs in 100% effluent
30.	Manganese (as Mn)	2 mg/l	2 mg/l	2 mg/l
31.	Iron (as Fe)	3 mg/l	3 mg/l	3 mg/l
32.	Vanadium (as V)	0.2 mg/l	0.2 mg/l	0.2 mg/l
33.	Nitrate Nitrogen	10 mg/l	—	20 mg/l

(1)	(2)	(3)	(4)	(5)
34.	Pesticides			
	(*i*) Benzene hexachloride	10 μg/l	—	10 mg/l
	(*ii*) Carboxyl	10 μg/l	—	10 μg/l
	(*iii*) DDT	10 μg/l	—	10 μg/l
	(*iv*) Endosulfan	10 μg/l	—	10 μg/l
	(*v*) Diamethoate	450 μg/l	—	450 μg/l
	(*vi*) Penitrothion	10 μg/l	—	10 μg/l
	(*vii*) Malathion	10 μg/l	—	10 μg/l
	(*viii*) Phorate	10 μg/l	—	10 μg/l
	(*ix*) Methyl parathion	10 μg/l	—	10 μg/l
	(*x*) Phenthoate	10 μg/l	—	10 μg/l
	(*xi*) Pyrethrums	10 μg/l	—	10 μg/l
	(*xii*) Copper oxychloride	9600 μg/l	—	9600 μg/l
	(*xiii*) Copper sulphate	50 μg/l	—	50 μg/l
	(*xiv*) Ziram	1000 μg/l	—	1000 μg/l
	(*xv*) Sulphur	30 μg/l	—	30 μg/l
	(*xvi*) Paraouat	2300 μg/l	—	2300 μg/l
	(*xvii*) Proponil	7300 μg/l	—	7300 μg/l
	(*xviii*) Nitrogen	780 μg/l	—	780 μg/l

8.3. Dilution in Rivers and Self Purification of Natural Streams

When sewage is discharged into a natural body of water, the receiving water gets polluted due to waste products, present in sewage effluents. But the conditions do not remain so for ever, because the natural forces of purification, such as *dilution, sedimentation, oxidation-reduction in sunlight*, etc., go on acting upon the pollution elements, and bring back the water into its original condition. This automatic purification of polluted water, in due course, is called the *self-purification phenomenon*. However, if the self-purification is not achieved successfully either due to too much of pollution discharged into it or due to other causes, the river water itself will get polluted, which, in turn, may also pollute the sea where the river outfalls.

The **various natural forces of purification** which help in effecting self-purification process are summarised below :

1. Physical forces are :
   (*i*) *Dilution and dispersion,*
   (*ii*) *Sedimentation,* and
   (*iii*) *Sunlight (acting through bio-chemical reactions).*
2. Chemical forces aided by biological forces (called bio-chemical forces) are :
   (*iv*) *Oxidation (Bio),*
   (*v*) *Reduction.*

These forces are described below :

(*i*) **Dilution and Dispersion.** When the putrescible organic matter is discharged into a large volume of water contained in the river-stream, it gets rapidly dispersed and diluted. The action, thus, results in diminishing the concentration of organic matter, and thus reduces the potential nuisance of sewage.

When sewage of concentration C_S flows at a rate Q_S in to a river stream with concentration C_R flowing at a rate Q_R, the concentration C of the resulting mixture is given by

$$C_S Q_S + C_R Q_R = C(Q_S + Q_R)$$

or
$$C = \frac{C_S Q_S + C_R Q_R}{Q_S + Q_R} \quad \text{...(8.1)}$$

This equation is applicable separately to concentrations of different impurities, such as, oxygen content, BOD, suspended sediments, and other characteristic contents of sewage.

(*ii*) **Sedimentation.** The settleable solids, if present in sewage effluent, will settle down into the bed of the river, near the outfall of sewage, thus, helping in the self purification process.

(*iii*) **Sunlight.** The sun light has a bleaching and stabilising effect of bacteria. It also helps certain micro-organisms to derive energy from it, and convert themselves into food for other forms of life, thus absorbing carbon dioxide and releasing oxygen by a process known as **photosynthesis.** The evolution of oxygen in river water due to sunlight will help in achieving self-purification through oxidation, as given below :

(*iv*) **Oxidation.** The oxidation of the organic matter present in sewage effluent, will start as soon as the sewage outfalls into the river water containing dissolved oxygen. *The deficiency of oxygen so created, will be filled up by the atmospheric oxygen.* The process of oxidation will continue till the organic matter has been completely oxidised. This is the most important action responsible for affecting self purification of rivers.

(*v*) **Reduction.** Reduction occurs due to hydrolysis of organic matter settled at the bottom either chemically or biologically. Anaerobic bacteria will help in splitting the complex organic constituents of sewage into liquids and gases, and thus paving the way for their ultimate stabilization by oxidation.

The various factors on which these natural forces of purification depend are : (*a*) temperature ; (*b*) turbulence ; (*c*) hydrography such as the velocity and surface expanse of the river-stream ; (*d*) available dissolved oxygen, and the amount and type of organic matter present, (*e*) rate of reaeration, etc.

Besides affecting the dilution and sedimentation rates, the *temperature* also affects the rate of biological and chemical activities, which are enhanced at higher temperatures and depressed at lower temperatures. The dissolved oxygen content of water, which is very essential for maintaining aquatic life and aerobic conditions (so as to avoid the anaerobic decomposition and subsequent nuisance caused by the eruption of foul odours) is also influenced by temperature. *At higher temperatures, the capacity to maintain the D.O. concentration is low ; while the rate of biological and chemical activities are high, causing thereby rapid depletion of D.O.* This is likely to lead to anaerobic conditions, when the pollution due to putrescible organic matter is heavy.

The turbulence in the body of water helps in breaking the surface of the stream or lake, and helps in rapid re-aeration from the atmosphere. Thus, it helps in maintaining aerobic conditions in the river stream, and in keeping it clean. Too much of turbulence, however, is not desirable, because it scours the bottom sediment, increases the turbidity, and retards algae growth, which is useful in reaeration process. Wind and undercurrents in lakes and oceans cause turbulences which affect their self-purification.

The Hydrography affects the velocity and surface expanse of the river stream. High velocities cause turbulence and rapid reaeration, while large surface expanse (for the same cubic contents) will also have the same effects.

The larger the *amount of dissolved oxygen* present in water, the better and earlier the self-purification will occur.

The *amount and the type of organic matter and biological growth* present in water will also affect the rate of self-purification. Algae which absorbs carbon dioxide and gives out oxygen, is thus, very helpful in the self-purification process.

The *rate of reaeration i.e.* the rate at which the D.O. deficiency is replenished, will considerably govern the self-purification process. The greater is this rate, the quicker will be the self-purification, and there will be no chances of development of anaerobic conditions.

8.3.1. Zones of Pollution in a River-Stream. A polluted stream undergoing self-purification can be divided into the following four zones :

(*i*) *Zone of degradation* ;

(*ii*) *Zone of active decomposition* ;

(*iii*) *Zone of recovery* ; and

(*iv*) *Zone of cleaner* water.

These zones are discussed below :

(*i*) **Zonc of degradation or Zone of pollution.** This zone is found for a certain length just below the point where sewage is discharged into the river-stream. This zone is characterised by water becoming dark and turbid with formation of sludge deposits at the bottom. D.O. is reduced to about 40% of the saturation value*. There is an increase in carbon dioxide content; reoxygenation (*i.e.* re-aeration) occurs but is slower than de-oxygenation.

These conditions are unfavourable to the development of aquatic life ; and as such, algae dies out, but fish life may be present feeding on fresh organic matter. Moreover, certain typical bottom worms such as *Limondrilus* and *Tubifex* appear with *sewage fungi*, such as *sphaerotilusnatans*.

(*ii*) **Zone of active decomposition.** This zone is marked by heavy pollution. It is characterised by water becoming greyish and darker than in the previous zone. D.O. concentration falls down to zero, and anaerobic conditions may set in with the evolution of gases like methane, carbon dioxide, hydrogen sulphide, etc., bubbling to the surface, with masses of sludge forming an ugly scum layer at the surface. As the organic decomposition slackens due to stabilisation of organic matter, the re-aeration sets in and D.O. again rises to the original level (*i.e.* about 40%).

---

*Saturation value at 30°C = 7.63 mg/l. (Pl. see Appendix Tables A-3 given at the end of the book).

In this zone, bacteria flora will flourish. At the upper end, anaerobic bacteria will replace aerobic bacteria ; while at its lower end, the position will be reversed. Protozoa and fungi will first disappear and then reappear. Fish life will be absent. Algae and Tubifex will also mostly be absent. Larvae of *maggots* and *psychoda* (sewage fly) will, however, be present in all but the most septic sewage.

(*iii*) **Zone of recovery.** In this zone, the river stream tries to recover from its degraded condition to its former appearance. The water becomes clearer, and so the algae reappears while fungi decreases. B.O.D. falls down and D.O. content rises above 40% of the saturation value ; *protozoa, rotifers, crustaceans* and *large plants like sponges, bryozons*, etc. also reappear. Bottom organisms will include : tubifex, mussels, snails, etc. The organic material will be mineralised to form nitrates, sulphates, phosphates, carbonates, etc.

	← Zones of pollution →				
	Clear water	Zone of degradation	Zone of active decomposition	Zone of recovery	Zone of clearer water
Dissolved oxygen sag curve	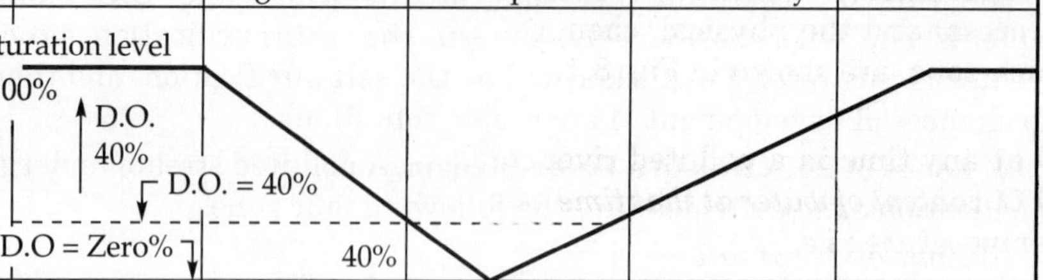				
Physical Indices	Clear water, no bottom sludge, no colour	Floating solids; bottom sludge present, colour getting turbid	Darker and greyish colour, evolution of gases like CH_4, CO_2 H_2S, etc. A lot of sludge coming to the surface forming an ugly scum layer at top	Turbid with bottom sludge	Clear water with no bottom sludge
Fish presence	Ordinary fish like game, pan, food & forage etc. present	Tolerant fishes like carp, buffalo, gary, etc. present	No fish present	Tolerant fish like carp, buff alo, etc. are present	Ordinary fish like game, pan, food, and forage, etc, present
Bottom Animals					
Algae & Protozoa etc. called plankton	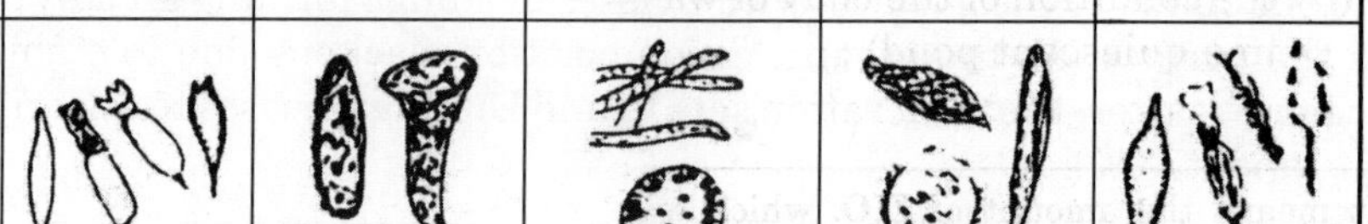				

Fig. 8.1. Showing Zones of Pollution along a River stream.

(*iv*) **Zone of cleaner water.** In this zone, the river attains its original conditions with D.O. rising upto the saturation value. Water becomes attractive in appearance, and Game fish (which requires atleast 4 to 5 mg/l of D.O.) and usual aquatic life prevails. Some pathogenic organisms may still, however, survive and remain present, which confirms the fact that "when once a river water has been polluted, it will not be safe to drink it, unless it is properly treated."

8.3.2. Indices of Self-Purification. The stage of self-purification process can be determined by the physical, chemical and biological analysis of the water. Colour and turbidity are the physical indices, while DO, BOD and suspended solids (SS) are the chemical indices which can mark the stages of purification. Moreover, the biological growth present in water can also indicate the stage of purification process, as different types of micro and macro organisms will exist in polluted water under different conditions, as discussed in the previous sub article.

The different zones of pollution (*i.e.* various stages in the self-purification process) and the physical, chemical and biological indices, characteristics of each zone, are shown in Fig. 8.1.

8.3.3. The Oxygen Deficit of a Polluted River-Stream. The oxygen deficit *D* at any time in a polluted river-stream is the *difference between the actual D.O. content of water at that time and the saturation D.O. content** at the water temperature ; *i.e.*

$$\left[\text{Oxygen deficit } (D) = \text{Saturation D.O.} - \text{Actual D.O.}\right] \quad ...(8.2)$$

In order to maintain clean conditions in a river-stream, the oxygen deficit must be nil, and this can be found out by knowing the rates of de-oxygenation and re-oxygenation.

De-oxygenation Curve. In a polluted stream, the D.O. content goes on reducing due to decomposition of volatile organic matter. The rate of de-oxygenation depends upon the amount of the organic matter remaining to be oxidised at the given time (*i.e.* L_t) as well as on the temperature of reaction (*i.e.* *t*). Hence, at a given temperature, *the curve showing depletion of D.O. with time, i.e.* deoxygenation curve (Refer curve 1 of Fig. 8.2) is similar to the first stage B.O.D. curve (Refer Fig. 7.3). It can also be expressed mathematically as per Eq. (7.12).

Re-oxygenation Curve. In order to counter-balance the consumption of D.O. due to de-oxygenation, atmosphere supplies oxygen to the water, and the process is called re-oxygenation. The rate at which the oxygen is supplied by the atmosphere to the polluted water depends upon :

(*i*) the depth of the receiving water (rate is more in a shallow depth) ;

(*ii*) the condition of the body of water (rate is more in a running stream than in a quiescent pond) ;

---

*It means the amount of D.O. which will be present in water if it is fully saturated with oxygen. The normal saturation D.O. value for fresh water varies between 14.6 mg/l to 7.6 mg/l for temperature varying between 0°C to 30°C. (Pl. see Appendix Table A-3) (*iii*) the saturation deficit or the oxygen deficit (*i.e.* the deficit of D.O. below the saturation value) ; and

(*iv*) the temperature of water.

Depending upon these factors, the rate of re-oxygenation can also be expressed mathematically and plotted in the form of a curve called re-oxygenation curve (Refer curve II—Fig. 8.2).

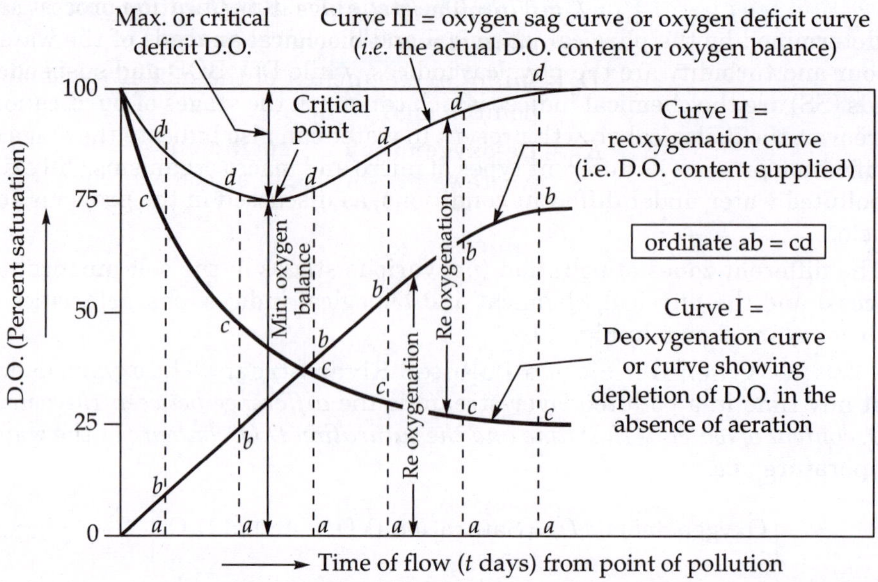

Fig. 8.2. Oxygen sag curve in a river receiving discharge of sewage.

Oxygen Deficit Curve. In a running polluted stream exposed to the atmosphere, the de-oxygenation as well as the re-oxygenation go hand in hand. If de-oxygenation is more rapid than the re-oxygenation, an oxygen deficit results.

[**Note.** If the D.O. content becomes zero, aerobic conditions will no longer be maintained and putrefaction will set in.]

The amount of resultant oxygen deficit can be obtained by algebraically adding the de-oxygenation and re-oxygenation curves (see curve III—Fig. 8.2). The resultant curve so obtained is called **the oxygen sag curve** or **the oxygen deficit curve.** From this curve, the oxygen deficit (D) and oxygen balance (*i.e.*, 100—D) per cent in a stream after a certain lapse of time, can be found out.

It can also be seen that when the de-oxygenation rate exceeds the re-oxygenation rate, the oxygen sag curve shows increasing deficit of oxygen ; but when both the rates become equal, the **critical point** is reached, and then finally when the rate of de-oxygenation falls below that of re-oxygenation, the oxygen deficit goes on decreasing till becoming zero.

The entire analysis of super-imposing the rates of de-oxygenation and re-oxygenation have been carried out mathematically, and the obtained results expressed in the form of famous **Streeter-Phelps equation** ; *i.e.*,

$$D_t = \frac{K_D \cdot L}{K_R - K_D}\left[(10)^{-K_D \cdot t} - (10)^{-K_R \cdot t}\right] + \left[D_0 \times (10)^{-K_R \cdot t}\right] \quad ...(8.3)$$

where D_t = the D.O. deficit in mg/l after t days.

L = *Ultimate first stage B.O.D. of the mix* at the point of waste discharge in mg/l.

D_0 = *Initial oxygen deficit* of the mix at the mixing point in mg/l.

K_D = **De-oxygenation coefficient** for the wastewater, which can be considered as equal to the *BOD rate constant** determined in the laboratory, through BOD tests performed at different times on BOD bottles, as discussed in article 7.4.1.9. Also, K_D varies with temp. as per Eq. (7.18), as :

$$K_{D(T)} = K_{D(20)}\left[1.047\right]^{T-20} \quad ...(i.e.\ \text{Eq. } 7.18)$$

The typical values of $K_{D(20)}$ vary between 0.1 to 0.2, generally taken as 0.1.

K_R** = **Re-oxygenation coefficient** for the stream. It can be determined by the field tests by using eqn. :

$$K_{R(20^\circ)} = \frac{3.9\sqrt{v}}{y^{1.5}} \quad ...(8.4)$$

where v = Av. stream velocity in m/s

y = Av. stream depth in m.

K_R varies with temperature as per the eqn. :

$$K_{R(T)} = K_{R(20)}\left[1.016\right]^{T-20^\circ} \quad ...(8.5)$$

where $K_{R(T)}$ is the K_R value at T°C and $K_{R(20)}$ is the K_R value at 20°C.

Typical values of $K_{R(20)}$ are given in Table 8.4.

---

*Deoxygenation coefficient for a river is, infact, a little more than the BOD rate constant, since BOD is exerted more rapidly in a river than that in a BOD bottle due to terbulent mixing in a river. However, for practical purposes, this value is taken to be the same, as determined in the laboratory.

**Certain authors use K_2 as the symbol for this constant.

Table 8.4. Values of Re-oxygenation Coefficient (K_R) at 20°C

S. No.	*Type of water body*	*Value of $K_{R(20)}$ per day*
1.	Small ponds and back waters	0.05—0.10
2.	Sluggish streams, large lakes and impounding reservoirs	0.10—0.15
3.	Large streams of low velocity	0.15—0.20
4.	Large streams of normal velocity	0.20—0.30
5.	Swift streams	0.30—0.50
6.	Rapids and waterfalls	Over 0.5

The oxygen deficit curve can be plotted easily with the help of Eq. (8.3), by using different values of t in days :

The critical time (t_c) after which the minimum dissolved oxygen occurs can be found by differentiating Eq. (8.3) and equating it to zero ; which on solving gives

$$t_c = \left[\frac{1}{K_R - K_D}\right] \log\left[\left\{\frac{K_D L - K_R D_0 + K_D D_0}{K_D L}\right\}\frac{K_R}{K_D}\right] \quad \text{...(8.6)}$$

and the critical or max. oxygen deficit is given as :

$$D_c = \frac{K_D L}{K_R}\left[10\right]^{-K_D . t_c} \quad \text{...(8.7)}$$

The constant $\frac{K_R}{K_D}$ is sometimes represented by f, called **self-purification constant**, the values of which are given in Table 8.5.

Table 8.5. Values of Self Purification Constant $\left(f = \frac{K_R}{K_D}\right)$

S. No.	*Type of water body*	*Value of f*
1.	Small ponds and back waters	0.5—1.0
2.	Sluggish streams, large lakes and Impounding reservoirs	1.0—1.5
3.	Large streams of low velocity	1.5—2.0
4.	Large streams of normal velocity	2.0—3.0
5.	Swift streams	3.0—5.0
6.	Rapids and waterfalls	Over 5.0

Using $\frac{K_R}{K_D}$ as f, the Eq. (8.6) becomes as

$$\boxed{t_c = \frac{1}{K_D(f-1)} \log\left[\left\{1 - (f-1)\frac{D_0}{L}\right\} f\right]} \quad \text{...(8.8)}$$

and equation (8.7) becomes

$$D_c = \frac{L}{f}\left[10\right]^{-K_D.t_c} \quad ...(8.9)$$

Taking log, we get

$$\log D_c = \log \frac{L}{f} - K_D \cdot t_c \quad ...(8.10)$$

Substituting the value of t_c from Eq. (8.8) in Eq. (8.10), we get

$$\log D_c = \log \frac{L}{f} - \frac{K_D \cdot 1}{K_D(f-1)} \log\left[f\left\{1-(f-1)\frac{D_0}{L}\right\}\right]$$

or

$$\log D_c = \log \frac{L}{f} - \frac{1}{(f-1)} \log\left[f\left\{1-(f-1)\frac{D_0}{L}\right\}\right]$$

or

$$(f-1)\left[\log \frac{L}{f} - \log D_c\right] = \log\left[f\left\{1-(f-1)\frac{D_0}{L}\right\}\right]$$

or

$$\log\left[\frac{\frac{L}{f}}{D_c}\right]^{f-1} = \log\left[f\left\{1-(f-1)\frac{D_0}{L}\right\}\right]$$

or

$$\boxed{\left(\frac{L}{D_c \cdot f}\right)^{f-1} = f\left[1-(f-1)\frac{D_0}{L}\right]} \quad ...(8.11)$$

This is the important first stage equation in which L is the B.O.D. of the mixture of sewage and stream, and f (K_D and K_R also) corresponds to the temperature of the mixture of sewage and stream at the outfall.

The above equations are of practical value in predicting the oxygen content at any point along a stream, and thus help us in estimating the degree of waste treatment required, or of the amount of dilution necessary, in order to maintain a certain D.O. in the stream.

A few examples showing the use of these eqns. have been solved below.

Example 8.1. *The sewage of a town is to be discharged into a river stream. The quantity of sewage produced per day is 8 million litres, and its B.O.D. is 250 mg/l. If the discharge in the river is 200 l/s and if its B.O.D. is 6 mg/l, find out the B.O.D. of the diluted water.*

Solution. Sewage discharge = Q_S

$$= \frac{8 \times 10^6}{24 \times 60 \times 60} \; l/s. = 92.59 \; l/s.$$

Discharge of the river = Q_R = 200 l/s

B.O.D. of sewage = C_S = 250 mg/l

B.O.D. of river = C_R = 6 mg/l.

Using equation (8.1), we have

B.O.D. of the diluted mixture

$$= C = \frac{C_S \cdot Q_S + C_R \cdot Q_R}{Q_S + Q_R}$$

or $$C = \frac{250 \times 92.59 + 6 \times 200}{92.59 + 200}$$

$$= 83.21 \text{ mg}/l. \quad \textbf{Ans.}$$

Example 8.2. *In the above example, what should be the river discharge, if it is desired to reduce the B.O.D. of diluted water to 20 mg/l.*

Solution. Here $C = 20$ mg/l

$\therefore$ $$20 = \frac{250 \times 92.59 + 6 \times Q_R}{92.59 + Q_R}$$

or $$Q_R = 1521 \; l/s. \quad \textbf{Ans.}$$

Example 8.3. *A city discharges 1500 litres per second of sewage into a stream whose minimum rate of flow is 6000 litres per second. The temperature of sewage as well as water is 20°C. The 5 day B.O.D. at 20°C for sewage is 200 mg/l and that of river water is 1 mg/l. The D.O. content of sewage is zero, and that of the stream is 90% of the saturation D.O. If the minimum D.O. to be maintained in the stream is 4.5 mg/l, find out the degree of sewage treatment required. Assume the de-oxygenation coefficient as 0.1, and re-oxygenation coefficient as 0.3.*

Solution. From the table given at the end of the book in Appendix A-3, the value of saturation D.O. at 20°C is found out as 9.17 mg/*l*.

D.O. content of the stream

= 90% of the saturation D.O.

$$= \frac{90}{100} \times 9.17 = 8.25 \text{ mg}/l.$$

D.O. of mix at the start point (*i.e.* at $t = 0$)

$$= \frac{8.25 \times 6000 + 0 \times 1500}{6000 + 1500} \qquad (\because \text{ D.O. of sewage is zero})$$

$$= 6.6 \text{ mg}/l$$

$\therefore$ D_0 = initial D.O. deficit

= [Saturation D.O. at mix. temp. – D.O. of mix.]

= 9.17 – 6.6 = **2.57 mg/l** (*Assume instantaneous mixing*)

Minimum D.O. to be maintained in the stream

= 4.5 mg/*l*.

$\therefore$ Max. permissible saturation deficit (*i.e.*, critical D.O. deficit)

$= D_c = 9.17 - 4.5$

= 4.67 mg/*l*.

Now, using equations (8.11), the first stage B.O.D. of mixture of sewage and stream (L) is given by

$$\left[\frac{L}{D_c f}\right]^{f-1} = f\left[1 - (f-1)\frac{D_0}{L}\right]$$

Substituting the values as :

$$D_0 = 2.57 \text{ mg/l} \quad \text{and} \quad D_c = 4.67 \text{ mg/l}$$

$$f = \frac{K_R}{K_D} = \frac{0.3}{0.1} = 3$$

we get
$$\left[\frac{L}{4.67 \times 3}\right]^{3-1} = 3\left[1-(3-1)\frac{2.57}{L}\right]$$

or
$$\left[\frac{L}{14.01}\right]^{2} = 3\left[1-\frac{5.14}{L}\right]$$

Solving by hit and trial, we get the value

$$L = 21.1 \text{ mg}/l$$

Now, using $Y_t = L\left[1-10^{-K_D \cdot t}\right]$, we have

Max. permissible 5 day B.O.D. of the mix (at 20°C)

$$Y_5 = 21.1\left[1-10^{-0.1\times 5}\right] \qquad \text{(where } K_D \text{ at 20°C} = 0.1)$$

$$= 14.43 \text{ mg}/l.$$

Now, using equation (8.1), we have

$$C = \frac{C_S Q_S + C_R Q_R}{Q_S + Q_R}$$

where C stands for concentrations of B.O.D.

Substituting the values, we get

$$14.43 = \frac{C_S \times 1500 + 1 \times 6000}{1500 + 6000}$$

where C_S will represent the permissible B.O.D.$_5$ (at 20°C of course) of the discharged wastewater.

Solving, we get

$$C_S = 68.16 \text{ mg}/l.$$

∴ Degree of treatment required (per cent)

$$= \frac{\text{Original B.O.D. of sewage} - \text{Permissible B.O.D.}}{\text{Original B.O.D.}} \times 100$$

$$= \frac{200 - 68.16}{200} = \frac{131.84}{200}$$

$$= 65.9\%. \quad \textbf{Ans.}$$

Example 8.4. *A city discharges 100 cumecs of sewage into a river, which is fully saturated with oxygen and flowing at the rate of 1500 cumecs during its lean days with a velocity of 0.1 m/sec. The 5-days BOD of sewage at the given temperature is 280 mg/l. Find when and where the critical D.O. deficit will occur in the downstream portion of the river, and what is its amount. Assume coefficient of purification of the stream (f) as 4.0, and coefficient of de-oxygenation (K_D) as 0.1.*

Solution. The initial D.O. of river

$$= \text{Saturation D.O. at the given temp.} = 9.2 \text{ mg}/l \text{ (say)}$$

D.O. of mix at $t = 0$ *i.e.*, at start

$$= \frac{9.2 \times 1500 + 0 \times 100}{1500 \times 100} \quad \textit{(assuming that D.O. of sewage is nil)}$$

$$= 8.62 \text{ mg}/l$$

Initial D.O. deficit of the stream

$$= D_0 = 9.2 - 8.62 = \mathbf{0.58 \ mg/l}$$

Also, 5-day BOD of the mixture of sewage and stream is given by

$$C = \frac{C_S Q_S + C_R Q_R}{Q_S + Q_R}$$

$$= \frac{280 \times 100 + 0 \times 1500}{100 + 1500}$$

$$= \frac{280 \times 100}{1600} = 17.5 \text{ mg/l.}$$

∴ 5 day BOD of mix at the given temp. $= Y_5 = 17.5$ mg/l

$$Y_5 = L\left[1 - (10)^{-K_D \times 5}\right] \text{ and } K_D = 0.1 \text{ (at 20°C)}$$

∴ The ultimate BOD of the mix (*i.e.* L)

$$= \frac{17.5}{0.684} = 25.58 \text{ mg}/l.$$

Now, using equation (8.11), we have

$$\left[\frac{L}{D_c \cdot f}\right]^{f-1} = f\left[1 - (f-1)\frac{D_0}{L}\right]$$

or

$$\left[\frac{25.58}{D_c \times 4}\right]^3 = 4\left[1 - \frac{3 \times 0.58}{25.58}\right]$$

or

$$D_c = 4.12 \text{ mg}/l. \quad \textbf{Ans.}$$

Now, from equation (8.8), we have

$$t_c = \frac{1}{K_D(f-1)} \log_{10}\left[f\left\{1 - (f-1)\frac{D_0}{L}\right\}\right]$$

or

$$t_c = \frac{1}{0.1(4-1)} \log_{10}\left[4 \times \left\{1 - \frac{3 \times 0.58}{25.58}\right\}\right]$$

$$= \frac{1}{0.3} \times 0.571 = \mathbf{1.905 \ days.}$$

Now, distance = Velocity of river × Travel time

= 0.1 m/sec × (1.905 × 24 × 60 × 60 sec)

= 16,460 m = **16.46 km**

Hence, the most critical deficit will occur after 1.905 days and at point 16.46 km downstream of the point of sewage disposal. **Ans.**

Example 8.5. *A town with a population of 30,000 has to design a sewage treatment plant to handle industrial as well as domestic wastewaters of the town. A sanitary survey revealed the following :*

Dairy wastes of 3 million litres per day with BOD of 1100 mg/l, and sugar mill waste of 2.4 million litres per day with BOD of 1500 mg/l are produced. In addition, domestic sewage is produced at the rate of 240 litres per capita per day. The per capita BOD of domestic sewage being 72 gm/day. An overall expansion factor of 10 per cent to be provided. The sewage effluents are to be discharged to a river stream with a minimum dry weather flow of 4500 litre per second and a saturation dissolved oxygen content of 9 mg/l. It is necessary to maintain a dissolved oxygen content of 4 mg/l in the stream. Determine the degree of treatment required to be given to the sewage. Assume suitable values of coefficients of de-oxygenation and re-oxygenation.

Solution. Per capita B.O.D. of the domestic sewage

$$= 72 \text{ gm/day} = 72 \times 1000 \text{ mg/day.}$$

The per capita sewage produced

$$= 240 \text{ litre/day.}$$

∴ B.O.D. per litre of the domestic sewage

$$= \frac{72 \times 1000}{240} \text{ mg/l.} = 300 \text{ mg/l.}$$

Amount of domestic wastewater produced per day

$$= 30{,}000 \times 240 \text{ litres} = 7.2 \text{ million litres}$$

Net B.O.D. of all wastewaters (*i.e.* domestic + industrial)

$$= \left[\frac{7.2 \times 300 + 3 \times 1100 + 2.4 \times 1500}{7.2 + 3 + 2.4}\right] = \mathbf{719\ mg/l.}$$

Total wastewater discharge

$$= \frac{\text{Vol. of wastewaters entering per day}}{\text{No. of secs in 1 day}}$$

$$= \frac{3 \text{ Ml} + 2.4 \text{ Ml} + 7.2 \text{ Ml}}{1 \times 24 \times 60 \times 60 \text{ sec}}$$

$$= \frac{12.6 \times 10^6}{24 \times 3600} \ l/s = 145.8 \ l/s$$

Total wastewater discharge with 10% expansion factor

$$= 1.1 \times 145.8 \ l/s = \mathbf{160\ l/s}$$

Initial D.O. of saturated stream water

$$= 9 \text{ mg}/l \ (\textit{i.e.} \text{ saturation D.O. as given})$$

∴ D.O. of mixture at $t = 0$ *i.e.*, at start point

$$= \frac{\text{D.O. of river} \times Q_R + \text{D.O. of sewage} \times Q_S}{Q_R + Q_S}$$

$$= \frac{9 \times 4500 + 0 \times 160}{4500 + 160}$$

(Assuming that the D.O. of wastewaters is nil)

$$= 8.69 \text{ mg}/l$$

∴ Initial D.O. deficit

$= D_0 = 9 - 8.69$ *(assuming instantaneous mixing)*

= 0.31 mg/l

Also, critical D.O. deficit, *i.e.* allowable max. D.O. deficit

$= D_c = 9 - 4.0$

= 5 mg/l

Now, using eqn. (8.11), we have

$$\left[\frac{L}{D_c \cdot f}\right]^{f-1} = f\left[1 - (f-1)\frac{D_0}{L}\right]$$

where $D_c = 5$ mg/l,
$D_0 = 0.31$ mg/l,
$K_D = 0.1$; $K_R = 0.3$; $f = 3$
(assumed values at mix. temp.)

$$\therefore \quad \left[\frac{L}{5 \times 3}\right]^2 = 3\left[1 - \frac{2 \times 0.31}{L}\right]$$

Solving by hit and trial,

L = **25.65 mg/l**

Max. permissible 5 day B.O.D. of mix at mix temp.

$$= Y_5 = L\left[1 - (10)^{-0.1 \times 5}\right]_{(K_D \text{ at mix temp. is assumed } = 0.1)}$$

$= 0.684\ L$

$= 0.684 \times 25.65$ = **17.54 mg/l**

Using eqn. (8.1) as

$$C = \frac{C_S Q_S + C_R Q_R}{Q_S + Q_R}$$

we get $$17.54 = \frac{C_S + 160 + 0 \times 4500}{160 + 4500}$$

where C_S = Max. permissible B.O.D.$_5$ of wastewaters

or C_S = 511 mg/l.

∴ Permissible B.O.D. of wastewaters

= 511 mg/l

Initial B.O.D. of city wastewaters

= 719 mg/l

∴ *Degree of treatment reqd.*

$$= \frac{719 - 511}{719} = \mathbf{28.93\%.} \quad \textbf{Ans.}$$

Example 8.6. *In the previous example, determine what should be the dilution ratio if no treatment was required, and thus determine the river discharge for such a condition.*

Solution. When no treatment is required, the value of max. permissible BOD_5 of wastewaters, *i.e.* C_S should be 719, Q_R can then be determined as :

$$17.54 = \frac{719 \times 160 + 0 \times Q_R}{160 + Q_R}$$

or $\quad 17.54\,[160 + Q_R] = 719 \times 160$

or $\quad 160 + Q_R = \dfrac{719 \times 160}{17.54} = 6559$

or $\quad Q_R = 6399\ l/s$ (say). **Ans.**

$$\text{Dilution ratio} = \frac{6399}{160} = 39.99 \text{ ; Say 40 times.}$$

Hence when the dilution ratio is 40 and the minimum river discharge is 6400 *l/s*, no treatment will be required. **Ans.**

[**Note.** Strictly speaking, when Q_R increases, D_0 will reduce, increasing Y_5, needing repeat of calculations, to obtain precise results. But the effect will be very small and on safer side, and hence is generally ignored.

Example 8.7. *A waste water effluent of 560 l/s with a BOD = 50 mg/l, DO = 3.0 mg/l and temperature of 23°C enters a river where the flow is 28 m³/sec, and BOD = 4.0 mg/l, DO = 8.2 mg/l, and temperature of 17°C. k_1 of the waste is 0.10 per day at 20°C. The vel. of water in the river downstream is 0.18 m/s and depth of 1.2 m. Determine the following after mixing of waste water with the river water :*

(i) Combined discharge ;

(ii) BOD ;

(iii) DO ; and

(iv) Temperature. *(Civil Services, 1981)*

Solution.

Particulars of Sewage thrown	*Particulars of River*
$Q_S = 560\ l/s = 0.56\ m^3/sec$	$Q_R = 28\ m^3/sec$
Concentrations (C_S)	**Concentrations (C_R)**
BOD = 50 mg/*l*	BOD = 4.0 mg/*l*
DO = 3.0 mg/*l*	DO = 8.2 mg/*l*
Temp. = 23°C	Temp. = 17°C.
k_1 at 20° = 0.1 per day	

(i) Combined discharge

$= Q_S + Q_R$

$= 0.56 + 28 =$ **28.56 m³/sec. Ans.**

Now, using equation (8.1), for conc. of mix as

$$C = \frac{C_S \cdot Q_S + C_R \cdot Q_R}{Q_S + Q_R}, \text{ we have}$$

(ii) BOD of mix

$$= \frac{50 \times 0.56 + 4.0 \times 28}{0.56 + 28} = \frac{140}{28.56} = \textbf{4.9 mg/l.} \quad \textbf{Ans.}$$

(iii) DO of mix

$$= \frac{3.0 \times 0.56 + 8.2 \times 28}{0.56 + 28} = \mathbf{8.098\ mg/l} \quad \textbf{Ans.}$$

(iv) Temp. of mix

$$= \frac{23 \times 0.56 + 17 \times 28}{0.56 + 28} = \mathbf{17.12°C.} \quad \textbf{Ans.}$$

Example 8.8. *125 cumecs of sewage of a city is discharged in a perennial river which is fully saturated with oxygen and flows at a minimum rate of 1600 cumecs with a minimum velocity of 0.12 m/sec. If the 5 day BOD of the sewage is 300 mg/l, find out where the critical DO will occur in the river. Assume :*

(i) the coefficient of purification of the river as 4.0,

(ii) the ultimate BOD as 125% of the 5 day BOD of the mixture of sewage and river water.

(iii) Saturation DO of river = 9.2 mg/l

Solution. *Saturation D.O. Concentration of the given river*

= **9.2.** (given in assumption *iii*)

The D.O. of the river at the mixing point after disposal of sewage (D)

$$= \frac{125 \times 0 + 1600 \times 9.2}{125 + 1600} = 8.53 \text{ mg/l}$$

Initial D.O. deficit $(D_0) = D_s - D$

= 9.2. – 8.53 = **0.67 mg/l.**

BOD_5 of the river at the mixing point after disposal of sewage (Y_5)

$$= \frac{125 \times 300 + 1600 \times 0}{125 + 1600} = \mathbf{21.74\ mg/l}$$

The *ultimate BOD of river (mix) at mixing point (L)*

= 125% BOD_5 [as per given in assumption (*ii*)]

= 1.25 × 21.74 = **27.17 mg/l.**

Now, using eqn. (7.16), we have

$$BOD_5 = L\left[1 - (10)^{-K_D \times 5}\right]$$

or $$21.74 = 27.17\left[1 - (10)^{-K_D \times 5}\right]$$

or $$0.8 = \left[1 - (10)^{-5K_D}\right]$$

or $$(10)^{-5K_D} = 0.20$$

or $$-5K_D \log 10 = \log 0.20$$

or $$K_D = \mathbf{0.14}.$$

Now, using eqn. (8.8) as

$$t_c = \frac{1}{K_D(f-1)} \log\left[\left\{1 - (f-1)\frac{D_0}{L}\right\} f\right], \text{ where } f = 4.0 \text{ (given)}$$

$$\therefore \quad t_c = \frac{1}{0.14(4-1)} \log\left[\left\{1 - (4-1)\frac{0.67}{27.17}\right\} 4\right] = \mathbf{1.354\ days.}$$

The distance along the river, where the critical D.O. deficit will occur

$= S =$ Velocity × Time

$= 0.12$ m/sec × (1.354 × 24 × 3600 sec)

= 14.04 km ; Say 14 km

Hence, critical D.O. deficit will occur at 14 km downstream of the sewage disposal point. **Ans.**

Example 8.9. *A wastewater treatment plant disposes of its effluents into a stream at a point A. Characteristics of the stream at a location fairly upstream of A and of the effluent are as below:*

Item	*Units*	*Effluent*	*Stream*
Flow	m^3/s	*0.20*	*0.50*
Dissolved oxygen	*mg/l*	*2.00*	*8.00*
Temperature	*°C*	*26*	*22*
BOD_5 at 20°C	*mg/l*	*40*	*3*

Assume that the deoxygenation constant K_1 at 20°C (base e) = 0.20 d^{-1} and the re-aeration constant K_2 at 20°C (base e) = 0.40 d^{-1} for the mixture. Equilibrium concentration of dissolved oxygen C_s for the fresh water is as follows:

Temperature °C	*18*	*20*	*22*	*23*	*24*	*25*	*26*
C_s (mg/l)	*9.54*	*9.17*	*8.99*	*8.83*	*8.53*	*8.38*	*8.22*

The velocity of the stream downstream of the point A is 0.2 m/s. Determine the critical oxygen deficit and its location.

[Use temperature coefficients of 1.04 for K_1 and 1.02 for K_2]

(Civil Services, 1989)

Solution. K_1 at 20°C (base *e*)

$= 0.2\ d^{-1} = 0.2$ per day

∴ K_D at 20°C (base 10)

$$= \frac{K_1}{2.3} = 0.434\ K_1$$

$= 0.434 \times 0.2$ per day = **0.087 per day.**

Similarly, K_R at 20°C

$= 0.434 \times 0.4\ d^{-1}$ = **0.174 per day.**

The formulas to be used in this question for converting K_D and K_R at any other temperature (T°C) will be

$$K_{D(T^\circ)} = K_{D(20^\circ)} \left[1.04\right]^{T^\circ - 20^\circ} \text{ ; and}$$

$$K_{R(T^\circ)} = K_{R(20^\circ)} \left[1.02\right]^{T^\circ - 20^\circ} \quad \text{(as per the given values)}$$

(*i*) *We will now determine DO, BOD and temperature of mixture as below:*

$$\text{DO of mixture} = \frac{\text{DO of sewage} \times Q_S + \text{DO of river} \times Q_R}{Q_S + Q_R}$$

$$= \frac{2 \times 0.20 + 8 \times 0.50}{0.20 + 0.50}$$

$$= \mathbf{6.29\ mg/l.}$$

BOD_5 of mixture
(i.e. 5 day BOD at 20°C)

$$= \frac{40 \times 0.20 + 3 \times 0.50}{0.20 + 0.50}$$

$$= \mathbf{13.57\ mg/l.}$$

Temperature of mixture

$$= \frac{26 \times 0.20 + 22 \times 0.50}{0.20 + 0.50} = \mathbf{23.14°C}$$

(ii) Ultimate BOD of mixture (L)

$$L = \frac{Y_5 (i.e.\ 5 \text{ day BOD of mixture at } 20°\text{C})}{1 - (10)^{-K_D \times 5}}$$

where K_D is at 20°C = 0.087 per day

$$= \frac{13.57}{1 - (10)^{-0.087 \times 5}} = \frac{13.57}{0.633} = \mathbf{21.45\ mg/l.}$$

(iii) Initial D.O. Deficit of mixture,

D.O. of mixture = 6.29 mg/*l*

Saturation D.O. at mixture temperature of 23.14°C
= 8.79 (interpolated from given values)

$\therefore \quad D_0$ = D.O. deficit
= 8.79 – 6.29 = **2.50 mg/l.**

(iv) Corrected values of K_D and K_R are :

$$K_{D(23.14°)} = K_{D(20°)} \left[1.04\right]^{T-20}$$

$$= 0.087\ [1.04]^{3.14} = \mathbf{0.098}$$

$$K_{R(23.14°)} = K_{R(20°)} \left[1.02\right]^{T-20}$$

$$= 0.174\ [1.02]^{3.14} = \mathbf{0.185}$$

(v) The time (t_c) after which critical D.O. deficit (D_c) occurs is given by eqn. (8.8) as

$$t_c = \frac{1}{K_D(f-1)} \log_{10} \left[\left\{ 1 - (f-1) \frac{D_0}{L} \right\} f \right]$$

where $K_R = 0.185$
$K_D = 0.098$

$$\therefore \quad f = \frac{K_R}{K_D} = \frac{0.185}{0.098} = 1.888$$

$$L = 21.45 \text{ mg}/l$$
$$D_0 = 2.5 \text{ mg}/l.$$

$$\therefore \quad t_c = \frac{1}{0.098(1.888-1)} \log_{10}\left[\left(1 - \frac{0.888 \times 2.5}{21.45}\right) 1.888\right]$$
$$= \frac{1}{0.098(0.888)} \times 0.228 = \mathbf{2.625 \text{ days.}}$$

(*vi*) Now, *Distance* = Velocity × Travel time

$$= 0.2 \text{ m/s} \times (2.625 \times 24 \times 60 \times 60 \text{ sec})$$
$$= \mathbf{45.36 \text{ km.}} \quad \textbf{Ans.}$$

(*vii*) D_c *is now given by eqn. (8.11) as*

$$\left(\frac{L}{D_c \cdot f}\right)^{f-1} = f\left(1 - (f-1)\frac{D_0}{L}\right)$$

$$\text{or} \quad \left(\frac{21.45}{D_c \times 1.888}\right)^{0.888} = 1.888\left(1 - \frac{0.888 \times 2.5}{21.45}\right)$$

$$\text{or} \quad \frac{21.45}{1.888\, D_c} = \left(1.692\right)^{\frac{1}{0.888}} = (1.692)^{1.126} = 1.808$$

$$\text{or} \quad D_c = \frac{21.45}{1.888 \times 1.808} = 6.28 \text{ mg}/l.$$

Hence, the critical D.O. deficit equal to 6.28 mg/l occurs at 45.36 km downstream of A, after 2.625 days. **Ans.**

Example 8.10. *A treated waste water is discharged at the rate of 1.5 m^3/sec into a river of minimum flow 5 m^3/sec. The temperature of river flow and waste water flow may be assumed as 25°C. The BOD removal rate constant K_1 is 0.12/d (base 10). The BOD_5 at 25°C of the waste water is 200 mg/l, and that of the river water upstream of the wastewater outfall is 1 mg/l. The efficiency of waste water treatment is 80%. Evaluate the following :*

(*i*) *BOD_5 at 25°C, if river water receives untreated waste water*

(*ii*) *BOD_5 at 25°C if river water receives treated waste water*

(*iii*) *ultimate BOD of the river water after it receives treated waste water.*

(*Civil Services, 1993*)

Solution.

Discharge of waste water $= Q_W = 1.5 \text{ m}^3/\text{s}$

Discharge of river $= Q_R = 5 \text{ m}^3/\text{s}$

Temperature $= T = 25°\text{C}$

$$K_{D(25°)} = K_1 = 0.12/d.$$

C_W = Conc. of BOD_5 for untreated waste water = 200 mg/l

C_R = Conc. of BOD_5 for river water = 1 mg/l

Using eqn. (8.1),

(*i*) Conc. of BOD_5 of the mixture if untreated waste water is discharged into the river

$$= C = \frac{C_W \cdot Q_W + C_R \cdot Q_R}{Q_W + Q_R} = \frac{200 \times 1.5 + 1 \times 5}{1.5 + 5}$$

$$= \textbf{46.92 mg/l.} \quad \textbf{Ans.}$$

(*ii*) *BOD_5 of the treated wastewater is given by*

$$C_{T.W.} = 20\% \text{ of the } BOD_5 \text{ of untreated wastewater}$$

($\because$ efficiency of wastewater treatment is 80%)

$$= 20\% \times C_W = 20\% \times 200 \text{ mg/l}$$

$$= 40 \text{ mg/l.}$$

BOD_5 of mixture if treated wastewater is discharged into the river

$$= C' = \frac{C_{T.W.} \cdot Q_W + C_R \cdot Q_R}{Q_W + Q_R}$$

$$= \frac{40 \times 1.5 + 1 \times 5}{1.5 + 5} = \textbf{10 mg/l.} \quad \textbf{Ans.}$$

(*iii*) BOD_5 of river water after it receives treated wastewater

$$= 10 \text{ mg/l (as computed above)}$$

Ultimate BOD of this mixture

$$= Y_u = L = ?$$

Using eqn. (7.16), we have

$$Y_{t_{(day)}} = L\left[1 - (10)^{-K_D \cdot t}\right]$$

or $$Y_5 = L\left[1 - (10)^{-0.12 \times 5}\right]$$

or $$10 = L\left[1 - (10)^{-0.6}\right]$$

or $$L = \textbf{13.35 mg/l.} \quad \textbf{Ans.}$$

8.4. Disposal of Wastewaters in Lakes and Management of Lake Waters

8.4.1. Lake Pollutants. Disposal of wastewaters in confined lakes or reservoirs, is much more harmful than its disposal in flowing streams and rivers. Water quality management in lakes in entirely different from that in rivers. It is infact the **phosphorous** (a nutrient largely contained in industrial as well as domestic waste waters), which seriously affects the water quality of lakes; and is hence considered as the **prime lake pollutant**. *Oxygen demanding wastes may be the other important lake pollutants.* The *toxic chemicals* from industrial waste waters may also sometimes very adversely effect some special classes of the lakes. However, phosphorous (a nutrient) constitute the most important lake pollutant, and needs special study in water quality management of lakes.

A study of the lake systems is essential to understand the role of phosphorus in lake pollution. The study of lakes is called **limnology**.

8.4.2. Stratification in Lakes. The water of a lake gets stratified during summers and winters, as discussed below :

During summer season, the surface water of a lake gets heated up by sunlight and warm air. This warm water being lighter, remains in upper layers near the surface, until mixed downward by turbulence from winds, waves, boats and other forces. Since such turbulence extends only to a limited depth from below the water surface, the top layers of water in the lake become well mixed and aerobic. This warmer, well mixed and aerobic depth of water is called ***epilimnion zone.*** The lower depth, which remains cooler, poorly mixed and anaerobic, is called the ***hypolimnion zone.*** There may also exist an intermediate zone or a dividing line, called ***thermocline or metalimnion***, as shown in Fig. 8.3 (*a*).

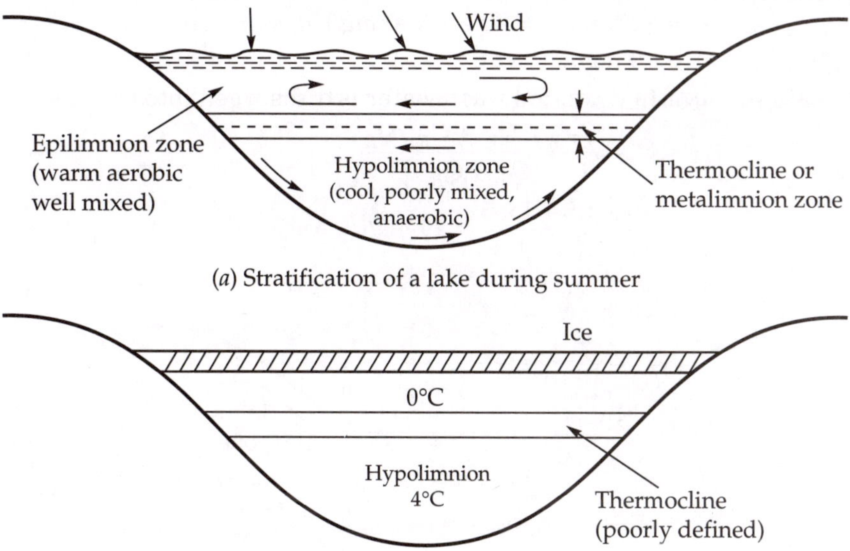

(*a*) Stratification of a lake during summer

(*b*) Stratification of a lake during winter

Fig. 8.3. Stratification of lakes.

The change from epilimnion to hypolimnion can be experienced while swimming in a lake. When you swim in top layers horizontally, you will feel the water warmer ; and if you dive deeper, you will find the water cooler. ***The change line will represent monocline.*** The depth of epilimnion zone depends upon the size of the lake for the same temperature changes. It may be as little as 1 m in small lakes and may be as large as 20 m or more in large lakes. This depth also depends upon the storm activity in the spring when stratification is developing. A major storm at the right time will mix the warmer water to a substantial depth and thus create a deeper epilimnion zone than its normal depth. *Once formed, lake stratification is very stable, and can only be broken by exceedingly violent storms.* As a matter of fact, as summer progresses, this stability increases, the epilimnion continues to warm, while the hypolimnion remains at a fairly constant temperature.

With the onset of winter season, the epilimnion cools, until it is more dense than the hypolimnion. The surface water then sinks, causing '**overturning**'. The water of the hypolimnion rises to the surface, where it cools and again sinks. *The lake, thus becomes completely mixed*, making it quite aerobic. In regions of freezing temperatures, when the temperature drops below 4°C, the above process of overturning (or *turn over*) stops, because water is most dense at this temperature. Further cooling or freezing of the water surface results in winter stratification, as shown in Fig. 8.3 (*b*).

With the passing of winters and commencement of spring season, the surface water again warms up and overturns, and lake becomes completely mixed. The lakes in regions of temperate climate will, therefore, have at least one, if not two, cycles of stratification and *turn-over* every year.

8.4.3. Biological Zones in Lakes. Lakes have been found to exhibit distinct zones of biological activity, largely determined by the availability of light and oxygen. The most important biological zones are :

(*i*) *Euphotic zone* ; (*ii*) *Littoral zone* ; and

(*iii*) *Benthic zone*.

These zones are shown in Fig. 8.4, and briefly discussed below :

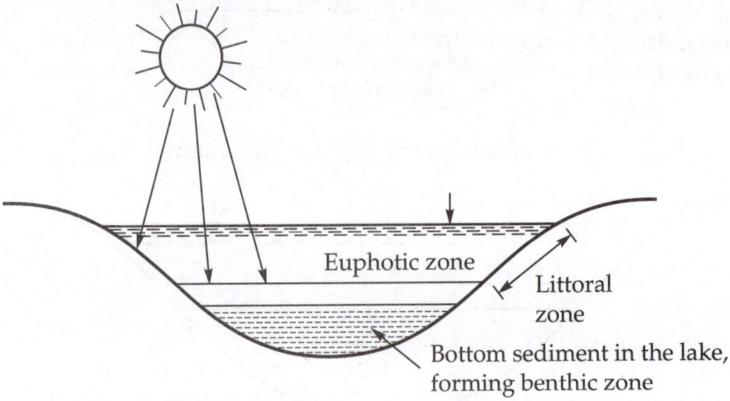

Fig. 8.4. Biological zones in a lake.

8.4.3.1. Euphotic zone. The upper layer of lake water through which sunlight can penetrate, is called the euphotic zone. All plant growth occurs in this zone. In deep water, algae grow as the most important plants, while rooted plants grow in shallow water near the shore.

The depth of the euphotic zone is reduced by the turbidity, which blocks sunlight penetration. In most lakes, the turbidity is due to algal growth ; although colour and suspended clays may substantially reduce sunlight penetration in some lakes. It is important to note that the bottom of euphotic zone only rarely coincides with the thermocline.

The depth of the euphotic zone can be approximated and measured by a simple device, called the **Secchi disk**, shown in Fig. 8.5. This disk is lowered into the lake water until the observer can no longer distinguish between the boundaries between its white and black quadrants. This depth, called the *Secchi disk depth*, does not correspond exactly to the depth of euphotic zone, but provides a good approximation of its extent. This Secchi disk depth also provides a measure of the aesthetic quality of water. *The greater is the secchi disk depth, the clearer is the water*.

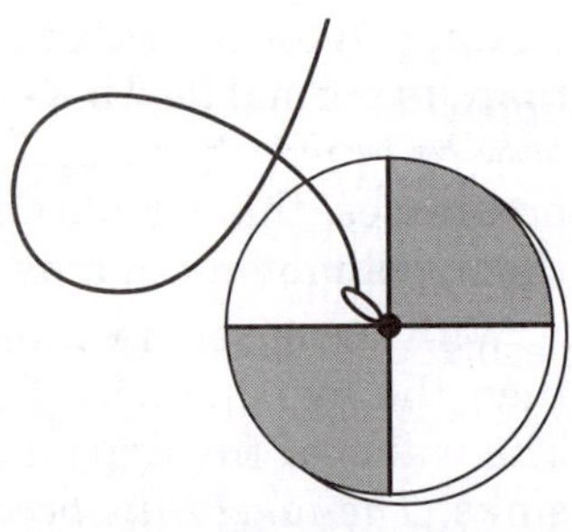

Fig. 8.5. Secchi disk.

8.4.3.2. Littoral zone. The shallow water near the shore, in which rooted plants grow, is called the littoral zone. The extent of the littoral zone depends on the slope of the lake bottom, and the depth of the euphotic zone. The littoral zone cannot extend deeper than the euphotic zone, as shown in Fig. 8.4.

8.4.3.3. Benthic zone. The bottom sediments in a lake comprises what is called the benthic zone. As the organisms living in the overlying water die, they settle down to the bottom, where they are decomposed by the organisms living in the benthic zone. Bacteria are always present in this zone. The presence of higher life forms, such as worms, insects and crustaceans however, depends upon the availability of oxygen at the lake bottom.

8.4.4. Productivity of a Lake. The productivity of a lake is defined as a measure of its ability to support a food chain. Since the algae forms the base of this food chain, which is required by the other forms of living organisms to thrive, its presence measures the lake productivity. Although, more productive lakes will have a higher fish population, yet since such a lake will have to support heavier algal growth, its water quality will be reduced, because of the undesirable changes that occur as algal growth increases. Moreover, due to reduced water quality, the most desirable fish which flourish in better quality waters, will be lost.

A lake's productivity level may, therefore, be determined by measuring the amount of algal growth that can be supported by the available nutrients. This productivity level of a lake is thus, reflective of the water quality of the lake. *As the productivity of a lake increases, its water quality reduces*. Because of the important role productivity plays in determining water quality, it forms a basis for classifying lakes.

Depending upon the increasing level of its productivity, the lakes may be classified as :

(*i*) *Oligotrophic lakes* ; (*ii*) *Mesotrophic lakes* ;

(*iii*) *Eutrophic lakes* ; and (*iv*) *Senescent lakes*.

There four types are discussed below :

8.4.4.1. Oligotrophic lakes. Oligotrophic lakes have a low level of productivity due to a severely limited supply of nutrients to support algal growth. The water of such a lake is, therefore, clear enough as to make its bottom visible upto considerable depth. In such a case, the euphotic zone often extends into the hypolimnion, which is aerobic. Oligotrophic lakes, therefore, support cold water game fish. An important example of such a lake is offered by the *Tahoe lake* on California-Nevada border in USA.

8.4.4.2. Mesotrophic lakes. The lakes having medium productivity levels, with medium growth of algae and turbidity, are usually classified as mesotrophic lakes. In such a lake, *although substantial depletion of oxygen may occur in the hypolimnion, yet it remains aerobic*.

8.4.4.3. Eutrophic lakes. Eutrophic lakes do have a high level of productivity, because of an abundant supply of algal nutrients. The flourishing growth of algae make the lake water to be highly turbid, so that the euphotic zone may extend only partially into the epilimnion. As the algae die, they settle to the lake bottom, where they are decomposed by benthic organisms. In a eutrophic lake, this decomposition is sufficient to deplete the hypolimnion of oxygen during summer stratification.

Because the hypolimnion is anaerobic during summer, such a lake will only support warm water fish. As a matter of fact, all types of cold water fish are driven out of the lake, before the hypolimnion becomes anaerobic, because they generally require dissolved oxygen levels of at least 5 mg/L. Highly eutrophic lakes may also have large mats of floating algae that typically impart unpleasant tastes and odours to the water.

8.4.4.4. Senescent lakes. These are very old shallow lakes, having thick organic sediment deposits at their bottoms. Rooted water plants abundantly grow in such shallow ponds, which ultimately become *marshes*.

8.4.5. Eutrophication of Lakes. Eutrophication is a natural process under which lakes get infested with algae and silt up gradually to become shallower and more productive through the entry and cycling of nutrients like carbon, nitrogen and phosphorus. The initially clear water of oligotrophic lakes, therefore, gradually turns into mesotrophic, eutrophic, and senescent stages, due to continuous entry of silt and nutrients.

This natural process of eutrophication infact can always get its carbon and nitrogen requirements from the atmospheric gases like CO_2 and NO_2 ; while the requirement of phosphorus is met by its presence in natural run off due to disintegration of rocks, which produce phosphorus. The increased phosphorus in lake water, entering either through the agricultural use of its drainage area or through the entry of domestic and industrial waste waters, will cause accelerated eutrophication of lakes, and is called the **cultural eutrophication.** The natural process of eutrophication thus, gets intensified by the entry of wastewater discharge into the lakes, causing permanent damage to its water quality and leading to its siltation. *Eutrophic lakes are, however, not necessarily polluted, but pollution contributes to eutrophication.*

The water quality management for a lake, therefore, aims at reducing its eutrophication to atleast at its natural level, by controlling and reducing the input of phosphorus in the lake water. Once the input of phosphorus, (which directly controls the production of *chlorophyll* and hence the algal development) is reduced, the phosphorus concentration will gradually fall down, as the existing phosphorous will get buried into the sediment or is flushed out of the lake along with the excess river flow feeding the lake.

Strategies suggested for reducing eutrophication by removal of phosphorus by its precipitation by addition of lime to the lake water ; or by dredging out the phosphorus rich sediment from the lake bottom, will not succeed until the entry of phosphorus is not curtailed. It, therefore, becomes imperative to stop the entry of sewage and industrial wastewaters, which largely contain phosphorus, into the lakes. Even the treated sewage will not be low in phosphorus, since phosphorus can be removed only by costly advanced methods of waste treatment. *The lake waters should, therefore, not be used even for discharge of treated sewage*. The phosphorus content of the domestic sewage can, however, be reduced by banning the use of phosphorus rich polyphosphates in detergents, which presently contribute heavy input of phosphorus in domestic sewage, as large as twice that contributed by human excreta. Several advanced countries have, therefore, banned the use of phosphates in detergents and soaps. India has yet to follow suit.

The continuous entry of seeping septic tank effluents from cottages and houses built adjoining the lakes, through the sub-soil towards the lake, will also cause phosphorus pollution in the lake, after the soil gets too saturated to absorb any further phosphorus, finally passing it on to the lake. The time it takes for phosphorus to break through to the lake depends on the type of soil, the distance to the lake, the amount of wastewater generated, and the concentration of phosphorus in the seeping wastewater. Entry of phosphorus through such sources should also be controlled to reduce cultural eutrophication of lakes, by collecting the effluents of septic tanks in sewers, to be carried up to the treatment plants, before disposing them off, safely.

The use of fertilizers in fields in the drainage area should also be controlled to reduce the entry of phosphorus through the surface run off flowing over such fields and finally entering the lakes. Treatment of such catchments to reduce soil erosion will also help in reducing phosphorous entry into the surface runoff.

8.5. Disposal of Wastewaters in Sea Water

Sea water normally contains 20% less oxygen than that contained in fresh water of a river stream. Moreover, sea water normally contains a large amount of dissolved matter. As such, the capacity of sea water to absorb sewage solids is not as high as that of fresh water of a stream. Moreover, sewage solids, when thrown into sea water, chemically react with the dissolved matter of sea water, resulting in precipitating some of the sewage solids, *giving a milky appearance to the sea water and forming* **sludge banks.** These sludge banks are undesirable, as they are likely to produce offensive hydrogen sulphide gas by reacting with the sulphate rich water of the sea.

As pointed out earlier, the oxygen content of sea water is less than that of fresh water, and also, its re-aeration is slower. However, since the sea contains too larger volumes of water, most of these deficiencies get removed, provided the sewage is taken deep into the sea and away from the coast line.

Since the specific gravity of sea water is greater than that of sewage, and temperature of sea water is lower than that of sewage, the lighter and the warmer sewage will rise up to the surface when thrown into the sea water. This will result in spreading of the sewage at the top surface of sea in a thin film or 'sleek'.

To prevent the backing up and spreading of sewage on the sea shore, the sewage should preferably be disposed off during low tides. Large sized tanks may, therefore, sometimes, be constructed to hold the sewage during high tides. Provision of a large sized sewer, grated with a non-return valve at the outfall end, is also an alternative to hold the sewage during high tides.

In all, the following points should be kept in mind while discharging sewage into the sea to avoid marine pollution :

(*i*) The sewage should be discharged in deep sea only.

(*ii*) In order to mix the sewage properly with sea water, the sewage should be released at a minimum depth of 3 to 5 metres below the water level, and by taking it sufficiently inside (about $1\frac{1}{2}$ km) from the shore line. This measure will prevent the sewage from accumulating on the shore, and thus preventing nuisance to baths and recreation centres on the shores.

(*iii*) Before deciding the position of outfall point, the sea currents, wind direction, velocity, etc. should be thoroughly studied. The point of disposal should be such that the sewage is taken away from the shore by the winds, and not brought back near to the shore.

(*iv*) The outfall sewer should be placed on a firm rocky foundation, and encased in thick stone masonry, so as to properly protect it from wave action, floating debris, etc.

The discharge of wastewaters into the sea should, however be controlled in respect of the quality of the effluents, by adhering to the prescribed standards. The standards prescribed by BIS under IS : 1968—1976 are shown in Table 8.6.

Table 8.6. BIS (ISI) Standards for Wastewater Effluents to be Discharged into Marine Coasts

S. No.	*Constituent Pollutant contained in the Wastewater Effluent*	*Tolerance Limit*
(1)	(2)	(3)
1.	BOD_5	100 mg/l
2.	COD	250 mg/l
3.	pH value	5.5 to 9.0
4.	Total suspended solids	100 mg/l
5.	Oil and grease	20 mg/l
6.	Fluorides (as F)	15 mg/l
7.	Ammoniacal Nitrogen (as N)	50 mg/l

Note. In addition to these major constituents, limits are *prescribed for many trace metals and non-metals, as given in Table 8.3 (col. 5).*

The standards prescribed for discharge of polluted waste waters into the marine coasts by GoI under Environmental (Protection) Rules, 1986 have already been given in (col. 5) of Table 8.3.

An **Indian example of discharge of city's sewage into the sea** can be seen in Mumbai city, where the *Love Grove Waste Water Treatment Facility Terminal* at Worli Mumbai, is disposing of the partially treated sewage at 3.5 km distance inside the sea, via. an underground *marine outfall* tunnel, at 65 m below the ground level, after removing the floating material and inorganic solids through the screens and aerated grit chambers. This STP (Sewage Treatment Plant) facility is designed for treating a peak flow of 1820 MLD. The sewage discharge from the area comprising of Church Gate on South to Dadar on North, Mumbai Harbour, Ray Road, Seweri on the East and Arabian Sea on the West, flows upto this STP site, and is lifted, at the Influent Pumping Station (IPS) containing seven duty pumps (with one additional standby unit), each having a designed capacity of 260 MLD each for a designed head of 12.5 m. Before being lifted, the sewage is made to pass through 2 No *trash racks* (size of openings = 150 mm) and 4 No. *coarse screens* (size of openings = 100 mm) installed upstream of the IPS. The lifted sewage is then passed through 8 No. *fine screens* (size of openings = 20 mm), which discharge the sewage to 4 No. *aerated grit chambers* of size 34.5 m × 7 m × 5.1 m. The grit chambers are designed as spiral flow aeration tanks, provided with air diffusion tubes placed on one side of the tank. The compressed air is discharged at a pressure of 70 kN/m^2 @ 1500 m^3/hr. The compressed air and the geometry of the tank induces a "spiral roll" that provides centrifugal force to assist in the removal of inorganic materials of size 0.2 mm in diameter or larger, with a sp. gravity of 2.65, thereby keeping organic matter of lower sp. gravity in suspension. The roll also "sweeps" the settled grit into a bottom hopper or collection trough, from where it is periodically removed.

The outflowing screened and degritted sewage from the grit chambers flows to the sea through *marine outfall conduit*, normally by by-passing the *Effluent Pumping Station (EPS)*. At higher discharges from the IPS, sewage from the grit chambers spills over to the EPS, containing 5 centrifugal pumps of 520 MLD capacity each at 5.8 m head. As the water level rises in the EPS wet well, the pumps at the EPS are turned on. Once the EPS is discharging into the channel leading to *marine outfall*, the flow in the by-pass channel will backup, closing the by-pass flap gates. All flow will then be pumped through the EPS.

The sewage coming out of the EPS channel or the by-pass channel now enters into a *drop shaft* on shore, from which it flows into a 3 km long underground R.C.C. circular tunnel of 3.5 m diameter, constructed at an upward slope under the sea bed at a depth of about 65 m below the ground level. This underground tunnel is named as the *marine outfall*.

The outfalling 240 m end length of this tunnel is provided with 10 No riser pipes, each of 1 m diameter, capped just above the sea bottom with circular caps containing horizontal outlet ports, the elevation of which is about 1 m above the sea bottom above the layer of sediments suspended by waves. There are 10 outlet ports of size varying from 220 mm diameter to 280 mm diameter on each riser. As such, there are 100 outlets on 10 risers. They are at a depth of about 9 m from the mean sea level to maximise dilution in initial discharges, as they rise to the surface. At this selected discharge point in the sea, the direction of currents is such that the plume of sewage is taken away from the shore side to towards the deep sea side, and spreading the sewage on a wide area.

DISPOSAL ON LAND

8.6. Disposal of Sewage Effluents on Land for Irrigation

In this method, the sewage effluent (treated or diluted) is generally disposed of by applying it on land. The percolating water may either join the water-table, or is collected below by a system of under-drains. This method can then be used for irrigating crops.

This method, in addition to disposing of the sewage, may help in increasing crop yields (by 33% or so) as the sewage generally contains a lot of fertilising minerals and other elements*. However, the sewage effluent before being

*Fertilising elements like nitrogen, phosphate, potash, etc. do help in increasing the yields of cash crops like cotton, sugarcane, plantain, etc. However, crops to be eaten raw or which do not have any skin to be removed before eating, should not be grown on sewage farms, although this is seldom practised by the farmers.

used as irrigation water, must be made safe. In order to lay down the limiting standards for sewage effluents, and the degree of treatment required, it is necessary to study as to what happens when sewage is applied on to the land as irrigation water.

When raw or partly treated sewage is applied on to the land, a part of it evaporates, and the remaining portion percolates through the ground soil. While percolating through the soil, the suspended particles present in the sewage are caught in the soil voids. If proper aeration of these voids is maintained, the organic sewage solids caught in these voids get oxidised by aerobic process. Such aeration and aerobic conditions will more likely prevail, if the soil is sufficiently porous and permeable (such as sands and porous loams). However, if the land is made up of heavy, sticky and fine grained materials (such as clay, rock, etc.), the void spaces will soon get choked up, and thus resulting in non-aeration of these voids. This will lead to the developing of non-aerobic decomposition of organic matter, and evolution of foul gases. Moreover, excessive clogging may also result in ugly ponding of sewage over the farm land, where mosquitoes may breed in large number, causing further nuisance.

Application of too strong or too heavy load of sewage will also similarly result in the quick formation of anaerobic conditions. The greater is the sewage load, more likely it will be for the soil to get clogged. Hence, if the sewage load is reduced either by diluting it or by pre-treating it, it may be possible to avoid the clogging of the soil pores. The degree of treatment required will, however, considerably depend upon the type of the soil of the land. *If this soil, to be irrigated, is sandy and porous, the sewage effluents may contain more solids and other wastes, and thus requiring lesser treatment, as compared to the case where the soil is less porous and sticky*.

The pretreatment process may be adopted by larger cities which can afford to conduct treatment of sewage ; whereas the dilution technique may be adopted by smaller cities. When sewage is diluted with water for disposal for irrigation, too large volumes of dilution water are generally not needed, so as not to require too large areas for disposal. The extent of land area required for disposing a certain volume of sewage effluent can be worked out from the values given in Table 8.7.

Table 8.7. Recommended Doses* for Sewage Farming in India

Type of soil	*Doses of sewage in cubic metres per hectare per day*	
	Raw sewage	*Settled sewage*
Sandy	120—150	220—280
Sandy loam	90—100	170—220
Loam	60—80	110—170
Clayey loam	40—50	60—110
Clayey	30—45	30—60

*These quantities will be governed by the requirements of crops grown. These quantities can also be increased by laying a system of under-drains, and thus increasing the rate of percolation. The land area will, however, then act like a crude intermittent sand filter. But in such a case, the effluents from the under-drains will then also have to be discharged some where.

Example 8.11. *A town having population of 40,000 disposes sewage by land treatment. It gets a per capita assured water supply from waterworks at a rate of 130 l/day. Assuming that the land used for sewage disposal can absorb 80 m^3 of sewage per hectare per day, determine the land area required, and its cost at the rate of ₹ 25,000 per hectare. Make suitable assumptions where needed.*

Solution.

Population = 40,000

Rate of water supply = 130 l/day/person

Total water supplied per day

= 40,000 × 130 l

= 52,00,000 lites = 5,200 cu. m.

Assuming that 80% of this water appears as sewage, we have

The quantity of sewage produced per day

= 0.8 × 5200 = 4160 cum.

∴ Area of land reqd. for disposing sewage

$$= \frac{4160}{80} = 52 \text{ hectares.}$$

Providing 50% extra land for rest and rotation, we have

The total land area required

= 1.5 × 52 = 78 hectares. **Ans.**

Cost of land involved

= ₹ 25,000 × 78 = ₹ 19,50,000. **Ans.**

Example 8.12. *A town disposes sewage by land treatment. It has a sewage farm of area 150 hectares. The area included an extra provision of 50% for rest and rotation. The population of the town being 50,000 and rate of water supply 140 litres per capita per day. If 75% of the water is converted into sewage, determine the consuming capacity of the soil.*

Solution. Quantity of water produced per day

= 50,000 × 140 litres/day

= 70,00,000 l/day = 7,000 cu-m/day.

Quantity of sewage produced

= 0.75 × 7000 = 5,250 cu-m/day.

Area of farm land provided

= 150 hectares including 50% additional reserve for rest and rotation.

∴ Area provided for immediate need

$$= \frac{150}{1.5} = 100 \text{ hectares.}$$

100 hectares is capable of passing 5250 cum per day.

∴ Consuming capacity of soil

$$= \frac{5250}{100}$$

= 52.5 cu-m per hectare per day. **Ans.**

8.7. Quality Standards For Wastewater Effluents to be Discharged on Land For Irrigation

The Bureau of Indian Standards (BIS), previously known as Indian Standard Institution (ISI), has vide its Code IS : 3307-1965, laid down the tolerance limits for various polluting characteristics/constituents of wastewater effluents for their discharge on land for irrigation. These standards are based upon the quality of irrigation water required by the crops, and thus limit the concentrations of pollutants contained in sewage or industrial liquid wastes, which may prove harmful to the crops. The prescribed BIS standards are shown in table 8.8.

Table 8.8. BIS (ISI) Standards of Wastewater Effluents to be Discharged on Land for Irrigation

S. No.	*Characteristic/constituent of Effluent wastewater*	*Tolerance limit as per IS : 3307-1965*
(1)	(2)	(3)
1.	BOD_5 at 20°C	500 mg/l*
2.	pH value	5.5 to 9.0
3.	Total Dissolved Solids (TDS)	2100 mg/l
4.	Oil and grease	30 mg/l
5.	Chlorides (as Cl)	600 mg/l
6.	Boron	2 mg/l
7.	Sulphates	1000 mg/l
8.	Percentage of Sodium with respect to total content of Sodium, Calcium, Magnesium and Potassium	60%
9.	Radioactive Materials	
	(*i*) α-emitters	10^{-9} μC/ml
	(*ii*) β-emitters	110^{-8} μC/ml

In order to make them legally enforceable, GOI has notified the standards of polluted effluents for discharge on land (for irrigation) under Environment (Protection) Rules 1986. These notified standards are shown in Table 8.9.

Table 8.9. Standards for Discharge of Polluted Effluents on Land for Irrigation under Environment (Protection) Rules, 1986

S. No.	*Characteristics of effluent i.e. pollutant of waste water*	*Standards under Environment (Protection) Rules, 1986*
(1)	(2)	(3)
1.	Colour and odour	All efforts should be made to remove colour and unpleasant odour as far as practicable
2.	BOD_5 at 20°C	100 mg/l

(*Contd.*)

*Law enforcing Pollution Control Boards are generally following an upper limit for BOD_5 of 100 mg/l, as shown in Table 8.9.

(1)	(2)	(3)
3.	Suspended Solids (SS)	200 mg/l
4.	pH Value	5.5 to 9.0
5.	Oil and grease	10 mg/l
6.	Arsenic (As)	0.2 mg/l
7.	Cyanide (as CN)	0.2 mg/l
8.	*Radioactive materials*	
	(*a*) Alpha emitters	10^{-8} μC/ml
	(*b*) Beta emitters	10^{-7} μC/ml
9.	Bio-assay test	90% survival of fish after 96 hours in 100% effluent
10.	*Pesticides*	
	(*i*) Benzene hexachloride	10 μg/l
	(*ii*) Carbaryl	10 μg/l
	(*iii*) DDT	10 μg/l
	(*iv*) Endosulfan	10 μg/l
	(*v*) Diamethoate	450 μg/l
	(*vi*) Penitrothion	10 μg/l
	(*vii*) Malathion	10 μg/l
	(*viii*) Phorate	10 μg/l
	(*ix*) Methyl parathion	10 μg/l
	(*x*) Phenthoate	10 μg/l
	(*xi*) Pyrethrums	10 μg/l
	(*xii*) Copper oxychloride	9600 μg/l
	(*xiii*) Copper sulphate	50 μg/l
	(*xiv*) Ziram	1000 μg/l
	(*xv*) Sulphur	30 μg/l
	(*xvi*) Paraouat	2300 μg/l
	(*xvii*) Proponil	7300 μg/l
	(*xviii*) Nitrogen	780 μg/l

The effluent irrigation method for disposal of sewage can be **favourably adopted under the following conditions** :

(*i*) When some natural rivers or water courses are not located in the vicinity, the land treatment is the only alternative left, and has to be adopted.

(*ii*) When irrigation water is scarcely available, the use of sewage for irrigating crops is a good alternative.

(*iii*) When large areas of open land are available, broad irrigation may be practised over it with the help of sewage effluents, and good returns can be earned by raising cash crops. Crops like wheat, cotton, sugarcane, plantain, grasses, fodder, coconut, orange trees. etc. have been successfully grown with advantage on sewage farms.

(*iv*) The method of effluent irrigation will prove useful in areas of low rainfall, as this will help in maintaining good absorption capacity of the soil.

(*v*) The area for land treatment or sewage farming should preferably be porous, such as sandy, loamy, or alluvial soils, or soft moorum. It should not be made of heavy retentive soils like clay, etc., which prevent easy aeration of the soil voids, and thus creating anaerobic conditions.

(*vi*) This method of disposal of sewage, poses problems during the periods when no irrigation water is required for the crops especially during rains. This method is, therefore preferred when sewage can be diverted to some river streams (flowing high during rainy season).

(*vii*) This method is preferred in areas of low water-table, where rate of percolation may be quite high.

8.8. 'Effluent Irrigation' and 'Sewage Farming'—Difference Thereof

Although, outwardly, both these terms are used as synonyms to each other, yet there is one basic difference between them. This difference is that : *in 'effluent irrigation' (or broad irrigation), the chief consideration is the successful disposal of sewage, while in 'sewage farming', the chief consideration is the successful growing of the crops.*

Hence, in *broad irrigation*, the raw or settled sewage is discharged on a vacant land, which is provided underneath, with a system of properly laid under-drains. These under-drains, usually, consist of 15 to 20 cm dia porous tile pipes, laid open jointed at a spacing of 12 to 30 m. The effluent collected in these drains after getting filtered through the soil pores is generally small (as a large quantity gets evaporated) and well stabilised, and can be easily disposed into some natural water courses, without any further treatment.

In case of *sewage farming*, however, the stress is laid upon the use of sewage effluents for irrigating crops and increasing the fertility of the soil. The pre-treatment of sewage, in removing the ingredients which may prove harmful and toxic to the plants is, therefore, necessary in this case.

However, in general, for all practical purposes, both these terms are used as synonyms, and both means : use of sewage effluents for irrigating crops.

8.9. Methods of Applying Sewage Effluents to Farms

The sewage effluents can be used for irrigating farms exactly in the same manner as irrigation water is used for farming. The various techniques that are employed for irrigating crops are :

(1) *Surface Irrigation called Broad Irrigation*. In this method, sewage is applied in different ways, on to the surface of the land. Depending upon the mode of application, it can be of different types, such as :

(*i*) Free flooding ;
(*ii*) Border flooding ;
(*iii*) Check flooding ;
(*iv*) Basin flooding ; and
(*v*) Furrow irrigation method.

(2) *Sub-surface Irrigation*. In this method, sewage is supplied directly to the root zone of crops, through a system of properly laid open-jointed pipes. Sewage, as it flows through these pipes, percolates through the open joints, and is distributed in the surrounding area by the action of capillarity.

(3) *Sprinkler or Spray Irrigation*. In this method, sewage is spread over the soil through nozzles, which are fitted at the tips of pipes carrying sewage

under pressure. The process, being costly, is not preferred in India, although it gives very good results, like those of a natural rainfall.

All these methods of irrigation have been thoroughly discussed in "Irrigation Engineering and Hydraulic Structures" (Chapter 1) by the same author ; and can be referred to in special needs.

8.10. Sewage Sickness

When sewage is applied continuously on a piece of land, the soil pores or voids may get filled up and clogged with sewage solids retained in them. The time taken for such a clogging will, of course, depend upon the type of soil* and the load present in sewage. But when once these voids are clogged, free circulation of air will be prevented, and anaerobic conditions will develop within the pores. Due to this, the aerobic decomposition of organic matter will stop, and anaerobic decomposition will start. The organic matter will thus, of course, be mineralised, but with the evolution of foul gases like hydrogen sulphide, carbon dioxide and methane. This phenomenon, of soil getting clogged, is known as *sewage sickness of land* in question.

In order to prevent the sewage sickness of a land, the following **preventive measures** may be adopted :

(*i*) *Primary Treatment of Sewage*. The sewage should be disposed of, only after primary treatment, such as screening, grit removal, and sedimentation. This will help in removing settleable solids and reducing the B.O.D. load by 30% or so ; and as such, soil pores will not get clogged, quickly.

(*ii*) *Choice of Land*. The piece of land used for sewage disposal should normally be sandy or loamy. Clayey lands should be avoided.

(*iii*) *Under-drainage of Soil*. The land on which the sewage is being disposed of, can be better drained, if a system of under-drains (*i.e.* open joined pipes) is laid below, to collect the effluent ; and this will also minimise the possibilities of sewage sickness.

(*iv*) *Giving Rest to the Land*. The land being used for disposal should be given rest, periodically, by keeping some extra land as reserve and standby for diverting the sewage during the period the first land is at rest. Moreover, during the rest period, the land should be thoroughly ploughed, so that it gets broken up and aerated.

(*v*) *Rotation of Crops*. Sewage sickness can be reduced by planting different crops in rotation instead of growing single type of a crop. This will help in utilising the fertilising elements of sewage and help in aeration of soil.

(*vi*) *Applying shallow Depths*. The sewage should not be filled over the area in larger depths, but it should be applied in thin layers. Greater depth of sewage on a land does not allow the soil to receive the sewage satisfactorily, and ultimately results in its clogging.

A sewage-sick land can be improved and made useful by thoroughly ploughing and breaking the soil, and exposing it to the atmosphere.

---

*As discussed earlier, porous sandy soils will clog less as compared to heavy clayey soils.

8.11. Crops Grown in Sewage Farms and Their Hygienic Aspect

Infact, the city sewage is screened and sedimented before discharging it for irrigation. Neither biological treatment is given to it, nor it is disinfected. Due to this fact, the sewage effluents used for irrigation, normally remain contaminated. And since in all the broad irrigation methods (except furrow irrigation method), sewage applied for irrigation will come in direct contact with the plants, it may endanger the life and health of the people, if at all the crops grown are eaten raw, as the sewage is a potential carrier of germs and pathogenic bacteria, likely to cause diseases like typhoid, cholera, dysentery, etc. From public health point of view, therefore, **no** *crops which are to be eaten raw or without any skin removal, and which come in direct contact with the sewage, should be grown on sewage farms.*

Green vegetables, potatoes, etc. should, therefore, never be grown in such farms. However, crops like cotton, groundnut, grains, tobacco, sugarcane, etc. may be raised successfully on a sewage farm. Grass and fodder may also be raised, provided however, care is taken to see that the cattle do not graze on sewage wet fodder.

In any case, the crops grown on sewage farms should be properly labelled and marked*, so that the public knows about their source, and is cautious in their use.

It is suggested that, if at all vegetables are grown and sold in the market without any control, as is happening in Delhi (the capital city of India), the public should take preventive measures by keeping the raw vegetables submerged for at least 10 minutes in 1 in 400 water solution of bleaching powder, and then thoroughly wash them with water, before use. The use of potassium permanganate for disinfecting vegetables is not recommended, as by its use, the vegetables tend to loose freshness and crispness.

The health of the farmers working in the fields, is also an important factor. They should also take adequate precautions by wearing gum boots and gloves, as otherwise, the larvae present in sewage effluent may get into their body, through direct contact, causing diseases, most commonly the intestinal worms.

8.12. Dilution Method *Vs.* Land Disposal Method for Disposal of Sewage

(*i*) For disposal by dilution, large volumes of natural clean waters (with nil or very low B.O.D.) are required ; whereas, for land disposal, large areas of land, preferably with sandy soils, are required.

(*ii*) The cost of land, in land disposal method, is generally very high, specially in cities and urban areas, which are generally situated near rivers or oceans, and thus suggesting the choice of dilution method. Whereas, in the rural areas, where the amount of sewage produced is less, land being available at lesser cost, no easy water source being available, the land disposal method is a better choice.

(*iii*) When the cost of land is high in land disposal method, some return may be available due to sewage farming ; but then good management is also

*Practically, no farmer does so, as it will reduce the cash value of his crop.

required. Dilution method of disposal, however, when possible, is a simple method, and does not require too much of management.

(*iv*) Land disposal method requires either no pre-treatment of sewage or only preliminary treatment ; whereas, dilution method of disposal, generally requires either full or at least partial treatment.

(*v*) Dilution method of sewage disposal requires nil or small head pumping, because rivers flow through the lowest contours, while land disposal requires high head pumping, thus making the land disposal method costlier.

(*vi*) Land disposal method is generally found to be a better choice in hot climatic areas. This is because of the fact that in hot areas, D.O. contents of natural waters are reduced, while fish and aquatic life need higher D.O. concentrations, thus rendering rivers (particularly those with small dry weather flows) susceptible to too much pollution and unfit for throwing untreated or less treated sewage. A lot of pretreatment, if required, will make the dilution method costlier.

(*vii*) Land disposal saves the inland rivers from getting polluted by sewage, and returns to the land the fertilising elements, which were once drawn from it. Dilution in river water will naturally cause downstream river water to become polluted, and an unavoidable health hazard. The towns and the cities situated downstream of the disposal point (particularly those near the zone of degradation and zone of active decomposition) cannot utilise river water without thorough and costlier treatment. Sometimes, the river water may get so much polluted that it may become almost impossible to purify it within the normal economics. Due to the pollution of the river, fish life may be lost, resulting in a good loss to the economy. Moreover, the recreational uses of the river will be gone, if it becomes polluted.

For most of the Indian towns and cities, the land disposal method is a better choice, because of the following reasons :

(*i*) Due to hot climate, D.O. contents of river waters are low, while the fish and aquatic life requires higher D.O. concentrations for their survival, thus rendering the dilution method more difficult and prohibitive.

(*ii*) There are only a few coastal towns in India, which have strong tidal currents moving in the forward directions, and the necessary depth of water at the point of disposal, thus, prohibiting the disposal of sewage in the sea on large scale.

(*iii*) Most of the inland rivers, flowing in the vicinity of our cities and towns, do have very small amounts of dry weather flows in them. Most of them run dry during summer season. Even the perennial rivers like Yamuna, have very low dry weather flows in them. Further, these very rivers are the only sources of water supply to these cities and towns. All these factors make it almost imperative upon us, as not to throw away our sewage in rivers, but to throw it away on land. Even if it is decided to be thrown into the rivers, a lot of pre-treatment will be required, which will make it very costly.

(*iv*) Except for a few major cities, the water supplies in India are very low (90 to 120 litres per capita per day), thus resulting in the production of highly concentrated sewage, which on travelling in our hot climate, becomes very stale and septic by the time it reaches the disposal point, thus prohibiting the use of dilution method for its disposal.

PROBLEMS

1. Enumerate the two general methods adopted for sewage disposal ; and discuss their merits, and demerits explaining the conditions favourable for their adoption.

2. What are the factors affecting self-purification of polluted streams ? What measures would you recommend to control stream pollution in India.

3. What is meant by "Environmental Pollution" ? Describe what happens when untreated sewage from a town is discharged into a nearby stream.

4. Sewage disposal systems are to be provided for :

(*i*) an isolated residential building with ten users ;

(*ii*) a small town of 2000 persons located on the bank of a river ;

(*iii*) a town with 10,000 persons located on the bank of a small river.

Describe the possible methods of sewage disposal for each case and bring out the advantages and disadvantages of the various methods listed.

5. Write a detailed note on land treatment of sewage dealing with the chemical as well as the engineering aspects of the process.

6. What do you understand by self-purification property of a stream ? Explain the factors affecting this property.

7. (*a*) A district town with a population of 60,000 has to discharge its sewage into a small stream. Explain various considerations involved in determining the degree of treatment required.

(*b*) In what process of treatment, do you observe the following phenomenon, and why :

(*i*) sewage sickness ;

(*ii*) sludge bulking.

Mention the remedial measures that are normally taken to correct the situation.

8. (*a*) What is sewage farming ? What are its advantages over the method of disposal of sewage by dilution ? What crops should be grown on such farms ? What precautions should be taken in its operations to prevent health hazards either to the farm workers or to the consumers using the produce ?

(*b*) Calculate the area of sewage farm and type of crop to be grown if waste flow is 5 Mlpd, (nitrogen required may be taken as 40 kg/ha/crop and nitrogen available in waste as 20 ppm).

9. (*a*) Write a brief note on the use of sewage effluent for irrigation. Mention the places (Indian cities)*, where such uses of sewage are being made on a large scale.

(*b*) A town with a population of 25,000 is to be provided with a sewage treatment plant to handle both its industrial as well as domestic wastes. A sanitary survey has revealed the following facts :

Dairy wastes of 2.7×10^8 litre/day with BOD of 1200 mg/l and sugar mill waste of 2.2×10^6 litre/day with B.O.D. of 800 mg/l are produced. In addition, domestic sewage produced at the rate of 200 litres/capita/day, the per capita BOD of sewage being 76 gm/day. An expansion factor of 10% is to be provided. The effluent from the treatment plant is to be discharged into a stream with a minimum flow of 7000 litres/sec and a dissolved oxygen content of 8 mg/l. It is necessary to maintain a D.O. content of 4.0 mg/l. Find the degree of treatment required for the plant.

[**Hint.** Follow example 8.5]

10. (*a*) Explain self-purification of streams and indicate how sunlight helps in such purification.

(*b*) Write a short note on "Minimum D.O. content in polluted stream for survival of aquatic life".

---

*Bangalore, Delhi, Ahmedabad and Pune, all using partly treated *i.e.* sedimented sewage for irrigating crops.

11. (*a*) Explain the importance of study of re-oxygenation and de-oxygenation in problems of stream sanitation.

(*b*) The sewage discharge of a city is 85 m^3/sec in the river having a minimum discharge of 930 l/sec with a velocity of 0.12 m/sec. The B.O.D. at 20°C of the sewage is 325 mg/l. The B.O.D. of the river is zero. Determine the quantity and point of critical D.O. deficit.

12. (*a*) Explain the difference between the dilution process, if the waste water effluents are disposed in stream water and sea water.

(*b*) Explain clearly the methods, problems, and limitations of land disposal of sewage.

13. (*a*) Explain how domestic sewage is different from industrial waste.

(*b*) Write a detailed explanatory note on the self-purification of streams.

(*c*) What are the different indices that can be used to assess the self purification obtained above ? Which of them is most significant and why ?

14. 0.04 cumecs of settled sewage having a 5 day B.O.D. of 120 ppm and containing 4.6 ppm of D.O. are discharged into a large normal stream carrying an average flow of 1.02 cumecs with a 5 day B.O.D. at 20°C of 9 ppm, and containing 6 ppm of D.O. what is the value of maximum deficit that is likely to occur downstream, assuming the temp. at 20°C throughout, $k_D = 0.1$, $k_R = 0.24$, and saturation D.O. at 20°C = 9.2 ppm. If the minimum D.O. is to be kept at 5 ppm, what percentage of removal is necessary in a treatment plant ?

15. (*a*) What factors are considered in estimating the amount of dilution required for disposal of sewage in bodies of water ? What dilution ratios are usually, employed in India while disposing of domestic sewage in inland surface streams, which are sufficiently clean ?

(*b*) Find out the flow in stream required per 1000 population in disposing of sewage from a residential town in India with the following data :

Average stream temperature = 30°C.

5 day BOD of sewage at 30°C = 300 ppm.

Average sewage flow = 135 lphd.

Value of de-oxygenation and reoxygenation constant of the stream at 30°C are 0.15 and 0.27, respectively.

Minimum D.O. concentration to be provided in stream = 4 ppm.

16. Write short notes on :

(*i*) Oxygen sag curve ;

(*ii*) Broad Irrigation and Effluent Irrigation ;

(*iii*) Sewage sickness ;

(*iv*) Sewage treatment required *Vs* Available dilution ratio for its disposal ;

(*v*) Self-purification of streams, and self-purification constant ;

(*vi*) Dilution method *Vs* Land disposal method for sewage disposal.

9

Treatment of Sewage

9.1. Classification of Treatment Processes

Sewage, before being disposed of either in river streams or on land, has generally to be treated, so as to make it safe. The degree of treatment required, however, depends upon the characteristics of the source of disposal, as discussed in the previous chapter.

Sewage can be treated in different ways. Treatment processes are often classified as :

(*i*) *Preliminary treatment* ;

(*ii*) *Primary treatment* ;

(*iii*) *Secondary* (or *Biological*) *treatment* ; and

(*iv*) *Complete final treatment,* as discussed below :

9.1.1. Preliminary Treatment. Preliminary treatment consists solely in separating the floating materials (like dead animals, tree branches, papers, pieces of rags, wood, etc.), and also the heavy settleable inorganic solids. It also helps in removing the oils and greases, etc. from the sewage. This treatment reduces the *BOD* of the wastewater, by about 15 to 30%. The processes used are : **Screening** for removing floating papers, rags, clothes, etc. ; **Grit chambers or Detritus tanks** for removing grit and sand ; and **Skimming tanks** for removing oils and greases.

9.1.2. Primary Treatment. Primary treatment consists in removing large suspended organic solids. This is usually accomplished by sedimentation in **Settling basins.**

The liquid effluent from primary treatment, often contains a large amount of suspended organic material, and has a high BOD (about 60% of the original).

Sometimes, the preliminary as well as primary treatments are classified together, under primary treatment.

The organic solids, which are separated out in the sedimentation tanks (in primary treatment), are often stabilised by anaerobic decomposition in a *digestion tank* or are *incinerated*. The residue is used for land fills or soil conditioners.

9.1.3. Secondary Treatment. Secondary treatment involves further treatment of the effluent, coming from the primary sedimentation tank. This is generally accomplished through biological decomposition of organic matter, which can be carried out either under aerobic or anaerobic conditions. In these biological units, bacteria will decompose the fine organic matter, to produce clearer effluent.

The treatment reactors, in which the organic matter is decomposed (oxidised) by aerobic bacteria are known as ***aerobic biological units ;*** and may consist of : (*i*) **Filters** (*intermittent sand filters* as well as *trickling filters*) ; (*ii*) **Aeration tanks,** with the feed of *recycled* **activated sludge** (*i.e.* the sludge, which is settled in secondary sedimentation tank, receiving effluents from the aeration tank)* ; (*iii*) **Oxidation ponds and Aerated lagoons.** *Since all these aerobic units, generally make use of primary settled sewage, they are easily classified as secondary units.*

The treatment reactors, in which the organic matter is destroyed and stabilised by anaerobic bacteria, are known as ***anaerobic biological units*** and may consist of **Anaerobic lagoons, Septic tanks, Imhoff tanks,** etc. Out of these units, only anaerobic lagoons, make use of primary settled sewage, and hence, only they can be classified under secondary biological units. *Septic tanks and Imhoff tanks, using raw sewage, are, therefore, not classified as secondary units.*

The effluent from the secondary biological treatment will usually contain a little BOD (5 to 10% of the original), and may even contain several milligrams per litre of DO.

The organic solids/sludge, separated out in the primary as well as in the secondary settling tanks, will be disposed of by stabilising them under anaerobic process in a **Sludge digestion tank.**

9.1.4. The Final or Advanced Treatment. This treatment is sometimes called **tertiary treatment,** and consists in removing the organic load left after the secondary treatment, and particularly to kill the pathogenic bacteria. This treatment, which is normally carried out by **chlorination,** is generally not carried out for disposal of sewage in water, but is carried out, while using the river stream for collecting water for re-use or for water supplies. It may, however, sometimes be adopted, when the outfall of sewage is very near to the water intake of some nearby town.

The sewage treatment is, therefore, usually confined up to secondary treatment only. Well in fact, various physical, chemical and biological processes are available for treatment, depending upon the particular requirements. The choice of the treatment methods, depends on several factors, including the disposal facilities available. *Actually, the distinction between primary, secondary and tertiary treatment is rather arbitrary, since many modern treatment methods incorporate physical, chemical and biological processes in the same operation.*

The various treatment units, which may be used for treating sewage, and the extent of BOD, solids and bacteria removed by them, have been summarised in Table 9.1.

The individual operations of these units may be combined in different ways, depending upon the topography and other local needs, so as to create different types of treatment plants.

---

*Aeration tank(s), with feed of active sludge, along with secondary settling tank, etc., constitute an **Activated sludge plant.**

Table 9.1 Types of Treatments Units Employed in Sewage Treatment, Their Functions and Efficiencies

S. No. (1)	*Types of Treatment* (2)	*Purification effected* (3)	*Process or unit employed* (4)	*BOD removal as percentage of original* (5)	*Removal of suspended and dissolved solids as percentage of original* (6)	*Removal of bacterial load as percentage of original* (7)	*Disposal of residuals* (8)
1.	Preliminary Treatment	(*a*) Removal of *floating materials* like dead animals, tree branches, pieces of rags, wood and other large sized floating materials.	Coarse and fine **screens** of different designs	5—10	2—20	10—20	Screenings can be disposed of easily, either by burial or burning.
		(*b*) Removal of *heavy settleable inorganic solids.*	**Grit chambers** or *Detritus tanks*	10—20	20—40	10—20	The grit can be easily disposed of either by burial or burning for raising low lying areas.
		(*c*) Removcal of *fats and greases.*	**Skimming tanks** or *Vacuators.*	20—30	20—40	10—20	The skimmings contain unstable volatile organic materials, and have to disposed of by first stabilising them in digestion tanks by anaerobic process.
2.	Primary treatment	Removal of *suspended settleable organic solids*	(*i*) **Sedimentation tanks** *or*	30—35	60—65	25—75	Sludge containing organic material has to be stabilised first, in digestion tanks, and the digested material is then used as a manure or soil builder.

(1)	(2)	(3)	(4)	(5)	(6)	(7)	(8)
			(*ii*) *Septic tanks* (used only for small colonies or in rural areas). *or*	20—30	40—60	25—75	The effluents are generally disposed of for sewage farming on land. These units combine sludge digestion along with sedimentation in these tanks
			(*iii*) *Imhoff tanks* (seldom used in modern days).	30—40	60—65	25—75	
3.	Secondary or Biological Treatment	Removal of fine suspended non-settleable solids and colloidals, including dissolved organic matter	(*i*) *Chemical flocculation and sedimentation* (seldom used in sewage treatment) *or*	50—85	70—90	40—80	Sludge containing organic material has to be stabilised in digestion tanks, and the residue is used as manure or soil builder.
			(*ii*) *Intermittent sand filters*, followed and preceded by sedimentation (seldom used in modern days) *or*	90—95	85—95	95—98	Sludge containing organic material has to be stabilised in digestion tanks, and the residue is used as manure or soil builder.
			(*iii*) **Conventional low rate trickling filters,** followed and preceded by sedimentation. *or*	90—95	70—92	90—95	Sludge containing organic material has to be stabilised in digestion tanks, and the residue is used as manure or soil builder.
			(*iv*) **Modern high rate trickling filters,** followed and preceded by sedimentation. *or*	65—95	65—92	80—95	Sludge containing organic materials has to be stabilised in digestion tanks and the residue is used as manure or soil builder.

(1)	(2)	(3)	(4)	(5)	(6)	(7)	(8)
			(*v*) **Activated sludge treatment** in *Aeration tank* and *secondary settling tank*	75—95	85—90	90—98	Sludge containing organic materials has to be stabilised in digestion tanks, and the residue is used as manure or soil builder.
			or				
			(*vi*) **Oxidation ponds**	85—90	85—90	90—98	Effluent are generally disposed of by using them for irrigation.
4.	Final or Tertiary Treatment	Removal of Pathogens and the remaining very fine dissolved organic matter	*Chlorination* (seldom used for sewage treatment)	100%	100%	100%	—

Fig. 9.1 shows diagrammatic sketches of some **standard types of sewage treatment plants**, which give full treatment, and may be adopted under different conditions. The first plant [Fig. 9.1 (*a*)] consists of *Imhoff tanks* and *low rate trickling filters,* and may be adopted for very small towns, although the use of Imhoff tanks has become old and obsolete these days. The second plant [Fig. 9.1 (*b*)] consists of *sedimentation tanks* and *high rate trickling filters,* and is suitable for cities of small and medium sizes. The third type of treatment plant [Fig. 9.1 (*c*)] consists of a *sedimentation tank* and an *activated sludge treatment plant,* and is suitable for larger cities, where continuous attendance and supervision is possible.

These three different types of treatment plants contain different combinations of treatment units. These combinations can also be changed, and some other combination made, depending upon the local needs. The topography and geology will influence the choice of the units to be adopted. For example, the trickling filter plant consumes high head and requires steep ground slope to avoid pumping ; Imhoff tanks are vary deep and can not be constructed economically under adverse underground conditions. Similarly, the size of the plant will also influence the type of units to be housed in it. For example, smaller sized plants cannot house such units which require continuous skilled attendance and supervision, and as such, they should house such units, which do not employ complicated mechanical equipment.

All these treatment units, will now be discussed in details.

SCREENING

Screening is the very first operation carried out at a sewage treatment plant, and consists of passing the sewage through different types of screens, *so as to trap and remove the floating matter,* such as pieces of cloth, paper, wood, cork, hair, fibre, kitchen refuse, fecal solids, etc. present in sewage. These floating materials, if not removed, will choke the pipes, or adversely affect the working of the sewage pumps. Thus, the main idea of providing screens is to protect the pumps and other equipments from the possible damages due to the floating matter of the sewage.

Screens should preferably be placed before the *grit chambers* (described in the next article). However, if the quality of 'grit' is not of much importance, as in the case of land fillings, etc., screens may even be placed after the grit chambers. They may sometimes be accommodated in the body of the grit chambers themselves.

9.2. Types of Screens, their Designs and Cleaning

Depending upon the size of the openings, screens may be classified as coarse screens, medium screens, and fine screens.

(*i*) **Coarse screens** are also known as **Racks,** and the spacing between the bars (*i.e.* opening size) is about 50 mm or more. These screens do help in removing large floating objects from sewage. They will collect about 6 litres of solids per million litre of sewage. The material separated by coarse screens, usually consists of rags, wood, paper, etc., which will not putrefy, and may be disposed of by *incineration, burial,* or *dumping.*

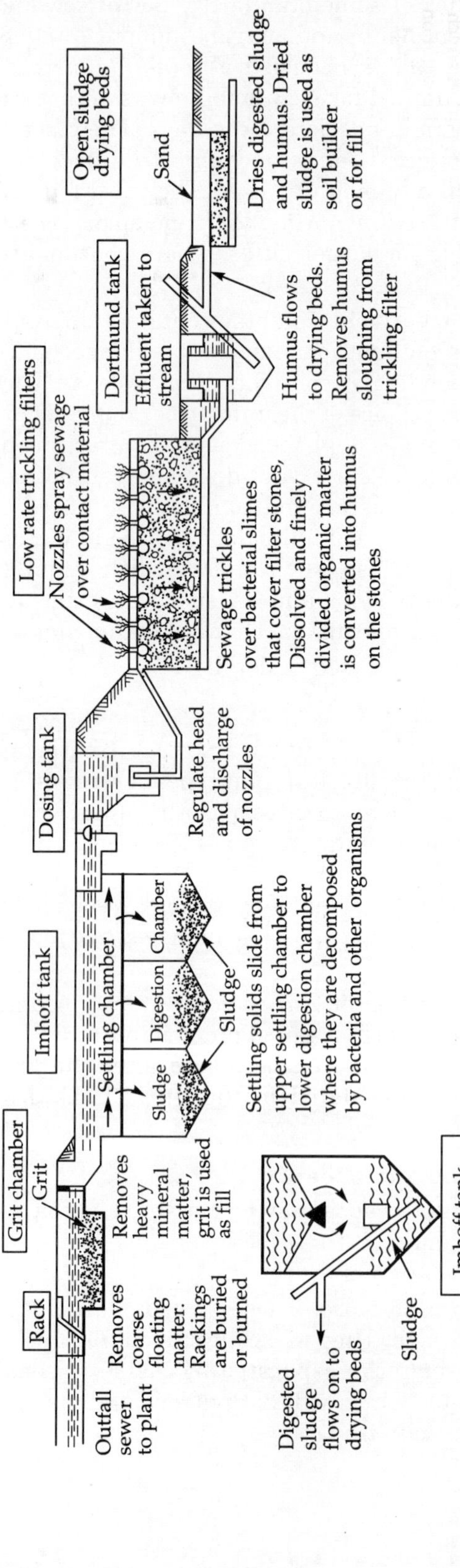

Fig. 9.1 (*a*) Flow diagram of a sewage treatment plant, which may to used for very small towns (use of Imhoff tank, through, has become old and obsolete)

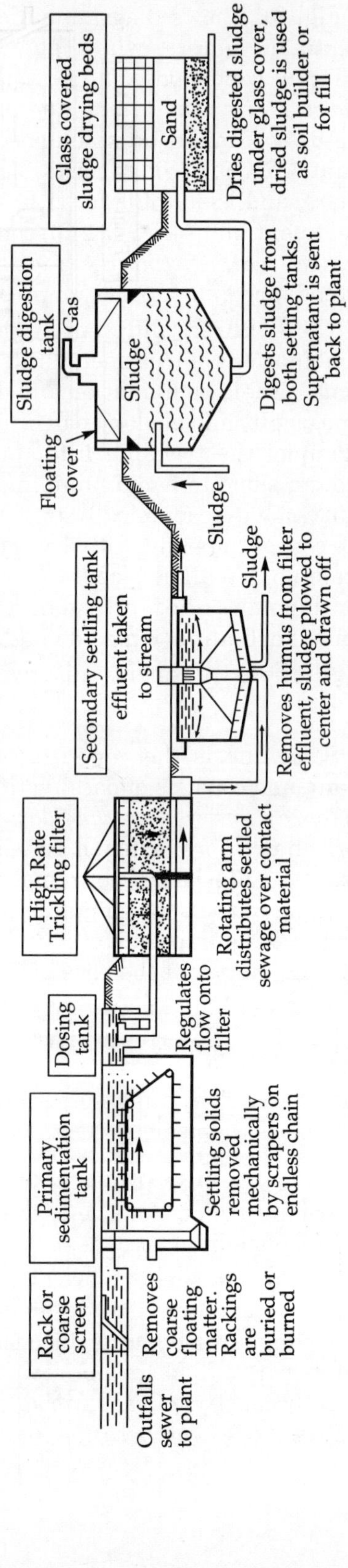

Fig. 9.1 (*a*) Flow diagram of a sewage treatment plant useful for small and medium sized cities

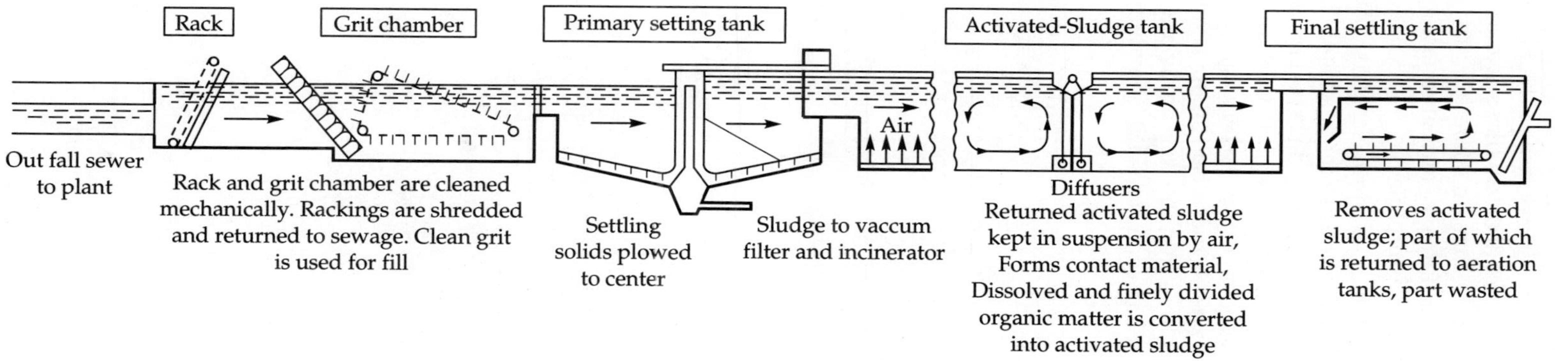

Fig. 9.1 (*c*) Flow diagram of sewage treatment plant useful for large sized cities

(*ii*) In **medium screens,** the spacing between the bars is about 6 to 40 mm. These screens will ordinarily collect 30 to 90 litres of material per million litre of sewage. The screenings usually contain some quantity of organic material, which may putrefy and become offensive, and must, therefore, be disposed of by *incineration,* or *burial* (not by dumping).

Rectangular shaped coarse and medium screens are now-a-days widely used at sewage treatment plants. They are made of steel bars, fixed parallel to one another at desired spacing on a rectangular steel frame, and are called **bar screens.** The screens are set in a masonry or R.C.C. chamber, called the *screen chamber*. Now-a-days, *these screens are generally kept inclined at about 30 to 60° to the direction of flow*, so as to increase the opening area, and to reduce the flow velocity ; and thus making the screening more effective. A most commonly used hand cleaned bar type screen is shown in Fig. 9.2.

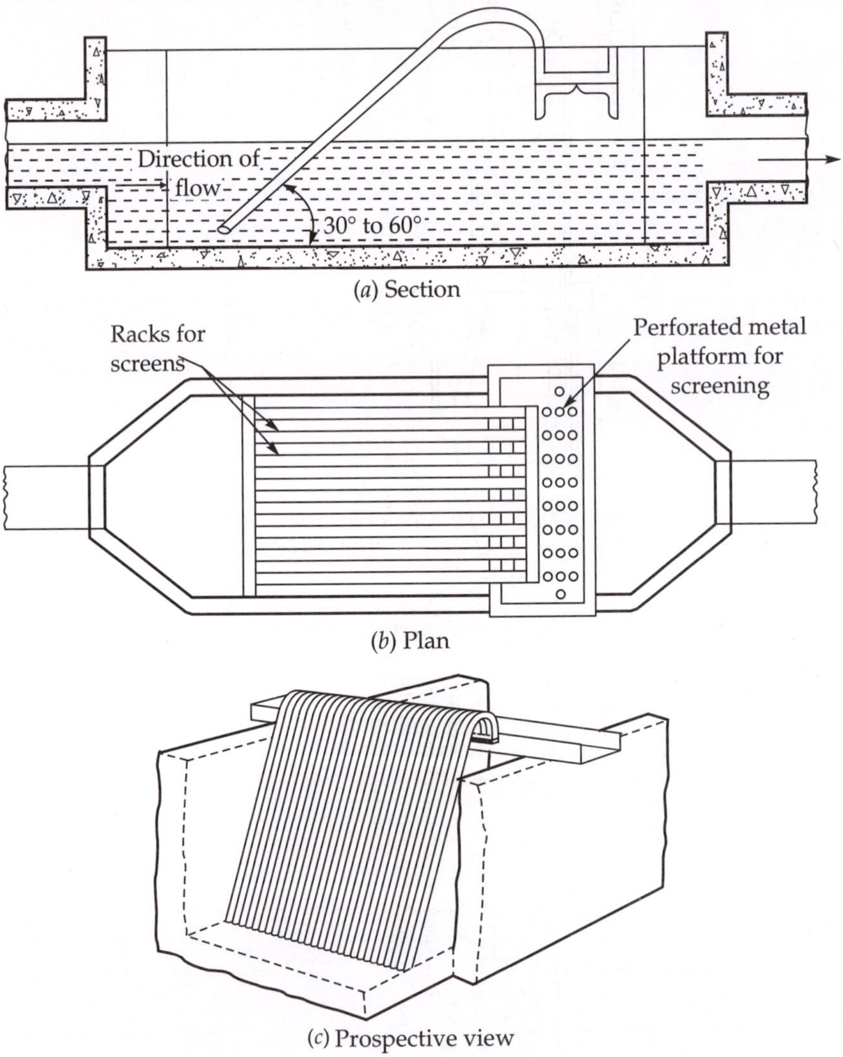

(*a*) Section

(*b*) Plan

(*c*) Prospective view

Fig. 9.2. Fixed Bar type Hand Cleaned Coarse or Medium Screen or Rack.

While designing the screens, clear openings should have sufficient total area, so that the velocity through them is not more than 0.8 to 1 m/sec. This limit, placed on velocity, limits the head loss through the screens, and, thus, reduces the opportunity for screenings to be pushed through the screens.

The material collected on bar screens can be removed either manually or mechanically. Manual cleaning is practised at small plants with hand operated **rakes**.* The inclined screens help in their cleaning by the upward stroke of the rake. Large plants, however, use mechanically operated rakes, which move over the screens, either continuously or intermittently.

The cleaning of screens by rakes will be hindered by cross bars, if at all provided. They are, therefore, generally avoided.

Screens are sometimes classified as *fixed* or *movable*, depending upon whether the screens are stationary or capable of motion. ***Fixed screens*** are permanently set in position. ***Movable screens*** are stationary during their operating periods, but they can be lifted up bodily and removed from their positions for the purpose of cleaning. A common movable bar medium screen is a 3-sided cage with a bottom of perforated plates. It is mainly used in deep pits ahead of pumps.

(*iii*) **Fine Screens** have perforations of 1.5 mm to 3 mm in size. The installation of these screens prove very effective, and they remove as much as 20% of the suspended solids from sewage. These screens, however, get clogged very often, and need frequent cleaning. They are, therefore, used only for treating the industrial wastewaters, or for treating those municipal wastewaters, which are associated with heavy amounts of industrial wastewaters. These screens will considerably reduce the load on further treatment units.

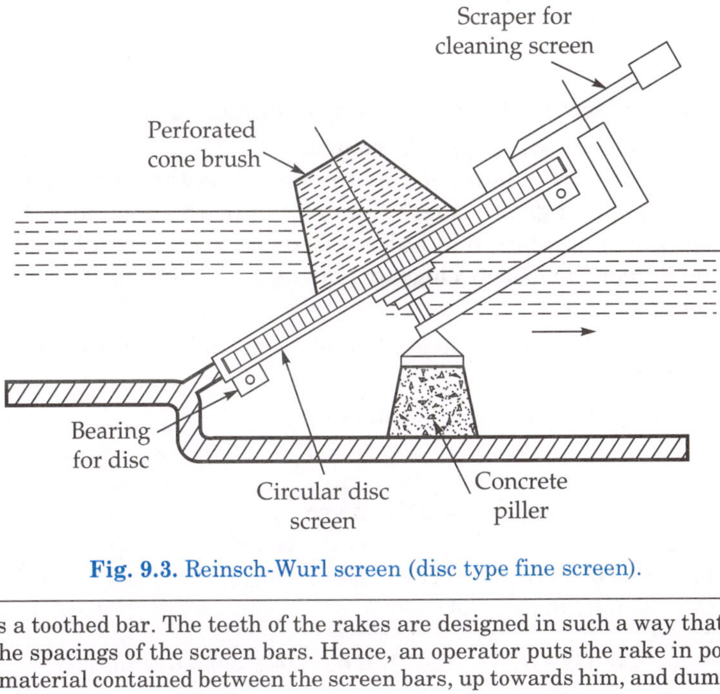

Fig. 9.3. Reinsch-Wurl screen (disc type fine screen).

---

*A **rake** is a toothed bar. The teeth of the rakes are designed in such a way that they fit in between the spacings of the screen bars. Hence, an operator puts the rake in position, and drags the material contained between the screen bars, up towards him, and dumps it on the floor.

Brass or Bronze plates or wire mesh are generally used for constructing fine screens. The metal used should be resistant to rust and corrosion.

The fine screens may be **disc** or **drum** type, and are operated continuously by electric motors. Fig. 9.3 shows a typical disc type of fine screen, which is cleaned by a cone brush.

Example 9.1. *Estimate the screen requirement for a plant treating a peak flow of 60 million litres per day of sewage.*

Solution. Peak flow = 60 M*l*/day

$$= \frac{60 \times 10^6}{1000} \text{ cu-m/day}$$

$$= \frac{60{,}000}{24 \times 60 \times 60} \text{ cu-m/sec} = 0.694 \text{ m}^3\text{/sec.}$$

Assuming that the velocity through the screens (at peak flow) is not allowed to exceed 0.8 m/sec, we have

The net area of screen openings required

$$= \frac{0.694}{0.8} \text{ m}^2 = 0.87 \text{ m}^2.$$

Using rectangular steel bars in the screen, having 1 cm width, and placed at 5 cm clear spacings, we have

The gross area of the screen required

$$= \frac{0.87 \times 6}{5} = 1.04 \text{ m}^2$$

Assuming that the screen bars are placed at 60° to the horizontal, we have

The gross area of the screen needed

$$= \frac{1.04}{\frac{\sqrt{3}}{2}} = \frac{1.04 \times 2}{\sqrt{3}} = 1.2 \text{ m}^2.$$

Hence, a coarse screen of 1.2 m^2 area is required. **Ans.**

While designing the screen, we have also to design its **cleaning frequency.** The cleaning frequency is governed by the head loss through the screen. The more the screen openings are clogged, more will be the head loss through the screen. Generally, not more than half the screen clogging is allowed. To know whether the screen has been clogged and needs cleaning, we can check or measure the head loss.

The head loss through the cleaned screen and half-cleaned screen, can be computed as follows :

Velocity through the screen = 0.8 m/sec.

Velocity above the screen

$$= \frac{0.8 \times 5}{6} \text{ m/sec} = 0.67 \text{ m/sec}$$

Head loss through the screen

$$= 0.0729\,(V^2 - v^2) \qquad ...(9.1)$$

$$= 0.0729\,(0.8^2 - 0.67^2) = 0.0134 \text{ ; say } \mathbf{0.013\ m.}$$

When the screen openings get half clogged, then

The velocity through the screen

$$= v = 0.8 \times 2 = 1.6 \text{ m/sec.}$$

$\therefore$ Head loss $= 0.0729\,(1.6^2 - 0.67^2) = 0.1538$; **say 0.15 m.**

This shows that when the screens are totally clean, the head loss is negligible *i.e.* about 1.3 cm only ; whereas, the head loss shoots up to about 15 cm at half the clogging. The screens should therefore be cleaned frequently, as to keep the head loss within the allowable range. **Ans.**

9.3. Comminutors

Comminutors or *Shredders* are the patented devices, which break the larger sewage solids to about 6 mm in size, when the sewage is screened through them. Such a device consists of a revolving slotted drum, through which the sewage is screened (Fig. 9.4). Cutters mounted on the drum, shear the collected screenings against a comb, until they are small enough to pass through 5 mm to 10 mm wide slots of the drum. These are usually arranged in pairs to facilitate repairs and maintenance. Comminutors are of recent origin,

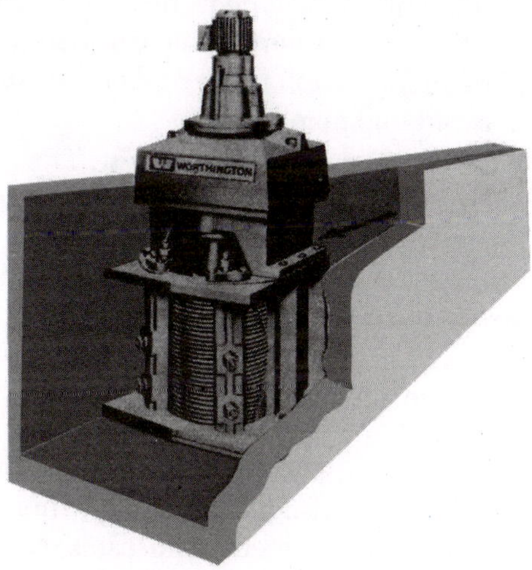

(*a*) Photo view of a comminutor

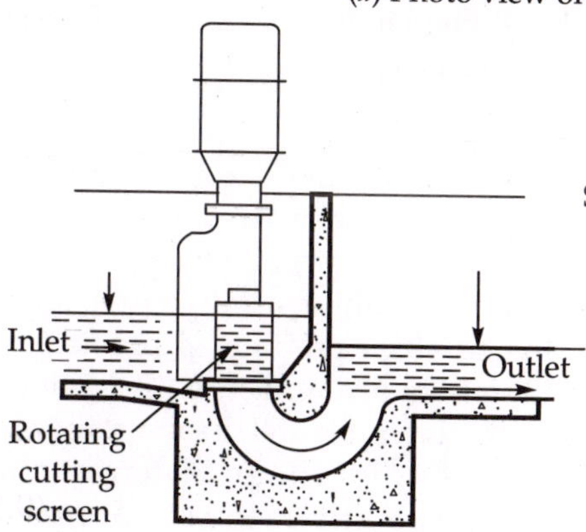

(*b*) Comminutor-section

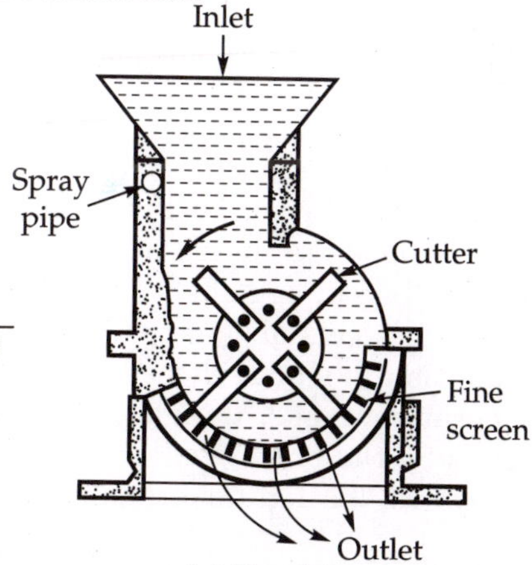

(*c*) Shredder

Fig. 9.4

and eliminate the problem of disposal of screenings, by reducing the solids to a size which can be processed elsewhere in the plant. They should always be preceded by grit chambers to prevent their excessive wear.

Such devises are used only in developed countries like USA, and generally not adopted in our country.

9.4. Disposal of Screenings

The material separated by screens is called the screenings. It contains 85 to 90% of moisture and other floating matter. It may also contain some organic load which may putrefy, causing bad smells and nuisance. To avoid such possibilities, the screenings are disposed of either by burning, or by burial, or by dumping. The dumping is avoided when screenings are from medium and fine screens, and are likely to contain organic load, as pointed out earlier. The screenings may also sometimes be broken up by a comminutor and then passed on to the grit chamber.

Burning of the screenings is done in the incinerators, similar to those used for burning garbage. The process of burning is called **Incineration.** The screenings are first dried with sun's heat by spreading on ground or by compressing through hydraulic or other presses, so as to reduce the moisture content to about 60%. The incineration is carried out at temperatures of about 760 to 815°C. This will avoid bad smells.

The screenings may also be disposed of by *burial*. The process is technically called **Composting.** In this process, the screenings are buried in 1 to 1.5 m deep trenches, and then covered with 0.3 to 0.45 m of porous earth. In due course of time, oxidation-reduction of screenings will take place, and the contents can be used as manure.

Another method of disposing of the screenings is by **dumping** them in low lying areas (away from the residential areas) or in large bodies of water, such as sea. *Dumping in sea* will be suitable only where strong forward currents do exist to take the dumped material away from the shore line. The *dumping on land* for raising low lying areas is also adopted only when screenings are from the course screens, and not from the medium or fine screens, and as such not containing much organic load.

Digestion of screenings along with the sewage sludge in a sludge digestion tank has also been tried, but not found successful.

GRIT REMOVAL BASINS

Grit removal basins, such as **Grit chambers** or **Grit channels** or **Detritus tanks** are the sedimentation basins placed in front of the wastewater treatment plant to remove the inorganic particles (specific gravity about 2.65), such as sand, gravel, grit, egg shells, bones, and other nonputresible materials that may clog channels or damage pumps due to abrasion, and to prevent their accumulation in sludge digestors. Grit chambers are, infact, nothing but like sedimentation tanks, designed to separate the intended heavier inorganic materials by the process of sedimentation due to gravitational forces, and to pass forward the lighter organic material*. The grit chambers are usually

*The organic material is not allowed to settle in this process, as otherwise, the organic matter gets entangled with the inorganic matter, causing septicity of sewage and requiring unnecessary labour and expenses for disposal or removal.

placed after the fine screens, to remove particles of size up to 0.2 mm or so. Many engineers, however, prefer to place them before the fine screens, as to avoid silting of the screen chambers.

Many a times, the grit chambers are **aerated** by providing outside air through compressors. The diffused air creates spiral current within the grit, as to help in its settlement.

Since the grit chambers are sedimentation or settling basins, we will discuss theory of sedimentation first, before coming to the design of grit chambers.

9.5. Settling of Discrete Particles (Type I* Sedimentation or Settling) in Waste Waters or Raw Waters

Sedimentation is the physical separation of suspended material from a water or a wastewater by the action of gravity. It is a common operation for water treatment and found in almost all waste water treatment plants. It is less costly than many other treatment processes.

The very fundamental principle underlying the process of sedimentation is that the organic matter present in sewage is having specific gravity greater than that of water (*i.e.* 1.0). *In still sewage, these particles will, therefore, tend to settle down by gravity ; whereas, in flowing sewage, they are kept in suspension, because of the turbulence in water. Hence, as soon as the turbulence is retarded by offering storage to sewage, these impurities tend to settle down at the bottom of the tank offering such storage.* This is the **principle behind sedimentation.**

The design of sedimentation basins is thus, totally dependent upon the **settling velocity** of the sewage solids. The design of an ideal settling basin is based on the removal of all particles that have a settling velocity greater than a specified settling velocity.

The settling velocity of a discrete** solid particle is mathematically computed and analysed on the basis of the theory of sedimentation, which is discussed below :

9.5.1. Theory of Type I Settling. The settlement of a particle in water brought to rest is opposed by the following factors :

(*i*) *The velocity of flow* which carries the particle horizontally. The greater the flow area, the lesser is the velocity, and hence more easily the particle will settle down.

(*ii*) *The viscosity of water* in which the particle is travelling. The viscosity varies inversely with temperature. Warm water is less viscous and, therefore, offers less resistance to settlement. However, the temperature of water cannot be controlled to any appreciable extent in "water or wastewater purification processes" and hence this factor is generally ignored.

(*iii*) *The size shape and specific gravity of the particle.* The greater is the specific gravity, more readily the particle will settle. The size and shape of the

---

**Type I sedimentation refers to discrete particles settling.

'Discrete**' or 'granular' particles are those which do not change their size, shape and weight. The other type of particles called "**flocculated**" particles are those which change their size, shape and weight, and thus loose their identity.

particle also affect the settling rate. For example, the weight and volume of the spherically shaped particle, varies with the cube of its diameter $\left(\text{volume being equal to } \frac{\pi d^3}{6}, \text{ where } d \text{ is the dia}\right)$ or its size ; and its area varies with the square of the diameter $\left(\text{area being equal to } \frac{\pi d^2}{4}\right)$. Hence, very small sized particles will settle very slowly. It, therefore, clearly follows that the shape and size of the particles do affect their setting velocities.

The settling velocity of a spherical particle is expressed by **Stoke's law,** which takes the above three factors into account. The final Stoke's equation, is expressed as :

$$\boxed{v_s = \frac{g}{18}(G-1)\,\frac{d^2}{\nu}} \quad \text{for } d < 0.1 \text{ mm} \qquad \text{...(9.2)}$$

[for viscous flow and small sized particles, represented by $R_e < 0.5$]

where v_s = velocity of settlement of particle (assumed to be spherical) in m/sec.

d = diameter of the particle in m.

G = sp. gravity of the particle.

ν = kinematic viscosity of water in m²/sec*, and is equal to $\frac{\mu}{\rho_w}$, where μ = absolute or dynamic viscosity of water in N sec/m²

R_e = Reynold Number

$= \frac{v_s \cdot d}{\nu}$

Derivation of Stoke's Law. When a solid particle settles down in water, its downward settlement is opposed by the drag force offered by the water. The effective weight of the particle (*i.e.* actual weight—buoyancy) causes the particle to accelerate in the beginning, till it attains a sufficient velocity (v_s) at which the drag force becomes equal to the effective weight of the particle. After attaining that velocity, the particle falls down with that constant velocity (V_s). Now, the drag force offered by the fluid is given by Newton's law, as

$$\text{Drag force} = C_D A.\, \rho_w \cdot \frac{v^2}{2} \qquad \text{...(1)}$$

where C_d = Coefficient of drag

A = Area of particle

---

*1 m²/sec = 10^4 Stoke.

1 centistoke = 10^{-2} Stoke.

ρ_w = Density of water

v = velocity of fall

[**Note.** This drag force increases with the increasing velocity, till it becomes equal to the effective weight of the particle ; and at that time, v becomes equal to v_s.]

The effective weight of the particle

$$= \text{Total weight} - \text{Buoyancy}$$

$$= \frac{4}{3} \cdot \pi r^3 \gamma_s - \frac{4}{3} \pi r^3 . \gamma_w$$

[∵ weight = volume × unit weight]

where r = radius of particle

γ_s = unit weight of particle

γ_w = unit weight of water.

$$= \frac{4}{3} \pi r^3 \left[\gamma_s - \gamma_w \right] \qquad ...(2)$$

Eqs. (1) and (2) will become equal when v becomes equal to v_s in Eq. (1).

$$\therefore \quad C_D . A . \rho_w . \frac{v_s^2}{2} = \frac{4}{3} \pi r^3 \left[\gamma_s - \gamma_w \right]$$

But $A = \pi r^2$

$$\therefore \quad C_D . \pi r^2 . \rho_w . \frac{v_s^2}{2} = \frac{4}{3} \pi r^3 \left[\gamma_s - \gamma_w \right]$$

or
$$v_s^2 = \frac{4}{3} \cdot \frac{(r) . (\gamma_s - \gamma_w) . 2}{\rho_w . C_D}$$

or
$$v_s^2 = \frac{\frac{4}{3}\left(\frac{d}{2}\right)(\gamma_s - \gamma_w) . 2}{\rho_w . C_D}$$

or
$$v_s^2 = \frac{\frac{4}{3}(\gamma_s - \gamma_w) . d}{\rho_w . C_D} \qquad ...(3)$$

Now, $\gamma_s = \rho_s . g$

and $\gamma_w = \rho_w . g$

$$\therefore \quad \gamma_s - \gamma_w = g(\rho_s - \rho_w)$$

$$= g . \rho_w \left(\frac{\rho_s}{\rho_w} - 1 \right)$$

$$= g \, \rho_w . (G - 1)$$

Eq. (3) then becomes

$$v_s^2 = \frac{4}{3} \cdot \frac{g . \rho_w (G-1) \, d}{\rho_w . C_D}$$

or $$v_s^2 = \frac{\frac{4}{3} \cdot g \cdot (G-1)\, d}{C_D} \qquad \text{...(9.3)}$$

The coefficient of drag (C_D), has been found for a *viscous flow and small particles* (size $d < 0.1$ mm), to be equal to $\frac{24}{R_e}$ ...(9.4*a*)

where R_e is the particle Reynold number

$$= \left(\frac{v_s \cdot d}{\nu}\right)$$

∴ Eq. (9.3) then becomes,

$$v_s^2 = \frac{\frac{4}{3} \cdot g \cdot (G-1) \cdot d}{\frac{24}{R_e}}$$

or $$v_s^2 = \frac{4}{3} \cdot g\,(G-1) \cdot d \cdot \frac{R_e}{24} = \frac{g}{18}\,(G-1)\, d \cdot \frac{\nu_s \cdot d}{\nu}$$

or $$v_s = \frac{g}{18}\,(G-1)\,\frac{d^2}{\nu} \text{ for } d < 0.1 \text{ mm}$$

which is the requisite equation 9.2.

The above Stoke's equation is valid for particles of size less than 0.1 mm ; in which case, the viscous force predominates over the 'inertial force, leading to what is known as *laminar* or *streamline settling*.

If, however, the settling particles are larger than 0.1 mm, the nature of settling tends to become *turbulent*, with a *transition zone* in between. It has been established that *turbulent settling* occurs for particle size greater than 1.0 mm, whereas settling remains *transition settling* for particle sizes between 0.1 mm to 1.0 mm.

The relation between coefficient of drag (C_D) and R_e for these three types of settling are as follows :

(*a*) *For streamline* or *laminar settling* ($d < 0.1$ mm)

Here $R_e < 1$; and

$$C_D = \frac{24}{R_e} \qquad (\textit{i.e. Eq. 9.4a})$$

(*b*) *For transition settling* (d between 0.1 mm and 1.0 mm)

Here $1 < R_e < 10^4$*

$$C_D = \frac{24}{R_e} + \frac{3}{R_e^{0.5}} + 0.34 \qquad \text{...(9.4bi)}$$

or $$\boxed{C_D = \frac{18.5}{R_e^{0.6}}} \qquad \text{...(9.4bii)}$$

*According to Newton, this value of R_e varies between 10^3 to 10^4. However, for sureset development of turbulent conditions, a higher value of 10^4 is considered here.

(c) *For turbulent settling* ($d > 1.0$ mm)

Here $R_e > 10^4$

$$C_D = 0.4 \qquad \text{...(9.4bc)}$$

Grit particles, generally, lie between 0.1 mm and 1 mm sizes, and hence, undergo settling which lies in transition settling zone. In such a case, therefore, the settling velocity equation (9.3) becomes :

$$v_s^2 = \frac{\frac{4}{3} \cdot g \cdot (G-1) \cdot d}{C_D} \qquad \text{where } C_D = \frac{18.5}{R_e^{0.6}}$$

or
$$v_s^2 = \frac{4}{3} g \cdot (G-1) \cdot d \times \frac{R_e^{0.6}}{18.5}$$

or
$$v_s^2 = \frac{g}{13.88} (G-1) \cdot d\, R_e^{0.6}$$

or
$$v_s^2 = \frac{g}{13.88} (G-1) d \cdot \left(\frac{v_s \cdot d}{\nu}\right)^{0.6}$$

or
$$v_s^{1.4} = \frac{g(G-1)}{13.88} \frac{d^{1.6}}{\nu^{0.6}}$$

or
$$v_s = \left[\frac{g(G-1)}{13.88} \frac{d^{1.6}}{(\nu)^{0.6}}\right]^{1/1.4}$$

or
$$v_s = \left[\frac{g(G-1)}{13.88} \cdot \frac{d^{1.6}}{(\nu)^{0.6}}\right]^{0.714} \qquad \text{...(9.5)}$$

for d lying between 0.1 mm and 1.0 mm.

For *turbulent settling*, the equation (9.3) reduces to :

$$v_s^2 = \frac{4g}{3} (G-1)\, d \cdot \frac{1}{C_D} \quad \text{where } C_D = 0.4 \text{ (eq. 9.4c)}$$

or
$$v_s^2 = \frac{4g}{3 \times 0.4} (G-1) \cdot d = 3.33g\,(G-1)d$$

or
$$v_s = 1.8\sqrt{gd(G-1)} \quad \text{for } d > 1 \text{ mm} \qquad \text{...(9.6)}$$

This equation is known as Newton's equation for turbulent settling.

The above formulas represent the theoretical settling velocities of discrete spherical particles. The actual settling velocities in the sedimentation basins will be much less than those calculated by these formulas, because of : the non-sphericity of the particles, the upward displacement of the fluid caused by the settling of other particles, and convection currents.

Based upon experiments, Hazen has formulated a table (Table 9.5) giving the values of the settling velocities (popularly called (*hydraulic settling values*) for different sized particles in still liquids at 10°C.

The above experimental values have also been expressed in mathematical form as **modified Hazen's equation for transition zone,** given by :

$$v_s = 60.6\,(G-1)\,d \cdot \left(\frac{3T+70}{100}\right) \qquad \text{...(9.7)}$$

for particles between 0.1 and 1 mm

Table 9.5. Hydraulic Settling Values in mm/sec in still liquids at 10°C*

Type of material	*Dia of particles in mm*	*Settling velocity in mm/sec*	
		Particles of Sp. gravity 2.65	*Particles of Sp. gravity 1.20*
(1)	*(2)*	*(3)*	*(4)*
Course sand	1.00	100	12.0
	0.80	83	9.6
	0.60	63	7.2
Medium sand	0.50	53	6.0
	0.40	42	4.8
	0.30	32	3.6
Fine sand	0.20	21	2.4
	0.15	15	1.8
	0.10	8	1.2
Very fine sand	0.08	6	0.53
	0.06	3.8	0.30
	0.05	2.9	0.21
Silt	0.04	2.1	0.13
	0.03	1.3	0.076
	0.02	0.62	0.034
	0.015	0.35	0.019
	0.010	0.154	0.0084
Fine silt	0.008	0.098	0.0054
	0.006	0.055	0.0030
	0.005	0.0385	0.0021
Clay	0.004	0.0247	0.0013
	0.003	0.0138	0.00076
	0.002	0.0062	0.00034
	0.0015	0.0035	0.00019
Fine clay	0.001	0.00154	0.000084
	0.0001	0.0000154	0.00000084
Colloidal clay	0.0001		

*__Note.__ At higher temperatures such as 26°C (*i.e.* average temperature prevailing in our country), the value of settling velocity will be about 50% more than these values.

The above equation yields the following :

For inorganic solids,

$G = 2.65$; and therefore,

Settling velocity for inorganic solids

$$v_{s(in)} = 60.6d\,(1.65)\,.\,\frac{3T+70}{100}$$

or $$\boxed{v_{s(in)} = d(3T + 70)} \qquad ...(9.8)$$

Similarly, settling velocity for *organic matter* (for which $G = 1.2$)

$$v_{s(0)} = 60.6d\,(0.2)\left(\frac{3T+70}{100}\right)$$

or $$\boxed{v_{s(0)} = 0.12d\,(3T + 70)} \qquad \text{...(9.9)}$$

Equation (9.7) above computes the actual settling velocity for non-spherical particles.

Infact, Hazen has originally given the following equations.

Hazens Original Equation for Settling Velocity

$$v_s = 418\,(G-1)\,d\,.\left(\frac{3T+70}{100}\right) \qquad \text{...(9.7a)}$$

This equation was computing the theoritical values of settling velocity for discrete spherical particles in a suspension. However, since the suspended matter is seldom spherical, and may be rod like or disc-like spheroidal shapes, they settle more slowly than the spheres of equivalent volumes. Also, frictional drag changes with the orientation of the particles relative to the direction of settling. The actual settling velocity being less than the theoretical one, is given by modified Hazen's equation, i.e., equation (9.7)

9.6. Grit Chambers or Grit Channels

As stated earlier, the *grit chambers* or **grit channels**, as they are usually called, are the sedimentation basins placed usually after the fine screens and certainly, before the primary sedimentation tank. *The grit chamber removes the inorganic grit, such as sand, gravel, and other mineral matter that has a nominal diameter of 0.15 to 0.20 mm or more.* Actually, grit will also include smaller mineral particles that may settle, as well as non-putrescible organic matter, such as rags, coffee grounds, vegetable cuttings, ash clinker, wood pieces, and tea leaves, etc. Even though, some of the grit components, such as coffee grounds are organic, they are essentially non-biodegradable over time spans involved here in grit collection and disposal.

The amount of grit collected is a function of the removal device, its operation, and the quantity of grit in the sewage, and therefore, it varies over a wide range. The grit quantity may vary between 0.004 – 0.037 $m^3/1000\ m^3$ of sewage for separate sewage system ; while this may range between 0.004 – 0.180 $m^3/1000\ m^3$ for combined sewage system.

Generally, grit channels are designed to remove all particles of higher specific gravity of 2.65 or so, with a nominal diameter of 0.20 mm and more, having settling velocity of about 21 mm/s (at 10°C) ; although some grit removal channels are designed to remove particles above 0.15 mm size, having settling velocity of about 15 mm/s (at 10°C).

It is not at all desirable to remove any organic matter in the grit chambers, because no further treatment of removed grit is provided. The grit chamber must hence, be designed to scour the lighter organic particles, while the heavier*

*The organic material is not allowed to settle in this process, as otherwise, the organic matter gets entangled with the inorganic matter, causing septicity of sewage and requiring unnecessary labour and expenses for disposal or removal.

grit particles remain settled.

Grit channels can be divided into the following two general types :

(*i*) *horizontal flow type (non-aerated)* ; and (*ii*) *aerated*

In the **horizontal flow type**, the flow passes through the channel in a horizontal direction. The chamber (or channel) is designed to give a horizontal straight line flow velocity, which is kept constant over varying discharge.

The constant velocity is achieved by providing a *velocity control section, such as a proportional flow weir* at the effluent end of a rectangular channel ; or a *parabolic (or V) shaped channel*, having varying width with depth, as discussed in article 9.7.

The **aerated type of grit channel** consists of a spiral flow aeration tank, where the spiral velocity is controlled by the dimensions/shape of the chamber and the quantity of air supplied to the unit. This type of grit channels are discussed in article 9.8.

9.7. Constant Velocity Horizontal Flow Grit Channels

Such a grit channel is an enlarged channel or a long basin, in which the cross-section is increased, so as to reduce the flow velocity of sewage to such an extent that the heavy inorganic materials do settle down by gravity, and the lighter organic materials remain in suspension, and, thus, go out along with the effluent of the grit basin. *The important point in the design of the grit basins is that the flow velocity should neither be too low as to cause the settling of lighter organic matter, nor should it be so high as not to cause the settlement of the entire silt and grit present in sewage.* The flow velocity should also be enough to scour out the settled organic matter, and reintroduce it into the flow stream. Such a **critical scouring velocity** is, infact, given by the *modified Shield's formula*, which states that

$$\text{Critical scour velocity} = V_H = 3 \text{ to } 4.5\sqrt{gd(G-1)} \qquad \text{...(9.10)}$$

For grit particles of 0.2 mm (*d*), the above formula gives critical velocity values of 0.17 to 0.26 m/sec. This fixes the limits for *optimum flow velocity* for design of grit basins. *In practice, a flow velocity of about 0.25 to 0.3 m/sec is adopted for the design of grit basins.*

In order to prevent large increase in flow velocity at peak hours due to increased discharge, and thus, to avoid the scouring of the settled grit particles from the bottom, it is preferable to design the grit chambers for D.W.F. (Dry weather flow), and to provide additional units for taking increased discharge at peak hours. If, however, a single unit is to be designed, or there are large variations in discharge, then the grit chamber is designed for generating optimum velocity at peak discharge and a **velocity control section**, such as a properly designed modified weir, called a **proportional flow weir** or a **sutro weir** (Fig. 9.5*a*), is provided at the lower (effluent) end of the rectangular grit channel, which helps in varying the flow area of the section in direct proportion to the flow, *and thus, helps to maintain a constant velocity in the channel (within the permissible limits of ± 5 to 10% over the designed value), even at varying discharges.*

A **channel with a varying cross-section geometry** (Fig. 9.5*b*) is an alternative to providing a modified weir. The channel may discharge into a **parshall flume** (Fig. 9.7) to measure the discharge accurately. In this case, there is a rectangular control section at the end of the channel, but the channel width varies as a function of water depth, as shown in Fig. 9.5 (*b*) and 9.8 to keep the velocity constant at varying discharge.

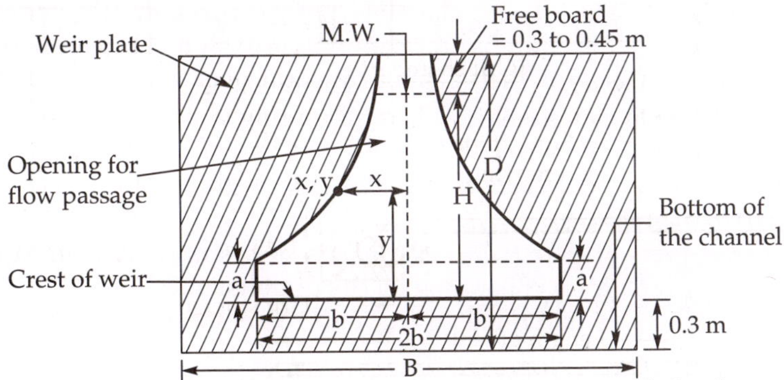

(*a*) Properly designed proportioning flow weir provided at the effluent end of a rectangular grit channel

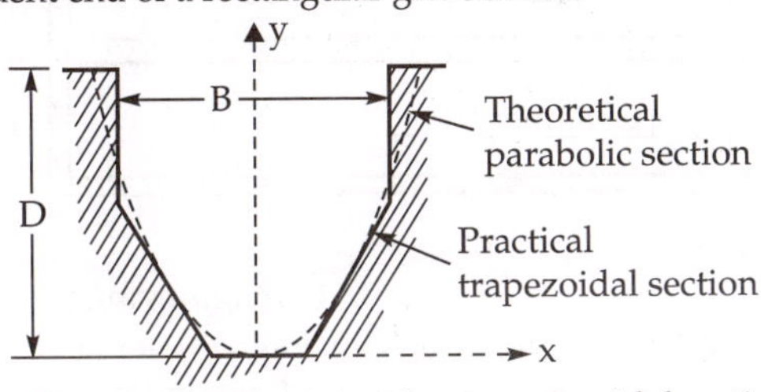

(*b*) Parabolic channel section (approximated to a trapozoidal section) helping in producing constant velocity with varying discharge, since the channel width varies with water depth

Fig. 9.5. Velocity control sections for horizontal grit channels.

9.7.1. Design of a Rectangular Grit Chamber provided with a Proportioning Flow Weir at Effluent End. The *depth* and *detention time* provided for a grit basin are inter dependent, and are based on the considerations of settling velocity of inorganic particles through water. A detention time of about 40 to 60 seconds (1 minute) is generally sufficient for a water depth of about 1 to 1.8 m.* After fixing the water depth and the detention time, we can easily design the dimensions of a rectangular chamber, as its length will then be equal to velocity × detention time.

As stated earlier, generally two to three separate chambers in parallel (as shown in Fig. 9.6) should be provided ; one to pass the low flow, and the other to pass (along with the first of course) the high flow. This will also help in

---

**Settling velocity of inorganic matter* is given by equation (9.8) as,

$$V_s = d(3T + 70).$$

For $T = 25°C$, and $d = 0.2$ mm, we get

$$V_s = \frac{0.2}{10 \times 100}(3 \times 25 + 70) = 0.029 \text{ m/sec.}$$

$\therefore$ Detention time in secs $= \dfrac{\text{Depth}}{\text{Settling velocity}} = \dfrac{\text{Depth}}{0.029} = 34$ Depth in m.

$\therefore$ Detention time $= 34D$ (where D is the depth in m).

$\therefore$ For depth of 1 m, detention time = 34 secs (≈ 40 secs).

and for a depth of 1.8 m, detention time = 34 × 1.8 = 61 secs (≈ 60 secs).

manual cleaning of the chambers, as one unit can work, while the other is shut down for cleaning.

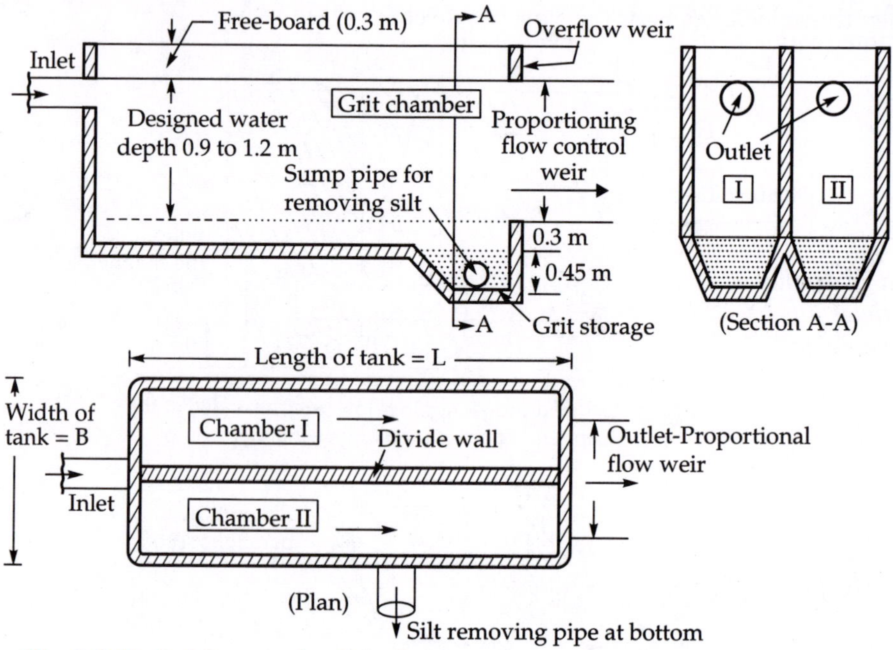

Fig. 9.6. Typical Rectangular Grit chamber (Not showing Scraper arrangement).

The grit chambers can be cleaned periodically at about 3 weeks interval, either manually, mechanically or hydraulically. Hand cleaning is done only in case of smaller plants (of capacity less than about 4.5 million litres per day), while mechanical or hydraulic cleaning is adopted for larger plants. In *manual cleaning*, grit is removed manually by shovels, etc., while in *mechanical cleaning*, grit is removed with the help of machines ; and in *hydraulic cleaning*, grit is removed by the force of water-jet directed from a central point and removed through the pipes in the side walls or bottom of the chamber.

The removed grit may contain some organic matter, and can be washed prior to its disposal, if necessary, by using certain patented machines, and the wash water returned to the plant influent. Washed grit may still contain about 1 to 5% of putrescible organic matter.

The silt and grit, etc. removed by the grit chambers can be easily disposed off either by burial or burning (incineration) or for raising low lying areas by dumping. It cannot be used for preparation of concrete, as it contains sufficient organic matter.

9.7.7.1. Design of proportioning flow weir. The modified weir shape w.r. to Fig. 9.5 (*a*) can be designed by using the following equation

$$x = \frac{2 . B . V_h}{C_d \cdot \sqrt{2g}\, \pi . \sqrt{y}} \qquad ...(9.11)$$

where B = width of channel

V_h = horizontal flow velocity

C_d = coefficient of discharge ≈ 0.62

x and y are coordinates on weir profile, shown in Fig. 9.5 (*a*).

The bottom portion of the weir is slightly modified to a rectangle of width $2b$ and height a. The recommended minimum value of a is 0.025 m (25 mm) for domestic wastewaters. The value of b can be computed from Eq. (9.11) for assumed value of a, by using $y = a$ in the above equation, *i.e.*

$$b = \frac{2 . B . V_h}{C_d . \sqrt{2g}\, \pi . \sqrt{a}} = \frac{2B . V_h}{0.62 \times \sqrt{2 \times 9.81} \times 3.14 \times \sqrt{0.025}}$$

or $$\boxed{b = 1.467 . B . V_h} \qquad ...(9.12)$$

Example 9.2 (*a*) *A rectangular grit chamber is designed to remove particles with a diameter of 0.2 mm, specific gravity 2.65. Settling velocity for these particles has been found to range from 0.016 to 0.022 m/sec, depending on their shape factor. A flow through velocity of 0.3 m/sec will be maintained by proportioning weir. Determine the channel dimensions for a maximum wastewater flow of 10,000 cu m/day.*

Solution. Let us provide a rectangular channel section, since a proportional flow weir is provided for controlling velocity of flow.

Now,

Horizontal velocity of flow = V_h = 0.3 m/sec.

Settling velocity is between 0.016 to 0.022 m/sec, and hence let it be 0.020 m/sec.

Now, Q = velocity × cross-section

or $Q = V_h \times A$

where Q = 10,000 cu m/day

$$= \frac{10000}{24 \times 60 \times 60} \text{ m}^3/\text{s} = 0.116 \text{ m}^3/\text{s}$$

$\therefore$ $0.116 = 0.3 . A$

$$A = \frac{0.116}{0.3} \quad \text{or} \quad A = 0.385 \text{ m}^2.$$

Assuming a water depth (H) of 1 m above the crest of the weir, which is kept at 0.3 m above the channel bottom, we have the width (B) of the basin as

$1 \times B = 0.385$

or B = 0.385 m ; **say 0.4 m.**

Overall depth of grit chamber (D)

= Water depth above the crest of weir + 0.3 m

+ Free board of 0.45 m

= 1.0 m + 0.3 m + 0.45 m = **1.75 m**

Now, settling velocity

V_s = 0.02 m/sec

$$\therefore \quad \text{Detention time} = \frac{\text{Water depth in the basin}}{\text{Settling velocity}} = \frac{1}{0.02} = 50 \text{ secs.}$$

$\therefore$ Length of the tank = V_h × Detention time = 0.3 × 50 m = 15 m.

The length of the channel is increased beyond theoretical value by a factor of 10—50% to account for non-idealities in the flow and settling of particles. Hence, using about 30% (say) increased length, we get $L = 20$ m.

Hence, use a rectangular tank, with dimensions :

$$\left.\begin{aligned} \text{Length } (L) &= 20 \text{ m} \\ \text{Width } (B) &= 0.4 \text{ m} \\ \text{Depth } (D) &= 1.75 \text{ m} \end{aligned}\right\} \textbf{ Ans.}$$

9.7.2. Design of Parabolic or V-shaped Grit Chamber provided with a Parshall Flume

9.7.2.1. Parshall flume. A parshall flume, also called a **venturi flume,** is a horizontally constricted vertical throat provided in a rectangular section of an open channel, as shown in Fig. 9.7. Such a venturi-flume, as we know, can be used as a the best discharge measuring device, as it involves negligible head loss, giving smooth hydraulic flow and preventing deposition of solids. A parshall flume can also work under submerged conditions for certain limits. These *limits of submergence* are : 50% in case of 0.15 m throat width, and 70% for wider throat widths upto 1 m. Another advantage of a venturiflume is, that : one control section can be installed for 2 to 3 grit chambers. All treatment plants are often equipped with such discharge measuring devices, to exercise control on flow. Some existing STP's have venturimeters in the discharge piping from the wet well, but these are not recommended because of potential operation and maintenance problems.

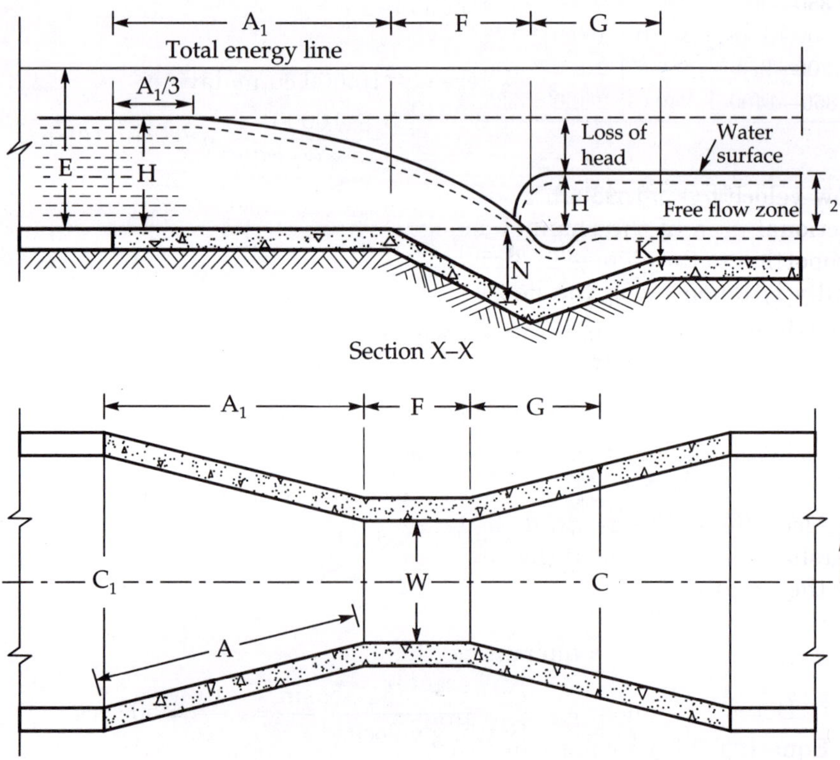

Fig. 9.7. The Parshall flume.

The discharge passing through a parshall flume of the type shown in Fig. 9.7, is related to the water head (*i.e.* upstream water depth) by the formula :

$$Q = 2.264\, W \,.\, (H_a)^{3/2} \quad \text{...(9.13)}$$

where W = The width of the throat in m

Q = Flow in m^3/sec

H_a = Depth of flow in upstream leg of flume at one third point in m.

The above equation is applicable to flumes of 0.3 m to 3 m in width.

Typical design dimensions for parshall flumes are given in table 9.2.

Table 9.2. Standard Dimensions of Parshall flume (w.r. to Fig. 9.7) in cm.

Flow range max. in M.L./sec	*Throat width W*	*A*	*A_1*	*C*	*C_1*	*F*	*G*	*K*	*N*
Upto	7.5	46.0	45.0	17.5	25.5	15.0	60.0	2.5	5.6
5—30	15.0	61.0	60.0	31.5	39.1	30.0	60.0	7.5	11.3
30—45	22.5	86.5	85.0	37.5	56.6	30.0	75.0	7.5	11.3
45—170	30.0	135.0	132.2	60.0	83.1	60.0	90.0	7.5	22.5
170—250	45.0	142.5	139.7	75.0	101.0	60.0	90.0	7.5	22.5
250—350	60.0	150.0	147.2	90.0	118.8	60.0	90.0	7.5	22.5
350—500	90.0	165.0	161.9	120.0	154.7	60.0	90.0	7.5	22.5
500—700	120.0	180.0	176.6	150.0	190.6	60.0	90.0	7.5	22.5
700—850	150.0	210.0	206.0	210.0	262.5	60.0	90.0	7.5	22.5
850—1400	240.0	240.0	235.3	270.0	334.4	60.0	90.0	7.5	22.5

9.7.2.2. Parabolic grit channel. In order to maintain a constant horizontal flow velocity (V_h) through a grit channel, we have to ensure that the cross-sectional area of the channel changes with the changed discharge in direct proportion to the change in discharge. Thus, if x is taken along the width side, and y is taken along the depth side of a channel x-section, then the cross-section must be such that

$$V_h = \frac{Q}{\int_0^y x \,.\, dy} = \text{constant} = C_1$$

or

$$Q = C_1 \,.\, \int_{y=0}^{y=y} x\, dy \quad \text{...(9.14)}$$

Also, the discharge through the control section such as a sutro weir or a parshall flume, placed at the down end of the grit channel, in general, is given by the equation

$$Q = C_2 \,.\, y^n \quad \text{...(9.15)}$$

where n is the discharge coefficient of the control section ; *i.e.* 1.5 for parshall flume, and 1 for proportional flow weir.

Equating Q from equations (9.14) and (9.15), we get

$$C_1 \int_0^y x \,.\, dy = C_2 \,.\, y^n$$

or $$C_1 \,.\, x \,.\, y = C_2 \,.\, y^n \quad \text{or} \quad y^{n-1} = \frac{C_1}{C_2} \,.\, x$$

or $$y^{n-1} = C' \,.\, x \qquad ...(9.16)$$

For a parshall flume, $n = 1.5$,

$$\therefore \quad y^{1.5-1} = C' \,.\, x \quad \text{or} \quad y^{1/2} = C' \,.\, x$$

or $$y = C \,.\, x^2 \qquad ...(9.16a)$$

Hence, when parshall flume is provided at the effluent end of a channel, then the channel section should be governed by equation $y = C \,.\, x^2$ (x representing width, and y representing depth) ; which is a parabolic section. *Such a parabolic section can be approximated to a trapezoidal section.* The method of designing the dimensions of a parabolic grit channel is explained in example 9.2 (*b*).

If a rectangular channel is used instead of a parabolic channel, then its width is constant ;

i.e. $x = k$. Then, using equation (9.16), we have

$$y^{n-1} = C' \,.\, x$$
$$= C' \,.\, k \quad \text{or} \quad y^{n-1} = C''$$

or $$n - 1 = 0 \qquad \text{or} \qquad n = 1.$$

Hence, for a rectangular channel section, we need a control section, such as a **proportional flow weir**, *whose discharge equation is of the form*

$$Q = K \,.\, y \qquad ...(9.16b)$$

where K is a constant.

Example 9.2 (*b*) *Design a parabolic (trapezoidal) grit chamber for a horizontal velocity of 25 cm/sec and a flow which ranges from a minimum of 25000 cu-m/day to a maximum of 1,00,000 cu-m/day. Average flow is 62500 cu-m/day.*

Solution. Let us adopt 4 grit channels, each designed to carry discharges as :

(*i*) *Minimum discharge* $= \frac{25000}{4} = 6250$ cu m/day

(*ii*) *Normal maximum discharge* $= \frac{1,00,000}{4} = 25000$ cu m/day

(*iii*) *Average discharge* $= \frac{62500}{4} = 15625$ cu m/day

(*iv*) Let each unit be designed to carry a *peak discharge* of 1.33 times the normal max. discharge, say

$$= 1.33 \times 25000 = 33,325 \text{ cu-m/day}.$$

Let a vertically controlled flume be used to maintain constant velocity. The flow in the control section is assumed to be at critical depth.

Let us now design the parabolic channel X-section :

For a parabolic channel,

Area of cross-section $= A = \frac{2}{3}$ B.D.

where B = Top width

D = Depth.

Since $A = \dfrac{Q}{V_h}$; and $V_h = 0.25$ m/sec,

(for all discharges), we can easily work out, A values corresponding to peak, maximum, average, and minimum discharges.

$$\therefore A \text{ for } Q_{peak} = (A_{peak}) = \frac{Q_{peak}}{0.25} = \frac{33325}{24 \times 60 \times 60} \times \frac{1}{0.25} \text{ m}^2$$

$$= \frac{0.386}{0.25} = 1.54 \text{ m}^2. \qquad ...(i)$$

Similarly,

$$A \text{ for } Q_{max} = (A_{max}) = \frac{25000}{24 \times 60 \times 60} \times \frac{1}{0.25} = \frac{0.289}{0.25} = 1.16 \text{ m}^2 \qquad ...(ii)$$

$$A \text{ for } Q_{av.} = (A_{av.}) = \frac{15625}{24 \times 60 \times 60} \times \frac{1}{0.25} = \frac{0.181}{0.25} = 0.72 \text{ m}^2 \qquad ...(iii)$$

$$A \text{ for } Q_{min} = (A_{min}) = \frac{6250}{24 \times 60 \times 60} \times \frac{1}{0.25} = \frac{0.072}{0.25} = 0.29 \text{ m}^2. \qquad ...(iv)$$

At maximum discharge

Let us limit the maximum width of the channel to 1.5 m at $Q_{max.}$ Then, we have

$$A_{max} = \frac{2}{3} B \,.\, D_{max} = \frac{2}{3} \times 1.5\, m \times D_{max} \qquad ...(v)$$

Equating eqn. (*ii*) and (*v*), we have

$$1.16 = \frac{2}{3} \times 1.5 \,.\, D_{max}$$

or
$$D_{max} = \frac{1.16 \times 3}{2 \times 1.5}$$

or
$$D_{max} = 1.16 \text{ m.}$$

The total energy upstream of control section is given by

$$E_1 = D + \frac{V_1^2}{2g} \qquad ...(vi)$$

$$\approx D \text{ (neglecting velocity of approach in channel)}$$

or
$$E_1 = D \qquad ...(vii)$$

Hence at maximum discharge

$$E_1 = D_{max} = 1.16 \text{ m} \qquad ...(viii)$$

Total energy at critical point *i.e.* at the point of jump formation

$$= E_c = y_c + \frac{V_c^2}{2g} + H_L \qquad ...(9.17)$$

where H_L is the energy or head loss in jump.

Assuming that the head loss in control section (H_L) is 10% of the velocity head in the control section, we have

$$H_L = 0.1\,\frac{V_c^2}{2g}$$

$$\therefore \quad E_c = y_c + \frac{V_c^2}{2g} + 0.1\,\frac{V_c^2}{2g} \qquad \text{...(9.17}a\text{)}$$

But, in the control section, at critical depth, we know

$$y_c = \frac{V_c^2}{g} \qquad \text{...(9.18)}$$

$$\therefore \quad E_c = \frac{V_c^2}{g} + \frac{V_c^2}{2g} + 0.1\frac{V_c^2}{2g}$$

or $$E_c = \frac{3.1}{2}\cdot\frac{V_c^2}{g}$$

or $$E_c = 1.55\,\frac{V_c^2}{g} \qquad \text{...(9.17}b\text{)}$$

Using Bernoulli's theorem,

Total energy at *u*/s point in channel

= Total energy at critical point in control section

or $$E_1 = E_c = 1.55\,\frac{V_c^2}{g}$$

But $E_1 = D$ (from eq. *vii*)

$$\therefore \quad \boxed{D = 1.55\,\frac{V_c^2}{g}} \qquad \text{...(9.19)}$$

Using the value of D_{max}, as equal to 1.16 at maximum discharge, we have

$$\therefore \quad V_c \text{ at } Q_{max} = \sqrt{\frac{1.16 \times 9.81}{1.55}} = 2.71 \text{ m/sec.}$$

Also $$y_c \text{ at } Q_{max} = \frac{V_c^2 \text{ at } Q_{max}}{g} = \frac{(2.71)^2}{9.81} = 0.74 \text{ m.}$$

The discharge through the control section is :

$$Q = (W \,.\, y_c)\, V_c \qquad \text{...(9.20)}$$

where W is the throat width ;
and $W \,.\, y_c$ is the flow area of the throat.

or $$W = \frac{Q}{y_c \,.\, V_c} \qquad \text{...(9.20 }a\text{)}$$

$$\therefore \quad W = \frac{Q_{max}}{y_c \,.\, V_c (\text{both at } Q_{max})}$$

$$= \frac{0.289}{0.74 \times 2.71} = 0.144 \text{ m}$$

Hence, use throat width $W = 0.144$ m.

For other flow conditions. Using the above used two formulas, we have

$$y_c = \frac{V_c^2}{g} \quad \text{...(i.e. eqn. 9.17)}$$

Also $$W \cdot y_c \cdot V_c = Q \quad \text{...(i.e. eqn. 9.19)}$$

$$\therefore \quad V_c = \frac{Q}{W \cdot y_c} \quad \text{or} \quad y_c = \frac{\left(\frac{Q}{W \cdot y_c}\right)^2}{g}$$

or $$g \cdot y_c \cdot y_c^2 W^2 = Q^2$$

or $$y_c^3 = \frac{Q^2}{g \cdot W^2} \quad \text{or} \quad y_c = \sqrt[3]{\frac{Q^2}{gW^2}} \quad \text{...(9.21)}$$

Knowing Q and W, we can find y_c at different discharges.

From eqn. (9.13) derived above, we have

$$D = 1.55 \cdot \frac{V_c^2}{g}$$

or $$D = 1.55\, y_c \quad \text{...(9.22)}$$

Knowing y_c at different discharges, we can find D at different discharges.

Then finally, we have for a parabolic section

$$\frac{2}{3} \times B \times D = A$$

$$\therefore \quad B = \frac{1.5A}{D} \quad \text{...(9.23)}$$

Knowing A at various discharges, already computed, we can find B at different discharges, because D is known at different discharges.

The values of B are, thus, calculated for other discharges, as shown in table 9.3.

With B and D values, computed in col. (6) and (7) of the above table, we can draw the parabolic section, which is approximated to a practical trapezoidal section, as shown in Fig. 9.8.

Table 9.3

Cond.	*Q* *m³/day*	*Q in* *m³/sec*	*A* *m²*	$y_c = \sqrt[3]{\frac{Q^2}{gW^2}}$ *where W = 0.144*	$D = 1.55\, y_c$	$B = \frac{1.5A}{D}$
(1)	(2)	(3)	(4)	(5)	(6)	(7)
Peak	33,325	0.386	1.54	0.90	1.40	1.65*
Maximum	25000	0.289	1.16	0.74	1.15	1.51
Average	15625	0.181	0.72	0.54	0.84	1.29
Minimum	6250	0.072	0.29	0.29	0.45	0.97

*Limited to 1.51 m.

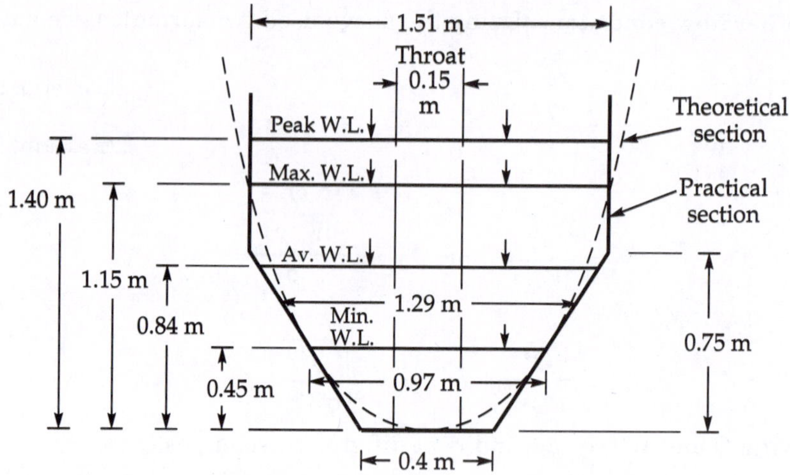

Fig. 9.8. Grit channel section for example 9.2 (*b*).

9.8. Aerated Grit Channels

Aerated grit channels are like *spiral flow aeration tanks**, usually designed to provide detention periods of about 3 minutes at the maximum rate of flow. The rectangular tank cross-section is also similar to that provided for spiral circulation in activated-sludge aeration tanks*, except that a grit hopper of about 0.9 m depth with steeply sloping sides is located under the air diffusers, as shown in Fig. 9.9 (*a*). The air diffusers are located at about 0.45 m to 0.60 m above the normal bottom of the tank.

The spiral roll or *helical flow* of waste water in the grit chamber (Fig. 9.9 (*b*)) 'drives' the grit into the hopper. The shearing action of the air bubbles is supposed to strip the inert grit of much of the organic material that adheres to its surface.

The performance of an aerated grit chamber is a function of the *roll velocity* and *detention time*. If the velocity of roll is too high, grit will be carried through the chamber ; and if it is too low, organic matter may also settle down along with the grit. The roll velocity can, however, be easily controlled by adjusting the air feed rate, which may range between 0.15 to 0.45 cubic metre per minute of air per metre of tank length (typical value being 0.30 m^3/min per m length).

Liquid detention time is kept in the design at about 3 minutes at maximum flow. Length to width ratios may range from 2.5 to 7.0 with depths of the order of 2 to 5 m, with width-depth ratio varying between 1 : 1 to 5 : 1 (typical value being 2 : 1).

Grit accumulation in the chamber varies largely, depending on whether the sewer system is of a combined type, or of a separate type ; and also on the efficiency of the chamber. For combined systems, 90 m^3 of grit per M.cum of sewage is not uncommon, while in a separate system, one might expect something less than 30 m^3/M-cum of sewage. Normally, the grit is buried in a sanitary landfill site, being free of organic matter.

---

*Explained in article 9.33.2.

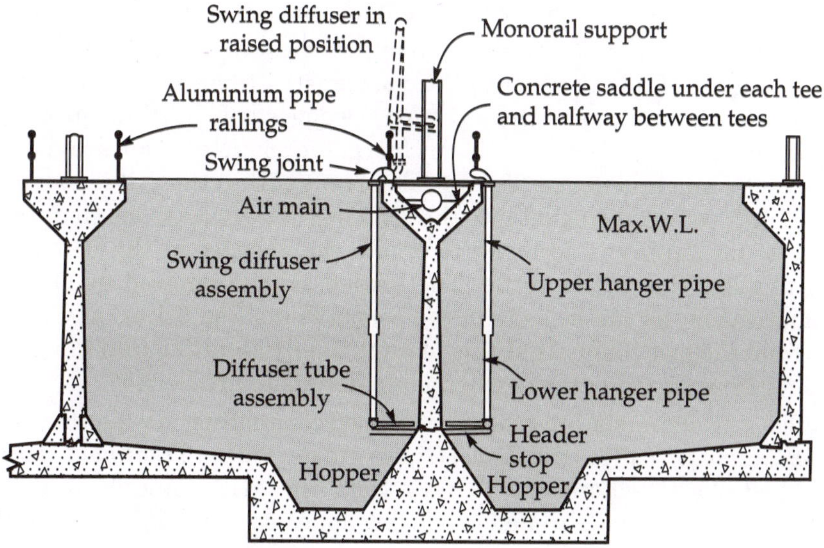

(*a*) Typical section through a two chambered aerated grit channel

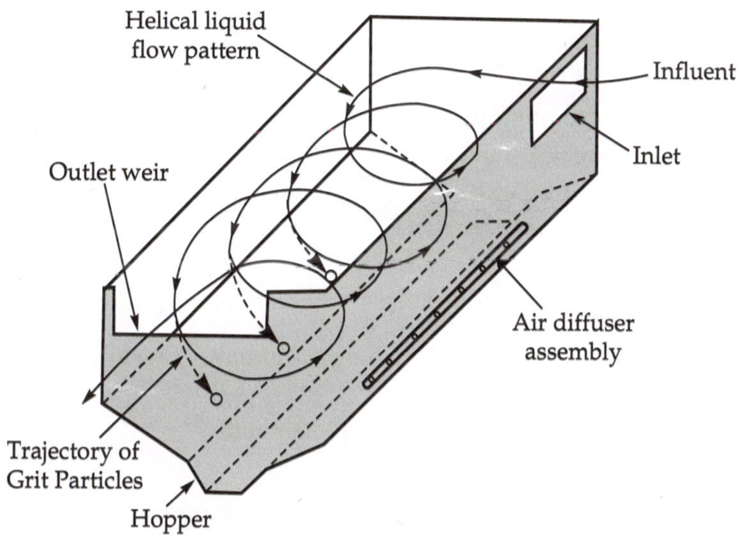

(*b*) Helical liquid flow pattern (spiral roll) in an aerated grit chamber

Fig. 9.9. Aerated grit chamber.

Many aerated chambers have been provided with mechanical means for grit removal in the form of *grab buckets* [Pl. see Colour Photo Fig. 9.1(*a*) under article 9.47] travelling on monorails, centred over the grit collection and storage trough. Other installations are equipped with chain-and-bucket conveyor, running the full length of the storage troughs, which move the grit to one end of the trough, and elevate it above the waste water level in a continuous operation. Screw conveyors, jet pumps, and air lift pumps have also been used for hydraulical removal of the grit.

Aerated grit channels have been used at 690 MLD *Ghatkopar Sewage Treatment Plant, Mumbai*, where the lifted sewage at the plant is first made to pass through the 4 No. semi automatic screens of 1600 × 2000 mm size to arrest the floating solids, and the screened sewage is made to flow through the 4 No. aerated grit channels to settle out inorganic particles of sp. gravity = 2.6 and of size 0.2 mm and above. The spiral current within the grit is created by the use of diffused air, being let into the grit channels through the compressor. The rate of air supply is adjusted to create the velocity in the bottom, low enough to allow the grit to settle. The settled grit is removed by Grab and Crane machinery, as can be seen in the colour Photo Fig. 9.1 (*a*) and (*b*). The sewage from the grit channels at this plant is finally admitted into the aerated lagoons for final treatment, as explained in details in article 9.47.

Aerated grit channels have also been used at another sewage treatment plant at Mumbai, called the *Love Grove Waste Water Treatment Facility Terminal at Worli*, where the outflow from the grit channels is directly discharged into the sea through a 3.5 km long *marine outfall* conduit, as explained in article 8.5.

Example 9.3. *Design an aerated grit chamber for treating municipal waste water with average flow rate of 0.5 m³/s (43.2 MLD). Assume the peak flow rate to be 3 times the average.*

Solution. Peak flow rate = $0.5\ m^3/sec \times 3 = 1.5\ m^3/s$

Assume average liquid detention time = 3 min = 180 sec.

$\therefore$ Aerator volume = $1.5\ m^3/s \times 180\ s = 270\ m^3$

In order to drain the channel periodically for routine cleaning and maintenance, *use two chambers*.

$$\therefore \quad \text{Volume of one aerated channel} = \frac{270\ m^3}{2} = 135\ m^3$$

To determine the *dimensions of the aerated channel*, assume depth of 3 m and width-depth ratio of 2 : 1.

$\therefore$ Width of channel = Depth × 2 = 3 m × 2 = 6 m

$$\therefore \quad \text{Length of channel} = \frac{135\ m^3}{3\ m \times 6\ m} = 7.5\ m$$

Increase the length by about 20% to account for inlet and outlet conditions.

$\therefore$ Provided length = 7.5 × 1.2 = **9 m.**

Hence, use 2 chambers, each of size 9 m × 6 m × 3 m depth. **Ans.**

Air supply requirement. Assume that $0.03\ m^3/min$ per m length may be adequate,

$$\text{Air required} = 0.03\ \frac{m^3}{min\ .\ m} \times 9\ m = \mathbf{0.27\ m^3/min.} \quad \textbf{Ans.}$$

Volume of grit produced daily. Assume that $50\ m^3$/M.cum of sewage of grit is produced by the incoming sewage, the daily grit volume produced

= Peak flow rate of sewage in m³/d × Grit produced per m³ of sewage

$$= 1.5\ \frac{\text{m}^3}{\text{s}} \times \left(24 \times 3600\ \frac{s}{d}\right) \times \frac{50\ \text{m}^3}{10^6\ \text{m}^3} = \mathbf{6.48\ m^3/d.}\quad \textbf{Ans.}$$

[**Note.** The grit handling facilities must be based on sustained peak flow rate. Hence, the arrangement for removal of grit @ 6.48 m^3/d must be provided.]

9.9. Detritus Tanks

Detritus tanks are nothing but rectangular grit chambers, designed to flow with a smaller flow velocity (of about 0.09 m/sec) and longer detention periods (about 3 to 4 minutes) so as to separate out not only the larger grit, etc., but also to separate out the very fine sand particles, etc. Due to this, a large amount of organic matter will also settle out along with the inorganic grit, sand, etc. This organic material is then separated from the grit by control of currents in the tank through baffles, or by controlled aeration of the flow through the tank. The rising air bubbles will then separate the lighter organic matter from the descending grit. The grit is removed continuously by means of scraper mechanism. All other details of detritus tanks remain the same as those of a rectangular grit chamber.

Example 9.4. *Design a suitable grit chamber cum Detritus tank for a sewage treatment plant getting a dry weather flow from a separate sewerage system @ 400 l/s. Assume the flow velocity through the tank as 0.2 m/sec ; and detention period of 2 minutes. The max. flow may be assumed to be three times of dry weather flow.*

Solution. The length of the tank

$$= \text{Velocity} \times \text{Detention time} = 0.2 \times (2 \times 60) = 24\ \text{m}.$$

Since the peak flow is three times the DWF, let us provide three detritus tanks, each designed for passing D.W.F.

∴ The discharge passing through each tank = 400 l/s = 0.4 m^3/sec.

$$\therefore\quad \text{Cross-sectional area required} = \frac{\text{Discharge}}{\text{Velocity}} = \frac{0.4}{0.2} = 2\ \text{m}^2.$$

Assuming the water depth in the tank to be 1.2 m, we have the width of the tank

$$= \frac{\text{Area of } X\text{-section}}{\text{Depth}} = \frac{2}{1.2} = 1.67\ \text{m} \text{ ; say } 1.7\ \text{m}.$$

Hence, use a Detritus tank with 24 m × 1.7 m × 1.2 m size.

At the top, a free-board of 0.3 m may be provided ; and at the bottom, a dead space depth of 0.45 m for collection of detritus may be provided.

Thus, the overall depth of the tank

$$= 1.2 + 0.3 + 0.45 = 1.95\ \text{m}.$$

The tank will be 1.7 m wide upto 1.5 m depth, and then the sides will slope down to form an elongated trough of 24 m length and 0.8 m width at the bottom with rounded corners, as shown in Fig. 9.10.

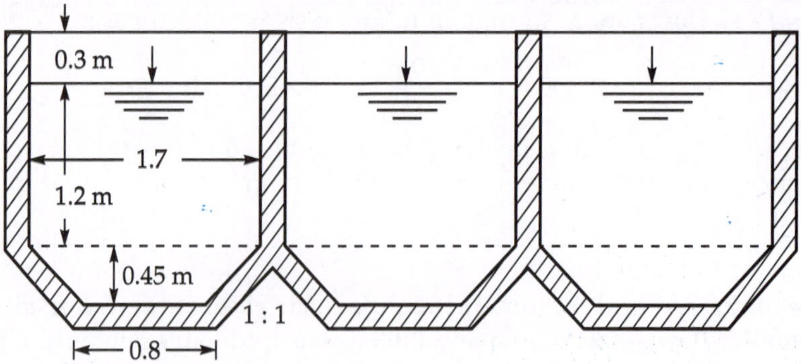

Fig. 9.10

TANKS FOR REMOVING OILS AND GREASE

9.10. Skimming Tanks

Skimming tanks are sometimes employed for removing oils and grease from the sewage, and placed before the sedimentation tanks. They are, therefore, used where sewage contains too much of grease or oils, which include fats, waxes, soaps, fatty acids, etc. These materials may enter into the sewage from the kitchens of restaurants and, houses, from motor garages, oil refineries, soap and candle factories, etc. They are, thus, normally present in large amounts in the industrial wastewaters.

If such greasy and oily matter is not removed from the sewage before it enters further treatment units, it may form unsightly and odorous scums on the surface of the settling tanks, or interfere with the activated sludge treatment process, and inhibit biological growth on the trickling filters.

These oil and greasy materials may be removed in a skimming tank, in which air is blown by an aerating device through the bottom. The rising air tends to coagulate and congeal (solidify) the grease, and cause it to rise to the surface (being pushed in separate compartments), from where it is removed.

The typical details of a skimming tank are shown in Fig. 9.11. It consists of a long trough shaped structure divided into two or three lateral compartments

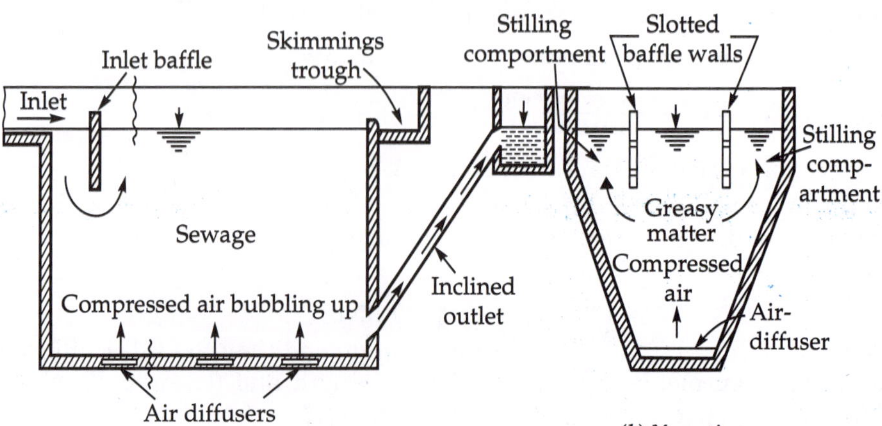

Fig. 9.11. Skimming Tank.

by means of vertical baffle walls (having slots in them) for a short distance below the sewage surface, as shown in Fig 9.11 (*b*). The baffle walls help in pushing the rising coagulated greasy material into both the side compartments (called stilling compartments). The rise of oils and grease is brought about by blowing compressed air into the sewage from *diffusers* placed at the bottom of the tank. The collected greasy materials are removed (*i.e.* skimmed off) either by hand or by some mechanical equipment. It may then be disposed of either by burning or burial.

Sewage enters the tank from one end, flows through longitudinally, and finally goes out through a narrow inclined duct, as shown. This is so narrow that the suspended heavier particles are carried up its slope and out of the tank. *A detention period of about 3 to 5 minutes is usually sufficient*, and the amount of compressed air required is about 300 to 6000 m^3 per million litres of sewage. The surface area required for the tank can be found out by using the formula :

$$A = 0.00622 \frac{q}{V_r} \qquad \text{...(9.24)}$$

where q = Rate of flow of sewage in m^3/day.

V_r = Minimum rising velocity of greasy material to be removed in m/minute

= 0.25 m/minute in most cases.

The efficiency of a skimming tank can be increased considerably (three to four times) by passing chlorine gas (2 mg per litre of sewage) along with the compressed air. Chlorine may also be added as a solution with the sewage discharge, just ahead of the air diffuser plates. The action of chlorine is to destroy the protective colloidal effect of protein, which holds the grease in emulsified form.

9.11. Vacuators

Grease can also be removed from the sewage by *vacuum floatation method*, by subjecting the aerated sewage to a vacuum pressure of about 0—25 cm of mercury for 10 to 15 minutes in a *vacuator*. This causes the air bubbles to expand and move upward through the sewage to the surface. The rising bubbles lift the grease and the lighter waste solids to the surface, where they are removed through skimming troughs. Heavier solids settle to the tank bottom, where they are collected and carried away for sludge treatment and disposal.

9.12. Disposal of Skimmings

The oil and greasy material removed as skimmings from the skimming tanks or vacuators can be disposed of either by burning or burial. It is generally too polluted to be of any economic use. However, it may sometimes be converted in to soap lubricants, candles and other non-edible products. It may sometimes be digested in digestors, which prove beneficial only if the mineral oils are less in amount, and vegetable and organic matters predominate. The latter digest easily, and produce gases of high fuel value.

9.13. Necessity and Use of Skimming Tanks in India

In hot countries like India, it is very difficult to skim out the oils and grease, because they do not coagulate and congeal (solidify due to cooling) easily. Moreover, in normal municipal wastewaters, they are generally found in very meagre quantities. They are, however, largely present in particular industrial wastewaters. Attempts are, therefore, generally made in India, to remove the greasy matters before they enter the municipal sewers, by constructing **oil and grease traps,** as discussed in chapter 5.

Skimming tanks are, hence, generally not provided here in India at our treatment plants. They may, however, be provided in particular industrial towns, or at colder hill stations like Shimla, Mussoorie, Nainital, etc., where the greasy materials may congeal easily.

SEDIMENTATION

9.14. Necessity of Sedimentation in Treatment of Waste Waters

As discussed in the previous pages, the screens and the grit chambers do remove most of the floating materials (like paper, rags, cloth, wood, tree branches, etc.) and the heavy inorganic settleable solids from the sewage. However, a part of the suspended *organic solids* which are too heavy to be removed as floating matters, and too light to be removed by grit chambers (designed to remove only the heavy inorganic solids of size more than 0.2 mm, and of sp. gravity 2.65), are generally removed by the sedimentation tanks. *The sedimentation tanks are thus designed to remove a part of the organic matter from the sewage effluent coming out from the grit chambers.*

In a complete sewage treatment, the sedimentation is, in fact, carried out twice ; once before the biological treatment (*i.e. primary sedimentation*) and once after the biological treatment (*i.e. secondary sedimentation*). When chemical coagulants are also used for flocculating the organic matter during the process of sedimentation, the process is called *chemical precipitation* or *sedimentation aided with coagulation*. This is generally not used in modern days, as discussed, later.

Other sewage treatment units which work on the principle of sedimentation are : *Septic tanks, Imhoff tanks*, etc. Septic tanks and Imhoff tanks combine sludge digestion with sedimentation, whereas the sludge deposited in primary as well as in the secondary settling tanks, is separately digested in the *sludge-digestion tanks*.

9.15. Sedimentation Tanks

The clarification of sewage by the process of 'sedimentation' can be affected by providing conditions under which the suspended material present in sewage can settle out. This is brought about in specially designed tanks called sedimentation tanks.

Out of the three forces which control the settling tendencies of the particles (enumerated earlier), the two forces *i.e.* the velocity of flow, and the shape and size of the particles, are tried to be controlled in these settling tanks. The third force *i.e.* the viscosity of sewage or the temperature of sewage is left uncontrolled, as the same is not practically possible.

The velocity of flow can be reduced by increasing the length of travel, and by detaining the particles for a longer time in the sedimentation basin. The size and the shape of the particles can be altered by the addition of certain chemicals in water. These chemicals are called *coagulants*, and they make the sedimentation quite effective leading to the settlement of even very fine and colloidal particles. However, their use is not made in "Plain Sedimentation" (or generally called "sedimentation"), but is being made in the process called "chemical precipitation" or "sedimentation with coagulation". This will be discussed later.

Sedimentation basins are thus designed for effecting settlement of particles by reducing the flow velocity or by detaining the sewage in them. They are generally made of reinforced concrete and may be rectangular or circular in plan. Long narrow rectangular tanks with *horizontal flow* (Fig. 9.12*a*) are generally preferred to the circular tanks with *radial* or *spiral flow* (Fig. 9.12*b*).

The capacity and other dimensions of the tank should be properly designed, so as to effect a fairly high percentage of removal of the suspended organic material. A plain sedimentation tank under normal conditions may remove about 60 to 65% of the suspended solids, and 30 to 35% of the BOD from the sewage*.

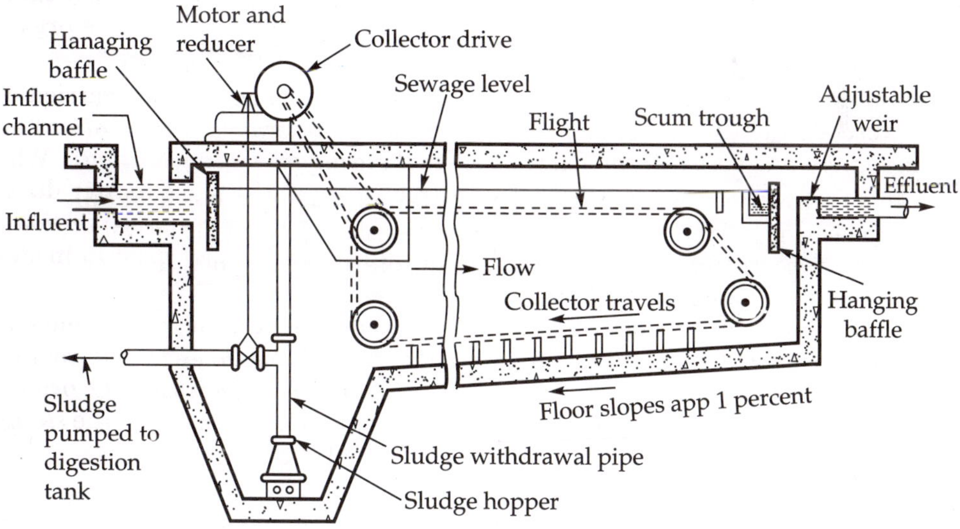

Fig. 9.12 (*a*) Rectangular sedimentation tank.

*The percentage removal of BOD is lesser than the percentage removal of the suspended solids, because more BOD is caused by the finer non settleable and dissolved solids which escape out through the effluent than due to the settleable solids which are removed in these tanks.

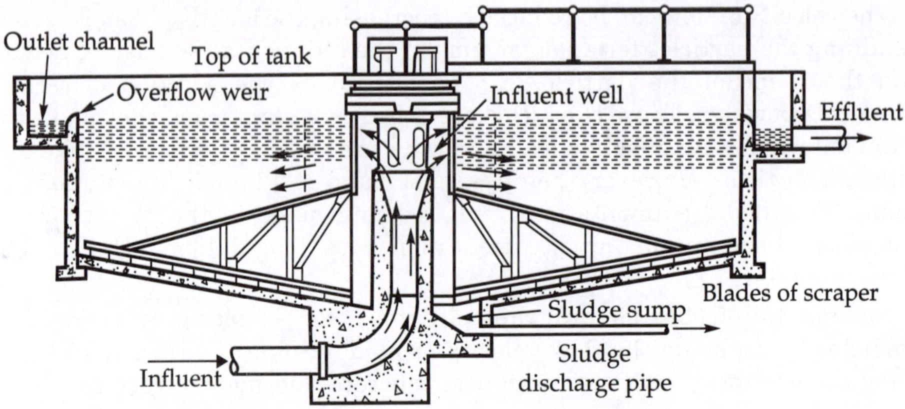

Fig. 9.12 (*b*) Circular sedimentation tank.

9.15.1. Types of Sedimentation Tanks. Sedimentation tanks may function either intermittently or continuously.

The Intermittent Settling tanks called ***quiescent type tanks*** are simple settling tanks which store sewage for a certain period and keep it in complete rest. After giving it a rest of about 24 hours, during which the suspended particles settle down to the bottom of the tank, the cleaner sewage from the top may be drawn off, and the tank be cleaned off the settled silt. The tank is again filled with raw sewage to continue the next operation. This type of tank, thus, functions intermittently, as a period of about 30 to 36 hours is required to put the tank again in working condition. This necessitates the commissioning of atleast two tanks. Such tanks are generally not preferred, because a lot of time and labour is wasted and more units are required, *They have, therefore, become completely obsolete these days.*

In a **continuous flow type** of a sedimentation tank, which is generally used in modern days, the flow velocity is only reduced, and the sewage is not brought to complete rest, as is done in an intermittent type. The working of such a tank is simple, as the water enters from one end, and comes out from the other end. The velocity is sufficiently reduced by providing sufficient length of travel. *The velocity is so adjusted that the time taken by the particle to travel from one end to another is slightly more than the time required for settlement of that particle.* The theory and design of such a tank is discussed below in details.

9.15.2. Design of a Continuous Flow Type of a Sedimentation Tank. In the theory which is applied to the design of such sedimentation basins, it is assumed that the sediment is uniformly distributed as the sewage enters the basin. In Fig. 9.13, let the water containing uniformly distributed sediment, enters the rectangular tank with a uniform velocity V. If Q is the discharge entering the basin, the flow velocity V is given by

$$V = \frac{Q}{BH} \quad \text{...(9.25)}$$

where B = Width of the tank or basin, and

H = Depth of sewage in the tank.

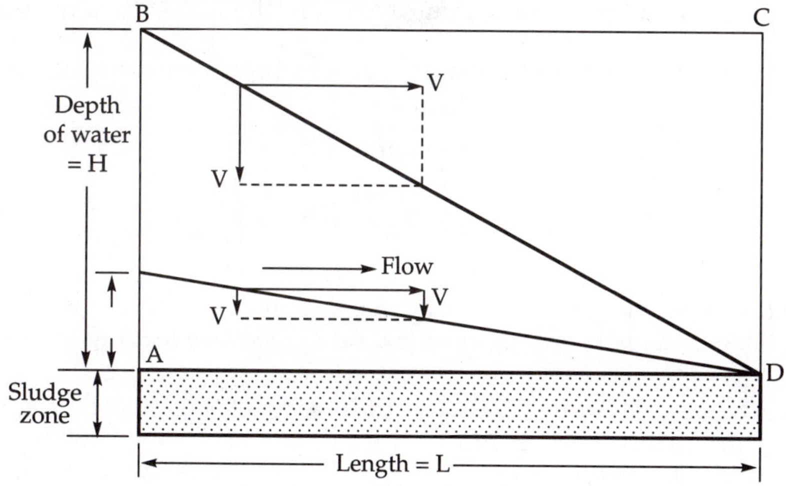

Fig. 9.13. Elevation or L-section of a rectangular sedimentation tank (Line diagram).

Now, every discrete particle is moving with a horizontal velocity V and a downward vertical velocity V_s. The resultant path is given by the vector sum of its flow velocity (V) and its settling velocity (V_s).

Now assuming that all those particles whose paths of travel are above the line BD, will pass through the basin, we have from geometric considerations :

$$\frac{V}{V_s} = \frac{L}{H} \qquad \text{...(9.26)}$$

or

$$V_s = \frac{V \cdot H}{L}.$$

But

$$V = \frac{Q}{BH} \qquad \text{from equation (9.25)}$$

$$\therefore \quad V_s = \frac{Q}{BH} \cdot \frac{H}{L} \quad \text{or} \quad V_s = \frac{Q}{BL} \qquad \text{...(9.27)}$$

It shows that all those particles with a settling velocity equal to or greater than Q/BL will settle down, and be removed. In other words, no particle having a settling velocity more than or equal to Q/BL will remain suspended in such a tank.

It is mentioned above, that a particle having settling velocity greater than or equal to Q/BL will be removed. In fact, it is the case when the particle entering at full height H of the tank is considered. Truly speaking, even the smaller particles having settling velocities lower than Q/BL will also settle down, if they happen to enter at some other height h of the tank. In that case, when the particles are entering at some other height h of the tank, all those particles having their settling velocities $\geq \frac{h}{H}\frac{Q}{BL}$ will settle down.

If y_0 represents the number of particles of a given size that are settled out, and y being the total number of particles of that size, then the percentage of that particular sized particles which shall be removed is y_0/y ; and is equal to h/H for an assumed uniform distribution of particles. Hence, if 70% of particles of a given size are proposed to be removed in a settling tank, then the settling velocity of that sized particles must be kept $\geq \frac{70}{100} \cdot \frac{Q}{BL}$. In other words

$\frac{Q}{BL}$ must be kept less than equal to

$$\frac{100}{70} \times \text{settling velocity of that sized particle} \qquad ...(9.28)$$

It, therefore, follows that the quantity Q/BL *i.e.* the discharge per unit of plan area is a very important term for the design of continuous flow type of settling tanks ; and is known as **overflow rate** or **surface loading** or **overflow velocity**.

Normal values of overflow rates ranges between 40,000 to 50,000 litres/sq. m/ day for plain* primary sedimentation tanks ; and between 50,000 to 60,000 litres/sq. m/day for sedimentation tanks using coagulants as aids ; and about 25,000 to 35,000 litres/sq. m/day for secondary sedimentation tanks.

Decreasing the overflow rate will lead to the settlement of even those particles which are having lower values of their settling velocities. *Hence, smaller particles will also settle down, if the overflow rate is reduced.* Further, with a given Q, the overflow rate can be reduced by increasing the plan area of the basin. It therefore, follows that an increase in the plan area (*i.e.* width × length) will increase the efficiency of sedimentation tank ; and *theoretically speaking, depth does not have any effect on the efficiency of sediment removal.* However, it is important for practical considerations, and also for making allowance for deposition of sludge and silt.

Usual values of effective depth (*i.e.* depth excluding the bottom sludge zone) *ranges between 2.4 to 3.6 m* (generally not exceeding 3 m).

Another important term, which is used in connection with the design of sedimentation basins, is its **detention time** or **detention period** or **retention**

---

*It is intended to remove 60 to 70% of the organic particles (sp. gr. 1.2) of size larger than 0.06 mm. The settling velocity for such particles is about 0.3 mm/sec or 1.2 m/hr. For affecting settlement, we have

$$\frac{Q}{BL} \leq \frac{100}{70} \times 1.2 \text{ m/hr (For affecting 70\% removal)}$$

or $$\frac{Q}{BL} = 1.7 \text{ m/hr.}$$

or $$\frac{Q}{BL} = 1.7 \text{ m}^3/\text{m}^2 \text{ plan area/hr.}$$

$$= 1700 \text{ litres/m}^2 \text{ of plan area/hr.}$$

$$= 1700 \times 24 \text{ litres/m}^2\text{/day}$$

$$= 40800 \text{ litres/m}^2\text{/day}$$

Similarly, for affecting 60% removal, it can be calculated that

$$\frac{Q}{BL} = 48{,}000 \; l/\text{m}^2/\text{day}$$

Hence, surface loading adopted is about 40,000 l/m^2/day to 50,000 l/m^2/day for removing 70% to 60% of organic particles. Lesser is this value, higher is the removal.

period. The detention time (t) of a settling tank may be defined as the average theoretical time required for the sewage to flow through the tank. It is, thus, the time that would be required for the flow of sewage to fill the tank, if there was no outflow. In other words, it is the average time for which the sewage is detained in the tank. Hence, it is the ratio of the volume of the basin to the rate of flow (*i.e.* discharge) through the basin.

∴ Detention time, t, for a Rectangular tank

$$= \frac{\text{Volume of the tank}}{\text{Rate of flow}}$$

$$= \frac{B.L.H.}{Q} \qquad \text{...(9.29)}$$

Similarly, the detention time for a circular tank

$$= \frac{d^2(0.011\,d + 0.785\,H)}{Q} \qquad \text{...(9.30)}$$

where d = Dia of the tank

H = Vertical depth at wall or Side water depth.

The detention time for a sewage sedimentation tank usually ranges between 1 to 2 hours*. The lower value of detention period (*i.e.* 1 hour) is generally adopted when the activated sludge treatment is used in secondary treatment after the sedimentation ; and the higher and more normal value (*i.e.* 2 hours) is generally adopted when the trickling filters are used as the secondary treatment.

Larger detention periods will result in higher efficiency ; but too long a period induces septic conditions, and should be avoided. However, if the secondary sedimentation is to be avoided, a longer detention period of about $2\frac{1}{2}$ hours to 3 hours may be adopted.

The width of the tank is normally kept at about 6 m, and not allowed to exceed 7.5 m or so. The length of the tank is generally not allowed to exceed 4 to 5 times the width. The cross-sectional area of the sedimentation tank is such as to provide a horizontal flow velocity of about 0.3 m/minute. The total amount of flow from the tank within 24 hours, generally equals the maximum daily flow of sewage.

The maximum dia of a circular tank may be kept 60 m or so.

9.15.3. Short Circuiting in the Sedimentation Tanks. For the efficient removal of sediment in the sedimentation tanks, it is necessary that the flow is uniformly distributed throughout the cross-section of the tank. If currents, on the other hand, permit a substantial portion of the water to pass directly through the tank without being detained for the intended time, the flow is said to be *short circuited. Properly designed inlets and outlets near the entrance and the exit may reduce the short circuiting tendencies, and distribute the flow more evenly*. Moreover, relatively narrow tanks are less affected by inlet and outlet disturbances, and by currents caused by breezes.

---

*The settling velocity of the particles to be separated being of the order of 0.3 mm/sec or 1.2 m/hr., means that a detention period of 2 hrs will give satisfactory removal of intended solids (larger than 0.06 mm or so) in a tank of 2.4 m depth.

But however, in actual practice, certain amount of short circuiting will always exist, and, therefore, the actual average time taken by a batch of water in passing through a settling tank (called **flowing through period**) will always be less than the detention period, which is the corresponding theoretical time. The ratio of the 'flowing through period' to the 'detention period' is called the *Displacement efficiency*.

∴ Displacement efficiency (%)

$$= \frac{\text{Flowing through period}}{\text{Detention period}} \quad ...(9.31)$$

Note. In order to counteract the effects of short circuiting, it may be necessary to keep a high detention period or a smaller surface loading than that obtained from the theoretical considerations for obtaining the desired results.

9.15.4. Constructional Details of the Sedimentation Tanks. The following features of such tanks need special attention :

(*i*) **Inlet and Outlet Arrangement.** In order to reduce short circuiting, and to distribute the flow uniformly, proper arrangement must be made for smooth entry of water. A most suitable type of an inlet for a rectangular settling tank is in the form of a channel extending to full width of the tank, with a submerged weir type baffle wall, as shown in Fig. 9.14.

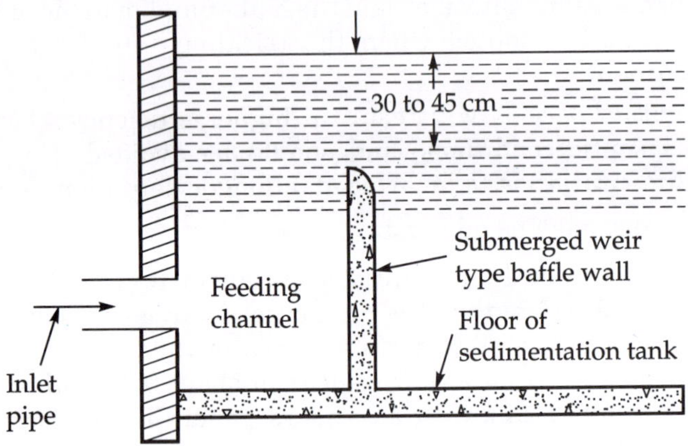

Fig. 9.14. Section of a 'Submerged type' or a 'Weir type' inlet.

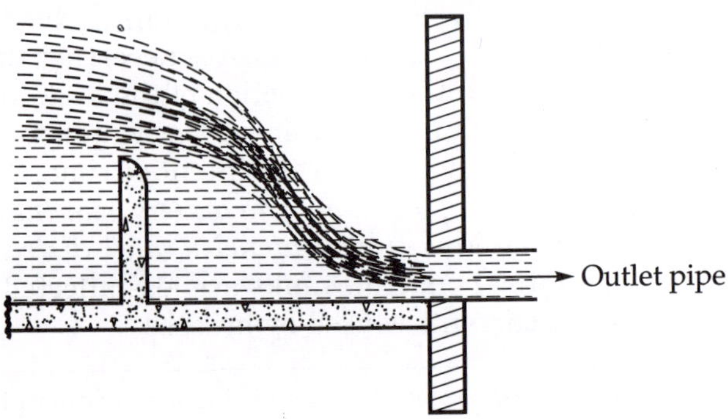

Fig. 9.15. Section of a 'submerged type' or 'weir type' outlet.

A similar type of outlet arrangement is also used these days. It consists of an outlet channel extending for full width of the tank and receiving the water after it has passed over a weir, as shown in Fig. 9.15.

(*ii*) **Baffles.** Baffles are required to prevent the movement of organic matter and its escape along with the effluent ; and to distribute the sewage uniformly through the cross-section of the tank, and thus to avoid short circuiting. Both inlets and outlets are, therefore, protected by hanging baffles, 0 to 90 cm in front of them, and submerged 45 to 60 cm below the flow line, as shown in Fig. 9.11. Some other arrangement of placing baffles may be designed, but it should not be too complicated. Moreover, too many baffles may result in concentration of currents and is as bad as too less baffles are.

(*iii*) **Skimming Troughs.** When the amount of oils and greasy matter present in a sewage is small, it is generally uneconomical to provide a separate skimming tank. In such cases, a scum trough is generally provided in the sedimentation tank itself, near its outlet end, as shown in Fig. 9.12 (*a*).

In manually operated tanks, the skimmings that float to the surface may be pushed into the trough by squeezes with hand. Whereas, in mechanically operated tanks (such as in Fig. 9.12 *a*), the skimmings may be pushed by the same scraper blades which collect the sludge while moving along the bottom, and push the skimmings into the end trough, when they move near the surface along with the endless chain to which they are attached.

(*iv*) **Cleaning and Sludge Removal.** The suspended organic solids contained in sewage, settle down at the bottom of the sedimentation tank, and have to be removed periodically. The removal of the deposited sludge before it becomes stale and septic is necessary not only because it reduces the capacity of the tank and its detention period, but also because it leads to the evolution of foul gases formed due to the anaerobic decomposition of the settled organic matter. The sedimentation tanks are, therefore, cleaned from time to time at frequent intervals, either manually or they are provided with mechanical arrangements for cleaning.

In manual cleaning, the tank is, first of all, put out of service (thus needing duplicate tanks), and the supply of sewage stopped. The already contained sewage is drained off till the depth remains about 30 cm or so. The deposited sludge is now stirred and removed as a slurry through a separate pipe provided with a gate valve at the bottom of the tank. *Pick axes* and *pharaohs* may be used for displacing hard deposits. The removed sludge is first taken to a pump-sump, and then pumped into a sludge digestion tank for further treatment.

Modern sedimentation tanks, however, are generally provided with mechanical cleaning devices. Say for example, in Fig. 9.12 (*a*), the sludge is scraped by scrapers and brought to the hopper at the inlet end, and is removed daily or often. The scrapers can work either continuously or at any desired intervals of time. Similarly, in a circular tank (Fig. 9.12 *b*), the sludge is scraped and brought to the centre, and likewise removed. For tanks without mechanical sludge removing equipment, an additional minimum depth of about 0.8 to 1.2 m should be provided for storage of settled material, and is called the *sludge zone.*

Example 9.5. *Design a suitable rectangular sedimentation tank (provided with mechanical cleaning equipment) for treating the sewage from a city, provided with an assured public water supply system, with a max. daily* *demand of 12 million litres per day. Assume suitable values of detention period and velocity of flow in the tank. Make any other assumptions, wherever needed.*

Solution. Assuming that 80% of water supplied to the city becomes sewage, we have the quantity of sewage required to be treated per day (*i.e.* max. daily).

$$= 0.8 \times 12 \text{ million litres} = 9.6 \text{ M. litres}$$

Now assuming the detention period in the sewage sedimentation tank as 2 hours, we have

The quantity of sewage to be treated in 2 hours *i.e.* the capacity of the tank required

$$Q = \frac{9.6}{24} \times 2 \text{ M. litres} = 0.8 \text{ M litres} = 800 \text{ cu. m.}$$

Now, assuming that the flow velocity through the tank is maintained at 0.3 m/minute ; we have

The length of the tank required

$$= \text{Velocity of flow} \times \text{Detention period}$$

$$= 0.3 \times (2 \times 60) \text{ m} = 36 \text{ m.}$$

Cross-sectional area of the tank required

$$= \frac{\text{Capacity of the tank}}{\text{Length of the tank}} = \frac{800}{36} \text{ m}^2 = \mathbf{22.2 \text{ m}^2.}$$

Assuming the water depth in the tank (*i.e.* effective depth of tank) as 3 m,

The width of the tank required

$$= \frac{\text{Area of } X\text{-section}}{\text{Depth}} = \frac{22.2}{3} = \mathbf{7.4 \text{ m.}}$$

Since the tank is provided with mechanical cleaning arrangement, no extra space at bottom is required for sludge zone.

Now, assuming a free-board of 0.5 m, we have

The overall depth of the tank = 3 + 0.5 = 3.5 m.

Hence, a rectangular sedimentation tank with an overall size of 36 m × 7.4 m × 3.5 m can be used.

[**Note.** This satisfies the requirements like : length $\not>$ 4 to 5 times the width ; and the width not more than 7.5 m or so ; the depth between 2.4 to 3.6 m, etc.]. **Ans.**

Alternatively,

Instead of assuming the depth, we may assume an overflow rate, say as 40,000 litres/m²/day.

$$\therefore \quad \frac{Q}{BL} = 40{,}000 \text{ litres/m}^2\text{/day}$$

But $\quad Q = 9.6 \text{ M. litres/day} = 9.6 \times 10^6 \text{ litres/day}$

$$\therefore \quad BL = \frac{Q}{40{,}000} = \frac{9.6 \times 10^6}{40{,}000} \text{ m}^2 = 240 \text{ m}^2$$

or $\quad B = \dfrac{240}{L} = \dfrac{240}{36} = 6.67 \text{ m} \; ; \; \textbf{Say 6.7 m.}$

$\therefore$ The depth required

$$= \frac{\text{Area of } X\text{-section}}{\text{Width}} = \frac{22.2}{6.7} = 3.3 \text{ m.}$$

Hence, we can alternatively use a tank of dimensions 36 m × 6.7 m × (3.3 + 0.5 m) overall depth *i.e.* 36 m × 6.7 m × 3.8 m size. **Ans.**

Example 9.6. *Design a circular settling tank unit for a primary treatment of sewage at 12 million litres per day. Assume suitable values of detention period (presuming that trickling filters are to follow the sedimentation tank), and surface loading.*

Solution. Assuming the normal detention period for such cases as 2 hr, and surface loading as 40,000 litres/sq. m/day ; we have

The quantity of sewage to be treated per 2 hours

$$= 12 \text{ M. litres} \times \frac{2}{24} = 1 \text{ M. litres} = 1000 \text{ m}^3.$$

$\therefore$ Capacity of tank = 1000 m^3.

Now, surface loading

$$= \frac{Q}{\text{Surface area of tank}} = \frac{Q}{\frac{\pi}{4}.d^2}$$

or $$40{,}000 = \frac{12 \times 10^6}{\frac{\pi}{4}.d^2}$$

where d is the dia. of the tank

or $$\frac{\pi}{4}.d^2 = \frac{12 \times 10^6}{40{,}000}$$

or $$d = \sqrt{\frac{300 \times 4}{\pi}} = 19.55 \text{ m } \textbf{Say 19.6 m.}$$

Now, effective depth of tank

$$= \frac{\text{Capacity}}{\text{Area of X-section}} = \frac{1000}{\frac{\pi}{4} \times (19.6)^2} = \frac{1000}{302}$$

$$= 3.2 \text{ m. (Say).}$$

Hence, use a settling tank with 19.6 m dia. and 3.2 m water depth (with free board of 0.3 m extra depth). **Ans.**

SEDIMENTATION AIDED WITH COAGULATION (Type II. Sedimentation)

9.16. Chemical Precipitation and Coagulation

Very fine suspend particles, present in wastewaters, which cannot be removed in plain sedimentation, may sometimes, be settled by increasing their size by changing them into flocculated particles. For this purpose, a chemical compound (like ferric chloride, ferric sulphate, alum, chlorinated copperas, etc.) called *coagulant,* is added to the wastewater, which on thorough mixing, forms a gelatinous precipitate, called ***floc***. The fine mud particles and other colloidal matter present in wastewater get absorbed in these flocs, forming the bigger sized flocculated particles. The process of addition and mixing of chemical is called coagulation. The coagulated sewage is then made to pass through sedimentation tank, where the flocculated particles settle down, and get removed.

The process of coagulation of sewage is similar to that of water, as described thoroughly in "Environmental Engg. Vol. I—Water Supply Engineering" (chapter 9). The purpose, principle, feeding and mixing of coagulants and their chemical reactions have been discussed at length in that chapter. **Since the coagulation of sewage is seldom adopted in modern days**, *we shall not repeat the entire coagulation process here in this volume.* The readers may, however, refer to volume 1, in special needs.

The characteristics and efficiency of the important coagulants used in sewage treatment are, however, given in Table 9.5.

9.17. Merits and Demerits of Coagulation Process in Sewage Treatment

As pointed out earlier, the coagulation process is generally not adopted in modern sewage treatment plants, mainly because of the following reasons :

(1) More advanced methods of sewage treatment based on biological actions are available these days, and they are preferred to coagulation.

(2) The coagulation process has various disadvantages, such as discussed below :

Disadvantages or Demerits of Coagulation in Sewage Treatment

(*i*) The biological secondary treatments used these days for treating sewage are complete in themselves, and do not require coagulation. Moreover, coagulation rather makes some of these processes more difficult.

(*ii*) The chemicals used in coagulation react with sewage, and during these reactions, they destroy certain micro-organisms, which are helpful in digestion of the sludge, thus creating difficulties in sludge digestion.

(*iii*) Cost of chemicals is added to the cost of sedimentation, without much use, and thereby making the treatment costlier.

(*iv*) The process of coagulation and subsequent sedimentation produces larger quantities of sludge than that produced in plain sedimentation, and thus adding to the problem of sludge disposal.

(*v*) The process of coagulation requires skilled supervision and handling of chemicals.

In view *of all these disadvantages, the coagulation of sewage has become obsolete these days. It may still, however, be adopted in certain special cases, such as* :

(*a*) For treating sewage from industries, using some specific chemicals in their processes.

(*b*) It is particularly advantageous, where there is large seasonal variation in sewage flow or as an emergency measure to increase the capacity of an overloaded plain sedimentation tank. It may, thus be of use at places, like hill stations, where flow may considerably go up during rush seasons.

The various advantages offered by the coagulation process in sewage treatment are :

(*i*) Sedimentation aided with coagulation produces better effluent with lesser BOD and suspended solids, as compared to plain sedimentation.

(*ii*) Coagulated settling tank requires less space than that required by an ordinary plain settling tank.

Table 9.5. Properties of Important Coagulants Used in Sewage Treatment

S. No.	*Name of coagulant*	*B.O.D. removed as percentage of total present*	*Suspended solids removed as percentage of total present*	*Dosage required in ppm*	*pH value required for proper functioning*	*Remarks*
1.	**Ferric chloride**	80—90	90—95	25—35	5.5 to 7.0	This cogulant is widely used for sewage treatment*, wherever, coagulation is adopted.
2.	**Ferric sulphate with lime**	60	80	35—40	8.0 to 8.5	Ferric sulphate has been found to be more effective than chlorinated copperas, if used in conjunction with lime. *Hence ferric chloride and ferric sulphate are mainly used, as the main coagulants in sewage, if at all coagulation is adopted.*
3.	**Alum**	60	80	40—90	6 to 8.5	It is generally not used in sewage** although used for treating water supplies on a large scale.
4.	**Chlorinated copperas**	70—80	80—90	35—80	5.5 to 7.0 and 9.0 to 9.5	This coagulant is effective for producing sludge for activated sludge process.

*Ferric chloride forms a dense heavy floc, which settles rapidly. The sludge formed is also not so bulky, and can be digested or dewatered easily. However, it is corrosive in nature and requires care in storing and handling.

**Alum, although cheap and easy to handle, yet is not used in sewage treatment, because it forms a spongy floc, which settles slowly. Moreover, the volume of sludge produced is also large.

(*iii*) Coagulation process may also remove the phosphates from sewage, which may help in controlling **eutrophication*** of river, into which the effluent is being discharged.

SECONDARY TREATMENT THROUGH BIOLOGICAL FILTRATION OF SEWAGE

(Aerobic Attached Culture)

9.18. Introduction to Sewage Filtration

The effluent from the primary sedimentation tank contains about 60 to 80% of the unstable organic matter originally present in sewage. This colloidal organic matter, which passes the primary clarifiers, without settling there, has to be removed by further treatment. This further treatment of sewage, called *secondary* or *biological treatment,* is carried out by changing the character of the organic matter, and thus converting it into stable forms (like nitrates, sulphates, etc.) by oxidation or nitrification.

The character of the organic matter in sewage may be changed by different methods, which are broadly classified into : (*i*) *filtration*, and (*ii*) *activated sludge process*. These processes help in changing the unstable organic matter into stable forms, and thus ensuring their removal, when the effluent from these units is settled in *secondary sedimentation tanks*. The sludge settled in secondary settling tanks may be disposed of in the manner explained earlier.

All the secondary treatment processes (i.e. filters as well as activated sludge process) are designed to work on aerobic bacterial decomposition. This is because of the fact that aerobic decomposition does not produce bad smells and gases, as are produced by the anaerobic decomposition, and also because aerobic bacteria are more active than anaerobic bacteria (*the rate of doing work by aerobic bacteria being three times that of the anaerobic type at 30°C*). Hence, in the biological treatment units, the main aim is to produce suitable environmental conditions, which may help in stimulating the growth of these aerobic bacteria.

The ***filter units*** used for secondary treatment consist of *open beds* of coarse aggregates over which sewage (*i.e.* effluent from the primary clarifier) is sprinkled intermittently. The necessary contact surface for the growth of aerobic bacteria is provided by the aggregates in the bed, and the aeration is provided by nature. The effluent from filter units is settled out in secondary clarifiers. On the other hand, an ***activated sludge treatment unit*** consists of *aeration tanks* of long detention period, in which the activated sludge mixed with the sedimented sewage (*i.e.* effluent from primary clarifiers) is agitated

---

*The excessive growth of algae and other aquatic plants in a river stream is called ***eutrophication***. The *nitrogen and phosphorous* present in the sewage and industrial effluents being discharged into the rivers, serve as fertilising elements, which in the presence of sun-light, stimulate the growth of algae and other aquatic plants.

Certain limited quantity of these plants is, of course, necessary in a river stream, as to serve as a source of food for fish. However, large growths are undersirable, as they tend to clog the flow and to form floating mats and to decrease water clarity. They are unsightly and become breeding places for mosquitoes, thus resulting in spread of malaria. The decomposition of algae may also lead to unpleasant odours and smells.

Contents of nitrogen and phosphorous are, therefore, controlled these days in the effluents being discharged in such infested streams.

and aerated. The sludge solids, thus kept in suspension will provide the necessary contact surfaces for the growth of aerobic bacteria. The mixed liquor after agitation is settled out in secondary clarifiers. *The sludge so settled in secondary clarifier is called the activated sludge as it contains a lot of oxygen, and is thus biologically active.* This sludge will be disposed of after digestion, although a small part of it will be taken to aeration tanks for keeping the activated sludge process going.

There are *various kinds of filters* that are commonly employed either singly or in combinations, for giving secondary treatment of sewage. All these filters try to oxidise the unstable organic matter into stable forms, mainly be aerobic decomposition. The different types of filters that may be used in sewage treatment are :

(1) *Contact beds* (used at very small plants, and have become almost obsolete these days).

(2) *Intermittent sand filters* (used at smaller plants).

(3) *Trickling filters* (most commonly used in modern days).

(4) *Miscellaneous types of filters* (used under special circumstances in particular projects).

All these different types of filters are described below in details :

9.19. Contact Beds for Biological Filtration of Sewage

In contact beds, also known as *contact filters,* sewage is kept in contact with the filtering media for a certain period (*i.e.* during first contact period). As the sewage percolates through the filtering media, fine colloidal organic matter is trapped in the voids between the particles of the filtering media. *Due to this, a film of organic matter is formed around the particles of the filtering media.* A large number of aerobic bacteria present in the sewage will get attached to the filter sand, and will, thus, be present in this biofilm. These bacteria will carry out the oxidation of the organic matter present in this film, by the process of aerobic decomposition. In order to make sufficient oxygen available for the growth of aerobic bacteria, the filter is left empty (without any sewage) for a certain period (*i.e.* second contact period). The filter, during this period, will absorb oxygen from the atmosphere, and will complete the oxidation process. This constitutes one cycle of operation. The stabilised organic matter in the first cycle will be washed away by the fresh sewage in the next cycle of operation.

9.19.1. Constructional Details. A typical plan and section of a contact beds unit is shown in Fig. 9.16. A contact bed is a water tight rectangular tank, filled with a filtering media, consisting of gravel ballast, or broken bricks or stones. The size of the media particles may vary between 20 to 40 mm. The depth of the filtering media varies between 1 to 1.8 metres (the common being 1.2 m). These tanks are generally dug below the ground level, and are provided with water-tight concrete lining, as shown. A siphonic dosing tank is generally provided to serve two or three contact beds. The dosing tank will receive sewage from the primary sedimentation tanks, and will supply it to the different contact beds, turn by turn, at regular intervals by the timed siphonic action. Ordinary tanks without siphons, and operated manually by valves may also be used. The sewage percolating through the filtering media is collected at the bottom by a system of suitably laid open-jointed under-drains.

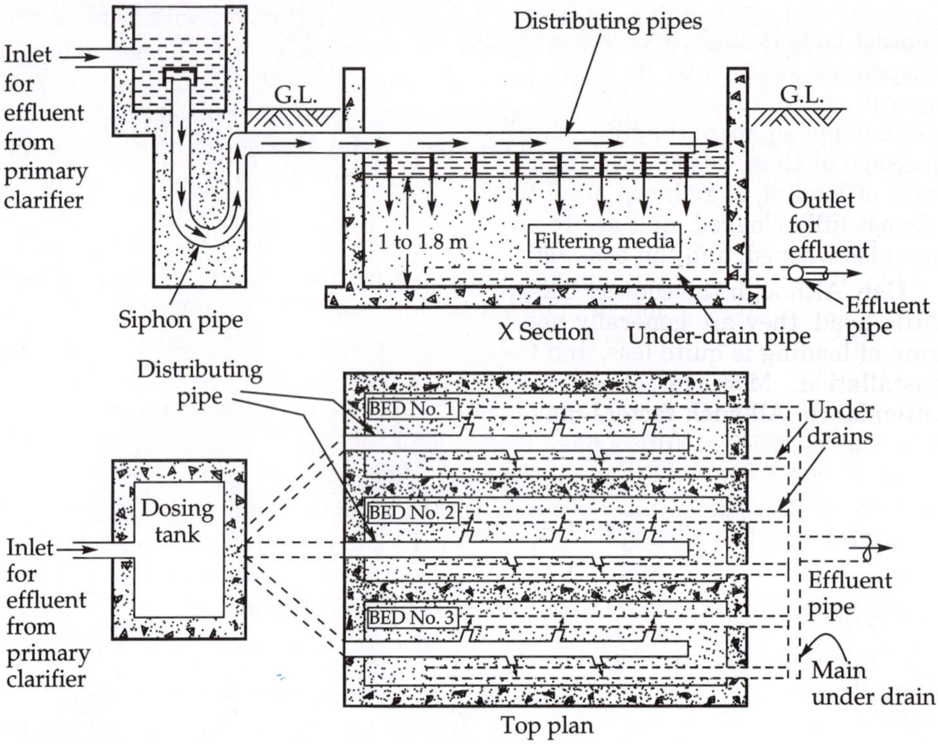

Fig. 9.16. Contact beds.

The effluent coming out from the effluent pipe may be taken to the secondary sedimentation tank for settling out the oxidised organic matter.

9.19.2. Operation. The sewage is, first of all allowed from the dosing tank, so as to fill the bed completely, and to stand to a depth of about 5 to 10 cm above their top surface. While doing so, the outlet valve is kept closed. This filling takes about 1 hour. The dosing tank supply is then closed, and the sewage is left over the bed as it is, for about two hours or so (called the first contact period). The colloidal matter present in sewage will be transferred to filter voids during this period.

The outlet valve is then opened, and the sewage is removed slowly. This emptying of the bed should be carried out slowly so that the colloidal matter trapped in the voids of the filter media does not get washed out with the down-moving sewage. The bed is now kept empty at rest for a period of about 4 to 6 hours (called the 2nd contact period or the empty period). During this period, the organic matter trapped in the voids get oxidised by aerobic bacteria in the presence of free oxygen that rushes into the filter voids from the atmosphere. This completes one full cycle of operation, which takes about 8 to 12 hours*. The stabilised organic matter from the voids is finally washed down the contact material in the next emptying of the tank.

Since the inflow of sewage is continuous, while the beds are to be operated intermittently, several units will have to be constructed, so that the sewage can be applied in turn to the different units. As pointed out earlier, a siphonic dosing tank is generally used for this purpose. Continuous attendance is, of course, necessary to operate the different units in succession.

*The contact beds are thus given two to three loadings (generally two) per day.

9.19.3. Performance and Rate of Loading. The effluent obtained from the contact beds is slightly turbid and odourless, as it removes about 80% of the suspended solids, and 60 to 75% of the BOD. But due to its intermittent operation, the rate of loading is very low ; and it should not exceed 110 litres per day per sq. m. of the filter for beds of 1.2 to 1 m depth. Moreover, with the passage of time, as the filtering media goes on clogging more and more, the rate of loading is reduced accordingly. After 4 to 5 years, the beds become almost fully clogged, and are removed, washed and dried. The under-drains are also washed, and the beds reinstated.

Use. Although the contact beds produce satisfactory effluent and consume little head, they are generally not adopted these days ; mainly because their rate of loading is quite less, and they thus, require large land areas for their installation. Moreover, their intermittent operation requires continuous attendance. All such factors make them unsuitable for use in modern days, when more efficient filters have been designed. They may, however, still be used for small isolated institutions, where the quantity of sewage to be treated is very small ; or at places where available head is limited, and further pumping is undesirable.

9.20. Intermittent Sand Filters for Biological Filtration of Sewage

Intermittent sand filters were the earlier forms of biological units of sewage treatment, and are now rarely used because of their large area requirement. They may still however, be used at certain places, to give highly nitrified and polished effluent, such as in case of treating sewage from tuberculosis hospitals.

Intermittent sand filters are more or less like contact beds, with the difference that the contact media here is finer than that in the contact beds ; and also, there is no concrete lining around the filter media, as was there in the contact beds.

9.20.1. Construction and Operation. The plan and section of a typical intermittent sand filter is shown in Fig. 9.17. It consists of an underground rectangular tank, 1 to 1.25 m deep, 1000 to 4000 sq. m. in plan, and made by excavating the earth without any lining on sides or bottom. The sand, (*i.e.* the filtering media) is filled in this tank to a depth of about 0.75 to 1.05 m.

The bottom of the tank is made sloping gently towards the under-drains, which are placed at about 9 m apart in trenches below the bottom of the sand. These under-drains are nothing but the vitrified tile pipes with open joints of about 1 cm, and surrounded by graded gravel.

A siphonic dosing tank is generally provided to serve two or three filter beds. The dosing tank will receive sewage from the primary sedimentation tank, and supply it to the different filter beds, turn by turn, at regular intervals, by the timed siphonic action. Ordinary tanks without siphons, and operated manually by valves, may also be used.

The sewage from the dosing tank is spread intermittently over the filter bed, through *distribution pipes*, which are provided with side openings, as shown. These distribution pipes are laid at about 9 to 18 m apart. The depth of sewage applied to the entire filter bed is about 5 to 7.5 cm. As the sewage percolates through the bed, the suspended organic matter present in sewage gets trapped in the voids between the particles of the filtering media. The organic matter is thus first removed by the straining action of the filter. This trapped organic matter is then acted upon by the aerobic bacteria which can flourish well in the presence of free oxygen rushing in from the atmosphere during the period the filter bed is kept at rest after the sewage dose has percolated down.

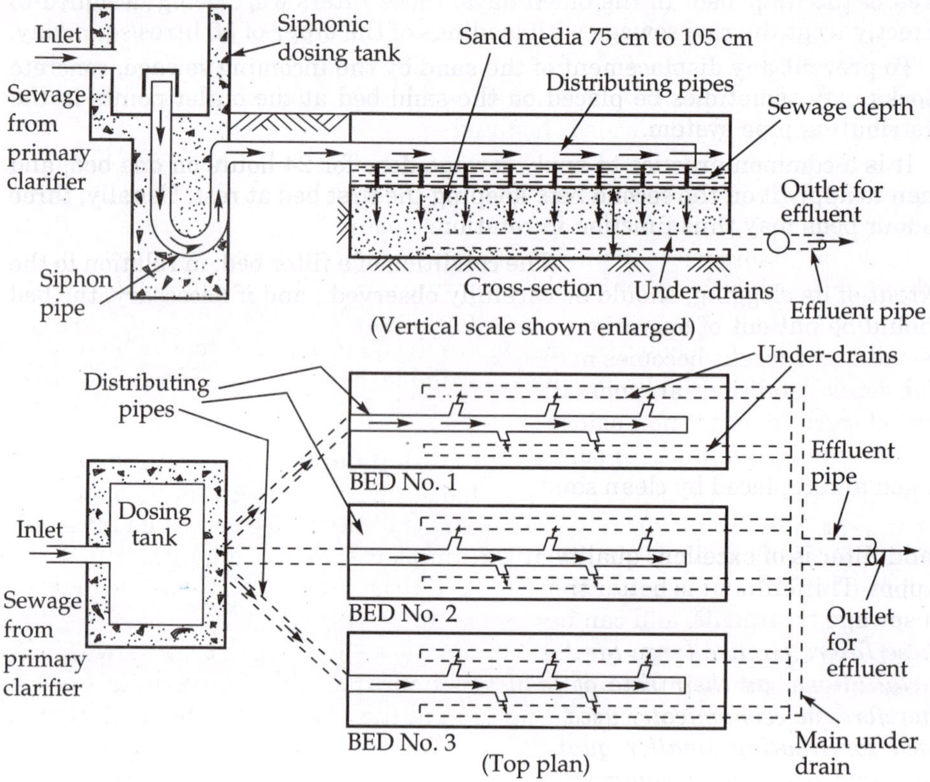

Fig. 9.17. Intermittent sand filters.

Design Criteria. As stated earlier, the intermittent sand filters are rectangular in plan with an area of about 1000 to 4000 sq. m, with length to width ratio, as about 3 to 4. The depth of sand varies between 0.75 to 1.05 m.

The sand used in the filtering media should be clean and sharp, with an effective size (D_{10}) of about 0.2 to 0.5 mm and uniformity coefficient $\left(\frac{D_{60}}{D_{10}}\right)$ ranging between 2 to 5. If the finer sand is used, the filter will be rapidly clogged, and if the coarser sand is used, the penetration of the solids will be too deep, and the distribution of sewage will be difficult. *The top 15 cm layer of this sand may be kept of finer variety than that of the rest, which is generally kept uniform in grain size. However, if different gradation of sand are used (thought not necessary), then the coarsest layers should be placed near the bottom, and the finest towards the top.* The finer the sand used, the purer and better effluent will be obtained, as more impurities and bacteria will be trapped, but the filter will clog quickly. In addition to this sand layer, a bottom layer of about 15 cm of gravel may also sometimes be placed, so as to facilitate the drainage of the effluent.

The rate of loading for the intermittent sand filters is less than that permitted for the contact beds (which contain coarser filtering media). The commonly adopted values of the hydraulic loading (*i.e.* the rate at which the sewage is applied) ranges between 80 to 110 litres per day per sq. m of plan area of the filter bed. In the olden days, these filters were being designed to directly treat the raw sewage with loadings of the order of 50 litres/sq. m/day.

To prevent any displacement of the sand by the incoming sewage, concrete blocks may sometimes be placed on the sand bed at the outlet points in the distributing pipe system.

It is a common practice to apply sewage dose for 24 hours on one bed, and then to apply it on the second bed, keeping the first bed at rest. Usually, three to four beds may thus function in rotation.

9.20.2. Cleaning of the Filter. The condition of a filter bed, in relation to the extent of its clogging, should be carefully observed ; and if necessary, the bed should be put out of operation, when it requires cleaning or rest. A complete rest of 1 to 4 weeks becomes necessary, if septic conditions prevail in the bed. The septic conditions are more likely to occur, when the top layers of the sand are clogged due to non-oxidation of some suspended solids, which are transferred into *humus*. In such case, the top 5 to 7.5 cm sand layer may be raked and replaced by clean sand.

9.20.3. Performance and Use. The effluent obtained from an intermittent sand filter is of excellent quality, with suspended solids < 10 ppm, and BOD < 5 ppm. This effluent is better than that obtained from any other ordinary type of sewage treatment, and can be disposed of even without dilution. *However, these filters require large land areas, and huge quantities of sand for their installations, as their rate of loadings are very low. These filters cannot, therefore, be economically used at larger plants. They may, however, still be used for treating smaller quantities of polluted sewages coming out from hospitals or isolated commercial and industrial institutions. In very cold climates, these filters may further pose troubles due to danger of sand surface freezing.*

9.21. Trickling Filters for Biological Filteration of Sewage

The **conventional trickling filters** and their improved forms, known as **high rate trickling filters** are now almost universally adopted for giving secondary treatment to sewage. These filters, also called as *percolating filters* or *sprinkling filters*, consist of tanks of coarser filtering media, over which the sewage is allowed to sprinkle or trickle down, by means of *spray nozzles* or *rotary distributors*. The percolating sewage is collected at the bottom of the tank through a well designed *under-drainage system.*

The decomposition of the organic matter and the resultant purification of the sewage is brought about by a population of micro-organisms. Micro-organisms and bacteria, which are naturally present in sewage, get attached to the filter media. Organic matter from the sewage influent is also adsorbed on the **biological film**, which is formed by the micro-organisms around the filtering media particles (*i.e.*, sand particles). In the outer portions of tnis film of biological mass or **slime layer**, the organic matter is degraded by the aerobic bacteria. As the micro-organisms grow, the thickness of the slime layer increases, and the diffused oxygen is consumed by the upper portions of the slime layer, thereby creating an anaerobic environment near the surface of the media particles. As the slime layer increases in thickness, the adsorbed organic matter is metabolised before it can reach the micro-organisms near the media face. This creates shortage of external source of organic carbon near the media face, due to which, the micro-organisms near the media face enter into an endogenous phase of growth, and lose their ability to cling to the media surface. The liquid sewage, exerting a ***shearing action***, then breaks up some biological mass (slime) from the media, but the new slime layer continues to grow. The continuing growth and break up of biological mass from this **biofilm** (slime layer), infact, creates a balance in the thickness of the formed biofilm. *The break up or detachment of the biomass (biological solids) from the slime layer is known* as **sloughing**.

The *sloughing*, infact, imparts turbidity to the filter effluent, and the sloughed material (biological solids) is separated from the treated sewage in the secondary settling tank.

The extent of *sloughing* is primarily a function of organic and hydraulic loading on the filter. The hydraulic loading accounts for shear velocities that are developed in the filter (which causes sloughing) ; while the organic loading accounts for the rate of metabolism in the slime layer (which loosens the biological mass in the slime layer, leading to sloughing).

In order to ensure the large scale growth of the aerobic bacteria in the biofilm, sufficient quantity of oxygen is supplied by providing suitable ventilation facilities in the body of the filter ; and also to some extent by the intermittent functioning of the filter.

The effluent obtained from the under-drains of the filter, must be taken to the secondary sedimentation tank for settling out the sloughed biomass. This secondary tank thus, separates the treated sewage from the biological solids. A portion of the filter effluent or the settled effluent is recycled, usually to dilute the incoming wastewater. The sewage influent entering the filter must, of course, be given pre-treatments, including screening and primary sedimentation.

9.21.1. Construction and Operation of Trickling Filters. Trickling filter tanks are generally constructed above the ground. They may either be rectangular or more generally *circular* [Refer Fig. 9.18 (*a*), (*b*) and (*c*)]. Rectangular filters are provided with a network of pipes having fixed nozzles, which spray the incoming sewage into the air, which then falls over the bed of the filter, under gravity. The circular filter tanks on the other hand, are provided with **rotary distributors** having a number of distributing arms (generally four arms are used). These distributors rotate around a central support either by an electric motor, or more generally by the force of reaction on the sprays. Such self-propelled reaction type of distributors (Fig. 9.19) are now-a-days preferred and used. The rate of revolutions vary from 2 RPM for small distributors to less than $\frac{1}{2}$ RPM for large distributors. The advantage of having two or more arms is not only to get reaction sufficient to rotate the entire mechanism but is also to pass the fluctuating demands by taking low flows in two arms, and the remaining two arms coming into operation only at the times of higher flows. The distributing arms should remain about 15 to 20 cm above the top surface of the filtering media in the tank.

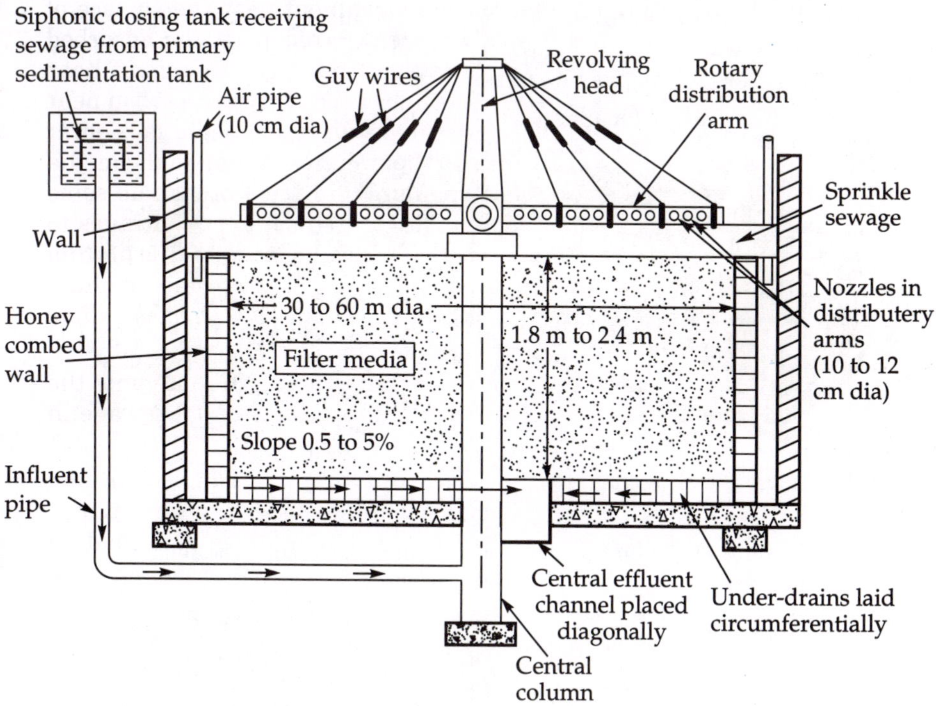

Fig. 9.18(a) Typical section of a conventional Circular-Trickling filter with Rotary distributors (vertical scale shown enlarged).

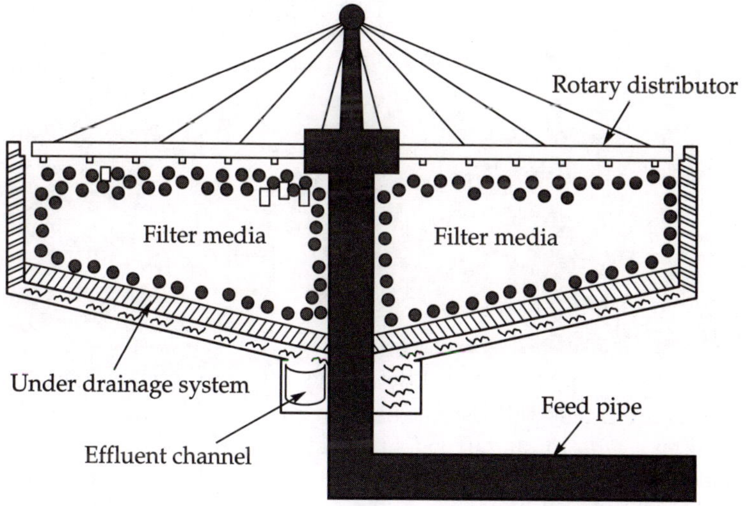

Fig. 9.18(b) Simplified Line Sketch of Trickling Filter.

There is an important difference between the action of rotary distributors and that of spray nozzles. With a rotary distributor, the application of sewage to the filter is practically continuous ; whereas with spray nozzles, the filter is dosed for 3 to 5 minutes, and then rested for 5 to 10 minutes before the next application. In any case, however, dosing tanks with siphons, receiving sewage from the primary clarifiers and supplying it to the filters at regular intervals, are used. The dosing tank for a filter with distributors will, however, be designed to have a smaller capacity (*i.e.* about 1 to 3 minutes detention capacity), as against a higher capacity (of about 5 to 10 minutes detention capacity) for filters with spray nozzles.

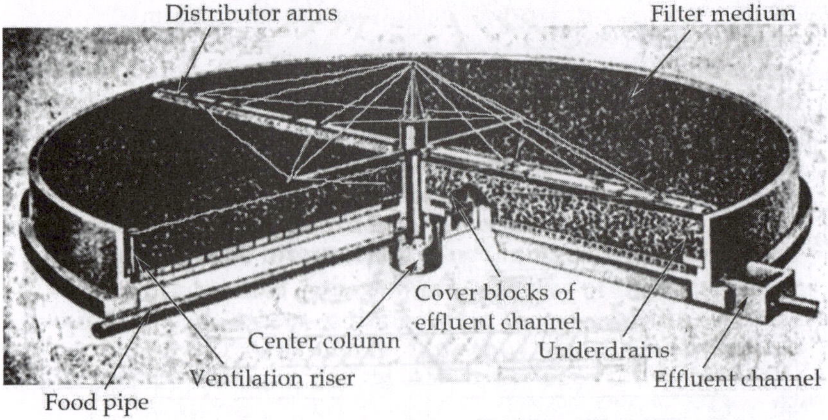

Fig. 9.18. (*c*) Photographic view of a conventional Circular Trickling filter with Rotary distributors.

Fig. 9.19. Photographic view of Rotary Distributors (Reactions type).

The **filtering media,** as pointed out earlier, consists of coarser materials like cubically broken stones or slag, free from dust and small pieces. The size of the material used may vary between 25 to 75 mm. The filtering material should be washed before it is placed in position. The quality of the stone used should be such as not to be easily affected by acidic sewage, and should be sufficiently hard (with hardness of about 12 on Brinels Hardness machine), and should have a minimum compressive strength (crushing strength) of about 100 N/mm^2 ($\approx$ 1000 kg/cm^2). Its resistance to freezing and thawing is another important property, especially for northern regions. Usually, stones from rocks like granite or limestone may be used.

Now a days, ***plastic media,*** instead of ***stone media,*** is also being used, since it ofters higher surface area per unit of volume. Such plastic media of 25 to 100 mm size are, in fact, being used in these modern days, in ***super-rate trickling filters***, made specially for the treatement of waste waters having very high BOD. *Since the plastic media have high specific surface area, high void space, and low mass, the resulting trickling filters can have much higher hydraulic loading rates as well as higher organic loading rates.*

The ***hydraulic loading rate**** for such super-rate filters can, infact, be as high as 40–200 m^3/m^2.d (normally 30–90 m^3/m^2.d) as compared to stone media high rate trickling filters having hydraulic loading rate of 10–40 m^3/m^2.d. Similarly, the ***organic loading rate***** for plastic media super rate filters can be as high as 1.0 to 6.0 kg BOD/m^3.d (normally 1 to 2 kg BOD/m^3.d) as compared to the organic loading rate of 0.37 to 1.85 kg/m^3.d for stone media high rate trickling filters.

The depth of the filtering media may vary between 2 to 3 metres. The filtering material may be placed in layers ; with coarsest stone used near the bottom, and finer material towards the top.

The walls of the filter tank are made *honey-combed* or otherwise provided with openings for circulation of air, all through. Sometimes, instead of constructing the supporting walls, the filtering material may be stacked above the ground with its natural angle of repose, so as to ensure better circulation of air from the surrounding atmosphere.

Sometimes, forced ventilation, by forcing the air vertically upwards through the filter by the use of fans or other mechanical equipment, may be used ; but it has not been found to increase the capacity of well-constructed trickling filters.

A satisfactory ventilation is achieved when properly designed **under drains,** having adequate openings, are provided under the filter bed. Besides ensuring satisfactory drainage, such drains, will also ensure satisfactory ventilation and aeration of the filter bed. Vitrified clay blocks (Fig. 9.20) are generally used as under-drains. These blocks have top openings of such size that the stone can be placed directly on them, and yet they furnish flow channels with sufficient capacity for the heaviest hydraulic loading. These blocks are laid on a reinforced concrete floor (about 10 to 15 cm thick), which is sloped at 0.5 to 5% $\left(i.e. \frac{1}{200} \text{ to } \frac{1}{40}\right)$ towards the main effluent rectangular channel. This **main effluent channel** may be provided *diagonally* adjoining the central column of

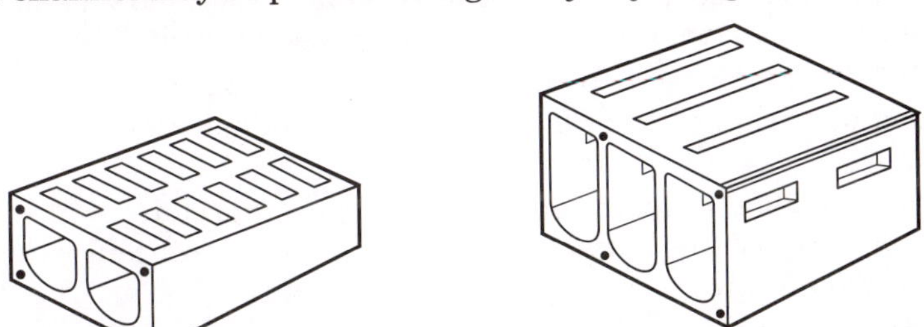

(*a*) Under-drain block for standard trickling filters

(*b*) Under-drain block for high rate trickling filters with heavy hydraulic loading

Fig. 9.20

*Explained in next article 9.21.4.

Also known as ***Volumetric loading rate.*

the distributor, as shown in Fig. 9.18 (*a*) and (*b*) ; or may be provided along the circular periphery of the filter. The flow in this channel has characteristics similar to the flow in a wash-water trough of a rapid sand filter used in water supplies. The slope of the channel should be sufficient to ensure a minimum flow velocity of about 0.9 m/sec. The depth and the width of this central channel should be such that it can carry the maximum flow entering the filter. The main effluent channel will receive lateral flows from the precast, perforated vitrified tile or concrete blocks. These under drains are laid at right angles to the main channel. Hence, if the main channel is laid diagonally, then these under drains will be laid circumferentially ; and if the main channel is laid along the outer circular periphery of the filter, then the under drains will be laid radially. The under drains will be designed to flow not more than ½ full to promote ventilation. The min flow velocity should be about 0.75 m/sec.

9.21.2. Types of Trickling Filters. Trickling filters can be broadly classified into :

(1) **Conventional trickling filters** or *Ordinary trickling filters* or **Standard rate** or **Low rate** trickling filters ;

(2) **High rate filters** or *High rate trickling filters.*

Strictly speaking, all what we have spoken so far is about conventional or standard rate trickling filters. The high rate filters of modern advancements, also function on the same lines, and are having the same constructional details, but with the difference that provision is made in them for *recirculation of sewage through the filter*, by pumping a part of the filter-effluent to the primary settling tank, and repassing through it and the filter. The high rate filters make it possible to pass sewage at greater loadings, thus requiring lesser space and lesser filter media. The process of 'recirculation' and its use in making high rate filters shall be discussed thoroughly, a little later, after we finish the general description of trickling filters, and the comparative characteristics of both types.

9.21.3. Merits and Demerits of Trickling filters. The various **advantages** of the trickling filters are :

(*i*) Rate of filter loading is high, as such requiring lesser land areas and smaller quantities of filter media for their installations.

(*ii*) Effluent obtained from the trickling filters is sufficiently nitrified and stabilised. They can remove about 75% of BOD and about 80% of suspended solids. The effluent can, therefore be easily disposed of in smaller quantity of dilution water.

(*iii*) Working of tricking filters is simple, and does not require any skilled supervision.

(*iv*) They are flexible in operation, and they can, therefore, withstand the application of variety of sewages having different concentrations and compositions. Even if they are over-loaded, they can recoupe after rest.

(*v*) They are self-cleaning.

(*vi*) Mechanical wear and tear is small, as they contain less-mechanical equipment.

(*vii*) Moisture content of sludge, obtained from trickling filters, is as high as 99% or so.

(*viii*) Trickling filters have been found to operate more efficiently in warm weather, and produce an effluent appreciably lower in BOD. Hence, they are of immense-use in hot countries like India.

The **disadvantages** of the trickling filters are :

(*i*) The head loss through these filters* is high, making automatic dosing of the filters necessary (through siphonic dosing tanks).

(*ii*) Cost of construction of trickling filters is high.

(*iii*) These filters cannot treat raw sewage, and primary sedimentation is a must.

(*iv*) These filters pose a number of **operational troubles** such as given below :

(*a*) ***Fly Nuisance.*** The filter fly, *psychoda*, which develops in the filter particles, may prove to be nuisance, as the same may be carried away into the habitation.

This problem may be controlled by flooding the filter with sewage for 24 hours or more. The flooding will destroy the larvae, and usually interfere slightly with the results of operation. However, two or more filters will be needed, in order that, while one is flooded, another is available to treat the sewage.

Another method of controlling fly nuisance is by using insecticides, such as D.D.T., chlordance, and benzene hexachloride, etc. in the filter plant. When using insecticides for fly control, the flies may sometimes develop resistance to one particular insecticide. In such a case, the insecticide used must be changed. An insecticide may be applied at the siphon chamber or at any other place in the plant that may be more convenient. The dosage of insecticide required is normally 1 to 2 mg/l based on the total daily sewage flow, but is applied over a period of about 2 hours. Application should be used on active content of the insecticide.

(*b*) ***Odour Nuisance.*** Odours, generally, do not prevail in trickling filters using rotary distributors ; but however, when fixed nozzles are used, H_2S and other odourous gases are frequently released from the sprays into the atmosphere. The usual remedy is to chlorinate the sewage to prevent formation of H_2S gas, or to neutralise that which is already formed.

Some odours may also be evolved due to sludge, and also due to anaerobic conditions which may prevail as and when the sewage becomes stale. The remedy is to chlorinate the sewage, and to keep sewage fresh by recirculation, as is done in rapid filters.

(*c*) ***Ponding trouble.*** Sometimes, the voids in the filter media gets clogged due to heavy growth of fungi and algae. This may result in ponding of the sewage over the filter bed. This trouble, posed by algae and fungi, can be controlled by chlorinating the sewage, which kills the algae, thus causing unloading of the accumulated material. Other methods of controlling algae are to add copper sulphate to the sewage, and resting the bed for some time.

Besides these external controlling factors, the algae is utilised by the worms and larvae particularly of the psychoda fly, which feed on the vegetable matter ; and in the process, bring about a biological change.

---

*The total head loss through such a filter will be equal to the head loss through the circular distributor [about 45 to 58 cm] plus the loss in the feed piping and in the siphon as well as the entrance loss. Generally, these extra losses will total to at least 15 cm.

Ponding at the surface is, therefore, most common during the winter months, when the worms and larvae are driven down from the surface by cold. Near spring time, however, they return to the surface, breaking up the vegetable growth in the filter, and causing heavy discharge of *humus** from the bed. This process is called *spring off loading*.

The humus so produced is easily settleable and is allowed to settle down in the secondary clarifier. Sometimes, separate tanks, similar to settling tanks, called *Humus tanks* may be used for this purpose. They are provided, particularly when the effluent from the filters is to be discharged into the rivers, and a high degree of purification is needed. Although humus is inert, yet it cannot be allowed to pass into the river streams, because ultimately it produces a large quantity of mud, and impedes the course of the river stream in question.

9.21.4. Design of Trickling Filters. The design of the trickling filter, primarily involves the design of the dia of the circular filter tank, and its depth. The design of the *rotary distributors* and *under-drainage system* is also involved in the filter design.

The design of the filter size is based upon the values of the **filter-loadings,** adopted for the design. This loading on a filter can be expressed in two ways :

(*i*) By the quantity of sewage applied per unit of surface area of the filter per day. This is called **hydraulic-loading rate** and expressed in million litres per hectare per day. The value of hydraulic loading for conventional trickling filters may vary between 22 to 44 (normally 28) million litres per hectare per day (as against the value of 1.1 ML/ha/day for intermittent sand filters). The hydraulic loading can still be increased to about 110 to 330 (normally 220) M.L/ha/day in the high rate trickling filters.

(*ii*) By the mass of BOD per unit volume of the filtering media per day. This is called **organic loading rate** or ***volumetric loading rate,*** and expressed in kg of BOD_5 per hectare metre of the filter media per day. The value of organic loading for conventional filters may vary between 900 to 2200 kg of BOD_5 per ha-m. This organic loading value can be further increased to about 6000—18000 kg of BOD_5 per ha-m in high rate trickling filters.

With an assumed value of organic loading (as between 900 to 2200 kg/ha-m), we can find out the total volume of the required filter, by dividing the total BOD_5 of the sewage entering the filter per day in kg, by the assumed value of the organic loading. The organic loading can thus, decide the volume of the filter.

The hydraulic loading, on the other hand, gives us the area of the filter required, when the total sewage volume entering the filter per day is divided by the hydraulic loading, (assumed between 22 to 44 M.*l*/ha/day).

Knowing the volume and area of the cylindrical filter, we can easily find out its dia and depth.

It may also be mentioned here that the filter dia and depth is designed for average value of sewage flow. The rotary distributors, under-drainage system, and other connected pipe lines etc. are, however, designed for peak flow, and of course checked for the average flow. Moreover, since the rotary distributors are available indigenously only up to 60 m in length, it is desirable to keep the dia of the filter tank up to a max. of 60 m. If the required filter dia is more than

*Humus is like flocculated floating matter present in the filter effluent.

60 m, then it is better to use more units of lesser dia. The design of rotary distributors, and under-drainage system is explained in the numerical example 9.7.

9.21.5. Performance of Conventional Filters and Their Efficiencies. The effluent obtained from a conventional trickling filter plant is highly nitrified and stabilised. The BOD is reduced to about 80 to 90% of the original value. The BOD left in the effluent is generally less than 20 ppm or so. The sludge obtained in the secondary clarifier is thick, with moisture content of about 92%. It is heavy and easily digestible. The filter is very flexible, and can even take intermittent shock loads without any detrimental effects. *Hence, a conventional or standard rate trickling filter plant is very useful to medium towns and industrial cities requiring full treatment of sewage.*

The efficiency of such a conventional filter plant can be expressed by the equation evolved by National Research Council of U.S.A., and given as :

$$\eta\ (\%) = \frac{100}{1 + 0.0044\sqrt{u}} \quad ...(9.32)$$

where η = Efficiency of the filter and its secondary clarifier, in terms of percentage of applied BOD removed.

u = Organic loading in kg/ha-m/ day applied to the filter (called unit organic loading).

This equation shows that the BOD removed by the filter plant depends upon the organic loading adopted. Greater is the loading, lesser is the efficiency, and thus lesser BOD is removed. This equation holds good when there is no recirculation. In case of recirculation, this equation modifies to as given in Eq. (9.34).

Example 9.7. (*a*) *Design suitable dimensions of a circular trickling filter units for treating 5 million litres of sewage per day. The BOD of sewage is 150 mg/l.*

(*b*) *Also design suitable dimensions for its rotary distribution system, as well as the under-drainage system.*

Solution. Total BOD present in sewage to be treated per day

$$= 5 \text{ ML} \times 150 \text{ mg/L} = 5 \times 10^6 \times 150 \text{ mg}$$
$$= 5 \times 150 \text{ kg} = 750 \text{ kg}.$$

Assuming the value of organic loading, say as, 1500 kg/hectare metre/day [*i.e.* between 900 to 2200 kg/ha-m/day], we have

The volume of filtering media required

$$= \frac{750}{1500} \text{ hectare-metre} = 0.5 \text{ ha-m} = 5000 \text{ m}^3.$$

Assuming the effective depth of filter, as, say 2 m, we have

$$\text{The surface area of the filter required} = \frac{5000}{2} \text{ m}^2 = 2500 \text{ m}^2.$$

Using a circular trickling filter of dia 40 m, we have the number of units required

$$= \frac{\text{Total area required}}{\text{Area of one unit}} = \frac{2500}{\frac{\pi}{4}(40)^2} \approx 2 \text{ Nos.}$$

Check for Hydraulic loading. The surface area of the filter bed required can also be worked out by assuming the value of hydraulic loading, say as, 25 million litres per hectare per day [*i.e.* between 22 to 44 ML/ha/day]

∴ Surface area required

$$= \frac{\text{Total sewage to be treated per day}}{\text{Hydraulic loading per day}}$$

$$= \frac{5\text{ ML}}{25\text{ ML/hr}} = \frac{5}{25}\text{ hectares} = \frac{5}{25} \times 10{,}000\text{ m}^2 = 2000\text{ m}^2.$$

The surface area chosen is 2500 m^2, which is greater than 2000 sq. m, and hence safe.

Hence, 2 units each of 40 m dia and 2 m effective depth (*i.e.* 2.6 m overall depth), can be adopted. An extra third unit as stand-by may also be constructed. **Ans.**

Design of Rotary Distributors

Rotary distributors, as said earlier, are to be designed for peak flow, which may be considered as 2.25 times the average flow.

∴ Peak sewage flow per day

$$= 2.25 \times 5\text{ M.L/day} = 11.25\text{ M.L/day}$$

$$= 11.25 \times \frac{10^6}{10^3} \times \frac{1}{24 \times 60 \times 60}\text{ cu-m/sec}$$

$$= 0.13\text{ m}^3/\text{sec}.$$

This flow is divided into two filter units ; and, therefore,

Flow through each unit at peak flow

$$= 0.065\text{ m}^3/\text{sec}.$$

Assuming that the velocity at peak flow is 2 m/sec through the central column of the distributor, we have

The dia of the central column

$$= \sqrt{\frac{0.065}{2} \times \frac{1}{\pi/4}} = 0.2\text{ m}.$$

Provide a *central column* of 0.2 m in dia, but check the velocity through the column at average flow, as it should not be less than 1 m/sec or so.

Check for flow velocity through the central column at average flow

Discharge through each unit at average flow

$$= \frac{5}{2}\text{ M.L. day} = 2.5\text{ M.L./day}$$

$$= 2.5 \times \frac{10^6}{10^3} \times \frac{1}{24 \times 3600}\text{ m}^3/\text{sec.} = 0.029\text{ m}^3/\text{sec.}$$

∴ Velocity at average flow

$$= \frac{0.029}{\frac{\pi}{4} \times (0.2)^2} = 0.92\text{ m/sec.}$$

Strictly speaking, since the velocity at average flow becomes slightly less than the minimum permissible of 1 m/sec, we should reduce the adopted dia slightly. Say, let us use 0.19 m dia, then

The velocity at average flow

$$= \frac{0.029}{\frac{\pi}{4} \times (0.19)^2} = 1.023 \text{ m/sec} > 1 \text{ m/sec}\,; \quad \therefore \text{ permissible}$$

The velocity at peak flow will then be

$$= \frac{0.065}{\frac{\pi}{4} \times (0.19)^2} = 2.29 \text{ m/sec.}$$

Hence, we may use a **central column** of 0.19 m in dia. If, however, the central column of 0.19 m in dia is not available, we may permit 0.2 m (8″) dia central column, as the velocity at average flow is quite near the allowable minimum value.

Design of Arms. Now, let us use rotary reaction spray type distributor with 4 arms.

Then, the discharge per arm $= \frac{0.065}{4}$ m³/sec. = 0.016 m³/sec.

Dia of filter used = 40 m.

$\therefore$ Arm length $\approx \frac{40-2}{2} = \frac{38}{2} = 19$ m.

We can use each arm of 19 m length with its size reducing from near the central column towards the end. The first two sections, each of 6 m length, and the third section of 7 m length, can be used.

The flow in the arms has to be adjusted in the proportion of the filter area covered by these lengths of arm. Therefore, the area covered by the different lengths of arm are calculated first.

Let A_1, A_2, and A_3 be the circular filter areas covered by each length of arm, starting from the central column. Allowing for 0.3 m dia in centre to be used for central column, etc., these areas would be :

$$A_1 = \pi(r_2{}^2 - r_1{}^2)$$
$$= \pi(6.15)^2 - (0.15)^2]$$
$$= \pi(37.82 - 0.0225) = 118.69 \text{ m}^2.$$
$$A_2 = \pi[(12.15)^2 - (6.15)^2]$$
$$= \pi(147.62 - 37.82) = 344.77 \text{ m}^2$$
$$A_3 = \pi[(20)^2 - (12.15)^2]$$
$$= \pi(400 - 147.62) = 792.5 \text{ m}^2$$

Total area of filter (A) = $\pi[(20)^2 - (0.15)^2]$ = 1256 m²*.

$\therefore$ Proportionate areas served by each section of arm (numbered from the centre towards the end) are worked out as :

$$\text{1st} = \frac{A_1}{A} = \frac{118.69}{1256} = 9.45\%$$
$$\text{2nd} = \frac{A_2}{A} = \frac{344.77}{1256} = 27.45\%$$
$$\text{3rd} = \frac{A_3}{A} = \frac{792.5}{1256} = 63.1\%$$
$$\underline{\underline{100\%}}$$

**Check.* A should be equal to $A_1 + A_2 + A_3$.

i.e., = 118.69 + 344.77 + 792.5 = **1255.96 ≈ 1256 m².**

Now, full discharge through an arm, *i.e.*, 0.016 m^3/sec, will flow through the first section, and this will go on reducing through the second and third sections.

(*i*) *Design of first section*

Discharge = 0.016 m^3/sec.

Assuming the velocity through the arm as 1.2 m/sec., we have

$$\text{The area required} = \frac{0.016}{1.2}\ \text{m}^2 = 0.0133\ \text{m}^2$$

$$\therefore\ \text{Dia required} = \sqrt{\frac{0.0133}{\pi/4}} = 0.13\ \text{m. say 13 cm, or } \mathbf{130\ mm.}$$

(*ii*) *Design of second section*

Discharge through the second section

$$= (100 - 9.45)\ \% \times 0.016$$

$$= \frac{100 - 9.45}{100} \times 0.016 = 0.0145\ \text{m}^3\text{/sec.}$$

$$\therefore\ \text{Area required} = \frac{Q}{V} = \frac{0.0145}{1.2} = 0.012\ \text{m}^2$$

$$\text{Dia required} = \sqrt{\frac{0.012}{\pi/4}} = 0.124\ \text{m} = \mathbf{124\ mm.}$$

(*iii*) *Design of third section*

Discharge through the third section

$$= (100 - 9.45 - 27.45)\ \% \times 0.016$$

$$= \frac{63.1}{100} \times 0.016 = 0.010\ \text{m}^3\text{/sec.}$$

$$\therefore\ \text{Dia required} = \sqrt{\frac{0.010}{1.2 \times \frac{\pi}{4}}} = 0.103\ \text{m} = \mathbf{103\ mm.}$$

Each arm length can thus be made of three sections, *i.e.* first 6 m from centre to be 130 mm dia, next 6 m of 124 mm dia, and the last 7 m of 103 mm dia. If economy is not much effected, and or if different sized pipes are difficult to join, then the entire arm length may be kept of 130 mm dia ; or the first 12 m may be kept of 130 mm dia, and the last 7 m of 103 mm dia. **Ans.**

Design of Orifices. Each arm section will be provided with different number of orifices, depending upon the discharge to be passed through each section.

Total discharge through each arm

$$= 0.016\ \text{m}^3\text{/sec.}$$

Assuming that 10 mm dia orifices are provided with coefficient of discharge (C_d) being 0.65, we have

The discharge through each orifice with an ***assumed water head, causing flow, as 1.5 m,***

$$= C_d \cdot A \cdot \sqrt{2gh}$$

$$= 0.65 \times \frac{\pi}{4} \times (0.01)^2 \times \sqrt{2 \times 9.81 \times 1.5}$$

$$= 2.768 \times 10^{-4}\ \text{m}^3\text{/sec.}$$

Total No. of orifices through each arm

$$= \frac{\text{Total discharge through each arm}}{\text{Discharge through each orifice}} = \frac{0.016}{2.768 \times 10^{-4}} \approx 58.$$

Hence, use a total of 58 orifices in the arm ; the number of orifices in each section of the arm is given as :

Number of orifices through the first section $= \frac{9.45}{100} \times 58 \approx 5$

Number of orifices through the second section $= \frac{27.45}{100} \times 58 \approx 16$

Number of orifices through the third section $= \frac{63.1}{100} \times 58 \approx 37.$

Spacing of orifices can be

In the first section, 5 number in 6 m length, *i.e.* @ $\frac{6}{5}$ = 1.2 m c/c.

In the second section, 16 numbers in 6 m length, *i.e.* @ $\frac{6}{16}$ = 0.375 m c/c.

In the third section, 37 numbers in 7 m length, *i.e.* @ $\frac{7}{37}$ = 0.189 m *i.e.* 18.9 cm c/c.

Design of Under-drainage System

Total discharge through each filter unit at peak flow = 0.065 m^3/sec.

Let us design the under-drainage system with a central rectangular channel, fed by radial laterals discharging into the channel. The under-drain block lengths, containing semi-elliptical openings, can be used as laterals.

The size and slope of the rectangular effluent channel should be such as to allow, say a velocity of 1 m/sec through it (min. = 0.9 m/sec).

$$\therefore \quad \text{Area of channel} = \frac{\text{Discharge}}{\text{Velocity}} = \frac{0.065}{1} \text{ m}^2 = 0.065 \text{ m}^2$$

Using 0.225 m width, we have

$$\text{its depth} = \frac{0.065}{0.225} = 0.288 \text{ m ; say } 0.3 \text{ m.}$$

The slope of the bed of this channel S, is given by :

$$Q = \frac{1}{N} \,.\, A \,.\, R^{2/3}.\, S^{1/2}$$

where N = Manning's coefficient = 0.018 (say)

$A = 0.225 \text{ m} \times 0.3 \text{ m} = 0.0675 \text{ m}^2$

$$R = \frac{A}{P} = \frac{0.0675}{(0.225 + 0.3 + 0.3)} = \frac{0.0675}{0.825} = 0.082$$

$$\therefore \quad 0.065 = \frac{1}{0.018} \times 0.0675 \times (0.082)^{2/3}.\, S^{1/2} = 0.706\sqrt{S}$$

or $$\sqrt{S} = \frac{0.065}{0.706} = \frac{1}{10.86}$$

or $$S = \frac{1}{117.9}\text{ ; say 1 in 100.}$$

Hence, use a central effluent channel, 0.225 m (9″) in width and 0.3 m (12″) in depth-below the bottom level of laterals. The channel may be laid at a slope of 1 in 100. **Ans.**

The under-drain block lengths can be placed in rows, discharging into the effluent channel. ***The under-drains may be laid circumfrentially at a slope of 1 in 40 if the main effluent channel is provided diagonally, [otherwise they may be laid radially if the main effluent channel is laid along the outer periphery of the filter].***

Let us use 10 cm (4″) dia, semi-circular under-drain blocks. These laterals should be designed to run less than $\frac{1}{2}$ full, so as to ensure proper ventilation. Let us assume that laterals run at a depth of say $0.3D$, where D is the dia of circle of which the lateral section is a semi circle. Then from Table 4.8, we have

$$q = 0.196\ Q$$

$$a = 0.252\ A$$

Also $$Q = \frac{1}{N} . A . R^{2/3} . S^{1/2}$$

∴ Q through a circular sewer of dia = D = 0.1 m (10 cm) is given as :

$$Q = \frac{1}{0.013} \times \frac{\pi \times (0.1)^2}{4} \times \left(\frac{0.1}{4}\right)^{2/3} . \frac{1}{\sqrt{40}} = 0.00815 \text{ m}^3\text{/sec.}$$

∴ $$q = 0.196\ Q = 0.196 \times 0.00815 = 0.0016 \text{ m}^3\text{/sec.}$$

Discharge through the filter = 0.065 m^3/sec.

Discharge through each lateral = 0.0016 m^3/sec.

∴ Number of laterals required = $\frac{0.065}{0.0016}$ = 40 (say).

So use 40 laterals in all, laid circumfrentially in the circular filter tank of 40 m in dia @ about 1 m spacing to discharge in the central effluent channel below.

Now, velocity through the laterals at peak flow

$$= \frac{q}{a} = \frac{0.0016}{0.252 \times \frac{\pi}{4} \times (0.1)^2} = 0.81 \text{ m/sec.} > 0.75 \text{ (required.)}$$

Velocity at average flow :

$$q_{\text{at average flow}} = \frac{0.0016}{2.25} = 0.00071 \text{ m}^3\text{/sec.}$$

∴ $$\frac{q_{av}}{Q} = \frac{0.00071}{0.00815} = 0.087$$

For $\frac{q}{Q}$ of 0.087, from table 4.8, $\frac{d}{D} = 0.2$ and $\frac{a}{A} = 0.143$

∴ $$a_{av.} = 0.143\ A$$

$$v_{av} = \frac{q_{av}}{a_{av}} = \frac{0.00071}{0.143 \times \frac{\pi}{4} \times (0.1)^2}$$

$$= 0.632 \text{ m/sec} > 0.5 \text{ m/sec (required). O.K.}$$

Hence, use 40 semi circular laterals of 10 cm dia. each, discharging into the effluent channel and laid at a slope of 1 in 40. **Ans.**

Example 9.8. *The sewage is flowing @ 4.5 Million litres per day from a primary clarifier to a standard rate trickling filter. The 5-day BOD of the influent is 160 mg/l. The value of the adopted organic loading is to be 160 gm/m³/day, and surface loading 2000 l/m²/day. Determine the volume of the filter and its depth. Also calculate the efficiency of this filter unit.*

Solution. Total 5-day B.O.D. present in sewage

$$= \frac{160 \times 4.5 \times 10^6}{10^3} \text{ gm/day} = 7{,}20{,}000 \text{ gm/day}$$

Volume of the filter media required

$$= \frac{\text{Total B.O.D}}{\text{Organic loading}}$$

$$= \frac{7{,}20{,}000 \text{ gm/day}}{160 \text{ gm/m}^3\text{.day}} \text{ m}^3 = \frac{7{,}20{,}000}{160} \text{ m}^3$$

$$= 4{,}500 \text{ m}^3. \quad \textbf{Ans.}$$

Surface area required for the filter

$$= \frac{\text{Total flow}}{\text{Hydraulic loading}} = \frac{4.5 \times 10^6 \text{ l/d}}{2000 \text{ l/m}^2\text{.d}}$$

$$= \frac{4.5 \times 10^6}{2000} \text{ m}^2 = 2.25 \times 10^3 \text{ m}^2 = 2250 \text{ m}^2.$$

∴ Depth of the bed required

$$= \frac{4500}{2250} \text{ m} = \textbf{2 m.} \quad \textbf{Ans.}$$

Efficiency of the filter is given by Eq. (9.32) as

$$\eta = \frac{100}{1 + 0.044 . \sqrt{u}}$$

where u = organic loading in kg ha-m/day.

Organic loading, u = 160 gm/m³/day (given)

Now, 1 hectare-m = 10^4 m². m = 10^4 m³.

$$\therefore \quad u = \frac{160}{1000} 10^4 \text{ kg/ha-m/day} = 1600 \text{ kg/ha-m/day.}$$

Hence, $$\eta = \frac{100}{1 + 0.0044\sqrt{1600}}$$

$$= \frac{100}{1 + 0.176} = \frac{100}{1.176} = 85.03\%. \quad \textbf{Ans.}$$

9.21.6. Comparison of Conventional and High Rate Trickling Filters. This comparison is given in Table 9.6.

Table 9.6. Conventional Vs. High Rate Trickling Filters

S. No.	*Characteristics*	*Conventional or Standard rate filters*	*High rate filter*
(1)	(2)	(3)	(4)
1.	Depth of filter media	Varies between 1.6 to 2.4 m.	Varies between 1.2 to 1.8 m.
2.	Size of filter media	25 to 75 mm.	25 to 60 mm.
3.	Land required	More land area is reqd. as the filter loading is less	Less land area is reqd. as the filter loading is more.
4.	Cost of operation	It is more for treating equal quantity of sewage	It is less for treating equal quantity of sewage.
5.	Method of operation	Continuous application, less flexible, and requires less skilled supervision,	Continuous application, more flexible, and more skillful operation is required.
6.	Type of effluent produced	The effluent is highly nitrified and stabilised, with BOD in effluent $\leq$ 20 ppm or so.	The effluent is nitrified up to nitrite stage only and is thus less stable, and hence it is of slightly inferior quality. BOD in effluent $\geq$ 30 ppm. or so.
7.	Dosing interval	It generally varies between 3 to 10 minutes. The sewage is generally not applied continuously but is applied at intervals.	It is not more than 15 seconds, and the sewage is thus applied continuously.
8.	Filter loading values (*i*) *Hydraulic loading*	Varies between 22 to 44 M.L per hectare per day	Varies between 110 to 330 M.L per hectare per day.
	(*ii*) *Organic loading*	Varies between 900 to 2200 kg of BOD_5 per hectare-metre of filter media per day.	Varies between 6000 to 18,000 kg of BOD_5 per hectare metre of filter media per day.
9.	Recirculation system	Not provided generally.	Always provided for increasing hydraulic loading.
10.	Quality of secondary sludge produced	Black, highly oxidised with slight fine particles.	Brown, not fully oxidised with fine particles.

9.22. Recirculation of Treated Sewage and its Use in High Rate Trickling Filters

Recirculation of sewage is an essential and important feature of high rate filters. The recirculation consists in returning a portion of the treated or partly treated sewage to the treatment process. Usually, the return is from the secondary settling tank to the primary settling tank, or to the dosing tank of the filter, as shown in Fig. 9.21. Sometimes, the effluent from the filter itself, before it enters the secondary clarifier, may be sent back to the primary clarifier.

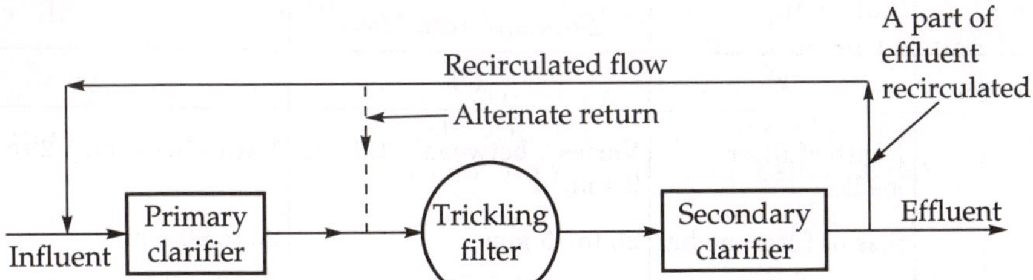

Fig. 9.21. Single stage commonly adopted Recirculation Process.

In some other cases, and to obtain better efficiency, two stage recirculation process may be adopted. A two stage recirculation process consists of having two filters arranged in series, as shown in Fig. 9.22. Various other combinations are possible.

Recirculation improves the operating results of filters, because of the following reasons :

(*i*) Recirculation allows continuous dosing of the filters, irrespective of the fluctuations in flow.

(*ii*) Recirculation equalises and reduces loading, thereby increasing the efficiency of the filter.

(*iii*) Recirculation provides longer contact of the applied sewage with the bacterial film on the contact media, thereby seeding it with bacteria, and accelerating the biological oxidation process.

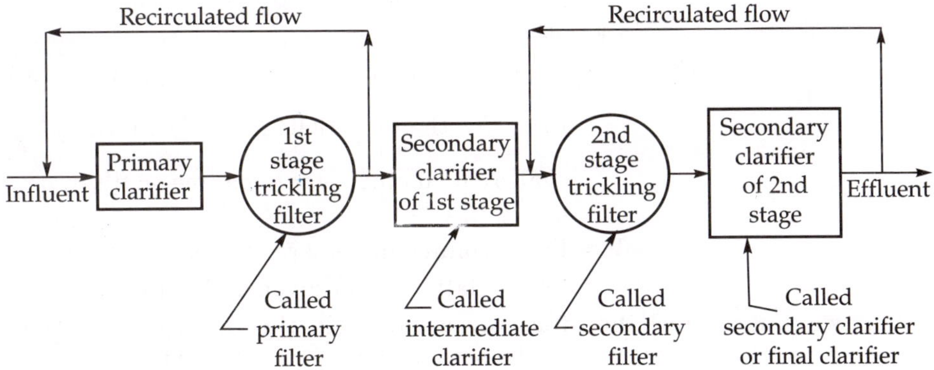

Fig. 9.22. Two stage commonly adopted Recirculation Process.

(*iv*) The influent remains fresh all the time, and also helps in reducing odours. The fly nuisance is also comparatively less.

It may, however, be noted that although the effluent would greatly be reduced in its BOD content because of the recirculated flow, yet a large volume of sewage through the filter tends to wash off the filter before nitrification has had time to take place, resulting in loss of nitrates in the effluent, thereby slightly lowering the quality of the effluent. *For this reason, a high rate filter plant with single stage recirculation may not show as good results as those obtained from a conventional trickling filter plant.* For comparable or better results, two high rate filters are constructed and operated in series, as shown in Fig. 9.22. As pointed out earlier, various recirculation schemes may be used, the most common being to pump back a part of the effluent from each filter to the influent of the same filter.

9.22.1. Efficiency of High Rate Filters. The efficiency of high rate filters depend upon the volume of the recirculated flow (in comparison to the volume of raw sewage) and also upon the organic loading.

The ratio of the Volume of sewage recirculated (*R*) to the Volume of raw sewage (*I*), *i.e.* $\frac{R}{I}$ is called **recirculation ratio,** and is an important feature in obtaining the efficiency of the filter plant (or to work out the required degree of treatment for obtaining a certain efficiency). The recirculation ratio (*R*) is connected to another term, called **recirculation factor** (*F*) by the relation

$$F = \frac{1+\frac{R}{I}}{\left[1+0.1\frac{R}{I}\right]^2} \quad ...(9.33)$$

The recirculation factor (*F*) also represents the number of effective passages through the filter. Thus, when there is no recirculation and $\frac{R}{I}$ is zero, *F* is unity.

The efficiency of the single stage high rate trickling filter can then be worked out by using the equation,

$$\eta(\%) = \frac{100}{1+0.0044\sqrt{\frac{Y}{V.F}}} \quad ...(9.34)$$

where *Y* = the total organic load in kg/day applied to the filter, *i.e.* the total BOD in kg. The term $\frac{Y}{V.F}$ is also called unit organic loading on filter, *i.e.*, *u*.

V = Filter volume in hectare-metres.

F = Recirculation factor given by Eq. (9.33).

In a two stage filter, the efficiency achieved in the first stage will be obtained as per Eq. (9.34) ; and in the second stage, it is obtained as :

Final efficiency in the two stage filter

$$= \eta'_{(\%)} = \frac{100}{1+\frac{0.0044}{1-\eta}\sqrt{\frac{Y'}{V'.F'}}} \quad ...(9.35)$$

where *Y'* = Total BOD in effluent from first stage in kg/day.

V' = Volume of second stage filter in ha-m.

F' = Recirculation factor for the second stage filter.

η' = Final efficiency obtained after two stage filtration.

These equations are very important, as they form the basis of designing high rate filters, as will become clear when we solve examples 9.9 to 9.14.

9.22.2. Effects of Recirculation on Sizes of Treatment Units. Recirculation through the primary sedimentation tanks requires extra capacity in these tanks, because the flow passing through them is increased, and under some conditions, the size of the secondary sedimentation tanks may also have to be increased. The volume of the filter, no doubt, will not be affected by recirculation, but the distributor should have the capacity to handle the increased flow.

If recirculation is from the effluent of the secondary sedimentation tank to the inlet of the primary sedimentation tank, then both the tanks, will have to be made larger. For example, with a flow of 4 million litres per day and a recirculation ratio of 0.5, both tanks will have to be designed, so as to provide the required detention period and overflow rate for a flow of 6 million litres per day, and the volume will thus have to be 50% greater. However, if the recirculation is from the effluent of the filter (before entering the secondary clarifier) the secondary clarifier need not be larger, since the detention period and overflow rate remain constant.

9.22.3. Types of High Rate Filters. There are three principal proprietory types of high rate filters. Their patented proprietory names are : (1) *Biofilters*, (2) *Accelo filters*, and (3) *Aero filters*. The patent for two of them is for the method of application of sewage to the filter. Although in practice, through cross-licensing and expiration of patents, these features no longer control the use of high rate filters.

These three important types of filters are discussed below in brief.

(1) **Bio-filters.** The bio-filters are comparatively shallow filters with 1.2 m to 1.5 in depth (the depth is kept less on the consideration that the main action of treatment is involved in the upper surface layers of the filter). The filter utilises recirculation of a portion of the filter effluent to the primary settling tank for a second passage through the filter. If additional treatment is necessary to lower the BOD content in the effluent, such as in the case of strong sewages, a second stage filter may be provided. Also the quality of final effluent can also be changed by altering the loading rate and the recirculation ratio. Fig. 9.23 shows the various arrangements of Bio-filters utilising one stage or two stage treatment.

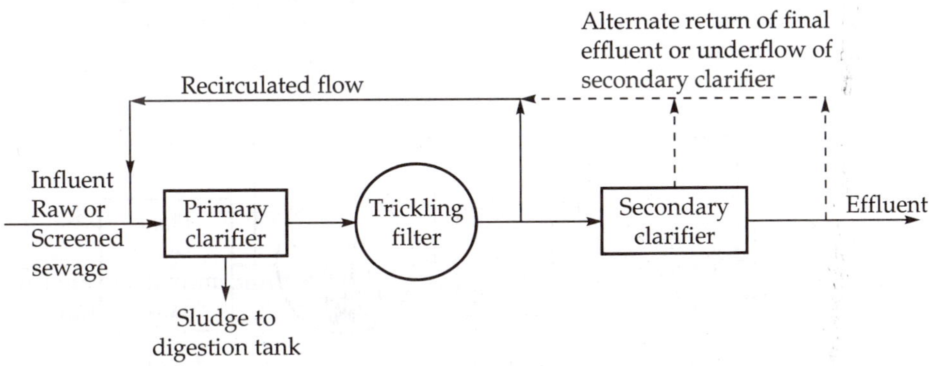

(a) Single stage complete treatment (biofilter)

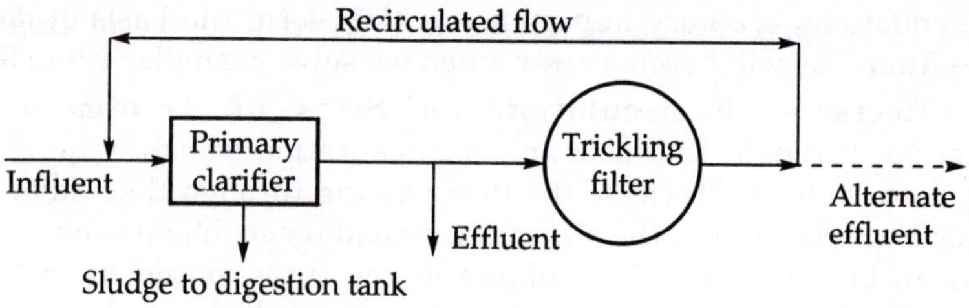

(*b*) Single stage intermediate treatment (biofilter)

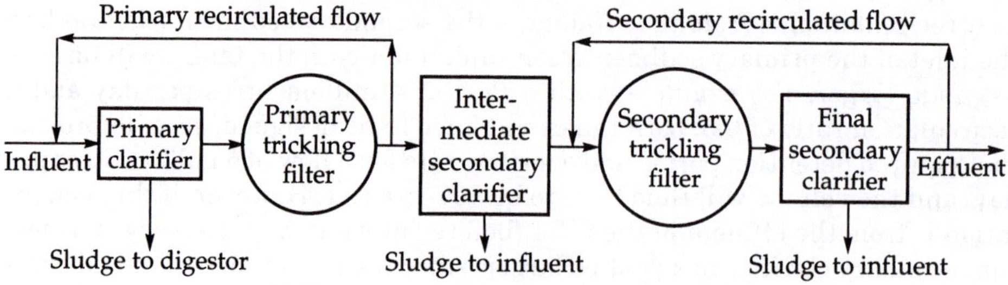

(*c*) Two stage complete treatement (biofilter)

Fig. 9.23. Flow diagrams of Bio-filter plants.

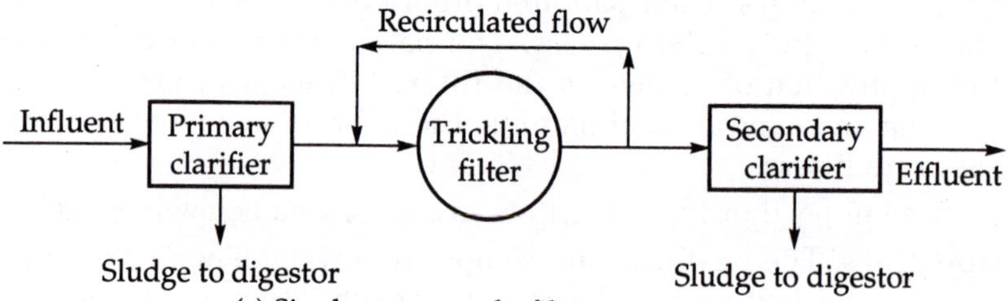

(*a*) Single stage accelo-filter arrangement

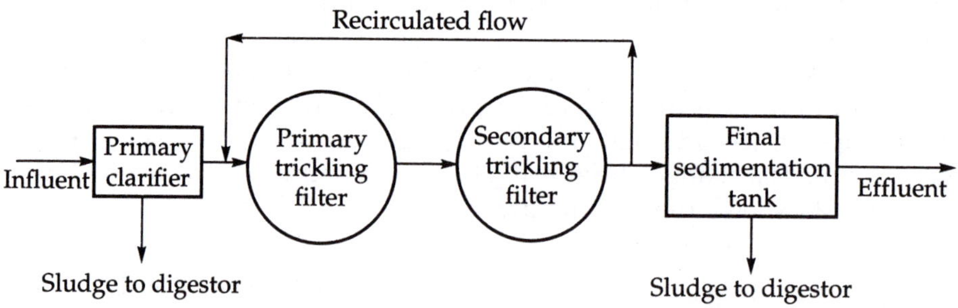

(*b*) Two stage (with recirculation from 2nd stage to the 1st stage) accelo filter

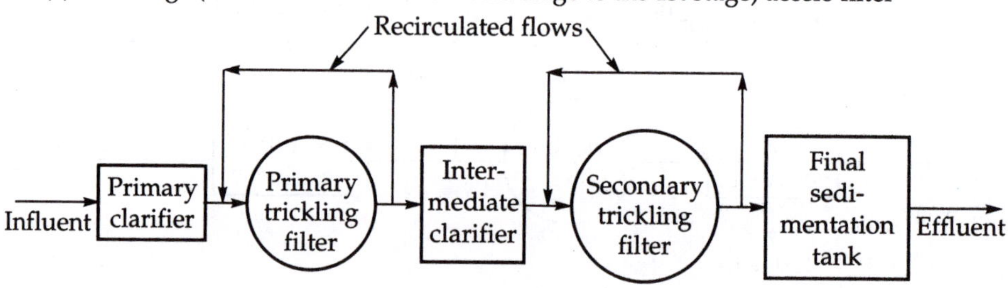

(c) Two stage (with direct recirculation within each stage) accelo-filter

Fig. 9.24. Flow diagrams for Accelo-filters.

These filters are capable of giving any desired degree of treatment by appropriate selection of flow diagram and recirculation ratio. Organic loading adopted, normally, ranges between 9,000 to 11,000 kg of BOD_5 per hectare metre per day. The total hydraulic loading may range between 110 to 330 million litres per day per hectare.

(2) **Accelo-filters.** These filters are normally 1.8 to 2.4 m deep, and utilises the direct recirculation of unsettled filter effluent to the distributor feed. Various arrangements, such as shown in Fig. 9.24, may be used. The applied loadings usually vary between the same limits as for the bio-filters.

(3) **Aero filters.** In the aero-filters, a continuous, uniform and low rate of application of sewage over the filter media is maintained. Whereas, in the bio-filters and accelo-filters, the recirculation is used for more intimate contact of sewage with the biological film of organic matter formed within the particles of the filter media ; in the aero-filters, on the other hand, the same results are obtained by the use of special type of distributor designed for frequent "rain drop" applications. Recirculation is used only during periods of low sewage flow ; and then only in amounts necessary to ensure operation of the distributor. If additional treatment is required to lower the BOD of the effluent, a second stage filter may also be used ; flow diagrams used are shown in Fig. 9.25.

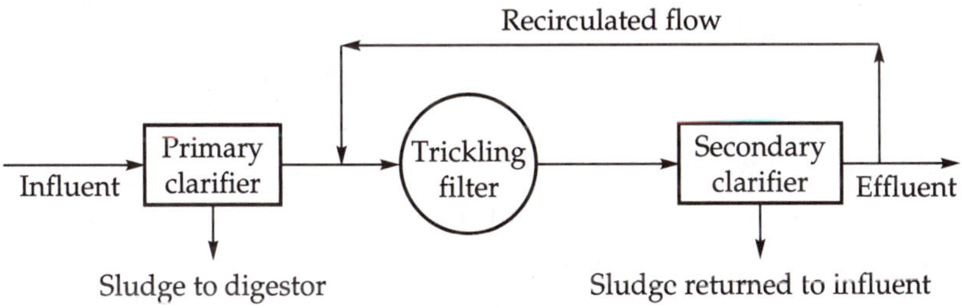

(*a*) Single stage Aero-filter

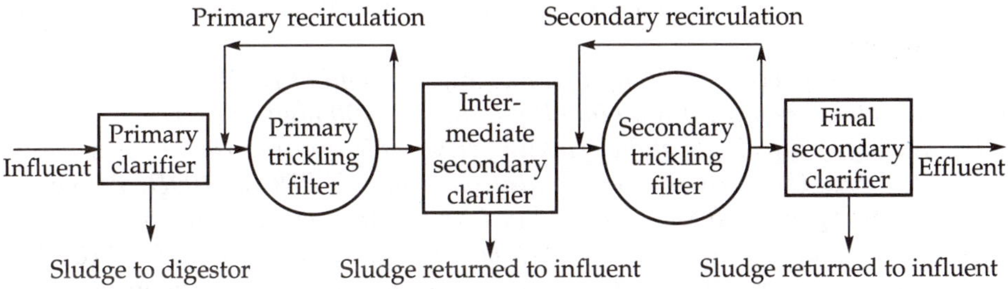

(*b*) Two stage Aero-filter

Fig. 9.25. Flow diagrams for Aero-filters.

Aero-filter beds are usually more than 1.8 m deep. The recommended organic loading may range between 11000 to 12000 kg of 5 day-BOD per hectare metre per day. The hydraulic dose, for proper functioning of filters, should not be less than 150 million litres per hectare per day (which may need recirculation during low flow rates).

9.22.4. Super Rate Trickling Filters. Super rate trickling filters, also called ***Roughing filters*** or ***Biotowers,*** are the modern advancements, generally used for treating waste waters having very high BOD. Since the power required in trickling filters per unit of BOD removal is less as compared to Activated Sludge Plant (ASP), these are becoming popular these days. They are used ahead of the existing trickling filter or ASP, and are generally constructed above ground. The diameter of the biotowers can vary from 3 m to 70 m. The walls of the biotowers can be made from RCC ; or when modular plastic media is used, the walls can be made from the plastic, since there is no hydrostatic pressure on the walls. Air blower may be provided in addition to natural air draft in biotower to enhance oxygen resources of the system to handle higher organic loading rates.

Design values of hydraulic loading rates, organic loading rates, and other parameters for low rate, high rate, and super rate trickling filters are given below in table 9.6 (*a*).

Table 9.6. (a) Design Values of Parameters for Trickling Filters

Parameter	*Low rate trickling filter*	*High rate trickling filter*	*Super rate roughing filter*
Hydraulic loading, $m^3/m^2.d$	1–4	10–40	40–200
Organic loading or volumetric loading kg BOD/m^3.d	0.11–0.37	0.37 to 1.85	1.0–6.0
Depth, m	1.5–3.0	1.0–2.0	4–12
Recirculation ratio	0	1–4	1–4
Power requirement, kW/10^3 m^3	2–4	6–10	10–20
Dosing intervals	Less than 5 min.	15 to 60 seconds	Continuous
Sloughing	Intermittent	Continuous	Continuous
Effluent quality	Fully nitrified	Nitrified only at low loading	Nitrified only at low loading

Example 9.9. *A town having a population of 30,000 persons is producing the following sewages :*

(*i*) *Domestic sewage @ 120 l.p.c.d. having 200 mg/l of BOD.*

(*ii*) *Industrial sewage @ 3,00,000 l.p.d. having 800 mg/l of BOD.*

Design high rate single stage trickling filters for treating the above sewage. Assuming that the primary sedimentation removes 35% of BOD. Allow an organic loading of 10,000 kg/ha-m/day (excluding recirculated sewage). The recirculation ratio is 1.0 ; and the surface loading should not exceed 170 M.l./ha/day (including recirculated sewage). Also determine the efficiency of the filter and the BOD of the effluent.

Solution. Quantity of domestic sewage produced per day

= 120 × 30,000 litres/day = 3.6 M.*l*/day.

BOD for domestic sewage = 200 mg/l.

∴ Total BOD of domestic sewage per day

= 3.6 × 200 kg/day = 720 kg/day ...(*i*)

Quantity of industrial sewage produced per day

= 3,00,000 litres.

BOD of industrial sewage = 800 mg/l

∴ Total BOD of industrial sewage

$$= \frac{3{,}00{,}000 \times 800}{10^6} \text{ kg/day}$$

$$= 240 \text{ kg} \quad \text{...(ii)}$$

Total BOD of domestic as well as industrial sewage per day

$$= 720 + 240 = 960 \text{ kg/day}$$

Out of this BOD, 35% is already removed in primary clarifier.

∴ BOD to be removed by filter unit

= 960 × (0.65) = **624 kg/day.**

Volume of filter media required

$$= \frac{\text{Total BOD removed}}{\text{Organic loading}} = \frac{624}{10{,}000} \text{ ha-m} = \frac{624}{10{,}000} \times 10^4 \text{ m}^3$$

= **624 m³.**

Now, the total volume of sewage flowing

$$= (3.6 \times 10^6 + 3{,}00{,}000) \text{ litres/day}$$

$$= (3.9 \times 10^6) \text{ litres/day} = 3.9 \text{ M}l\text{./day}$$

A recirculation ratio of 1 means that the volume of recirculated sewage

$$(R) = \text{Original volume} = 3.9 \text{ M.}l\text{./day} \quad \text{...(i)}$$

Total volume (*i.e.*, original + recirculated)

$$= 2 \times 3.9 \text{ M.}l\text{/day} = 7.8 \text{ M.}l\text{./day}$$

∴ Filter area required

$$= \frac{\text{Total flow volume}}{\text{Surface loading}} = \frac{7.8 \text{ Ml/d}}{170 \text{ Ml/ha.d}}$$

$$= \frac{7.8}{170} \text{ hectares} = \frac{7.8}{170} \times 10^4 \text{ m}^2 = 458.8 \text{ m}^2.$$

∴ Dia of the filter tank required

$$= \sqrt{\frac{458.8 \times 4}{\pi}} = 24.17 \text{ m.}$$

Hence, use, say, 24 m dia tank with area as = 452.16 m². **Ans.**

∴ Depth of filter media required

$$= \frac{\text{Volume of filter media}}{\text{Surface area}} = \frac{624 \text{ m}^3}{452.16 \text{ m}^2} = 1.38 \text{ m.} \quad \textbf{Ans.}$$

Efficiency of this filter is given by Eq. (9.34) as

$$\eta = \frac{100}{1 + 0.0044\sqrt{\dfrac{Y}{V.F.}}}$$

where Y = Total organic load, *i.e.*, total BOD applied to filter in kg/day = 624 kg per day.

V = Volume of filter in ha-m

$$= \frac{624}{10{,}000} = 0.0624 \text{ ha-m}$$

F = Recirculation factor as given by Eq. (9.33) as :

$$= \frac{1+\frac{R}{I}}{\left[1+0.1\frac{R}{I}\right]^2} \quad \text{where } \frac{R}{I} = 1 \text{ (given)}$$

$$= \frac{1+1}{(1+0.1)^2} = \frac{2}{(1.1)^2} = \frac{2}{1.21} = 1.65$$

or
$$\eta = \frac{100}{1+0.0044\sqrt{\frac{624}{0.0624 \times 1.65}}} = \frac{100}{1+0.0044\sqrt{\frac{10^4}{1.65}}} = 74.5\% \quad \textbf{Ans.}$$

BOD of the effluent left

$$= \frac{(100-74.5)}{100} \times 624 \text{ kg/day}$$

$$= \frac{25.5}{100} \times 624 = 159.12 \text{ kg}$$

Total volume of effluent = 3.9 M *l*/day

∴ BOD concentration in the effluent

$$= \frac{159.12 \times 10^6}{3.9 \times 10^6} \text{ mg}/l = 40.8 \text{ mg/l.} \quad \textbf{Ans.}$$

Example 9.10. *Determine the size of a high rate trickling filter for the following data :*

(*i*) *Sewage flow = 4.5 Mld ;*

(*ii*) *Recirculation ratio = 1.5 ;*

(*iii*) *BOD of raw sewage = 250 mg/l ;*

(*iv*) *BOD removal in primary tank = 30% ;*

(*v*) *Final effluent BOD desired = 30 mg/l.* (*A.M.I.E. 1974*)

Solution. Quantity of sewage flowing into the filter per day = 4.5 M.*l*/day.

BOD concentration in raw sewage = 250 mg/*l*.

∴ Total BOD present in raw sewage = 4.5 Ml × 250 mg/l = 1125 kg.

BOD removed in primary tank = 30%

BOD left in the sewage entering per day in the filter unit

= (1125) 0.7 = 787.5 kg.

BOD concentration desired in final effluent = 30 mg/*l*.

∴ Total BOD left in the effluent per day = 4.5 × 30 kg. = 135 kg.

∴ BOD removed by the filter = 787.5 – 135 = **652.5 kg.**

∴ Efficiency of the filter

$$= \frac{\text{BOD removed}}{\text{Total BOD}} \times 100 = \frac{652.5}{787.5} \times 100 = 82.85\%.$$

Now, using equation (9.34), we have

$$\eta = \frac{100}{1 + 0.0044\sqrt{\dfrac{Y}{V.F.}}}$$

where η = 82.85%

Y = Total BOD in kg = 787.5 kg.

$$F = \frac{1 + \dfrac{R}{I}}{\left(1 + 0.1\dfrac{R}{I}\right)^2} \quad ; \text{ where } = \frac{R}{I} = 1.5$$

(given)

$$\therefore \quad F = \frac{1 + 1.5}{[1 + 0.1 \times 1.5]^2} = \frac{2.5}{(1.15)^2} = \frac{2.5}{1.322} = 1.89$$

$$\therefore \quad 82.85 = \frac{100}{1 + 0.0044\sqrt{\dfrac{787.5}{V \times 1.89}}}$$

or $$1 + 0.0044 \,.\, \sqrt{\frac{416.6}{V}} = 1.207$$

or $$\sqrt{\frac{416.6}{V}} = \frac{0.207}{0.0044} = 47.05$$

or $$\frac{416.6}{V} = 2213.3$$

or V = 0.188 hectare-m. = **1880 m³**

Assuming the depth of the filter as 1.5 m, we have

The surface area required

$$= \frac{1880}{1.5} \text{ m}^2 = 1253 \text{ m}^2$$

$\therefore$ Dia of the circular filter required

$$= \sqrt{1253 \times \frac{4}{\pi}} = \mathbf{40\ m.}$$

Hence, use a high rate trickling filter with 40 m dia., 1.5 m deep filter media, and with recirculation (single stage) ratio of 1.5. **Ans.**

Example 9.11. *Determine the size of a high rate trickling filter for the following data :*

Flow	*= 4.5 Mld*
Recirculation ratio	*= 1.4*
BOD of raw sewage	*= 250 mg/l*
BOD removed in primary clarifier	*= 25%.*
Final effluent BOD desired	*= 50 mg/l.*

Calculate also the size of the standard rate trickling filter to accomplish the above requirement. *(Calcutta University 1967)*

Solution. Total BOD present in raw sewage per day

= 4.5 Ml × 250 mg/l = 1125 kg.

BOD removed in the primary clarifier = 25%.

∴ BOD entering per day in the filter units

= 0.75 × 1125 kg = 843.75 kg.

Permissible BOD concentration in the effluent = 50 mg/*l*.

∴ BOD allowed to go into the effluent

= 50 mg/l × 4.5 Ml = 225 kg.

∴ BOD removed by the filter per day = 843.75 – 225 = 618.75 kg.

$$\text{Efficiency of the filter} = \frac{\text{BOD removed}}{\text{Total BOD entering}} \times 100 = \frac{618.75}{843.75} \times 100 = 73.3\%.$$

Now, efficiency of the filter is given by Eq. (9.34) as

$$\eta = \frac{100}{1 + 0.0044\sqrt{\dfrac{Y}{V.F}}}$$

where Y = Total BOD applied to the filter per day in kg.

= 843.75 kg

F = Recirculation factor

V = Vol of filter in ha.m.

$$= \frac{1 + \dfrac{R}{I}}{\left[1 + 0.1\dfrac{R}{I}\right]^2}\ ; \text{ where } \frac{R}{I} = 1.4 \text{ (given)}$$

$$= \frac{1 + 1.4}{(1 + 0.1 \times 1.4)^2} = \frac{2.4}{(1.14)^2} = \frac{2.4}{1.3} = \mathbf{1.85}$$

$$\therefore \qquad 73.3 = \frac{100}{1 + 0.0044\sqrt{\dfrac{843.75}{V \times 1.85}}}$$

$$\text{or} \quad 1 + 0.0044 \,.\, \sqrt{\frac{456}{V}} = \frac{100}{73.3} = 1.364$$

$$\text{or} \quad \sqrt{\frac{456}{V}} = \frac{0.364}{0.0044} = 82.78$$

$$\text{or} \quad \frac{456}{V} = 6853$$

$$\text{or} \quad V = \frac{45.6}{6853} \text{ hectare-m.} = \frac{456}{6853} \times 10^4 \text{ m}^3 = \mathbf{665.4\ m^3.}$$

Using 1.5 m depth of the filter, we have

$$\text{Area required} = \frac{665.4}{1.5} = 413.6 \text{ m}^2$$

∴ Dia of the filter tank required

$$= \sqrt{\frac{413.6 \times 4}{\pi}} = 23.8 \text{ m.} \quad \textbf{Ans.}$$

For an equivalent standard rate filter ; F = 1.

$$\therefore \qquad 73.3 = \frac{100}{1 + 0.0044 \, . \sqrt{\dfrac{843.75}{V}}}$$

or $$1 + 0.0044 \sqrt{\frac{843.75}{V}} = \frac{100}{73.3} = 1.364$$

or $$\sqrt{\frac{843.75}{V}} = \frac{0.364}{0.0044} = 82.73$$

or $$\frac{843.75}{V} = 6843$$

or $$V = \frac{843.75}{6843} \text{ ha-m} = 0.1233 \text{ ha-m.} = 1233 \text{ m}^3$$

$$\left[\because \text{ ha-m} = 10^4 \text{ sq. mm.} = 10^4 \text{ m}^3\right]$$

Using depth of filter as 1.5 m, we have

$$\text{Surface area required} = \sqrt{\frac{1233}{1.5}} = 822 \text{ m}^2$$

∴ Dia of the filter tank required

$$= \sqrt{\frac{822 \times 4}{\pi}} = 32.4 \text{ m.} \quad \textbf{Ans.}$$

Example 9.12. *A single stage filter is to treat a flow of 3.79 M.l.d. of raw sewage with BOD of 240 mg/l. It is to be designed for a loading of 11086 kg of BOD in raw sewage per hectare metre, and the recirculation ratio is to be 1. What will be the strength of the effluent, according to the recommendations of the National Research Council of U.S.A.*

Solution. Total BOD present in raw sewage

$$= 3.79 \text{ Ml} \times 240 \text{ mg/l} = 909.6 \text{ kg}$$

Now, filter volume required

$$= \frac{\text{Total BOD in raw sewage in kg}}{\text{Given BOD loading rate of } 11{,}086 \text{ kg / ha-m}}$$

$$= \frac{909.6}{11086} \text{ ha-m} = 0.082 \text{ ha-m.}$$

Now, assuming that 35% of BOD is removed in primary clarifier, we have

The amount of BOD applied to the filter

$$= 0.65 \times 909.6 \text{ kg} = 591.24 \text{ kg.}$$

Now, using equation (9.34), we have

$$\eta = \frac{100}{1 + 0.0044 \, . \sqrt{\dfrac{Y}{V \, . \, F}}}$$

where Y = Total BOD applied to the filter in kg = 591.24 kg

∴ V = Vol. of the filter in ha-m. = 0.082 ha-m.

$$F = \frac{1+\dfrac{R}{I}}{\left(1+0.1\dfrac{R}{I}\right)^2}\ ; \quad \text{where} \quad \frac{R}{I} = 1$$

$$\therefore \quad F = \frac{1+1}{(1+0.1)^2} = \frac{2}{1.21} = 1.65.$$

$$\therefore \qquad \eta = \frac{100}{1+0.0044\sqrt{\dfrac{591.24}{0.082 \times 1.65}}} = 77.47\%.$$

∴ The amount of BOD left in the effluent

= 591.24 [1 – 0.7747] kg. = 133.21 kg.

∴ BOD concentration in the effluent

$$= \frac{\text{Total BOD}}{\text{Sewage volume}} = \frac{133.21 \times 10^6}{3.79 \times 10^6} \text{ mg}/l = 35.15 \text{ mg}/l. \quad \textbf{Ans.}$$

Example 9.13. *It is proposed to use a two stage plant instead of the single stage plant in example 9.12. The total volume of filter medium remains the same as was in one filter, i.e. 0.082 ha-m, and each filter is to contain half of this material, and the recirculation ratio is to be 1 for each filter. Determine the BOD of the plant effluent.*

Solution. For each filter $F = 1.65$.

For the first stage filter, the efficiency is given by

$$\eta = \frac{100}{1+0.0044\sqrt{\dfrac{Y}{V.F}}}$$

where Y = Total BOD applied to filter

= 591.24 kg (from previous example)

$$V = \text{Volume of filter} = \frac{0.082}{2} = 0.041 \text{ ha-m}$$

$$\therefore \qquad \eta = \frac{100}{1+0.0044\sqrt{\dfrac{591.24}{0.041 \times 1.65}}} = 70.85\%$$

∴ Percentage of BOD removed in first stage filter = 70.85%.

∴ Amount of BOD left in the effluent from that filter

= 591.24 [1 – 0.7085] = 172.32 kg.

For the second stage filter, the efficiency is given by equation (9.35), *i.e.*,

$$\text{The efficiency} \qquad \eta' = \frac{100}{1+\dfrac{0.0044}{1-\eta}\sqrt{\dfrac{Y'}{V'.F'}}}$$

where Y' = 172.32 kg

V' = 0.041 ha-m.

F' = 1.65.

η = 0.7085

$$\therefore \qquad \eta' = \frac{100}{1+\dfrac{0.0044}{1-0.7085}\sqrt{\dfrac{172.32}{0.041\times 1.65}}} = \frac{100}{1+\dfrac{0.0044}{0.2915}\sqrt{\dfrac{172.32}{0.041\times 1.65}}}$$

$$= 56.76\%$$

∴ The amount of BOD left in the effluent from the plant

$$= 171.9\left[\frac{100-56.76}{100}\right] \text{kg.} = 74.33 \text{ kg.}$$

∴ BOD concentration in the effluent

$$= \frac{\text{Total BOD}}{\text{Sewage volume}} = \frac{74.33\times 10^6}{3.79\times 10^6} \text{ mg}/l = 19.61 \text{ mg}/l. \quad \textbf{Ans.}$$

Example 9.14. *The design flow of sewage is 3.8 million litres per day, and the BOD of the raw sewage is 300 mg/l. Design a single stage Bio filter to produce an effluent having a BOD of 45 mg/l or less.*

Solution. Total BOD present in raw sewage per day

$$= 3.8 \times 300 \text{ kg.} = 1140 \text{ kg.}$$

Assuming that 35% of this BOD is removed in the primary sedimentation tank, we have

The total daily BOD applied to the filter = 0.65 × 1140 kg = 741 kg.

Now, the total daily BOD present in the effluent (permissible maximum)

$$= 3.8 \times 45 \text{ kg.} = 171 \text{ kg.}$$

∴ Total daily BOD to be removed by the filter = 741 – 171 = 570 kg.

$$\therefore \quad \text{Efficiency of the filter} = \frac{570}{741} \times 100 = 76.92\%.$$

Assuming an organic loading of say 10,000 kg/ha-m/day [*i.e.*, between 9,000 to 14,000], we have

$$\text{Volume of filter required} = \frac{\text{Total daily BOD removed}}{\text{Organic loading}}$$

$$= \frac{570}{10{,}000} \text{ ha-m.} = 0.057 \text{ ha-m.} = 570 \text{ m}^3$$

Now, using equation (9.34), we have

$$\eta = \frac{100}{1+0.0044\,.\sqrt{\dfrac{Y}{V.F}}}$$

where Y = Total daily BOD applied to filter in kg
= 570 kg.
V = Volume of filter = 0.057 ha-m.
and η = 76.92% (worked out earlier)

$$\therefore \qquad 76.92 = \frac{100}{1+0.0044\,.\sqrt{\dfrac{570}{0.057\,F}}}$$

$$\text{or} \qquad 76.92 = \frac{100}{1+0.44\dfrac{1}{\sqrt{F}}}$$

or $$1 + \frac{0.44}{\sqrt{F}} = \frac{100}{76.92} = 1.3$$

or $$\frac{0.44}{\sqrt{F}} = 0.3$$

or $$F = \left(\frac{0.44}{0.3}\right)^2 = 2.15$$

or $$F = \frac{1+\frac{R}{I}}{\left(1+0.1\frac{R}{I}\right)^2}$$

or $$2.15 = \frac{1+\frac{R}{I}}{\left(1+0.1\frac{R}{I}\right)^2}$$

This gives $\frac{R}{I} \approx 1.47$ (Solving by trial)

Hence, the recirculation ratio used in the bio filter will be 1.47 or say 1.5. **Ans.**

(**Note.** The ratio can be reduced by reducing the organic loading.)

Example 9.15. *Explain Eckenfelder trickling filter equation.*

Determine the BOD_5 of the effluent from a low rate trickling filter that has a diameter of 35 m and a depth of 1.5 m, if the hydraulic loading is 1900 m^3/day and the influent BOD_5 is 150 mg/l. Assume the rate constant as 1.89 d^{-1} and η = 0.67. (*Engineering Services, 1994*)

Solution. Eckenfelder has developed an equation for measuring the performance of trickling filters, on the basis of rate of waste removal. His final equation which helps to compute the BOD removed by the filter, is given as :

$$\frac{Y_t}{Y_0} = \left[e\right]^{-\frac{KD}{Q_L{}^n}} \qquad \ldots(9.36)$$

where Y_0 = BOD_5 of the influent entering the filter, in mg/L

Y_t = BOD_5 of the effluent getting out of the filter, in mg/L

K = Rate constant per day

D = Depth of filter in m

Q_L = Hydraulic loading rate per unit area of filter in m^3/day/m^2

$= \frac{Q}{A}$

The values given in the question are

Y_t = BOD_5 of the effluent = ?

Y_0 = BOD_5 of the influent = 150 mg/L

D = Depth of filter = 1.5 m

K = Rate constant per day = 1.89 d^{-1}

$\eta = 0.67$

Q_L = Hydraulic loading rate in m³/d.m²

$$= \frac{1900 \text{ m}^3/\text{d}}{\text{Area of filter}}$$

$$= \frac{1900}{\frac{\pi}{4} \times (35)^2} \text{ m}^3/\text{d.m}^2 = \mathbf{1.976\ m^3/d.m^2}$$

Substituting the above values in Eqn. (9.36), we get

$$\frac{Y_t}{150} = [e]^{-\frac{1.89 \times 1.5}{(19.76)^{0.67}}}$$

$$= (e)^{-1.796} = \frac{1}{(e)^{1.796}} = \frac{1}{6.027}$$

or $$Y_t = \frac{150}{6.027} = \mathbf{24.89\ mg/l} \text{ say } 25 \text{ mg/l}$$

Hence, the BOD_5 of the filter effluent = 25 mg/L. **Ans.**

9.23. Other Miscellaneous Types of Filters

Besides the ordinary sand filters, contact beds, and both types of trickling filters (standard rate as well as high rate), other types of filters, such as, Dunbar filters, magnetic filters, rapid sand filters, etc. may also sometimes be used under special circumstances. These three types of filters are briefly noted below :

(*i*) **Dunbar Filters.** Dunbar filters are similar to intermittent sand filters in so far as their construction and operation is concerned, except that, here the filter media is graded from bottom to top from 150 to 1 mm size. At each dose of sewage, the filter is flooded, which then slowly percolates down. The filtering process is continued till the filter gets clogged. The surface of the filter is then cleaned.

Thus, surface loading may be kept at about 25 million litres per hectare of surface area per day (*i.e.* 2500 M.*l*/m²/day). The BOD removed is upto 85% or so.

(*ii*) **Magnetic Filters.** In this type of filter, a layer of crushed magnetic ore of iron is provided in about 80 mm thickness, and is supported on a non-magnetic metal wire screen. Sewage is filtered through the magnetic layer, which removes the impurities purely by mechanical straining action. Filter is regularly cleaned, as and when the loss of head becomes more. These filters are not in common use these days.

(*iii*) **Rapid Sand Filters.** Rapid sand filters are like rapid gravity filters used in treating raw public water supplies. They may be adopted in treatment of sewage for reclaiming the used water, and then to utilise the reclaimed water for gardening, etc. However, such filters are generally not adopted, as they get clogged very soon.

SECONDARY SEDIMENTATION

9.24. Secondary Settling Tanks or Humus Tanks

It was pointed out earlier also that a filter, infact, does remove *only a very small percentage* of the finely divided suspended organic matter present in sewage, while *the majority of this organic matter undergoes a change of character, only* due to the biological oxidation and nitrification taking place in the filter. Due to this aerobic oxidation and nitrification that takes place in the filter, the organic solids are converted into coagulated suspended mass, which is heavier and bulkier, and would thus settle down by gravity, if allowed to do so in sedimentation tanks. The effluent of the filter is, therefore, passed through a sedimentation tank, called the *Secondary clarifier** or the *Secondary settling tank* or the *Humus tank**.

A secondary clarifier used for settling the effluents from the filters is similar to a primary settling tank, with a detention period of 1.5 to 2 hours, and over-flow rate ranging between 40,000 to 70,000 litres per sq. m. of plan area per day. Ordinary rectangular tanks with horizontal flow [Fig. 9.12 (*a*)] or circular tanks with spiral flow [Fig. 9.12 (*b*)] may be used. Sometimes, a third type, *i.e. hopper bottomed vertical flow type* tank (such as shown in Fig. 9.26), called **Dortmund tank,** is also used, especially for the secondary clarifier. They are circular tanks with deep conical hoppers, with dia as well as height ranging between 7.5 to 9 m. They are very costly, and as such, not generally used these days, and ordinary rectangular tanks with horizontal flow are generally preferred even for secondary clarification.

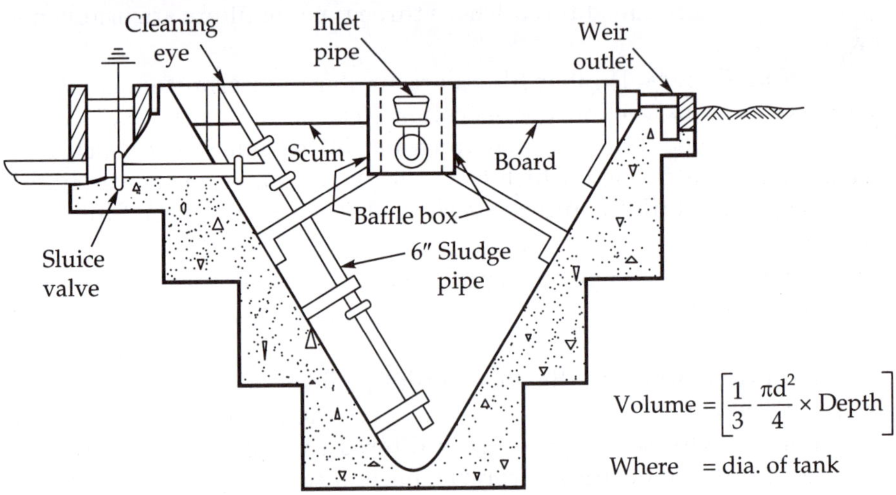

Fig. 9.26. Hopper-bottomed conical vertical flow type Sedimentation tank, called Dortmund tank.

---

*A ***scondary clarifier*** is a general term used for the tank, which is used for settling the effluent from a filter or from an activated sludge process. However, the word ***Humus tank*** specifically refers to the secondary settling tank, used for settling the filter effluent, because humus is produced in the filters only.

The secondary clarifiers provided for settling the effluent from the high rate filters (provided with recirculation of its filtered-settled sewage), will have larger capacities, so as to provide adequate settling time for the total rate of flow, and in addition, should be provided with necessary recirculation devices.

The sludge settled at the bottom of the humus tanks should be continuously and rapidly removed, and sent to the digestion tanks, either directly or through primary clarifiers.

Example 9.16. *Design percolating filters with conical humus tanks for treating 7 million litres of sewage per day.*

Solution. We will design conventional trickling filters, with a hydraulic loading, say, 28 M.l/ha/day [*i.e.* between 22 to 44].

$\therefore$ Surface area of the filter bed required $= \frac{7}{28}$ ha $= \frac{7}{28} \times 10^4$ m^2 = 2500 m^2

Using two equal sized circular tanks, we have the dia of each tank

$$= \sqrt{\frac{2500}{2} \times \frac{4}{\pi}} = 39.9 \text{ m ; say 40 m.}$$

Hence, use two trickling filters, each of 40 m dia. A third extra unit as stand-by unit, may also be provided. The depth of the filter provided may be kept 2 m or so. **Ans.**

Design of Conical Humus Tank

Assuming the detention period as 1.5 hours, we have the capacity of the tank required

$$= \frac{7}{24} \times 1.5 \text{ M.}l. = 0.437 \text{ M.}l. = 437 \text{ cu. m}$$

The volume of each conical tank, such as shown in Fig. 9.26, is given by

$$V = \left[\frac{1}{3} \cdot \frac{\pi d^2}{4}\right] \times \text{Depth} \quad ...(9.37)$$

Using 4 Nos tanks, with 7.5 m depth of each tank, we have the vol. of each tank $= \frac{437}{4}$ m^3 = 109.25 m^3.

$\therefore$ Surface area of each tank $= \frac{109.25}{7.5}$ m^2 = 14.57 m^2

$\therefore$ Dia of each tank (d) is given by

$$\frac{1}{3} \cdot \frac{\pi d^2}{4} = 14.57$$

or $$d = \sqrt{\frac{14.57 \times 12}{\pi}} = 7.46 \text{ m ; say 7.5 m.}$$

Hence, use 4 tanks with dia 7.5 m and height also as 7.5 m. One additional tank as stand-by unit may also be provided ; thus, using 5 tanks in all. **Ans.**

DIGESTION AND DISPOSAL OF PRIMARY AND SECONDARY SLUDGE

9.25. Sludge and Its Moisture Content

The sludge, which is deposited in a primary sedimentation tank, is called the *Raw sludge* ; and the sludge which is deposited in a secondary clarifier is called the *Secondary Sludge*. **Raw sludge** is odourous, contains highly putrescible organic matter, and is, thus, very very objectionable. **Secondary sludge** is also putrescible, though a little less objectionable.

The sludge withdrawn from the bottom of the sedimentation basins must, therefore, be stabilised before its final disposal. In addition to its putrescibility, another problem posed by the sludge is its high moisture content. In case of raw sludge, the moisture content is about 95% ; in case of secondary sludge from a trickling filter plant, it is about 96 to 98% ; and in case of secondary sludge from an activated sludge treatment plant*, it is about 98% to over 99%. *The sludge containing high moisture content becomes very bulky, and difficult to handle*. For example, sludge with 95% moisture, contains 5 litres of solid matter in 100 litres of sludge. Similarly, the sludge with 90% moisture will contain 10 litres of solid matter in 100 litres of sludge, *i.e.* 5 litres of solid matter in 50 litres of sludge. *Thus, the sludge with 90% moisture will be half in quantity as compared to that of the sludge with 95% moisture, for the same volume of solids*. Similarly, a sludge with 99% moisture contains 1 litre of solids in 100 litres of sludge, whereas, a sludge with 95% moisture contains 5 litres of solids in 100 litres of sludge, *i.e.* 1 litre of solids in 20 litres of sludge.

Hence, the sludge with 99% moisture will be $\frac{100}{20} = 5$ times more bulky than the sludge with 95% moisture, for storing the same solid content.

It, therefore, follows that if the moisture content of the sludge is reduced, its volume will go on decreasing. If moisture content is reduced to about 70 to 80%, the sludge becomes viscous ; and at about 10% moisture content, it becomes dry, and assumes powder form. The complete moisture can, however, be removed only by special treatments, since the water is so tenaciously held in the sludge.

Example 9.17. *A sedimentation tank is treating 4.5 million litres of sewage per day containing 275 ppm of suspended solids. The tank removes 50% of suspended solids. Calculate the quantity of sludge produced per day in bulk and weight, if (a) moisture content of sludge is 98% ; (b) moisture content of sludge is 96%.*

Solution. Volume of sewage treated = 4.5 M-litres/day

Since suspended solids amount to 275 mg/*l*, we have the mass of suspended solids present in sewage

$$= \frac{275 \times 4.5}{10^6} \times 10^6 \text{ kg/day} = 1237.5 \text{ kg/day}$$

*Described after Sludge Digestion.

Since 50% of solids are removed in sedimentation tank, we have

The mass of solids removed in sedimentation tank

$$= 1237.5 \times \frac{50}{100} = 618.75 \text{ kg/day.}$$

(*a*) *When moisture content of sludge is 98%,* then 2 kg of solids (dry sludge) will make

= 100 kg of wet sludge

∴ 618.75 kg of solids (dry sludge) will make

$$= \frac{100}{2} \times 618.75$$

= 30937.5 kg = 30940 kg (say).

Hence, wet sludge or *sludge* produced per day

= 30,940 kg = **30.94 tonnes. Ans.**

Assuming the sp. gravity of wet sludge (sludge) as 1.02, we have

Unit wt. of sludge = $1.02 \times 1 \text{ t/m}^3 = 1.02 \; t/\text{m}^3$

[∵ Unit wt. of water = 1 t/m^3]

∴ Vol. of wet sludge produced per day

$$= \frac{\text{Wt.}}{\text{Unit wt.}} = \frac{30.94}{1.02} = 30.33 \text{ m}^3$$

∴ Vol. of sludge (when its m.c. is 98%) = **30.33 m³. Ans.**

(*b*) *When moisture content is 96%*, then

4 kg of solids will make

= 100 kg of wet sludge.

∴ 618.75 kg of solids will make

$$= \frac{100}{4} \times 618.75 \text{ kg of wet sludge} = 15{,}470 \text{ kg (say) of wet sludge}$$

= 15.47 tonnes of wet sludge.

Hence, wt. of sludge (when its m.c. is 96%) = **15.47 tonnes. Ans.**

If sp. gravity of sludge is 1.02, then

$$\text{Vol. of sludge (when its m.c. is 96\%)} = \frac{15.47}{1.02} \text{ m}^3 = 15.17 \text{ m}^3.$$

Hence, the vol. of sludge at 96% m.c. = **15.17 cu. m. Ans.**

Note. ***It shows that the sludge is reduced to half its volume when its m.c. is lowered from 98% to 96%.***

Example 9.18. *There is a sewage sludge with volume containing a certain moisture content p_1 (per cent). What will be the volume of this sludge if its moisture content is reduced to p (per cent).*

Solution. Let the given sewage contains solids = W kg. Let its volume be V_1 at a moisture content of p_1 (per cent), and V at a moisture content of p (per cent).

At moisture content of p_1, we have

$(100 - p_1)$ kg of solids will make

= 100 kg of wet sludge.

W kg of solids will make

$$= \frac{100 \,.\, W}{(100 - p_1)} \text{ kg of wet sludge.}$$

or Wt. of sludge produced $= \dfrac{100 \,.\, W}{(100 - p_1)}$ kg

$\therefore$ If γ_s is the unit wt. of sludge, in kg/m^3, then

Vol. of sludge produced $= \dfrac{100 \,.\, W}{(100 - p_1)} \cdot \dfrac{1}{\gamma_s}$ m^3.

or

$$V_1 = \frac{100 \,.\, W}{100 - p_1} \cdot \frac{1}{\gamma_s}. \quad \text{...}(i)$$

At moisture content of p (per cent), similarly, we have

Vol. of sludge produced $(V) = \dfrac{100 \,.\, W}{100 - p} \cdot \dfrac{1}{\gamma_s}$ m^3

or

$$V = \frac{100 \,.\, W}{100 - p} \cdot \frac{1}{\gamma_s} \quad \text{...}(ii)$$

From equation (*i*), we have

$$W = \frac{(100 - p_1)\, V_1 \,.\, \gamma_s}{100}. \quad \text{...}(iii)$$

From equation (*ii*), we have

$$W = \frac{(100 - p)\, V \,.\, \gamma_s}{100}. \quad \text{...}(iv)$$

Equating (*iii*) and (*iv*), we get

$$\frac{(100 - p_1)\, V_1 . \gamma_s}{100} = \frac{(100 - p)\, V . \gamma_s}{100}$$

or

$$\boxed{V = V_1 \left[\frac{(100 - p_1)}{(100 - p)}\right]. \quad \textbf{Ans.}} \quad \text{...(9.38)}$$

Example 9.19. *The moisture content of a sludge is reduced from 95 to 90% in a sludge digestion tank. Find the percentage decrease in the volume of sludge.*

Solution. Using Eqn. (9.38), we have

$$V = V_1 \left[\frac{100 - p_1}{100 - p}\right]$$

or

$$V = V_1 \left[\frac{100 - 95}{100 - 90}\right] = V_1 \times \frac{5}{10} = \frac{V_1}{2}$$

Thus, the volume at 90% moisture will be half of that at 95% moisture. Hence, the percentage decrease in moisture will be 50%. **Ans.**

9.26. Sludge Digestion Process

As pointed out earlier, the sludge withdrawn from the sedimentation basins contains a lot of putrescible organic matter, and if disposed of without any treatment, the organic matter may decompose, producing foul gases and a lot of nuisance, pollution, and health hazards. In order to avoid such pollutions, the sludge is, first of all, *stabilised by decomposing the organic matter under controlled anaerobic conditions**, and then disposed of suitably after drying on drying beds, etc. The process of stabilisation is called the *sludge digestion* ; and the tank where the process is carried out is called the *sludge digestion tank*. In a sludge digestion process, the sludge gets broken into the following three forms :

(*i*) **Digested sludge.** It is a stable humus like solid matter, tary black in colour, and with reduced moisture content, and, is therefore, having reduced volume (about $\frac{1}{3}$ times the undigested sludge volume). Moreover, the quality of digested sludge is much better than that of the undigested sludge, and it is free of pathogenic bacteria which are killed in the digestion process. It may still, however, contain cysts and eggs of bacteria, protozoa and worms.

(*ii*) **Supernatant liquor.** It includes the liquified and finely divided solid matter, and is having high BOD (about 3000 ppm).

(*iii*) **Gases of decomposition.** Gases like methane (65 to 70%), carbon dioxide (30%), and traces of other inert gases like nitrogen, hydrogen sulphide, etc. are evolved. They may be collected (particularly the methane which has a high calorific value) and used as a fuel.

The sludge gas, having 70% methane, has a fuel value of about 5800 kilo calorie/cu. m (*i.e.* 650 Btu per cu. ft. app.). The amount of gas produced, on an average, is about 0.9 cu. m. per kg of volatile solids reduced in digestion. The gas produced thus varies with the sewage produced, and works out to about 14 to 18 litres per capita per day (usually 17 *l*/*c*/d).

The digested sludge is *dewatered, dried up, and* **used as fertiliser** ; while the gases produced are also **used for fuel** or for driving gas engines. The supernatant liquor contains about 1500 to 3000 ppm of suspended solids ; and is, therefore, re-treated at the treatment plant along with the raw sewage.

9.27. Stages in the Sludge Digestion Process

Three distinct stages have been found to occur in the biological action involved in the natural process of sludge digestion. These stages are :

(*i*) *Acid fermentation ;*

(*ii*) *Acid regression ;* and

(*iii*) *Alkaline fermentation.*

These stages are briefly summarised here :

(*i*) **Acid Fermentation Stage or Acid Production Stage.** In this first stage of sludge digestion, the fresh sewage-sludge begins to be acted upon by *anaerobic and facultative bacteria,* called **acid formers**. These organisms solubilize the organic solids through *hydrolysis*. The soluble products are then fermented to volatile acids and organic alcohols of low molecular weight like propionic acid, acetic acid, etc. Gases like methane, carbon dioxide and

---

*40 to 60% of the organic solids are converted by bacteria into carbon dioxide and methane gases. The organic matter which remains, is chemically stable and practically odourless, and contains 90 to 95% moisture.

hydrogen sulphide are also evolved. *Intensive acid production makes the sludge highly acidic, and lowers the pH value to less than 6. Highly putrefactive odours are evolved during this stage, which continues for about 15 days or so* (at about 21°C). BOD of the sludge increases to some extent, during this stage.

(*ii*) **Acid-Regression Stage.** In this *intermediate stage,* the volatile organic acids and nitrogenous compounds of the first stage, are attacked by the bacteria, so as to form acid carbonates and ammonia compounds. Small amounts of hydrogen sulphide and carbon-dioxide gases are also given off. The decomposed sludge has a very offensive odour, and its pH value rises a little, and to be about 6.8. The decomposed sludge, also, entraps the gases of decomposition, becomes foamy, and rises to the surface to form scum. *This stage continues for a period of about 3 months or so* (at about 21°C). BOD of the sludge remains high even during this stage.

(*iii*) **Alkaline Fermentation Stage.** In this *final stage* of sludge digestion, more resistant materials like proteins and organic acids are attacked and broken up by anaerobic bacteria, called **methane formers,** into simple substances like ammonia, organic acids and gases. *During this stage, the liquid separates out from the solids, and the digested sludge is formed.* This sludge is granular and stable, and does not give offensive odours. (It has a musty earthy odour). This digested sludge is collected at the bottom of the digestion tank, and is also called **ripened sludge.** Digested sludge is alkaline in nature. The pH value during this stage rises to a little above 7 (about 7.5 or so) in the alkaline range. *Large volumes of methane gas (having a considerable fuel value) alongwith small amount of carbon dioxide and nitrogen, are evolved during this stage. This stage extends for a period of about one month or so (at about 21°C). The BOD of the sludge also rapidly falls down during this stage.*

It is, thus, seen that several months (about $4\frac{1}{2}$ months or so) are required for the complete process of digestion to take place under natural uncontrolled conditions at about 21°C. This period of digestion is, however, very much dependent upon the temperature of digestion, and other factors. If these factors are controlled, quicker and effective digestion can be brought about, as discussed below.

9.28. Factors Affecting Sludge Digestion and Their Control

The important factors which affect the process of sludge digestion, and are, therefore, controlled in a digestion tank, are :

1. *Temperature ;*
2. *pH value ;*
3. *Seeding with digested sludge ; and*
4. *Mixing and stirring of the raw sludge with digested sludge.*

Besides these important factors, certain other minor conditions like quality of water supply ; presence of copper, fluorides, and radio active substances, etc., may also affect the rate of digestion, but not to any appreciable extent. The important factors which are largely responsible for controlling the rate and effectiveness of sludge-digestion are discussed below :

(1) **Temperature.** The process of digestion is greatly influenced by temperature ; *rate of digestion being more at higher temperatures* and *vice-versa*. The effect of temperature on digestion period is shown in Fig. 9.27. In this figure, two distinct temperature zones are indicated ; *i.e.*

(*i*) ***Zone of Thermophilic Digestion.*** In this zone of high temperature, digestion is brought about by heat loving *thermophilic* organisms. The temperature in this zone ranges between 40 to 60°C. The optimum temperature in this zone is about 54°C, and at this temperature, the digestion period can be brought down to about 10—15 days only. However, thermophilic range temperatures are generally not employed for digesting sewage sludge, owning to odours and other operational difficulties.

(*ii*) ***Zone of Mesophilic Digestion.*** In this zone of moderate temperature, digestion is brought about by common *mesophilic* organisms. The temperature in this zone ranges between 25 to 40°C. *The optimum mesophilic temperature is about 29°C ; and at this temperature, the digestion period can be brought down to about 30 days.*

Hence, it can be concluded that the sludge can be quickly digested, if the temperature in the digestion tank is kept high. The best results are obtained

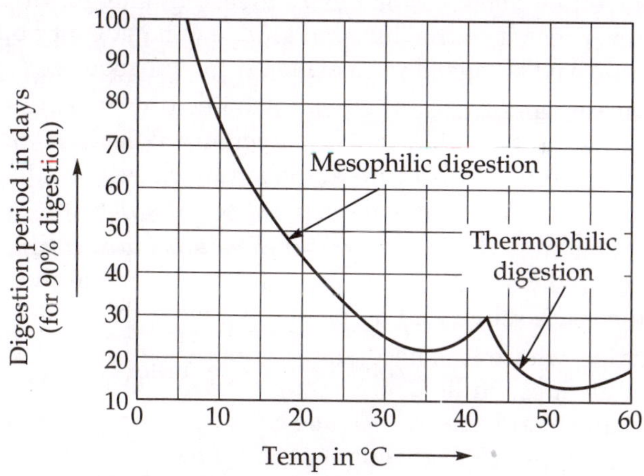

Fig. 9.27. Showing effect of temperature on sludge digestion period.

at about 29°C (*i.e.* the ***optimum mesophilic temperature***) when about 90% of digestion takes place in about 30 days. But it may, however, be pointed out that it is difficult to control temperature in practice, as it mainly depends upon the prevailing local climatic conditions. Although external heating devices may sometimes be employed to control temperature in the digestion tanks, especially in cold countries.

(2) **pH Value.** It was pointed out earlier that during the digestion process, a lot of volatile organic acids are formed, as an intermediate step, in the breakdown of organic material. These volatile acids are then converted into methane gas by a specialised group of strictly anaerobic and slow growing bacteria, called **methane formers.** If the methane formers are not operating properly, an accumulation of volatile acids may occur, causing the pH to drop to a value as low as 5.0, *which will suppress further bacterial action.* Hence,

during digestion, care must be taken to keep the acidity well under control, so that the pH during the digester start-up does not go below 6.5 or so, and thus to see that alkaline conditions (with optimum pH about 7.2 to 7.4) may prevail ultimately, in the final stage of digestion.

The acidity increases, (*i*) with the overdosing of raw sludge ; (*ii*) with the over withdrawal of digested sludge ; and (*iii*) with the sudden admission of industrial wastes. *The remedy in such cases is to add hydrated lime in doses of 2.3 to 4.5 kg. per 1000 persons to the raw sludge.* The weight of raw sludge to be added daily, for the maintenance of optimum value of pH, should also be limited to 3 to 5 per cent of the weight of the digested sludge removed.

(3) **Seeding with the Digested Sludge.** When a sludge digestion tank is first put in operation, it is highly beneficial to *seed* it with the digested sludge from another tank. Without seeding, it may take a few months to get a tank operating properly. Proper seeding will help attain quick balance conditions of reaction.

(4) **Mixing and Stirring of the Raw Sludge with the Digested Sludge.** Incoming fresh raw sludge should be thoroughly mixed with the digested sludge, by some effective method of agitation, so as to make a homogenous mass of raw as well as digested (or partly digested) sludge. In this way, the bacterial enzymes present in the digested sludge will get every opportunity to get mixed with the raw sludge, and to attack it for subsequent decomposition.

The mixing of raw and digested sludge achieved by stirring the sludge in the sludge digestion tank by slow moving mechanical devices ; or the gases of decomposition may be used to set up agitation by circulating from bottom to top of the tank and *vice versa*, by means of a pumping device.

Excessive stirring may produce harmful effects, as it may kill the bacteria. The proper stirring however, results in even distribution of incoming sludge, breaks and reduces the scum, and helps in increasing the production of gases.

In cold countries, where it is necessary to heat the digestion tanks, so as to maintain optimum mesophilic temperature (about 29°C), the stirring may help in transmitting heat from the heating coils to the tank contents ; and thus to attain uniform temperature throughout the tank.

9.29. Sludge Digestion Tank or Digestors

(Aerobic Suspended Culture)

9.29.1. Constructional Details. A typical sludge digestion tank is shown in Fig. 9.28. It consists of a circular R.C.C. tank with hoppered bottom, and having a fixed or a floating type of roof* over its top. The raw sludge is pumped into the tank, and when the tank is first put into operation, it is seeded with

---

***Floating type of roof** rises or falls with the pressure of gas, and it keeps out air successively. It may also be enlarged to work as a gas holder. Such type of roof is useful for small installations.

Fixed type of roof is used when separate gas holders are provided. The arrangement in this case should be so made that when the working level of the tank falls due to the withdrawal of supernatant liquor or the digested sludge, the gas from gas holders returns to the tank. Such an arrangement will ensure positive pressure of gas inside the digestion tank. Such type of roofs are used for larger plants.

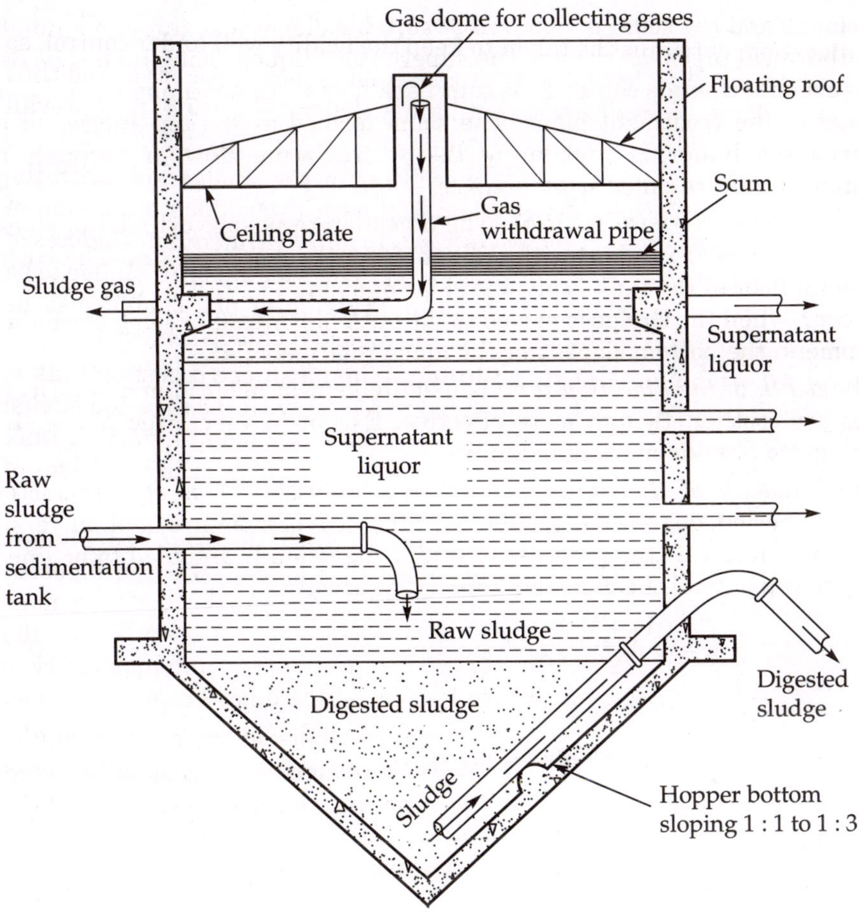

Fig. 9.28. Cross-section of a typical sludge digestion tank.

the digested sludge from another tank, as pointed out earlier. A screw pump with an arrangement for circulating the sludge from bottom to top of the tank or *vice versa* (by reversing the direction of rotation of the screw) is commonly used, for stirring the sludge. Sometimes, power driven mechanical devices may be used for stirring the sludge, although these are not very popular, at present.

In cold countries, the tank may have to be provided with heating coils through which hot water is circulated in order that the temperature inside the tank is maintained at optimum digestion temperature level.

The gases of decomposition (chiefly methane and carbon dioxide) are collected in a gas dome* (in smaller tanks) or collected separately in gas holders (in larger tanks) for subsequent use. The digested sludge which settles down to the hoppered bottom of the tank is removed under hydrostatic pressure, periodically, once a week or so. The supernatant liquor lying between

*A **gas dome** is made of suitable metal, and is cylindrical in shape. It is fixed in the roof of the digestion tank, along with various accessories such as, gas meter, pressure relief valve, etc., Gas is taken off from gas dome, and is then stored in gas holders.

the sludge and the scum is removed at suitable elevations, through a number of withdrawal pipes, as shown. The supernatant liquor, being higher in BOD and suspended solids contents, is sent back for treatment along with the raw sewage in the treatment plant. The scum formed at the top surface of the supernatant liquor is broken by the recirculating flow or through the mechanical rakers called *scum breakers*.

9.29.2. Design Considerations. The digestion tanks are cylindrical shaped tanks (*i.e.* circular in plan) with *dia* ranging between 3 to 12 m. The bottom hoppered floor of the tank is given a slope of about 1 : 1 to 1 : 3 (*i.e.* 1 H : 3 V). However, when the sludge is moved to the outlet by means of some mechanical equipment, the bottom slopes may be made relatively flat.

The *depth of the digestion tank* is usually kept at about 6 m or so. Deeper tanks are costlier, though more effective. Except in very large plants, it is usual not to provide more than 2 units.

The *capacity of the digestion tank* is a function of sludge production, digestion period, degree of digestion required, loss of moisture, and conversion of organic matter. If the progress of sludge digestion is assumed to be linear, then the capacity of the digestion tank (V) is given as :

$$V = \left(\frac{V_1 + V_2}{2}\right) t \qquad \text{...(9.39)}$$

where V = Vol. of the digestor, m^3

V_1 = Raw sludge added per day, m^3/d

V_2 = Equivalent digested sludge produced per day on completion of digestion, $m^3/d \approx \frac{V_1}{3}$

t = Digestion period, d.

When the daily digested sludge could not be removed (even though digestion gets completed) due to the factors, such as monsoon season, winter season, etc. ; then separate capacity for its storage should be provided in the tank. This capacity eventually amounts to $V_2 . T$, where T is the no. of days for which the digested sludge is stored, and is called the ***monsoon storage.*** The total digestor volume is then given as :

$$V = \left(\frac{V_1 + V_2}{2}\right) t + V_2 . T \qquad \text{...(9.39 a)}$$

However, when the change during digestion is assumed to be parabolic* rather than linear, the average volume of digesting sludge will be $\left[V_1 - \frac{2}{3}(V_1 - V_2)\right]$ and the required capacity of the digestor will be given as :

$$V = \left[V_1 - \frac{2}{3}(V_1 - V_2)\right] t \qquad \text{...(9.40)}$$

without monsoon storage

*Parabolic variation is more closed to realistic happenings in the tank, though it gives lesser capacity. Hence, in order to be on conservative side, certain people calculate the capacity on linear basis.

or $$V = \left[V_1 - \frac{2}{3}(V_1 - V_2)\right] t + V_2 \,.\, T \qquad \text{...(9.40a)}$$

with monsoon storage

The capacity of the sludge digestion tank calculated above, may also be modified for the following additional factors :

(*i*) amount of sludge withdrawn and its interval ;

(*ii*) provision of adequate free-board at top ;

(*iii*) storage for winter ;

(*iv*) type of pre-treatment given to fresh sludge ;

(*v*) collection of gas ; etc.

Depending upon all these variable factors, digestion tanks are also designed on per capita basis. The capacity provided per capita may range between 21 to 61 litres per capita (taking into account the usual one month digestion period) as shown in Table 9.7.

Table 9.7. Design Capacities of the Digestion Tanks on Per Capita Basis

S. No.	*Type of raw sludge to be digested*	*kg. of volatile solids present per cu. m. of sludge per month*	*Per capita capacity in cum/capita*
1.	Primary sludge	8	0.021
2.	Mixture of primary sludge and secondary sludge from trickling filters (humus tanks)	7.36	0.036
3.	Mixture of primary sludge and secondary activated sludge	5.76	0.061
4.	Chemically coagulated sludge	—	0.056

As pointed out earlier, the amount of the *sludge gas produced* in the digestion tank, ranges between 14 to 28 litres per capita per day (usually 17 litres per capita per day or 900 litres per gm of volatile solids digested, is quite common). The gas collected may be utilised for operating gas engines, and for heating sludge to promote quick digestion. However, the gas collection and its utilisation is found fruitful only in case of large plants serving more than 50,000 persons or so, so as to produce at least about 800–1000 cu. m. of gas per day.

Example 9.20. *Design a digestion tank for the primary sludge with the help of following data :*

(*i*) *Average flow = 20 Mld ;*

(*ii*) *Total suspended solids in raw sewage = 300 mg/l ;*

(*iii*) *Moisture content of digested sludge = 85%.*

Assume any other suitable data you require. (*A.M.I.E. 1974*)

Solution. Average sewage flow = 20 M.*l*.d.

Total suspended solids = 300 mg/*l*

∴ Mass of suspended solids in 20 Ml of sewage flowing per day

$$= \frac{300 \times 20 \times 10^6}{10^6} \text{ kg} = 6000 \text{ kg/day.}$$

Assuming, that 65% solids are removed in primary settling tanks, we have

Mass of solids removed in the primary settling tank

$$= 65\% \times 6000 \text{ kg/day} = 3900 \text{ kg/day.}$$

Assuming that the fresh sludge has a m.c. of 95%, we have

5 kg of dry solids will make = 100 kg of wet sludge

∴ 3900 kg of dry solids will make

$$= \frac{100}{5} \times 3900 \text{ kg of wet sludge per day}$$

$$= 78{,}000 \text{ kg of wet sludge per day.}$$

Assuming the sp. gravity of wet sludge as 1.02 (*i.e.* Density = 1020 kg/m^3), we have

The volume of raw sludge produced/day

$$= V_1 = \frac{78000}{1020} \text{ m}^3\text{/day} = 76.47 \text{ m}^3\text{/day.} \qquad ...(i)$$

The volume of the digested sludge (V_2) at 85% m.c. is given by the formula (9.38) as

$$V_2 = V_1 \left[\frac{100 - p_1}{100 - p_2}\right]$$

or $$V_2 = V_1 \left[\frac{100 - 95}{100 - 85}\right] \quad \text{or} \quad V_2 = V_1 \left[\frac{5}{15}\right]$$

or $$V_2 = \frac{1}{3} \times V_1 = \frac{1}{3} \times 76.47 \text{ cu. m./day} = 25.49 \text{ m}^3\text{/day.}$$

Now, *assuming the digestion period as 30 days*, we have the capacity of the required digestion tank, given by Eqn. (9.40) as :

$$\text{Capacity} = \left[V_1 - \frac{2}{3}(V_1 - V_2)\right] t$$

or $$= \left[76.47 - \frac{2}{3}(76.47 - 25.49)\right] 30$$

$$= \left[76.47 - \frac{2}{3} \times 50.98\right] 30 = 1274.5 \approx \mathbf{1275 \text{ cu. m.}}$$

Now, providing 6.0 m depth of the cylindrical digestion tank, we have

$$\text{Cross-sectional area of the tank} = \frac{1275}{6} = 212.5 \text{ m}^2$$

∴ Dia of tank $= \sqrt{\dfrac{212.5}{\pi/4}}$ m = **16.45 m ; say 16.5 m.**

Hence, provide a cylindrical sludge digestion tank (typical section shown in Fig. 9.28) 6 m deep and 16.5 m diameter, with an additional hoppered bottom of 1 : 1 slope for collection of digested sludge. **Ans.**

Example 9.21. *Raw waste water is entering a treatment plant and contains 250 mg/l suspended solids. If 55% of these solids are removed in sedimentation.*

(a) Find the volume of raw sludge produced per million litre of waste water. Assume that the sludge has a moisture content of 96%, and specific gravity of solids is 1.2.

(b) Find the unit weight of raw sludge.

(c) If 45% of raw sludge is changed to liquid and gas in the digestion tank, find the volume of digested sludge per million litre of waste water. Assume that the moisture content of the digested sludge is 90%.

Solution. (*a*) Suspended solids in waste water = 250 mg/l

Since 55% of these solids are removed in sedimentation, we have

The solids removed in sedimentation as sludge

$$= 55\% \times 250 \text{ mg/l} = 137.5 \text{ mg/l}.$$

If vol. of waste water is 1 million litre, then solids removed as sludge

$$= \frac{137.5}{10^6} \times 10^6 \text{ kg} = 137.5 \text{ kg}.$$

Sludge produced will, thus, have 137.5 kg solids, and the rest will be water. Now, since the moisture content of sludge is 96%, we have

4 kg of solids will produce 100 kg of wet sludge, by joining with 96 kg of water.

∴ Water contained in 4 kg of solids = 96 kg

∴ Water contained in 137.5 kg of solids $= \frac{96}{4} \times 137.5$ kg = 3300 kg.

Hence, *volume of sludge produced per million litre of waste water*

$$= \frac{\text{Mass of solids}}{\text{Density of solids}} + \frac{\text{Mass of water}}{\text{Density of water}}$$

$$= \left[\frac{137.5}{1.2 \times 1000} + \frac{3300}{1000}\right] \text{ cu-m}$$

[∵ Density of solids = sp. gravity of solids × Density of water = 1.2 × 1000 kg/m^3]

$$= 0.115 + 3.3 = 3.415 \text{ cu m}.$$

Hence, vol. of sludge produced per million litre of waste water

= **3.415 cu m. Ans.**

(*b*) Density of raw sludge

$$= \frac{\text{Mass of solids} + \text{Mass of water}}{\text{Volume of sludge}}.$$

$$= \frac{137.5 + 3300}{3.415} \text{ kg/m}^3 = \frac{3437.5}{3.415} \text{ kg/m}^3 = \mathbf{1007 \text{ kg/m}^3}. \quad \textbf{Ans.}$$

(*c*) 45% of raw sludge is changed to liquid and gas, means that 45% of solids are consumed (digested).

∴ Mass of dry solids left in the digested sludge

= (100 – 45) % of total solid

$$= \frac{55}{100} \times 137.5 \text{ kg} = 75.625 \text{ kg.}$$

Since digested sludge contains 90% m.c., we have

The vol. of digested sludge

$$= \left[\frac{\text{Mass of solids left in digested sludge}}{\text{Density of solids}}\right] + \left[\frac{\text{Mass of water}}{\text{Density of water}}\right]$$

$$= \left[\frac{75.625}{1.2 \times 1000} + \frac{75.625 \times \frac{90}{100}}{1000}\right] \text{m}^3 = 0.063 + 0.068 = 0.131 \text{ cu m}$$

Hence, the vol. of digested sludge per million litre of waste water

= **0.131 cu-m. Ans.**

Example 9.22. *The sewage of a certain town contains 600 ppm of suspended matter. Assuming that 55% of this is settled down in plain sedimentation tank, and the sludge collected has a water content of 95%, calculate its quantity per million litre, both in bulk and weight. Assume sp. gr. 1.02.*

Solution. Suspended matter in sewage = 600 ppm = 600 mg/l.

For 1 million litre of sewage, we have the suspended matter

$$= \frac{600}{10^6} \times 10^6 \text{ kg} = 600 \text{ kg.}$$

Now, 55% of this matter is settled as sludge, and therefore quantity of sludge solids

$$= 0.55 \times 600 = 330 \text{ kg.}$$

The sludge is having 95% m.c., which means 5 kg of dry solids will make 100 kg of wet sludge.

∴ 5 kg of dry solids make = 100 kg of sludge

$$\therefore \quad 330 \text{ kg of dry solids make} = \frac{100}{5} \times 330 = 6600 \text{ kg of sludge.}$$

Hence, the mass of sludge formed per million litre of sewage = 6600 kg. **Ans.**

$$\therefore \quad \text{Volume of sludge} = \frac{\text{Mass of sludge}}{\text{Density of sludge}} = \frac{6600 \text{ kg}}{1020 \text{ kg/m}^3}$$

[∵ Density of sludge = Sp. gr. × Density of water = 1.02 × 1000 kg/m³ = 1020 kg/m³].

$$= 6.47 \text{ m}^3.$$

Hence, the vol. of sludge formed per million litre of sewage

= **6.47 cu-m. Ans.**

Example 9.23. *Design a sludge digestion tank for 40,000 people. The sludge content per capita per day is 0.068 kg. The moisture of the sludge is 94%. The sp. gravity of the wet sludge is 1.02 and 3.5 per cent of the digestor volume is daily filled with the fresh sludge, which is mixed with the digested sludge.*

Solution. Dry sludge content produced by 40,000 persons

= 0.068 × 40,000 kg = 2,720 kg/day

94% moisture content means that 6 kg of dry sludge will produce 100 kg of wet sludge.

$\therefore$ 6 kg of dry sludge produces wet sludge = 100 kg

2720 kg of dry sludge produces wet sludge $= \dfrac{100}{6} \times 2720$

= 45333 kg = 45.3 t/day

Volume of wet sludge produced

$$= \frac{\text{Mass of sludge}}{\text{Density of sludge}}$$

$$= \frac{45.3}{1.02} \text{ m}^3\text{/day}$$

[$\because$ Density of sludge in t/m^3 = Sp. gr. × Density of water (which is 1 t/m^3) = 1.02 × 1 = 1.02 t/m^3]

= 44.4 m^3/day.

44.4 m^3 of fresh sludge is added to the tank daily, to fill 3.5% of the digestor capacity.

$\therefore$ $\dfrac{3.5}{100}$ × Capacity of digester

= Volume of fresh sludge produced daily = 44.4 m^3

or Capacity of digester required

$$= \frac{44.4 \times 100}{3.5} = 1268.9 \text{ cu-m.}$$

Providing 30% additional capacity for fluctuations, we have

The required digester capacity

= 1268.9 × 1.3 = 1650 cu-m (say).

Now, providing 6 m depth of the cylindrical digestion tank, we have

The cross-sectional area of the tank

$$= \frac{1650}{6} = 275 \text{ m}^2$$

$$\therefore \quad \text{Dia of tank} = \sqrt{\frac{275}{\pi/4}} = \sqrt{\frac{275}{0.785}} = 18.7 \text{ m.}$$

Hence, provide a cylindrical sludge digestion tank (typical section drawn in Fig. 9.28), 6 m deep and 18.7 m in diameter, with an additional hoppered bottom of 1 : 1 slope for collection of digested sludge. **Ans.**

Example 9.24. *A waste water plant produces 1000 kg of dry solids per day at a moisture content of 95 percent. The solids are 70 percent volatile with a sp. gr. of 1.05, and the remaining are non-volatile with specific gravity of 2.5. Find the sludge volume after digestion which reduces volatile solid content by 50 percent and decreases the moisture content by 90 percent.*

(*Engineering Services, 2001*)

Solution. Total solids produced = 1000 kg (dry mass)

Volatile solids = 70% total solids = 700 kg.

Non-volatile solids = 30% T.S. = 300 kg.

Volatile solids removed in digestion

= 50% volatile solids = 50% × 700 kg = 350 kg.

Volatile solids left in digested sludge = 350 kg.

Non-volatile solids in digested sludge = 300 kg (as in original sludge)

∴ Mass of solids in digested sludge = 300 + 350 = 650 kg.

Mass of water in wet digested sludge = 90%

10 kg of solids make = 100 kg of wet sludge

∴ 10 kg of solids contain = 90 kg of water

or 650 kg of solids contain = $\frac{90}{10} \times 650$ = 5850 kg of water

Density of volatile solids = $\rho_w \,.\, S_s = 1000 \frac{\text{kg}}{\text{m}^3} \times 1.05 = 1050 \text{ kg/m}^3$

Similarly density of non-volatile solids = 1000 × 2.5 = **2500 kg/m³**

∴ Vol. of volatile solids in wet sludge = $\frac{\text{Mass}}{\text{Density}} = \frac{350}{1050} \text{ m}^3 = \mathbf{0.33 \text{ m}^3}$

Vol. of non-volatile solids in wet sludge = $\frac{\text{Mass}}{\text{Density}} = \frac{300}{2500} \text{ m}^3 = \mathbf{0.12 \text{ m}^3}$

Vol. of water in wet sludge = $\frac{\text{Mass}}{\text{Density}} = \frac{5850}{1000} \text{ m}^3 = \mathbf{5.85 \text{ m}^3}$

∴ Total volume of wet sludge = 0.33 + 0.12 + 5.85 = **6.30 m³. Ans.**

Example 9.25. *A trickling filter plant treats 1500 cum per day of sewage with a BODs of 220 mg/l and SS of 250 mg/l. Estimate the total solid production assuming that primary clarification removes 30% of BOD and 60% of influent solids. Take the solid production in the trickling filter as @ 0.5 kg/kg of the applied BOD.* (*Engineering Services, 2001*)

Solution. Solids removed in primary clarification units of the trickling filter plant

= 60% of SS applied

= 60% × 250 mg/l = 150 mg/l = 150 g/m³

= $\frac{150 \times 1500}{1000}$ kg/day (for 1500 cum of sewage)

= 225 kg/day ...(*i*)

BOD_5 removed in primary treatment = 30%

Balance BOD_5 applied to filters

= 70% × 220 mg/l = 154 mg/l

= 154 g/m³ = $\frac{154 \times 1500}{1000}$ kg/day (for 1500 cum of sewage)

= 231 kg/day.

Solids production in filters

= 0.5 kg/kg of BOD applied

= 0.5 × 231 kg/day = **115.5 kg/day** ...(*ii*)

Total solids produced daily in treating 1500 cum of sewage

= (*i*) + (*ii*)

= 225 + 115.5 = **340. 5 kg/day. Ans.**

9.29.3. Estimated Gas Production. If it is possible to analyse the wastewater to be treated, and to determine the characteristics of the sludge, its gas producing ability may be estimated. When such data cannot be obtained, the following approximate values may be used as a basis for estimating the amount of gas produced by digestion :

About 60% of the suspended solids of sewage are removed by sedimentation ; 75% by chemical coagulation and settling ; and 90% by complete treatment, such as by the activated sludge or the trickling filters, preceded and followed by sedimentation.

About 70% of the suspended solids in the sewage are volatile, and the reduction of the volatile matter in sludge, is about 65%. In digestion, the amount of gas produced is about 0.6 cu. m per kg of volatile matter present in the sludge, or is about 0.9 cu-m per kg of volatile matter reduced. The gas produced usually contains 65% methane, 30% carbon dioxide, and trace amounts of other gases. The heat content of methane is approximately 36000 kJ/m^3* (8800 kC/m^3). The procedure in using this information is exhibited in the following examples :

Example 9.26. *A sewage containing 200 mg/l of suspended solids is passed through primary settling tanks, trickling filters, and secondary settling tanks. How much gas will probably be produced in the digestion of sludge from one million litres of sewage ?*

Solution. Total suspended solids in sewage = 200 mg/*l*.

Assuming 90% removal of suspended solids in complete treatment, we have

The suspended solids removed

= 90% × 200 mg/*l*. = 180 mg/*l*

Assuming volatile solids to be equal to 70% of suspended solids, we have

Volatile solids removed

= 70% × 180 mg/*l* = 126 mg/*l*.

Now, assuming that the volatile solids (matter) is reduced by 65% in the sludge by digestion, we have

Volatile solids reduced = 65% × 126 mg/*l* = 81.9 mg/*l*.

∴ Volatile matter reduced per million litre of sewage = 81.9 kg.

Now, assuming that 0.9 cu. m. of gas is produced per kg of volatile matter reduced, we have

The gas produced per million litre of sewage

= 0.9 × 81.9 cu.m. = 73.71 cu.m.

= 73,710 litres. **Ans.**

Example 9.27. *A sewage containing 200 mg/l of suspended solids is passed through primary settling tank. The solids from the primary settling tank are digested to recover the gas. Find the likely volumes of methane and carbon*

---

*1 kJ = 1.0551 BTU (British Thermal Unit)

4.12 kJ = 1 kC (kilo Calorie).

dioxide produced in the digestion of the sludge from 10,000 m^3 of sewage. Calculate the fuel value of the gas produced. State clearly the assumptions made. *(A.M.I.E. 1975)*

Solution.

Total suspended solids in sewage = 200 mg/*l*.

Assuming that 60% of suspended solids are removed in the primary settling tank, we have

The suspended solids removed as sludge

= 60% × 200 mg/*l* = 120 mg/*l*.

Now, assuming that the volatile solids present are 70% of the suspended solids, we have

∴ The volatile solids removed

= 70% × 120 mg/*l* = 84 mg/*l*.

Further, assuming that the volatile matter is reduced by 65% in sludge digestion, we have

Volatile matter reduced = 65% × 84 mg/*l* = 54.6 mg/*l*

Hence, volatile matter reduced in 10,000 cu. m. of sewage

$$= 54.6 \times \frac{(10{,}000 \times 1000)}{10^6} \text{ kg} = 546 \text{ kg}.$$

Now, assuming that 0.9 cu. m. of gas is produced per kg of volatile matter reduced, we have

Total quantity of gas produced

= 0.9 × 546 cu. m. = **491.4 cu. m. Ans.**

Assuming that the produced gas contains 65% methane and 30% carbon dioxide, we have

Methane produced = 0.65 × 491.4 cu m = **319.41 cu m. Ans.**

Carbon dioxide produced = 0.30 × 491.4 cu m = **147.42 cu m. Ans.**

Now, assuming that the methane in the sludge gas has a fuel value of 36000 kJ/m^3, we have

The fuel value = 36,000 × 391.41 kJ = 11.50 kJ

Now, assuming a boiler efficiency of 80%, we have the amount of heat that can be furnished by the boiler

= 80% × 11.50 MkJ = **9.2 MkJ. Ans.**

$$= \frac{9.2}{4.12} \text{ MkC} = \textbf{2.233 Million kilo Calorie. Ans.}$$

9.29.4 Two Stage Digestion. While treating sewage on a large scale, two stage digestion of sludge is generally adopted instead of a single stage digestion. In two stage digestion, two digestion tanks, called *primary and secondary digestors*, are used.

Sludge is, first, admitted into the primary digestion tank (or primary digestor) and is kept there for a period of about 7 to 10 days. The gas produced is collected in this unit. The partly digested sludge and supernatant liquor from this primary digestor are then transferred to the secondary digestion tank (or secondary digestor), where they are kept for a period of about 20 days or so. The digested sludge and supernatant liquor from the secondary digestor are then finally removed, and disposed of suitably in a sanitary manner.

Two stage digestion gives the following *advantages* compared to single stage digestion :

(*i*) Two stage digestion is an effective method of preventing (or reducing greatly) any tendency for the sludge to *short-circuit*, as may happen in a single stage digestor. The quality of supernatant liquor produced is, therefore, of much better quality here.

(*ii*) Only the primary digestor is provided with heating, stirring, gas collection arrangements ; while the secondary unit merely acts as a closed settling tank, so as to produce clear supernatant liquor, thus affecting economy in construction.

(*iii*) Two stage digestion offers the freedom from large scum formations in any of the digestion tanks.

(*iv*) It has been estimated that the total cost with two stage digestion may be less than that for two tanks operated in parallel.

9.30. Disposal of Digested Sludge

The digested sludge from the digestion tank contains a lot of water, and is, therefore, first of all, **dewatered** or **dried up,** before further disposal either by burning or dumping. In India, the water of sludge is removed by drying the sludge on *drying beds* ; whereas, in western countries, dewatering is generally done by *vacuum filters* or by using *high speed centrifuges*. These methods of dewatering the sludge, are discussed below :

9.30.1. Dewatering, Drying and Disposal of Sludge by Sludge Drying Beds. Drying of the digested sludge on open beds of land (called sludge drying beds) is quite suitable for hot countries like India, and is discussed below :

Sludge drying beds are open beds of land, 45 to 60 cm deep, and consisting of about 30 to 45 cm thick graded layers of gravel or crushed stone varying in size from 15 cm at bottom to 1.25 cm at top, and overlain by 10 to 15 cm thick coarse sand layer. Open jointed under-drain pipes (15 cm in dia) @ 5 to 7 m c/c spacing are laid below the gravel layer in valleys, as shown in Fig. 9.29, at a longitudinal slope of about 1 in 100. The beds are about 15 × 30 m in plan, and are surrounded by brick walls rising about 1 metre above the sand surface, as shown.

The sewage sludge from the digestion tank is brought and spread over the top of the drying beds to a depth of about 20 to 30 cm, through *distribution troughs* (Fig. 9.29) having openings of about 15 cm × 20 cm at a distances of about 2 m or so.

A portion of the moisture, drains through the bed, while most of it is evaporated to the atmosphere. It usually takes about two weeks to two months, for drying the sludge, depending on the weather and condition of the bed.

The required area for sludge drying beds normally ranges between 0.05 to 0.2 sq. m per capita, as shown in Table 9.8. In some instances, the required area is reduced by building a glass roof over the beds, to give protection against rain and snow.

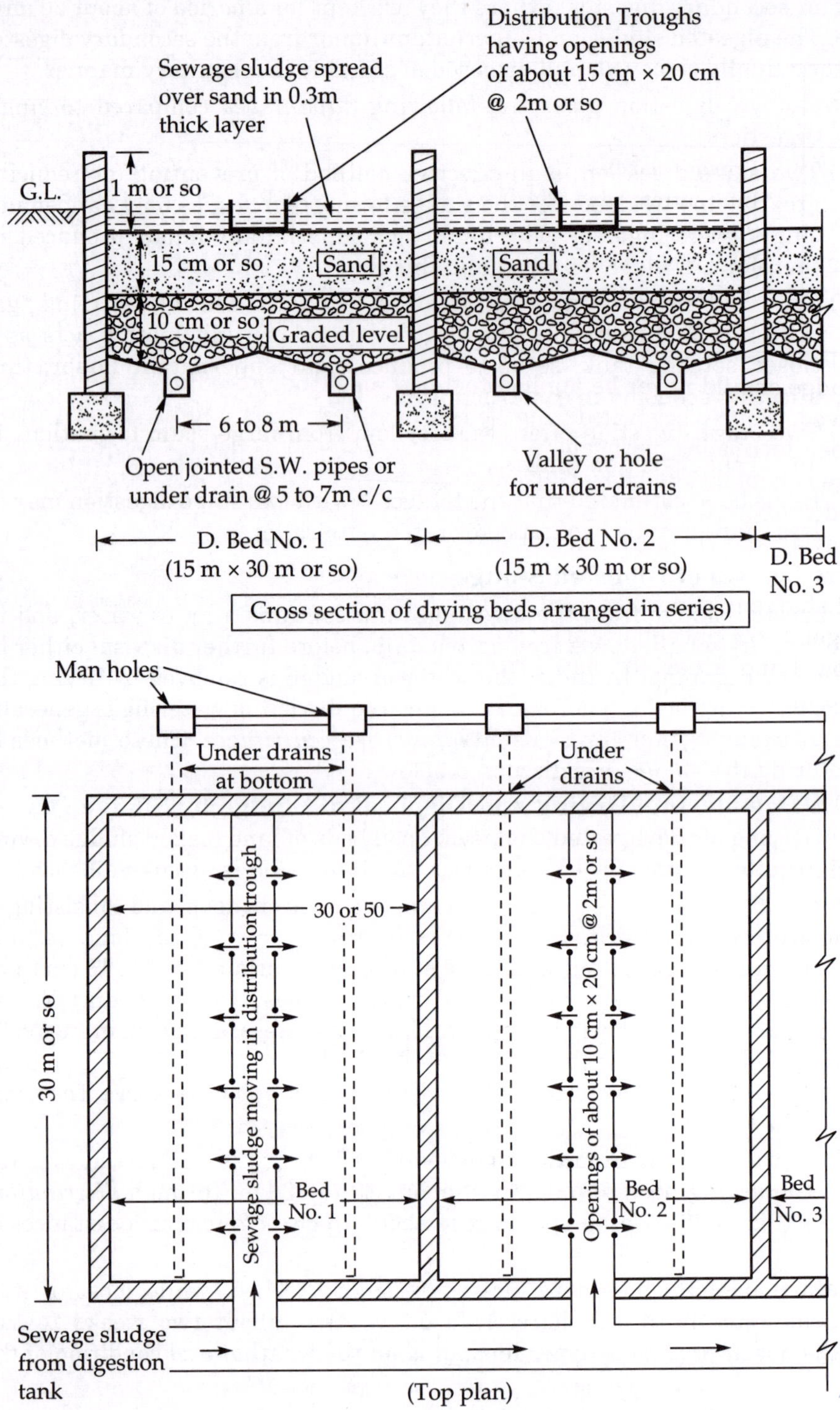

Fig. 9.29. Sludge drying beds.

Table 9.8. Approximate area of Drying beds for different types of sludges

Sl. No.	*Source of sludge*	*Area in m²/capita*	
		Open beds	*Covered beds*
1.	Primary clarifiers	0.1	0.08
2.	Intermittent sand filters	0.1	0.08
3.	Standard rate trickling filters	0.12	0.10
4.	High rate trickling filters	0.15	0.12
5.	Activated sludges	0.18	0.14
6.	Coagulated sludges	0.20	0.15

Sludge should never be applied to a bed until the preceding dose has been removed. Hence, several drying beds will generally be required, with their number increasing with an increase in the number of days for which the sludge is kept on the beds. Normally, sludge is removed from the beds after a period of about 7—10 days ; as within this period, about 30% of the moisture goes away and the surface of sludge gets cracked. The sludge cakes are then removed by spades, and they are dumped into a pit for further drying. The dried sludge is generally used as manure in our country, as it contains 1.7% nitrogen, 1.5% phosphoric acid, and 0.5% potash. It may also be used for filling up low lying areas. It may sometimes be disposed of by burning (*i.e.* by incineration).

Example 9.28. (*a*) *Calculate the area of land required for drying the sludge from the digestion tank for 40,000 population, designed in Example 9.23.*

(*b*) *Also design the dimensions of beds.*

Solution. (*a*) The volume of wet sludge from the sewage of 40,000 population was worked out as 44.4 m³/day. Let it be spread in 22.5 cm thick layer (*i.e.* between 20 to 30 cm thick layer) on under-drained sand beds, then

The area of beds required

$$= \frac{44.4}{0.225} \text{ m}^2 = 197.3 \text{ m}^2/\text{day}.$$

Under tropical Indian conditions, the beds get dried out in about 10 days, and hence taking 2 weeks as average drying time including wet days of rainy season, we can utilise the same bed $= \frac{52^*}{2}$ = 26 times in an year*.

∴ Area of bed required

$$= \frac{197.3 \times 365}{26} = 2770 \text{ m}^2 \text{ (say)}.$$

Making 100 per cent allowance for space for storage, repairs, and resting of beds, etc., we have

The total area of beds required

$$= 2 \times 2770 \text{ m}^2 = 5540 \text{ m}^2 = \textbf{0.554 hectares.} \quad \textbf{Ans.}$$

---

*52 weaks are there in an year.

(*b*) Now, using 15 × 30 m sized beds, we have the No. of beds required

$$= \frac{5540}{15 \times 30} = 12.3$$

So let us use 14 No. of beds :

Then, Area of each bed $= \frac{5540}{14} = 395.7 \text{ m}^2$

Using 15 m width of each bed

$$\text{Length of each bed} = \frac{395.7}{15} = 26.4 \text{ m.}$$

Hence, use 14 beds, of size 15 m × 26.4 m in plan. The beds should be provided with under-drains and side walls, with typical section and plan, as shown in Fig. 9.29. **Ans.**

9.30.2. Mechanical Methods of Dewatering Sludge. In developed countries, digested sludge (sometimes even raw sludge), is often dried or dewatered by mechanical means, such as by vacuum filtration or by high speed centrifuges.

In **vacuum filtration** process, the sludge is first mixed with a coagulant such as ferric chloride, and then conveyed to a vacuum filter, consisting of a hollow rotating drum, covered with a replaceable filter cloth. The drum rotates partly submerging into the sludge. The vacuum created by a pump within the drum draws the moisture from the sludge through the cloth. The sludge cake which is formed on the outside of the drum is removed by a scraper, as the drum rotates. A typical vacuum filter is shown in Fig. 9.30.

Fig. 9.30. Photographic View of a typical Vacuum filter for sludge dewatering.

High-speed centrifuges are also used for drying of raw or digested sludges, and are becoming more popular because of small area requirements. These methods may remove about 50% moisture.

Vacuum filtration or centrifugation of raw sludge is often adopted in situation where sludge is to be disposed of by incineration (*i.e.* burning). These mechanical methods of drying are generally used when the available area is less than that required for sludge dryidge beds, or where the climates are too cold or at places where rains are frequent as not to permit natural drying, or as a preliminary to heat drying for making fertiliser.

9.30.2.1. Elutriation of sludge. Before the sludge is dewatered by vacuum filtration method, it is generally *elutriated*. Elutriation is the process of *washing the sludge water,* to remove the organic and fatty acids from it. Sometimes, before dewatering, the sludge is *thickened* or *conditioned,* and then elutriated.

During the sludge digestion process, the volatile acids, alcohols, and organic acids are formed, which if not removed, will interfere with coagulation process during dewatering by vacuum filtration. If elutriation is done before dewatering, it will reduce the quantity of coagulants considerably.

Sludge elutriation is carried out in sludge or multiple tanks by washing the sludge with water. During washing, the solids are continuously kept in suspension by air or by mechanical action.

Single stage, or *multistage*, or *counter-current washing* may be employed as one of the three-methods for washing. The choice of any of these methods will depend upon the availability of water, because *single stage elutriation* will require 2.5 times the water required for two-stage elutriation, and 5 times that required for counter-current washing, for the same alkalinity reduction.

In the elutriation process, sludge and water are mixed in a chamber fitted with mechanical devices, keeping it for 20 seconds (*i.e.* detention period). The sludge is then settled in settling tanks, and excess water is decanted. The maximum surface loading on settling tank may be 40 m^3/m^2/day with a detention period of 4 hours.

Counter-current elutriation is generally carried out in twin tanks, similar to the sedimentation tanks, in which sludge and water enter at opposite ends. Piping and channels are so provided that wash water entering the second stage tank comes first in contact with sludge already washed in the first stage. The quantity of wash water required is about 2 to 3 times the quantity of sludge elutriated.

9.30.3. Disposal of Dewatered Sludge. The dewatered sludge obtained from mechanical devices in western countries, is generally **heat dried,** so as to produce fertilizers. As a matter of fact, the mechanical dewatering removes only about 50% of the moisture, and hence the mechanically dewatered sludge is actually heated, so as to fully remove the moisture from it. The dry residue is used as manure. This method is looked upon more as a method of producing fertilizer rather than as a method of sludge disposal, because if this method is adopted only for sludge disposal, it proves to be extremely costly, and thus feasible only for rich countries.

The wet sludge, after mechanical dewatering, is sometimes, directly disposed of either in sea or in underground trenches, or burnt, as discussed below :

9.30.3.1. Disposal by Dumping into the Sea. The dewatered wet sludge may, sometimes, be discharged at sea from hopper barges or through outfall sewers. This method can, however, be adopted only in case of cities situated on sea shores, and where the direction of the normal winds are such as to take the discharged sludge into the sea, away from the shore line.

9.30.3.2. Disposal by Burial into the Trenches. In this method, the digested sludge, without dewatering is run into trenches, which are 0.9 m wide × 0.6 m deep, and rectangularly spaced at 1 to 1.5 m apart in parallel rows. When the sludge has dried to a firm state, it is covered at top with a thin layer of soil. After about a month, the land is ploughed up with powdered lime and planted with crops.

9.30.3.3. Disposal by Incineration. The dewatered wet sludge produced in waste water treatment plant may also be disposed of by burning, in suitably designed incinerators, when sufficient space is not available for its burial near the plant site, or the sludge cannot be dried and used as manure. The following types of incinerators (furnaces) have primarily been designed and used for incinerating wet sludges :

(1) *Multiple-hearth furnace* ;

(2) *Fluid-bed furnace* ;

(3) *Flash-type of furnace* ; and

(4) *Infra-red (Electric or Radiant heat) furnace*.

Out of the above types, *multiple hearth furnace* is most widely used for sludge incineration in countries like USA ; although India does not incinerate the sludges of its waste water treatment plants, and generally uses them as manure, after drying in the sludge drying beds.

The above mentioned types of incineration systems are, however briefly discussed below :

(1) *Multiple Hearth Incinerators*. The **multiple hearth furnace** was originally developed for ore-roasting in the early 1900s. In 1930, it was adopted for sludge incineration. In USA, about 400 such furnaces are in existence, most of which are used for sludge incineration of waste water treatment plants.

A multiple-hearth furnace is a vertical structure, consisting of a series of circular refractory hearths, located one above the other, as shown in Fig. 9.31. The structure consists of a steel shell, lined internally with refractory. The hearths are self supporting, each acting as an arch supported from the refractory lining of the furnace shell. The hearths may be numbered as 1, 2, 3, 4, ... from top to the bottom. There may be five to nine hearths in a typical furnace and (commonly 6) ; they are normally built in diameters ranging from 3 to 7.5 m, or so.

Wet sludge (cake) is fed by gravity from the top of the furnace, by placing it on the top-most hearth, which passes downward, over the successive hearths, as explained below :

A vertical shaft is positioned in the centre of the furnace. Rabble arms are attached to the central shaft above each hearth, as shown. The central shaft rotates and drives the rabble arms. A series of rabble teeth on each rabble arm wipes the sludge across each hearth. The central shaft is driven at a speed of $\frac{3}{4}$ to 2 rpm by a variable speed driving mechanism provided at the bottom of the furnace. A bull and pinion gear system is normally used to drive the central shaft. *Alternative hearths have a large annular opening between the hearth and the central shaft. These are called* **in-hearths.** *The teeth on the rabble arms will rabble the sludge to the centre of these hearths, "in" towards the centre of the* *furnace, where sludge will drop off the edge of the refractory, landing on the hearth below, an* **out-hearth.** The out-hearths allow the sludge to rabble "out" away from the centre of the furnace. A series of '*dropholes*', shown in Fig. 9.31, are placed equidistant around the periphery of the out-hearths, to allow the sludge to drop down to an in-hearth, before it reaches the vertical furnace wall.

The rising hot air and flue gases of the furnace will dry and burn the wet sludge during its downward journey through the furnace. The sludge will eventually be dried due to evaporation of moisture at the upper hearths ; while burning (combustion) of sludge will occur in the middle hearths. In the lower hearths, the burnt sludge (ash) will get cooled, passing its heat to the flue gas. The dried ash is finally taken out and collected at the bottom of the furnace. The exhausted flue gas will escape out from near the top of the furnace, as shown. 10 to 20% of the non-combustible component of the sludge will also escape with the off-gas, and may need passing through an **After-burner** (*subsidiary combustion chamber*) to ensure its full combustion, and also through an *air-pollution control equipment* (**scrubber**) to catch the harmful particulate matter before discharging into the open atmosphere through a chimney stack.

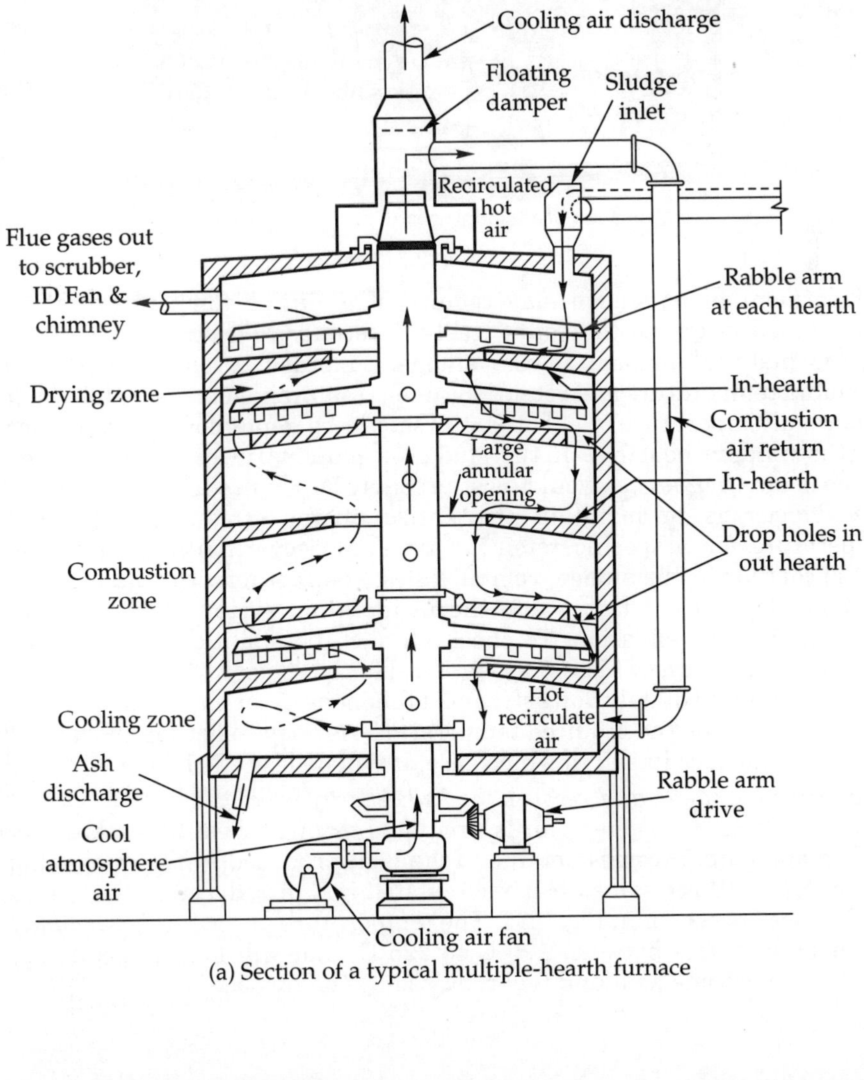

(a) Section of a typical multiple-hearth furnace

(b) Photoview of a Multiple-hearth incinerator installation

Fig. 9.31

Treatment plant sludge may require a minimum temperature of about 750°C and a retention time of at least 0.5 seconds to destroy the majority of organic material present in the sludge. In order to ensure that such a minimum temperature and retention time is maintained in a multiple-hearths furnace, a temperature of about 900°C should be kept above the two burning hearths (middle hearths). If the sludge is too dry (in excess of 25% solids content), an *After burner* would most probably be required. The purpose of the *After burner* is to maintain the entire off-gas stream at a controlled temperature for a specific retention time. Moreover, with a high solid or grease content in the sludge, combustibles would otherwise volatalize above the burning zone, and may exit the incineration system unburnt, causing smoke and odour emission. To check all such possibilities, an *After burner* becomes necessary, which may preferably be placed separate from the incinerator, immediately downstream of the incinerator exit, with one or more burners to provide the required temperature ; or may sometimes be made as a part of the furnace itself ; in which case, it will be less efficient.

In order to *recycle and recover the heat energy* produced by the burning of the sludge, cold atmospheric air is circulated through the central shaft, which while moving up, becomes hot due to the heat produced by the combustion of the sludge ; and the same air is recirculated and introduced at the bottom of the furnace, as shown in Fig. 9.31. The temperature of the '*sludge combustion air*', usually varies between 120°C to 225°C, and will help in reducing the quantum of outside heat and fuel required by the furnace.

The multiple-hearth furnace is maintained at a *negative pressure*, as to avoid leakage of hot gases from "*out*" of the furnace, but "*into*" the furnace. As a matter of fact, several leakage points exist in a multiple hearth furnace, such as at the access/observation doors (normally two per hearth), at the sludge inlet and outlet mechanisms, and at the top and bottom central shaft seals. Although the system is designed to minimise leakage, yet some leakage will occur. If draft (negative pressure) is maintained in the system by an **I.D. fan** (*Induced Draft fan*), then naturally, the leakage will occur 'into' the furnace, and hot gases will not leak out of the furnace.

The *Induced-Draft fan* (*I.D. fan*) is connected to the system after the *scrubber* (*i.e.* wet scrubber, called the cyclone scrubber). The scrubber removes majority of the particulate matter remaining in the off-flue gas stream, and discharges it into a spent water stream. The gas stream is also cooled to approximately 50°C in the scrubber, and is made to pass through an I.D. Fan, and is finally directed through a **noise silencer** to an **exhaust chimney** (stack) for discharging the off-gas into the open atmosphere.

The most important operating problem in a multiple-hearth furnace occurs when the furnace goes positive ; *i.e.* when it losses draft. When this occurs, then naturally, hot dirty gases will find their way out, through the leakage points of the furnace, and such an uncontrolled leakage may cause serious damage to equipment and personnel damage to the persons working in and around the furnace.

(2) *Fluid-Bed Incinerators*. Fluid-bed or **Fluidized bed incinerator** systems are comparatively of modern origin, and consist of a furnace (reactor) having silica *sand* filled in its bottom portion (above the wind box—to be used for burning of fuel), as shown in Fig. 9.32, which figure shows the typical section of such a furnace. Complete details of installation of such an incinerator system are shown in Fig. 9.33.

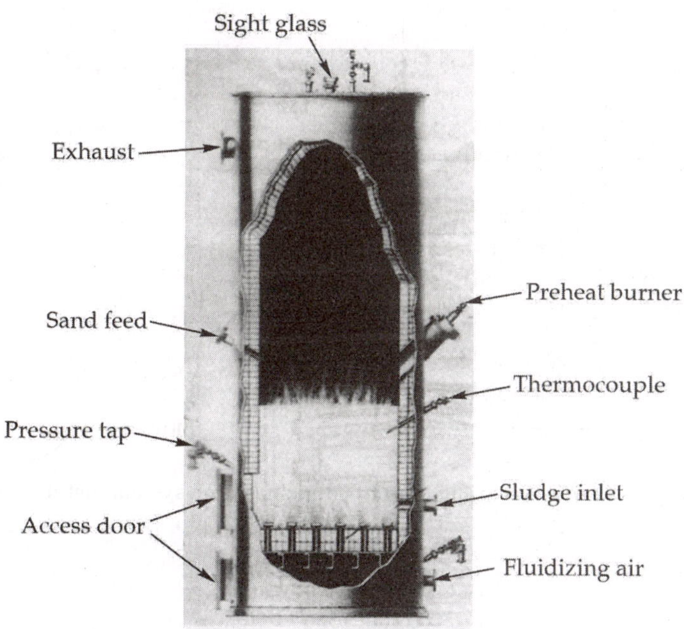

Fig. 9.32 (a) Prospective view of a typical fluidized bed furnace (Reactor).

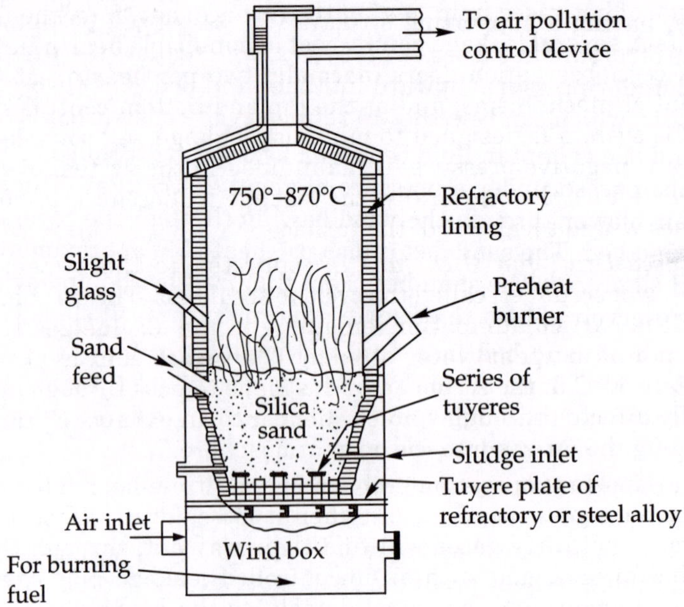

Fig. 9.32 (b) Section of a fluidized bed furnace

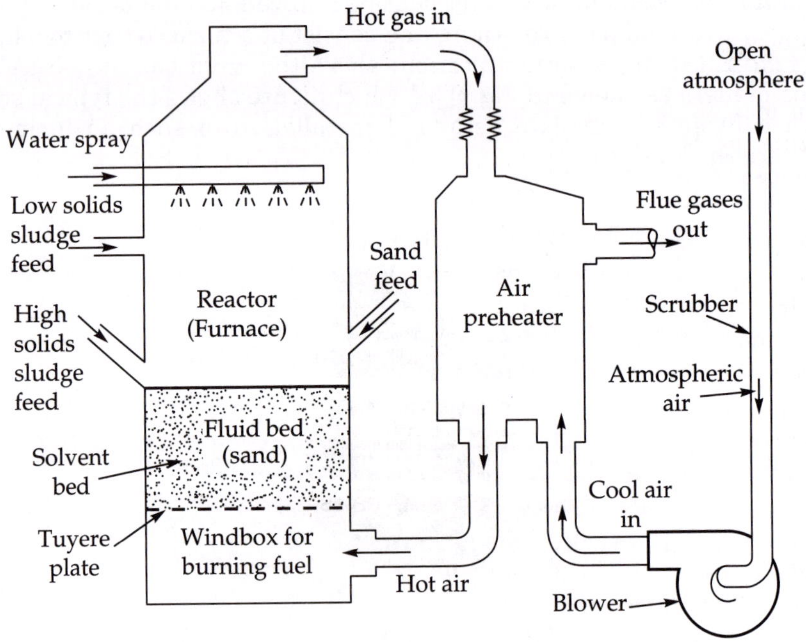

Fig. 9.33. A conventional fluid-bed incineration system installation.

The furnace consists of a vertical cylindrical steel shell, lined with refractory, having a supporting structure in its bottom portion, to hold the sand. This structure has a series of *tuyeres*, which allow the passage of air (to be passed under pressure) upward into the sand bed. The tuyeres rest on the *tuyeres plate*, which can be either in refractory or constructed of steel alloy.

The sand bed is kept in fluid condition by creating turbulence by an upflow of air, to be passed under a pressure of about 25-35 kPa*. The air is passed from an air blower through the wind box, up through the tuyeres plate, and into the sand bed. The sand bed is also pre-heated to approximately 650°C by using fuel oil or gas. The sand-bed, kept fluidized by the upmoving air, thus, acts as a reservoir of heat to promote uniform burning of sludge solids.

The depth of sand bed in a fluid-bed incinerator may usually vary from about 0.6 m to 2.4 m, or so. The depth of sand bed should, however, be maintained at a minimum value designed to be consistent with complete combustion and minimum excess air.

When dewatered sludge or waste to be burnt in introduced in the hot sand bed, rapid combustion occurs, maintaining an operating temperature of about 800°C (The operating temperature increases from the pre-heating temperature due to heat produced by the burning of sludge). Ash and water vapour, derived from the burning of the wet sludge, along with some sand, finally escape out with the flue gas coming out of the furnace. This off-gas is taken to a wet scrubber, where the ash and particulate matter is removed in a waste water-stream, and the cleaner air is discharged into the open atmosphere through a chimney (stack).

In most of the designs, the system is made air-tight and leak-proof, as a *positive pressure system*. This feature is useful in applications, where the furnace is required to operate on a non-continuous basis. However, if leakages cannot be prevented from the system, then the system will be made to have a *negative pressure*, by connecting the off-gas stream to an I.D. fan, after it comes out of the scrubber, as is done in a multiple-hearth furnace.

Since a certain amount of sand goes out with the off-gas stream, its deficiency will have to be occasionally made up in the furnace bed @ about 5% of bed volume at every 100 hours of incineration operation.

With the burning of the sludge waste, certain burnt ashes, salts, etc. may remain entrapped in the sand bed. These accumulated residuals will tend to increase the sand-bed volume. Some of the waste residues, particularly those from the metal salts, may also cause the sand bed particles to grow in size, and eventually to combine, or to cause *agglomeration* of particles. Clinkers may, thus, also be formed, which will reduce the fluidization of the bed, leading to bed-seizure. The **bed agglomeration**, can, thus, cause problems in the fluidized bed furnaces. The bed-agglomeration can, however, be detected through an increase in the differential pressure across the bed. Cleaning or replacement of sand bed, or addition of additives which may reduce agglomeration, may be used as remedies to the bed-agglomeration problems.

---

*1 kPa = 102 kg/m^2 = 0.102 m of water head.

Since a fluid-bed incinerator offers a high degree of turbulence and large surface area for heat transfer, made available on the inert bed material, besides offering a continuous exposure of the heated sand to the waste, it results in highly efficient incineration with low excess air levels and a minimal temperature gradient through the bed.

The residence time of fluid-bed incinerator is in a high range from 5 to 8 seconds or more at a temperature of 750° to 870°C, or so, with a slight *positive pressure*. This long residence time further ensures a high *Destruction and Removal efficiency (DRE)** for the organic waste, and produces an inert ash, which can easily be disposed of in unhazardous land fills.

Due to its capacity to easily incinerate organic and hazardous wastes, a fluid-bed incinerator is not only useful for incinerating waste sludges of municipal sewage treatment plants, but is also useful for incinerating hazardous wastes of industrial plants and sometimes of hospitals and biomedical institutions.

(3) *Flash Type of Furnace*. A **flash type of incinerator** consists of a tower, which is heated in the beginning by burning fuel. The wet sludge is then thrown from the top of the tower, and the super-heated gases come up from the bottom of the tower. The rising hot gases remove the moisture from the sludge, and the water vapour will pass along with the hot gases. The dried sludge is collected at the bottom, and can be used as burning material for the tower, so as to reduce the cost of the external fuel required by the tower.

(4) *The Infra-red Furnaces*. The **infra-red (i.e. Electric or Radiant heat) furnace** (Fig. 9.34) is a conveyor belt system, passing through a long refractory-lined chamber. An induced draft fan (I.D. fan) maintains a negative pressure throughout the system. Combustion air is introduced at the discharge end of the conveyor belt. Air picks up heat from the hot burning sludge, as sludge and air, travel counter to each other.

Supplemental heat is provided by electric infra-red heating elements within the furnace above the belt (not shown in figure). Cooling air is injected into the incinerator chamber to prevent local hot spots in the immediate vicinity of the heaters, and is used as secondary combustion air within the furnace. The furnace is designed to provide and maintain a temperature of about 900°C in the space above the travelling conveyor.

Wet sludge cake is fed by gravity to the belt, which immediately gets levelled to a depth of approximately 2.5 cm due to motion of the belt. There is no other sludge-contact or sludge-handling mechanism on most of these units.

***Destruction and Removal efficiency** (DRE) for an incinerator burning hazardous waste, for each hazardous constituent, is defined as :

$$\textbf{DRE} = \frac{W_{in} - W_{out}}{W_{in}} \times 100\%$$

where W_{in} = mass feed rate of one POHC (Principal organic hazardous constituent) in the waste stream.

W_{out} = mass emission rate of the same POHC present in the exhaust emissions prior to release to the atmosphere.

The belt speed and travel is selected to provide burn-out of the sludge without agitation. This feature results in relatively low level of particulate emissions.

Fossil fuel or electrical energy may be used to provide a supplemental fuel for start up of furnace, or to provide heat required to maintain combustion temperatures.

These furnaces provide a number of advantages, such as : (*i*) *low particulate matter emissions* ; (*ii*) *possible use for intermittent or infrequent loadings* ; (*iii*) *non-production of additional flue gas by the heat generated by electric power / elements* ; etc.

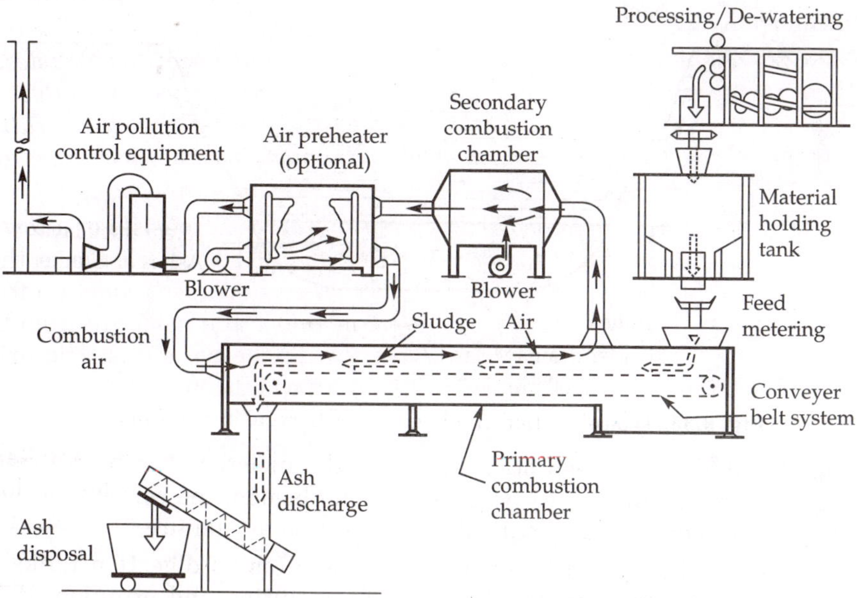

Fig. 9.34. Flow chart of an infra-red incineration system.

Disadvantages of incinerators. Incinerators are usually avoided for burning waste sludges of waste-water treatment plants, so long as alternative methods of drying them to manure, or by burial, are available ; since they not only cause air-pollution but also consume a lot of fuel, and require continuous monitoring and upkeep. Their initial installation costs are also very high, besides the recurring costs.

The use of incinerators for burning municipal refuse (solid waste), and for burning hazardous hospital wastes, have further been discussed in articles 12.3.2 and 17.11.5, respectively.

9.31. Use of Lagoons for Disposal of Raw Sludge

This method is, sometimes, used at smaller places for disposing of raw sludge without digestion. In this method, the raw sludge is kept at rest in a large shallow open pond, called a *lagoon*. The detention period is 1 to 2 months, and may extend up to 6 months. During its detention in the lagoon, the sludge undergoes anaerobic digestion thereby getting stabilised. Due to this anaerobic decomposition of sludge, foul gases will be evolved from a lagoon ; and hence the lagoons, should be located away from the town, and direction of the common winds should be such that the smells are not carried towards any localities.

A typical section for a lagoon is shown in Fig. 9.35. It is a shallow pit, 0.6 to 1.2 m deep, formed by excavating the ground. At the bottom of this pit, a 15 cm thick layer of ashes or clinker is placed. Agricultural tile drains of about 10 cm diameter are laid at bottom as *under-drains*. These are placed at about 3 m centre to centre spacings. Banks are formed on both sides of the pit from the excavated earth, as shown.

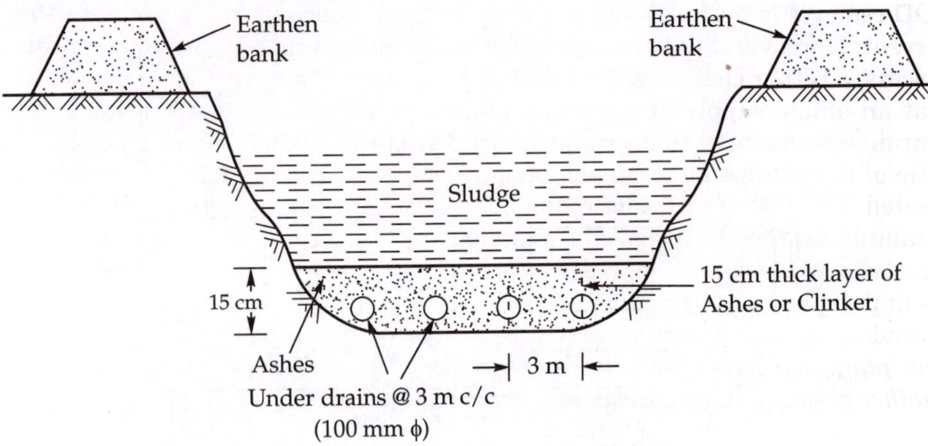

Fig. 9.35. Typical section of a Lagoon for disposal of raw sludge.

After the sludge has been stabilised, and the moisture has been drained away or evaporated during its detention in the lagoon, the contents are dug out to about half of their original volume, and used as manure.

This method of sludge disposal is quite cheap (as no digestion tanks are required), but the greatest drawback is the evolution and eruption of foul gases, polluting the environment. Its use is, therefore, restricted only to non residential areas.

SECONDARY TREATMENT THROUGH ACTIVATED SLUDGE PROCESS

(Aerobic Suspended Culture)

9.32. Definition of Activated Sludge Process

The activated sludge process provides an excellent method of treating either raw sewage or more generally the settled sewage. The sewage effluent from primary sedimentation tank, which is, thus normally utilised in this process, is mixed with 20 to 30 per cent of own volume of *activated sludge*, which contains a large concentration of highly active aerobic micro-organisms. The mixture enters an *aeration tank*, where the micro-organisms (coated around the sludge solids) and the sewage, are intimately mixed together, with a large quantity of air for about 4 to 8 hours. Under these conditions, the moving* organisms will oxidise the organic matter, and the suspended and colloidal matter tend to coagulate and form a precipitate, which settles down readily in the secondary settling tank. The settled sludge (containing micro organisms)

*The organisms coated around the sludge solids are kept moving by agitation/mixing.

called *activated sludge*, is then recycled to the head of the aeration tank, to be mixed again with the sewage being treated. New activated sludge is continuously being produced by this process, and a portion of it being utilised and sent back to the aeration tank, whereas the excess portion is disposed of properly along with the sludge collected during primary treatment, after digestion.

The effluent obtained from a properly operated activated sludge plant is of high quality, usually having a lower BOD than that of a trickling filter plant. BOD removal is up to 80—95 per cent, and bacteria removal is up to 90—95 percent. Moreover, land area required is also quite less. But, however, in this process, a rather close degree of control is necessary in operation to ensure (*i*) that an ample supply of oxygen is present ; (*ii*) that there is intimate and continuous mixing of the sewage and the activated sludge ; and (*iii*) that the ratio of the volume of activated sludge added to the volume of sewage being treated is kept practically constant. Moreover, there is the problem of obtaining activated sludge, at the start of a new plant. Hence, when a new plant is put into operation, a period of about 4 weeks may be required to form a suitable return sludge, and during this period, almost all the sludge from the secondary sedimentation tank will be returned through the aeration tank. *A new plant may also sometimes be seeded with the activated sludge from another plant, so as to quickly start the process in the new plant.*

9.33. Various Operations and Units of an Activated Sludge Plant

9.33.1. Primary Treatment Units of an Activated Sludge Plant. A typical flow diagram for a conventional activated sludge plant is shown in Fig. 9.36. As pointed out earlier, the removal of grit and larger solids by *screening in grit chambers* and *primary sedimentation tanks is generally* considered necessary before *aeration*. The pre-removal of these settleable solids is helpful in preventing deposits on aeration devices, and thereby not reducing their efficiencies. Moreover, such materials, if not pre-removed, may settle down in

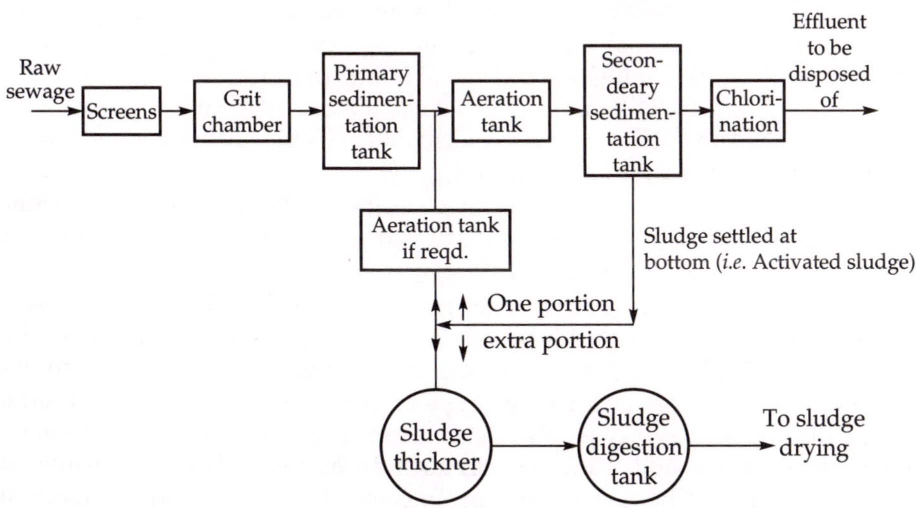

Fig. 9.36. Flow diagram for a conventional activated sludge plant giving high degree of treatment.

the aeration tank, and by decomposition, interfere with the treatment process. *Accordingly, grit removal, screening, and primary sedimentation are considered necessary for a conventional activated sludge process.*

Since, in this process, it is very much desirable to keep the sewage as fresh as possible, a *somewhat shorter detention period is provided in the primary sedimentation tanks than is required for most other treatment processes.* This period of primary detention may vary with the size of the plant and the characteristics of sewage, but tank size will generally provide an overflow rate of about 40,000 litres per sq-m of plan area per day. For a depth of about 2.4 m, the detention time will be about 1.4 hours.

9.33.2. Aeration Tanks of an Activated Sludge Plant. From the primary sedimentation tank, the sewage flows to the aeration tank, and is mixed with the activated sludge. The aeration tanks (or aeration chambers, as they are sometimes called) are normally rectangular tanks, 3 to 4.5 m deep, and about 4 to 6 m wide. The length may range between 20 to 200 m, and the detention period between 4 to 8 hours for municipal sewages. Air is continuously introduced into these tanks.

Methods of Aeration. There are two basic methods of introducing air into the aeration tanks, *i.e.*

(1) *Diffused air aeration or Air diffusion ; and*

(2) *Mechanical aeration*

(3) Sometimes, a combination of both may also be used which may then be called as *combined aeration.*

These methods are discussed below :

(1) Diffused Air Aeration. In the diffused air aeration method, compressed air under a pressure of 35 to 70 kN/m^2 (0.35 to 0.7 kg/cm^2), is introduced into the aeration chamber, through diffusion plates or other devices, called *diffusers.* The main criteria for selection of a particular diffuser is that it should be capable of diffusing air in small bubbles, so as to provide the greatest possible efficiency of **aeration.** Porous plates and porous tubes, made of quartz or crystalline alumina (Aluminium oxide) are generally used as **diffusers.** Plates are generally square in shape with dimensions of 30 cm × 30 cm, and they are usually 25 mm thick. These plates are fixed at the bottom of aeration tanks. Tube diffusers are generally 60 cm long with internal dia of 75 mm and thickness of wall equal to 15 mm. These tubes are suspended in the aeration tank, and can be taken out for cleaning, without emptying the tank. The effective areas for the above standard plate and tube diffusers work out to 780 cm^2 and 1160 cm^2, respectively.

Two types of aeration tanks are generally used. In one design, the tank is formed into a succession of ridges and furrows (Fig. 9.37), and air is forced upward through diffuser plates placed at the bottom of the furrows. Such a tank is called the **ridge and furrow type** of aeration tank. Another popular design is the **spiral flow type** of aeration tank (Fig. 9.38). In this tank, air is introduced near the side of the tank in such a way that spiral flow results in the tank, as shown. The compressed air, in such a tank, can be supplied either through a plate diffuser or a tube diffuser, although tube diffusers are most widely used. This type of tank requires small quantity of compressed air at low pressure. Spiral motion set up by the compressed air, released through the tube diffusers, causes the required aeration.

Quantity of air required in diffused aerators. In order to determine the capacity of the air compressor, it is necessary to determine the quantity of air that will be required. On an average, it may be assumed that about 4000 to 8000 m^3 of free air will be required per million litres of sewage being treated. But this provision depends on the strength of sewage and various other factors. With respect to the BOD removal, the usual rate adopted is 100 m^3/day of air per kg of BOD removed. Since only about 5% of oxygen in the air is actually involved in the biochemical action, the modern practice in design is to rationally calculate the oxygen requirement of the given waste-water and select aerator accordingly, as discussed in article 9.40.

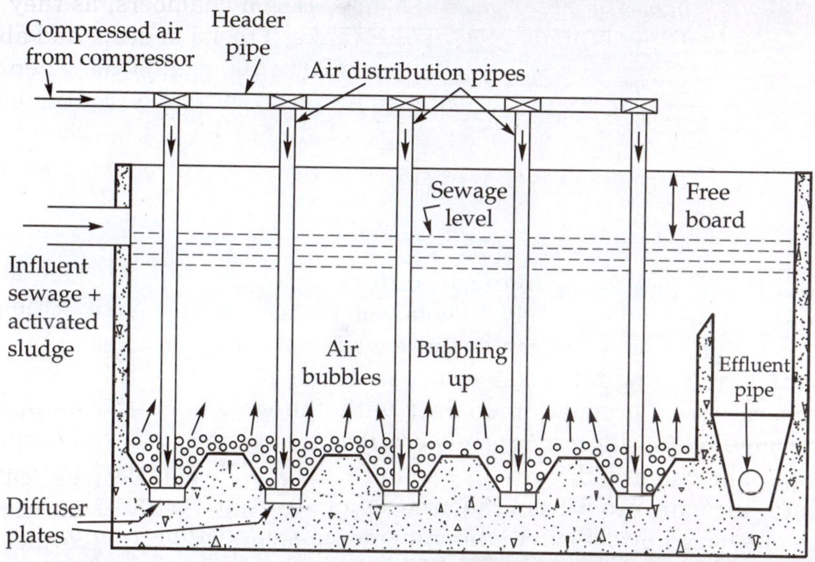

Fig. 9.37. L-section of a ridge and furrow type of an aeration tank (using compressed air).

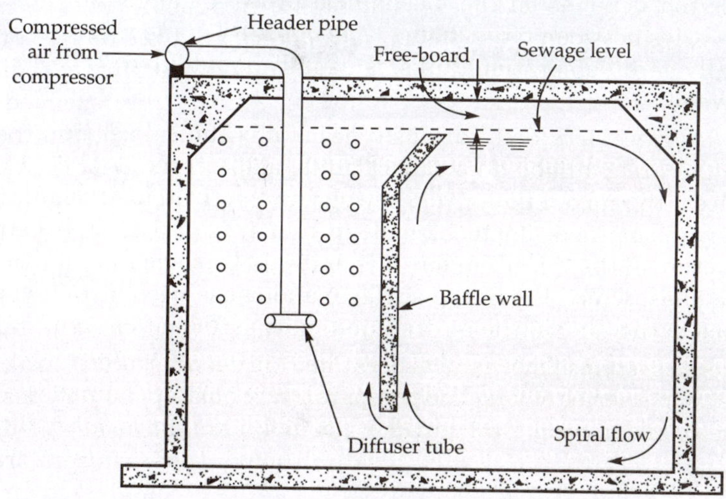

Fig. 9.38. Cross-section of a spiral flow type of an aeration tank (using compressed air).

Volume of returned activated sludge. The volume of returned activated sludge from secondary clarifier to the aeration tank, mainly depends upon the extent of BOD desired to be removed. It is usually expressed as percentage of flow of sewage as $\frac{Q_R}{Q}$, where Q_R is the returned sludge rate in m^3/d, and Q is the sewage inflow rate in m^3/d. Its rational computation is discussed in article 9.37 while its simple variation with the extent of desired BOD removal is shown in table 9.9 below :

Table 9.9. Volume of Returned Activated Sludge

Extent of BOD removal desired in ppm	*Quantity of returned sludge as percentage of sewage flow*
150	25
250	30
300	35
400	40
500	48
600	53

(2) Mechanical Aeration. In the air-diffusion method, as pointed out above, a lot of compressed air (90 to 95%) gets wasted, as it simply escapes through the tank without giving oxygen to the sewage ; although it helps in bringing about the required agitation of sewage mixture. In order to affect economy, atmospheric air is brought in contact with the sewage in the mechanical aeration method. In this *mechanical aeration method*, the sewage is stirred up by means of mechanical devices, like paddles, etc. (called **surface aerators**) ; so as to introduce air into it from the atmosphere by continuously changing the surface of sewage by the circulation of sewage from bottom to top. The only important requirement in this method is to have thorough agitation of sewage*, so as to bring it in intimate contact with the atmosphere. The *aeration period* depends on the mechanical process adopted for agitation. It generally varies between 6 to 8 hours. *The quantity of the returned sludge,* in mechanically aerated aeration tanks, is usually about 25 to 30 per cent of the flow of sewage.

Types of Surface Aerators. Various types of mechanical aerators have been designed, and are available in compact units, as patented devices. In most of these devices, the mixed liquor (*i.e.* sewage and activated sludge) is made to move slowly in channels, 0.9 to 1.2 m deep, and 1.2 to 1.5 m wide. During this forward motion of the mixed liquor, it is agitated by some mechanical means, such as paddles, which break the surface, and cause wave action exposing a new surface to the air, till the entire liquid gets sufficient oxygen.

In **Haworth system** of mechanical aeration, 1 m deep aeration tank is divided by thin walls in a series of long and comparatively narrow channels (70 m × 1.5 m plan area), and gives a total travel of about 1.5 km. At about midway of its

*If agitation of sewage is not done, the surface film of sewage, of course, will get saturated with the oxygen, but the oxygen so obtained is not conveyed to the body of sewage.

length, two rows of paddles, revolving at 1.5 r.p.m. are provided with a horizontal shaft crossing each channel for the aeration of sewage. The liquid is imparted a velocity sufficient to keep the sludge in suspension. The detention period of these tanks is about 15 hours, and the returned sludge ranges between 15 to 20% of the sewage flow. This method is generally used at Sheffield in England ; and is hence also known as *Sheffield system of mechanical aeration.*

Another system of mechanical aeration, called **Hartley system,** is similar to the above Haworth system, with the modification that here the paddles are kept inclined at some angle with the vertical, and are fixed at the end of the channel. The diagonal baffles are provided across the channel to maintain spiral flow.

In a third method, known as **Simplex aeration method,** a tank square in plan with hopper bottom (Fig. 9.39) is generally used. Sometimes, a rectangular tank may also be used ; but in that case, it must be divided suitably, into square units. At the centre of the tank, a hollow *uptake tube* is suspended from the top with a distance of about 15 cm from the bottom of the tank. A steel cone with spiral vanes is provided at the top of the uptake tube, and it is driven by a motor placed at the top of the tank. The cone is revolved at a high speed (60 r.p.m.) which sucks the mixed liquor through the uptake tube by creating suction at the bottom, and sprays it at the surface towards the sides in the tank. Numerous air bubbles are formed in this process, which bring about satisfactory aeration of sewage. Sewage also gets thoroughly mixed up with the activated sludge during its downward journey.

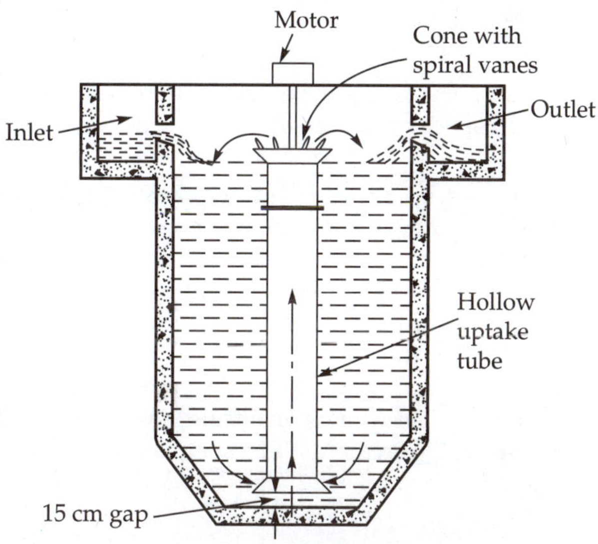

Fig. 9.39. Section of a simplex aerator.

The mechanical aerators have gained considerable popularity in recent years. Their simplicity and ease of maintenance has resulted in their increasing popularity, especially for smaller plants. They are, however, sensitive even to slight variations in water levels.

(3) Combined Aeration. In combined aerators, the diffused-air aeration as well as the mechanical aeration are combined together in a single unit. A well known type of such an aerator unit is called *Dorroco aerator*, and is shown in Fig. 9.40. In this type of an aerator, the aeration of sewage is achieved by diffusing air through bottom diffuser plates, as well as by rotating paddles @ 10 to 12 r.p.m., as shown. Spiral motion, so set up, brings about the required aeration.

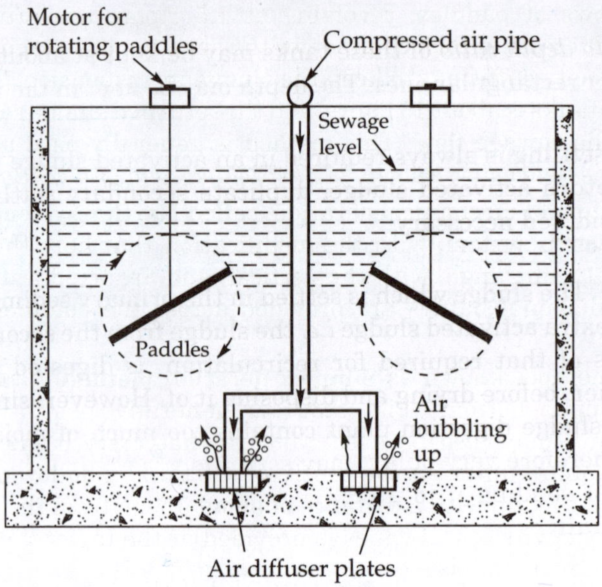

Fig. 9.40. Section of a Dorroco aerator or Imhoff aerator.

Such an aerator is very efficient, detention period is smaller (3 to 4 hours), and requires less amount of compressed air as compared to the diffused-air aeration.

9.33.3. Secondary Sedimentation Tank of an Activated Sludge Plant. From the aeration tank, the sewage flows to the final sedimentation tank. This tank will normally be of the general type (like a primary tank), with certain modifications. Since, there are no floating solids here, provisions for the removal of scum or floatage are not needed. The suspended particles in the aeration tank effluent are light in weight, and are thus, markedly influenced by currents. Therefore, in these secondary settling tanks, a considerable length of overflow weir is desirable, to reduce the velocity of approach. Good design, should provide a *weir overflow rate,* not exceeding 150 m^3/day per lineal metre of weir. This value is based on average flow of sewage and not to the mixed liquor flow.

Solids loading is another important factor, which governs the design of secondary basin. This is because of the fact that ***in the secondary tank, hindered settling occurs,*** and hence the settling velocity of discrete particles may not govern its design, as in the case of a primary sedimentation tank. The *solids loading rate* based on mixed liquor flow to the settling tanks, may be kept at about 100—150 kg/m^2 per day at average flow, and should not exceed 250 kg/m^2 per day at peak flows. Such rates ensure adequate sludge thickening and concentrated sludge returns.

The surface area for activated sludge settling tanks should be designed for both overflow rate and solids loading rate, and larger value adopted.

The *detention period* for such a sedimentation tank may be kept between $1\frac{1}{2}$—2 hours, as the same is usually found to give optimum results.

The *length to depth ratio* of these tanks may be kept at about 5 for circular tanks, and 7 for rectangular ones. The depth may be kept in the range of 3.5 to 4.5 m.

Since final settling is always required in an activated sludge plant, so as to provide the return activated sludge, duplicate secondary settling tanks are generally considered necessary.

9.33.4. Sludge Thickener and Sludge Digestion Tank of an Activated Sludge Plant. The sludge which is settled in the primary sedimentation tank as well as the extra activated sludge *i.e.* the sludge from the secondary settling tank in excess of that required for recirculation, is digested as usual and described earlier, before drying and disposing it of. However, since the sludge obtained in a sludge digestion plant contains too much of moisture (98% to 99%), and is therefore very bulky, may sometimes be reduced in its moisture content by first sending it to a **sludge thickener unit,** also called a **sludge concentrator unit.**

Sludge thickening is a procedure used to increase the solids content of the sludge by removing a portion of the liquid fraction. To illustrate, if waste activated sludge, which is typically pumped from the secondary tanks with a solids content of 0.8% (say), can be thickned to a solids content of 4% (say), then a five fold decrease in sludge volume is achieved.*

The thickening of sludge can be carried out in three types of thickening units ; *i.e.*,

(*i*) *gravity thickeners* ;

(*ii*) *floatation thickeners* ; and

(*iii*) *centrifugal thickeners*.

The **gravity thickeners** are the simplest and most commonly adopted units, for thickening sludge. Such a thickener consists of a small circular open tank, similar to a conventional sedimentation tank except that it is deeper to accommodate a greater volume of sludge, and has a heavier raking mechanism. **Slowly rotating rake mechanism,** like *deep trusses* (Fig. 9.41) or *vertical pickets*, are provided to stir the sludge gently, opening up channels for water to escape, thereby promoting densification of sludge. Incoming flow enters from behind an inlet well in the centre of the tank, and is directed downward. The supernatant liquor continuously overflows a peripheral weir, while underflow of thickened sludge is drawn from a bottom sump in the tank.

Depending upon the characteristics of the sludge, these tanks may be designed for **typical surface loading @ 15 to 35 m^3 per day of sludge per m^2 of tank area.** They can also be designed based on the kg wt. of solids present in the

---

*Solids content of 0.8% means that 0.8 kg of solids is contained in 100 kg of wet sludge. Therefore, 4 kg of solids will be contained in 500 kg of this wet sludge of 0.8% solids.

As against this, if the solids content of the sludge in 4%, it means that 4 kg of solids will be contained in 100 kg of sludge. Hence, the sludge weight or volume will decrease from 500 kg to 100, if its solids content increases from 0.8% to 4%.

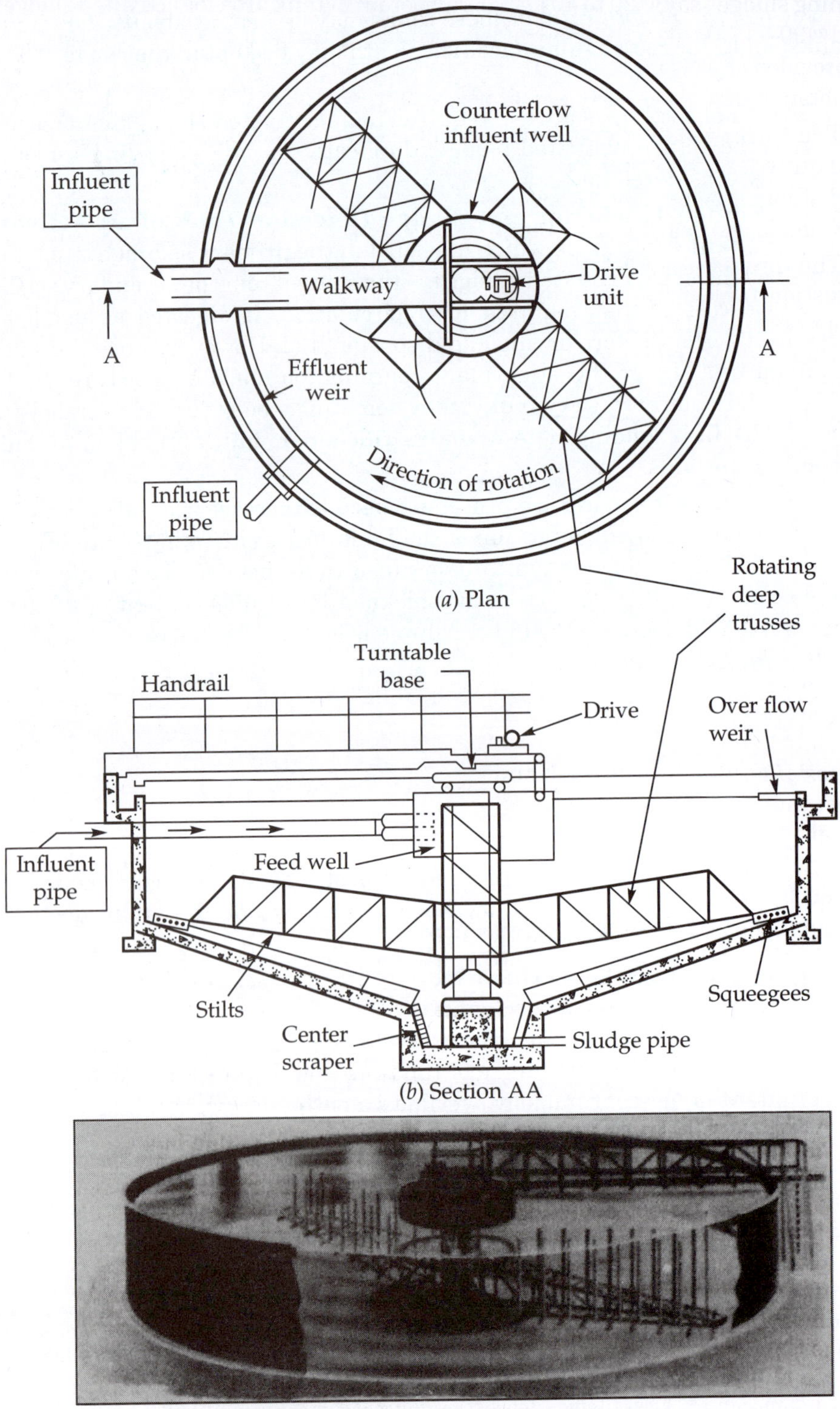

(*a*) Plan

(*b*) Section AA

(*c*) Photoview of a gravity sludge thickener

Fig. 9.41. A typical gravity sludge thickener.

coming sludge ; say @ 20 to 40 kg per day per m^2 of tank area for activated sludge ; and @ 90—120 kg per day per m^2 of tank area for primary sludge (if at all a thickener is provided or used for that also). The capacity of the tank would be finally decided to obtain a **detention period of 3 to 4 hours,** thereby fixing its depth.

The thickened sludge, having reduced moisture content of 90—93% (say), and quite less in volume (say $\frac{1}{2}$ of original), is finally sent to sludge digestion, and sludge drying. The supernatant liquor, having properties similar to raw sewage, is sent back to join the plant influent.

The process of sludge digestion, using a sludge thickener before the digestion tank, helps in reducing the capacity of the digestion tank ; and by proper control of digestion conditions, lower detention periods are used, which further reduce their capacities, and the rate of digestion is also made high. Such a digestion, which is used in modern large sized plants, is called **high rate digestion.**

It may also be pointed out that in an activated sludge treatment plant, special provision must be made for removing the activated sludge quickly from the secondary settling tank (and for resending it to the head of the aerator), so as to prevent it from becoming anaerobic.

9.34. Bulking and Foaming Sludge in an Activated Sludge Treatment Plant

Foam formation and poorly settling sludge are the two most common problems of the activated sludge process. A sludge that exhibits poor settling characteristics is called a *bulking sludge. Filamentous micro-organisms* have been found to be responsible for a bulked sludge. Large surface area to volume ratios of these micro-organisms retard their settling velocities. *Fungi* are the most familiar filamentous micro-organisms. The vegetative structure of most fungi is composed of filaments, which actually contain a number of nuclei. Fungi, however, are not commonly significant in wastewater treatment. Some other bacteria, particularly of type *021 N* bacteria are among the most frequently reported filamentous micro-organisms found in bulking sludges. These bacteria are usually found to develop in activated sludge systems which are characterised by the low or variable nutrient concentrations. These bacteria may live on variety of carbon and nitrogen sources present in the activated sludge system.

Organic acids form an important class of carbon sources for the growth of filamentous sulphur bacteria. These bacteria, however, do not develop well at low pH values. The other types of bacteria which are most commonly found in foaming and bulking sludges of activated plants are : *Nocardia amorae, Microthrix parvicella, N. amarae* like organisms, *N. pinesis* like organisms, and *type 0092*, etc. *Nocardia growth is supported by high sludge ages, low F : M ratios, and higher waste temperatures*. The most successful methods to control these organisms, as per latest research, are :

(*i*) *reduction of the sludge age* to less than 6 days* ; and

(*ii*) *chlorination of return activated sludge.*

---

*Pl. see the article 9.35.4

Foam removal is also a logical and beneficial control measure, since *Nocardia filaments* are usually concentrated in foam compared to the mixed liquor.

Classical control measures like *adjustment of the F : M* (Food : Micro-organism) ratio**, *raising or lowering D.O.*, or *applying a disinfectant (chlorine)* are found to help in controlling filamentous bacteria. *Carbohydrate-rich waters* are more prone to give rise to filamentous populations. Exclusion of such wastewaters may, therefore, sometimes help in controlling sludge bulking.

Nutrient deficiencies as said above, cause growth of filamentous micro-organisms. One of the first check made on a bulking activated plant, therefore, is to measure concentrations to determine whether they are sufficient for the amount of sludge being produced. If not, the problem is readily solved by addition of nutrients to the waste water.

Rising Sludge. Another problem that is encountered in the secondary clarifiers of an activated sludge plant is the **rising sludge** caused by *denitrification** in the secondary clarifier. Denitrification results in the formation of nitrogen gas bubbles, which lifts up the settled sludge, thereby deteriorating the quality of the clarified effluent.

Denitrification in the secondary clarifiers become a distinct possibility when the activated sludge process is nitrifying** the ammonia rich wastewaters, sending high nitrate concentrations of 6-8 mg/l in the influent of its secondary clarifier. High temperatures (above 20°C) do accelerate the rate of denitrification. Hence, when the nitrates concentration in the influent of the secondary clarifier of an activated plant exceeds 6-8 mg/l and temperatures are more, then enough nitrogen gas (due to denitrification) would be produced in the secondary clarifier during its usual 1 hour detention time, causing *rising sludge.* Plants in areas of warm climates will be susceptible to this problem ; and the only practical solution to be problem is to denitrify the effluent of the activated sludge process before allowing its entry into the secondary clarifier.

DESIGN CONSIDERATIONS INVOLVED IN AN ACTIVATED SLUDGE PLANT

9.35. Aeration Tank Loadings

The important terms which define the loading rates of an activated sludge plant, include :

(*i*) *Aeration Period* (*i.e.* Hydraulic Retention Time—H.R.T.)

(*ii*) *BOD loading per unit volume of aeration tank (i.e. Volumetric loading)* ;

(*iii*) *Food to Micro-organism Ratio (F/M Ratio)* ; and

(*iv*) *Sludge age.*

All these loading terms are discussed below in details.

9.35.1. The Aeration Period or H.R.T. The aeration period (*t*) empirically decides the loading rate at which the sewage is applied to the aeration tank. For continuous flow aeration tank, this value is determined in the same

* Pl. see article 9.35.3.

*Reconversion of nitrates into free nitrogen, as shown in Fig. 7.1 (*a*).

**Converting ammonia into nitrates.

manner as it is determined for an ordinary continuous sedimentation tank, as per equation (9.29) as :

$$\text{Detention period} \quad (t) = \frac{\text{Volume of the tank}}{\text{Rate of sewage flow in the tank}} \qquad ...(9.41)$$

or $$t = \frac{V \text{ in m}^3}{Q \text{ in m}^3/\text{day}}$$

or $$t = \frac{V}{Q} \text{ day} \qquad ...(9.41a)$$

or $$t = \frac{V}{Q} \,.\, 24 \text{ hour} \qquad ...(9.41)$$

where t = aeration period in hours
V = Volume of aeration tank
Q = Quantity of wastewater flow into the aeration tank, *excluding the quantity of recycled sludge.*

9.35.2. Volumetric BOD Loading. Another empirical loading parameter is *Volumetric loading,* which is defined as the BOD_5 load applied per unit volume of aeration tank. This loading is also called **organic loading.**

∴ Volumetric BOD loading or Organic loading

$$= \frac{\text{Mass of BOD applied per day to the aeration tank through influent sewage in gm}}{\text{Volume of the aeration tank in m}^3}$$

$$= \frac{Q \,.\, Y_0 (\text{gm})}{V(\text{m}^3)} \qquad ...(9.42)$$

where Q = Sewage flow into the aeration tank in m^3.
Y_0 = BOD_5 in mg/l (or gm/m^3) of the influent sewage.
V = Aeration tank volume in m^3

This loading is quite similar to the BOD loading rate per cum of filter volume, as used and adopted in biological filtration, and discussed earlier in article 9.21.4 (*ii*).

9.35.3. Food (F) to Micro-organisms (M) Ratio. F/M ratio is an important rational organic loading rate, adopted for an activated sludge process. It is a manner of expressing BOD loading with regard to the microbial mass in the system. The BOD load applied to the system in kg or gm is represented as food (F), and the total microbial suspended solids in the mixed liquor of the aeration tank is represented by M.

∴ F/M ratio

$$= \frac{\text{Daily BOD load applied to the Aerator System in gm}}{\text{Total Microbial mass in the system in gm}}$$

...(9.43)

If Y_0 (mg/*l*) represents the 5 day BOD of the influent sewage flow of Q m³/day, then eventually,

The BOD applied to the Aeration system = Y_0 mg/*l* or gm/m³

∴ *BOD load applied to the aeration system* = $F = Q \,.\, Y_0$ gm/day ...(*i*)

The total microbial mass in the aeration system (*M*) is computed by multiplying the average concentration of solids in the mixed liquor of the aeration tank, called **Mixed Liquor Suspended Solids (MLSS),** with the volume of the aeration tank (*V*).

$$\therefore \quad M = \text{MLSS}^* \times V$$
$$= X_T \, . \, V \qquad \text{...(ii)}$$

where $\mathbf{X_T}$ **is MLSS* in mg/l**

Dividing (*i*) by (*ii*), we get

$$\text{F/M ratio} = \frac{F}{M} = \frac{Q \, . \, Y_0}{V \, . \, X_t} \qquad \text{...(9.44)}$$

F/M ratio for an activated sludge plant is the main factor controlling BOD removal. *Lower the F/M value, the higher will be the BOD removal in the plant*. The F/M ratio can be varied by varying the MLSS concentration in the aeration tank.

9.35.4. Sludge Age. The sludge age is an operation parameter related to the F/M ratio. *It may be defined as the average time for which particles of suspended solids remain under aeration. It, thus, indicates the residence time of biological solids in the system*. While aeration period (*i.e.* liquid retention time) may be as short as 3 to 30 h, the residence time of biological solids in the system is much greater, and is measured in days.

While sewage passes through the aeration tank only once and rather quickly, the resultant biological growths and the extracted waste organics (solids) are repeatedly recycled from the secondary clarifier back to the aeration tank, thereby increasing the retention time of solids. This time is called **Solids Retention Time (SRT) or Mean Cell Residence Time (MCRT) or Sludge Age.**

The most common method of expressing sludge age, usually represented by θ_c in days, is to express it as the ratio of the mass of MLSS in the aeration tank relative to the mass of suspended solids leaving the system per day.

---

*MLSS represents the *mixed liquor suspended solids* (*i.e.* total suspended solids) in the aeration tank, and may be represented by X_T.

MLSS is generally taken as a measure of *active biological mass* (i.e., micro-organisms or bacteria) in the system. However, the MLSS *will contain not only the active micro-organisms but also dead cells as well as inert inorganic matter derived from the influent sewage. For this reason, the term MLVSS (mixed liquor volatile suspended solids) is also sometimes used and may be preferred to MLSS, as it eliminates the effect of inorganic matter.*

Hence, MLSS wherever used can be read/replaced by MVLSS.

The total solids in the mixed liquor of the aeration tank (*i.e.* mix of influent sewage + returned sludge), called MLSS, can infact be measured by filtering a known volume of mixed liquor sample, in the same manner as the suspended solids are measured in wastewater.

Evidently MLSS being TSS (Total suspended solids) will be retained on the filter, as these are not dissolved solids, such as NaCl. Hence the retained solids on filter, when measured in *mg* and divided by the volume of liquor sample, will give the value of MLSS in mg/L.

Some of the MLSS, as stated above, is inorganic material and sometimes this biologically non-active mass may be quite large. Hence, MLVSS, being the fraction of MLSS that volatizes at 550°C, may be taken as the measure of active mass; if it can be measured, without much problem.

MLVSS is usually found to be around 75% to 80% of MLSS.

∴ Sludge age (θ_c)

$$= \frac{\text{Mass of suspended solid (MLSS) in the system } (M)}{\text{Mass of solids leaving the system per day}} \quad ...(9.45)$$

For a *conventional activated sludge plant*, with the flow (Q), concentrations of solids (X_T), and BOD_5 (Y_o), as marked in Fig. 9.42, we can easily write :

(*a*) **Mass of solids in the reactor**

$$= M = V \times (\text{MLSS})$$
$$= V \,.\, X_T \quad ...(9.46)$$

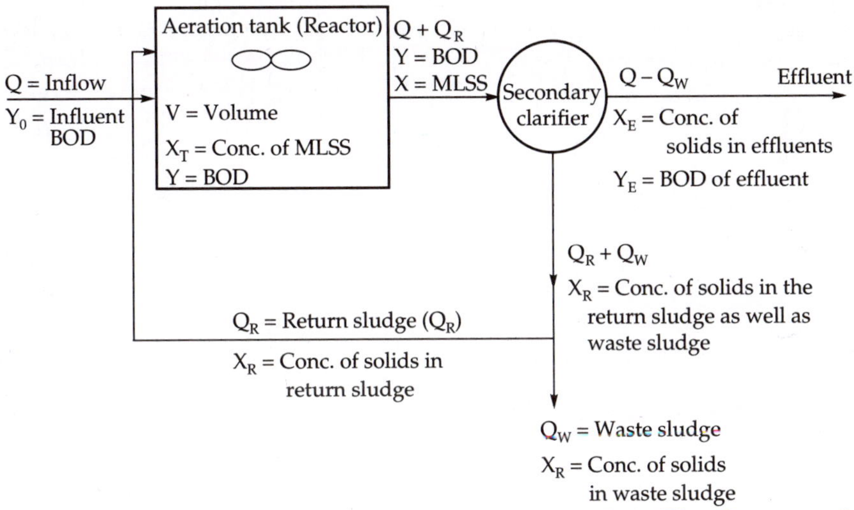

Fig. 9.42. Flow chart of conventional activated sludge plant (Plug flow reactor).

where X_T is MLSS in the aeration tank (in mg/*l*).

(*b*) (*i*) mass of solids removed with the *waste sludge* (*underflow*) per day

$$= Q_w \,.\, X_R \quad ...(i)$$

(*b*) (*ii*) mass of solids removed with the effluent per day

$$= (Q - Q_W)X_E \quad ...(ii)$$

∴ (*b*) **Total solids removed from the system per day**

$$= (i) + (ii)$$
$$= Q_W \,.\, X_R + (Q - Q_W)X_E \quad ...(9.47)$$

Now, Eq. (9.45) can thus be represented as :

$$\text{Sludge age} = \theta_c = \frac{V \,.\, X_T}{Q_W \,.\, X_R + (Q - Q_W)X_E} \quad ...(9.48)$$

where X_T = Concentration of solids in the influent of the Aeration Tank, called the MLSS, *i.e. Mixed Liquor Suspended Solids,* in mg/*l*.

V = Volume of Aerator

Q_W = Volume of waste sludge per day.

X_R = Concentration of solids in the return sludge or in the waste sludge* (both being equal) in mg/*l*

*also called ***underflow.***

Q = Sewage inflow per day

X_E = Concentration of solids in the effluent of secondary tank in mg/l

When the value of X_E (suspended solids concentration in the effluent of activated sludge plant) is very small, then the term $(Q - Q_W)\, X_E$ in Eq. (9.48) can be ignored, leading to :

$$\theta_c = \frac{V . X_T}{Q_W . X_R} \qquad \text{...(9.48 } a\text{)}$$

In addition to using sludge retention time (θ_c) as a rational loading parameter, another rational loading parameter which has found wider acceptance is the ***specific substrate (BOD) utilization rate (U) per day,*** and is defined as :

$$U = Q . \frac{(Y_0 - Y_E)}{V . X_T} \qquad \text{...(9.49)}$$

Under steady state operation, the mass of wasted activated sludge is further given by

$$Q_W . X_R = \alpha_y . Q\,(Y_0 - Y_E) - K_e . X_T . V \qquad \text{...(9.50)}$$

where α_y = max. **yield coefficient for conversion of BOD into bacterial cells in mg of MLSS per mg of BOD required**

$$= \frac{\text{microbial mass synthesised}}{\text{mass of substrate (BOD) utilised}}$$

K_e = **Endogenous respiration rate constant**, (per day), also called ***Decay rate of bacteria*** (per day). It may also be represented by K_d.

The values of α_y and K_e are found to be constant for municipal waste waters, their typical values being

α_y = 1.0 w.r.to TSS (*i.e.* MLSS) ; and 0.6 w.r.to VSS (*i.e.* MLVSS)
K_e = 0.05 to 0.06 (per day)

From the above equations, we can also work out as :

$$\frac{1}{\theta_c} = \alpha_y U - K_e \qquad \text{...(9.51)}$$

where U is the specific substrate (BOD) utilization rate per day.

Since both α_y and K_e are constants for a given wastewater, it becomes necessary to define either θ_c or U.

Eq. (9.51) is plotted in Fig. 9.43 for typical values of α_y = **0.5** and $\boldsymbol{K_e}$ = **0.06** (per day) for municipal wastewaters.

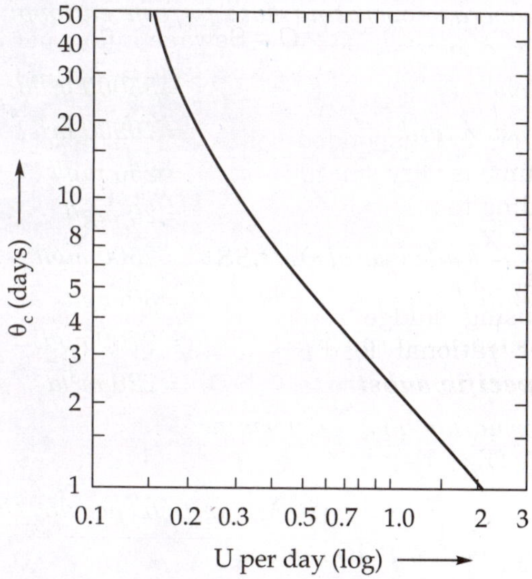

Fig. 9.43. Relationship between SRT (θ_c) and specific substrate (BODs here) utilization rate (U) for $\alpha_y = 0.5$ and $K_e = 0.06d^{-1}$.

Note : *Young sludge is indicated by the excessive white billowy (swelling and surging) foam, and low MLSS concentration, with high food to microorganism (F/M) ratios. On the other hand over abundance of dark brown and greasy foam is an indicator of old sludge, which also has high MLSS concentration and low F/M ratio.*

The θ_c value adopted for the design, controls the effluent quality and settleability and drainability of the biomass. Other parameters which are affected by the choice of θ_c values, are oxygen requirement, and quantity of waste activated sludge. Fig. 9.44 gives θ_c values as a function of temperature for 90—95% reduction of BOD of municipal wastewaters.

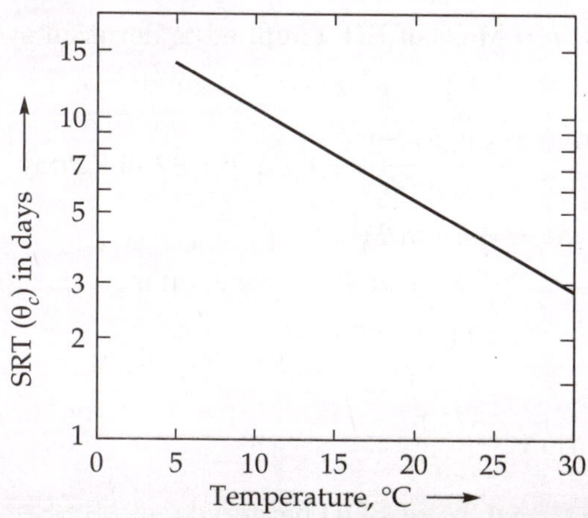

Fig. 9.44. SRT as a function of aeration basin temperature for 90-95% BOD removal.

Typical values of different types of loadings for a conventional as well as other types of Activated sludge plants are given in Table 9.10.

Example 9.29. *An average operating data for conventional activated sludge treatment plant is as follows :*

(*1*) *Wastewater flow*	= *35000 m*^{3}*/d*
(*2*) *Volume of aeration tank*	= *10900 m*^{3}
(*3*) *Influent BOD*	= *250 mg/l*
(*4*) *Effluent BOD*	= *20 mg/l*
(*5*) *Mixed liquor suspended solids (MLSS)*	= *2500 mg/l*
(*6*) *Effluent suspended solids*	= *30 mg/l*
(*7*) *Waste sludge suspended solids*	= *9700 mg/l*
(*8*) *Quantity of waste sludge*	= *220 m*^{3}*/d.*

Based on the information above, determine:

(*a*) *Aeration period (hrs)*

(*b*) *Food to microorganism ratio (F/M) (kg BOD per day/kg MLSS)*

(*c*) *Percentage efficiency of BOD removal*

(*d*) *Sludge age (days).* (*G.A.T.E., 1993*)

Solution. Given values are symbolised as :

$$Q = 35000 \text{ m}^3/\text{d} \; ; \qquad V = 10900 \text{ m}^3$$
$$Y_0 = 250 \text{ mg/l} \; ; \qquad Y_E = 20 \text{ mg/l}$$
$$X_T = 2500 \text{ mg/l} \; ; \qquad X_E = 30 \text{ mg/l}$$
$$X_R = 9700 \text{ mg/l} \; ; \qquad Q_W = 220 \text{ m}^3/\text{d}$$

These values are now used to calculate the desired factors, as below :

(*a*) Aeration period (t) in hr is given by Eq. (9.41) as

$$t = \frac{V}{Q} \cdot 24 = \frac{10,900}{35,000} \times 24 = 7.47 \text{ h} \; ; \text{ say } 7.5 \text{ h.} \quad \textbf{Ans.}$$

(*b*) *F/M* ratio $\quad F$ = Mass of BOD applied to aeration system

$$= Q \cdot Y_0 = 35000 \times 250 \text{ gm/day}$$

$$= \frac{35000 \times 250}{1000} \text{ kg/day} = \mathbf{8750} \text{ kg/day}$$

M = Mass of MLSS

$$= V \cdot X_T = 10900 \text{ m}^3 \times 2500 \text{ mg/l } (i.e. \text{ gm/m}^3)$$

$$= \frac{10900 \times 2500}{1000} \text{ kg} = \mathbf{27,250 \text{ kg}}$$

$$\therefore \qquad F/M \text{ ratio} = \frac{8750}{27,250}$$

$$= \mathbf{0.32} \text{ kg BOD per day/kg of MLSS.} \quad \textbf{Ans.}$$

(*c*) *Percentage efficiency of BOD removal*

$$= \frac{\text{Incoming BOD} - \text{Outgoing BOD}}{\text{Incoming BOD}}$$

$$= \frac{250 - 20}{250} \times 100\% = \frac{230}{250} \times 100\% = \mathbf{92\%}. \quad \textbf{Ans.}$$

(*d*) Sludge age in days (θ_c) is given by Eq. (9.48) as

$$\theta_c = \frac{V \,.\, X_T}{Q_W \,.\, X_R + (Q - Q_W) \,.\, X_E}$$

$$= \frac{27250 \text{ kg}}{(220 \text{ m}^3/\text{d} \times 9700 \text{ mg/l}) + (35000 \text{ m}^3/\text{d} - 220 \text{ m}^3/\text{d})\ 30 \text{ mg/l}}$$

$$= \frac{27250 \text{ kg}}{\frac{220 \times 9700}{1000} \text{ kg/d} + (35000 - 220)\frac{30}{1000} \text{ kg/d}}$$

$$= \frac{27250}{2134 + 1043.4} = \frac{27250}{3177.4} = \mathbf{8.58 \text{ days}}. \quad \textbf{Ans.}$$

9.36. Sludge Volume Index (S.V.I.)

The term *sludge volume index* or ***sludge index*** is used to indicate the physical state of the sludge produced in a biological aeration system. It represents the degree of concentration of the sludge in the system, and hence decides the rate of recycle of sludge (Q_R) required to maintain the desired MLSS and *F/M* ratio in the aeration tank to achieve the desired degree of purification.

S.V.I. is defined as the volume occupied in ml by one gm of solids in the mixed liquor after settling for 30 minutes, and is determined experimentally.

The standard test, which is performed in the laboratory to compute SVI of an aeration system involves collection of one litre sample of mixed liquor from the aeration tank from near its discharge end in a graduated cylinder. This 1 litre sample of mixed liquor is allowed to settle for 30 minutes and the settled sludge volume (V_{ob}) in ml is recorded as to represent sludge volume. This volume V_{ob} in ml per litre of mixed liquor will represent the quantity of sludge in the liquor in ml/l.

The above sample of mixed liquor, after remixing the settled solids, is further tested in the laboratory for MLSS by the standard procedure adopted for measuring the suspended solids in sewage. Let this concentration of suspended solids in the mixed liquor in mg/*l* be X_{ob}. Then SVI is given by the equation

$$\text{SVI in ml/g} = \frac{V_{ob}\,(\text{ml/l})}{X_{ob}\,(\text{mg/l})} \times 1000 \left(\frac{\text{mg}}{\text{g}}\right) = \frac{V_{ob}}{X_{ob}} \times 1000 \text{ ml/g}$$

or

$$\boxed{\text{SVI} = \frac{V_{ob}}{X_{ob}} \times 1000 \text{ ml/g}} \qquad ...(9.52)$$

The usual adopted range of SVI is between 50—150 ml/gm and such a value indicates good settling sludge.

Note. When the given SVI value in ml/gm is divided by $10^3 \times 10^3$ (*i.e.* 10^6), we will get SVI value in l/mg. SVI value in l/mg will therefore be $\frac{\text{SVI}}{10^6}$.

SVI value in mg/l will, thus, be given by $\frac{10^6}{\text{SVI}}$.

9.37. Sludge Recycle and Rate of Return Sludge

The MLSS concentration in the aeration tank is controlled by the **sludge recirculation rate** and the sludge settleability and thickening in the secondary sedimentation tank. The relationship between *sludge recirculation ratio* or *recycle ratio* $\left(\frac{Q_R}{Q}\right)$ with X_T (MLSS in tank) and X_R (MLSS in return or waste sludge) is given as :

$$\frac{Q_R}{Q} = \frac{X_T}{X_R - X_T} \quad \text{...(9.53)}$$

where Q_R = Sludge recirculation rate in m³/d
X_T = MLSS in the aeration tank in mg/L
X_R = MLSS in the returned or wasted sludge in mg/L

The settleability of sludge, as stated in the previous article, is determined by sludge volume index (SVI), which is determined in the laboratory.

If it is assumed that the sedimentation of suspended solids in the laboratory is similar to that in the sedimentation tank, then

$$X_R = \frac{10^6}{\text{S.V.I.}} \quad (\textit{i.e.}\ \text{SVI value in mg/l}) \quad \text{...(9.54)}$$

Eq. (9.53) then becomes

$$\frac{Q_R}{Q} = \frac{X_T}{\left(\frac{10^6}{\text{S.V.I.}} - X_T\right)} \quad \text{...(9.55)}$$

Values of ***Return sludge ratios*** or ***Recycle ratios*** $\left(\frac{Q_R}{Q}\right)$ adopted in different types of Activated sludge systems are shown in table 9.10. Its value for conventional sludge plant varies between 0.25 to 0.50.

The return sludge has always to be pumped, and the pump capacity should be designed for a minimum return sludge ratio of 0.50 to 0.75 for large plants and 1.0 to 1.5 for smaller plants, irrespective of the theoretical requirement. The required capacity should be provided in multiple units to permit variation of return sludge ratio, as found necessary during the operation of the plant.

9.38. Wasting of Excess Sludge (Q_w)

We know that the sludge generated in the aeration tank has to be partly discharged and wasted out of the plant to maintain a steady level of MLSS in the system. *The excess sludge quantity will increase with the increasing F/M ratio, and decrease with temperature.* In the case of domestic sewage, Q_w will be about 0.50—0.75 kg per kg BOD removed for the conventional sludge plants (having F/M ratio varying between 0.4 to 0.3).

Excess sludge may be wasted either from the sludge return line, or directly from the aeration tank as mixed liquor. The latter procedure is usually preferred, since the concentration of suspended solids will then be fairly steady in the waste discharge making the control easy.

In conventional plants, the wasted sludge is taken directly to a sludge thickener and digestor, or to the primary settling tank for its disposal along with the primary sludge. In extended aeration plants, however, the excess sludge is directly taken to the sludge drying beds.

9.39. Modifications of the Basic Activated Sludge Process

In the **basic activated sludge process,** also called ***conventional aeration process,*** the recirculated activated sludge is added to the inlet end of the aeration tank as a single dose. The ***regime flow employed in the aeration tank is plug flow and not mixed flow.***

Plug flow implies that the sewage moves down progressively along the aeration tank, essentially unmixed with the rest of the tank contents. The other type of flow regime, called complete mixed flow, involves the rapid dispersal of the incoming sewage throughout the tank, and is adopted in the extended aeration process and also in complete mix process (described a little later).

In a conventional aeration tank (of plug flow type), the F/M ratio and the oxygen demand will be the highest at the inlet end, and will then progressively decrease. In the complete mix system on the other hand, the F/M ratio and oxygen demand will be uniform throughout the tank.

The plug flow regime is achieved in such an activated process by employing a long and narrow configuration of the aeration tank with length equal to 5 to 50 times the width. The sewage and the mixed liquor are let in at the head of the tank and withdrawn at its end. Because of the plug flow regime, the oxygen demand at the head of the aeration tank is high and then tapers down. However, air is supplied in the process at a uniform rate along the length of the tank. This leads to either oxygen deficiency in the initial zone or wasteful application of air in the subsequent reaches.

The conventional system is always preceded by primary settling. The plant itself consists of an aeration tank, a secondary settling tank, a sludge return line and an excess sludge waste line leading to digestor. The BOD removal in this process is 85—92%.

The main **limitations of the conventional system** are that : *(i) the aeration tank volume requirement is high ; and (ii) there is a lack of operational stability at times of excessive variation in the rate of inflow or its BOD strength.* In order to overcome such difficulties posed by a conventional system plant, and to meet specific treatment objectives, several modifications of the conventional system have been suggested by modifying the process variables. The important modified processes are :

(*i*) *Tapered aeration process* ;

(*ii*) *Step aeration process* ;

(iii) *Contact stabilisation process* ;
(iv) *Complete mix process* ;
(v) *Modified aeration process* ;
(vi) *Extended aeration process* ; *and*
(vii) *Activated aeration process.*

Inspite of its various limitations, the conventional system for historical reasons, is the most widely used type of the activated sludge process. Plants up to 300 MLD capacity have been built in India. In addition to conventional activated sludge plants, the **complete mixed plants** and the **extended aeration plants** have also been found a wider acceptance in modern days, particularly for obtaining high BOD removals in smaller capacity plants.

We will now describe the above mentioned modifications of the basic activated sludge process.

9.39.1. Tapered Aeration Process. This process involves a very little modification of the conventional process, and ensures higher air supply at the inlet and in the initial length of the tank, as compared to the downstream length. The process is surely based on the fact that as the mixed liquor progresses through the aeration tank, its air requirement goes on reducing. ***Therefore, in a tapered aeration plant, compressed air is supplied at higher rates near the inlet end of the tank, and is gradually decreased as sewage moves towards the outlet end of the tank. Such a process therefore helps us in ensuring optimal application of air in the aeration tank.***

Ordinarily, 45% of air is supplied to the first one-third length of the tank, 30% to the second one-third length of the tank, and the rest 25% to the remaining one-third length of the tank.

No. of diffuser plates are thus varied accordingly. *Such a modification to the conventional activated plants using diffused air aeration, has now-a-days become a common feature, and is invariably adopted in all modern designs.* The loading parameters of such a plant do not materially differ from a conventional one, and are given in Table 9.10.

9.39.2. Step Aeration Process. In the step aeration process, the sewage is introduced along the length of the aeration tank in several steps, while the return sludge is introduced at the head, as shown in Fig. 9.45.

Such an arrangement results in a uniform air requirement along the entire length of the tank, and hence the uniform air supply of the conventional plants, can be efficiently used. The process enables an appreciable reduction in the aeration tank volume, without lowering the BOD removal efficiency. Step aeration method has considerable capacity to absorb shock organic loadings. The method has found application for larger plants of capacities up to about 1000 MLD.

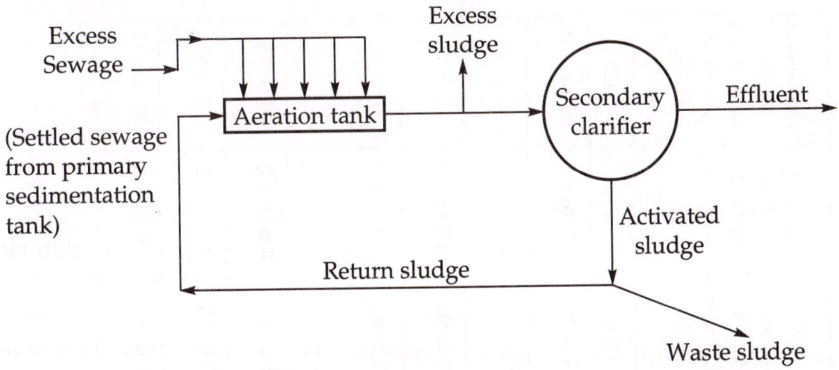

Fig. 9.45. Flow chart of step-aeration process.

The loading parameters of a such a plant are given Table 9.10.

9.39.3. Contact Stabilisation Process or Biosorption Process. This process has been designed for treating colloidal wastewaters.

In this process, the sewage and recycled or returned sludge are mixed and aerated for a comparatively shorter period of 0.5 to 1.5 hour in a special mixing tank, called *contact tank*. This mixing will allow the suspended and dissolved organic matter to be sorbed to the activated sludge floc. The sorbed organics and flocs are removed in the secondary settling tank, where the effluent from the contact tank enters. These settled sorbed organics and flocs are then transferred to a sludge aeration tank (called stabilisation tank or **aerodigestor**) where the organics are stabilised over a period of about 3 to 6 hours before it is fed back into the contact aeration tank. The stabilised sludge is then mixed with the influent wastewater again, and the process is repeated. The flow diagram for this process is shown in Fig. 9.46, and the loading parameters are given in Table 9.10.

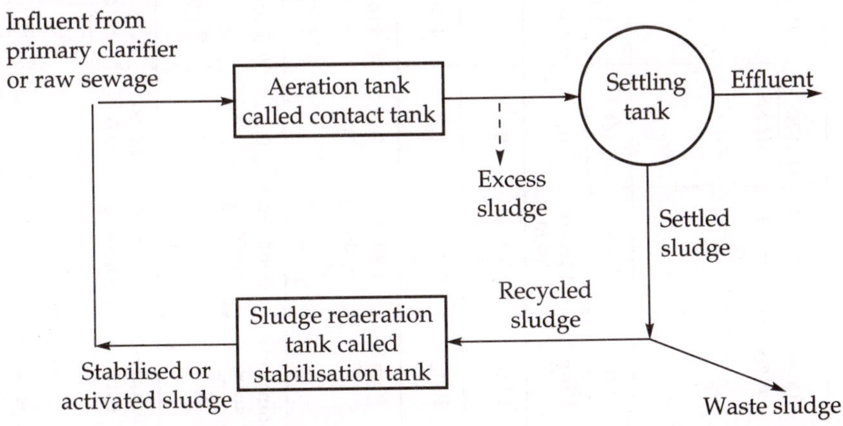

Fig. 9.46. Contact-stabilisation process.

The contact stabilisation process is quite effective in the removal of colloidal and suspended organic matter, but it is not very effective in removing soluble organics. The method is well suited for the treatment of fresh domestic sewage, containing only a low percentage of soluble BOD.

Table 9.10. Characteristics and Design Parameters of Different Activated Sludge Systems

Process type	*Flow regime*	*MLSS mg/l* (X_T)	$\frac{MLVSS++}{MLSS}$	*F/M*	*HRT hrs*	*Volume-tric Loading kg* BOD_5 *per* m^3	*SRT (days)* θ_c	$\frac{Q_R}{Q}$ *Return sludge ratio*	*BOD removal Percent or Efficiency (%)*	*kg of* O_2 *reqd. per kg* BOD_5 *removed*	*Air/Oxygen require-ment in* m^3 *per kg of* BOD_5 *removed* ÷
(1)	(2)	(3)	(4)	(5)	(6)	(7)	(8)	(9)	(10)	(11)	(12)
Conven-tional	**Plug**	**1500 to 3000**	**0.8**	**0.4 to 0.3**	**4 to 6**	**0.3 to 0.7**	**5 to 8**	**0.25 to 0.5**	**85 to 92**	**0.8 to 1.0**	**40 to 100**
Tapered aeration	Plug	1500 to 3000	0.8	0.4 to 0.3	4 to 6	0.3 to 0.8	5 to 8	0.25 to 0.5	85 to 92	0.7 to 1.0	50 to 75
Step aeration	Plug	2000 to 3000	0.8	0.4 to 0.3	3 to 6	0.7 to 1.0	5 to 8	0.25 to 0.75	85 to 92	0.7 to 1.0	50 to 75
Contact stabi-lisation	Plug	1000 to 3000* 3000 to 6000**	0.8	0.5 to 0.3	0.5 to 1.5* 3 to 6**	1.0 to 1.2	5 to 8	0.25 to 1.0	85 to 92	0.7 to 1.0	50 to 75÷
Complete mix.	**Complete mix.**	**3000 to 4000**	**0.8**	**0.5 to 0.3**	**4 to 5**	**0.8 to 2.0**	**5 to 8**	**0.25 to 0.8**	**85 to 92**	**0.8 to 1.0**	**50 to 75**
Modified aeration	Plug	300 to 800	0.8	3.0 to 1.5	1.5 to 3	1.2 to 2.4	0.2 to 0.5	0.05 to 0.15	60 to 75	0.4 to 0.6	25 to 50
Extended aeration	**Complete mix.**	3000 to 5000	**0.5 to 0.6**	**0.18 to 0.10**	**12 to 24**	**0.2 to 0.4**	**10 to 25**	**0.5 to 1.0**	**95 to 98**	**1.0 to 1.2**	**100 to 135**

* in contact aeration tank

**in sludge aeration tank

÷divided equally between contact aeration tank and sludge reaeration tank.

++Mixed Liquor Volatile Suspended Solids.

Compared to the conventional system, the contact stabilisation process has greater capacity to handle shock organic loadings, because of the biological buffering capacity of the sludge reaeration tank. The process also presents greater resistance to toxic substances in the sewage as the biological mass is exposed to the main stream of sewage containing the toxic constituents only for a short time.

The air requirements of the process are the same as for the conventional system, the air supply being divided equally between the contact aeration tank and the sludge re-aeration tank. However, the total aeration tank volume required for both the aeration tanks is only about half of the volume reqd. in the single conventional sludge aeration tank. The process therefore presents an effective method of uprating the existing conventional plants, where sewage characteristics are satisfactory. Moreover, the total aeration time is considerably reduced, and the plant capacity is thereby increased.

The process has found application in medium sized plants with capacities up to 40 MLD.

9.39.4. Complete Mix Process. Complete mix activated sludge plants or *continuous flow stirred Tank Reactors* were developed particularly for smaller cities, where the hourly variations in sewage were quite high, and as such, conventional plants were experiencing serious problems of biological instability. They are now-a-days quite popular.

In such a plant, the plug flow regime of a conventional plant is replaced by a completely mixed flow regime. The flow chart for such a system is shown in Fig. 9.47.

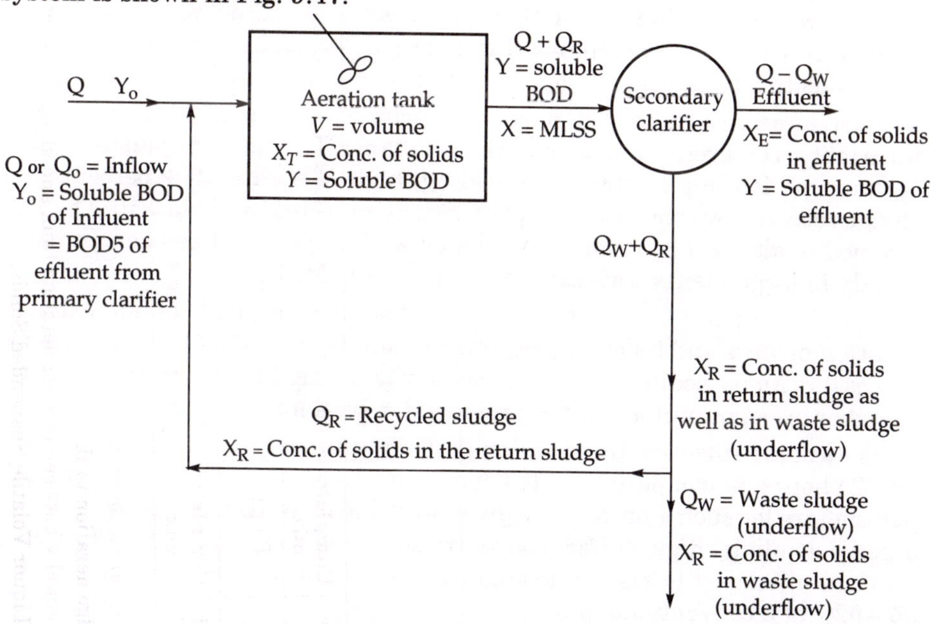

Fig. 9.47. Flow chart of a complete mix plant or a Continuous flow stirred tank plant. In such a plant, mechanical aerators are installed in the centre of a square aeration tank. Such aerators must have adequate mixing capacity to ensure thorough mixing of sewage and return sludge. The reactor contents are completely mixed, and it is assumed that there are no micro-organisms in the wastewater effluent.

The parameter Y_o and Y are the *soluble food concentration* or *soluble BOD values* in the influent and reactor, respectively.

In designing such a plant, it is assumed :

(1) That the influent and effluent biomass (micro-organisms) concentrations are negligible compared to biomass at other points in the system.

(2) The influent food concentration *i.e.*, BOD (Y_o) is immediately diluted to the reactor concentration Y, because of the complete mix regime.

(3) All reactions occur in the reactor, *i.e.*, neither the biomass (bacteria) production nor food (BOD) utilization occurs in the clarifier (secondary settling tank).

The loading parameters of such a plant are given in Table 9.10.

The complete mix plant possesses capacity to hold much higher MLSS concentration level in the aeration tank ; say 3000 to 4000 mg/l as against 1500 to 3000 mg/l of a conventional plant. *This helps in adopting smaller volume for the aeration tank. The plant has an increased operational stability at shock organic loadings, and also increased capacity to treat toxic biodegradable wastewaters like phenols. Such a plant is less liable to upsets by slugs of flows of toxic wastewaters.*

Such plants have been widely used for smaller plant capacities of less than 25 MLD or so, particularly for the towns where municipal and industrial wastewaters flow together.

9.39.5. Modified Aeration Process. When an intermediate quality of effluent containing higher BOD is permissible, such as at a place where effluent is to be used for farming, this modified aeration plant may be adopted, as it leads to substantial savings in construction and aeration costs.

Such a process does not need any primary sedimentation tank, as is invariably required in a conventional plant. The process ensures short aeration period, high volumetric loading, high *F/M* ratio, low percentage of sludge return, low concentration of MLSS, as reflected in Table 9.10. The BOD removal is also low, say only 60—75% or so. The process has been employed mainly in large plants with capacities above 200 MLD.

9.39.6. Extended Aeration Process. The flow scheme of an extended aeration process and its mixing regime are similar to that of the complete mix process. Primary sedimentation is frequently avoided in this process, but grit chamber or comminutor is often provided for screenings.

As its name suggests, the aeration period is quite large and extends to about 12—24 hours, as compared to 4 to 6 hours in a conventional plant. The loading parameters for such a process are given in Table 9.10. The process permits low organic loading, high MLSS concentration, and low F/M ratio. The BOD removal efficiency is also quite high, to say about 95—98% as compared to 85—92% of a conventional plant.

The air or oxygen requirement, is of course quite high, which increases the running cost of the plant considerably. The plant, however, offers another advantage, as no separate sludge digestor is required here, because the solids undergo considerable endogenous respiration and get well stabilised over the long detention periods, adopted in the aeration tank. The sludge produced is, thus, capable to be directly taken to the sludge drying beds. Also, the excess sludge production is minimum. *The operation is also simpler due to the elimination of primary settling and separate sludge digestion.* The typical plant based on this principle is known as an **oxidation ditch**, and is described separately under article 9.46.

Such a process is quite suitable for small communities having sewage flows of less than 4 MLD or so.

9.39.7. Activated Aeration Process. In the activated aeration process, a pair of conventional activated sludge plants operating in parallel are used. Excess activated sludge from the secondary settling tank of one unit will supply to the aeration tanks of both units, and the sludge from the second unit is pumped to final disposal. This arrangement, although no modification of the conventional process in the strict sense, reduces the construction costs, and probably operating costs, but does not modify the operating results.

9.40. Size and Volume of the Aeration Tank

Eqns. (9.48 *a*) and (9.50) can also be combined to yield

$$\boxed{V \,.\, X_T = \frac{\alpha_y \,.\, Q(Y_0 - Y_E)\theta_c}{1 + K_e \,.\, \theta_c}} \qquad \text{...(9.56)}$$

This equation can be used to calculate the *volume of the aeration tank* (V) for an assumed value of X_T (MLSS concentration in aeration tank) and a selected value of θ_c. *Alternatively*, the tank volume can be determined from Eq. (9.45) for an assumed value of F/M ratio and tank MLSS (X_T).

It can be seen that economy in reactor volume can be achieved by assuming a higher value of X_T. However, it is seldom taken to be more than 5000 g/m^3 (*i.e.* mg/L). A common range is between 1500 to 3000 g/m^3 or mg/l (see Table 9.10). *Considerations which govern the upper limit are :*

(*i*) initial and running cost of sludge recirculation system to maintain a high value of MLSS ;

(*ii*) limitations of oxygen transfer equipment to supply oxygen at required rate in a small reactor volume ;

(*iii*) increased solids loading on secondary clarifier which may necessitate a larger surface area to meet limiting solid flux ;

(*iv*) design criteria for the tank and minimum H.R.T. (t) for the aeration tank for stable operation under hydraulic surges.

Except in the case of extended aeration plants and completely mixed plants, the aeration tanks are designed as *long narrow channels*. This configuration is achieved by the provision of *round-the-end baffles* in small plants when only one or two tank units are proposed ; and by constructing long and narrow rectangular tanks with common intermediate walls in large plants when several units are proposed.

In extended aeration plants (other than oxidation ditches) and in complete mix plants, the tank shape may, however, be kept *circular* or *square* when the tank capacity is *small*, and rectangular with several side inlets and equal number of side outlets when the plant capacity is large.

The **width and depth of aeration channel** for conventional plants, depend upon the type of aeration equipment used. The depth controls the aeration efficiency and usually ranges from 3 to 4.5 m. The higher value of depth of 4.5 m is found to be more economical for plants of more than 50 MLD capacity. Beyond 70 MLD, duplicate units are preferred. The width controls the mixing and is usually kept between 5 to 10 m. Width-depth ratio should be adjusted between 1.2 to 2.2. The length should not be less than 30 m and not ordinarily longer than 100 m in a single section length before doubling back. The horizontal velocity should be around 1.5 m/min. *Excessive width may lead to settlement of solids in the tank.* Triangular baffles and fillets are used to eliminate dead spots and induce spiral flow in the tanks. The free-board in the tank is generally kept between 0.3—0.5 m.

While designing the aeration tanks, due consideration should also be given to the need of emptying them for maintenance and repair of aeration equipment. Intermediate walls should be designed for empty conditions on either side. The method of dewatering should be considered in the design and provided for during construction.

The inlet and outlet channels of the aeration tank should be designed for empty conditions on either side. The method of dewatering should be considered in the design and provided for during construction. The inlet and outlet channels should be designed to maintain a minimum velocity of 0.2 m/s to avoid deposition of solids. The channels or conduits and their appurtenances should be sized to carry the maximum hydraulic load to the remaining aeration tank units when any one unit is out of operation.

The inlet should provide for free fall into aeration tank when more than one tank unit or more than one inlet is proposed. The free fall will enable positive control of the flows through the different inlets. Outlets usually consist of free fall weirs. The weir length should be sufficient to maintain a reasonably constant water level in the tank. When multiple inlets or multiple tanks are involved, the inlets should be provided with valves, gates or stop planks to enable regulation of flow through each inlet.

9.41. Oxygen Requirement of the Aeration Tanks

Oxygen is required in the activated sludge process in the aeration tank for oxidation of part of the influent organic matter, and also for endogenous respiration of the micro-organisms in the system.

The total oxygen requirement may be computed by using the equation

$$O_2 \text{ required} = \left[\frac{Q(Y_0 - Y_E)}{f} - 1.42\, Q_W \,.\, X_R\right] \text{gm/day} \qquad ...(9.57)$$

$$\text{where } f = \frac{BOD_5}{BOD_u} = \frac{\text{5 day BOD}}{\text{Ultimate BOD}} \approx 0.68$$

1.42 = oxygen demand of biomass in gm/gm.

The above formula represents the oxygen demand for carbonaceous BOD removal and does not account for nitrification. The extra requirement of oxygen for nitrification is theoretically found to be 4.56 kg O_2/kg NH_3—N oxidised to NO_3—N.

The total oxygen requirement per kg BOD_5 removed for different activated sludge processes are given in table 9.10 in col. (11). *The amount of oxygen required for a particular process will increase within the range shown in the table, as the F/M value decreases.*

The aeration facilities of the activated sludge plant shall be designed to provide the calculated oxygen demand of the waste water against a specific level of D.O. in the wastewater. The aeration devices, besides supplying the required oxygen demand, shall also provide adequate mixing or agitation, so that the entire MLSS present in the aeration tank will become available for the biological activity. The recommended D.O. concentration in the aeration tank is in the range of 0.1 to 1.0 mg/L for conventional activated plants ; and is in the range of 1—2 mg/L for extended aeration type of activated plants ; and shall be above 2 mg/L when nitrification is required in the activated sludge plant.

Aerators are rated on the basis of the amount of oxygen (kg) that they can transfer to the tap water *under standard conditions* at 20°C, 760 mm Hg barometric pressure, and zero D.O., per unit of energy consumed.

The *oxygen transfer capacity (N) under field conditions* can be calculated from the *standard oxygen transfer capacity) (N_s)* by the formula :

$$N = \frac{N_s \,.\, (D_s - D_L)\,(1.024)^{T-20°}\,.\,\alpha}{9.17} \qquad \text{...(9.58)}$$

where N = Oxygen transferred under field conditions in kg O_2/kWh (or MJ*).

N_s = Oxygen transfer capacity under standard conditions in kg O_2/kWh (or MJ*).

D_s = Dissolved oxygen-saturation value for sewage at operating temperature.

D_L = Operation D.O. level in aeration tank, usually 1 to 2 mg/L

T = Temperature in degree C.

α = Correction factor for oxygen transfer for sewage, usually 0.8 to 0.85.

Oxygen may be supplied either by *surface aerators* or by *diffused air aeration systems* employing *fine or coarse diffusers*. In India, *surface aerators* are preferred because of easier maintenance. The oxygen transfer capacities of surface aerators, and fine and coarse diffused air systems, under standard conditions, lie between 1.2—2.4, 1.2—2 and 0.6—1.2 kg O_2/kWh, respectively.

Example 9.30. *Design a conventional activated sludge plant to treat domestic sewage with diffused air aeration system, given the following data :*

Population	*= 35,000*
Average sewage flow	*= 180 lpcd*
BOD of sewage	*= 220 mg/l*
BOD removed in primary treatment	*= 30%*
Overall BOD reduction desired	*= 85%.*

*1 watt = 1 Joule/sec = 3600 J/h

1 kWh = 3600 kJ = 3.6 MJ.

Solution. Daily sewage flow

$= Q = 180 \times 35000$ l/day $= 6300$ m³/day.

BOD of sewage coming to aeration

$= Y_0 = 70\% \times 220$ mg/l = **154 mg/l**

(∵ 30% BOD is removed in primary settling)

BOD left in effluent $= Y_E = 15\% \times 220$ mg/l = 33 mg/l

(∵ Overall 85% BOD removal is desired)

∴ BOD removed in activated plant

$= 154 - 33 =$ **121 mg/l**

∴ Efficiency required in Activated plant

$$= \frac{121}{154} = 0.79$$

From table 9.10, for efficiency of 85–92%, we use F/M ratio as 0.4 to 0.3, and MLSS between 1500 to 3000 for conventional activated plant. Since efficiency required is on lower side, we can use moderate figures for F/M ratio and MLSS.

So let us adopt F/M = 0.33

Similarly adopt MLSS (X_T) = 2000 mg/l

Using equation (9.44), we have

$$\frac{F}{M} = \frac{Q \, . \, Y_0}{V \, . \, X_T}$$

where $F/M = 0.33$ (assumed)

$Q = 6300$ m³/day

$Y_0 = 154$ mg/l = 154 gm/m³

$X_T = 2000$ mg/l (assumed)

$$\therefore \quad 0.33 = \frac{6300 \times 154}{V \times 2000}$$

∴ V = Volume of aeration tank

$$= \frac{6300 \times 154}{2000 \times 0.33} = \mathbf{1470\ m^3.}$$

(*i*) ***Check for Aeration period or H.R.T. (t)***

Using Eq. (9.41), we have

$$t = \frac{V}{Q} \times 24 \text{ h} = \frac{1470}{6300} \times 24 \text{ h}$$

= 5.6 h (within the limits of 4 to 6 h) O.K.

(*ii*) ***Check for S.R.T. (θ_c)***

From equation (9.56), we have

$$V \, . \, X_T = \frac{\alpha_y \, . \, Q(Y_0 - Y_E)\theta_c}{1 + K_e \, . \, \theta_c}$$

where $V = 1470$ m³

$X_T = 2000$ mg/L

α_y = yield coefficient = 1.0 w.r.t. TSS or MLSS (as defined under Eq. 9.50)

$Q = 6300\ m^3/d$

K_e = Endogeneous respiration rate constant = $0.06d^{-1}$. (as defined under Eq. 9.50).

Y_0 = BOD of influent in aeration tank = 154 mg/L

Y_E = BOD of effluent = 33 mg/L

Substituting the values, we get

$$1470 \times 2000 = \frac{1.0 \times 6300(154 - 33)\theta_c}{1 + 0.06 \times \theta_c}$$

or
$$1 + 0.06\theta_c = \left(\frac{1.0 \times 6300 \times 121}{1470 \times 2000}\right)\theta_c = 0.275\ \theta_c$$

$$1 + 0.06\theta_c = 0.259\ \theta_c$$

or
$$1 = (0.259 - 0.06)\ \theta_c$$

or
$$1 = 0.199\ \theta_c$$

$$\theta_c = \frac{1}{0.199} = 5.02 \text{ days} \approx \textbf{5 days}$$; O.K., as it lie between 5 to 8 days. (Table 9.10)

(*iii*) ***Check for volumetric loading***

Using equation (9.42), we have

$$\text{Volumetric loading} = \frac{Q \,.\, Y_0}{V} \text{ gm of BOD/m}^3 \text{ of tank vol.}$$

$$= \frac{6300 \times 154}{1386} \text{ gm/m}^3 = 700 \text{ gm/m}^3 = 0.7 \text{ kg/m}^3$$

(within the permissible range of 0.3—0.7 kg/m³) ; O.K.,

(*iv*) ***Check for Return sludge ratio***

Using equation (9.55), we have

$$= \frac{Q_R}{Q} = \frac{X_T (i.e.\ \text{MLSS})}{\dfrac{10^6}{\text{SVI}} - X_T}$$

where SVI = 100 ml/gm (assumed-since this value should be in the range of 50-150. Pl. see under Eq. 9.52)

X_T = 2000 mg/l

or
$$\frac{Q_R}{Q} = \frac{2000}{\left(\dfrac{10^6}{100} - 2000\right)}$$

= 0.25 (*i.e.* within the prescribed range of 25 to 50%) ; which is O.K.

We will, for conservative purposes, however provide 33% return sludge.

The adapted SVI with this return sludge ratio is then computed as :

$$0.33 = \frac{200}{\left(\frac{10^6}{SVI} - 2000\right)}$$

or $$\frac{10^6}{SVI} - 2000 = \frac{2000}{0.33} = 6060$$

or $$SVI = \frac{10^6}{8060} \approx \mathbf{125}\text{, which is O.K.}$$

The **sludge pumps** for bringing recirculated sludge from the secondary sedimentation tank will thus have a capacity = 33% × Q = 33% × 6300 m³/d = **2100 m³/d. Ans.**

Tank Dimensions. Adopt aeration tank of depth 3 m and width 4.5 m. The total length of the aeration channel required

$$= \frac{\text{Total volume required}}{B \times D} = \frac{1470}{4.5 \times 3} \text{ m}$$

$$= 108.9 \text{ m ; say } 111 \text{ m.}$$

Provide a continuous channel, with 3 aeration chambers, each of 37 m length. Total width of the unit, including 2 baffles each of 0.25 m thickness = 3 × 4.5 m + 2 × 0.25 = 14 m. Total depth provided including free-board of 0.6 m will be 3 + 0.6 = 3.6 m.

Overall dimensions of the Aeration tank will be 37 m × 14 m × 3.6 m. **Ans.**

Rate of Air Supply Required. Assuming the air requirement of the aeration tank to be 100 m³ of air per kg of BOD removed*, we have

Air required *i.e.* blower capacity

$$= 100 \times \frac{121 \times 6300}{1000} \text{ m}^3\text{/day}$$

$$= 121 \times 630 \times \frac{1}{24 \times 60} \text{ m}^3\text{/min}$$

$$= \mathbf{53 \text{ m}^3\text{/min.}} \quad \mathbf{Ans.}$$

*Let **standard diffuser plates** of 0.3 m × 0.3 m × 25 mm size, releasing 1.2 m³ of air / min / m² with 0.3 mm pores may be used.* Then, the total No. of plates required

$$= \frac{53}{1.2 \times 0.3 \times 0.3} = 491 \text{ ; say } \mathbf{500.}$$

Let us now provide more plates in the initial length of the tank ; say provide 45% of plates in the 1st $\frac{1}{3}$ length ; 30% of plates in the 2nd $\frac{1}{3}$ length ; and 25% in the last $\frac{1}{3}$ length.

∴ Plates to be provided in the 1st chamber = 45% × 500 = 225

Plates to be provided in the 2nd chamber = 30% × 500 = 150

Plates to be provided in the 3rd chamber = 25% × 500 = 125

The **diffuser plates** are now adjusted as to ensure a minimum clear distance

*See col. (12)—Table 9.10, it is between 40—100 m³/kg.

of 0.9 m along the channel length, to avoid interference from the rising streams of bubbles. The minimum centre to centre spacing of plates along channel length, will therefore be 1.2 m, and hence in a channel of 37 m length, we would be able to accommodate only about $\left[\frac{37}{1.2}(i.e.\text{ No. of spacings} - 1)\right]$ = 31 – 1 = 30 plates. In order to provide a total of 225 plates in this length of 37 m, we will have to use $\frac{225}{30} \approx 8$ plates in each row, placed side by side along the width of the chamber. Hence, use 8 plates placed along the width of each chamber, constituting as one row, and provide 30 such rows @ 1.2 m c/c distance with total 240 plates in total.

In the second chamber of 37 m length, we have to provide only 150 plates, and hence c/c spacing of rows (8 plates in each row)

$$= \frac{37}{\frac{150}{8}} = \frac{37 \times 8}{150} = 1.97 \text{ m}\ ; \text{ say @ 2 m c/c.}$$

In the 3rd chamber, No. of plates required is only 125 ; and hence spacing of rows in this chamber $= \frac{37}{\frac{125}{8}} = \frac{37 \times 8}{125} = 2.37$ m ; say 2.4 m.

Hence adopt 8 plates in each row ; and rows are placed @ 1.2 m c/c in the 1st chamber, @ 2.0 m c/c in 2nd chamber ; and @ 2.4 m c/c in the 3rd chamber. **Ans.**

Design of Secondary sedimentation Tank. Adopting a surface loading rate of 20 m^3/day/m^2 at average flow of 6300 m^3/day, we have

(*i*) Surface area required

$$= \frac{6300}{20} \text{ m}^2 = 315 \text{ m}^2$$

Adopting a solids loading of 125 kg/day/m^2 for MLSS of 2000 mg/*l*, we have

(*ii*) the surface area required

$$= \frac{6300 \times 2000}{1000} \times \frac{1}{125} \text{ m}^2 = 100.8 \text{ m}^2$$

The higher surface area of 315 m^2 is adopted.

Adopting a circular tank,

$$\text{dia of tank} = \sqrt{\frac{315 \times 4}{\pi}} = 20 \text{ m.}$$

Weir loading for a circular weir placed along the periphery of the tank having length 20 π will be

$$= \frac{6300}{20\pi} \text{ m}^3\text{/day/m.} = 100.3 < 150 \text{ ; O.K.}$$

Note. If weir loading exceeds the permissible value, we may provide a trough instead of a single weir at the periphery.

Hence, provide 20 m dia secondary settling tank. **Ans.**

Design of Sludge Drying Beds. In order to design sludge drying beds, the quantity of **excess wasted sludge** will be calculated by using the equation (9.48 *a*) as :

$$\theta_c = \frac{V . X_T}{Q_W . X_R}$$

or $$5.02 = \frac{1470 \times 2000}{Q_W . X_R}$$

or $$Q_W . X_R = \frac{1470 \times 2000}{5.02} \text{ gm/d} = 586000 \text{ g/d} = 586 \text{ kg/d}$$

For 10 kg/m³ S.S. concentrating in secondary sludge, (*i.e.* $X_R = 10$ kg/m³),

the excess secondary sludge volume = $Q_w = \frac{586}{X_R} = \frac{586 \text{ kg/d}}{10 \text{ kg/m}^3}$ = **58.6 m³/d.**

Note. This secondary sludge volume of 58.6 m³/d shall be taken to sludge drying beds, along with the primary sludge. The volume of primary sludge can be calculated if the concentration of suspended solids in sewage is known along with knowing the degree of removal of suspended solids in primary settling. Since S.S. of sewage is not given in this question, the quantity of primary sludge cannot be worked out ; and hence the design of sludge drying beds cannot to be done with the given data.

Example 9.31 on designing an activated-sludge reactor. *An activated-sludge system is to be used for secondary treatment of 8000 m³/d of municipal wastewater. After primary clarification, the BOD$_5$ is 150 mg/L, and it is desired to have not more than 5 mg/L of soluble BOD in the effluent. A completely mixed reactor is to be used, and pilot-plant analysis has established the following kinetic values: α_y = 0.5 kg/kg, K_e or K_d = 0.05 d^{-1}. Assuming an MLSS concentration of 2800 mg/L and an underflow concentration of 8000 mg/L from the secondary clarifier, determine (1) the volume of the reactor, (2) the volume of solids that must be wasted each day, and (3) the recycle ratio.*

Assume sludge age θ_c = 12 days.

Solution. Q or Q_o = 8000 m³/d (given)

Y_o = BOD$_5$ of effluent from primary clarifier = 150 mg/L (given)

Y_E = Soluble BOD of effluent from ASP (Activated Slude Plant)

= 5 mg/L (given)

α_y = 0.5 kg MLSS/kg of BOD (given)

K_e = 0.05/*d* (given)

X_T = 8000 mg/L (given)

θ_c = 12 d (given)

V = ?

Q_W = ?

Use Eq. (9.56) to compute V, as

$$V . X_T = \frac{\alpha_y . Q (Y_o - Y_E) \theta_c}{1 + K_e . \theta_c} \qquad \text{...(Eq. 9.56)}$$

$$\therefore \quad V \,.\, X_T = \frac{\alpha_y \; Q\,(Y_o - Y_E)\,\theta_c}{1 + K_e \,.\, \theta_c}$$

Substituting values, we get,

$$V \times 2800 \; \frac{\text{mg}}{\text{L}} = \frac{0.5 \times 8000 \dfrac{m^3}{d} (150 - 5) \dfrac{mg}{L} \times 12d}{1 + \dfrac{0.05}{d} \times 12d}$$

$$\text{or} \;\; V \times 2800 = \frac{4000 \dfrac{m^3}{d} \times 145 \times 12d}{1 + 0.6}$$

$$\text{or} \qquad V = \frac{4000 \times 145 \times 12}{1.6 \times 2800} \text{ m}^3 = 1554 \text{ m}^3$$

Hence, Vol. of reactor = $\boldsymbol{V}$ **= 1554 m³** **Ans.**

2. From Eq. (9.48*a*), we have

$$\theta_c = \frac{V . X_T}{Q_W \;.\; X_R} \qquad \text{...(Eq. 9.48}a\text{)}$$

where Q_W = Vol. of sludge wasted per day = ?

X_R = MLSS in return sludge or in underflow

= 8000 mg/L (given)

θ_c = 12 d

$$\therefore \qquad 12\, d = \frac{1554 \text{ m}^3 \times 2800 \dfrac{\text{mg}}{\text{L}}}{Q_W \times 8000 \dfrac{\text{mg}}{\text{L}}}$$

or $\quad \boldsymbol{Q_W = 45.33 \; m^3/d}$ **Ans.**

3. Recycle Ratio is given as :

$$\frac{Q_R}{Q} = \frac{X_T}{X_R - X_T} \qquad \text{...(Eq. 9.53)}$$

$$\text{or} \qquad \frac{Q_R}{Q} = \frac{2800 \text{ mg/L}}{(8000 \text{ mg/L} - 2800 \text{ mg/L})} = \frac{2800}{5200}$$

= **0.54 Ans.**

Example 9.32. *A municipality has directed to upgrade its primary wastewater treatment unit to a secondary unit that can meet an effluent standard of 20 mg/L BOD_5 and 20 mg/l total suspended solids. They have selected a completely mixed activated sludge system. BOD_5 of total suspended solids is 63% of TSS concentration. Estimate the required volume of aeration tank. The following data is available from existing primary plant :*

Flow = 0.150 m^3/s, BOD_5 = 80 mg/l.

Assume the following values for half velocity constant = 95 mg/l of BOD_5; Maximum growth rate constant = 2.5/day; Decay rate of micro-organism = 0.050/day; Yield coefficient = 0.50 mg VSS per mg BOD_5 removed; MLVSS = 2000 mg/l. *(Engineering Services, 2020)*

Solution. Let the secondary clarifier produces an effluent with 20 mg/L of TSS (max permissible)

The BOD casued by TSS in the effluent = 63% × 20 mg/L = 12.6 mg/L

Total max allowed BOD in the effluent = 20 mg/L

Hence, the treated liquid should have soluble BOD $\not>$ 20 – 1.26 = **7.4 mg/L**

Monod, a French biochemist worked out a mathematical equation on the basis of mathematical modelling to relate microbial growth rates in an aqueous environment to the concentration of a limiting nutrient. His basic equation (Eq. 28.2) was further expanded for a Complete Activated Sludge process, to produce the following equation.

$$Y_E = \left[\frac{K_s(1 + K_e \,.\, \theta_c)}{\theta_c(\mu_m - K_e) - 1}\right] \qquad ...(9.59)$$

where Y_E = Soluble BOD in the affluent of the plant = 7.4 mg/L (calculated above)

μ_m and K_s are empirical (experimental) coefficients to the Monod equation and are defined below:

μ_m is the maximum value of μ, *where μ is the **growth rate of a considered bacteria** or micro-organism.* This value μ_m is known as *maximum growth rate constant (per day).* Its usual value is 2.5/day.

Its given value in question is also 2.5/day

K_s = ***Half saturation constant*** = concentration of substrate (Y) in mg/L, when μ is equal to 0.5 μ_m. It, thus represents the substrate (Y) concentration of one half of the maximum growth rate (in mg/L). Its usual value is 100 mg of BOD/L.

Its given value in question is 95 mg/L of BOD.

k_e = *Decay rate of bacteria constant or Endogenous respiration (or decay) rate constant = 0.05/d (given)*

θ_c = Sludge age or Solids retention time in days = ?

It can be seen that only θ_c is unknown in this Equation (i.e., Eq. 9.59), as Y_E (effluent soluble BOD from activated sludge plant (ASP) is known or can be measured. This value of θ_c can be used in Eq. (9.56) to compute V.

Now, using Eq. (9.59), we have

$$Y_E = \left[\frac{K_s(1 + K_e \,.\, \theta_c)}{\theta_c(\mu_m - K_e) - 1}\right] \qquad \text{...(Eq. 9.59)}$$

Using values defined below Eq. (9.59), we get,

$Y_E = 7.4$ mg/L

$K_s = 95$ mg/L

$K_e = 0.05$/d

$\mu_m = 2.5$ /d

$\theta_c = ?$

$$\therefore \qquad 7.4 = \left[\frac{95(1 + 0.05\,\theta_c)}{\theta_c(2.5 - 0.05) - 1}\right]$$

$$\text{or} \qquad 7.4 = \frac{95(1 + 0.05\,\theta_c)}{2.45\theta_c - 1}$$

$$\text{or} \qquad \frac{7.4}{95} = \frac{1 + 0.05\,\theta_c}{2.45\,\theta_c - 1}$$

or $0.078\,(2.45\,\theta_c - 1) = (1 + 0.05\,\theta_c)$

or $0.191\,\theta_c - 0.078 = 1 + 0.05\,\theta_c$

or $\quad 0.141\theta_c = 1.078$

$$\text{or} \qquad \theta_c = \frac{1.078}{0.141} = \mathbf{7.65d}$$

Now, using Eq. (9.58), we have

$$V X_T = \frac{\alpha_y \,.\, Q\,(Y_o - Y_E)\,\theta_c}{1 + K_e \,.\, \theta_c} \qquad \text{...(Eq. 9.56)}$$

where $\theta_c = 7.65$ days (calculate above)

$\alpha_y = 0.5$ mg MLSS/mg BOD

$Y_0 =$ Soluble influent BOD to ASP $\approx$ BOD_5* of the effluent from the primary treatment = 80 mg/L (given)

$Y_E = 7.4$ mg/L (calculated above)

$K_e = 0.05$/d (given)

---

* Note BOD_5 is generally equated to soluble BOD, because BOD_5 represents oxygen demand by most common wastewater compounds, such as sugars, fatty acids, starches, etc. What remains in the BOD bottle after 5 days tends to be *insoluble BOD* due to insoluble organic compounds and solids.

There are two forms of soluble BOD *i.e.*, (*i*) nitrogenous BOD, and (*ii*) carbonaceous BOD.

$$X_T = 2000 \text{ mg/L (given)}$$
$$Q = 0.150 \text{ m}^3\text{/s (given)}$$

$$\therefore\ V \times 2000 \text{ mg/L} = \frac{0.5 \times 0.15\text{m}^3\text{/s } (80 - 7.4) \text{ mg/L} \times 7.65\ d}{(1 + 0.05/d \times 7.65d)}$$

$$\text{or} \quad \text{V} \times 2000 = \frac{0.5 \times 0.15 \times 72.6 \times 7.65 \times (24 \times 3600)\ \text{m}^3}{(1 + 0.3825)}$$

$$\text{or} \quad V = 1301.6 \text{ m}^3 \quad \textbf{Ans.}$$

9.42. Advantages and Disadvantages of an Activated Sludge Plant

The chief advantage of the activated sludge process is that it offers secondary treatment with minimum area requirements, and an effluent of high quality is obtained. The conventional process was somewhat difficult to operate and needed a lot of supervision. However, the modifications described earlier, have made the process less difficult to operate than formerly, with the result that most secondary treatment plants being installed today are of these types.

Compared to the trickling filter plant, the capital cost of an activated sludge plant is less, but the operating cost is more mainly because of high power consumption for operating air compressors and the sludge circulation pumps. *Generally, the power requirement in an activated sludge plant varies between 55 to 110 H.P. per million litres of sewage.*

Loss of head through the plant is also quite less, and there is no fly or odour nuisance here, as is there in a trickling filter plant. However, if there is a sudden increase in the volume of sewage or if there is a sudden change in the character of sewage, adverse effects are produced on the working of the process, producing inferior effluent. Moreover, the quantity of sludge obtained is larger, and needs suitable thickening and disposal.

9.43. Activated Sludge Process Vs Trickling Filter Process and the Choice of One

As discussed earlier, the conventional biological (secondary) treatment of sewage is usually carried out, either by using *trickling filters* or by using an *activated sludge process*. The basic difference between an activated sludge process and the action involved in a trickling filter is that ; whereas in a trickling filter, the bacterial film coating the grains of the filter media is *stationary* and likely to become clogged after sometime ; in the activated sludge process, on the other hand, the finer suspended organic particles of sewage (settled as activated sludge) are themselves coated with the bacterial film, which is kept *moving* by the constant agitation. *In the activated sludge process, therefore, the sludge flocs are coated with bacteria, and they act like free moving organisms,* which are being continuously swept through the sewage, and which in their search for food and work, oxidise the organic matter present in sewage *in a much more efficient way* than that carried out in a *filter* by the bacteria coated around the particles of filter media. *As such, it can be stated that an activated sludge process is more efficient than a trickling filter.*

The quality of the effluent obtained in a conventional activated sludge plant is also better than that of a trickling filter plant. But since a conventional activated sludge plant requires a lot of skilled attendance and supervision during its operation, the modified activated sludge processes are generally used these days (as discussed earlier).

Moreover, the modern plants are operated more rapidly at higher rates (than those adopted in conventional plants) with shorter detention periods in the aeration tanks (2 to 3 hours), smaller amounts of air (3.13 m^3/m^3 of sewage treated), and lesser amount of returned sludge (10 to 25% of sewage flow). Such a **high rate activated sludge plant*** can produce sufficiently good quality effluent (removing 70 to 80% of suspended solids, and 80 to 85% of BOD from sewage), with the results lying in between those obtained by a conventional activated sludge plant and a high rate trickling filter plant ; although the available finances and the technical development of a country is another major guiding factor.

In addition to giving a slightly better effluent, other **advantages offered by an activated sludge plant** are :

(*i*) lesser land area is required ;

(*ii*) the head loss in the plant is quite low ;

(*iii*) there is no fly or odour nuisance ;

(*iv*) capital cost is less ; and

(*v*) greater flexibility of operation, permitting control on the quality of effluent, is possible.

The **disadvantages of the activated plant**, on the other hand, are also laudable, as :

(*i*) high cost of operation, with greater power consumption ;

(*ii*) a lot of machinery to be handled ;

(*iii*) the sudden change in the quantity and character of sewage may produce adverse effects on the working of the process, thus producing inferior effluent ;

(*iv*) *bulking of sludge* is a common trouble, which has to be controlled, especially when industrial wastewaters with high carbohydrate content or antiseptic properties are present. In any case, if such bulking is not there, comparatively larger sludge volume is to be handled ; and

(*v*) the quantity of returned sludge has to be adjusted every time, as and when there is a change in the quantity of sewage flow, thus making the operation a little cumbersome.

Since these advantages and disadvantages weigh against each other, it is therefore, suggested that before recommending any of these two methods, the following factors must be thoroughly studied.

(*i*) availability of land for installing the treatment ;

(*ii*) availability of suitable method of sludge disposal ;

---

*The sludge settled in secondary tanks in a high rate activated process is much dense than the conventional activated sludge, and therefore, settles much faster. The surface loading in secondary clarifiers can, therefore be increased to 50,000 l/m^2/day or so. However, the sludge becomes septic too early, and must be immediately removed.

(*iii*) cost and availability of power ;

(*iv*) availability and cost of machinery and its spare parts required for installing the plant ;

(*v*) degree of flexibility required in the operation of the process ;

(*vi*) the quality and quantity of sewage, and the chances of variation in its quality and quantity ; and wear and tear of civil works and machinery employed in the process.

Moreover, detailed estimates for both types of treatments should also be prepared for a particular project. The pros and cons of both methods for that project should also be reviewed and thoroughly considered. A final decision is then taken, keeping in consideration the economy as well as the comparative merits and demerits of each method, and their effects on the desired aims. Normally, *it is found that for towns or small cities with medium sized plants, trickling filters are better ; whereas in big cities with large sized plants, the activated sludge plant is better*.

SECONDARY TREATMENT THROUGH ROTATING BIOLOGICAL CONTACTORS

(Aerobic Attached Culture)

9.44. Rotating Biological Contactors (RBCs)

The Rotating Biological Contractor's method of secondary waste-water treatment has been recently developed and does not fit precisely in to either the trickling filter or the activated sludge categories, but does employ principle common to both of them.

A **rotating biological contactor** (RBC) is a cylindrical media made of closely mounted thin flat circular plastic sheets or discs of 3 to 3.5 m in diameter, 10 mm thick, and placed at 30 to 40 mm spacings mounted on a common shaft. Thinner materials can be used by sandwitching a corrugated sheet between two flat discs and welding them together as a unit, as shown in Fig. 9.48.

Fig. 9.48. Rotating Biological Contactors placed in series.

The R.B.C.'s are usually made in up to 8 m length, and may be placed in series or parallel in a specially constructed tank(s), through which the wastewater is allowed to pass. The RBC's are kept immersed in wastewater by about 40% of their diameter. The RBC's are rotated around their central horizontal shaft, at a speed of 1—2 rpm by means of power supplied to the shaft. Approximately 95% of the surface area is thus alternately immersed in the wastewater and then exposed to the atmosphere above the liquid.

When the process is operated, the microorganisms of the waste-water begin to adhere to the rotating surfaces and grow there, untill the entire surface area of the discs gets covered with 1 to 3 mm layer of biological slime. As the discs rotate, they carry a *film of wastewater* into the air, where it trickles down the surface of the discs, absorbing oxygen. As the discs complete their rotation, this film mixes with the wastewater in the tank, adding to the oxygen of the tank and mixing the treated and partially treated wastewater. As the *attached microorganisms* pass through the tank, they absorb other organics for breakdown. The excess growth of microorganisms is sheared from the discs, as they move through the wastewater tank. The dislodged organisms are kept in suspension by the moving discs. This suspended growth finally moves down with the sewage flowing through the tank to a downstream settling tank for removal. The effluent obtained is of equal or even better quality than what is obtained from other secondary treatments. The quality of the effluent can further be improved by placing several contractors in series along the tank. The method can thus provide a high degree of treatment, including biological conversion of ammonia to nitrates.

As is evident, a given set of discs (*i.e.* an RBC) serves the following purposes :

(*i*) They provide media for build up of *attached microbial growth*.

(*ii*) They bring the growth of microbes in contact with the waste-water.

(*iii*) They aerate the wastewater and the suspended microbial growth in the wastewater tank.

In this process, the attached growths are similar in concept to a trickling filter, except that here the microorganisms are passed through the wastewater, rather than the wastewater passing over the microbes, as happens in a trickling filter. This method realises some of the advantages of both the trickling filter and the activated sludge process.

The sludge produced in the process contains about 95—98% moisture, and may amount to about 0.4 kg per kg of BOD_5 applied. The theoretical model of the process is similar to that for trickling filter, *but actual design is still empirical* and based on the data from the successful working plants and as developed by the process manufacturers.

The *hydraulic loading* rates may vary between 0.04—0.06 m/day, and *organic loading rates* between 0.05—0.06 kg BOD_5/m^2 per day, based upon the disc surface area. Sloughing of biological solids is more or less continuous and the effluent contains a relatively constant concentration. The solids settle well and clarifier surface overflow rates of about 33 m^3/m^2 per day are reported to be satisfactory.

AEROBIC STABILISATION UNITS

(Aerobic Suspended Culture)

9.45. Oxidation Ponds and Stabilisation Ponds

Stabilisation ponds are open flow-through earthen basins, specifically designed and constructed to treat sewage and biodegradable industrial wastewaters. Such ponds provide comparatively long detention periods, extending from a few days to several days, during which time the wastes get stabilised by the action of natural forces.

Stabilisation ponds may be classified as *aerobic*, *facultative* or *anaerobic*, depending upon the mechanism of purification.

In a *totally aerobic pond*, the stabilisation of wastes is brought about by aerobic bacteria, which flourish in the presence of oxygen. The oxygen demand of such bacteria in such a pond is met by the combined action of algae and other microorganisms, (such as bacteria and protozoa*), called **algal photosynthesis**, or **algal-symbiosis.** In this symbiosis, the algae (which are microscopic plants) while growing in the presence of sunlight, produce oxygen by the action of *photosynthesis* ; and this oxygen is utilised by the bacteria for oxidising the waste organic matter. The end products of the process are carbon dioxide, ammonia and phosphates, which are required by the algae to grow and continue to produce oxygen.

In an *anaerobic pond*, however, the stabilisation of waste is mainly brought about by the usual anaerobic conversion of organic wastes to carbon dioxide, methane, and gaseous end products, with eruption of foul odours and pungent smells.

In a *facultative pond*, the upper layers work under aerobic conditions, while the anaerobic conditions prevail in the bottom layers. *The upper aerobic layer of the pond acts as a good check against the evolution of the foul odours from such a pond.* The treatment affected from such a pond is comparable to that of conventional secondary treatment processes.

The totally aerobic ponds will have to be of very small depths, such as below 0.5 m and still then may require occasional stirring, to prevent anaerobic conditions in the settled sludge. *It is, therefore practically difficult to construct and use such purely aerobic ponds.* The facultative ponds, with depths varying between 1.0 m to 1.5 m, are thus most widely used for treatment of sewage. Deeper ponds in the depth range of 2.5 m to 4 m may also sometimes be constructed to work anaerobically.

Treatment ponds have been used to treat wastewaters for many years, especially for small communities.

The term **oxidation pond** was originally referred to that stabilisation pond which received partially treated sewage ; whereas the pond that received raw sewage was used to be called a **sewage lagoon** ; but in recent years, the term *oxidation pond* has been widely used as a collective term for all types of ponds, and most particularly the *facultative stabilisation ponds*.

The results of the oxidation pond treatment are : the oxidation of the original organic matter and the production of algae, which are discharged with

*Aeorbic bacteria are most commonly present, and protozoa may be present under certain loading conditions.

the effluent. This results in a net reduction in BOD, since the algae are more stable than the organic matter in wastewater, and degrade slowly in the river stream into which the effluent is discharged.

The oxidation ponds, throwing their effluents in rivers, just upstream of some lakes or reservoirs, are therefore, generally not preferred, as the discharged algae may settle in the reservoirs, and cause anaerobic decomposition and other water quality problems. *However, the effluents from oxidation ponds can be easily used for land irrigation,* particularly at places, where they cannot be discharged into river streams.

9.45.1. Constructional Details. A typical plan and section of an oxidation pond is shown in Fig. 9.49. It is an earthen pond, dug into the ground, with shallow depth. The pond should be atleast 1.0 m deep to discourage growth of aquatic weeds, and should not exceed 1.8 m or so, as otherwise the pond may turn into a deeper anaerobic pond rather than remaining facultative in character without giving foul odours. *The* **detention time** *in the pond is usually 2 to 6 weeks,* depending upon sun light and temperature. In cold countries, higher figure is to be adopted.

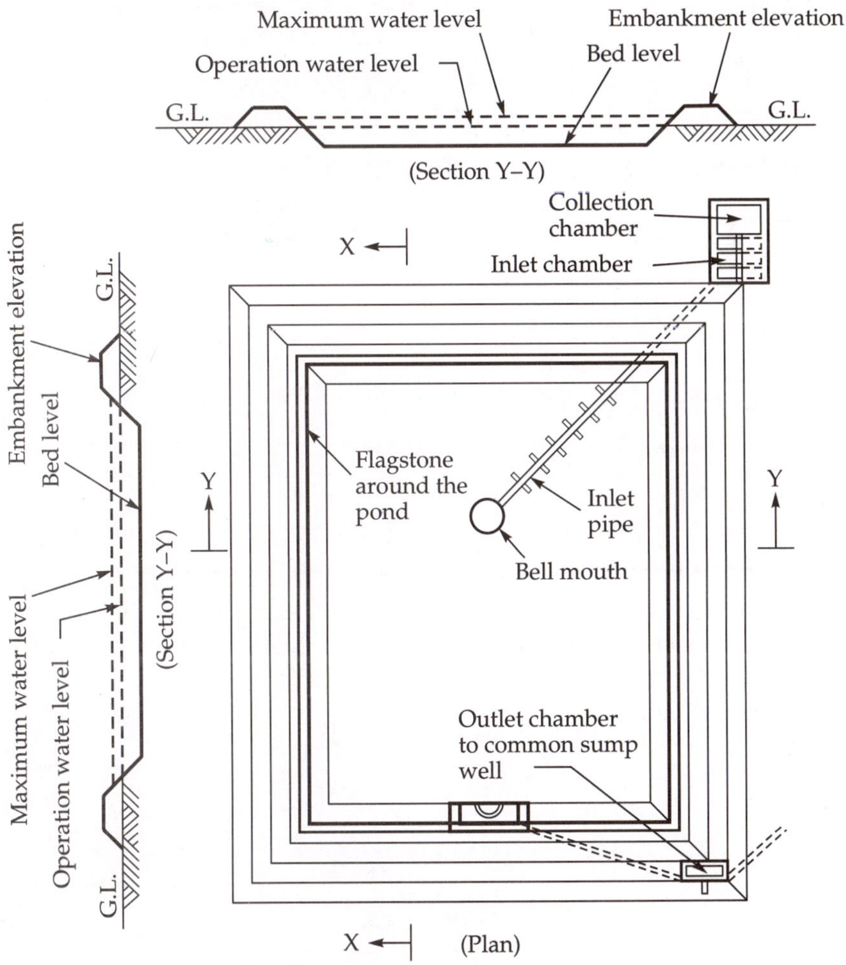

Fig. 9.49. Oxidation Pond.

Better efficiency of treatment is obtained, if several ponds are placed in series, so that the sewage flows progressively from one to another unit, until it is finally discharged.

The effluent should be entered at centre above or below the liquid surface for a uniform mixing of the influent with the oxygen saturated pond liquor. The outlet should provide withdrawal of effluent from different depths of pond.

9.45.2. Design Criteria. The surface area of the tank may be worked out by assuming a suitable value of **organic loading,** which may range from 300—150 kg/hectare/day* or so in hot tropical countries like India to about 90—60 kg/hectare/day or so for colder countries situated at higher latitudes. Each unit may have an area ranging between 0.5 to 1 hectare.

The length of the tank may be kept at about twice the width. The depth may be kept between 1 to 1.5 m. A free-board of about 1 m may also be provided above a capacity corresponding to 20—30 day of detention period. It is found that with the above assumed organic loading, and a per capita daily BOD production of about 0.08 kg, one hectare land area will suffice for

$$\frac{300}{0.08} = 3750, \text{ to } \frac{60}{0.08} = 750 \text{ persons.}$$

IS 5611 recommends the following values of BOD loadings over different parts of the country, depending upon temperature, which inturn, depends on the latitude of the place, as given in Table 9.11.

Table 9.11. BOD Loadings for Stabilisation Ponds

Latitude °N	*BOD_5 Loading in kg/ha/d*
8	325
12	300
16	275
20	250
24	225
28	200
32	175
36	150

The **detention period** for the stabilisation pond as stated above, varies from 7 days to 42 days or so. It can be roughly estimated by using the formula :

$$\text{Detention period in days} = \frac{1}{K_D} \log_{10}\left(\frac{L}{L-Y}\right) \quad ...(9.60)$$

where L = the BOD of the effluent entering the pond

Y = the BOD removed ; say 90% of L or 95% of L, etc.

K_D at 20°C is approximately 0.1 per day, and at other temperatures. it can be determined by eqn :

*5 day BOD of course.

$$K_{D(T^\circ)} = K_{D(20^\circ)}\,[1.047]^{T-20^\circ} \qquad \text{...[Eqn. 7.18]}$$

[**Note.** Since K_D value increases with temperature, the detention period will decrease with the increase in temperature.]

9.45.3. Results Obtained. Properly operated ponds may be as effective as trickling filters in reducing the BOD of sewage. The BOD removal is upto 90%, and coliform removal is upto 99% or so.

9.45.4. Removal of Settled Sludge. During the designed detention period in an oxidation pond, the organic solids will either be oxidised or will settle down. The average sludge accumulation observed in several ponds is about 2 to 5 cm depth per year. At a max. of this rate, sludge removal is thus required, once in 6 years for 1.2 m deep ponds, and once in 12 years for 1.5 m deep ponds, so as to ensure a minimum liquid depth of 0.3 m all the time.

9.45.5. Advantages and Disadvantages. Oxidation ponds are quite suitable in hot dry countries like India, and places where 200 or more sunny days are expected per year ; especially for small cities or towns where large land areas are cheaply available at about 0.5 to 1 km away from the habitation. The *biggest advantage* of an oxidation pond treatment is that it is very cheap, the capital cost being 10 to 30% of that of the conventional plant (using trickling filters or activated sludge process). Their maintenance cost is also very minor, and no skilled supervision is required at any stage of construction or operation. In addition, they are quite flexible, and do not get upset due to fluctuations in organic loading.

The *main disadvantage* of oxidation ponds, however, is the nuisance due to mosquito breeding and bad odours. To avoid mosquito breeding, the banks of the ponds should be kept clear of any grasses and bushes, etc. Similarly, to avoid bad odours, the oxidation ponds should be located sufficiently far from the residential areas. Odours may also be kept under control by avoiding the over-loading. However, when a pond gets over-loaded, the algae growth may be stimulated by adding sodium nitrate, which is both a plant food and an oxidising agent.

Example 9.33. *Design an oxidation pond for treating sewage from a hot climatic residential colony with 5000 persons, contributing sewage @ 120 litres per capita per day. The 5-day BOD of sewage is 300 mg/l.*

Solution. The quantity of sewage to be treated per day

$$= 5000 \times 120 = 6{,}00{,}000 \text{ litres}$$

$$= 0.6 \text{ M. litres} = 600 \text{ cu. m.}$$

The BOD content per day

$$= 0.6 \text{ Ml} \times 300 \text{ mg/l} = 180 \text{ kg}$$

Now, assuming the organic loading in the pond (in hot climates) as say 300 kg/hectare/day, we have

The surface area required

$$= \frac{180 \text{ kg/d}}{300 \text{ kg/ha.d}} = \frac{180}{300}\text{ ha} = \frac{180}{300} \times 10^4 \text{ m}^2 = 6{,}000 \text{ m}^2.$$

Assuming the length of the tank (L), as twice of its width (B), we have

$$2.B^2 = 6000$$

or $$B = \sqrt{3000} = 54.7 \text{ m} \text{ ; } \textbf{say 55 m.}$$

Use $L = \frac{6000}{55} \approx 110$ m

Using a tank with effective depth as 1.2 m ; we have

The provided capacity = 110 × 55 × 1.2 = 7260 m^3.

Now, Capacity = Sewage flow per day × Detention time in days.

∴ Detention time in days

$$= \frac{\text{Capacity in cu. m.}}{\text{Sewage flow per day in cu. m/day}}$$

$$= \frac{7260}{600} = 12.1 \text{ days ; say 12 days.}$$

Hence, use an oxidation pond with length = 110 m ; width = 55 m ; and overall depth = (1.2 + 1) = 2.2 m ; and a detention period of 12 days. **Ans.**

Design of Inlet Pipe. Assuming an average velocity of sewage as 0.9 m/sec, and daily flow for 8 hours only,

$$\text{Discharge} = \frac{600}{8 \times 60 \times 60} \text{ cumecs.}$$

∴ Area of inlet pipe required

$$= \frac{\text{Discharge}}{\text{Velocity}} = \left(\frac{600}{8 \times 60 \times 60}\right)\frac{1}{0.9} \text{ m}^2$$

$$= \frac{1}{7.2 \times 6} \text{ m}^2 = \frac{1}{43.2} \text{ m}^2 = 232 \text{ cm}^2$$

∴ Dia of inlet pipe

$$= \sqrt{\frac{4 \times 232}{\pi}} = 17.2 \text{ cm ; say 18 cm.} \quad \textbf{Ans.}$$

Dia of outlet pipe may be taken as 1.5 times that of the inlet ; say 27 cm. **Ans.**

Example 9.34. *Stabilisation ponds for a town of 3000 population are provided to operate in series. The larger cell has an area of 60,000 m^2, and the smaller one 30,000 m^2. The average daily waste flow is 900 m^3/day containing 200 kg of BOD (222 mg/l).*

(*i*) *For series operation, calculate the BOD loadings based on both the total pond area and the larger cell only.*

(*ii*) *Estimate the number days of winter storage available between 0.6 m and 1.5 m water levels, assuming an evaporation and seepage loss of 2.5 mm of water per day.*

Solution. (*i*) (*a*) ***BOD loading based on total pond area***

Total pond area of both cells joined in series

$$= 60{,}000 \text{ m}^2 + 30{,}000 \text{ m}^2 = 90{,}000 \text{ m}^2 = 9 \text{ hectares.}$$

Total BOD per day = 200 kg/day.

BOD loading in kg/ha/day

$$= \frac{200}{9} \text{ kg/day/ha} = 22.2 \text{ kg/ha/day.} \quad \textbf{Ans.}$$

(*i*) (*b*) ***BOD loading based on area of larger cell only***

Area of larger cell $= 60{,}000\ m^2 = 6$ ha.

BOD = 200 kg/day

BOD loading in kg/ha/day

$$= \frac{200}{6} \text{ kg/day/ha} = 33.3 \text{ kg/ha/day.} \quad \textbf{Ans.}$$

(*ii*) To calculate the number days of storage between *WL* 0.6 m and 1.5 m, we have

Depth available for storage = 1.5 – 0.6 = 0.9 m.

Total area $= 90{,}000\ m^2$

∴ Volume of storage available = 90,000 × 0.9 = **81,000 m³.**

Daily inflow of sewage = 900 cu. m/day

The sewage volume, which percolates and evaporates daily

$$= 2.5 \text{ mm depth}$$

$$= \frac{2.5}{10} \times \frac{1}{100} \text{ m} \times \text{surface area of tanks.}$$

$$= \frac{2.5}{1000} \text{ m} \times 90{,}000\ m^2 = 225\ m^3.$$

∴ Net effective daily inflow of sewage

$$= (900 - 225)\ m^3 = 675\ m^3/\text{day}.$$

∴ Winter storage available as days

$$= \frac{\text{Vol. of storage in } m^3}{\text{Daily net sewage inflow in } m^3/\text{day}}$$

$$= \frac{81{,}000}{875} \text{ days} = \textbf{120 days.} \quad \textbf{Ans.}$$

9.46. Oxidation Ditches (Pasveer Type) or Extended Aeration Lagoons

The *conventional activated sludge plant* has been modified to eliminate the primary sedimentation tank, and sludge digestion tank, in a process, called *Extended Aeration*, which aims at providing an aeration tank with a *longer aeration time*. The flow chart for such a process, is shown in Fig. 9.50.

A common application of this process has been made in **Pasveer ditches,** which have been widely used in Europe and elsewhere, and also in "*package plants*", available for small installations.

Such on *oxidation-ditch-plant* has been found to be economical, up to say 1.5 lakh population, compared to the conventional *activated sludge plant* or *tricking filter plant.*

Such a plant, also proves better than an ordinary simple oxidation pond, in the sense that it requires only about one-fifth of the land area of that required for an oxidation pond.

The construction of an *oxidation ditch plant*, generally involves the construction of a number of ditch channels, placed together side by side,

---

*The depth can be increased now a days to 2.5—4 m, by using *vertical axis rotor* in place of a horizontal axis cage rotor.

having depth of about 1 to 1.5 m*. The width of the ditches is limited to the type and availability of the aeration rotors used, and may vary between 1 to 5 m.

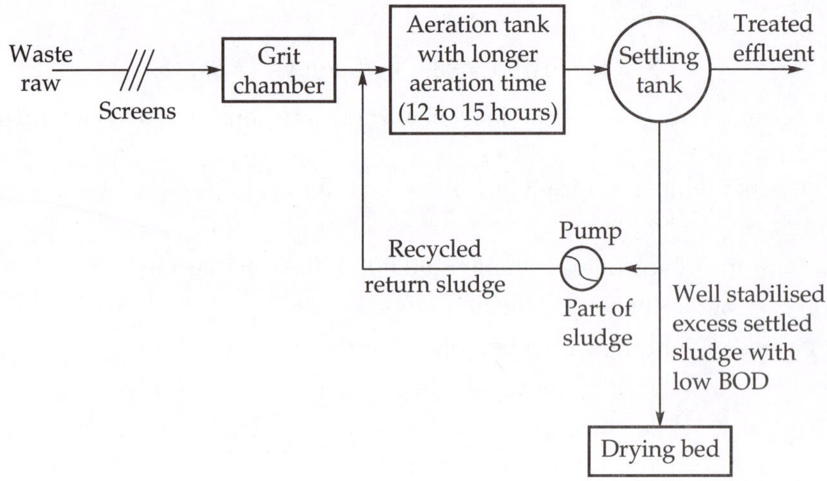

Fig. 9.50. Extended Aeration Process.

The length of the ditches is not too important, but some effort may be made to keep the perimeter of the walls to minimum, when two or more ditches are used side by side. The length may vary from, say, 150 m to 1000 m, or so.

The **oxidation ditches** may be constructed either in earthwork with earthen embankments or in brick or stone masonry with vertical walls. Brick or masonry walls are preferred, because they occupy lesser land area, and don't get eroded. Whether they are constructed in earthwork or masonry, *water-tightness is essential*, to ensure the desired immersion of the rotor.

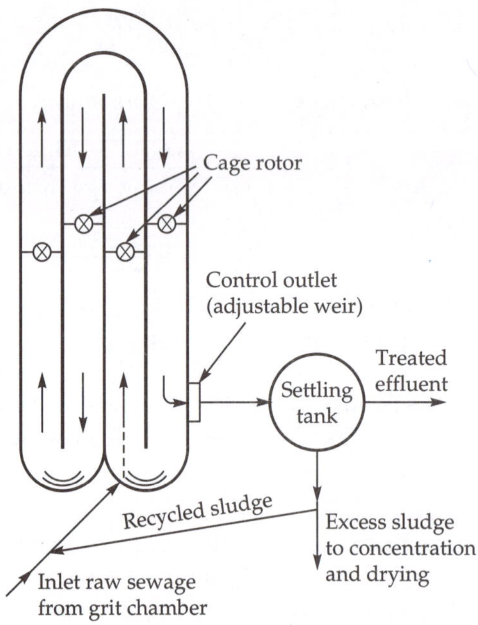

Fig. 9.51. Plan view of a typical Oxidation ditch.

Each ditch channel is usually equipped with a special type of horizontal axis rotor, which serves the purpose of agitating and circulating the sewage, and thereby oxygenating the same. The rotor, not only gives oxygen supply to the sewage, but also helps in keeping the sewage-solids in suspension (which act as carriers of aerobic bacteria), by imparting sewage a velocity of more than 0.3 m/sec. The thoroughly aerated sewage is then settled in a settling tank by stopping the rotors for about 2 hours, and then taking out the clear supernatant liquor at the time of no flow (*i.e.* during night hours in case of small communities). A part of the settled sludge is used for recirculation, and the excess settled sludge, which is well stabilised due to long detention periods (characterised by low BOD), can be easily dried on sand beds, and disposed off suitably. A typical arrangement of an oxidation ditch is shown in Fig. 9.51. A photo view of such a plant, using a number of ditch channels, as constructed at Baroda in India, is also shown in Fig. 9.52.

Fig. 9.52. Photoview of oxidation ditches constructed at Baroda (Gujarat) in 1975—the largest such ditches of India when constructed.

Design Considerations. The volume of the aeration channel required under Indian climates should be such as to give a detention period of 12 to 15 hours, or 800 to 2500 litres per kg of BOD load present in the sewage admitted. The concentration of the suspended solids in the mixed liquor should be high, say about 4000 to 5000 mg/l. The volume of the ditch approximates 120 to 150 cum per metre length of the rotor. The length of the standard horizontal axis cage type rotor used is 1 m. The rotor is in the form of a cylindrical cage of 0.7 m in diameter, and is rotated at a speed of 75 RPM, while dipping 10 cm into the liquid.

Results obtained. The quality of the effluent obtained is quite good, with suspended solids removal at about 95% and BOD removal at about 98%. The power consumed in the process, is however, more than what is required in an ordinary activated sludge process (*i.e.* about 4.4 kW per kg of BOD admitted). However, the extra cost due to this factor is compensated by elimination of primary settling tank and anaerobic digestion tank. This method however, requires skilled supervision.

The method can be advantageously used for treating small quantities of wastewaters at places where sufficient land is not available for constructing large sized oxidation ponds ; but where sufficient funds and trained personnel are available. This method is preferably used for treating wastewaters from industries, like pharmaceuticals, canning, dairy, etc., located in densely populated areas.

9.47. Mechanically Aerated Lagoons

A mechanically aerated lagoon is a deeper oxidation pond, with oxygen introduced by mechanical aerators rather than relying on the photosynthetic oxygen production alone. *As these ponds are deeper than the oxidation ponds, and as they are artificially aerated, less detention time and areas are required.*

The depth of these basins, normally ranges between 2.4 to 3.6 m, and detention time between 4 to 10 hours. The land area required is about 5 to 10% of that required for an equivalent oxidation pond. With the above detention period of 4 to 10 hours, the efficiency obtained ranges between 65 to 90%. The aerated lagoons are frequently used for treating industrial waste waters as well as city sewage.

A simple flow chart of such a lagoon is shown in Fig. 9.53.

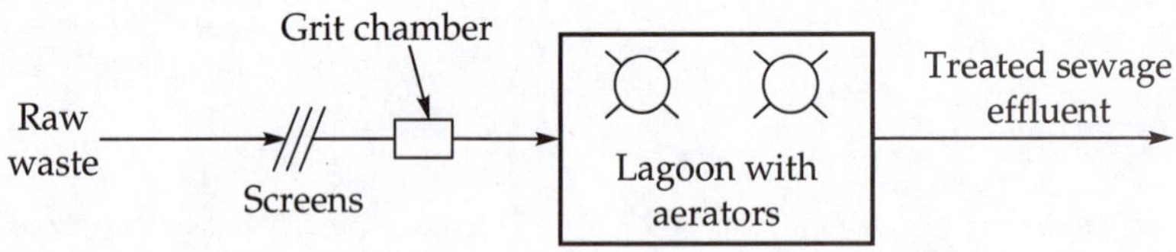

Fig. 9.53. Flowsheet for a mechanically aerated lagoon.

In India, aerated lagoons have been used in Mumbai city at *Ghatkopar Sewage Treatment Plant* of 690 MLD (150 MGD) capacity. At this plant, the city's sewage from Ghatkopar, Kurla, Vikhroli, Deonar area is received at the plant through a tunnel to the wet well of Influent Sewage Pumping Station (IPS), where 5 No. sewage pumps of 138 MLD (30 MGD) capacity each, are installed. The sewage is pumped from the wet well to the inlet of S.T.P. The lifted sewage is made to pass through 4 No. semi automatic screens of 1600 × 2000 mm size to arrest floating solids and to protect downstream units. The screening materials are disposed off by dumping in low lying area. The screened sewage is made to pass through 4 No. **aerated grit chambers** to settle out inorganic particles of sp. gravity = 2.6 and of size 0.2 mm and above. The spiral current within the grit is created by the use of defused air, being let into the grit channels through a compressor. The air supply rate is adjusted to create the velocity in the bottom, low enough to allow the grit to settle. The settled grit is removed by **Grab and Crane machinery** (Pl. see **Colour Photo Fig. 9.1** of the aerated grit chamber) and disposed off regularly. Each of the four rectangular grit channels, at its end length, are provided with a parshall flume section chamber, to measure discharge.

The sewage from the grit chamber through partial flume chamber finally enters into 4 Nos. of **aerated lagoons** (Pl. see **Colour Photo Fig. 9.2** of the aerated lagoon). There are 4 such ponds, each of size 230 m × 160 m × 4.5 m with 16 floating aerators of 50 HP installed in each pond. The detention period in each aerated lagoon is 43 hour (1.8 day). *The aeration increases D.O. in sewage and accelerates microbiological process of bacteria.* The settled sludge is required to be removed once in 6-8 years. The final effluent is let out into a creek.

The characteristic parameters of the influent and effluent sewage at this STP are as follows :

Parameters of Influent Sewage

BOD_5 = 170-200 mg/L

COD = 300 mg/L

TSS = 260 mg/L

Photo Fig. 9.2. Photoview of one of 4 aerated oxidation lagoons (ponds) at Ghatkopar Sewage Treatment Plant, Mumbai

Photo Fig. 9.1 (a) Enlarged photoview of two of the four grit channels (entry side) provided after the fine screens at the Ghatkopar Sewage Treatment Plant, Mumbai.

Photo Fig. 9.1 (b) Photoview of the two of the four grit channels (outfall end side) provided after the fine screens at the Ghatkopar Sewage Treatment Plant, Mumbai.
Grab and Crane machine arrangement provided to periodically remove the settled grit can be seen at the top of the photo.

Parameters of Treated Effluent Sewage

BOD_5 = 30-40 mg/L

COD = 50-60 mg/L

TSS = 30 mg/L

ANAEROBIC STABILISATION UNITS

9.48. Anaerobic and Facultative Stabilisation Ponds

As stated under article 9.45, the deeper stabilisation ponds, usually operating under the action of anaerobic bacteria, are called the *anaerobic stabilisation ponds*. These ponds generally receive untreated sewage, and are deep enough to restrict the penetration of light to reduce the algae growth to a negligible level. The photosynthesis, which primarily causes stabilisation of wastes in an oxidation pond, can be reduced by decreasing the surface area and increasing the depth of the pond. Deeper ponds, usually in the depth range of 2.5 to 4.0 m are, therefore, sometimes used for anaerobic decomposition of wastes. The anaerobic conditions are further ensured by applying BOD load that exceeds oxygen production from photosynthesis.

Anaerobic treatment of complex wastes involves *two distinct stages*. In the first stage, known as ***acid fermentation,*** complex organic materials are broken down mainly to short-chain acids and alcohols. In the second stage, known as ***methane fermentation,*** these materials are converted to gases, primarily methane and carbon dioxide. The proper design of anaerobic ponds must result in environmental conditions favourable to methane fermentation.

Anaerobic ponds may primarily be used as a pre-treatment process, and are particularly suited for the treatment of high temperature, high strength industrial wastewaters. However, they have been used successfully to treat municipal wastewaters as well. *Such ponds are, however, generally avoided, unless the process is used in a close bodied tank, due to evolution of foul odours in the process of anaerobic decomposition of wastes.*

9.49. Septic Tanks

A septic tank may be defined as *a primary sedimentation tank**, with a longer detention period (12 to 36 hours, against a period of 2 hours in an ordinary sedimentation tank), and *with extra provisions for digestion of the settled sludge.* Since the digestion of the settled sludge is carried out by anaerobic decomposition process, *the septic tank unit is generally classified under the units which work on the principle of anaerobic decomposition.* Moreover, due to anaerobic decomposition of the settled sludge, foul gases will be evolved in the tank, and as such, a septic tank will be a completely covered tank, provided with a high vent shaft for escape of gases.

A septic tank, is thus, a horizontal continuous flow type of a sedimentation tank, directly admitting raw sewage, and removing about 60 to 70% of the dissolved matter from it. The effluent from such a tank will be sufficiently foul in nature, and will have to be disposed of either for sub-surface irrigation or in cess-pools or soak pits, or to be treated in trickling filters before disposing it of

*Horizontal continuous flow type of sedimentation tank, of course.

in water courses, as will be discussed a little later. *The sludge settled at the bottom of the tank, and the oils and greasy matter rising to the top surface of sewage as scum, are allowed to remain in the tank for a period of several months*, during which they are decomposed by the anaerobic bacteria to form gases and liquids by the process of sludge digestion. The resultant sludge so obtained, will consequently be much less than that obtained in ordinary settling tanks (the quantity is reduced because of digestion). The digested sludge from the tank is periodically (at intervals of 6 to 12 months, and in no case exceeding 3 years) removed, and taken to a nearby sump, from where it may be pumped up to the ground, and disposed of in a sanitary manner, preferably after drying it.

Septic tanks are generally provided in areas where sewers have not been laid and for catering to the sanitary disposal of sewage produced from isolated communities, schools, hospitals, hotels, other public institutions, etc.

9.49.1. Construction Details. The plan and section of a typical septic tank are shown in Fig. 9.54. The tank is designed to prevent direct currents between the tank inlet and outlet. This ensures effective sedimentation, and is achieved by using pipe tees with submerged ends as inlet and outlet, as shown in Fig. 9.54 or alternatively, a better arrangement is to provide hanging baffles in front of inlet and before the outlet, as was shown in sedimentation tank of Fig. 9.12(*a*). The T-pipe or the baffle at the outlet will also help in retaining the *scum* (*i.e.*, oily and greasy matter floating to the surface, due to evolution of gases produced by anaerobic decomposition of sludge at bottom) in the tank.

It may be noted that holding of scum in the tank not only frees the effluent from it, but is also desirable to hold back odours and to create a sort of heat insulation which aids the bacterial action. The tank is generally covered at top with a R.C.C. slab. Manhole covers are provided in the top slab, so as to permit inspection and maintenance. Sometimes, stairs or cast iron steps may also be provided to go into the tank, if needed. The R.C.C. cover prevents the escape of foul gases and odours from the tank on to the ground ; prevents the chances of accidents due to falling of children or animals : and prevents wind agitation, thus keeping the sewage warm, and helping in bacterial activity. The foul gases produced by the anaerobic decomposition of deposited sludge are separately removed through a vent pipe projecting sufficiently high above the ground, so as not to cause any nuisance to the neighbouring areas. The sludge deposited and digested, is removed at an interval of 6 to 12 months, into a side sump-well through a pipe under hydrostatic head or with the help of a portable pump. In rural areas, and in case of small septic tanks, the sludge may be removed by manual labour. The sludge from the sump well is then lifted up by pumps, etc. ; and disposed of suitably after drying, as pointed out earlier.

9.49.2. Design Considerations. The septic tanks will be designed like ordinary sedimentation tanks with the following data :

(*i*) ***Capacity of Septic Tanks.*** The volume of liquid which a septic tank can accommodate is called its capacity. A septic tank should be capable of storing the sewage flow during the detention period, and an additional volume of sludge for 6 months to 3 years, depending upon the periodicity of cleaning.

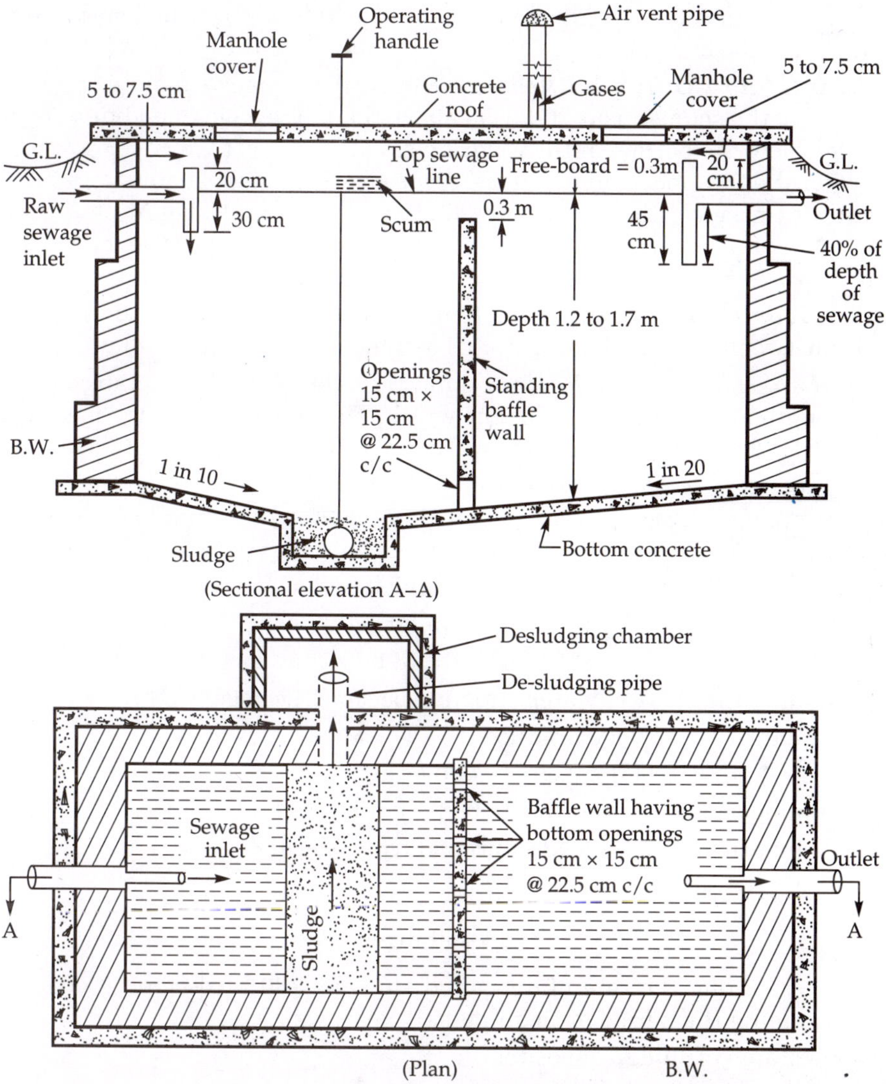

Fig. 9.54. Septic Tank.

If only water closets are connected to the septic tank, the sewage flow will be about 40—70 litres/capita/day. But when sullage is also discharged in to the septic tank, the sewage flow may be as much as 90 to 150 litres/capita/day. Where a large quantity of sludge is expected, it may separately be disposed of in the manner as the effluent from the septic tank.

The sludge and the scum will accumulate in the tank at varying rates, depending upon the characteristics of raw sewage, and the efficiency of the sedimentation and digestion. *The rate of accumulation of sludge has been recommended as 30 litres/person/year.*

The minimum capacity of a septic tank for about 8 to 10 persons, may be kept 2,250 litres (when all liquid wastes are discharged into the tank) and 1,400 litres (when only water closet wastes are discharged).

Free-board. A free-board of about 0.3-0.5 m may be provided above the top sewage line in the tank. This will help accommodate the scum in the tank.

(*ii*) ***Inlet and Outlet baffles,*** *etc.* The baffles or tees should extend up to top level of the scum (about 20-22 cm above the top sewage line), but must stop a little below the bottom of the covering slab (by at least 7.5 cm or so) so as to allow for the free-movement of gases. Moreover, the inlet should penetrate by about 30 cm below the top sewage line, and the outlet should penetrate to about 40% of the depth of sewage. Further, the outlet invert level should be kept 5 to 7.5 cm below the inlet invert level.

(*iii*) ***Detention Period.*** The detention period for a septic tank generally varies between 12 to 36 hours, but is commonly adopted as 24 hours.

(*iv*) ***Length to Width Ratio.*** Septic tanks are usually rectangular with their length at about 2 to 3 times the width. The width should not be less than 90 cm. The depth of the tank generally ranges between 1.2 to 1.8 m.

Example 9.35. *Design the dimensions of a septic tank for a small colony of 150 persons provided with an assured water supply from the municipal head-works at a rate of 120 litres per person per day. Assume any data, you may need.*

Solution. The quantity of water supplied

= Per capita rate × Population

= 120 × 150 litres/day = 18,000 *l*/day.

Assuming that 80% of water supplied becomes sewage, we have

The quantity of sewage produced

= 18,000 × 0.8 = 14,400 *l*/day.

Assuming the detention time to be 24 hours, we have

The quantity of sewage produced during the detention period (*i.e.*, the capacity of the tank)

$$= 14,400 \times \frac{24}{24} = 14,400 \text{ litres.}$$

Now assuming the rate of deposited sludge as 30 litres/capita/ year ; and also assuming the period of cleaning as 1 year, we have

The volume of sludge deposited

= 30 × 150 × 1 = 4,500 litres.

∴ Total required capacity of the tank

= Capacity for sewage + Capacity for sludge

= 14,400 + 4,500 = 18,900 litres = **18.9 cu-m.**

Assuming 1.5 m as the depth of the tank, we have

The surface area of the tank

$$= \frac{18.9}{1.5} \text{ m}^2 = 12.6 \text{ m}^2.$$

If the ratio of the length to width is kept as 3 : 1, we have

$$3 \,.\, B^2 = 12.6$$

or $$B = \sqrt{\frac{12.6}{3}} = \sqrt{4.2} = 2.05 \text{ m ; say } \mathbf{2.1 \text{ m.}}$$

∴ Provide width = 2.1 m ; and

Provide length of the tank = 6 m.

∴ Area of cross-section provided

$$= 6 \times 2.1 = \mathbf{12.6\ m^2} \text{ (same as required)}$$

Thus, the dimensions of the septic tank will be

$$6 \text{ m} \times 2.1 \text{ m} \times (1.5 + 0.3) \text{ m overall depth}$$

[0.3 m used as free-board]

Hence, use a tank of size 6 m × 2.1 m × 1.8 m. **Ans.**

9.49.3. Disposal of the Effluent from the Septic Tank. The effluent coming out from a septic tank is no better than the effluent of an ordinary sedimentation tank. It contains large amount of putrescible organic matter (200 to 250 mg/*l*), and its BOD is quite high (100 to 200 mg/*l*). This effluent should, therefore, be disposed of carefully, so as to cause minimum nuisance or risk to the health of the people.

The following three methods of disposal of septic tank effluent are usually adopted :

(*i*) *Soil absorption system* ;

(*ii*) *Biological filters* ; and

(*iii*) *Upflow anaerobic filters.*

These methods are discussed below :

9.49.3.1. Soil absorption system. The soil absorption system involves the disposal of effluent on land, and can be adopted only when sufficient land is available and the soil is sufficiently porous as to give a *percolation rate* not exceeding 60 minutes (per cm).

The **percolation rate** of a soil or a ground is defined as *the time in minutes required for seepage of water through that ground by 1 cm.* Higher percolation rate naturally would reflect less porous soil.

In case, sufficiently porous ground is not available, the effluent of the septic tank shall be subjected to a secondary treatment either in a *biological filter* or in an *upflow anaerobic filter*, before its final disposal either in a city sewer or drain, or for gardening purposes.

The **soil absorption system** may be of the following two types :

(*a*) *seepage pit* or *soak pit* ; and

(*b*) *dispersion trench.*

Both the above systems are explained below :

(*a*) ***Disposal in Soak Pits.*** A soak pit is a circular covered pit, through which the effluent is allowed to be soaked or absorbed into the surrounding soil. The soak pit (or sewage pit) may either be filled with stone aggregate as shown in Fig. 9.55 (*a*) or may be kept empty as shown in Fig. 9.55 (*b*). When the soak pit is empty, the pit is lined with brick, stone or concrete blocks with dry open joints. In addition to this, this lining is supported below the inlet level by at least 7.5 cm thick backing of coarse aggregate, as shown. The lining above the inlet level should be plastered with cement mortar, as shown. However, when the soak pit is filled with stone or brick aggregate, no lining is required except for a top masonry ring constructed to prevent damage by

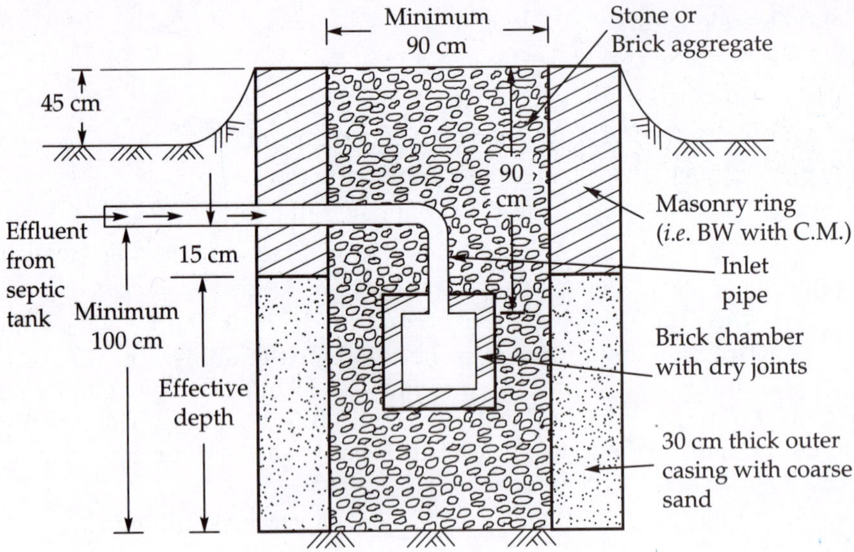

Fig. 9.55. (*a*) Unlined soak pit filled with stone or brick aggregates.

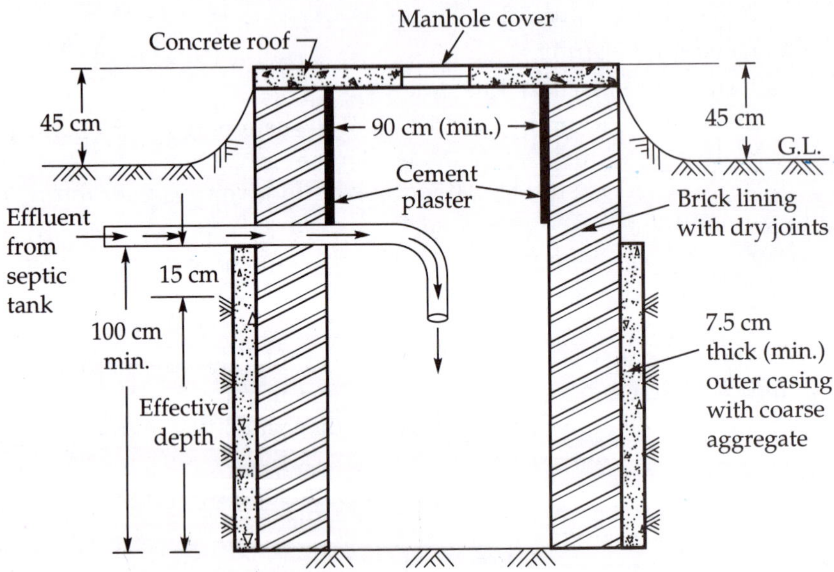

Fig. 9.55. (*b*) Lined soak pit.

flooding of the pit by surface run off [Fig. 9.55 (*b*)]. The inlet pipe may be taken down to a depth of 0.9 m from the top as an anti-mosquito measure.

For disposal in soak pits, the underground soil should be highly porous with percolation rate not exceeding 30 minutes.

(*b*) ***Disposal in Absorption Trenches.*** In this method, the septic tank effluent is allowed to enter into a masonry chamber (called distribution box), from where it is uniformly distributed through an underground network of open jointed pipes into absorption trenches, called the **dispersion trenches**, as shown in Fig. 9.56.

The suspended organic matter present in the effluent will be absorbed in the absorption trenches, which are filled with gravel and well graded aggregate, as shown (Fig. 9.56). The clearer water will seep down to the water-table. Plants are usually grown on the top of the absorption trenches, which get their irrigation water requirement fulfilled by capillarity from the seeping water in the absorption trenches.

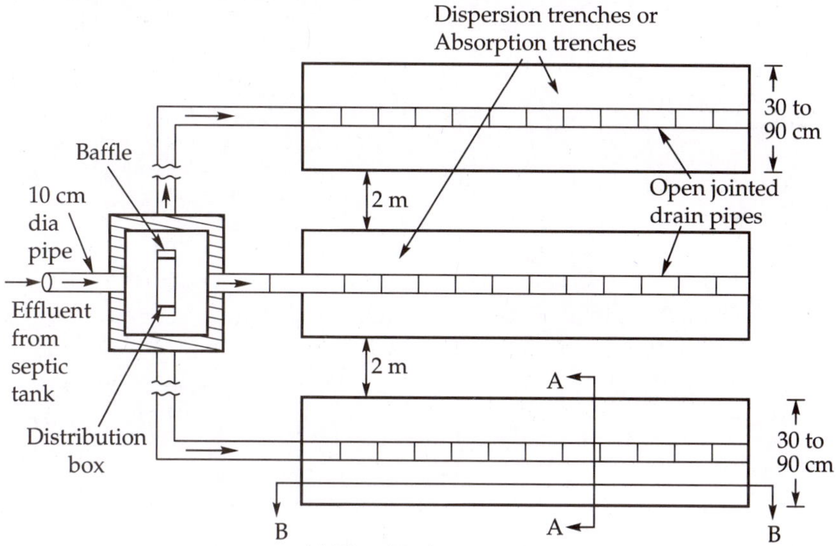

(*a*) Plan (Underground)

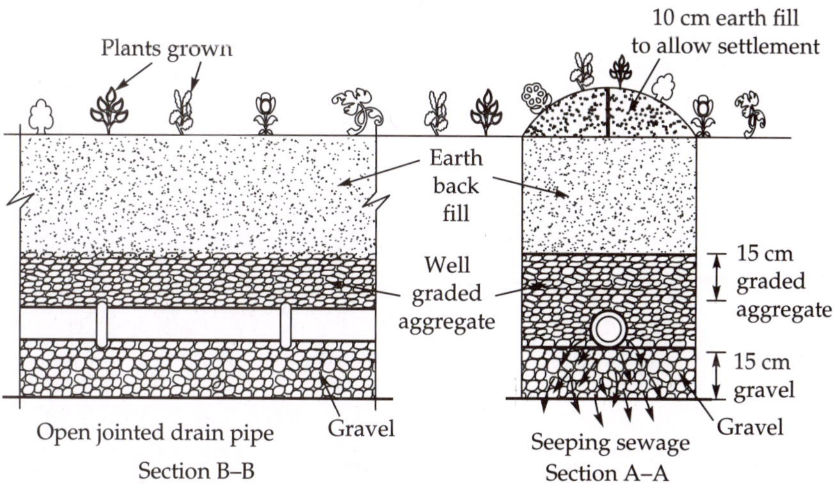

(*b*) Section of Dispersion Trench

Fig. 9.56. Absorption trench method for disposal of septic tank effluent.

Dispersion trenches are not recommended in areas where fibrous roots of trees or vegetation are likely to penetrate the system and cause blockages.

Dispersion trenches may be adopted on soils having percolation rate not exceeding 60 minutes.

The minimum absorption area of the pits or trenches required, shall be worked out on the basis of the max. allowable rate of effluent application (q), which can be computed by using the empirical equation

$$q = 130\sqrt{t} \qquad ...(9.61)$$

where q = max. rate of effluent application in l/m^2/day of leaching surface

t = standard percolation rate in minutes

The standard percolation rate or percolation rate of the ground or soil is determined in a square or a circular test hole at the level of the proposed soak pit or trench. The test hole width (or die) may be 10 to 30 cm. The rate is defined as the time in minutes required for the water to fall 1 cm. This rate is determined for the drop that occurs in the final 10 minutes of the test during a total 1 hour period of the test. At the start of the test, 30 cm water depth is impounded for 24 hour in the test hole (which is filled with 5 cm depth of gravel) to allow full swelling of soil to approximate its condition during rainy season.

The value of max. allowable rate of effluent application, as obtained from Eq. (9.60) which is given by "Manual on Sewage and Sewage Disposal" Govt. of India, is generally found to be on higher side ; and hence BIS has, vide IS : 2470 (Part II) 1985, given another empirical equation for determining this rate, as :

$$q = \frac{204}{\sqrt{t}} \qquad ...(9.62)$$

where q and t have the same meaning as in Eq. (9.61).

The allowable rates of effluent application for certain selected values of percolation rates are given in table 9.12.

Table 9.12. Allowable Rates of Effluent Applications to Soil Absorption Systems

Standard Percolation rate in minutes	*Maximum Rate of Effluent Application in $l/m^2/d$.*
1 or less	204
2	143
3	118
4	102
5	90
10	65
15	52
30	37
45	33
60	26

Note 1. The absorption area for a dispersion trench is the trench bottom area.

Note 2. The absorption area for soak pits is the effective side wall area, effective depth being measured from 150 mm below invert level of inlet pipe to the bottom of the pit (Fig. 9.55).

Note 3. If the percolation rate exceeds 30 minutes, the soil is unsuitable for soak pits. If the percolation rate exceeds 60 minutes, the soil is unsuitable for any soil absorption system.

Location of Sub-Surface Absorption System. A sub surface dispersion system shall not be installed closer than 18 m from any source of drinking water, such as a well, to mitigate the possibility of bacterial pollution of water supply. It shall also be as far away from the nearest habitable building as is economically feasible but not closer than 6 m, to avoid damage to the structures. The actual distance, however, shall be based on the soil conditions in relation to both percolation and bearing capacity. Care should be taken to see that the ground below the adjacent building is not likely to be affected by the effluent seeping into the soil.

In limestones or creviced rocks, the soil absorption system is not recommended, as there may exist channels in such formations, which may carry contamination over long distances.

9.49.3.2. Biological filters. Biological filters are suitable for treatment of septic tank effluent where the soil is relatively impervious (percolation rate > 60 minutes), or in water-logged areas, or where limited land area is available. In a biological filter, the effluent from septic tank is brought into contact with a suitable medium, the surfaces of which become coated with an organic film. The film assimilates and oxidises much of the polluting matter through the agency of micro-organisms.

The biological filter requires ample ventilation and an efficient system of under-drains leading to an outlet, as discussed earlier.

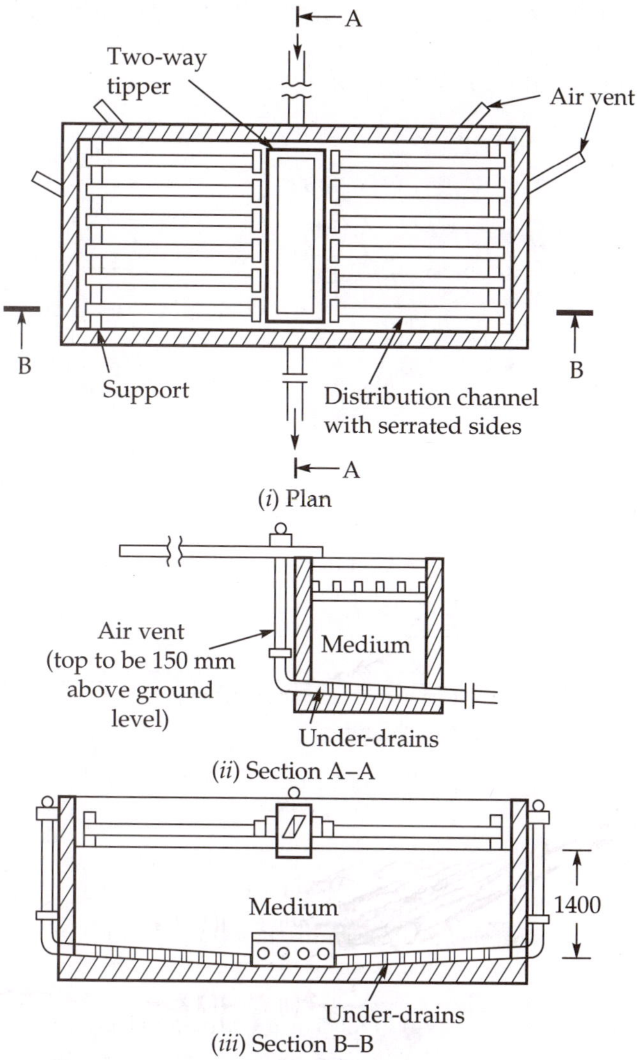

Note. Flexible joints may be required on inlet or outlet connections, where rigid pipes are used.

Fig. 9.57. (*a*) Rectangular biological filters.

The depth of medium should be 1.4 m but never less than 0.9 m. The medium should be retained in position by walls of adequate trench. The filter should have a concrete floor, with a system of under-drains laid on it and consisting of half channels laid upside down and open-jointed, or special tiles discharging to the outlet.

Biological filters may either be rectangular or circular as shown in Figs. 9.57 (*a*) and (*b*) respectively. A series of fixed channels or rotating arm distributors may be used for distributing the effluent on the media. Adequate ventilation of biological filters is essential. Air vents communicating with the floor level of the filter should be provided. Where the filter is below the ground level, the vent pipes from the ends of the under-drains should be carried to 150 mm above the ground level outside the filter. Normally the filter should not be covered, but wire netting may be used to prevent falling of leaves fouling the surface of the filter or blocking the ends of the vent pipe.

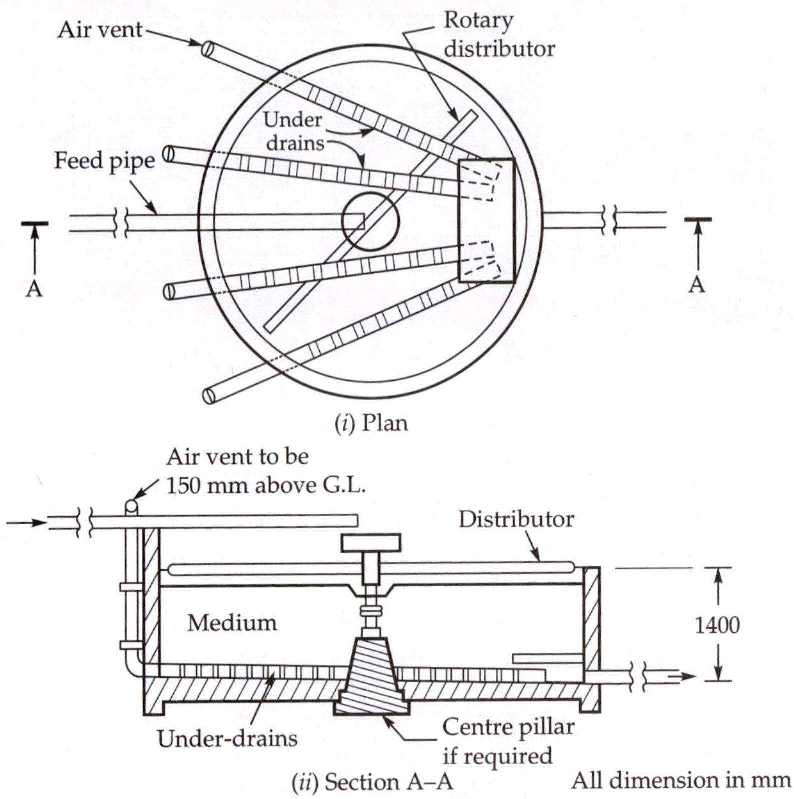

Note. Flexible joints may be required on inlet or outlet connections, where rigid pipes are used.

Fig. 9.57 (*b*) Circular biological filters.

It is essential that the *volume of the filter medium* is sufficient to allow for sewage flow which occurs with small installations, such variation being more pronounced if the smaller number of persons are served. For populations of upto 10 persons, the volume should be 1 m^3 of medium per head ; for resident population of over 10 and up to 50 persons, 0.8 m^3 of medium per head ; and for population of over 50 person the volume of media may be taken @ 0.6 m^3/head.

If pumping of the septic tank effluents forms part of the scheme, recirculation of final effluent to dilute the septic tank effluent may be introduced to reduce the volume of the filter.

The filter effluent is finally discharged into the city surface drain or sewer, or is evenly distributed over a grass plot from the system of channels. Where the effluent is likely to contaminate the water course, the effluent should be adequately disinfected.

9.49.3.3. Upflow anaerobic filters. The upflow type of filters (reverse filter) operating under submerged condition, as discussed in details in article 9.52, are also used for giving secondary treatment to the effluent of septic tank, before its final disposal. Such filters like biological filters, are used in areas which donot permit land disposal, such as in areas of high water-table, or in comparatively less porous areas, or at places where sufficient land is not available.

In such a filter, the septic tank effluent is introduced from the bottom and the microbial growth is retained on the stone media making possible higher rates and efficient digestion. The capacity of the unit is 0.04 to 0.05 m^3 per capita, or 1/3 to 1/2 the liquid capacity of the septic tank it serves. BOD removals of 70% can be expected and the effluent is clear and free from odour and nuisance.

Single and double chambered type of rectangular upflow anaerobic filters, as shown in Figs. 9.58 (*a*) and (*b*) respectively, are generally used in conjunction with septic tanks.

In a ***single chambered rectangular upflow tank*** [Fig. 9.58 (*a*)], the chamber is packed with coarse material and the size of the packing media should be 20 mm. The stone media rests on a perforated concrete false bottom slab. The effluent from the septic tank enters the bottom of the filter chamber through a 150 mm dia pipe, and is distributed upward through the media from a perforated slab at the bottom.

The vertical inlet is fitted with a tee at the bottom, one branch of which leads to the filter and the other branch is kept plugged while the filter functions. The plug can be removed to facilitate emptying into an adjoining chamber and cleaning the filter where required. The effluent from the top of the bed is allowed to escape over a *V*-notch. The sill level is kept 150 mm above the top of the medium.

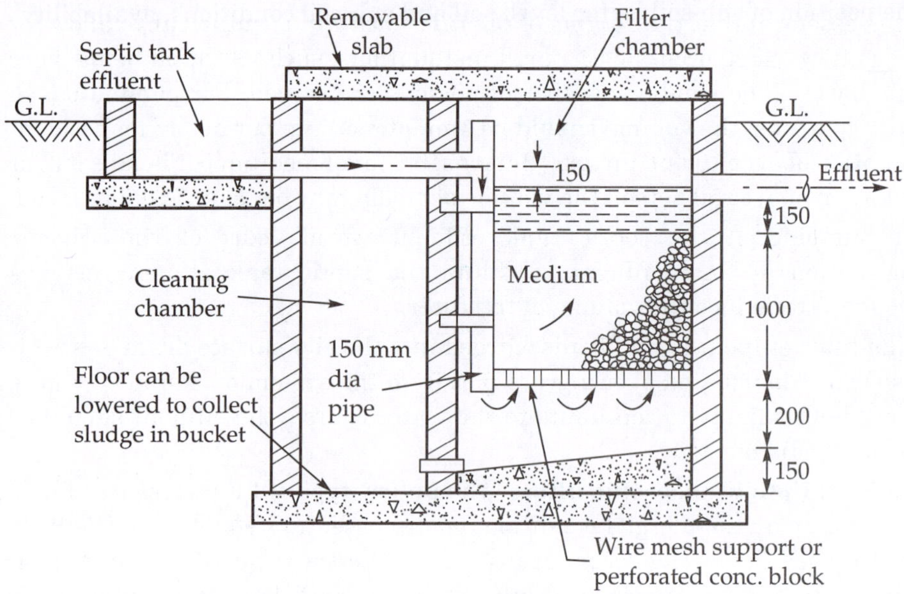

Fig. 9.58. (*a*) Single chambered rectangular upflow anaerobic filter.

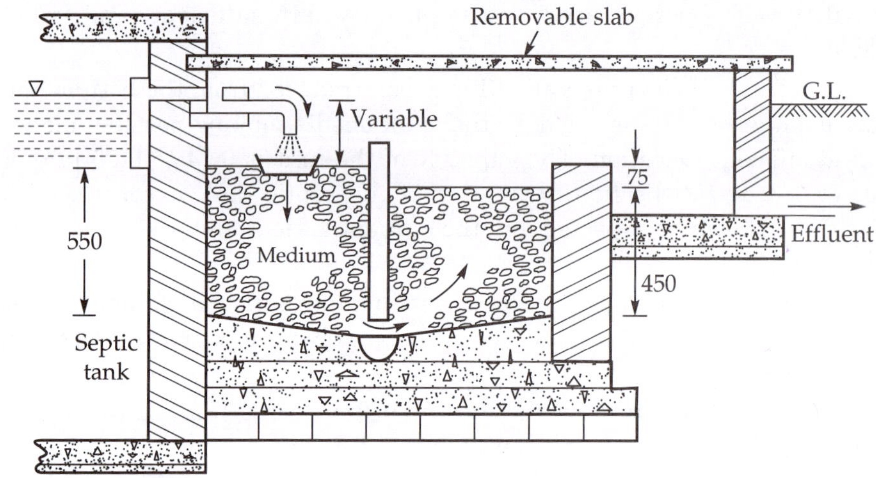

Fig. 9.58. (*b*) Double chambered rectangular upflow anaerobic filter.

In a ***double-chambered rectangular type,*** the filter consists of two inter-connected compartments. [Fig. 9.58 (*b*)]. The first chamber is filled to a depth of 0.55 m with 20 mm coarse medium. The second chamber is filled to a depth of 0.45 m with 20 mm size coarse aggregate. The septic tank effluent falls through a perforated tray over the medium in the first compartment and enters the second compartment directly from the bottom. The effluent passes up through the medium in the second chamber and escapes over a *V*-notch placed 75 mm above the top of the medium. By this arrangement, the time of travel of the effluent through the filter is lengthened.

9.49.3.4. Adoption of a particular method of disposal of septic tank effluents. The use of a particular method of disposal will usually be governed by the position of sub-soil water level, soil and sub-soil conditions, availability of sufficient land, etc. The recommended method of disposal for site conditions is shown in table 9.13.

Table 9.13. Recommended Method of Disposal for Effluents of Septic Tanks

Position of Subsoil water level (SSWL) below the ground level	*Soil and Subsoil Condition*		
	Porous soil with percolation rate		*Dense and clayey soils with percolation rate exceeding 60 min.*
	not exceeding 30 min.	*exceeding 30 min. but not exceeding 60 min.*	
≤ 1.8 m	*Dispersion trench* loca- ted partly or fully above the ground level in a mound.	*Dispersion trench* loca- ted partly or fully above ground level in a mound.	*Biological filter* partly or fully above ground level with under drains; or *Upflow anaerobic filter*. The final effluent is discharged into a city sewer or a drain,, or used for gardening purposes.
> 1.8 m	*Soak pit* or *Dispersion trench*	*Dispersion trench*	*Sub-surface biological filter* with under-drains or *Upflow anaero-bic filter*. The final effluent is discharged into a city sewer or used for gardening purposes.

Example 9.36. (*a*) *Design a septic tank for the following data :*

No. of people = *100*

Sewage/capita/day = *120 litres*

De-sludging period = *1 year*

Length : width = *4 : 1.* (*A.M.I.E., Section B, 1976*)

(*b*) *What would be the size of its soak well if the effluent from this septic tank is to be discharged in it. Assume percolation rate through the soak well to be 1250* $l/m^3/d$.

Solution. Quantity of sewage produced per day = 12,000 litres/day.

Assuming the detention period to be 24 hours, we have

The quantity of sewage produced during the detention period, *i.e.* the capacity of the tank

$$= 12{,}000 \times \frac{24}{24} = 12{,}000 \text{ litres.}$$

Now assuming the rate of sludge deposit as 30 litres/capita/year, and with the given 1 year period of cleaning, we have

The quantity of sludge deposited

$$= 30 \times 100 \times 1 = 3{,}000 \text{ litres.}$$

Total required capacity of the tank

$$= 12{,}000 + 3{,}000 = 15{,}000 \text{ litres.} = 15 \text{ cu-m.}$$

Assuming the depth of the tank as 1.5 m, the cross-sectional area of the tank

$$= \frac{15}{1.5} = 10 \text{ m}^2$$

Using $L : B$ as 4 : 1 (given), we have

$$4B^2 = 10$$

or $$B = \sqrt{2.5} = 1.5 \text{ m.}$$

$$L = 4 \times 1.5 = 6 \text{ m.}$$

∴ The dimensions of the tank will be 6 m × 1.5 m × (1.5 + 0.3 m) as overall depth with 0.3 m free-board. Hence, use a tank of size 6 m × 1.5 m × 1.8 m. **Ans.**

(*b*) *Design of Soak-well*

The soak-well or soak-pit can be designed by assuming the percolating capacity of the filtering media as 1250 litres per cu-m per day (as given).

Sewage outflow = 12000 *l*/d

Percolation rate = 1250 $l/\text{m}^3/\text{d}$

∴ Volume (of filtering media) required for the soak-well

$$= \frac{12000 \; l/\text{d}}{1250 \; l/m^3/\text{d}} = 9.6 \text{ m}^3$$

If the depth of the soak well is taken as say 2 m, then

Area of soak-well required

$$= \frac{9.6}{2} = 4.8 \text{ m}^2.$$

∴ Dia of soak-well required

$$= \sqrt{\frac{4.8 \times 4}{\pi}} = 2.47 \text{ m ; } \textbf{say 2.5 m.} \quad \textbf{Ans.}$$

Example 9.37. *Estimate the size of a septic tank (length to width ratio 2.25, liquid depth 2 m with 300 mm freeboard), desludging intervals in years, and the total trench area (m^2) of the percolation field, for a small colony of 300 people. Assume water supply of 100 litres per capita per day, waste water flow at 80% of water consumption, sludge production of 0.04 cubic metre per capita per year, and the retention time of 3 days at start up. Desludging is done when the tank is one-third full of sludge. A percolation test indicated an allowable hydraulic loading of 100 L per square metre per day.* (*GATE—1995*)

Solution.

(*i*) Size of the tank = ?

(*ii*) Desludging interval in years = ?

(*iii*) Total trench area of perpocolation fields in m² = ?

Given : $$\frac{L}{B} = 2.25$$

$$D_w = 2 \text{ m}$$

Free-board = 0.3 m

Population = 300

Water supply = 100 L/c/day
Waste flow = 80% of water supplied
Sludge production = 0.04 m^3/c/year
Retention time = 3 days
Using the above data, we have
Water supply to the colony = 100 L/c/d × 300 persons = 30,000 L/d
Sewage produced per day = 80% × 30,000 L/d = **24,000 L/d.**
Sewage produced in 3 days (*i.e.* during retention period at start up)

= 3 × 24,000 L/d = 72,000 L = **72 m³.**

Deslugding is done when the tank is filled up to $\frac{1}{3}$ rd of its capacity (C).

Hence, sludge volume collected is $\frac{C}{3}$.

$\therefore$ Capacity (C) = Max. Sewage volume retained + Sludge volume retained

or
$$C = 72 \text{ m}^3 + \frac{C}{3}$$

$\therefore$
$$\frac{2}{3} C = 72 \text{ m}^3$$

$$C = 72 \times \frac{3}{2} = \mathbf{108\ m^3}$$

Hence, the capacity of the tank = **108 m³**

But $C = L \times B \times D_w$

$= (2.25B)(B)(2 \text{ m}) = 108 \text{ m}^3$

$\therefore$ $B^2 = \frac{108}{4.5} = 24$

$B = \mathbf{4.9\ m}$

$L = 2.25 \times 4.9 = \mathbf{11.1\ m}$ (say)

(*i*) Hence, the **tank size**

= **11.1 m × 4.9 m × (2 + 0.3 m) depth. Ans.**

Sludge volume removed in desludging = $\frac{C}{3} = 36 \text{ m}^3$.

Sludge produced per year = $0.04 \frac{\text{m}^3}{\text{capita.year}} \times 300$ persons

= 12 m^3/year

36 m^3 of sludge will therefore be produced in

$= \frac{1}{12} \times 36$ years = 3 years.

(*ii*) Hence, **desludging interval = 3 years. Ans.**

Hydraulic loading of percolation trench

= 100 L/m^2/day.

Outflowing sewage per 1 day = 24 m^3 = **24,000 L/d.**

(*iii*) ∴ *Trench area reqd.* $= \dfrac{24{,}000 \text{ L/d}}{100 \text{ L/m}^2\text{/d}} = \mathbf{240 \ m^2}$. **Ans.**

Example 9.38. *Design the absorption field system for the disposal of septic tank effluent for a population of 100 persons with sewage flow rate of 135 lpcd. The percolation rate for the percolation test carried out at the site of the absorption field may be taken as 3 minutes.* (*Civil Services, 1997*)

Solution. Total sewage flow

$$= 135 \text{ lpcd} \times 100 \text{ persons} = 13{,}500 \text{ l/day}$$

For dispersion trench (Fig. 9.56), the maximum rate of effluent application, is given by BIS Eq. (9.61) as :

$$q = \frac{204}{\sqrt{t}} \text{ l/d/m}^2, \text{ where } t \text{ is in minutes} = 3 \text{ min}$$

$$\therefore \qquad q = \frac{204}{\sqrt{3}} = 117.78 \text{ l/d/m}^2$$

$$\therefore \quad \text{Area of Trench required} = \frac{13{,}500}{117.78} \text{ m}^2 = 114.62 \text{ m}^2$$

Using 0.75 m bottom width of the trench, the length of the trench

$$= \frac{114.62}{0.75} \approx 153 \text{ m.}$$

Depending upon the availability of space, we may use say 3 no. trenches, each of length $\dfrac{153}{3} = 51$ m, as per Fig. 9.56. **Ans.**

9.49.4. Advantages and Disadvantages of Septic Tanks. They are enumerated below :

Advantages

(*i*) Septic tanks can be easily constructed, and do not require any skilled supervision during construction. Moreover, there is no maintenance problem (except periodical cleaning) as there is no moving part in it.

(*ii*) Their cost is reasonable compared to the advantages and sanitation they offer in rural or semi-urban areas, where no sewage system has been laid.

(*iii*) An excellently functioning septic tank can considerably reduce the suspended solids and BOD from sewage.

(*iv*) The sludge volume to be disposed of is quite less, as compared to that in a normal sedimentation tank. The quantity is reduced due to digestion taking place in the tank itself. The reduction in volume is about 60%, and reduction in weight is about 30%.

(*v*) The effluent from the septic tank can be disposed of on land in a soak-pit or a cess-pool, without much trouble.

(*vi*) They are best suited for isolated rural areas, and for isolated hospitals, buildings, etc.

Disadvantages

(*i*) If the tank is not properly functioning, which happens many a times, then the effluents will be very foul, dark, and even worser than the influent.

(*ii*) They require too large sizes for serving many people.

(*iii*) Leakage of gases from the top cover of septic tank may cause bad smells and environmental pollution.

(*iv*) Periodical cleaning, removal and disposal of sludge remains a tedious problem.

(*v*) The working of a septic tank is unpredictable and non-uniform.

9.50. Imhoff Tanks

An Imhoff tank is an improvement over septic tank, in which the incoming sewage is not allowed to get mixed up with the sludge produced, and the outgoing effluent is not allowed to carry with it large amount of organic load, as in the case of a septic tank. An Imhoff tank is, infact, a two storeyed tank, and was designed by a German scientist Mr. Karl Imhoff. They are sometimes also known as *Two-storey Digestion Tanks.*

9.50.1. Constructional Details. The typical details of an Imhoff tank are shown in Fig. 9.59. It is a double chamber rectangular tank. The upper chamber is called the *sedimentation chamber* or *flowing through chamber,* through which the sewage flows at a very low velocity ; and the lower chamber is the *digestion chamber,* in which the sludge gets digested due to anaerobic decomposition.

The solids of the slowly moving sewage, settling down to the bottom of the sedimentation chamber, through the sloping bottom sides of the sedimentation chamber (slope being 1.25 vertical to 1 horizontal, as shown) will slide down into the digestion chamber through an *entrance slot* at the lowest point of a sedimentation chamber. The slot (with minimum width as 15 cm) is trapped or overlapped in such a way that the gases generated in the digestion chamber cannot enter the sedimentation chamber, and thus avoiding direct contact of sewage with the foul gases, and its consequent pollution.

A Gas vent also called *Scum chamber* is also provided above the digestion chamber and along side the sedimentation chamber to take care of the gases escaping to the surface. The chief gas is methane, having a considerable fuel value, and may, therefore, be separately collected for use.

In order to prevent the particles of sludge or scum from entering into the sedimentation chamber from the digestion chamber, the scum and sludge must be maintained at least 45 cm above and below the slots, respectively. This free or clear zone is called the *neutral zone*.

The digestion chamber (*i.e.* the lower chamber) is divided into a number of (three to four generally) interconnected compartments as shown. The bottom of each digestion compartment is made up in the form of an *inverted cone* or *hopper* with sides sloping 1 : 1, so as to concentrate the sludge at the bottom of the hopper. The digested sludge from the bottom of the hoppers is removed periodically (after 1 to $1\frac{1}{2}$ months, depending upon the temperature of sludge) through the cast-iron de-sludging pipes provided in each compartment. The sludge is removed with flow under hydrostatic pressure of 1.2 to 1.8 m. Moreover, all the sludge is not removed, and only the bottom layers which are

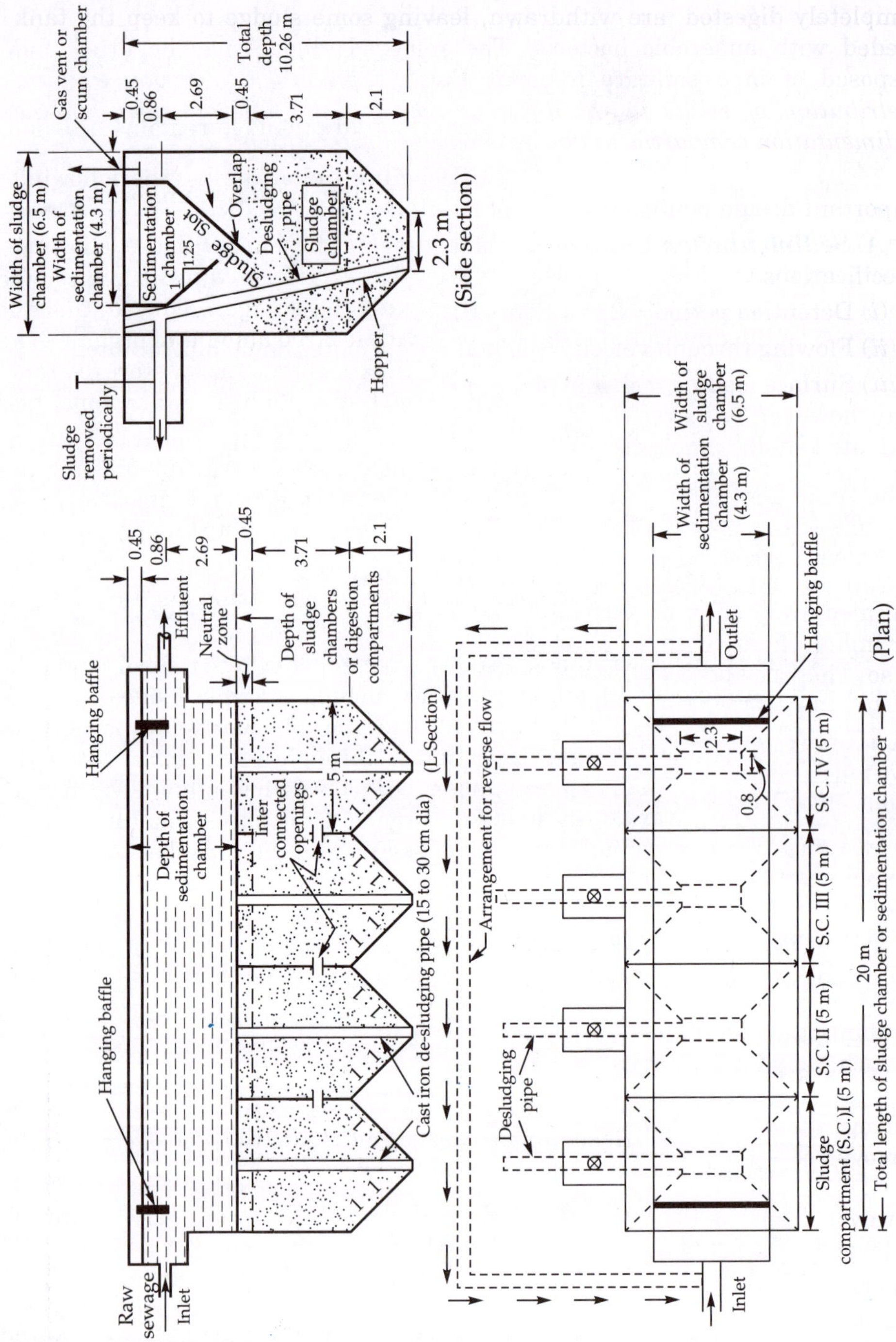

Fig. 9.59. Imhoff tank.

completely digested, are withdrawn, leaving some sludge to keep the tank seeded with anaerobic bacteria. The removed sludge may be dried and disposed of in a sanitary manner. Further, *in order to ensure uniform distribution of solids in the different hoppers, the flow of sewage in the sedimentation compartment above, is reversed intermittently.*

9.50.2. Design Considerations. In designing Imhoff tanks, the following important design points may be kept in mind :

(*A*) ***Sedimentation Chamber.*** It is rectangular in shape with the following specifications :

(*i*) Detention period = 2 to 4 hours (usually 2 hours)

(*ii*) Flowing through velocity= should not be more than 0.3 m/minute.

(*iii*) Surface loading = should not exceed 30,000 litres/m^2 of plan area/day. It may, however, be increased to about 45,000 l/m^2/day for effluent coming from activated sludge plant, or where recirculation is adopted.

(*iv*) Length of the tank should preferably not exceed 30 m or so, as to provide good sludge distribution. Length to width ratio may vary between 3 to 5.

(*v*) Depth of this chamber should as far as possible be kept shallow, so as to permit sliding of the solids up to the slot before reaching the end of the sedimentation. In practice, a total depth of 9 to 11 m has been found sufficient for Imhoff tanks ; with the depth of sedimentation chamber as about 3 to 3.5 m or so. The free-board provided may be about 45 cm.

(*B*) ***Digestion Chamber.*** This chamber is generally designed for a minimum capacity of 57 litres per capita*. But in warmer climates, where shorter periods between sludge withdrawals are possible, it may be reduced to about 35 to 40 litres per capita.

(*C*) ***Gas Vent or Scum Chamber.*** The surface area of the scum chamber should be about 25 to 30 per cent of the area of the horizontal projection of the top of the digestion chamber. Sufficient area for escape of gases is necessary, so as to prevent troubles due to foaming. Moreover, the width of a vent should be 60 cm or more.

9.50.3. Advantages and Disadvantages of Imhoff Tanks. Imhoff tanks combine the advantages of both the septic as well as sedimentation tanks, and as such, they find use in case of small treatment plants requiring only primary treatment. They are quite economical, and do not require skilled attention during operations. The results obtained are quite good, with 60 to 65% removal of solids, and 30 to 40% removal of BOD. Moreover, there is no problem of sludge disposal, as in the case of sedimentation tanks. They suffer, however, from the following ***drawbacks :***

(*i*) Depth of tank is more, which may make the constructions costlier, especially in hard rocks or quick sands. At such places, these tanks may, thus become uneconomical.

---

*This is larger than that provided in separate digestion tanks because of the lack of control on temperature of digestion (since there is no heating arrangement in Imhoff tanks), and inadequate mixing of the raw and the digested sludge (since there are no mechanical devices installed in Imhoff tanks for this purpose as is done in sludge digestion tanks).

(*ii*) Imhoff tanks may give out offensive odours, when improperly operated.

(*iii*) They are unsuitable and do not function properly where sewage is highly acidic in character.

(*iv*) These tanks have a tendency to foam or boil. This may cause the scum to go up to the top of the tank, and it may also force the sludge particles to enter the sedimentation chamber through the slot. The foaming may, thus, adversely affect the quality of the effluent.

(*v*) There is no adequate control over their operation. This makes them unsuitable for use in large treatment plants, where separate sludge digestion tanks are preferred in addition to sedimentation tanks. Imhoff tanks are, therefore, useful only for small cities and institutions, where it is not possible or economical to install separate sludge digestion tanks. **Mostly, however, they have become obsolete these days.**

Example 9.39. *Design an Imhoff tank to treat the sewage from a small town with 30,000 population. The rate of sewage may be assumed as 150 litres per head per day. Make suitable assumptions, wherever needed.*

Solution.

Design of Sedimentation Chamber

The sewage discharge per day

$$= 30{,}000 \times 150 = 4.5 \text{ M. litres/day} = 4500 \text{ cu-m/day}$$

Assuming a detention period of sewage in the sedimentation chamber as 2 hours, we have

The volume of sewage entering in two hours, *i.e.* the capacity of the sedimentation chamber

$$= 150 \times 30{,}000 \times \frac{2}{24} \text{ litres} = 3{,}75{,}000 \text{ litres} = 375 \text{ cu-m.}$$

Assume an effective depth of 2.2 m (effective depth includes part of the bottom sloping walls of the chamber) and a width of 4.3 m (say).

The length of the sedimentation chamber

$$= \frac{375}{2.2 \times 4.3} = 39.64 \text{ m ; say 40 m.}$$

This length is too large for a single tank. So let us adopt two tank units, each of length 20 m and width 4.3 m ; then

$$\frac{L}{B} = \frac{20}{4.3} = 4.65$$

which is between 3 to 5 and, therefore, satisfactory.

Now, discharge passing through each unit

$$= \frac{1}{2} \text{ of the total discharge} = \frac{4.5}{2} \text{ M. litres/day} = 2.25 \text{ M. litres/day}$$

Check for velocity

Length of tank = Velocity × Detention time

$\therefore$ 20 m = Velocity in m/min × (2 × 60 min.)

∴ Velocity in m/min

$$= \frac{20}{2 \times 60} = 0.17 \text{ m/min} < 0.3 \text{ m/min};$$

and, therefore, Safe.

Check for Surface Loading

$$\text{Surface loading} = \frac{Q}{BL} = \frac{2.25 \times 10^6}{4.3 \times 20} = 26{,}162\ l/\text{m}^2/\text{day}$$

which is less than 30,000 l/m^2/day ; and, therefore, satisfactory. Hence, the dimensions chosen can be accepted.

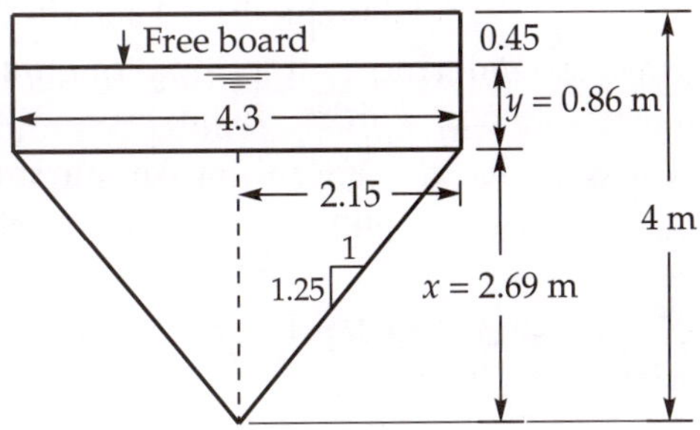

Fig. 9.60

Now let us decide the depth of the rectangular and sloping portions of the sedimentation chamber with its effective depth as 2.2 m.

With 4.3 m width and bottom sides sloping at 1 H : 1.25 V, the height of the sloping bottom = x = 1.25 × 2.15 = 2.69 m (Fig. 9.60).

Now, with effective depth of 2.2 m, the height of the vertical portion below the liquid surface (y) is given by

$$y = 2.2 - \tfrac{1}{2}\,(2.69)^*$$

$$= 2.2 - 1.345 = 0.855 \text{ ; say } 0.86 \text{ m.}$$

Adding 0.45 m for the free board, the total depth of the sedimentation chamber up to bottom at the entrance of the slot

$$= 0.45 + 0.86 + 2.69 = 4.00 \text{ m.}$$

Design of Gas Vent and Neutral Zone

Provide a neutral zone of 0.45 m below this depth of 4 m. The tank, in general, is of 20 m length, but below this 4.0 m depth, it shall be divided into a number of compartments, say 4, each of length

$$\frac{20}{4} = 5 \text{ m.}$$

The area of gas vent has now to be provided on both sides of the sedimentation chamber. This width should be about 25 to 30% of the total width of the tank. Using an overall width of 6.5 m, the total width of the gas vent (*i/c* both sides of sedimentation chamber), assuming 15 cm thickness of chamber walls

$$= 6.5 - 4.3 - 2 \times 0.15 = 1.9 \text{ m};$$

*Effective depth of triangular portion will be half, to make it equivalent to a rectangular section.

This is about $\frac{19}{6.5} \times 200 = 29.23\%$ of the total width, and therefore, OK (between 25 to 30%). Hence, 0.95 m width of gas vent will be provided on either side of the sedimentation chamber.

Design of Digestion Chamber

Assuming the capacity of the digestion chamber @ 40 litres/ capita, we have

The capacity of the digestion chamber

$$= 30{,}000 \times 40 = 12 \times 10^5 \text{ litres} = 1200 \text{ cu-m.}$$

Now, considering four compartments or units per tank (8 units in both tanks with 6.5 m width), we have

The capacity of each unit or compartment

$$= \frac{1200}{8} = 150 \text{ cu-m.}$$

Now, assume the depth of each hopper as 2.1 m, side slopes 1 : 1, and bottom section as

$$(6.5 - 2 \times 2.1 = 2.3 \text{ m}) \times (5 - 2 \times 2.1 = 0.8)$$

Capacity of each hopper

$$= \frac{h}{3}\left[A_1 + A_2 + \sqrt{A_1 A_2}\right]$$

where $h = 2.1$ m

$A_1 = 6.5 \times 5 \text{ m} = 32.5 \text{ m}^2$

$A_2 = 2.3 \text{ m} \times 0.8 \text{ m} = 1.84 \text{ m}^2$ [See Fig. 9.61]

$\therefore$ Capacity of each hopper

$$= \frac{2.1}{3}\left[32.5 + 1.85 + \sqrt{32.5 \times 1.84}\right]$$

$$= 0.7[32.5 + 1.84 + 7.73] = 0.7[42.07] = \mathbf{29.45\ m^3.}$$

Balance capacity to be provided by rectangular portion of section 6.5 m × 5 m

$$= 150 - 29.45 = 120.55 \text{ m}^3.$$

$\therefore$ Height of this portion

$$= z = \frac{120.55}{6.5 \times 5} = 3.71 \text{ m.}$$

$\therefore$ Total height of digestion chamber including neutral zone

$$= 0.45 + 3.71 + 2.1 = 6.26 \text{ m}$$

Total height of tank from top to bottom

= Height of sedimentation chamber + Height of sludge chamber

= 4 + 6.26 = **10.26 m.**

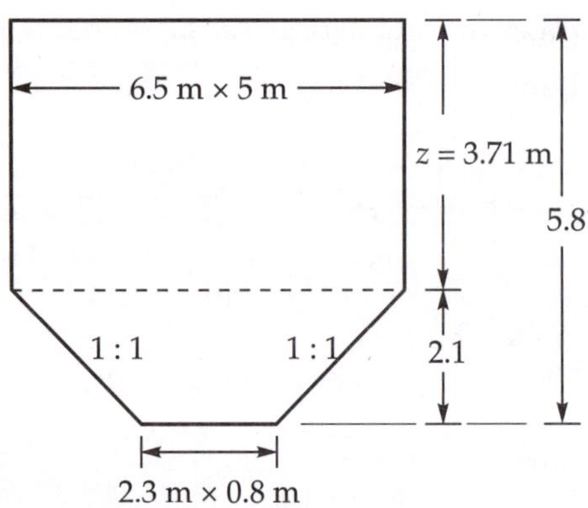

Fig. 9.61

This height is well within the practical limits (of 9 to 11 m) and hence the design is O.K. The plan, *L*-section and cross-section of the tank with these dimensions have been shown in Fig. 9.59. **Ans.**

Example 9.40. *Design an Imhoff-tank to treat the sewage from a small town with 30,000 population, assuming that the suspended solids in the influent sewage are 200 ppm. Water content of sludge is 97 per cent. Design the tank for three months sludge storage. Rate of sewage is 135 litres per head per day.*

(*Engg. Services, 1968*)

Solution.

Design of Sedimentation Chamber

The sewage discharge per day = 30,000 × 135 litres

= 4.05 M-litres/day = 4050 cu. m/day

Assuming a detention period for sewage in the sedimentation chamber as 2 hours, we have

The volume of sewage entering in two hours, *i.e.* the capacity of the sedimentation chamber

$$= 4050 \times \frac{2}{24} \text{ cu-m.} = 337.5 \text{ cu-m.}$$

Assuming an effective depth of 1.5 m (effective depth includes part of the bottom sloping walls of the chamber) and a width of 4.5, we have

The length of the sedimentation chamber

$$= \frac{\text{Capacity}}{\text{Area of } X\text{-section}} = \frac{337.5}{4.5 \times 1.9} = 39.5 \text{ m ; say 40 m.}$$

This length is too long for a single tank. So let us adopt two units, each of length 20 m, and width 4.5 m. Then $\frac{L}{B} = \frac{20}{4.5}$ = 4.42, which is between 3 to 5, and hence satisfactory.

Now, discharge passing through each unit

$= \frac{1}{2}$ of the total discharge

$$= \frac{4.05 \text{ M. litres per day}}{2} = 2.025 \text{ M. litres per day.}$$

Check for velocity in the sedimentation tank

Length of the tank = Velocity × Detention time

or $20 = V \times (2 \times 60)$

or $V = \frac{20}{2 \times 60}$

= 0.167 m/minute < 0.3 m/minute ;

and therefore satisfactory.

Check for surface loading in the sedimentation tank

$$\text{Surface loading} = \frac{Q}{BL} = \frac{2.025 \times 10^6}{4.5 \times 20}$$

= 22,500 litres/m^2 of plan area/day

= < 30,000 ; and, therefore, satisfactory.

Hence, the dimensions chosen are satisfactory and acceptable.

Now, let us decide the depth of rectangular and triangular portions of the sediment chamber.

With 4.6 m width and bottom sides sloping at 1 H : 1.25 V, the height of the sloping triangular portion,

$$x = 2.25 \times 1.25 = 2.81 \text{ m.} \quad \text{(Fig. 9.62)}$$

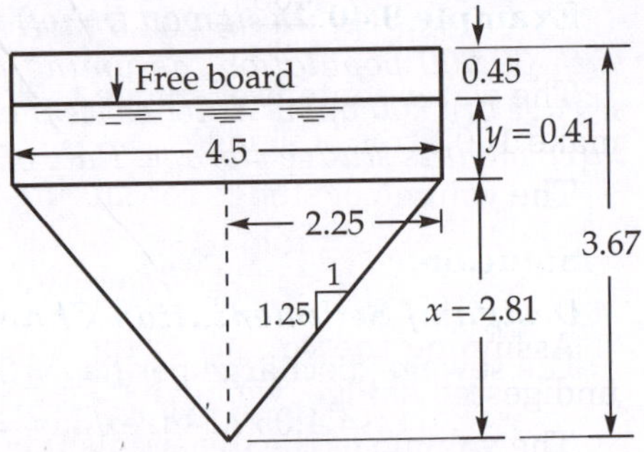

Fig. 9.62

Now, with effective depth of 1.9 m, the height of vertical portion below the liquid surface, *i.e.* y, is given as :

$$y = 1.9 - \frac{1}{2}(2.81)^* = 1.9 - 1.405 = 0.405 \text{ m ; } \textbf{say 0.41 m.}$$

Assuming 0.45 m free-board above the top liquid line, the total depth of the sedimentation chamber upto the bottom at the entrance of the slot

$$= 0.45 + 0.41 + 2.81 = 3.67 \text{ m.}$$

Design of Gas Vent and Neutral Zone

Provide a neutral zone of 0.45 m below this depth of 3.67 m. The tank in general is of 20 m length, but below this 3.67 m depth it shall be divided into a number of compartments, say 4, each of length $\frac{20}{4} = 5$ m.

The area of gas vent has now to be provided on both sides of the sedimentation chamber. This width should be about 25 to 30% of the total width of the tank. Using an overall width of 6.5 m, the total width of gas vent (including both sides of sedimentation tank), and assuming 15 cm thickness of chamber walls

$$= 6.5 - 4.5 - 2 \times 0.15 = 1.7 \text{ m.}$$

This is about $\frac{1.7}{6.5} \times 100 = 26.2\%$ of the total width, and therefore, O.K. (it is between 25 to 30%). Hence $\frac{1.7}{2} = 0.85$ m width of gas vent, will be provided on either side of the sedimentation chamber.

Design of Digestion Chamber

The suspended solids present in sewage are 200 ppm. Hence in one tank in which the volume of sewage entering is @ 2.025 M. litre/day, the volume of sewage solids

$$= 2.025 \times 10^6 \times \frac{200}{10^6} \text{ litres/day} = 405 \text{ litres/day.}$$

Assuming 50% solid removal (given), the volume of sewage solids to be removed in each tank

*Effective depth of triangular portion will be half, to make it equivalent to a retangular section.

$$= \frac{405}{2} = 202.5 \text{ litres/day.}$$

The water content of sludge is 97%, which means that 3 litres of solids will make 100 litres of sludge.

The volume of sludge containing 202.5 litres of solids

$$= \frac{100}{3} \times 202.5 \text{ litres/day} = 6750 \text{ litres/day.}$$

Assuming the volume of the digested sludge to be one third of that of the undigested sludge, we have

The volume of the digested sludge

$$= \tfrac{1}{3} \times 6750 = 2250 \text{ litres/day.}$$

The average of the undigested (new) and the digested sludge volumes in the tank, stored during 3 months (*i.e.* 3 × 30 days) is given by

$$C = \frac{(V_1 + V_2)^*}{2} \times t$$

$$= (6750 + 2250)\ \tfrac{1}{2} \times (30 \times 3 \textit{ i.e. } \text{No. of days})$$

$$= 9000 \times \tfrac{1}{3} \times 90 = 4{,}05{,}000 \text{ litres} = \mathbf{405\ cu\text{-}m.}$$

Since the digestion chamber has 4 compartments (per tank) each of 5 m length,

The sludge capacity required per compartment

$$= \frac{405}{4} = 101.25 \text{ cu-m.}$$

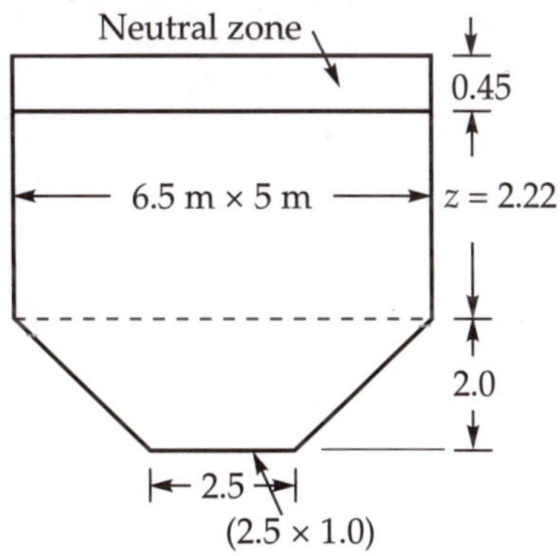

Fig. 9.63

Now, assume the depth of each hopper as 2.0 m, side slopes 1 : 1 and bottom section as (6.5 – 2 × 2.0 = 2.5 m) × (5 – 2 × 2 = 1 m)

Capacity of each hopper

$$= \frac{h}{3}\left[A_1 + A_2 + \sqrt{A_1 A_2}\right]$$

where h = 2.0 m

A_1 = 6.5 m × 5 m = 32.5 m^2

A_2 = 2.5 m × 1 m = 2.5 m^2

∴ Capacity of each hopper

$$= \frac{2}{3}\left[32.5 + 2.5 + \sqrt{32.5 \times 2.5}\right]$$

$$= \frac{2}{3}\ [35 + 9] = \frac{88}{3} = 29.3 \text{ m}^3.$$

Balance capacity to be provided by rectangular portion

$$= 101.25 - 29.3 = 71.95 \text{ m}^3.$$

---

*This may also be computed using parabolic average instead of st. line average by using $C = V_1 - \frac{2}{3}(V_1 - V_2)$ as explained under "Sludge Digestion Tank".

Height of rectangular portion

$$z = \frac{71.95}{6.5 \times 5} = \mathbf{2.22\ m.}$$

∴ Total height of the digestion tank including neutral zone

$= 0.45 + 2.22 + 2.0 = 4.67$

∴ Total height of the Imhoff tank from top to bottom

$= 3.67 + 4.67 = 8.34$ m, which is within limits, and so O.K.

The tank cross-section is shown in Fig. 9.64.

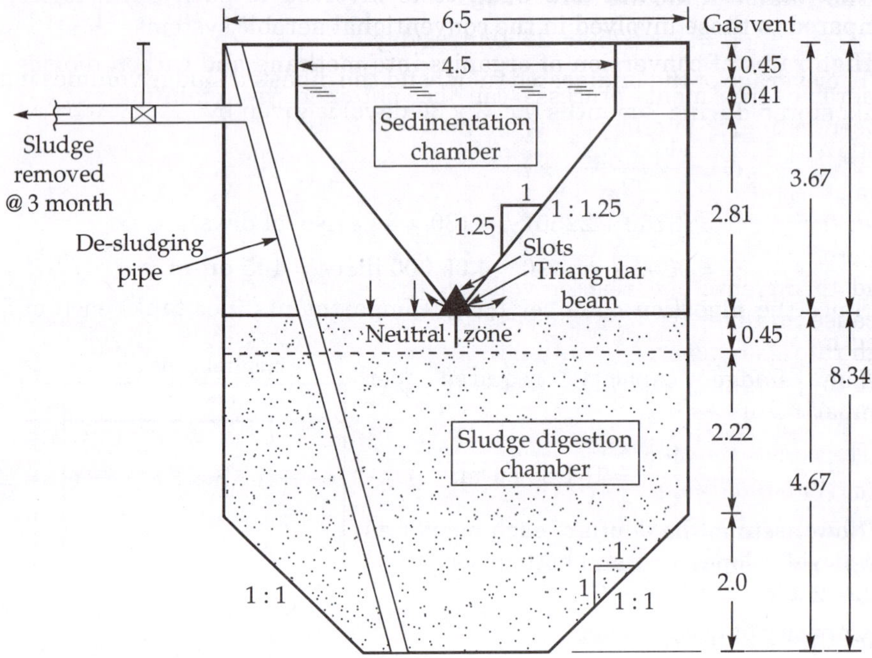

Fig. 9.64. Cross-section of Imhoff Tank of example 9.38 (2 Tanks of 20 m length etc.)

9.51. Clarigesters

Clarigesters are small patented circular Imhoff type double storey tanks, without bottom hoppers, and fitted with mechanical sludge and scum breaking equipment. Since the hoppers are omitted in them, their depth is generally less than 6 m or so, due to which they can be easily and economically constructed in areas where deep excavations are difficult.

These tanks function exactly like Imhoff tanks, with their upper compartment settling the sewage, and passing the sludge to the lower compartment. The sludge is digested in the lower compartment, and removed under hydrostatic pressure, from time to time.

Such small compact units are quite useful for treating limited sewage from small and isolated colonies, institutions, hospitals, etc. Effluents from such tanks are foul, and should be disposed of carefully, like those from Imhoff tanks.

9.52. High Rate Anaerobic Systems

Although the anaerobic treatment of wastewaters has been adopted since the beginning of the century, yet conventionally, the anaerobic process is considered to be a slow process, requiring **digestors of large hydraulic retention time (HRT)**. In recent years, however, *high rate anaerobic systems* have been devised and constructed to treat concentrated industrial wastewaters and for direct treatment of municipal wastewaters.

Application of anaerobic treatment technology to municipal wastewaters is quite significant for a developing country like India, because of high energy savings and low capital and OMR costs involved in such technologies, as compared to those involved in the conventional aerobic systems.

High rates of conversion of organics into methane and carbon dioxide, by anaerobic treatment process, can be obtained by maintaining a high concentration of microbes in a reactor, and preventing them from escaping with the effluent. This concept is expressed by **Sludge Retention Time (SRT)**, being the *ratio of the mass of biological solids in the system to that escaping from the reactor*. Maximal SRT is therefore, desirable for the stability of an anaerobic system and for minimising sludge production. This will however, tend to increase the reactor volume, which will increase the capital cost, because minimal HRT minimises the reactor volume. Other requirements of high rate systems are : (*i*) intimate contact between the incoming wastewater and the detained biological solids in the reactor, and (*ii*) maintenance of sufficiently warm temperatures.

The **various high rate anaerobic systems,** that have been devised, include :

(*a*) *Anaerobic contact (AC) process ;*

(*b*) *Anaerobic filters (AF) ;*

(*c*) *Anaerobic fixed films (AFF) reactors* ;

(*d*) *Fluidized bed (FB) reactor ; and*

(*e*) *Upflow Anaerobic Sludge Blanket (UASB) reactor.*

The basic configurations of these systems are shown in Fig. 9.63 (*a*) to (*e*).

All the above systems are briefly discussed below except the last one, which is discussed in more details, due to its successful working.

In any event, these reactors usually provide a little bit of incomplete treatment, bringing down the BOD and S.S. by 50 to 70% only. Their effluents may therefore need post treatment either by *aerobic filters*, *aerated langoons*, etc. ; or by *holding pond* of 1 day detention time, followed by *fish pond / aqua culture pond*.

9.52.1. Anaerobic Contact (AC) Process. The system involves a closed *stirred* tank reactor, in which the biomass leaving with the reactor effluent, is settled in a sedimentation tank and is recycled to the stirred tank, as shown in Fig. 9.65 (*a*), to increase SRT. Continuous and complete mixing in the reactor is not recommended, since this may adversely affect the settling characteristics of the sludge. On the other hand, inadequate mixing may lead to the formation of dead zones inside the reactor. This process has been generally used for treatment of industrial wastewaters.

9.52.2 Anaerobic Filter (A.F.). In an anaerobic filter, a *stationary filter media*, consisting of crushed rock of 15—25 mm size, or plastic beads, etc. are packed in a closed tank, and the wastewater is entered from the bottom to move up the packing media, as shown in Fig. 9.65 (*b*). The sludge is entrapped in the void space between the packing material. The effluent is recycled to increase SRT of the reactor. The reactor is operated as an upflow submerged bed reactor.

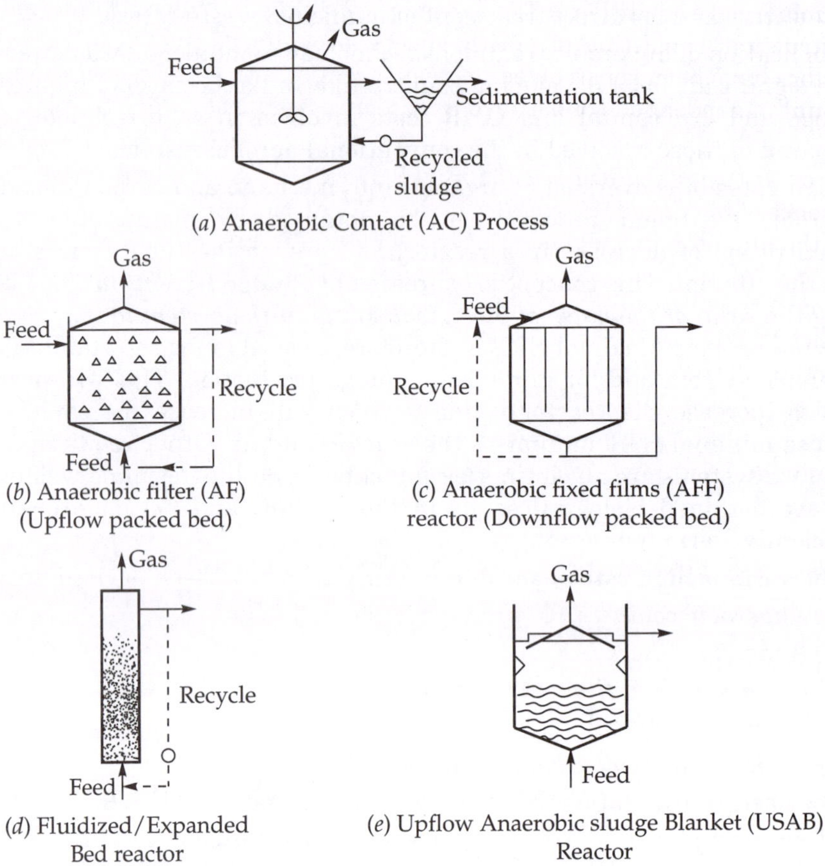

Fig. 9.65. Basic Configurations of High Rate Anaerobic Systems.

A number of such filters have been constructed for treatment of low strength municipal wastewaters. 65 to 75% efficiency in removal of BOD and SS, at volumetric organic loading of the order of 0.3-1.2 kg of COD*/m^3.d, has been obtained.

---

*The organic matter in wastewater can be expressed in terms of BOD or COD. In anaerobic treatment systems, however, the COD value is finding greater usage, which lends itself directly to mass balance calculations. Reduction in COD for municipal wastewaters would normally correspond to approximately equivalent amount of reduction in ultimate BOD, since non-biodegradable solids are negligible in such wastewater.

SRT, which is a more rational design parameter, is however, difficult to be calculated for anaerobic reactors. For AC and UASB, it ranges between 15 to 30 days ; while for other systems, it is estimated to be about 100 days or more giving them greater operational stability.

9.52.3. Anaerobic Fixed Films (AFF) Reactors. Like an anaerobic filter (AF), this process is also characterised by the presence of a *stationary packing material* in the reactor, but here the aim is to avoid entrapment of suspended solids in the voids of the packing material. In order to prevent accumulation of solids in the reactor, the AFF reactor is worked with *downflow mode*, as shown in Fig. 9.65 (*c*). The reactor may be operated either in submerged or unsubmerged condition. The reactor packing is usually of modular construction, consisting of plastic sheets, providing a high void ratio. Such reactors have been constructed to treat high strength wastes, containing high amount of suspended solids, such as screened manures.

9.52.4. Fluidized and Expanded Bed Reactors. A *fluidized bed reactor* as well as an *expanded bed reactor* [Fig. 9.65 (*d*)] are both characterised by the presence of *mobile packing material*, such as sand, clay, coal, etc. The organics and the microbes in the reactor get attached to these particles, with which the reactor tank is filled only to a part height. The wastewater is entered from the bottom of the tank with an upflow velocity to move (fluidize) the media particles, which act as *mobile biomass carriers*, causing digestion of the organics present in the wastewater.

In a *fluidized bed reactor*, the overflow velocity is kept higher, causing the entire bed material to move with the up moving sewage, thereby distributing the sludge, virtually almost over the entire reactor volume. Whereas, in an expanded bed reactor, the upflow velocity is kept lower, thereby confining the sludge mostly in the lower part of the reactor. The *expanded bed reactors*, therefore, do not aim at complete fluidization of the bed, permitting the use of lower upflow pressure on the wastewater, thereby consuming lesser electrical power. These reactors can be used for treating industrial as well as municipal wastewaters.

9.52.5. The UASB Reactor. The Upflow Anaerobic Sludge Blanket (UASB) reactor [Fig. 9.65 (*e*)] maintains a high concentration of biomass through the formation of highly settleable microbial sludge aggregates. The wastewater flows upwards through a layer of very active sludge to cause anaerobic digestion of organics of the wastewater. At the top of the reactor, three phase separation between gas-solid-liquid takes place. Any biomass leaving the reaction zone is directly recirculated from the settling zone. The process is suitable for both *soluble wastewaters* as well as *wastewaters containing particulate matter*.

The large scale adoption of this technique for treating municipal wastewaters is comparatively of recent origin, although the system was developed as late as in the year 1979 by a Netherland's scientist, Mr. Gatze Lettinga. The system, however, has proved very promising, and several such treatment plants have come into existence in countries like China, India, Pakistan, Bangladesh, and some Latin American countries. As far as India is concerned, this simple technique was first of all adopted in the year 1989, when a demonstration sewage treatment plant (STP) of 5 MLD capacity was installed at Kanpur (U.P.) under the *Ganga Action Plan*. This has been followed by the installation of a 36 MLD plant at Kanpur (U.P.), and 14 MLD plant at Mirzapur (U.P.). These plants have been successfully working since their commissioning. Many more such plants are proposed to be commissioned for treating municipal wastewaters of towns in Haryana and U.P.

The *X*-section of a typical UASB reactor is shown in Fig. 9.66. This reactor consists of an upflowing treatment tank, provided with a feed inlet distribution system at the tank bottom. A gas-solid-liquid separator (GSS) device is provided at the top to help provide a quiescent zone at the top of the reactor.

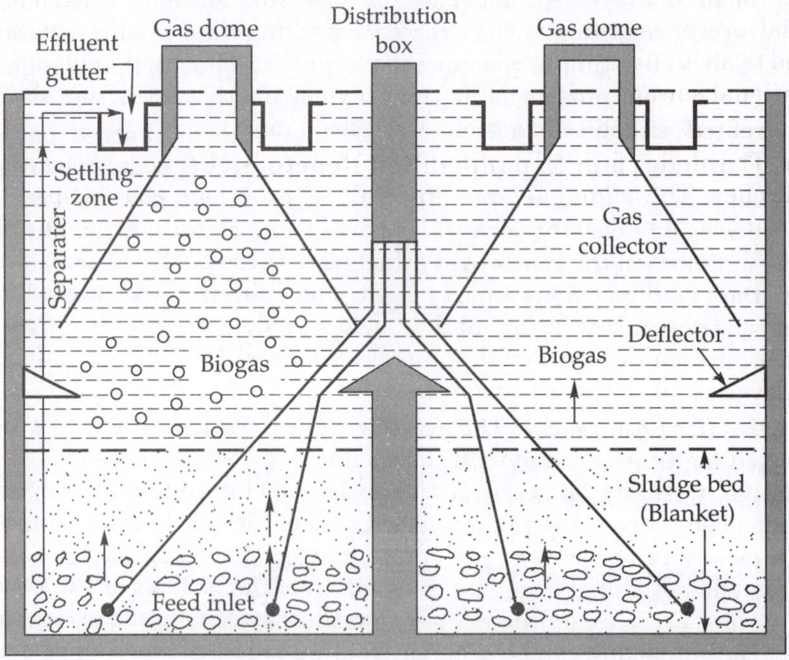

Fig. 9.66. X-section of a typical UASB reactor.

The wastewater enters the tank from the bottom, and flows upward through the sludge bed, which gets formed during the process itself. The sludge bed develops micro-organisms capable of flourishing in an oxygen deficient environment. The **sludge bed (blanket)** traps the suspended organics of the upmoving wastewater. The suspended solids trapped in the sludge bed are degraded by the anaerobic and anaerobically working facultative bacteria, producing methane and carbon dioxide (*i.e.* biogas, which is a mixture of 65—70% methane, and 30—35% carbon dioxide). The biogas produced during the anaerobic decomposition helps in providing gentle mixing and stirring of the biomass, thereby increasing the efficiency of decomposition, reducing the BOD and suspended solids of the wastewater.

The methane or biogas is collected at the top of the tank in a *gas collector*, from where it can be withdrawn for use as a by-product ; while the water sludge mixture is made to enter a settling tank where the sludge settles down and flows back into the bottom of the reactor. The sludge will show good settling properties after an initial start up period, followed by granulation, forming a sludge blanket or sludge bed in the lower part of the reactor.

The UASB, evidently operates as a **suspended growth system**, with no packing material in the reactor. In this system, the microbes attach themselves to each other, or to small particles of the suspended matter of sewage, to form granules or agglomerates and ultimately the sludge bed (blanket). The gas produced in the process causes sufficient agitation to keep the sludge fully mixed.

Retention of the bacteria containing sludge in the reactor is one of the most important features of the UASB process. The bacteria in the sludge continue to perform their function of treating the incoming effluent. The continuous bacterial presence and activity enables retention time in the reactor to be reduced to about 6-8 hours, as compared to atleast 30 hours, that is required in conventional sewage treatment systems.

The treated effluent is collected in gutters, and discharged out of the reactor. The sludge is periodically shifted into the drying beds, to be used as a soil enricher. The methane generated can be used as a gas for domestic or industrial use. It may also be used for generation of electricity for running the plant, after appropriate dehydration and cleaning.

This process can be reactivated even after the plant remains shut down for days or months, or after power breakdowns and interruptions in wastewater supply.

Like other high rate anaerobic systems, the various advantages offered by UASB system are summarised below :

Advantages of UASB System. The various advantages offered by UASB system over the Conventional Aerobic systems are given below :

(*i*) The space requirement of the system is quite comparable to that of an Activated sludge system ; *i.e.* about 0.5 acres per MLD, as compared to 2.5 acres per MLD required for Oxidation ponds, and 1.5 acres for Aerated lagoons.

(*ii*) The capital cost investment of such a plant is about ₹ 20 lakh/MLD as compared to about ₹ 35 lakh/MLD for an Activated sludge plant, ₹ 7.5 lakh/MLD for Oxidation ponds, and ₹ 15 lakh/MLD for Aerated lagoons.

(*iii*) The system requires lesser and simpler electromagnetic parts as compared to the ones required in an Activated sludge plant, leading to lower O and M (Operation and Maintenance) Costs.

(*iv*) Electricity consumption in this system, like all anaerobic systems, is quite low, and the system is quite capable of withstanding long power failures.

(*v*) The sludge production in this system is low, and the produced sludge is having quick dewatering characteristics.

(*vi*) The system enables quicker sludge digestion, as compared to the conventional digestors.

(*vii*) Biogas is produced in the system as a by-product, which can be used to produce electricity to run the system.

Limitations or Drawbacks of UASB System. The various drawbacks of the UASB system as compared to the Conventional Aerobic system are given below :

(*i*) The system helps to lower only two parameters of wastewaters; *i.e.* (*a*) BOD ; and (*b*) Suspended solids. Eventually, the system does not help in the removal of toxic pollutants, like heavy metals, which may be present in some of the wastewaters. The USAB system will therefore have to be supported by subsidiary disposal system to remove the toxic pollutants, if present in the wastewater.

(*ii*) Like all other anaerobic high rate systems, UASB reactor also requires larger quantity of organic matter as compared to the aerobic reactors, because the growth of aerobic bacteria per unit of organic matter is about 10—20 times the growth of anaerobes. In order to support microbial growth and metabolism in UASB system, therefore, 20 to 30 times more of organic matter has to be metabolised, as compared to that in Aerobic systems. For the success of UASB, it therefore becomes necessary to ensure the presence of at least 10% of suspended solids in the wastewater. This requirement factor can not always be met by all types of wastewaters.

(*iii*) Some of the wastewaters may contain minerals, which may interfere with the efficiency of the anaerobic microbes. The system also does not respond well to the wastewaters of tanneries, which contain more than 500 auxiliary chemicals, offering varying response to the UASB technology.

(*iv*) The acids produced during the breakdown of organic matter in a UASB reactor, may cause corrosion of the reactor.

(*v*) The efficiency of BOD and S.S. removal is a little bit low, as compared to that in an Activated sludge plant. With generally adopted organic loadings of 1.0—2.0 kg COD/m^3.d in UASB reactors, the achieved efficiency varies between 50 to 70% only. The effluent BOD of municipal wastewaters treated in UASB reactor system, will therefore be higher.

Say for example, the effluent BOD may be about 50 mg/l for influent BOD of 200 mg/l. For concentrated influents, the effluents BOD may still be higher. Direct disposal of effluent containing such high BOD may not always be permissible. Depending upon the situation, the effluent from a UASB system may have to be given further aerobic treatment in Aerated lagoons, or Oxidation ponds, or Filters. Where, enough space is not available, the post treatment may consist of using a *holding pond* of 1 day detention time followed by **fish pond/aqua culture pond.**

(*vi*) Pre-treatment of wastewater with *screening* and *grit removal*, are usually found necessary for direct anaerobic treatment.

(*vii*) The system responds well in high temperature climate areas, because the activity of methanogenic bacteria is strongly influenced by temperature, which approximately doubles for every 10°C rise in temperature in the range of 18°C to 38°C. However, high micro-organism concentration in high rate anaerobic reactors like a UASB, compensates the decreased activity of the anaerobic organisms at the lower temperatures.

(*viii*) The methanogenic bacteria do require iron, cobalt, nickel and sulphide, in addition to nitrogen and phosphorous. These elements are generally present in municipal wastewaters, but may have to be added to anaerobically treat some specific industrial wastewaters, which may have deficiency of these elements.

Conclusions. With all the above drawbacks, the UASB system has responded well for the treatment of high strength soluble wastewaters, like those from *municipalities* and industries like *food processing, edible oils, distilleries, wineries, breweries, dairies, cheese processing, organic chemicals*, etc. It has been opined that the installation of a UASB plant should be made a mendatory requirement of developing every new colony or township, having population of more than 5000 or so.

The UASB technology has, however, not become very popular so far, because there is a lack of awareness of this technology. Even the engineers of the Public Health Engg. Deptts. are usually unaware of this technology, and the adopted technologies at different locations are being widely decided either by the funding agencies (such as World Bank) ; or by the contractor firms, who undertake execution of plants on turn-key basis including the cost of design and technology, rather than deciding the technology on simplicity and cost benefit considerations. So much so that even in Delhi, the capital city of India, two extremely costly *biological filter* plants*, each of 10 MLD capacity and each costing about ₹ 4.5 crores, have recently been constructed to treat the sewage/sullage flows of Delhi Gate and Sen Nursing Home nalls, which shall not only require very high power consumption, but shall also require very high O and M expenses. Although of course, the adopted technology at these two plants shall produce effluent of low BOD (≈ 10 mg/l) and utilise lesser space (about 1.25 acres for 10 MLD plant), yet the adoption of such costly methods of sewage treatment must not remain undebatable. Be that as it may, many many more STP's are still needed even in this capital city of India, since at present about 630 MLD** of sewage is joining the Yamuna river, without being subjected to any kind of treatment.

CHLORINATION OF SEWAGE

9.53. Disinfection of Sewage by Using Chlorine

Sewage may be disinfected by adding chlorine to it, either before the treatment or after the treatment, as is done for disinfecting water supplies.÷

When chlorine is added to the treated sewage as a final step in its treatment, it is called **post-chlorination.** This may be adopted where an effluent, low in bacterial count, is necessary. The disinfecting action of chlorine will, no doubt, reduce the bacteria, but, its oxidising action will also reduce the BOD to some extent.

Pre-chlorination of sewage before it enters the sedimentation tank may help in controlling odours, may prevent flies in trickling filters, and may assist in the removal of grease.

When chlorine is added to raw sewage in proper quantities, it will destroy the bacteria, which may breakdown the sulphur compounds in the sewage, and reduce the production of hydrogen sulphide. For this reason, chlorine is sometimes added into the main collecting sewers to prevent the destructive action of hydrogen sulphide on concrete pipes.

*In biological filters, the wastewater is filtered downward through a fully submerged bed of small stones, which help to form biofilms, and air is forced into the bed. No settler is used, but periodic back washing is required.

**At present, Delhi generates about 1900 MLD of sewages against an installed treatment capacity of 1270 MLD, leaving a gap of 630 MLD. The entire existing capacity is also not up to the desired secondary standards.

÷Please see "Environmental Engg. Vol. I—"Water Supply Engg." by the same author.

Chlorine may sometimes be added to sewage, both at the beginning as well as at the end of the treatment process, and this is known as **split-chlorination.**

The amount of chlorine required for chlorinating sewage may vary considerably*, but is always much greater than that required in water purification. Hence, it is a very costly process, and as such, generally not adopted in normal sewage treatment processes. Most of our sewage treatment plants are, therefore, generally not equipped with chlorination facilities, and even those plants which are provided with **chlorinators** often use them intermittently to relieve over-loads during peak flows. Chlorination may, however, be adopted, when needed, to prevent contamination of beaches, shelfish beds, and water supplies.

PROBLEMS

1. Give a list of methods available for treatment of sewage for both rural and urban conditions.

2. Give the flow diagram for 'the activated sludge process', and describe the working of the activated sludge plant.

3. Describe the aim of the following sewage treatment methods, using one or more appropriate key words given below :

(*i*) Screen ;

(*ii*) Activated sludge process ;

(*iii*) Sedimentation ;

(*iv*) Grit chamber.

Key words : Removal of—organic, inorganic, suspended or settleable solids ; organic, inorganic dissolved solids ; pathogenic, micro-organisms ; large floating solids.

4. (*a*) BOD removal by sedimentation of sewage can at best be 40 percent. Why ?

(*b*) What is the necessity of maintaining constant velocity in grit channel ?

5. Differentiate between aerobic and anaerobic treatment of sewage, giving major end products. Name one treatment method in each category.

6. Give the approximate reasonable values in percentage of the BOD removal of the following unit processes :

(*i*) Plain sedimentation ;

(*ii*) Activated sludge plus final sedimentation ;

(*iii*) Trickling filters plus final sedimentation ; and

(*iv*) Imhoff tank.

7. Give the approximate reasonable values of the detention time in hours of the sedimentation tanks, as noted below :

In waste treatment, (*i*) final sedimentation tank after activated sludge ; and (*ii*) primary settling tank before activated sludge treatment.

8. Classify the following under aerobic or anaerobic processes :

(*i*) Sludge digestion tanks ;

(*ii*) Trickling filters ;

---

*Pre-chlorination dose may be as high as 25 mg/l, while post-chlorination, usually, requires atleast 3 mg/l.

(*iii*) Activated sludge treatment ;

(*iv*) Intermittent sand filters ;

(*v*) Seepage tile bed following the septic tank ;

(*vi*) Lower compartment of Imhoff tank.

9. Indicate in appropriate space what effect an increase in variable A will have on variable B.

S. No.	*Variable A*	*Variable B*	*Increase / Decrease / No Effect*
(*i*)	Depth of tank	Overflow rate	
(*ii*)	Temperature of water	Settling velocity	
(*iii*)	Oxygen content	Rate of absorption of oxygen	
(*iv*)	Water temperature	Head loss in filter	

10. Discuss anaerobic digestion of sludge under the following heads :

(*i*) Definition and composition of sludge ;

(*ii*) Fundamental mechanism of anaerobic digestion ;

(*iii*) End products and their disposal.

11. Design an oxidation pond for treating domestic sewage of 10,000 persons supplied with 200 litres per capita water per day. The BOD and the suspended solids are each of 300 mg/l.

Permissible organic loading for the pond is not less than 500 kg/ha/day. The detention period is not to exceed 6 days. Assume width of the pond to its length as 1 : 2 and the operational depth as 1.2 m. Assume any other suitable data.

12. With the aid of sketches, describe the principles involved in the design and construction of any *three* of the following :

(*a*) Grit chamber ;

(*b*) Sludge digestion tank ;

(*c*) Imhoff tank ;

(*d*) Septic tank ;

(*e*) Percolating filter.

13. (*a*) Explain with a neat sketch the working of a percolating filter (trickling filter). What is the principles on which it works ?

(*b*) Compute the diameter of a circular trickling filter for 250 users. Dry weather flow is 120 *litres per capita* per day. Rate of filtration of trickling filter may be taken as 10 million litres per hectare/metre.

14. Design a septic tank for 200 users. Water allowance is 120 litres per head per day. Detention period may be taken as 8 hours. Draw a neat dimensioned sketch of the septic tank you design.

15. What do you understand by digestion of sewage sludge ? Give a neat sketch of separate digestion tank and explain its working.

16. What is meant by activated sludge ? Describe with sketches the treatment of sewage by activated sludge process. Mention the advantages and disadvantages of this system.

17. Give by a line diagram the outline of the method of purifying sewage in different steps from outfall sewer to the effluent discharge points in the river.

18. (*a*) Compare and contrast septic tank with Imhoff tank and clarigester, in scope, function and performance.

(*b*) Design an Imhoff-tank to treat the sewage from a small town with 30,000 population, assuming that the suspended solids in the influent sewage are 200 ppm, water content of sludge is 97 per cent. Design the tank for three month sludge storage. Rate of sewage is 137 litres per head per day.

19. Write short notes on five of the following, illustrating your answers with neat sketches where needed :

(*i*) Grit chambers ;

(*ii*) Cesspool ;

(*iii*) Dispersion trenches ;

(*iv*) Imhoff tanks ;

(*v*) Sludge digestor tank ;

(*vi*) Septic tank ;

(*vii*) Oxidation ditch ;

(*viii*) Oxidation pond ;

(*ix*) Sludge drying beds.

20. (*a*) Distinguish between primary treatment and secondary treatment. Draw the flow diagram for sewage treatment using activated sludge process.

(*b*) Design standard rate trickling filter for the following :

Average incoming flow = 350 m^3/hr.

BOD of primary effluent = 210 ppm.

No. of units = 4.

21. (*a*) What is the object of sewage treatment ? Distinguish between fresh sewage, stale sewage, and septic sewage.

(*b*) Explain how you would decide the type of treatment to be given for a particular sewage.

22. Design a septic tank for the following data :

No. of people = 100

Sewage/capita/day = 120 litres

De-sludging period = 1 year

Length : width = 4 : 1.

23. (*a*) Distinguish between low rate (standard) trickling filters and high rate filters.

(*b*) Design an activated sludge plant with diffuser plates, given the following particulars :

Average inflow = 380 m^3/hr.

Amount of return sludge = 25%

Aeration time = 6 hours.

24. (*a*) Compare and contrast septic tank with Imhoff tank in scope, function and performance.

(*b*) Design a lagoon to serve for 5000 Persons at 100 l.p.c.d., B.O.D. of raw sewage is 200 ppm.

25. (*a*) Answer with respect to standard rate trickling filter :

(*i*) Size of filter media ; (*ii*) Depth of filter ;

(*iii*) Hydraulic loading ; (*iv*) B.O.D. loading ;

(*v*) Effluent characteristics.

(*b*) Mention the operational troubles of a standard rate trickling filter and their remedies.

26. (*a*) Mention the various methods of disposal of effluent from septic tank. Describe one of them in detail.

(*b*) Mention merits and demerits of Imhoff tank.

(*c*) Calculate the volume and number of aeration tanks and the rate of air supply for the following data of the activated sludge unit ;

Population	= 35,000
Average sewage flow	= 180 lpcd.
BOD of raw sewage	= 220 mg/l
BOD removed in primary treatment	= 30%
Overall BOD reduction desired	= 85%.

27. Differentiate between the following pairs :

(*a*) Step aeration and tapered aeration in activated sludge process ;

(*b*) Oxidation pond and oxidation ditch.

28. (*a*) Define sludge volume index. What is its importance in sewage treatment ?

(*b*) Sketch and explain the types of aeration systems in an activated sludge process.

29. (*a*) Discuss the effects of pH and temperature on sludge digestion.

(*b*) Mention the composition and uses of sludge gas.

(*c*) How will you compute the quantity of sludge gas and its fuel value.

30. (*a*) Discuss how a standard rate trickling filter differs from a high rate onc.

(*b*) Design the activated sludge units for the following data :

Population served	= 50,000
Average sewage flow	= 180 l.p.c.d.
B.O.D. of raw sewage	= 200 mg/l
Raw sewage suspended solids	= 300 mg/l
B.O.D. removal in primary treatment	= 35%
Overall B.O.D. reduction desired	= 80%.

31. (*a*) Prove that the area and overflow rates rather than the detention period govern the design of a settling tank. (*Engg. Services, 1976*)

(*b*) In a continuous flow settling tank 3 m deep and 60 m long, what flow velocity of water would you recommend for effective removal of 0.025 mm particles at 25°C. The sp. gravity of particles is 2.65 and kinematic viscosity n for water may be taken as 0.01 cm^2/sec.

32. (*a*) Why coagulants are used in the sewage treatment ? Name a few common coagulants used. Describe a method of application of any one of the coagulants.

(*b*) Calculate kg of alum needed per day if alum dose is 28.6 mg/l and the flow is 18 Ml/day.

33. **Enumerate the various methods that can be used to dispose of the digested sludge. Discuss in details the working of any of these two important methods, which are useful in Indian conditions.**

34. **(*a*) Comment on the statement that the 20 days BOD is an approximation of the ultimate BOD.**

(*b*) (*i*) Discuss the mechanism of working of a stabilisation pond.

(*ii*) Design a stabilisation pond to treat the night soil from a population of 10,000. The design temperature is 18°C. Give sketches where necessary.

—10

Process Design of a Complete Sewage Treatment Plant

In our previous chapter, we have discussed the designs of various types of treatment units, which are used for treating municipal wastewaters (sewage). In this chapter, we will give the typical design and detailing of the hydraulic design of a ***complete sewage treatment plant (STP).*** The S.T.P. which we have chosen has been designed for treating a part of the sewage discharge of the *Dalhousie town* (a hill station in Himachal Pradesh State of India near Dharamsala).

10.1. Hydraulic Design and Detailing of a Complete Sewage Treatment Plant of 1.5 MLD (Average Annual Flow) to be Constructed at Block I of Dalhousie Township

10.1.1. Basic Data for the Proposed S.T.P. The following basic data was first of all worked out and stipulated for the proposed sewage treatment plant, on the basis of population, per capita sewage produced, quality of sewage produced, and standards of effluent specified :

Average flow = *1.25 MLD = 1250* m^3/day

Peak flow = *3.125 MLD = 3125* m^3/day

Sewerage Characteristics	*Influent*	*Effluent*
BOD, 20°C	*295 mg/l*	*20 mg/l*
Suspended Solids	*4653 mg/l*	*30 mg/l*
PH	*Around 7.5*	*Around 7.5*

G.L. near plant (av) R.L. 1860 m above MSL formation level

TWL in inlet chamber : 1863.50 m

TWL in outlet channel of FST : 1861.30 m.

10.1.2. Choosing the Type of Secondary Treatment. Since the desired quality of the treated effluent was of a high standard, and the available land area for construction of plant was limited, it was decided to design the S.T.P. using ***activated sludge process*** for the secondary treatment instead of using the trickling filters.

Moreover, considering the various advantages offered by the extended aeration process, specially for smaller plants up to 4 MLD capacity, as described in article 9.38.5, it was decided to use **extended aeration type of activated sludge plant**, which eliminates primary settling tank as well as sludge digestion tank.

10.1.3. Treatment Units. The proposed S.T.P. will hence contain the following units, as discussed in the previous chapter :

1. *Inlet channel* (chamber)
2. *Screen chamber*
3. *Grit chamber*
4. *Aeration tanks* for activated sludge treatment
5. *Secondary clarifier* ; and
6. *Sludge drying beds.*

The positioning of these units are shown in location plan, **Chart Fig. 10.1 at the end of the book.**

The hydraulic designs of the above mentioned treatment units is now carried out below :

(1) Design of Inlet Chamber

TWL in inlet chamber = 1863.50

Peak flow = 3125 m^3/day = 0.038 m^3/sec

Plan dimension of inlet chamber = 1.3 × 1 m

Free-board = 1.0 m

Top of Inlet chamber = RL 1864.50 m

Size of the Bye-Pass chamber by the side of Inlet Chamber

= 1.3 × 1 m (minimum size)

The outflow from the inlet chamber shall be taken to the screen chamber. A bye-pass channel 0.60 m × 1 m or 400 mm ϕ pipe shall be provided from the bye-pass chamber up to final effluent channel to meet with any exigencies of the S.T.P.

(2) Design of Screen Chamber

2 Nos. screen chambers/channels shall be provided as per sound engineering practice.

The flow from the inlet chamber to the screen channels shall be controlled by C.I. penstock gates.

Q_{max} = 0.038 m^3/sec

Assumptions :

Shape of Bar = M.S. Flats

Size = 10 mm × 50 mm (10 mm facing flow)

Clear spacing between the bars = 20 mm

Inclination of bars with horizontal = 80° (Cleaning-manually)

Assuming velocity normal to screen = 0.8 m/sec

At peak flow, net inclined area required $= \frac{0.038}{0.8} = 0.047$ sqm

Gross inclined area = 0.047 × 1.5 = 0.070 sqm.

Gross vertical area required = 0.070 × sin 80° = 0.069 sqm.

Provide submergence depth = 0.3 m

Width of channel $= \frac{0.069}{0.3} = 0.23$ m ; provide 0.30 m

Check velocity in duct $= \dfrac{0.038}{0.30 \times 0.3} = 0.42$ m/sec

(Approach velocity u/s of screen) > 0.4 m/sec

Provide 20 bars of 10 mm × 50 mm at 20 mm clear spacing

Screen chamber shall be 60 cm wide.

U/s of screens, 2 Nos. C.I. penstock gates shall be provided (one for each channel).

Min. drop of 150 mm shall be provided in the bed of screen channel.

Size of the Penstock gates :

2 Nos. of 350 × 450 mm size be provided.

(3) Design of Grit Chamber

Flow from screen channels shall be taken into Grit Chamber, provided in duplicate. 2 no. C.I. gates, one each at Inlet and Outlet, are provided for each Grit Chamber.

One more C.I. Gate shall be provided at the inlet to bye-pass channel in between two Grit Chambers.

Design flow $= \dfrac{2.5 \times 1.25}{2} = 1.56$ MLD = 1560 m^3/day

Surface loading = 1100 m^3/sqm/day

To account for turbulence and short circuiting, reduce the Surface loading to about 800 m^3/sqm/day

Area required $= \dfrac{1560}{800} = 1.95$ sqm.

Provide 1.70 m dia chamber, or 1.70 × 1.70 m square chamber.

Detention time = 60 sec.

Volume $= \dfrac{1560 \times 60}{24 \times 3600} = \dfrac{156}{144} = 1.08$ cum

Liquid depth $= \dfrac{1.08}{1.95} = 0.554$ m

Size of the Grit Chamber

= 1.70 m × 1.70 m (or 1.70 m dia)

× 1.2 m (*i.e.* 0.554 + 0.6 FB = 1.2 m) depth

Check for Horizontal Velocity

Cross sectional area of Grit Chamber

= 1.7 × 0.554 = 0.942 sqm

Velocity $= \dfrac{1560}{1.7 \times 0.554 \times 24 \times 3600} = 0.0192$ m/sec.

= 1.9 cm/sec < 18 cm/sec. Hence O.K.

Grit generation = 0.05 m^3 per 1000 m^3 of sewage flow (Assume)

Even though the grit is continuously raked, still 8 hrs. grit storage is provided for average flow.

Storage Volume required $= \dfrac{1250 \times 8}{24} \times \dfrac{0.05}{1000} = 0.028\ m^3$

Grit storage area = $\frac{\pi}{4} \times 1.7^2 = 2.27$ sqm.

Grit storage depth = $\frac{0.028}{2.27} = 0.012$ m

Total liquid depth = 0.554 + 0.012 = 0.566 ; say 0.6 m

Provide grit chamber of size : 1.7 × 1.7 × (0.60 + 0.60 FB) depth, or

= 1.7 m × 1.7 m × 1.2 m size.

Outflow from grit chamber shall be carried to the aeration tanks through a 600 mm wide R.C.C. channel provided with ***fine bar screen (manually operated).*** The clear spacing between the bars shall be 10 mm.

Note. *Parshall Flume and measuring arrangement has not been proposed for two reasons viz :*

(*a*) *The measuring instruments are rarely working satisfactorily at any plant.*

(*b*) *The plant is very small, to economise the cost.*

(4) Design of Aeration Tanks

No. of tanks = 2

Average flow to each tank = $\frac{1.5 \text{ MLD}}{2} = 0.625$ MLD

or $Q = 625$ m³/day.

The total BOD entering S.T.P. = 295 mg/L

Assuming that negligible BOD is removed in screening and grit chamber (since it mainly removes inorganic solids), the BOD of sewage coming to aeration tank.

$= Y_0 = 295$ mg/L.

BOD left in the effluent

$= Y_E = 20$ mg/L

∴ BOD removed in activated plant

= 295 – 20 = 275 mg/L

∴ Minimum efficiency required in the activated plant

$= \frac{275}{295} = 93\%$; O.K.

since the adopted extended aeration process can remove BOD up to 95—98%.

Volume of the Aeration Tank can be designed by assuming a suitable value of MLSS and θ_c (or F/M ratio)

Let us assume MLSS

= **3000 mg/L** ; (between 3000—5000 mg/L)

F/M ratio = **0.12** (between 0.18 to 0.10)

Now, using equation (9.45), we have

$$\frac{F}{M} = \frac{Q}{V} \cdot \frac{Y_0}{X_T}$$

where $Q = 625$ m³/day

$$V = ?$$
$$Y_0 = 295 \text{ mg/L}$$
$$X_T = 3000 \text{ mg/L (assumed)}$$
$$F/M = 0.12 \text{ (assumed)}$$

$$0.12 = \frac{625}{V} \times \frac{295}{3000}$$

or $$V = \frac{625 \times 295}{0.12 \times 3800} = 512 \text{ m}^3.$$

Aeration tank dimensions. Let us adopt an aeration tank of liquid depth 3.5 m and 9 m width. Then,

The length of the tank

$$= \frac{V}{B.D} = \frac{512}{9 \times 3.5} = 16.25 \text{ m ; say 16 m}$$

∴ Volume (V) provided = 16 m × 9 m × 3.5 m = **504 m³.**

(*i*) *Check for aeration period or H.R.T. (t)*

$$t = \frac{V}{Q} \,.\, 24 \text{ h (From eqn. 9.41)}$$

$$= \frac{504}{625} \times 24 \text{ h} = 19.35 \text{ h, say 20 hr ; OK,}$$

since it lies between 10 to 25 h

(*ii*) *Check for volumetric loading*. From equation (9.43), we have volumetric loading

$$= \frac{Q \,.\, Y_0}{V} \text{ gm of BOD}_5\text{/m}^3 \text{ vol. of tank}$$

$$= \frac{625 \times 295}{504} \text{ gm/m}^3$$

$$= 366 \text{ gm/m}^3$$

$$= 0.36 \text{ kg/m}^3. \text{ OK,}$$

since it should lie between 0.2 to 0.4

—See col. (7) of Table 9.10

(*iii*) *Check for return sludge ratio (for SVI ranging between 50—150 ml/gm).*

Using equation (9.54 *a*), we have

$$\frac{Q_R}{Q} = \frac{X_T}{\left(\dfrac{10^6}{\text{SVI}} - X_T\right)}$$

Using SVI = 100 ml/gm ; and X_T = 3000 mg/L, we have

$$\frac{Q_R}{Q} = \frac{3000}{\left(\dfrac{10^6}{100} - 3000\right)} = \frac{3000}{7000}$$

= 0.43 ; but it should be within 0.5 to 1.0

(please see col. (9) Table 9.10)

Hence, we will provide 50% sludge recirculation, giving SVI = 111, OK.

(*iv*) *Check for S.R.T.* (θ_c). From Eqn. (9.53), we have

$$V \,.\, X_T = \frac{\alpha_y \,.\, Q \,.\, (Y_0 - Y_E) \,.\, \theta_c}{1 + K_e \,.\, \theta_c}$$

where α_y = 1.0 (constant for municipal sewage w.r.t. to MLSS)

$K_e = 0.06\ d^{-1}$ (constant for municipal sewage)

Y_0 = 295 mg/L

Y_E = 20 mg/L

V = 504 m^3

X_T = 3000 mg/L

Q = 625 m^3/d

$$504 \times 3000 = \frac{1.0 \times 625 (295 - 20)\theta_c}{1 + 0.06\,\theta_c}$$

or $1 + 0.06\,\theta_c = 0.114\,\theta_c$

or $0.054\,\theta_c = 1$

$$\theta_c = \frac{1}{0.054} = 18.6 \text{ days ; OK}$$

since it is between 10 to 25 days, please see col. 8 Table 9.10.

The adopted tank size is thus O.K. Hence, adopt an aeration tank having an overall size of 16 m × 9 m × (3.5 + 0.6) m overall depth, with 0.6 m of free-board.

The outlet weir shall be of adjustable type.

The effluent from the aeration tank will be taken to the final (Secondary) clarifier. The inflow to the secondary clarifier shall be by means of 250 mm ϕ C.I. pipes, which will give a velocity of 0.78 m/sec at peak flow.

Aerators Sizing :

BOD_5 applied to each tank = 295 mg/l

Average flow in each tank = 625 m^3/day

BOD_5 to be removed in each tank = 625 × 0.295 = 184.375 kg/day

= 8 kg/hr (say)

Oxygen requirement = 1.2 kg/kg BOD applied

Peak oxygen demand = 125%

Oxygen transfer capacity of the aerator in standard conditions

= 1.9 kg/kWh = 1.41 kg/HP/hr

Oxygen transfer capacity of aerators at field conditions

= 0.7 × 1.41 = 0.98 kg/HP/hr

Oxygen to be applied in each tank

= 1.2 × 8 × 1.25 = 12 kg/hr

H.P. of aerators required

$$= \frac{12}{0.98} \text{ H.P.} = 12.24 \text{ ; say 14 H.P.}$$

Provide 2 Generators, each of 7 HP.

Check for Mixing Consideration

As per practice, power required for mixing = 0.02 kW/m^3

Volume of each A.T. = 504 m^3

SHP required = 0.02 × 504 kW = 10.08 kW

Providing 2 aerators, and considering gear efficiency as 97%

H.P. of each aerator required = $\frac{10.08}{2 \times 0.97}$ = 5.2 HP

Considering a power margin of 25% on motor rating

Motor H.P. required = 5.2 × 1.25 = 6.50 ; say 7

Provide 2 no. of 7 H.P. motors/aerators in each tank.

The dimension of the tank selected on the basis of suitable aerator motor is the zone of influence.

The zone of influence for each aerator will be about 11 m sq. with 3.5 m depth. The suitable aerator be selected, and accordingly, the depth and size of aeration tank adjusted. The submergence of the aerator will be between 75 mm to 115 mm.

(5) Design of Secondary Clarifier

Secondary Clarifier :

No. of secondary clarifier = 1 No.

Average flow = 1250 m^3/day

Recirculated flow, say 50% = 625 m^3/day

Total inflow = 1875 m^3/day

Provide hydraulic detention time = 2 hrs.

Volume of tank (exclusive of hopper portion) = 1875 × $\frac{2}{24}$ = 156.25 m^3

Assume liquid depth = 3.5 m

Area (superficial) = $\frac{156.25}{3.5}$ = 44.64 m^2

Surface loading rate of average flow = 15 m^3/m^2/day

Surface area to be provided = $\frac{1250}{15}$ = 83 m^2

(provide area greater of two *i.e.* 83 m^2)

Dia of Circular tank (d)

$$d = \sqrt{\frac{83 \times 4}{\pi}} = 10.28 \text{ m ; say 11 m.}$$

Actual area provided = 95 sq. m.

Check for Weir Loading :

Average flow = 1250 m^3/day

Weir loading = $\frac{1250}{\pi \times 11}$ = 36.2 m^3/day/m ; O.K. as it is less than 185 m^3/day/m

Provide a peripheral launder

Check for Solids Loading :

Recirculated flow = 625 m^3/day

Average flow = 1250 m^3/day

MLSS in the tank = 3000 mg/l

Total solids inflow = (1250 + 625) × 3 = 5625 kg/day

Solids loading = $\frac{5625}{83}$ = 67.80 kg/day/m^2

Provide a clarifier of 11 m dia having liquid depth as 3.5 m

Hopper slope shall be 1 in 12.

Free-board will be 0.3 m

Sludge will be withdrawn from the clarifier through a C.I. pipe. The sludge will be taken to the return sludge pump house. The treated effluent from the secondary clarifier can be disposed of in the nearby nallah valley.

Return Sludge Pump House :

Total return flow = 625 m^3/day = 0.434 m^3/min

Detention time = 15 min

Volume of wet well = 0.435 × 15 = 6.5 m^3

Provide wet well = 2.5 m × 1.5 m × 1.8 m SWD

Provide dry well = 2.5 m × 2.5 m

Size of annexe control room = 2.5 m × 2.5 m

Provide 2 Nos. pumps, each of 0.625 MLD capacity in the dry well for returning the sludge to the aeration tank.

The return sludge pipe line shall be 150 mm ϕ.

Excess sludge shall be taken to the sludge drying beds by providing necessary fittings in the main return sludge header. The sludge feed line to the sludge drying beds shall be 150 mm ϕ C.I. pipe.

(6) Design of Sludge Drying Beds

Sludge applied to drying beds @ 100 kg/MLD

Sludge applied = 125 kg/day

Sp. gravity = 1.015

Solid contents = 1.5%

Volume of sludge = $\frac{125}{1.5\%} \times \frac{1}{1000 \times 1.015}$ = 8.2 m^3/day

Considering monsoon etc. total no. of cycle in one year = 33

Period of each cycle $\frac{365}{33}$ = 11 days

Volume of sludge/cycle = 8.2 × 11 = 90.2 m^3

Spreading a layer of 0.3 m/cycle area of beds required = $\frac{90.2}{0.3}$ = 300.67

Provide 4 beds of 1.2 m × 7 m

Thus, providing = 336 sqm. area

Filtrate Pump House and Sump

Actual BOD_5 20°C removed per day = $1250 \times \dfrac{(295-20)}{1000}$ = 343.25 kgm.

Excess Wasted Sludge. From equation (9.49 *a*), we have

$$\theta_c = \frac{V \cdot X_T}{Q_W \cdot X_R}$$

$$\therefore \qquad 18.6d = \frac{504 \text{ m}^3 \times 3000 \dfrac{\text{gm}}{\text{m}^3}}{Q_W \cdot X_R}$$

or $\qquad Q_W \cdot X_R = \dfrac{504 \times 3000}{18.6}$ = 81,290 g/d = 81.3 kg/d

Thus, excess sludge produced = 81.3 kg/d.

Assuming the excess sludge to contain 1% solids and specific gravity = 1.015, we have

Volume of excess sludge

$$= \frac{81.3}{1\% \times (1000 \times 1.015)} \text{ m}^3/\text{d}$$

$$= \frac{81.3 \times 100}{1000 \times 1.015} = 8 \text{ m}^3/\text{d} = 0.34 \text{ m}^3/\text{hr}.$$

Taking detention time as 8 hrs.

Volume of wet well = 8 × 0.34 = 2.64 m^3 for 1% concentration.

Provide liquid depth = 1.0 m

Area required for 1% concentration of solids = $\dfrac{2.64}{1.0}$ = 2.64 m^2

Dia of wet well = $\sqrt{\dfrac{2.64 \times 4}{\pi}}$ = 1.83 m ; Provide 2.0 m ϕ

Hence, provide 2 m ϕ of the wet well for (worst condition) filtrate sump with 1.2 m liquid depth below incoming pipe line. This will be covered on top and a 4 m high pump room shall be constructed over it. 2 No. vertical non-clog pumps, each of 1 H.P. capacity, shall be provided in this pump house. The filtrate from the pump house shall be pumped to the distribution chamber upstream of the Aeration Tank through 100 mm ϕ C.I. pipe line.

The details of the designed units are shown in **Chart Fig. 10.2 at the end of the book.**

—11

Treatment of Industrial Wastewaters

11.1. Introduction

Wastewaters obtained from industries are generally much more polluted than the domestic or even commercial wastewaters. Still, however, several industrialists try to discharge their effluents into our natural river streams, through unauthorised direct discharges. Such a tendency on the part of the industries may pollute the entire river water to a grave extent, thereby making its purification almost an impossible task. Sometimes, the industries discharge their polluted wastewaters into municipal sewers, thereby making the task of treating that municipal sewage, a very difficult and a costly exercise.

The industries are, therefore, generally prevented by legal laws, from discharging their untreated effluents. It, therefore, becomes necessary for the industries to treat their wastewaters in their individual treatment plants, before discharging their effluents either on land or lakes or rivers, or in municipal sewers, as the case may be.

11.2. Methods of Treating Industrial Wastewaters

Industrial wastewaters, as pointed out above, usually contain several chemical pollutants and toxic substances in too large proportions. *The characteristics of the produced wastewater will usually vary from industry to industry, and also vary from process to process even in the same industry.* Such industrial wastewaters cannot always be treated easily by the normal methods of treating domestic wastewaters*, and certain specially designed methods or 'sequence of methods' may be necessary.

In order to achieve this aim, it is generally always necessary and advantageous to isolate and remove the troubling pollutants from the wastewaters, before subjecting them to usual treatment processes. *The sequence of treatment processes adopted should also be such as to help generate useful by-products.* This will help economise the pollution-control measures, and will encourage the industries to develop such treatment plants.

---

*The normal biological treatment methods for sewage are dependent upon the bacterial activity within the sewage ; and as the toxic chemicals present in the industrial wastewaters may hinder or destroy the bacterial activity, these normal methods may not suffice, unless medified and/or supplemented by additional techniques.

Depending upon the *quantum, concentration, toxicity* and *presence of non-biodegradable organics* in an industrial wastewater, its treatment may consist of any one or more of the following processes :

(1) *Equalisation* ;

(2) *Neutralisation* ;

(3) *Physical treatment*

(4) *Chemical treatment* ; and

(5) *Biological treatment.*

These processes are briefly discussed below :

11.2.1. Equalisation. Equalisation consists of holding the waste-water for some pre-determined time in a continuously mixed basin, to produce a uniform wastewater. Such an arrangement will, of course be necessary when the wastewater produced by the industry varies in characteristics and quantity over the entire day.

11.2.2. Neutralisation. Neutralisation means neutralising the excessive acidity or alkalinity of the particular wastewater, by adding alkali or acid, respectively, to the wastewater. This may be achieved either in the ***equalisation tank,*** where possible, or a ***separate neutralisation tank*** may be used.

11.2.3. Physical Treatment. Physical treatment consists of separating the suspended inorganic matter by physical processes, like *sedimentation* and *floatation.*

(*i*) **Sedimentation.** Sedimentation as we know, is employed to separate the heavier settleable solids, and hence sedimentation tank may be provided only when the wastewater contains a high percentage of such heavy inorganic solids.

(*ii*) **Floatation.** Floatation consists of creation of fine air bubbles in the waste tank, by introduction of air into the tank from the bottom. The rising air bubbles, attach themselves to the fine suspended particles, increasing their buoyancy, and finally lifting them to liquid surface for consequent removal by **skimming.**

11.2.4. Chemical Treatment. Chemical treatment is often necessary before the biological treatment, though sometimes, it may not be required at all. Sometimes, it may however, serve as the final stage of treatment*.

Since chemical treatment is a costly and expensive exercise, care should be taken to see, if it could be avoided altogether, to achieve our required goals.

The chemical treatment is used to recover the dissolved organic matter from the wastewater, and may consists of one or more of the following processes :

(*i*) *reverse osmosis* or *hyper filtration* ;

(*ii*) *electrodialysis* ;

---

*Chemical treatment was largely used in England during 1870 to 1930, when it was replaced by the advent of biological treatment methods. Since 1970 or so, it has again come up for isolated uses on specific industrial wastewaters, and is increasingly being developed for separation of impurities as byproducts.

(iii) *chemical oxidation* ;

(iv) *chemical coagulation* or *chemical precipitation* ;

(v) *adsorption* ;

(vi) *deionisation* ;

(vii) *thermal reduction* ; and

(viii) *air-stripping.*

All these processes are briefly pointed out below :

(i) **Reverse Osmosis.** In reverse osmosis treatment process, the wastewater containing dissolved salts are filtered through semipermeable membranes at a pressure higher than the osmotic pressure, as discussed in article 9.29.6(3)—Vol. I. Such a treatment requires pre-treatments like : (a) *activated carbon adsorption* ; or (b) *chemical precipitation* followed by some kind of *filtration.*

(ii) **Electrodialysis.** In electrodialysis treatment process, dissolved salts from wastewater are separated by passing an electric current through the wastewater tank, installed with 'ion exchange membranes', as discussed in article 9.29.6(2)—Vol. I. This treatment process also requires some pretreatment, as is required in reverse osmosis process.

(iii) **Chemical Oxidation.** In chemical oxidation, chemicals like *chlorine* and *ozone* are used to reduce substances like *ammonia* and *cyanide*, etc. from the wastewater, in addition to reducing BOD load on biological treatments.

(iv) **Chemical Coagulation.** Chemical coagulation, as we know, is adopted in treating raw water supplies, and helps in sedimentation of unsettleable micro and colloidal impurities, which get absorbed in the gelatinous flocs, formed by the chemical reactions between coagulants, or between the coagulant and the alkalies present in raw water.

However, in treatment of industrial waste-waters, such chemicals are generally added to help out the precipitation of dissolved and troublesome impurities, such as *phosphorous* or *phosphoric compounds,* etc. This process is then popularly known as **chemical precipitation.**

The *alum*, or ferric salts like *ferric chloride* or *ferric sulphate* may be used to directly *precipitate out phosphorous*, as shown below :

$$Al^{+++} + H_nPO_4^{---} \rightleftharpoons \underset{\text{Al. Phosphate}}{AlPO_4} \downarrow + nH^+ \quad \text{...(11.1)}$$

$$F^{+++} + H_nPO_4^{---} \rightleftharpoons \underset{\text{Ferric Phosphate}}{FePO_4} \downarrow + nH^+ \quad \text{...(11.2)}$$

Sometimes, *lime* is also used to precipitate out phosphorous ; wherein lime will initially react with natural bicarbonate alkalinity to precipitate $CaCO_3$, and then excess lime will react with phosphorous to precipitate *hydroxylapatite*, as given in equation (11.3) below :

$$10\,Ca^{++} + 6\,PO_4^{--} + 2OH^- \rightleftharpoons \underset{\text{Hydroxylapatite}}{Ca_{10}(PO)_6(OH)_2} \downarrow \quad \text{...(11.3)}$$

Chemical precipitation, though introduces additional chemicals in wastewater and produces larger quantity of sludge, yet is generally adopted these days for removing organic compounds like *phosphorous*, and also sometimes for removing nutrients like *nitrogen from* specific industrial wastewaters. Toxic metals may also sometimes be precipitated out by using and developing appropriate chemical processes.

(*v*) **Adsorption.** Adsorption is generally used to remove non-biodegrable organics (like synthetic detergents), colour and odour from the wastewater. The process involves passing the pre-treated wastewater through the beds of activated carbon, as discussed in article 9.25 (2) of Vol. I.

(*vi*) **Deionisation.** Deionisation involves a similar passage of wastewater through the beds of synthetic ion exchange resins, where some undesirable cations or anions of the wastewater get exchanged for sodium or hydrogen ions of the resin.

(*vii*) **Thermal Reduction.** Thermal reduction requires burning and consequent oxidation of some toxic and refractive substances, like organic cyanide, which may be present in certain specific industrial wastewaters.

(*viii*) **Air-stripping.** In air-stripping, the liquid waste is poured down through a packed tower, equipped with an air blower at the bottom. This method is a modification of *aeration*-process, used for removal of gases from the wastewater.

11.2.5. Biological Treatment. Biological treatment of industrial wastewaters is necessary, when they contain large quantities of biodegradable substances. Such biological treatment may be used either with or without *acclimatisation*.

Laboratory tests on wastewater for determining its $\frac{\text{BOD}}{\text{COD}}$ ratio, will help in determining the type of treatment required. Say for example, if this ratio is more than 0.6, the wastewaters are biologically treatable without acclimatisation ; and if the ratio is less than 0.6 and upto 0.3, then acclimatisation is needed for biological treatment ; and if the ratio is less than 0.3, biological treatment may not be necessary.

Acclimatisation consists of the gradual exposure of the waste-water in increasing concentration to the seed or initial micro-biological population under a controlled condition.

The criteria for selecting a particular *conventional biological treatment process**, may differ for different types of industrial wastewaters. The system parameters for a given type of industrial wastewater may be determined by laboratory experiments.

In the absence of any actual test data, the performance data of a similar type of industrial wastewater may be used for design.

---

*Conventional biological treatment may consists of : (*a*) *aerobic units like : cheaper stabilisation ponds or aerated lagoons*, or costlier units like *trickling filters or activated sludge plant* ; and (*b*) *anaerobic units like anaerobic lagoons and septic or imhoff tanks.*

It has also been observed that most of the industrial wastewaters do not contain sufficient nutrients for micro-biological growth ; and hence nutrients like *urea* (containing nitrogen), *super phosphates* (containing phosphorous) etc. may have to be added to the reactors. For **balanced growth of micro-organisms** in a biological treatment reactor, the ratios of BOD : Nitrogen : Phosphorous should be 100 : 5 : 1 for aerobic systems, and 100 : 2.5 : 0.5 for anaerobic systems.

In certain cases, **special types of micro-organisms** are found to cause better biological oxidation, since commonly available micro-biological population may fail to achieve oxidation.

Acute *Toxic wastewaters* must be handled with special care and specific treatments. Say for example, toxic wastes, like cyanides, formaldehydes, phenols, etc. may need acclimatised-growth of special type of bacteria. Toxic metals, like copper, zinc, chromium etc. may need pre-separation chemically, as otherwise, they may interfere with the biological oxidation by tying up the enzymes, essentially required for microbial growth.

Segregation of strong wastewaters from weak wastewaters may also sometimes help in reducing problems in industrial wastewater treatment plants.

In conclusion, it can be stated that the selection of a sequence of particular treatment processes depends on the characteristics of the wastewaters, and also upon the permissible requirements of the effluents.

The treatment processes needed for different industries will, therefore, generally vary from one industry to another.

The attempts made for treating industrial wastewaters, simply by biological methods, though cheaper, have generally failed in the absence of making efforts for pre-recovery of certain pollutants or chemicals from the wastewater by using chemical methods. However, use of simple biological methods have fairly succeeded in industries, like *fruit processing, dairies, slaughter houses, and textiles.*

Physical and chemical treatments prior to biological treatment, for separation of troubling pollutants, like chromium, arsenic, cyanide, mercury, several nitrogenous substances, lignin, etc., is very necessary and important for industries like *fertilizers, dyes and pigments, pesticides, electroplating, paper and pulp*, etc. Suitable pre-treatments *to the wastewaters of such industries, before subjecting them to the biological treatment, is thus, the prime requirement for designing and planning the treatment plant for a particular industry.*

Possible large scale re-use of the treated water in the industry is another important factor, which must be considered, while deciding the sequence of treatment processes for a particular industry. Such a possible reuse, if can be made possible, will help in large scale economy in the industry.

The **pollution characteristics of certain typical Indian industries** along with suggested treatments with **flow charts**, are broadly reflected in Tables 11.1 and 11.2 so as to serve as a first hand guide to the design engineers. For detailed planning and design of the treatment works, however, the design engineer will have to be well conversant with the detailed operations and stages involved in the given industry, study the characteristics of the wastewaters generated at different places in the industry, and to finally work out the most satisfactory and economical sequence of treatment processes for that industry, using his intuition, intelligence, experience, and of course, ingenuity.

Table 11.1. Pollution Characteristics of Certain Typical Indian Industries

Sl. No. (1)	*Industry Type* (2)	*Pollution Characteristics* (3)	*Suggested Treatments* (4)
1.	**Dairy** *(Milk processing industry)*	(*i*) moderate BOD (800—1000 mg/l) (*ii*) $\frac{\text{BOD}}{\text{COD}} \approx 0.6$ (*iii*) heavy oil and grease (200—300 mg/l) (*iv*) high dissolved solids (1000 mg/l or so) (*v*) high suspended solids (800 mg/l or so) (*vi*) high nitrogen *Very strong, though bio-degradable.*	Aerobic Biological treatment, either in the form of *High rate trickling filters* or *Activated sludge plant* will generally suffice for this industry. Still however, due to intermittent nature of wastewater discharge, an *Equilisation tank* with or without *aeration* may preferably be used. In unadvanced locations, *Oxidation ditch* or *Aerated lagoon* may also help ensure good disposal.
2.	**Distillery** and *Brewery wastes.* (*Wines ; alcohols,* and *brandy producing industry*)	(*i*) extremely high BOD (12000—40000 mg/l) (*ii*) very high COD (28000—73000 mg/l) (*iii*) high chlorides and sulphates—dissolved solids (7000—10000 mg/l) (*iv*) highly coloured-brownish yellow.	*Brewery wastewaters,* being less strong, can be generally treated by two stage aerobic biological treatment units like *High rate trickling filters,* after *screening* and *neutralisation* (Fig. 11.1). But yeast sludge *Distillery wastewaters* generally make it highly polluted, as shown in col (3) ; thereby necessitating *two stage treatment* ; consisting of *biological anaerobic* treatment followed by *aerobic* treatment (Fig. 11.2)

(*Contd.*

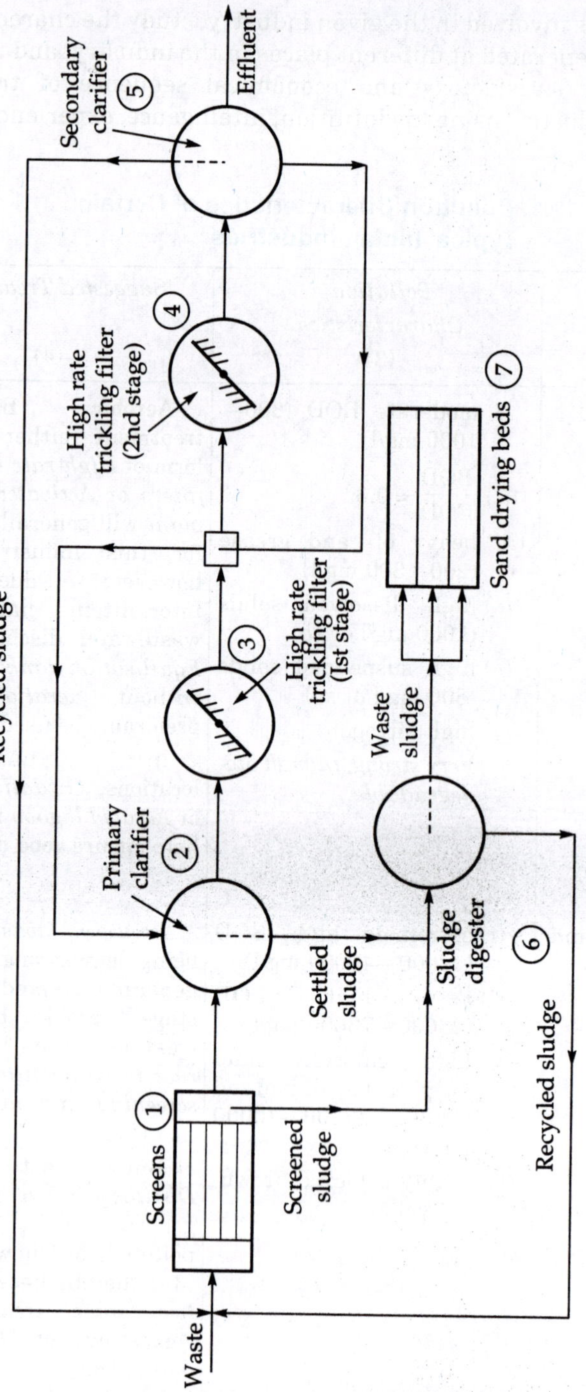

Fig. 11.1. Flow chart for treating liquid wastes of a Brewery Unit—*Two Stage Aerobic Treatment suggested ; 90—94% BOD removal.*

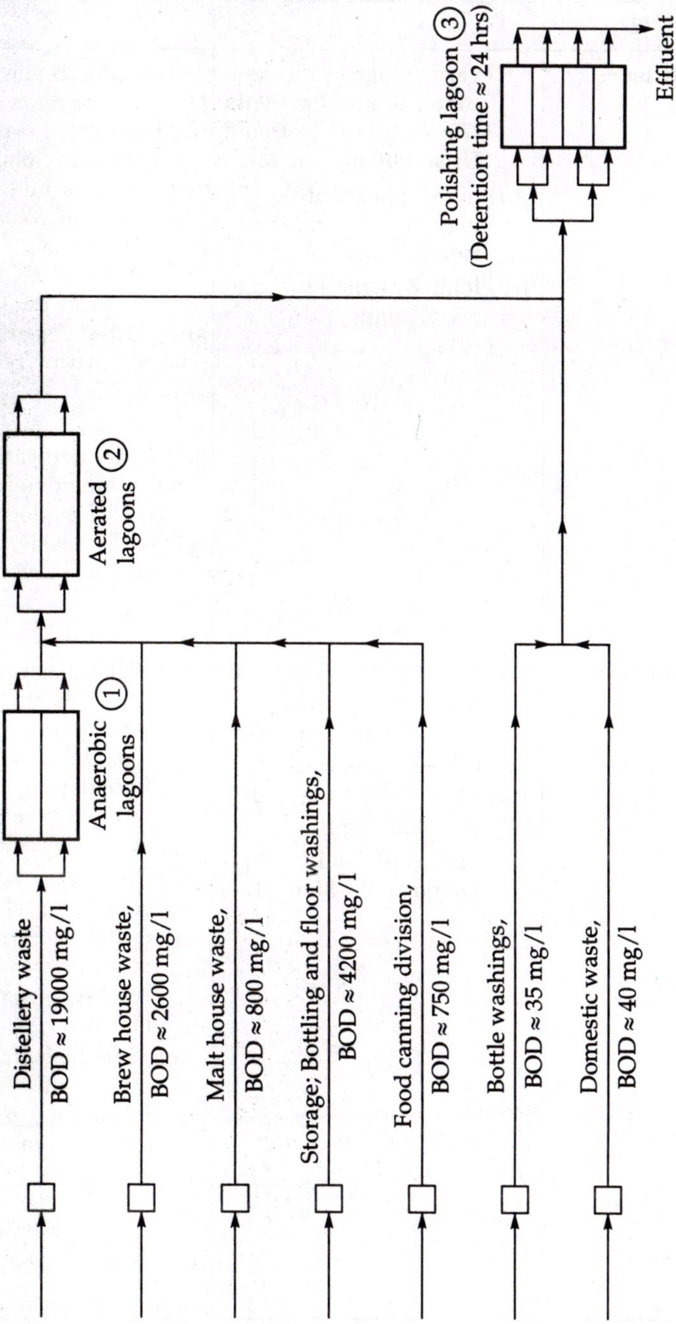

Fig. 11.2. Flow chart for treating wastewaters of a Complex Distillery Plant, containing brewery, distillary and fruit carving divisions—*Both Aerobic and Anaerobic Treatments suggested.*

(1)	(2)	(3)	(4)
3.	**Fertilisers**	(*i*) very high nitrogen content as Ammonia (700 mg/l or so) and Urea (600 mg/l or so) ; (*ii*) high phosphates (75 mg/l or so) and Fluorides (15 mg/l or so) ; (*iii*) high Arsenic (1.5 mg/l or so) ; and (*iv*) Oils	Oils are removed in a *Gravity separator* ; arsenic is segregated after its concentration ; phosphates and fluorides are segregated and chemically coagulated by lime ; ammonia is then removed by *Lagooning* or by using an *Air-stripped tower* packed with red wool stakes. Biological nitrification and denitrification to remove ammonia can also be used. Urea does not get removed easily. Flow diagram is shown in Fig. 11.3.
4.	**Oil Refineries**	(*i*) very high free oils (2000—3000 mg/l). (*ii*) high emulsified oils (90—120 mg/l). (*iii*) high H_2S and RSH (10—220 mg/l). (*iv*) high phenolic compounds (12—30 mg/l)	(*a*) Physical separation of free oils in *Gravity separators*. *Air floatation* with or without flocculation may also be tried to remove oils in some special cases. (*b*) Chemical treatment consisting of *Coagulation*, *Flocculation* and *Sedimentation* to remove emulsified oils, etc. (*c*) Final biological treatment, such as *Stabilisation pond* or *Aerated lagoon*, or *Trickling filter*, or *Activated sludge process*. Flow diagram is shown in Fig. 11.4.

(*Contd.*)

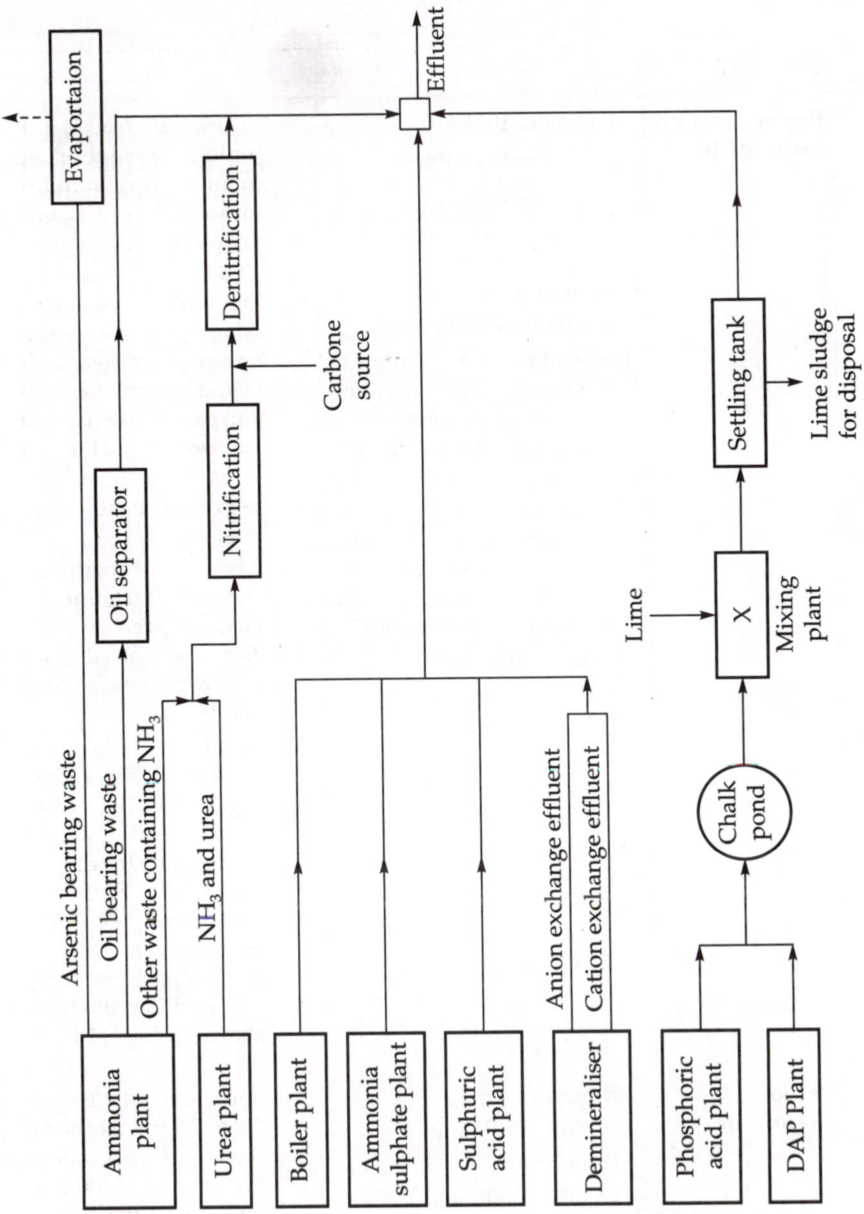

Fig. 11.3. Flow chart for treating wastewaters from a Complex Fertiliser Plant.

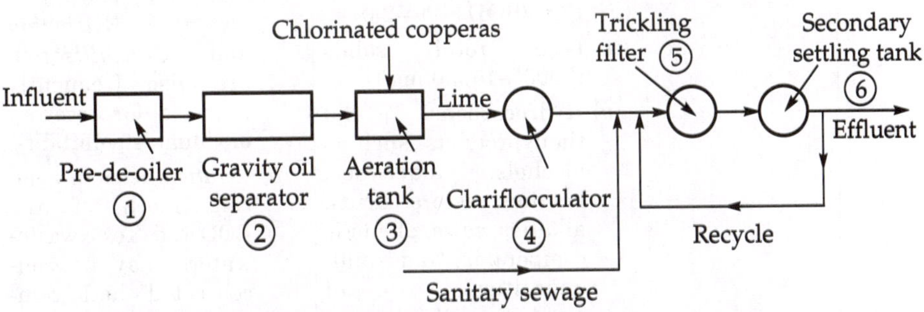

Fig. 11.4. Flow chart for treating Oil-Refinery (Petroleum) wastewaters—*Only Aerobic treatment suggested after oil segregation and aeration.*

(1)	(2)	(3)	(4)
5.	**Paper and Pulp Mills**	(*i*) moderate BOD (150—1250 mg/l) (*ii*) Low $\frac{\text{BOD}}{\text{COD}}$ ratio (0.25—0.20) (*iii*) highly coloured (4000—8000 units) (*iv*) acidic for sulphite plants,, and alkaline for other processes ; (*v*) Lignin, which is highly resistant to biological treatment, is present in paper mill waste-waters, especially in plants based on "Sulphite-technology". (*vi*) Low nitrogen.	(*a*) *Chemical Recovery of lignin* (produced in plants using sulphite process), as a useful by product. (*b*) Chemical treatment for colour removal. Massive **lime treatment** process is widely used for this purpose. **Activated carbon** at pH lower than 3 is also found to serve as good colour remover. (*c*) Physical treatment like *Mechanically cleaned circular clarifier* is used for removing suspended solids. (*d*) Final Biological treatment for reducing BOD. Both conventional or low cost processes (such as *Lagoons, Stabilisation ponds*, etc.) can be used. The process also helps in reducing colour. Flow diagram shown in Fig. 11.5.
6.	**Petro-chemicals**	(*i*) high BOD (600—5800 mg/l) (*ii*) high COD (580—5000 mg/l) (*iii*) COD/BOD (1.4—2.0) (*iv*) high total solids (2000—15000 mg/l), (*v*) Hydrocarbons and their products, such as alcohols, aldehydes, ketones, esters, acids, alkalies, salts, phenolic compounds, and small quantities of emulsified oils.	(*a*) Recovery of byproducts by methods, such as *Stripping, Scrubbing, Adsorption*, and *Distillation*. *Oxidation*, *Extraction* and *Crystallisation* are also frequently required for recovery of valuable pollutants. (*b*) *Aeration* helps in reducing much of the BOD load. Toxic waste-waters may be concentrated and removed by *Incineration*.

(*Contd.*)

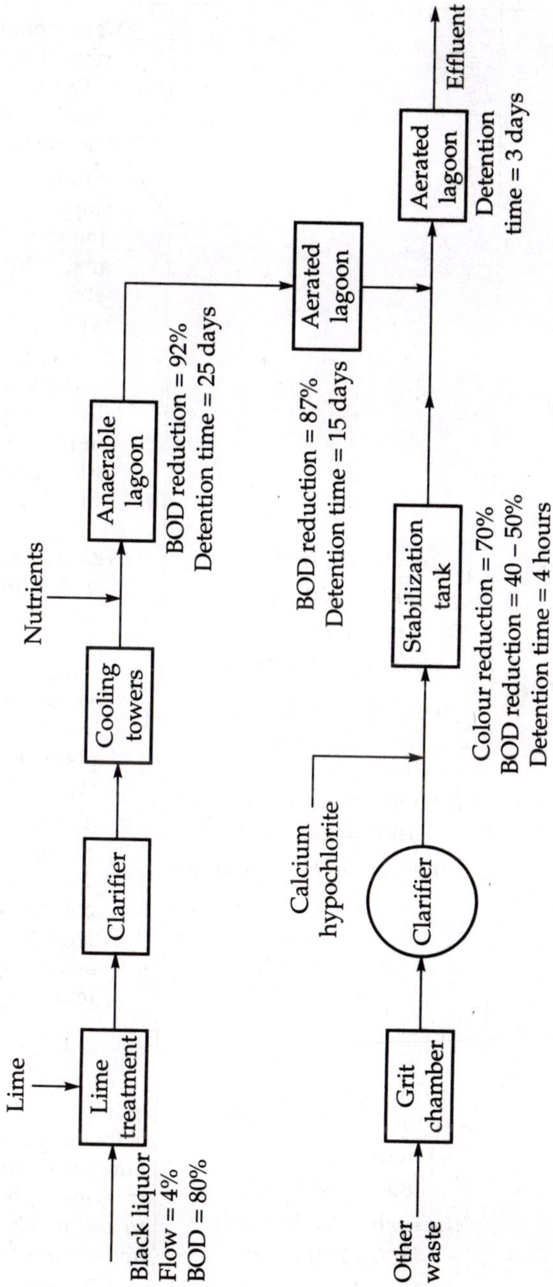

Fig. 11.5. Flow chart for treating wastewaters of a typical Pulp Mill.

(*Contd.*

(1)	(2)	(3)	(4)
			(*c*) To remove colloids and to control pH, *Chemical treatment* is used previous to biological treatment. (*d*) *Final Biological treatment* preceded by Sedimentation, Screening, Oil separation, etc. All modes of biological units, sometimes succeeded by secondary units may have to be used. Flow sketch shown in Fig. 11.6.
7.	**Pharmaceuticals** (*a*) *Antibiotics* (*b*) *Synthetic drugs*,	(*i*) moderate BOD (500—1000 mg/l) (*ii*) very high COD (5000—7000 mg/l) (*iii*) very low $\frac{\text{BOD}}{\text{COD}}$ ratio (0.10—0.13) (*iv*) high total solids (1000—5000 mg/l) (*v*) either acidic or alkaline	(*a*) *Equalisation, Neutralisation with lime, Clarifier, Anaerobic digestion + conventional Aerobic biological processes* are used in a typical antibiotic plant (Fig. 11.7). (*b*) For synthetic drug plants, very careful planning is required due to excessive pollution potential. Highly specialised sequence of operations necessary.
8.	**Sugar Industry**	(*i*) moderate BOD (300—2000 mg/l) (*ii*) high COD (600—4400 mg/l) (*iii*) high volatile solids (400—2200 mg/l) (*iv*) low pH (4.6—7.1)	Anaerobic treatment using *Digestors* or *Lagoons* is found to be very effective and economical. Biological treatment, however, needs to be preceded by *Screens* and *Grease traps*. Where sufficient land is available, two stage biological treatment with *Anaerobic lagoons followed*

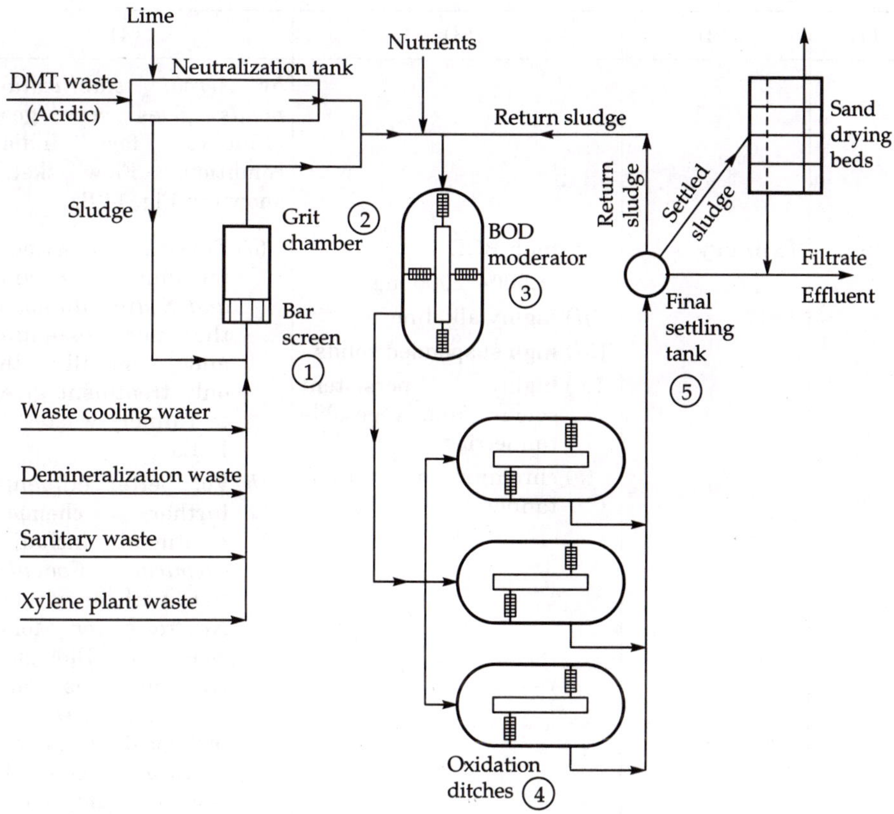

Fig. 11.6. Flow chart for treating Aromatic wastewaters of a Petrochemical Complex.

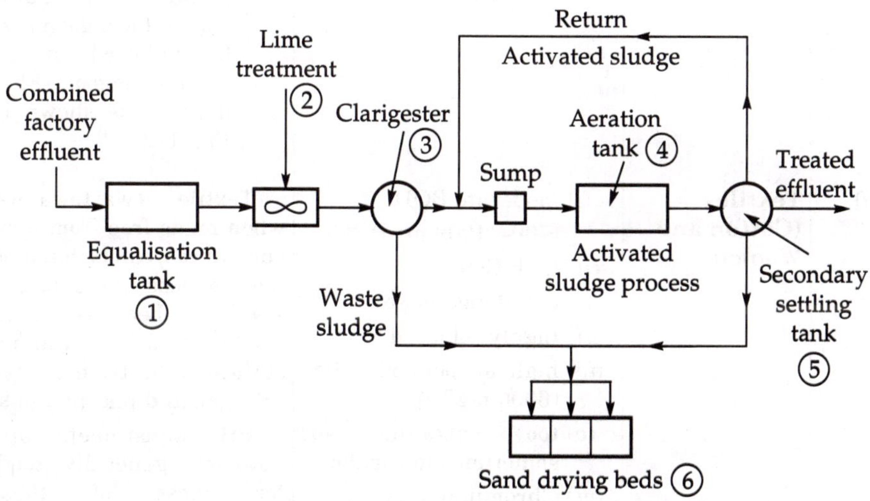

Fig. 11.7. Flow chart for treating wastewaters from a Pharmaceutical Plant, producing combined antibiotic and other chemical wastes—*Only Aerobic activated treatment after lime treatment is suggested.*

(1)	(2)	(3)	(4)
			by *Aerobic stabilisation ponds* gives very good effluents for Indian conditions. Flow sketch shown in Fig. 11.8.
9.	**Tannery**	(*i*) high BOD (1.500—2000 mg/l) (*ii*) highly alkaline (*iii*) high suspended solids (*iv*) highly persistent colour from vegetable tanne-ries (*v*) chromium in chrome tanneries.	(*a*) Physical treatment including *Screening and Sedimentation* is the most essential, and generally the only treatment given to Tannery wastes, in India. (*b*) For better effluents, further chemical treatment including *chemical Coagulation*, with or without *Neutralisation*, followed by Biological treatment is also required. Low cost biological units like **Lagoons** can be cheaply and easily used for smaller and isolated plants. To obtain better results, Anaerobic lagoon may be followed by an Aerated lagoon. Flow diagram is shown in Fig. 11.9.
10.	**Textiles (Cotton and Woolen)**	(*i*) moderate BOD (700—1000 mg/l) (*ii*) high COD (1400 mg/l or so) (*iii*) highly coloured (*iv*) high suspended solids (6000 mg/l or so) (*v*) too alkaline or sometimes too acidic (*vi*) Chromium (10—15 mg/l)	Textile wastewaters, when made free from toxic substances, can be handled more or less like domestic sewage, because such wastewaters contain sufficient nutrients like nitrogen and phosphorous. PH adjustments are, however, generally reqd. for most of these wastewaters.

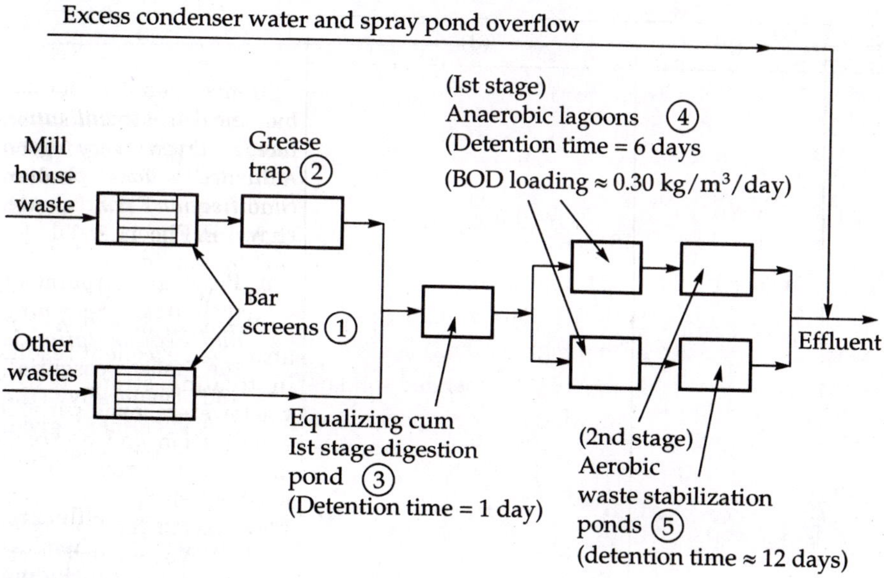

Fig. 11.8. Flow chart for treating wastewaters of a Sugar Mill—*Anaerobic treatment followed by Aerobic treatment is suggested.*

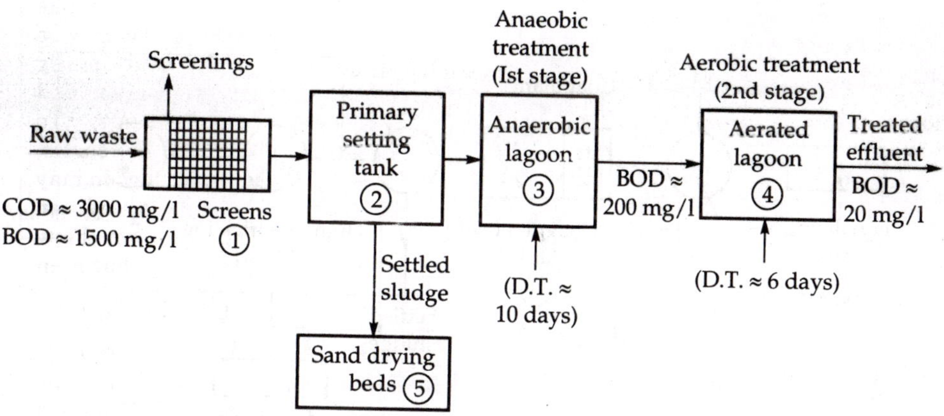

Fig. 11.9. Flow chart for treating Tannery wastewaters—*Anaerobic treatment followed by Aerobic treatment is suggested.*

(1)	(2)	(3)	(4)
			Conventional aerobic biological treatment units, like, *Trickling filters*, *Activated sludge process*, *Stabilisation ponds*, etc. have all been used for textile wastewaters, and found quite effective. *Extended aeration* has also given excellent results in treating strong textile wastewaters, even without equalisation and pretreatment. It also removes the need of sludge digestion. Flow sketch for Cotton as well as Woolen textile industries are shown in Fig. 11.10 (*a*) and (*b*), respectively.

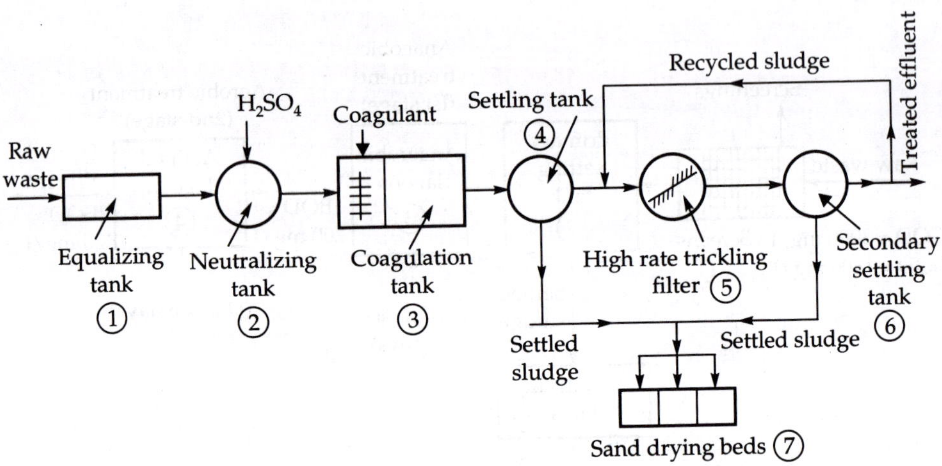

Fig. 11.10. (*a*) Flow chart for treating Cotton Textile mill wastewaters—*mostly treated like domestic sewage.*

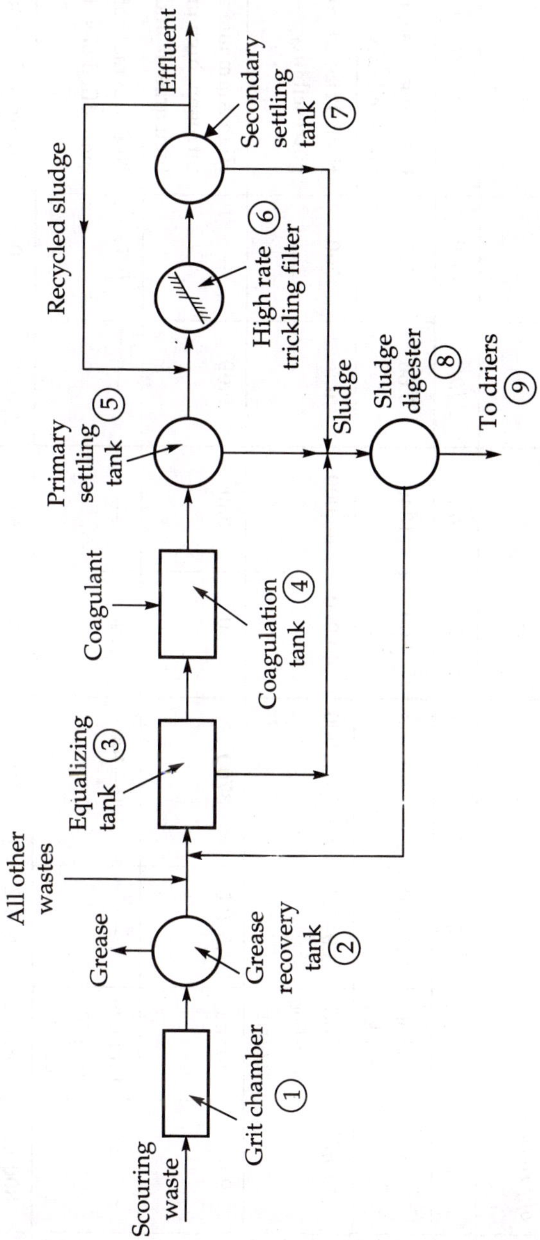

Fig. 11.10. (*b*) Flow chart for treating Woolen mill wastewaters—*mostly treated like domestic sewage.*

Table 11.2. Characteristics of Waste-Waters from Selected Industries in India

S. No.	*Name of Industry*	*Amt. of waste-water produced per unit of produc-tion*	*Unit of produc-tion*	*pH value*	*Suspended solids in mg/l*	*Typical BOD_5 value in mg/l*	*Typical COD value in mg/l*	*Typical $\frac{BOD_5}{COD}$ ratio*	*BOD load per kg of product*	*Other important char-acteristics*
(1)	*(2)*	*(3)*	*(4)*	*(5)*	*(6)*	*(7)*	*(8)*	*(9)*	*(10)*	*(11)*
1.	Dairy	7 m^3	m^3 of milk	7.2–8.0	700	800	1340	0.66	5.7	Oil and grease, ready putrescibility
2.	Distillery	14 m^3	m^3 of rectified spirit	3.9–4.3	4500—12000	40,000	80,000	0.5	560	High TDS, Cl, SO_4 and putrescibility
3.	Fertilizer (Ammonia)	9 m^3	tonne	7.5–9.5	3700	30	330	0.09	—	High Ammonia and urea nitrogen, high phosphates, high arsenic and oil
4.	Oil Refinery	1.7 m^3	tonne of oil pro-cessed	—	200—400	200	—	—	0.34	High quantity of free and emulsified oils, H_2S, etc.
5.	Pulp and Paper	300 m^3	tonne of paper	6.5–8.2	1000—3000	160	610	0.26	50	Lignin, colour, and suspended fibre

(1)	*(2)*	*(3)*	*(4)*	*(5)*	*(6)*	*(7)*	*(8)*	*(9)*	*(10)*	*(11)*
6.	Petro-chemical (Primary)	0.2–5.4 m^3	tonne	5.3–10.0	—	100–1000	500–3000	0.2–0.33		Phenol, oil and pH
7.	Phar maceutical (Synthetic drugs)			6.5–7.0	—	500–1000	5000–7000	0.10–0.13		Either acidic or alkaline, high total solids, highly specialised sequence of operation necessary
8.	Sugar mills	3 m^3	tonne of cane crushed	4.6—7.1	220—800	300—2000	600—4400	0.7—0.5		Low pH, high volatile solids
9.	Tannery	3.2 m^3	tonne of hide	8—9	1000—2000	1700	3500	0.48	5.4	$Cr^{+3} \approx$ 0.4 kg/100 kg of hide, Cl $\approx$ 4600 mg/l, tanin, colour
10.	Textiles (cotton)	25 m^3	1000 m of cloth	7—10	375	350	525	0.66	8.75	Colour, TDS $\approx$ 300 mg/l, % sodium $\approx$ 0.9
11.	Viscous Rayon	570 m^3	tonne	2.8—4.1	200	—	210	—	—	Zinc $\approx$ 6 to 18 mg/l, poly sulphides

11.3. Thermal Pollution

Certain industries release hot waters directly into the natural streams or lakes, thereby polluting their water supplies due to *warming*. Although small amount of heat released into a water body may not have any serious repercussions, yet large quantities may certainly kill heat-intolerant plants and water animals. Elimination of heat intolerant species would lead to flourishing of *heat tolerant species,* which are usually *less desirable species*.

The warming of water also lowers the DO (dissolved oxygen) and increases its BOD, and at the same time increases the metabolic rate of aquatic organisms. Since metabolism requires oxygen, certain species may be entirely eliminated if the temperature of the water body rises by 10°C or so. Heated waters may also interfere with the movement of migratory fish and their spawning process ; and similarly, too warm waters may not allow hatching of trout eggs. Warming of water also increases the susceptibility of aquatic organisms to parasites, toxins and pathogens, due to increased BOD.

The hot water discharges in water bodies, thus, drastically alter the aquatic ecosystem, thereby affecting the entire aquatic food chain. Excessive cold discharges in water bodies from industries may also similarly cause drastic changes in the biota of the waterbody. *Such pollution of water, which is caused by its abnormal warming or cooling, is known as thermal pollution.*

Sharp and sudden changes of temperature in a water body may even lead to large scale death of fish and other organisms, that fail to escape. Such an incident of sudden and heavy deaths of aquatic life is called a **thermal shock.** Evidently, a *thermal shock* may be caused by a new industry, which on a fine day, starts releasing its hot water from its cooling process ; or a shut industry, which restarts its operation after remaining idle for a long time.

Thermal pollution is largely caused by the industries, which release hot waters from their cooling processes. The worst offending industries in this respect are :

nuclear power plants, steel mills, oil-refineries, paper mills, etc.

Thermal pollution can be controlled by constructing *cooling ponds* for collecting hot water for its cooling before its release into the nearby water body. **Cooling towers** are also constructed and used in modern days to dissipate heat energy before releasing the wastewater to avoid its hot discharge into the water body.

—12

Excreta Disposal in Unsewered and Rural Areas

12.1. General Introduction

Sewers, as we know, have not been laid in large part of India, due to non-availability of enough funds. The priority was first given to the piped potable water supply, education, industrialisation, etc. in comparison to sewerage. Hence, ever after more than 60 years of independence, we find the existence of unsewered trans-Yamuna colonies even in the capital of this country (*i.e.* in Delhi), where a large population is living. When the capital city of our country is having large population living with conservancy system of sanitation, one can easily imagine the fate of our town and village folks. Most of these places are without sewers, and are provided with conservancy system of sanitation.

Even the unsewered colonies can be divided into two types, *i.e.* (*i*) those which are provided with organised municipal water supplies ; and (*ii*) those which are not provided even with such water supplies. Most of our cities and towns fall under the first category ; whereas, most of our villages and other isolated dwellings, institutions and camping sites fall under the second category. In the absence of a sewerage system, various methods of the conservancy system are used to dispose of the sullage and human excreta. Due to nonavailability of large amounts of water, the waste obtained from such communities is generally small. Particularly, in the second category, only the human excreta requires disposal ; whereas, in the first category, some amount of sullage will also need disposal.

Sullage in such areas is generally taken along side drains flowing along houses, and then finally discharged into some open surface drain or just left unattended into some outside pond, depression, etc. The human excreta, however is burried underground through the *latrines* **(privies)** of different designs. Sometimes, such latrines directly discharge their wastes into deep underground pits, or sometimes the wastes are collected by sweepers, carted manually as head load or in carts, and finally buried outside the town, and thus, composted.

Where comparatively, larger amount of water is available due to the availability of an organised municipal supply, however, **septic tanks** are constructed, provided land is also available. The sewage from individual houses, may also sometimes be disposed of for farming by subsurface irrigation. The design and construction details of septic tanks, alongwith the methods adopted for disposal of its effluents, have already been discussed in details in article 9.45. The sub-surface irrigation methods for disposal of sewage or the septic tank effluents, have also been discussed in that article, and hence need not be repeated here.

We will, however discuss in this chapter, the various types of latrines, which have been designed and developed for use in such unsewered areas.

12.2. Privies of a Conservancy System of Sanitation

The different types of privies, which are used in such unsewered areas, are briefly discussed below :

12.2.1. Removable Receptacle Privy. This is the oldest and primitive type of a privy, still being used in India at various places, including various villages and smaller towns.

Fig. 12.1 shows the typical sketch of such a latrine. It essentially consists of a removable metal container placed below the simple squatting seat. The excreta is collected from this container daily by sweepers, and carted away for further disposal by composting, etc.

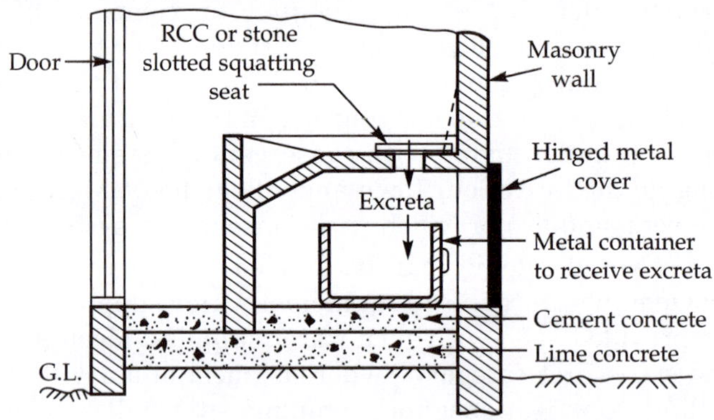

Fig. 12.1. Removable Receptacle Privy.

12.2.2. Bore-hole Privy (Dry Latrine or Soak Pit Privy). This type of a latrine consists of a 40 cm dia bore hole, normally drilled 4 to 6 m deep (but certainly 1 m above the water-table, to avoid pollution of ground water). The hole may either be lined or unlined from inside, but should preferably be lined.

The bore hole (or sometimes even a pit may be used), is connected at its top to the squatting seat. Either an ordinary slotted R.C.C. seat, as used in a removable receptacle privy, may be used here ; or to make things more sanitary and aesthetic, a pan and a trap arrangement may be used. Both these arrangements are shown in Fig. 12.2 (*a*) and (*b*) respectively.

In this type of latrine, the excreta directly goes into the pit, and gets digested in due course, while the liquid portion gets absorbed. The construction of such a latrine is simple and cheap. Since no human element is required for disposal of excreta, it is a much better type of a privy than the one described earlier.

The smell and fly nuisance will, however remain in the simple squatting plate type, which will also be eliminated in the improved type provided with pan and trap arrangement, which of course will require flushing cistern and water.

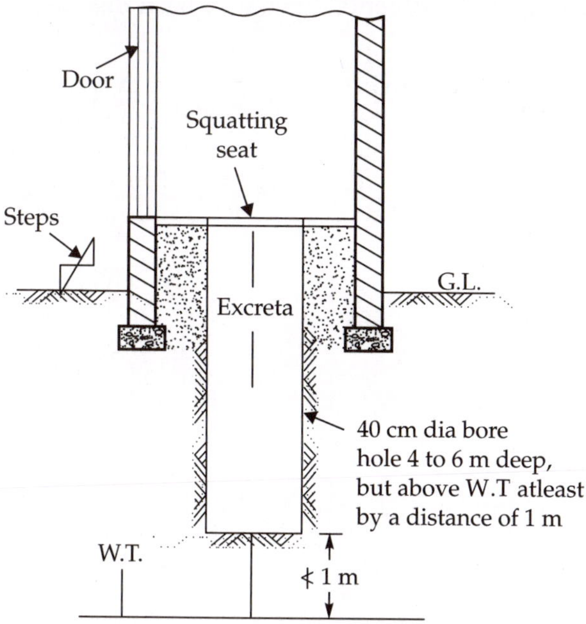

(*a*) Bore hole latrine with squatting plate

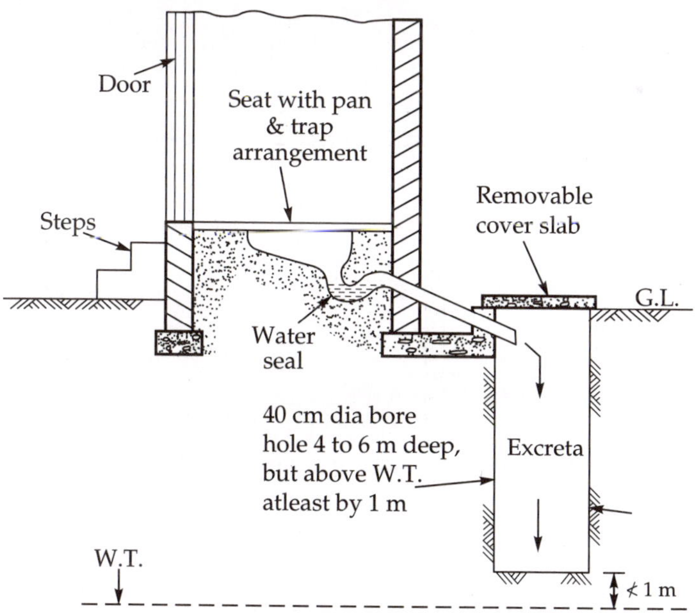

(*b*) Bore hole latrine with a pan and a trap arrangement

Fig. 12.2. Bore hole Latrines (Dry Latrines).

The biggest drawback with this type of latrine, however, is that the bore hole gets filled up after sometime, and a new bore hole will need to be excavated, thereby requiring shifting of the entire superstructure and accessories. Moreover, in areas of high water-table, the depth and hence the capacity of the bore-hole will be very limited, needing frequent shifting. In such areas, the ground water is also susceptible to contamination.

12.2.2.1. Concrete vault privy. This in an improved form of bore hole privy, for areas of high water-table or in sandy soils. In sandy soils, the construction of a pit or a bore hole will be very difficult, and in areas of high water-table, there is a likelihood of its getting contaminated by bore hole privies. Hence, in such circumstances, a concrete vault privy is generally adopted.

A concrete vault privy (Fig. 12.3) essentially consists of a water-tight concrete vault (chamber) constructed under-ground. Squatting pan with compartment is placed over the concrete vault as shown. When the vault is filled up, it is emptied and cleaned through the manhole cover. To prevent bad odours, generally soil is thrown daily after use, over the excreta. Squatting pan should also be constructed in such a way that no water can enter the vault.

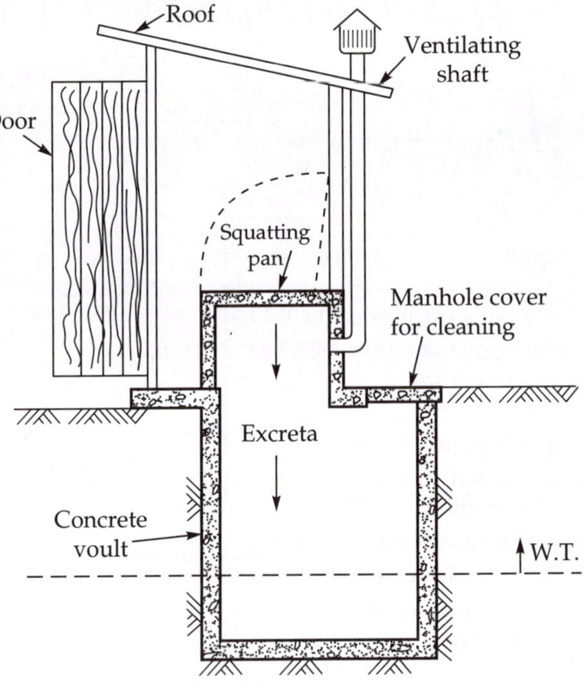

Fig. 12.3. Concrete Vault Privy.

12.2.3. Aqua Privy (Wet Latrine). The dry latrines, described above, are either temporary in nature or need frequent cleaning. The aqua privy, on the other hand, is an improved form of privy, and is a permanent arrangement. *Such type of privies are very convenient and ideal for factories, hill stations, isolated colonies, villages, etc.*

They are, however more costly and require certain good amount of attendance for obtaining good results.

An aqua privy (Fig. 12.4) essentially consists of three interconnected underground masonry chambers, as shown. The end of the squatting pan is taken about 10 cm below the flow line in the digestion chamber (A) to provide water seal. Human excreta (stools and urine) directly falls into the chamber (A), and is decomposed under anaerobic conditions. Gases formed during the decomposition are allowed to escape into the atmosphere through the vent pipe provided in the second chamber (B).

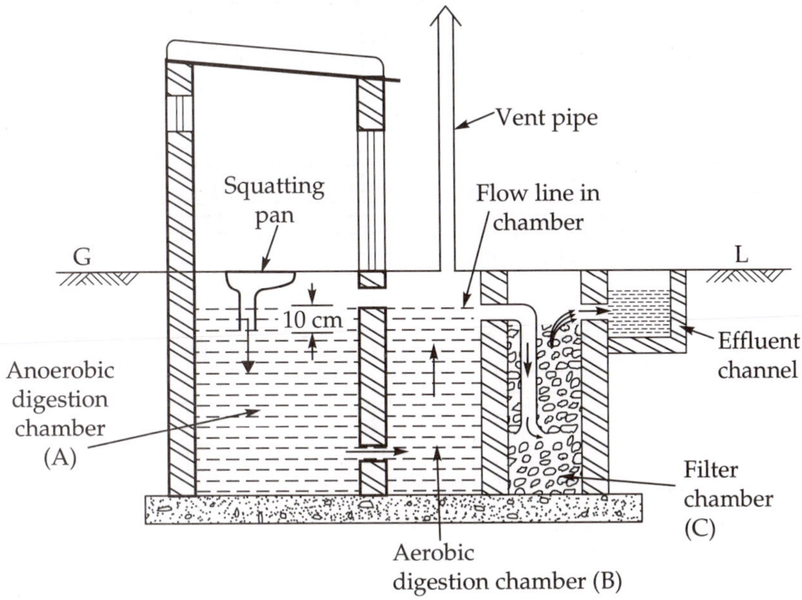

Fig. 12.4. Aqua Privy (wet latrine).

Since the second chamber (B) is connected to the atmosphere, the aerobic digestion takes place in this chamber. The digested sludge gets settled in both these chambers [*i.e.* (A) and (B)], and the effluent from the second chamber flows to the third chamber (C) through a pipe whose outlet is taken down, upto somewhat near the bottom of the third chamber. Chamber (A) and (B) are of equal size and the third chamber (C) is of smaller size, and is filled with clinker to act as filter media. Effluent, while rising up through the clinker, gets purified, and can be used for gardening or farming.

Such a privy was considered to be a very useful and economical method of disposal of human excreta, till the advent of *Sulabh Sauchalayas*, for unsewered and rural areas.

12.2.4. Sulabh Shauchalaya (Hand-flush Water Seal Latrine). A sulabh shauchalaya (Fig. 12.5) is a modern answer to the old conservancy system of dry latrines, since it completely avoids the carriage of night soil as head load. Such a latrine is also considered to be a better choice than a septic tank latrine, since it neither involves periodic manual cleaning as is involved for a septic tank, nor it emits foul orders as are emitted by the vent pipe of a septic tank. Additionally, a sulabh shauchalaya requires about $\frac{1}{5}$ times the water required by a septic tank privy. Moreover, a sulabh shauchalaya can be constructed very easily by an ordinary mason, and is quite economical (costing about ₹ 5000–8000 only).

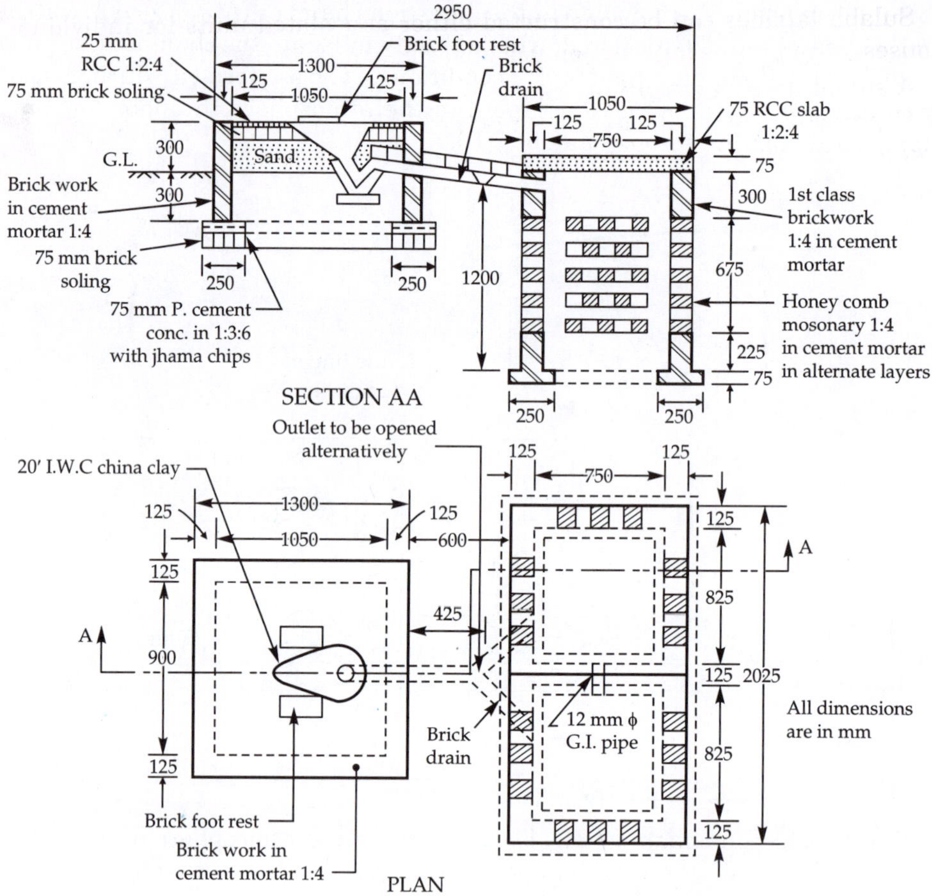

(*a*) Plan and section

(*b*) Photographic view

Fig. 12.5. Sulabh shauchalaya with rectangular tanks.

Sulabh latrines can be constructed either as isolated units for individual houses, or in groups for common use of a colony or for public conveniences.

A sulabh latrine essentially consists of a specially designed 500 mm long *W.C. pan*, made of moulded mosaic or china clay, mounted on a 200 mm *water seal with groove* as shown in Fig. 12.6 (*a*) and (*b*).

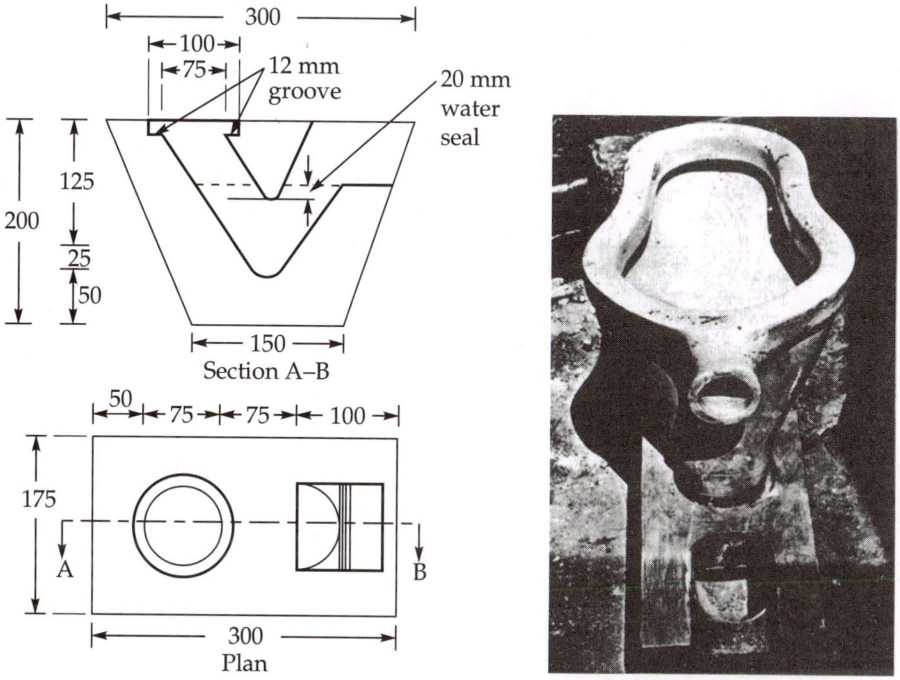

(*a*) Sketch and plan (*b*) Photographic view

Fig. 12.6. Sulabh pan mounted on sulabh seal.

This pan is called a *sulabh pan*, and the seal is called a *sulabh seal*. The sulabh pan, mounted on the sulabh seal, is fixed in a raised pedestal to provide the W.C. seat, as shown in Fig. 12.5. The W.C. is connected to the two *underground tanks* by covered *drain channels*, as shown in Fig. 12.5 and 12.7.

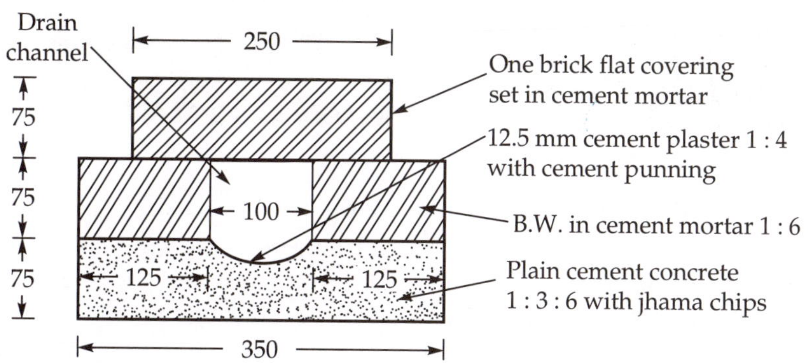

Fig. 12.7. Section of the drain channel connecting the w.c. pan with the tanks.

The two underground tanks are either circular (1 m dia × 1 m deep) or rectangular (1 m × 1 m × 1.2 m deep), and are used in turn. The rectangular tanks require lesser space, as the tanks can be adjusted side by side, as shown in Fig. 12.5. The use of circular tanks, is shown in Fig. 12.8 (*a*) and (*b*).

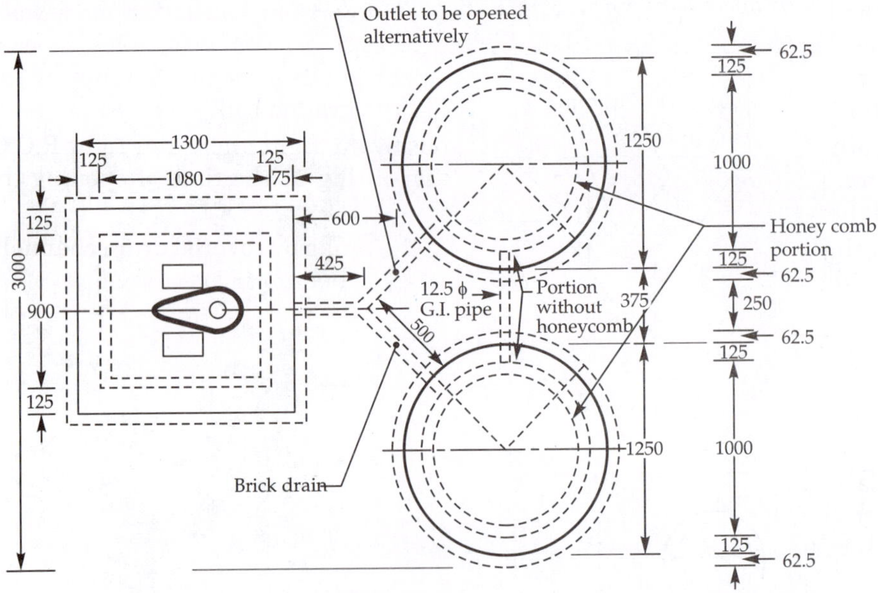

(*a*) Plan view

(*b*) Photographic view

Fig. 12.8. Sulabh shauchalaya with circular tanks.

Both the tanks are interconnected by a 12 mm dia C.I. pipe near their top levels, so that the gases can pass from one tank to another. The walls of both the chambers are made of 125 mm thick brick masonry with 1 : 4 cement mortar on a 250 mm wide footing. The roof level of tanks are kept slightly higher than the adjacent ground level, so that the rain water may not enter into these tanks. Three of the four walls of each tank are made pucca in bottom

225 mm depth and top 300 mm height ; while the remaining height is made in *honey-combed brick work* (Jalli). The 4th side of each tank, which is adjoining each other, is made solid in brick work to avoid leakage of effluent from one tank to the other.

The bottom base of both the tanks are kept earthen, so that all the sewage water leaches out. The contact of earth with excreta also helps in easy decomposition and transformation of the excreta into organic manure by soil bacteria, which are found in abundance in the ground soil.

Both the tanks are covered at their tops by air-tight and water tight R.C.C. covers, which are provided with hooks for lifting at the time of opening the tank for removal of manure.

Although the W.C. is connected to both the tanks by covered drain channels, but only one of the two connections is to be opened for use at a time. In other words, only one of the two tanks shall receive the night soil at a time.

Scientific experiments have shown that the excreta starts decomposing as soon as it comes into contact with the earth, and its quantity goes on reducing as the *anaerobic decomposition* proceeds. Millions of anaerobic bacteria, thus, decompose this excreta, ultimately transforming it to *manure*.

Each tank will serve a family of 5 members for a period of 5 years ; after which time, the firstly used tank shall be closed by blocking its inlet drain, and the second tank will be put into use by opening its inlet drain. 2 years after the stoppage of the use of the first tank, the manure can be taken out preferably in summer season for remaking the first tank ready for reuse.

In this way, the tanks can be used alternatively for a total life period of about 100 years.

PROBLEMS

1. Enumerate and discuss the various types of privies that are used in conservancy system of sanitation. Draw neat sketches for any two types of these.

2. Differentiate between 'dry latrine' and 'wet latrine' drawing neat sketches of both. Also write down their merits and demerits in relation to their use in Indian conditions.

3. Discuss the various methods that are used to dispose of human excreta in areas where sewers have not been laid. Illustrate your answer with Indian examples.

4. What are sulabh shauchalayas ? Why are they finding increasing use in developing countries like India ? Draw a neat sketch to illustrate the working of a sulabh shauchalaya.

5. Differentiate between a septic tank privy and a sulabh shauchalaya, and discuss their merits and demerits.

6. Write short notes on :

(*i*) Aqua privy ;
(*ii*) Concrete vault privy ;
(*iii*) Bore hole privy ;
(*iv*) Sulabh shauchalaya.

—13

Sewage Collection from Houses and Buildings

13.1. General Introduction

The sewage produced in houses and buildings has to be conveyed and connected to the municipal sewers by the owners of the houses. The provision and construction of an efficient **plumbing system,** for collection and movement of the sewage produced in the building, till it is carried and discharged into the nearest municipal sewer, is an important aspect of building construction.

Many of us, who are living in areas provided with sewers, might have experienced the bad or improperly designed plumbing works, which everyday poses problems, in either the kitchen drain or the wash basin or the water-closet getting choked, or the foul smells escaping from somewhere or the other. A properly designed and carefully constructed plumbing system in buildings, is therefore, absolutely necessary to avoid such frequent troubles.

13.2. General Principles Governing the Design of a Sanitary Plumbing System

The various **sanitary fittings** provided in buildings, such as *wash basins, bath tubs or bath rooms, kitchen sinks, urinals, latrines (with their flushing cisterns)*, etc. do contribute to the sewage, and hence a *system of pipes* should be laid and connected with the sanitary fittings in such a way that the sewage (or sullage) automatically flows by gravity through this system, and ultimately discharges into the municipal sewer. Such a plumbing system, should be designed keeping into consideration the following ***general principles :***

(1) For better maintenance and easy repairs, the house sewer pipes should preferably be laid by the side of the building rather than in walls or underground. But in modern days in small houses, they are generally embedded in walls ; and in large buildings, kept outside and accommodated in shafts, to avoid their bad appearance.

(2) Drains should be laid straight between the manholes (inspection chambers). All sharp bends and junctions, which tend to become dead-ends for collection of sewage solids, should be avoided except through manholes.

(3) The entire system should be properly ventilated from the start to the end.

(4) The *house* sewer should be connected to the public sewer, keeping the outfall level of the house sewer sufficiently higher than the water level of the public sewer* to avoid back flow and inefficient building drainage.

(5) The plumbing system should contain enough number of *traps*, at suitable points for its efficient functioning or to avoid evolution of foul smells.

(6) The *house sewer* should be disconnected from the *public sewer* by the provision of an ***intercepting trap,*** so as not to allow foul gases from public sewer to enter the house sewer.

(7) Joints of sewers should be water-tight, and should be properly checked so. Any leaking joint, especially when the pipe is buried in the wall or floor, will later-on pose serious problems, needing dismentalling of walls and floors.

(8) Lateral sewers should be laid at sufficient slopes (gradients), so as to develop self cleansing velocities.

(9) The layout of house plumbing system should, as far as possible, be such as to permit easy cleaning in case of blockade, or repair in case of leakage, or additions if additional sanitary fittings are provided on a future date.

(10) High quality sewer pipes should be used in the system.

(11) Possibilities of formation of ***air-locks, siphonage, undue deposits,*** etc. should be properly studied, and remedies should be accommodated in the design to avoid them.

(12) The size of lateral sewers should be such that they will not overflow at the time of maximum discharge.

(13) Rain water from roofs or open courtyards should not be allowed to mix with house sewage or sullage, but should be separately discharged on road or street surface, for *catch basins or inlets* to convey it to the storm water drains, etc. The rain water is best used for rain water harvesting.

13.3. Functions and Types of Traps being used in Sanitary Plumbing Systems

13.3.1. Definition. Traps may be defined as fittings, placed at the ends of the soil pipes or the sullage pipes (waste pipes)** to prevent the passage of foul gases from the pipes to the outside. This is possible because traps do enclose or maintain *water seal* between the pipe and the outside. This water depth does not allow gases to escape to the outside of the pipe. The efficiency and effectiveness of a trap will depend upon the depth of the water seal. Greater is this depth, more effective the trap will be. This water seal generally varies from 25 mm to 75 mm ; 50 mm being quiet common in most of the traps.

13.3.2. Qualities. A good trap should possess the following *qualities* :

(*i*) It should provide sufficient water seal (50 mm—or so) with large surface area ;

(*ii*) Its interior should be smooth, so as not to obstruct flow, and the trap should thus be self cleansing ;

---

*The public sewers are of course, normally laid quite deep, compared to the general ground level of the area.

Soil pipes** are the pipes which carry the night soil, and ***sullage pipes are the pipes which carry the sullage from bathrooms and kitchens.

(*iii*) It should be provided with an *access door* for cleaning ; and

(*iv*) It should be made of some non-absorbent material.

13.3.3. Types. *Depending upon their* **shapes,** the traps may be of three types, *i.e.*

(*i*) *P-trap* ;

(*ii*) *Q-trap ; and*

(*iii*) *S-trap*.

These three types of traps are shown in Figs. 13.1 and 13.2. A trap essentially consists of a U-tube, which retains water, acting as a seal, between the foul gases (inside the pipe) and the outside atmosphere. They are largely used for baths, sinks, and laboratories. In all such needs, they are made with enlarged mouth, so that the waste pipe may be thoroughly flushed out.

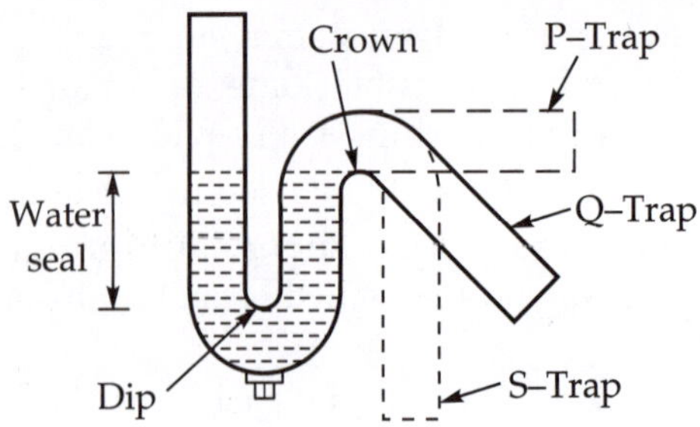

Fig. 13.1. P, Q and S Traps shown together.

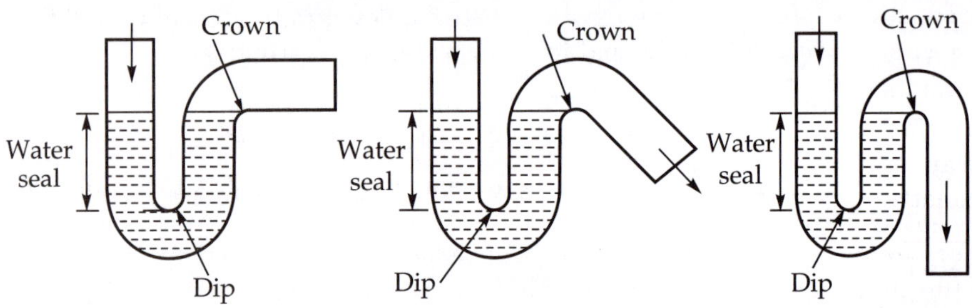

(*a*) P-Trap (*b*) Q-Trap (*c*) S-Trap

Fig. 13.2. P, Q and S Traps, shown separately.

Depending upon their use, the traps may again be of three types *i.e.*

1. *Floor trap* ;
2. *Gully trap* ; and
3. *Intercepting trap*.

These three different types of traps are briefly described below :

(1) Floor Traps. These traps are generally used to admit waste water (sullage) from the floors of rooms, kitchens, baths, etc. into the said room drain (sullage pipe). These are invariably provided with cast iron or galvanised or stainless steel gratings (*Jallis*) at the top, so as to prevent the entry of solid and larger sticky matter, into the drain pipe, to avoid frequent blockade. A commonly used patented name of such a trap is **Nahni trap.** A floor trap is shown in Fig. 13.3.

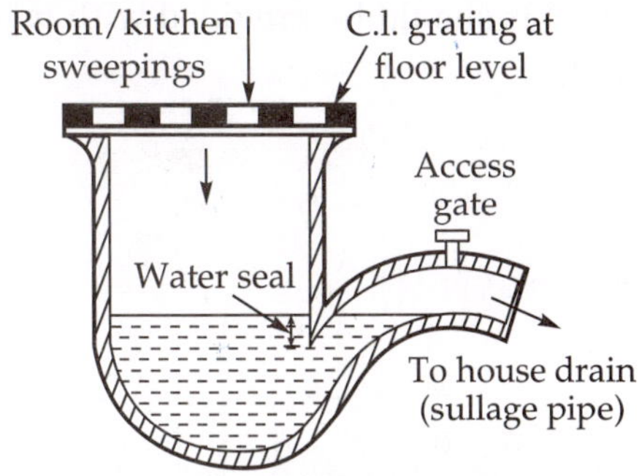

Fig. 13.3. Floor Trap.

(2) Gully Traps. A gully trap or a **gully** (Fig. 13.4) is often provided at the junction of a room or a roof drain and the other drain coming from bath, kitchen, etc. The foul sullage from baths, will enter through the side inlet (called **back inlet**), and the unfoul room washings or rain water from roof or courtyard will enter from the top.

The rain water pipes or sullage pipes discharging into drains, are often connected to them through such traps. Gully traps may either have a S-trap or a P-trap. The water seal is usually 50 mm to 75 mm deep. The top of the trap is covered by a C.I. grating to exclude the entry of coarser materials to avoid blockade.

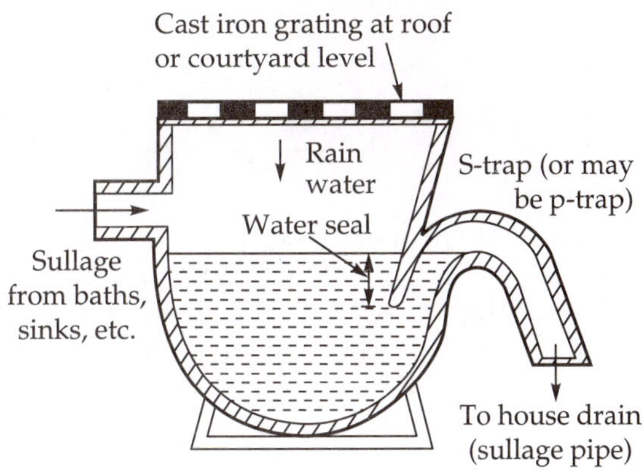

Fig. 13.4. Gully Trap.

(3) Intercepting Traps. An intercepting trap (Fig. 13.5) is often provided at the junction of a house sewer and a municipal sewer, so as to prevent the entry of the foul gases of the municipal sewer, into the house drainage system. This trap at such a junction is often provided in a small man-hole (Fig. 13.6) constructed just near the house, either outside in the street or in a corner inside the house boundary. This trap is provided near its top with an access gate or a plug, called *cleaning eye*, for removing silted matter from inside the trap incase of blockade. It has a high depth of water seal, say about 100 mm. It is interesting to note the merits and demerits of an interceptor, which are given below :

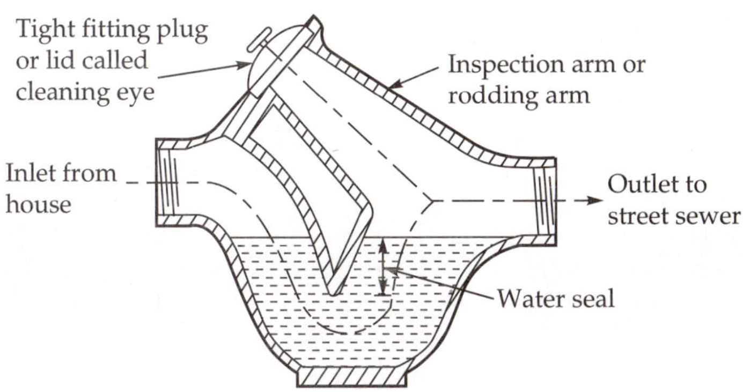

Fig. 13.5. An Intercepting Trap.

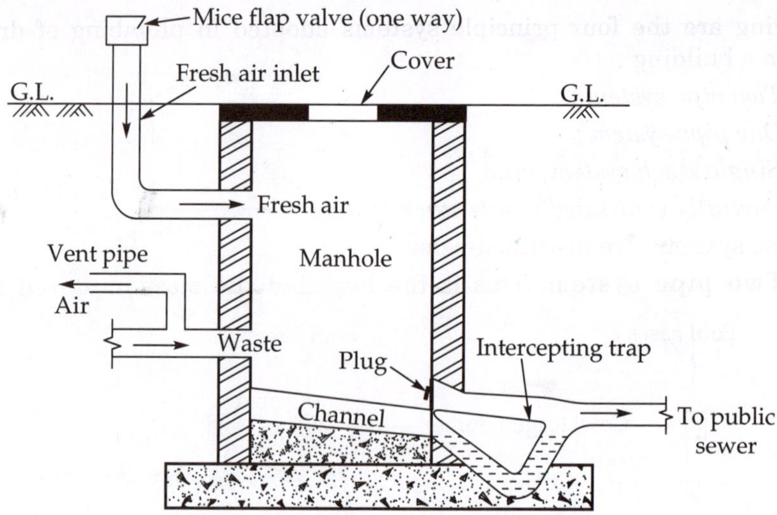

Fig. 13.6. Manhole with Intercepting trap.

Merits of interceptors. (*i*) Foul gases of public sewer cannot pass through the interceptor, and hence prevented from entering the house drainage system. If the interceptor is not provided, these gases will enter the vent pipes of the house drainage system, and spread around in the surrounding atmosphere, causing serious air pollution.

(*ii*) Harmful pathogenic bacteria contained in the public sewers are thus prevented from entering the house drains, due to the presence of the interceptor.

(*iii*) Properly designed and constructed interceptors can quickly remove the foul matter of the house drains into the public sewer.

Demerits of interceptors. (*i*) If the discharge from house drains is small, the solid heavy matter may be retained in the trap, and may start decomposing, producing foul gases. The basic purpose of interceptors of preventing foul gases, will then no longer be served.

(*ii*) If the lid or the plug is not fitted properly, or is broken, foul gases from public sewer will do enter the house drain.

(*iii*) Cleaning through the inspection arm of the trap is not easy.

(*iv*) Interceptor itself forms an obstruction to the normal flow of sewage.

(*v*) Omission of interceptors from house drainage is found not to present too serious a difficulty or a problem.

(*vi*) Presence of interceptors installed by house owners, is found to seriously affect the ventilation of public sewers, as in such cases, the foul gases of public sewers will find an outlet only through the *ventilating columns,* which are provided at the head of every branch sewer and at other key points in the city sewerage system. Hence, if interceptors are allowed, then public sewers will need greater ventilation arrangements, and hence involving more expenditure, consequently leading to greater taxes on the public.

In view of this difficulty, the city municipality itself decides whether to allow the house owners to install or not to install interceptors.

13.4. Systems of Plumbing

Following are the four principle systems adopted in plumbing of drainage work in a building :

(1) *Two pipe system* ;

(2) *One pipe system* ;

(3) *Single stack system* ; and

(4) *Partially ventilated single stack system.*

These systems are discussed below :

(1) Two pipe system. This is the best and the most improved type of

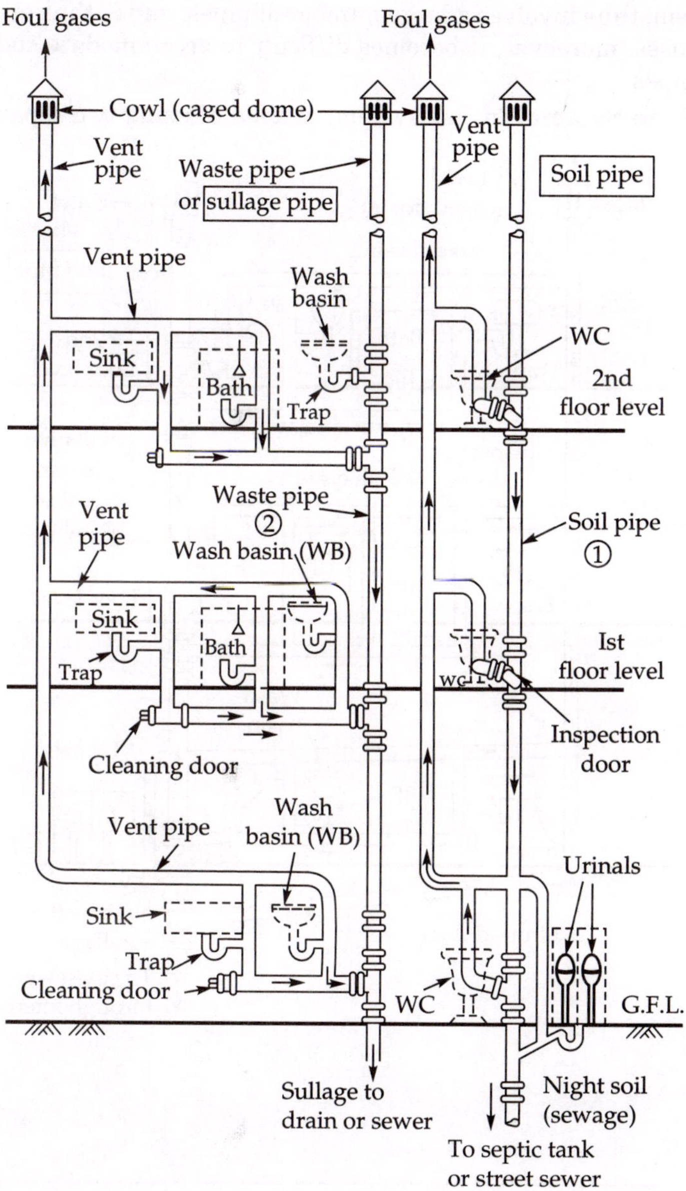

Fig. 13.7. Two pipe system of Plumbing.

system of plumbing. In this system, two sets of vertical pipes are laid, *i.e.*, one for draining night soil, and the other for draining sullage. The pipes of the first set carrying night soil are called ***soil pipes,*** and the pipes of the second set carrying sullage from baths etc., are called ***sullage pipes or waste pipes.***

The soil fixtures, such as latrines and urinals, are, thus, all connected through branch pipes (laterals) to the *vertical soil pipe* ; whereas, the sludge fixtures such as baths, sinks, wash basins, etc. are all connected through branch pipes to the vertical *waste pipe*. The soil pipe as well as the waste pipe, are separately ventilated by providing separate vent pipes or antisiphonage pipes*, as shown in Fig. 13.7. This arrangement, thus requires four pipes.

This system, thus involves a large number of pipes, and is thus quite costly. In small houses, moreover, it becomes difficult to accommodate such a large number of pipes.

(2) One Pipe System. In this system, instead of using two separate pipes

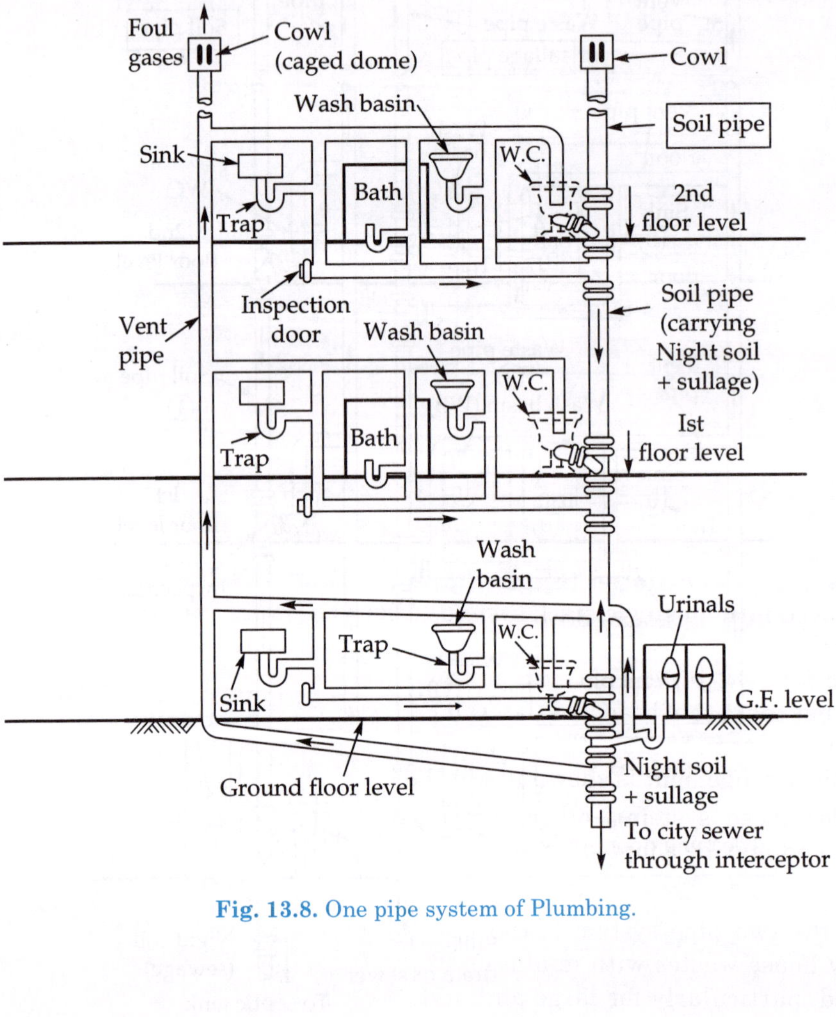

Fig. 13.8. One pipe system of Plumbing.

*Described in Article 13.7.7.

(for carrying sullage and night soil, as is done in the above described two pipe system), only one main vertical pipe is provided, which collects the night soil as well as the sullage water from their respective fixtures through branch pipes. This main pipe is ventilated in itself by providing cowl at its top, and in addition to this, a separate vent pipe, however, is also provided, as shown in Fig. 13.8. This system, thus, has two pipes, instead of four pipes of the two pipe system.

(3) Single Stack System. This system is a single pipe system without providing any separate ventilation pipe. Hence, it uses only one pipe, which carries the sewage as well as the sullage, and is not provided with any separate vent pipe, except that it itself is extended upto about 2 m higher than the roof level and provided with a cowl, for removal of foul gases, as shown in Fig. 13.9.

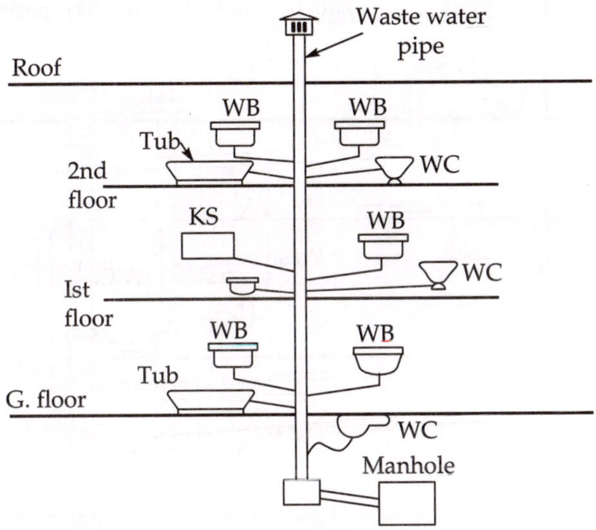

Fig. 13.9. Single stack system of Plumbing.

(4) Partially Ventilated Single Stack or Single Pipe System. This is an improved form of single stack system in the sense that in this system, the traps of the water closets are separately ventilated by a separate vent pipe called ***relief vent pipe.*** This system, thus, uses two pipes as in a single pipe system, but the cost of branches (laterals) is considerably reduced compared to single pipe system, because the sullage fixtures are not connected to the vent pipe. This arrangement is shown in Fig. 13.10.

Besides these systems, other combinations and permutations are also possible and may be adopted by some people.

13.4.1. Choice of a Particular System of Plumbing. As pointed out earlier, the two pipe system is the best system for efficient conveyance of sanitary house wastes with minimum use of traps, and is therefore, largely favoured, particularly for large and multistoreyed buildings. This system, however, requires a large number of pipes and their connections, and is hence

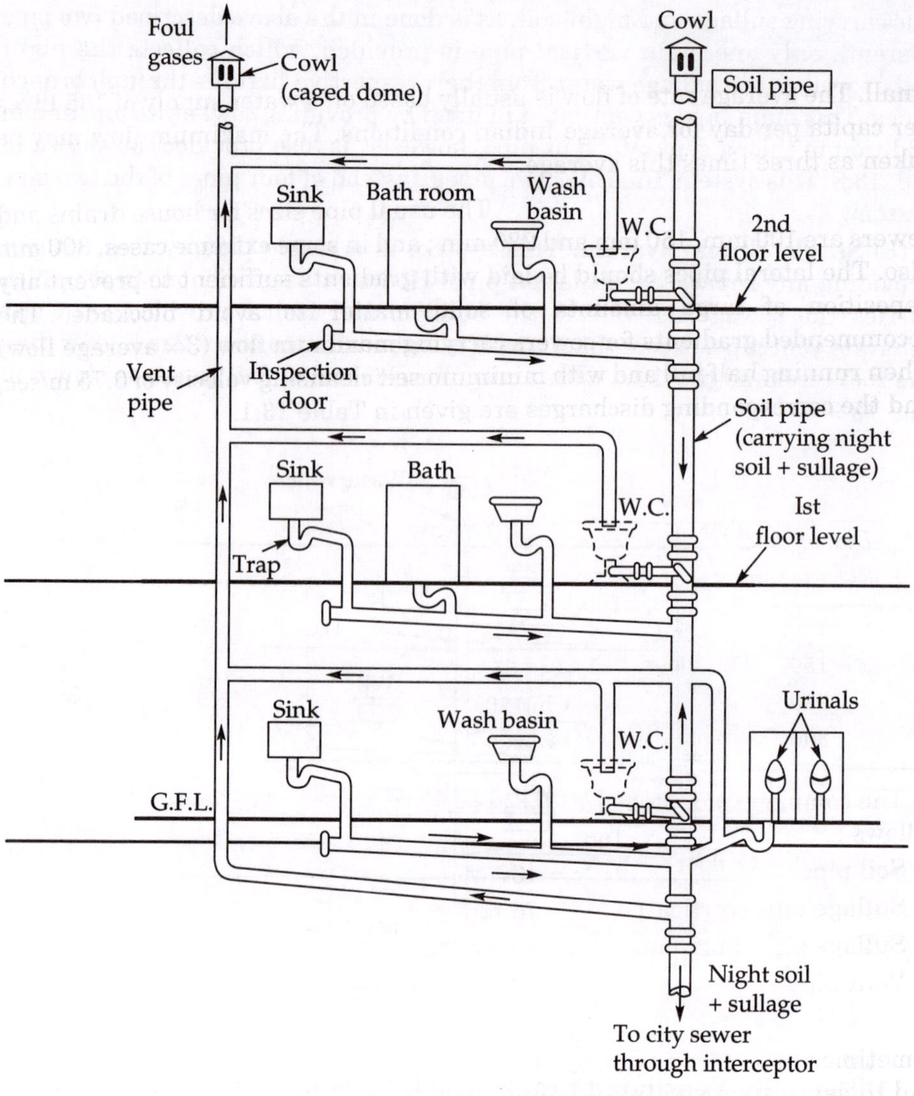

Fig. 13.10. Partially ventilated single pipe system.

costly. Moreover, it is also difficult to find suitable place for accommodating so many pipes in small houses and buildings. In that case, one pipe system is more economical and easy to accommodate, but requires sufficient safe guard, to make the drainage effective, in the form of proper ventilation, adequate water-seals, and proper connections between the sanitary fixtures and the soil pipe. In multi-storeyed buildings, moreover, use of one pipe system, generally makes it imperative to place the lavatory blocks of various floors one above the other.

13.5. Sewerage Plans of Buildings and Design of Sewer Pipes

13.5.1. Quantity of Flow. The flow in the drainage pipes of a house sewerage system, is irregular in nature and is not continuous. The quantity is also small. The average rate of flow is usually based on a water supply of 135 litres per capita per day for average Indian conditions. The maximum flow may be taken as three times this average.

13.5.2. Pipe Sizes and Gradients. The usual pipe sizes for house drains and sewers are 100 mm, 150 mm and 230 mm ; and in some extreme cases, 300 mm also. The lateral pipes should be laid with gradients sufficient to prevent any deposition of large amounts of solid matter to avoid blockade. The recommended gradients for sewers carrying maximum flow (3 × average flow) when running half full and with minimum self cleansing velocity of 0.75 m/sec, and the corresponding discharges are given in Table 13.1.

Table 13.1. Recommended Gradient and Discharge Capacity of House Sewers (Pipes)*

Dia. of Pipe sewer in mm	*Gradient*	*Discharge in m^3/minute (m^3/day)*
100	1 in 35	0.18 (259)
150	1 in 65	0.42 (605)
230	1 in 120	0.93 (1339)
300	1 in 200	1.66 (2390)

The commonly adopted sizes for house drains, for average condition, are as follows :

Soil pipe = 100 mm ϕ
Sullage pipe (vertical) = 75 mm ϕ
Sullage pipe (horizontal) = 32 mm to 50 mm ϕ
Vent pipe = 50 mm ϕ

13.5.3. Sewerage Plans of Buildings. It is generally convenient and sometimes necessary to prepare the detailed plans of house sewerage system and to get it approved from the municipalities. While preparing such plans, the following points need attention :

(*i*) The site plan of the building should be drawn to a convenient scale, and positions of gully traps (G.T.) should be marked on it, as shown in Fig. 13.11.

(*ii*) The L-section of the proposed sewer (drain) should be drawn to a convenient scale. Generally, the L-sections of drains larger than 150 mm ϕ are only drawn. Such an L-section should show distances, ground levels, invert levels, depths of cutting, sizes of manholes, sizes and gradients of pipes, etc., as shown in Fig. 13.12.

(*iii*) The position of public sewer should be clearly marked on the site plan as well as on the sewer L-section. *It is advisable to join the house sewer to a manhole on public sewer line, if site conditions permit.*

*As per IS : 1742–1960.

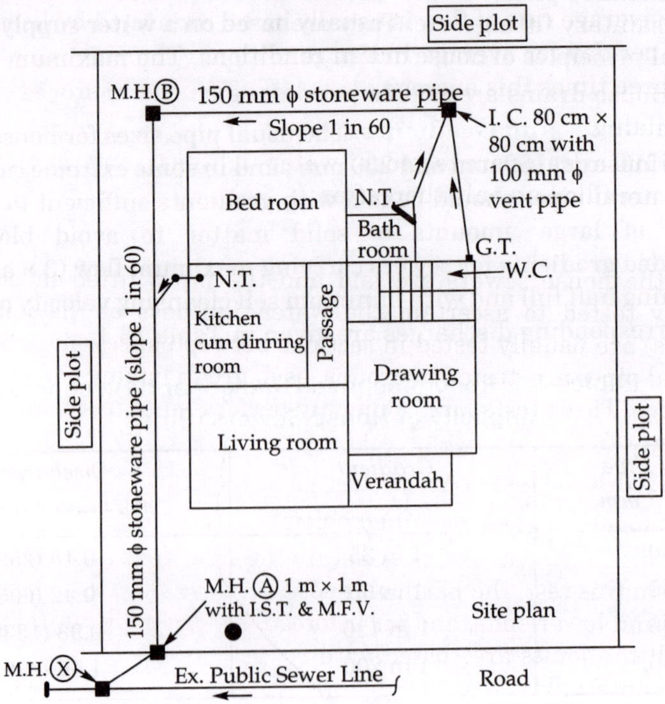

Fig. 13.11. General Sewerage Plan of a Building.

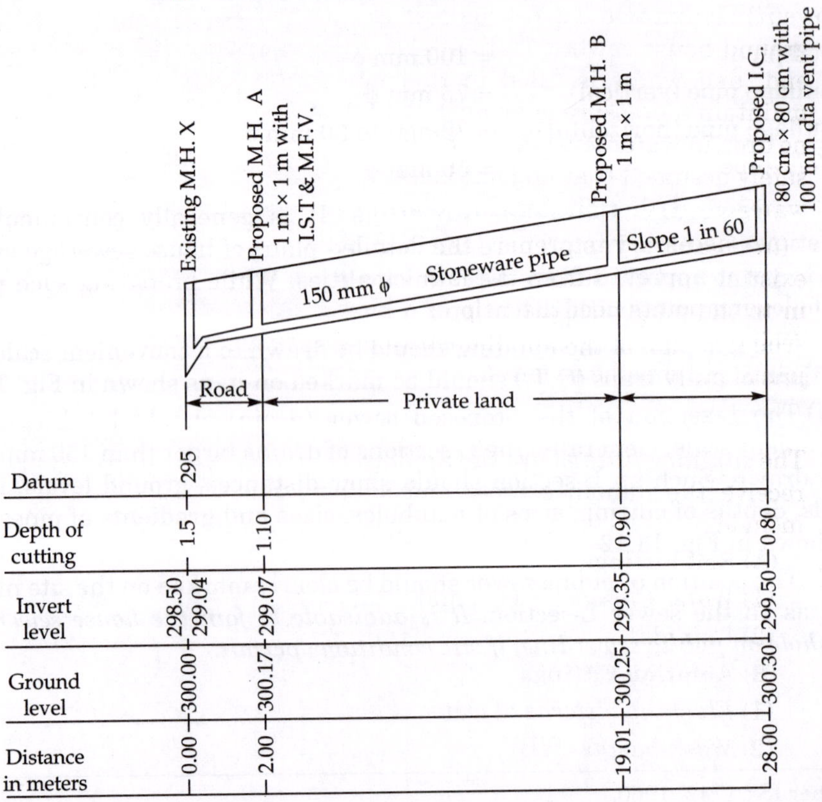

Fig. 13.12. L-Section of a House Sewer.

(*iv*) Detailed plans of different floors of building show the positions of floor traps, sanitary fittings, etc. They are to be connected by keeping in mind the general principles of house drainage.

(*v*) House drains are laid as per details on the approved sewerage plan of the building.

(*vi*) House drains are tested for water-tightness, and the trenches in walls or floors, are filled up only thereafter.

13.6. Testing of House Sewers

After the house sewers are laid and the pipes fitted in position, they are usually tested to ascertain the water-tightness of the joints. The sewers (drains) are usually tested in sections between successive chambers, and the vertical pipes are tested in groups, each group having one common inspection chamber. Three tests, are commonly performed to test house drains ; *viz.*

(*i*) Air test ;

(*ii*) Hydraulic test or Water test ; and

(*iii*) Smoke test ; as explained below ;

(*i*) **Air Test.** Air test is generally used to test underground and vertical pipes. In this test, the particular section of the pipe to be tested, is plugged at upper and lower ends, and air is forced into the pipe with the help of a pump, etc. All the joints are coated with soap solution. The escape of air from the leaky joints will then form bubbles at the joint, which can be visually seen, and hence repaired.

(*ii*) **Hydraulic Test.** This test is most commonly used for testing underground house sewers. In this test, the drain (sewer) is plugged at the lower end, and water is filled in the pipe through the upper end manhole, to maintain a pressure of about 1 to 1.5 m depth of water. If there is found to be no drop in water level in this manhole in about 10 minutes time, then it can be safely assumed that no joint is leaking. However, if there is some drop in the water head, then the leaking joints have to be watched visually, and repaired.

(*iii*) **Smoke Test.** This test is generally performed to detect leakage of existing vertical sullage pipes or rain water pipes. In this test, smoke is forced in to the pipe with the help of a blower. Leakage can be easily detected by seeing or smelling over the emitted smoke at the leaky joints. The leaking joints can then be repaired.

13.7. Sanitary Fittings and Other Accessories

The *sanitary fittings* are the appliances used in a house plumbing system, to receive the human excreta and sullage waste of the house. Such fittings include :

(A) Soil Fittings

(1) *Water Closets* (W.C.) ; and

(2) *Urinals.*

(B) Ablution Fittings

(1) *Flushing cisterns of water closets* ;

(2) *Wash basins* (WB) ;

(3) *Baths ; and*

(4) *Kitchen sinks.*

All these fittings are briefly discussed below :

13.7.1. Water Closets (W.C.).

13.7.1. Water Closets (W.C.). A water-closet is a water-flushed plumbing fixture, designed to receive the human excreta directly, and is connected to the soil pipe by means of a trap. Sometimes, the term water-closet is applied not only to represent the particular sanitary fitting, but to the entire compartment or the room (toilet), in which this fixture is placed.

Two types of water-closets are in common use these days :

(*i*) *The Squatting* or the *Indian type* ; and

(*ii*) *The Pedestal or the European type.*

These types are discussed below :

(*i*) **The Squatting or the Indian Type W.C.** The Indian type W.C. has a squatting *pan* 45 cm to 63 cm in overall length, and 45 to 50 cm in height. The pan is fitted either with a *P*-trap, or *S*-trap, which is then connected to the soil pipe [Fig. 13.13 (*a*)].

It is made of vitreous china or porcelain, and the pan (seat) and trap are in two different pieces and joined together, as shown in Fig. 13.13 (*a*). The trap has an opening for fixing an *anti-siphonage pipe* or a vent pipe. The pan has a *flushing rim* to spread the flush water. The excreta does not fall directly into

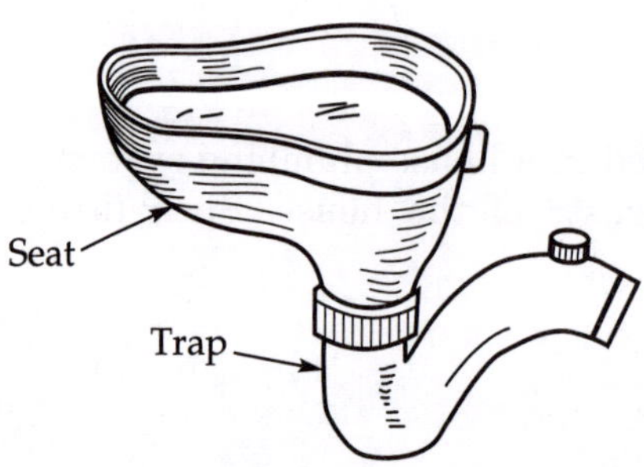

Fig. 13.13. (*a*) Indian Type W.C. Pan and Trap.

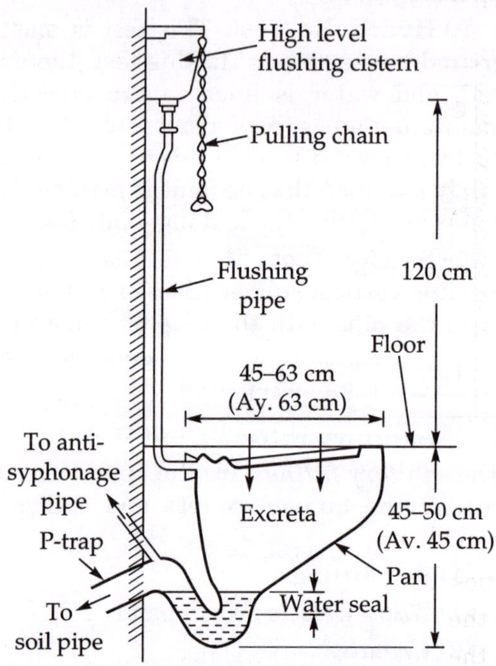

Fig. 13.13. (*b*) Section of an Indian type W.C. provided with a Flushing Cistern.

the trap, and needs to be forced into it and into the soil pipe, by flushing the pan with water released under pressure from the flushing cistern. The excreta may therefore become foul or get decomposed if not properly flushed. This W.C. requires 5-10 litres of water for proper flushing.

The pan is fixed in a squatting or *sitting* position at the floor level or at slightly raised level, and is usually provided with a pair of foot-rests for the convenience of defecating in a sitting position. A section of such a W.C. is shown in Fig. 13.13 (*b*).

This type of W.C. is largely used in Indian houses.

(*ii*) **The Pedestal or the European Type W.C.** The European type, sometimes called ***wash-down type of W.C.*** is mostly used in houses of western countries or in high class buildings and hotels in India. In modern days, however, most of city homes are also fitted with such European W.C.s. It is provided with a wide flushing rim and a 50 mm trap. It is one piece construction, in which the *pan* and the *trap are not separate,* as is there in an Indian type. It is provided with an inlet or a *supply horn* for connecting to the flushing pipe. The W.C. may be provided with a *P-trap* or a *S*-trap, as desired, depending upon the outlet existing in the *wall* or the floor, respectively.

These types of water-closets require less space than the Indian type, and can be located along with the bath in the same room as it won't look bad. They are normally flushed with a low level cistern. A section and a perspective view of such a water-closet is shown in Fig. 13.14 (*a*) and (*b*), respectively.

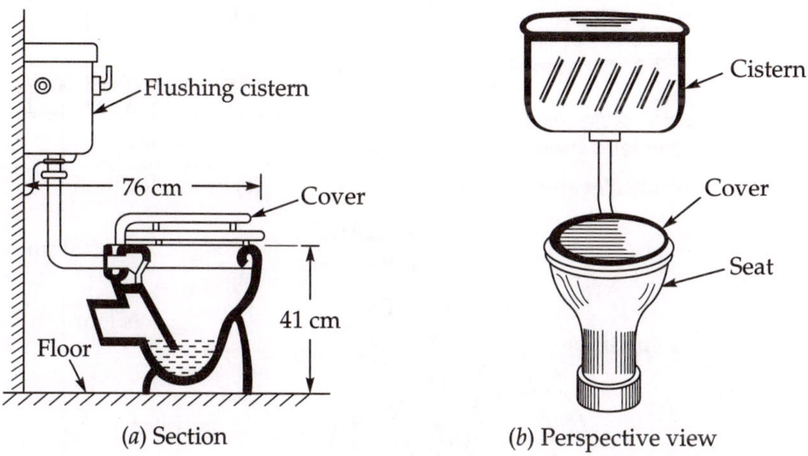

Fig. 13.14. Pedestal or European type W.C.

13.7.2. Urinals. Urinals can be of different forms and shapes. The common varieties are :

(*i*) the *bowl type* or *the basin type* ; and

(*ii*) the *stall type* or the *slab type*.

The bowl type is generally used for private houses or buildings, and the stall type is generally used for public buildings and municipal toilets. A third type, called *squatting plate type,* is used specifically in ladies toilets, only in high classes.

The **Bowl or Basin type urinals** are shown in Fig. 13.15. This type of urinals are of one piece construction ; each urinal being provided with two fixing holes on each side, for fixing on the wall. This type of urinals are more costly but better in appearance and convenient to use, as compared to the *slab type*. They are made of glazed earthenware or vitreous china clay. They are usually flushed with siphonic type of flushing cisterns, which act automatically at desired intervals of time, say 10 to 15 minutes or so.

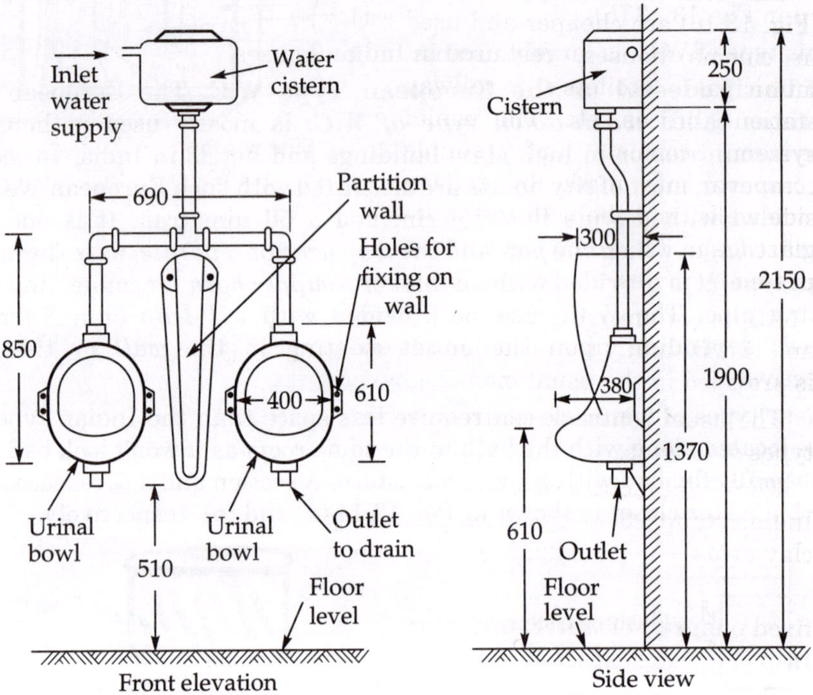

(*a*) Side elevation (*b*) Front elevation

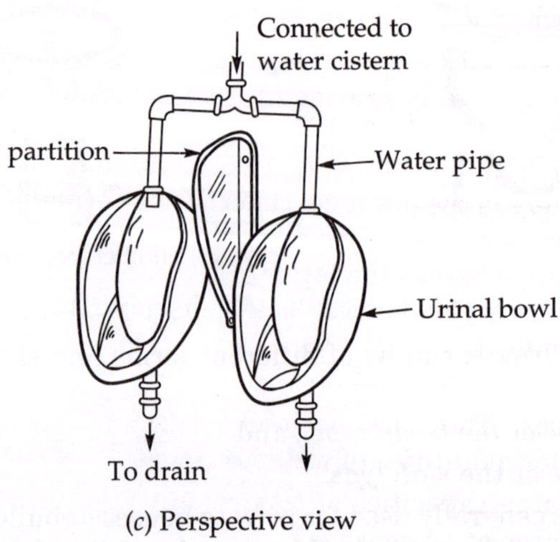

(*c*) Perspective view

Fig. 13.15. Bowl or Basin type urinals (Two units placed side by side).

At the bottom of the bowl shaped vessel (*i.e.* urinal), an *outlet* horn is provided for connecting the urinal to the soil pipe through a floor trap (**Nahni trap**).

The **Stall or Slab type urinals** (Fig. 13.16) are cheaper and used in cinema houses, restaurants, municipal toilets, railway stations, offices, etc. The urinal system consists of simple compartments made by raising side walls, and the wall infront is glazed, against which the urine is passed by males.

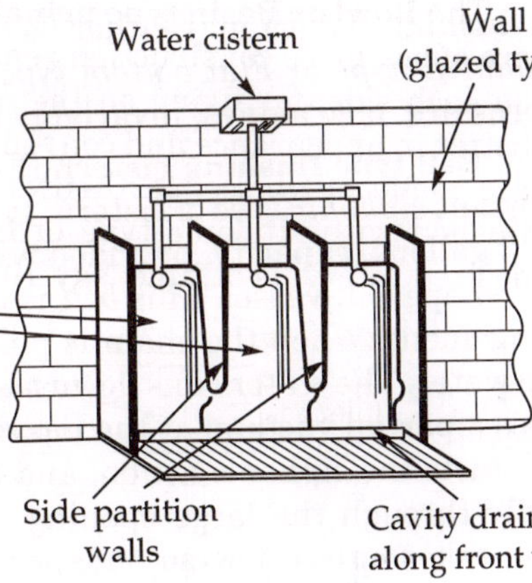

Fig. 13.16. Perspective view of Stall or Slab type of urinals (3 units placed side by side).

At the bottom, a drain is made along the front wall in which the urine flows, and this is finally connected to the soil pipe through a floor trap. The flushing is provided in the usual manner, by a cistern.

The usual centre to centre spacing of units is kept at about 60 cm in both types of *urinals.*

The manufacture of glazed earthenware urinals in India is governed by Indian standard specification code : IS : 771—1963 ; and that of vitreous china clay urinals by I.S. : 2556 (Part VI)—1967.

13.7.3. Flushing Cisterns. A flushing cistern is a device which releases a fixed quantity of water, under pressure, so as to flush and clean the pan and trap of a water-closet or a urinal.

High level cisterns, are those which operate with a minimum height* of 1.25 m ; and the **Low level cisterns** are those which operate with a maximum height of 30 cm.

In Indian type water-closets, high level cisterns fixed at about 2 m above the floor level are generally provided ; whereas, in European type water closets, low level cisterns fixed at about 60 cm above the floor level are generally provided. The high level flushing cisterns used in India are normally made of cast iron ; whereas the low level cisterns are normally made of glazed vitreous ware. The former type, normally used in Indian homes, are operated by pulling a chain, whereas the latter type, usually used in public buildings or in modern western styled homes, looking decent in appearance, are operated by turning a handle.

Urinals are normally provided in India only in public buildings and not in private houses. They are normally flushed by providing cisterns, which release smaller quantities of water, automatically, at fixed interval of time.

The discharge capacities of cisterns are usually 5, 10 and 15 litres. They should discharge at an average rate of 5 litres in 3 seconds. Their construction and design in India, is governed by the BIS code IS : 774—1964.

**i.e.* the distance between the bottom or underside to the cistern and the top of the pan.

A flushing cistern may either be of *siphonic type* (without any valves) or of *mechanical type* or *Flat bottom type* (provided with valves). Siphonic type of cisterns are, now a days, invariably used every where.

The **Bell type flushing cistern** (Fig. 13.17) is a well known example of the important *siphonic type* of cistern. It consists of a cast iron box of 10 to 15 litres (2 to 3 gallons) capacity, provided with a central outlet *stand pipe*, covered by a *bell* shape C.I. vessel. This *bell* is connected to a chain through a lever arm arrangement. When the chain is pulled, the bell receives a jerk and splashes some water. The water runs down the pipe, takes away some air with it, which causes a partial vacuum at the top of the bell shaped vessel (*i.e.* at its crown). This starts the siphonic action, and all the water inside the cistern is sucked rapidly through the large opening near the base of the bell. This emptying action takes only a few seconds, causing powerful flush in the water-closet below.

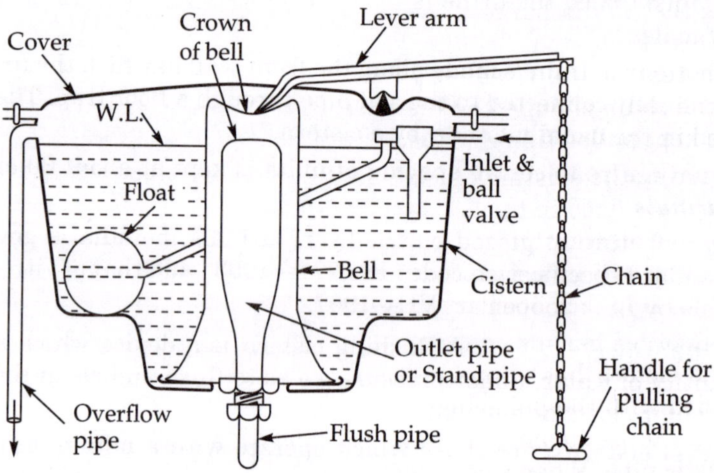

(*a*) Section

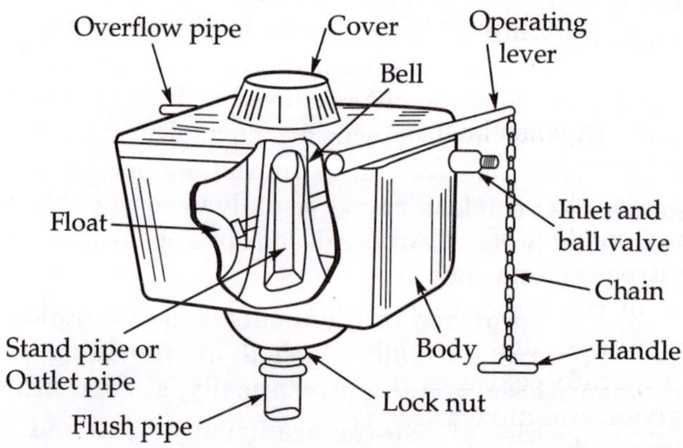

(*b*) Perspective view

Fig. 13.17. Bell type Flushing Cistern.

In this type of cistern, the chain should be released immediately after being pulled, to force the water out of the flush pipe. If this precaution is not taken, the partial vacuum developed by pulling of the chain may be destroyed by the entry of air from the flush pipe.

The cistern is supplied with water through a *ball valve arrangement*, so that the water inlet gets closed when the cistern is full with water upto the water line, and gets open when it is empty permitting entry of water. It is also provided with an overflow pipe, to drain out excess incoming water, if the ball cock gets out of order, so as to avoid spilling of water over the cistern in the toilet.

It is evident that this type of a cistern, releases a fixed quantity of water, even if some one simply urinates in the toilet, which could be washed down with a much smaller quantity of water. The problem becomes acute in areas where there is a shortage of water. There has, thus, been some demand in the past to find ways to reduce the quantity of water consumption. C.B.R.I. Roorkee has consequently developed **dual flushing cistern,** which allows the *fractional* or the *full* discharge of the cistern at a time. All the existing cisterns can be converted into dual flushing cisterns, by making arrangements to cut the vacuum seal at the fixed water level. Hence, to obtain the fractional discharge, the chain is pulled and left ; whereas, for obtaining full discharge, the chain is kept pulled till full capacity is discharged.

For converting the existing cisterns into this type of dual cisterns, a 6 mm dia rubber tube is fixed in the bell at a fixed height. The other end of the tube is connected to a plastic pipe with a stop cock at the end, to control the entry of air. The cistern will give full discharge with stop cock closed, and partial discharge with stop cock open.

C.B.R.I. Roorkee has also developed an automatic flushing cistern for the urinals. This cistern has eliminated the use of copper fittings, which were being provided with the public urinal cisterns, and which are not only costly but also liable to be stolen. The new developed fittings consist of a U-tube, made of plastic pipe. When the water level in the cistern reaches the level of the bend, the siphonic action takes place and the water present in the cistern rushes to the urinals. This new automatic flushing cistern can be cheaply manufactured and easily fixed in position.

13.7.4. Wash Basins. A wash basin is a vessel used for washing hands or brushing teeth, etc. in a standing position. It is made of glazed earthenware or vitreous china clay. It is provided with one or two water taps at either ends. Sometimes, and in high class buildings, separate cold and hot connections are given, and mixed water comes out through a mixer arrangement. The hotness of the coming water can be adjusted upto the

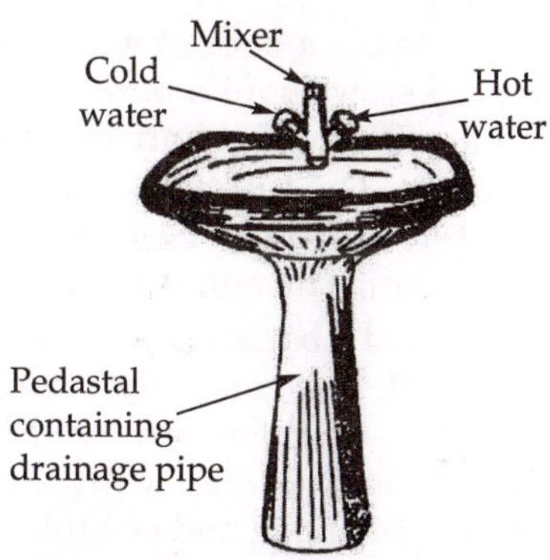

Fig. 13.18. Perspective view of a wash basin supported on a *Pedestal* provided with hot and cold mixer arrangement.

required extent, by opening the cold and hot taps to the desired extent. Perspective view of such a modern wash basin, provided with a mixer, is shown in Fig. 13.18.

The wash basins may either be supported on brackets fixed on wall, or supported on **pedestals.** Fig. 13.19 shows the former type and Fig. 13.18, shows the latter type.

The wash basins may either be of flat back pattern for mounting on walls, or **angle back** for fixing at the junction of two walls. Their usually available sizes are given in Table 13.2.

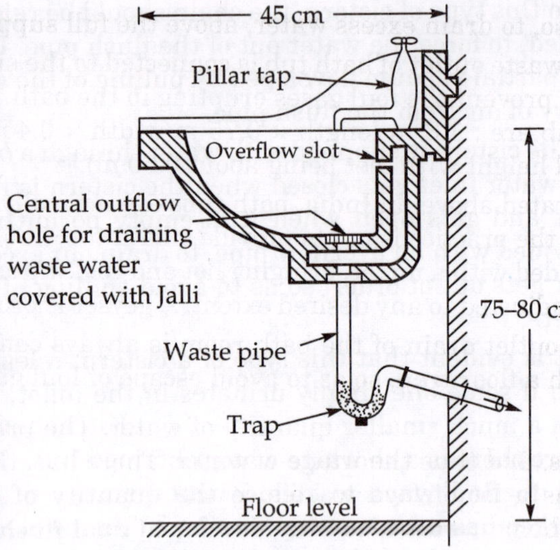

Fig. 13.19. Section of a wash basin (Flat back).

Table 13.2. Commonly available Sizes of Wash Basins

Pattern	*Size*
Flat back	650 × 450 mm 550 × 400 mm
Angle back	600 × 480 mm 400 × 400 mm

A central hole covered with a jalli is provided in the wash basin for draining the used water to the sullage pipe through a trap. For cleaning and for holding water in the basin, sometimes, a rubber plug attached with a chain secured by a stay is used to close this central outlet hole.

13.7.5. Baths and Bath Tubs. Bath rooms often provide waste water, either when one takes a bath or when one washes clothes. In normal Indian homes, bath is taken without a *bath tub*, either under a *shower* or by using a bucket and a mug. Western countries, however, use a bath tub, in which one sits down or lies down. The tub can be filled with water, which can be drained out by opening the outlet plug.

A section of a bath tub is shown in Fig. 13.20. It is generally made of vitreous earthenware or of marble. It is provided with an outlet, 4 to 8 cm dia. and one inlet pipe for filling it. In modern days, two taps are provided, so as obtain hot and cold supplies both, It is provided with an overflow

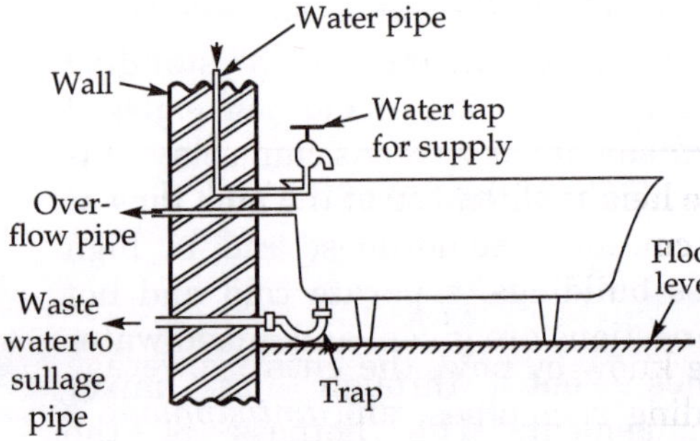

Fig. 13.20. A section through a bath tub.

pipe also, to drain excess water, above the full supply level.

The waste outlet of bath tub is connected to the sullage pipe through a trap, so as to prevent the foul gases erupting in the bath. The usual dimensions of a bath tub are : 1.8 m length × 0.75 m width × 0.45 m depth near waste pipe (overall height with feet being about 0.60 m).

As stated above, in India, bath tubs are normally not used. Even in modern homes, the practice is not to provide a bath tub, but to provide a shower, which is provided with a mixer, bringing hot and cold water, the temperature of which can be adjusted to any desired extent. A geyser is provided for hot water supply.

The outlet drain of the bath room is always connected to the sullage pipe through a floor trap, so as to avoid escape of foul gases in the bath room.

13.7.6. Kitchens and Kitchen Sinks. Kitchen sinks are used to clean the utensils, and thus the waste water obtained from them is normally foul, as it contains a lot of food wastes. Kitchen sinks are normally used in western styled homes, and a normal Indian home uses the kitchen floor itself for washing utensils. In that case, the outlet drain of the kitchen room is provided with a floor trap, so as to avoid evolution of foul gases in the kitchen.

Kitchen sinks are rectangular shallow vessels, as shown in Fig. 13.21. It is a one piece construction provided with or without a rim. It is made of marble or vitreous earthenware or vitreous china clay or stainless steel. The floor of the sink is given a slope towards the waste outlet. The sinks are provided with circular waste hole. All the kitchen sinks are provided with a draining board, which is fixed on the right side of the user. Weir type overflow slots are also provided in some sinks. The usually available sizes of sinks are given in Table 13.3.

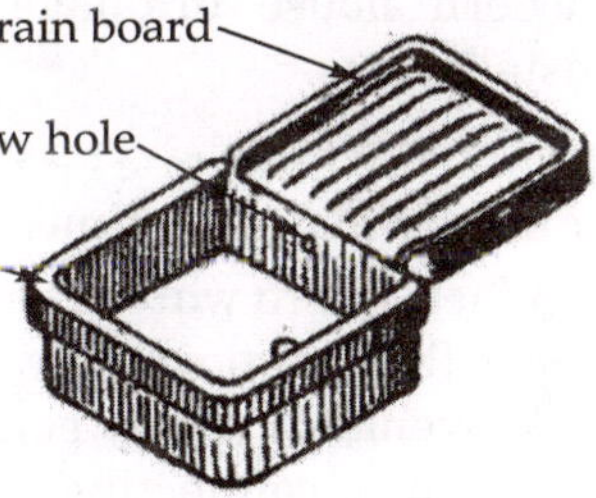

Fig. 13.21. Kitchen sink.

Table 13.3. Usual Dimensions of Kitchen and Laboratory Sinks

	Available size
Kitchen Sinks	600 × 450 × 150 mm 600 × 450 × 250 mm 750 × 450 × 250 mm
Laboratory Sinks	400 × 250 × 150 mm 450 × 300 × 150 mm 500 × 350 × 150 mm 600 × 400 × 200 mm

The height of the top of the sink floor above the room floor should normally be 90 cm.

13.8. Ventilation of House Drains

As we know by now, the entire sewerage system of a house is ventilated by providing *vent* pipes, and *antisiphonic* pipes*, and sometimes by *fresh air*

*Explained in article 13.8.1.

inlet. Such a ventilation system is required to serve the following purposes :

(*i*) to relieve the pressure of foul gases developed in house drains, which may otherwise break the water seals of the shallow seal traps ;

(*ii*) to prevent breakage of trap's seals by siphonic action,* and

(*iii*) to dilute the foul gases in the drains and to reduce their obnoxious effects, in case such gases happen to escape out into the house due to some reason or the other.

The *ventilation* is usually achieved by providing a **fresh air inlet,** connected to the house drain through the lower-most manhole (inspection chamber), as shown in Fig. 13.6. Fresh air from the atmosphere will enter through this inlet into the manhole, and finally goes out though the vent pipes provided in conjunction with soil or waste pipe, as shown in Fig. 13.6. This air along with foul gases will finally escape out from the cowl provided at the head of the vent pipe, at least 2 m above the roof level.

A *mica flap valve* is provided at the head of the fresh air inlet, with slits for admitting fresh air, which is so hinged at the top that it easily opens inside to admit fresh air, but it closes with the slightest back pressure of gases from the manhole, to avoid the escape of foul gases in the street or the house courtyard.

Such a ventilating pipe installation should normally be provided in an modern house drainage system, though may be avoided for very small installations.

13.8.1. Antisiphonage Pipes. Water seals of traps in multi-storeyed buildings or houses may sometimes get broken due to siphonic action, as explained below:

When waste water is suddenly discharged from a sanitary fixture A on the upper floor, it moves down rapidly through the soil (or the waste) pipe ; and in its movement, it may suck some air from the lateral pipe connecting the soil pipe with the fixture at the lower floor, as shown in Fig. 13.22. This sucked air causes siphonic action, resulting in the flow of water from the trap of the fixture to the soil pipe and thus, breaking its water seal.

To overcome this difficulty, a separate pipe of smaller diameter is attached to the traps, which connects the trap with the vent pipe, as shown in Fig. 13.22. This pipe is known as *anti siphonage pipe* and it supplies air to the short branch pipe of the lower fixture, at the time of suction ; otherwise also, it acts as a vent pipe connection of the lower fixture. This pipe, thus will normally serve as a *vent pipe*, and as an antisiphonage pipe in case suction takes place, as explained above. Hence, some people get confused when *antisiphonage pipes* or *vent pipes* are sometimes described separately and sometimes written as one and the same thing.

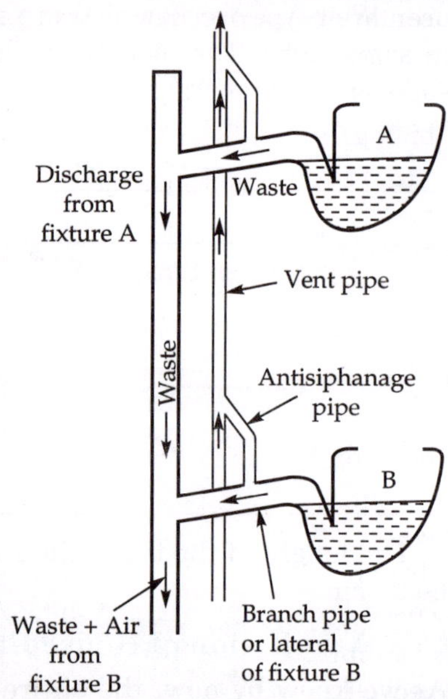

Fig. 13.22. Induced Siphonage

13.9. Wastewater Recycling in Buildings

In water scarce urban areas having large building complexes or institutions, *wastewater recycling* can be resorted to, to reduce the consumption of fresh treated water supplies. In waste water recycling, the less polluted sullage is collected from the building(s), and is treated for resupply for *non-potable uses*, such as for *gardening, construction, car washing, toilet flushing*, and in *scenic waters* and *fountains*, etc. The fresh water is then used only for **potable uses**, such as for *drinking, washing*, etc.

The waste water recycling can be resorted to, throughout the year, to provide large scale saving of fresh water, or to reduce the daily requirement of fresh treated water. Reusing waste water also means cleaner rivers. Although the waste water recycling appears to be a costly exercise, yet actually, it works out to be cheaper, since the cost of production of treated fresh water is 4-5 times the cost of production of reusable waste water.

Contrary to the popular belief, recycling of waste water is a tried and tested method. Its technology for domestic units is readily available in the market. *A two tier structure*, comprising laying of two pipe system for draining sewage and sullage in a building, is ideal for practising water recycling. In this two pipe system, as explained earlier, one pipe system collects sewage from water-closets, and connects the same to the city's sewage system. The other set of pipes connects kitchens and bath rooms together to bring down the polluted sullage discharge into an R.C.C. sullage collection tank, from where it is passed on to the mini treatment plant installed for the purpose of treating the same. Such a mini treatment plant treats the sullage in the following four stages :

(*i*) **Stage 1.** This stage consists of providing preliminary treatment to remove coarse solids and other large sized materials.

(*ii*) **Stage 2.** This stage involves primary treatment, consisting of sedimentation, to remove settleable organic and inorganic solids. A part of biological oxygen consuming material is also removed in this process.

(*iii*) **Stage 3.** This stage provides secondary treatment to oxidize the residual and suspended organic solids by aerobic microorganisms (principally bacteria).

(*iv*) **Stage 4.** The water after secondary treatment from 3rd stage is finally disinfected before reuse, by chlorine.

The entire technique is shown in Fig. 13.23.

Compact waste treatment units based on the above principles, are nowadays being provided by companies, like Ion Exchange, etc.

Another method that is in use involves *plant-based treatment* of waste water. Here, waste water is filtered through a bed of plants. The only drawback of this system is that it requires larger space, which may be a very costly proposition in a big metro city like Delhi or Bombay.

Some institutions and hotels in Delhi are already practising waste water recycling to generate water for non-potable uses. Hyatt Regency, a five star hotel, for example, is recycling its waste water from laundry and kitchen for

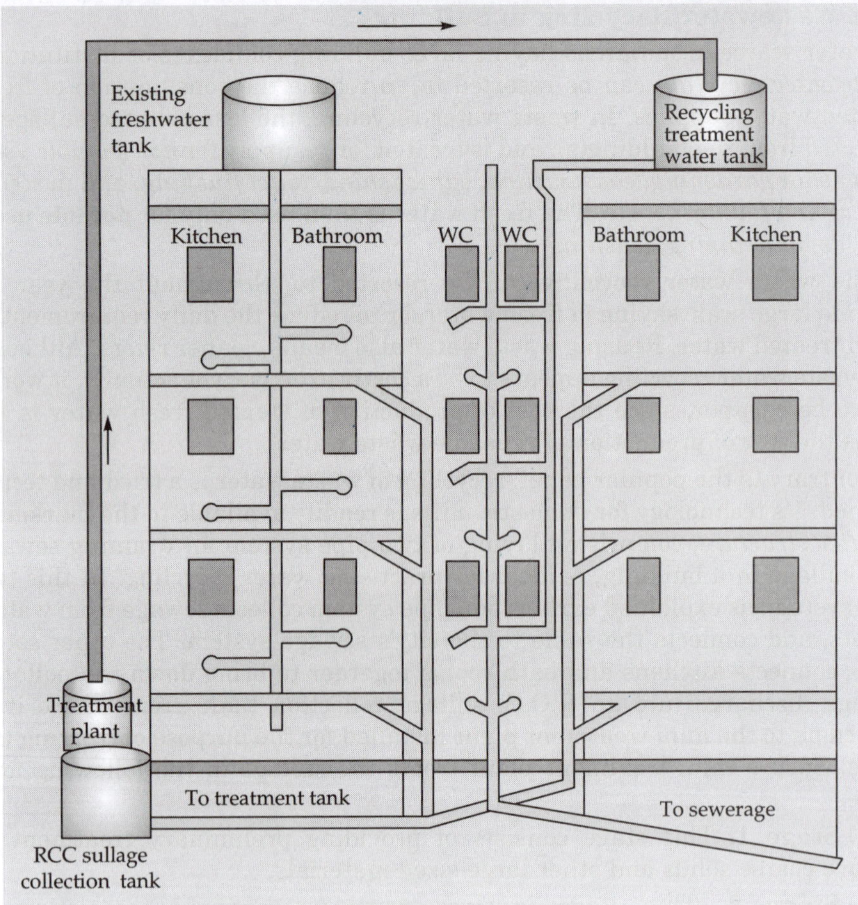

Fig. 13.23. Line diagram for Recycling of waste water from a building using two pipe system.

gardening purposes. Fig. 13.24 shows a photo view of the waste water treatment plant of Hyatt Regency hotel in Delhi. Certain other institutions like the Imperial hotel at Janpath in New Delhi, the Maurya Sheraton hotel at Diplomatic Enclave in New Delhi, the famous Netaji Subhash Institute of Technology (previously known as DIT) at Dwarka in West Delhi, are also recycling their waste water for horticulture purposes. The India Habitat Centre at Lodi Road, New Delhi, is also expected to follow suit.

As a matter of fact, the Union Ministry of Urban Affairs and Poverty Alleviation, GoI, has issued a notification on July 28, 2001, vide which not only *rain water harvesting* has been made compulsory for all the new buildings in Delhi, but has also made it compulsory for all the buildings and institutions to develope *waste water recycling for horticulture purposes*, in case their outflow discharge is 10,000 litres or more. Though this regulation is binding on only upcoming buildings and not the existing sprawl, yet it can go a long way in helping the capital city of NCT of Delhi to tide over its water shortages, provided the civic bodies implement this by law, effectively.

Fig. 13.24. Photoview of Waste water Recycling treatment plant at Hyatt Regency hotel in New Delhi.

PROBLEMS

1. What are the principles adopted in providing an efficient and economical system of drainage in buildings ?

Design a toilet block with modern conveniences for an office with 200 officials, of whom about a quarter are women, and indicate in plan various amenities you would provide.

Draw a cross-section of the water closets in a multi-storeyed building showing all the connecting pipes and fixture. The cross-section may be limited to the top two floors.

2. (*a*) Sketch an Indian type water closet, with associated fittings and fixtures, and explain its working.

(*b*) Critically discuss one and two pipe systems of plumbing.

3. (*a*) Differentiate between single pipe and single stack plumbing system.

(*b*) What is an intercepting trap, and when is it used in house plumbing system ? Explain your answer with neat sketch.

4. Draw a neat sketch showing water and sanitary fittings required for a building, and prepare a list of plumbing materials.

5. (*a*) With the help of sketches, describe the following sanitary fittings :

(*i*) Indian type water closet ; and

(*ii*) Wash basin.

(*b*) Write short notes on.

(*i*) Automatic flushing tank ; and

(*ii*) Intercepting trap.

6. What is meant by ventilation of house sewers, and how is it achieved ? Also discuss the uses of antisiphonage pipes in multistoreyed blocks.

7. (*a*) Name and explain with sketches, the different types of traps used in house sewer connections. What is the patented name of famous trap, used in India in kitchens or bath drains.

(*b*) Show by sketches the urinals generally used in high class public buildings and in municipal street toilets, in India.

8. Enumerate the various sanitary conveniences required in a public building. Indicate the minimum requirements of such conveniences for school of 500 boys and 50 girls, according to Indian standards.

9. Describe 'one pipe', two pipe', and 'single stack' system of plumbing for buildings, and state their merits and demerits. Which one of these would you recommend for the toilet blocks of students hostel in the campus of your university and why ?

10. Draw a typical floor plan of a house, showing the position of different sanitary fittings, traps, pipes, inspection chambers, etc., and indicate their approximate sizes. Draw a sketch to show its connection with the city sewer under the street, assuming that normal necessary slopes are available.

11. (*a*) What are traps, and why are they provided ? How can the water seal in traps be broken ? What methods are to be employed to maintain the seal ?

(*b*) When can traps be omitted in a certain section of a house drain ? How will you connect this section to the main drain ? Illustrate your answer with sketches.

12. Why it is necessary to provide 'ventilation' of the house drainage system ? Draw a sketch to show the typical ventilation arrangements including inlet, outlet, etc. to a house drainage system, and explain its working.

13. What is an intercepting trap, and where is it used in a house drainage system ? Explain its location with a neat sketch. Can you avoid the use of such a trap in a house drainage system ? And if yes, what will be the problems that may arise due to its absence.

14. Write short notes on any *three* of the following :

(*i*) Gully trap and S-trap ;

(*ii*) Anti-siphonage pipe ;

(*iii*) Automatic flushing cistern ;

(*iv*) One pipe system of plumbing ;

(*v*) Ventilation of house sewers ;

(*vi*) Intercepting trap, and its merits and demerits ;

(*vii*) Soil pipe, waste pipe, vent pipe and anti-siphonage pipe ; and

(*viii*) Testing of house drains.

14

Disposal of Municipal and Industrial Solid Wastes

14.1. Definition, Introduction, and Necessity of Safe Disposal of Solid Wastes

The term **solid waste** includes all those solid and semi-solid materials that are discarded by a community. The solid waste generated through domestic and commercial activities is classified as *municipal solid waste (MSW),* and is also called **'refuse'.** The solid waste generated by industries is known as *industrial solid waste.* Solid waste may also be generated by agricultural activities, a large portion of which may also become a part of the municipal solid waste. The *biomedical waste* from hospitals and nursing homes, similarly, finds entry into the municipal solid waste, though is supposed to be disposed off separately as a *hazardous biomedical waste.* The animal excreta (gobar) and human excreta, also, does not stand included in the municipal solid waste, though a portion of this type of semi solid waste may also find entry into the municipal solid waste. In general, therefore, the solid wastes are usually divided into the following two categories ; *i.e.*

(*i*) *municipal solid waste* ; and

(*ii*) *industrial solid waste.*

The municipal solid waste is disposed of by the municipal corporations and such local bodies ; while the industrial solid waste is disposed off by the respective industries.

The **safe disposal of solid wastes** of a society was not a serious problem, as long as the population was small and the land available for assimilation of wastes was large. However, in olden societies, no proper collection and safe disposal practices for getting rid of the solid wastes were in vogue. The society's solid wastes used to be dumped on roads, unpaved streets, and vacant plots of land. Such practices prevailing in medieval towns, served to swell the population of rodents (like rats) and flies, which functioned as the carriers of germs and pathogenic bacteria. The outbreak of plagues starting with the *Black Death*–which killed 50% of European population in the fourteenth century, and claimed many more lives in subsequent years by repeated outbreaks, provide a grim testimony to the then prevailing standards of sanitation. Improper or poor collection and disposal of society's refuse also caused the outbreak of what was widely thought to be the pneumonic plague in September 1994 in Surat City of Gujarat State of India.

In addition to causing possible outbreaks of diseases and epidemics, the improper management of solid wastes causes adverse effects on ecology. For example, dumping of solid wastes at a particular waste dumping site, may lead to the formation of *leachates* during rains, which may seep down and contaminate the ground water. This happens when the dumped refuge contains non-biodegradable and carcinogenic substances, such as plastics, unused medicines, paints, insecticides, sanitary napkins, etc. which may start troubling on coming in contact with the rain water seeping through the dumped garbage, producing a coloured liquid called **leachate.** This highly poisonous and polluted leachate, containing organic compounds like chlorinated hydrocarbons such as benzene, toluene, xylene, etc. ; and toxic elements such as arsenic, copper, uranium, etc. is likely to seep to the underground water-table to pollute the ground water, leading to diseases like cholera, typhoid, polio, etc.

Although the natural processes possess enough capacity to dilute, disperse, degrade, absorb, or otherwise dispose off the unwanted materials (wastes), but that capacity has now come under tremendous stress due to the enormous quantities of wastes being generated by the modern society. India, similarly, faces an uphill task of controlling pollution of all kinds, such as water pollution, air pollution, and land pollution, as may be caused by the solid wastes.

MUNICIPAL SOLID WASTE (MSW)

14.2. Composition and Quantity of the Generated Municipal Solid Waste (MSW) or Refuse

The municipal solid waste (MSW) is a heterogeneous mixture of various kinds of solid wastes-which are not transported with water as sewage, and may include biodegradable (putrescible) food wastes called *garbage*, and the non-putrescible solid wastes like paper, glass, rags, metal items, etc., called *rubbish.*

The **garbage** includes all sorts of *putrescible (bio-degradable) organic wastes* obtained from kitchens, hotels, restaurants, etc. All waste food articles, vegetable peelings, fruit peelings, etc., are thus, included in this term. These wastes are organic in nature, and thus, likely to decompose quickly, producing foul odours and health hazards. They may also result in breeding of flies, mosquitoes, insects, etc. Hence, garbage must be disposed of, properly and quickly. When it is scientifically processed and *composted,* then it is possible to obtain valuable products, like grease, hog wood, fertiliser, etc. from the garbage. The density of garbage usually varies between 450 to 900 kg/m^3.

The **rubbish** can include a variety of materials, which may either be *combustible* (such as paper, plastic, textile, etc.) or *incombustible* (such as broken glass, crockery, metal, masonry, etc.). Most of these types of wastes are discarded on a regular basis from homes, offices, and small commercial establishments. The density of rubbish usually varies between 50 to 400 kg/m^3.

In addition, the **incombustible ashes** from household hearths (*chulhas*), called *ashes,* may also stand included in the MSW, although its quantity is getting reduced in modern urban homes, due to increasing use of cooking gas, kerosene oil stoves, and lesser use of cooking coal or firewood. The density of ashes generally varies between 700 to 850 kg/m^3.

Besides the above three important components, the municipal solid waste may also include the fine dust, silt and sand, obtained from **street sweepings**.

In addition to the above four usual components of refuse, certain **special wastes** such as construction debris, abandoned appliances and automobiles, etc. may also be collected at sporadic intervals from different places.

Not included in the components of municipal waste described above, there are many other solid wastes that are normally not the responsibility of the municipal bodies. These may include ashes from thermal power plants, sludges from water and waste-water treatment plants, wastes from animal feed lots (cattle dung), mine tailings, biomedical wastes, and other industrial solid wastes. All such wastes require separate arrangements for their disposal.

Although the amount of municipal solid waste is usually very small (say as less as up to 5% of the total solid wastes of all kinds), but it receives the greatest attention, because of the adverse effects its improper disposal can have on public health and on water supplies from both the surface water as well as ground water sources.

Besides the above technical classification of the components of the municipal solid waste, based on the type of generated waste, the refuse may also be classified depending on its source, as : (*i*) **house refuse** ; (*ii*) **street refuse** ; and (*iii*) **trade refuse**. All these terms are quite apparent, and known to all of us, and hence need no explanation.

The **quantity of municipal solid waste (MSW)** produced by a society depends upon the living standards of its residents. The degree of commercialisation, urbanisation, and industrialisation, infact, has resulted in a vast increase in the amount of refuse generated per person. For example, the average per capita daily refuse production in United States is as high as about 2.8 kg ; whereas, the figure is only about 0.5 kg in India. That is why, a city like New York with a population of about 20 million produces MSW of about 50,000 tonnes/day ; while, a city like Delhi with about 15 million population produces MSW of about 6000 tonnes/day.

With the modernisation and industrialisation, new types of solid wastes have come in abundance. Say for example, cans, plastic bottles and containers, plastic toys and appliances, tyres and many other items are now being used and frequently discarded on a large scale. Packaging foils and films, cardboard boxes, and several such items are nowadays increasingly being used.

All such factors have increased the MSW by a large extent.

Although only a few reliable statistics are available on generation of solid wastes in India at national level, yet CPCB has estimated that the daily per capita solid wastes generation in India ranges from about 0.1 kg in small towns to about 0.5 kg in larger cities.

The quantity of refuse produced in a city not only depends upon the type of the city and on the living standards of the residents, but also depends upon the season. Say for example, in India, average summer refuse is about 25% higher than the yearly average, due to larger carriage and consumption of fruits, like mangoes, melons, etc.

The **average composition of refuse** (by weight) is estimated to be as shown in table 14.1.

Table 14.1. Average Composition and Other Properties of MSW (Refuse)

Sl. No.	*Item / Constituent*	*App. Av. composition (% age by weight at disposal site)*	
		For a typical Indian city	*For a typical city of U.S.A.*
1.	Garbage	45%	20%
2.	Rubbish (paper, glass, rags, etc.)	15%	50%
3.	Ashes	15%	15%
4.	Fine dust, silt and sand	25%	15%
	Density	400 to 600 kg/m³	100 to 250 kg/m³
	Calorific value* in kilo. Joule/kg**	5000–6600	15,000
	*Heat, which garbage can generate. kJ/kg = 0.243 kC/kg.		

As can be seen from table 14.1, the presence of rubbish in Indian refuse, as compared to that in USA, is very small, as in India, large quantities of papers, cardboards, plastics, synthetic polymers, rags, etc. are picked up and removed enroute, by the rag pickers (men, women and children) before the refuse reaches the disposal site. This, infact, reduces the calorific value of the Indian refuse. On the other hand, the quantity of garbage in U.S.A. is very small, because of the use of garbage grinders, and use of tinned and ready-made packed foodstuffs.

The density of Indian refuse is generally higher than that of the developed countries, and hence the Indian refuse can be carried efficiently and economically by mechanical transport (carrying more weight for the same volume), for land fillings. But since the calorific value of the Indian refuse is much smaller, and its moisture content is high, it cannot be easily burnt or incinerated, and hence the incineration method of refuse disposal, is not so suitable for India, as it is for the developed countries like U.S.A.

As pointed out earlier, although the biomedical wastes from hospitals and nursing homes are required to be collected and treated separately, but in most cities and towns of India, such wastes largely continue to form a part of MSW. However, after the notification of *Biomedical wastes (Management and Handling) Rules, 1998* enacted under the *Environment Protection Act 1986*, a great emphasis is being laid and pressure is being mounted on hospitals and nursing homes of the country to develop combined treatment plants or individual treatment and disposal plants to collect, treat, and dispose of their wastes, either individually or collectively, in accordance with the provisions of these notified rules.

As a matter of fact, rules have also been notified for proper management of municipal solid waste under the *Environment (protection) Act, 1986,* so as to legally enforce proper collection and disposal of municipal solid wastes. These rules are called the *Municipal Solid Wastes (Management and Handling) Rules 2000,* and are discussed in article 14.8.2.

14.3. Collection and Transportation of Municipal Solid Waste (MSW)

In India, the refuse is generally collected in individual houses in small containers, and from there, it is collected by sweepers in small hand-driven lorries/carts, and then dumped into the masonry chambers constructed by the municipalities along roadsides. The refuse is finally carted away by municipal trucks, for further disposal during some day time. *The methods adopted here are highly unsatisfactory, and need tremendous improvements and changes.*

Infact, the house sweepers and street scavengers, do not bother much for carrying refuse properly, and they go on scattering it here and there, while carrying it up to the municipal chambers. Many of them do not bring hand lorries even. Even at the entrance point of the municipal refuse chamber, they just throw the refuse, scattered alround in and out of its gate. The street animals, would further scatter it, leading to all round scattering of the refuse, almost everywhere, resulting in highly insanitary conditions and health hazards.

Similarly, the municipal trucks, do not generally clear and clean the refuse chambers properly, and the residual refuse remains dumped for a long time, resulting in its decomposition and evolution of obnoxious gases, and consequently causing health hazards. Even municipal trucks, while carting refuse, are generally not closed-bodied, and hence go on throwing refuse on way, and also giving pungent smells. This is happening so, even in Delhi, the capital of India.

The process of refuse collection and its carriage, therefore, needs vast changes in this country. The refuse, should, therefore, infact be collected by municipal trucks directly from the houses. *Roadside refuse collection masonry chambers, need complete elimination.* Municipal trucks should be completely closed, and should visit homes and houses, twice a day, once in the morning and once in the evening, to collect household refuse and street sweepings.

The transporting trucks (vehicles) should also be of high quality, of special design and be properly maintained. They should be strong, durable, and water-tight, and be made of stainless steel with smooth interior, having rounded corners and edges, for facility of cleaning. They should have a low loading line, say upto about 1.5 m, so that minimum of time and effort is required in filling them. They should have a cover, which should be made as a part of the body with hatches, which can be opened during collection. Mechanical devices should be installed in these vehicles, for lifting the body to the sides or back, or for pushing the refuse out, so that they can be quickly and easily emptied.

TERI (*The Energy Research Institute,* previously known as Tata Energy Research Institute) has, infact, carried out a survey of the efficiency of the presently prevailing system of collection of refuse in India, wherein the following conclusions have been drawn.

It takes anywhere from 3 to 7 days for the solid wastes to be disposed of from the time it is generated. The collection, transportation and disposal of MSW in India are totally labour intensive activities, because modern automated systems are not used. The number of municipal workers employed on these jobs are also short, because of the general policy of the governments to restrict employment. As a result, on an average, less than three–fourth of the MSW is collected.

Not only the trucks deployed for cartage of MSW are of bad quality, their numbers are also very much less than the actual requirement. Considering an average per capita production of MSW of 0.5 kg, the expected mass of daily Municipal solid waste produced per million of population would be 5,00,000 kg with an average density of MSW as 500 kg/m^3. The volume of MSW produced per million population would be 1000 m^3. Assuming that a truck makes three trips a day, the minimum carrying capacity of deployed trucks should be about 333 m^3/ million population. A survey carried out by TERI, however, has shown that about 70% of the surveyed 44 No. of Indian cities, do have less than the minimum reqd. trucks (total capacity being less than 300 m^3/million population), and only 6.8% of the cities had more than 400 m^3/million population capacity (*i.e.,* capacity of total deployed trucks), as shown in table 14.2.

Table 14.2. Available Transport Capacity in 44 Cities of India (1996)

Available Capacity in m^3/million population	*Cities (%) Total 44 cities surveyed*
< 100	4.5
100–200	34.6
200–300	29.6
300–400	25.0
> 400	6.8
	100.00

A part of the uncollected and untransported MSW, normally finds its way into sewers; some of it is eaten by cattle ; while some portion of it continues to lie scattered for a few days before sweepers collect it together and burn it in the open.

14.4. Separation for Recycling and Reuse of Plastics, Paper, and Glass from the Municipal Solid Waste (MSW)

In a developing country like India, papers, plastics, glasses, rubbers, ferrous nonferrous metals, and other materials that can be recycled, are salvaged from the municipal solid wastes to produce low-cost products extensively used for lower-income groups of society. Recycling has been receiving increasing attention worldwide over the fast few years. Though the consumption of goods in India is very low compared to that in the developed world, recycling has always been a common practice in India.

Selling old newspapers, magazines, books, empty bottles of glass and plastic, metal cans, discarded appliances, etc. to the *Kabaris,* is a very common practice even amongst the richest homes in our country. Till a few years ago, the emphasis in India was on 'recover and reuse' which has now changed to 'recover and recycle'. The change from reuse to re-cycle has occurred due to changing lifestyles and attitudes of people to the kinds of products they use. Of all the wide varieties of recyclable material present in Indian household waste, it is the paper and the plastic that get maximum attention, because the two together form the single largest component of household wastes. The recycling of plastics, papers and glasses that are mostly adopted in India is briefly discussed below.

14.4.1. Recycling of Waste Plastics. Over the last four decades (1961-2000), the production and consumption of plastics in India has increased by more than about 100 times, increasing from the meagre 0.02 kg/capita/annum to about 2.0 kg/capita/annum, as shown in table 14.3.

Table 14.3. Annual Per Capita Consumption of Plastics in India

Year	*Annual per capita consumption of plastics in kg*
1960-61	0.02
1984-85	0.64
1986-87	0.71
1988-90	1.00
1990-91	1.06
1993-94	1.42
1994-95	1.62
1999-2000	2.00

Consequently, the share of plastics in Indian MSW has risen rapidly from about 0.7% in 1971 to about 5-10% in 2000. Nearly, all plastic waste is sold in India to *kabariwallahas,* who form the first link in the chain of recycling plastic. Small scale plastic recycling units have mushroomed in metro cities. These units are, infact, using outdated technology and poor quality controls, causing several adverse environmental impacts. These units are producing wide range of plastic products like chappals, mugs, utensils, toys, etc. These units, although not complying with industrial area regulations and environmental laws, but are saving the scarce and valuable petro-based virgin raw material. Unfortunately, without any government support and specifications regarding the quality of the raw materials to be used for recycling, plastic waste recycling continues to be a disorganised and technically backward sector, with some of the units working in residential areas without even complying with the water pollution and air pollution control laws.

Overall, the annual consumption of plastic in India is estimated to be about 2 kg per person (2000), as compared with the world's average consumption of about 20 kg/person, and that of the developed-world of 80 kg/person. Moreover, India recycles about 40% of its plastic waste, as compared to 15 to 20% in the developed world. The developed world is, thus, primarily responsible for throwing or burning the non-bio-degradable harmful plastic waste, with India contributing to it only by a very small extent. However, given the current trends, with plastic consumption having been doubled in India during 1990-2000, the use of plastics in India is found to be increasing by a large extent ; and considering the scarcity of petroleum products, there is an urgent need to organize the plastic recycling industry.

14.4.2. Recycling of Waste Papers. Over the past few decades, India has realised the disadvantages of cutting trees for manufacturing paper ; and accordingly, the focus has changed from the paper made from *wood pulp* to the paper made from the *agricultural residues based pulp* and on paper made from *old waste paper*, as shown in table 14.4.

Table 14.4. Raw Material Inputs Used in India for Manufacturing Paper (% of Total Production) Over the Last Decades

S. No.	*Raw material used*	*Year* 1970	1985	1989	1992	1994/95
1.	Wood-based pulp	95	65	60	48	37.6
2.	Agro-based pulp	5	22	25	29	31.4
3.	Waste paper	Nil	13	15	22	31.0

It can be seen from table 14.4 that whereas in 1970, 95% of our total paper production was being derived from wood-pulp by cutting trees ; in 1994-95, only 37.6% of our total paper production was being made by using wood pulp. Similarly, whereas no old waste paper was being used for manufacturing new paper in 1970, about 31% of our total paper production was being obtained from recycled paper waste in 1994–95. The increasing use of waste paper for manufacturing new paper indicates very good progress made by India, to save our valuable forest resources. However, many of our paper mills that use waste paper prefer to use imported waste paper, because domestic waste paper is of poorer quality, given its unorganised collection. So much so that about two thirds of waste-paper-based Indian paper mills use imported waste paper, because of its better quality in terms of fibre-strength, and also due to inadequate domestic supply.

Each tonne of imported waste paper can, infact, produce as much as 0.9 tonne of recycled paper, which is considerably higher than the figure of 0.4 tonne for agro-pulp, and 0.5 tonne for wood-pulp. Moreover, the waste-paper-based production of new recycled paper is far less polluting. The BOD values of the waste water effluent discharged from a wastepaper mill is approximately 25 mg/L, which is far below the value of 150 mg/L for agro-based-paper mill and of 100 mg/L for wood-based paper plant. *Not only a wastepaper based paper mill produces far less polluting waste water, it also requires far less energy than the energy required by paper mills based on other raw materials.*

With the current stress on liberalization, industrial growth, and changing life styles, consumption levels for paper are growing in India. That is why, the annual per capita consumption of paper in India has steadily risen from about 3.6 kg in 1986 to about 6 kg in 2004-05.

As a matter of fact, the consumption of paper in India has increased due to its increasing use for packaging. While the share of the packaging paper in India has been increasing over the last four decades, its conventional cultural share (for writing and printing) has shown a decline. That is why, the increasing percentage of paper has been observed in the municipal solid waste

(MSW), since the packaging paper is generally not collected for recycling, although most of the old newspapers, magazines, books etc. do enter the recycling chain. At present, about 8 to 12% of the household waste consists of paper, which ends up being used for landfill along with most other types of wastes, because it is primarily the packaging paper, a large portion of which is not sorted and separated. Yet, however, a small percentage of this—what can be recycled and has not been soiled by other wastes—is still recovered by the vast network of rag pickers working especially in large towns and cities, and is finally used for recycling.

Unfortunately, we have not been able to increase the recovery, sorting and bailing of waste paper even in our metropolitan cities in India, since the recovery and recycling of paper has received little support, either institutional or governmental. In the absence of financial incentives and the latest technology, paper recycling industry in India remains an undeveloped industry.

14.4.3. Recycling of Glasses. Glass is the third important component of municipal solid waste that is normally recycled. It is usually picked up in our country by the rag pickers, and finally reaches the glass industry through dealers. The waste glass is generally used in manufacturing glass containers and hollow wares, for which the largest use (more than 99%) of glass is made in India. The proportion of waste glass in the raw material varies from 25% to 90%.

14.5. Disposal of Municipal Solid Waste (MSW)

The municipal solid waste, as stated earlier, is collected in India from houses, roads, streets, markets, and from the commercial office buildings by the private or municipal sweepers and scavengers in hand lorries, and carried and dumped into the municipal refuse chambers (popularly called *dhalaos*). During the time the refuge is discarded and thrown away, and till the time it is loaded in trucks for transportation, a lot of recyclable materials like plastics, papers and glasses are removed by the rag pickers. After the wastes are loaded in trucks, it is taken for its final disposal either for dumping on low levelled land, or for *sanitary land filling,* or for *composting*, or for *incineration,* etc. As far as India is concerned, the MSW is normally brought and dumped into the designated land fill site, normally a low-lying area on the outskirts of the city. As per the available data, about 14,30,000 tonnes (1.43 million tonne) of garbage is generated daily in Indian cities. Out of which, about 35,600 tonnes (about 25% of the total) is processed through scientific methods like composting and other methods like pelletization, vermin-composting, etc. The remaining about one lakh tonne of garbage is dumped into unmanaged open land fill sites, or allowed to remain in the open urban areas or dumped in the backyards of smaller towns, suburbs and villages. All in all, only about 90% of total generated garbage is disposed of, while about 10% continue to litter in the streets and open vacant land plots.

The choice of a site for land filling is more a matter of what is available, then what is suitable. Only a few cities in India follow good scientific practices, such as organized tipping of wastes, using mechanised equipment for levelling and compacting the wastes, and covering the top layers with good earth before compacting further, as are required in *sanitary land filling. Most of the land filling practices adopted in the country are, thus, confined only to carrying and dumping the waste rather than adopting the engineered land filling technique, which in future is going to be adopted more and more, with advancing technologies and shortage of land filling sites.* In general, the following scientifically managed methods can be used for disposal of municipal solid waste (refuse) :

(1) *Sanitary land-filling,* also called *Controlled tipping* ;

(2) *Shredding and pulverisation ;*

(3) *Composting* including *Vermicomposting* ;

(4) *Incineration* and *Thermal pyrolysis* ; and

(5) *Barging it out into the sea.*

Out of these methods, only the methods given at S. No. (1) and (5) above, are complete in themselves, while all others require further disposal of some left out residues or by-products.

All these methods are described below :

14.5.1. Disposal of MSW (Refuse) by Sanitary Land Filling or Controlled Tipping Method. In this method of refuse disposal, refuse is carried and dumped into the low lying area (earmarked as the land fill site) under an *engineered operation, designed and operated in an environmentally sound manner*, as not to cause any public nuisance or hazards to public health or safety.

In this method, the refuse is dumped and compacted in layers of about 0.5 m thickness, and after the days work-when the depth of filling becomes about 1.5 m, it is covered by ***good earth*** of about 15 cm thickness. This cover of good earth is called the ***daily cover.*** Since the refuse is well compacted with bulldozers, trucks, rollers, etc. and is well covered daily with good-earth, it does not cause any public nuisance, like scattering of wind blown litter, and evolution of unpleasant odours and foul smells, as may be caused by an ordinary dumping of refuse on land.

The filling of refuse is actually done in sanitary land filling by dividing the entire land-fill area into smaller portions, called **cells**. These cells are initially filled with daily compacted refuse of about 1.5 m depth, inturn. After filling all the cells with first lift, the second lift is laid in about 1.5 m height, and covered with good earth cover of about 0.15 depth, called the ***intermediate cover.*** After all the cells have been filled up with second lift, the third and more lifts can be piled up in about 1.5 m depth each, all laid over by the intermediate earth covers, turn by turn. The process will continue till the top most lift is piled up, over which the ***final cover*** of good earth of about 0.6 m depth shall be

---

*****Good earth*** is an ideal soil material, which is a combination of approximately 50% clay-silt and 50% sand. Such a ***sandy-loam*** soil mixture is porous, compacts well, and is not subjected to cracking upon drying. Clay alone, if used, will crack on drying, giving access to insects and rodents, and hence not used for cover.

laid, and well compacted, to prevent the rodents from burrowing into the surface. This filling operation is illustrated in Fig. 14.1. A **cap-system*** may also be installed over the top of the final cover.

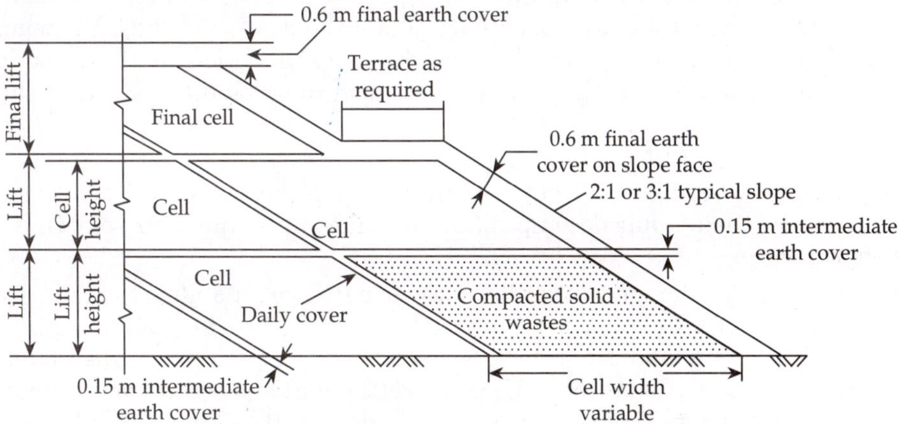

Fig. 14.1. Sectional view of a sanitary land fill.

With the passage of time, the filled up refuse will get stabilised due to the decomposition of organic matter and subsequent conversion into stable compounds. *The land filling operation is essentially a biological method of waste treatment, since the waste is stabilised by aerobic as well as anaerobic bacterial processes.*

Initially, the bacterial decomposition occurs under the aerobic conditions, because a certain amount of air is trapped within the landfill. However, the oxygen in the trapped air is soon exhausted within a few days and the long term decomposition occurs under anaerobic conditions.

The entire period of refuse stabilisation can infact, be divided into five distinct phases : (*i*) During the first phase of operation, aerobic bacteria and fungi, which are dominant, deplete the available oxygen to effect oxidation of organic matter. As a result of aerobic respiration, the temperature in the fill

---

*The *cap system* or the **cap**, which is usually provided at the top of the sanitary land fill, after filling the same to full height, to avoid formation of leachates, consists of the following layers.

45 cm thick *compacted clay layer* (permeability less than 10^{-5} cm/s) is provided at the bottom, which is overlain by a 15 cm thick *drainage layer* consisting of sand (having permeability greater than about 10^{-2} cm/s) ; above the second layer of sand, is laid a *geotextile filter membrane* to protect the drainage layer from being clogged. Finally, 30 cm thick *soil layer* is laid at the top to promote vegetative cover and grassing. This top most soil layer is called the *vegetative layer*. This entire arrangement is known as the cap, and is provided over the final cover of about 0.6 m thickness.

The purpose of the cap is to prevent the rain water from reaching the solid waste (to avoid formation of *leachate*) which otherwise, is drained through the drainage layer. *Gas wells* may also be installed through this cap, which may be located in the top earth cover, and will extend up through the cap, up to the atmosphere, like chimneys.

If the sanitary land fill site is covered with a cap, it should be ensured that sufficient moisture is present in the waste fill, since moisture is necessarily required in inaerobic decomposition of the waste, which the waste has to undergo after the initial oxygen trapped in the voids is consumed in the aerobic decomposition.

increases. (*ii*) In the second phase, anaerobic and facultative bacteria develop to decompose the organic matter ; and H_2 and CO_2 gases are thus evolved through acidogenic activity. (*iii*) In the third phase, methanogenic bacteria develop to cause evolution of methane gas. (*iv*) In the fourth phase of decomposition, the methanogenic activity gets stabilised. (*v*) In the fifth stage, the methanogenic activity subsides, representing depletion of the organic matter ; and ultimately, the system returns to aerobic conditions within the land fill.

For better biological degradation, the moisture content of the dumped material should be high, say not less than 60% or so, which is sometimes maintained by the aerobic decomposition brought out by fungi, or sometimes by sub-soil water.

The refuse, in managed landfills, may usually get stabilised, generally within a period of 2 to 4 months, and settle down by 20—40% of its original height. The filled up land can, infact, be used for developing some green land, parks, or other recreational spots. Unequal settlement and odour trouble, may however, be there, and hence, normally, for the first 1—2 years, the land is grassed or planted, fenced, and left out as reserved green land. This can, on a later date, be preferably used for developing some regular play grounds, or picnic spots. Such sites may also sometimes be used for constructing houses ; though they are not generally preferred, because such constructions may prove to be costly due to their deeper foundations for avoiding unequal settlements. Such houses, may further pose problems like those of bad odours and cracks in walls and plasters, on a later date.

This method of refuse disposal is very suitable to the heavier type of Indian refuse, and also to the rural communities, hostels, camps, etc. Hence, it is widely adopted in our country. So much so, that about 90% of Indian refuse is disposed of in this manner.

The various advantages and disadvantages of this method of refuse disposal are given below :

Advantages : (*i*) This method is most simple and economical. No costly plant or equipment is required in this method, as is required in other methods of incineration or pulverization.

(*ii*) Separation of different kinds of refuse, as required in incineration method, is also not required in this method.

(*iii*) There are no residues or byproducts left out/evolved in this method, and hence no further disposal is required ; this being a complete method in itself.

(*iv*) Low lying water-logged areas and odd quarry pits can be easily reclaimed and put to better use. The mosquito-breeding places are also, thus, eliminated.

Disadvantages : (*i*) Low lying depressions or dumping sites may not always be available ; or even if they are available today, they may ultimately become scarce or unavailable in future, since the production of solid waste is a continuous process.

(*ii*) There is a continuous evolution of foul gases near the fill site, especially during the times the refuse is being dumped there. These gases may often be explosive in nature, and are produced by the decomposing or evaporating organic matter. These gases, known as **land fill gases,** become a serious environmental problem at sanitary land fill sites. These gases need be estimated, and properly disposed off as discussed in article 14.5.1.1.

(*iii*) Since the dumped garbage may contain harmful and sometimes carcinogenic non-bio-degradable substances, such as plastics, unused medicines, paints, insecticides, sanitary napkins, etc., they may start troubling on a later date, particularly during rainy season, when excess water seeping through the area, may come out of the dump, as a coloured liquid, called **leachate.** This highly poisonous and polluted leachate, containing organic compounds like chlorinated hydrocarbons, benzene, toluene, xylene, etc. ; is likely to seep to the underground water-table, to contaminate the ground water, leading to diseases, like cholera, typhoid, polio, etc. In order to avoid such harmful effects, the *leachates* may have to be scientifically assessed, collected, and disposal of, as discussed in article 14.5.1.2.

14.5.1.1. Gas production and its control at properly managed sanitary land fill sites. The rate of decomposition of solid waste in land fills, as measured by gas production, reaches a peak within the first 2 years, and then tapers off, continuing in many cases, for periods as large as up to 25 years or more.

The total volume of gases released during anaerobic decomposition can be estimated in a number of ways. If all the organic constituents in the refuse (with the exception of plastics, rubber and leather) are represented with a generalised formula of the form $C_a H_b O_c N_d$, then the total volume of gas can be estimated by using the equation (14.1), given below, with the assumption of complete conversion to carbon dioxide and methane :

$$C_aH_bO_cN_d + \left(a - \frac{b}{4} - \frac{c}{2} + \frac{3}{4}d\right)H_2O \longrightarrow \left(\frac{a}{2} + \frac{b}{8} - \frac{c}{4} - \frac{3d}{8}\right) CH_4 \uparrow$$
$$+ \left(\frac{a}{2} - \frac{b}{8} + \frac{c}{4} + \frac{3d}{8}\right) CO_2 \uparrow + d \,.\, NH_3 \uparrow \qquad ...(14.1)$$

Under ideal conditions, the gases generated from a landfill should either be vented out to the atmosphere ; or in larger landfills, it may be collected and supplied to houses for warming or cooking, or used for production of energy. In most cases, over 90% of the gas volume produced from the decomposition of solid wastes, consists of methane and carbon dioxide. Although most of the methane escapes to the atmosphere, both methane and carbon dioxide have been found in large concentrations, of say up to 40%, at lateral distances of up to 120 m from the edges of the landfills. Hence, if the gases are allowed to be released to the atmosphere in an uncontrolled manner (as happens in most of the cases in India), then methane may accumulate below buildings, or in enclosed spaces, or close to the sanitary land fill site (upto about 200 m), since the sp. gravity of methane gas is less than that of air. This methane collection may prove to be dangerous, as it may develop an explosive mixture with air, even when methane is present in 5-15% by volume.

On the other hand, carbon dioxide gas tends to move towards the bottom of the landfill, since carbon dioxide is about 1.5 times as dense as air, and is 2.8 times as dense as methane. As a result, the concentration of CO_2 in the lower portions of the landfills may remain high for years. Ultimately, because of its density, CO_2 will also move downward through the underlying formation, until it reaches the ground water. Because CO_2 is readily soluble in water, it usually lowers the pH of ground water, which, in turn, may increase the hardness and mineral content of the ground water, through the solubilization of calcium and magnesium carbonates.

The movement of gases in landfills, should therefore, be controlled with properly designed engineering methods, such as by constructing vents and barriers, and by gas recovery, as discussed below.

(*i*) ***Control of Gas Movement with Vents and Barriers.*** The lateral movement of gases produced in a landfill can be controlled by installing gas *vents*, made of materials that are more permeable than the surrounding soil. Typically, as shown in Fig. 14.2 (*a*) to (*c*), gas vents are constructed of gravel.

The gravel layer of a cell vent is laid directly above the daily cell cover. The spacings of **cell vents** [Fig. 14.2 (*a*)] depends on the width of the waste cells, but usually varies from 18 to 60 m. The thickness of the gravel layer should be such that it will remain continuous, even though there may be differential settling. A depth of about 0.3 m to 0.45 m is usually recommended.

In the **trench vent**, a trench is dug as deep as the solid wastes and filled with gravel, as shown in Fig. 14.2 (*b*).

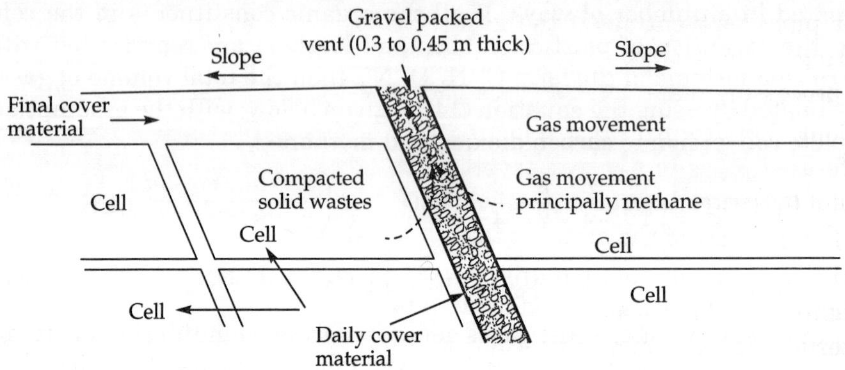

(*a*) Cell vent

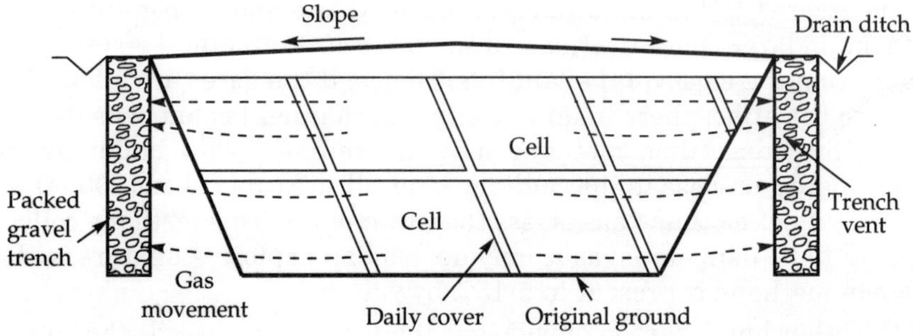

(*b*) Trench vents

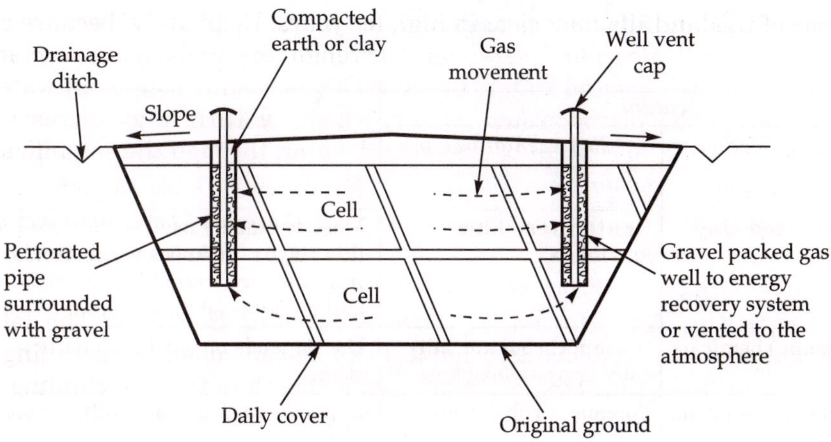

(*c*) Well vents

Fig. 14.2. Typical gravel gas vents installed in sanitary landfills for venting landfill gases.

A **well vent**, consisting of a perforated pipe surrounded with gravel, may also be installed through the waste fill, as shown in Fig. 14.2 (*c*), to control the lateral movement of gases. Such pipe vents will collect and convey the gases to riser pipes for venting. If the gases are to be collected for recovery for its heating value, then the riser pipe may be connected to a main for pumping, to the processing plant.

Control of downward movement of gases can be accomplished by installing perforated pipes in a gravel layer at the bottom of the landfill. If the gases cannot be vented laterally, it may be necessary to install *gas wells* and to vent the pumped gas to the atmosphere.

The movement of landfill gases through adjacent soil formations (neighbours lands) can be controlled by erecting **vertical well barriers** of materials that are more impermeable than the soil (Fig. 14.3). Some of the **landfill sealants** that are available for this use, are shown in table 14.5. Out of these available sealants, the *compacted clay* is most commonly used.

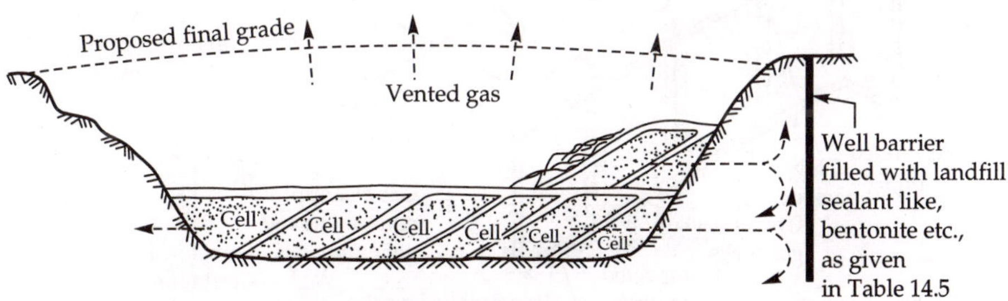

Fig. 14.3. Barrier wells or clay barriers.

Table 14.5. Landfill Sealants for the Control of Gas and Leachate Movement

Sealant		*Remarks*
Classification	*Representative types*	
Compacted soil		Should contain some clay or fine silt.
Compacted clay	**Bentonites,** illites, kaolinites	Most commonly used sealant for landfills ; layer thickness varies from 0.15 to 1.2 m ; layer must be continuous and not allowed to dry out and crack.
Inorganic chemicals	Sodium carbonate, silicate, or pyrophosphate	Use depends on local soil characteristics.
Synthetic chemicals	Polymers, rubber latex	Experimental, use not well established.
Synthetic membrane liners	Polyvinyl chloride, butyl rubber, hypalon, polyethylene, nylon-reinforced liners	Expensive, may be justified where gas is to be recovered.
Asphalt	Modified asphalt, asphalt-covered polypropylene fabric, asphalt concrete	Layer must be thick enough to maintain continuity under differential settling conditions.
Others	Gunite concrete, soil cement, plastic soil cement.	

(*ii*) ***Control of Gas Movement by Gas Recovery.*** In large sized landfills, the gases evolved due to decomposition of refuse may be collected through installing **gas recovery wells,** as shown in Fig. 14.4. The recovered gas may be

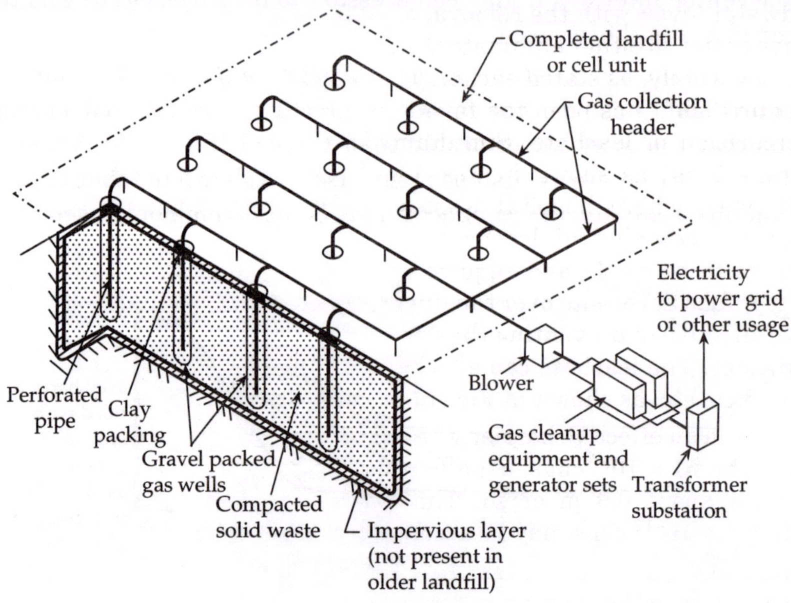

Fig. 14.4. Gas Recovery Well system used for the recovery of gases from large sized landfills.

used for generation of electric power or may be supplied to nearby houses for domestic use. Although gas recovery systems have been installed in some advance countries in some large municipal landfills, yet the economics of such operations are not well defined. The cost of cleaning of the generated gas and that of the processing equipment may limit the recovery of landfill gases, especially from small landfills.

14.5.1.2. Controlling the movement of leachate in properly managed sanitary landfill sites. Under normal conditions in hazardous landfills, leachate is found at the bottom of the landfills. From there, its downward movement occurs through the underlying strata, although some lateral movement may also occur, depending upon the characteristics of the surrounding soil.

The rate of downward seepage from the bottom of the landfill can be estimated by Darcy's law, by assuming that the material below the landfill to the top of the water-table is saturated, and that a small layer of leachate exists at the bottom of the fill. Under these conditions, the discharge rate of leachate per unit area is equal to the value of K (coefficient of permeability) expressed in m/day. This value will indicate the maximum amount of seepage* that would be expected, and this value can be used for designing a suitable system for controlling or collecting the leachate.

As leachate percolates through the underlying soil strata of the landfill, many of the chemical and biological constituents originally contained in it, will be removed by the filtering and adsorptive action of the soil strata. The degree of removal of pollutants from the leachate will usually depend on the characteristics of the soil strata, especially its clay content. However, because of the inherent potential risk involved in allowing leachate to percolate to the groundwater (even with the removal of some of its organics and chemicals), it is always better to either eliminate the production of leachate, or to collect and treat it separately, as stated earlier. The use of **clay liners or synthetic liners like geotextiles**** has been the favoured method for reducing or eliminating the percolation of leachate. Simultaneous use of clay lining and synthetic membrane liner, as shown in Fig. 14.5 may be required for all hazardous landfill sites, where hospital wastes are dumped frequently, and where chances of production of leachates are high. The use of synthetic liners may prove to be costly and require care as not to get damaged during the filling of refuse at the site over the liner, although such a lining layer proves to be quite effective in preventing the downward movement of the leachate to the groundwater. The leachate can also be collected over this layer through open jointed pipes, etc. as shown in Fig. 14.5.

For it to be an effective barrier, the thickness of clay liner with permeability not more than 1×10^{-7} cm/s, usually provided at the bottom of the fill, should be at least about 0.9 m or so. The synthetic liner with permeability not exceeding 1×10^{-12} cm/s may be acceptable for the upper liner.

---

*Under normal conditions, the actual rate would be less than the theoretical value (*i.e.* K m/d), because the soil column below the landfill would not be saturated.

**For detailed description of Geotextiles, please refer "Soil Mechanics and Foundation Engineering" by the same author.

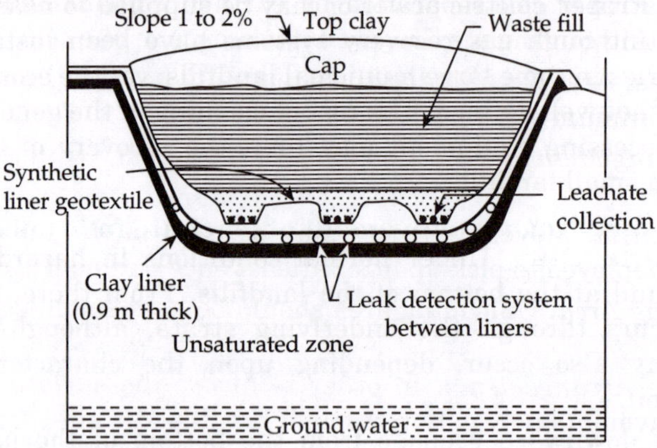

Fig. 14.5. Simultaneous use of clay lining and synthetic lining in hazardous landfills to prevent and control movement of leachate to ground water.

Another important method to control the production of leachate is to provide an impervious clay layer over the top of the fill, which should be appropriately sloped (1 to 2 percent) and provided with adequate surface drainage. This will prevent the infiltration of surface water, which is the major contributor to the total volume of the leachate.

14.5.1.3. Factors to be considered in the design and operation of a sanitary landfill site. Usually, the following factors must be considered for properly designing and managing a sanitary landfill site :

(i) *Access to the site*. The sanitary land fill site must be provided with suitable access roads for easy plying of trucks, carrying the city's solid waste (refuse) to the site.

(ii) *Cell Design and construction*. The design for piling the waste at the landfill site will largely depend upon the terrain and position of water-table, etc. of the landfill site, as well as on the type and quantum of daily waste required to be dumped at the filling site. Circumstances under which the different methods of land filling are to be adopted have already been discussed along with the arrangements required to be made for collecting gas and leachate, if required. Depending on the specific requirements, the arrangements may be designed and adopted to suit the specific needs of the given site.

(iii) *Cover Material*. Approximately 1 m^3 of cover earth (good earth) will be required to cover every 4 to 6 m^3 of refuse. Clayey soils would be preferable as covers or sealants. Such clayey earth should be available near the sanitary fill site, as otherwise, the same will have to be transported over distances. Bringing of such earth will have to be identified and arranged in advance, for properly managing a landfill site. *Use of on site earth by excavation, may be maximised to avoid transport of outside earth* by preferring the trench system of land filling.

(iv) *Equipment Requirements*. Heavy earth moving machinery to pile and compact the waste and the good earth will be required at the land fill site, and this requirement may vary with the size of the landfills.

(v) *Fire Prevention*. Non potable water outlets, if provided at site to extinguish fires at the landfill sites, must be clearly marked and distinguished. Proper cell separation by earthen embankments will surely help in preventing continuous fire, even if it occurs in one cell.

(*vi*) *Land Area*. Area for filling at the site should be large enough to hold all wastes for a minimum of 1 year but preferably for 5 to 10 years.

(*vii*) *Land filling Method*. Selection of method will vary with the terrain and available cover.

(*viii*) *Litter control*. Moveable fences may be used at the unloading areas ; crews may be deployed to pick up litter atleast once per month, or as required.

(*ix*) *Unloading Area*. Unloading area should be kept small, generally under 30 m.

(*x*) *Drainage Arrangements*. In order to divert the surface run off (from rains, etc.) to avoid its percolation through the filled up site, *drainage ditches* shall be installed, and final top earth cover shall be smoothly finished with 1 to 2% slope to avoid ponding of water.

(*xi*) *Gas Recovery or Prevention of Gas Movement*. Arrangements for prevention of gas movement or gas recovery may be suitably designed and made, as discussed in article 14.5.1.1.

(*xii*) *Protection of underground water*. Underground springs, if any, may be suitably diverted. Sealant barrier for leachate control may be installed. Wells for gas and groundwater monitoring may also be installed.

(*xiii*) *Communications Facilities*. Telephones may be installed at the site to ensure proper communication in case of emergencies, like fires, etc.

(*xiv*) *Operational Records*. Proper personnel may be positioned to receive the city's refuse and solid waste. Tonnages, transactions, and billing may be installed, if any disposal fee is to be charged.

(*xv*) *Employees Facilities*. Restrooms and drinking water should be provided.

(*xvi*) *Days and Hours of Operation*. Usual practice is to adopt 5 to 6 days a week, with 8 to 10 h/day. Night shifts are also provided in big cities, since carriage of refuse is usually done during night hours.

(*xvii*) *Equipment Maintenance*. A covered shed or workshop at the site may be installed for field maintenance of equipment.

14.5.2. Disposal of MSW (Refuse) by Shredding and Pulverisation. The size and volume reduction of municipal solid waste (MSW) is accomplished by the physical processes of *shredding* and *pulverisation*. **Shredding** refers to the actions of cutting and tearing ; whereas, **pulverisation** refers to the actions of crushing and grinding. Shredding and pulverising may help in reducing the overall volume of the original MSW, by as much as 40%. The shredding and pulverising not only helps in reducing the volume of MSW, but also helps in changing the physical character of the waste, which becomes practically odourless and unattractive to the insects.

The pulverised refuse, though contains fertilizing elements like potash, phosphorus, and nitrogenous materials, yet cannot be suitably used as manure. It has, therefore, to be further disposed of either by filling in trenches, or is digested in *open windrows* or *closed digestors,* as discussed in article 14.5.3.

The pulverisation of MSW is usually achieved in a **hammer mill,** where the raw solid waste is cramped with a force, sufficient to crush or tear individual components of the waste. Impact is provided by several hammers that rotate at high speeds (up to 1500 revolutions/minute) around a central horizontal or vertical shaft. The mill helps to produce a uniform or homogeneous mass of solid waste. Such a hammer mill proves to be a very versatile device for size reduction of solid wastes, because it accepts almost any type of solid waste material (except of course, very bulky or dense ones, such as tree stumps, or engine blocks, etc.). A hammer mill, thus, makes it possible to reduce the size of the various components of solid waste material, to uniform fragments of size 25 to 50 mm or so.

In order to further reduce the waste volume, and facilitate handling of the pulverised MSW, it is usually compacted to form *rectangular blocks* or *bails.* MSW bails are typically of about 1.5 m^3 in size and weigh roughly 1 kN. The pulverised MSW is converted into bails by compacting the pulverised MSW under high pressures (of about 700 kPa) in either *vertical or horizontal presses.* This process is called *bailing of MSW.* The bails are frequently wrapped with steel wire to help retain their rectangular shape during handling. They may also be coated and stabilised with portland cement or hot asphalt, or tied with metal bands, depending upon the intended use, or the method of disposal. If moisture content and compaction pressures are high enough, the bails may retain their shape without being wire-wrapped or encased. Semi automatic horizontal presses can bail up to 36 kN of pulverised waste per hour. Volume reduction due to pulverising and bailing together, may occur by as much as 90% of the original waste volume.

The bailing process, thus, has significant advantages in decreasing the waste volume, and in providing an easy way for handling waste as blocks rather than as litter, thereby reducing the litter nuisance of scattering, etc.

14.5.3. Disposal of MSW (Refuse) by Composting. Composting of refuse is a biological method of decomposing solid wastes. This decomposition can be affected either under *aerobic conditions*, or under *anaerobic conditions*, or both. The final end product, is a manure, called the *compost* or *humus*, which is in great demand in European countries as fertilizer for farms.

Basically, composting is considered to be an aerobic process, because it involves piling up of refuse and its regular turning, either manually or by mechanical devices, so as to ensure sufficient supply of air and oxygen during its decomposition by bacteria, fungi and other microorganisms, like antinomycetes. Initially, the process starts with the *mesophilic bacteria*, which oxidise the organic matter (in the refuse) to carbon dioxide and liberate heat. The temperature rises to about 45°C, and at this point, the *thermophilic bacteria* take over and continue the decomposition. During this phase, the temperature further rises to about 60°C, which has to be maintained for at least 3 days in order to destroy pathogenic bacteria. This temperature control is crucial, because optimal decomposition occurs between 55 and 60°C, but if the temperature exceeds 60°C, decomposition slows down. In about 4 to 5 weeks, the easily biodegradable fraction gets consumed and the temperature of the compost mass starts falling. Complete stabilisation occurs after the compost is allowed to *cure* for another 2 to 8 weeks. During the active early decomposition phase, the **thermophilic bacteria** (mainly *Bacillus, Clostridium* and *Pseudomones*) act as the principal decomposers ; while **fungi** (such as *Mucor, Penicillium* and *Aspergillus*) are more active during the curing stage. The entire composting, thus, gets completed in about 3–4 months time. Volume reductions of the original organic material of up to 50% are achieved under ideal conditions. The finally produced compost usually, has earthy smell and a dark brown colour.

Moisture content of the compost mass should, however, be controlled to ensure optimum aerobic decomposition, because excessive moisture will make it difficult to maintain aerobic conditions, while deficient moisture inhibits biological life. A moisture content of about 55% should be established, so that aerobic biological activity may proceed at an optimum rate. It may, therefore, become necessary to add water to the compost mass during its turning, for maintaining satisfactory moisture content.

The Municipal Solid Wastes (Management and Handling) Rules, 2000 has laid down the following specifications (Table 14.6) for compost quality, to ensure its safe application.

Table 14.6. Composition/Quality Standards for Compost

S. No.	*Parameter*	*Concentration not to exceed in mg/kg of dry compost, except for pH value and C/N ratio*
1.	Arsenic	10.00
2.	Cadmium	5.00
3.	Chromium	50.00
4.	Copper	300.00
5.	Lead	100.00
6.	Mercury	0.15
7.	Nickel	50.00
8.	Zinc	1000.00
9.	C/N ratio	20–40
10.	pH value	5.5–8.5

In India, the composting is practised in rural areas on the mixture of night soil and refuse. Two methods, which are generally adopted here, are :

(*i*) *Indore* process* ; and

(*ii*) *Bangalore* process*, as discussed below :

Indore method of composting uses manual turning of piled up mass (refuse + night soil), *for its decomposition under aerobic conditions*. In this method, layers of vegetable wastes and night soil are alternatively piled in depths of about 7.5 to 10 cm each, to a total depth of about 1.5 m in a trench ; or above the ground to form a mound called a *windrow*.

A **windrow** is a long mound or stack of the organic MSW (mixed with cattle dung and human excreta if needing disposal) dumped on land in a height of

* Names of Indian cities.

about 1.5 m to 2 m, usually about 2.5 m to 3 m wide at the base. Most *windrows* are conical in cross-section and about 50 m in length. The composting waste is aerated by periodically turning the waste mix in the windrow, or in the trench, as the case may be. The manual turning with a *pitchfork* can be adopted at smaller installations ; while at larger plants, mechanical devices like self-propelled *overcab loaders, rotary ploughs,* etc. may be used to turn the refuse once or twice per week, which serves to introduce oxygen and to control the temperature. *Bulldozers* and *Packlain machines* are used for reshaping the stacks and for forming new ones. The moisture content of the turning mass is maintained at about 55% for getting **optimal decomposition** of the waste mass. This process of turning is continued for about 4 to 5 weeks, during which time, the readily biodegradable organics are consumed. The waste compost mass is finally allowed to cure for another 2 to 8 weeks without any turning. The entire composting process, thus, takes about 3-4 months time to complete, after which the compost becomes ready for being taken out for use or for sale.

The **Bangalore method of composting,** on the other hand, involves anaerobic decomposition of wastes ; and does not involve any turning or handling of the mass, and is, hence, cleaner than the Indore method. *This method is, therefore, widely adopted by municipal authorities throughout India*. The refuse and night soil, in this method, are, therefore, piled up in layers in an underground earthen trench (about 10 m × 1.5 m × 1.5 m). This mass is covered at its top by layer of earth of about 15 cm depth, and is finally left over for decomposition.

Within 2 to 3 days of burial, intensive biological action starts taking place, and organic matter begins to be destroyed. Considerable heat gets evolved in the process, which raises the temperature of the decomposing mass to about 75°C. This heat prevents the breading of flies by destroying the larvae. After about 4 to 5 months (depending upon the season), the refuse gets fully stabilised and *changes into a brown coloured odourless innocuous powdery mass*, called **humus.** This humus is removed from the trenches, sieved on 12.5 mm sieves to remove stones, broken glass, brickbats, etc., and then sold out in the market as manure. The empty trenches can again be used for receiving further batches of refuse.

The initial **C-N ratio*** and moisture content of the compost heap are the two important controlling factors in the success of anaerobic digestion, which

---

Carbon-Nitrogen ratio** ; *i.e.* C/N of the input material in the compost heap is an important factor for the bacterial activity to continue, since the bacteria use nitrogen for building their cell structures (as proteins) and carbon for food (as energy). The anaerobic bacteria, developing in this digestion, use up carbon about 30-50 times faster than they use up nitrogen. Hence, for proper development of anaerobic digestion, C/N ratio of the digestive material should be between 30 to 50 for optimum digestion (i.e. carbon should be 30-50 times more than the nitrogen***).

However, when there is too much of carbon, *i.e.* C/N is higher than the optimum, then nitrogen will be used up and carbon left over, thereby leaving the digestion of organic matter incomplete. The digestion will, thus, remain inefficient. Similarly, when there is too much of nitrogen, *i.e.* C/N ratio is lower than the optimum, then the carbon will soon get exhausted and fermentation stop, leaving nitrogen in the digestor, which will combine with hydrogen to form ammonia (NH_3). This can kill or inhibit the growth of bacteria, especially the methane producers. ***The anaerobic digestion will hence, require an optimum C/N ratio of about 30-50.***

finally produces a compost free from pathogens and contains 1% N, 1.1% P (as P_2O_5), and 1.5% K (as K_2O) on dry basis, thus proving to be a valuable nutrient for the soils, along with producing biogas as a by-product.

Fully Mechanised Composting Plants involve mechanised *receipt, segregation*, and *pulverization* of the refuse, before its digestion in closed digestors. Closed digestors for digestion of pulverised waste are, however, a costly proposition, and are hence being used only in advanced developed countries of Western Europe, USA, Japan, Israel, etc., Rotterdam, in Holland (Netherland)*, for example, is having a major composting plant to complement its waste management programme. Although, the same interest in composting exists in third world countries like India, but in these developing countries, the digestion is usually not done in the costly *closed digestors*, but is carried out in *open windrows* by turning the compost mass by mechanical devices.

The procedures adopted for composting in mechanised plants are briefly discussed below :

(*a*) *Receipt of Refuse*. The refuse is received at the plant @ 2 to 6 tonnes per vehicle, and storage capacity at the plant should be about 25 to 50% of the daily arrival of refuse at the plant, pending segregation and grinding.

(*b*) *Segregation of Refuse*. Dry refuse is allowed to move along a *belt conveyor*, whereby the ferrous (iron) matter is removed by *magnetic separators* ; and other non-compostable organic materials, such as papers, rags, card-boards, non-ferrous metals, etc. are also separated out, for recycling and reuse. The separation of such material can be done manually as well as by *rotary* and *shaker screens*.

(*c*) *Grinding or Pulverizing of refuse*. The remaining refuse is now pulverised by using different equipments, like *hammer mill*, *rasping machine, dano grinder*, etc.

(*d*) *Digestion or Fermentation of refuse*. The above refuse is finally digested under controlled conditions of temperature and moisture, in mechanical digestors. *Vertical or horizontal cylinders,* or *Silo type digestors* are the common types of **closed digestors.** Such closed digestors are the most hygienic and require minimum space, though costly, and hence largely adopted in developed countries. *Digestion period* in such digestors normally vary between 2 to 5 days for refuse containing low cellulose or low $\frac{C}{N}$ ratio ; and 7 to 9 days for refuse having more quantities of cellulose and higher $\frac{C}{N}$ ratio.

Fig. 14.6 shows such a closed digestor plant, where the refuse and garbage is processed in four stages ; *viz.* (*i*) is taken to a *grinder* ; (*ii*), then to a *bucket conveyor* ; (*iii*) for segregation ; and (*iv*) finally to a closed digestor containing rotating mechanism for thorough mixing of the refuse for its aerobic digestion under controlled conditions of temperature and moisture.

---

*Solid waste composting technique has been used in countries like Holland (Netherland) since as early as the 1930s.

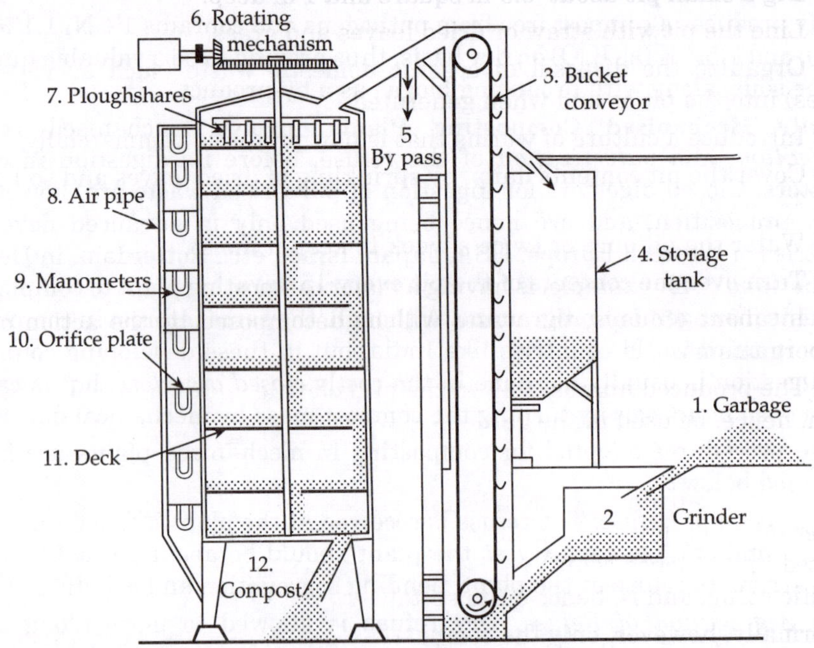

Fig. 14.6. Diagram of a large scale modern Refuse Digestor Plant at Michigan (U.S.A.).

Such closed digestors, as stated earlier, may prove to be costly, and hence uneconomical for composting large quantities of refuse, particularly for developing countries, like India. In under-developed and developing countries, the refuse is segregated and pulverised by using different techniques, and finally digested in *open windrows*. The digestion process in open windrows, as explained earlier, will further digest and reduce the already reduced quantity of waste (through pulverisation) by about 50%. This finally stabilised humus can be collected, dried, sieved, and sometimes enriched by the addition of chemical nutrients like nitrogen and phosphorous (*i.e.*, Urea and DAP, etc.), packed and sold in the market as *manure* or *soil-conditioner*. The compost (humus) can also be granulated or pellatised for sale, instead of selling it in powder form.

The largest potential market for compost is the agriculture sector. It may also be used for land reclamation, and in landscaping.

14.5.3.1. Vermi-composting. Vermi-composting uses the natural composting process of decomposition of biodegradable organic matter by the soil bacteria-as in ordinary composting technique described earlier, but takes the assistance of **cultured earth worms,** that are now produced commercially. These *earth worms* do help in quicker decomposition of the organic matter. The method helps in adopting the composting technique in individual bungalows and institutions, to dispose of the domestic waste, and more particularly for disposing of the yard and garden wastes, particularly the leaves and grass clippings, which cannot be thrown away with MSW in countries like USA. The various steps involved in applying the vermi-composting technique at individual domestic level are summarised below :

(*i*) Dig a small pit-about 0.5 m square and 1 m deep.

(*ii*) Line the pit with straw or dried leaves and grass.

(*iii*) Organize the disposal of organic domestic waste (such as vegetable wastes) into the pit as and when generated.

(*iv*) Introduce a culture of worms that is now produced commercially.

(*v*) Cover the pit contents daily, by sprinkling of dried leaves and soil every day.

(*vi*) Water the pit once or twice a week to keep it moist.

(*vii*) Turn over the contents of the pit every 15 days.

(*viii*) In about 45 days, the waste will be decomposed by the action of the microorganisms.

(*ix*) The produced humus (soil) in the pit is fertile and rich in soil nutrients. It can, hence, be used in the garden.

14.5.4. Disposal of MSW (Refuse) by Incineration and Thermal Pyrolysis*. Burning of refuse at high temperatures in furnaces, called *incinerators,* is quite a sanitary method of refuse disposal, and is widely adopted in developed countries like USA, where the collected refuse is of high calorific value, and is, hence quite suitable to burning.

Normally, however, only the combustible matter, such as garbage, rubbish and dead animals are burnt, and the incombustible matter like broken glass, chinaware, metals, etc. are either left unburnt or are separated out for *recycling* and *reuse,* before burning the solid wastes. Prior separation of such materials will reduce the load on the furnaces, and shall more than compensate the cost of this separation.

The left out ashes and clinkers** from the incinerators along with the non-recycled incombustible materials may, however, measure as much as 10 to 25% of thc original waste, which in any case, shall have to be disposed of either by sanitary land filling, or in some other productive manner. Say for example, the clinkers can be used as aggregates for making low grade concrete or as road metal, and the ashes can be used for making bricks.

The heat produced during burning of the refuse is used in the form of steam power for running turbines to generate electricity. The quantity of steam generated as a byproduct in this method, shall, however, be uncertain, as it will depend on the moisture and calorific value of the refuse. Hence, the power so generated will fluctuate and will need stand-by diesel units to obtain continuous and assured power.

Incinerators, besides being useful for burning city refuse, are also quite useful for burning the hazardous solid and liquid wastes of the hospitals and the industries. *Rotary kiln type* of incinerators and *liquid waste incinerators* are widely used for disposal of hazardous biomedical and industrial wastes. Incinerators are also used for disposing of the wet waste sludge of waste-water

---

***Thermal pyrolysis** is the burning of refuse in the presence of very sparce air or oxygen, so that the wastes donot completely burn, but only smoulder to produce charcoal like product.

**Clinkers are large fused masses of incombustible materials.

treatment plants, where sufficient land for converting it into manure or burying it underground for land filling, is not available.

Conventional Incinerator for Incinerating Municipal Solid Waste. The basic arrangement of a conventional incinerator, which is widely used for incinerating *municipal solid waste,* is shown in Fig. 14.7.

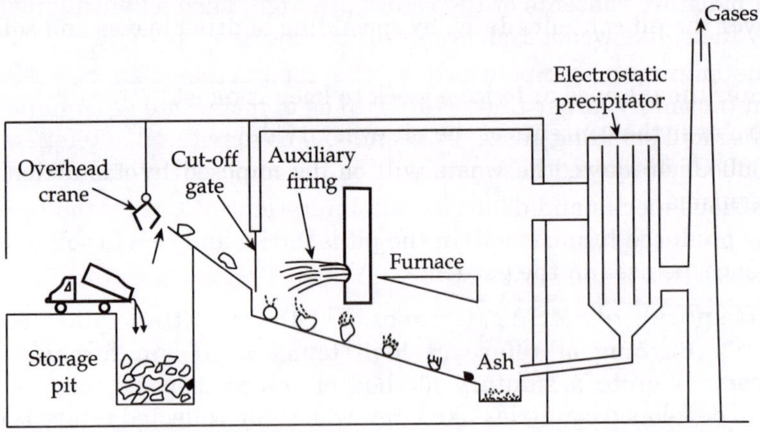

Fig. 14.7. Schematic sketch of a conventional incinerator (travelling grate type), to be used for incinerating municipal solid waste (refuse).

Since the solid wastes reaching the incinerator plant are generally quite wet, inspite of their being of high calorific value, it is generally found necessary to dry them out before burning. Conventionally, auxiliary fuel is used for initial drying of these wastes. Electrostatic precipitators are also installed in the incineration plant to reduce air pollution, caused by the escaping furnace gases, as shown in the figure. Still, however, large scale air pollution, particularly due to the emissions of **dioxins**, remains a serious problem with the incinerators. Moreover, the produced fly ash from the incinerators is often found to be highly contaminated with substances like lead, and, hence, has to be treated as a *hazardous waste*, and be treated before being disposed of in *hazardous waste land fills.*

Due to air pollution problems, the incineration plants should generally be located in sparsely populated areas and on the leeward side of the city, so that any foul or combustible gases, either eminating from the burning or due to any bad working of the plant, may not adversely affect the residents on a larger scale.

Large sized incinerators are called **destructors**, and they can burn 100 to 150 tonnes of refuse per hour. A destructor consists of a *furnace chamber, combustion chamber, expansion chamber,* and a *tall chimney* (25 m or so in height). The ancillary works consist of *ash-pit, charging apparatus, forced draft apparatus, pre-heating arrangements, steam generating apparatus, etc.*

While operating the incinerators, the following points should be carefully observed :

(*a*) The charging should be thorough and rapid.

(*b*) Each batch of refuse entering the furnace should be well mixed, and the proportion of the fuel in the charge be adjusted to provide complete combustion, and proper temperature.

Refuse containing 80% garbage and 20% rubbish, will normally burn without any auxiliary fuel, if air supplied for combustion is pre-heated to about 150°C. Whereas, the refuse containing 50 to 60% garbage and 40 to 50% rubbish will burn satisfactorily, without any pre-heated air. If the percentage of garbage is less than 50%, flashy fire may result.

When moisture contents of the refuse are high, such as during monsoons, auxiliary fuels, like wood, coal, or oil will be required.

(*c*) The minimum temperature in the combustion chamber should be sufficient (normally larger than 670°C), so as to incinerate all organic matter and oxidise foul smelling gases. If steam is to be generated, a temperature of about 1000°C is required to be produced in the combustion chamber.

The various merits and demerits and limitations of this method are given below :

Merits. (*i*) This is the most sanitary method of refuse disposal, and ensures complete destruction of pathogenic bacteria and insects.

(*ii*) There is no odour trouble or dust nuisance.

(*iii*) Some cost can be recovered by selling the steam power and clinkers.

(*iv*) The disposal site can be conveniently located within the city near the outskirts, and transportation problems sorted out easily.

(*v*) It requires very less space for refuse disposal.

Demerits and Limitations. (*i*) It is a very costly method, and requires a lot of technical know-how.

(*ii*) Solid wastes to be burnt should have a high calorific value.

(*iii*) Smoke, odour, and ash nuisance may result due to the improper and incompetent operation of the plant, particularly if substances like plastics, giving high calorific value to the wastes, are present in the wastes.

(*iv*) Transport vehicles are required in slightly large numbers, as there may occur delays in their emptying near the incinerators.

Use of incinerators in India. Incinerators for waste disposal, as stated earlier, have been widely constructed and used in developed countries like U.S.A. So much so that more than 100 such plants are working in USA alone. Developing countries like India, however, do find this method to be a costly affair, and the one needing a lot of technical know-how. Still, however, a medium sized incineration plant to dispose of 300 tonnes of daily refuse, was installed in Delhi (near Timarpur) at a cost of about ₹ 18 crores. This plant was set up by BHEL (Bharat Heavy Electricals Limited) with Danish assistance (extended in the form of interest free investment to be returned over 35 years). The plant was designed to produce 3.75 Mega watt (MW) of electricity ; out of which, about 3 MW power was to be utilised by the Wazirabad Water-Works, and the rest was to be consumed in running the plant itself.

The commissioning of the plant, infact, was completed by the end of year 1986 ; but the operational tests found the choking of the apron conveyor, due to settlement of silt and sand on the apron conveyor*. This problem arose due to

*Which conveys the residual burnt garbage on vibrating conveyor for final disposal.

the presence of huge quantity of silt and sand found in the refuse of Delhi city. Devising of remedial measures to overcome this trouble delayed the operation of the plant by about 2 years. *The plant has, however, still generally remained out of order, mainly because the calorific value of the city refuse is found to be hardly 600—800 kilo cal / kg* (2500–3300 kJ/kg), *while the plant is designed to burn wastes with calorific value of 1500 kilo cal / kg* (6200 kJ/kg) *or more.*

This North Delhi plant is having a huge refuse bunker, 21 m high, 9.5 m long and 19 m wide, in which four trucks could unload garbage at a time, and refuse housed for two-three days. Two large mechanical polygrabs, each having a capacity to lift 1.6 tonnes of garbage at any time, are used to hold the garbage, lifted to the bunker with the help of cranes operating from the top, and unloaded through funnel-shaped hoppers, directly into the furnace. The furnace consists of a grate made of small metallic slabs which could be moved up and down, and also sideways by hydraulic power. Through this churning, sufficient air would pass through even the smallest bits of the garbage, and complete combustion would thus take place. There are three such grates through which the garbage would pass. The garbage from the grates would go for further combustion into the rotary kiln. Bulky and carbonaceous wastes requiring high retention time would be completely burnt here. The hot flue gases, being generated during the combustion, would heat the water in the boiler, which inturn, would generate steam to drive the turbines and the generators, for producing electricity. To make this plant pollution free, the gases are to be passed through an electrostatic precipitator before being released from the chimneys. The dust level of the emitted gas would be 150 mg/cum, and the total quantity emitted through chimneys would be 24000 cum/hour.

14.5.4.1. Pyrolysis. Upon heating in closed containers in oxygen free atmosphere, most of the organic substances of solid waste can be split through a combination of *thermal cracking* and *condensation reactions* into gaseous, liquid and solid fractions. This process is known as *pyrolysis* or *thermal pyrolysis*.

In contrast to the combustion process which is highly *exothermic* (releasing heat on burning in the presence of oxygen), the pyrolysis is highly **endothermic** (consuming heat). That is why, this process is also known as **destructive distillation.**

When the organic solid waste is pyrolysed, we obtain the following three types of products at different stages or temperatures :

(*i*) *a gas stream*, which primarily contains hydrogen, methane, CO, CO_2 and other gases, depending upon the organic character of the solid waste being pyrolysed ;

(*ii*) *a liquid fraction*, consisting of a tar and/or an oil stream, which is a liquid at room temperature and is found to contain chemicals such as acetic acid, acetone, and methanol ; and

(*iii*) *a solid fraction*, consisting of *charcoal* like product of almost pure carbon plus any inert material that may have entered the process.

The respective quantums of three end products (*i.e.* gas, oil and charcoal) is found to depend upon the temperature of pyrolysis. Under conditions of maximum gasification, the energy content of the resulting gas is found to be about 26000 kJ/m^3 ; and that of the resulting oil to be about 23000 kJ/kg.

Pyrolysis may be used for reducing the quantities of sludge produced in a water or a wastewater treatment plant, before their ultimate disposal by methods like *landfill* and *land application*.

14.5.5. Disposal of MSW (Refuse) by Barging it Out into the Sea. This method had been used in the past to dispose of refuse by throwing it away into the sea, after carrying it at reasonable distance from the coast (say 16 to 20 kilometres inside the sea) on barges. The sea depth at such disposal point should not be less than 30 m or so, and the direction of the currents should be such as not to bring it back towards the shore. This method may, however, have a limited use, and that too only in a few coastal towns :

This method, however, proves to be quite cheap and simple, but possesses the following **disadvantages** :

(*i*) The bulky and lighter parts of the refuse do not settle down, remain floating, and tend to return to the shores, especially during high tides ;

(*ii*) This method requires ships or barges for taking the refuse into the interior of the sea, the movement of which may be difficult during monsoons and stormy weathers. During such periods, the refuse will either have to be collected or disposed of by some other method ;

(*iii*) Even inspite of the best care and efforts, some refuse, in this method, returns to the shores, spoiling their beauty.

(*iv*) The dumping of refuse in sea may cause pollution of sea, leading to large scale fish kills, if the refuse happens to contain any toxic material.

Due to such disadvantages and due to general ban imposed on dumping of wastes into the sea under the International Treaty of *Law of the Sea* (described in article 14.3.2), this method has now become obsolete.

14.6. Collection and Disposal of Solid Wastes from High-Rise Apartment Buildings

The basic composition of the solid wastes (refuse) generated by a high-rise apartment building is no different than that of solid wastes produced from independent residential homes, since the residents of high-rises in a given area are of the same types, as residents of independent houses, producing the same type of refuse. However, the composition varies from country to country. Say for example, in developed countries like USA, the major part of the solid waste consists of paper and plastics; whereas in a country like India, papers and plastics are usually sold to 'kawariwalas', and hence solid wastes from Indian houses contain very less amount of papers and plastics, while the major part consists of biodegradable type of organic waste including food and kitchen waste (particularly called *garbage*), besides containing small quantity of *rubbish* (paper, glass, rags, etc.), ashes, fine dusts, etc. with very small amount of waste paper and broken plastic items. In multi-storeyed apartments, however, lawn and garden waste from common green area, and sand and dust from street sweepings also find entry, which may generally be absent in wastes from independent houses.

For high-rise residents, disposing of wastes can be a little bit complicated affair, as compared with residents from single family units or independent houses, from where scavengers usually pick up refuse. In many high-rises today, for example, all residents have to carry their household trash to a central location, usually in the basement of the building, where it is collected in roll-off or standard refuse containers. This common disposal method, still practised in many of the old taller storeyed buildings even in a developed city like New York (USA) is the most labour intensive for the residents.

Another more popular disposal method for high-rise waste is the chute system. Using the principle of gravity as the heart of its technology, a **garbage chute system** is simply a series of vertical long down tubes or tunnels that are constructed in a building to carry down the wastes to a central collection *community bin* in a lower level of the building, either at ground floor or basement. Drawers that provide access to the chute are located on every floor and make for less walking for the residents for disposal of their house wastes.

Although particular features of different chute systems vary, a ***compactor**** is usually attached to the collection bin, in all modern developed countries, to *densify* the incoming wastes in order to reduce its volume and to maximize space. The reduction in volume of the generated wastes by the use of a compactor, do help in easy handling and transportation of the wastes to the recycling or disposal centres.

When compactors are used, carriage of refuse through haulers (trucks and trolleys) may be done twice or thrice a week, instead of daily necessary carriage when no compactors are used. The transportation of refuse through trucks and trolleys in case of high-rise building complexes have largely been privatised at several places, as against the collection of refuse by municipal trucks from independent homes.

---

*Very few high-rises in India do use compactors to reduce the volume of generated wastes. A **compactor** reduces large amounts of wastes to smaller units by using power driven rams to crush the wastes. These machines dispose of solid wastes, such as metals, plastics, papers, cotton, clothes, card board boxes, etc. Compactors can press these waste materials into bales that can be easily transported to disposal facilities or recycling centres. A compactor, infact, has five basic components; *i.e. ram*, *cylinder*, *power unit*, *electrical panel box*, and a *body*. The body is a steel structure that houses all of the other parts. The waste material is loaded into the body in a compartment, called a ***charge box***, which has reinforced sides able to withstand the pressure from compaction. The ram, a steel device with a heavy face plate, moves back and forth through the charge box, and forces the material into a compaction container. The cylinder is a hydraulically operated piston and rod that moves the ram. The power unit contains the hydraulic machinery that forces the piston in the cylinder to move. The electrical panel box contains the fuses, relays, motor starter, transformer, and switches that operate the compactor.

Segregation for Recycling of residential complex solid wastes is another important issue that is now-a-days gaining popularity, particularly in developed and developing countries. Individual high-rise complexes have also started segregating (for recycling) of their wastes in several cities particularly in developed countries. Say for example, in New York city (USA), which began the first *high-rise apartments recycling programme* in 1985, mandatory high-rise recycling has now been implemented in most parts of the city to help achieve a state recycling goal of more than 40% by the end of decade (2010).

Just as the types of high-rises vary, so do the different ways in which people recycle in them. Many high-rises simply set up centrally located recyclable bins next to regular trash containers or even trash chute doors. A few high-rises ask their residents to place their recyclables in blue boxes and leave them right outside their apartment doors. These boxes are then collected by the building maintenance staff or by the haulers themselves.

In Vancouver B.C. (Canada) which boasts of 91% participation rate for recycling programme in multi-family units, apartment residents are given blue tote-like bags to carry their recyclables to centrally located material-specific carts, from where they are transported to a recycling centre.

Some high-rises are now turning to a new technology to solve their recycling needs. The Park Sutton Apartments in Silver Spring in USA has recently installed an *automated segregation system,* called the ***automated recycling system*** from a company called Hi-Rise Recycling Systems, Inc (Miami), which allows the residents to source separate two to four different types of recyclables using the existing floor trash chutes. Such a new recycling system is very convenient for residents, since they do not have to go out of their way to carry their recyclable materials to a central place. *By simply pressing a button for a certain type of waste material, residents can drop their different waste collections down their trash chutes and into various revolving material specific, roll away containers.* Sales of such modern systems are on the rise. Although these new automated segregation recycling systems, which allow separate collection of different types of wastes (such as biodegradable, papers, plastics, and other non biodegradables) in different containers, without inconveniencing the residents to carry them to the central location, they are costly. The system may cost as much as $200 per unit or more. This, however, is one time cost for large scale convenience. Owner residents (condominiums) are more inclined to incur this one time cost as compared to tenant residents (renters). With the development of such modern methods, apartment owners are more motivated to segregate for recycling of their wastes. Segregation for recycling of wastes also helps in reducing the load on the sanitary land fill sites, since the quantity of refuse being taken for dumping at such sites gets considerably reduced, because the papers and plastics are separately recycled, and organic and garden wastes are generally treated in in-house compost plants installed within the high-rise building complexes.

When the refuse of a city is primarily organic waste (garbage) such as in Indian cities, *eco-friendly solid waste management systems* are preferably being developed. Say for example, Kochi City Corporation in Kerala state of India, has decided in 2007, to develop such a waste management system, which separately treats the segregated biodegradable wastes through the ***aerobic microbial composting system.*** The system requires the residents to implement a bio-bin system for collecting biodegradable and non-biodegradable wastes separately in two different containers. The biodegradable wastes of a building complex are separately collected and processed at a suitable place in the complex by using naturally occurring microbes through a scientific process called the *bio-trigger mechanism.* Consequently, the natural biodegradation happens and the garbage turns into useful organic fertilizer. The microbial solution is sprayed on the garbage in the compost chamber to initiate the bio-trigger mechanism, which takes about 10–15 days to convert garbage into decomposed organic matter. Such a system, after being tried successfully on trial basis in certain residential complexes, has been adopted on a large scale in many other apartment complexes, and as of now approximately 20% of the biodegradable wastes in Kochi is being treated at the source itself.

In order to overcome the initial reluctance and to achieve a certain level of acceptance, a campaign was infact initiated initially by housewives by way of audio-visual presentations in different residential complexes. After successful implementation in a few apartment complexes, convincing others was comparatively easy, and 70% of apartments in Kochi are now covered by this scheme. In all, about 230 apartment complexes have been covered with installation of such composting facilities in cities around Trivandrum, Calicut, Trichur, and Kochi.

The microbial solution being used in the process, is being procured by Kochi Corporation from an approved service provider, certified by the Govt. of Kerala for the *Clean Kerala Mission*. The service provider procures the solution from Mumbai, and supplies to Kochi Corporation, which now-a-days is also promoting an NGO, having a product manager, and supervisors with other workers employed as a part of a team to carry out the work of developing such systems in more and more complexes. This helps to provide employment to rural women on a daily basis between 10 am to 3 pm with a salary package of ₹ 3500-4000/- depending upon the number of flats in a complex. These workers are infact employed by residential associations. Supervisors and Executives employed by Kochi Corporation regularly visit complexes as per stipulated schedules. The Kochi Corporation has also hired Executives and Auditors to monitor whether the system is working properly, and if there are any complaints.

Under the scheme, flat residents collect their house wastes into two different coloured plastic containers, provided to them by the operators. The housewives, thus, separately collect their household biodegradable and non-biodegradable wastes into these separate colour coded containers, from where they are collected by municipal workers, who transfer the biodegradable wastes to the in-house compost chamber/plant, and take away the non-biodegradable waste separately through haulers to a transfer station or a waste disposal facility centre or a land fill site, etc.

After the success of this project, Kochi Corporation is also addressing *plastic wastes* separately to overcome 'plastic menace'. A *plastic shredding unit* has been constructed at Ravipuram by Corporation of Kochi and CREDAI Kochi. Plastic collected from apartments are shredded and stored for using it along with bitumen for ***polymarized tarring of roads*** in Kochi city. The plastic when used for tarrying (carpetting) helps in enhancing the life of roads. Shredded plastic has been utilised for road paving on Manappatti Parambu road, Kaloor and Manimala road, Edapally. These roads bear the distinction of first *polymarized road tarring* done in Kerala. CREDAI Kochi has donated a shredding machine for this project. The technology to use shredded plastic for bitumen carpetting of roads has been provided by Bharat Petroleum (BP).

INDUSTRIAL SOLID WASTES

14.7. CLASSIFICATION OF INDUSTRIAL SOLID WASTES, AND EXTENT OF GENERATION OF THESE WASTES IN INDIA

The solid wastes produced by industries can be broadly divided into the following two categories :

(1) *Non-hazardous solid wastes* ; and

(2) *Hazardous solid wastes.*

Both these types of solid wastes are discussed below :

14.7.1. Non-hazardous Solid Wastes. The non-hazardous solid wastes generated by various industries can be further sub-divided into the following two categories.

(*i*) *Biodegradable wastes* ; and

(*ii*) *Non-biodegradable wastes.*

(*i*) **Biodegradable wastes.** The major industries in urban areas that generate substantial amounts of biodegradable non-hazardous solid wastes are : fruit processing, cotton mills, paper mills, sugar mills, textile factories, etc.

(*ii*) **Non-biodegradable wastes.** The non-biodegradable non-hazardous solid wastes of industries are usually referred to as the **industrial solid wastes.** They are usually produced by industries like thermal power plants—which produce coal ash ; integrated iron and steal plants—which produce blast furnace slag and steel melting slag ; nonferrous industries like aluminium, zinc, and copper—which produce red mud and tailings ; fertilizer and allied industries—which produce phospho-gypsum. Together, these industries produce huge amounts of wastes in India, as shown in table 14.7.

Table 14.7. Annual Generation of Industrial Solid Wastes in India in Million Tonnes

S. No.	*Type of Solid Waste*	*Quantity generated in million tonnes (Mt)*
1.	Fly ash	45.0
2.	Slag	8.5
3.	Phospho-gypsum	5.0
4.	Lime sludge	4.0
5.	Red mud	3.5
Source : Confederation of Indian Industries 1996 and TERI 1996.		

14.7.2. Hazardous Solid Wastes. Some of the wastes generated by industries are deemed to be 'hazardous wastes', because they contain substances that are toxic to plants, animals, and humans, or are flammable, corrosive, explosive, or highly reactive chemically.

The major industries that produce hazardous wastes include metals, chemicals, drugs and pharmaceuticals, leather, pulp and paper, electroplating, refining, pesticides, dyes, rubber goods, etc. Table 14.8 shows the amount of wastes generated by such industries per unit output.

Table 14.8. Quantum of Hazardous Wastes Produced from Select Industries (in tonnes per tonne of production)

S. No.	*Industry / Industrial produce*	*Solid waste generation (tonne / tonne of product)*
(i)	Caustic soda	0.03
(ii)	Drugs and pharmaceuticals	0.04
(iii)	Dye and dye intermediates	1.36
(iv)	Fertilizer	0.085
(v)	Inorganic chemicals	0.4
(vi)	Organic chemicals	0.15
(vii)	Pesticides	0.07
(viii)	Petrochemicals	1.38
(ix)	Refinery	1.015
(x)	Textile processing	0.02/1000 m (2 grams/metre)
Source : Confederation of Indian Industries, 1996		

Wastes which are classified to be 'hazardous' under the "*Hazardous Wastes (Management and Handling) Amendment Rules 2002* are listed in table 14.10. Waste substances exceeding their prescribed concentration limits becoming hazardous, as specified in these Rules, are also shown in table 14.11.

It is estimated that currently the Indian industrial sector generates about 100 million tonnes of non-hazardous solid wastes, and 2 million tonnes of hazardous wastes, per year.

14.8. Storage, Transportation, and Disposal of Industrial Solid Wastes

Management of industrial wastes have generally remained somewhat neglected in India. Of late, slag from steel industry and fly ash-from thermal power plants have been utilised by cement industry. However, only 1—2% of fly ash and 30—60% of blast furnace slag are being utilised by cement industry.

The disposal of huge fly ash has always remained a big headache in India. The use of fly ash bricks and use of fly ash in land reclamation projects have, however, now been made compulsory by the Ministry of Environment and Forests GoI, in all areas located within 100 km of an existing coal or lignite based thermal power plant*. Under these regulations, every construction agency carrying out building works has to use fly ash bricks in all building works up to 100% wef 31.8.05 for places located within 50 km of thermal power plants, and 50% and 100% wef 31.8.05 and 31.8.07 respectively, for places located within 100 km of the thermal power plants.

Some **serious drawbacks** noticed in collection, storage, transport, and disposal of industrial solid wastes in India are listed below :

(*i*) No municipal corporation (or any other public body) is assigned the duty of safe collection, transportation, and disposal of industrial wastes. The responsibility lies totally with the industrial units and is largely neglected.

(*ii*) Wastes are shovelled into storage containers and loaded into lorries manually.

(*iii*) Useful items from the wastes are salvaged by factory workers, who usually handle the wastes without gloves.

(*iv*) Wastes piled against a wall or on open ground is a common sight in factories,

(*v*) Concrete bays or disused drums are often used for storage.

(*vi*) Wastes are rarely covered.

(*vii*) Wastes are transported in open trucks, which often spill the wastes.

(*viii*) Drivers are not advised about the precautions to be taken.

Industrial solid wastes pose a serious threat to the environment in India, because their disposal is largely uncontrolled. The responsibility of their safe disposal lies solely on the industrial units ; and when there exists an enormous number of small scale industrial units even in unconfirmed areas, then how can one expect these units to ensure safe disposal of their wastes ?

---

*These instructions are contained in the MoEF notification No. SO 979 (E) dated 27.08.2003, amending the earlier notification No. SO 763 (E) dated 14.09.1999, issued under Section 3(2) (1) (d) of the Environment (Protection) Act, 1986 and Section 5 (3) (d) of the Environment (Protection) Rules, 1986.

Even where the laws have been enacted to obtain compliance for safe disposal of the specific industrial wastes, the actual compliance is much less than required. Majority of the industrial units, thus, continue to throw and dump their solid wastes on public land, or make them a part of the municipal solid wastes. The practices, which are being adopted by certain select industries, for treating and disposing of their solid wastes, are shown in table 14.9, which shows the clear cut highly unsatisfactory prevailing position in this respect in India.

Table 14.9. Existing Methods of Waste Disposal in India for Certain Select Industries

S. No.	*Industry*	*Disposal of wastes*
(*i*)	Petrochemicals	All sludge and solid wastes from a plant are collected and sent to a disposal site outside the plant.
(*ii*)	Paints, varnish, lacquers	Solid waste is collected at point of generation in barrels and given to a contractor for disposal. Date expired paints are sold off. Waste treatment sludge is disposed of along with other solid wastes.
(*iii*)	Dye intermediates (naphthalene-based), H acid, G acid, peracids	Manually collected and dumped in low-lying areas
(*iv*)	Ethanolamine	Put in drums and stored underground, or incinerated
(*v*)	Orthochlorine	Burnt in open pits

Highly toxic wastes are sometimes destroyed by controlled incineration, but more commonly, small quantities of sludges are burnt along with factory garbage in open pits, Secured land-filling of containers of wastes, again, is not as common as uncontrolled dumping, which is the most common practice. Industrial waste is disposed along with municipal solid waste in industrial agglomerations.

EXISTING LEGISLATION TO MANAGE SOLID WASTES

14.9. Existing Laws to Regulate and Control the Disposal of Municipal and Industrial Solid Wastes in India

Management of municipal solid waste is essentially the duty and responsibility of municipal committees and corporations in India, and the constitution and powers of municipal corporations are a subject included in the 'State List' in the Seventh Schedule of the Indian Constitution. As a result, India, Until recently, did not have any Central act, specifically dealing with the management of solid wastes. Though each State has its own Municipal Act, by and large, these acts have been framed on similar lines. The municipal corporations have a wide range of subjects to look after : health and sanitation, primary education, construction and maintenance of roads and bridges, health-care, registration of births and deaths, inspection and licensing of food, and so on. *Management of solid wastes has, therefore, been low on the priority list of municipal authorities and seldom funded adequately.*

Although, most of the municipal acts have all the necessary legal provisions to (*i*) take action against the civic officials if they fail to provide a clean and safe environment to the general public ; and (*ii*) to penalize those who spoil the environment ; yet in the absence of proper enforcement of all these laws, municipal and industrial wastes have been disposed of with little regard to the environment, resulting in serious environmental problems. The situation has deteriorated to the point where 'public-interest litigations' are being filed in the Supreme Court to legally compel the corporations to provide effective civic amenities.

In order to bring the overall deteriorating environmental degradation under control, the Govt. of India, enacted *the Environment (Protection) Act, 1986* under Articles 253 of the Indian Constitution. This article empowers the Central Govt. to make any law for implementing any international treaty, agreement, or decisions taken in any international conference. The enactment of this Central law to protect environment could become possible only after decisions to protect the environment in general, were taken at the United *Nations Conference on Human Environment* held in Stockholm, in June 1972.

Rules for disposal of municipal solid wastes and for hazardous industrial wastes have been inacted in India and are discussed below in brief.

14.9.1. The Municipal Solid Wastes (Management and Handling) Rules, 2016. The rules for the management and handling of municipal solid wastes, under the *Environment (Protection) Act, 1986,* were for the first time, notified by the Ministry of Environment and Forests, Govt. of India on *03.10.2000.* These rules were called **Municipal Solid Waste (Management and Handling) Rules 2000,** and were notified to provide some kind of responsibility on the municipal bodies for proper collection and disposal of municipal solid wastes.

Schedule 1 to these Rules provided target dates to municipal authorities for : (*i*) setting up waste processing and disposal facilities (31.12.2003) ; (*ii*) improving existing land fill sites as per provisions of these rules (31.12.2001) ; (*iii*) identifying land-fill sites for future use and making these sites ready for operation (31.12.2002) ; and (*iv*) for constant monitoring of the performance of waste processing and disposal facilities (once in six months). Although all the target dates have elapsed, but no compliance is visible even in our metro cities, what to talk of small cities, towns, and villages.

Under these Rules, Secretary Urban Development of the concerned State is the nodal agency—who is given the overall responsibility for enforcement of the provisions of these rules in metro cities. The Dy. Commissioners or the District Magistrates of the concerned districts are further responsible for implementing the provisions of the rules in their respective districts. The CPCB, SPCBs and SPCCs are supposed to monitor the compliance of the standards laid down in these rules regarding ground water, ambient air, leachate quality, compost quality, including incineration standards, as specified under these rules. The SPCBs and SPCCs shall also grant authorisation to the municipal authorities for setting up any waste processing disposal facility, including land fills. The proposals of local authorities shall be examined by SPCBs and SPCCs by taking into consideration the views of other agencies like the State Urban Dev. Deptt. ; the Town and Country Planning Deptt., the Air Port or Air Base authorities, the Ground Water Board, etc., prior to issuing the authorization.

In *Schedule II*, these Rules further lay down guidelines on collection, segregation, storage, transportation, processing, and disposal of MSW. Specifications for land fill sites are given in *Schedule III*, whereby standards for ground water quality and ambient air quality in the vicinity of land fill sites are also laid down. *Schedule IV* to these rules, provide standards for composition of compost (to ensure its safe use) ; standards for disposal of treated leachates in public sewers and for land disposal (to avoid contamination of disposal sources) ; and operating as well as emission standards for incinerators.

These Rules however appear to be more like a BIS Code, than an anti-pollution law. No remedial measures or punitive actions are specified in these rules for violations of the rules, or for not making necessary arrangements for safe collection and disposal of the society's solid wastes.

The above mentioned rules of 2000, have however, recently been replaced with new set of rules, called ***"Municipal Solid Wastes (Management and Handling) Rules, 2016"***, which have been notified by the Ministry of Environment, Forests and Climate Change (MoEF CC), GoI, on ***08.04.2016.***

Even these new rules fail to incentivise or impose a strict penalty in case of poor or non-implementation. The rules have not pushed for decentralised management of solid wastes, but have encouraged centralised treatment, such as waste to energy, the present state of which is not good in the country. Also, the informal sector has been considerably neglected in these new rules.

According to Union Minister of State for Environment, Forests and Climate Change GoI, 62 million tonnes of waste is generated annually in the country at present, out of which 5.6 million tonnes is plastic waste, 0.17 million tonnes is biomedical waste, hazardous waste generation is 7.90 million tonnes per annum, and 15 lakh tonnes is e-waste. *Out of the total generated waste, only about 75-80 percent of the municipal waste gets collected and only 22-28 percent of this waste is processed and treated.*

The new rules are now applicable beyond municipal areas and have included urban agglomerations, census towns, notified industrial townships, areas under the control of Indian Railways, airports, special economic zones, places of pilgrimage, religious and historical importance, and State and Central Government organisations in their ambit.

Major highlights of the new Solid Waste Management Rules, 2016

1. Segregation at source. The new rules have mandated the source segregation of waste in order to channelise the waste to wealth by recovery, reuse and recycle. Waste generators would now have to segregate waste into three streams–***Biodegradables, Dry*** (plastic, paper, metal, wood, etc.) and ***Domestic Hazardous waste*** (diapers, napkins, mosquito repellants, cleaning agents etc.) in suitable different coloured bins, before handing them over to the authorised rag-pickers or waste collectors.

Institutional generators, market associations, event organisers and hotels and restaurants have been directly made responsible for segregation and sorting the waste and manage in partnership with local bodies. In case of an event, or gathering of more than 100 persons at any licensed/unlicensed place, the organiser will have to ensure segregation of waste at source and handing over of segregated waste to waste collector or agency, as specified by the local authority.

All hotels and restaurants will also be required to segregate biodegradable waste and set up a system of collection to ensure that such food waste is utilised for composting/biomethanation. The rules mandate that all resident welfare and market associations and gated communities with an area of above 5,000 sq m will have to segregate waste at source into material like plastic, tin, glass, paper and others; and hand over recyclable material either to authorised waste-pickers and recyclers or to the urban local body.

2. Collection and disposal of sanitary waste. The manufacturers or brand owners of sanitary napkins are responsible for awareness for proper disposal of such waste by the generator and shall provide a pouch or wrapper for disposal of each napkin or diaper along with the packet of their sanitary products.

3. Collect Back scheme for packaging waste. As per the rules, brand owners who sale or market their products in packaging material which are non-biodegradable, should put in place a system to collect back the packaging waste generated due to their production.

4. User fees for collection. *The new rules have given power to the local bodies across India to decide the user fees. Municipal authorities will levy user fees for collection, disposal and processing from bulk generators. As per the rules, the generator will have to pay "User Fee" to the waste collector and a "Spot Fine" for littering and non-segregation, the quantum of which will be decided by the local bodies.*

Also, the new rules have mentioned about the integration of rag pickers, waste pickers, and kabadiwalas from the informal sector to the formal sector by the state government.

The rules also stipulate zero tolerance for throwing; burning, or burying the solid waste generated on streets, open public spaces outside the generator's premises, or in the drain, or waste bodies.

5. Waste processing and treatment. As per the new rules, it has been advised that the bio-degradable waste should be processed, treated and disposed of through *composting* or *bio-methanation* within the premises as far as possible, and the residual waste shall be given to the waste collectors or agency as directed by the local authority. The developers of Special Economic Zone, industrial estate, industrial park shall earmark at least 5 per cent of the total area of the plot or minimum 5 plots/sheds for recovery and recycling facility.

Waste processing facilities will have to be set up by all local bodies having a population of 1 million or more, within two years. For census towns with a population below 1 million or for all local bodies having a population of 0.5 million or more, common, or stand-alone sanitary landfills will have to be set up in three years time. Also, common, or regional sanitary landfills to be set up by all local bodies and census towns with a population under 0.5 million will have to be completed in three years.

Also, the rules have mandated bio-remediation or capping of old and abandoned dump sites within five years.

6. Promotion of use of compost. As per the rules, the Department of Fertilisers, Ministry of Chemicals and Fertilizers should provide market development assistance on city compost and ensure promotion of co-marketing of compost with chemical fertilisers in the ratio of 3-4 bags is to 6-7 bags by the fertiliser companies to the extent compost is made available for marketing to the companies. Also, the Ministry of Agriculture should provide flexibility in Fertiliser Control Order for manufacturing and sale of compost, propagating use of compost on farm land, set up laboratories to test quality of compost produced by local authorities or their authorised agencies.

7. Promotion to waste to energy. Rules of 2016 also emphasise promotion of waste to energy plants. The rules mandate all industrial units using fuel and located within 100 km from a solid waste-based *Refuse-Derived Fuel (RDF)* plant, to make arrangements within six months from the date of notification of these rules to replace at least 5 per cent of their fuel requirement by RDF so produced.

The rules also direct that non-recyclable waste having calorific value of 1500 K.cal/kg or more, shall be utilised for generating energy either through RDF not disposed of on landfills and can only be utilised for generating energy either or through refuse derived fuel or by giving away as feed stock for preparing refuse derived fuel. High calorific wastes shall be used for co-processing in cement or thermal power plants.

As per the rules, the Ministry of New and Renewable Energy Sources should facilitate infrastructure creation for Waste to Energy plants and provide appropriate subsidy or incentives for such Waste to Energy plants. The Ministry of Power should fix tariff or charges for the power generated from the Waste to Energy plants based on solid waste and ensure compulsory purchase of power generated from such Waste to Energy plants by discoms.

8. Revision of parameters and existing standards. As per the new rules, the landfill site shall be 100 metres away from a river, 200 metres from a pond, 500 metres away from highways, habitations, public parks and water supply wells, and 20 km away from airports/airbase. Emission standards are completely amended and include parameters for dioxins, furans, reduced limits for particulate matters from 150 to 50. Also, the compost standards have been amended to align with Fertiliser Control Order.

9. Management of waste in hilly areas. As per the new rules, construction of landfills on hills shall be avoided. Land for construction of sanitary landfills in hilly areas will be identified in the plain areas, within 25 kilometres. However, transfer stations and processing facilities shall be operational in the hilly areas.

10. Constitution of Central Monitoring Committee. The government has also constituted a Central Monitoring Committee under the chairmanship of Secretary, MoEF&CC to monitor the overall implementation of the rules. The Committee comprising of various stakeholders from the Central and state governments will meet once a year to monitor the implementation of these rules.

11. Author's comments. It will take almost 4-5 years to see the drastic change in the waste management regimes in the country. The SWM Rules, 2016 diminish hopes in pushing for adoption of a decentralised mechanism for solid waste management. *However, it would be challenging* to *see how segregation at source shall work on the ground.* A massive awareness compaign in association with communities, NGOs, students and other stakeholders needs to be planned to push for better implementation of these rules. The Rules need to focus on making solid waste management a people's movement by taking the issues, concerns and management of solid waste to citizens and grass-roots.

14.9.2. Plastic Waste Management Rules, 2016

Huge generation of plastic waste in India, as in other world countries, is causing a great menace of its disposal. The usually adopted method of dumping it along with other municipal solid waste, causes serious problems, as the plastic waste is not biodegradable.

Single use thin plastic carry bags, and other such waste plastic items like tetra packs, chips packs, ketch up pouches, straws, coffee stirrers, water bottles, tooth paste tubes, etc. do cause serious problems, as most of such waste find its way into waterways, causing their blockages and floods.

Since the use of plastic in our lives is increasing by leaps and bounds, we do generate as much as about 15000 tonnes of plastic waste everyday; out of which, only about 9000 tonnes is collected daily and processed, while the remaining about 6000 tonnes/day is not being collected. This huge quantity of plastic waste, mostly containing *single use plastic,* goes into the environment, polluting land, air and water clogging our waterways including oceans, and it takes an average of one thousand years to decompose completely.

Considering the large scale damage caused by the uncollected and and non-recycled plastic waste, the Indian Government introduced the first legislation to control and manage plastic waste in India in the year 2011, and the law named ***"Plastic Waste (Management and Handling) Rules, 2011***, was notified, which has now been superceded by the *Plastic Waste Management Rules, 2016*. ***The new rules aim to :***

(*a*) Increase minimum thickness of plastic carry bags from 40 to 50 microns and stipulate minimum thickness of 50 micron for plastic sheets also to facilitate collection and recycle of plastic waste;

(*b*) Expand the jurisdiction of applicability from the municipal area to rural areas, because plastic has reached rural areas also;

(*c*) To bring in the responsibilities of producers and generators, both in plastic waste management system and to introduce collect back system of plastic waste by the producers/brand owners, as per extended producers responsibility;

(*d*) To introduce collection of plastic waste management fee through pre-registration of the producers, importers of plastic carry bags/multi-layered packaging and vendors selling the same for establishing the waste management system;

(*e*) To promote use of plastic waste for road construction as per Indian Road Congress guidelines or energy recovery, or waste to oil etc. for gainful utilization of waste and also address the waste disposal issue; to entrust more responsibility on waste generators, namely payment of user charge as prescribed by local authority, collection and handling over of waste by the institutional generator, even organizers.

In addition, the expected outcomes from the new rules includes:

(*i*) Increase in the minimum thickness of carry bags and plastic sheets from 40 to 50 micron. This regulation is likely to increase the cost by about 20%. Hence, the tendency to provide free carry bags will come down, and collection by the waste-pickers may also increase to some extent.

(*ii*) **Collect back system.** The producers, importers and brand owners who introduce the plastic carry bags, multi-layered plastic sachet, or pouches, or packaging in the market within a period of six months from the date of publication of these rules, need to establish a system for collecting back the plastic waste generated due to their products. They shall work out modalities for waste collection system based on Extended Producers Responsibility and involving State Urban Development Departments, either individually or collectively, through their own distribution channel or through the local body concerned. This plan of collection has to be submitted to the State Pollution Control Boards while applying for consent to Establish or Operate or Renewal. The producers/brand owners whose consent has been renewed before the notification of these rules shall submit such plan within one year from the date of notification of these rules, and implement within two years thereafter.

The introduction of the collect back system of waste generated from various products by the producers/brand owners of those products will improve the collection of plastic waste, its reuse/recycle.

(*iii*) **Phase out of manufacture and use of non-recyclable multi-layered plastic.** Manufacture and use of non-recyclable multi-layered plastic if any should be phased out in two years' time.

(*iv*) **Responsibility of Waste Generator**. All institutional generators of plastic waste, shall segregate and store the waste generated by them in accordance with the Solid Waste Management Rules, and handover segregated wastes to authorized waste processing or disposal facilities or deposition centers, either on its own or through the authorized waste collection agency.

All waste generators shall pay such user fee, or charge, as may be specified in the bye-laws of the local bodies for plastic waste management, such as waste collection, or operation of the facility thereof, etc.

Every person responsible for organising an event in open space, which involves service of food stuff in plastic, or multi-layered packaging, shall segregate and manage the waste generated during such events, in accordance with the Solid Waste Management Rules.

(*v*) **Responsibility of local bodies and Gram Panchayat**. The local bodies shall be responsible for setting up, operationalization and coordination of the waste management system and for performing associated functions.

(*vi*) **Responsibility of Retailers and Street Vendors**. Retailers or street vendors shall not sell, or provide commodities to consumers in carry bags, or plastic sheet, or multi-layered packaging, which are not manufactured and labelled or marketed, as prescribed under these rules.

Every retailer, or street vendor, selling or providing commodities in, plastic carry bags or multi-layered packaging or plastic sheets, or like, or covers, made of plastic sheets which are not manufactured, or labelled, or marked, in accordance with these rules, shall be liable to pay such fines, as specified under the bye-laws of the local bodies.

(*vii*) **Pre-registration Fee**. The shopkeepers and street vendors willing to provide plastic carry bags for dispensing any commodity shall register with local body. The local body shall, within a period of six months from the date of final publication of these rules on the Official Gazette of India notification of these rules, by notification, or an order under their appropriate state statute or byelaws shall make provisions for such registration on payment of plastic waste management fee of minimum Rs. 48,000/- @ 4,000/- per month. The concerned local body may prescribe higher plastic waste management fee, depending upon the production, or sale capacity. The registered shopkeepers shall display at prominent place that plastic carry bags are given on payment.

Only the registered shopkeepers, or street vendors shall be eligible to provide plastic carry bags for dispensing the commodities.

The local body shall utilize the amount paid by the customers for the carry bags exclusively for the sustainability of the waste management system within their jurisdictions.

The introduction of provision to collect fee from the producers, importers of plastic carry bags/multi-layered packaging and vendors selling the same, will strengthen the financial status of local authorities and improve Plastic Waste Management System.

(*viii*) **Reuse of Plastic Waste.** The options on reuse of plastic in various applications namely, road construction, waste to oil, waste to energy will enhance the recycling of plastic.

(*ix*) **Land for Waste Management Facility**. The responsibility to provide land for establishing waste management facility has been made to the Department with business allocation of land allotment in the State Government. This would eliminate the issue of getting land for the waste management facilities.

The ministry of Environmental, Forest and Climate Change, GOI, had initially notified the Recycled Plastic Manufacture and Usage Rules in 1999, which was mainly on manufacturing and usage of Plastic carry bags. It is specified that the minimum thickness of plastic bags should be of 20 microns. The Plastic Waste (Management and Handling) Rules, 2011, laid down certain conditions for manufacturing, stocking, sale and use of plastic carry bags and sachets, which were required to be monitored and implemented by the State Pollution Control Boards/Municipal Authorities. It specified that the minimum thickness of plastic bags should be of 40 microns. This was to facilitate its collection and recycle. However, the implementation of these rules was not so effective because the ambit of these rules was limited to notified municipal areas whereas today, the plastic has reached to our rural area also. There were no provisions of responsibility of waste generators. The rules did not address the promotion of conversion of waste to useful resources. Though, it provided for *Extended Producers Responsibility* for the establishment of waste management system, pricing of carry bags etc. those were not exercised by the local authorities, as its was simply left at the discretion of municipal authorities.

To implement these rules more effectively and to give thrust on plastic waste minimization, source segregation, recycling, involving waste pickers, recyclers and waste processors in collection of plastic waste and adopt polluter pays principle for the sustainability of the waste management system, the Central Government reviewed the existing rules and drafted revised rules. Stakeholders' consultation meets were organized in New Delhi, Mumbai and Kolkata wherein major Associations, industrial units, experts in various fields were invited. Consultative meetings with relevant State governments, State Pollution Control Boards were also held on these draft Rules. The suggestions/objections (238 in number) were received on these draft rules and have been examined by the Working Group. Based on the recommendations of the Working Group, now, the Plastic Waste Management Rules, 2016 have come into force.

Plastic has multiple uses and the physical and chemical properties lead to commercial success. However, the indiscriminate disposal of plastic has become a major threat to the environment. In particular, the plastic carry bags are the biggest contributors of littered waste and every year millions of plastic bags end up into the environment vis-à-vis soil, water bodies, water courses, etc. and its takes an average of one thousand years to decompose completely. Therefore, to the *address the issue of scientific plastic waste management, new regulations namely, the Plastic Waste (Management and Handling) Rules, 2011 were notified in 2011, which included plastic waste management.*

14.9.3. Hazardous and other Wastes (Management and Trans-boundary Movement) Rules 2016. Rules for handling and management of hazardous wastes were first of all notified by MoEF, GoI on **28.04.1989** under the *Environment Protection Act, 1986*. These rules were subsequently revised and renotified on 06.01.2000, 23.5.2003, 24.09.2008, and finally on **04.04.2016.**

The latest rules of the year 2016 have been notified by the *Ministry of Environment Forests and Climate Change,* GoI, vide GSR No. 395 (E) Dt. 04.04.2016, and can be copied from their web site in 68 pages, if specifically required.

These rules have been divided into the following chapters:

Chapter-I: Preliminary

Chapter-II: Procedure for Management of Hazardous and other Wastes

Chapter-III: Import and Export of Hazardous and other Wastes.

Chapter-IV: Treatment, Storage and Disposal Facility for Hazardous and other Wastes

Chapter-V: Packaging, Labelling, and Transport of Hazardous and Other Wastes.

Chapter-VI: Miscellaneous.

The above chapters also refer to various schedules, as pointed out below:

Schedule I. Schedule I to these rules lists the various hazardous wastes. **128 types of wastes generated from 38 types of industrial operations have been listed here.**

Schedule II to these rules, gives a list of different kinds of waste substances with concentration exceeding the specified values, which have been categorised as hazardous, under class A, class B, class C (class C1 to C13), depending upon their reducing hazard.

Schedule III to the rules further lists the various types of hazardous wastes, the imports and exports of which have some sort of restrictions, needing or not needing prior informed consent of the competent authority.

Schedule IV gives the list of commonly recyclable hazardous wastes; while **Schedule V** provides specifications for used oil suitable for recycling.

Schedule VI *to these rules provides list of hazardous and other wastes which are prohibited for import.*

Schedule VII lists the various authorities and their respective duties under this Act.

Schedule VIII provides the list of various documents required for verification by customs for import of wastes specified in part D of Schedule III.

In the end, these rules provide formats of 12 Nos. of Forms, which are required to be filled for seeking various permissions or for providing requisite information to competent authorities under these rules.

These rules shall apply to the management of hazardous and other wastes as specified in the schedules, but are not applicable to the following types of wastes, where other specific laws have been inacted:

(*a*) Waste water and exhaust gases as covered under the provisions of (*Prevention and Control of Pollution*) *Act 1974* and Air (*Prevention and Control of Pollution Act 1981*) and rules made there under and as amended from time to time.

(*b*) Wastes arising out of the operation from ships beyond five kilometres of the relevant baseline, as covered under the provisions of *Merchant Shipping Act, 1958* (44 of 1958) and the rules made there under, and as amended for time to time.

(*c*) Radioactive wastes as covered under the provisions of *Atomic Energy Act, 1962* (33 or 1962) and rules made there under, and as amended from time to time.

(*d*) Biomedical wastes covered under the *Biomedical Wastes (Management and Handling) Rules,* 1998, made under the Act (33 of 1962) and as amended from time to time.

(*e*) Wastes covered under the *Municipal Solid Wastes (Management and Handling) Rules, 2000,* made under the Act and as amended from time to time.

Any person or establishment generating or handling any of the listed hazardous wastes under *the Hazardous Wastes (Management and Transboundery Movement) Rules 2016*, is called an occupier or an operator. Such an ***occupier*** or an ***operator*** is supposed to obtain an *authorisation* to handle such wastes, from the concerned *State Pollution Control Board* (SPCB) in case of a State, or from the concerned *State Pollution Control Committee* (SPCC) in case of UT's—including NCT of Delhi. The authorisation granted to an occupier or an operator to handle a given hazardous waste can be cancelled, if the said occupier or operator fails to fulfil the conditions laid down in the authorisation so granted, to take adequate steps while handling hazardous wastes to contain contaminates, limit accidents, and to provide adequate training and equipment to the staff to handle the hazardous wastes, as not to cause adverse effects on humans and the environment in general.

Under these Rules, the hazardous wastes shall be collected, treated, stored, and disposed of only in such facilities as may be authorised by SPCB or SPCC. An occupier or recycler handling hazardous wastes, not having hazardous waste treatment and disposal facility of his own, can be allowed to become a member of a *common treatment, storage and disposal facility* existing in the area. Such an occupier will send his wastes to this common facility for proper treatment and disposal, as otherwise, his authorisation is liable to be cancelled.

These Rules further provide *guidelines for packaging, labelling and transport of hazardous wastes*. The occupier or operator of any waste facility is further supposed to identify, design and set up proper treatment and disposal site/facility. *Guidelines are also given for imports and exports of hazardous wastes,* listed in Schedule 3 to these Rules.

The occupier, transporter and operator of a facility is further liable for damages caused to the environment due to improper handling and disposal of hazardous wastes. Provision of imposition of fines for violations of these rules is also made in these rules. *However, no provision is made for any kind of imprisonments under these rules for violations of the rules. Evidently, these rules only provide a sort of guidelines for obtaining licenses and authorisation, rather than providing any deterrents against defaulters. That is why accidents continue to occur frequently, and hazardous wastes are occasionally continued to be mishandled and thrown at public places to become a part of MSW.* Some hue and cry is, of course, made when some serious accident occurs, but after a few days things return to their ordinary normal course.

INTERNATIONAL TREATY ON IMPORTS AND EXPORTS OF HAZARDOUS WASTES

14.10. Basel Convention

The *Basel Convention* provides international guidelines to control the transboundary movements of hazardous wastes between different countries. It is an international treaty designed to reduce the movements of hazardous wastes between nations and specifically to prevent transfer of hazardous wastes from *developed* to *less developed countries* (*LDCs*). This treaty, however, does not cover the movement of radioactive wastes. This convention is also intended to minimize the amount of toxicity of the generated wastes, to ensure their environmentally sound management as closely as possible to the source of generation and to assist the less developed countries in their environmentally sound management of the hazardous and other wastes they generate.

This convention was opened for signatures on March 22, 1989, and came into force on **May 5, 1992,** 90 days after the ratification by atleast 20 signatory States (nations). India signed the treaty on June 24, 1992. There are ***173 parties to this Convention****. Afghanistan, Haiti and U.S.A. have signed the convention but have not yet ratified it.

14.10.1. History and Necessity of Basel Convention. The necessity for enforcing an international legislation on the subject arose because the developed countries were found to be moving their hazardous wastes to the less developed countries (LDCs) on a large scale at a price, to avoid incurring of huge costs on their safe treatment and disposal in their respective nations. The globalization of shipping made the transboundary movement of wastes more accessible, and people in many less developed nations (including India) were quite eager to receive such wastes for recycling or recovery of costly components, parts, precious, metals, etc., to make profits in the trade, irrespective of the pollution caused to the people and the environment by their crude recovery methods.

The trade in hazardous wastes, consequently, increased to a large extent, causing large scale health hazards in LDCs and pollution of the global environment in general.

One of the incidents, which led to the creation of the Basel Convention, was the ***Khian Sea Waste Disposal Incident,*** in which a ship carrying incinerator ash from the city of Philadelphia in the USA, after having dumped half of its load on the beach in Haiti, was forced away where it sailed for many months, changing its name several times. Unable to unload the cargo at any port, the crew is believed to have dumped much of it at sea. Similarly, in 1988, 5 ships from Italy were found to have transported 8000 barrels of hazardous waste to the small town of **KOKO** in Nigeria, in exchange for 100 U.S. dollar monthly rent, which was paid to a Nigerian resident for use of his farmland.

The adoption of such practices by the various companies of the developed nations to move their industrial wastes to the less developed countries,

---

*A complete list of parties to the convention and their ratification status can be seen on the Basel Secretariat's web page (http.//www.basel.int/ratif/convention.htm.)

created a lot of criticism from the environmentalists, NGOs, and the Governments of the less developed nations. It was also stressed by the environmental experts that environmental pollution caused anywhere over the globe will ultimately affect the entire global environment in general, and that it is important to safely treat and dispose of the hazardous waste near the point of its production rather than transporting it over long distances to other places, irrespective of the high costs involved in safe treatment and disposal of such wastes in the developed countries themselves.

The practice of transboundary movements of their wastes by the developed countries was termed as ***"Toxic colonialism"***, by many developing countries. This forced the developed world to sit and work out a treaty to put guidelines and some controls on the transboundary movement of the hazardous wastes from one country to another. This finally resulted in signing and ratification of the Basel convention.

14.10.2. Definitions of Hazardous Waste under the Basel convention. A waste will be termed as **hazardous waste** under the Basel convention, if it falls within the listed categories in Annex-1 of the convention (Table 14.13), and it does exhibit one of the hazardous characteristics such as being explosive, flammable (liquids or solids), oxidizing, poisonous infectious, corrosive, or toxic. The other way that waste may fall under the scope of the convention is if it is defined as or considered to be a hazardous waste under the laws of either the exporting country or the importing country, or of the countries of transit. **Other Wastes** are listed in Annex-II as household wastes and residue that comes from incinerating household wastes.

Table 14.13. Annex 1 of the Basel Convention Listing the Hazardous Wastes to be Controlled under the Convention

Category	*Type of Waste*
(1)	(2)
	Waste Streams
Y1	Clinical wastes from medical care in hospitals, medical centres and clinics.
Y2	Wastes from the production and preparation of pharmaceutical products.
Y3	Waste pharmaceuticals, drugs and medicines.
Y4	Wastes from the production, formulation and use of biocides and phytopharmaceuticals.
Y5	Wastes from the manufacture, formulation and use of wood preserving chemicals.
Y6	Wastes from the production, formulation and use of organic solvents.
Y7	Wastes from heat treatment and tempering operations containing cyanides.
Y8	Waste mineral oils unfit for their originally intended use.
Y9	Waste oils/water, hydrocarbons/water mixtures, emulsions.
Y10	Waste substances and articles containing or contaminated with polychlorinated biphenyls (PCBs) and/or polychlorinated terphenyls (PCTs) and/or polybrominated biphenyls (PBBs).

(1)	(2)
Y11	Waste tarry residues arising from refining, distillation and any pyrolytic treatment.
Y12	Wastes from production, formulation and use of inks, dyes, pigments, paints lacquers, varnish.
Y13	Wastes from production, formulation and use of resins, latex plasticizers, glues/adhesives.
Y14	Waste chemical substances arising from research and development or teaching activities which are not identified and/or are new and whose effects on man and/ or the environment are not known.
Y15	Wastes of an explosive nature no subject to other legislation.
Y16	Wastes from production, formulation and use of photographic chemicals and processing materials.
Y17	Wastes resulting from surface treatment of metals and plastics.
Y18	Residues arising from industrial waste disposal operations.
	Wastes having metals as Constituents
Y19	Metal carbonyls.
Y20	Beryllium; beryllium compounds
Y21	Hexavalent chromium compounds
Y22	Copper compounds
Y23	Zinc compounds
Y24	Arsenic; arsenic compounds
Y25	Selenium; selenium compounds
Y26	Cadmium; cadmium compounds
Y27	Antimony; antimony compounds
Y28	Tellurium; tellurium compounds
Y29	Mercury; mercury compounds
Y30	Thallium; thallium compounds
Y31	Lead; lead compounds
Y32	Inorganic fluorine compounds excluding calcium fluoride
Y33	Inorganic cyanides
Y34	Acidic solutions or acids in solid form
Y35	Basic solutions or bases in solid form
Y36	Asbestos (dust and fibres)
Y37	Organic phosphorus compounds
Y38	Organic cyanides
Y39	Phenols; phenol compounds including chlorophenols
Y40	Ethers
Y41	Halogenated organic solvents
Y42	Organic solvents excluding halogenated solvents
Y43	Any congenor of polychlorinated dibenzo-furan
Y44	Any congenor of polychlorinated dibenzo-p-dioxin
Y45	Organohalogen compounds other than substances referred to in this Annex (*e.g.,* Y39, Y41, Y42, Y43, Y44).

14.10.3. General Obligations of the Parties to the Basel Convention

The obligations of the Parties to this convention, as specified in Article 4 of the convention are listed below:

1. (*a*) Parties exercising their right to prohibit the import of hazardous wastes or other wastes for disposal shall inform the other Parties of their decision pursuant to Article 13.

(*b*) Parties shall prohibit or shall not permit the export of hazardous wastes and other wastes to the Parties which have prohibited the import of such wastes, when notified pursuant to subparagraph (*a*) above.

(*c*) Parties shall prohibit or shall not permit the export of hazardous wastes and other wastes if the State of import does not consent in writing to the specific import, in the case where that State of import has not prohibited the import of such wastes.

2. Each Party shall take the appropriate measures to:

(*a*) Ensure that the generation of hazardous wastes and other wastes within it is reduced to a minimum, taking into account social, technological and economic aspects;

(*b*) Ensure the availability of adequate disposal facilities, for the environmentally sound management of hazardous wastes and other wastes, that shall be located, to the extent possible, within it, whatever the place of their disposal;

(*c*) Ensure that persons involved in the management of hazardous wastes or other wastes within it take such steps as are necessary to prevent pollution due to hazardous wastes and other wastes arising from such management and if such pollution occurs, to minimize the consequences thereof for human health and the environment;

(*d*) Ensure that the transboundary movement of hazardous wastes and other wastes is reduced to the minimum consistent with the environmentally sound and efficient management of such wastes and is conducted in a manner which will protect human health and the environment against the adverse effects which may result from such movement;

(*e*) Not allow the export of hazardous wastes or other wastes to a State or group of States belonging to an economic and/or political integration organization that are Parties, particularly developing countries, which have prohibited by their legislation all imports or if it has reason to believe that the wastes in question will not be managed in an environmentally sound manner, according to criteria to be decided on by the Parties at their first meeting;

(*f*) Require that information about a proposed trans-boundary movement of hazardous wastes and other wastes be provided to the States concerned, according to Annex V A, to state clearly the effects of the proposed movement on human health and the environment;

(*g*) Prevent the import of hazardous wastes and other wastes if it has reason to believe that the wastes in question will not be managed in an environmentally sound manner;

(*h*) Co-operate in activities with other Parties and interested organizations, directly and through the Secretariat, including the dissemination of information on the transboundary movement of hazardous wastes and other wastes, in order to improve the environmentally sound management of such wastes and to achieve the prevention of illegal traffic.

3. The Parties consider that illegal traffic in hazardous wastes or other wastes is criminal.

4. Each Party shall take appropriate legal, administrative and other measures to implement and enforce the provisions of this Convention including measures to prevent and punish conduct in contravention of the Convention.

5. ***A Party shall not permit hazardous wastes or other to be exported to a non-Party or to be imported from a non-Party.***

6. The Parties agree not to allow the export of hazardous wastes or other wastes for disposal within the area south of 60° South latitude, whether or not such wastes are subject to transboundary movement.

7. Furthermore, each Party shall:

(*a*) Prohibit all persons under its national jurisdiction from transporting or disposing of hazardous wastes or other wastes, unless such persons are authorized or allowed to perform such types of operations;

(*b*) Require that hazardous wastes and other wastes that are to be the subject of a transboundary movement be packaged, labelled and transported in conformity with generally accepted and recognized international rules and standards in the field of packaging, labelling and transport, and that due account is taken of relevant internationally recognized practices;

(*c*) Require that hazardous wastes and other wastes be accompanied by a movement document from the point at which a transboundary movement commences to the point of disposal.

8. Each Party shall require that hazardous wastes or other wastes to be exported, are managed in an environmentally sound manner in the State of import or elsewhere. Technical guidelines for the environmentally sound management of wastes subject to this Convention shall be decided by the Parties at their first meeting.

9. Parties shall take the appropriate measures to ensure that the transboundary movement of hazardous wastes and other wastes only be allowed if:

(*a*) The State of export does not have the technical capacity and the necessary facilities, capacity or suitable disposal sites in order to dispose of the wastes in question in an environmentally sound and efficient manner; or

(*b*) The wastes in question are required as raw material for recycling or recovery industries in the State of import; or

(*c*) The transboundary movement in question is in accordance with other criteria to be decided by the Parties, provided those criteria do not differ from the objectives of this Convention.

10. The obligation under this Convention of States in which hazardous wastes and other wastes are generated to require that those wastes are managed in an environmentally sound manner may not under any circumstances be transferred to the States of import or transit.

11. Nothing in this Convention shall prevent a Party from imposing additional requirements that are consistent with the provisions of this Convention and are in accordance with the rules of international law, in order to protect human health and the environment.

12. Nothing in this Convention shall affect in any way the sovereignity of States over their territorial sea established in accordance with international law and the sovereign rights and the jurisdiction which States have in their exclusive economic zones, and their continental shelves in accordance with international law, and the exercise by ships and aircraft of all States of navigational rights and freedoms as provided for in international law and as reflected in relevant international instruments.

13. Parties shall undertake to review periodically the possibilities for the reduction of the amount and/or the pollution potential of hazardous wastes and other wastes which are exported to other States, in particular to developing countries.

14.10.4. Basel Ban Amendment. After the initial adoption of the Convention, some LDCs and environmental organizations argued that it did not go far enough. Many nations and NGOs argued for a total ban on shipment of all hazardous waste to LDCs. In particular, the original Convention did not prohibit waste exports to any location except Antarctica but merely required a notification and consent system known as "prior informed consent" or PIC. Further, many waste traders sought to exploit the good name of recycling and began to justify all exports as moving to recycling destinations. Many believed a full ban was needed including exports for recycling. These concerns led to several regional waste trade bans, including the ***Bamako Convention.***

Lobbying at the 1995 Basel conference by LDCs, Greenpeace and key European countries such as Denmark, led to a decision to adopt the **Basel Ban Amendment to the Basel Convention.** *Not yet in force, but considered morally binding by signatories, the Amendment prohibits the export of hazardous waste from a list of developed (mostly OECD) countries to developing countries. The Basel Ban applies to export for any reason, including recycling.* An area of special concern for advocates of the Amendment was the sale of ships for salvage, shipbreaking. The Ban Amendment was strenuously opposed by a number of industry groups as well as nations including Australia and Canada. This Ban Amendment is still not ratified. The European Union, however, fully implemented the Basel Ban in its Waste Shipment Regulation (EWSR), making it legally binding in all EU member states. Norway and Switzerland have similarly fully implemented the Basel Ban in their legislation.

ENVIRONMENTAL IMPACTS OF SOLID WASTES

14.11. Impacts of Improper Disposal of Solid Wastes on Human Health and Environment

Improper removal and disposal of solid wastes is a proven cause of health hazards, which may even spread epidemics, like plague, and water borne diseases. The main diseases that may spread out and infect people, arises from the breeding of flies and rats. It has been estimated that in warm climates, exposed garbage may generate as many flies as up to 25,000 per m^2 of garbage dump per week. The refuse dumps also serve as a source of food for rats and small rodents, which quickly proliferate and spread to neighbouring areas.

The flies, as you all know, are basically responsible for spreading various kinds of diseases, as they carry the human feces and other infections to food or water, which finally infect the humans, and even animals. The most common cause for bacillary dysentery, amoebic dysentery and diarrhoeal diseases in India is the transmission of these disease vectors by flies. The rats, similarly infect us by direct bite and spread various diseases, like plague, endemic typhus, salmonellosis, and trichinosis, besides destroying the stored grains, clothes, and other household items.

Apart from causing diseases, for which flies and rats act as carriers, the handling and transfer of biological waste pose an infection threat to the workers, as well as to those who come in contact with such infected workers. The infections to the workers may pass from the waste through direct contact with the waste, through infection of sores or through vectors.

The hazardous wastes, thus, prove injurious to human health. Some wastes may produce acute ill-effects, while others may cause diseases after prolonged period of exposure. Improper disposals of municipal solid wastes have resulted in spread of epidemics, causing large scale human deaths, through contamination of water supplies or crops.

The environmental damage caused by solid wastes mostly pertains to aesthetics, and is a cause of choking of the storm water drains-leading to rain water stagnation and flooding in case of heavy rains, as happened in Mumbai city in 2005. Uncontrolled dumping of urban refuse eventually destroys the beauty of the country side, as the wastes are taken away by high winds to far and wide. The garbage dumps lying unattended for a number of days may also contaminate the ground water, as the rain water seeping through the garbage may become poisonous leachate, as it may leach the toxic heavy metallic compounds from the solid wastes. Such leachate may also contaminate surface water sources, as it may flow along with rain runoff and join the surface sources. In addition, uncontrolled burning of solid wastes may cause air pollution.

PROBLEMS

1. (*a*) Define 'municipal solid waste' and explain the necessity of its safe disposal.

(*b*) Explain the term 'refuse', and give its composition and classification. Describe various methods employed for the collection and disposal of the refuse.

2. (*a*) Describe and explain the importance of 'recycling and reuse' in disposal of solid waste (MSW) of a society. Which items are generally recycled from the MSW ?

(*b*) Write a brief note on the status of collection and transportation of municipal solid waste (MSW) in India.

3. Enumerate the various methods which can be used for disposal of municipal solid waste, and explain the two most widely adopted methods in India.

4. What do you understand by 'sanitary land filling' and how is it practised ? Draw a neat sketch and explain the filling process adopted in such filling practices. In this context, clearly explain the use of intermediate and final earth covers, and of the cap system. Also explain the digestion process involved in such land fills, and their adverse environmental impacts.

5. (*a*) What do you understand by the term 'leachate' ? What problems are posed by leachates and how are they overcome ?

(*b*) Differentiate between 'composting' and 'vermi-composting', and explain their utilities in disposal off society's solid wastes.

6. Explain the term 'composting'. Give the different types of compostings in use, and describe with the aid of sketches, their working.

7. Differentiate between refuse and garbage.

Mention the most significant property of the city refuse which guides the adoption of each of the following methods of refuse disposal.

(*i*) Incineration (*ii*) Sanitary land fill (*iii*) Composting.

[**Ans.** (*i*) high calorific value ; (*ii*) high density ; (*iii*) high organic garbage content.]

8. (*a*) Tabulate the quantitative and qualitative characteristics of refuse from an Indian town.

(*b*) Write short notes on :

(*i*) Indore method of refuse disposal ;

(*ii*) Bangalore method of refuse disposal ;

(*iii*) Incineration method of disposal of MSW ;

(*iv*) Mechanised composting plants ;

(*v*) Vermi-composting.

9. Explain and describe in details the incineration method of disposal of municipal solid waste (MSW), drawing a line diagram of a conventional incinerator which is usually adopted for the disposal of MSW. List out the various adverse environmental impacts of such a plant.

10. Explain and describe the aerobic as well as the anaerobic methods of composting of the organic solid wastes of a society, giving details of the digestion process of each method and their environmental impacts.

11. (*a*) What is the importance of C–N ratio and moisture content of a compost heap mass of the organic solid waste to be used for anaerobic composting process ?

(*b*) Briefly describe the working of a fully mechanised composting plant.

(*c*) Define and explain the term 'thermal pyrolysis,' and its use.

12. (*a*) How had the solid wastes of a society been disposed off in the sea, and why and how the practice has been phased out ?

(*b*) Differentiate between the hazardous and non-hazardous industrial solid wastes, giving approximate quantities of generation of various types of industrial solid wastes in India.

(*c*) How are industrial solid wastes disposed off in India, and what are their drawbacks.

13. (*a*) Explain the salient provisions of the "Municipal Solid Wastes (Management and Handling) Rules 2016", indicating their limitations.

(*b*) Explain the impacts of improper disposal of solid wastes on human health and the environment in general.

14. Explain the salient provisions of "Hazardous and other Wastes (Management and Trans Boundary Movement) Rules, 2016", of India.

15

Disposal of Environmentally Hazardous Biomedical, Radioative, and E-Wastes

BIOMEDICAL WASTES

15.1. Biomedical Wastes and Their Impacts on Health and Environment

The wastes (solid wastes as well as liquid wastes) produced by hospitals, nursing homes, clinics, research laboratories, diagonostic centres, veternary hospitals, etc., are potential sources of health hazards, and need to be specially treated and disposed of.

Approximately 1.45 kg of waste is generated per patient per day in Indian hospitals ; whereas it is much higher in developed countries. In western-countries, approximately 15-20% of this total waste is found to be hazardous and infected, whereas the percentage of hazardous and infected waste in India is usually much higher, since we donot have proper waste segregation practices in all the hospitals except in some big hospitals.

The greatest risk of clinical waste is from the *infectious* and *sharp components* like needles, syringes, etc. Health care workers and people associated with the hospital waste disposal, often get needle prick injuries and can even get AIDS/HIV or other types of infections, like Hepatitis B & C, through skin route. In one survey, it has been found that every second nursing staff gets needle prick or cut during their work every week, which may increase their chances of getting infection, if proper handling of waste is not ensured. *In addition to its infectivity, the medical waste is highly toxic and may possess variable radioactivity*.

15.2. Legislative Laws on Management of Biomedical Waste in India

Keeping in view its potential hazardous effects on humans-living in the surrounding of the medical wastes, the Indian Union Ministry of Environment and Forests* had initially notified *"Biomedical Waste (Management and Handling) Rules 1998"* on 20-7-1998, which have further been superceded with ***Bio-medical Waste Management Rules 2016,*** made effective wef **28.03.2016.** These rules have been notified under the *Environment (Protection) Act 1986.*

Under these rules, every health care institute, big or small, will have to make arrangements to ensure that their produced waste is handled and managed without causing any adverse health effects on human beings and the environment. In case of non-compliance with these rules, the occupier, *i.e.* the head of such Institution, shall be liable to be punished under the Environment (Protection) Act, 1986.

*This Ministry has now been renamed as the **Ministry of Environment, Forest and Climate Change.**

In order to safely and economically dispose of the medical wastes, it is necessary to collect the wastes by suitable segregation, as to label the same into *hazardous* and *non-hazardous* categories. These hazardous and non-hazardous wastes can then, be suitably transported and treated for final disposal.

The various steps involved in safe collection and disposal of medical wastes are discussed in article 15.3A to 15.8 after discussing the New Biomedical Waste Management Rules, 2016 in article 15.3.

15.3. Biomedical Waste Management Rules, 2016 (BMW Rules 2016)

15.3.1. Definitions under BMW Rules 2016. These rules shall apply to all persons who generate, collect, receive, store, transport, treat, dispose, or handle bio medical waste in any form including hospitals, nursing homes, clinics, dispensaries, veterinary institutions, animal houses, pathological laboratories, blood banks, ayush hospitals, clinical establishments, research or educational institutions, health camps, medical or surgical camps, vaccination camps, blood donation camps, first aid rooms of schools, forensic laboratories and research labs.

These rules shall however, not apply to:

(*a*) radioactive wastes as covered under the provisions of the Atomic Energy Act, 1962(33 of 1962) and the rules made there under;

(*b*) hazardous chemicals covered under the Manufacture, Storage and Import of Hazardous Chemicals Rules, 1989 made under the Act;

(*c*) solid wastes covered under the Municipal Solid Waste (Management and Handling) Rules, 2000 made under the Act;

(*d*) the lead acid batteries covered under the Batteries (Management and Handling) Rules, 2001 made under the Act;

(*e*) hazardous wastes covered under the Hazardous Wastes (Management, Handling and Transboundary Movement) Rules, 2008 made under the Act;

(*f*) waste covered under the e-Waste (Management and Handling) Rules, 2011 made under the Act; and

(*g*) hazardous micro organisms, genetically engineered micro organisms and cells covered under the Manufacture, Use, Import, Export and Storage of Hazardous Microorganisms, Genetically Engineered Micro organisms or Cells Rules, 1989 made under the Act.

Definitions. (*a*) Under these rules, an ***authorised person*** is defined as the person who is authorised by the prescribed authority to generate, collect, receive, store, transport, treat, process, dispose, or handle biomedical waste in accordance with these rules and the guidelines issued by the Central Government or the Central Pollution Control Board, the case may be

(*b*) **"biological"** means any preparation made from organisms or micro-organisms or product of metabolism and biochemical reactions intended for use in the diagnosis, immunisation or the treatment of human beings or animals or in research activities pertaining thereto;

(*c*) **"bio-medical waste"** means any waste, which is generated during the diagnosis, treatment or immunisation of human beings or animals or research activities pertaining thereto or in the production or testing of biological or in health camps, including the categories mentioned in Schedule I appended to these rules;

(*d*) **"occupier"** means a person having administrative control over the institution and the premises generating bio-medical waste, which includes a hospital, nursing home, clinic, dispensary, veterinary institution, animal house, pathological laboratory, blood bank, health care facility and clinical establishment, irrespective of their system of medicine and by whatever name they are called;

(*e*) **"operator** of a common bio-medical waste treatment facility" means a person who owns or controls a ***Common Bio-medical Waste Treatment Facility (CBMWTF)*** for the collection, reception, storage, transport, treatment, disposal or any other form of handling of bio-medical waste;

(*f*) **"prescribed authority"** means the State Pollution Control Board in respect of a State, and Pollution Control Committees in respect of Union territories;

3. Duties of the Occupier. It shall be the duty of every occupier to:

(*a*) take all necessary steps to ensure that bio-medical waste is handled without any adverse effect to human health and the environment and in accordance with these rules;

(*b*) make a provision within the premises for a safe, ventilated and secured location for storage of segregated biomedical waste in colored bags or containers in the manner as specified in Schedule I, to ensure that there shall be no secondary handling, pilferage of recyclables or inadvertent scattering or spillage by animals; and the bio-medical waste from such place or premises shall be directly transported in the manner as prescribed in these rules to the common bio-medical waste treatment facility or for the appropriate treatment and disposal, as the case may be, in the manner as prescribed in Schedule I;

(*c*) pre-treat the laboratory waste, microbiological waste, blood samples and blood bags through disinfection or sterilisation on-site in the manner as prescribed by the World Health Organisation (WHO) or National AlDs Control Organisation (NACO) guidelines, and then sent to the common bio-medical waste treatment facility for final disposal;

(*d*) phase out use of chlorinated plastic bags, gloves and blood bags within two years from the date of notification of these rules;

(*e*) dispose of solid waste other than bio-medical waste in accordance with the provisions of respective waste management rules made under the relevant laws and amended from time to time;

(*f*) not to give treated bio-medical waste with municipal solid waste;

(*g*) provide training to all its health care workers and others, involved in handling of bio medical waste at the time of induction and thereafter at least once every year and the details of training programmes conducted, number of personnel trained and number of personnel not undergone any training shall be provided in the Annual Report;

(*h*) immunise all its health care workers and others, involved in handling of bio-medical waste for protection against diseases including Hepatitis B and Tetanus that are likely to be transmitted by handling of bio-medical waste, in the manner as prescribed in the National Immunisation Policy or the guidelines of the Ministry of Health and Family Welfare issued from time to time;

(*i*) establish a Bar-Code System for bags or containers containing bio-medical waste to be sent out of the premises or place for any purpose within one year from the date of the notification of these rules;

(*j*) ensure segregation of liquid chemical waste at source and ensure pre-treatment or neutralisation prior to mixing with other effluent generated from health care facilities;

(*k*) ensure treatment and disposal of liquid waste in accordance with the Water (Prevention and Control of Pollution) Act, 1974 (6 of 1974);

(*l*) ensure occupational safety of all its health care workers and others involved in handling of bio-medical waste by providing appropriate and adequate personal protective equipments;

(*m*) conduct health check up at the time of induction and at least once in a year for all its health care workers and others involved in handling of bio-medical waste and maintain the records for the same;

(*n*) maintain and update on day to day basis the bio-medical waste management register and display the monthly record on its website according to the bio-medical waste generated in terms of category and colour coding as specified in Schedule I;

(*o*) report major accidents including accidents caused by fire hazards, blasts during handling of bio-medical waste, and the remedial action taken and the records relevant thereto, (including nil report) in Form I to the prescribed authority **and also** along with the annual report;

(*p*) make available the annual report on its web-site, and all the health care facilities shall make own website within two years from the date of notification of these rules;

(*q*) inform the prescribed authority immediately in case the operator of a facility does not collect the bio-medical waste within the intended time or as per the agreed time;

(*r*) establish a system to review and monitor the activities related to bio-medical waste management, either through an existing committee or by forming a new committee, and the Committee shall meet once in every six months and the record of the minutes of the meetings of this committee shall be submitted along with the annual report to the prescribed authority and the healthcare establishments having less than thirty beds shall designate a qualified person to review and monitor the activities relating to bio-medical waste management within that establishment and submit the annual report;

(*s*) maintain all record for operation of incineration, hydro or autoclaving etc., for a period of five years;

(*t*) existing incinerators to achieve the standards for treatment and disposal of bio-medical waste as specified in Schedule II for retention time in secondary chamber and Dioxin and Furans within two years from the date of this notification.

4. Duties of the Operator of a Common Biomedical Waste Treatment and Disposal Facility: It shall be the duty of every operator to:

(*a*) take all necessary steps to ensure that the bio-medical waste collected from the occupier is transported, handled, stored, treated and disposed of, without any adverse effect to the human health and the environment, in accordance with these rules and guidelines issued by the Central Government or, as the case may be, the central pollution control board from time to time;

(*b*) ensure timely collection of bio-medical waste from the occupier as prescribed under these rules;

(*c*) establish bar coding and global positioning system for handling of bio-medical waste within one year;

(*d*) inform the prescribed authority immediately regarding the occupiers which are not handing over the segregated bio-medical waste in accordance with these rules;

(*e*) provide training for all its workers involved in handling of bio-medical waste at the time of induction and at least once a year thereafter;

(*f*) assist the occupier in training conducted by them for bio-medical waste management;

(*g*) undertake appropriate medical examination at the time of induction and at least once in a year and immunise all its workers involved in handling of bio-medical waste for protection against diseases, including Hepatitis B and Tetanus, that are likely to be transmitted while handling bio-medical waste and maintain the records for the same;

(*h*) ensure occupational safety of all its workers involved in handling of bio-medical waste by providing appropriate and adequate personal protective equipment;

(*i*) report major accidents including accidents caused by fire hazards, blasts during handling of bio-medical waste and the remedial action taken and the records relevant thereto, (including nil report) in Form I to the prescribed authority **and also** along with the annual report;

(*j*) maintain a log book for each of its treatment equipment according to weight of batch; categories of waste treated; time, date and duration of treatment cycle and total hours of operation;

(*k*) allow occupier, who are giving waste for treatment to the operator, to see whether the treatment is carried out as per the rules;

(*l*) shall display details of authorisation, treatment, annual report etc on its web-site;

(*m*) after ensuring treatment by autoclaving or microwaving followed by mutilation or shredding, whichever is applicable, the recyclables from the treated bio-medical wastes such as plastics and glass, shall be given to recyclers having valid consent or authorisation or registration from the respective State Pollution Control Board or Pollution Control Committee;

(*n*) supply non-chlorinated plastic coloured bags to the occupier on chargeable basis, if required;

(*o*) common bio-medical waste treatment facility shall ensure collection of biomedical waste on holidays also;

(*p*) maintain all record for operation of incineration, hydro or autoclaving for a period of five years; and

(*q*) upgrade existing incinerators to achieve the standards for retention time in secondary chamber and Dioxin and Furans within two years from the date of this notification.

15.3.2 Treatment and Disposal of Biomedical Waste under BMW Rules 2016. (1) Biomedical waste shall be treated and disposed in accordance with **Schedule I**, and in compliance with the standards provided in **Schedule-II** by the health care facilities and common bio-medical waste treatment facility.

(2) Occupier shall hand over segregated waste as per the Schedule-I to common bio-medical waste treatment facility for treatment, processing and final disposal:

Provided that the lab and highly infectious bio-medical waste generated shall be pre-treated by equipment like autoclave or microwave.

(3) No occupier shall establish on-site treatment and disposal facility, if a service of' common bio-medical waste treatment facility is available at a distance of seventy-five kilometres.

(4) In cases where service of the common bio-medical waste treatment facility is not available, the Occupiers shall set up requisite biomedical waste treatment equipment like incinerator, autoclave or microwave, shredder prior to commencement of its operation, as per the authorisation given by the prescribed authority.

(5) Any person including an occupier or operator of a common bio medical waste treatment facility, intending to use new technologies for treatment of bio medical waste other than those listed in Schedule I shall request the Central Government for laying down the standards or operating parameters.

(6) On receipt of a request referred to in sub-rule (5), the Central Government may determine the standards and operating parameters for new technology which may be published in Gazette by the Central Government.

(7) Every operator of common bio-medical waste treatment facility shall set up requisite biomedical waste treatment equipments like incinerator, autoclave or microwave, shredder and effluent treatment plant as a part of treatment, prior to commencement of its operation.

(8) Every occupier shall phase out use of non-chlorinated plastic bags within two years from the date of publication of these rules and after two years from such publication of these rules, the chlorinated plastic bags shall not be used for storing and transporting of bio-medical waste and the occupier or operator of a common bio-medical waste treatment facility shall not dispose of such plastics by incineration and the bags used for storing and transporting biomedical waste shall be in compliance with the Bureau of Indian Standards. Till the Standards are published, the carry bags shall be as per the Plastic Waste Management Rules, 2011.

(9) After ensuring treatment by autoclaving or microwaving followed by mutilation or shredding, whichever is applicable, the recyclables from the treated bio-medical wastes such as plastics and glass shall be given to such recyclers having valid authorisation or registration from the respective prescribed authority.

(10) The Occupier or Operator of a common bio-medical waste treatment facility shall maintain a record of recyclable wastes referred to in sub-rule (9) which are auctioned or sold and the same shall be submitted to the prescribed authority as part of its annual report. The record shall be open for inspection by the prescribed authorities.

(11) The handling and disposal of all the mercury waste and lead waste shall be in accordance with the respective rules and regulations.

15.3.3. Segregation, packaging, transportation and storage. (1) No untreated bio-medical waste shall be mixed with other wastes.

(2) The bio-medical waste shall be segregated into containers or bags at the point of generation in accordance with Schedule I prior to its storage, transportation, treatment and disposal.

(3) The containers or bags referred to in sub-rule (2) shall be labeled as specified in **Schedule IV.**

(4) Bar code and global positioning system shall be added by the Occupier and common bio-medical waste treatment facility in one year time.

(5) The operator of common bio-medical waste treatment facility shall transport the bio-medical waste from the premises of an occupier to any off-site bio-medical waste treatment facility only in the vehicles having label as provided in part 'A' of the Schedule IV along with necessary information as specified in part 'B' of the Schedule IV.

(6) The vehicles used for transportation of bio-medical waste shall comply with the conditions if any stipulated by the State Pollution Control Board or Pollution Control Committee in addition to the requirement contained in the Motor Vehicles Act, 1988 (59 of 1988), if any or the rules made there under for transportation of such infectious waste.

(7) Untreated human anatomical waste, animal anatomical waste, soiled waste and, biotechnology waste shall not be stored beyond a period of forty-eight hours:

Provided that in case for any reason it becomes necessary to store such waste beyond such a period, the occupier shall take appropriate measures to ensure that the waste does not adversely affect human health and the environment and inform the prescribed authority along with the reasons for doing so.

(8) Microbiology waste and all other clinical laboratory waste shall be pre-treated by sterilisation to Log 6 or disinfection to Log 4, as per the World Health Organisation guidelines before packing and sending to the common bio-medical waste treatment facility.

Table 15.1. Schedule I: Under BMW Rules 2016

Biomedical wastes categories and their segregation, collection, treatment, processing and disposal options

Category	Type of waste	Type of Bag or container to be used	Treatment and Disposal options
(1)	(2)	(3)	(4)
Yellow	(*a*) **Human Anatomical Waste:** Human tissues, organs, body parts and fetus below the viability period (as per the medical Termination of Preganancy Act 1971, amended from time to time)	**Yellow** coloured non-chlorinated plastic bags	Incineration or Plasma Pyrolysis or deep burial*

	(b) Animal Anatomical Waste: Experimental animal carcasses, body parts, organs, tissues, including the waste generated from animals used in experiments or testing in veterinary hospitals or colleges or animal houses.		
	(c) Solid Waste: Items contaminated with blood, body fluids like dressings, plaster casts, cotton swabs and bags containing residual or discarded blood and blood components.		Incineration or Plasma Pyrolysis or deep burial* In absence of above facilities, autoclaving or micro-waving/ hydroclaving followed by shredding or mutilation or combination of sterilization and shredding. Treated waste to be sent for energy recovery.
	(d) Expired or Discarded Medicines: Pharmaceutical waste like antibiotics, cytotoxic drugs including all items contaminated with cytotoxic drugs along with glass or plastic ampoules, vials etc.	**Yellow** coloured non-chlorinated plastic bags or containers	Expired cytotoxic drugs and items contaminated with cytotoxic drugs to be returned back to the manufacturer or supplier for incineration at temperature > 1200°C or to common bio-midical waste treatment facility or hazardous waste treatment, storage and disposal facility for incineration at > 1200°C or Encapsulation or Plasma Pyrolysis at > 1200°C. All other discarded medicines shall be either sent back to manufacturer or disposed by incineration.
	(e) Chemical Waste: Chemicals used in production of biological and used or discarded disinfectants.	**Yellow** coloured containers or non-chlorinated plastic bags	Disposed of by incineration or Plasma Pyrolysis or Encapsulation in hazardous waste treatment, storage and disposal facility.
	(f) Chemical Liquid Waste: Liquid waste generated due to use of chemicals in production of biological and used or discarded disinfectants, Silver X-ray film developing liquids, dis-	Seperate collection system leading to effluent treatment system	After resource recovery, the chemical liquid waste shall be pre-treated before mixing with other wastewater. The combined discharge shall conform to the discarge norms given in Schedule-III.

	carded Formalin, infected secretions, aspirated body fluids, liquid from laboratories and floor washings, cleaning, house-keeping and disinfecting activities etc.		
	(g) Discarded linen, mettresses, beddings contaminated with blood or body fluid.	Non-chlorinated **yellow** plastic bags or suitable packing material	Non-chlorinated chemical disinfection followed by incineration or Plazma Pyrolysis or for energy recovery. In absence of above facilities, shredding or mutilation or combination of sterilization and shredding. Treated waste to be sent for energy recovery or incineration or Plazma Pyrolysis.
	(h) Microbiology, Biotechnology and other clinical laboratory waste: Blood bags, Laboratory cultures, stocks or specimens of microorganisms, live or attenuated vaccines, human and animal cell cultures used in research, industrial laboratories, production of biological, residual toxins, dishes and devices used for cultures.	Autoclave safe plastic bags or containers	Pre-treat to sterilize with non-chlorinated chemicals on-site as per National AIDS Control Organisation or World Health Organisation guidelines thereafter for Incineration.
Red	**Contaminated Waste (Recyclable)** *(a)* Wastes generated from disposable items such as tubing, bottles, intravenous tubes and sets, catheters, urine bags, syringes (without needles and *fixed needle syringes*) and vaccutainers with their needles cut) and gloves.	**Red coloured** non-chlorinated plastic bags or containers	Autoclaving or micro-waving/hydroclaving followed by shredding or mutilation or combination of sterilization and shredding. Treated waste to be sent to registered or authorized recyclers or for energy recovery or plastics to diesel or fuel oil or for road making, whichever is possible. ***Plastic waste should not be sent to landfill sites.***

White (Translucent)	**Waste sharps including Metals:** Needles, syringes with fixed needles, needles from needle tip cutter or burner, scalpels, blades, or any other contaminated sharp object that may cause puncture and cuts. This includes both used, discarded and contaminated metal sharps.	**Puncture proof,** Leak proof, tamper proof containers	Autoclaving or Dry Heat Steri lization followed by shredding or mutilation or encapsulation in metal container or cement concrete; combination of shredding cum autoclaving; and sent for final disposal to iron foundries (having consent to operate from the State Pollution Control Boards or Pollution Control Committees) or sanitary landfill or designated concrete waste sharp pit.
Blue	**(a) Glassware:** Broken or discarded and contaminated glass including medicine vials and ampoules except except those contaminated with cytotoxic wastes.	**Cardboard boxes** with **blue** colored marking	Disinfection (by soaking the washed glass waste after cleaning with detergent and Sodium Hypochlorite treatment) or through autoclaving or microwaving or hydroclaving and then sent for recycling.
	(b) Mettallic Body Implants	Cardboard boxes with **blue** colored making	

***Disposal by deep burial is permitted only in rural or remote areas where there is no access to common bio-medical waste treatment facility. This will be carried out with prior approval from the prescribed authority and as per the Standards specified in Schedule-III. The deep burial facility shall be located as per the provision and guidelines issued by Central Pollution Control Board from time to time.**

Schedule II under BMW Rules, 2016. Standards have been laid down under this schedule for:

(*i*) *Incineration* and their emission standards ; (*ii*) *Autoclaving* (*iii*) *Microwaving;* (*iv*) *Chemical disinfection;* (*v*) *Dry heat sterilization; and Standards for Liquid westes.* These standards can be found under these BMW (Rules) 2016 booklet or on website ***mpcb.gov.in***, by the needy organisation or engineers employed by them:

Schedule III under BMW Rules, 2016. This schedule lists the prescribed authorities and their duties under the BMW Rules, 2016. The details can also be found on the website of the MoEFCC, on site ***mpcb.gov.in>pdf BMW-Rules 2016(2)***

Schedule IV under BMW Rules 2016

LABLE FOR BIO-MEDICAL WASTE CONTAINERS OR BAGS

HANDLE WITH CARE

CYTOTOXIC HAZARD SYMBOL

HANDLE WITH CARE

Part B

LABEL FOR TRANSPORTING BIO-MEDICAL WASTE BAGS OR CONTAINERS

Day .... Month ....

Year ....

Date of generation ....

Waste category Number ....

Waste quantity ....

Sender's Name and Address

Phone Number ....

Fax Number ....

Contact Person ....

Receiver's Name and Address:

Phone Number ....

Fax Number ....

Contact Person ....

In case of emergency please contact:

Name and Address:

Phone No.

Note: Label shall be non-washable and prominently visible.

15.3.4. Conclusive Comments on Biomedical Waste Management Rules, 2016. The above described new bio-medical waste management rules are likely to change the way India used to manage these wastes earlier. Under the new rules, the coverage has been increased and provides for pre-treatment of lab wastes, blood samples, etc. It also mandates ***bar code system*** for proper control, and has simplified categorisation and authorisation in comparison to previous rules.

The major salient features of BMW Management Rules, 2016 include the following:

(*a*) The ambit of the rules has been expanded to include vaccination camps, blood donation camps, surgical camps or any other healthcare activity;

(*b*) Phase-out the use of chlorinated plastic bags, gloves and blood bags within two years;

(*c*) Pre-treatment of the laboratory waste, microbiological waste, blood samples and blood bags through disinfection or sterilisation on-site in the manner as prescribed by WHO or NACO;

(*d*) Provide training to all its health care workers and immunise all health workers regularly;

(*e*) Establish a ***Bar-Code System*** for bags or containers containing bio-medical waste for disposal;

(*f*) Report major accidents;

(*g*) Existing incinerators to achieve the standards for retention time in secondary chamber and Dioxin and Furans within two years;

(*h*) ***Bio-medical waste has been classified into 4 categories intead 10 to improve the segregation of waste at source;***

(*i*) Procedure to get authorisation simplified. Automatic authorisation for bedded hospitals. The validity of authorization synchronised with validity of consent orders for Bedded ***Health Care Facilities*** (HCFs). One time Authorisation for Non-bedded HCFs;

(*j*) The new rules prescribe more stringent standards for incinerators to reduce the emission of pollutants in environment;

(*k*) Inclusion of emissions limits for Dioxin and furans;

(*l*) State Government to provide land for setting up common bio-medical waste treatment and disposal facility;

(*m*) ***No occupier shall establish on-site treatment and disposal facility, if a service of common bio-medical waste treatment facility is available at a distance of seventy-five kilometre.***

(*n*) Operator of a common bio-medical waste treatment and disposal facility to ensure the timely collection of bio-medical waste from the HCFs and assist the HCFs in conduct of training.

Proper and scientific treatment and disposal of bio-medical waste through segregation, collection, and treatment in an environmentally sound manner, is likely to minimise the adverse impact on health of workers, petients, visitors, and on the environment in general. The hospitals are required to ensure proper disposal of their bio-medical waste either directly or through common bio-medical waste treatment and disposal facilities. The hospitals serving more than 1000 patients or more per month are now required to obtain authorisation and segregate bio-medical waste into 4 categories (against 10 categories, earlier) to be contained into different coloured bags. At present, our country is having 198 ***common bio-medical waste treatment facilities (CBMWF)*** in operation, and 28 are under construction. 21870 ***Health Care Facilities*** (HCFs) have their own treatment facilities; and 1,31,837 HCFs are using the CBMWF. ***The total bio-medical waste generation in the country is estimated to be 484 Tonnes per day (TPD) from 1,68,869 healthcare facilities, out of which 447 TPD is treated.***

The quantum of bio-medical waste generated is estimates to be 1-2 kg per bed per day in a hospital, and 600 gm per day per bed in a clinic. 85% of the hospital waste is non-hazardous; 15% is infectious/hazardous. Mixing of the hazardous waste with non-hazardous waste results into contamination and makes the entire waste hazardous. Hence, there is a necessity to segregate and treat. Improper disposal increases the risk of infection; encourages recycling of prohibited disposables and disposed drugs; and develops resistant to micro-organisms.

After describing the various statutory requirement of BMW Rules 2016 in the above article 15.3, we will now explain the various other technical terms and methods which are used in collection, storage, transportation, and treatment of biomedical wastes.

15.4. Colour Coding of Biomedical Wastes and Their Collection in Different Coloured Bins or Bags

The various types of medical wastes are divided into different categories, and collected in different coloured bins or containers, as to help in segregation of hazardous and non-hazardous wastes, needing different types of treatments. The process is known as **waste segregation**. The correct classification for collection of different waste items in different coloured dust bins, is to be done by the various hospitals, and it is the duty of the hospital doctors and nurses to ensure correct disposal of different waste items in their respective dust bins or containers. The system of using different coloured bins and bags to collect different types of solid medical wastes, is known as **colour coding**. Such a system, eventually helps in separately collecting the non-hazardous medical waste items, such as the uncontaminated packaging materials like plastics, paper, card-board, food and kitchen waste, garden waste, etc. Such non-hazardous waste can, then, easily and economically be disposed of by *landfill*, but only if it is free of biomedical waste that would cause adverse effects on human health and environment.

Proper segregation of wastes helps in the following ways :

(*i*) It minimises the amount of potentially hazardous waste that requires the specialised and costly treatment and disposal ;

(*ii*) It facilitates proper *packaging* and *labeling* of wastes ;

(*iii*) It reduces occupational health and safety risks to the health care workers and rag-pickers ;

(*iv*) It improves infection control within the hospital, and

(*v*) It helps in establishing uniform waste management practice, and to comply with the the national laws and legislations.

According to the 2016 Indian Rules notified for the disposal of biomedical wastes, such wastes are to be segregated in the following coloured bins or containers.

(*i*) ***Yellow bins or containers*** shall be used to store human/animal anatomical waste, solid waste, expired medicines, chemical waste, body fluids, etc.

(*ii*) ***Red bins or containers*** shall be used to store contaminated plastic bag, bottles, pipes or containers, etc.

(*iii*) ***White bins or containers*** shall be used to store scalpels, blades, needles, syringes and sharp metal items, etc.

(*iv*) ***Blue bins or containers*** shall be used to store broken glessware or metallic body implant, etc.

The use of these different coloured containers for storage of different types of biomedical wastes and their treatment and disposal are already given in details in table 15.1.

15.4A. Collection of Sharp Wastes

In collection of various types of biomedical wastes, special attention is required to be given to the **sharp wastes**, as they constitute a unique category of medical wastes, and always pose a risk factor of many infectious pathogens. Sharps may be defined as objects that are capable of puncturing or cutting due to any points or proturbences, and may include : syringes, needles, scalpels, blades, pasteur pipettes, and broken glasses or plastics. As a matter of fact, 98% of all health facilities generate sharp wastes.

During medical procedures, health care workers get a large number of injuries due to sharps. Similarly, a significant amount of injuries are caused by sharp wastes to the solid waste workers, who handle their collection.

To ensure their safe handling, manufacturers must develop mechanisms to render the syringes 'non-sharp', immediately after use. This will reduce the needle stick injury, by preventing the punctures. Self-sheathing syringes, self-blunting blood collection needles, and even needleless syringes are, hence, gaining popularity.

No loose sharp wastes should be disposed of into plastic bags, as these bags can get punctured or cut by the sharps, which may cause injuries to the waste handlers, or the content of the bag may spill during transportation. The sharp wastes should, therefore, be placed in puncture-resistent containers or should be *mutilated* at the point of generation by some *needle destroyer*.

15.5. Labelling for Identification of Bio-medical Wastes

Labelling is essential for the correct identification and safe management of medical wastes. Labelling makes the identification and handling of different types of medical wastes, easier. It will also warn the workers, patients, and the public about the existence of the wastes and their potential health hazards.

All labelling and sign posting should follow the international symbols and colour coding. All the infected, soiled, pathological, human and sharp wastes, should be marked with *biohazard symbol* in black colour, representing *cytotoxic wastes*. The ***symbol of cytotoxic hazard***, *i.e. C* in a triangle (△C) is used for this labelling. Labelling of the wastes at the generation point should be in the form of tag or adhesive label, attached to the collection bag or container, prior to it being collected by the cleaning staff. This ***waste tagging*** *system* will allow *waste audits* conducted at the treatment/disposal site, to identify those areas that are in compliance or non-compliance with the required hospital waste management practices. If needed, a corrective action can then be easily initiated. It would be better, if the bags and bins provided are already labelled with the appropriate hazard symbol.

15.6. Storage of Bio-medical Wastes

According to the rules, the infectious waste cannot be stored for more than 48 hours. However, hospital wastes may need to be stored, if immediate treatment and disposal cannot be done. These wastes should be *refrigerated on site* to prevent rotting and evolution of offensive smells. These wastes may then the carted and stored in off-site storage facilities, built in R.C.C. in notified areas, away from the general public. These storage houses must be properly designed to allow proper outflow of wash down water and rain water.

15.7. Transportation of Bio-medical Wastes to the Treatment and Disposal Sites

Wastes in hospitals should be frequently lifted from well marked designated collection points in hospital wards or rooms. Open trolleys are often used to collect wastes from individual bins. In all such cases, each bin and each trolley must be throughly cleaned and disinfected with 1% solution of bleaching powder (hypochlorite) at least once a week. The movement of the sanitation staff should be prevented or minimised through the designated clean areas of the hospital. Automated waste handling and transport system is much better than the manual ones. The vehicles used for waste transport to the disposal sites should not be used for any other purpose. Such vehicles should be labeled and designated for the purpose by the authorities.

15.8. Treatment and Disposal Methods for Biomedical Wastes

All the plastic disposable items contained in medical wastes are vulnerable for reuse without proper sterilization, unless destroyed completely. Several cases of repacking, supply and sale of syringes, catheters, etc. just after washing, or after low temperature autoclaving in unauthorised factories, have been reported in the recent past in the country and even in the capital city of Delhi. This practice is highly risky for the health of patients, for whom these *used and repacked items* are used. It has, therefore, been made mandatory to mutilate and shred such items at the generation point itself. The wastes should also be disposed of in a manner which does not cause any harm to the people or their environment.

The various methods which may be used for disposing of the hazardous wastes from hospitals include :

(1) *Chemical disinfection* ; **(Chemical process)**

(2) *Autoclaving* ;
(3) *Hydroclaving* ;
(4) *Microwaving* ; and
(5) *Incineration*
} **(Thermal processes)**

In addition to the above methods, **land filling** may also sometimes be used for disposal of some particular types of biomedical wastes, as stated in table 15.1. The landfill site should, however, always be away from the residential areas, as it could be dangerous for the community health. Landfill or burial of hospital wastes may pose the following problems :

(*i*) difficulty posed in safe transport
(*ii*) difficult availability of waste landfill site
(*iii*) potential source of air, soil and groundwater contamination
(*iv*) health threat to the community
(*v*) potential risk to the scavenger-animal birds and rag pickers.

Due to the above demerits, bio-medical wastes are generally not taken for burial, but are usually disposed of by **thermal processes**, like *autoclaving, hydroclaving, microwaving,* and *incinerating*.

Chemical disinfection followed by *mutilation* or *shredding* may also sometimes be adopted for disposal of the waste items like syringes, needles, globes, etc.

All these techniques are described below in brief.

15.8.1. Chemical Disinfection. In this method, the waste is disinfected by using chemicals like *chlorine compounds*, such as bleaching powder, etc. Other chemicals like iodine, alcohols, phenolic compounds, hexachlorphene, formaldehydes, etc. may also sometimes be used. However, 1% solution of hypochlorite (bleaching power) is most commonly used. All such chemicals shall be used in aqueous solutions, in which the waste items shall be kept submerged. Many of the waste items may even need shredding before disinfection, as to help in providing sufficient contact between the waste and the disinfectant. The disinfected waste will usually be shredded.

In this method, the shredded wet wastes containing disinfectant chemicals will need final disposal. The method can hence be used for disposing solid waste items like tubes, catheters, blood or urine bags, globes and sharps like needles, syringes etc. which can be buried after disinfection and shredding ; or for disposing certain types of liquid wastes which can be discharged into the sewers after disinfection.

15.8.2. Autoclaving. Autoclaving (**steam sterilization**) is a ***low heat thermal process****, and is designed to bring steam into direct contact with the wastes, in a controlled manner and for sufficient duration to disinfect the wastes. The three basic types of steam autoclave systems are :

(*i*) *gravity system*

(*ii*) *pre-vacuum system* ; and

(*iii*) *retort system*.

These three types of autoclave systems are briefly indicated below :

(*i*) **Gravity type autoclaves.** They are those in which pressure of steam alone is used to evacuate air from the treatment chamber. They operate with steam temperatures of about 121°C. These systems require a cycle time of approximately 60-90 minutes in order to achieve full steam penetration into the most densely packed waste loads.

(*ii*) **Pre-vacuum type autoclaves.** They are those which evacuate air from the treatment chamber using vacuum pumps. This enables them to reduce cycle time to about 30-60 minutes, as the time to heat the air within the chamber is eliminated. These systems operate at about 132°C.

(*iii*) **Retort type autoclaves.** They consist of large volume treatment chambers designed for much higher steam temperatures and pressures ; and hence their cycle times can be substantially lower than those of the other systems.

After autoclaving the wastes, they are shredded and disposed of suitably. The items which can be disposed of in this manner are indicated in table 15.1.

15.8.3. Hydroclaving. An innovation of the autoclave is the hydroclave. Here, indirect heating is done by providing steam into the outer jacket of a double walled container, while the waste inside the inner container is turned on by a suitable mechanism. This causes the waste to be fragmented and continuously tumbled against the hot vessel walls. The moisture content of the waste changes into steam, and the vessel starts to pressurise. In the absence of enough moisture in the waste to pressurise the vessel, a small amount of

---

****Incineration***, on the other hand, is a ***high heat thermal process.***

steam is added until the desired pressure is reached. The treatment time is hardly 15 minutes at 132°C or 30 minutes at 121°C. In the process, sufficient sterilization occurs, and the resultant waste is fragmented and dehydrated with reduction in volume and weight.

15.8.4. Microwave Treatment. In the thermal treatment systems described above, the waste is heated externally ; while in microwave, heating occurs inside the waste material. This process involves pre-shredding the waste, injecting it with steam, and heating it for 25 minutes at 25°C, under a series of microwave units.

Microwave radiation is designated as that portion of the electromagnetic radiation spectrum lying between the frequencies of 3000 MHz to 3,00,000 MHz and the microbial inactivation occurs as a result of thermal effect of radiation, and not from any intrinsic non-thermal property.

Microwaving, however, cannot be used for human organs or tissue wastes. This technique does not reduce the volume of waste, also.

15.8.5. Incineration. Incineration systems involve high temperature combustion, under controlled conditions, to convert wastes containing infections and pathological material to inert mineral residues and gases. The following two types of incinerators are generally used for treatment of hazardous liquid and solid wastes, like those of hospitals, industries, etc.

(*i*) *Liquid waste incinerators* ; and

(*ii*) *Rotary kiln type of incinerators.*

Almost 90% of all the hazardous waste incinerator installations belong to the above two categories. The rest about 10% are the *fluidized beds* and *starved air/pyrolysis* systems, which are primarily used for incinerating sludge wastes of sewage treatment plants and city refuse, respectively.

The *liquid waste incinerators* are used for burning liquid wastes. The line between liquid and non-liquid waste is now well defined ; *since a material is considered liquid, if it can be pumped to a burner and atomized, i.e., fired in suspension. Atomization is a function of nozzle type and liquid viscosity*. Such an incineration system will involve injecting the liquid waste at 350 to 700 kPa through an atomizing nozzle into the combustion chamber (furnace). The liquid waste, passing through the nozzle, ejects in the form of fine droplets, which volatilizes in the hot gas stream, oxidising the gas. Efficient destruction of liquid hazardous wastes requires minimising unevaporated droplets and unreacted vapour. Temperature, residence time, and turbulence are optimised to increase destruction efficiency. The liquid flow injections may be horizontal, vertical, or tangential in different incinerator designs. Vertical units are less likely to experience ash, build up. Tangential units have a much higher heat release and generally superior mixing.

The *rotary kiln incinerators* are the most versatile of all the types of incinerators, as they can be used to burn solid, liquid, gaseous, and sludge* wastes. Throughout the world, more rotary kiln incinerators are used for destruction of non-liquid hazardous wastes than any other incinerator ;

***Sludges** are materials with semi-liquid consistency that comprise small particles within a liquid matrix, as opposed to **slurries,** which are large particles in a liquid medium (usually water).

although of course, the kilns can be designed to incinerate aqueous* as well as non-aqueous wastes.

Considering their wide spread use for hospital waste incineration, we herein describe, the conventional rotary kiln type of incinerators, and the variations of the kiln design, in greater details.

15.8.5.1. Conventional rotary kiln incinerators and their variations. A typical arrangement of a conventional rotary kiln system is shown in Fig. 15.1 (*a*), while Fig. 15.1 (*b*) shows the photoview of such a conventional rotary kiln.

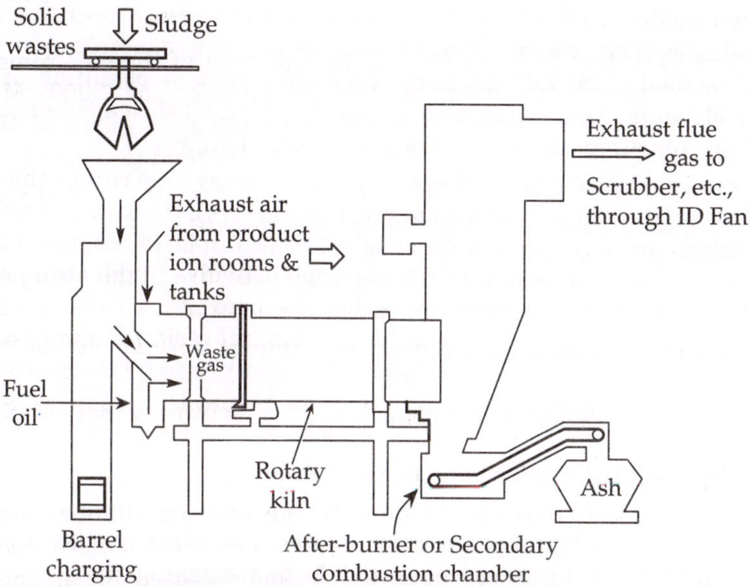

(*a*) Typical section of a rotary kiln system

(*b*) Photoview of a conventional rotary kiln

Fig. 15.1

*A waste with atleast 60% water content is known as an **aqueous waste.**

The rotary kiln is a horizontal cylinder, lined inside with refractory. The kiln is capable to rotate about its longitudinal axis @ $\frac{3}{4}$ to 4 rpm. The waste is deposited in the kiln at one end, and the waste burns out to ash by the time it reaches the other end.

The entire system assembly consists of a waste feed, an *air injection*, the *kiln* itself, an *afterburner* (a secondary combustion chamber), and an *ash collection* system. The gas discharge from the afterburner is directed to an *air emission control system*, to reduce air pollution impacts.

As shown in Fig. 15.1 (*a*), there are a number of areas within the system, where leakages can occur. The feeding ports, for example, cannot be completely sealed ; the kiln seals are also the areas of potential leakage. Similarly, although the ash system is usually provided with a water seal, but with dry ash collection, there will always be some leakage.

In order to ensure that the leakage occurs *into* the system rather than *out of* the system (*i.e. from the kiln to the outside atmosphere*), the kiln is maintained with a negative pressure (draft). An **induced draft fan** is, hence, provided into the system, and is sized to maintain a negative pressure in the entire system, as to ensure leakage into the system, rather than out of it.

The conventional kiln is mounted on its frame *at a slight incline from the horizontal plane*, at about 2 to 4% slope $\left(\frac{1}{4} \text{ to } \frac{1}{2} \text{ inches per foot of length}\right)$, with the higher end at the feed end of the kiln.

Some other designs, particularly adopted for treating liquid wastes, may have a zero or slightly negative rake (slope) with lips at the input and discharge ends. Such kilns ore operated in the slagging mode*, with the internal kiln geometry designed to maintain a pool of molten slag between the kiln lips.

Most kiln designs use smooth refractory lining on the kiln interior. Some designs, particularly those for the processing of granular material (dirt or powders) may have internal vanes or paddles to encourage motion along the kiln length, and to promote turbulence of the feed. Provision of baffles need a lot of care, since with certain material consistencies, such as soils with 10-20% moisture content, baffles may lead to retard the movement of material through the kiln.

The kiln is supported by atleast two trunnions. One or more sets of trunnion rollers are idlers. Kiln rotation can be caused by a set of powered trunnion rollers ; or by gear drive around the kiln periphery ; or through a chain, driving a large sprocket around the body of the kiln.

A source of heat like gas or diesel is required to bring the kiln up to the operating temperature, and to maintain its temperature during incineration of the waste feed. Supplemental fuel is normally injected into the kiln through a conventional burner or a ring burner, when gas fuel is used.

---

*A kiln designed to operate at very high temperature as to produce ash in molten form is called a slagging kiln, as exaplained a little later in this article.

There occurs a number of variations in kiln design, including the following common ones :

(*a*) *Parallel (i.e. Co-current)* or *Counter current flow* types ;

(*b*) *Slagging* or *Non-slagging* types ;

(*c*) *Refractory*, or *Bare wall* types.

The most commonly used kiln design, called the **conventional kiln**, is a *parallel flow, non-slagging, refractory lined system.*

The differences between various types of kilns, mentioned above, are clarified below :

(*a*) ***Parallel and Countercurrent Kilns.*** When the gas flow through the kiln is in the same direction as the waste flow, the kiln is said to have *parallel* or *co-current flow*, as indicated in the conventional kiln in Fig. 15.1 (*a*). The burner(s) is placed at the front of the kiln, *i.e.* the face of the kiln, from which the air gas mixture originates. The waste is introduced in this type from the front end of the kiln.

A **counter-current kiln** is generally used when an **aqueous waste** is to be incinerated. With *counter current flow*, the gas flows opposite to the direction of the flow of the waste. Waste is introduced at the end of the kiln far from the burner. The gases exiting from the kiln will then dry the aqueous waste, and its temperature will drop. If aqueous waste were dropped into the kiln with parallel flow, water would be evaporated at the feed end of the kiln. The feed end would then be the end of the kiln at the lowest temperature, which would require much longer kiln for burning of the wastes.

Wastes containing a light volatile fraction (containing greases, for example), should use a kiln with parallel flow. These volatiles will likely be released from the feed, immediately upon entering the kiln. The parallel kiln provides residence time that a counter-current kiln would not, for the burn-out of these volatiles.

(*b*) ***Slagging and Non-slagging Kilns.*** When a kiln, like the conventional one, is designed to produce ash in dry or powder form, it is known as a **non-slagging kiln.** However, with higher temperatures, the dry ash can be converted into molten ash ; and hence, when a kiln uses such high temperatures, then it is termed as a *slagging kiln*.

At temperatures in the range of 1100—1200°C, the ash starts to deform ; and as the temperature increases, the ash will melt. The temperatures at which these things will actually happen will depend upon the chemical constituents present in the ash. It is also a function of the presence of oxygen in the furnace. When a kiln is designed to operate at such high temperatures as to generate and maintain molten ash during operation, then it is known as a ***slagging kiln***. *Operation in a slagging mode provides a number of advantages over non-slagging operation*. However, if slagging occurs in a non-slagging kiln (designed as such), then it is undesirable and must be eliminated.

The construction of *slagging kiln* is more complex than that of a non-slagging one, requiring a lip at the kiln exit to contain the molten material. A non-slagging kiln, on the other hand, will have a smooth transition with no impediments to the smooth discharge of ash. For this reason again, the non-slagging kiln is given a raking (2 to 4% slope to horizontal), while a slagging kiln is either kept horizontal or is given a negative rake (*i.e.* the surface of the kiln at the feed end is kept lower than the kiln surface at the discharge end), as to permit accumulation of more slag in the kiln than with zero or positive rake.

A *slagging kiln* can be used to destroy *metal drums* and *salt laiden wastes*, which cannot be incinerated in a non-slagging kiln, since they need higher temperatures for melting ; and the dropping of the drums in a conventional kiln may damage its refractory lining, since molten slag to absorb the impact will not be available, as will be available in a slagging kiln. The temperature in a slagging kiln, as stated above, must be very high to maintain the ash in the molten-slag. Temperatures as high as 1400-1550°C are quite common for slagging kilns ; while a non-slagging kiln usually operates at temperature below 1100°C.

The kiln will burn out solids and will volatilize the organics. All the organics will generally not be incinerated in the kiln, and will require that a high temperature be maintained at a specific residence time for destruction. For this purpose, a *secondary combustion chamber*, also called an *Afterburner*, is normally placed immediately downstream of the kiln, as shown in Fig. 15.1 (*a*). It is a stationary piece of equipment that is designed to maintain the temperature of the gas stream exiting the kiln at a pre-selected temperature of destruction for a specific period of time. The released volatiles, infact, will exit the kiln in the flue gas and will enter the After-burner. The After-burner will normally contain at least one burner to provide the supplemental fuel required for burn out of the organics in the gas stream. In some of the kiln system designs, the volatiles released from the kiln, may have enough heating value, and they may not require any external source of supplemental fuel for complete combustion. In these designs, the Afterburner does not require supplemental fuel, and it acts as an extension of the kiln, providing the residence time for burn-out of the organics. This secondary combustion chamber would be initially heated by hot gases from the kiln.

The Afterburner required for slagging kiln can often be much smaller than that required for a non-slagging kiln. Similarly, the other downstream equipment, such as the exhaust gas scrubber or the induced draft fan, can be reduced in a slagging kiln, since a slagging kiln represents a more stable and a predictable system, because it will maintain a relatively more constant temperature profile under rapid changes in the kiln loading, due to the presence of massive melt in it.

The tumbling action of a rotating kiln encourages the release of particles to the gas stream. Thus, from 5 to 25% of the non-volatile solids in a feed stream may become air-borne with the use of a conventional non-slagging kiln. Since the molten slag in a slagging kiln acts, as does the fluid ash in a P.C. (pulverized coal) burner, the slag will absorb particulate matter from the gas stream. *The particulate emissions from a slagging kiln will, therefore, be reduced to about* $\frac{1}{4}$ *to* $\frac{3}{4}$ *of that produced by a non-slagging kiln. On the other hand,* NO_x *emissions may be more in a slagging kiln*, since such NO_x generation is usually insignificant so long as the process temperature does not exceed 1100°C or so. At 1450°C, the generation of NO_x is almost ten times as great as its generation at 1000°C.

The molten slag from a slagging kiln is finally dropped in a wet sump, where it immediately hardens into a granular material, called the **frit**. The frit has the appearance of dark glass. The ash from a non-slagging kiln, on the other hand, is collected wet or dry. *Thus, the end product of a slagging kiln can be much easily disposed of, as it is much less leachable than the ash of a non-slagging kiln.*

Refractory linings for slagging kilns will, however, need to be much more durable and costly, as to withstand higher temperatures and impact and corrosive action of waste drums, etc.

The destruction of organic compounds in a rotary kiln furnace is achieved by a combination of high temperature and residence time. Generally, the higher is the temperature, the lower will be the residence time required for destruction of wastes. For this reason, the residence time for a slagging kiln will usually be less than that for a non-slagging kiln, since temperature in a slagging kiln will be higher than that in a non-slagging kiln.

The *residence time*, or the **waste retention time** of a particular kiln, depends upon its geometry and kiln speed, and is connected by the equation :

$$t = \frac{0.19}{D.S.N} \qquad ...(15.1)$$

where t = mean residence time in minutes

L = Length of the cylindrical kiln in m. The typical values may range between 3 to 10 m

D = Dia of the cylindrical kiln in m. Typical values may range from 1.5 m to 4 m

S = Kiln rake or slope in m per m length of kiln

N = Rotational speed of the kiln in revolutions per minute (rpm). Typical values of rotational speed may range from 0.5 to 2.5 cm/sec.

From the above equation, it can be easily stated that for a given rotary kiln with given values of $\frac{L}{D}$ and S, the solid residence time (t) will be inversely proportional to the kiln speed. Thus, by doubling the speed of the kiln, the residence time can be reduced to half.

15.8.5.2. Modern Advancements in Incineration. In order to incinerate the hazardous organic liquid wastes, the modern research has led to the development of some such systems, called the **Plasma Arc Systems**. Such a system uses extremely high temperatures developed within a plasma stream to destroy the hazardous organic matter. Such an incinerator, which can even be made mobile on a self contained trailor, is preferably used to destroy leachates, or such highly polluting organic liquid wastes. These incinerators are generally not used for destroying solids, but may eventually be used to destroy sludge wastes, containing fine solid particles.

Steam reforming and **Electrical heating types** of incinerators are also being developed in modern days for treating hazardous wastes.

15.8.5.3. Air Polluting Effects of Incinerators. While selecting biomedical wastes for incineration, care must be taken to exclude all the chlorinated plastic materials or products, since they liberate harmful and carcinogenic gases, which may pollute the environment, and affect human and animal health. Chlorinated plastics are often used in medical equipments ; and after being used, they become heavily contaminated with free flowing body fluids. The burning of such plastics cause an increased risk of production of dioxine(s) & furnas-which are carcinogenic, and harmful chlorine compounds, which get liberated more when the waste is burnt on low temperatures.

Due to their potential of causing serious air pollution problems, many people do oppose the installation of incinerators, particularly within the hospital premises, although incineration has been found to be the most effective method of destroying infection and toxic materials, as well as for volume and weight reduction of waste. The incinerator system also helps to produce energy and electric power which itself can be used to run the incineration plant. The incineration plant can, however be used with great advantage, if an off-site comprehensive system of waste treatment is developed for a city, where the plant is installed away from the residential areas, and the waste is carried upto the plant through careful and safe methods in closed containers.

15.8.6. Minimising Biomedical Waste. Good bio-medical waste management desires that the waste should be reduced and minimised by its *re-use and recycling*. The use of disposable items should therefore, be minimised, and the use of long lasting reusable items like syringes, etc. with proper sterilization, should be promoted, except where their use becomes absolutely essential in performance of necessary procedures. Reduction in waste could ultimately be beneficial for the environment.

15.8.7. Human Resource Management Issues in Disposal of Biomedical Wastes. The management of a hospital or a nursing home must develop a commitment for safe disposal of its wastes. A detailed plan should be prepared indicating each step of waste management, extending from procurement of materials, waste generation, collection, handling, storage, transport, treatment, and disposal. Biomedical waste management requires system approach involving all procedures necessary for waste management that should minimise the risk to health and environment. At the same time, it should not put extra burden on health care staff in their duties. *Devoted and trained personnel become more important than the technology*. Almost any treatment and disposal which is operated by well trained and devoted staff can eventually provide more protection for the staff, patients, and the community in general, than an expensive or a sophisticated system which is managed by staff having carefree altitude and who donot understand the risk and importance of their contribution and duties.

15.8.7.1. Training and Research. Each person in every health care establishment should be made aware of the potential risk of mishandling of the hospital wastes. The management of medical wastes requires diligence and care from the chain of people, starting from doctors, nurses, sanitary staff, labourers and persons who dispose of the residue after final treatment.

The staff in the hospital must be trained to properly segregate and collect the hazardous wastes separately, at the point of its generation. The staff should also be trained in proper packaging and labeling of various types of wastes. Colour coded bags and containers may be adopted for such segregation of wastes, as discussed earlier. The staff should also be advised to minimise the use of disposable items, which automatically will reduce the quantity of waste.

15.8.7.2. Community Awareness. People visit hospitals from all sections of the society, starting from the very poor to the very rich, and from the illiterate to the highly educated. Many of them donot have any knowledge about the risk involved in the hospital environment. Patients and their relatives go on changing everyday. Education of general public, therefore, becomes an important issue in proper management of biomedical wastes. Through press and TV advertisements, the general public should be taught to behave and take care of the wastes being produced by them during their stay in the hospital with their patient relative. Health care staff alone cannot look to the sanitation aspects, unless the people co-operate, either through education or through compulsions.

Implementation of the environmental laws and safe practices is not only the responsibility of the government alone, but active participation of the people is essential. National level educational programme for the masses about proper waste management is hence called for, in any good waste management system. The public awareness programme should be continuous and consistent, and should not last for a limited time. It should also be wide spread over the country, as a whole.

15.8.8. Conclusions. Environmental pollution has become a major concern with respect to the present as well as the future life on our planet. Health care facilities do contribute significantly to this pollution. It is the legal duty of the management of every health care institution to ensure safe disposal of their wastes, as not to cause any adverse impacts on human health or the environment. The authorities in the government or Pollution Control Boards are responsible to ensure the implementation of the legal laws and the statues, concerning such important pollution issues.

RADIOACTIVE WASTES

15.9. Radioactive Isotopes or Radionuclides

An atom, as you know, is the smallest unit of an element. An atom, in most simple terms, is made up of a nucleus containing protons and neutrons, surrounded by the revolving electrons. The total mass of an atom is contained only in the nucleus, since the electrons are considered weightless, though they occupy a large space to increase the volume of the atom.

The chemical properties of an atom are governed by the number of protons, which also indicates the **atomic number** of that element. *This means that the atomic number (i.e., the number of protons) contained in one atom of an element, will determine as to what that element is. Eventually, a change in the number of protons (i.e., atomic number) will change the very element itself.*

In the nucleus of an ordinary atom, the number of neutrons usually equals to the number of protons. Such a balanced nucleus of an atom of an element, containing equal number of protons and neutrons, is a **stable nucleus,** and the atoms of most of the elements do occur in such stable forms. The sum of the number of protons and the number of neutrons in the nucleus is known as the **atomic mass number** (which approximately equals the **atomic mass** in amu*). However, atoms of *radio-active elements* may release or absorb some neutron(s), and to become unstable isotopes, called **radioactive isotopes** or **radioactive nuclides.** The radioactive isotope of a radioactive element is, thus, formed due to the excess or shortage of neutrons in its atoms, as compared to the number of neutrons contained in the balanced atoms of that element. Evidently, the number of neutrons in an atom of such an element shall not be equal to the number of protons, but will be more or less. This changed number of neutrons (from the balanced atom) will eventually change the atomic mass number, *which shows that the isotopes of an element will have different atomic mass (i.e., atomic mass number) but the same atomic number.* For example, a normal carbon-12 atom has 6 protons and 6 neutrons along with 6 revolving electrons ; but its radioactive isotope, carbon-14 has 8 neutrons instead of the normal 6.

Atoms of such radioactive isotopes, due to having disproportionate number of neutrons, possess **unstable nucleus,** which tries to become stable by disintegration. This disintegration can take place in different ways, with the release of radiations of different types (*i.e.,* α, β and γ radiations). It can, hence, be concluded that the radioactive nuclides, when present in the natural environment, will disintegrate, releasing **ionised radiations** or rays**. These emitted radiations (alpha particles, beta particles, and gamma rays) are very powerful, and contain enormous energy. Some of these rays are even capable of penetrating through thick steel sheets.

Radioactive nuclides, thus, continue to disintegrate (or decay) with the passage of time, releasing *radioactive radiations* or *radioactivity.* One of the important parameters to characterise the radio-nuclides is their **half life,**

---

*∵ mass of 1 proton = 1.0088925 amu (amu is **atomic mass unit**)

mass of 1 neutron = 1.0088665 amu

1 amu = 1.6606×10^{-24} gm.

When atoms of radioactive isotopes disintegrate, **ionising particles (alpha and beta particles) are given off, and in some cases, *ionising electromagnetic radiations* (gamma rays) are also evolved. The most important characteristic of these radioactive radiations is their ionising nature ; *i.e.,* they possess enough energy to expel the electrons from the atoms or the molecules on which they strike. This **ionising characteristic** (*i.e.,* the capability to produce an ion or to excite a molecule) of radioactive radiations causes damage to the living matter.

which is the time required for 50% of the total atoms to decay. The half life periods (usually denoted as $T\frac{1}{2}$) of radioactive nuclides can vary from a few seconds to thousands of years. Accordingly, radioactive nuclides with higher half life periods may continue to release harmful radioactive radiations for a very long period of time, and may, hence, continue to cause hazards to biological life for very long periods.

Most of the present day man-made nuclear programmes are based on **nuclear fission,** in which process, the heavier radio isotope such as Uranium-235 (or say Plutonium-239) is bombarded with neutrons, resulting in its split in to two smaller and lighter atoms (lighter because the atomic mass of the original isotope will be equal to the sum of the atomic masses of both the fission fragments). *These smaller atoms of fission products are beta or gamma emitters only, since they are too small to emit alpha particles.* This splitting is accompanied with the release of neutrons and energy, as explained in article 15.10. These newly released neutrons, in turn, attack the fresh Uranium atoms, causing their splitting, with the release of more neutrons and energy. In this way, a chain reaction starts, which is self-sustaining with respect to neutrons, and lasts until the whole Uranium is exhausted. In this manner, a tremendous amount of cumulative energy is generated. This huge generated heat energy is exploited in a closed **nuclear reactor** to produce steam for running turbines to produce nuclear power, or to use this energy for any other beneficial uses in a controlled manner. However, when these reactions undergo in an uncontrolled manner in the open, they cause devastations, as in the case of **nuclear weapons.**

The elements or their isotopes which emit characteristic radioactive radiations are called radioactive elements, or radio-nuclides, as stated above. More than 42 elements have been identified as having naturally occurring radioactive isotopes, and more than 1100 radioactive isotopes (usually called radioisotopes) are known. **Out of these 1100 known isotopes, only about 60 are found in nature, while others may be man made.** A list of a few commonly occurring radionuclides is given in table 15.2.

Table 15.2. Commonly Occurring Radionuclides

Radio Isotope	*Emitted Particles*	*Half-life period*
Ameticium (Am)-237	Alpha	433 yr
Ameticium (Am)-243	Alpha	7,370 yr
Bismuth (Bi)-210	Beta	***5 days***
Carbon (C)-14	Beta	5730 yr
Cobalt (Co)-60	Beta	5.27 yr
Cesium (Cs)-135	Beta	3×10^6 yr
Cesium (Cs)-137	Beta	30.17 yr
Curium (Cm)-245	Alpha	8500 yr
Iodine (I)-129	Beta	1.6×10^7 yr
Iodine (I)-131	Beta	8 days
Krypton (Kr)-85	Beta/gamma	10.7 yr

Contd...

Lead (Pb)-210	Beta	22.3 yr
Polonium (Po)-210	Alpha	138.4 yr
Plutonium (Pu)-239	Alpha	24,000 yr
Plutonium (Pu)-240	Alpha	6570 yr
Radium (Ra)-226	Alpha	1630 yr
Radon (Rn)-222	Alpha	3.82 days
Strontium (Sr)-90	Beta	28.8 yr
Technetium (Te)-99	Beta	2.14×10^5 yr
Thorium (Th)-230	Beta	80,000 yr
Tin (Sn)-126	Beta	10^5 yr
Tretium (H)-3	Beta	12.33 yr
Uranium (U)-234	Alpha	**2.45×10^8 yr**
Uranium (U)-235	Alpha	7.04×10^8 yr
Uranium (U)-238	Alpha	4.47×10^8 yr
Xenon (Xe)-133	Beta/gamma	5.25 days

15.10. Nuclear Fission and Generation of Nuclear Power

15.10.1. Nuclear Fission. Tremendous heat energy can be obtained from certain fissionable radio-isotopes, like Uranium-235, Plutonium-239, or Uranium-233, which are arranged in such a way in a nuclear reactor, as to get a self-sustaining chain reaction. The process involved is known as *nuclear fission.* The process of **nuclear fission** is accomplished by bombarding the nuclei of some fissionable heavy radio-isotope (commonly Uranium-235) with neutrons. This action causes the uranium nuclei to split into smaller lighter nuclei **(fission fragments)** and to emit neutrons and heat energy. In one splitting of an atom of U-235, atleast one of the produced neutrons must remain available to produce another fission, instead of escaping from the assembly, or being used up in some other nuclear reaction. This newly available neutron, ejected from the first splitting of an atom of U-235, will bombard the nuclei of the second uranium atom, splitting it into fission products with ejection of two or three neutrons. These newly ejected neutrons will again split other atoms of the U-235. This will create a chain reaction to split various atoms of U-235.

In such a fission reaction of U-235, the nucleus of a U-235, atom will capture a neutron (introduced into the reactor from outside neutron producing mass), making the nucleus of this atom unstable, causing it to break. Since the atomic mass number (*A*) of a neutron is 1, the sum of the atomic mass numbers of the product species (*fission fragments*) must be 235 + 1 = 236. One possible fission reaction is shown below :

$$\underset{\text{Uranium-235}}{{}^{235}\text{U}} + \underset{\text{Neutron}}{{}^{1}n} \longrightarrow \underbrace{{}^{95}\text{M}_0 + {}^{139}\text{La}}_{\text{Fission fragments}} + \underset{\text{Neutron}}{2\,.\,{}^{1}n} + \text{Heat} + \text{gamma rays} \quad ...(15.2)$$

The *fission fragments* are, infact, simply lower elements (having lower atomic mass numbers). There are commonly two fission fragments from each nucleus with release of energy of the order of 200 M. eV*, shared between them. *The uranium nucleus, infact, does not split into the same two fragments* *each time. The break up is far from symmetrical and can occur in more than 30 different ways.* The most commonly produced isotopes are grouped around the mass numbers 95 and 139. The fragments formed from the fission process have very large neutron-to-proton ratios, highly unstable. Many more splittings will, hence, occur, before a stable nucleus is finally achieved. These successive decays or disintegrations will give rise to a decay chain.

The process of capturing of a neutron by an atom of U-235 and its fission (splitting) happens very very quickly. The time taken in splitting of one atom is of the order of pico-seconds (10^{-12} sec). The fission of a single atom of U-235 releases as much as 200 M eV of energy*. Since the finite used amount of U-235 will have enormous number of its atoms (1 gm of U-235 will have 2.55×10^{21} atoms), the generated heat energy will also be tremendous. An incredible amount of heat energy, is thus, released by the fission, and is accompanied with the release of gamma radiation. The two new atoms of the fission fragments, later in stabilisation stage, release beta and gamma emissions of their own, as well.

15.11. Radioactive Disintegration and Radioactive Radiations

The above explained disintegration of a radio-nuclide, often called as the *radioactive disintegration,* may be defined as the spontaneous break up of the nucleus of an atom of a radioactive element or its isotope. The break up of nucleus itself is important here, as compared to that of the entire atom or the molecule, as may happen in other types of hazardous wastes. In dimensions of radioactivity, the nucleus of a radioactive atom itself, is too large, making it unstable. This unstable nucleus breaks up, and the atom changes to the atom of some other element, which is lower in mass. The radiations or the radioactivity generated due to such break up of the nucleus of the radioactive atom, which is found to be extremely hazardous to life, may be of three types ; *i.e.*, : (*i*) *alpha particles* ; (*ii*) *beta particles* ; and (*iii*) *gamma radiations,* as discussed below :

(*i*) **Alpha Radiations.** The release of alpha radiations from the nucleus of a radioactive element is equivalent to the release of positively charged helium nucleus, consisting of *two neutrons* and *two protons.* This release causes the parent atom to loose four *atomic mass,* (since an **atomic mass** is the sum of the number of protons and the number of neutrons in the nucleus) ; while the loss caused in *atomic number* in such a break up will be equal to 2 only (because **atomic number** is equal to the number of protons in the nucleus).

Disintegration of Uranium-238 (atomic number 92) to release alpha particles, given by eq. (15.3) can be quoted as an example of this type of break up of a radioactive element, which is explained below :

When U-238 releases the alpha particles, the atomic number decreases by 2, and atomic mass (or atomic mass number) decreases by 4. The atomic number of Uranium is 92. From the periodic table (given at the end of the book as Appendix Table A-8), the atom with an atomic number of 90 is *Thorium*

---

*M.eV (million electron volt) = 10^6 eV

where **1 eV = 1.602 × 10^{-19} Joules** = 1.602×10^{-12} ergs

$\therefore$ 1 MeV 1.602×10^{-13} J.

(Th). Hence, the new element formed will be Thorium. This reaction equation is written as :

$$\underset{\text{Uranium-238}}{{}^{238}U_{92}} \xrightarrow[\text{(break up)}]{\substack{\text{radioactive}\\\text{disintegration}}} \underset{\text{Thorium-234}}{{}^{234}Th_{90}} + \underset{\text{Alpha paticles}}{{}^{4}He_{2}} \quad ...(15.3)$$

where ${}^{238}U_{92}$ = Uranium having atomic number = 92, and atomic mass number = 238

${}^{234}Th_{90}$ = Thorium having atomic number = 92 – 2 = 90 ; and atomic mass number = 238 – 4 = 234

${}^{4}He_{2}$ = Helium having atomic number = 2 and atomic mass number = 4

Likewise, Plutonium-239 (atomic number 94) disintegrates to Uranium-235 (atomic number 92) with the release of an alpha particle, as per equation (15.4) given below :

$$\underset{\text{Plutonium-239}}{{}^{239}Pu_{94}} \xrightarrow[\text{(break up)}]{\substack{\text{radioactive}\\\text{disintegration}}} \underset{\text{Uranium-235}}{{}^{235}U_{92}} + \underset{\text{Alpha particle}}{{}^{4}He_{2}} \quad ...(15.4)$$

Likewise, Radium-226 is transformed into Radon-222, as per equation (15.5) :

$$\underset{\text{Radium-226}}{{}^{226}Ra_{88}} \xrightarrow[\text{(break up)}]{\substack{\text{radioactive}\\\text{disintegration}}} \underset{\text{Radon-222}}{{}^{222}Rn_{86}} + \underset{\text{Alpha particle}}{{}^{4}He_{2}} \quad ...(15.5)$$

[**Note.** *In the above equations, the pre-exponent is the atomic mass number, and the subscript is the atomic number.*]

Alpha particles released in the above equations, are *slow moving ionising particles,* having an average velocity almost 10 times lesser than that of light. These particles can travel not more than 10 cm in air, and have a weak penetration power. They can be stopped even by a thin sheet of paper. *Since the alpha particles cannot penetrate the barrier of skin, they also have insignificant effects on the body from outside. However, they can be extremely dangerous when emitted by the ingested radioactive alpha emitters, since they cause an intense ionisation in the biological molecules.*

(ii) **Beta Radiations.** Beta radiations are the *release of electrons* from a radioactive nucleus, caused by the break up of neutron into a proton and an electron. Due to this break up of one neutron, the nucleus will have one increased proton, and hence one increased positive charge or one increase in its atomic number. Since the mass of the released electron is very small, the reaction will not change the atomic mass, which equals to the number of protons and number of neutrons (here the total number of neutrons and protons remains unchanged, since one neutron is changed into one proton with the release of one electron). However, due to change in charge (increased charge), the parent atom changes to another atom of higher atomic number. *Being an electron, beta radiation has a negative charge.*

Disintegration of Strontium-90 (having atomic number = 38) to produce Yttrium-90 with the release of a beta particle, given by eq. (15.6) can be quoted as an example of this type of disintegration of a radioactive element :

$$\underset{\text{Strontium}}{{}^{90}Sr_{38}} \xrightarrow[\textit{i.e. break up}]{\text{disintegration}} \underset{\text{Yttrium}}{{}^{90}Y_{39}} + \underset{\substack{\text{Beta} \\ \text{particle} \\ \text{(high vel.} \\ \text{electron)}}}{\beta} \qquad ...(15.6)$$

Likewise, Xenon-139 (with atomic number 54) is converted into Cesium-139 (with atomic number 55), after emitting a beta particle, as indicated by equation (15.7).

$$\underset{\text{Xenon-139}}{{}^{139}Xe_{54}} \xrightarrow[\text{disintegration}]{\text{radioactive}} \underset{\text{Cesium-139}}{{}^{139}Cs_{55}} + \underset{\substack{\text{Beta} \\ \text{particle}}}{\beta} + \underset{\substack{\text{Gamma} \\ \text{ray}}}{\gamma} \qquad ...(15.7)$$

Gamma rays may or may not be released with beta particles in this type of disintegrations. Beta particles are the released high velocity electrons having velocity ranging from 30% to 99% of the speed of light, which is more than the spread of electrons moving in an electric current. *The penetration power of beta particles varies with their velocities, and may range from a few millimetres to a centimetre under the skin. However, because of their lower mass, beta particles possess a lower ionising power than alpha particles.* Hence, beta particles, though can prove to be hazardous to body as an external radiation, but may not cause that much harm as may be caused by alpha particles—if produced internally from an ingested radioactive element.

(*iii*) **Gamma Radiations.** Gamma radiation or **gamma ray**, as it is also called, has no charge or mass, being simply an electromagnetic radiation that travels at the speed of light. Gamma rays are normally produced where an electron in an atom of a radioactive nuclide, moves from a higher energy level (*excited state*) to the lower stable energy level. An analogous phenomenon also occurs in a **nuclear reaction ;** *wherein, a nucleus in an excited state releases the gamma rays when it transforms into a more stable lower form.* Gamma radiation may accompany either alpha or beta radiation. **X-rays are also a sort of gamma rays**.

Gamma rays are, hence, defined as *very high frequency photons.*

Being an electromagnetic radiation, gamma radiation may be thought of as a wave or a massless *photon*, having a very short wave length in the range of 10^{-3} to 10^{-7} μm. Because of their shorter wave lengths, gamma rays possess very high energy, and greater penetrating power, which increases with the decrease in wavelength. *Gamma radiation has got a high capability of ionisation, and are highly damaging to biological life.*

15.12. Impact of Radioactivity on Biological Environment

When the radioactive radiations strike a particular material or body, they cause ionisation of the atoms or the molecules of that material, or of any material present in that body.

For example, when +vely charged alpha particles pass through an absorbing material, their positive charge attracts the electrons of the atoms in the molecules of that material, leading to a rise in their energy levels ; which, in many cases, forces the electrons completely out of the atoms, thus, bringing about the ionisation in the molecules.

Similarly, when the –vely charged beta particles pass through the electron orbits of the atoms of the molecules of the absorbing material, the force of repulsion between the two, increases the energy level of the electrons, pushing them to higher energy orbits or completely out of the atom. Beta ionisations are usually encountered less frequently than the alpha ionisations.

The ionisation caused by the fast moving highly energetic gamma rays involves a direct collision with the orbital electrons, due to which, the electrons are ejected free of their respective atoms. The net result of radiation on the molecules and atoms is to form positive ions and free electrons. Gamma radiations, especially those having very high energy, can even be absorbed by the nucleus, leading to the release of neutrons.

Any atom or molecule in a biological body can be ionised by such radioactive radiations, when they happen to strike that biological body (such as human body). Since the body of a living being (*i.e.,* biological life) normally contains huge amount of water-constituting more than 80% of most tissues, it has the maximum chance of being ionised by radioactive radiations. The ionisation of the water of the various tissues of the biological body, causes formation of various radicals and compounds, which possess great properties of oxidation and reduction, and can readily interact with the biological molecules, bringing about tissue damage, death, or even mutagenic effects. The equations showing ionisation of water and associated reactions are indicated and explained below :

$$H_2O + \text{radioactive radiation} \longrightarrow \underset{\substack{+\text{ vely charged} \\ H_2O}}{H_2O^+} + \underset{\substack{\text{free} \\ \text{electron}}}{e^-}$$

$$H_2O + e^- \longrightarrow \underset{-\text{ vely charged } H_2O}{H_2O^-}$$

$$H_2O^+ \longrightarrow H^+ + OH^-$$

$$H_2O^- \longrightarrow H^- + OH^-$$

The hydroxyl ion (OH^-) may, on loosing an electron due to radiation, exists as hydroxyl **free radical** (OH′) – which is short of one electron. The hydroxyl radical (OH′) is a highly reactive radical*, in search of one electron, thereby acting as a strong *oxidising agent.* Similarly, the Hydrogen free radical (H′) wants to lose its sole electron to form a stable hydrogen ion (H^+), and, therefore, acts as a strong *reducing agent.* The radiation, thus, leads to formation of these OH′ and H′ radicals, along with formation of some other radicals like HO_2 and H_2O_2, as per reactions indicated below :

---

*A **radical** in chemistry, as you may be knowing, is a group of two or more charged or neutral atoms that have atleast one unpaired electron. Molecular oxygen, O_2, for example, has two unpaired electrons, and is a common radical. Most radicals are extremely, reactive, and combine with other atoms or radicals to form compounds or ions. However, some radicals, called **free-radicals,** may exist for relatively short times, unbound to any other group. Radicals play a major role in certain chemical reactions of commercial or medical significance.

$$OH' + OH' \longrightarrow H_2O_2'$$
$$H' + O_2 \longrightarrow HO_2'$$
$$HO_2 + H' \longrightarrow H_2O_2'$$

All these free radicals, (*i.e.*, OH′, H′, HO_2' and H_2O_2') cause many oxidising and reducing reactions to occur in the biological molecules of a living body, with the further formation of more free radicals and oxidised molecules, as indicated below :

$$RH + HO_2' \longrightarrow R' + H_2O_2'$$
$$RH + HO_2' \longrightarrow RO + H_2O'$$
$$RH + OH' \longrightarrow R' + H_2O'$$

Besides, these free radicals may also interact with the double bonds, hydrogen bonds, and sulpha-hydroxyl group (SH) present in proteins, DNA, RNA, and other bio-molecules.

In certain cases, the free neutrons when absorbed in the body can transmit some elements into others, as indicated below :

$$^{31}P + \underset{\text{Neutron}}{n} \longrightarrow {}^{28}Al + a + y$$
$$^{28}Al \longrightarrow {}^{28}Si + b$$

The net result of all such reactions and interactions leads to deactivation of enzymes and destruction of nucleic acid and other biologically active molecules in living tissues, which ultimately may lead to cell division, disruption of cell membranes, and overall damage to the biological tissues, causing death and mutagenic effects.

It can, hence, be concluded that the ionization produced by radiation disrupts the normal chemical processes of living cells, causing the cells to grow abnormally or to die.* The seriousness of the biological damage depends on the density of the ions formed. Cells can, infact, repair the damage produced by only a few ions, but they generally do not recover from a high density of ionisation.

The various types of ionizing radiations, thus, produce different degrees of biological damage. The cell damaging capacity of radiations are, hence, compared by using a term called **the relative biological effectiveness (RBE)** or **the Quality factor (Q).** The RBE of a particular radiation, thus, indicates the extent to which it damages the body cells, compared with equal doses of other ionizing radiations. For example, dense track of ions formed by alpha particles causes about 20 times as much damage as is being caused by a thin ion trail generated by electrons. Thus, alpha particles will have an RBE or Q of 20, if electrons have an RBE of 1**.

The amount of radiation dose absorbed by a substance (including a body) is measured in unit, called **rad.** A **rad** is defined as a dose of $\frac{1}{100}$ joule of radiation energy per kilogram mass of material absorbing the radiation. **The**

*cells from tissues.

**The factor Q or *RBE* for X-rays, γ-rays, and electrons is 1 ; whereas for α particles, it is 20.

new S.I. unit of absorbed radiation is Gray (Gy), which replaces rad, and is defined as that dose of ionizing radiation that imparts one Joule of energy to 1 kg of absorbing material.

Naturally 1 Gy = 100 rad. ...(15.8)

Another unit, called **rem,** measures the biological damage produced in the human body by different types of ionizing radiations. The number of rems of radiation equals the number of rads multiplied by the RBE of the radiation involved. The new S.I. unit replacing rem is *Sievert (Sv),* which equals the radiation dose in gray multiplied with the RBE of the radiation involved.

Evidently, 1 Sivert = 100 rem ...(15.9)

The rem or sievert, thus, directly indicates the damaging capacity of a dose of a particular type of radiation. A person can receive a radiation dose of up to 25 rem without showing any immediate effects. A dose of 100 rem (*i.e.,* 1 Sv) may produce **radiation sickness.** People exposed to doses between 300 to 800 rem (*i.e.,* 3 to 8 Sv) experience severe radiation sickness Radiation dose of more then about 800 rem (8 Sv) are always fatal.

Over a life time, a person receives 7 to 14 rem of radiation exposure from *natural sources of radiations*, such as from the radioactive rocks and minerals in the Earth and from the sun and other objects in space, including cosmic rays, which probably arise from the sun or even beyond it. A single exposure of 5 to 75 rem of radiation, usually produces a few observable symptoms. Vomiting, fatigue, and loss of appetite may, however, accompany radiation exposures of 75 to 200 rem, and recovery may take a few weeks. Severe changes in blood cells and hemorrhage may occur with exposures of more than about 300 rem. *Above 600 rem, additional symptoms include loss of hair, and loss of body's ability to fight infections, usually resulting in death.*

Other adverse effects of radiation may become evident late in life. These include a shortened life span and an increased probability of developing cancers and cataracts. Radiation can also damage a person's genes. Damaged genes can transmit harmful traits to the person's off-springs.

On the community level, especially in aquatic ecosystems, fish eggs have been found to be extremely sensitive to even low levels of radioactivity. Fish species, particularly those with low fecundity, are badly affected by the radioactivity present in water. Fish and other biological life (including animals, plants, and humans) are also susceptible to mutational effects, leading to birth of abnormal organisms. In general, fishes are the most sensitive organisms in water to radioactive radiations (*i.e.,* irradiation), followed by crustaceans, mollusks, algae (except cyanophyceac), and protozoa.

Besides the radiations emitted by the natural sources in the environment-exposing humans to such radiations, a smaller amount may come from the testing of nuclear weapons. Certain amount of exposure may also come from medical and dental X-rays ; while a certain amount may come from items, such as colour television sets, smoke detectors, airport baggage X-ray machines, etc. The operation of nuclear power plants account for only a tiny amount of total normal radiation exposure. In USA, the average annual radiation dose received by a person has been estimated to be about 180 milli rem (1 milli rem $= \frac{1}{1000}$ rem = 10^{-5} Sv).

The nuclear fall-outs from the explosions of nuclear weapons and accidental discharges from the nuclear power reactors, as well as the radioactivity emanating from the solid and liquid wastes of nuclear power plants, may cause very harmful radioactive exposures on humans and other animals. Three isotopes produced from fission reactions taking place in the nuclear reactors and atomic explosions, and, thus, occurring in the nuclear fallouts and in the wastes of nuclear reactors, have been particularly signaled out for their adverse effects on human health. Strontium (Sr)-90, Cesium (Cs)-137, and Iodine (I)-131 are these three highly dangerous radioactive isotopes. **Strontium-90** is made even more dangerous than Cesium-137 due its ability to behave like calcium, and getting incorporated into the bones. As such, the milk produced by animals into whose bodies Sr-90 has been concentrated, may become radioactively contaminated.

Iodine-131 is also a highly dangerous isotope, as this element can enter the food chain at any level, and become concentrated in human body, consuming such contaminated food. Once within the body, it can damage white blood cells, bone marrow, spleen lymph nodes, and can cause lung tumours, skin cancers, sterility, defective eye sights, etc. The accidental discharges of such radio-isotopes from nuclear power stations due to causes like over-heating, melting and vaporisation of the fuel, etc., should, therefore, need to be prevented by providing very high safety levels.

A summary of exposure to organs from various radio-nuclides* is given in table 15.3, which also shows the probable means of exposure.

Table 15.3. Type of Exposure from Typical Radio-Nuclides in the Environment

Radio nuclide	*Half-life*	*Critical organ*	*Form*	*Probable cause of exposure*
H(T)-3	12.3 y	Total body	HTO	Water intake
Mn-54	300 d	Total body	Insol. particle	External
Fe-55	2.6 y	Lung	Insol. particle	Inhalation
Zn-65	245 d	Total body	Sol. particle	Dietary intake
Kr-85	10.8 y	Total body	Gas	Immersion
Sr-89	52 d	Bone	Sol. particle	Dietary intake
Sr-90	28 y	Bone	Sol. particle	Dietary intake
Y-91	59 d	Lung	Insol. particle	Inhalation
Zr-95	65 d	Total body	Insol. particle	External
Nb-95	35 d	Total body	Insol. particle	External

Contd...

*Either produced from radioactive testing and nuclear power stations, or occurring in natural environment.

Ru-103	40 d	Total body	Insol. particle	External
Ru-106	1.0 y	Total body	Insol. particle	External
Sb-125	2.7 y	Total body	Insol. particle	External
I-131	8.1 d	Thyroid	Vapour particle	Dietary intake
Xe-133	5.3 d	Total body	Gas	Immersion
Cs-134	2.0 y	Total body	Sol. particle	Dietary intake
Cs-137	30 y	Total body	Sol. particle	Dietary intake
Ba-140	12.8 y	Total body	Insol. particle	External
Ce-144	28.4 d	Total body	Insol. particle	External
Pm-147	2.6 y	Lung	Insol. particle	Inhalation
Th-232	1.4×10^{10} y	Lung	Insol. particle	Inhalation
U-233	1.6×10^{5} y	Lung	Insol. particle	Inhalation
U-235	7×10^{8} y	Lung	Insol. particle	Inhalation
U-238	4.5×10^{9} y	Lung	Insol. particle	Inhalation
Pu-238	86 y	Lung	Insol. particle	Inhalation
Pu-239	2.4×10^{4} y	Lung	Insol. particle	Inhalation
Pu-240	6.6×10^{3} y	Lung	Insol. particle	Inhalation
Pu-241	13 y	Lung	Insol. particle	Inhalation

Whether the radio-isotopes are natural or man-made, their danger to the environment depends on the following factors :

1. Physical properties, such as half-life ($T\frac{1}{2}$), type of emissions emitted, and energy of emission.
2. Their ability to enter the food chain.
3. Their tendency to become concentrated in animal tissues.

Some of the physical properties of a few radio-nuclides are listed in table 15.4, together with the approximate energies of emission.

Table 15.4. Energy Emission Levels of Certain Radio Nuclides

Radio nuclide	*Half life* $\left(T\frac{1}{2}\right)$	*Energy of emission E (MeV)*	*Type of emission*
Carbon-14	5,700 y	0.15	Beta
Phosphorus-32	14 days	1.7	Beta
Iodine-131	8 days	0.8, 0.08	Beta
Strontium-90	28 y	0.5	Beta
Cesium-137	30 y	0.5	Beta
Plutonium-238	86 y	5.5	Alpha
Plutonium-239	24,000 y	5.1	Alpha
Uranium-235	7×10^{8} y	204	Beta/gamma

It can be observed from this table that the high energy of emission with the presence of alpha particles make Plutonium-238 (with half life of 86 years) particularly toxic. These particles base particularly dense trails of ionization inside living cells, which might take up the isotope.

In order to ensure that such radiations do not cause harmful effects on the body, it is necessary that the quantum of these radiations, to which the body is subjected to, be restricted to safe standards. *International Commission on Radiological Protection,* by their research and past experience on the use of X-rays and radium, has stipulated the limiting radiation exposures per year per normal person.

These *standard values of annual permissible radiation exposures to different organs of human body* are given below in table 15.5.

Table 15.5. Annual Permissible Radiation per Year per Person

S. No.	*Part of the body*	*Annual permissible radiation per year per person*
1.	Blood forming organs, and genades like ovaries, testies, etc.	5 rem*
2.	Bone, skin, and thyroid	30 rem
3.	All other organs of the body	15 rem
4.	Hands, forearms, feet, and ankles	75 rem
	For pregnant ladies	
	(*i*) during the initial 2 months of pregnancy	< 1 rem
	(*ii*) during the remaining 7 months	≥ 1 rem

British Medical Council (BMRC) has specified a dose, called the **Emergency Reference Level (E.R.L.),** which has been defined, as the concentration of radioactivity, which different parts of the body can tolerate, varying from 10 to 60 rem. This *reference level* is a guide to initiate counter measures like *evacuation,* etc., as and when the radioactivity in a particular area is detected to be more than the reference level. Other counter measures, like giving anti-atomic drugs to remove the bad effects of radioactivity may also sometimes be adopted ; say for example, contamination of Iodine-131 may be removed by treating the affected people with ordinary iodine tablets**.

Larger concentrations of radioactivity not only cause adverse impacts on humans and other animals & birds, but it also affects the trees and plants. Deadly effects of gamma radiations have been observed, leading to complete destruction of trees-subjected to gamma rays for a period of 6 month, in an experimental verification. Since radioactivity causes adverse effects on all forms of biological life, it becomes imperative for man to watch and control the presence of such dangerous radiations in the environment.

Fortunately, radioactivity released in the natural environment, under normal conditions, known as the **back ground radiation,** is found to be hardly 100—200 milli rem, a value much lower than the tolerance limits. However, in case of accidental nuclear discharges, anti-radioactivity measures may have to be initiated, as pointed out above.

---

*It would be worthwhile to mention here that an ordinary chest X-ray subjects our body to a radiation of 1 rad, which equals to 1 rem for X-rays, since Q for X-rays is equal to 1.

** Such iodine tablets help in removing Iodine-131 from our body.

15.13. Sources of Radioactivity in the Environment

The sources of radioactivity can be broadly divided into the following two categories :

(1) *Natural sources* ; and

(2) *Man-made sources.*

These are discussed below :

15.13.1. The Natural Sources of Radioactivity. The natural sources of radioactivity include the following sources :

(*i*) cosmic rays ;

(*ii*) radioactive rocks and minerals contained in the body of the Earth, or present in the environment ;

(*iii*) radionuclides present within the bodies of the organisms.

(*i*) **The cosmic rays** are of extra-terrestrial origin that originate from the sun and other objects beyond, in outer space. These high-energy particles include electrons, protons, helium nuclei, hydrogen nuclei, and the nuclei of such heavier elements as carbon and oxygen. Cosmic rays travel at tremendous speeds with many rays travelling near the **speed of light-which is 2,99,792 km/sec.** Large quantities of the cosmic rays originate from exploding stars, called *supernovae,* and extremely dense stars known as *pulsars.* The sun releases cosmic radiation during spectacular radiations called *solar flares.* These cosmic rays, which originate in space, are also called **primary cosmic rays or primaries.** Every second, about three to six cosmic rays strike each square inch (6.25 cm^2) of the Earth's upper atmosphere. However, only a few of them reach the Earth's surface. Some cosmic rays, which otherwise would enter the Earth's atmosphere, are reflected back into the space by the magnetic field of the Earth. The Earth's magnetic field can be pictured as a system of imaginary lines, called *flux lines,* that curve from the Earth's north magnetic pole to the south magnetic pole. Only particles with extremely high energies can penetrate the magnetic field near the equator, because they have to cut across the flux lines. Near the poles, even rays with low energies can travel along the flux lines and enter the atmosphere. Thus, the quantity of cosmic rays that enter the atmosphere is lowest at the equator, and increases towards the poles.

The cosmic rays entering the atmosphere collide with the gas molecules present in the upper atmosphere, bringing about intense ionization in gases. The collision of primary cosmic rays with the atomic nuclei (oxygen and nitrogen nuclei) in the atmosphere also causes the formation of **secondary cosmic rays or secondaries.** When a primary cosmic ray as well as the atomic nucleus it strikes, break into these smaller particles (called secondaries), energy is released. Some of this energy is converted into matter, creating more secondaries. A number of new particles collide again, resulting in even more particles. These collisions produce a shower of secondaries that includes all types of subatomic particles. Secondaries lose energy to the atmosphere, and so only a small fraction of them reach the Earth's surface. Still however, every minute, an average of about six particles, strike each square inch (6.25 cm^2) of the Earth's surface.

The secondary cosmic rays mainly composed of neutrons, mesons, and gamma rays. Eventually, a complex mixture of particles reaches the Earth as cosmic rays. These particles also form substantial quantities of ^3H (tritium) and ^{14}C (carbon-14) in the atmosphere. Such radioactive substances, generated naturally in the Earth's atmosphere by cosmic rays, are washed out slowly, or absorbed in the precipitation to reach water bodies. The gaseous radio-nuclides from the atmosphere can also be transported to the water bodies by direct chemical reaction to form H_2O, or by their simple dissolution in water on coming in contact with the water bodies due to their movement by winds.

(*ii*) The **radionuclides contained in the Earth's body** (*i.e.*, in the **lithosphere** and **hydrosphere**), or those handled on the Earth, provide another important source of natural radiations. *All the elements above the atomic number of 82 (i.e., Pb_{82}) are radioactive in nature and emit a variable quantity of radiation.* The most abundant naturally occurring radionuclides on the Earth are thorium, uranium, and potassium-40. Soils, rocks and even building materials contain small quantities of ^{40}K (*i.e.,* K-40), and tiny amounts of Uranium and its daughter products.

(*iii*) Amongst the **radionuclides naturally present within the bodies of living organisms,** potassium-40 (K-40) is the most important, besides others like tritium (H-3), carbon-14, radium, polonium (especially contained in bones), and radon (especially contained in lungs).

The average annual radiation released by all these natural sources like cosmic rays, and radionuclides present internally in the body or existing externally on Earth, is estimated to be about 100 milli rem, though there may be variations up to the tune of 400% in different parts of the world, depending upon the altitude and local concentration of the natural radioactivity. These radiations caused by natural sources are called the **background radiation.**

15.13.2. Man-made Sources of Radioactivity. Man-made radiations may enter into the environment through various sources, including appliances of daily use, like colour TV sets, luminous clocks, watches, and even telephone dials, etc. ; all of which may contribute to about 1% of the total background radiation received by a person. Besides, humans are significantly exposed to radiations received from medical use of X-rays and radioactive isotopes ; and from use of X-rays at baggage check points at airports, etc. A small amount of radioactivity is also introduced into the environment by the use of radionuclides in industry, medicine and research. However, the most important man-made sources of radioactivity are the *nuclear explosions* and the *nuclear power plants*. Both these important man-made sources of radioactivity are briefly discussed below :

15.13.2.1. Nuclear Explosions and Nuclear fallouts as sources of radioactivity. The nuclear explosions are carried out by man for testing of nuclear weapons, and for beneficial uses like mining and excavation of harbours, canals, etc. Such nuclear explosions release huge quantity of radio-nuclides along with heat in the open environment. These radioisotopes fused with dust and dirt rise in the atmosphere along with hot gases, and then fall down slowly on to the surface of the Earth by gravity. This is known as

radioactive fall-out. The radionuclides in this fall-out differ from those in other radioactive wastes in that they remain fused with iron, silica and dust ; and become insoluble. They look like marble particles of different colours, having different sizes, varying from a few microns to as big as snow particles.

The radio-nuclides resulting from nuclear explosions may include those that originate directly from fuel as fission products (such as Sr-90 and Cs-137, etc.), and those that are formed by the absorption of released neutrons by the ambient elements and molecules (such as Fe-59, Mn-54, Co-60, Zn-65, etc.).

In the nuclear fall-out resulting from a nuclear explosion, while the larger sized particles start reaching the Earth within a few hours by gravity, the smaller sized particles stay in the atmosphere for quite some times, till swept down by rain or snow. These smaller sized particles, usually adhere strongly to the leaves of vegetation, from where they are passed on to the herbivores, and finally to man through milk or meat. The most significant food chain transfer in this manner has been reported for Sr-90 and Cs-137, as pointed out earlier.

In aquatic ecosystems, several fall-out nuclides, such as Fe-59, Mn-54, Co-60, Zn-65, Ce-144, Rn-160, and Zn-95 form complexes with organic detritus, from where they are readily transferred to the aquatic organisms.

15.3.2.2. Nuclear power plants as sources of radioactivity. The nuclear power plants use the heat energy released from the fission reaction of the nuclear fuel ; uranium being the most commonly used fuel. The uranium to be used in a nuclear power plant has to be prepared fit for such a use, through several processes, right from its mining to its inception into the nuclear reactor. Besides, the spent materials obtained from the reactors are also reprocessed to recover unburnt uranium, plutonium, and some other important fission isotopes, which can be used in medicine, research, etc.

The entire operation, starting from the mining of the fuel to the final disposal is called the *uranium fuel cycle*—when uranium fuel is used in the nuclear reactor ; or the *nuclear fuel cycle*—when any other fuel such as thorium or plutonium is used as a fuel.

At almost every stage of the ***nuclear fuel cycle***, liquid, gaseous and solid radioactive wastes are generated. These wastes have a great potential to contaminate the environment, and, hence, have to be disposed of very carefully, to avoid harmful effects on biological life, and the environment in general.

The important processes involved in a **uranium fuel cycle** are shown in Fig. 15.2. These processes start from mining of the ore, followed by its milling, and then uranium concentration into its oxide (U_3O_8) by chemical extraction process, leaving behind the crushed material containing several radioactive daughter isotopes of uranium, called ***Uranium mill tailings***. A low activity liquid waste is produced from chemical extraction of uranium, whereas, the uranium tailings cause radioactive pollution of air due to wind action, and pollution of water by run off, seepage, and leaching. After the concentration of uranium into its oxide (U_3O_8), it is converted by chemical conversions to uranium hexa-fluoride (UF_6), a form of uranium suitable for its enrichment and final conversion to uranium or its oxide for use as fuel.

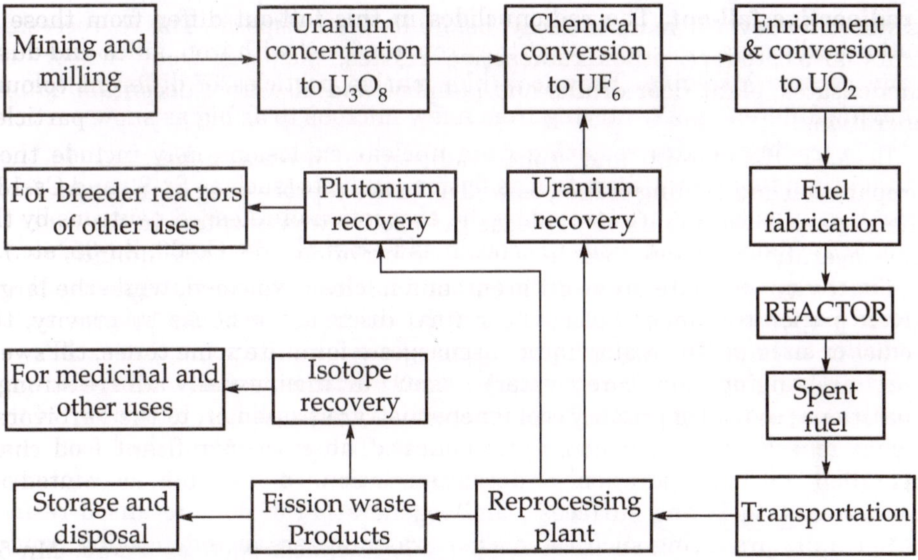

Fig. 15.2. Uranium fuel cycle.

The uranium obtained from its ore like pitchblende, as you know, is mainly U-238, with only 0.7% of U-235, which is not enough to work as a nuclear fuel. To make this uranium fit for use as fuel, the concentration of U-235 has to be increased from about 0.7% to about 2–3% in the uranium fuel, to obtain the final useful uranium, called the *enriched uranium.* The enriched uranium is finally converted into uranium or its oxide (UO_2) by chemical processes. This material is, then, powdered and fabricated into pellets to be used in the reactors.

The chemical conversion processes for enrichment of uranium and fuel fabrication process, produce a small quantity of gaseous radioactive waste, and significant quantities of liquid and solid wastes of low radioactivity, which need suitable disposal to avoid water pollution.

The fabricated fuel pellets are assembled in the fuel rods, which along with the control rods are loaded in the reactor core immersed in water. In the beginning, ^{235}U is attacked by the neutrons, resulting in its fission by release of neutrons and energy. Some of the released neutrons are absorbed by ^{238}U, converting it into ^{239}Pu – that acts like ^{235}U and breaks down into smaller elements (fission fragments) with the release of further neutrons and energy. This ultimately results to the formation of a mixture of isotopes, numbering nearly 80, and usually ranging from atomic number 30 to 63. A small quantity of some heavier elements is also present in the mixture. The fuel once introduced into the reactor core, runs for about 2-3 years, after which it needs replacement with new fuel. The *spent fuel,* called ***irradiated fuel,*** is stored for a few months at the reactor site in water ponds, and then transported to the *reprocessing units*.

In a radioactive fuel reprocessing plant, uranium and plutonium are chemically separated from the spent fuel, together with some other important radio-nuclides. The remaining solid mixture, which is still highly radioactive, is stored for several years to decay, before its final disposal. The reprocessing units also produce other radioactive wastes, such as from evaporated condensates, plant floor drains and laundry, and also from decontamination operations.

Besides the liquid and solid radioactive wastes generated in the preparation and loading of fuel, as well as in the reprocessing of the spent fuel, the nuclear power plants also release radioactive waste-waters in their day to day operations. These waste-waters, containing variable quantities of radioactivity, eminate from different units. These waste-waters should be given proper treatment before their final discharge into the nearby water bodies or on land. The major liquid discharges originate during core exchanges (*i.e.,* exchanging the core water) ; routine maintenance and clean-up operations ; control of primary coolant chemistry ; samplings ; blow down of the steam generators ; expansion water obtained when the plant goes from cold operation to hot operation ; decontamination of clothing of workers, components, tools and surfaces ; and regeneration of demineralised resins. Other units producing such radioactive waste-waters include cooling ponds, laundries & bathrooms, and radiochemical laboratories. These waste-waters, usually, have low levels of radioactivity.

Accidental discharges of radioactive radiations from accidental explosions in nuclear reactors may also lead to sudden rise of radioativity concentrations in the environment, which may prove quite harmful not only to humans, but also to plants, animals and birds. The most prominent example of release of such dangerous radioactive radiations is offered by the ***chernobyl atomic accident*** that occurred in Russia on April 26, 1986. As a matter of fact, the chernobyl nuclear reactor accident was the second such nuclear accident. Prior to this accident, the ***three mile island (TMI) nuclear accident has occurred*** on March 28, 1979 near Middletown Pennsylvania, U.S.A., because of a series of minor errors that started with a malfunctioning valve in the nuclear reactor, which was compounded with the fact that no one discovered it for a long time. The fuel rods started to melt, but the whole situation was finally brought under control after two weeks and no casualities occurred, except that some radioactive gas and water was discharged into a nearby river. The ***chernobyl nuclear accident of 1986,*** however, proved a disaster, as it resulted into a full scale explosion, due to the over-heated melted uranium reacting with steam. *The roof of the reactor was blown off releasing radioactive radiations 200 times more than those released by the Hiroshima and Nagasaki nuclear bombings together—in the second world war**. This explosion had scattered radio-active radiations over a vast area, as far as Europe. So much so that the radiation levels in Scotland (UK) were found to be 10,000 times more than the normal, for a while. Though the immediate explosion killed only around 30 people, yet the large scale *genetic effects* of the resulting radio-activity shall continue for generation, and some estimates put the death toll of clean-up workers alone at 7,500.

---

*In the second world war, U.S.A. has dropped atomic bombs on Japan, killing more than 2 lakh people, forcing Japan to surrender. The first bombing on Hiroshima was done on 05.03.1945 ; while Nagasaki was bombed three days later.

Inspite of several measures taken to avoid the long range genetic effects of radioactivity in these areas, and more than 50 lakh people being constantly undergoing regular medical check ups, the crisis of confidence is growing up, as birth defects in farm animals and increased ill health among children and villagers, are being reported.

The accidents in nuclear reactors or nuclear stations differ from the explosions of atomic bombs, since a nuclear reactor cannot explode with uncontrolled chain reaction like an atomic bomb. This is because of the fact that uranium fuel in an atomic bomb is much more enriched than the uranium fuel present in a regular commercial nuclear reactor. What happens in a nuclear reactor is, that due to insufficient coolant, the temperature becomes so high in the core of the reactor that the fuel rods actually melt to liquid state. In a complete reactor melt down, the extremely hot (about 2700°C) molten uranium fuel rods melt through the bottom off the reactor, and actually sink as deep as 50 ft into the earth, beneath the power station. The molten uranium reacts with the ground water, producing large explosions of radio-active steam and debris that would affect nearby towns and population centres.

15.14. Classification of Radioactive Wastes and Their Disposal

The radioactive wastes are usually classified into the following three categories, depending upon their radioactivity levels :

(*i*) *High Level Wastes* (HLW) ;

(*ii*) *Intermediate Level Wastes* (ILW) ; and

(*iii*) *Low Level Wastes* (LLW).

The *high level wastes* are those which have radioactivities measured in *curies* per litre* ; *intermediate level wastes* may have radoactivities measured in *millicuries per litre* ; and the *low level wastes* have their radioactivities in *microcuries per litre*. The intermediate level category, however, is usually ignored, and only High Level and Low Level radioactive wastes are referred to.

15.14.1. High Level Radioactive Wastes and Their Disposal. High level solid wastes are produced in nuclear power plants in the form of used or spent fuel. This irradiated nuclear fuel is in solid form, consisting of small fuel pellets in long metal tubes. The spent fuel is produced after about 2-3 years, when the reactor fuel is changed by fresh fuel. This spent fuel is usually stored for about 3-4 months at the reactor site, and later transported to the *reprocessing units*. This temporary storage of the spent fuel for a few months at the reactor site helps in decaying those radionuclides that do have short half lives. The transportation of this solid waste to the reprocessing plant has to be done with extreme care in shielded casks.

***Curies** (Ci) or **Becquerel** (Bq) represent the No. of disintegrations per second of the given radioactive material, and hence the radioactivity of a material is measured in these units. The Bq represents one disintegration per second (dps) ; while Ci represents 3.7×10^{10} disintegrations per second.

$\therefore$ $1 \text{ Ci} = 3.7 \times 10^{10} \text{ Bq}$

$\therefore$ $1 \text{ Pico Curie} = 1 \text{ P.Ci} = 10^{-12} \text{ Ci} = 3.7 \times 10^{-2} \text{ Bq} = 0.037 \text{ Bq}$

or **1 Bq (S.I. Unit) = 27 pCi.**

The high level solid waste eminating from the nuclear reactor, usually contains plutonium 239, which is formed when uranium-238 (the principal component of nuclear fuel)* captures a neutron. Since plutonium-239 has a very long half life of about 24000 years, the nuclear waste poses great difficulties in disposal, since it is likely to remain a great radioactive hazard for a very long period, unless plutonium-239 is separated out from this waste. Although, it has been possible to separate the plutanium-239 isotope in the reprocessing plant, and then to use it for breeder reactors of other uses (Pl see Fig 15.2), yet this technology is just not easily available, as is the case with the technology of uranium-enrichment (when content of U-235 has to be increased from about 0.7% as available in the ore – to about 2 to 3%). The nuclear club countries can not easily dispense with these technologies, since plutonium-239 is a critical ingredient in the production of ***nuclear bombs*** (or atomic bombs) ; and hence the world could be in the danger of proliferation of nuclear weapons, if the technology for isotope separation falls into the wrong hands. The plutonium waste, due to its long half life, thus, always poses great problems of disposal, in nuclear power generation.

When the technology of isotope separation is available, then the used fuel from the nuclear reactor is taken to the reprocessing plant, where uranium and plutonium isotopes along with some other radio-nuclides are separated chemically from the spent fuel. The recovered uranium isotopes are recycled to make the fuel for the nuclear power reactor, while plutonium-239 isotope is used as a fuel for breeder reactors of other uses. Other recovered radio-nuclides are also sent for medical and other uses.

Even after certain isotopes are recovered from the spent fuel, the left out mixture still contains high levels of radioactivity, and represents a *high level solid waste*. Moreover, during various stages of the reprocessing plant, a lot of high level radioactive waste-water is produced. This *high level liquid waste* need careful handling and treatment for ultimate conversion to solid form for final disposal. Before they are solidified, these high level waste-waters are usually stored in underground multi-layered thick concrete tanks for about 5 years.

As these waters have high temperatures and remain boiling, cooling is provided by coils for condensation of the vapours of the boiling waste-waters.

Extreme care is taken to prevent leakage from these underground tanks, called ***holding depositories (vaults).*** The liquid wastes, after being stored in the underground tanks for about 5 years, are solidified by using advanced technological methods. These waste-waters are usually calcined in a fluidised bed process, and the calcinated powder is then incorporated into glass. The spray calcination processed liquid waste is atomised by spraying steam or compressed air in a stainless steel column, which is kept at about 600 to 800°C. The calcine so formed is then melted with borosilicate glass to be converted into solid mass.

All the high level radioactive solid wastes are kept in especially designed containers and transported for burial to the permanent deep disposal sites under the ground. The favoured sites are some special geological formations like bedded salt, basalt, gneiss, or schist, where it is believed that these wastes

*The fissils isotope U-235 which provides the energy in the nuclear fuel is only about 2-3% of the fuel mass, and the rest of the fuel is the non-fissils U-238.

will maintain their integrety, and will remain in place for geological periods of time.

As a matter of fact, different sites are identified for holding high level and low level radioactive wastes. The only site identified for long time holding of high level wastes in USA is in the *Yucca Mountains in the Nevada desert.**

This lone high level waste site of USA is designated as a monitored, *retrievable storage facility,* which implies that the waste is to be retrieved after a long period of time, when its radioactivity dies down and reaches a safe level. Some high level solid wastes, having comparatively lesser half life (say only a few centuries) can, sometimes, be kept in safe leak proof containers, and disposed of by shallow burial or sea dumping.

Every year, the world's 442 nuclear power plants and reactors** produce about 12000 tonnes of high level radioactive waste, which is needed to be disposed of safely. The nuclear power production is further likely to grow in future, which will further increase the production of nuclear waste.

15.14.2. Low Level Radioactive Wastes and Their Disposal. Low level radioactive waste-waters are produced frequently in *day to day operations of nuclear power plants,* in various operations, such as during : core water exchanges, maintenance and clean up operations, samplings, blowing down of steam generators, washings of contaminated tools, components, clothings, etc. Low level radioactive solid and liquid wastes are also produced when the uranium is extracted and prepared for nuclear fuel. The *mill tailing wastes* are the residues, remaining after the processing of natural ore to extract uranium and thorium. The radioactivity in the resulting solid and liquid wastes is generally high enough, but is lower than those of high level wastes. These intermediate level wastes are produced in large quantities, and finally disposed of with low-level wastes after giving some treatment.

Low level wastes are also produced from *research laboratories, hospitals, diagnostic centres, army ammunition plants,* etc. The low level wastes, besides contaminated waste-waters, may contain solid wastes like protective clothings, tools, filters, rags, medical tubes, and many such waste items.

Very low level radioactive waste-waters are usually produced in high volumes, which do not permit their complete containment, and are, hence, usually disposed of into the open environment including water bodies, with or without any treatment. Care must, however, be taken to see that the discharged radio-active isotopes do not concentrate in the food chain, or raise the background radiation levels to any appreciable value.

The low-level radioactive waste-waters are generally treated by the usual waste-water treatment methods, such as *flocculation, precipitation, filtration, ion exchange, and thermal evaporation*. The *precipitation* of radio-nuclides can be achieved by using aluminium salts, iron salts, tannic acid with lime, phosphate with lime, ferrocyanides, and excess lime-soda ash. While the decanted effluent is disposed of in any manner, the radioactive sludge so formed is sent to the special burial sites.

---

*This site was selected in 1987 after hectic search for 47 years, and was approved as the final disposal site for spent nuclear fuel, by the House of Representatives, USA.

**USA alone is having around 102 nuclear reactors.

The *ion-exchange* procedure can be followed in case the volume of waste-water is small, and the total solids content is up to 1000 mg/L, and when a simultaneous final polishing of wastes is also required.

The *thermal evaporation* involves continuous evaporation of waste-waters in evaporators, made of special stainless steel. The condensed vapour (condensate) is discharged as usual, while the concentrated radionuclides remaining in the evaporators are taken to the burial sites.

The concentrates or sludges obtained from the above mentioned conventional treatments of low level radioactive waste-waters can, sometimes, be disposed of in hydro-fractures in the ground by mixing the waste with cement, etc. In this technique, a well is drilled to the desired geological formation and provided with a casing. The casing is perforated at desired levels. The pressure is applied through the perforated casing, as to fracture the desired geological formation. The radioactive waste sludge, mixed with cement, fly ash, attapulgite, illite, beta gluconolactone and tributyl phosphate is then injected in to the fractured rocks, which spreads in the rock fractures in thin sheets and forms a dense solid for permanent retainment of wastes under the ground at selected isolated sites.

In addition to the above mentioned conventional methods used for reducing the volumes of low activity waste-waters, various new methods are being developed for *volume reduction* of intermediate level waste-waters, such as those eminating from army ammunition plants. Such highly contaminated waste-waters can now be treated to remove **depleted uranium (DU)** through a newly discovered process, called the **Meta Lock process,** developed by KEECO, an environmental company of USA that offers water management systems based on its silica micro encapsulation process. *Meta-Lock* is highly specific towards radioactive ions and is able to reduce high volumes of diluted radioactive effluents to a small quantity of radioactive sludge and non-radioactive water. The volume of radioactive material is reduced by 95-96%. The process involves a calcium and silica based powder composed of various chemically active groups that result in the encasing of the radioactive contaminants into an inert sludge, when brought into contact with moisture or waste water. The process is controlled by using pH as the primary process parameter. Water treated by Meta-Lock has a concentration of uranium below 1 ppm (instrument detection limit) and is also non-radioactive. The universal adoption of the process, however, is still under initial stages.

Invariably, all such new processes are very complex and expensive.

Low level or Intermediate level wastes may, for safe disposal, be kept in containers, which can be stored in *water-tight R.C.C. vaults*, which are properly designed from engineering considerations, and are constructed either above the ground or below the ground. *Before being placed in containers, the radioactive wastes must, however, be reduced in volume by techniques like evaporation (for liquid wastes), and incineration and compression (for solid wastes).*

The important low level waste depository sites in USA are : *Maxey Flats-Kentucky, Sheffield-Illinois, West Valley—New York*, etc.

Sometimes, the radioactive liquid wastes (low level wastes of course) eminating from the nuclear boilers, etc., have been discharged by man into the open sea, although the emphasis is always on *recycling* and *containment in containers placed in safe deposit vaults*, as stated earlier. While discharging such waste waters in oceans, one should, however, atleast make sure that the discharged isotopes do not settle down and get trapped in the sea bed, as may happen with plutonium-239, which gets bound in the sediment. Moreover, the radioactivity concentrations in such water bodies, should be continuously watched to keep their radio-activity levels at much-much below the limiting standards.

15.14.3. Disposal of Nuclear Wastes in India. The position of disposal of nuclear wastes in India is, though not very acute at present, but is liable to become so in future, as our nuclear programme gets momentum. We have three sources of nuclear wastes ; *i.e.* (*i*) *the nuclear mines* ; (*ii*) *the nuclear power plants* ; and (*iii*) *the nuclear research reactors*.

Our nuclear mines for extracting uranium ore are located at *Jaduguda* village in Bihar State. Once the uranium ore is extracted at these mines, it is sent to a facility centre in Hyderabad for processing to secure uranium. Earlier, the waste water from the processing plant was being dumped in a nearby pond, but since it contaminated some nearby wells and residents protested, the waste is now returned to Jaduguda mines and dumped inside the mine campus.

The wastes from our nuclear power plants are presently being dumped in each power plant. The spent fuels, from which weapon grade uranium and plutonium are produced after reprocessing, are also being stored at the plants site, until it is removed for reprocessing. *This waste is kept for years in chilled water, to cool*. The nuclear waste after reprocessing has to be suitably disposed of. According to one estimate, India has so far produced about 17000 tonnes of high level waste from reprocessing on spent nuclear fuel.

Our nuclear research reactors, which are controlled by the **Bhabha Atomic Research Centre (BARC) Trombay** are also producing nuclear wastes.

Inspite of India producing a lot of nuclear waste, it still needs a safe site to store the nuclear wastes generated by its vast establishment, but the problem is that no country in the world has found a totally safe fool-proof way to keep the deadly waste buried safely. The Bhabha Atomic Research Centre is presently having a *solid waste storage surveillance facility* in operation at Tarapur, Maharashtra. This facility is used for the interim storage of solidified high level radioactivity waste products, produced in the *waste immobilisation plant* at the Tarapur nuclear power station. A *waste immobilisation plant is also under construction at Trombay, and work has begun on another plant at Kalpakkam, Tamil Nadu*.

In its search for a safe underground storage facility for the radioactive wastes, BARC had once (in 1990s), examined the possibility of using an abandoned gold mine in Kolar Gold Fields in Karnataka (one of the deepest mines in the world). It even set up a laboratory to simulate the high temperature conditions, which are likely to be generated by the nuclear wastes. But due to the presence of underground water, the site was abandoned. The BARC is still searching for a deep site, within the thick granite rock deposits (upto 1 km thick) of Rajasthan, in barren land of the desert village, *Sanawada*, just a few kilometres from Indias nuclear testing site, *Pokhran*. After several failures, this place has been selected as the most suitable place for setting up a laboratory in preparation for the site, which could be used to store nuclear wastes from various nuclear units. The granite rock samples upto 150 m depth, which were excavated/drilled in the area were, infact, throughly studied for heat resistant and hydrological characters before declaring this area as a fit place for developing the country's nuclear waste depository. Such surveys and testing is essential, since the rocks of the depository have to withstand the high temperatures of radioactive wastes, and the area should not have any ground water source, as not to cause its contamination by the wastes. Keeping all such factors into view, the Indian Govt. in August 1999, has declared that the country's nuclear waste would be deposited in parts of Rajasthan and Madhya Pradesh's granite belts. The selection of the site in village Sanawada near Pokhran, was, however, opposed by the local NGOs and the villagers, and the drilling/sampling work in second phase below 150 m depth, was suspended by the Govt. The notification of safe depository for Indian high level radioactive wastes, thus, still continues to be a dream.

Disposal of nuclear wastes may, thus, continue to be a serious problem with every country, including India, since no country wants a disposal site on its land.

An international effort is, therefore, on for cooperative disposal. In July 2004, the Russian President, Mr. Vladimir Putin, called for regional disposal sites, which suggestion was readily accepted by the *International Atomic Energy Association (IAEA)*. However, no country, except Russia, offered such a dumping site on their land. Russia offered to store spent fuels of other countries at its *Mayak Chemical Combine Facility* in the Southern Urals. The cost demanded for such an offer was, however, exhorbitantly high. There were no takers for the offer, since Russia through this offer, hoped to secure 20 billion dollars for modernising its nuclear establishments. No country has accepted this offer so far.

The requirement of such storage facilities will however, increase in future, as the USA's initiative for dismantling of nuclear weapons gets momentum. Keeping this in mind, the Russian Parliament, in the year 2001, has legalised the import of nuclear spent fuel. Russia is also intending to import and store such fuels in future, at a heavy price, of course. The Russia is also busy in searching new sites in its country to offer storage facilities for spent nuclear fuel, in future, on a large scale.

Krasnokamensk city, 7000 km east of Moscow, has been suggested as a new possible site for major spent fuel repository. Russia intends to earn huge money through such storage facilities. USA alone, is going to spend about $450 million under its *Global Threat Reduction Initiative (GTRI)*, which envisages call back of all the highly enriched uranium, which was supplied by the former USSR to 17 nations. The spent fuel of the uranium will be retrieved by the year 2010. Both will be shipped to Russia for storage—a huge project, on which USA will spend about $450 million. USA, infact, especially after the terrorist attack of 9/11 on its World Trade Towers, is very keen to trace and retrieve all the nuclear fuel that was supplied by the USSR ; a daunting task indeed.

15.14.4. Discharge of Low Level Radioactive Wastes in to the Sea. Inspite of the fact that more and more emphasis is now being laid on reprocessing and concentrating the radioactive wastes for their safe disposal in expensive containers to prevent it from reaching the open environment, yet because of the high cost of chemical reprocessing and non-availability of technology, smaller amounts of radioactive wastes have customarily been released into the sea in carefully controlled amounts off the west coast of Britain. For example, the radio isotopes like ^{95}Zr, ^{95}Nb, ^{106}Ru, ^{137}Cs, ^{144}Ce, $^{238,239,240}Pu$, have been released in fresh sea over the past thirty five years. The amounts of these isotopes are, however, limited in accordance with the rules of *International Commission for Radiological Protection (ICRP)*. The amounts must be low enough to fulfill the following *objectives* :

(*i*) Regardless of cost, the public must never be exposed to levels above the ICRP recommended limits.

(*ii*) Regardless of cost, the entire population must not receive radiations more than 1 rem per person over 30 years.

(*iii*) Keeping the cost in mind, doses should be kept as far below the mandatory limits as possible.

When radioactive waste is discharged into the sea, one must make sure that the discharged isotopes do not become trapped or concentrated in the sea bed, as happens with Plutonium-239, almost whole quantity of which, gets bound in sea bed near the point of outfall, as shown in table 15.6.

Table 15.6. Concentration of Plutonium-235 in Sea Bed

Distance from outfall in km	*Concentration of Plutonium-235*		*Concentration factor in sediment*
	in water pCi/l	*in sediment pCi/g*	
1	0.48	36	7×10^4
9	0.46	12	3×10^4
110	0.04	0.04	0.1×10^4

The discharge of Plutonium-235 must, therefore, be absolutely minimised, or completely stopped.

It is possible to monitor the amount of radioactive isotopes in the sea by making use of *biological monitoring organisms*, which live attached to the sea bed for long periods in the same area. In Finland, the authorities are studying *mussels* (*mytilus*) inorder to obtain an early warning of any environmental danger from their nuclear power programme.

The back ground radiation in the sea also needs observations, as it is significantly affected by nuclear waste discharge. The normal level of this background radiation is about 350 milli rem per year. In the Irish sea near the Windscale reprocessing plant, the background radiation has been observed to be as high as about 600 milli rem per day. Controlling the background radiation in the sea becomes very important for fish and sea animals, since higher radiation levels (such as about 10 rem) may bring about mutation or genetic changes ; and still higher levels (of the order of 100 rem) may lead to large scale mortality.

The legislation which controls the handling and disposal of radioactive wastes in India, as stated in article 14.8.1, is the *Atomic Energy Act, 1962*.

E-WASTE

15.15. Definition of E-Waste

Electronic waste, popularly known as *E-waste*, can be defined as electronic and electrical equipments/products (including the connecting power plugs and batteries), which have become obsolete due to

(*i*) changes in fashion, style and status ; and

(*ii*) nearing the end of their useful life.

15.16. Classification of E-Waste

E-waste encompasses ever growing range of obsolete electronic devices, such as computers, servers, main frames, monitors, TVs and display devices, telecommunication devices such as cellular phones and pagers, calculators, audio and video devices, printers, scanners, copiers and fax machines besides refrigerators, air conditioners, washing machines, and microwave ovens. E-waste also covers recording devices such as DVDs, CDs, floppies, tapes, printing cartridges, military electronic waste, automobile catalytic converters, electronic components such as chips, processors, mother boards, printed circuit boards (PCB's), industrial electronics such as sensors, alarms, sirens, security devices, automobile electronic devices, etc.

15.17. Adverse Health and Environmental Impacts of E-waste and its Improper Disposal

Electronic and electrical equipments are made up of several components, many of which contain toxic substances, like *lead, chromium, mercury, beryllium, cadmium, acids, and plastics,* etc. These toxic substances can have highly adverse impacts on human health and the environment, if not handled properly. *Often these hazards arise due to improper recycling and rudimentary processes used for disposal of E-waste.*

For example, *Cathode Ray Tubes* (*CRTs*) of Television sets have high content of carcinogens such as *lead*, *cadmium, barium*, *phosphor* and *other heavy metals.* When disposed carefully in a controlled environment, they do not pose any serious health or environmental risk. However, breaking, recycling or disposing CRTs in an uncontrolled environment without the necessary safety precautions can result in harmful side effects for the workers and release toxins into the soil, air and groundwater.

Similarly, improper breaking or burning of *printed circuit boards* (*PCBs*) and switches may lead to the release of *mercury*, *cadmium* and *beryllium* which are highly toxic to human health.

Another dangerous process is the recycling of components containing hazardous compounds such as halogenated chlorides and bromides used as flame-retardants in *plastics*, which form persistent dioxins and furans on combustion at low temperatures (600-800°C). **Copper**, which is present in printed circuit boards and cables, acts as a catalyst for dioxin formation when flame-retardants are incinerated. The PVC sheathing of wires is highly corrosive when burnt and also induces the formation of dioxins. A study on burning printed wiring boards in India showed alarming concentrations of dioxins in the surroundings of open burning places, reaching 30 times the Swiss guidance level.

Land-filling of e-waste, one of the most widely used methods of disposal, is prone to hazards because of leachate which often contains heavy water resources. Even state-of-the-art landfills which are sealed to the long-term are not fully safe. Older landfill sites and uncontrolled dumps, however, pose a much greater danger of releasing hazardous emissions. Mercury, Cadmium and Lead are among the most toxic leachates. Mercury, for example, will leach when certain electronic devices such as circuit breakers are destroyed. Lead has been found to leach from broken lead containing glass, such as the cone glass of cathode ray tubes from TVs and monitors. When brominated flame retarded plastics or plastics containing cadmium are landfilled, both PBDE and cadmium may leach into soil and groundwater. In addition, landfills are also prone to uncontrolled fires which can release toxic fumes.

An overview of the toxic substances released along with their adverse health effects, due to improper operations involved in rudimentary disposal methods generally adopted for disposal of E-Waste, are shown in table 15.7.

Table 15.7. Environmental and Occupational Hazards posed by Disposal of Certain Categories of E-waste

S. No.	*Component*	*E-waste product and operation involved in disposal*	*Adverse Health Effects*
(1)	(2)	(3)	(4)
1.	*Lead*	Mechanical breaking of Cathode Ray Tubes (CRTs) and Removal of solder from microchips, releasing lead as powder and fumes.	A neuro-toxin that affects the kidneys and the reproductive system. High quantities can be fatal. It affects mental development in children.
2.	*Chromium*	Used to protect metal housings and plates in a computer from corrosion.	Inhaling hexavelant chromium or chromium VI can damage liver and kidneys, and cause bronchial maladies, including asthmatic bronchitis and lung cancer.
3.	*Beryllium*	Found in switch boards and printed circuit boards.	It is a carcinogen and causes lung diseases.
4.	Mercury	Released while breaking and burning of circuit boards and switches.	Damages brain and kidneys, Impairs foetus growth and harm infants through mother's milk. Mercury in water bodies can form methylated mercury through microbial activity, which is toxic and can enter human food chain through aquatic life forms.

Contd...

(1)	(2)	(3)	(4)
5.	*Cadmium*	It is released as powder while crushing and milling of plastics, CRTs and circuit boards.	A carcinogen, long term exposure causes itai-itai, which affects kidneys and softens bones. It may be released with dust, entering surface water and groundwater.
6.	*Plastics*	Found in circuit boards, cabinets and cables. Burning PVC a component of plastics also produces dioxins. Even the dust on computer cabinets contains BFR.	They contain carcinogens. *BFRs* or *Brominated Flame Retardants* give out carcinogenic brominated dioxins and furans. Dioxins can harm reproductive and immune systems.
7.	*Acids*	Sulphuric and hydrochloric acids are often used to separate metals from *circuit boards.*	Fumes contain chlorine and sulphur dioxide which cause respiratory problems. They are corrosive to the eye and skin.

15.18. Menace of E-Waste in India

India produces about 4,00,000 tonnes of electronic waste each year of its own. In addition, undefined quantities of E-waste are received in this country through legal or illegal transboundary movements from abroad, particularly in the form of computer waste, as reusable components.

With the extensive and increasing use of computers, mobile phones and other electronic equipments, and people dumping old obsolete goods for new improved ones, the amount of E-waste generation in this country has been steadily increasing. So much so that India generated about 3,30,000 tonnes of E-waste in 2007, which is equivalent of 110 million laptops. About 10% of the E waste generated every year is recycled and the remaining is refurbished.

Due to the lack of government legislations on E-waste and the absence of any standards for their handling, recycling or disposal, these toxic hi-tech products mostly end up in the hands of *informal operators* for recycling and recovery of metals and other costly biproducts, through rudimentary and risky processes, under unhygienic uncontrolled and open environment. During these processes, open burning of plastics and use of acids for recovery of metals, are often used, causing a lot of air pollution through the evolution of dioxines and discharge of acidic and other wastes into the rivers or in landfills. Seelampur colony in Delhi and Moradabad town in U.P. have become the hubs of such informal operators who are breaking and burning various electronic parts to extract metals, causing a lot of air, water and noise pollution, besides endangering the health of workers, who are involved in these manual operations.

Inspite of the fact that such a huge quantity of E-waste is being generated or handled in the country, there does not exist many organised *formal operators*, having proper handling and recycling facilities. The first such facility has of course been created by a company called the *Attero Recycling*, which has established its recycling plant near Roorkee. The company claims it to be a ₹ 35 crore plant, having recycling technology, claimed to have been developed in their laboratories. There are various units in its plant, like *shredder*, *separator*, *smeltor*, and *electro-refinery*, with research and development facility. This plant can process up to 36000 tonnes of E-waste in a year, though at present, it is getting only 600 tonnes of E-waste. Table 15.8 highlights the major differences between an informal recycler and a formal organised recycler in their operations involved in dismantling and disposal of a discarded computer.

Evidently, the cost incurred by an organised formal recycler will be much higher than that incurred by an informal recycler, while the pollution and hazard potential of a formal operator is much less than that of an informal operator. Unless the Government comes out with a law, which forces the informal operators to unite, organise and form societies to create factories and plants for safe, proper and scientific disposal of E-waste, the imminent threat to human health and environment will not abate.

Table 15.8. Informal Vs Formal Recyclers in their Operations Adopted for Disposal of a Dismantled Computer

Note : Dismantling a computer into its basic parts is not Hazardous. Then begins the dirty work	
Informal Recycler / Operator	*Formal Recycler / Operator*
• Cathode ray tubes' (CRTs) are broken down manually to separate its components—glass, metal and copper. The glass, comprising lead, is sold to bakeries or bangle makers. Since it retains heat, the glass goes into the base of ovens. Phosphors, if inhaled can be toxic. The old CRTs are also sometimes sold to non-branded television makers.	• Components of cathode ray tubes are separated by heating in a closed chamber, which sucks out phosphors from the components. They are then crushed in shredder machines. The glass that contains lead is sold to companies that manufacture CRT's.
• Circuit boards have gold-plated brass pins, microchips and condensers. Heating separates these components. Fumes released during heating are toxic. Gold-plated brass pins are soaked in acid to recover the gold and brass separately. Microchips and condensers are heated in big containers filled with acid to extract metallic parts.	• Circuit boards are crushed in shredder machines. They are sent to approved smelters abroad, where after smelting, at 1200°C, the metals in the circuit board collect together. The plastic gets burnt. Since smelting is carried out in closed chambers at high temperature, it is not hazardous. The metals—lead, copper, nickel, tin, gold, silver, palladium—are than separated by electro-refining.
• No safety precautions are followed. Informal recyclers paid ₹ 200-300 daily in Seelampur; ₹ 100-150 in Moradabad to their workers.	• Protective equipments—gloves, masks, shoes, caps—are provided to employees. ₹ 5,000 per month paid to unskilled workers.
• Minimal capital investment required. Cost includes price of purchased e-scrap, bribes given to transfer it across state borders, Cost incurred on set up and shops, and rent for the work space.	• Investment for a dismantler is about ₹ 30 lakh and for a recycling plant, it may be as high as about ₹ 25 crore.

Reacting to the above necessity of urgent action needed on behalf of the government, the Ministry of Environment and Forests, GoI, has notified the ***E-waste (Management and Handling) Rules, 2011,*** on May 12, 2011, one year after the Draft Rules were published for public response on May 14, 2010. These rules have come in to effect wef May 01, 2012, one year after they were notified. The one year gap was given to the various stakeholders, including manufacturers, collectors, dismantlers and recyclers, a chance to put their mechanisms of e-waste management in place.

These Rules fix responsibilities of the Producers, Collection centres, Consumers or Bulk Consumers, Dismantlers, Recyclers separately. These rules also provide f*or seeking Authorization and Registration for handling E-wastes* from the State Pollution Control Boards or Pollution Control Committees in U.Ts. State Pollution Control Boards and Committees are given powers to suspend or cancel such authorizations. These rules also describe the procedure for gant of registration

These rules also have chapters on Storage of e-wastes; Reduction in the use of hazardous materials in the manufacture of electrical and electronic equipments, etc.

Certain salient provisions of these Rules are shown in Table 15.9.

Table 15.9. Salient Provisions of the E-waste (Management and Handling) Rules, 2011

- **Producers extended responsibilities** include collecting e-wastes generated from the end of life of their products, and to ensure that such e-wastes are channelized to registered dismantlers or recyclers. Producers, shall hence ensure such collection and channelization by their authorised collecting agencies or by their collection centres, and shall finance and organise a system to meet the costs involved in the environmentally sound management of e-wastes generated from the 'end of life' of its own products and historical wastes available on the date from which these rules come in to force. The financing arrangement shall be tranparent. The producers may choose to establish such a system either indivisually or by joining a collective scheme. They shall also create public awareness of such facilities and collection centres.
- Producers need to comply with the threshold limits specified for the use of certain hazardous substances like Lead, Mercury, Cadmium, Chromium, Polybrominated biphenyl or diphenyl ethers, in the production of electrical and electronic items. However, manufacturing of electrical and electronic equipment for the use of defence and other suh strategic applications are excluded from this provision.
- Every dismantler and recycler shall have to be registered with State Pollution Contol Board or Pollution Contol Committee.
- State Pollution Control Boards or Committees, responsible for grant of authorization, monitoring compliance of authorization and registration conditions, will take action against violations of rules. However, there is an emphasis on the need to hear out the other party-the violator-before any action is taken against that party. The Central Pollution Control Board shall monitor the compliance conditions stipulated for granting registration.

These rules, however, lack in listing out environment friendly technologies to be used while disposing of e-wastes. This is primarily because such

*Complete text can be seen on website:

http://moef.nic.in/downloads/rules- and regulations/1035 e_eng.pdf.

technologies are in the evolving and experimental stages. It is also expected that the producers(manufacturers) of electrical and electronic equipments & components shall also gear up to bring changes in the compositions of their products to reduce the use of harmful metals and substances, and will certainly try to develop their own methodologies for recycling and reusing their own obsolete products to make gains.

Reputed companies like "DELL" clims to have already established a thorough mechanism for dealing with its own electronic wastes, even in the absence of any legislative compulsions. In 2006, Dell started a programme in India, which involved picking up the unit that needs to be recycled from the consumer. In another effort to encourage recycling of personal computers, Dell has launched a discount coupon where consumers can send their old computers to the firm for free recycling and redeem a coupon of ₹ 1000/- on the purchase of their next Dell computer. All such measures need to be initiated by even smaller companies of electrical and electronic components to encurage general public to return their obsolete items at some token price to the producers rather than selling them over to the kwariwallas.

PROBLEMS

1. (*a*) Define and explain the term 'biomedical wastes.'

(*b*) How are bio-medical wastes endanger biological life including the humans ?

(*c*) How are bio-medical wastes managed successfully, as to avoid their harmful effects on humans ?

2. (*a*) Explain the term 'colour coding' as is used in relation to the biomedical wastes, and how does it help in safe disposal of bio-medical wastes ?

(*b*) How are sharp wastes taken care of in safe collection of bio-medical wastes.

(*c*) How are hazardous biomedical wastes labelled for easier identification ?

(*d*) How are biomedical wastes stored on site till they are transported for treatment ?

3. Enumerate and explain the various methods which are used for safe disposal of hazardous solid and liquid wastes from hospitals and nursing homes.

4. Write brief notes on the following :

(*i*) Autoclaving method of disposal of biomedical wastes.

(*ii*) Microwaving method of disposal of biomedical wastes.

(*iii*) Hydorclaving Vs. Microwaving treatment methods for disposal of biomedical wastes.

5. How are hazardous solid and liquid wastes disposed of by incineration ? Name and describe the working of the two widely used types of incinerators which are used for incinerating the liquid and solid wastes.

6. (*a*) Differentiate between solid wastes and liquid wastes.

(*b*) How are liquid wastes incinerated ?

(*c*) Name and describe the working of a conventional rotary kiln incinerator, which is used for disposal of solid hazardous wastes : Can such an incinerator be designed to incinerate aqueous wastes, and if yes, then how ?

7. Draw a neat sketch and explain the working of a conventional kiln incinerator ; and indicate and describe possible variations.

8. (*a*) Differentiate and explain between the :

(*i*) Parallel and counter current flow types of kiln incinerators ;

(*ii*) Slagging and non-slagging types of kiln incinerators.

(*b*) Write down an equation which can be used to compute the waste retention time of a rotary kiln of an incinerator.

(*c*) Discuss the modern advancements in incineration.

(*d*) Describe the air polluting effects of incinerators which are used for disposal of hospital wastes.

9. (*a*) Discuss the importance of human resource management in the disposal of biomedical wastes.

(*b*) How can you minimize biomedical wastes ?

(*c*) Enumerate the various methods which are used for disposal of biomedical wastes, and what are the environmental impacts of each of these methods of disposal.

10. Write brief notes on :

(*i*) Colour coding and proper collection of hospital wastes.

(*ii*) Methods of disposal of hospital wastes.

11. (*a*) Define the following :

(*i*) Atom (*ii*) Atomic number of an element

(*iii*) Atomic mass number of an element (*iv*) Radioactive isotopes

(*v*) Half life of radio-nuclides.

(*b*) Explain the various types of radioactive radiations. How are they evolved and what are their properties ? How do these radiations impact the biological environment ?

12. (*a*) Define and explain 'radionuclides', indicating their characteristic features.

(*b*) What is nuclear fission ? How does it help in producing nuclear electricity ?

13. (*a*) Define and explain radio-activity and radioactive radiations.

(*b*) What are the old and new S.I. units of absorbed radiation, and how are they related ?

(*c*) How is the damage produced in human body by radioactive radiations, measured ? Indicate the relation between the old and new units which are used to indicate such a damage.

(*d*) What types of diseases are caused by radioactive radiations in human beings, and how and when such exposures occur ?

14. Enumerate and briefly explain the various sources of radioactivity in the environment.

15. (*a*) How are radioactive wastes generated in production of nuclear power ? Explain the uranium fuel cycle in this context.

(*b*) How are radioactive wastes classified depending upon their levels of radioactivity, and how are these different level wastes disposed off ?

16. Differentiate between high level and low level radioactive wastes, indicating the methods of safe disposal of both these types of wastes. How are radioactive wastes being disposed off in India ?

17. Write notes on the following :

(*i*) Radioactive isotopes, their characteristics, and life spans.

(*ii*) Radioactive disintegration and radio-active radiations.

(*iii*) Storage of high level radioactive wastes.

(*iv*) Impacts of radioactivity on biological environment.

(*v*) Basel convention.

18. Define and explain the following terms :

(*i*) Relative Biological effectiveness (RBE) of a radiation

(*ii*) Annual permissible radiation exposure on humans, and Emergency Reference level

(*iii*) Cosmic rays

(*iv*) Back ground radiations

(*v*) Holding depositories and vaults.

19. Write a brief note on E-waste management.

16

Ventilation of Buildings for Controlling Indoor Air Pollution

Environment, as we all understand, is nothing but *our surroundings*, which can be badly affected by *smokes, smells, dusts, gases, oxygen deficits, noises,* and *vibrations,* etc. When such *substances* or *actions* hazardously affect our environment, we call them as **pollutants,** and the process whereby the surroundings get adversely affected, is known as **environmental pollution.**

A public health engineer, who is responsible for removing all kinds of wastes of a society, is evidently responsible for cleaning the environment from such pollutants, and, thus, to ensure a healthy and wholesome surroundings, to ensure health and happiness for the people.

Environmental sanitation, which evidently means *cleaning of the environment,* therefore, becomes the major task of a public health engineer, and this task primarily, includes ; *collection and disposal of refuse and sewage from houses, buildings, and other public areas,* the subject which has already been dealt in the earlier chapters of this book. *Provision of sufficient and wholesome air to the buildings and residents* for controlling **indoor air pollution,** is also included as a work of environmental sanitation, and hence, usually, included in the subject of 'Public Health Engineering' ; more so, because it is the wholesome air, on which depends the health of the public. Under this context, we are adding this chapter on *Ventilation of Buildings,* although strictly speaking, the topic should normally be covered under 'Building Construction'.

However, before we discuss design aspects governing ventilation of buildings, we shall first describe the harmful effects caused by indoor air pollution and its present status.

16.1. Sources, Effects and Status of Indoor Air Pollution

In a developing country like India, **the most important source** of indoor air pollution is combustion of domestic fuel (such as *cow dung, wood,* and *crop residues)* used for cooking, on which 80% (1991 census) of our population relies. It has further been estimated by *Indian Council of Medical Research* (ICMR) New Delhi, that globally 30 lakh people die every year due to air pollution, out of which 18 lakh people die due to indoor air pollution in developing countries. In India alone, 5.89 lakh people die annually due to indoor air pollution (4.96 lakh in rural areas and 0.93 lakh in urban areas).

The indoor air pollution has infact, been found to be much worse than the outdoor air pollution, since a pollutant released indoor is thousand times more likely to reach the lungs than a pollutant released outdoors. It is generally the women and young children and infants who face the maximum adverse effects of indoor air pollution. In poor households, there is no separate kitchen, and people usually stay in the same room where they cook, or burn fuel to heat during winters. Moreover, women who work on cooking, and their young children particularly the infants, always necessarily stay at the place of cooking or burning of traditional fuel. This exposes them to continuous indoor air pollution over long hours. This proves worst for the children (up to the age of about 5 years), whose lungs are in the developing stages.

The Environment Health Scientists of California have, infact, strongly linked the indoor air pollution with acute respiratory infections and chronic obstructive lung disease in children below five years of age. Cross-country analysis by Anita Zaidi of Agha Khan hospital in Islamabad (Pakistan) also shows a strong link between traditional biomass fuel use and infant mortality. The experts have concluded *that traditional fuel emits large quantities of dangerous pollutants and are often burnt in poorly ventilated conditions. Hence they are responsible for substantial ill health in the country.*

Promoting the use of cleaner fuels, improved stoves and better ventilation of homes and kitchens should therefore be given top priority to reduce indoor air pollution. Moreover, children and pregnant women, who are most susceptible to ill effects of indoor air pollutions, need to be protected on top priority from indoor air pollution.

16.2. Definition of Ventilation

Ventilation may be defined as the art of *supplying air* to a given space, and also includes the art to *remove* the old vitiated air from that space.

In order to understand as to how a particular space is to be provided with *sufficient and wholesome air* for its residents, it is necessary to understand as to how a particular space gets affected by the manner in which it is occupied by humans or other living creatures.

16.3. Effects of Occupancy of a Space

When an enclosed space is inhabited by humans and/or animals, etc., the following effects are produced :

(*i*) *Oxygen content* of the space *reduces* ;

(*ii*) *Carbon dioxide content* of the space *increases* ;

(*iii*) *Temperature* of the space increases ;

(*iv*) *Humidity* of the space increases ; and

(*v*) *Organic matter, and odours* in the space *increases* ; as discussed below :

(*i*) Fresh air usually contains 21% of oxygen (by volume)* ; whereas, the exhaled air contains only about 17% oxygen. The continuous respiration will, therefore, continuously reduce the available oxygen in an enclosed space. More is the number of occupants respirating in the given space, more deficiency of oxygen will be created. Such oxygen deficits cause headache, nausea, anoxenia, vertigo, etc. *In order to avoid such oxygen deficits in a given building inhabited by people, it is necessary to ensure supply of sufficient quantity of fresh air through ventilation.*

---

*The purest sea or mountain contains 20.999% of O_2 ; and this oxygen content in an over-crowded hall may fall down to as low as 20.65%.

(*ii*) Similarly, fresh air normally contains about 0.04% of carbon dioxide ; whereas the exhaled air contains as high as 50% of carbon dioxide. On an average, about 17 litres per hour of CO_2 is exhaled by an adult human being through respiration.

CO_2 is not toxic, and its smaller concentrations upto about 0.04% rather stimulates respiration ; but the higher concentrations of about 1.5% cause nausia, depression, and headache. When CO_2 concentration rises to about 2.5% or so, it may even extinguish a burning candle. Still higher concentrations of the order of 5% may cause fainting and prove fatal.

To avoid such CO_2 excesses in a building, it is necessary to provide sufficient entry of fresh air, through proper ventilation.

(*iii*) A person, on an average, produces about 300 B.Th.U. (≈ 69 kC) of heat per hour. A large part of this heat is lost by leakage through the rooms. Inspite of such leakage, the room temperature may considerably rise, if it is occupied by many persons. This is because of the reason, that about 38 B.Th.U. (8.7 kC) of heat can raise the temperature of about 38 cum-sized room (4 m × 3 m × 3 m, approximately) by about 1°C.

The heat so produced by occupants of the room is either to leak out to the atmosphere, or is to be ultimately absorbed by their bodies. And since a body temperature is to be maintained at 37°C (98.4° F), this extra heat has to be continuously dissipated by the occupants themselves. Higher outside temperatures retard this dissipation, reduces leakage, and thereby causing discomfort to the occupants. If things become severe, it may lead to heat exhaustion, heat cramp, and finally heat stroke. *Continuous supply of fresh air through proper ventilation is, therefore, very necessary to avoid heat effects.*

(*iv*) *Moisture*, in the form of *water vapour,* is released by inhabitants of a building @ about 1.08 kg per day (45 gm per hour). This increases the humidity of the room/building, thereby decreasing evaporation and cooling of the bodies of the inhabitants, and thus causing them discomfort. This moisture, supported by the dust particles, also acts as carrier of germs and pathogenic bacteria from one person to another, causing water borne diseases.

(*v*) Organic matter and odours are released by human beings from their skins, clothings, and mouths. The increase of these in an enclosed space, occupied by several persons, may cause nausia, headache, and may even aggravate ones existing illness.

In fact, the water vapour (*moisture*), released by the inhabitants of a building, is the biggest carrier of bacteria and germs, and therefore is responsible for many airborne diseases like influenza, flu, viral fever, tuberculosis, etc. The *dust particles* present in the room, also help the water vapour in carting organic matter, odours, and diseases from one person to another.

Even serious diseases like scarlet fever, whooping cough, measles, and lung diseases, etc. get spread through crowded buildings and shops. In order to avoid such bad effects of occupancy of a building, it is absolutely necessary to design it in a manner to ensure proper and sufficient ventilation, thereby avoiding all the above bad effects of contamination of existing air, by ensuring entry of sufficient new outside air, free from increased CO_2, reduced O_2, increased moisture, temperature and odours, etc.

16.4. Purpose of Ventilation

Ventilation, as stated earlier, is meant for supply of fresh air, and to replace the old hot used up (exhausted) air. The ventilation ensures the removal of bad effects of occupancy of an enclosed space :

(*i*) by providing necessary oxygen to remove oxygen deficit caused by respiration ;

(*ii*) by removing and diluting CO_2 in the air ;

(*iii*) by lowering down the temperature by removing hot used up air and replacing it by *colder* fresh air ;

(*iv*) by reducing humidity ; and

(*v*) by reducing body odours.

16.5. Extent of Ventilation Required and Ventilation Standards

In olden days, it was thought that the poisonous CO_2 released by inhabitants is mainly responsible for causing pollution in houses and other public buildings. It was also thought that CO_2 content increasing beyond 0.06% in the room, would cause very harmful effects. Accordingly, the ventilation standards were framed on limiting the CO_2 of the used up air to 0.06%, as against the normal content of 0.04% of fresh air.

This purification standard of 0.06% of CO_2 means that the air gets contaminated when its CO_2 content increases from 0.04% to 0.06%. In other words, addition of 0.02% of CO_2 in the air by respiration will contaminate the air, which further means that the addition of 0.02 cum of CO_2 will contaminate the air of a room of 100 cum capacity.

Moreover, since an adult person releases 17 litres (0.017 cum) of CO_2 per hour, we can, conclude that 0.02 cum of CO_2 will be released by an adult in

$\frac{0.02}{0.017} \times 60$ min = 70.6 minutes. Hence, an adult person, if kept in a closed room of 100 cum volume, will contaminate its air in 70.6 minutes.

If we consider a room of an average size 10 ft × 10 ft × 10 ft, *i.e.* 1000 ft^3 = 28 cum volume, we find that such a room of 28 cum capacity will be contaminated by a single adult in

$$= 70.6 \text{ min.} \times \frac{28}{100} = 20 \text{ min.}$$

In other words, a room of 28 cum volume will be contaminated by the presence of a single individual in 20 minutes, and hence air of this room will need to be *changed* at every 20 minutes ; *i.e.* 3 changes per hour will be required, *with total air requirement of 3 × 28 = 84 cum per hour.*

This will, however, be true only if there is no automatic ventilation through the loose door and window fittings, and also the ventilation that takes place when the doors of a room are occasionally opened. Due to such *automatic ventilations,* the requirement of air will get reduced. On the other hand, the requirement of air will increase if the number of inhabitants is more than 1, or if the room volume occupied per person is less than 28 cum. Moreover, this requirement of air is based upon the assumption that 0.06% CO_2 content will start causing discomfort to the inhabitants, needing air replacement ;

whereas, infact, in these days, it has been established that CO_2 contents upto 1% or so, can be easily withstood. Due to these reasons, the air requirement was considered much more in olden times than in modern days. The air requirement of as high a value as 50 cum per hour per person* was not considered infrequent in olden days. This requirement has, nowadays, not only been toned down to about 15—30 cum/hr, but rather the entire concept of ventilation has undergone a change.

Now-a-days, it has been established that maximum air change is required not for keeping CO_2 under control, but is to ensure proper heat dissipation and cooling of the human body, as explained below :

[The metabolism in all living beings produces heat, which is partly (20%) consumed in their different physical and chemical activities. The remaining 80% heat, has to be dissipated through *conduction, convection*, and *evaporation*, so as to maintain the *thermal equilibrium of the body*.

The blood in the body carries the heat to fine capillaries near the skin, and from there, it dissipates into the atmosphere by *conduction*. Due to this, the temperature of the surrounding air rises, which sets up the *convection currents. Fresh air comes nearby, and carries away heat*. If the rate of its passage is slow, the body feels discomfort ; whereas if this passage is accelerated through fans, etc. more relief is secured.

During summer season, the temperature difference between the body (at 37°C) and the surrounding air, becomes very low, and the heat gradient becomes very flat, and hence the rate of conduction falls down. Mechanical ventilation then becomes most essential.

When the surrounding temperature increases even beyond the body temperature, as happens in a tropical country like India, where 40 to 45°C temperature is quite common during summers, both conduction and convection stops functioning. The evaporation of the body sweat can only cause cooling. The rate of this evaporation also reduces when humidity is high. Hence, on a hot and humid day, body feels more discomfort. The comfort can then be increased by increasing the air movement by using fans, and also by increasing the surface area of evaporation by removing clothings. Lungs also help in removal of heat through convection and evaporation. In hot weather, deeper and frequent breathing expels more heat.

It thus becomes evident that ventilation** is mainly required to control the body heat, and not to overcome CO_2 alone. The rate of ventilation required for body cooling exceeds the rate required for removing other bad effects of occupancy., such as decrease of O_2, increase of odours etc.

Hence, the air changes are required and provided these days on the basis of body cooling alone, and not on the consideration of CO_2.

On this consideration and in actual practice, *the fresh air is supplied at the rate of 15 to 30 cum per hour per person,* depending upon the type of building. When the number of occupants cannot be easily determined, the rate of air supply may be based upon the number of air changes to be provided.

Table 16.1 gives some common accepted standards in this regard.

---

*As against the theoretical value of 84 cum worked out above, considering no automatic ventilation.

**Strictly speaking, mechanical ventilation.

Table 16.1. Requirements of Air Supply and Air Changes with Mechanical Ventilation

S. No.	*Type of Buildings*	*Air required per person per hour in cum.*	*No. of air changes required per hour*
1.	Kitchens	60	10 to 40
2.	Factories and Workshops :		
	(*i*) *Workshops with flumes*	30—40	20 to 30
	(*ii*) *Closely occupied*	20—30	6 to 8
	(*iii*) *Open type*	15—30	1 to 4
3.	Hospitals :		
	(*i*) *Operation theatres*	15—40	10
	(*ii*) *X-ray rooms*		10
	(*iii*) *Treatment rooms*		6
	(*iv*) *Open wards*,		3
4.	Lavatories and Laboratories	30—40	6 to 12
5.	Assembly Halls, Crowded dining places, Public halls, Lecture rooms, Meeting rooms, etc.	35—40	6 to 10
6.	Restaurants, Shops, and Canteens	30	6
7.	Schools (Class rooms)	20—25	3 to 6
8.	Living rooms and Bed rooms	15—25	2 to 5
9.	General offices and Libraries	15—30	3

16.6. Systems of Ventilation

A *good ventilation system* should generally fulfil the following **requirements** :

(*i*) It should admit sufficient quantity of fresh air, and remove the requisite used up or vitiated air.

(*ii*) Admitted air should be properly controlled with respect to its quantity as well as velocity of movement.

(*iii*) The system should be capable of changing the old air thoroughly, without leaving any stagnant pockets in the room.

(*iv*) Should avoid *draughts,* for which maximum permissible velocity of admitted air should not exceed 15 m/min *i.e.* 0.25 m/sec.

(*v*) The system should admit clean and humid air.

(*vi*) The system should also be capable of controlling the temperature of admitted air.

The **ventilation systems can be broadly divided into two categories** ; *viz.*

(*i*) *Natural ventilation* ; and

(*ii*) *Artificial mechanical ventilation.*

16.6.1. Natural Ventilation. Natural ventilation is based upon providing suitable openings in a room, at lower levels for admitting free atmospheric air, and also at upper levels for removing the warmer and lighter used-up air. Doors and windows near the floor level, thus, admit fresh air, and ventilators near the ceiling, take out the vitiated air from a room.

In order to ensure privacy of a room, windows are generally provided at about 0.75 to 0.9 m above the floor level, for admitting fresh outside air into the room. The size and the number of the windows provided, will depend upon the size of the room, number of occupants, the purpose and use of the room, etc. Besides admitting fresh air, the windows help in admitting natural day light.

On minimum side, a window area of 0.052 m^2 per person* should generally be provided, so as to ensure admission of atleast 28 cum— (*i.e.* 1000 cft) of air per hour with a velocity not grater than 9 m/min*. Another recommendation is to provide about $\frac{1}{10}$ to $\frac{1}{15}$ th of the floor area in the living rooms for windows. Every room should preferably be provided with atleast 2 windows, and at least one of them should face open space or a varandah. Kitchens must be provided with more window area.

Provision of **deflectors** (also called **fan lights**) of 30 cm height at the bottom or top of a window, *opening inward*, permits the ventilation of the room, even when windows are closed, as shown in Fig. 16.1 (*a*) and (*b*).

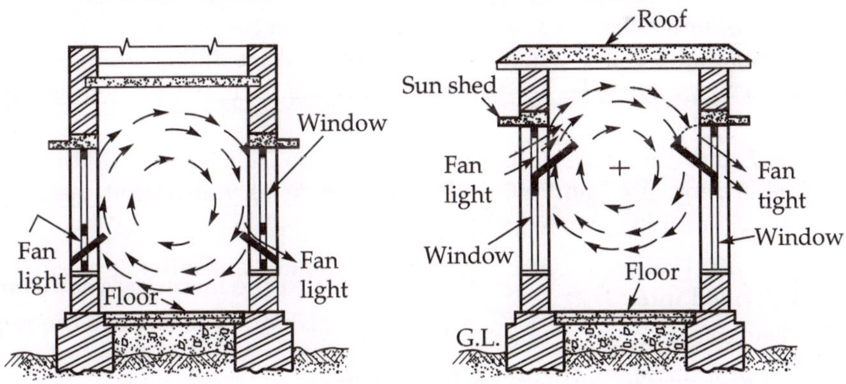

(*a*) Fan lights at the bottom of windows (*b*) Fan lights at the top of windows

Fig. 16.1. Fan lights located in the windows at bottom or at top to ensure natural ventilation.

In case of sloppy roofs, **ridge ventilators** may be provided, as shown in Fig. 16.2. Such ventilators are useful in taking out used vitiated air from large halls.

$$\frac{*\frac{28}{60}\text{ cum/min}}{9\text{ m/min}} = \frac{28}{60 \times 9}\text{ m}^2 = 0.052\text{ m}^2.$$

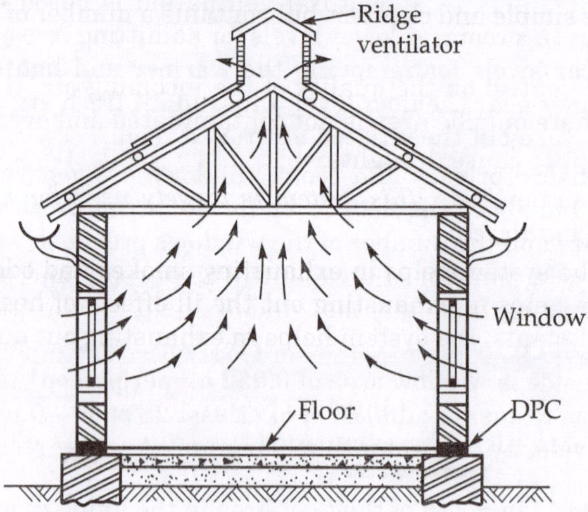

Fig. 16.2. Ridge ventilator.

In hot summer months, during day time, hot outside air may be warmer than the inside room air, and the ventilators may then reverse their functions. In other words, the bottom openings may start letting out the room air, whereas the top openings may start admitting outside air to fill up the partial vacuum created thereby. The inside of room will, therefore, become worse, unless the admitted air is cooled down by some method. *Khas* curtains may, therefore, be hanged at the roof level ventilators to cool the air by evaporation. But during night time, when outside temperature falls, all windows and ventilators may have to be kept open for allowing them to function in a normal manner.

In a hot and humid climate like that of India, natural ventilation fails to provide the requisite comfort, and hence artificial ventilation methods are adopted.

16.6.2. Mechanical Ventilation. The artificial ventilation system can be broadly divided into :

(1) The extraction or *vacuum system* ;

(2) The propulsion or *plenum system* ; and

(3) The *air conditioning system*.

The *air conditioning* is sometimes dealt separately, and not considered as a part of mechanical ventilation.

16.6.2.1. The Vacuum system is one in which the used-up vitiated air is thrown out to the outside atmosphere, by means of suitable exhaust fans or blowers, installed near the top ventilators. This exhaustion of the warmer air from the room, causes lower pressure inside the room, thereby permitting inward leakage of new fresh atmospheric air through the doors, windows, and walls.

Various types of fans are used for this purpose. They are usually of the rotary type, fixed in a convenient part of the ductwork, and driven usually by electric motors, though diesel and steam power may be used when electricity is not available.

This system is simple and cheaper, but contains a number of demerits, such as :

(*i*) there is no control on the quality of the incoming air. It can, hence, be installed only where outside area is not contaminated and over crowded.

(*ii*) It is also apt to cause draughts.

Inspite of these demerits, *this system is largely used for kitchens, public halls, industrial plants, etc.*

In kitchens, the system helps in exhausting smokes and odours ; in public halls, the system helps in exhausting out the ill effects of heavy occupancy ; and in industrial plants, the system helps in exhausting out dusts and fumes, etc.

16.6.2.2. The Plenum system involves *forcing* or *pumping in fresh air*, and causing the vitiated air to be exhausted out either by itself or through an exhaust fan placed at the outlet. The outlet exhaust fan is of smaller power than the inlet force fan.

Due to larger rate of incoming air, a slight pressure is created, which allows an outward leakage of vitiated air through the outlets.

Provision of a *cooler* at a window, with or without an exhaust fan at a ventilator, provides an example of such a plenum system of ventilation, largely adopted these days.

In larger buildings, such a plenum system may consists of distributing the incoming forced air at different points in the building through a system of ducts and grills.

Advantages. (*i*) This system enables us to control the quality, humidity, and temperature of the incoming air. And hence, this system is largely adopted these days for cinema houses, theatres, and even for individual homes.

(*ii*) It obviates the risk of draughts, because of outward leakage, and by making outlets less powerful than the inlets.

Moreover, such a system may either work *normally upward* with entry at floor level and exhaust at ventilator level ; or may work *downward*, with exhaust at floor level and entry at upper level.

Disadvantages. (*i*) The downward system, when adopted by providing air inlet ducts at roof level, is opposed to natural laws, and necessitates vitiated air to be rebreathed, causing discomfort.

(*ii*) The normal upward system when adopted, becomes costlier, as entry of air is provided at floor level under the seats, as is provided in some cinema halls, etc.

(*iii*) The air ducts require careful design in order to prevent draughts. This may involve using separate *dampers* for each room. The movement of air at the inlet as well as at the outlet needs to be properly controlled. This arrangement is quite costly. These objections may be overcome by (*i*) employing the upward system ; or (*ii*) by air-conditioning the buildings.

16.6.3. The Air Conditioning System. The atmosphere in the modern cities contains highly polluted smokes, fumes, dirts, germs, and bacteria, etc. In such environments, natural ventilation or even ordinary mechanical ventilation for bringing the outside polluted air into the building, will not serve us any useful purpose. In such conditions, it is necessary to completely control the temperature and quality of the outside air, before it is admitted into the room, and also to remove the heavy vitiated air at the optimum rates.

All these problems have been solved in these modern days, by air-conditioning the buildings, in which all the doors, windows, and ventilators *are kept closed.*

The *air-conditioning* may be defined as *the process of controlling the temperature, humidity and distribution of air in a building, with simultaneously removing the dirt, bacteria and toxic matter from the air*. Air-conditioning, thus, provides a comfortable and whole-some ventilation to the buildings.

However, as the atmospheric conditions vary, the requirements of occupants also vary with respect to season. In other words, during summer season, an air-conditioning system will be required to produce cool air inside the building ; and in winter season, it will be required to produce warmer air in the building. Accordingly, an air-conditioning system is required to work two ways.

We will now briefly discuss as to how a summer Air-conditioner and a winter Air conditioner, work.

(1) Summer air-conditioner. In summer season, the external atmospheric temperature is high, and the hot air has to be cooled before it can be distributed in the building. *During the process of cooling,* however, *the humidity of this air increases*, because at lower temperatures with the *same amount of moisture,* the relative humidity increases. Hence, after cooling, it becomes necessary to reduce the humidity of this air, by drying it through a *dehumidifier*. A dehumidifier is in the form of a hygroscopic substance like ammonia, calcium chloride, etc., or the air is cooled and dried through the *process of condensation*.

The clean, cooled, and dried air is then finally forced out into the space to be conditioned. A line diagram of this system is shown in Fig. 16.3 (*a*).

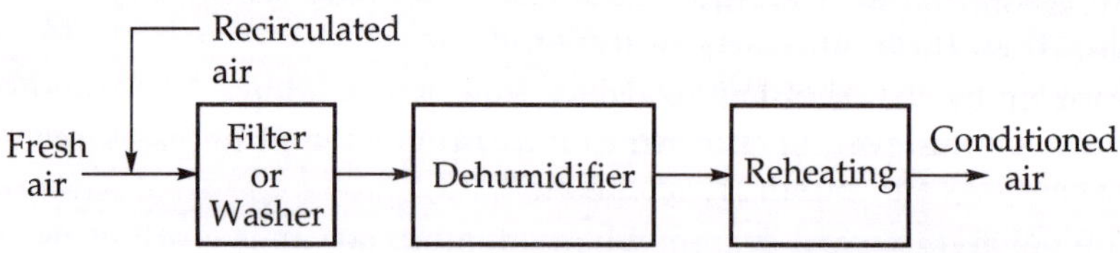

Fig. 16.3. (*a*) Summer A.C.—Line Diagram.

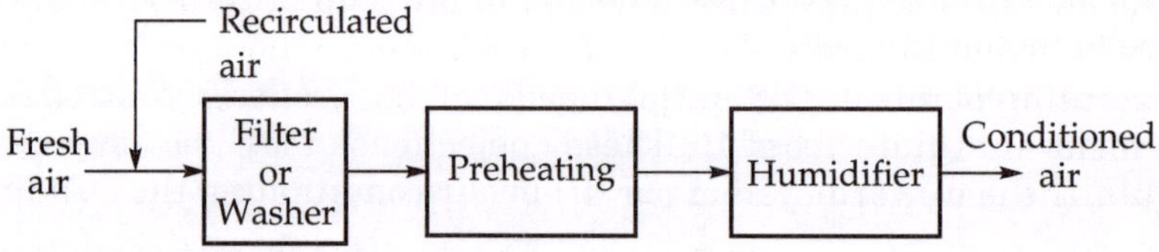

Fig. 16.3. (*b*) Winter A.C.—Line Diagram.

The various steps involved in this conditioning are briefly given below :

(*i*) **Filtration.** Fresh outside air as well as the vitiated reused air (*recirculated air*) is, first of all, passed through a **filter,** so as to remove the dust and other soot particles from it. *Dry filters* and *viscous filters* are the two types of filters, used for the purpose. **Dry filters,** made of close grained felt or cloth or jute are more commonly used. As a modern advancement on filter technique, **electric precipitators** are also, now-a days, sometimes used, though only rarely and particularly where absolutely dust free air is required, as for hospitals, etc. Such precipitators work on the principle of producing *high-intensity ionizing field* with a secondary field, where the dust gets precipitated and thus, removed.

(*ii*) **Cooling and Drying (Dehumidification).** This can be done by first lowering the temperature of the filtered air below the dew point, condensing out necessary amount of moisture, and then reheating the same with dry heat upto the desired temperature. Cooling of the air may be done in several ways ; two of which are (*i*) cooling by *mechanical refrigeration* : and (*ii*) cooling by *water spray method.*

Mechanical refrigeration cools the air like a household refrigerator, and is more suitable to tropical countries, where temperatures are to be lowered substantially. This method helps in cooling of air without wetting the same, with the help of a *compressor* and *metal coils* (tubes).

A *liquid refrigerant,* such as **freon,** enters the metal coils under high pressure, as soon as the machine is switched on. This liquid absorbs the heat from the coils and gets evaporated, thereby extracting heat from the metal coils and causing cooling of the coils. This gas refrigerant is then converted into liquid by the *condensor coils* and the *compressor.* This liquid again evaporates by absorbing heat from the coils, causing further cooling of the coils. The process continues till the cooling is caused upto the desired level.

The hot air coming in contact with these cooled metal coils, also gets cooled easily, upto the desired level.

This type of plants are known as **refrigeration plants.**

In the second type of plants, which cause cooling of the incoming air by **water spray method,** water is first of all, cooled either with the help of a refrigerant or by ice, and this cooled water is then sprayed on the hot air coming out of the filters. The air, thus, gets cooled by evaporation. This method increases the humidity of the air, and is generally not found suitable for humid climates.

Both the cooling methods, however, are used to cool the air upto *dew point,* so as to extract some desired moisture from it, and the balance air is reheated upto the desired temperature* for final distribution through fans and ducts in the space to be conditioned.

Refrigerating plants are often rated on ton basis. 1 M-ton being equivalent to heat removal at the rate of 100 C.H.U. (*i.e.* 180 B.Th.U.) per minute. As a rough rule, the power required to drive refrigerating plants is 1 B.H.P. per M. Ton.

---

*In summer, the most comfortable effective temperature is found to be 21.7°C ; whereas in winter, it lies near 18.9°C.

Individual air conditioning machines, called *room air conditioners*, are fitted on the windows of rooms ; whereas, for larger spaces, *central air-conditioning* may be adopted.

In a **room air conditioner,** the conditioned air comes out through a grill opening, as it is forced out by a fan. On the other hand, in **central air conditioning,** the conditioned air is produced by keeping the plant in one separate room, and *distribution ducts* are carried to individual rooms, and at various points in a big hall, for distributing the conditioned air, through fans and blowers, as shown in Fig. 16.4.

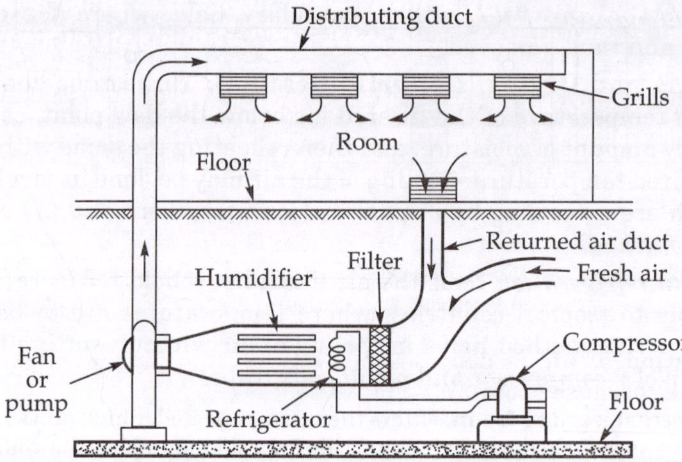

Fig. 16.4. Central system of air conditioning.

Recirculation of Air. During summer season, the temperature of room air is lower than the temperature of the outside air. And since an air conditioner has to cool down the hot air, it should be economical for it to cool the room air rather than the outside atmospheric air, if it becomes available to it. Moreover, in any case, the vitiated room air has to be removed by it.

On these two considerations, air conditioners are designed to extract and avail the vitiated room air at optimum rates, and to take from the outside atmosphere, only that much quantity of air, as is necessary to prevent concentration of odours. This air, which is withdrawn by an air conditioner from the conditioned room, and is recirculated through the filters and other cooling units, is known as ***circulated air,*** and the process is called ***recirculation of air.***

(2) **Winter air-conditioner.** In winter season, the external atmospheric temperature is low, and this cold atmospheric air has to be heated up before it can be dispersed in the conditioned room, as shown in the line diagram of a winter AC in Fig. 16.3 (*b*) During this process of pre-heating*, however, the humidity of the air *reduces*, because at higher temperatures with the same amount of moisture, the relative humidity decreases. Hence, after heating, *it becomes necessary to increase the humidity of this air,* by carrying it through a **humidifier,** where air, may be mixed with the water vapour, by throwing

---

*Preheating in a winter air conditioner may be done by admitting the air through a heater, or by passing it over the coils heated by hot circulating water.

water over the air with the help of spray nozzles. This humidified air is then passed through **eliminator plates,** so as to remove the water droplets, if any, in the humidified air.

The clean, hot and optimally humid air is finally forced out into the space to be conditioned.

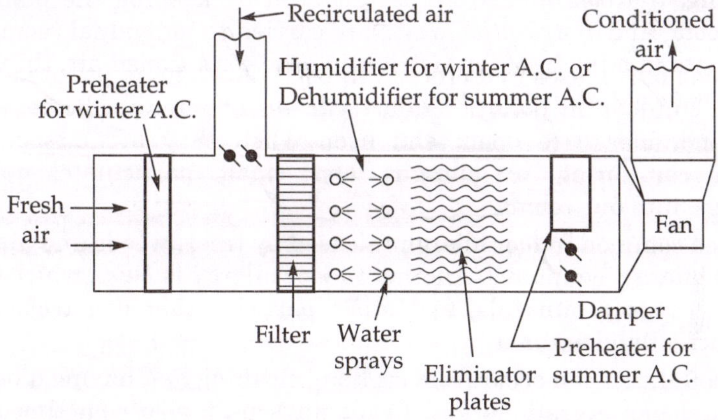

Fig. 16.5. Flow diagram of a complete two way air conditioner.

Recirculation of air is done in the winter A.C. also, as is done in the summer A.C., because room air is to be heated up to lesser extent than the outside air, which causes economy.

The flow diagram of a complete two way, air conditioner is shown in Fig. 16.5.

Example 16.1. *A copy machine is to be installed in a small room (volume 19.82 m³). There is a concern that ozone level inside the room would be high during operation of the machine. To maintain the proper air quality inside the room ventilation (namely, air changes @ 30 per hour) is recommended. Determine the capacity of exhaust fan.*

Solution. No. of air changes required per hour = 30

Volume of air in the room needing to be changed = 19.82 m^3

Total volume of air required to be changed per hour

$$= 30 \times 19.82 \text{ m}^3 = 595 \text{ m}^3/\text{hr}.$$

Hence, the minimum required capacity of the exhaust fan

$$= \mathbf{595\ m^3/hr.}\quad \textbf{Ans.}$$

Example 16.2. *In a small hall (volume 500 m^3), occupants (capacity 50 persons) are smoking @ 1.2 cigarettes per head per hour. To maintain proper ventilation, fresh air is pumped @ 1250 m^3/hr and the stale air is pumped out at the same rate. Estimate the concentration of formaldehyde inside the hall. Assume the process to be ready and complete mixing; cigarette contains 1.30 mg of formaldehyde per cigarette?*

Solution. 50 occupants consume cigarettes @ 1.2 No. per hr per head

Total No. of cigarrettes consumed per hour = 50 × 1.2 = 60

Since each cigarette contains 1.3 mg of formaldehyde the total formaldehyde produced per hour by 60 cigarettes

$= 60 \times 1.3 \text{ mg} = \mathbf{78\ mg}$

The 78 mg/hr of formaldehyde gets mixed in 1250 m³/hr of fresh air being injected in the room. Hence, the concentration of formaldehyde in the room air

$$= \frac{78 \text{ mg}}{1250 \text{ m}^3} = \mathbf{0.624\ mg/m^3}. \quad \textbf{Ans.}$$

16.7. Room Air Purifiers

The air inside our houses and offices can be 10-300% more polluted than the outside air. Outdoor air pollution infact, can easily enter our home, when the windows and doors are open; and even when they are closed, natural ventilation can bring in allergns and small particulates–under 2.5 micrometres, into our rooms.

Pollen is a common indoor allergn released by trees, weeds and plants that can trigger allergy symptoms. Another indoor allergy is *Dust*, which contains *dust mites*, outdoor minerals, and other particles that can trigger severe allergic reaction and asthma.

Indoor pollutants can occur from cooking, cleaning and having a household pet. New indoor decorations and furnishings may also sometimes release gases and volatile organic compounds (VACS).

To prevent bacteria and viruses from growing, the humidity levels of our homes should be 40-60%. A well humidified room makes the air more comfortable to breathe, and helps keep our skin free from drying out.

In order to avoid contracting bad health effects of polluted indoor air, *room air purifiers,* of various sizes and makes, have now-a-days been released in the market, and are becoming popular for common use, particularly in a city like Delhi, where the outdoor air is excessively bad, almost throughout the winter months. The room air purifiers, which are commonly available at reasonable rates in the markets these days, are from: Companies like *Atlanta, Philips, Kent, Godrej, Eureka Forbes, Hindware,* HSIL Ltd. (making Moonbow brand air purifiers) etc; etc.

Most of these room air purifiers work on the technology to suck and filter the room air through a series of filters, such as: (*i*) *prefilter*; (*ii*) *HEPA filter*; (*iii*) *Activated carbon filter*; (*iv*) *UV lamp*; and (*v*) *Refreshing energizers*, as shown in Fig. 16.6.

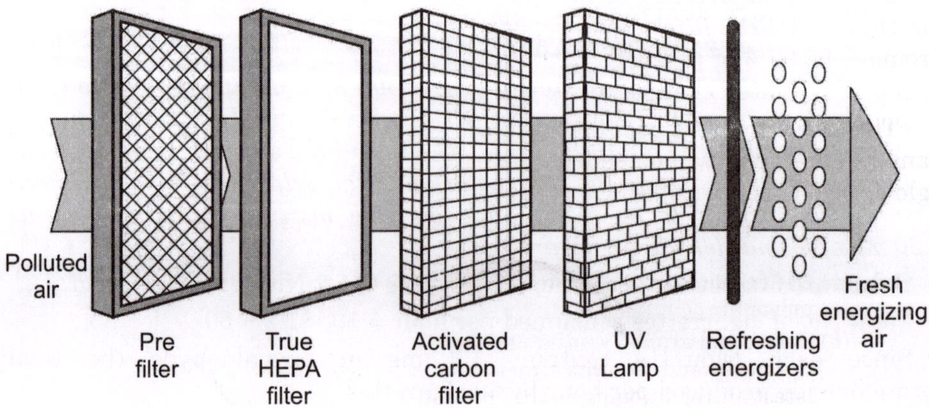

Fig. 16.6. Five stage air purification in a typical room air purifier.

The above indicated *five stage purification process* helps to capture indoor particulate pollutants and carcinogenic ultra-fine particles. The cleaned air, obtained after passing the above filtration stages, is finally released from the purifier unit (Photo Fig. 16.7) back into the room. As a result of this purification process, air-borne pollutants like dust, smoke, pollen, pet dander, mold and mildew spores, bacteria, viruses, and other particulates as small as 0.3 microns are usually removed. Some of the designs do display real time values of PM 2.5, as existing in the room, prior to starting the unit, as well as during its operation.

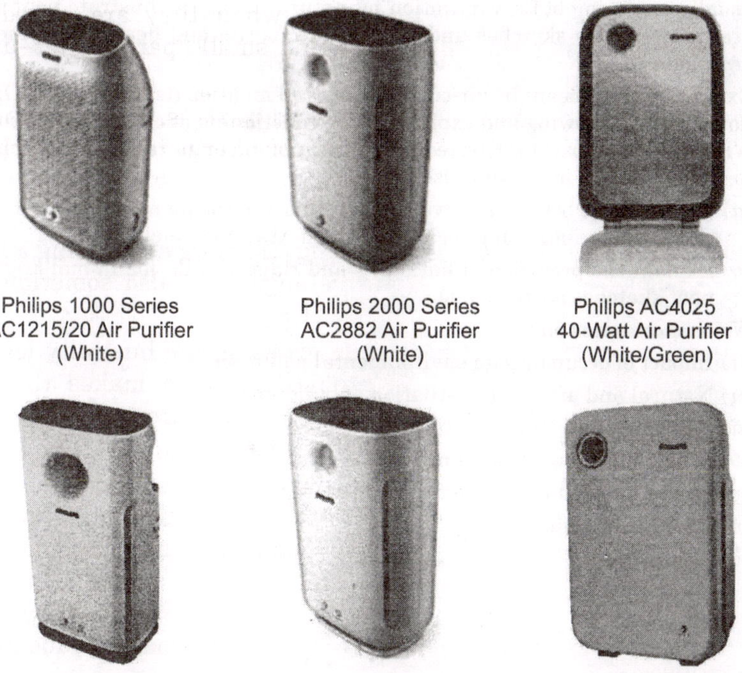

Philips 1000 Series AC1215/20 Air Purifier (White)

Philips 2000 Series AC2882 Air Purifier (White)

Philips AC4025 40-Watt Air Purifier (White/Green)

Philips 3000 Series AC3256 Aera Sense Air Purifier (White)

Philips 2000 Series Aera Sense AC 2887 Air Purifier (White)

Philips AC4012/10 36-Watt Tulip Air Purifier (White)

Fig. 16.7. Photoview of a typical room air purifer of Philips Co. in different capacities.

The most important role in the above indicated five stage filtration process is that of *HEPA filter*, that removes micro-scopic particles. It can usually remove 99.97% of particles large than 0.3 μm. Further to capture gases and odours, activated charcoal filter is used in conjunction.

The various filters used in these purifiers evidently need regular cleaning and replacement, when needed. Some units and design do have an indicator to glow, when such a cleaning or replacement is required.

PROBLEMS

1. (*a*) Define and explain indoor air pollution. What adverse effects are caused by indoor air pollution ?

   (*b*) Define and explain ventilation. How is it achieved in public buildings ? Discuss the merits and demerits of an ordinary mechanical system of ventilation and the latest air conditioning method of ventilation.

2. What are the ill effects of occupancy in an enclosed space ? How do these effects come about ? How does ventilation help in alleviation of these effects ? How do these effects affect humans ?

3. Indicate the number of air changes, per hour recommended for the following occupancies :
(1) General offices ; (2) Restaurants and Canteens ; (3) Lavatories ; (4) Assembly Halls ; (5) Factories and workshops ; (6) Living rooms and residences ; (7) Kitchens ; (8) Cinema Halls. Indicate or justify your answer with a single line reason for adopting high or low values for these cases.

4. Why is it necessary to provide ventilation in buildings ? What arrangements are usually made to obtain ventilation by natural means ? Illustrate your answer by drawing suitable sketches and explain the merits and demerits of the different arrangements.

5. Explain what is meant by air-conditioning and mention its advantages. Draw a neat flow diagram showing and explaining the functioning of a summer air conditioner. What additional parts will be required in your summer air conditioner if it is proposed to make it useful for winter also ?

6. (*a*) Differentiate between the vacuum system and the plenum system of ventilation of buildings, indicating their relative merits and demerits.
(*b*) Explain the provision of fan lights and ridge ventilation in buildings, drawing neat sketches for the same.

7. Write short notes on :
(*i*) Impact of occupancy on environmental pollution
(*ii*) Natural and artificial ventilation of buildings
(*iii*) Fan lights and Ridge ventillators
(*iv*) Air conditioning of Bulldings
(*v*) Indoor air pollution, its effects and status.
(*vi*) Room air purifiers.

17

The Environment, Its Pollution, and Present Status

17.1. Biosphere and Environment

From our knowledge of Geology, we know that the solid Earth and its interior is known as **lithosphere,** and the gaseous layers surrounding the Earth upto a distance of about 500 km, composes the **atmosphere.** The atmosphere is further sub-divided into (*i*) *troposphere* ; (*ii*) *stratosphere* ; and (*iii*) *thermosphere,* depending upon the distance of the gaseous layers from the surface of the Earth, as shown in Fig. 19.2.

The entire collection of water over the Earth as well as inside the Earth is called the **hydrosphere.**

A relatively narrow belt of lithosphere and atmosphere, a little below and above the surface of the land, and in water and air, which largely contains living organisms (such as plants and animals, including human life), is called the biosphere. **Biosphere** *is, therefore, that particular zone on Earth, where the lithosphere, the hydrosphere, and the atmosphere come into contact with one another.* It is, in fact, that portion on the Earth, where alone, life is in existence.

There is a continuous exchange of matter between these three elements (*i.e.* lithosphere, hydrosphere, and atmosphere). Say for example, plants draw their food from the nutrients and moisture found in the soil layers of the lithosphere. Similarly, plants use carbon dioxide from the atmosphere, and sunlight for their growth. Dead plants and animals are decomposed by bacteria and become soil nutrients. Some of this may be dissolved by the running water and added to the hydrosphere. The evaporation of water from the hydrosphere, and subsequent condensation of water vapour in the atmosphere and consequent rainfall on the Earth, provide water supply for the organisms in the biosphere.

17.2. Physical and Biological Environment

The four major elements, *i.e.* (*i*) *land*, (*ii*) *water*, (*iii*) *air*, and (*iv*) *living organisms* (plants and animals), together ; constitute what is known as **environment** or **ecosystem.**

Environment can be further sub-divided into :

(*i*) *Physical environment* ; and

(*ii*) *Organic* or *Biological environment.*

The non-living (abiotic) components like land, water, and air together, infact, forms one group of environment, called the **physical environment** ;

whereas, the *living organisms* (*biotic component*), form another group of environment, called the **biological environment.**

While the physical environment (land, water and air) is essential for the existence of life in various forms ; the biological environment provides the necessary food, so very essential for the sustenance of life on the Earth.

17.3. Ecosystem and Ecological Balance of Nature

The various **living organisms*** constituting the biological environment, are, infact, dependent upon each other for their survival, and each, inturn, depends on the physical environment of the area in which it lives. **Ecosystem,** as stated earlier, is the physical environment together with the organisms, which live therein. **Ecology** is the science which deals with the inter-relationships between the various organisms and their relationship with the physical environment.

Plants growing on land and in water provide food for *herbivorous animals* (generally smaller and weaker, plant-eating organisms including elephant) ; and these herbivorous animals, inturn, become the food for the *carnivorous animals* (flash eating animals, like lions) as well as for the *omnivorous animals* (*i.e.* animals who eat plant food as well as meat). The relative number of these different types of organisms in the biosphere is, however, such that there is no scarcity of food for any organism. The smaller organisms are much more in number, and their growth and reproduction is also much faster than those of larger organisms, which are fewer in number and reproduce slowly. This ensures the availability of sufficient food for larger animals.

In the natural environment, in the biosphere, there, infact, exists a perfect balance or equilibrium between the various organisms, and this is known as **ecological balance.** *In this equilibrium state, the relative numbers of different organisms in a particular environment remain constant.* This ecological balance may, however, get disturbed, when *changes take place in the natural environment***, which may consequently change the relative numbers of the different organisms in the biosphere. If the numerical ratio balance between the different organisms is disturbed, then naturally, there becomes a dearth of food for certain particular organisms, which may ultimately lead to large scale mortality of those particular organisms. *A new equilibrium is finally re-established under the changed conditions.* In this period of readjustment, evidently, certain old species get extinct, and new species may be born. This process of evolution of new species and extinction of old species is a continuous process.

The present existing pattern of organisms in the biosphere has been reached as a result of gradual evolution and extinction over several million years of earth's history. During the changing physical environment, various organisms adapted themselves, and survived ; but still however, various

---

*It includes all types of plants and animals, starting from the minutest microscopic size like bacteria, fungi, etc. to huge trees, whales, and elephants. *Strictly speaking, man is also a type of animal, who was formed during a particular type of physical environment, in the process of extinction of old species and evolution of new species.*

***i.e.* changes in physical factors, such as in temperature, precipitation, evaporation, light intensity, salinity of sea waters, etc.

particular species which could not tolerate the changing environment, died out, and became extinct. New species were also born in the new changing environment. *Man, infact, came into existence, as a result of large scale environmental changes that took place about one to two million years ago.*

17.4. Impact of Man on Biosphere

After the arrival and reproduction of man on the earth, a large scale impact has been caused on the biosphere, due to his unchecked actions. Say for example, *large scale deforestation* of forests for residential and agricultural land uses, has changed the habitat of organisms living in the forests. His *hunting of animals*, has led to the extinction of certain animal species. He has also developed new types of domesticated animals as well as plants to serve his own needs. His using pesticides and insecticides in the agriculture farms has also affected the relative proportion of various organisms in the biosphere. Such continuing imbalances in the biosphere, if not checked, will certainly prove disastrous to the very existence of man himself, because *it is the biosphere, on which depends, the life as well as the progress of human civilisation**.

17.5. Pollution and Conservation of Environment

The development of civilisation and rapid industrialisation by man has caused a great damage to the ecosystem. Things have worsened because no attention or a very little attention has been paid towards protecting the environment, while executing industries and other developmental projects.

Associated with any development, there is bound to be some amount of environmental degradation. An ecological survey, and effective measures for protecting the environment, are therefore, essentially required, before any developmental project is undertaken. But unfortunately, while developing industries or commercial or even urban properties, we have not bothered to look at the environmental degradation, likely to be caused by those establishments, either through our ignorance or through our sheer greed, for not spending any money on things which do not immediately affect us, individually. *There are, infact, people among us who believe that there is more money in destroying the environment (such as those who fell trees, and kill wild animals, unauthorisedly), rather than in conserving it. Tomorrow is not their immediate concern.*

Several examples of man's interference with the natural equilibrium may be mentioned. The excessive use of coal, petroleum, and natural gas for industries, automobiles, and power generation, has created enormous problems of pollution of environment.

Pollution of air is strikingly marked in the industrial and congested cities of U.S., Europe, Japan, and even in India, causing acid rains in many regions.

---

*All the food needs of man are met by plants and animals. Plants also provide fibres, such as cotton, jute, medicinal herbs, dyes, etc. Timber, resins, latex, gums, etc. are also obtained from forests. Plants and animals, thus, provide raw materials for various industries. Even the most essential power resources like coal, mineral oils, etc. are the products of the biosphere of earlier geological periods.

The *smokes* from factories, coke ovens and furnaces, steam engines, etc. ; *exhaust fumes* from automobiles, power plants, etc., injurious *chemical fumes* from oil-refineries, zinc refineries, chemical industries, metallurgical plants, iron and steel plants, incineration plants, etc. ; evolution of radioactive gases and suspended radioactive dusts from atomic explosions and accidental discharges from nuclear reactors, etc. have polluted the air *locally*, *regionally*, as well as *globally*, to such an extent that special steps are now essentially required at several places (with international cooperation) for reducing such air pollutions.

For example, the increase in CO_2 content of the environment has been responsible for gradual heating up of our globe, by a process called ***green-house effect***. The CO_2 layer, infact, acts like a glass cover used for a *green-house**, which allows the outside heat to enter the green house, but does not allow the inside heat to go out. *Similar to a glass house, CO_2 layer is transparent to short wave radiation from the sun, but absorbs the longer wave radiation from the Earth.* The net result is gradual heating up of the Earth. It is feared that by the year 2100, CO_2 in our atmosphere will be doubled, leading to 4°C rise in the world temperature. Such continuous warming of Earth may cause the glaciers to recede and ice to melt at poles. This may cause a rise in sea level by about 0.65 m, which ultimately may submerge most of our islands and coastal cities.

The salty sea water spreading to land may also lead to infertility of soil and spoil the underground water. Such global warming may also trigger severe climate changes, which will be discussed in detail in our next chapter.

Depletion of ozone in the stratosphere due to increasing release of chlorofluorocarbons (CFCs) due to human activities is another dreaded aspect of the global extent of environmental pollution. The presence of ozone in the upper atmosphere (*i.e.* stratosphere) in fact, cuts off the most harmful ultraviolet radiations (released by the rays of the sun) from reaching the surface of the Earth. Any appreciable reduction in ozone layer is likely to allow entry of harmful U.V. rays of sun, causing serious diseases like cataracts and skin cancers in humans. All such regional and global impacts caused by the pollution of our air environment, shall be discussed in detail in our next chapter.

Pollution of water is another aspect of environmental pollution. The waters of rivers, lakes, and oceans are, now-a-days being polluted on a large scale, by the outflow of effluents from factories and industries. Water pollution also results from the disposal of soiled urban wastes, such as *plastics***, *rubbers***, paper, untreated wastewaters and sewage, etc. Water pollution also interferes with the growth of organisms living in the water bodies, thus, retarding the natural purification process caused by such organisms.

---

*Glass or plastic covered green houses are generally used in cold countries to provide moderately heated environment to plants in nurseries or for growing out of season fruits and vegetables.

**Plastics and rubbers are very notorious wastes, as they cause cancers, unless properly collected, recycled, and reused.

All such environmental pollutions, lead to the spread of diseases through polluted air and water. The use of insecticides and pesticides has led to the concentration of DDT and other harmful chemicals in plants and animal food products, which when consumed by man, cause diseases.

In order to check such drastic ill-effects of environmental pollution on the health of the people as well as on the plants and the animals, it is extremely necessary that we wake up from our long slumber, and initiate some drastic remedial measures to protect our natural wholesome environment. Legislative laws and *their effective implementation, supplemented with spread of environmental education and awareness,* is most important to stop the greedy and/or ignorant people from spoiling our environment.

17.6. Status of Administrative Control on Environment in India

In our country, no attention was paid for controlling the environmental effects of developmental projects, almost till the year 1968 or so. It was in the 4th five year plan period (1968—1973), when, for the first time, environmental aspects were introduced for harmonious development.

In the year 1970, the Government of India appointed a committee under the chairmanship of Pandit Pitamber Pant (a member of the planning commission), which prepared a country report for presenting in the U.N. Conference on Human Environment, in 1972. Soon thereafter, a National Committee on Environmental Protection and Coordination was set up in the Deptt. of Science and Technology, for advising the GoI on environmental matters.

The 5th five year plan (1973-1977) continued to stress upon the environmental considerations. During this plan period, a Central law was enacted, under the name of **Water (Prevention and Control of Pollution) Act, 1974***. Under this Act, *The Water (Prevention and Control of Pollution) Rules, 1975* were notified by GoI on 27.2.1975. This law was meant for *checking and preventing water pollution*, which had become quite prominent at the time of enactment of this law, at several places in the country.

As a result of this legislative enactment of 1974, a Central board, called *Central Board for the Prevention and Control of Water Pollution,* was constituted for *monitoring* and *detecting* pollutions of water bodies, and for initiating remedial measures, including prosecutions in courts, so as to prevent the influential industrialists and municipalities from throwing their industrial wastes and sewages into our precious water bodies.

This act of 1974 also stipulated that the various States of India will separately constitute such pollution control boards in their respective States, and the Central board will serve as a watchdog body to advice and watch the performance of State boards, besides exercising its separate authority in respect of union territories, which are directly governed by the Centre. This Act, however, was mode applicable only in 12 States, which authorised the Union Govt. to do so, since water was a State subject under the Indian Constitution.

---

*Salient provisions of this Act are given in "Ecology and Environmental Studies" by the same author.

Another important Act, called the **Water (Prevention and Control of Pollution) Cess Act, 1977***, was also passed by the parliament. Under this Act, *The Water (Prevention and Control of Pollution) Cess Rules 1978* were notified by GoI on 24.7.1978. This law has proved quite effective in reducing the quantities of industrial wastes, as the act promotes *recycling* and *reuse* of the wastewaters. This Act was made applicable to all the States of the country, unlike the water pollution Act.

Protection of environment was further stressed in the 6th five year plan (1980—1985), which contained a separate chapter on 'Environment and Development'. During this plan period, a separate *Department of Environment* was set up on November 1, 1981, at the level of central cabinet. Air pollution was also recognised, and a Central legislation, called **Air (Prevention and Control of Pollution) Act, 1981***, was enacted. Under this Act, *The Air (Prevention and Control of Pollution) Rules* 1982 were notified by GoI on 18.11.1982. The Water Pollution Control Boards were given the additional charge of looking after air pollution control also.

A third act, called **The Environment (Protection) Act, 1986*** was subsequently promulgated by the Parliament after the occurrence of the *Bhopal gas tragedy*. This act extends to whole of India, and the Central Government has been empowered for taking any measures, which in its opinion, are necessary for improving and protecting the environment. Under section 9 of this act, it has also been made obligatory on the part of industries to prevent and mitigate environmental pollution, which may be caused due to any accident or unforeseen act at their industry. Under section 10 and 11 of this act, Government officers, empowered by the Central Government, may now enter the premises of the industries and collect samples, and/or carry out any reasonable action deemed fit for environmental protection.

Severe fines and penalties, including imprisonments upto 5 years have been prescribed for failures and continued failures on the part of industries failing to comply with the act, under section 5 of the act.

Under this important Environment (Protection) Act 1986, GoI has notified the following Rules :

(i) *The Environmental (Protection) Rules, 1986*
(ii) *Hazardous Wastes (Management and Handling) Rules, 1989*
(iii) *The Manufacture, Storage and Import of Hazardous Chemicals Rules, 1989*
(iv) *The Rules for the Manufacture, Use Import, Export and Storage of Hazardous Micro-organisms / Genetically Engineered Organisms or Cells, 1989*
(v) *The Chemical Accidents (Emergency Planning, Preparedness and Response) Rules, 1996*
(vi) *The Biomedical Waste (Management and Handling) Rules, 1998 & 2016*
(vii) *Plastic Waste Management Rules, 1998 & 2022*
(viii) *The Noise Pollution (Regulation and Control) Rules, 2000*
(ix) *The Ozone Depleting Substances (Regulation and Control), Rules, 2000*

---

*Salient provisions of this Act are explained in "Ecology and Environmental Studies" by the same author.

(*x*) *The Municipal Solid Wastes (Management and Handling) Rules, 2000*
(*xi*) *The Batteries (Management and Handling) Rules, 2001.*

Note : All the above Rules are available in compiled form together in "Universal's Environment & Pollution Laws", 2004 ed ; published by Universal Law Publishing Co. Pvt. Ltd, Delhi.

Moreover, in order to check the emitted smokes from badly maintained automobiles, containing too much of lead, carbon monoxide and particulate matter, a fourth legislation, called **Motor Vehicles Act, 1988,** has been passed by the parliament. The implementation of exhaust standards framed under *Central Motor Vehicles Rules 1989,* was to come into force w.e.f. 1-7-1989. However, due to non-availability of monitoring equipments, like *smoke-meters* and *gas-analysers,* for checking the quality of exhaust emissions, and also due to political and other influences, this act largely remains on the statue books of the country.

Even after fifteen years of the enactment of this legislation, nothing serious is being done to check and prevent the polluting vehicles, barring a few pollution checks carried out on private cars, and that too only in the capital city of Delhi. The authorities are thus, simply ralaxing, blaming interference from the politicians and the courts.

In order to prevent loopholes in the effective implementation of these environmental laws, the Government is now thinking to constitute special *environmental courts,* for speedy trial of the offenders of anti-pollution laws ; and also to carry out annual *environmental audits, to* easily detect disobedience of such laws by individual industries. *The earlier it is done-the better.*

17.7. Status of Water Pollution—Monitoring and Control in India

According to the assessment made on the subject, there are 14 river systems in India, which cover 83% of its physical area, 80% of its population, and 85% of its surface run off.

In order to monitor the water pollution in our country, it therefore becomes essential to watch out the waters of all these 14 river basins.

Several bodies have been engaged in the task of collecting and analysing water samples from these river basins, so as to work out the **water quality profiles*** for all the river basins.

State Governments collect water samples from about 50 river stations for analysis in the laboratories of CWC (Central water Commission). Central Board for Prevention and Control of Water Pollution has also established a

**The water quality profile for our most important Ganga river* has shown that the water quality of the Ganga is generally good in *West Bengal* (except for a small reach below Calcutta), as well as in *Bihar*. In U.P., the water quality is also found to be good, between Rishikesh and its confluence with river Kali near Kannauj ; with BOD < 2 mg/l and DO > 8—9 mg/*l* ; *but the river water is found to be highly polluted at the city of Kanpur* (with BOD between 10 to 55 mg/*l*, and DO between 4 to 6 mg/*l*). At Dalmau (50 km downstream of Kanpur), the water quality is again found to have considerably improved, the pattern is repeated at the cities of Allahabad. Varanasi, and Patna, where the river is highly polluted, followed by clear water stretches beyond about 50 km from each city. *Untreated wastes from human settlements are found to be the major cause for deterioration in the water quality near the cities.*

The above profile of Ganga was drawn in the year 1981.

monitoring network for sweet water bodies on a *three tier basis* (*i.e.* 40 stations under the *Global Environmental Monitoring Systems,* 61 stations under *Monitoring Indian National Aquatic Resources*, and additional 120 stations under *Minimal Action Programme*).

On the basis of assessments made so for, it has been established that in our country, about 85% of wastewater (by volume) is being generated by the domestic and *household sector, whereas, the balance about 15% is being contributed by the industries. In terms of BOD, domestic as well as industrial sectors are both contributing about 50% pollution load, each.*

In view of the above findings, it becomes abundantly clear that per-treatment of domestic as well as industrial wastewaters before disposal should receive our highest priority, if at all we want to control water pollution, and preserve public health, since more than 80% of diseases in our country are found to be water borne. Besides this, situations peculiar to our country like *Kumbh Mela*, which is found to increase the river coliforms from a few hundred to several thousands, need to be properly watched, and remedial measures devised to avoid heavy and large scale infections to the devotees taking bath in the holy rivers.

In order to monitor the water quality of surface sources w.r.t. different possible uses, and for their proper classification, the BIS **(Bureau of Indian Standards) has vide IS 2296-1982, classified the waters into the following five categories :**

1. **Category A** : Surface waters which can be used as drinking water sources without conventional treatment but after disinfection.
2. **Category B** : Surface waters for outdoor bathing.
3. **Category C** : Surface waters for use as drinking water sources with conventional treatment followed by disinfection.
4. **Category D** : Surface waters for use for fish culture and wild life propagation.
5. **Category E** : Surface waters for irrigation, industrial cooling or *control waste disposal.*

The BIS standards in respect of the physical and chemical, characteristics of all these categories of waters are shown in table 17.1. The physical, chemical and bacteriological standards laid down by the 'Central Board for Control and Prevention of Water Pollution' for these five categories of waters are shown in table 17.2.

The existing status of water quality at various monitoring stations on our rivers have shown the water quality to be either of category C or below.

The river *Ganga* is found to be polluted in stretches of 480 km, out of its total length of 2035 km.

The river *Yamuna* is found clean only in stretches of 522 km, out of its total length of 1044 km.

The quality of water at *Godavari* river at Mancherial and Polaveram, the *Krishra* at Vijaywada, the *Tunghabhadra* at KRS dam and Sathayalam, is however, found to be confirming to the desired levels.

The general water quality of *all the rivers* in Gujarat State is below category C due to high concentration of total coliforms in these rivers.

Table 17.1. BIS (ISI) Water Quality Standards for Classifying Surface Water Sources

S. No.	Characteristic	*Tolerance Limit*				
		Class A	*Class B*	*Class C*	*Class D*	*Class E*
(1)	(2)	(3)	(4)	(5)	(6)	(7)
1.	pH value	6.5—8.5	6.5—8.5	6.5—8.5	6.5—8.5	6.5—8.5
2.	Dissolved oxygen, mg/l, min.	6	5	4	4	—
3.	BOD (5-days at 20°C, mg/l, min.	2	3	3	—	—
4.	Total coliform organism, MPN/100 ml, max.	50	500	5000	—	—
5.	Colour, Hazen units, max.	10	300	300	—	—
6.	Odour	10	300	300	—	—
7.	Taste	Tasteless	—	—	—	—
8.	Total dissolved solids, mg/l, max.	500	—	1500	—	2100
9.	Total hardness (as $CaCO_3$), mg/l, max.	300	—	—	—	—
10.	Calcium hardness (as $CaCO_3$), mg/l, max.	200	—	—	—	—
11.	Magnesium hardness (as $CaCO_3$), mg/l, max.	100	—	—	—	—
12.	Copper (as Cu), mg/l, max.	1.5	—	1.5	—	—
13.	Iron (as Fe), mg/l, max.	0.3	—	0.5	—	—
14.	Maganese (as Mn), mg/l, max.	0.5	—	—	—	—
15.	Chlorides (as Cl), mg/l, max.	250	—	600	—	600
16.	Sulphates (as SO_4), mg/l, max.	400	—	400	—	1000
17.	Nitrates (as NO_3), mg/l, max.	20	—	50	—	—
18.	Fluorides (as F), mg/l, max.	1.5	1.5	1.5	—	—
19.	Phenolic compounds (as C_6H_5OH), mg/l, max.	0.002	0.005	0.005	—	—
20.	Mercury (as Hg), mg/l, max.	0.001	—	—	—	—

(Contd.)

(1)	(2)	(3)	(4)	(5)	(6)	(7)
21.	Cadmium (as Cd), mg/l, max.	0.01	—	0.01	—	—
22.	Salenium (as Se), mg/l, max.	0.01	—	0.05	—	—
23.	Arsenic (as Se), mg/l, max.	0.05	0.2	0.2	—	—
24.	Cyanide (as CN), mg/l, max.	0.05	0.05	0.05	—	—
25.	Lead (as Pb), mg/l, max.	0.1	—	0.1	—	—
26.	Zinc (as Zn), mg/l, max.	15	—	15	—	—
27.	Chromium (as Cr^{6+}), mg/l, max	0.05	—	0.05	—	—
28.	Anionic detergents (as MBAS) mg/l, max.	0.2	1	1	—	—
29.	Polynuclear aromatic hydrocarbons, (as PAH), mg/l	0.2	—	—	—	—
30.	Mineral oil, mg/l, max.	0.01	—	0.1	0.1	—
31.	Barium (as Ba), mg/l, max.	1	—	—	—	—
32.	Silver (as Ag), mg/l, max.	0.05	—	—	—	—
33.	Pesticides	Absent	—	Absent	—	—
34.	Alpha emitters, μC/ml, max.	10^{-9}	10^{-9}	10^{-9}	—	—
35.	Beta emitters, mC/ml, max.	10^{-8}	10^{-8}	10^{-8}	10^{-8}	10^{-8}
36.	Free ammonia (as N), mg/l, max.	—	—	—	1.2	—
37.	Electrical conductance at 25°C, mhos, max.	—	—	—	1000×10^{-6}	2250×10^{-6}
38.	Free carbon dioxide (as CO_2), mg/l, max.	—	—	—	61	—
39.	Sodium absorption ratio	—	—	—	—	26
40.	Boron (as B), mg/l, max.	—	—	—	—	—
41.	Percent sodium, max.	—	—	—	—	—

Table 17.2. Water Quality Standards for Surface Water Sources as per CPCB (New Delhi)

Category or class	*Designated Best use*	*Parameters affecting*	*Quality Criteria*
(1)	(2)	(3)	(4)
A.	Drinking Water Source, without conventional treatment but after disinfection	1. Coliform MPN 2. Turbidity 3. Colour 4. BOD 5. DO 6. Toxicants (including pesticides) 7. Plate Count 8. Floating Matter 9. Taste or Odour	< 50/100 < 10 units < 10 units < 2 mg/*l* > 6 mg/*l* No acute toxicity to be present < 50/100 ml Absent Not perceptible
B.	Bathing, Swimming and Recreation	1. Coliform MPN 2. Turbidity 3. Colour 4. BOD 5. DO 6. Toxicants (including pesticides) 7. Floating Matter 8. Taste or Odour	< 500/100 ml < 25 units < 10 units < 3 mg/*l* > 5 mg/*l* No acute toxicity to be present Not noticeable Not perceptible
C.	Drinking Water Source after conventional treatment	1. Coliform MPN 2. Colour 3. BOD 4. DO 5. Toxicants	< 5000/100 ml < 25 ml < 3 mg/*l* > 4 mg/*l* No acute toxicity to be present
D.	Propagation of Wild Life Fisheries	1. Coliform MPN 2. BOD 3. DO 4. Toxicants	< 5000/100 ml < 6 mg/*l* > 4 mg/*l* No acute toxicity to be present
E.	Irrigation, Industrial Cooling and Controlled waste disposal	1. TDS 2. (Ca + Mg) 3. Sodium Ratio 4. Chlorides 5. Boron	< 1000 mg/*l* < 100 mg/*l* < 0.5 < 250 mg/*l* < 2 mg/*l*

The BOD level at Ahmedabad on the *Sabarmati* ranged between 1.7 and 285 mg/l with an average value of about 70 mg/l, while the acceptable limit is less than 3 mg/l.

On an average, the water temperature of *rivers in Kerala* State were observed to be on the higher side than the other rivers of the southern region. The concentration of TKN was also observed to be comparatively higher in the rivers of Kerala as compared to the other rivers.

The water quality of *Subarnarekha* at Ranchi is worse than at Jameshedpur. The probable reason for bad quality of water at Ranchi is low flow.

17.8. Status of Air Pollution—Monitoring and Control in India

Air pollution was not very visible in our country, only upto a few decades ago. However, due to rapid urbanisation, commercialisation, and industrialisation, it has now acquired a threatening posture, although of course, the problem has become gigantic only in some of our big cities and towns, where either the industries are concentrated or there are too many automobiles.

In order to legally control air pollution, a Central legislation was passed by Indian parliament, and this act, called **"The Air (Prevention and Control of Pollution) Act 1981"***, came into force from May 16, 1981. This act is applicable to whole of the country, unlike the water pollution act of 1974, which is applicable only in 12 States, which had adopted it.

Under the above act, the *Central Board for Control and Prevention of Water Pollution,* popularly called the **Central Pollution Control Board (CPCB)** has been given the additional duties for looking after the administration of this air pollution control act, 1981, also. States where water pollution control boards were earlier constituted, have been asked to give air pollution control works also to these boards, as additional duties. However, States, where no such water pollution control boards were constituted earlier, have been asked to constitute separate *air pollution control boards.*

The CPCB, thus, has to exercise Central control on State Pollution Control Boards, and to act as State Pollution Control Boards in Union Territories, including NCT of Delhi.

As provided in this Act, the CPCB has formulated and notified *the Air (Prevention and Control of Pollution) Rules, 1982* wef. 18.11.1982. Most of the States have also notified their Boards and Rules. Some of the States have even notified their *air pollution control areas. Ambient air quality standards* have also been evolved for monitoring air quality, and notified under Environment (Protection) Act, 1986, and the Rules framed thereunder.

As a matter of fact, a large scale work has been done on air pollution by the *National Environment Engineering Research Institute Nagpur* (*NEERI*), under a **national air monitoring programme,** suggested and assisted by *World Health Organisation* (WHO), which had chosen our country as nodal point in South-East Asia for air monitoring in the year 1978. The main purpose of this programme was to alert the public and the authorities about abnormal increase in pollution levels, and to ultimately activate the local agencies to take up the programme into their hands.

Under the above mentioned National Air Monitoring programme, NEERI conducted the first survey of prevailing air quality in 9 major cities of India, in the year 1981. The results obtained in this survey are shown in Table 17.3.

---

*Amended by **The Air (Prevention and Control of Pollution) Amendment Act, 1987.**

This survey considered the following four **air pollution parameters :**

(*i*) *Total suspended particulate matter* (TSPM)* in microgram per cubic metre of air (μg/m^3) ;

(*ii*) *Sulphur dioxide* (SO_2) in μg/m^3 ;

(*iii*) *Sulphation rate* (SR) as $MgSO_3$/100 cm^2/day ; and

(*iv*) *Dust fall* (DF) in *t*/km^2/month.

Table 17.3. Annual Survey of Air Quality for Indian Cities for the year 1981

S. No.	*City*	*TSPM μg/m^3 (annual average)*	*SO_2 mg/m^3 (annual average)*	*SR (Sulphation rate $MgSO_3$"/100 cm^2/day (annual average)*	*DF (Dust fall) t/km^2/month (annual average)*
1.	Ahmedabad	217	31	0.10	91
2.	Bombay	184	38	0.45	11
3.	Calcutta	522	900	0.45	34
4.	Delhi	433	31	0.20	31
5.	Hyderabad	143	30		—
6.	Jaipur	308	6	0.18	19
7.	Kanpur	432	13	0.19	34
8.	Madras	155	8	0.33	—
9.	Nagpur	199	12	0.22	34

The above survey clearly established that the air of most of the Indian cities were found to be highly polluted as for as total suspended matter was concerned, since its existing average annual value was found to be much more than the national standard of 140 μg/m^3 (annual average). The existence of SO_2, however, was found to be exceeding the maximum tolerable value of 80 μg/m^3, only at Calcutta. The dust fall rate was also found to be quite high at Calcutta, Delhi, Kanpur and Nagpur. Realising this air pollution in the country, *the Air (Prevention and Control) Act 1981* was passed by Indian Government and made effective wef May 16, 1981, although, initially, it virtually remained on the statue books of the country, without any effective implementation.

Due to non-implementation of effective anti air pollution measures, the air pollutants continued to increase in various cities of India, as were detected in a survey conducted in 1989, the results of which are shown in table 17.4.

Considering the harmful effects of NO_x in air, this survey also included measurements of NO_x, which was found to be exceeding the tolerable limit for urban areas of 60 μg/m^3 in almost all the 5 cities surveyed, with Kolkata and Ahmedabad showing maximum pollution. This, survey also reflected excessive particulate pollution in Delhi and Kolkata.

---

*It includes **natural particulates** like *pollen, spores, bacteria, Viruses, protozoa, fungi, plant fibers, rusts, volcanic dusts,* etc. as well as **anthropogenic particulates** like fly ash, smoke soot particles, metallic oxides and salts, oily and tarry droplets, acid droplets, silicates and other inorganic dusts and metallic fumes, etc.

Table 17.4. Air Quality Data of Certain Indian Cities for the Year 1989

S. No.	*City*	*TSPM* $\mu g/m^3$			*SO_2* $\mu g/m^3$			*NO_x* $\mu g/m^3$		
		Min.	*Max.*	*Mean*	*Min.*	*Max.*	*Mean*	*Min.*	*Max.*	*Mean*
1.	Delhi	56	**1062**	**331**	*	42	13	18	56	33
2.	Madras	76	302	169	*	54	7	1	69	22
3.	Bangalore	21	478	141	7	104	27	8	70	23
4.	Ahmedabad	72	514	232	1	123	2	50	**188**	50
5.	Kolkata	**90**	**1093**	**554**	**5**	**230**	**33**	88	**363**	**109**

The excessive (serious) presence of TSPM (Total suspended particulate matter) in Bombay, Delhi and Calcutta was subsequently confirmed in a survey carried out by WHO in the year 1995, the results of which are shown in table 17.5. This survey placed **Bangkok** at the top of the 12 most polluted cities of the world, with three Indian cities Bombay, Kolkata and Delhi getting 3rd, 4th and 5th place, as shown in table 17.5.

Table 17.5. Most Polluted Cities of the World (WHO—1995)

S. No.	*Name of the city*	*Pollution Problem in respect of*					
		SO_2	*TSPM*	*NO_2*	*CO*	*Pb*	*O_3*
1.	**Bangkok**	L	S	L	L	M	L
2.	Beijing	S	S	L	—	L	M
3.	Bombay	L	S	L	L	L	L
4.	Kolkata	L	S	L	—	L	—
5.	Delhi	L	S	L	L	L	—
6.	Karachi	L	S	—	—	S	—
7.	Los Angeles	L	M	M	M	L	S
8.	Manila	L	S	—	—	M	—
9.	Moscow	L	M	M	M	L	—
10.	New York	L	L	L	M	L	M
11.	Rio de Janeiro	M	M	—	L	L	—
12.	Tokyo	L	L	L	L	—	S
S = Serious, M = Moderate, L = Low, — = Data not available							

With continuing advancements, it has further been realised that the smaller sized respirable particles present in the total suspended particulate matter (TSPM) are extremely harmful to human health**, and hence, needs to be monitored separately. Accordingly, smaller sized particles upto 10 micron in size, known as **Respirable suspended particulate matter (RSPM)** or **PM 10÷** are now a days also monitored in urban air, along with total suspended particulate matter (TSPM).

---

* Below detection limits.

** These can trigger lung cancer, respiratory and heart problems.

÷Even smaller particles of upto 2.5 micron size, called **PM 2.5,** are also, nowadays, being monitored in developed countries, due to their excessive health hazards.

According to our *National Ambient Air Quality Standards,* the average annual value of RSPM should not exceed 60 μg/m^3 ; whereas the limiting value of TSPM (average annual) is 140 μg/m^3.

The CPCB (*Central Pollution Control Board*) is monitoring RSPM since 1999 only.

The **CPCB in a report released in January 2003,** has presented a very grim picture of air pollution prevailing in Indian cities. This report has ranked 29 cities according to RSPM (PM10) levels, recorded during the year 2000, as shown in Fig. 17.1. According to this report, Ahmedabad air is the most noxious, followed by Kanpur, Solapur, Luknow and Delhi, in order. The small particulate levels (PM10 *i.e.* RSPM) here are 3-4 times the annual standard of 60 μg/m^3. This study has shown that 14 Indian cities are critically polluted with RSPM, exceeding 1.5 times the annual standard, and merely 3 cities are at moderate level. As compared to this critical existence of RSPM in Indian cities, PM 10 levels in USA are found to be far below 40 μg/m^3 in 90% of its cities.

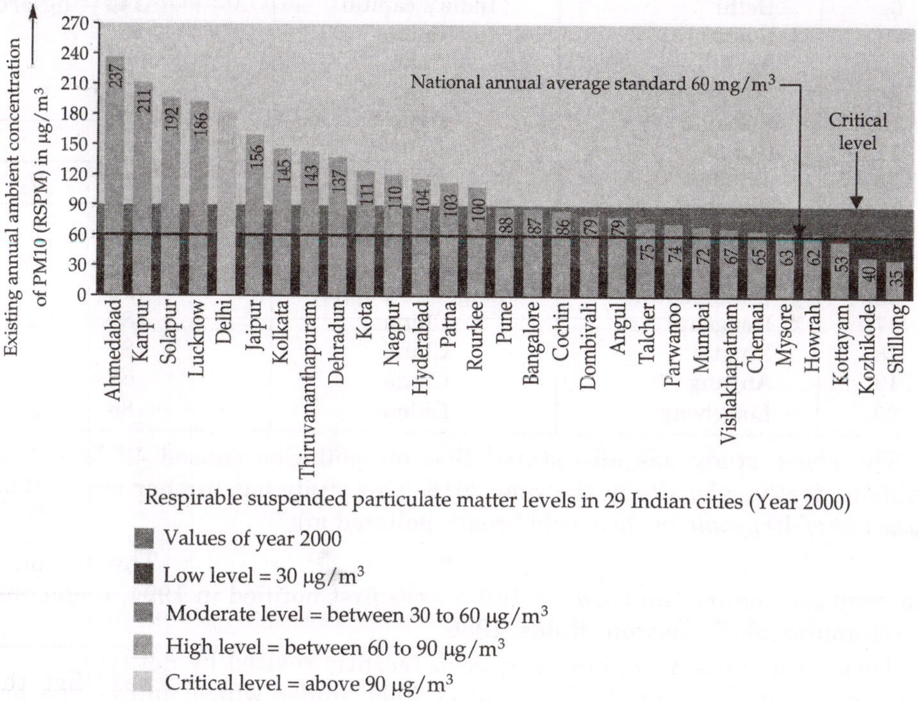

Fig. 17.1. Annual RSPM profile for 29 Indian cities for the year 2000.
Source : Computed on the basis of National Ambient Air Quality Status-1999 *(NAAQMS/2000-2001) and* National Ambient Air Quality Status-2000 *(NAAQMS/22/2001-02), Central Pollution Control Board, New Delhi.*

The excessive presence of TSPM and more so of the RSPM (PM 10) in Indian cities, is certainly causing a very adverse impact on the health of its residents, causing various diseases to residents, as shown in **Fig. 17.2.**

The Latest available report of WHO, based on the data of year 2016, has given results for 3000 cities of 103 countries. In this study, released on 02-05-2018, Delhi has been ranked at 6th place amongst the most polluted cities of the world, with PM 2.5 level of 143 μg/m^3. Not only this, Kanpur city of India has been ranked at No. 1, being the most polluted city of the world with PM 2.5 level of 173 μg/m^3. Surprisingly, 14 Indian cities have been ranked at S. No. 1 to 14 out of the top 20 most polluted cities of the world, as shown in table 17.6. All this data gives a very grim picture of the severe air pollution prevailing in India. *Maria Neira*, WHO's head of public health, said India should follow the example of China, which had made big improvements in air quality.

Table 17.6. List of 20 most polluted cities of the world (WHO's report of 2018).

S.No.	*Name of City*	*Name of Country*	*PM 2.5 level * μg / m³*
1.	Kanpur	India	173
2.	Faridabad	India	172
3.	Varanasi	India	151
4.	Gaya	India	149
5.	Patna	India	144
6.	Delhi	India's capital	143
7.	Lucknow	India	138
8.	Agra	India	131
9.	Muzaffarpur	India	120
10.	Srinagar	India	113
11.	Gurgaon	India	113
12.	Jaipur	India	105
13.	Patiala	India	101
14.	Jodhpur	India	98
15.	Baoding	China	93
16.	Ulaanbaatar	Mangolia's Capital	92
17.	Hengshui	China	87
18.	Xingtai	China	87
19.	Anyang	China	86
20.	Liaocheng	China	86

The above study has also stated that air pollution caused 42 lakh (4.2 million) deaths, globally in the year 2016. This study has further added that *nine out of 10 people* in the world breath polluted air.

17.8.1. National Ambient Air Quality Standards in India. The national ambient air quality standards in India were first notified in 1994, under the Environmental (Protection) Rules, 1986.

These standards have, however, been recently revised by notifying much tougher air quality standards on 16.11.2009, under which, unlike the old norms, the residential and industrial areas will have the same standards. Additionally, as compared to the regulation of only five pollutants under the old standards, now 12 pollutants are required to be monitored. The highly dangerous pollutants, like *PM 2.5, ozone gas* (ground level ozone), *lead, benzene, benzo(a) pyrene, arsenic and nickel* have been included for regulation under the new norms. The old pre-revised standards which still continue to be in practice are given in table 17.6; while the new revised standards are shown in Table 17.7.

Inspite of the fact that these standards have been notified under the legally enforceable Environment Protection Act, these cannot be legally enforced,

*Annual mean, μg/m^3.

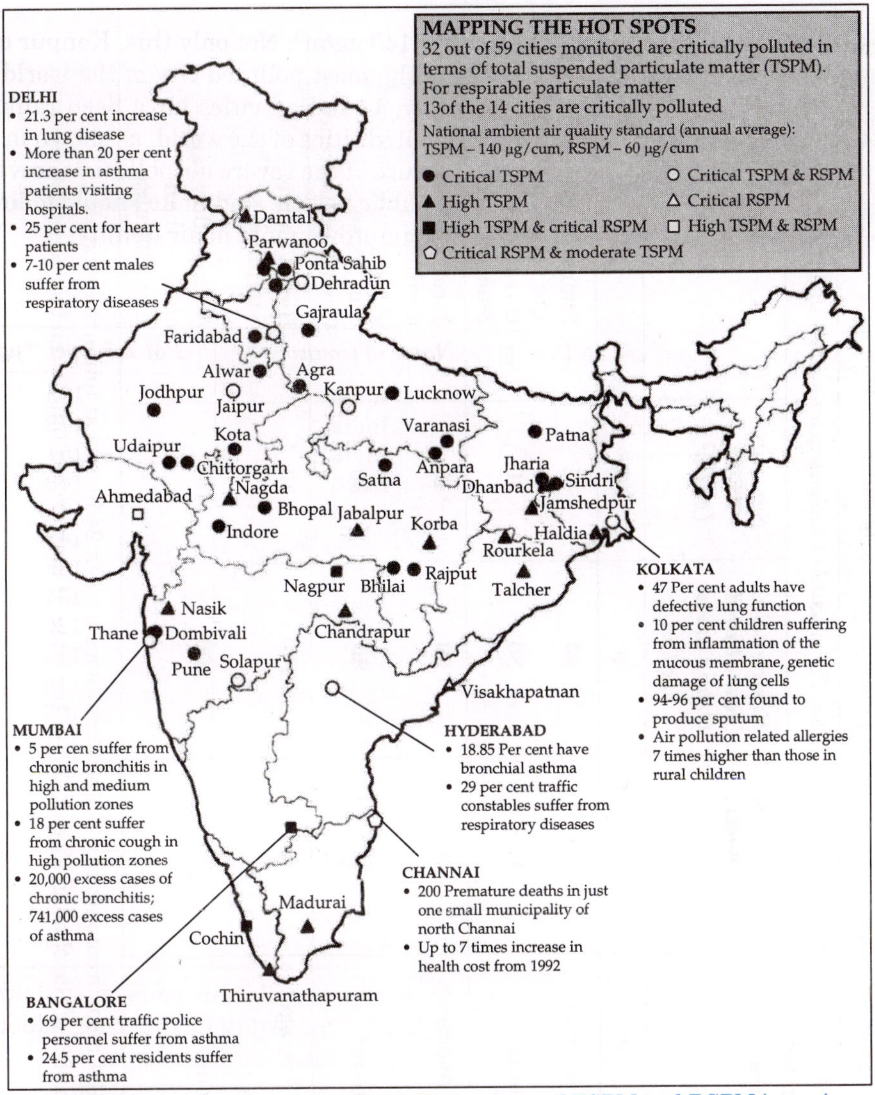

Fig. 17.2. Figure showing Critical and High values of TSPM and RSPM in various cities of India with Reported Adverse Health Effects on Residents.

because the public cannot file suit against the State for any defiance of its provisions, as is possible in United States. These standards are, hence, meant only for guidance and awakening the Governments in power, to take actions for reducing concentration of industries in critically polluted areas. That is why, the MoEF has notified these standards without preparing for their implementation strategy.

What to talk of implementation of these revised tougher standards, the Indian Governments have failed even to implement and enforce the pre-revised relaxed air quality standards. It can, infact, be observed that all the five metropolitan cities of India, with the exception of Madras, *do exceed* even the old relaxed standards in SPM pollution ; and Kolkata and Ahmedabad exceed limiting values in SO_2 and NO_2 pollution also. Other important cities, like Hyderabad, Jaipur, Kanpur, Nagpur, etc. are also full of dust, smoke, and other pollutants.

Table 17.6. Pre-Revised National Ambient Air Quality Standards (NAAQS) of India

S. No.	*Pollutants*	*Time weighted Average*	*Max. Permissible Concentrations in Ambient Air For*			
			Industrial Areas ($\mu g/m^3$)	***Residential, Rural and Other areas*** ($\mu g/m^3$)	*Sensitive Areas* ($\mu g/m^3$)	*Method of measurement*
(1)	(2)	(3)	(4)	(5)	(6)	(7)
(1)	Sulphur Dioxide (SO_2)	Annual Average*	80	**60**	15	— Improved West and Gaeke method
		24 hours**	120	**80**	30	— Ultraviolet Fluorescence
(2)	Oxides of nitrogen as NO_2	Annual Average*	80	**60**	15	— Jacab and Hochheiser modified (Na-Arsenite) method
		24 hours**	120	**80**	30	— Gas Phase Chemiluminescence
(3)	(*i*) Total Suspended Particulate Matter (TSPM)	Annual Average*	360	**140**	70	— High Volume Sampling
		24 hours**	500	**200**	100	— Average flow rate not less than 1.1 m^3/minute

* Annual Arithmeti mean of minimum 104 measurements in a year, taken twice a week 24 hourly at uniform interval.

** 24 hourly/8 hourly values shall be met 98% of the time in a year ; 2% of the time, it may exceed but not on two consecutive days.

(Contd.)

(1)	(2)	(3)	(4)	(5)	(6)	(7)
	(*ii*) Respirable Particulate Matter (size ≤ 10 μm) (RSPM or PM10)	Annual Average*	120	**60**	50	— Respirable particulate matter sampler
		24 hours**	150	**100**	75	
(4)	Lead (Pb)	Annual Average*	1.0	**0.75**	0.50	— AAS method after sampling using EMP 2000 or equivalent-filter paper.
		24 hours**	1.5	**1.00**	0.75	
(5)	Carbon Monoxide (CO)	8 hours**	5000	**2000**	1000	— Non dispersive, infrared spectroscopy
		1 hour	10,000	**4000**	2000	

* Annual Arithmeti mean of minimum 104 measurements in a year, taken twice a week 24 hourly at uniform interval.

** 24 hourly/8 hourly values shall be met 98% of the time in a year ; 2% of the time, it may exceed but not on two consecutive days.

Notes.

1. National Ambient Air Quality Standard : The levels of air quality necessary with an adequate margin of safety, to protect the public health, vegetation and property.
2. Whenever and wherever two consecutive values exceeds the limit specified above for the respective category, it shall be considered adequate reason to institute regular/continuous monitoring and further investigations.

Table 17.7. New-Revised National Ambient Air Quality Standards (NAAQS) of India
(Revised vide notification No. GSR 826(E) Dated Nov. 16, 2009)

S. No.	*Pollutant*	*Time weighted Average*	*Concentration in Ambient Air*		
			Industrial, Residential, Rural and other areas	*Ecologically Sensitive Area (notified by Central Govt.*	*Method of measurement*
(1)	(2)	(3)	(4)	(5)	(6)
(1)	Sulphur Dioxide (SO_2), μg/m^3	Annual* 24 hours**	50 80	20 80	— Improved West and Gaeke — Ultraviolet Fluorescence
(2)	Nitrogen Dioxide (NO_2), μg/m^3	Annual* 24 hours**	40 80	30 80	— Modified, Jacab & Hochheiser (Na-Arsenite) — Gas Phase Chemiluminescence
(3)	Particulate Matter (size less than 10 μg/m^3) or PM10, μg/m^3	Annual* 24 hours**	60 100	60 100	— Gravimetric — TOEM — Beta attenuation
(4)	Particulate Matter (size less than 2.5 μm) or PM2.5, μg/m^3	Annual* 24 hours**	40 60	40 60	— Gravimetric — TOEM — Beta attenuation
(5)	Ozone (O_3), μg/m^3	8 hours** 1 hour**	100 180	100 180	— UV photometric — Chemiluminescence — Chemical Method

* Annual Arithmetic mean of minimum 104 measurements in a year at a particular site taken twice a week 24 hourly at uniform intervals.

** 24 hourly or 08 hourly or 01 hourly monitored values, as applicable, shall be complied with 98% of the time in a year ; 2% of the time, they may exceed the limits but not on two consecutive days of monitoring.

(Contd.)

(1)	(2)	(3)	(4)	(5)	(6)
(6)	Lead (Pb), μg/m^3	Annual* 24 hours**	0.50 1.0	0.50 1.0	— AAS/ICP method after sampling on EPM 2000 or equivalent filter paper — ED-XRF using Teflon filter
(7)	Carbon Monoxide (CO), mg/m^3	8 hours** 1 hour**	02 04	02 04	— Non Dispersive, infrared (NDIR) — Spectroscopy.
(8)	Ammonia (NH_3), μg/m^3	Annual* 24 hours**	100 400	100 400	— Chemiluminescence — Indophenol blue method
(9)	Benzene (C_6H_6), μg/m^3	Annual*	05	05	— Gas chromatography based continuous analyzer — Adsorption and Desorption followed by GC analysis
(10)	Banzo (*a*) Pyrene (BaP)-particulate phase only, ng/m^3	Annual*	01	01	Solvent extraction followed by HPLC/GC analysis
(11)	Arsenic (As), ng/m^3		06	06	— AAS/ICP method after sampling on EPM 2000 or equivalent filter paper
(12)	Nickel (Ni), ng/m^3		20	20	— AAS/ICP method after sampling on EPM 2000 or equivalent filter paper

* Annual Arithmetic mean of minimum 104 measurements in a year at a particular site taken twice a week 24 hourly at uniform intervals.

** 24 hourly or 08 hourly or 01 hourly monitored values, as applicable, shall be complied with 98% of the time in a year ; 2% of the time, they may exceed the limits but not on two consecutive days of monitoring.

Note. "Whenever and wherever monitoring results on two consecutive days of monitoring exceed the limits specified above for the respective category, it shall be considered adequate reason to institute regular or continuous monitoring and further investigation." the Rules made thereunder, for various atmospheric pollutants.

Note. The various air quality surveys have established that the suspended particles of waste matter in the urban environment in this country are found to be 2 to 5 times more than that in the advanced western countries.

The *urban city air pollution* has been primarily *caused* by *automobiles, industries, thermal power plants,* and *burning of domestic coal or firewood.*

Say for example, Delhi alone releases about 2500 tonnes per day of pollutants into the atmosphere, consisting of about 12% of CO, 10% of NO, 70% of HC, and 8% of SO_2. About 64% of this total pollution load is caused by vehicular emissions, 16% by thermal plants, 12% by industry, and 8% by domestic sector. As a matter of fact, the pollution caused by thermal power plants has largely been reduced over the past years, due to the installation of Electro Static Precipitators (ESPs) at the city's three power plants (at Badarpur, Indraprastha and Rajghat), which have reduced the particulate matter from 720 tonnes in 1987 to 302 tonnes in 1992. These plants, however, continue to emit huge quantities of harmful SO_2 in the atmosphere. Pollution caused by industries has, however, not shown any disproportionate increase over the years due to installation of pollution control devices by about 400 large and medium sized industrial units. More than 10,000 small units, however, continue to cause air as well as water pollution.

The *vehicular pollution* has, however, increased quite disproportionately and alarmingly over the past years, to a level of about 1600 tonnes in 2005 from 870 tonnes in 1987. The total number of vehicles in Delhi has also gone up to a frightening figure of about 44 lakh in 2005 as against 39.4 lakh in 2003, 5.62 lakh in 1981 and 2.35 lakh in 1975, with continuing daily addition of about 500-700 new private vehicles*. Two wheelers constitute about 61% of total motor vehicles registered in Delhi, and hence constitute one of the major sources of air pollution. Not only the number of vehicles is high, but many of them are old and poorly maintained, emitting about four times more CO than those emitted in advanced countries. Installation of *catalytic convertors* in all the new motor cars to be registered in Delhi, has therefore, been made a mandatory requirement in this capital city.

All these factors have made Delhi, as the 5th most polluted city of the entire world. With such a high degree of prevailing air pollution, with excessive particulate matter and CO in the air, Delhi, the capital of India, has become a very unhealthy place to live, with different kinds of viruses always infecting its residents, and the incidence of bronchitis being 6—14 times more than the average elsewhere in India. The situation is, infact, no good in other metropolitan cities of India, with Bombay and Calcutta gaining 3rd and 4th place among the 12 most polluted cities of the world (Pl. refer table 17.5) in over a few past years. The situation is, infact, becoming bad to worse with every passing day.

17.8.2. Air Quality Index (AQI). Air quality Index is a number used by the air quality monitoring agencies to inform the public about the quality of air in their vicinity. It thus, informs as to how much polluted the air is, and how much polluted it is forecast to become. It is evidently an indicator of air quality, based on pollutants that have adverse effects on human health and environment.

As AQI value increases, the public is likely to suffer more adverse health effects of polluted air. Different countries have their own air quality indices, corresponding to their different national air quality standards.

The *Air Quality Index* is also referred to as the ***Air Quality Health Index*** (such as in Hong Kong), or the ***Air Pollution Index*** or the *Pollutant Standard Index*.

The term Air Quality Index (AQI) made its debut in USA in 1968; and since then many countries have developed their own methodologies and standards to compute and specify their respective indices.

Computatiozuality index requires consideration of the levels of the specified air pollutants as well as their averaging periods, both. The various air pollutants as

---

*The total number of vehicles in Delhi has further gone up to *105.64 lakh as on May 25, 2017* (66.49 lakh motorcycles and scotters, 31.73 lakh cars, and 7.46 lakh being all other type of vehicles), with an average daily addition of more than 1.6 lakh new vehicles every year. This is more than the combined total number of vehicles in Mumbai, Madras and Kolkata.

Proposed AQI Chart in India

AQI Category, Pollutants and Health Breakpoints								
AQI Category (Range)	PM_{10} (24hr)	$PM_{2.5}$ (24hr)	NO_2 (24hr)	O_3 (8hr)	CO (8hr)	SO_2 (24hr)	NH_3 (24hr)	Pb (24hr)
Good (0–50)	0–50	0–30	0–40	0–50	0–1.0	0–40	0–200	0–0.5
Satisfactory (51–100)	51–100	31–60	41–80	51–100	1.1–2.0	41–80	201–400	0.5–1.0
Moderately Polluted (101–200)	101–250	61–90	81–180	101–168	2.1–10	81–380	401–800	1.1–2.0
Poor (201–300)	251–350	91–120	181–280	169–208	10–17	381–800	801–1200	2.1–3.0
Very poor (301–400)	351–430	121–250	281–400	209–748	17–34	801–1600	1200–1800	3.1–3.5
Severe (401–500)	430+	250+	400+	748+	34+	1600+	1800+	3.5+
Note: Colours shown above are specific to each category of AQI								

AQI Category	AQI value	Associated Health Impacts
Good	(0–50)	Minimal impact
Satisfactory	(51–100)	May cause minor breathing discomfort to sensitive people.
Moderately polluted	(101–200)	May cause breathing discomfort to people with lung disease such as asthma, and discomfort to people with heart disease, children and older adults.
Poor	(201–300)	May cause breathing discomfort to people on prolonged exposure, and discomfort to people with heart disease.
Very poor	(301–400)	May cause respiratory illness to the people on prolonged exposure. Effect may be more pronounced in people with lung and heart diseases.
Severe	(401–500)	May cause respiratory impact even on healthy people, and serious health impacts on people with lung/heart disease. The health impacts may be experienced even during light physical activity.

Coloured Fig. 17.1. (a) Proposed Air Quality Index (AQI) Chart in India. Categories in Index are based on specified concentrations of Eight Air Pollutants; and Associated Health Impacts of Each category.

Global AQI Chart

AIR QUALITY INDEX | Recorded around 4pm (IST) on Nov 18, '14

India
Delhi | Anand Vihar: 399 ●; RK Puram: 298 ●; Punjabi Bagh: 182 ●

China
Beijing: 177 ●; **Shanghai:** 157 ●; **Guangzhou:** 155 ●

Japan
Tokyo | Nishitokyo: 53 ●; Nerima: 38 ●; **Tokorozawa:** 34 ●

Brazil
Sao Paulo | Americana: 52 ●; Paulinia: 17 ●

Malaysia
Kuala Lumpur: 32 ●
Perai: 52 (Moderate) ●

Thailand
Bangkok: 47 ●

France
paris: 58 ●

UK
London: 69 ●

US
New York | Utica: 56 ●; Albany County: 40 ●

AQI	AIR POLLUTION LEVEL	HEALTH IMPLICATIONS
0-50	Good	Quality satisfactory; pollution poses little or no risk
51-100	Moderate	Quality acceptable. However, some pollutants may moderately affect a very small number of people who are unusually sensitive to air pollution
101-150	Unhealthy for sensitive groups	Sensitive people may have trouble; others unlikely to be affected
151-200	Unhealthy	Everyone may get affected; sensitive people may have more trouble
201-300	Very unhealthy	Health warnings of emergency levels for all
300+	Hazardous	Everyone may experience more serious health effects

Coloured Fig. 17.1. (b) Global AQI Chart. The observed values dated Nov 18, '14 shows that air Quality in Delhi (India) is extremely poor.

well as their averaging periods, that need to be included in computing the AQI, are thus, needed to be specified by a country for its AQI values and standards.

The regulating agency takes real-time air quality data from its AQI monitoring sites to produce AQI readings for each location. AQI readings are reported to the public and news media at suitable time intervals, such as hourly reporting in Toronto (Canada), and daily reporting (as specified in India in the newly launched AQI concept).

Determining AQI. At the end of the specified period, such as one hour in Canada, the concentration of each pollutant, measured at each site, is converted into a number ranging from zero upwards using a common scale or index. The calculated number for each pollutant is referred to as a **sub-index**. At a given site, the highest sub-index for any given hour becomes the AQI reading for that hour. The index is a relative scale, in that, the lower the index, the better the air quality.

AQI Monitoring in India. *The Ministry* of *Environment, Forests and Climate Change** has launched the *National Air Quality Index* in New Delhi on September 17, 2014, under the '*Swachh Bharat Abhiyan*'. It is outlined as 'One Number-One Colour-One Description' for the common man to judge the air quality within his vicinity.

The computation of the newly launched AQI shall be based on eight air pollutants, namely PM_{10}, $PM_{2.5}$, NO_2, O_3, CO, SO_2, NH_3 and Pb**, for which short time (upto 24 hourly averaging period) National Ambient Air Quality standards are prescribed. Based on the measured ambient concentrations, corresponding standards and likely health effects, a sub-index is calculated for each of these pollutants. The worst sub-index reflects the final AQI value.

Based on pollutants concentrations shown in coloured Fig. 17.1 (*a*), there will be six AQI categories in India, namely (*i*) *Good*, (*ii*) *Satisfactory* (*iii*) *Modernately Polluted*; (*iv*) *Poor*; (*v*) *Very poor*; and (*vi*) *Severe*. The associated health impacts of each category have also been stipulated, as shown in coloured Fig. 17.1 (*a*). Coloured Fig. 17.1 (*b*) shows the global AQI chart and the observed values of AQI in certain World Cities, including Delhi. The different colours shown in the figures are specific to each category.

PROBLEMS

1. "Biosphere is an ecosystem", discuss critically this statement, explaining the essential components of the biosphere.
2. (*a*) Define Ecology and Ecosystem.
   (*b*) Explain the Ecological balance of nature.
3. (*a*) Define and explain the biotic and abiotic components of an ecosystem.
   (*b*) Explain in brief the impact of humans on the biosphere.
4. Write an essay on pollution and conservation of environment.
5. Explain the status of administrative control on environment in India.
6. What is the status of water pollution in India, and how is it monitored and controlled ?
7. Write a note on the status of air pollution in India, and explain as to how it is monitored and controlled.
8. Mention the national ambient air quality standards of important air pollutants prescribed in India under the Environment (Protection) Act, 1986, for different categories of areas.

---

*This is the new name given by the new BJP Govt. at the Centre (headed by Mr. Narendra Modi, the honourable Prime Minister of India) to the Ministry of Environment and Forests.

**Cancer causing substances like Benzene, which is found to be very high in cities like Delhi due to emissions from vehicle exhausts and vapours from petrol pumps, etc. is likely to be included at a later stage.

—18

Engineering of Air Pollution, Its Control, and Monitoring

18.1. Definition of Air Pollution

Air is one of the most important constituents of our environment. An average human being requires about 12 kg of air each day, which is nearly 12 to 15 times greater than the amount of food consumed. Eventually, even a small concentration of pollutants present in the air becomes more harmful to human health, in comparison to similar concentrations of pollutants present in food. The clean and pure air, free from outside solid, liquid or gaseous polluting substances (called pollutants), is evidently very essential for human health and survival. Any change in the natural and normal composition of the air (either qualitative or quantitative), that may adversely affect the living system, particularly the human life, invariably causes air pollution.

Air pollution is, therefore, defined as the presence of any solid, liquid, or gaseous substance (including noise) present in the atmosphere in such concentrations that may or tend to be injurious to human beings, or other living creatures, or plants, or property, or enjoyment. The solid, liquid, or gaseous substances, which when present in air, cause harmful effects on the abiotic and biotic components of our environment, are eventually called the **air-pollutants.** When the quantum of air pollutants exceed the self cleansing properties of the ambient air, and start causing harmful effects on the human health and his surrounding abiotic world, then the air is said to be polluted.

18.2. Natural and Man-made Air Pollution

Air pollution, may be caused by the naturally occurring events, like *volcanoes*—which release huge amounts of ash, dust, sulphur, and other gases into the air-atmosphere ; or by *forest fires*—that may occasionally be caused by lightening ; etc. In addition, the air pollution may be caused by human activities, such as *burning of fossil fuels, intentional burning of forests* to clear land for urbanisation or agriculture, etc. *Whereas, the air pollutants caused by the natural hazardous events tend to remain in the atmosphere for a short time ; the air-pollutants released by human activities may continue to stay in the air environment for long periods, and may even lead to permanent atmospheric changes.* One of the reasons for this is the fact that the natural hazardous events causing air pollution do occur very infrequently ; while the man-made releases of air pollutants is an ongoing continuous phenomena on daily basis.

Since the air pollution caused by the natural hazardous events is very infrequent and is automatically taken care of by the environment, we generally ignore this type of air pollution, and whenever we talk of air pollution, we always mean the air pollution caused by the human activities.

18.3. History and Causes of Air Pollution

The air pollution on Earth originated when the man started using firewood for cooking and heating purposes. Hippocrates, a well-known ancient physician, who practised medicine on the Greek Island of Cos, at about 400 BC and is considered to be the father of the medicine for his 80 written medical works, had written about the air pollution prevailing even in those ancient times. With the discovery and increasing use of coal, air pollution has gone on becoming more and more pronounced, especially in urban areas. Smoke pollution in London became so pronounced at about 1200 AD, that King Edward I was prompted to order the first anti-pollution law in the year 1273 to restrict people from using coal for domestic heating. In the year 1300, another law was enacted in London, banning the use of coal, and prescribing capital punishment for defiance of this law. Inspite of such harsh measures, air pollution assumed gigantic proportions in London during the *Industrial Revolution*, due to widespread use of coal in industries. So much so that a major disaster caused by '**London smog**' occurred in **December** 1952, which caused the death of about 4000 people due to accumulation of air pollutants over the London city for five consecutive days. By the middle of 19th century, barks of many trees in Europe, were blackened by smoke, due to which, the *peppered moths* (insect worms living on trees, like caterpillars) of all colours other than the black coloured varieties, were easily picked up by birds, as they failed to camouflage on black tree bark, as they did when on a clean lichen-covered tree. Thus, although the normal pepper-patterned moths continued to survive in clean non-industrial areas, only black-coloured moths could survive in industrial areas, as they alone could successfully camouflaged. Thus, with the increased air pollution, only black coloured moths could survive, and other species started to become extinct. *This is a classic example of air pollution, leading to adaptations and environmental imbalances caused in various biological species.*

Air pollution further began to increase in the beginning of the 20th century with the development of transportation systems, and large scale use of petrol and diesel in automobiles. The use of coal and diesel for running trains, use of petrol or diesel for running auto vehicles, and use of aviation fuel for running aeroplanes, have largely contributed to air pollution in modern days.

Air pollution caused by the emissions of *smokes* from factories, coke ovens, furnaces, steam engines, etc. ; *exhaust fumes* from automobiles, power plants, etc. ; injurious *chemical fumes* from oil-refineries, zinc refineries, chemical industries, metallurgical plants, iron and steel plants, incineration plants, etc. ; evolution of *radioactive gases* and *suspended radioactive dusts* from atomic explosions and accidental discharges from nuclear reactors, etc. have polluted our natural air environment to such an extent that special steps are now required at several places for reducing such air pollutions.

With the development of industries and increasing number of automobiles in its various towns and cities, India is now no longer left with its natural clean air. Air pollution disasters have, therefore, been occurring from time to time in India, as they used to occur in the presently developed countries in their initial years of development. The greatest Indian industrial disaster, leading to serious air pollution, took place in Bhopal on 03.12.1984, where the extremely poisonous Methyl Isocyanide (MIC) gas was accidentally released from the pesticide manufacturing plant of Union Carbide, a **U.S.** MNC.

18.4. Various Pollutants Causing Pollution of Air

The atmospheric air may contain hundreds of air pollutants from the natural or the anthropogenic (manmade) sources. All these pollutants, which are emitted directly from the identifiable sources, either from the natural hazardous events like dust storms, volcanoes, etc. ; or from human activities like burning of wood, coal, oil in homes or industries or automobiles, etc. ; are called the **primary pollutants.** The following five primary pollutants contribute to about 90% of the global air pollution :

The Important Primary air pollutants are :

1. *Oxides of sulphur*, particularly the *sulphur dioxide* (SO_2) ;
2. *Oxides of carbon* like carbon monoxide (CO) and carbon dioxide (CO_2), particularly the *carbon monoxide* (CO) ;
3. *Oxides of nitrogen*, like NO, NO_2, NO_3 (expressed as NO_x) ;
4. *Volatile organic compounds, mostly hydrocarbons* ; and
5. *Suspended particulate matter* (SPM).

Certain less important primary pollutants are *hydrogen sulphide (H_2S)* ; *hydrogen fluoride (H_2F) and other fluorides* ; *methyl and ethyl mercaptans*, etc.; which are usually rarely found in our general atmosphere, although if present, may prove quite harmful.

These primary pollutants often react with one another or with water vapour, aided and abetted by the sunlight, to form entirely a new set of pollutants, called the **secondary pollutants.** These secondary pollutants are the chemical substances, which are produced from the chemical reactions of natural or anthropogenic pollutants or due to their oxidation, etc., caused by the energy of the sun. These new pollutants are often more harmful than the original basic chemicals that produce them.

The important secondary pollutants are :

(*i*) *Sulphuric acid* (H_2SO_4) ;

(*ii*) *Ozone* (O_3) ;

(*iii*) *Formaldehydes* ; and

(*iv*) *Peroxy-acyl-nitrate* (PAN) ; etc.

H_2SO_4 is formed by the simple chemical reaction between SO_2 and H_2O vapour, and is a much more toxic pollutant than SO_2, having far reaching effects on environment, since it causes *acid rains* (discussed in detail in article 19.1).

Other secondary pollutants like ozone, formaldehyde, PNA, etc. are formed by the ***photochemical reactions,*** caused by sun light between two primary pollutants. Say for example, O_3 is formed due to the photochemical reaction between hydrocarbons (HC) and nitrogen oxide (NO). Similarly, aldehydes may be formed by photochemical oxidation of hydrocarbons in the atmosphere.

All these major air pollutants are now discussed below in details, mentioning their hazardous effects on human body, alongwith their sources of origin.

18.4.1. Oxides of Sulphur. Sulphur dioxide (SO_2) is the basic air pollutant amongst all the oxides of sulphur. SO_2 is an irritant gas, and when inhaled, affects our mucous membranes. It increases the *breathing rate* and *causes oxygen deficits* in the body, leading to *bronchial-spasms* in some of the affected persons. *Patients of asthma are very badly affected by this pollutant.*

Some quantity of atmospheric sulphur dioxide (SO_2) may oxidise to form sulphur trioxide (SO_3), which when inhaled, may dissolve in the body fluids to form sulphuric acid (H_2SO_4), which is a very strong corrosive acid. SO_3, thus, causes high and worse irritation even at lower concentrations, leading to *severe broncho-spasm*.

SO_2 is also responsible for causing acidity in fogs, smokes, and in rains, and hence is the major source of corrosion of buildings and metallic objects.

SO_2 mainly originates in the atmospheric air from the : *refineries* and *chemical plants, smelting operations,* and *burning of fossil fuels. Thermal power plants* may emit SO_2 quantities, as high as $\frac{1}{10}$th of the coal burnt by them. *Open burning of garbage* as well as *municipal incineration plants* may also emit sulphur dioxide in the air.

The new revised Indian ambient air quality standards (Table 17.7) prescribe the permissible maximum annual average concentration of SO_2 for residential areas, as equal to 50 μg/m^3, which approximates to 0.019 ppm* at 20°C.

*The concentrations of suspended particles and gaseous pollutants are generally expressed as micrograms per cubic metre (μg/m^3) at atmospheric temperature and pressure. The reason for expressing the pollutant's presence *as mass per unit volume of air is that : living things breathe air by volume but* are *affected by the mass of the pollutants absorbed by them.*

When the concentration of a pollutant is expressed in cum per million cum of air, we call it as ppm. The relation between ppm and μg/m^3 is given by the formula :

$$1\ \mu g/m^3 = \frac{1\ \text{ppm} \times \text{molecular mass of pollutant}}{\text{L/mol of pollutant at the given temperature and pressure } (V_2)} \times 10^3\ \text{L/m}^3.$$

The value of L/mol (*i.e.* volume in litres occupied by one molecule of pollutant gas) at 0°C and 1 atmospheric pressure *i.e.*, 760 mm Hg) is equal to 22.4. At other values of temperature and pressure, its value is governed by the **Combined gas law** as :

$$\frac{P_1V_1}{T_1} = \frac{P_2V_2}{T_2}$$

Contd...

where P_1 = 760 mm Hg

V_1 = Vol. at 0°C and 1 atm pressure = 22.4 L/mol

T_1 = 0°C = 273°K.

The value of V_2 at the given temperature (T_2) and pressure (P_2) can be easily determined and used.

Thus, at 20°C and at 1 atmospheric pressure,

$$\frac{760 \text{ mm of Hg} . V_1}{273 \text{ K}} = \frac{760 \text{ mm of Hg} . V_2}{(273 + 20) \text{ K}}$$

or $$V_2 = \frac{293}{273} \times V_1 = \frac{293}{273} \times 22.4 \text{ L/mol}$$

$$= \mathbf{24.04 \text{ L/mol.}}$$

$$\therefore \quad 1 \ \mu\text{g/m}^3 \text{ of } SO_2 = \frac{1 \text{ ppm} \times 64 \text{ g/mol}}{24.04 \text{ L/mol}} \times 10^3 \text{ L/m}^3$$

$$50 \ \mu\text{g/m}^3 \text{ of } SO_2 = \frac{SO_2 \text{ in ppm} \times 64 \times 10^3}{24.04}$$

or $$SO_2 \text{ in ppm} = \frac{50 \times 24.04}{64 \times 10^3} = 0.019 \text{ ppm.}$$

18.4.2. Carbon Monoxide (CO). Carbon monoxide is a colourless, odourless, and toxic gas, produced when organic materials like natural gas, coal, or wood are incompletely burnt. Vehicular exhausts are the single largest source of carbon monoxide, as the number of vehicles have increased and is increasing by leaps and bound over the world. Poorly maintained vehicles and those having inefficient engines, without being fitted with adequate pollution control devices, release greater amounts of carbon monoxide. Carbon monoxide possesses about 200 times affinity for blood-haemoglobin (H_b) than oxygen. Eventually, when inhaled, CO replaces O_2 from the haemoglobin, and forms what is known as carboxy-haemoglobin ($CO.H_b$). This carboxy-haemoglobin is of no use for respiratory purposes ; and hence, CO inhalation impairs normal oxygen transport carried out by the blood. Low levels of CO inhalations produce symptoms like headache, dizziness, reduction in reaction time, etc. Higher levels of CO inhalations may cause nausea, heart palpitations, difficulty in breathing, etc. In extreme cases, when about half of the haemoglobin of the blood is used up in forming carboxy-haemoglobin, death becomes a certainty.

Persons dying of carbon monoxide inhalations exhibit characteristic bright pink colour of the flesh due to the presence of pink coloured carboxy-haemoglobin in their bloods.

Presence of carbon monoxide in pregnant woman's blood has been attributed as a possible cause of still births and deformed offsprings.

Fortunately, however, carbon monoxide (CO) is not a permanent pollutant, since natural processes do convert carbon monoxide to other compounds that are not harmful. Therefore, the atmosphere can be cleared of carbon monoxide, if no new carbon monoxide is introduced into the atmosphere.

In cities having higher numbers of automobiles and industries, however, CO is found in as high concentrations as up to 60 mg/m³ (54 ppm) with still higher concentrations in tunnels, garages, and near the road intersections or the running automobiles, as against the maximum permissible concentration of about 10 mg/m^3.

The new revised national ambient air quality standards in India prescribe the maximum permissible concentration of CO, on hourly weighted average basis, as equal to 4 mg/m^3 for residential areas, as shown in Table 17.7.

18.4.3. Oxides of Nitrogen (NO_x). Atmospheric nitrogen may combine with oxygen at high temperatures, as generated during fuel combustion, to form nitric oxide (NO). The nitric oxide (NO) at low levels is relatively harmless, but at high concentrations may cause asphyxiation and respiratory discomfort, because it combines about 3 lakh times faster with haemoglobin than oxygen. Larger concentrations may cause quick deaths.

The nitrogen oxide further readily combines with the atmospheric oxygen, to form nitrogen dioxide (NO_2), which again is a very harmful compound for human health, as it is responsible for causing *eye and nasal irritations, respiratory discomforts,* and even deaths, as caused by NO. Prolonged exposures to somewhat higher concentrations of NO_2 exceeding the permissible concentration, has also been attributed to having caused serious diseases like *emphysema* (swelling due to air in body tissues), increased susceptibility to pneumonia, and even lung cancer.

Under the new revised national ambient air quality standards prescribed in India (Table 17.7), the maximum average annual concentrations of oxides of nitrogen as NO_2 for residential areas is 40 μg/m^3, which approximates to 0.021 ppm.

Very high concentrations of oxides of nitrogen, such as about 40 to 50 mg/m^3 of NO_2, may cause great respiratory discomfort, leading to quick deaths. For example, many deaths are reported to have occurred in a fire in a clinic in Clevland in USA in May 1929 due to evolution of NO_2 from the burning of X-ray films.

The oxides of nitrogen originates into the atmosphere from automobile exhausts, incineration plants, furnace smokes, etc ; all involving high temperatures due to combustion of fuel.

Oxides of nitrogen become more significant, as they are involved in the formation of secondary air pollutants, such as ozone.

18.4.4. Hydrocarbons (HC). Hydrocarbons are the group of compounds consisting of carbon and hydrogen atoms. They are either evaporated into the atmosphere from the petroleum fuel supplies (such as petrol, diesel, etc.) or are emitted out in the automobile exhausts as the remnants of petroleum fuel that did not burn completely. The hydrocarbons may, therefore, also be contained in the smokes of incinerators using petroleum fuel for burning. They may also be contained in the fumes of oil refineries.

Hydrocarbons are washed out of the air when it rains, and run into surface water. Their presence in water causes an oily film on the water surface.

In low concentrations and when washed away in quick succession by rains, they do not generally cause much harm ; but at high concentrations, they may cause irritation of eyes and respiratory tract, and atleast eight of them have been responsible for causing body cancers.

The hydrocarbons are generally divided into two categories, *i.e.*

(*i*) *aliphatic group* of hydrocarbons ; and

(*ii*) *aromatic group* of hydrocarbons.

The **aliphatic hydrocarbons** include : alkanes (methanes), alkenes (olefins), and alkynes. Out of these three varieties of aliphatic hydrocarbons, only the **alkenes (olefins) are found to be harmful,** as they are unsaturated and highly reactive in the atmosphere through photochemical reactions. Alkanes (methanes) are simply inert hydrocarbons and do not react photochemically. The third variety *i.e.*, the alkynes, though quite reactive, is generally not found present in the atmosphere, and hence is of no importance to us in the air pollution studies.

Due to this **photochemical reactionary property** of the **alkenes (olefin) type of hydrocarbons**, it has become very important to study and monitor the presence of such hydrocarbons in the air.

The presence of hydrocarbons in the city air, is, therefore, attracting a lot of attention, because the olefin type of hydrocarbons react with other pollutant gases, forming new pollutants, which are even more harmful than the individual original pollutants.

Hydrocarbons are chiefly released into the atmosphere by automobile exhausts. Paraffins and olefins have been found to be the chief hydrocarbons present in large quantities in the air of Los Angeles city of U.S.A., where heavy number of automobiles ply every day.

The specified *standard for non-methane types of aliphatic hydrocarbons under U.S. Ambient Air Quality Standards is 160* $\mu g/m^3$ (0.24 ppm).

The benzene and its related poly nuclear group of aromatic hydrocarbons are also considered to be very important in air pollution studies, because many of these compounds have been shown to be carcinogenic (likely to produce cancers). Increase in lung cancers in cities have been blamed on these hydrocarbons, caused by *automobile exhaust emissions.* Out of many varieties of such hydrocarbons, **Benzo (a) pyrene** has been found to be the most carcinogenic hydrocarbon, followed by *Benzo* (*e*) *acephenanthrylene*, and *Benzo* (*j*) *flouroranthene.* Various other varieties, with less carcinogenic properties are also known.

The Ambient Air Quality Standards adopted by the *Environment Protection Agency* (EPA) of USA, does not specify any particular standard for these aromatic hydrocarbons. However, considering their harmful potential, the new revised air quality standards of India has imposed limits on annual concentrations of Benzene and Banzo (a) Pyrene for not to exceed 5 $\mu g/m^3$ and 1 ng/m^3, respectively. (Pl. see Table 17.7)

Aldehydes and **ketones** may also be considered under hydrocarbons, because they are formed by the photochemical oxidation of hydrocarbons, as secondary pollutants in the atmosphere, although they may also be released by automobiles and incinerators along with hydrocarbons. Substances like **formaldehydes** cause *irritation of eyes, skins* and *lungs* ; and hence, may be quite injurious to health.

The presence of hydrocarbons in the environment may prove to be quite hazardous, and hence needs to be properly monitored.

18.4.5. Suspended Particulate Matter (SPM). The particulate matter in air may occur largely in *solid form* as particles of dust, smoke, fume, etc. ; and also in *liquid form* as mist and fog. The particles larger than a molecule but small enough to remain suspended in air are called **aerosols.**

The meanings of various types of solid and liquid particles, constituting **total suspended particulate matter (TSPM)** in air, are indicated below in table 18.1.

Table 18.1. Types of Particulates in Suspended Matter

S. No.	*Term*	*Meaning*	*Examples*
(*A*)	**Liquid particles**		
1.	*Mist*	Aerosols consisting of liquid droplets.	Sulphuric acid mist.
2.	*Fog*	Aerosols consisting of water droplets.	
(*B*)	**Solid particles**		
1.	*Dust*	Aerosols consisting of solid particles that are blown into the air, or are produced from larger particles by grinding them down.	Dust storm.
2.	*Smoke*	Aerosols consisting of solid particles or a mixture of solid and liquid particles produced by chemical reaction, such as by fires.	Cigarette smoke, smoke from burning of garbage, etc.
3.	*Fumes*	Generally means the same as smoke, but often used to indicate aerosols produced by condensation of hot vapour of metals.	Zinc/lead fumes.

The various types of suspended particles in air w.r. to their relative sizes are shown in Fig. 18.1.

The suspended particulate matter in the atmosphere is a variable component, and is introduced either through natural phenomena like winds, volcanic eruptions, pollens and spores, decomposing particles of organic matter, etc., or through human activities like mining, burning of fossil fuels, industrial processes, etc.

The suspended particulate matter in air may prove to be harmful to human health, inspite of the fact that the human respiratory system has a number of mechanisms for protecting the lungs from the entry of particles from air along with respiration. Infact, the bigger particles (> 10 μ) can be trapped by the hairs and sticky mucus in the lining of the nose (Pl. see Fig. 18.2). Smaller suspended particles of up to 10 microns (μ)* can, although reach tracheo-bronchial system, but get trapped there in the mucus. They are sent back to the throat by beating of hair like cilia, from where they can be removed by spitting or swallowing. However, very small suspended particles may still reach the lungs, and damage the lung tissues, causing diseases like asthma, bronchitis, and even lung cancer, when such particles bring with them toxic and carcinogenic pollutants attached to the surfaces of the particles.

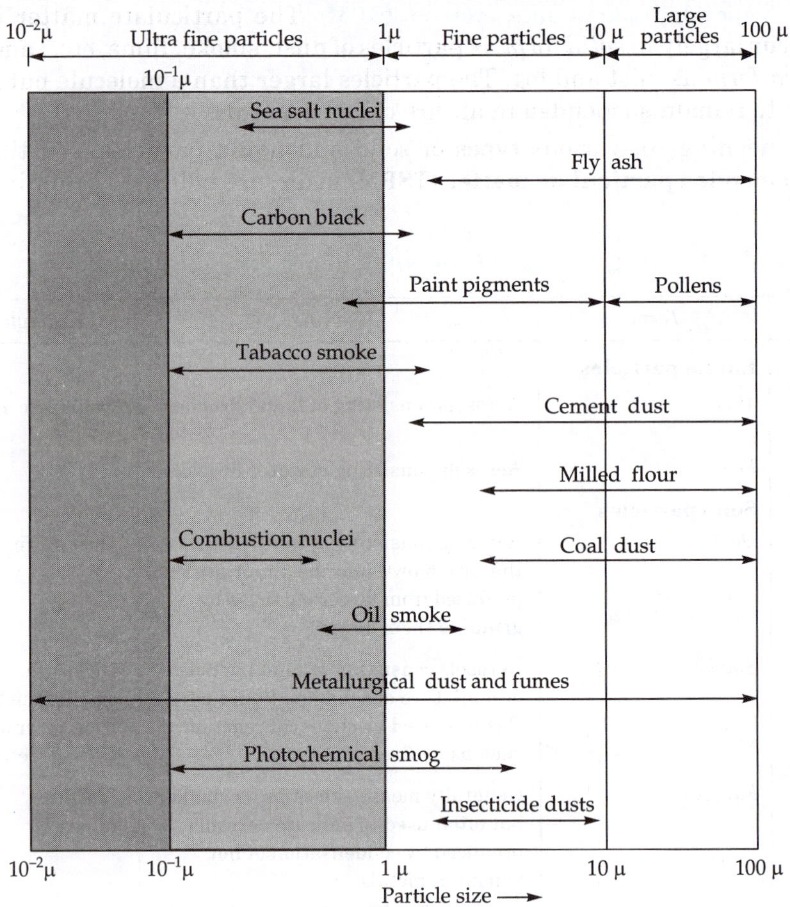

Fig. 18.1. Various types of suspended particles in air w.r. to their relative sizes.

During air pollution monitoring, it, therefore, becomes necessary as to not only study and control the total suspended particulate matter (TSPM), but also the more harmful smaller **Respiratory suspended particulate matter (RSPM).** The smaller sized particles of up to 10 micron in size are designated as **RSPM** or **PM 10.**

The old national ambient air quality standards (in India) prescribed the maximum annual average concentrations of TSPM = 140 $\mu g/m^3$; and of RSPM = 60 $\mu g/m^3$, for residential areas, as shown in table 17.6.

Further advancements in air pollution studies have also shown that the particles which are smaller than 10 micron in size, prove to be much highly dangerous to human health. Such smaller particles of sizes up to 2.5 micron, called **PM 2.5,** are hence being also monitored in modern days. Considering their importance, the revised national ambient air quality standards of India (Table 17.7) have specified max. annual concentrations of PM 10 and PM 2.5 to be 60 $\mu g/m^3$ and 40 $\mu g/m^3$, respectively.

Allergic agents. Various microscopic particulate substances, called **aero-allergens,** may also be present in air, and may cause allergic reactions in sensitive human bodies.

The reactions to such substances occur in our body mainly in the skin and the respiratory tract. Sneezing is one symptom of allergy, followed by *skin troubles and/or bronchitis, asthma*, etc.

Such organic allergens have been largely found to originate from living things like plants, animals, etc., although, however, finely powdered industrial materials may also sometimes cause allergic reactions in sensitive persons.

Pollen grains and fungal spores from local plants are one of the worst allergens. About 20 micro metre in diameter, *ragweed pollen* is usually deposited within 90 m of the parent plant, and hayfever and asthma sufferers coming within that range are liable to suffer severe allergic reactions and asthma attacks.

The high degree of allergy caused by *pollens and spores*, originating from the local plants, makes our beautiful city of Bangalore, a dreaded place for asthmatics and allergic persons. Instances have even been cited, where not only the asthmatics have had asthma attacks on them in Bangalore, but even normal persons who had never exhibited asthma tendencies earlier, got attacks of asthma in this city of fine climate. The *Asthma-Research Society* has infact, identified 75 types of air-borne pollens and 120 types of spores here ; among which the pollen of *parthenium plant* was found to be the highest (41%), followed by *grass pollen* (28.8%), and *pollens of cassia species* (11.8%).

Besides the pollens and spores, other important allergens are : *animal hairs, furs, feathers, dusts, spices, cotton flakes, flour, tobacco,* etc. besides certain chemical compounds. Say for example, SO_2 and compounds of cobalt and beryllium are found to cause allergic reactions in certain sensitive persons. Similarly, commercial fur dye—*paraphenylenediamine* is an allergic agent, capable of causing dermatitis in addition to bronchial asthma. Factories processing the castor beans for oil extraction, release a powdery material into the air, which is a strong allergen.

18.4.6. Lead is another major air pollutant that remains largely unmonitored, and is emitted by vehicles. High lead levels have been reported in the ambient air in our metro cities. Leaded petrol is the primary source of air-borne lead emissions in Indian cities. The compulsory use of unleaded petrol in cities like Delhi, has, however, reduced the lead menace from the atmospheric air here in this city. The revised air quality standards of India (Table 17.7) have prescribed a general annual concentration of 0.50 $\mu g/m^3$.

18.4.7. Photochemical Smog and Production of PAN and Ozone. Smog is a mixture of smoke and fog. It can be of two types ; *i.e.* photochemical and coal induced. The atmospheric conditions favourable for formation of *photochemical smog* are : the presence of high concentrations of hydrocarbons and nitrogen oxides in the atmosphere ; (*ii*) air stagnation, as not to cause dispersal of air pollutants ; and (*iii*) abundant sunlight. The formation of photochemical smog is restricted to highly motorised cities. The photochemical smog was first observed in Los Angeles (USA) in mid 1940 ; and since then, this phenomenon has been detected in most of the major metropolitan cities of the world. Smog is caused by the interaction of some hydrocarbons and oxidants like SO_2 or NO_x, under the influence of sunlight, giving rise to the formation of dangerous ***peroxy-acetyl-nitrate (PAN).*** The main constituents are nitrogen oxide, **PAN**, hydrocarbons, carbon monoxide, and large quantity of **ozone or ground level ozone. While PAN in itself is an extremely harmful gas, the ozone gas is also quite harmful to the respiratory system,** as it may enter deeper into the lungs (lower lobes) than even the oxides of sulphur. At concentration of about 0.3 ppm, cough and chest discomfort is increased. People suffering from chronic respiratory diseases are particularly susceptible.

The new revised national ambient air quality standards of India prescribe the maximum permissible concentration of O_3, on hourly weighted average basis, as equal to 180 $\mu g/m^3$ for residential areas, as shown in Table 17.7.

A very prominent example of occurrence of photochemical smog is offered by the **London smog,** which occurred over the city of London during 5th and 8th December, 1952, in which the residents were chocked and more than 4000 people died due to suffocation. It is understood that this incident occurred due to a combination of fog and smoke pollution containing sulphur dioxide released due to large scale burning of coal and oil, domestically as well as in fossil fuel power plants. At the time of this occurrence, the environment became stable without much wind movement, since an *anticyclone* reached London from the North-West and became stationary, with the result that the whole of London came under the pool of cold stagnant air, with a warm dense layer of air lying over the cooler air below, causing *subsidence inversion* [explained in article 18.6.2].

18.5. Effects of Air Pollution

One of the most obvious local effects of particles in the atmosphere is a *reduction in visibility*. Soot and smoke is usually quite apparent in highly industrialised and motorised cities. The reduction in visibility often causes slowdown of air-traffic, needing costly instrument-guided landing systems.

The presence of gases in the atmosphere has been causing the most prominent effect of producing *objectionable odour, taste,* or *obvious corrosive or chemical effects on objects.* Due to their chemical and corrosive effects, the air pollutants have been damaging the exteriors of buildings and cars by breaking down the paint emulsions, causing heavy economical losses. The air pollutants have been quickly spoiling the fabric and leather dresses worn by human beings, along with causing skin damages on their exposed body parts. Several monuments and old beautiful buildings made in exposed stone and marbles, have been badly spoiled and damaged by the acidic fumes. In addition to such apparent damages caused to the various objects, air pollution causes several adverse health effects on humans as well as plants and animals, besides affecting the goods and services, as discussed below :

18.5.1. Effects of Air Pollution on Human Health. The adverse effects of air pollution on human health has remained the prime consideration in air-pollution studies and research for obvious reasons. Although initially, the specific pollutants generating the observed adverse health effects, could not be clearly identified, but sufficient information could be gathered to implicate certain pollutants as the significant contributors to those adverse effects and diseases caused in humans. The research was, therefore, concentrated on those clearly identifiable pollutants.

As indicated earlier, the upper human respiratory system (nasopharyngeal) is quite efficient in filtering the larger particles of air we breath. Particles smaller than 5 microns (5 μm), however, can penetrate to the lungs, and be deposited in the alveoli. Fig. 18.2 shows the percentage of particles of various sizes deposited in the respiratory system, using an aerodynamic model.

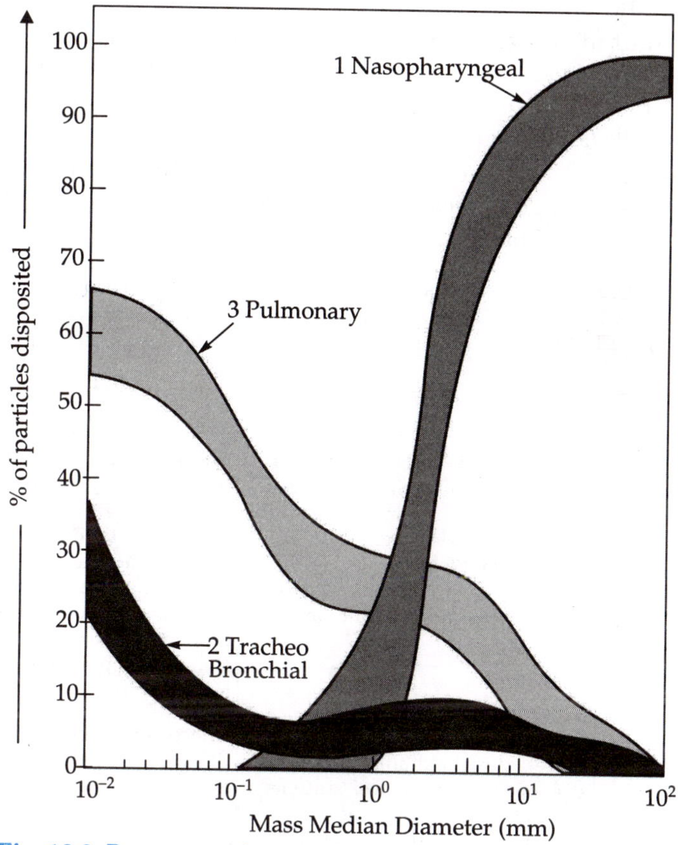

Fig. 18.2. Percentage of particles of different sizes deposited in the *respiratory tract*—consisting of (*i*) nose and throat region—called *nasopharyngeal* ; (*ii*) tubes leading to lungs—called *tracheobronchial* ; and (*iii*) lung regions—called pulmonary.

It can be seen from this figure that the smaller is the size of the particles, the greater percentage of the same will get deposited in the lung (pulmonary) ; whereas the larger sized particles will remain in the upper respiratory regions. This research led to the start of monitoring of smaller sized particles of sizes up to 10 micron (PM 10), called the Respiratory suspended particulate matter (RSPM) in 1987 in the United States, as compared to the previous monitoring of only the total suspended particulate matter (TSPM), as it was recognised that particles smaller than 10 micron are better indicator of potential adverse health effects of particulate air pollution. For example, cigarette smoke particles are smaller than 1 micron, and, hence, more than 25% of them get deposited in the lungs (Pl. see Fig. 18.2).

Some particles in the air are particularly damaging, because they absorb or adsorb gases, which cause more intense irritation locally. Gases also penetrate into the deepest lung pockets. Both particles and gases entering the body through the respiratory system can affect the gastro-intestinal system. Some chemicals such as lead, can enter the human blood stream either from the digestive system (ingestion) or by passing through the lung membranes (the respiratory system). Some other particles like tritium, systematic pesticides or herbicides, and a few other chemicals can even enter the blood stream through the skin.

Each pollutant affects the human body differently. The time of exposure is also an important consideration, as short term exposures to high concentrations of pollutants may cause *acute effects* ; while the long term exposures may cause *chronic effects.*

A brief description of the health effects caused by the different types of air-borne pollutants, though indicated while describing different pollutants, is again summarised in table 18.2.

Table 18.2. Effects of Air-Pollutants on Human Health

S. No.	*Pollutant*	*Characteristics if any*	*Source*	*Health Effects*
(1)	(2)	(3)	(4)	(5)
Sus-pen-ded	Particulate matter.	Solid particles like dust, smoke and fumes ; liquid particles like mist and fog.	Dust 1. storms ; cigarette smoke ; smoke from burning of garbage, fossil fuel ; and fumes like those of zinc or lead, etc.	Effects on breathing and respiratory systems, aggrevation of existing respiratory and cardiovascular diseases, alteration of body's defence systems against foreign materials, damage to lung tissues, carcinogenic effects, and premature mortality (still births). Elderly people and children are most sensitive.
2.	Sulphur dioxide (SO_2)	Colourless gas ; Taste threshold at about 0.3 ppm ; and Odour thres-hold at about 0.5 ppm.	Combustion of oil, and coal in power stations or automobiles.	Effects on breathing, respiratory illness, breakdown of lung defences, aggrevation of existing respiratory and cardiovascular diseases and death. Asthmatics and those suffering from chronic lung and cardiovascular diseases are sensitive to SO_2 exposures. Elderly people and children are greatly affected. It may lead to photochemical smog in some areas by oxidising the hydrocarbons.
3.	*Carbon monoxide gas* (CO)	Colourless, tasteless, odourless gas at atmospheric concentrations.	Incomplete combustion of coal and oil (diesel, petrol, etc.) fuels.	The health threat of CO is maximum to those having cardiovascular disease, because it reduces oxygen delivery to organs and tissues. At elevated concentrations, CO impairs visual perception, manual dexterity, and mental ability. Under short term exposure, it causes drowsiness and headaches. It also leads to formation of photochemical smog in some areas.

Contd...

(1)	(2)	(3)	(4)	(5)
4.	*Nitrogen oxides* (NO_x)	NO is a reddish-brown highly reactive gas. Odour threshold is at about 0.2 ppm.	High temperature combustion in automobiles (cars, etc.) and to some extent in thermal power stations.	NO plays a major role in tropospheric ozone formation. NO_2 irritates the lungs, causes bronchitis and pneumonia, lowers resistance to respiratory infections. Asthmatics are most susceptible, and increases susceptibility to viral attacks. It also plays a major role in tropospheric ozone formation.
5.	*Lead*	Colourless vapour. Consumed through inhalations. Lead compounds may also be consumed with food.	The major source is leaded petrol used by cars. Ingestion and inhalation may also occur from food, water, soils or dust.	High lead exposures can cause seizures, mental retardation, and behavioral disorders. Foetuses, infants and children are especially susceptible to low doses, resulting in disorders of central nervous system. Lead uptakes may be a factor in high blood pressure and heart disease.
6.	*Ozone* (O_3)	Colourless gas. Threshold odour is about 0.3 ppm.	A secondary pollutant produced by photochemical pollution, being the largest constituent of photochemical smog along with PAN, etc.	Ozone reduces lung function, and is associated with coughing, sneezing, chest pain, and pulmonary congestion. It may affect all healthy people as well as the people with impaired respiratory systems.
7.	Carbon dioxide gas (CO_2)	Colourless gas found in air.	Combustion of coal, petrol and diesel.	Increasing concentrations over the years cause green house effect, leading to global warming and climate changes.
8.	Nuclear waste	Invisible radio-active emissions and gases.	Nuclear power plants, nuclear weapon testing, wars, etc.	Causes radioactivity contaminations of areas ; cancers, mutations, deaths.

18.5.2. Effects of Air-pollution on Plants. The adverse impacts of air pollution are not limited to human health alone, but plants and animals are also detrimentally affected by air pollutants. The most prominent air pollutant, which causes severe damage to the plants is **fluorine.** Fluorine is emitted from factories manufacturing aluminium, glass, phosphate fertilizers, etc ; and in some clay baking operations. The concentrations of fluorine in excess of about 0.3 $\mu g/m^3$ cause *Phyto-toxicological effects* on plants. Frequently, the plant damage is observed on the fruit or on the flower, either of which significantly lowers the value of the crop. The plants are, infact, adversely affected by fluorine at much lower concentrations than what may be harmful to human health.

The most obvious damage caused by the general air pollutants like SO_2, H_2F, HCl, Cl_2, O_3, NO_x, NH_3, Hg, H_2S, H_2N, PAN, herbicides, smog, etc., to the plants and vegetation occurs in the leaf structure ; as the pollutants clog the stomata of the leaf, thereby reducing the intake of CO_2, which adversely affects the photosynthesis. These adverse impacts may range from reduction in growth rate to the total death of the plant.

Different plants and animals, however, have different susceptibilities to air pollutants. For example, *sugar maple* can tolerate relatively higher concentrations of sulphur dioxide alone, but it is susceptible to damage under the combined exposure of SO_2 and O_3 together ; *white pine* on the other hand is very sensitive to damage from either pollutant alone.

Plants also absorb low concentrations of pollutants like fluorine, and heavy metals like arsenic, lead, mercury, cadmium, etc ; which prove very harmful to cattle and animals grazing on these vegetations, even when the plants may not even show any signs of damage.

18.5.3. Effects of Air-Pollutants on Live-stock Animals. The adverse health effects of air-pollutants on farm animals occur, when animals eat the plants, grasses, fodder, and other vegetation, which has been contaminated by the air pollutants. Important contaminates that affect the live-stock are : (*i*) fluorine ; (*ii*) arsenic ; and (*iii*) lead. These pollutants originate either from the industries situated nearby ; or from dusting and spraying. Out of these contaminants, **fluorine** proves to be the worst pollutant, since cattle and sheep are found to be more susceptible to consumption of fluorine. The cattle grazing on vegetation that has been contaminated with fluorine may develop ***fluorosis,*** even when the contaminated plants do not show signs of damage. Symptoms of advanced fluorosis in animals include : lack of appetite, general ill-health due to malnutrition, lowered fertility, reduced milk production, and growth retardation.

Arsenic from dusts and insecticide sprays falling on the plants can similarly be accumulated by the plants, and when such contaminated vegetation is eaten by cattle, they may suffer from ***arsenic poisoning,*** with leading symptoms like salivation, thirst, vomiting, uneasiness, feeble and irregular pulse and respiration.

Lead contamination of the atmosphere occurs on account of various industries, such as smelters, coke ovens, and other coal-based industries. The lead fumes absorbed by the vegetation from the atmospheric air, contaminates the vegetation, which accumulates lead within itself. The lead-contaminated vegetation when eaten by live stock animals may cause ***lead poisoning,*** which may cause symptoms like inability to stand and staggering, prostration, etc ; besides complete loss of appetite, paralysis of digestive tract, and diarrhoea.

18.5.4. Effects of Air-Pollutants on Materials and Services. Air pollutants cause deteriorating effects on metallic surfaces, glass surfaces, painted surfaces, building stones, rubber surfaces, etc. along with damaging the paper and fabric.

Oxides of sulphur and nitrogen react in the atmosphere with water vapour to form acidic fumes, which attack and damage the metal surfaces, a problem that has been particularly acute for the computer, switch gear, and communications industries.

Fluorine is highly reactive, and at high atmospheric concentrations, may even cause etching of glass on windows, etc.

Hydrogen sulphide (H_2S) in the ambient air reacts with lead oxide in white paint, to form lead sulphate, due to which white painted surfaces in doors, windows, walls in buildings tend to acquire brownish tint overnight.

Accelerated aging of rubbers and synthetic fabrics due to exposure to the atmospheric oxidants have been causing very quick cracking of rubber tyres, and ageing and discolouration of synthetic fabrics. This was, infact, actually observed in 1940s in Los Angeles city of USA, where the air became so polluted that people started complaining for formation of cracks on their car tyres within a year, and the clothes worn by them started to wear out quickly along with showing discolouration in a short time.

Another peculiar impact of air pollution was noticed on electric cables and electricity poles, resulting in power leakage from high voltage electric cables. This was caused due to deposition of particles, which are good conductors of electricity, on insulators on electrical poles, resulting in leakage from high tension lines. In places like Windsor and North York, Ontario, it, therefore, becomes necessary to pressure-wash the insulators during extended dry periods, as to prevent electricity leakages, which in extreme cases, may even cause fires.

It is extremely difficult to estimate the financial loss caused by air pollution due to accelerated deterioration of materials and aesthetic items, such as the exteriors of the buildings, statues, monuments, or horticulture plantings ; or the cost of materials or service loss. It is, however, generally agreed that the annual damage from the atmospheric pollution runs into billions of dollars.

18.5.5. Effects of Air Pollutants on Aquatic Life. Air pollutants mixing up with rain can cause acidity in fresh water lakes, affecting aquatic life, especially the fishes. Some of the fresh water lakes have experienced total extinction of fish life.

18.5.6. Effects of Air Pollutants on Global and Regional Environment. Air pollutants may cause acid rains, ozone depletions, and global warming, thereby adversely effecting the environment at regional or global levels. These issues are discussed in the next chapter.

PREDICTING CONCENTRATIONS OF AIR POLLUTANTS

18.6. Dispersion of Air Pollutants into the Atmosphere

When once a pocket of smoke, containing air pollutants, is released into the atmosphere from a source like an *automobile* or a *factory chimney*, it gets *dispersed* into the atmosphere into various directions depending upon the *prevailing winds and temperature and pressure conditions* in the environment.

From our knowledge of meteorology and hydrology, we know that the temperature conditions of the environment are defined by a technical term, called *lapse rate.* For easier understanding, we will discuss lapse rate here, before discussing its effects on the *dispersion of the pocket of polluted smoke* (called **plume**).

18.6.1. Lapse Rate. In the *troposphere*, the temperature of the *ambient* (surrounding) *air* normally decreases with an increase in the altitude (height). *This rate of change of temperature is called the lapse rate.* This rate will differ from place to place, and from time to time even at the same place. Hence, the prevailing lapse rate at a particular time and the particular place, which can be determined by sending up a balloon equipped with a thermometer and a self recording mechanism, is known as the *prevailing lapse rate,* or the ***ambient lapse rate,*** or the **environmental lapse rate (ELR)**.

The graphical illustration of the lapse rate is shown in Fig. 18.3.

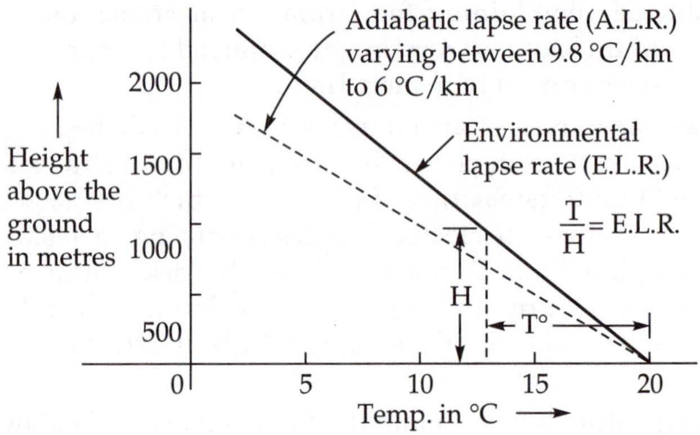

Fig. 18.3. The change of temperature with height in the environment, called E.L.R.

Under the prevailing environmental conditions, when a parcel of air, which is hotter and lighter than the surrounding air, is released, then naturally it tends to rise up, until of course, it reaches to a level, at which its own temperature and density becomes equal to that of the air surrounding it, at that height. Hence, when a pocket of artificially heated air (*i.e. automobile exhaust* or *stack gas*) is emitted into the environment, it rises up, *expands**, becomes lighter, and gets *cooled.* The rate at which the temperature decreases, as this parcel gains height (*i.e.* lapse rate), may be considerably different from the environmental lapse rate (ELR) of the air through which the smoke parcel moves. *Hence, it is very necessary to differentiate between the environmental lapse rate and the internal temperature change which occurs within the rising parcel of air gases.*

This internal decrease of temperature (cooling) with height, which occurs in the rising parcel of air mass, can be theoretically calculated, by assuming the cooling process to be *adiabatic***. In other words, as the air parcel moves up, its temperature decreases as its own heat energy is expanded due to increase in the volume of the air parcel. Using the *law of conservation of energy* and *gas laws,* therefore, it has been possible to mathematically calculate *this rate of decrease of temperature with height, called* **adiabatic lapse rate,** as shown for dry air in article 18.6.2.

---

*As we go up, the atmospheric *pressure reduces,* because the weight of air above us becomes less due to the reduced height of atmospheric air above us. Due to reduction in pressure, *the volume* of rising air *increases,* which means that the air *expands*.

***i.e.* a process occurring without the addition of outside heat or loss of its own internal heat.

Dry air, expanding and cooling adiabatically cools @ 9.8°C per km, and it is called the **dry adiabatic lapse rate.** In saturated (wet) air, this rate is calculated to be 6°C per km, and is known as **wet adiabatic lapse rate.**

Since a rising parcel of emitted smokes, will, normally, neither be fully dry nor fully saturated, the ***moist actual adiabatic lapse rate*** **(ALR)**, representing cooling of the emitted smokes, will be somewhere between the dry adiabatic rate (9.8°C/km) and wet adiabatic rate (6°C/km). This adiabatic lapse rate (ALR) is marked on the environmental lapse rate (ELR) diagram— Fig. 18.3, in a dotted line, for comparison of the two lapse rates. Depending upon the relative positions of the two lines (*i.e.* ALR line and ELR line) on the graph sheet, the ***stability of the environment*** is determined.

The three major relative positions of ELR line with reference to ALR line, are discussed below.

(*a*) **When the ELR (say 15°C per km) is more than the ALR (say 8′C per km)** as shown in Fig. 18.4 (*a*), then the environment is said to be **unstable.** In such a case, the rising parcel of air will always remain warmer than the surrounding environment. T*his is so because, as we go up, the environment is getting cooler more quickly than the rising parcel of lighter air, and hence the rising parcel of air will always remain warmer than the environment. The reverse is also true*, and hence a descending parcel of heavier air will always remain cooler than the surrounding air, because as we move down, the environment is getting warmer more quickly than the parcel of air.

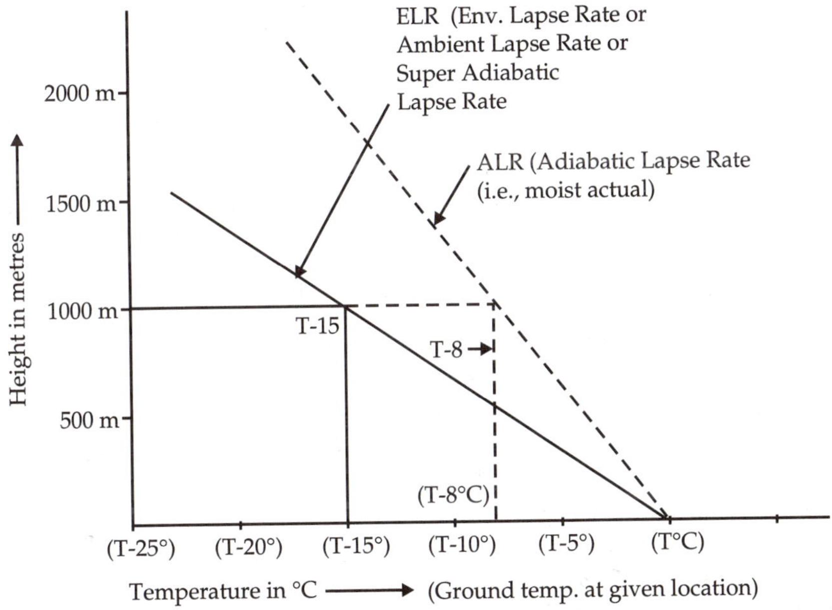

Fig. 18.4 (*a*). Unstable Environment.

It, therefore, follows that in such a case, when the environment lapse rate is more than the adiabatic lapse rate, a rising parcel of warmer lighter air will continue to lift up ; whereas a parcel of heavier cooler air will continue to come down. *In other words, the parcel of air will continue to accelerate in the direction of displacement.* ***In such circumstances, the environment is***

unstable, and the dispersion of pollutants will be rapid due to marked vertical mixing of the air, although, however, the high degree of turbulence may even sometimes bring the smokes touch the ground, under the pressure of the downward moving heavier air.

The prevailing environmental lapse rate (ELR) in such a condition, is known as **super adiabatic lapse rate,** as it is more than the adiabatic lapse rate.

(*b*) In the reverse case, **when ELR** (say 5°C per km) **is less than the ALR** (say 8°C/km), as shown in Fig. 18.4 (*b*), then the environment is said to be **stable,** and this prevailing environmental lapse rate is called the ***sub-adiabatic lapse rate*** (as it is less than the adiabatic lapse rate).

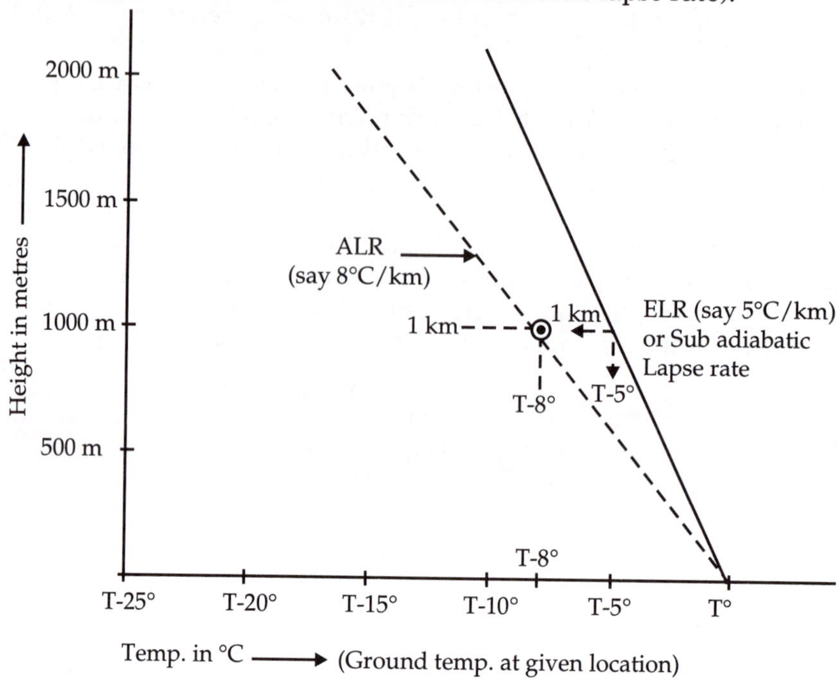

Fig. 18.4 (*b*) Stable Environment.

(*c*) The third case would be the one, when ELR equal the ALR, and both the lines coincide. The environment in such a case has **neutral stability**.

The ***super adiabatic conditions,*** thus. prevail when the air temperature drops more than the ALR (whose maximum value is 9.8°C/km). The ***sub-adiabatic conditions*** prevail when the air temperature drops at rate less than the ALR.

A special case of sub-adiabatic conditions is the ***temperature inversion,*** when the air *temperature actually increases with altitude and a layer of warm air exists over a layer of cold air.*

All these cases are shown together in Fig. 18.4(*c*). Super-adiabatic conditions are infact, unstable and favour dispersion of pollutants; while Sub-adiabatic conditions are stable and result in poor dispersion. Inversion causes more pollution, as it does not allow lifting of ground pollution.

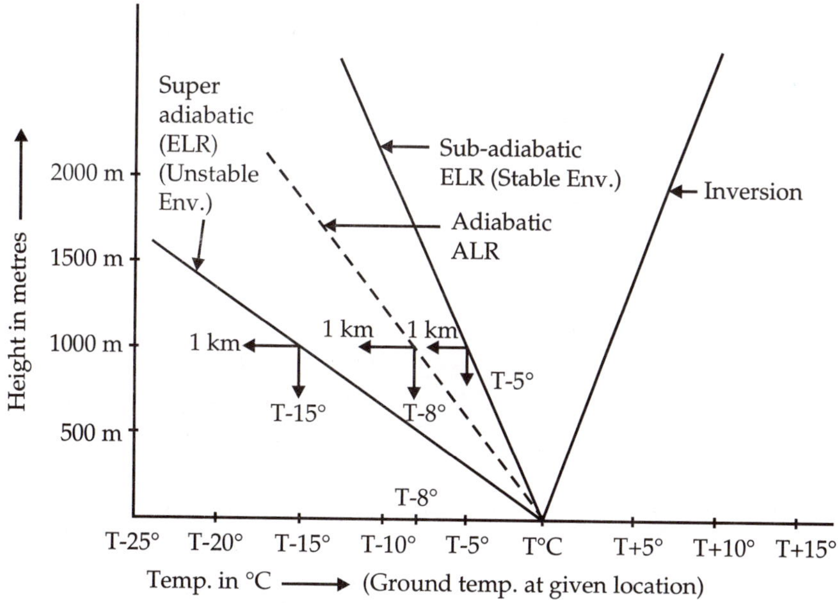

Fig. 18.4(c). Stability of the Environment.

18.6.2. Mathematical Expressions for Dry Adiabatic Lapse Rate. Lapse rate, as stated above, is defined as the rate of change of temperature (T) with height (z). The factor $\frac{dT}{dz}$ eventually represents the lapse rate. The lapse rate (LR) is usually represented by the symbol γ or α or T. We will, however, indicate Dry Adiabatic Lapse Rate as DALR, and Wet Adiabatic Lapse Rate as WALR. The mathematical equation for DALR is derived below :

A mathematical relation between the temperature (T) and altitude or height (z) can be obtained by considering the air to be an ideal gas, as discussed below :

(*i*) Let us first consider the *hydrostatic equilibrium of the atmosphere,* and derive the equation of hydrostatic equilibrium.

(i) Equation of Hydrostatic Equilibrium:

Consider a thin vertical slice of the atmosphere of cross-sectional area A, which starts at height z above the ground level and extends to a height of $z + dz$. The upward force exerted on this slice from the gas below is $P_{(z)} \cdot A$, where $P_{(z)}$ is the air pressure at height z. Likewise, the downward force exerted by the gas above the slice is $P_{(z+dz)} \cdot A$. The net upward force is clearly $[P_{(z)} - P_{(z+dz)}] \cdot A$. In equilibrium, this upward force must be balanced by the downward force due to the weight of the slice, which equals to $(A \cdot dz) \cdot (r_g)$ (*i.e.* vol. × unit wt.), where r is the density of the gas and g is the acceleration due to gravity. Eventually,

$$[P_{(z)} - P_{(z+dz)}]\, A \uparrow = [A \,.\, dz \,.\, \rho \,.\, g] \downarrow$$

or $$[P_{(z)} - P_{(z+dz)}]\, A = -[A \,.\, dz \,.\, \rho \,.\, g]$$

(–ve sign for reverse directions of forces)

Change in pressure $[P_{(z)} - P_{(z+dz)}]$ can also be expressed by dP.

$\therefore$ $$dP = -dz \cdot \rho \cdot g$$

or $$\frac{dP}{dz} = -\rho g \quad ...(18.1)$$

Now, let us use the *perfect gas law* as below :

The **Perfect Gas Law*** states that

$$PV = n \cdot R_u \cdot T \quad ...(18.2)$$

where P = Pressure in Pascal, P_a (N/m^2) of the gas

V = Vol. in m^3 of the gas

n = No. of moles of gas

T = Absolute temp. of gas in kelvin (K)

= °C + 273

R_u = ***The universal gas constant***

= **8.314 Nm/mol. K**

The Eq. (18.2) can further be modified to mass form implace of number of moles (n), since.

$$\text{No. of moles of gas } (n) = \frac{\text{Mass of gas } (m)}{\text{Molecular mass of gas } (M_0)}$$

or $$n = \frac{m}{M_0} \quad ...(18.3)$$

But mass = Volume × Density

or $$m = V \cdot \rho \quad (18.4)$$

or $$n = \frac{V \cdot \rho}{M_0} \quad (18.5)$$

---

*The various gas laws are

(*i*) ***Boyle's law:*** $V \propto \frac{1}{P}$

[Vol. of a gas of a given quantum is inversely proportional to its pressure at a constant temperature]

(*ii*) ***Charle's law:*** $V \propto T$

[Vol. of a gas of a given quantum is directly proportional to its temperature at a given pressure]

(*iii*) ***Avogadro's law:*** $V \propto n$

[Vol. of a gas is proportional to its amount *i.e.* number of moles (n) at a given pressure and temperature]

Combining all the above laws, we get the ***Perfect gas law*** as

$$V \propto \frac{n \cdot T}{p}$$

or $$PV \propto n \cdot T$$

or $$PV = n \cdot R_u T$$

where R_u ***is universal gas constant.***

Substituting the above value of n in Eq. (18.2), we get

$$PV = \frac{V\rho}{M_0} R_u \,.\, T$$

or
$$P = \rho\left(\frac{R_u}{M_0}\right).\, T \qquad \text{...(18.6)}$$

where $\frac{R_u}{M_0}$ is a new constant called the ***gas constant*** or the ***specific gas constant.*** Let us represent it by the symbol R for air (or dry air). The value of R for dry air is found = 287 J/kg.K. The value of R for water vapour, represented by R_w = 462 J/kg.K.

Eq. (18.6) then becomes

$$P = \rho \,.\, R \,.\, T$$

or
$$\rho = \frac{P}{R \,.\, T} \qquad \text{(18.7)}$$

Substituting this value of ρ (*i.e.* Eq. 18.7) in Eq. (18.1), we get

$$\frac{dP}{dz} = -\rho g = -\frac{P}{R \,.\, T} \cdot g$$

$$\frac{dP}{dz} = -\left|\frac{Pg}{R \,.\, T}\right|$$

or
$$\frac{dP}{P} = \left[\frac{-g}{R \,.\, T}\right] dz \qquad \text{...(18.8)}$$

The above equation is the equation of hydrostatic equilibrium.

Now, we can consider the adiabatic atmosphere :

(*ii*) ***Equation of the Adiabatic Atmosphere.*** Imagine a packet of air which is being swirled around in the atmosphere. It would be expected to remain at the same pressure as its surroundings, otherwise it would be mechanically unstable. Moreover, the air packet moves so quickly that it cannot exchange heat with its surroundings, mainly because air is a poor conductor of heat. Eventually, the air in the packet remains in adiabatic state (*i.e. it neither receives heat not lose its own heat*). In a *steady-state* atmosphere, as the air packet moves upwards, it expands due to reduced pressure, and cools adiabatically, its temperature always remains the same as that of its immediate surroundings. This means that we can use the *adiabatic gas law* to characterize the cooling of the atmosphere with increasing altitude. The most common form of the ***adiabatic law for adiabatic expansion of gases,*** from thermodynamics, is given as :

$$P \, . \, \rho^{-k} = \text{Constant} \qquad ...(18.9)$$

where $k^* = \dfrac{C_p}{C_v}$ = Ratio of sp. heats

$= \dfrac{\text{Sp. heat for constant pressure}}{\text{Sp. heat for constant volume}}$

But $\rho = \dfrac{P}{RT}$ (*i.e.* Eq. 18.7)

Eq. (18.9) then becomes

or $$P \, . \left(\frac{P}{RT}\right)^{-k} = \text{Constant}$$

or $$(P)^{1-k} \, . \, T^k = \text{Constant} \qquad ...(18.10)$$

Eq. (18.10) is an important useful form of adiabatic law.

Rearranging the above equation, we get

$$(P)^{1-k} = \text{Constant (say } C) \times T^{-k}$$

Differentiating, we get

$$(1-k)\,(P)^{(1-k)-1}\, dP = C \times -k \, . \, T^{-k-1} \, dT$$

or $$(1-k)\,(\rho)^{-k} \, . \, dP = C(-k)\, T^{-k-1}$$

Substituting the value of C as $(P)^{1-k} \, . \, T^k$, we get

$$(1-k)\,(P)^{-k} \, . \, dP = (P)^{1-k} \, . \, T^k \, (-k)\, T^{-k-1} \, dT$$

or $$dP = \frac{(P)^{1-k}}{(P)^{-k}} \left(\frac{-k}{1-k}\right) T^{-k-1+k} \, . \, dT$$

or $$dP = P\left(\frac{k}{k-1}\right) T^{-1} = P\left(\frac{k}{k-1}\right)\frac{1}{T} \, . \, dT$$

or $$\boxed{\frac{dP}{P} = \left(\frac{k}{k-1}\right)\frac{dT}{T}} \qquad ...(18.11)$$

Combining Eq. (18.11) with the equation of hydrostatic equilibrium (*i.e.* Eq. 18.8), we get the DALR as $\dfrac{dT}{dz}$ as :

$$\left(\frac{k}{k-1}\right)\frac{dT}{T} = \left[\frac{-g}{R \, . \, T}\right] dz$$

or $$\boxed{\frac{dT}{dz} = -\frac{g(k-1)}{k \, . \, R}} \qquad (18.12)$$

---

*Two specific heats are defined for gases ; *i.e.* one for constant pressure (C_p) and one for constant volume (C_v).

The ratio of specific heats (k) = $\dfrac{C_P}{C_v}$ equals to 1.66 for ideal monoatomic gas, and it is equal to **1.4 for air**, which is a diatomic gas.

This is the required final mathematical equation for computing the dry adiabatic lapse rate (DALR).

Using the above equation and substituting values of k and R for air as:

$$k = \frac{C_p}{C_v} = 1.4 \text{ for air}$$

$$R = \text{Specific gas constant for air}$$

$$= \frac{R_u}{M_0} = \frac{\text{Universal gas constant}}{\text{Molecular mass of air}}$$

where $R_u = 8314$ J/kilomole. K
$M_0 = 28.964$ kg/kilomole

$$\therefore \quad R = \frac{8314}{28.964} \text{ J/kg.K}$$

$$= 287 \text{ J/kg.K}$$

$\therefore$ Dry adiabatic Lapse Rate is hence given as :

$$\frac{dT}{dz} = \frac{-g\,.\,(1.4-1)}{1.4 \times 287 \text{ J/kg.K}} = \frac{-9.81\,(\text{m/sec}^2)\left(\dfrac{0.4}{1.4}\right)}{287 \text{ J/kg.K}}$$

or

$$\frac{dT}{dz} = (-)\,9.77 \times 10^{-3}\,\frac{\text{kg}}{\text{J}}\,.\,\text{K}\left(\frac{\text{m}}{\text{sec}^2}\right)$$

or

$$\frac{dT}{dz} = (-)\,9.77 \times 10^{-3}\,.\,\frac{\text{kg}\,.\,\text{K}}{\text{N.m}}\left(\frac{\text{m}}{\text{sec}^2}\right) \qquad [\because \; 1 \text{ J} = 1 \text{ N.m}]$$

$$= -\,9.77 \times 10^{-3}\,\frac{\text{kg}\,.\,\text{K}\,.\left(\dfrac{\text{m}}{\text{sec}^2}\right)}{\text{kg}\,\dfrac{\text{m}}{\text{sec}^2}\,.\,\text{m}} \qquad \left[\because \; N = \text{kg}\frac{\text{m}}{\text{sec}^2}\right]$$

$$= -\,9.77 \times 10^{-3} \text{ K/m}$$

$$= (-)\,9.77 \text{ K/km} \approx 9.8 \text{ K/km} \approx 9.8 \text{ K/km} = 9.8°\text{C/km}$$

[–ve sign indicates that T decreases as Z increases]

Hence, the dry adiabatic lapse rate is app. equal to 9.8°C/km.

The final equation for DALR *i.e.* Eq. (18.12) can be further simplified as:

$$\frac{dT}{dz} = (-)\,\frac{g\,(k-1)}{k\,.\,R} \qquad (i.e. \text{ Eq. } 18.12)$$

or

$$\frac{dT}{dz} = (-)\,g\,.\,\frac{\left(\dfrac{C_p}{C_v} - 1\right)}{\dfrac{C_p}{C_v}\,.\,R} \qquad \left(\because \; k = \frac{C_p}{C_v}\right)$$

or

$$\frac{dT}{dz} = -\,g\left(\frac{C_p - C_v}{C_p\,.\,R}\right) \qquad ...(18.13)$$

But we know that

$$\boxed{C_p - C_v = R} \quad \text{...(18.14)}$$

$$\therefore \quad \frac{dT}{dz} = -g \cdot \frac{R}{C_p \cdot R} = \frac{-g}{C_p}$$

or $$\boxed{\frac{dT}{dz} = \frac{-g}{C_p}} \quad \text{...(18.15)}$$

Hence, the dry adiabatic lapse rate can also be represented by equation (18.15) in addition to being represented by Eq. (18.13). The value of DALR when computed by Eq. (18.15) works out as:

$$\text{DLRL} = \frac{-g}{C_p} \quad (\textit{i.e.}\ \text{Eq. 18.15})$$

where C_p = Sp. heat for dry air in J/kg.K
= 1005 J/kg K (Appendix Table A-5)

$$= \frac{-9.81\ \text{m/sec}^2}{1005\ \text{kg} \cdot \frac{\text{m}}{\text{sec}^2} \cdot \text{m/kg.K}} \quad \left[\because \ \text{J} = \text{Nm} = \text{kg}\frac{\text{m}}{\text{sec}^2} \cdot \text{m}\right]$$

$$= \frac{-9.81}{1005} \cdot \text{K/m}$$

$$\approx -9.8\ \text{K/km} = -9.8°\text{C/km}.$$

18.6.3. Negative Lapse Rate and Inversion. In an unusual case, when the temperature of the environment (*i.e.* ambient air) increases with altitude, then the lapse rate becomes *inverted* or *negative* from its normal state. Negative lapse rate curve would then be of the type shown in Fig. 18.5, in comparison to the normal lapse rate line.

Negative lapse rate occurs under conditions, usually referred to as **inversion,** *a state in which the warmer air lies over the colder air below.* Such temperature inversions represent a *highly stable environment.*

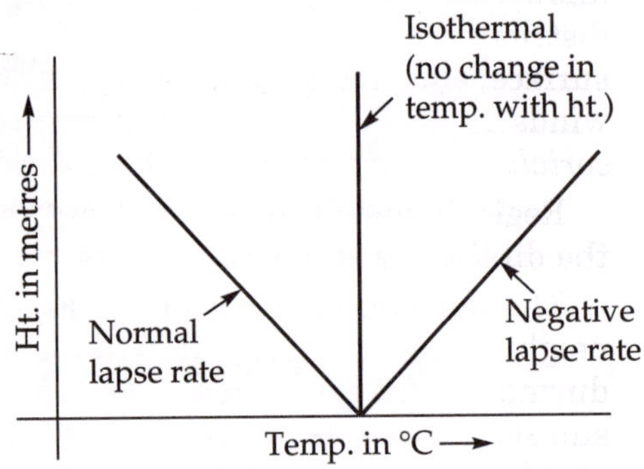

Fig. 18.5

There are two types of *inversions ; viz.*

(*i*) *radiation inversion* ; and

(*ii*) *subsidence inversion.*

(*i*) The **radiation inversion** is a phenomenon occurring from the unequal cooling rates for the Earth and the air above the Earth. In other words, *when* *the Earth cools rapidly and more quickly than the air above it**, then naturally the temperature in the environment will be less at the Earth and will increase above it*, causing negative lapse rate and inversion conditions.

This type of inversion may extend a few hundred metres into the friction layer, and is characteristically a nocturnal phenomenon that is likely to break up easily with the rays of the morning sun. Such an inversion in the environment, helps in formation of fog when air is wet, and simultaneously catches gases and particulate matter, as it stops their upward lifting, thereby creating concentration of pollutants in our close environment. This type of inversion is more common in winter than in summer, because of the longer nights. Valley areas may also have such inversions frequently, because of the absence of horizontal movement of air due to surrounding high ground.

(*ii*) The **subsidence inversion** is usually associated with a high pressure system, and is caused by the *characteristic sinking* or *subsiding motion of air in a high pressure area* surrounded by low pressure area (*i.e.* **anti-cyclone**). The air, circulating around the stationary high pressure, descends gently @ about 1000 m per day. As the air sinks, it is compressed and gets heated to form a warm dense layer over the cooler air below. Such inversion layers may be formed from the ground surface to around 1600 m or so. Such an inversion layer, by stopping the upward movement of polluting smokes, will cause the concentration of the pollutants in our immediate environment. When the thickness or height of this inversion layer is less than 200 m or so, extreme pollution would occur. Such an inversion may be more dangerous than the radiation inversion, and may occur at modest altitudes and may often remain for several days.

Sometimes, both the radiation as well subsidence inversions may occur simultaneously, causing what is known as **double inversion**.

.6.4. Impact of Winds on Dispersion of Pollutants. The moving air is known as *wind*. Such a movement in the air is caused by the unequal distribution of atmospheric temperature and pressure over the Earth's surface, and is largely influenced by the rotation of the earth. The direction of winds is always from high pressure areas to low pressure areas, but the *coriolis force* tends to deflect the air currents out of these expected patterns.

Regional and local geographical and topographical features may also affect the direction and speed of winds.

The quicker heating and cooling of the Earth as compared to the neighbouring sea, may also cause the flow of *sea breezes* from sea to land during day time, and flow of *land breezes* from land to sea during nights after sun set, respectively. *Such a wind pattern may also contribute to air pollution problems.* Los Angeles city of U.S.A., frequently experiences this wind pattern, which carries the air pollutants towards the sea in the evenings and nights, and brings them towards the city in the morning, as the sun advances.

In the *friction layer*** at the Earth's surface, winds are generally *gusty* and *changeable*, primarily due to *locally generated mechanical or thermal turbulence*.

---

*As may happen at nights, when the Earth may lose heat by radiation, and thereby cooling the surrounding air.

**Zone of atmosphere beneath 700 to 1000 m above the Earth's surface.

Wind speed is usually measured by an **anemometer** at a height, say Z_0. Knowing the wind velocity (u_0) at anemometer height (Z_0), we can work out the velocity u at any other height Z by using the formula

$$u = u_0\left[\frac{Z}{Z_0}\right]^k \quad ...(18.16)$$

where k is a constant $\approx \frac{1}{9}$ for large lapse rates, and $\frac{1}{3}$ for marked inversions, average normal value being $\frac{1}{7}$.

The direction and speed of surface winds primarily govern the drift and diffusion of *polluted gases* and *particulate* emissions from automobiles and factories, etc. emitted near the ground levels. *The higher the wind speed at or near the point of emission, the more rapidly the pollutants would be carried away from the source.* The pollutants so dispersed, will not exist at the same concentration, *but will rapidly be diluted with greater and greater volumes of air.*

On the other hand, when wind speeds are low, the pollutants tend to concentrate near the area of their emission ; and the longer the duration of such light winds, the larger will be the concentration of pollutants.

Gustiness, which is directly proportional to the wind speed, is another important characteristic of surface winds and determines the extent to which the pollutants are mixed and diluted with the ambient air.

Wind speed, can, thus, be related to the concentration of pollutants, *both being inversely proportional to each other*. This simple relationship is, however, complicated by various other factors, like *atmospheric turbulence and stability, geographical barriers in the flow of winds, presence of moisture in the environment, etc.*

18.6.5. Lapse Rates and Dispersion of Air Pollutants. We have pointed out in the previous article that the diffusion of pollutants into the environment is governed by the environmental lapse rate as well as the adiabatic lapse rate. *By comparing these two lapse rates, it is possible to predict to some extent, as to what will happen to gases emitted from a source ; the emitted gases being known as a plume, and their source of origin as a stack.*

Typical types of environmental conditions, characterised by different relative positions of environmental lapse rate and adiabatic lapse rate lines, which are generally encountered in the lower atmosphere (less than 300 m above the ground) are shown in Fig. 18.6 (*a*) to (*g*). The manner in which the emitted plume behaves under each of these seven conditions are also shown in these figures, and explained below :

(*a*) **Looping plume.** Looping plume [Fig. 18.6 (*a*)] has a wavy character and *occurs in super-adiabatic environment* ; which produces highly *unstable* atmosphere, because of rapid mixing. During the high degree of turbulence, the dispersion of plume would be rapid, yet higher concentrations near the ground may occur due to turbulence, before the dispersion is finally completed. Hence, in areas where environment is generally super-adiabatic, higher stacks may be needed to prevent premature contact of pollutants with the ground. Such conditions will then ensure a very good dispersion of pollutants ; but automobile exhausts cannot be dispersed well, because they are released at lower levels.

(*b*) **Neutral plume.** Neutral plume is the upward vertical rise of the plume from the stack, as shown in Fig. 18.6 (*b*), which occurs when the environmental lapse rate is equal to or very near to the adiabatic lapse rate. The upward lifting of the plume will continue till it reaches an air of density similar to that of the plume itself.

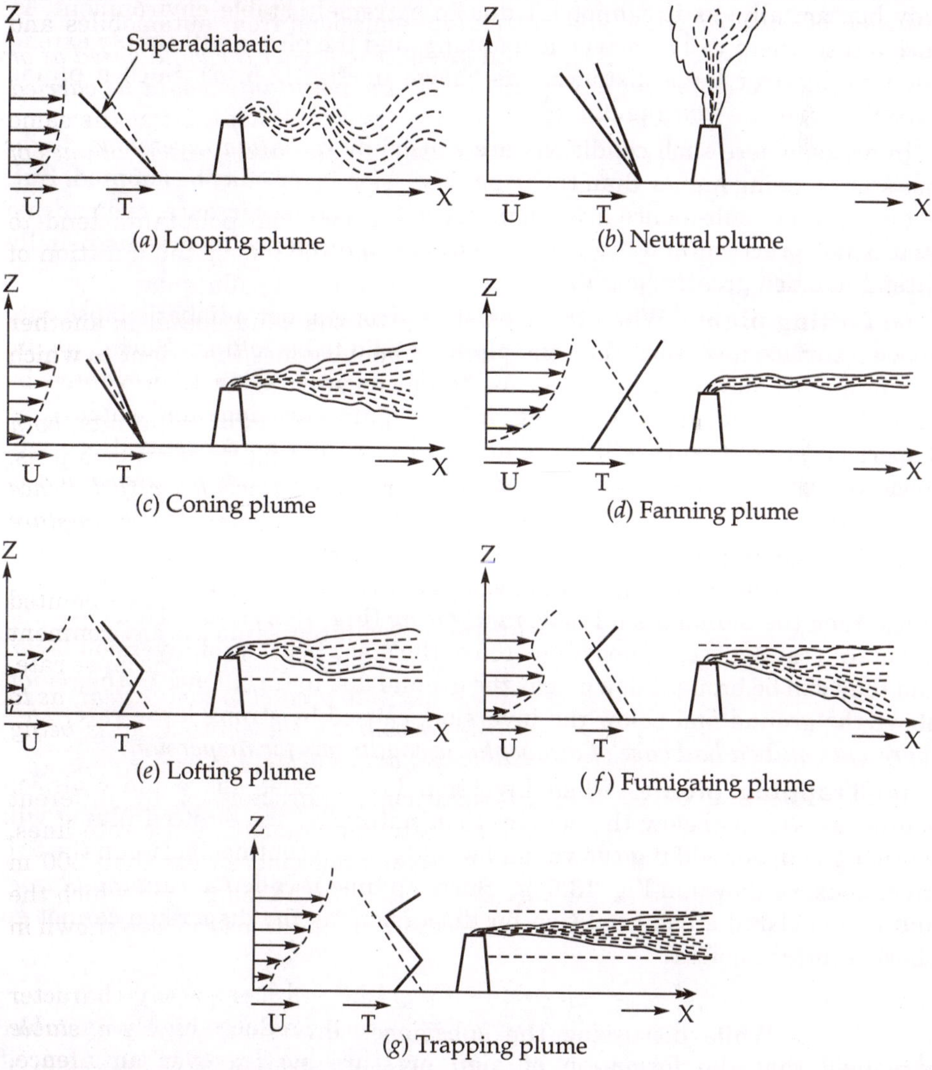

Fig. 18.6. Different types of plume behaviours.

(*c*) **Coning plume.** The neutral plume tends to *cone* as shown in Fig. 18.6 (*c*), when the wind velocity is greater than 32 km/hr, and when cloud cover blocks the solar radiation by day and terrestrial radiation by night.

Coning plume also occurs under *subadiabatic conditions* (*i.e.*, when environmental lapse rate is less than the adiabatic lapse rate). Under such conditions, the environment is slightly stable, and there is a limited vertical mixing, thereby *increasing the probability of air pollution in the area.* The plume dispersion is known as coning, because the plume makes a cone like shape about the plume line, as shown in Fig. 18.6 (*c*).

(*d*) **Fanning plume.** Under extreme *inversion conditions*, caused by negative environmental lapse rate, from the ground and upto a considerable height, extending even above the top of the stack, the emission will spread only horizontally, as it cannot lift due to extremely stable environment. In such a case, there will be no vertical mixing, and the plume will simply extend horizontally over large distances, as shown in Fig. 18.6 (*d*). Such a plume pattern is called a *fanning plume.*

In areas, where such conditions are caused by *radiation inversions,* high-rise stacks, rising higher than the usual inversion layer, may be adopted. But, in areas, where subsidence inversions are of frequent occurrence, even such a step is not practical and economical, because subsidence inversions usually extend to much greater heights.

(*e*) **Lofting plume.** When there exists a strong super adiabatic lapse rate above a surface inversion, then the plume is said to be '*lofting*'. Such a plume has minimum downward mixing, as its downward motion is prevented by inversion, but the upward mixing will be quite turbulent and rapid. The dispersion of pollutants will therefore, be rapid, and no concentrations will touch the ground. *Hence, this would be the most ideal case for dispersion of emissions.*

(*f*) **Fumigating plume.** When an inversion layer occurs at a short distance above the top of the stack, and super adiabatic conditions prevail below the stack, then the plume is said to be *fumigating* (Fig. 18.6 *f*). In such a case, the pollutants cannot escape above the top of the stack because of inversion layer, and they will be brought down near the ground due to turbulence in the region above the ground and below the inversion, caused by strong lapse rate. *This represents quite a bad case of atmospheric conditions for dispersion.*

(*g*) **Trapping plume.** When inversion layers exist above the emission source, as well as below the source, then naturally, the emitted plume will neither go up, nor will it go down, and would remain confined between the two inversions, as shown in Fig. 18.6 (*g*). Such a plume is called a *trapping plume,* and is considered a bad condition for dispersion, as the dispersion cannot go above a certain height.

18.6.6. Impact of Atmospheric Pressure on Dispersion of Air Pollutants. While discussing the subsidence inversion, we have earlier explained that the formation of high pressure system over an area, if continues for several days, could build up serious air pollution problems, due to formation of inversion conditions. Hence, *high pressure systems (anticyclones),* which are accompanied by *clear skies, light winds,* and *atmospheric stability, may prove to be bad for dispersion of pullutants.*

On the other hand, *low pressure systems* (**cyclones**)*, *which are associated with highly unstable atmospheric conditions, generally lead to good mixing and rapid dispersion of pollutants*. Hence, such cyclonic atmospheric conditions, which are usually accompanied by rains and storms, prove to be better for dispersion of pollutants.

However when a *warm front* overtakes a low pressure cell, conflicting influences are produced. Initially, warm front will reduce the pollutant load mainly due to storm activity along its leading edge ; but as the warm front develops, more stable conditions will form with an accompanying increase in the air pollution potential.

18.6.7. Impact of Moisture and Precipitation on Dispersion of Air Pollutants. The moisture content, and the *form* in which it is present in the atmosphere, may considerably affect the quality of air at a particular region.

The presence of water vapour (humidity) in the air affects the air quality, primarily by *blocking* and obstructing the solar radiation reaching the ground, and also the heat radiation reflected from the surface. Humidity also leads to formation of fogs, and increases the Earth's corrosive action of air pollutants.

Excessive moisture in the atmosphere will finally lead to rains, which are helpful in improving the quality of the ambient air, because they wash down the pollutants to the Earth, to be ultimately drained out with rain-run off.

The process of removal of atmospheric SO_2 through rain, may, however, cause problems due to reaction of SO_2 with water, forming H_2SO_3 or H_2SO_4, leading to fall of **acid rain,** which increases the rate of corrosion where air pollutants are present : and in addition, decreases the pH of rivers and streams, adversely influencing the algae and plant life of such water bodies.

18.7. Predicting Pollutants' Concentrations Through Dispersion Models and Equations

Mathematical approach has been applied to the problem of dispersion of pollutants into the atmosphere, although of course, the process of dispersion is quite complicate, and depends upon several meteorological factors, as discussed earlier. Moreover, the dispersion is generally three dimensional, which makes the mathematical solution a little more difficult. However, with the present day advent of computer modelling techniques, the problems can be solved with a little ease in these modern days.

Several empirical equations have been developed by several investigators. These equations or models do estimate the concentrations of pollutants in the plume at any distance x, y, and z, in horizontal down-wind direction, horizontal cross-wind direction, and vertical direction, respectively.

All such equations are based on **Fick's law of turbulent diffusion**. Assuming K_x, K_y and K_z as ***eddy diffusivity coefficients*** in three directions, and applying *continuity principle*, one gets :

$$\frac{\partial c}{\partial t} = \frac{\partial}{\partial x}\left(K_x \cdot \frac{\partial c}{\partial x}\right) + \frac{\partial}{\partial y}\left(K_y \cdot \frac{\partial c}{\partial y}\right) + \frac{\partial}{\partial z}\left(K_z \cdot \frac{\partial c}{\partial z}\right) + Q_{x,\, y,\, z} \qquad \text{...(18.17)}$$

*Low pressure at centre, surrounded by high pressure at outside.

where $Q_{x, y, z}$ represents the source or sink, expressed in appropriate terms.

While this equation will be comprehensive and general, it would not be amenable to direct integration, and can only be solved on a computer. Moreover, K_x, K_y, and K_z will also vary with location and time both, which will make the solution more difficult and complex.

On the basis of the above fundamental law, and based on the knowledge of the shape of the concentration distribution for a continuous single emission source, a simple statistical equation has been developed by Gaussian to compute turbulent transport of pollutants. The equation, known as **Gaussian Distribution***, is given as :

$$C_{x, y} = \frac{Q}{\pi u \sigma_z . \sigma_y} (e)^{-\frac{1}{2}\left(\frac{H^2}{\sigma_z^2} + \frac{y^2}{\sigma_y^2}\right)}$$

or
$$C_{x, y} = \frac{Q}{\pi u \sigma_z . \sigma_y} (e)^{-\frac{1}{2}\frac{H^2}{\sigma_z^2}} . e^{-\frac{1}{2}\frac{y^2}{\sigma_y^2}} \quad \text{...(18.18)}$$

where C = the concentration of pollutant in gm/m^3

Q = the pollutant emission rate in gm/sec

u = mean wind velocity in m/sec

x and y = downwind and cross-wind horizontal distances, respectively in m

σ_y = Plume's Standard deviation in cross-wind direction in m

σ_z = Plume's Standard deviation in vertical direction in m

H = Effective height of stack.

when concentration is required only along x-direction, *i.e.* in the downwind horizontal direction along the centre line of the plume, then naturally $y = 0$. Then, equation (18.18) becomes

$$C_{x, 0} = \frac{Q}{\pi u . \sigma_z \sigma_y} (e)^{-\frac{1}{2}\left(\frac{H^2}{\sigma_z^2}\right)} \quad \text{...(18.19)}$$

when the smokes are emitted at ground level, the effective stack height (H) is zero, then the above equation (18.19) gets further simplified, as

$$C_{x, 0} = \frac{Q}{\pi u . \sigma_z . \sigma_y} \quad \text{...(18.20)}$$

Values of σ_y and σ_z are not only a function of downwind distance (x) but are also a function of atmospheric stability. Values of σ_y and σ_z, for various values of x and different types of atmospheric stabilities, are given in Figs. 18.7 and 18.8 respectively.

*Representing the concentration of a gas aerosol < 20 μ in size.

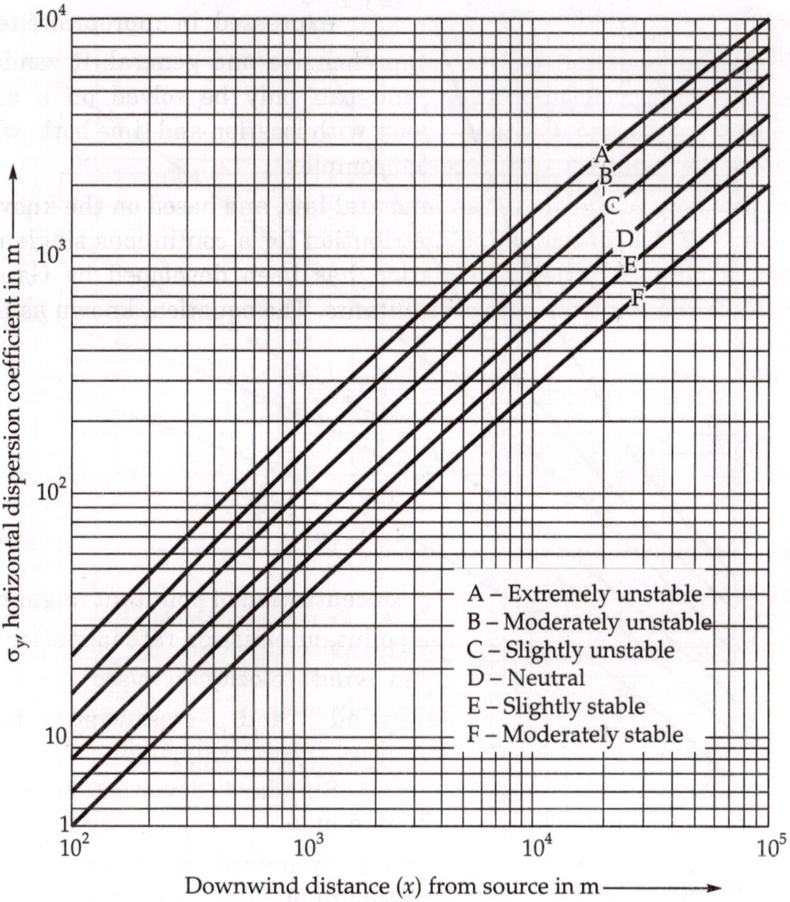

Fig. 18.7. σ_y *Vs x* **for different atmospheric stabilities.**

The generalised environment stability categories, mentioned in Figs. 18.7 and 18.8, such as *A*, *B*, *C*, *D*, *E* and *F* are detailed in Table 18.3.

Table 18.3. Pasquill Stability Types

Surface wind speed in m/sec	*Day*			*Night*	
	Incoming solar radiation			*Mostly overcast*	*Mostly clear*
	strong	*moderate*	*weak*		
< 2	*A*	*A—B*	*B*	—	—
2	*A—B*	*B*	*C*	*E*	*F*
4	*B*	*B—C*	*C*	*D*	*E*
6	*C*	*C—D*	*D*	*D*	*D*
> 6	*C*	*D*	*D*	*D*	*D*
	Category *A*—extremely unstable *B*—moderately unstable *C*—slightly unstable *D*—neutral *E*—slightly stable *F*—moderately stable				

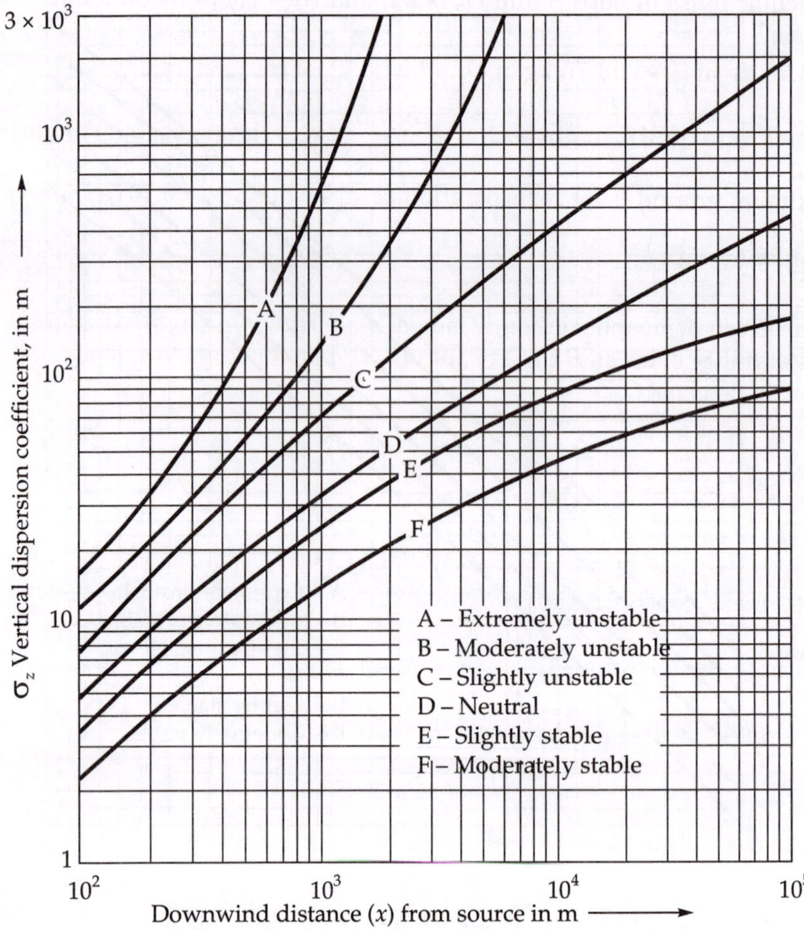

Fig. 18.8. s_z *Vs x* for different atmospheric stabilities.

The maximum ground level concentration occurs where $\sigma_z = 0.707H$, provided $\dfrac{\sigma_z}{\sigma_y}$ is constant with downwind distance x.

Example 18.1. *A coal-fired thermal power plant burns 6.25 tonnes of coal per hour, and discharges the combustion products through a stack having an effective height of 80 m. The coal has a sulphur content of 4.7%, and the wind velocity at the top of the stack is 8.0 m/sec. Atmospheric conditions are moderately to slightly unstable. Determine the maximum ground-level concentration of SO_2 and the distance from the stack at which this maximum occurs.*

Solution. *Emission rate for SO_2 :*

Coal burnt per hour = 6.25 t = 6250 kg

Sulphur content of coal = 4.7%

$\therefore$ Sulphur produced per hr $= 6250 \times \dfrac{4.7}{100}$ kg = 293.75 kg

Now, $S + O_2 = SO_2$

the molecular mass of both S and O_2 is 32, and they combine on a one to one mass basis.

$\therefore$ 293.75 kg of S + 293.75 kg of O_2 = 587.5 kg of SO_2 per hour

Emission rate of SO_2 in gm/sec $= \dfrac{587.5 \times 1000}{60 \times 60}$ gm/sec = 163.19 gm/sec.

The maximum ground level concentration occurs where $\sigma_z = 0.707\, H$, provided $\dfrac{\sigma_z}{\sigma_y}$ is constant with x.

For the given atmospheric conditions of *moderately to slightly unstable*, we have the stability class as *B* to *C*. (Table 18.3). For conservative designs, let us assume it to be class *C*.

The max. ground level concentration would occur where

$$\sigma_z = 0.707\, H = 0.707 \times 80 = 56.6 \text{ m}.$$

From Fig. 18.8, we can find that σ_z will reach a value of 56.6 m at x = 850 m (for *C* class of course).

Hence, max. concentration would occur at x = 850 m, provided $\dfrac{\sigma_z}{\sigma_y}$ is constant upto this distance. It can be confirmed from the chart (Fig. 18.8) that for class *C* conditions, $\dfrac{\sigma_z}{\sigma_y}$ is constant for distances (x) up to 1 km from the stack.

Hence, x_{max} = 850 m.

Now, to determine concentration at x = 850 m, we use the eqn. (18.19), as

$$C_{(x,\,0)} = \frac{Q}{\pi u \,.\, \sigma_z \,.\, \sigma_y} (e)^{-\frac{1}{2}\left(\frac{H^2}{\sigma_z^2}\right)}$$

From Fig. 18.7, σ_y = 88 m at x = 850 m (*C* class)

$$\therefore \quad C_{max} = \frac{163.19}{3.14 \times 8 \times 56.6 \times 88} (e)^{-\frac{80^2}{2(56.6)^2}} \text{ gm/m}^3$$

$$= 4.8 \times 10^{-4} \text{ gm/m}^3 = 480\ \mu \text{ gm/m}^3. \textbf{ Ans.}$$

Example 18.2. *From the data given in example 18.1, determine the ground level concentrations at a distance of two km downwind at : (a) The centre line of the plume ; and (b) at a crosswind distance of 0.5 km on either side of the centre line.*

Solution. (*a*) Concentration at x = 2 km along centre line of plume, means y = 0 and x = 2 km. This concentration is given by equation (18.19) as :

$$C_{(x,\,0)} = \frac{Q}{\pi u \,.\, \sigma_z \,.\, \sigma_y} (e)^{-\frac{1}{2}\frac{H^2}{\sigma_z^2}}$$

where σ_z = 130 (from Fig. 18.6 for x = 2 km and *C* class)

σ_y = 220 (from Fig. 18.7 for x = 2 km and *C* class)

$$\therefore \quad C_{(2,\,0)} = \frac{163.19}{3.14 \times 8 \times 130 \times 220}(e)^{-\frac{80^2}{2 \times (130)^2}} \text{ gm/m}^3$$

$$= 2.27 \times 10^{-4} \times (e)^{-0.189} \text{ gm/m}^3$$

$$= 2.27 \times 10^{-4} \times \frac{1}{(e)^{0.189}} \text{ gm/m}^3$$

$$= 2.27 \times 10^{-4} \times \frac{1}{1.208} \text{ gm/m}^3$$

$$= 1.878 \times 10^{-4} \text{ gm/m}^3 = \mathbf{187.8\ \mu\ gm/m^3.\ Ans.}$$

(*b*) Concentration at x = 2 km and y = 0.5 km (*i.e.*, 500 m) is given by eqn. (18.18) as :

$$C_{(x,\,y)} = \frac{Q}{\pi u \,.\, \sigma_z \,.\, \sigma_y} . (e)^{-\frac{H^2}{2 . \sigma_z^{\,2}}} . e^{-\frac{y^2}{2 . \sigma_y^{\,2}}}$$

$$= \frac{163.19}{3.14 \times 8 \times 130 \times 220}(e)^{-\frac{80^2}{2 \times (130)^2}} \times (e)^{-\frac{(500)^2}{2(220)^2}} \text{ gm/m}^3$$

$$= (1.878 \times 10^{-4}) . (e)^{-2.583} \text{ gm/m}^3$$

$$= 1.878 \times 10^{-4} \times 0.0755 \text{ gm/m}^3$$

$$= 0.142 \times 10^{-4} \text{ gm/m}^3 = \mathbf{14.2\ \mu.\ gm/m^3.\ Ans.}$$

18.7.1. Effective Height of a Stack. The value of H used in equation (18.18) is the effective height of the stack (chimney) and not its actual height. This effective height consists of actual height (h) plus the height (Δh) to which the plume rises above the stack before levelling out, as shown in Fig. 18.9.

$$\therefore \quad H = h + \Delta h \qquad \text{...(18.21)}$$

where h = actual height of stack in m

Δh = Plume height in m.

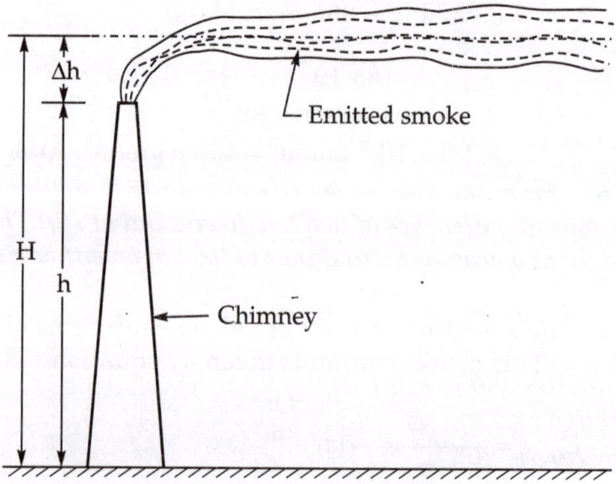

Fig. 18.9. Effective stack height.

There exists several equations for calculating the plume height Δh ; out of which, ***Holland's equation*** (18.22) is often used and is given as :

$$\Delta h = \frac{v_s \, . \, D}{u}\left[1.5 + 2.68 \times 10^{-3} \text{ P.D.}\left(\frac{T_s - T_a}{T_s}\right)\right] \qquad ...(18.22)$$

where Δh = rise of plume above the stack in m
v_s = stack gas velocity *i.e.* efflux emission velocity from the stack in m/sec
D = inside exit dia. of stack in m
u = wind speed in m/sec
P = atmospheric pressure in milli-bars
T_s = stack gas temperature in °K
T_a = air temperature in °K.

Equation (18.22) is quite suitable for computing Δh from neutral conditions. For unstable conditions, the above value of Δh should be increased by 10 to 20%, and for stable conditions, it should be decreased by 20 to 10%.

Another frequently used equation for computing Δh is given by *Davidson and Bryant,* as :

$$\Delta h = D \, . \left(\frac{v_s}{u}\right)^{1.4} . \left(1 + \frac{T_s - T_a}{T_s}\right) \qquad ...(18.23)$$

All the terms used in this equation have the same meaning as for equation (18.22).

The *Bureau of Indian Standards* (BIS), earlier known as ISI, has, through their code No. IS : 8829—1978 suggested the following empirical formulas for computing plume rise (Δh) :

(*a*) *For hot effluents with heat release of the order of 10^6 cal / sec or more :*

$$\Delta h = 0.84\,(12.4 + 0.09\,h)\,\frac{Q_H^{\;1/4}}{u} \qquad ...(18.24)$$

where Q_H = heat release in calories per second,
h = height of the chimney in m
u = wind velocity in m/sec

(*b*) *For not very hot releases, and which can be counted as momentum sources above :*

$$\Delta h = \frac{3v_s \, . \, D}{u} \qquad ...(18.25)$$

where v_s and D have the same meaning as in Eq. (18.22).

Example 18.3. *Determine the effective height of a stack, with the following given data :*

(*a*) *Physical stack is 180 m tall with 0.95 m inside diameter.*

(*b*) *Wind velocity is 2.75 m/sec.*

(*c*) *Air temperature is 20°C.*

(*d*) *Barometric pressure is 1000 millibars.*

(*e*) *Stack gas velocity is 11.12 m/sec.*

(f) Stack gas temperature is 160°C.

Solution. The given data is symbolised as below :

$h = 180$ m

$D = 0.95$ m

$u = 2.75$ m/sec

$T_a = 20'C = 20 + 273 = 293$ K

$P = 1000$ millibars

$v_s = 11.12$ m/sec

$T_s = 160'C = 160 + 273 = 433$ K.

Using equation (18.22), we have

$$\Delta h = \frac{v_s \cdot D}{u}\left[1.5 + 2.68 \times 10^{-3}\ PD\left(\frac{T_s - T_a}{T_s}\right)\right]$$

$$= \frac{11.12 \times 0.95}{2.75}\left[1.5 + 2.68 \times 10^{-3} \times 1000 \times 0.95 \times \frac{433 - 293}{433}\right]$$

$$= \frac{11.12 \times 0.95}{2.75}\left[1.5 + \frac{2.68 \times 0.95 \times 140}{433}\right] = 8.92 \text{ m}$$

or $\Delta h = 8.92$ m

H = Effective height of stack

$\therefore$ $= h + \Delta h$

$= 180 + 8.92 =$ **188.92 m. Ans.**

18.7.2. Design of Stack Height. The height of a chimney can, of course, be designed by utilising empirical equations connecting maximum concentrations of pollutants with stack height, and limiting the max. concentrations to permissible values. Such a thing would be possible, if there is only one stack, emitting effluents. There would, however, be several stacks in the area, which may cummulatively discharge their effluents, and hence, have to accounted for, properly.

Moreover, the meteorological conditions in the area would be variable throughout the year.

The emissions from tall stacks may also cause the fallout of contaminants at far off distances from the stacks.

Positions and locations of nearby buildings may also cause mechanical turbulence, and thus, bringing the plume to the ground level, especially when the stack is downwind of the building and wind speeds are high. To avoid this problem, stack height should be 2 to $2\frac{1}{2}$ times the height of nearby structures.

Moreover, the dispersion equations are generally derived for flat level topographic areas, and provision must be made for irregular terrain, when such areas are encountered.

Considering the Gaussian Plume Model with a number of assumptions, and such variable factors, *Central Board for Prevention and Control of Water Pollution, New Delhi,* has regulated the minimum chimney height for Indian conditions, by the following simple equations, under its *Emission Regulations*

*(July 1984) Part I** as well as under its publication titled *"A Method to Determine the Minimum Chimney Height"** :

(1) $$h = 74\,(Q_P)^{0.27} \qquad ...(18.26)$$

where h = chimney height in m

Q_P = Particulate matter emission in tonne/hr

(2) $$h = 14.(Q_S)^{1/3} \qquad ...(18.27)$$

where Q_s = SO_2 emission in kg/hr.

The maximum of the two heights calculated by the above equations, should be considered for adoption. The Board has further specified that the calculated chimney height for a given particulate matter and SO_2, by the above two equations, should be further subjected to the following **minimum values** :

(*i*) For Chimneys adopted *for industries in general* (except thermal power plants) ...30 m

(*ii*) *For Thermal Power Plants*

(*a*) Above 200 MW and below 500 MW capacity ...220 m

(*b*) Above 500 MW capacity ...275 m

As per these standards, the minimum stack was to be 30 m, but later on, the Board, has relaxed these provisions for chimneys of Boilers and Diesel Generator Sets to lower minimum values, as per their latest guidelines contained in "Emission Regulations Part IV". According to these provisions, *the minimum stack height for a boiler, generating steam @ less than 2 t/hr will be 9 m ; and that for a boiler producing steam @ 30 t/hr, it will be 30 m.* Intermediate minimum values for intermediate capacities are specified between 9 to 30 m.

Similarly, for *diesel generator sets* of different capacities (KVA) ; the minimum stack height to be kept is only 1.5 to 3.5 m more than the height of the building, and is to be worked out by using the equation

$$H = H' + 0.2\ \text{KVA} \qquad ...(18.28)$$

where H = stack height

H' = height of the building *KVA* is the capacity of generator.

Example 18.5. *An industry utilises 0.3 Ml of oil fuel per month. It has also been estimated that for every 1 Ml of fuel oil burnt in the factory, per year, the quantities of various pollutants emitted are given as :*

Particulate matter = 2.9 t/yr

SO_2 = *60 t/yr*

NO_x = *8 t/yr*

HC = 0.4 t/yr

CO = 0.5 t/yr.

Calculate the height of the chimney required to be provided for safe dispersion of the pollutants.

---

*These publications are available for sale at the Board's office at Privesh Bhawan, Sastri Nagar (Trans Yamuna area), Delhi.

Solution. Using equation (18.26), we have

h = height of chimney

$= 74\,(Q_P)^{0.27}$

where Q_P = emission of particulate matter in t/hr.

In the given case, the *particulate matter emission*

= 2.9 t/yr/Ml of fuel burnt

= 2.9 × (12 × 0.3) t/yr

= 10.44 t/yr (∵ fuel burnt per yr. = 12 × 0.3 Ml)

Now, $Q_P = \dfrac{10.44}{300 \times 24}$ t/hr (assuming 300 working days in the yr with 24 hr working)

or $Q_P = 1.45 \times 10^{-3}$ t/hr.

Using equation (18.26), we have

$h = 74\,(Q_p)^{0.27}$

$= 74 \times (1.45 \times 10^{-3})^{0.27}$ m

$=$ **12.67 m** ...(*i*)

It means that only 12.67 m high chimney is required for effective disposal of particulate matter.

Let us now calculate the height of the chimney required for effective disposal of SO_2 by using equation (18.27) as :

$h = 14\,(Q_S)^{1/3}$

where $Q_S = SO_2$ in kg/hr.

SO_2 emission = 60 t/yr/Ml of oil burnt

$= \dfrac{(60 \times 1000)}{(300 \times 24)}$ kg/hr/Ml of oil

$= \dfrac{60 \times 1000}{300 \times 24} \times (12 \times 0.3)$ kg/hr

= 30 kg/hr.

∴ $h = 14\,.\,(30)^{1/3}$ = **43.45 m ; say 43.5 m.**

Since the above requirement of 43.5 m is more than the minimum requirement of 30 m, 43.5 m high chimney is required for effective disposal of pollutants. **Ans.**

Note. Since the concentrations of NO_x, HC, and CO are generally much less than that of SO_2 in all industries, as in the above example, these pollutants have not been specified by the Board for considerations of chimney heights, and hence are ignored.

Example 18.6. *A coal fired 1000 MW power plant is operating around 38% efficiency. The ash and sulphur content in the coal used respectively are 35% and 3%, and the calorific value of 21 MJ per kg of coal. Find the emission rate of SO_2 from the plant.*

Solution. The electrical energy produced by 1000 MW power plant

= 1000 MJ/s (million joule per second)

Since the efficiency of the power plant is 38%, the heat energy required to be produced to generate 1000 MJ/s of electrical energy

$$= \frac{1000 \text{ MJ/s}}{0.38} = 2632 \text{ MJ/s}$$

Consumption of coal in producing 2632 MJ/s of heat energy

$$= \frac{2632 \text{ MJ/s}}{\text{Calorific value of coal in MJ/kg}} = \frac{2632 \text{ MJ/s}}{21 \text{ MJ/kg}} = 125.33 \text{ kg/s}$$

Sulphur content of consumed coal, *i.e.* sulphur produced per second

$$= 3\% \times 125.33 \text{ kg/s} = 3.76 \text{ kg/s}$$

Now, $S + O_2 = SO_2$

The molecular mass of both S and O_2 is 32, and they combine on a one to one basis:

$\therefore$ 3.76 kg of S + 3.76 kg of O_2 = 7.52 kg of SO_2/sec

$\therefore$ Emission rate of SO_2 in kg/sec = **7.52 kg/s. Ans.**

Example 18.7. *A coal fired 1000 MW power plant generates daily huge quantum of flyash, bottom slurries, besides particulate matter present in the burnt coal. Estimate the quantum of bottom ash, flyash and particulates generated daily by the plant. Also, suggest eco-friendly method of disposal of the flyash and bottom ash. Following information are available: Calorific value of the coal (ash content 35%) burnt 21 MJ / kg; 60% of ash discharged as flyash, and 75% of flyash and 90% of the bottom ash generated may be recovered; overall efficiency of the plant is 40%.*

Solution. Since 1 watt (W) is equal to 1 Joule/sec (1 J/s), we have

1000 MW power plant equals production of energy

$$= 1000 \text{ MJ/s}$$

Since the efficiency of the power plant is 40%, the total heat energy utilised by the power plant

$$= \frac{1000 \text{ MJ/s}}{0.4} = 2500 \text{ MJ/s}$$

Consumption of coal having calorific value of 21 MJ/kg

$$= \frac{2500 \text{ MJ/s}}{21 \text{ MJ/kg}} = 119.05 \text{ kg/s}$$

Since the coal used has an ash content of 35%, the total ash produced by consumption of 119.05 kg/s of coal

$$= 119.05 \text{ kg/s} \times 0.35 = 41.667 \text{ kg/s}$$

Hence, total ash produced per day

$$= 41.667 \times (24 \times 3600) \text{ kg/d} = 3.6 \times 10^6 \text{ kg/d} = \mathbf{3600 \text{ t/d.}} \quad \textbf{Ans.}$$

The best eco-friendly use of flyash generated by the plant can be made for manufacturing flyash bricks.

Example 18.8. *During rush hour on a busy road crossing, nearly 1200 vehicles ply per hour at an average speed of 20 kmph. Of these, about 70% cars use leaded petrol. The average fuel consumption is one litre for an average of 8 km of travel. Considering that 70% of the lead present in the fuel is emitted in the form of particulate aerosol, find the emission rate of lead aerosol in the ambient air. (Given : concentration of lead in the fuel 0.4 μg / L ; assume bright day with light wind).*

Solution. No. of vehicles passing per hour an the road

= 1200

Since 70% cars use leaded petrol, the no. of lead emitting vehicles passing per hour on the road

= 1200 × 70% = 840

Average fuel consumption per hour by each vehicle

$$= \frac{1 \text{ Litre}}{8 \text{ km}} \times 20 \text{ km} = 2.5 \text{ litres (L)}$$

Total fuel consumption by 840 vehicles per hour

= 840 × 2.5 L = 2100 L/h

Since the concentration of lead in fuel = 0.4 μg/L

The total lead contained in 2100 L/h of used fuel

= 0.4 μg/L × 2100 L/h

= 840 μg/h

Since 70% of lead present in the fuel is emitted as particulate aerosol, the lead aerosol released in air

= 840 μg/h × 0.7 = **588 μg/h**

Hence, 588 μg/h of lead is released in air as particulate aerosol. **Ans.**

AIR POLLUTION CONTROL

The atmosphere, like a river, do possesses self-cleansing properties, which continuously clean and remove the pollutants from the atmosphere under natural processes. So long as the pollutants discharged by man into the environment, is lower than the natural cleansing capacity of the environment, we live happily without any air pollution problem. But as and when the discharged pollutants exceed the natural cleansing capacity, our environment becomes polluted, giving us diseases, spoiling our clothes, plants, eatables, buildings, etc. Attempts are then made to reduce the emission of pollutants from the automobiles or factories by adopting mechanical means, or by using high rise chimneys for better dispersion and dilution of pollutants over a longer range of environment. The natural self-cleansing process of the environment, and the engineering measures adopted to artificially clean the industrial and vehicular gases, before they are emitted into the environment, are discussed in the following articles.

18.8. The Natural Self-Cleansing Properties of the Environment

The various natural properties, which continuously clean the environment, automatically, are :

(*i*) *dispersion* ;

(*ii*) *gravitational settling* with or without *flocculation* of particles ;

(*iii*) *absorption* including *washout* and *scavenging* ;

(*iv*) *rainout* ; and

(*v*) *adsorption*.

All these processes, naturally occur periodically in the ambient air, thereby removing or reducing the various pollutants entering the air. These processes are briefly discussed below :

18.8.1. Dispersion. Dispersion of pollutants by winds reduces the concentration of air pollutants at one place, although in the strict sense, it does not remove them from the environment as a whole. This mechanism is therefore, a *diluting* mechanism only.

Ironically, we have been stressing extremely hard on providing long chimneys for causing effective dispersion of pollutants, although however, in the strict sense, these long chimneys are only a means of spreading our own pollutants around our neighbours. The dispersion through long chimneys, therefore, helps us in diluting the air pollutants near their sources only, and does not reduce their long term undesirable effects on the community as a whole. For example, it has been estimated that SO_2 emitted in other countries, particularly in Great Britain, leads to *acid rains* in Swedon, whose environment is found to contain 15 to 50% of transported outside SO_2.

18.8.2. Gravitational Settling. Gravitational settling is the most important natural mechanism, under which large heavy particles from the ambient air settle down on buildings, trees and other objects. This generally happens for the particles which are larger than 20 μm in size.

This process also helps in removing **flocculated particles** formed by uniting of smaller particles over larger particles, till a floc particle, large and heavy enough to settle out under gravity, is formed.

18.8.3. Absorption. In the natural **absorption process,** the *gaseous* as well as *particulate* pollutants from the air get collected in the rain or mist, and may settle out with that moisture. This phenomenon takes place below the *cloud level*, when falling raindrops absorb pollutants, and is also known as **washout** or **scavenging.** The process, however, does not help in removing particles smaller than 1 μm in size.

The gaseous pollutants are removed in dissolved state with moisture, either with or without chemical changes.

18.8.4. Rainout. Rainout is the process involving precipitation above the cloud level, where submicron particles present in the atmosphere in the clouds, serve as *condensation nuclei,* around which drops of water may form, and fall out as raindrops. This phenomenon helps in increased rainfall and fog formation in urban areas, containing huge quantities of such particles, rising high above the cloud level.

18.8.5. Adsorption. Adsorption is the phenomenon in which the gaseous, liquid, or solid pollutants present in the ambient air are kept *attracted, generally electrostatically*, by a surface, where they are concentrated and retained. Natural surfaces, such as soils, rocks, leaves, blades of grass, buildings, and other objects can adsorb and retain pollutants. The particles may come in contact with such surfaces either by *gravitational settling* (as stated earlier) or by *inertial impaction,* under which the pollutants are carried to such surfaces by winds. *Impaction* is particularly effective for particles of size 10 to 15 μm.

18.9. Dilution Method for Controlling Air Pollution from Stationary Sources (Factories)

As stated in the earlier paragraph, the emitted smokes can be spread over a larger area, through the use of **high rise chimneys,** thereby transporting the pollutants over larger distances, and, thus, reducing the pollution near the emission source.

This method is largely adopted in developing countries, because here, the pollution is generally confined over smaller environment near cities and industrial towns only. The neighbouring environment, which is free from emissions, is thus made to share some of the pollutants burden, thereby causing somewhat equitable distribution of the pollutants. Nevertheless, this method only reduces the concentration of pollutants at particular place(s), rather than reducing or removing the pollution load from the total environment, as a whole.

18.10. Controlling Air Pollution from Stationary Sources by Installing Engineering Devices

In order to reduce the pollution load entering the environment from stationary sources, several measures may be taken ; out of them, replacement of burning fuel by electricity or solar energy is by far, the best method, as it will eliminate the very production of pollutants in the combustion process. Besides, this, we can use better quality of fuels and efficient engines, for reducing pollution loads from emissions. Say for example, LPG (liquid petroleum gas) and LNG (liquefied natural gas) may be used in industries in place of coal, as they will produce much less pollutants in the emissions. Replacement of old obsolete processes in industries with the new efficient processes may also lead to reduced pollution emissions. Besides such innovations and precautions, certain **mechanical devices** may be installed in the industrial processes, which may help in reducing the emission of pollutants.

Such mechanical devices, are generally divided into two categories ; *i.e.* (*i*) *those devices which help in reducing particulate* matter ; and (*ii*) *those devices which help in reducing gaseous pollutants.* These devices are discussed below:

18.10.1. Control of Particulate Pollutants in Industries. The important devices, which are used to control particulate matter, are :

(*a*) *Gravitational settling chambers.*

(*b*) *Centrifugal collectors* including *cyclone collectors* and *dynamic precipitators* ; and

(*c*) *Wet Scrubbers* (collectors) including *spray towers, wet cyclone scrubbers* and *venturi-scrubbers.*

(*d*) *Electrostatic precipitators* ; and

(*e*) *Fabric filters.*

The sizes of particles removed by these mechanical devices, their working principles, comparative merits, demerits, and uses are summarised in table 18.4. Their detailed designs are, however, beyond the scope of this book.

Table 18.4. Comparative Study of Various Mechanical Devices Used in Particulate Emission Control

S. No.	*Name of Mechanical Device*	*Minimum particle size that can be removed in μm*	*Efficiency %*	*Merits*	*Demerits*	*Principle of Working and Uses*
(1)	(2)	(3)	(4)	(5)	(6)	(7)
1.	**Gravitational Settling Chambers** (Fig. 18.10)	> 50 μm	< 50%	Simple to design and maintain, and low pressure loss.	Requires lar-ger space, and collection efficiency is also low. Only lar-ger sized particles are separated out.	The emitted smokes, when made to pass through a settling chamber (Fig. 18.10), drop some of their larger sized particles in the chamber, under Stoke's law. The largest size particle (*d*) that can be removed with 100% efficiency in such a chamber of length *L* and height *H* is given by equation $$d = C \cdot \sqrt{\frac{18\mu \cdot v_h \cdot H}{g \cdot L \cdot \rho_p}} \quad ...(18.29)$$ where v_h = horizontal velocity of gas passing through the chamber, between 0.5 to 2.5 m/sec. ρ_p = density of particles removed. μ = viscosity of air, at the given temperature in kg/m. sec C = correction factor for existing non-quiescent conditions in the gas flow through the chamber, generally taken at equal to 2.

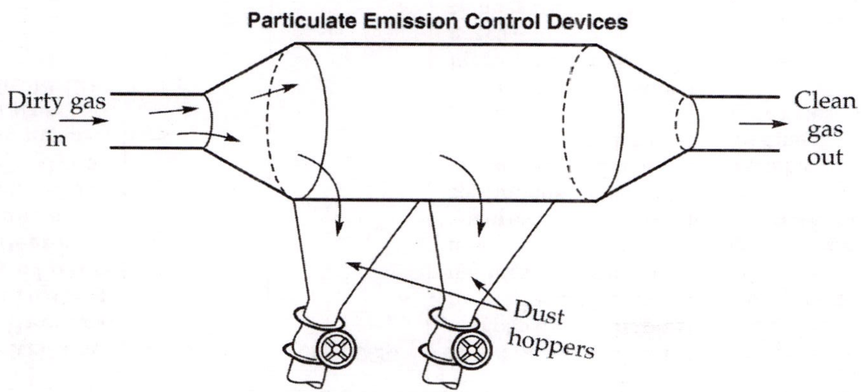

Fig. 18.10. Gravitational settling chamber.

Contd.

(1)	(2)	(3)	(4)	(5)	(6)	(7)
2.	**Centrifugal Collectors** including **Cyclone collector** or *separator* (Fig. 18.11) and **Dynamic precipitator** (Fig. 18.12)	5—25 μm	50 – 90%	Relatively inexpensive, simple to design and maintain ; requires less floor area ; ensures dry continuous disposal of collected dusts ; low to moderate pressure loss (2.5 to 20 cm) ; can handle large volumes of gases at temperatures upto 90°C. Cyclones are widely used in industries producing larger quantities of gas containing larger sized particles, like, Cement and Fertiliser plants, Petroleum refineries, Asphalt, mixing plants, Grain mills, Cotton gins, etc.	Requires much head room ; collection efficiency is low for smaller particles, quite sensitive to variable dust loadings and flow rates.	A **cyclone collector** (Fig. 18.11) is a specially designed closed chamber, in which the velocity of the inlet gas is transformed into spinning vortex, and the particles from the gas are thrown out under the centrifugal force. The particles thrown out on the walls of the chamber, slide down to the hopper, and, are thus removed. Its efficiency depends on the generated centrifugal force, which in turn, depends on mass of the particles (M_p), inlet gas velocity (v_i), and radius of cyclone (R), and is given by the equation F_c = centrifugal force generated $= M_p \cdot \frac{v_i^2}{R}$ ...(18.30) A **dynamic precipitator** (Fig. 18.12) imparts centrifugal force to the entering gas with the help of rotating vanes, and is, thus, about 7 times more effective than an ordinary Cyclone. Such a unit can work as an exhaust fan as well as a dust collector. They are widely used in Ceramics, Food and Pharmaceutical, and Wood working industries. They cannot, however, handle wet fibrous material, which tend to accumulate on the moving vanes. A photoview of such a device is shown in Fig. 18.12.

Contd.

Particulate Emission Control Devices

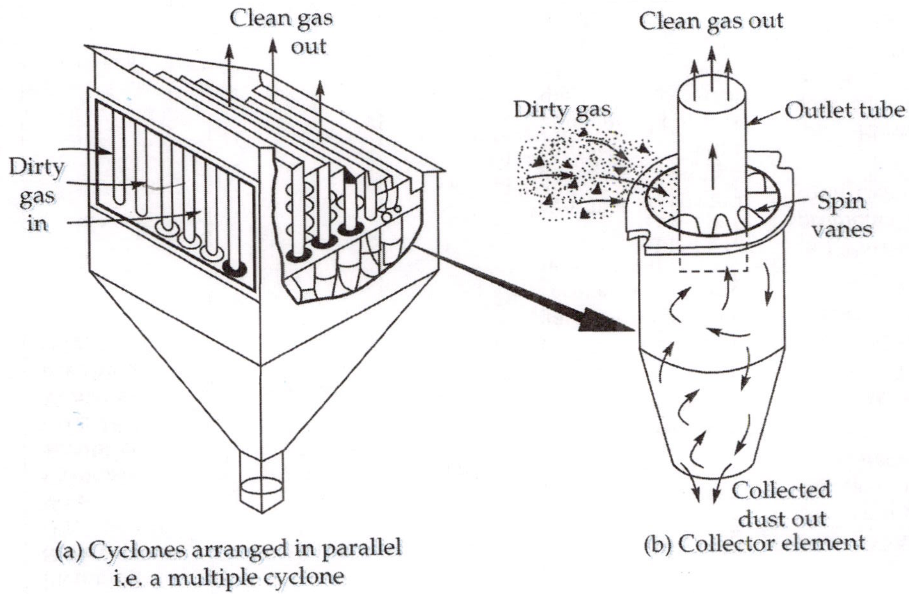

Fig. 18.11. Cyclone collector (Cyclone Separator).

Fig. 18.12. Photoview of a Dynamic Precipitator.

(1)	(2)	(3)	(4)	(5)	(6)	(7)
3.	**Wet Scrubbers** or **Collectors,** including ***Spray tower*** (Fig. 18.13) **Wet-Cyclonic scrubber** (Fig. 18.14) **Venturi-scrubber** (Fig. 18.15)	> 10 μm > 2.5 μm > 0.5 μm	< 80% < 80% < 99%	(*i*) The *Spray tower* and *Venturi scrubber* can be made to remove gaseous pollutants also, along with removing particulate matter. (*ii*) Hot gases can be cooled down. (*iii*) Corrosive gases can be recovered and neutralised. (*iv*) The separated gases through contact with aqueous chemicals may produce such useful byproducts, as chemicals and fertilisers.	(*i*) A lot of waste-water needing disposal may be produced. (*ii*) Maintenance cost is high, when corrosive materials are collected. (*iii*) Wet outlet gases cannot rise high from the stack. (*iv*) Poses freezing problems in cold countries. (*v*) Plume may sometimes be visible in the sky due to the presence of water vapour.	In these devices, the flue gas is made to push up against a down falling water (liquid) current. The particulate matter mixes up with water droplets and, thus, falls down and gets removed. Water solutions, when replaced with other aqueous chemical solutions, like lime, potassium carbonate, slurry of MnO and MgO, etc. do help in removing gaseous pollutants also from the flue gases.

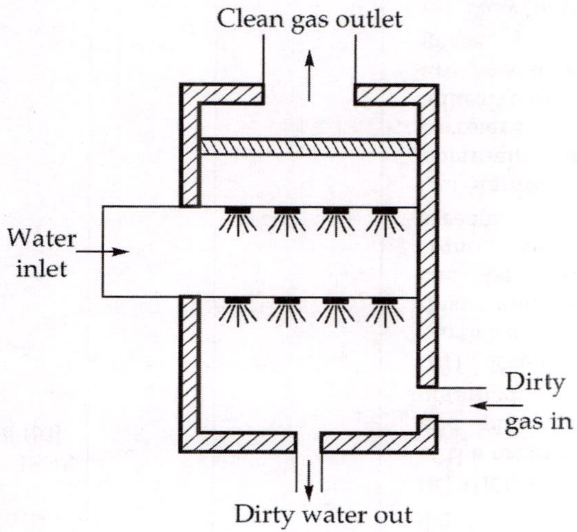

Fig. 18.13. Spray tower (Also used for removing gaseous pollutants).

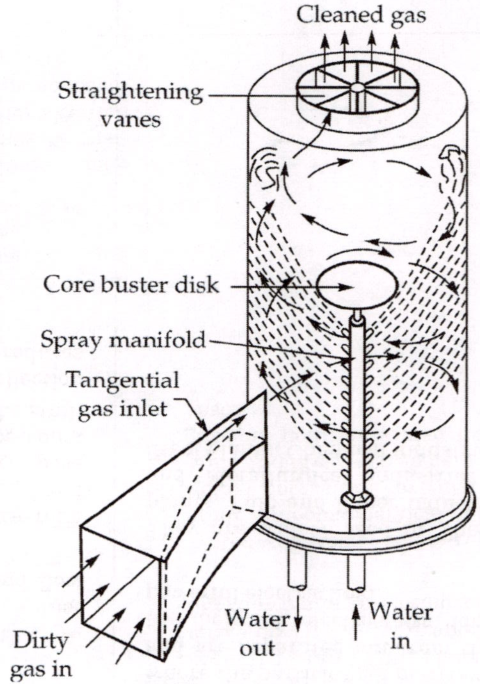

Fig. 18.14. Wet-Cyclone scrubber.

(1)	(2)	(3)	(4)	(5)	(6)	(7)
4.	**Electro-static Precipit-ators** [Fig. 18.16 (*a*) and (*b*)]	> 1 μm	95—99%	(*i*) Particles may be collected wet or dry. (*ii*) High efficiency of the order of 99% and plus can be obtained. (*iii*) Even small particles, such as dust and smoke can be removed from the air stream. (*iv*) Maintenance is nominal, unless corrosive and adhesive materials are present in flue gases. (*v*) They contain a few moving parts. (*vi*) They can be operated at high temperatures (upto 300—450°C), and pressures, and their power requirement is low.	(*i*) Higher initial costs. (*ii*) Sensitive to variable dust loadings and flow rates. (*iii*) They use high voltage, and hence may pose risk to personal safety of the staff. (*iv*) Collection efficiency reduces with time.	In electrostatic precipitators, the flue gas is made to pass through a highly ionised zone, where the particles get electrically charged and are separated out from the gas, with the help of electrostatic forces in the powerful electric field. They are widely used in Thermal power plants, Pulp and Paper industries, Mining and Metallurgical industries, Iron and Steel plants, Chemical industries, etc.

(Contd.)

Particulate Emission Control Device

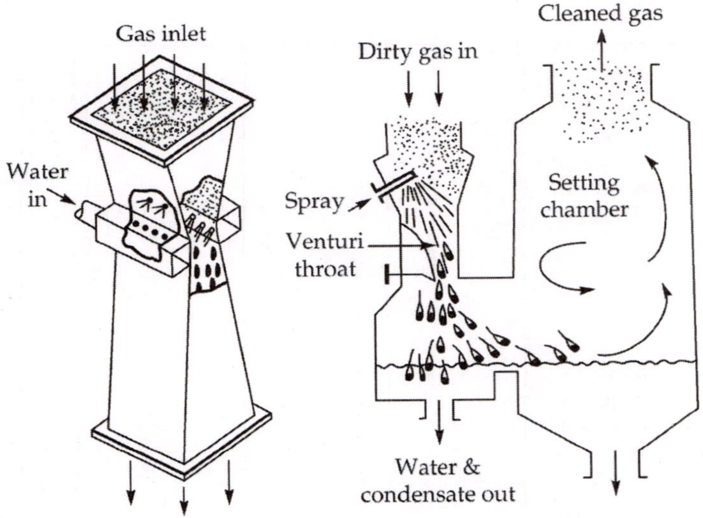

Fig. 18.15. Venturi-scrubber (also used for removing gaseous pollutants).

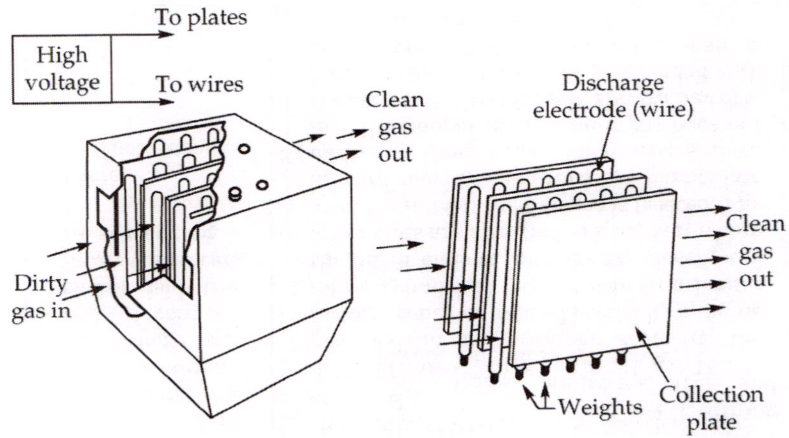

Fig. 18.16. (*a*) High Voltage Electrostatic Precipitator (plate type).

(1)	(2)	(3)	(4)	(5)	(6)	(7)
5.	**Fabric Filters** (Fig. 18.17)	< 1 μm	> 99%	(*i*) Fabric filters can give high efficiency, and can even remove very small particles in dry state. (*ii*) Performance decrease becomes visible, giving pre-warning.	(*i*) High temp. gases need to be cooled to the range 100—450°C, within which, the filters are stable. (*ii*) The flue gases must be dry, as otherwise, there is a risk of condensation inside the filter, which can cause clogging. (*iii*) The fabric is liable to chemical attack.	In such a system, the flue gas is allowed to pass through a woven or felted fabric, which filters out the particulate matter and allows the gas to pass. Small particles are retained on the fabric, initially through interception and electro-static attraction ; and later on, when a dust mat is formed, the fabric starts collecting particles more efficiently. A **baghouse** filter unit, (Fig. 18.17), provided in an ordinary room of the factory, contains several vertically hanging fabric cylindrical bags (1.8 to 9 m long) ; the upper ends of the bags are closed, and lower ends are attached to a hopper, where also, the inlet of the flue gas is located. The upward moving gas drops out particulate matter in these bags, which settles down into the hopper, and cleaner gas goes out through the fabric filters. The framework, housing the hanging bags, is provided with an automatic shaking device for cleaning the bags of the collected dust.

Particulate Emission Control Device

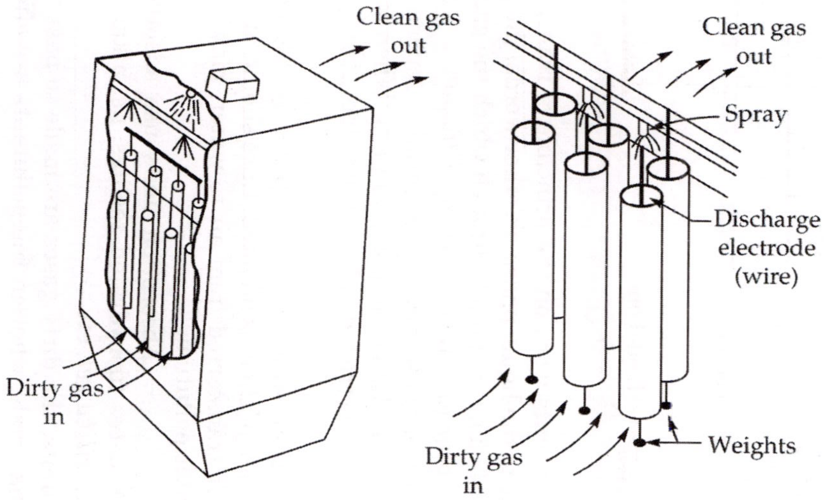

Fig. 18.16. (*b*) High Voltage Electrostatic Precipitator (Tube type).

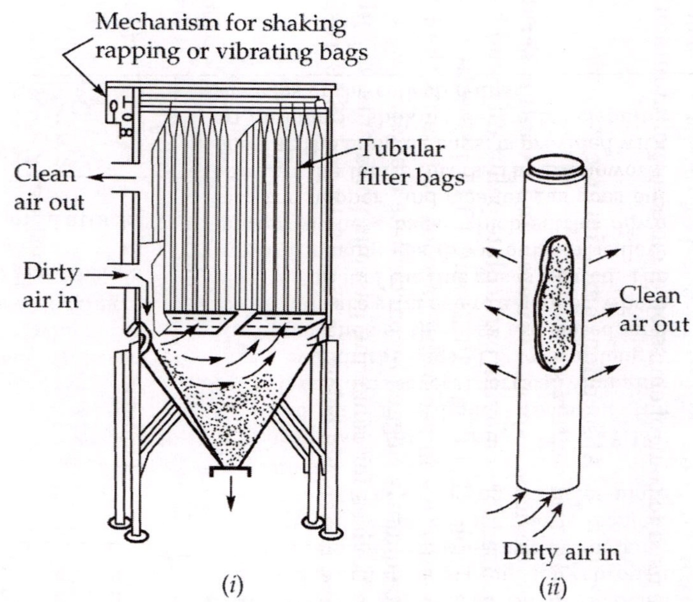

Fig. 18.17. Bag house fabric filter unit.

18.10.2. Control of Gaseous Pollutants in Industries. The *important devices*, which are used to control emission of gaseous pollutants in industries, are :

(1) Absorption Units. Absorption units include : *spray towers, plate towers, packed towers,* and *venturi scrubbers.*

Out of these devices, **spray towers** and **venturi-scrubbers** can simultaneously be used for removing particulate pollutants, and have already been described in Table 18.4 (Figs. 18.13 and 18.15 respectively). Scrubbers are generally less effective in removing gaseous pollutants than towers, but then, towers get frequently clogged by particulate matter.

Other effective devices for removing gaseous pollutants, *i.e.* **plate towers** and **packed towers,** are shown below, in Fig. 18.18 and 18.19, respectively.

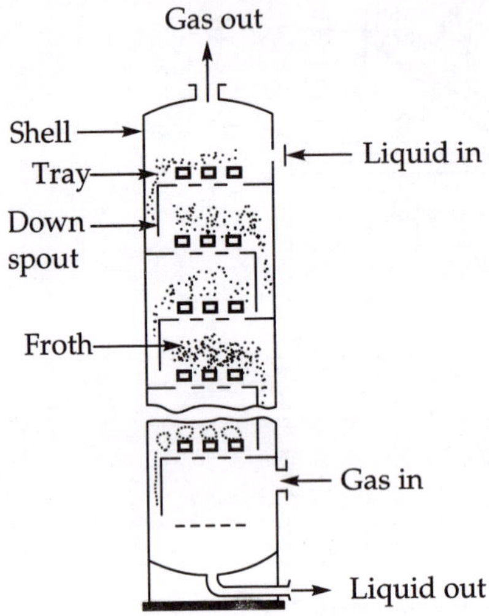

Fig. 18.18. Plate tower.

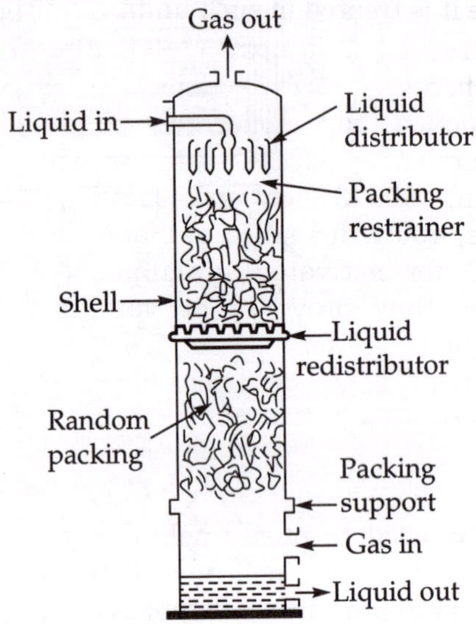

Fig. 18.19. Packed tower.

These absorption units work on the principle of transfer of the pollutants from the gas phase to the liquid phase. In other words, the pollutants from the dirty gas, get absorbed in the liquid, through which the gas is made to pass, in these units. Such absorption takes place by *diffusion* as well as *dissolution*.

The effectiveness of these devices, naturally, depends upon the solvent (liquid), through which the gas is made to pass. When water is used as the solute, the removal is restricted only to a few inorganic gases, such as NH_3, Cl_2, and SO_2.

All such abssorbent units must be properly designed for the given conditions, before one of them is adopted. Care should also be taken to ensure that the pollutants transferred from air to water, may not cause heavy and uncontrollable water pollution.

(2) Adsorption Units. Absorption units use adsorbents like *activated carbon, molecular sieves* (dehydrated zeolites, *i.e.* alkali or metal silicates), *activated alumina, silika gel*, etc.

In adsorption units, the dirty gases are made to pass through the beds of such adsorbent materials, wherein the pollutants are effectively caught and removed. **Activated carbon beds** can very effectively catch hydrocarbons, H_2S and SO_2.

One special form of molecular sieve can also capture NO_2. A schematic sketch of such a unit is shown in Fig. 18.20.

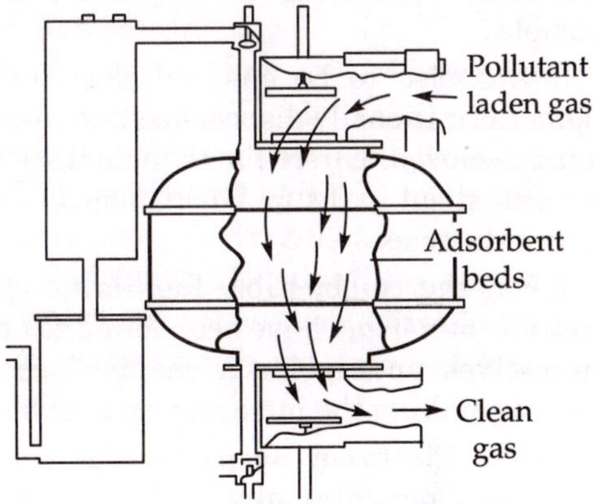

Fig. 18.20. Schematic sketch of an adsorption tower unit.

The above adsorbents, except activated carbon, have the drawbacks of preferably capturing water, before catching any of the gaseous pollutants ; and hence, water may have to be removed from the gas, before it is treated in such units.

Moreover, almost all the adsorbents are subject to destruction at moderately high temperatures (150°C for active carbon, 600°C for molecular sieves, 400°C for silica gel, and 500°C for activated alumina). Hence, they prove to be very inefficient for purifying industrial gases at such high temperatures.

(3) Combustion or Incineration Equipments may be used to purify polluted gases, when the pollutants in the gas streams are oxidisable to an inert gas. Pollutants, like hydrocarbons and carbon monoxide (CO) can be easily burnt, oxidised, and removed in such equipments.

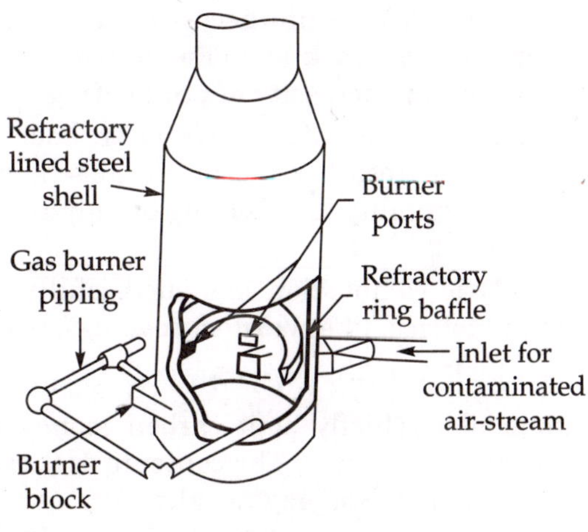

Fig. 18.21. Direct flame Incineration.

Both, *Direct flame combustion* by **afterburners** (Fig. 18.21), and *catalytic combustion* (Fig. 18.22), have been used in commercial applications. **Catalytic incineration** can be used when combustible materials in the waste polluted gas, are too low to make direct-flame incineration feasible. This generally happens when the energy of the polluted gas is lesser than 3.7 MJ/m^3. At all

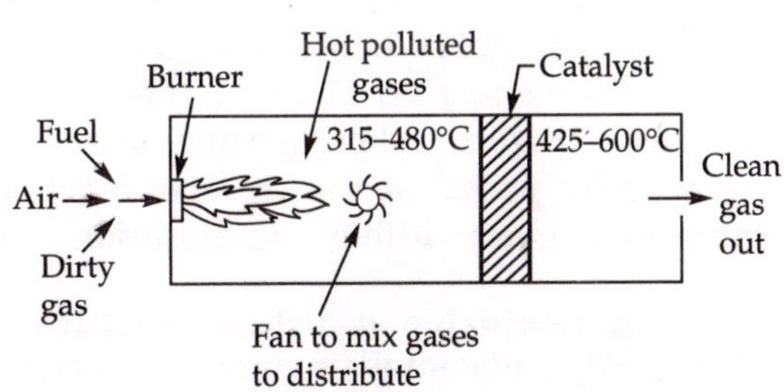

Fig. 18.22. Catalytic Incineration.

such energy contents, the gas burning becomes self supporting after initial ignition, thereby making possible the use of direct flame incinerators. Secondary supporting fuel may also sometimes be used to make such a method feasible.

It has also to be ensured that the final combustion product from the incinerator is easily disposable, and less toxic than the original gas pollutants, being removed. Direct flame incineration has been successfully used to purify the industrial gases in *Varnish cooking, Meat smoke houses,* and *Paint bake oven* industries.

When the combustible fuel value of the polluted gases are lower, some *catalytic materials* have been found to accelerate the rate of oxidation, without themselves undergoing a chemical change, thus reducing the incineration time, or making the incineration feasible even with low-energy polluted gases.

Costly **Platinum or palladium compounds*** have often been used as catalysts. Conventionally, such a catalytic compound is placed in a bed similar to absorption beds, and the supporting lattice is made of ceramic.

The old catalytic systems were effective only when the hot gases were being passed through them ; and hence, pre-heating by per-heaters (as shown in Fig. 18.22) was an absolute necessity. But in modern days, even cold catalytic systems, which work at ambient temperatures,* have been designed, thus eliminating the necessity of pre-heating.

Catalytic incinerators have often been successfully used to control SO_2, NO_x, CO, hydrocarbons*, etc. Besides being costly, their major drawback is their susceptibility to poisoning by sulphur and lead compounds, even in trace amounts.

Catalytic combustion has successfully been used in purifying emissions from industries, like, *Varnish cooking, Asphalt oxidation, Printing press,* etc.

18.11. Controlling Air Pollution from Automobiles

Automobiles, chiefly emit carbon monoxide (CO), hydrocarbons (HC), and nitrogen oxides (NO_x). The contents of these pollutants in the smoke, emitted by the automobiles, may usually vary, as :

CO	—	0.5 to 6.4%
HC	—	300 to 1000 ppm
NO_x	—	500 to 3000 ppm,

depending upon the type and condition of the engine of the automobile.

An ideal internal combustion engine (I.C. engine), when gets *stoichiometric mixture of Air and Fuel* (14.7 : 1 ratio)** burns them up completely to CO_2, water and nitrogen, without producing any CO. But in actual practice, two and three wheeler *petrol engines* receive *high fuel content mixture* (12 : 1 ratio), during cold start, idling, acceleration, and high speed cruising ; and

*__Vanadium pentoxide__ at high temperatures, has been used vary effectively, for removing SO_2 ; **platinum metals** have been used for removing NO_x ; *activated alumina impregnated with a metallic compound* is found useful in removing hydrocarbons ; palladium II (Pd II) and Cu (II) catalysts have been developed to effectively oxidise CO to CO_2 at ambient temperatures.

**14.7 parts of Air and one part of Fuel (petrol or diesel), giving A.F. ratio of 14.7.

similarly *diesel engines* receive *low fuel content mixture* (16 : 1 ratio) when running at a medium speed. So in practice, the combustion of fuel in automobiles, is never complete. Naturally then, the exhaust gases from such engines, invariably contain products of incomplete combustion ; *i.e.* CO, HC, H_2, hundreds of oxygenated hydrocarbons, as well as small fractions of nitrogen oxides. These pollutants are highly dangerous to the overall environment, and to the life in general ; *and hence the automobiles are nowadays seen as a symbol of technological menace.*

The emissions from petrol engines of two, three and four wheelers (including cars, having 4 stroke engines), contain heavier concentrations of HC and CO ; whereas, the four stroke diesel engines of diesel vehicles (buses and trucks), contain heavier concentrations of NO along with thick smoke and particles. The auto emissions also contain gaseous pollutants, like SO_2, and *lead compounds,* especially when **lead containing fuel** is used, as in India*.

In order to reduce **lead menace,** efforts are now on, in India, to reduce lead content in the gasoline, as it will take quite sometime before we are able to switch over completely to lead-free gasoline. Similarly, SO_2 emissions are caused in auto exhausts, because sulphur, or 'compounds of sulphur' are invariably present in the mineral oil. Efforts are also on, to reduce sulphur content, to as low a level as possible, in the gasoline.

Besides the above improvements required in the gasoline, the upkeep and maintenance of the auto engines, is also of utmost importance. The proper *tuning of the engine and carburettor* is very essential, in order to control pollutant emissions. *Central Motor Vehicles Rules 1989*, framed under *Motor Vehicles Act, 1988*, has therefore, stipulated permissible auto-emission levels. The maximum permissible CO emission, *while idling* is limited to 3.0% for cars, and 4.5% for two and three wheelers. Such stipulation for improved tuning and adjustment of the engines is only an initial step to limit the vehicular pollution ; as *infact, there is no other choice left now, but to change the basic design of our automobiles, incorporating catalytic convertors in their exhaust pipes,* and to adopt *multi-point fuel injection* (MPFi) *system* with dual intake valves to limit the fuel intake exactly to the needs of the engine, with no excess fuel usage at any point.

The **catalytic convertors** are usually made of noble metals, like *platinum, palladium*, etc., and help in oxidising CO and HC into their final end product of CO_2, and also to reduce NO into nitrogen. These noble metal catalysts are highly active, and resist sulphur poisoning.

They may be made in pellets, or may be in the form of a monolithic one piece metal.

A catalytic convertor is generally placed inside the tail exhaust pipe of the automobile, so as to pass through it the *'partially oxidised emissions*, before they are let out into the atmosphere.

In earlier years, two different catalytic metal beds were being used for oxidising CO and HC, and for reducing NO ; but in modern days, catalysts have been designed, which catalyse both sets of reactions. These catalysts, popularly called **three way catalysts,** (P_t, P_d and R_h), have simplified the dual bed to a single bed catalytic convertor.

---

*Tetra Ehyl Lead (TEL) is still being used in India **as octane booster ;** whereas, in advanced countries, like USA, Japan, etc., **TEL** has been replaced by (*i*) **MTBE** (Methyl Tertiary Butyl Ether) which does not contain any lead ; and (*ii*) less toxic **MMT** (Methyl cyclopentadienyl Manganese Tricarbonyl.)

Table 18.5. Emission Norms of Pollutants for Petrol and Diesel Vehicles (All Figs in gm.km)

S.No.	*Emission Norms*	*CO*		*HC*		*Nox*		*HC + NOx*		*PM10*	
		Petrol	*Diesel*	*Petrol*	*Diesel*	*Petrol*	*Diesel*	*Petrol*	*Diesel*	*Petrol*	*Diesel*
1.	**India-2000** (*Euro-I*). Inforced in Europe in 1993, and in India wef **2000**.	2.72	2.72	0.97	—	—	0.97	—	0.97	—	0.14
2.	BSII (*Euro-II*). Enforced in Europe in 1996, in Indias metro cities wef. April 2001. Extended to whole of India wef April 2005.	2.20	1.00	0.50	—	—	0.7-0.9	—	0.7-0.9	—	0.10
3.	**BS-III** (*Euro-III*). Enforced in Europe in 1996; in NCR and 13 cities of India wef **April 2005.** Extended nation wide in India wef **October 2010.**	2.30	0.64	0.20	—	0.15	0.50	—	0.56	—	0.05
4.	**BS-IV** (*Euro-IV*) Enforced in Europe in 2005; in NCR and 13 more Indian cities wef **April 2010**. Extended to whole of India wef **April 2017.**	1.0	0.50	0.10	—	0.08	0.25	—	0.30	—	0.025
5.	**BS-V** (*Euro-V*). Enforced in Europe in 2009. (To be skipped in India)	1.0	0.50	0.10	—	0.06	—	—	—	—	—
6.	**BS-VI*** (Euro-VI), enforced in Europe in 2014. Proposed to be enforced in India wef **April 2020** and in phases.	1.0	0.50	0.10	—	0.06	0.06	—	0.17	0.005	0.005

*Not yet defined. Hence the specified limits of pollutants are shown based on Euro-VI norms only.

Presently, installation of such three way catalytic convertors (Fig. 18.23) is compulsory for all automobiles plying on the roads of USA and Japan. The auto manufacturers in India, had however, been avoiding the incorporation of catalytic convertors in the vehicles, primarily due to the high cost of noble metals.

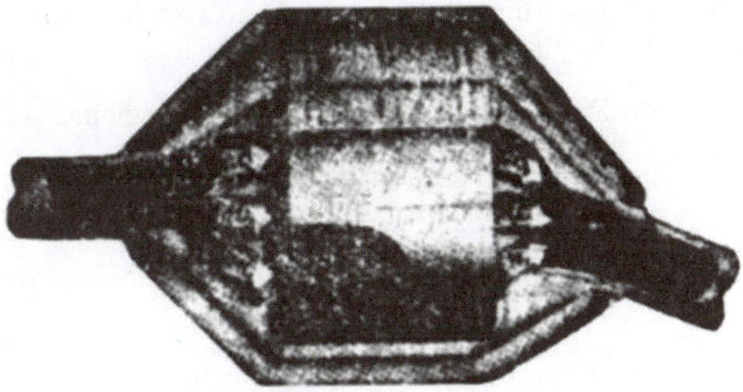

Fig. 18.23. Photoview of a typical Three way Catalytic Converter.

Be that as it may, the Indian car manufacturers have now started fitting catalytic converters in the new cars and the lead free petrol required to run such vehicles is now being made available by refineries by upgradation of thier processes.

In order to control air pollution caused by plying of vehicles, pollution control norms for petrol as well as diesel vehicles are being enforced throughout the world. These **pollution control norms** are known as **Euro-I to VI** in Europe, and as **Bharat Stage (BS)-I to VI** in India. These norms specify the maximum permissible limits of pollutants like CO, HC, NO_x and PM (Particulate Matter), as shown in table 18.5, for petrol and diesel vehicles both.

Note: (*i*) CO and HC emissions are infact more evident in petrol vehicles; while CO, NO_x and PM are emitted laragely by diesel vehicles.

(*ii*) It may also be pointed out that to upgrade the emission norms, the fuel (Petrol or diesel) also needs be upgraded by refineries. For example, BS-IV compliant fuels contain 50 ppm sulphur, while BS-VI grade fuel will have 10 ppm sulphur. Oil refineries in India, have, infact, earlier invested over ₹ 55,000 crore for production and supply of BS-III/IV fuels; while another ₹ 80,000 crore will have to be invested to further upgrade to BS-VI compliant fuels.

PROBLEMS

1. (*a*) Define air pollution, and differentiate between natural and man-made air pollutions.

   (*b*) Describe the historical background and causes of air pollution.

   (*c*) How does air pollution affects the different zones of the atmosphere ?

2. (*a*) Differentiate between the primary and secondary air pollutants, and enumerate the various such pollutants of both the categories, explaining each in details.

3. (*a*) What is the difference between RSPM and TSPM in relation to air pollution ?

   (*b*) What is photochemical smog and how is it formed ?

   (*c*) How does air pollution affects human health ?

4. (*a*) Enumerate the different major air pollutants, their characteristics, sources, and health effects on human beings in a tabular form.

   (*b*) How and which of the air pollutant(s) adversely affect the plant and animal life on Earth ?

5. (*a*) Describe the harmful effects of important air pollutants on materials and services, as well as on aquatic life.

(*b*) Enumerate the important local, regional and global impacts of air pollution.

6. (*a*) Differentiate between the environmental lapse rate (ELR) and adiabatic lapse rate (ALR). Also differentiate between ALR and Dry & wet ALRs.

(*b*) Explain the stability of the ambient environment w.r. to relative positions of ELR line and ALR line.

7. (*a*) Differentiate and explain the terms : unstable, stable, and neutral environment.

(*b*) Explain, negative lapse rate and temperature inversion conditions. How does inversion conditions affect pollution levels of the immediate environment of humans ?

8. Show that the lapse rate for dry adiabatic atmosphere can be given by

$$-\frac{dT}{dz} = \frac{g(k-1)}{kT}$$ where the notations have their usual meaning.

9. (*a*) Differentiate between radiation inversion and subsidence inversion, bringing out their impacts on the resulting air pollution effects.

(*b*) Enumerate the various factors which affect the dispersion of air pollutants into the atmosphere, and explain them briefly.

10. (*a*) What is double inversion ?

(*b*) How do winds impact the dispersion of pollutants into the ambient air environment.

(*c*) How and in what manner the environmental lapse rate (ELR) and adiabatic lapse rate (ALR) affect the dispersion of an air pollutant into the atmosphere. In this context, draw the various possible behaviours of the emitted plume.

11. (*a*) How will the existing atmospheric pressure and moisture prevalent in the atmosphere affect the dispersion of air pollutants into the ambient air atmosphere.

(*b*) Write down the Gaussian distribution equation which is commonly used to predict the concentration of an air pollutant evolving from a source (factory) at any downwind point, and the location of the maximum pollution.

12. (*a*) What is meant by the effective height of a chimney and how is it computed ? What is its use ?

(*b*) Write down the computing equations and other limiting standards which have been prescribed by CPCB for regulating the minimum chimney height of an air polluting source.

13. (*a*) Enumerate and describe the various natural properties which clean the ambient air (environment) over time by removing the various air pollutants.

(*b*) Enumerate the engineering measures that are adopted to reduce emissions of air pollutants from (*i*) factories ; and (*ii*) automobiles, giving brief descriptions of various such measures.

14. Enumerate and describe in brief (with neat sketches) the various types of engineering devices that are used to control emissions of particulate matter from factories (stationary sources)

15. Enumerate and describe briefly (with neat sketches) the various engineering devices that are used to control the emissions of gaseous air pollutants from industries.

16. Write a note on air pollution caused by automobiles and its control.

17. Write notes on any four of the following :

(*i*) Electrostatic precipitators (*ii*) Spray towers

(*iii*) Venturi-scrubbers (*iv*) Catalytic incineration

(*v*) Cyclone collectors (*vi*) Fabric filters

(*vii*) Gravitational settling chambers (*viii*) Dynamic precipators.

—19

Global Environmental Issues

Three important issues, which threaten the earth's biosphere, and are being widely discussed today on the world's environmental forums, are :

(1) *acid rains* ;

(2) *ozone depletion* and consequent development of what is called an *ozone hole* ; and

(3) *global warming.*

These three issues are described below :

19.1. Acid Rain—A Regional Environmental Challenge

Normal clean rainfall through unpolluted environment is very slightly acidic, with its pH value just less than 7. This slight acidity is caused by the equilibrium reaction between water and CO_2 present in air, forming a weak carbonic acid solution. However, when the environment is polluted with *primary pollutants* like SO_2 and NO_x gases, the resultant rainfall tends to become more and more acidic, with the increasing concentration of these pollutants. This extra acidity in rain water is caused due to the formation of the *secondary pollutants,* like : sulphuric acid (H_2SO_4) and *nitric acid* (HNO_3) due to the reaction of water vapour with SO_2 and NO_x gases, respectively. The greater is the concentration of these primary pollutants, the greater would be the acidity, and, hence, lesser the pH value of the resultant rain. *It has been specified that when the pH of the rain water falls to 5.6 or below, the rain is specifically termed as acidic (Pl. refer* **Colour Fig. 19.1**).

An acid rain does not spare any thing, which is exposed to it. What has taken mankind decades to build, and billions of years for the nature to evolve, is destroyed in a matter of few years-a *blink* in geological time scale.

Acid rain has the potential to *convert forests into wastelands* by damaging trees, plants and crops ; *acidify fresh waters* of lakes, rivers, reservoirs, springs, wells, and underground sources ; *induce respiratory illness* in humans ; *poison wild life ;* and *inflict irreversible damages on buildings* to deface the old treasured monuments, like the one shown in Fig. 19.1.

Acid rain is a versatile destroyer having regional implications at present, which may change to global ones, with the passage of time, if controlling measures are not taken. It has scourged much of the northern hemisphere in the last 30 years, and is now spreading to the industrialised regions of the south.

The primary pollutant, responsible for $\frac{2}{3}$rd of the acid rains, is SO_2, which is chiefly produced by the burning of coals and oils, largely used for generation of

Scale of Acidity (pH)

Delhi rain touched 6.1 in the 1980s
which shows a certain degree of acidity

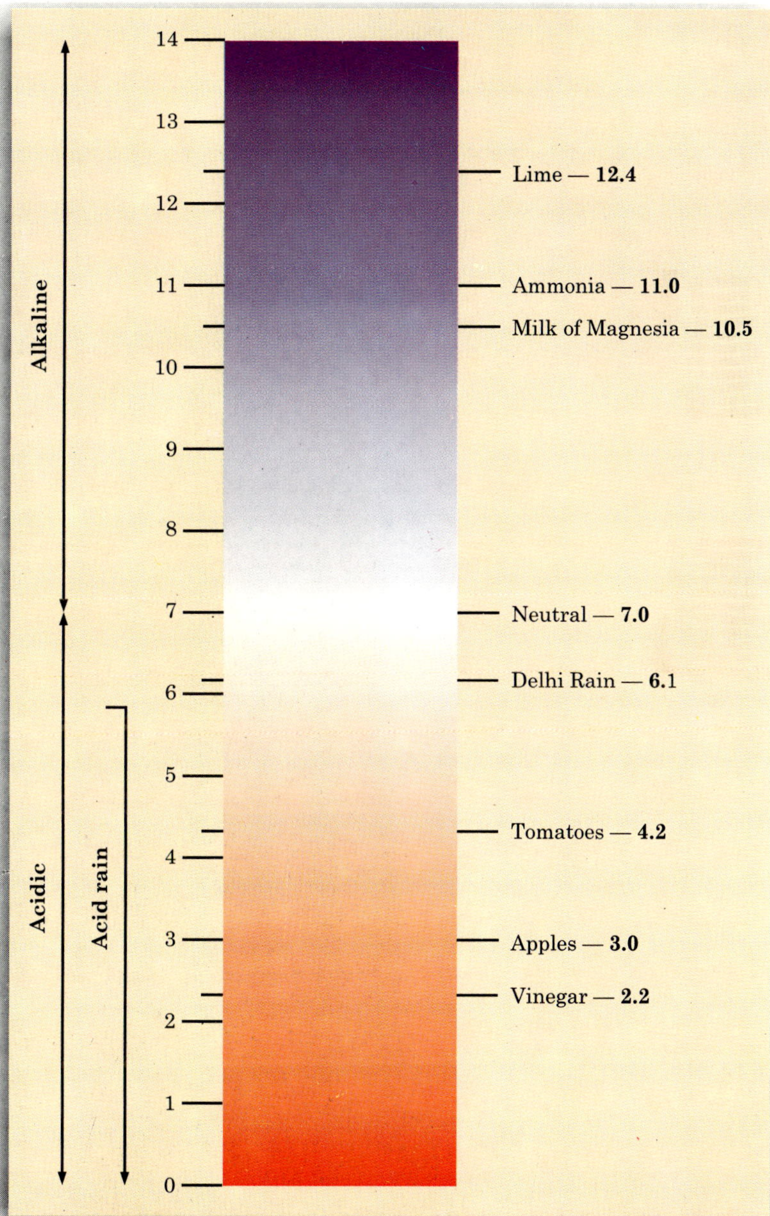

Coloured Fig. 19.1.

Coloured Fig. 19.2. On a war footing : a helicopter dumps lime into Lake Bergsjon in Sweden. Spraying was undertaken to neutralise the acidic nature of water which was destroying the lake.

heat and power in industries as well as in domestic uses ; besides the large scale use of petrol and diesel in automobiles. SO_2 pollutant, as stated above, finally leads to H_2SO_4 acidity

Similarly, nitrogen oxides (NO_x), another primary pollutant, responsible for $\frac{1}{3}$rd of acid rains, is chiefly produced by auto-emissions, nitrogenous fertilizers, fired forests, and grass lands, etc. NO_x pollutant, as stated above, finally leads to HNO_3 acidity.

Fig. 19.1. An ancient statue eroded by acid rains in Sweden.

Some of the industries, may release hydrogen chloride gas ($HCl\uparrow$) into the atmosphere, which may lead to HCl acidity ; but this is generally on a very small scale.

The extent of these primary pollutants, emitted into the atmosphere by the industries, is really staggering ; say for example, a single *nickel-copper smelting plant* in Canada is estimated to be throwing about 0.65 million tonnes of sulphur oxides (SO_x) each year. On an overall estimate, about 150 million tonnes of sulphur oxides and 40 million tonnes of nitrogen oxides, do enter annually into our environment from human actions alone, which approximately equals the natural emissions of these pollutants. The proportion of man-made emissions and the natural emissions is, however, as high as 9 : 1 in industrialised countries. Such huge man-made pollution is certainly going to pay us back in the same coin, and that is the story of the acid rain.

As stated above, rains having pH equal to or less than 5.6 (that are those which are about 20 or more times* acidic than a neutral rain having pH = 7), are specifically called **acidic rains.** And, you must believe us, when we tell you that rains, as acidic as 1,00,000 times more acidic than a neutral one (*i.e.* those having pH = 2) have occurred at several places, particularly in north-eastern US and Scotland ; causing extensive damage to the forests.

An idea of the large scale damage, that has occurred due to acid rains, can be had when we inform you that in *Germany, Switzerland* and *Sweden*, half of the natural forests are dead or dyeing ; over 60% of all trees in *Britain* show signs of decay ; 80% of all lakes in *Norway* and 20% in *Sweden* are acidified, and many voided of fish and aquatic life. Damages by acidic environments to the word's most treasured monumental buildings, such as in *Rome, Venice, Athens ; London, Oxford*, and even in *India*, are quite visible.

---

* $\frac{10^7}{10^{5.6}} = 19.95 \approx 20.$

Inspite of such large scale damage to the biosphere, we have not learnt lessons, simply because, there are no easy solutions to the problem. Moreover, the worst offenders, throwing SO_2 and NO_x in the air, are not always the worst sufferers, because these pollutants are, several times, carried away to the far off neighbouring countries, causing acid rains there. Say for example, SO_2 gases emitted by thermal power stations and other industries in Britain and West Germany, have been drifted away to northern European Countries like Sweden, Norway, etc. causing acid rains is these Scandinavian countries. So much so that an estimated 75% of acid rain in Sweden and 90% of acid rain in Norway have been caused due to air pollution from other countries. Similarly, millions of tonnes of industrial wastes are thrown into the atmosphere by USA every year. These are carried away by strong winds to Canada, where they precipitate as acid rain. Even within the USA, it has been estimated that more than 50% of the deposition of acid rain in Central Onterio is caused due to air masses passing over the major sulphur-emitting sources in the mid western States of USA like Ohio and Indiana.

The problem of acid rain, infact, was first of all detected in northern Europe, more than forty years ago, when the lakes of southern Sweden and Norway were found losing their fish population. In 70% of some 1500 lakes analysed in Norway, the pH values were found to be below 4.3, and they contained no fish at all. Soon scientist from other countries started reporting increased acidity in their water bodies. Nearly 200 lakes have since been found dead in North America and Canada, as the acid rains have killed fish, bacteria, and algae, causing complete collapse of the lakes into sterility, leaving crystal clear but ultimately dead lakes.

As far as India is concerned, the acid rains have so far prominently occurred in highly industrialised localities like Chembur in Maharashtra State, where acid rains occurred in 1970's and 1980's. The industries in this region had been using fuel with high sulphur content ; and when on Govt's intervention, the industries started using cleaner fuel, the acid rains stopped. Lower pH levels of rains have also been reported from areas in Delhi, U.P., Maharashtra, Madhya Pradesh, Tamil Nadu, and even Andaman islands. (Please see Fig. 19.2). Although , these rains cannot be strictly called acidic, since in most of the cases, the pH levels were higher than 5.6, yet the stage has been set for acid rains, since pH levels nearing 5.6 have often been recorded in the past, as shown in table 19.1.

Table 19.1. pH Data Collected in 1974-84 from Certain Indian Towns where Monitoring Stations were Located

S. No.	*Station*	*No. of observations*	*pH variations recorded*
1.	Allahabad	56	6.34–9.00
2.	Kodailkanal	62	5.18–8.70
3.	Minicoy	66	5.52–8.90
4.	Mohanbari	42	5.50–8.30
5.	Port Blair	48	5.65–8.90
6.	Pune	76	5.65–8.90
7.	Srinagar	88	6.15–8.40

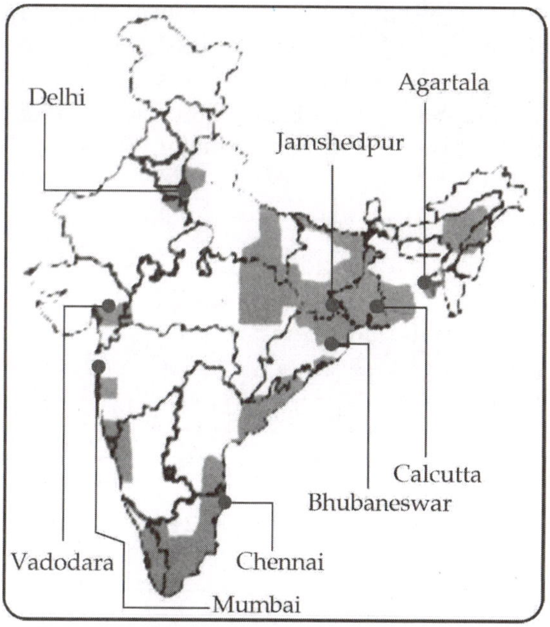

Fig. 19.2. Places where Acid Rains have occurred in India in the past.

Eventually, if the increasing air pollution emissions continue, particularly SO_2 emissions, primarily being caused by the use of diesel* vehicles, gen-sets, and other industries, then the days may not be too far off, when India may also face consequences of acid rains, which may cause irreparable damage to the country's biodiversity and even to the food chain.

Efforts have, however, started to provide *expensive scrubbers* in thermal power stations and other industries to check the emission of SO_2 and other gases ; to provide *three way catalyst convertors* in automobiles to check the emissions of both SO_2 and NO_x ; to protect forests and lakes by taking emergency measures, such as *spreading of lime* on their surface (**Colour Fig. 19.2**) ; by *fertilising the damaged trees* with calcium, potassium, magnesium, and so on.

All such corrective measures, of course, have a price, and are very costly, and may even cause other detrimental effects on the environment, such as the ones caused by the necessity of excessive quarrying of limestone mines for use in scrubbers. *The environmental gains of such corrective measures have, therefore, to be weighted and balanced against their costs and associated detriments.*

19.2. Ozone Depletion—The Most Dreaded Global Environmental Issue

Ozone depletion is the **most dreaded aspect of air pollution,** having wide spread implications, extending over the entire atmosphere. This problem is

*Indian diesel supplies contain about 0.5% sulphur by weight, except in Delhi and Agra where supplies contain about 0.25% sulphur. These sulphur contents are far higher than the sulphur levels prevailing in the diesel used in countries like Swedan (0.001%).

caused by the reduction of naturally available ozone layer in the atmosphere. Before we explain ozone depletion, we will like to summarise the contents of the *gaseous atmosphere,* that exists above the surface of the Earth.

The gaseous **atmosphere** extends to about 500 km beyond the Earth's surface, and consists of the following four major zones :

(*i*) *Troposphere ;*

(*ii*) *Stratosphere or Ozonosphere ;*

(*iii*) *Mesosphere ;* and] Ionosphere

(*iv*) *Thermosphere.*]

These four zones of the atmosphere are shown in Fig. 19.3, and explained below :

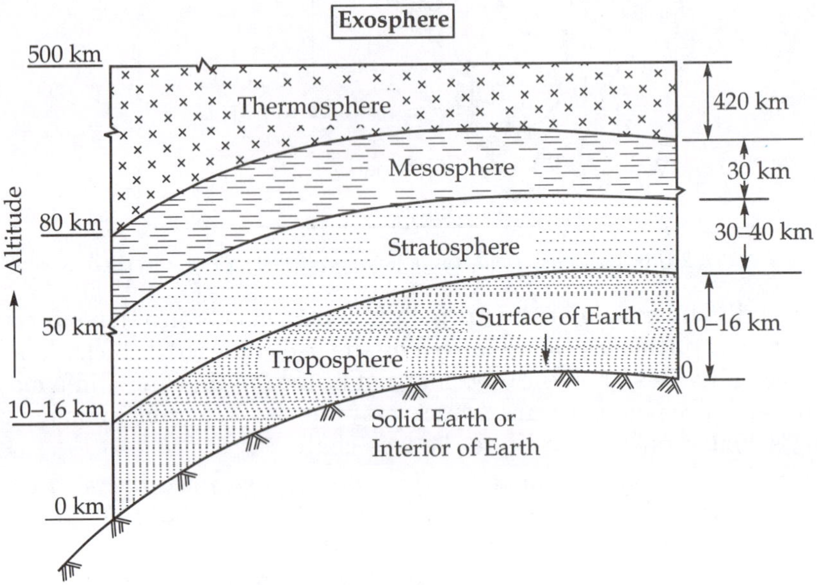

Fig. 19.3. Structure of the Atmosphere.

(*i*) **Troposphere.** It is the nearest gaseous zone extending from the surface of the Earth upto an average distance of about 11 km at the poles and 16 km at the equatorial regions.

Temperature decreases with height in the troposphere, since the main heat source is the solar radiation which is absorbed at the ground level. The rate of decrease of temperature, called the **lapse rate** is about 5° C/km. The troposphere contains most of the water vapour, clouds, and storms of the atmosphere. Winds tend to be the strongest at the **tropopause** (*i.e.*, the top level of troposphere, indicated by point T in Fig. 19.4), the level of the *jet streams.* This is also the level at which the jet aircrafts cruise. The troposphere contains about three-forth of the total weight of the atmosphere, and is responsible for most of the meteorological processes. Thus, it is the troposphere, in which winds are generated and clouds are formed, which cause precipitation.

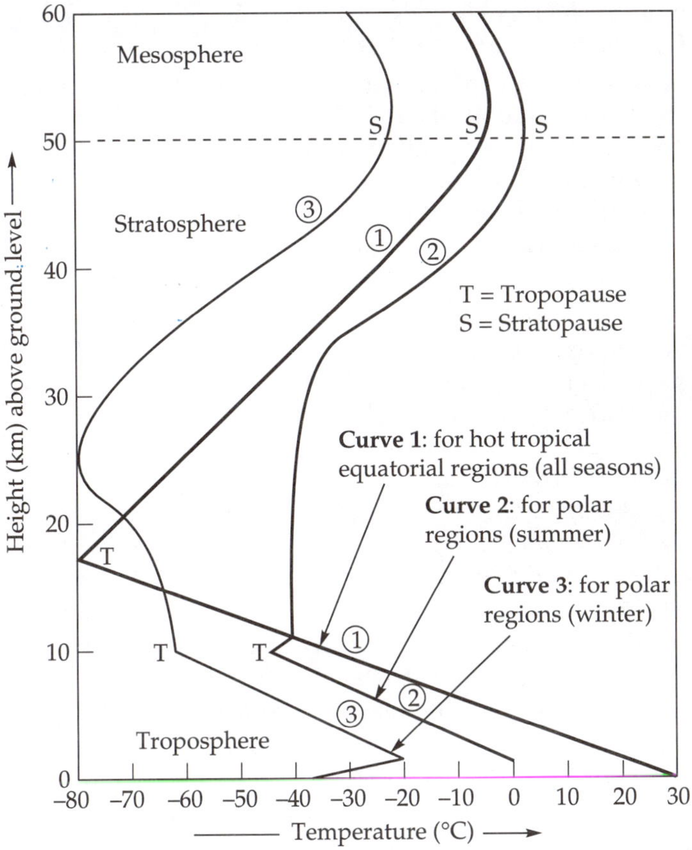

Fig. 19.4. Typical Temperature variation with height in a hot tropical region and polar region in summer and winter (T-marks the *tropopause* on each curve ; and S is the *stratopause*)

(*ii*) **Stratosphere.** Above the *tropopause* (at about 10–16 km altitude) lies another zone, called the *stratosphere,* extending up to about 50 km altitude, a point called the **stratopause** (represented by the latter S in Fig. 19.4). Between the tropopause (T) and strotopause (S), the temperature increases gradually with height, from about – 80°C to about 0°C or so. At the stratopause (S), the temperature, infact, is about as high as at the ground level. Its warmth is due to the absorption of ultraviolet radiation from the sun by oxygen and ozone. *Most of the world's ozone is found in the stratosphere, where it may exceed 5 ppmv (parts per million by volume). Hence, stratospheric air is lethal to human beings.* There is very little water vapour at these heights. Ozone (O_3) is, infact, formed in this zone due to the splitting of an oxygen molecule (O_2) into the atomic oxygen (O), and its subsequent union with an oxygen molecule (O_2) to form O_3 (Ozone). Ozone is, however, destroyed naturally by ultraviolet radiations, as it helps in preventing UV radiations from reaching the ground. ***In the stratosphere, ozone is, thus, constantly being formed and destroyed, and is, therefore, maintained in a rough equilibrium, under natural unpolluted environment.***

Since ozone is formed in the stratosphere, this zone is also sometimes called as the ozonosphere. ***The ozonosphere is of utmost importance to the life***

on the Earth, as it absorbs most of the U.V. radiations from the sun, which in its absence, would possibly have burnt all the life on Earth.

(*iii*) **Mesosphere.** The mesophere extends from the stratopause (S) at about 50 km altitude and at about 0°C to another temperature minimum of about – 110°C at about 80 km altitude. This top boundary of mesosphere is called the **mesopause**. The atmosphere in this zone is windy and turbulent, but there is usually too little water vapour for clouds to form.

(*iv*) **Thermosphere.** Above the mesopause (80 km altitude) the ***temperature increases*** indefinitely upward into the *thermosphere* (*i.e.* ***the hot upper atmosphere***), which zone may extend up to about 500 km altitude.

The mesosphere and the thermosphere zones contain many ions (electrically charged atoms and groups of atoms) and free electrons. Cosmic rays and radiations from the sun produces these ions. Both these zones together are, therefore, usually referred to as the **ionosphere.** *The ionosphere is a very light and rarefied mass, and as such, is very light in mass, despite having a large volume.* Different types of subzones, such as *D-region* (55-90 km altitude) ; *E-region* (90 to 145 km altitude) ; F_1*-region* (145 to 305 km altitude) and F_2*-region* (305 to 500 km altitude) have been differentiated within the ionosphere, depending upon the degree of ionisation.

The region of the atmosphere beyond about 500 km, is called the **Exosphere,** and till today, very little is known about it. It is perhaps a low density, ***high temperature region,*** with minimum atomic collisions.

As far as the chemical composition of the atmosphere is concerned, it has been found to be mainly a mixture of approximately 78% Nitrogen and 20% oxygen by volume. Other gases like carbon-dioxide, argon, neon, helium, ozone, and water-vapour will constitute the balance 2%. This composition is fairly constant in the regions of troposphere and a large part of the stratosphere. Dust particles and liquid droplets are also found present in the lower atmosphere and water vapour may be present in large variations.

The variation of temperature in the atmosphere is shown in Fig. 19.4.

Out of all the above described zones of the atmosphere, the second zone, i.e., the stratosphere, remains the most important to man, ***as it is the stratosphere which primarily contains ozone gas (O_3), chiefly in the layers between 25 and 40 km above the ground level.*** *This ozone layer cuts off short wave length radiations (called ultraviolet radiations) from reaching the surface of the Earth. These ultraviolet (UV) radiations, having wave lengths shorter than about 290 nm (neno metre)* are cut off by this ozone layer. These radiations, which are not allowed to cross below the ozone layer, produce warm conditions and raise the temperature in the stratosphere.* ***This existing ozone layer in the stratosphere, thus, helps in absorbing the harmful ultraviolet (UV) radiations released by the sun rays.*** *The sun light is thus, filtered out by the ozone molecules, before it reaches us on Earth.*

As a matter of fact, the ultraviolet radiations are extremely harmful to all kinds of life on Earth. Hence, if the filtering of the UV radiations is not done

---

*1 neno metre = 1 nm = 10^{-9} m.

effectively and fully due to shortage of ozone thickness in the stratosphere, then naturally, the harmful U.V. radiations are likely to reach our planet, causing several adverse impacts on humans, animals, as well as the plant life existing on Earth.

The most *dreaded adverse impact* of UV radiations is to cause mutation in DNA, leading to **skin cancers** including the deadly **melanoma.** So much so, that a *10% overall depletion of ozone is estimated to cause 26% increase in the incidence of non-melanoma skin cancers, world-wide.* UV radiations also *affect our eyes* leading to cataracts and ultimately to blindness, if not treated well in time. Excess UV radiations may also reduce the overall immune efficiency of our body, making us susceptible to infectious diseases. Moreover, UV radiations may interfere with the photosynthesis, leading to lower crop yields.

One of the most dangerous UV radiation, called UV-B, may even penetrate scores of metres into the oceans, killing *phytoplanktons* and *krills,* which constitute the very bottom of the ocean food chain, thereby causing food scarcity to larger animals like fishes, etc. and ultimately leading to shortage of marine food for humans.

Since 1985, it has been observed that the ozone layer is depleting in the stratosphere, due to its chemical action with chlorine atoms, being released by the dissociation of chlorofluorocarbons (**CFCs),** particularly in the presence of sunlight. The **halons,** which are compounds of bromine, also play similar role, as is played by the CFC's in depleting ozone ; but their use in industrial world is very limited. Hence, when we speak of ozone depletion, we generally speak of CFCs only.

The first appreciable ozone depletion was noticed in the **antarctic region** (south polar region) in 1985. This depletion has been quite noticeable, particularly in the *late winter and early spring.* As much as 50% ozone loss has been noticed at the worst periods, in the antarctic region. Countries of this region, like **Australia** and **Argentina** have recorded increased incidence of skin cancers, which are curable only if detected early. People in these countries are advised to avoid sun-bathing during the particular periods, when ozone depletions are appreciable. Even in **Newzealand,** school children are advised and urged to stay out of sun between 10 AM to 3 PM, and to wear hats and eat their lunches in the shades of trees, to avoid sun exposure.

Initially, it was thought that the ozone shortage may be existing only in the *colder antarctic region (south polar region)* ; but later on, ozone depletions were confirmed even in the *less cold arctic region (north polar region),* detecting an overall 4 to 8% ozone depletion over the northern hemisphere. Collected data further imply that the ozone layer over some regions, including the *northern-most parts of the U.S., Canada, Europe and Russia,* may deplete in the late winter and early spring, by as much as 40%. *If such a huge* **ozone hole** *does not open up in a particular year, it may open up, say a few years later. The threat is much worse than what was thought of, initially.* ***The countries, situated north of 60° N latitude, are most threatened by the likely ozone deficits.*** The countries, thus, likely to come in the above zone are : *Canada, British Isles, Scandinavia, the Netherland, large part of Belgium, Germany and Russia. All these counties may, thus, have to face*

ozone alerts *and stern warnings to wear sun glasses and sun screens, in the coming future.*

Evidently, the polar regions are more prone to ozone depletions, because **polar vortex*** collects dust particles, CFC's, etc. and the super frigid night air of the polar region, causes the formation of tiny ice crystals high up in the stratosphere. When the CFC's breakdown, the resulting chemicals cling to the crystals, where they can decompose further into ClO, etc., And finally, when the sun rises after the long winter night, its light triggers a wholesale demolition of ozone by chlorine monoxide.

Evidently, if the polar vortex breaks in a particular year, before the sun rises after month's of darkness to trigger the reaction, there will be no ozone hole in that particular year.

The extent of ozone in the stratosphere has usually been found to be of the order of 3.5 mm to 2.5 mm in thickness; which is usually expressed in the units of *Dobson.* 1 ***Dobson Unit*** (DU) = 0.01 mm. The usual ozone thickness measured in the stratosphere thus, varies from 350 DU to 250 DU. The maximum ozone thickness of about 500 DU, however, is found just above the equator, while it is found to be of the order of 250 – 350 DU (average value = 300 DU) in the adjoining regions located between 15° to 60° latitude.

This usually available ozone thickness of 350 – 250 DU is, however, found to have decreased up to as low a value as about 150 DU in upper latitude colder regions, in and around the poles. *This phenomenon is noticeably found in the southern hemisphere at the start of spring season after long winters in about September.*

The falling of ozone thickness to this low level of about 150 DU, indicates alarming shortage of ozone, resulting in the creation of **ozone hole.** The ozone hole has been getting bigger and bigger due to increasing concentrations of CFCs in the environment. The maximum area covered by the ozone hole (ozone concentration of the order of 150 DU) in the southern hemisphere was found to be about 29.4 million sq. km, and occurred in Sept. 2000.

Note. The most heartening news, however, is that after the world has initiated measures to control ozone depletion by reducing emissions of CFC's during the last 20 years, the maximum size of developed ozone hole has reduced to about 16.4 M.sq.km on Sept., 8, 2019 (minimum value during the 21st century). This is a huge reduction in the size of ozone hole from 29.4 M.sq.km to 16.4 M.sq.km in 19 years. ***A big achievement due to the impact of efforts made the by world community under the montreal protocol (explained in next article).***

Chlorofluoro carbons (CFCs), as pointed out above, are the chief agents of ozone destruction. They are largely used and subsequently released in modern world, in *refrigeration, air-conditioning, fire extinguishers,* as *cleaning solvents* in factories, as *blowing agents* to create certain kinds of *plastic foams,* and are spewed into the air in several countries as part of

---

*A huge air mass, trapped for months at a time due to the blowing of winds in circular pattern over the ocean (surrounding the continent) for a long winter season in a polar region.

aerosol sprays.

These *inert* CFC's do not disintegrate in the lower atmosphere, and do rise several kilometres up into the stratosphere, where they release *chlorine atoms* in the presence of sun light.

Each chlorine atom from CFC's then reacts with an ozone molecule, forming CIO (chlorine monoxide) as :

$$\underset{\substack{\text{Chlorine} \\ \text{atom from} \\ \text{CFC's}}}{Cl} + \underset{\text{Ozone}}{O_3} \longrightarrow \underset{\substack{\text{Chlorine} \\ \text{monoxide}}}{ClO} + \underset{\substack{\text{Oxygen} \\ \text{molecule}}}{O_2} \quad ...(19.1)$$

The ClO so formed, then reacts with another oxygen atom to form a new oxygen molecule and a chlorine atom. *The chlorine atoms, thus, replenished,* can go on to break apart thousands and thousands of more ozone molecules, leading to large scale ozone deficits.

Each atom of chlorine, infact, it is estimated, can destroy up to 1 lakh ozone molecules–at a faster rate than the gas is replenished naturally.

Unusually high concentrations of 1 to 1.5 ppb (parts per billion) of ClO*, found in the antarctic region, have confirmed that ozone depletion is certainly caused by CFCs. Naturally occurring nitrogen oxides in the environment, however, have been found to be helpful, as they tend to interfere and slow down the ozone-chlorine reactions.

What is most frightening about CFC's is that they remain in the atmosphere for decades, after they are first emitted. This long life of the already emitted CFC's poses a grave threat to the humanity, even if we reduce or stop their emissions in the coming future. But then, we have to wake up as quickly as possible, and stop CFCs usage, at least in the future.

Fortunately, India has thrown very less** CFC's in the world's environment, and the developed countries are the chief culprits on this account. Indians have started using the comforts of CFC's very late, and that is why, India initially refused to sign the **Montreal Protocol**÷, which aimed at complete phasing out of the CFC's by the year 2000, and to change over to the use of their costly substitutes. India's, well understood argument was that the developed countries, who had been largely benefitted in the past by the use of CFC's, must contribute and compensate the under-developed countries, for enabling them to change over to the use of costly CFC substitutes (which don't wipeout ozone). Why should poor nations like India, be forced to bear the cost of the mistakes committed by the developed countries, was the argument given by Indian politicians and diplomats. The Indian stand did force the developed world to agree to create a 240 million dollar special fund,

---

*World recorded concentration of ClO.

**Annual per capita CFC consumption in India had been only 0.0006 kg, as against a value of 1.75 kg in USA, and 1.1 kg in Europe.

÷Explained in the next article.

to help developing nations switch to CFC—free technology, based on HCFCs (Hydro Chlorofluoro carbons)*, patented by DUPONT of U.S. USA, though initially not agreeing to *ozone-aid,* had finally bowed down to contribute 25% of the above fund on the condition that this will not become a precedent in solving other environmental issues. India finally signed this treaty on **17.09.1992.**

19.2.1. Montreal Protocol—An International Treaty for Controlling Global Ozone Depletion. Considering the very fact that the ozone depletion has been proved to be the most dangerous and dreaded global environmental problem, the international community became active to provide some international legal bindings on various countries of the world to prevent them from producing and using Ozone Depleting Substances (ODSs). Keeping in view the harmful effects of CFC's and halons to the ozone layer, and the health threats of ozone depletions, as expressed and proved by the scientific community, the representatives of certain developed countries met in 1985 to adopt the **Vienna Convention**, which, established an international legal frame work for action. Two years later, in 1987, the international negotiators met again in Montreal, to adopt some legally binding commitment, wherein the **Montreal Protocol** was adopted on **16.09.1987.** This was followed by *Helsinki meet* in May 1989, and *London conference* (called **Saving the Ozone Layer Conference**) held in June 1990, to bring in certain amendments in the original protocol. The amendments made to the original protocol were called the ***London Amendments.***

In order to persuade more countries to join the Montreal Protocol, the London Conference was followed by *Copenhagen Meeting* held in November 1992. Based on the decisions taken, certain amendments were again made in the Montreal Protocol, which were called the ***Copenhagen Amendments.***

India signed the Montreal Protocol on 17.09.1992, about 2 months prior to the holding of the Copenhagen Meeting, when U.S.A. agreed to contribute 25% of the amount set out for creating a special fund worth 240 million dollars, to help developing nations switch to CFC-free technology based on the use of HCFCs (Hydro-chlorofluoro-carbons) patented by DUPONT of U.S.A.

In this *Copenhagen meeting*, the developed counties agreed to phase out the use of CFCs by 1995. It was also decided that there will be no compulsion of phase out time schedule for CFCs and Halons for the developing countries. *India, by then, has signed the Montreal protocol and became a party to this protocol on September 17, 1992. It was required to phase out CFCs and other ODSs completely by 2010.*

More amendments have also since been carried out twice in the Montreal Protocol, based on the meetings held in *Montreal in 1997,* and *Beijing in 1999.*

183 nations have infact become parties to the Montreal Protocol.

---

*HCFCs have an ozone depleting potential of 0.12 as against the ozone depleting potential of 1.0 for CFCs. Thus, HCFCs cause only 12% damage than that caused by the CFCs.

The salient points of the final revised Montreal Protocol are indicated below:

(*i*) The Protocol requires all parties (signatories to the treaty) to ban exports and imports of HCFCs (Hydrochloro-fluoro-carbons) *i.e.* substitutes to CFCs, to and from Non-Parties. This was adopted through Beijing Amendment effective from 25.2.2002.

(*ii*) Production and consumption of CFCs, halons, and other ozone depleting chemicals have been phased out in industrialised countries. A time schedule to eliminate the use of methyl bromide, a pesticide and agricultural fumigant, is in place. Developing countries have been given a different phase out schedule to give them enough time before phase out measures would apply to them. This has been done by recognising their need for industrial development and their relatively small production and use of ODSs ;

(*iii*) Developing countries have agreed to freeze most CFC's consumption as on July, 1, 1999 based on 1995-98 averages ; to reduce this consumption by 50% by Jan 1, 2005, and to fully eliminate these CFCs by Jan 1 , 2010. Other control measures apply to ODSs, such as halons, carbon tetrachloride, and methyl chloroform. For methyl bromide, used primarily as fumigant, developed countries are to freeze their consumption at 1995 levels and will eliminate all use by 2005; while the developing countries have committed to freeze consumption by 2002 (based on average 1995-98 consumptive levels) and will eliminate all use by Jan 2015. The control measures are summarised in table 19.2 below :

Table 19.2. Control Measures as Agreed to in the Montreal Protocol

Substance	*For Developed Countries*		*For Developing Countries*	
	Consumption Freeze	*Phase out*	*Consumption Freeze*	*Phase out*
Chlorofluoro-carbons (CFCs)	1 July 1989	1 Jan., 1996	1 July 1999	1 Jan., 2010
Halons	...	1 Jan., 1994	1 Jan., 2002	1 Jan., 2010
Other Fully Halogenated CFCs	...	1 Jan., 1996	...	1 Jan., 2010
Carbon Tetra-chloride	...	1 Jan., 1996	...	1 Jan., 2010
Methyl Chloroform	1 Jan., 1993	1 Jan., 1996	1 Jan., 2003	1 Jan., 2015
Methyl Bromide	1 Jan., 1995	1 Jan., 2005	1 Jan., 2002	1 Jan., 2015
Hydrochlorofluoro-carbons (HCFCs)	1 Jan., 1996	1 Jan., 2030	1 Jan., 2016	1 Jan., 2040

Through the Beijing amendment, it was also agreed that the consumption and production of bromo-chloro-methane, which is an industrial solvent and a fire extinguisher under the name of Halone-1011, shall also be monitored.

Impacts of Montreal Protocol. Nearly three decades after coming into being of the Montreal Protocol, a recent study of 2019 has shown a large shrinkage of the *seasonal ozone hole*, which has decreased from 29.4 M.sq.km in 2000 to 16.4 M.sq.km in 2019 (min. value since 2000), it shows that the *Montreal Protocol has shown its good and intended effects.*

19.3. Global Warming and Climate Change—A Global Environmental Challenge

During the past more than 40 years or so, Earth's climate has been noticed to be getting hotter and hotter, year after year. Say for example, in 1987, Stockholm had the warmest winter in 230 years ; France had the hottest May ever recorded ; Spain and Southern Italy sweltered under seemingly endless summer heat wave ; Britain baked in the summer and bashed in winter during an unusually warm dry air. In the U.S.A., when one of the century's worst heat wave gripped the East coast and dessicated the Midwest in 1988, the scientist declared that the *green house effect* is changing our climate.

As a matter of fact, **green house gases*** in the lower atmosphere, act like the glass of a green house, trapping some of the heat, as it radiates back from the Earth in to the space. Such gases and their heating effects, have infact, survived life on Earth, as otherwise, temperature on Earth could well have been below freezing, like that on Mars. Excessive quantities of these gases, evidently, may cause excessive heat, again to make life impossible, like that on Venus, where CO_2 is found 60,000 times than that on Earth, giving it as high a temperature as 425°C.

The continuous excessive burning of fossil fuels (coal, oil and natural gas) and large scale deforestation of forests (which absorb CO_2) by humans have increased the carbon dioxide gas in the atmosphere from an average annual value of about 316 ppmv (parts per million by volume) in 1959 to about 412 ppmv in 2019 giving an average increase of about 1.6 ppmv per year, during the 60 years period. The total increase of about 30% (in 60 years) gives an average *increase of about 5% per decade (10 years period)* or 0.5% per year.

As a matter of fact, the excessive burning of fossil fuels has not only increased the CO_2 gas in the atmosphere, but has also increased NO_x in the air. Similarly, large scale decomposition of organic matter in swamps, rice paddies, livestock yards, cattle raising, etc. is estimated to have increased methane (CH_4) gas in the atmosphere.

The presence of CFCs, another green house gas, has also increased due to our modern life style of using refrigerators, air-conditioners, sprays, etc.

Carbon dioxide* is considered to be the major green house gas, as it is responsible for about 60% of the total green house effect caused by all the green house gases. Other green house gases are : CH_4, NO_x *Chloro-flurocarbons, Per Fluoro Carbons, Sulphur Hexafluoride*, and some other 25 gases. Although CH_4 produces about 21 times more heat than CO_2, NO_x produces 390 times more heat than CO_2, and chlorofluorocarbons produces 3500—7300 times more heat than CO_2, yet **CO_2 is chiefly responsible for increasing the heat on Earth, because of its largest quantity (*i.e.* about 412 ppm (2019), i.e., 0.04% existing in Earth's air, as compared to other gases, which are only in very minor quantities.

Earth's air infact contains 78.09% nitrogen, 20.95% oxygen, 0.93% oxygen, and 0.04% carbon dioxide, and a variable amount of water vapour-on average about 1% at sea level and 0.4% over the entire atmosphere.

An alround increase in green house gases in the atmosphere, caused by large scale industrial developmental activities, has already increased the heating effect on Earth, which is further likely to increase with the passage of time, with its consequential adverse impacts of melting of glaciers, rising of sea levels, and large scale changes in climate around the world.

Even the Arctic climate in hotting up, causing Ilulissat glacier of Greenland to have shrunk several kms during the last about 50 years. The Arctic warming rate is about twice the global rate. Rising temperature put increasing pressure on certain plants and tree species, along with contributing to Greenland melting ice sheet. It has been estimated that in less than a century from now i.e., by start of 2022 century, Arctic ice could melt completely during the summer, threatening many species and life style of indigenous Inuit population.

The environmental scientists are, therefore, worried over the future prospects of global warming, and are making cautioning sounds to stop or to reduce activities, which may lead to emissions of increased green house gases into our biosphere.

Studies have, infact confirmed that the Earth's temperature, has increased by 0.8°C since 1880, *i.e.* the beginning of industrial revolution. Due to this heating, the water level of oceans has gone up by about 160 mm (16 cm) over its 1900 level. This rise is partly due to the thermal expansion of water volume contained in the world ocean and partly due to the melting of glaciers, giving an average sea level rise of 1.5 mm/yr over the last century. *This rise, however, is found to be about 1.7 mm / year during the last about 50 years, and about 3 mm / yr during the last decade.* This continuing and increasing global warming may also cause glaciers and ice caps of polar regions to melt, which will further contribute to the water volume contained in oceans. It has been estimated that uncontrolled global warming may cause the Earth's temperature to increase by 4 to 5°C during the next about 200 years (as against the present rise of 0.8°C above the pre-industrial years), which will melt all the polar ice caps and glaciers, ***leading to rise in sea levels by as much as about 7 metres (22 feet),*** submerging several island countries and low lying coastal regions.

It is estimated that *Low Elevation Coastal Zones* (LECZs) that are less than 10 metres above the current mean sea level include an area of about 2% of the world's total land area, and about 10% of the world's total population. Thus about 55% population of Vietnam, 46% of Bangladesh, 38% of Egypt, 26% Thailand, 18% of Philippines, 11% of China, and 6% of India is found to be living within 10 m LECZs. All these LECZ areas are threatened by global warming.

Globally, the year 2016 has been the warmest year on record (till 2019), followed by the year 2019, 2015, 2017, 2018, 2014, 2010, 2013, 2005 and 2009. Due to increased temperatures, many of the countries of the world have experienced increase in rainfall, particularly noticed in countries situated in mid high altitudes. While, some other countries like the parts of Asia and Africa have experienced lower rainfaill with increase in occurances of

*Indiginous people of northern Canada and part of Greenland and Alaska.

droughts. All such incidents clearly reflect that the ***Earth's climate is changing,*** which will make it more difficult for humans to survive. ***The nature, infact, is losing its equilibrium due to the imbalances created by unchecked human activities.***

Future impacts and climate changes have also been assessed by IPCC* on the basis of a series of experiments made by computer based global climate models. These also account for factors like future population growth, and energy use. *These studies and assessments have shown that the Earth's temperature is likely to rise by about 0.2°C/decade ; which rise can be reduced to 0.1°C/decade if the concentrations of all the GhGs is kept constant at the level of year 2000.* This eventually means that if GhGs emissions continue to evolve at the current rates, then a 2°C further rise in temperature is expected during the 21st century (2001–2100). Since the temperature has already risen by about 0.8°C from the pre-industrial temperatures, the Earth's temperature by the year 2100 may go up about 2.8°C and water level of oceans by about 60 cm.

The catastrophe likely to be caused by the rising sea levels may be gauged by the fact that more than half of the world's population lives within 60 km of the seas. They are likely to be seriously affected by the submerged land and ingress of salt water, due to rising sea levels. Some of the most vulnerable regions to these impacts are the Nile delta of Egypt, the Ganges and Brahmaputra deltas in India and Bangladesh, and many small island countries including the Marshal islands and the Maldives.

Human society is also likely to be seriously affected by the changing climate, which will increase floods and droughts. The rainfall is likely to increase in higher altitude areas and decrease at lower altitude levels. There will be increased frequency of droughts across the world. The frequency and intensity of tropical cyclones will increase with increase in the associated wind speeds and heights of storm surges. There will be more warm nights and fever cold ones.

The increasing retreat of glaciers will reduce the summer seasons streamflow in perennial rivers, causing shortage of water during summer seasons. A changing climate would bring about changes in the frequency and/or intensity of these extremes. The public health, to a large extent, depends upon factors such as safe drinking water, adequate food, secure shelter, and good climatic conditions, etc; and since all these factors are likely to be affected by the changing climate, it will seriously affect public health.

The changing climates may reduce the availability of clean water reqd. for drinking and washing during droughts as well as floods. Water may possibly be contaminated and sewage systems damaged. The risk of spread of infectious diseases such as diarrheal diseases is likely to increase. Food production may be adversely affected in vulnerable regions, directly due to warmer weather,

---

***IPCC (Intergovernmental Panel on Climate Change)** is an intergovernmental body constituted in 1988 by the *World Meteorological Organisation* (WMO) and the *United Nations Environmental programme* (UNEP), to assess the role of human activities in climate change. The IPCC is the co-winner of Nobel Peace Prize, 2007 alongwith Mr. Al Gore—a former Vice President of USA, for leading a campaign against global warming to create public awareness against its dangers, and its consequential climate changes.

and also indirectly through the increase in pests and plant or animal diseases. The local reduced food production may lead to starvation and malnutrition with long term health effects, especially for children. Food and water shortages may lead to conflicts in vulnerable regions.

Climate change related impacts on human health may lead to large scale migration and displacement of a large number of people, creating environmental refugees, thereby disturbing international peace. Changes in climate may also increase the spread of vector born species, such as mosquitoes, which inturn, may increase the spread of diseases, such as malaria, filariasis, dengue, yellow fever, etc. ***All in all, severe climate changes may adversely affect human health quite severally, leading to large scale death and destruction.***

The evil effects of the projected temperature increase of about 4°C by the year 2150, may be gauged by the simple estimate that about 1.6 million years ago, the Earth's temperature was about 5°C less than that of the present, and at that time—there used to be snow and snow all around. What will happen to our life, if the reverse happens, is any body's guess, but surely alarming.

In totality of the circumstances, and in view of the latest findings on melting of Arctic ice during the last years, it can be concluded that the build up of green house gases is surely a cause for caution to take effective measures for reducing emissions of green house gases into the atmosphere besides identifying and implementing engineering measures to save low island States, to counteract the impacts of rising sea levels.

There is certainly no denying the fact that there exists an urgent necessity of reducing our consumption of fossil fuels, which even otherwise, is an exhaustible resource. What will happen after about 200-500 years, when all stocks of these fossil fuels get exhausted, is any body's guess. Another interesting question that one may ask in this context, is : as to what will happen to the Earth's temperature when we completely stop burning fossil fuels, as and when they become unavailable ; will the old cold icy conditions return on the Earth ?

19.3.1. Kyoto Protocol—An International Treaty to Control Global Warming and Climate Change. In order to discuss global warming and to initiate measures for reducing the green house gases in the atmosphere, an international conference was held during December 1 to 10, 1997, in Kyoto, an ancient capital of Japan.

On the last day of this conference an agreement was drawn on **11.12.1997**, which was committed by the delegates of several countries. Under this agreement, many industrialised countries committed to a legally binding overall reduction of 5.2% in green house gases emissions below 1990 levels, by 2008–2012. Under this agreement, Japan was to reduce its emissions of green house gases by 6%, and the European Union (EU) by 8% by 2008–2012. It would be pertinent to mention here that India and China categorically rejected the ideas of controlling their emissions of green house gases by preventing their growths and industrial developments. Both these countries suggested that the issue was too complex to be rushed through. USA also refused to accept this protocol.

Although several countries agreed and signed the initial agreement, called the **Kyoto Protocol**, it could not come into force for about 8 years, since it could not be ratified by the minimum required 55 countries. Inspite of the fact that United States, which alone accounted for the world highest per capita annual emission of green house gases (primarily carbon dioxide gas)*, did not ratify the Kyota Protocol, the protocol came into force wef **16.2.2005,** 90 days after the Russian ratification.

This protocol was drafted keeping in view the fact that in todays carbon based, gas-guzzling global economy with few real alternatives, the fear is that reducing carbon emissions even by 5% might have drastic economic consequences.

The flexible mechanism that was built into the Kyota protocol is two fold ; viz (i) it is a case of Joint Implementation (JI), where the industrialised "Annex 1" countries can trade emissions (known as carbon trading) with each other ; and (ii) there is a provision of Clean Development Mechanism (CDM), whereby the industrialised nations can claim certified emission reductions from developing countries. The provision of CDM is built to encourage the industrialised nations to transfer technology to the developing countries that would reduce carbon emissions in those countries.

To explain as to how this is possible under CDM, let us consider an industrialised nation 'A', who finds that it is not able to meet its emission reduction target. It would then invest in a project in a developing country 'B', in a way that the project would reduce carbon emissions to a greater degree than in the normal course in country 'B'. What it means is that a high emission establishment working in country 'B', would be replaced by a low emission establishment with the financial and technological assistance provided by the industrialised country 'A'. The carbon saved in this manner can be claimed by country 'A' as its own emission reduction. In this way, the country 'A' can meet its target of reducing carbon emissions at a cheaper cost, since the developing economies tend to be cheaper.

The said trading of carbon dioxide emissions has come to be known by the name of **carbon trading.** ***One carbon credit*** or ***carbon point*** or ***certified emission reduction (CER)*** is equivalent to the reduction of one tonne of carbon dioxide gas (CO_2). These carbon points are being earned by organisations

*Although the total annual CO_2 gas emission (2020) is the highest for China, yet the **per capita** annual emission of CO_2 (2020) is highest for USA, as exhibited in Table 19.3. below : India's total annual CO_2 emission is 4th highest, while its annual per capita emission is quite low.

Table 19.3. 2020 Annual CO_2 Emissions of Certain Top Countries of the World.

S. No.	*Country*	*Annual CO_2 emission in (million tonne)*	*Population in Million (M)*	*Per capita CO_2 emission in tonne/c*
1.	China	**10,668**	1,402	7.6
2.	USA	4,713	331	**14.24*** (highest)
3.	European Union (27 countries)	2,540	445	5.71
4.	India	2,442	1,380	1.77
5.	Russian Federations	1,577	144	10.95
6.	Japan	1,031	126	8.18
	Whole world	34,600	7,753	4.47

*Qatar has the highest per capita value of 30 t/c

located in Third world countries, and traded with the developed countries at a good enough price. *The price of one carbon credit in August 2008 in the international market was as much as 12-15 Euros,* which went down to about 5 to 6 Euros in December 2011, and further to about 2.15 Euro (2.4 US dollar) in Feb. 2019 due to uncertain future of the CDM route, ***which is likely to end on 31.12.2020 with the ending of Kyoto protocol.***

Since the developed countries, who have to reduce their emissions, cannot do so of their own, they buy such certified carbon credit points from the developing countries under kyota protocol. *In this way, carbon credit points have been traded like the shares or units of a stock exchange.*

Many organisations and companies located in the Third world countries are, hence, earning carbon points by shifting to environment-friendly technologies and practices, such as by developing forests, by reducing use of cement in concrete by using fly ash ; or by adopting non-polluting energies like wind or solar energy to replace the thermal power that uses fossil fuels ; or by providing mass transportation to replace the individual cars ; or by adopting regenerative technology in trains that may save as much as 30% energy each time a break is applied ; or by producing power from wastes by using the green house gases (CH_4 and CO_2) produced from a sanitary landfill for power generation ; or by using energy efficient appliances including the use of LED's bulbs (Light Emitting Diode) in place of normally used incandescent lamps (bulbs) etc.

Several companies, corporations and Private Equity (PE) investors such as Goldman Sachs, DEG Germany, Basemer Venture Partners, IDFC Private Equity, Olympus Capital, etc.) have entered or are trying to enter into such ventures to earn money and help reducing environmental pollution. However, the process of applying for and obtaining sanctions for the *carbon credit projects* is a time consuming process, because, first of all, a project needs approval from the *Designated National Authority* (DNA), which in India, is the Ministry of Environment and Forests (MoEF). The proposal is then to be sent to an international body—the *UNFCC's CDM Methodology Panel*, which would ensure that the project protocols follow the approved methodology. Following this approval, the *CDM Executive Board* has to approve the project for its registration. Even after the registration of a project as a CDM project, the verification and auditing by independent agencies called the *Designated Operational Entities (DOEs)* will continue during the implementation phase of the project. The DOEs are the accredited agencies that monitor the projects, like the ISO (*International Standards Organisation*) and certify that the carbon savings have indeed taken place.

The CDM route, thus, has surely been earning profits for developing countries, but is a complex, costly, and time consuming process. Moreover, yield has decreased over the years as ***the countries have not agreed to continue with the Kyoto Protocol, and it will now end on 31.12.2020, thus bringing an end to the carbon trading.***

Even before the expiry of the first phase of Kyota Protocol, the world has been trying to negotiate an alternative treaty or agreement to replace the Kyota Protocol. Several conferences and meetings were held, during 2009 to 2015 to come to some conclusion. Ultimately, an agreement called ***Paris Agreement*** was drawn and agreed to in the 21st yearly session of conference of Parties (COPs) to the 1992 United Nations Framework Convention on climate change (UNFCCC) held in Paris during November 30, 2015 to December 12, 2015. This conference was the 11th session of meeting of Parties to the 1997 Kyota Protocol.

This agreement got final acceptance and approval of the world communities on 04.11.2016, 30 days after it was ratified by atleast 55 countries accounting for 55% of global green house gas emissions. This could become possible after the agreement was ratified by India representing 4.1% of total emission and also by EU countries (representing 12.1% of total emissions, on October 2, 2016 and October 5, 2016 respectively.

After having been accepted and approved on 04.11.2016 for replacing Kyota Protocol wef. 01.01.2021, the various nations are supposed to work out Action Plans for implementation of the Paris Agreement. ***The Paris climate change agreement will finally come into force for all countries wef January 01, 2021.***

In the 12 page document of the Paris Agreement, the members agreed to reduce their carbon output as soon as possible, and do their best to keep global warming to well below 2°C from the time fossil fuel were used (1880) and persuing efforts to limit the temperature increase to 1.5°C above the pre-industrial levels (1880).

Under this Paris Agreement, each country that ratifies the agreement, is required to set a target for emission reduction, but the amount will be voluntary.* There will be neither a mechanism to force a country to set a target by a specific date, nor any enforcement measures, if a set target is not met. There will only be a **"name and shame" system**, or **"name and encourage plan".**

Aims of Paris Agreement. The aims of this agreement are described in article 2, "enhancing the implication" of the UNFCCC through:

(*a*) Holding the increase in the global average temperature to well below 2°C above pre-industrial levels and to pursue efforts to limit the temperature increase to 1.5°C above pre-industrial levels, recognizing that this would significantly reduce the risks and impacts of climate change;

(*b*) Increasing the ability to adapt to the adverse impacts of climate change and foster climate resilience and low greenhouse gas emissions development, in a manner that does not threaten food production;

(*c*) Making finance flows consistent with a pathway towards low greenhouse gas emissions and climate-resilient developmet."

Countries furthermore aim to reach "global peaking of greenhouse gas emissions, as soon as possible".

---

*As on Febraury 2020, all UNFCC members have become party to this aggreement, except Iran and Turkey.

Countributions to be made by each Country. The contribution that each individual country should make in order to achieve the worldwide goal are determined by all countries individually and called **"nationally determined contributions" (NDCs)**. Article 3 requires them to be "ambitious", "represent a progression over time" and set "with the view to achieving the purpose of this Agreement". The contributions should be reported every five years and are to be registered by the **UNFCCC Secretariat.** Each further ambition should be more ambitious than the previous one, known as the principle of 'progression'. Countries can cooperate and pool their nationally determined contributions. The **Intended Nationally Determined Contributions** pledged during the **2015 Climate Change Conference** serve–unless provided otherwise–as the initial Nationally determined contributions.

The level of NDCs set by each country will set that country's targets. However the 'contributions' themselves are not binding as a matter of international law, as they lack the specificity, normative character, or obligatory language necessary to create binding norms. Furthermore, there will be no mechanism to force a country to set a target in their NDCs by a specific date and no enforcement if a set target in an NDC is not met. There will be only a "name and shame" system or as **János Pásztor**, the U.N. assistant secretary-general on climate change, told CBS News (US), a "name and encourage" plan.

Periodic Evaluation and Stock Taking: The implementation of the agreement by all members countries together will be evaluated every 5 years, with the first evaluation in the year 2023. The outcome is to be used as input for new nationally determined contributions of member states. The stock-taking will not be of contributions/achievements of individual countries, but a collective analysis of what has been achieved and what more needs to be done **Green Climate Fund**. Not actually a part of the Paris Agreement (and not legally binding) is a (non binding) plan to provide $ 100 billion a year in aid to the developing counries for implementing new procedures to minimize climate change with additional amounts to be provided in subsequent years. Accordingly, in early March 2016 the U.S.A gave $ 500 million grant to the **"Green Climate Fund"** as the first chunk of a $ 3 billion commitment, made at Paris climate talks".

Although the Paris Climate Change Agreement has been lauded by many, including French President Francis Hollande, and U.N. Secretary General Ban ki-moon, yet criticism has also surfaced. For example, Professor *James Hansen*, a former NASA scientist and a climate change expert, voiced anger about the fact that most of the agreement consists of "promises" or aims, and not committments.

Institutional asset owners associations and think-tanks such as the **World Pensions Council (WPC)** have also observed that the stated objectives of the Paris Agreement are implicitly *"predicated upon an assumption – that **member states of the United Nations**, including high polluters such as China, the US, India, Brazil, Canada, Russia, Indonesia and Australia, which generate more than half the world's greenhouse gas emissions, will somehow drive down their carbon pollution voluntarily and assiduously without any* *binding enforcement mechanism to measure and control CO_2 emissions at any level from factory to state, and without any specific penalty gradation or fiscal pressure (for example a carbon tax) to discourage bad behaviour.*

Be that as it may, this Agreement has confirmed the death of Kyota Protocol, thereby denying the large scale profits to India Under CDM route. However, India cannot forget that climate changes likely to be brought in by global warming will most badly impact their country due to increasing floods, cyclones, droughts, and water scarity. Rich countries can always undertake preventive and adaptation measures, but it will be the poorer countries who will be the worst sufferers. Hence, any global agreement, which aims to reduce global warming, must be welcomed by one and all, and particularly by the poorer and undeveloped countries.

It will be worthwhile to mention here that India has already submitted its "Nationally Determined Contribution (NDCs)"; i.e., its post 2020 promises, as follows :

India's Post 2020 Promises

(*i*) Reduce emission intensity or emissions per unit of GDP, by atleast 33% by 2030 compared with 2005 levels.

(*ii*) At least 40% of power to come from non-fossil fuel sources in 2030.

(*iii*) Additional carbon sink of about 2.5—3 billion tonnes of adding forest/ tree cover to be created.

PROBLEMS

1. Write detailed notes on any two of the following :

(*i*) Acid rains and their harmful effects on environment

(*ii*) Ozone depletion and it dreaded effects on biotic as well as abiotic world

(*iii*) Global warming and Green house gases

(*iv*) Montreal Protocol

(*v*) Kyoto Protocol

(*vi*) Carbon trading and Clean Development Mechanism (CDM).

(*vii*) Paris climate change agreement.

—20

Noise Pollution

20.1. Definition and Introduction

Sound in the environment is caused by vibrations in the air (or some other medium) that reach human ears and stimulate a sensation of hearing. When the sound becomes loud, or disagreeable, or unwanted, it becomes **noise.** Since the unwanted sound (*i.e.* noise) certainly produces several undesirable effects on our body health, it can be termed as an environmental pollutant. *The Air (Prevention and Control of Pollution) Act 1981* of India, includes noise as one of the air pollutant.

Noise can, therefore, also be defined as that *unwanted sound pollutant, which produces undesirable physiological and psychological effects in an individual, by interfering with one's social activities like work, rest, recreation, sleep, etc.*

It may also be noted that when noise is defined as unwanted sound, then certainly it becomes a subjective definition, independent of the loudness of the sound, since a particular loudness may not be liked by one person, whereas the same loudness may be quite pleasant to another person. A particular sound being enjoyed by one person can, thus, become a noise pollution to another person.

The noise, as an air pollutant, infact, differs from other pollutants in the sense that it is *transient* in nature, and is not a continuing or persisting phenomenon. When once the noise pollution stops, the environment becomes free of this pollutant, unlike other pollutants like gases and particulate matter, which continue to linger on, after once entering into the air atmosphere.

20.2. The Effects of Noise

The various effects of noise can be divided into the following categories :

1. *Noise induced annoyance*
2. *Noise induced diseases*
3. *Sleeplessness*
4. *Communication interference*
5. *Noise induced hearing loss*
6. *Effect of noise on wild life.*

These effects are discussed below :

(1) Noise Induced Annoyance. One of the most important effect of noise on humans is annoyance and irritation due to disturbance. A noise can be said to be annoying when a person does not like the same, and hence wants the sound to be put off, or wish to leave the noisy area at the earliest.

(2) Noise Induced Diseases. Noise may produce several undesirable physiological and psychological diseases in human beings. These effects may vary depending upon the physical and mental condition of an individual. The diseases caused may include: *anxiety, tenseness, nervousness, headaches, fatigue, nausea, insomnia, high blood pressure, high pulse rate, greater perspiration, gastric secretions,* etc.

(3) Sleeplessness. The noise may induce sleep disturbances including *shorter sleep durations, more frequent awakenings,* etc. Sleep disturbances due to noise may depend upon the characteristics of the noise, such as its frequency, loudness (intensity), and continuity/intermittency.

(4) Communication Interference. Noise can badly disturb communication when a person is speaking on telephone; or when individuals are talking face to face. This may require greater speaking effort on the part of the speaker, and much more attentive care on the part of the listener, than what would have been needed, otherwise.

(5) Noise Induced Hearing Loss. Exposure of human ears to intense noise for a long enough duration may cause damage to the inner ear, thereby decreasing ones ability to hear. On lower scale, the damage may cause the individuals to complain of tinnitus in the ears, while longer exposure may lead to deafness/permanent loss of hearing.* It may be caused due to complete destruction of the organ of corti that transmits sound from ear to brain. WHO data suggests that around 5% of the school children suffer from varying degree of deafness. According to U.S. Public Health Service, more than 7 million people in USA are working where noise levels are high enough to damage hearing. No such data is, however, available for India, although noise is a major health hazard in industry, at airports, at railway stations, and at highways.

(6) Effect of Noise on Wild Life. Wildlife, like humans, is also badly affected by noise. Health of several zoo-animals, particularly those of deers, rhinos, lions, etc. are adversely affected by noise. They become inactive and dull under noisy conditions. Several migratory birds have stopped resting in habitats close to noisy cities. Musk-oxen, kangaroos, rats, grizzly bears have been noticed to have left their noisy habitats to move to calmer places. Such a change of habitat of wild life may sometimes alter their food habits, health, and mating behaviour, bringing about environmental imbalances.

The basic reason as to why 'noise' has so far not been regarded as a killer pollutant is the fact that so far noise has generally not exceeded to such levels, as to cause drastic adverse impacts on human beings. Except in industrial areas and big cities, noise pollution is not prominent, and hence has generally remained ignored from being treated as an environmental pollutant. However, with the increasing industrialisation and commercialisation of our society, and with consequential increase in noise producing auto-mobiles and machines, noise is becoming a slow poison, and is becoming more and more prominent as a pollutant. *It of course remains an undisputed fact that several technological advancements, such as rapid transportation, labour saving machines, and other recreational devices have been associated with increased*

*Permanent hearing loss is known as ***"boiler maker's deafness".***

noise production and consequential environmental pollution. But then the question arises as to whether we should accept the increased noise as a price for progress, or should shun the new technological advances all together to lead secluded life, or should devise and implement noise controlling techniques to avoid or reduce the adverse impacts of noise. So far we have been accepting the first choice, but we cannot continue with it much longer. The second choice of course may not be liked by anyone in the modern world, and, therefore, we are left with no other alternative but to choose the third choice of devising techniques for abatement and control of adverse impacts of noise. Before we discuss the suitable technology available to control different types of indoor and outdoor noises, it is essential to first understand the *properties of sound* including its tolerance levels, and the *sources of noise* including the degree of noise produced by all such sources.

20.3. Characteristics of Sound and its Propagation

By our knowledge of physics, we are aware that sound is produced in the environment by the vibrations of solid objects, which produce pressure waves, causing alternating pressure changes in the air.

This pressure wave causes perticles in the surrounding air medium to have vibrational motion. As the particles vibrate, they move nearby particles, transmitting the sound further through the medium.

The pressure waves caused by the vibrations will cause the surrounding air to undergo compression, then rarefaction, again compresson, then rarefaction, and so on.

Such alternating compression and rarefaction of the surrounding air produces sound waves, which propagate in the form of sinusoidal path, as shown in Fig. 20.1 (a) and (b).

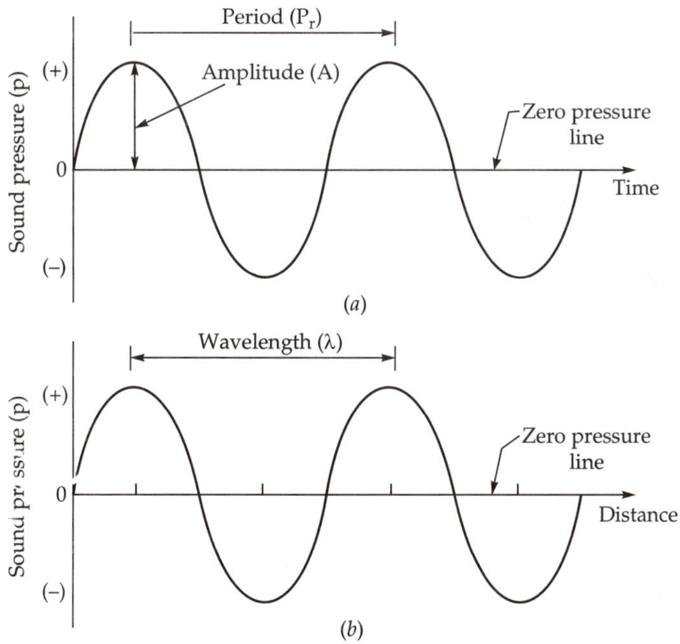

Fig. 20.1. Typical sinusoidal sound waves produced by alternating compression and rarefaction of air molecules.

As indicated in Fig. 20.1 (*a*), the time between the successive peaks or troughs of oscillation is called **the period** (P_r), and its inverse, which represents the number of times a peak arrives in one second, is called the **frequency** (f). Hence

$$f = \frac{1}{P_r} \qquad \text{...(20.1)}$$

Moreover, as shown in Fig. 20.1 (*b*), the distance between successive peaks or troughs is called the **wavelength** (λ), which is related to frequency (f) by the relation

$$\lambda = C \cdot \frac{1}{f} \qquad \text{...(20.2)}$$

where C = the velocity of sound wave.

20.4. Sound Pressure

The amplitude (A) of the sound wave is the height of the peak sound pressure, measured above or below the zero pressure line* . The equivalent pressure of such a sine wave is represented by root mean square pressure (p_{rms}) as

$$p_{rms} = \sqrt{p_{(t)}^2} = \sqrt{\frac{1}{T} \cdot \int_0^T p_{(t)}^2 \, dt} \qquad \text{...(20.3)}$$

where $p_{(t)}$ = Pressure at any time t.

The r.m.s. sound pressure is, thus, obtained by adding the squares of amplitude values at small time intervals, and dividing the total by the averaging time, and finally taking the square root of the total. This works out equal to $\frac{A}{\sqrt{2}}$ for a regular sine wave.

All sounds produce pressure fluctuations in the air over the atmospheric pressure, which are very very small, incomparison to the atmospheric pressure.

The average atmospheric pressure of air is 760 mm (0.76 m) of Hg at 0°C. This is a pressure of 1,01,325 Pa (Pascal) *i.e.*, approximately 10^{-5} Pa. Incomparison to such a large value of atmospheric pressure, the sound pressure or the sound amplitude may be much smaller, say of the order of 20 micro pascal (20×10^{-6} Pa) to about 60 Pascal.

Note. *The lowest sound pressure possible to hear is approximately* 20×10^{-5} *Pa (20 micro pascal) or* 20×10^{-11} *times of an atmosphere. The maximum sound pressure which may start causing disturbance and pain to normal human ear, is about 60 Pa (or* 60×10^{-5} *atmosphere].*

The sound pressure is a dynamic pressure, (which changes continuously), while the ambient atmospheric pressure is a static pressure. The sound pressure diagram is shown in Fig. 20.2.

---

*The amplitude, thus, represents sound pressures, and hence *sound pressure* may be referred to as *pressure amplitude*. ***The larger the pressure amplitude, the louder the sound will be.***

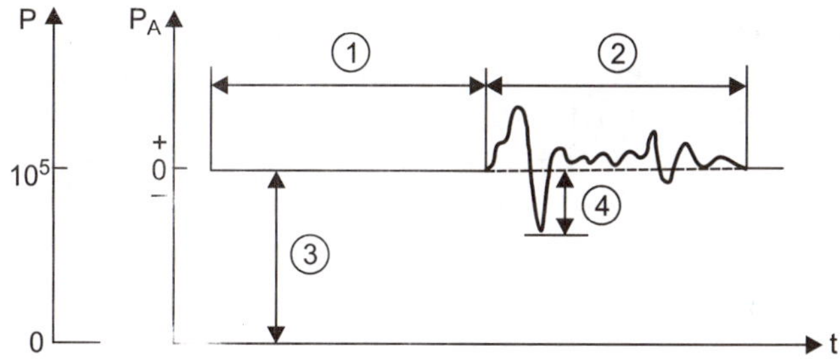

Fig. 20.2. Sound Pressure Diagram

Sound pressure diagram :

1. Silence
2. Audible sound
3. Atmospheric pressure (P_A) as datum in Fig. 20.2
4. Sound pressure

Sound pressure denoted by p is defined as :

$$p_{\text{total}} = p_{\text{stat}} + p$$

where p_{total} is the total pressure; and p_{stat} is the static pressure.

We will again explain more details about sound pressure in article 20.7.

20.5. Sound Power or Power of Sound

The power of sound or *the power of a source of sound* is the amount of energy released by the sound source per unit time. It is thus, the energy which is trasmitted by a sound wave in the direction of its propagation.

Since energy is measured in joules, the unit of power is Joules per second (J/s). This unit of power is also called ***watt***, in honour ***Mr. James Watt***, the 18th century developer of the steam engine.

Sound power is neither room dependent nor distance dependent. It is a properly of the sound source and equals the total power emitted by that source in all directions. Sound power passing through an area is sometimes called *sound flux* or *acoustic flux* through that area.

Sound Power Level (SWL) or L_W is indicated in decibels (db) as :

$$L_W = 10 \log_{10} \left(\frac{W}{W_0} \right) \text{ in dB*} \qquad \text{...(20.4)}$$

where W = Sound power emitted by a source in all directions in watts or joules/sec (J/s)

W_0 = the reference sound power, which is equal to 10^{-12} W (*i.e.*, ***1 pico watt or pW), which is the lowest sound, which a person of excellent hearing can hear.***

$$\therefore \quad L_W = 10 \log_{10} \left(\frac{W}{10^{-12}} \right) \text{ in dB} \qquad \text{...(20.5)}$$

*dB is decibel and equals one tenth of bel.

20.6 Sound Intensity

In addition to sound pressure or pressure amplitude, the sound intensity is another important term which is used to measure sound. It is defined as the sound power averaged* over the time, per unit area normal to the direction of propagation of the sound wave. Intensity and power of a sound wave are related by the equation

$$I = \frac{W}{a} \text{ (watt/m}^2\text{)} \qquad ...(20.6)$$

where I = Intensity of sound wave in watt/m^2

W = Power of sound wave in watts (averaged over the time)

a = A unit area ⊥ to the direction of wave motion.

Intensity of sound (I) at a distance r from a point source of sound located on ground, having power W, is hence given as :

$$I = \frac{W}{\text{Surface area of the sound propagation}}$$

Since for a sound source located on ground, the area of propagation of sound is equal to the surface area of a hemisphere (upper half sphere) which will be equal to $2\pi r^2$. The value of sound intensity I will than be given as:

$$I = \frac{W}{2\pi r^2} \text{ (W/m}^2\text{)} \qquad ...(20.7)$$

where r = distance of the receiving point from the source of power, W(watt), located on ground

Sound intensity (I) is also related to r.m.s. sound pressure by the equation

$$I = \frac{p_{rms}^2}{\rho . C} \qquad ...(20.8)$$

where p_{rms} = r.m.s. sound pressure in pascals (Pa)

ρ = Density of air, in which sound wave is travelling in kg/m^3. Its standard value for dry air at 15°C at sea level is equal to 1.225 kg/m^3.

C = Velocity of sound wave in air is m/s. Its value is 335 m/s (1100 ft/sec) on a standard day at sea level static conditions.

Using these usual values of ρ and C, we get

$$I = \frac{p_{rms}^2}{1.225 \times 335} = \frac{p_{rms}^2}{410}{}^{*}$$

---

*Weighted average.

** It can, thus be seen that sound intensity varies with the square of the sound pressure (rms of course).

or $$I = \frac{p_{rms}^2}{410} \quad \text{...(20.9)}$$

The density of air (ρ) and the speed of sound (*C*) used in the equation (20.8), infact, can be known, if the temperature of air is known. In other words, given the temperature and pressure of air, the density of air can be determined from the standard tables ; and the speed of sound in air at 101.325 kPa may be determined by the equation

$$C = 20.05 \sqrt{T} \quad \text{...(20.10)}$$

where *T* is the absolute temperature in kelvin (K)

C is in m/s.

20.6.1. Sound Intensity Level. It is a log quantity of the intensity of a sound relative to a standard reference value. The reference value is 1 $pW/m^2 = 10^{-12}$ W/m^2. Sound intensity level (L_I) is hence given as :

$$L_I = 10 \log_{10} \left(\frac{I}{I_o}\right) \text{dB} \quad \text{...(20.11)}$$

where I = sound intensity in W/m^2

I_o = standard reference value

$= 1\ pW/m^2 = 10^{-12}\ W/m^2$

$(\because pW = \text{pico watt} = 10^{-12}\ W)$

or $$L_I = 10 \log_{10} \left(\frac{I}{10^{-12}}\right) \text{dB} \quad \text{...(20.12)}$$

20.7. Sound Pressure Level (SPL)

The sound pressure of the faintest sound that can be heard by a normal healthy individual is about 20 micro-pascal (μ-Pa). On the other hand, the loudest sound produced by a Saturn rocket at the lift off stage is about 200 Pa. This astronomical variation in sound pressure (varying from 20 μ-Pa to 200 Pa) is usually avoided by expressing sound pressure on a scale based on the log of the ratio of the measured sound pressure and a reference standard pressure.

The sound pressure level is thus expressed by an equation as

$$L_p \text{ in dB} = 10 \log_{10} \left(\frac{p_{rms}}{p_{ref}}\right)^{2*} \text{(dB)} \quad \text{...(20.13)}$$

The reference standard quantity is taken to be equal to 20 μPa in the above equation. Thus

Sound pressure level (L_p) in dB is thus given as :

$$= L_p = 10 \,.\, \log_{10} \left(\frac{p_{rms}}{20\ \mu\text{Pa}}\right)^{2*} \text{ where } p_{rms} \text{ is in } \mu\text{Pa} \quad \text{...(20.14)}$$

or $$\boxed{L_p \text{ in dB} = 20 \log_{10} \left(\frac{p_{rms}}{20\ \mu\text{Pa}}\right)} \quad p_{rms} \text{ is in } \mu\text{Pa} \quad \text{...(20.15)}$$

---

*Since sound measuring instruments measure the r.m.s. pressure, the sound pressure is computed by squaring their r.m.s. values.

The sound pressure levels so measured are reported as dB *re* : 20 μPa.

Note. ***All the sound pressure meters record sound pressure level in dB implace of sound pressure in Pascal, due to too wide variation in Pascal readings.***

20.7.1. Sound Pressure Level (L_p) at a distance r from the Power Source.

L_p at a distance r from the sound source located on ground is given by the Eq.

$$L_p \text{ in dB} = \left[L_W - 10\log_{10}(2\pi r^2)\right] \text{ (dB)} \quad ...(20.16)$$

where $2\pi r^2$ represents the surface area of a hemisphere on which sound propagates.

Note 1. It can be found from Eq. (20.15) that an increases of 20 dB in sound pressure level will correspond to the sound pressure (p_{rms}) or the loudness of sound, increasing by 10 times (Pl. see Fig. 20.2).

Note 2. One dB is the faintest sound which can be perceived by a human ear, and the maximum sound which can be tolerated by human ear is about 180 dB.

The co-relation of sound intensity level and sound pressure level is shown in Table 20.1.

Table 20.1. Co-relation of Sound Intensity and Sound Pressure with the type of sound source.

Sound Pressure Level dB (A)	*Sound Pressure level in μ Pa i.e. μ N/m²*	*Sound Intensity Level in W/m²*	*Sound source*
0	20	1×10^{-12}	Threshold of hearing
20	20×10^1	1×10^{-10}	Whisper
40	20×10^2	1×10^{-8}	Public Library
60	20×10^3	1×10^{-6}	Normal conversation at 1 m distance
80	20×10^4	1×10^{-4}	Loud voice
100	20×10^5	1×10^{-2}	Siren at 5 m distance
120	20×10^6	1	Pneumatic drill
140	20×10^7	100	Jet aircraft or Artillery fire

Fig. 20.3 shows some common sound pressure values (in μPa) as well as their corresponding sound pressure level values (dB) with the types of sources producing such sound levels.

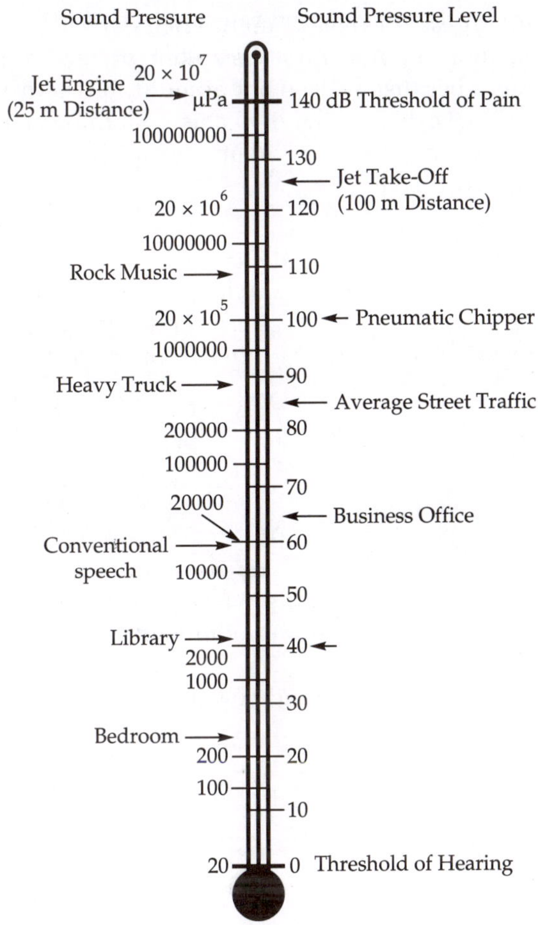

Fig. 20.3. Relative scale of sound pressure levels

20.7.2. Addition of Sound Pressure Levels. The addition of sound pressure levels in dB cannot be done by simple arithmetic addition, because of the log scale involved. Hence, if you consider 50 decibel noise and want to add another 50 decibel noise, it will not make up 100 decibel noise, but will make up only 53 decibel noise, as calculated below :

$$50 \text{ decibel} = 20 \log_{10}\left(\frac{p_{rms}}{20}\right) \text{ where } p_{rms} \text{ is in } \mu\text{Pa}$$

$$\therefore \quad \left(\frac{p_{rms}}{20}\right) = \text{Antilog}\left(\frac{50}{20}\right) = 316.227$$

or $$p_{rms} = 6324.55 \ \mu\text{Pa}.$$

$\therefore$ 50 decibel + 50 decibel in r.m.s.

$$= \sqrt{(6324.55)^2 + (6324.55)^2} = 8944.26 \ \mu\text{Pa}$$

$$= 20 \log_{10}\left(\frac{8944.26}{20}\right) \text{dB} = \mathbf{53 \ decibel.} \quad \textbf{Ans.}$$

Although *'sound pressure level'* mentioned above, gives a physical measurement of sound, yet it cannot be taken at face value as a perfect indication of loudness, because it does not account for the frequency of sound which has a lot to do with how loud a noise, sounds to human ear. The frequency of noise, thus, becomes an important factor in measuring a sound, and is explained in article 20.8.

Example 20.1. *The sound power from a voice shouting is 0.002 W. What is the Sound Power Level? What are the Sound Intensity, Sound Intensity Level, the Sound Pressure and the Sound Pressure Level at a distance of 10 metres from the source? Assume that sound radiates from the source in all directions.*

(Engineering Services, 2020)

Solution. (*i*) *Sound Power Level (SWL) or L_W is given in decibels (dB) by Eq. (20.5) as :*

$$L_W = 10 \log_{10}\left(\frac{W}{10^{-12}}\right) \text{ dB} \qquad \text{...(Eq. 20.5)}$$

where W = Sound power emitted by a sound source in all directions in watts or Joules/sec (J/s) 10^{-12} is the value of P_o *i.e.*, the reference sound power.

Here, W = 0.002 W.

$$\therefore \quad L_W = 10 \log_{10}\left(\frac{0.002}{10^{-12}}\right) \text{ dB} = 10 \log_{10}\left(\frac{2\times10^{-3}}{10^{-12}}\right)$$

$$= 10 \log_{10}(2\times10^{-3}\times10^{12}) \text{ dB} = 10 \log_{10}(2\times10^{9})$$

$$= 10 \times 9.3 \text{ dB} = \mathbf{93\ dB} \quad \textbf{Ans.}$$

(*ii*) *Sound Intensity at 10 m from the source.*

Sound intensity by a sound source on ground at a distance of r metre from the source is given by Eq. (20.7) as :

$$I = \frac{\text{Power of the source } (W)}{2\pi r^2} \qquad \text{....(Eq. 20.7)}$$

$$\therefore \quad I = \frac{0.002\ W}{2\pi\times(10\text{ m})^2} = \frac{0.002}{2\pi\times100}\text{ W/m}^2$$

$$= \mathbf{3.18\times10^{-6}\ W/m^2} \quad \textbf{Ans.}$$

(*iii*) *Sound Intensity Level (L_I)* in dB is given by Eq. (20.12) as :

$$L_I = 10 \log_{10}\left(\frac{I}{10^{-12}}\right) \text{ dB} \qquad \text{...(Eq. 20.12)}$$

But I = Sound intensity in W/m^2
= 3.18×10^{-6} W/m^2 (calculated above)

$$\therefore \quad L_I = 10 \log_{10}\left(\frac{3.18\times10^{-6}}{10^{-12}}\right) \text{ dB}$$

or $L_I = 10 \log_{10} (3.18 \times 10^6) = \mathbf{65\ dB}$ **Ans.**

(*iv*) *Sound Pressure Level (SPL)* or L_p at a distance of r from a sound source situated on ground is given by the Eq. (20.16) as :

$$L_p = L_W - [10 \log_{10} (2\pi\, r^2)] \text{ dB} \quad \text{...(Eq. 20.16)}$$

or $L_p = 93 - [10 \log_{10} (2\pi \times 10^2)]$

or $L_p = 93 - 10 \times 2.798 = 93 - 27.98 = 65.02$ dB **Ans.**

(*v*) *Sound Pressure* (p_{rms}) can be computed from Eq. (20.15) as :

$$L_p \text{ in dB} = 20 \log_{10} \left(\frac{p_{rms}}{20\ \mu\ \text{Pa}} \right) \quad \text{...(Eq. 20.15)}$$

But L_p = Sound pressure level, computed above = 65.02 dB

$$\therefore \quad 65.02 = 20 \log_{10} \left(\frac{p_{rms}}{20\ \mu\ \text{Pa}} \right)$$

$$\text{Antilog} \left(\frac{65.02}{20} \right) = \frac{p_{rms}}{20\ \mu\ \text{Pa}}$$

or $\dfrac{p_{rms}}{20\ \mu\ \text{Pa}} = 1782.38$

or $p_{rms} = 35{,}648\ \mu$ Pa

or $p_{rms} = 0.0356$ Pa **Ans.**

Check p_{rms} value from Eq. (20.9) as :

$$I = \frac{p_{rms}}{410}$$

or $3.18 \times 10^{-6} = \dfrac{p_{rms}^2}{410}$ where $I = 3.18 \times 10^{-6}$ W/m^2 (calculated above)

or $p_{rms} = \sqrt{3.18 \times 10^{-6} \times 410} = 0.036$ Pa

which tallies with the value computed above; and hence OK.

20.8 Frequency of Sound and Filtering Networks and Sound Meters

The frequency of sound is defined as the number of vibrations per unit time, and its unit is **hertz** (Hz). One hertz is equal to I vibration per second.

Although, humans can hear sound waves in the frequency range of 10 to 20,000 Hz*, but human ear is more sensitive to sounds of middle frequency (in the range of about *500 to 5000 Hz*) as compared to the sounds of lower and higher frequencies in this audible range. Hence, while measuring the loudness of a sound, *electronic filtering circuits* (called ***weighting networks***) are built into the sound pressure measuring meters, which filter out certain frequencies. Such circuits, therefore, permit the sound level meter to respond more to some frequencies than to others, with prejudice something like that of

*Many animals can hear sounds that are inaudible to human ear.

a human ear. Although three types of filtering circuits, *i.e.*, **A, B, and C types** of are often used in sound meters, but **A network** is generally used in our day to day sound measurements. Hence, the measured sound in dB (decibel) is written as dB(A). The *A-network largely filters out very low and very high frequencies below about 200 Hz and above about 10,000 Hz.*

The measured sound, or strictly speaking sound pressure level, is usually represented by the symbol L_p. Thus, L_p at a place can be measured in dB (A) units, and may be represented by L_A. Different values of L_p will be measured at different times of the day, since the sound will vary throughout the day.

Characteristics of three weighting networks. As a matter of fact, very low frequencies are quite severely filtered out by the *A* network ; moderately by the *B* network ; and hardly at all by the *C* network. Therefore, if measured sound level on *C* network is much higher than that on *A* network, much of the noise is probably of low frequency.

Fig. 20.4 shows the basic characteristics of the three weighting networks (*i.e. A, B* and *C*). When such a network is used, the sound level meter electronically adds or subtracts the number of *dB* shown at each frequency in table 20.2 from the actual sound pressure level at that frequency. It then sums up all the resultant numbers by log addition to give a single reading. The readings taken on different networks *A, B* and *C* are designated as *dB(A), dB(B), dB(C)* respectively. They may also be notated by L_A, L_B and L_C.

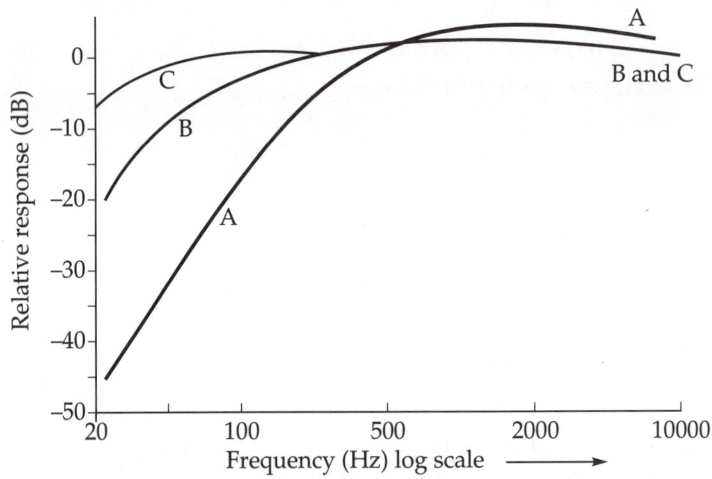

Fig. 20.4. Response characteristics of *A, B* and *C* types of weighting networks.

Readings taken when a network is in use are said to be **"sound levels"** rather than **"sound pressure levels".**

A perusal of the *dB* values of table 20.2 reflects that in low frequency range, the *dB* values to be subtracted are reducing from network *A* to *C*, and there is wide variation in their values. *Hence for such low frequency noise, the resulting values of* L_A, L_B *and* L_C *will differ by wide margins and will be reducing from A to C.* In other words $L_A > L_B > L_C$ by wide margins.

Table 20.2. Network Weighting Values of Sound Level Meters

Frequency (Hz)		*Curve A (dB)*	*Curve B (dB)*	*Curve C (dB)*
Low frequency range	10	– 70.4	– 38.2	– 14.3
	20	– 50.5	– 24.2	– 6.2
	50	– 30.2	– 11.6	– 1.3
	100	– 19.1	– 5.6	– 0.3
Mild-frequency range	200	– 16.9	– 2.0	0
	500	– 3.2	– 0.3	0
	1000	0	0	0
High frequency range	2,000	1.2	– 0.1	– 0.2
	5,000	0.5	– 1.2	– 1.3
	10,000	– 2.5	– 4.3	– 4.4
	20,000	– 9.3	– 11.1	– 11.2

Similarly, for *mild frequency noise*, these values will of course be in decreasing order, but with very little difference between them. On the other hand for *high frequency noise*, these values will be in increasing order, because the subtraction for *C* network is more than for *B* network, which is more than for *A* network. The difference will however be quite small.

A pair of such values as L_A = 94 dB ; L_B = 88 dB, and L_C = 80 dB, therefore, reflects *low frequency noise* ; and a pair of values such as L_A = 94 dB, L_B = 93 dB, and L_C = 92 dB, will thus reflect a *mild-frequency noise* ; and similarly, a pair of values like L_A = 96 dB, L_B = 97 dB, and L_C = 98 dB will reflect a *high frequency noise*.

20.8.1. Octave Band Analysis and Sound Spectra. A noise can be fully characterised by breaking it down into its *frequency components,* called *spectra*. Normal practice is to consider 8 to 11 octave bands. *An octave band is the frequency interval between a given frequency and twice that frequency.* For example, for a given frequency of 22 Hz, the first octave band is from 22 Hz to 44 Hz. A second octave band would be from 44 Hz to 88 Hz, and so on. The standard octave bands and their geometric mean frequencies are given in table 20.3.

Table 20.3. Standard Octave Bands

Octave frequency range (Hz)	*Geometric mean (centre band) frequency (Hz)*
22—44	31.5
44—88	63
88—175	125
175—350	250
350—700	500
700—1400	1,000
1400—2800	2,000
2800—5600	4,000
5600—11200	8,000
11200—22400	16,000
22400—44800	31,500

An octave band analysis is usually sufficient for community noise control and for identifying violators ; more refined analysis is required for corrective action and design. Such a refined analysis is usually provided by *one-third octave band analysis.* This improved resolution is generally sufficient for determining corrective action for community noise problems. More refined analysis like *narrow band analysis,* involving band widths down to 2 Hz, may, however, be justified only for instrument design and testing, or in trouble shooting industrial machine noise and vibrations.

A typical noise spectra of automobiles is shown in Fig. 20.5.

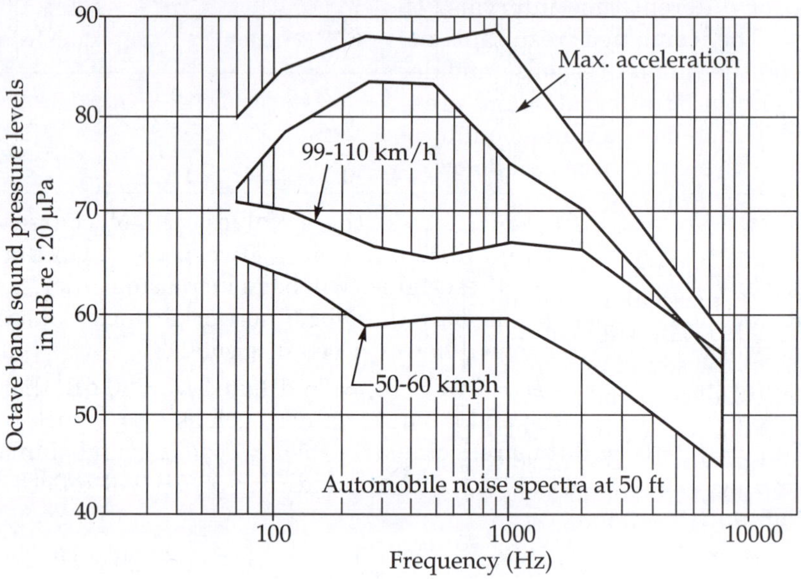

Fig. 20.5. Typical noise spectra of automobiles.

20.8.2. Averaging Sound Pressure Levels. The average value of the various recorded sound pressure levels (L_p) at a particular place over a given period cannot be computed by simple averaging due to log scale involved in their values. On the other hand, the following equation is used to compute average pressure level :

Average pressure level

$$= \overline{L}_p = 20 \log_{10} \frac{1}{N} \sum_{n=1}^{n=N} (10)^{L_{n/20}} \quad \text{...(20.17)}$$

where $\overline{L}_p$ = Average sound pressure level in dB *re* : 20 μPa

N = Number of measurement readings.

L_n = nth sound pressure level in dB *re.* 20 μPa.

n = 1, 2, 3 .... N.

Say for example, the average of 4 measurement readings recorded as 40, 50, 62 and 72 dB *re* : 20 μPa is computed to be 63 dB, inplace of straight arithmetic average value of 56 dB, as follows :

$$\sum_{n=1}^{n=4} (10)^{L_n/20} = [(10)^{40/20} + (10)^{50/20} + (10)^{62/20} + (10)^{72/20}]$$

$$= [100 + 316.23 + 1258.92 + 3981.07] = 5656.22$$

and $$\overline{L}_p = 20 \log_{10} \frac{1}{4} \times 5656.22 = \mathbf{63\ dB.} \quad \textbf{Ans.}$$

20.9. Noise Rating Systems

A noise may consist of *different types of sounds** with different pressure levels operating for different time intervals. The frequencies of these sounds may also vary. The combined resultant noise will infact be responsible for determining the human response and degree of annoyance caused by it. The combined impact of different sound pressures lasting for different periods is therefore attempted to be worked out by using some statistical measures, such as L_N and L_{eq} systems, discussed below :

The L_N concept. The parameter L_N is a statistical measure indicating how frequently a particular sound level is exceeded. The value of L_N will represent the sound pressure level that will exceed for N% of the gauging time. Say for example, the given 70 dB value of L_{60} will mean that the sound level will exceed 70 dB for 60% of the measuring time.

When L_N is plotted against N (where N = 1, 2, 3, ... 100%), a *cumulative distribution curve,* as shown in Fig. 20.6, is obtained.

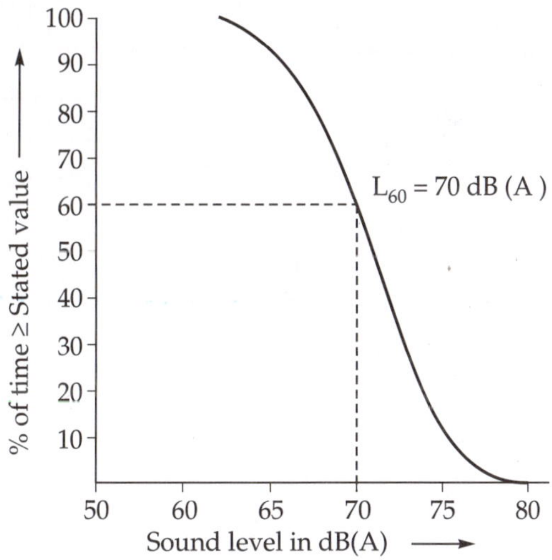

Fig. 20.6. Cumulative Distribution Curve.

The L_{eq} Concept. L_{eq} is that statistical value of sound pressure level that can be equated to any fluctuating noise level. Say for example, a sound of 80 dB lasting for 10 minutes, followed by a sound of 60 dB lasting for the next 80 minutes, and then followed by a sound of 100 dB for the next 5 minutes, will compose a fluctuating noise level, which can be represented by a certain dB

---

*Sound is usually divided into three types ; *viz* :

(*i*) *continuous* ; (*ii*) *intermittent ; and* (*iii*) *impulsive* ; as explained below :

(*i*) **Continuous noise** is an uninterrupted sound level that varies less than 5 dB during the entire period of observation. A running fan is an example of such a sound.

(*ii*) **Intermittent noise** is a noise which continues for more than 1 second and is then interrupted for more than 1 second. A drilling machine used by a dentist produces such type of sound.

(*iii*) **Impulse noise** is characterised by a change of sound pressure of atleast 40 dB within 0.5 second with a duration of less than one second. The noise produced from firing of a weapon would fall in this category.

value which is indicative of producing the same effect over the entire time period of 10 min + 80 min + 5 min = 95 min, as the original fluctuating noise. This value is called the ***Equivalent continuous equal energy level*** or the **Equivalent Noise Level** (L_{eq}).

Thus, L_{eq} is defined as the constant noise level, which, over a given time, expands the same amount of energy, as is expanded by the fluctuating levels over the same time. This value is expressed by the equation :

$$L_{eq} = 10 \log \sum_{i=1}^{i=n} (10)^{\frac{L_i}{10}} \times t_i \qquad \text{...(20.18)}$$

where n = Total number of sound samples

L_i = The noise level of any ith sample

t_i = Time duration of ith sample, expressed as fraction of total sample time.

Using the above equation, L_{eq} value for fluctuating noise level of 95 minutes indicated earlier (*i.e.* the one with 80 dB lasting for 10 minutes, followed by 60 dB for 80 minutes, followed by 100 dB for 5 minutes can be worked out as below :

$$\sum_{1}^{3} (10)^{\frac{L_i}{10}} \times t_i = \left[(10)^{\frac{80}{10}} \times \frac{10}{95} + (10)^{\frac{60}{10}} \times \frac{80}{95} + (10)^{\frac{100}{10}} \times \frac{5}{95}\right]$$

$$= 1.053 \times 10^7 + 0.842 \times 10^6 + 0.52632 \times 10^9$$

$$= 10^6[10.53 + 0.84 + 526.32] = 537.69 \times 10^6$$

$\therefore \quad L_{eq} = 10 \,.\, \log_{10} (537.69 \times 10^6) =$ **87.3 dB. Ans.**

The above fluctuating sound with its equivalent energy level (L_{eq}) is represented graphically in Fig. 20.7, which reflects that high occasional sounds do influence the resultant equivalent value by a large extent.

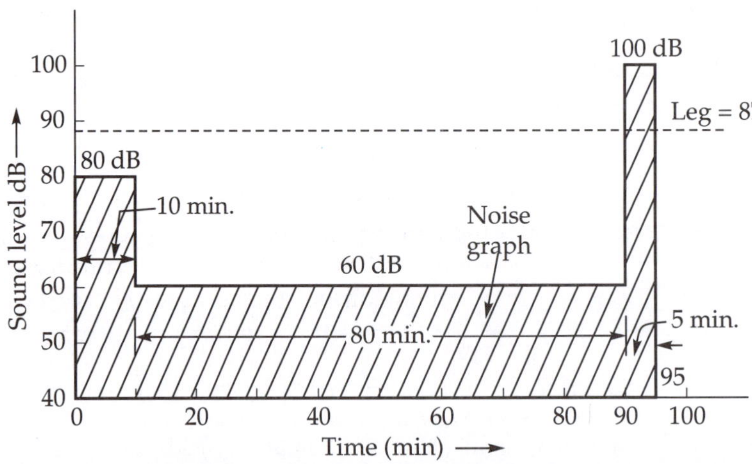

Fig. 20.7. Noise graph and its equivalent noise level.

The *equivalent noise level* (L_{eq}) constitutes an important parameter for evaluating the impact of fluctuating noises of all kinds, such as from aircraft, street and road traffic, rail traffic, industrial machines, sports stadiums, play grounds, etc.

Moreover, the duration in hours, over which L_{eq} is worked out for a given site, is further mentioned in bracket, such as $L_{eq(8)}$ which means that L_{eq} is based on 8-hour measurement ; when, however, no such time is mentioned, then L_{eq} always corresponds to one hour measurement.

The equivalent noise level (weighted on A network) during 24 hour time period with + 10 dB (*A*) correction applied to measurements recorded during night time of 22 hr to 07 hr, is usually represented by L_{dn}.

Example 20.2. *Traffic noise data are given in the table below. Compute* L_{eq}.

Time (in s)	*10*	*20*	*30*	*40*	*50*	*60*	*70*	*80*	*90*	*100*
Noise (dBA) L(t)	*71*	*75*	*70*	*78*	*80*	*84*	*76*	*74*	*75*	*74*

(*Engineering Services 2014*)

Solution. *Equivalent Noise Level* (L_{eq}) is expressed by equation (20.13) as :

$$L_{eq} = 10 \log_{10} \sum_{i=1}^{i=n} (10)^{\frac{L_i}{10}} . t_i \qquad \text{...(Eq. 20.18)}$$

Using the above equation, L_{eq} value for the given fluctuating noise level of 100 seconds is computed as under :

$$\sum_{1}^{10} (10)^{\frac{L_i}{10}} \times t_i = \Bigg[(10)^{\frac{71}{10}} \times \frac{10}{100} + (10)^{\frac{75}{10}} \times 10 + (10)^{\frac{70}{10}} \times \frac{10}{100}$$

$$+ (10)^{\frac{78}{10}} \times \frac{10}{100} + (10)^{\frac{80}{10}} \times \frac{10}{100} + (10)^{\frac{84}{10}} \times \frac{10}{100} + (10)^{\frac{76}{10}} \times \frac{10}{100}$$

$$+ (10)^{\frac{74}{10}} \times \frac{10}{100} + (10)^{\frac{75}{10}} \times \frac{10}{100} + (10)^{\frac{74}{10}} \times \frac{10}{100} \Bigg]$$

$$= \frac{1}{10} \big[1.2589 \times 10^7 + 3.1622 \times 10^7 + 1 \times 10^7 + 6.3096 \times 10^7 + 10 \times 10^7 + 2.5119 \times 10^7 + 3.9811 \times 10^7 + 2.5119 \times 10^7 + 3.1623 \times 10^7 + 2.5119 \times 10^7 \big]$$

$$= \frac{10^7}{10} [36.4098] = 36.4098 \times 10^6$$

$$\therefore \quad L_{eq} = 10 \times \log_{10} (36.4098 \times 10^6) = \mathbf{75.61\ dB.} \quad \textbf{Ans.}$$

20.10. Noise Level Standards

The prescribed noise standards generally put a limit either on L_{eq} or L_N values. Different such limiting values are then prescribed for different areas. Say for example, VIP and serene areas will impose lower limiting values as compared to those for commercial or industrial areas within a city. The usually prescribed limiting L_{eq} values for different areas of a city is of the order of 40 to 70 decibels.

Table 20.4 and 20.5 show two sets of noise standards, which have been laid down in USA. The first standards (table 20.4), developed by *U.S. Federal*

Highway Administration are the levels prescribed for newly planned highways through different types of areas. These levels are above those that would be absolutely harmless, but are below those of many existing highways.

Table 20.4. Noise Standards laid down by U.S. Federal Highway Administration for Highways in New Developments

Land use category	*Exterior design noise level dB(A)**		*Description of land use category*
	L_{eq}	L_{10}	
A	57	60	VIP and other protected areas, where serenity and quietness is to be maintained
B	67	70	Residences, hotels, schools, hospitals, picnic areas, play grounds, parks, etc.
C	72	75	Developed lands and properties not included in A and B category, such as commercial and industrial areas.
D	No limit	No limit	Undeveloped lands
E	52 (interior)	55 (interior)	Public meeting halls, schools, libraries, churches, hospitals, etc.

*Either L_{eq} or L_{10}, based on 1 hour samples, is to be used, but not both. [L_{10} will evidently mean that the indicated dB will be exceeded for only 10% of the measuring time of 1 hr].

The other set of standards (Table 20.5) has been evolved by *U.S. Environmental Protection Agency* and shows the limiting noise levels for different types of areas for indoor as well as outdoor noises, that should be maintained for adequately safeguarding public health over long periods (accounting annual average levels in L_{eq} values).

Table 20.5. Average yearly energy Levels (L_{eq}) identified as requisite to protect public health and welfare with an adequate margin of safety

Types of areas	*Measure*	*Indoor*			*Outdoor*		
		Activity interference	*Hearing loss consideration*	*To protect against both effects (b)*	*Activity interference*	*Hearing loss consideration*	*To protect against both effects (b)*
(1)	(2)	(3)	(4)	(5)	(6)	(7)	(8)
Residential with outside space and farm residences	L_{dn} $L_{eq(24)}$	45	 70	45	55	 70	55
Residential with no outside space	L_{dn} $L_{eq(24)}$	45	 70	45			
Commercial	$L_{eq(24)}$	(*a*)	70	70(*c*)	(*a*)	70	70(*c*)

Contd...

Inside transportation	$L_{eq(24)}$	(*a*)	70	(*a*)			
Industrial	$L_{eq(24)}(d)$	(*a*)	70	70(*c*)	(*a*)	70	70(*c*)
Hospitals	L_{dn} $L_{eq(24)}$	45	70	45	55	70	55
Educational area	$L_{eq(24)}$ $L_{eq(24)(d)}$	45	70	45	55	70	55
Recreational areas	$L_{eq(24)}$	(*a*)	70	70(*c*)	(*a*)	70	70(*c*)
Farm land and general unpopulated land	$L_{eq(24)}$				(*a*)	70	70(*c*)

Note. (*a*), (*b*), (*c*) and (*d*) are defined at the bottom of the table.

Code :

(*a*) Since different types of activities appear to be associated with different levels, identification of a maximum level for activity interference may be difficult except in those circumstances where speech communication is a critical activity.

(*b*) Based on lowest level.

(*c*) Based only on hearing loss.

(*d*) An $L_{eq(8)}$ of 75 dB may be identified in these situations so long as the exposure over the remaining 16 hour per day is low enough to result in negligible contribution to the 24-hour average, that is, no greater than an L_{eq} of 60 dB.

Note. Explanation of identified level for hearing loss. The exposure period that results in hearing loss at the identified level is a period of 40 years.

The acceptable outdoor noise levels in different types of residential areas, as well as the acceptable indoor noise levels for various types of buildings, as recommended by *Indian standard code* IS : 4954—1968, are given in table 20.6.

Table 20.6. Acceptable Noise Levels for Residential Areas, as per B.I.S. code 4954-1968

	Acceptable outdoor noise levels in residential areas			*Acceptable indoor noise levels for various types of buildings*	
S. No.	*Location*	*Noise levels dB(A)*	*S. No.*	*Location*	*Noise levels dB(A)*
1.	Rural areas	25—35	1.	Radio and TV studios	25—30
2.	Suburban areas	30—40	2.	Music rooms	30—35
3.	Urban Residential areas	35—45	3.	Hospitals, class rooms, auditoria	35—40
4.	Residential and business urban areas	40—50	4.	Apartments, hotels, homes Conference-rooms, small offices	35—40
5.	City areas	45—55	5.	Court rooms, private offices, libraries	40—45
6.	Industrial areas	50—60	6.	Large public offices, banks, stores, etc.	45—50
			7.	Restaurants	50—55

In order to legally enforce control on noise levels, the GoI has notified **the Noise Pollution (Regulation and Control) Rules 2000,** under *the Environment (Protection) Act, 1986*. Under these Rules, Ambient Air Quality Standards in respect of Noise, have been specified in respect of different types of area uses (such as industrial area, commercial area, residential area and silence zones), as shown in Table 20.7.

Table 20.7. Ambient Air Quality Standards in Respect of Noise as Specified under the Indias' Environment Protection Act, 1986

Area Code	*Category of Area / Zone*	*Limits in dB (A) L_{eq}**	
		Day Time	*Night Time*
(A)	Industrial Area	75	70
(B)	Commercial Area	65	55
(C)	Residential Area	55	45
(D)	Silence Zone	50	40

Note : 1. Day time shall mean from 6.00 a.m. to 10.00 p.m.

2. Night time shall mean from 10.00 p.m. to 6.00 a.m.

3. Silence zone is an area comprising not less than 100 metres around hospitals, educational institutions, courts, religious places or any other area which is declared as such by the competent authority.

4. Mixed categories of areas may be declared as one of the four above mentioned categories by the competent authority.

*dB(A) L_{eq} denotes the time weighted average of the level of sound in decibels on Scale A which is relatable to human hearing. Day time specified values will, thus, be weighted values over 6 am to 10 pm, and night time specified values will be weighed values over 10 pm to 6 am.

A "decibel" is a unit in which noise is measured.

"A", in dB(A) L_{eq}, denotes the frequency weighting in the measurement of noise and corresponds to frequency response characteristics of the human ear.

Leq : It is an energy mean of the noise level over a specified period.

The major highlights of the provisions of these Noise Pollution (*Regulation and Control Rules, 2000*) are indicated below :

- The State Government shall take measures for abatement of noise including noise emanating from vehicular movements and ensure that the existing noise levels do not exceed the ambient air quality standards (see Table 20.7) specified under these rules.
- All development authorities, local bodies and other concerned authorities while planning developmental activity or carrying out functions relating to town and country planning shall take into consideration all aspects of noise pollution as a parameter of quality of life to avoid noise menace and to achieve the objective of maintaining the ambient air quality standards in respect of noise.
- The noise levels in any area/zone shall not exceed the ambient air quality standards in respect of noise as specified in the Schedule (see Table 20.7).

- The authority shall be responsible for the enforcement of noise pollution control measures and the due compliance of the ambient air quality standards in respect of noise.

  (*Responsibility as to enforcement of noise pollution control measures*)
- *A person may, if the noise level exceeds the ambient noise standards by 10 dB(A) or more given in the corresponding columns against any area/ zone, make a complaint to the authority.*
- *The authority shall act on the complaint and take action against the violator in accordance with the provisions of these rules and any other law in force.*

[**Note:** *You don't need to measure the noise. It has to be beyond the 'tolerable' level to have irritated you.*]

Author's Note: Although quite stringent noise level standards have been prescribed by GoI, yet there is practically no enforcement of these standards even in the capital of Delhi, where the existing sound levels are as high as 83 dB during day time and 77 dB during night time, as compared to the highest permissible values of 75 and 70 dB, respectively.

20.11. Sources of Noise and their Noise Levels

Outdoor noise is usually produced by the following sources :

(1) *Traffic* including air traffic, road traffic, and sea-shore and inland water traffic ;

(2) *Industries* ; and

(3) *Other miscellaneous activities* ; as discussed below.

20.11.1. Traffic Noise. Traffic usually produces a lot of noise, particularly in big cities like Delhi, and is responsible for causing a lot of nuisance to the society. Thc amount and type of noise produced by traffic is largely dependent upon the type of traffic. Say for example, the noise produced will depend upon the type of aircraft or automobile used for the traffic. A jet aircraft produces more noise than a propeller type ; a diesel jeep or a truck produces more noise than a petrol car ; badly maintained automobiles produce more noise than new and good conditioned vehicles ; diesel and steam locomotives produce more noise than the electric trains ; badly designed and maintained rail tracks produce much more noise than those in good condition and constructed in deep cuts. Noise levels produced by different types of traffic are given in Table 20.8.

Table 20.8. Noise Levels of Different Sources of Traffic

S. No.	*Source of Noise*	*Noise level in dB*
1.	Air traffic	
	(*i*) *Jet aircraft* at take off stage at about 300 m	100—110
	(*ii*) *Propeller type aircraft* at take off stage at about 300 m	90—100
2.	Rail traffic (at about 30 m)	90—110
3.	Heavy road traffic (highway)	80—90
4.	Medium road traffic (main streets)	70—80
5.	Light road traffic (side streets)	60—70

In order to ensure that the vehicles on roads do not produce excessive noise, the Union Environment Ministry has prescribed noise standards for different types of vehicles, as shown in Table 20.9.

Table 20.9. GoI Noise Standards for Different Types of Vehicles

S. No. (1)	*Type of vehicles* (2)	*Noise level in dB* (3)
1.	Two wheelers	80
2.	Cars	82
3.	Passenger or commercial vehicles	
4.	(*i*) Upto 4 MT (Tonnes)	85
	(*ii*) Between 4 MT to 12 MT	89
	(*iii*) More than 12 MT	91

At present, the noise pollution level clearance for different brands of vehicles is only required at the manufacturer's level. However, by October 2004, all the 6 crore vehicles plying in the country were to meet the above norms or face prosecution. The owners of vehicles were to secure certificates that their vehicles meet the prescribed noise pollution norms.

However, there has been no implementation of these standards, since sufficient manpower and enough sound meters are not available as per the needs.

20.11.2. Industrial Noise. Noise is the essential by-product of industry ; its intensity and nature being dependent upon the type of the industry. Industrial noises are usually produced by rotating, reciprocating or any other types of machinery, or by high pressure high velocity gases, liquids or vapour involved in the industrial processes. The usual noise level of the industries is of the order of 60 to 95 decibels.

20.11.3. Noise Produced by Other Sources. Several other human activities, such as blaring of loudspeakers and sirens, shouting of hawkers, playing of children, general life and activity, ringing of temple and church bells, etc. do produce noises of different levels, tones and spectra.

20.12. Noise Abatement and Control

Exposure of mankind to the day by day increasing noise nuisance must be reduced and abated, if its adverse effects on human health are to be controlled. The society must therefore be protected from the harmful effects of noise by devising and implementing ways and means for the abatement of noises.

There are certain noises which can be kept under control by legal laws and ordinances, and there are others which have to be dampened and attenuated by the use of good technology and town planning. For example, the noises produced by motor vehicles can be controlled to some extent by proper maintenance of vehicles, which can be ensured only by prescribing maximum permissible noise levels for different types of automobiles, through Motor Vehicles Act. Similarly, industrial noises, can be brought under the control of Factories Act by specifying maximum permissible noise levels and other checks. Similarly, public blaring of loudspeakers and plying of radio sets at loud levels can be prevented by general legal laws of public nuisance or laws specifically made for noise pollution. Yet however, there exist several other noises which have become part and parcel of our modern life. All such noises are to be reduced by better design technology to be used in the modern day gadgets like fans, air-conditioners, washing machines, refrigerators, mixers and grinders, etc.

Another very important method for abating noise effects on mankind is to use **proper town planning techniques,** and thus, to ensure construction of houses and offices away from the major sources of noise. Proper segregation, **zoning** and separation of residential complexes from the commercial and industrial ones, by means of physical barriers, roads, railway lines, parks or green belts, do constitute an important aspect of such good town-planning.

The noises produced by automobiles and trains, being the biggest noise nuisance in a modern city life, can be abated by construction of walls on both sides of roads and railway lines. **Raising of such obstructions and barriers in between the noise sources (auto-mobiles) and residences,** may considerably reduce the noise levels reaching the residences. Attenuation of upto 15 decibels is possible in this manner. The extent of attenuation achieved by raising of such a barrier wall, is shown in Fig. 20.8.

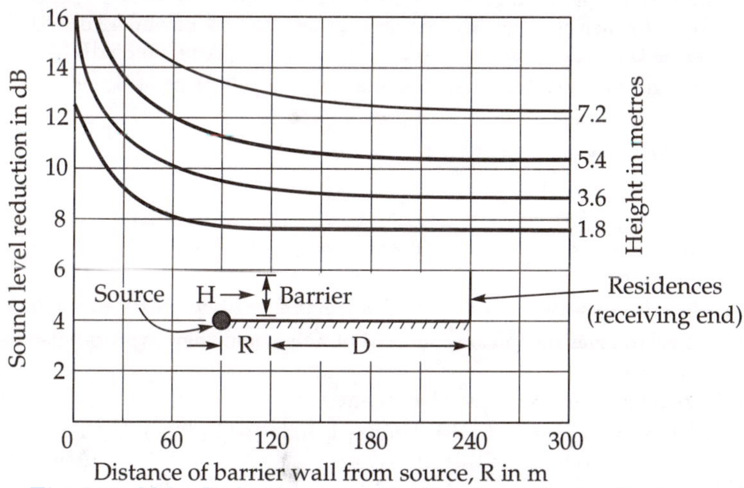

Fig. 20.8. Noise Reduction caused by Barrier wall raised between a noise source and the receiving end.

When $D \geq R$ and $R >> H$, the noise reduction may also be calculated by the equation* :

$$\text{Noise reduction (dB)} = 10 \log_{10}\left(\frac{20H^2}{\lambda R}\right) \quad \text{...(20.19)}$$

where H = Height of the barrier wall

λ = Wavelength of sound

D = Distance between barrier and the receiving point

*This equation is generally found to be giving higher value than that observed in actual life.

Raising of thick and high vegetation and tree growing along sides of roads and railway lines, offer cheaper barriers to cause such noise reductions.

Locating the noisy sources on the downwind side of the residences, may be another important consideration in good town planning, because noise will then travel farther in the downwind direction away from the residences.

Noise levels in residential buildings can be reduced to some extent by **offsetting the building from the main or street roads** by a suitable distance. The farther is this distance, the better will be the attenuation, because the intensity of noise gets reduced with increase in distance.

PROBLEMS

1. (*a*) Define 'noise' and explain as to how and why it should be regarded as an environmental pollutant.

(*b*) Differentiate between L_n, L_{eq} and L_{dn} in relation to expression of sound levels.

2. (*a*) Differentiate between the terms "sound pressure levels" and "sound levels", bringing out clearly the basic characteristics of the three types of weighting networks.

(*b*) 60 dB (A) *re* : 20 μPa noise is accompanied with another 60 dB (A) *re* : 20 μPa noise. What will be the total noise level. [**Ans.** 63 dB (A) *re* : 20 μPa]

3. (*a*) While recording *A*-weighted sound levels, 4 readings have been taken at a site at different times of a day. These readings are : 20, 56, 66 and 42 dB(A) *re* : 20 μPa. What is the average sound level ? [**Ans.** 56.85 dB (A) *re* : 20 μPa]

(*b*) Mention typical values of acceptable sound levels as per I.S. code for :

(*i*) Rural areas ;

(*ii*) Urban residential areas ;

(*iii*) City areas ; and

(*iv*) Industrial areas.

4. (*a*) Explain in brief the major factors and actions that may help in noise abatement in a modern society.

(*b*) Differentiate between *continuous* noise, *intermittent* noise, and *impulse* noise.

5. (*a*) Discuss in brief the various sources of noise, and their typical noise levels, in a modern society.

(*b*) Write a brief note on noise abatement.

6. (*a*) 50 dB (A) noise lasting for 55 minutes is followed by 90 dB (A) noise lasting for 5 minutes. What is L_{eq} of this noise ? [**Ans.** 79.2 dB (A)]

(*b*) Write brief but explanatory notes on any two of the following :

(*i*) Noise pollution, its causes, effects, and remedies,

(*ii*) Octave band analysis of noise and noise spectrum

(*iii*) Noise rating systems

(*iv*) Noise levels and their specified Indian standards.

—21

Environmental Impacts of Thermal Power Plants, Mining and River Valley Projects

21.1. Environmental Impact Assessment

Engineering projects involving development of thermal power, mining operations, and even river valley water resources developments, have been found to be causing certain adverse and negative impacts on our surrounding environment, which has forced us to make it compulsory to evaluate these adverse impacts in details, well before the project is cleared for execution. With this end in view, all project clearance cells do evaluate and examine the detailed environmental assessment report, which is prepared and submitted along with the DPR (Detailed Project Report) of every such project. Submission of such *Environmental Impact Assessments* or *Environmental Impact Statements (EIS)* have been made compulsory by the Indian Government for all projects, which are likely to cause harm to our surrounding environment. These statements are thoroughly examined by the concerned Ministries before giving *environmental clearance* to the project, without which, administrative and financial sanction to the project, will not be given.

All such impact assessments should thoroughly examine and discuss the various possible environmental damages, whether pertaining to *water pollution, air pollution, ground (land) pollution, noise pollution*, or any other kind of environmental pollution ; and their suggested remedial measures to prevent or to mitigate such hazardous environmental effects.

The environmental impacts which are likely to be caused by the three types of projects (*i.e.* thermal power plants, mining, and river valley development projects) are discussed here in the following articles, as they stand included in the syllabus of Civil Services Exam.

21.2. Environmental Impacts of Thermal Power Plants

Thermal power plants, generally use coal as the fuel for producing steam to run their turbines for production of electricity. Fuel oil is also sometimes used as fuel. Use of fuel gas is the most modern advancement in this field.

Fossil fuels, including **coal** as well as **fuel oil**, when used in the power plants, prove very harmful to the air environment, as they release very heavy amounts of pollutants into the atmosphere.

If coal is used, pollutants like *fly ash, sulphur dioxide*, and *nitrogen oxides*, are produced on a large scale. In case of oil, however, only *sulphur dioxide* and *nitrogen oxides* are produced, as the major pollutants. Coal, thus, proves to be a worse fuel, and is an important source responsible for particulate air pollution. The amount of fly ash and sulphur dioxide produced by the coal depends upon its quality, and sulphur and ash content in it.

It has also been observed that the fly ash from a thermal power plant falls out, upto large distances from the plant. Heavy amount of fly ash has been observed to be falling even upto distances as large as 3 to 6 km. The fallout rates are found to increase in dry windy seasons.

The fallout rates of fly ash at the thermal power plant at Korba in Madhya Pradesh State has been found to be as high as 300—550 t/km^2/month in the vicinity of the plant, about 130—550 t/km^2/month at distances of about 2.5 km, and about 60—450 t/km^2/month at distances of about 4 km.

Similarly, the Indraprastha and Rajghat power houses at Delhi near ITO, are responsible for polluting Delhi's air on a large scale. *Dust and fly ash* has been thrown by these plants even upto Karol Bagh, about 10 km away. However, the distances upto which, these dusts fall, have now been reduced with the commissioning of electrostatic precipitators at these plants, because the *precipitators catch the particulate matter*. Yet, however, the menace of disposal of huge quantity of fly ash caused a lot of worries to the Govt. of National Capital Territory of Delhi, which is now utilising the fly ash for making bricks.

In order to reduce the quantum of fly ash production and consequential air pollution, DVB (*Delhi Vidhut Board*) is also now planning to run these plants on fuel gas, in place of coal. A gas pipe line, existing between Hazira-Bijapur-Jagdishpur is being extended by the *Gas Authority of India*, for bringing fuel gas to these plants. This pipe line will also supply gas to an NTPC's power plant at Dadri, enroute to Delhi from Jagdishpur.

21.3. Emission Norms for Thermal Power Plants in India

The GoI has recently revised the emission norms for thermal power plants in India. As per the notification, dated December 7, 2015, the new standards for PM10, SO_2 and NO_x have been fixed by dividing the plants into three categories, as indicated in table 21.1.

These norms/standards have been fixed with the aim of reducing emission of PM 10 to 0.98 kg/kWh, SO_2 to 7.3 kg/kWh, and NO_2 to 4.8 kg/kWh, which in turn, will bring an improvement in the ambeint air quality (AAQ) in and around the thermal power plants.

The technology employed for the control of the proposed limits of sulphure dioxide (SO_2) and nitrogen oxides (NO_x) will also help in control of mercury emissions (at about 70–90%) as a co-benefit.

In addition to prescribing revised standards for emission of PM10, SO_2 and NO_x, the new notification also stipulates that the new power plants to be commissioned from January 1, 2017 will have to achieve **'zero waste water discharge'**, standard by putting up adequate captive infrastructure. All

Table 21.1. Revised Emission Norms for Thermal Power Plants (TPPs) in India*

S. No.	*Category of TPP*	*Revised Standards of Emission of Gases from TPPs*			
		PM 10 mg/m³	*SO_2 mg/m³*	*NO_x mg/m³*	*Mercury mg/m³*
1.	TPP installed before Dec. 31, 2003	100*	(*i*) 600 for plants having capacity of less than 500 MW	600	None
			(*ii*) 200 for plants having capacity of 500 MW or more	600	0.03
2.	TPP installed after Dec. 31, 2003 and upto Dec. 31, 2016	50*		300	0.03
3.	TPP to be installed from Jan. 1, 2017	30	100	100	0.03

incoming power plants will be given environmental clearance only after getting such committments under the new norms.

Standards have also been specified for water consumption as follows :

(*i*) All plants with once through cooling (OTC) shall install cooling towers (CTs) and achieve specific water consumption-max of 4 m^3/M.Wh within 2 year period.

(*ii*) All existing CT-based plants shall reduce specific water consumption up to max of 3.5 m^3/M.Wh within 2 years from the date of notification.

(*iii*) New plants to be installed after January 1, 2017 shall meet specific water consumption-max of 2.5 m^3/M.Wh and achieve zero liquid discharge.

21.3. Environmental Impacts of Mining

Mining, as we know, is the extraction of minerals and ores from the Earth. It is a vital sector of our economy, just next to agriculture.

Mining helps us to extract costly resources from the Earth, which we fashion into countless goods of our daily and industrial uses. Mining provides us *fuels* for our energy requirements, *metals* for making machines and engines, the *ores* for making chemicals and fertilisers, the *clays* for making crockery, the *rocks* and *stones* for our buildings and dams, the *gold, diamond* and other *precious stones* for our jewellery making, and what not. *Mining is, therefore, considered as an important industry, and it is necessary for any*

*Up till now *i.e.* **before the Revised norms**, PM 10 standard for TPPs with power generation capacity of more than 210 MW was 150 mg/Nm^3, and for TPPs having generation capacity of less than 210 MW, PM10 standard was 350 mg/Nm^3. No standards existed for SO_2, NO_x and Mercury.

nation to ensure that this industry continues to flourish and develop. The total number of mines, strictly speaking **open cast mines**, in our country is around 5600 ; out of which, about 500 mines are producing coal, and the remaining are in the non-coal sector.

In general, mining activities are accompanied by a *variety of environmental problems*. The process of environmental degradation, which starts with the extraction of minerals resulting in *land degradation and addition of pollutants in the air and water*, continues even as the extracted mineral ores are further processed and developed for their ultimate uses.

The *environmental impacts of mining* can be broadly divided into four parts, *viz.*

(1) *water pollution* ;

(2) *air pollution* ;

(3) *de-spoilation of land, including land subsidence, land degradation and deforestation* ; and

(4) *noise* and *ground vibrations* ; as discussed below :

(1) Water pollution. Mining causes water pollution in two ways. Firstly, mining causes discharge of acid-mine-waters into surface water bodies like ponds, rivers, etc. ; and secondly, mining adds toxic radioactive substances like selenium, nickel, fluorine, etc. to the water bodies.

Mining is also responsible for changing the hydrology of an area in many ways. Say for example, sometimes, mining may change the river course and discharge, thereby affecting the agriculture and flora and fauna of the area. Similarly, sometimes, mining activities, may lower the water-table of the area.

(2) Air pollution. Mining causes air pollution in two ways. Firstly, it adds *gaseous pollutants* to the air ; and secondly, it emits and adds *dust particulates* to the atmosphere.

The harmful **gaseous pollutants**, like *sulphur dioxide, oxides of nitrogen, carbon monoxide, etc.* are originated in mining activities from the *waste dumps* ; whereas, the **dust particulates** are produced during ore/coal handling, blasting, and transportation.

Mine fires, which are largely associated with the **underground mining** and also in solid waste disposals, also do pollute the atmosphere on a large scale through smokes, etc. Such pollution has been caused on a large scale near Jharia in Bihar State of our country.

(3) De-spoilation of Land. The land degradation effects of mining can be divided into three parts, as discussed below :

(*i*) ***Land Subsidence.*** Due to *underground mining*, the land surface subsides, which not only causes damage to buildings and surface drains ; but also, sometimes, damages the highways, bridges, water and sewage lines, etc. existing on the ground.

The incidents and threats of **land subsidence** have mainly been reported from areas of coal-mining ; and to a lesser extent, from the areas where salt and gypsum have been mined on a large scale. There have been a few reports of subsidence arising from the extraction of limestone and other minerals. In recent times, land subsidence incidents have also been reported in some areas, which have been subjected to excessive mining of oil and natural gas.

Apart from causing the loss of buildings, agricultural land and other property, land subsidence has also taken a toll of several lives in many disasters and near disasters. Say for example, an 21.01.1960, in South Africa, as much as 3 sq. km of Coalbrook colliery collapsed at a depth of 143 metres below the ground, killing 437 underground miners. The disaster was reportedly caused by the failure of one single pillar, setting off a chain reaction. Similarly, in 1970, 38 persons were killed in *Mefulira copper mines of Zambia*, when about $\frac{1}{2}$ M. cum of mud entered the mine because of subsidence.

In India, the problem of land subsidence caused by coal mining has become most acute in *Raniganj Jharia Coal belt in eastern coal fields*. Such problems have, infact arisen due to unscientific practices followed in underground mining operations. The subsidence, though is a relatively slow process, extending over decades, eventually leads to drastic accidents such as disaster floods, failure of dams and levees on the rivers that flow on the ground over the underground mines, collapse of buildings, destruction of sewers, drains and canals. *The water of the Barak river often fills the depressed area with flood water in the Jharia coal fields*. The **Chasnala tragedy** of the 1978 that took a toll of 376 lives, is an eye-opening example of this kind of a hazard. According to a report of the Director General of Mines Safety, substantial parts of *Asansol* and *Raniganj* are infact, facing serious threat of a collapse. Subsidence has already occurred in several localities of Asansol like Dhadka, Dhandadihi village near Parasea, Kendra, Dehergarh and Jamuria Bazar, etc. Some experts have pointed out that the town of Barkar is precariously perched on the water of Damodar river, as it is practically hollowed out by mining. The subsidence of coal fields, infact, has been occurring in this region since 1920's, or so. As the pillars of the abandoned mines give way either through erosion by water accumulation, or because of vibrations caused by over-head rail traffic, or from blasting in neighbouring mines, the surface land subsides. In recognition of the problem, *sand stowing* (hydraulic stowing of sand) was made a statutory requirement after the report of the coalfields committee in 1922. But its provisions were virtually flouted by the private mining contractors and companies for over more than 50 years, leading to such precarious conditions. The menace of land subsidence, has, thus been looming large in the Raniganj Jharia coal fields in India for several decades.

(ii) ***Land Degradation.*** Open cast mining removes the top soils and causes deep and large excavations, pits and cuts. As such, large areas become unusable and denude of vegetation, causing continuous soil erosion and its ill effects like silting and degradation of streams and water bodies. Additionally and most importantly, the waste dumps, created by the open cast as well as underground mining could lead to erosion and slips ***(land slides)***, if not attended and treated timely and properly. Such land slides and flowing debris may sometimes cause serious damage to life and property, and may sometimes result in disasters or virtual disasters.

Say for example, on 21.10.1966, a 250 m high rain soaked tip of the mining waste, standing on the valley side above the town of Aberfen in South Wales, collapsed, and while flowing downwards, it engulfed a row of houses and a school, killing 144 people including 116 school children. A former Headmaster had given repeated warnings of this danger, which were just ignored. Similarly, a huge rock fall was caused by the systematic quarrying and removal of slate from the valley side of a cliff in Elm village of Switzerland, leading to the collapse of more than 10 M. cum of rock just within 20 minutes, killing 115 people. Similarly, a rockfall occurred at Frank Alberto Canada in the year 1903, which killed 70 people and buried a part of the Frank town. During investigation, it was concluded that careless coal mining was primarily responsible for this disaster.

In India, a disastrous land slide occurred at Hanibal in Mizoram on 09.08.1992, which should open ones eyes to the hazards of indiscriminate mining and quarrying. This land slide occurred in an area of large scale quarrying work, and was reportedly caused by the indiscriminate use of explosives during mining and removal of supports from the mined area. The unscientific and indiscriminate mining operations, infact, are carried out unstopped, by the mining companies or contractors, who usually are influential people. *Such lease holders (miners) frequently undertake mining work without sufficient geological investigations, and ignore precautionary measures, like forming steps, minimising the use of dynamite (explosives), avoiding of under-cutting operations, reducing tree loss and planting more trees, or stabilising the land in other ways.* Such carefree attitude during mining of hills, often results in throwing of stones, dust and rubble at the villagers settled on the foot hills, destroying their fields, canals, forests and animals, besides causing imminent threats of some serious disasters like land slides and land subsidence.

(*iii*) ***Deforestation.*** Large scale felling of trees from the forest areas, may sometimes have to be adopted for extraction of valuable mineral deposits from those areas.

The thick tree cover to be felled in such operations, may either be found directly on the area to be mined or it may have to be destroyed for making in-roads in the mining area.

Moreover, during mining, dirt, dust and over-burden spreads on the nearby areas, thereby causing large scale damage to the nearby vegetations.

Several mining areas, such as near Mussorrie, are getting bereft of thick forests, posing tremendous environmental problems.

Large scale deforestation, besides causing oxygen deficiency, drastically changes the climatic conditions, like rainfall, temperature, humidity, etc. of the area ; which may result in large scale deaths of wild animals and birds living in that forest, causing imbalance to the natural ecology of the area.

(4) Noise and Ground Vibrations. The large scale noise caused by mining operations, and the vibrations induced into the ground, have been established to be a major environmental hazard, as it interferes with the hearing and speech communication, causes distraction, annoyance, and ultimately *hearing loss*. It not only damages the ears, but may even cause headache and high blood pressure. Heavy ground vibrations, introduced by heavy blasting of areas, may also damage the nearby structures, buildings, etc. ; besides causing irritation to the residents.

Besides the above direct environmental impacts, mining operations may also cause adverse social and visual effects.

21.4. Environmental Impacts of River Valley Projects

A multipurpose river valley project, involving storage of river water by the construction of a dam across the river, can cause several effects on the environment of the area.

Some of these impacts may adversely affect the ecology and environment, while most others may prove beneficial to the environment, as discussed below :

21.4.1. Negative Impacts. Among the *adverse negative impacts*, the following effects may be examined and highlighted, while formulating the *environmental impact statement* for the project :

(*i*) ***Loss of wild life habitat.*** Loss of wild life and possible extinction of some rare species of the flora and fauna of the area, likely to be caused by the submergence of the vast tract of the forested area, needs to be evaluated.

(*ii*) ***Loss of valuable forest land.*** Loss of valuable forest land and the consequent loss of wood, particularly fuel wood should be evaluated.

(*iii*) ***Loss of agricultural land.*** Loss of agricultural land due to submergence, and consequent loss of food and nonfood crops, and particularly those of vegetables of daily use, needs to be evaluated and analysed.

(*iv*) ***Loss of religious sites.*** Loss of religious sites like mosques, temples, etc. needing shifting or relocation, need to be identified.

(*v*) ***Loss of adventure sports and river rafting.*** Due to the construction of the dam, the original rapid flow of river may not remain available to the river-runners and rafters, thereby reducing their charm and adventure opportunities. Such impacts are to be analysed.

(*vi*) ***Displacement of people coming in the submergence zone of the dam.*** The people living in areas coming under the submergence zone of the dam will have to be shifted and relocated, which may lead to overall misery, chaos, and public distress. This factor is extremely important, and must be well planned with compassion and liberal spendings.

(*vii*) ***Growing pressure of civilization and industrialisation on nearby areas.*** The increase in population due to dam construction is certainly likely to occur, when once the project is completed, because the project will ensure availability of water, power, and flood free land, inducing development of industries and other human activities.

Even during the construction period, there will occur heavy human activities in the area, including movements of heavy machineries, trucks, heavy labour forces, etc.

All such activities are likely to cause degradation of the overall environment of the area, and such activities may sometimes cause irreparable harm to the animals, birds and local populace living in the nearby areas.

(*viii*) ***Post project effects, like salinity and water-logging of irrigated land.*** The water logging and consequent salinity of land, likely to be irrigated by the reservoir water, may be *caused* by over-irrigation in the irrigation command area of the project. This can however, be eliminated by proper control and conservative use of valuable water.

(*ix*) ***Reservoir induced seismicity.*** The dam reservoir may increase the susceptibility of the area to the earthquakes. Its possible effects need be examined even on the dam design.

(*x*) ***Adverse impacts to fisheries.*** Adverse impact may be caused to the fisheries, as the fish is not likely to find the original rapid flow and sea water environment in the river downstream, and the dam will also check and prevent the free up and down movements of the fish in the river. *Migratory anadromous fish*, like **Hilsa**, are surely going to be badly affected, and may perish in large numbers. Fish may also perish while passing through the turbines or through the sluiceways along with the water released downstream for power generation and other uses.

21.4.2. Positive Impacts. Among the *favourable positive impacts* of a dam project on environment, we may count and highlight the following beneficial effects :

(*i*) ***Net improvement in public health.*** Net improvement in public health is caused due to the availability of ample domestic water supplies, leading to overall sanitation, cleanliness, and better living conditions.

(*ii*) ***The overall increase in wood production.*** The overall increase in production of crops and wood is certain to occur after the implementation of the project, due to flourishing growth of trees and crops in the irrigated command area of the project. Such production may far exceed the loss of wood and crops, caused by the submergence of forests and agricultural land, probably 100 times greater. This may need evaluation.

(*iii*) ***Excellent habitat for fisheries and water liking birds.*** The lake created by the dam on its upstream side, usually shall provide an excellent habitat for development of fisheries, and bird sanctuaries, which may properly be planned and evaluated.

(*iv*) ***Development of tourism and recreation.*** Development of tourist spots can be made feasible by a dam reservoir. **Brindavan gardens** in Mysore is an excellent example of such facilities, which have been developed on the down-stream side of the Krishanaraja Sagar dam across the Krishna river. Boating facilities in the reservoir lake are also, sometimes, developed to increase and induce tourism and recreation, as to provide boost to the environment.

(*v*) ***Improved micro-climate.*** Improved micro climate is caused in the adjoining areas due to evaporation from the open water surface of the reservoir and from the irrigated command area of the project. This can be evaluated and stipulated among the positive impacts of the project.

(*vi*) ***Overall improved oxygen production.*** This is caused due to the increased photosynthetic rate from green crops and trees, likely to flourish in the irrigation command area of the project, and which is likely to far outweigh such losses from the inundation area.

(*vii*) ***Development of sanctuaries and wild life.*** Development of wild life sanctuaries becomes feasible, when once adequate sweet water becomes available round the year by the implementation of the dam project. Sanctuaries can then be well planned and developed for the overall growth of wild life and promotion of tourism. Five such sanctuaries have been planned in the execution of the Sardar Sarovar multipurpose dam project, as discussed in the next article.

21.4.3. Conclusions. *In totally, it can be stated that the multipurpose water resources projects, do not by themselves, cause any environmental degradation* and do generally justify their nature of being in **environmental-harmony**. They also provide benefits and prosperity, and make available food and power, so very necessary for the development of a country. However, *it is the haphazard and unplanned development and industrialisation, encouraged by the execution of such water-resources projects, which causes all round environmental degradation*, and this must be checked.

We must, therefore, make attempts to stop and prevent such unplanned growths of human and industrial developments, rather than stopping and preventing the execution of the water-resources projects, which are the backbone of existence and prosperity of a nation, and that of the entire community as a whole.

The evaluation of large scale inundation and satisfactory resettlement of the effected people, however, remain the important adverse environmental impact of a water resources project.

Be that as it may, the Environmental Impact Assessment for each project shall be prepared to thoroughly study and examine the adverse impacts of the project on the environment, and to device ways and means to reduce or minimise the environmental damages before a project is cleared from environment point of view.

In our next articles, we are giving a detailed description of the environmental impact studies of the Sarovar dam project, as to make this matter more elaborate.

ENVIRONMENTAL IMPACTS OF SARDAR SAROVAR DAM PROJECT—A CASE STUDY

21.5. Salient Features and Status of the Narmada Valley Development Plan including Narmada Sagar (Indira Sagar) and Sardar Sarovar Dam Projects

The *Narmada* and *Tapti* are the two major ***west flowing*** river systems of India, which fall into the Arabian sea. These two river systems have so far remained undeveloped, while all the major east flowing river systems like *Mahanadi, Godawari, Krishna* and *Kaveri* have largely been harnessed for multiple uses of their water resources. In order to harness the water of the Narmada river, a committee was appointed in 1965 to develop a master plan for the basin, but its recommendations were not accepted by the riparian States of Gujarat, Madhya Pradesh and Maharashtra.

Even prior to the appointment of the committee for preparing a master plan for the development of the river, the Govt. of Gujarat had conceived the construction of a major dam across Narmada river near Navagam village, which is now called the *Sardar Sarovar dam*. The Sardar Sarovar dam was thus, infact conceived in late forties, planned in fifties, and launched in early sixties. However, as pointed out above, the project could not be implemented due to inter-state dispute on sharing of waters of river Narmada between the riparian States of **Madhya Pradesh**, **Gujarat** and **Maharashtra**, through whose territories the river flows (or touches their boundaries). When the Govt. of India failed to resolve the dispute, it was referred to a Tribunal, appointed by GoI under the provisions of the *Inter-State Water Disputes Act, 1956. The Narmada Water Disputes Tribunal (NWDT)*, after deliberating the issue for about 10 years, announced its award in Dec. 1979, allocating 18.25 MAF *i.e.* about 67% of total dependable* river flow to Madhya Pradesh and 9 MAF *i.e.* about 33% to Gujarat. 0.25 MAF was allocated to Maharashtra and 0.5 MAF to Rajasthan for drinking purposes. The *Narmada Control Authority* was then constituted by GoI to overlook proper implementation of the award, to create necessary storages and releases by the respective States, and for a coordinated development of the Narmada valley, especially Sardar Sarovar and Narmada Sagar dam projects.

The Narmada river, as you may be aware, is the largest west flowing river of India. It rises in Madhya Pradesh State from Vindhyachal mountain ranges at an elevation of about 900 m, from near a place known as *Amarkantak*. Passing down, it flows through the cities of *Mandla, Jabalpur, Narshimpur, Barna, Hoshangabad* and *Omkareshwar* (all in Madhya Pradesh). *Between Hosangabad and Omkareshwar, the Narmada Sagar dam is proposed to be built at village punasa in Khandwa District of Madhya Pradesh State*.

After flowing through the Madhya Pradesh State, the river Narmada enters into the territory of Gujarat State, where *Sardar Sarovar dam is proposed to be built across it, at a place known as Kevadia near Navagam*. The river, after travelling for about 1312 km, ultimately falls into the Arabian sea through the Gulf of Khambhat (Combay) near Bharuch city. During its sojourn through the Gujarat State, the river on its left bank touches the Maharashtra State in some of its length, thereby making it an inter-state river between the States of Madhya Pradesh, Gujarat and Maharashtra.

The Narmada valley development plan includes the construction of 4 large dams ; *i.e.* (*i*) *Bargi dam* ; (*ii*) *Narmada Sagar dam* renamed as *Indira Sagar Dam*; (*iii*) *Maheswar dam* ; and (*iv*) *Sardar Sarovar dam*.

Out of these four mega dams, the *Bargi dam* was completed way back in the year 1988. The ***Narmada Sagar dam*** also called *Indira Sagar dam* consisting of a 653 m long 92 m high concrete gravity dam, could, however, be completed only in May 2005, after it faced a lot of hurdles in its construction from the Activists of *Narmada Bachao Andolon* (NBA).

Maheshwar dam located in Madhya Pradesh state, d/s of Narmada sagar dam and u/s of Sardar Sarovar dam, slated to provide 400 MW of power, has also been completed in the year 2015, after facing enormous hurdles in its construction and with changes in executing agencies, primarily due to opposition of NBA activists.

---

*75% dependable flow was worked out to be 28 MAF (MAF stands for million acre feet), which equals to 1230 M m^3 = 1.23 Bm^3.

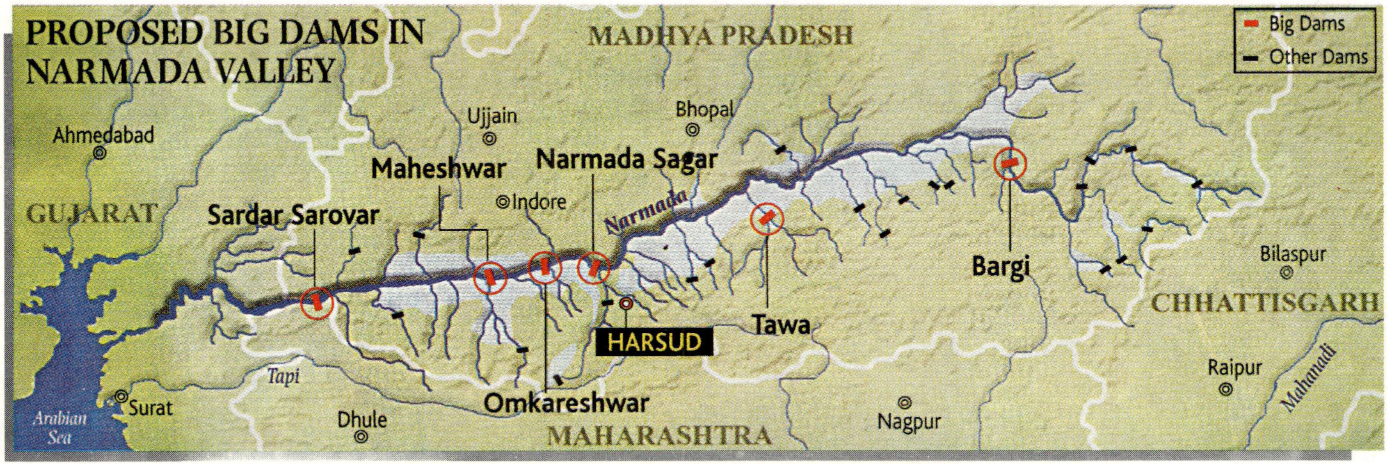

Coloured Fig. **Photo Fig. 21.1.** Proposed Dams in Narmada Valley.

Main Features of Sardar Sarovar Dam

	Feature		Value
1.	FRL		= RL 138.68 m
2.	MWL		= RL 140.20 m
3.	Length of main concrete gravity dam		= 1210.00 m
4.	Max. height above the deepest foundation level		= 163.0 m
5.	Catchment area of river above dam site		= 88,000 sq. km.
6.	Gross storage capacity	= 7.7 MAF	= 9497.07 M.cum.
	Live storage capacity	= 4.75 MAF	= 5859.80 M.cum.
7.	Length of reservoir		= 214.00 km
8.	Maximum width		= 16.10 km
9.	Average width		= 1.77 km
10.	Spillway gates		
	Chute spillway 7 Nos. (60' x 60')		= 18.3 m x 18.3 m
	Service spillway 23 Nos (60' x 55')		= 18.3 m x 16.75 m

Coloured Fig. 21.2. Sardar Sarovar Dam

Sardar Sarovar dam in Gujarat has also faced severe hurdles from the activists of the NBA from time to time, and has been completed in June 2017, with great efforts and huge cost over-runs. It has, infact, been raised to full designed crest level of 121.92 m, and FRL of 138.68 m (with installed gates). However, till the monsoon of 2016, the storage was permitted only up to its crest level of 121.92 m. **It was a only on 16.06.2017** that permission was given by the Narmadea Control Authority (NCA) to close all the 30 radial gates to permit water storage up to FRL of 138.68 m. This will, of course, cause more submergence in the upstream areas of Gujarat, Maharashtra and Madhya Pradesh, but will greatly help in irrigating additional perched area of Gujarat state, due to additional water storage of 3.48 MAF* *i.e.*, 0.428 Mham or 4.28 km^3.

In addition to these 4 mega dams, several smaller dams were recommended for construction under the Narmada Development Project. In all, 29 major dams, 135 medium dams and several minor dams were to be constructed. Of the 29 major dams, only one project, *i.e. Sardar Sarovar dam* was to be constructed in Gujarat State, and the rest 28 projects including the mega *Narmada Sagar dam* were in Madhya Pradesh State. An index plan of the Narmada Valley showing locations of proposed major dams is shown in **Colour Photo Fig. 21.1.**

Out of the 75% dependable flow of about 34.6 Bm^3 (28 MAF)** of the river Narmada, the Narmada Sagar dam project of height upto FRL 262.13 m, is proposed to have a live storage of 9.75 Bm^3 (gross storage equalling 12.22 Bm^3); while the **Sardar Sarovar dam (Colour Fig. 21.2)** is proposed to have a live storage of 5.8 Bm^3 (gross storage equalling 9.5 Bm^3).

Regulated releases from the Narmada Sagar dam reservoir will enable a large block of power to be generated through a cascade of dams at *Narmada Sagar, Omkareshwar, Maheshwar* and *Sardar Sarovar* to the extent of over 2000 MW (installed capacity of 3000 MW), and will supplement withdrawals for irrigation from Sardar Sarovar dam.

Omkareshwar and Maheshwar will be only lift dams without any significant storage ; while the *Narmada Sagar Project (NSP)* and *Sardar Sarovar Project (SPP)* will be multi-purpose projects. The benefits shall be shared by the States of Madhya Pradesh, Maharashtra, Gujarat and Rajasthan. Nearly half of the dependable flow of the Narmada river will be stored behind these two dams, but the quantum of regulated flows over the year will be nearly 2/3rd of the available dependable flow.

21.6. Environmental Impacts of Sardar Sarovar Project

The Sardar Sarovar and the Narmada Sagar dam projects were the first of the large multi-purpose water-resources projects of India, which were thoroughly studied for their environmental impacts, before granting them environmental clearance by the ministry of Environment and Forests-GoI, and the subsequent investment clearance by the Planning Commission GoI. These projects were cleared by the Ministry of Environment and Forests in

---

*MAF is Million Acre Feet.

**1 MAF = 1230 M m^3 = 0.123 M. ham = 1.23 Bm^3 (billion cubic metres).

1987, after thoroughly discussing and debating the possible environmental impacts of the projects (good as well as bad), with the suggested environmental improvement measures to be taken by the project implementing States, either prior to or during the execution of the projects. We shall herein discuss the various environmental impacts of the Sardar Sarovar project, in particular.

The various possible **adverse environmental impacts of the Sardar Sarovar Dam project**, including the post dam effects, can be discussed under the following heads :

(1) *Submergence causing loss of Agricultural and Forest land including displacement of people*

(2) *Loss of wild life (Flora and Fauna)*

(3) *Displacement of Tribals living in harmony with forests*

(4) *Submergence of the Ancient Monuments*

(5) *Water-logging and salinity of the land to be irrigated by the Dam Project*

(6) *Sedimentation rate affecting the life of the reservoir*

(7) *Seismic Effects of Reservoir*

(8) *Health aspects and possible increase in Malaria and other water related Diseases.*

All the above environmental impacts, in respect of Sardar Sarovar Dam project, are discussed below :

(1) Sumbergence Causing Loss of Agricultural and Forest Land including Displacement of People. As you know, the most important adverse environmental impact of a water-resources project, like the SSP and the NSP, relates to the *loss of agricultural and forest land*, coming under the submergence zone of the high dam, and the *consequent displacement of the* affected people. These adverse impacts will, in any case, have to be absorbed by adopting suitable mitigation measures, when other benefits of the project (like flood control, hydropower generation, storage of water for irrigation and water supply, etc.) outweigh these impacts and the execution of the project becomes much more useful and beneficial to the general public as a whole, as in the present case, when water supply to the thirsty Gujarat alone is enough to justify its execution.

However, in order to minimise the adverse effects of submergence, it becomes necessary to provide a *very liberal and compassionate rehabilitation and resettlement package* to the affected people, and to keep the submergence area of the dam to the minimum possible. In order to minimise the submergence zone of a dam, the dam height needs to be reduced. But then, the necessity of storage, puts a limit on the dam height, which cannot be reduced beyond a certain minimum value, under the considerations of optimum planning techniques, and requirements & availability of water supplies.

In the case of SSP, the dam height was earlier recommended to be upto FRL 152.44 m by the Khosla Committee in the year 1965. The State of Gujarat wanted to increase this recommended height of the dam, while the State of

Madhya Pradesh wanted it to be reduced. Even when the entire dispute was discussed and debated before the Narmada Water Disputes Tribunal, the State of Gujarat pleaded that it may be allowed to construct SSP with a height upto FRL 161.6 m to have a storage of 22.2 bm^3 ; *but the tribunal after thoroughly examining various alternatives, decided for a dam of height upto* ***FRL 138.68*** *m at this location, giving a live storage of 5.8 bm^3 only*. The Tribunal allowed this reduced height to reduce the submergence area, as to cause inconvenience to the minimum number of people. *This height of FRL 138.68 m was declared by the Tribunal as the height, which would cause less hardships but would give larger benefits*. Since the live storage of 5.8 bm^3 with the dam height of FRL 138.68 m, would not have been sufficient, the Tribunal also provided for the construction of upstream Narmada Sagar dam of 125 m height with FRL of 262.13 m in the State of Madhya Pradesh, from which regulated water releases would compensate the water needs of the downstream States of Gujarat and Maharashtra. The State of Gujarat accepted the decision of the Tribunal in the best of the spirits, and did not plead for any changes in its decision. Although the decisions of the Tribunal are final and binding on the riparian States, yet the State of Madhya Pradesh continued to create controversies, by delaying the resettlement of the affected persons in its State ; and by making appeals at political forums as well as at the Hon. Supreme Court of India, for reviewing the decision of the Tribunal, which though was final and binding till the year 2025. This inspite of the fact that the decision of the Tribunal cannot even be challenged in the Hon. Supreme Court in terms of the provisions of the Inter-State River Water Disputes Act.

With the above decided dam height of FRL 138.68 m of the Sardar Sarovar dam, the following submergences had been worked out (Table 21.2).

Table 21.2. Submergence Details of SSP (138.68 m FRL)

State	*Area of Submergence in hectares*				*No. of villages to be submerged*	
	Agriculture land	*Forest land*	*River bed / Waste land*	*Total land*	*Fully*	*Partly*
Gujarat	1877	4523	1,069	7,469	3	16
Madhya Pradesh	7883	2737	10,208	20,828	—	193
Maharashtra	1519	3459	1,592	6,570	—	36
Total	11279	10719	12,869	34,867	3	245

The submergence in Madhya Pradesh in terms of number of villages appears to be very high in terms of figures (193 No) but out of these 193 villages, there are 116 villages, whose agricultural land coming under submergence is hardly 20 ha (as against the average agricultural land per village equalling 350 ha or so). Thus, only 77 villages are left with 2100 families, where private land going under submergence is more than 20 ha in a village, as can be seen from the data given in Table 21.3.

Table 21.3. Submergence Details of SSP (138.68 m FRL) in Madhya Pradesh State Alone

Particulars	*No. of villages submerged*	*Private land coming under submergence in ha*	*% of total Agricultural land coming under submergence*	*Families affected*	*Remarks*
Total no. of villages affected	193	6700	12	14824	
Total no. of villages where less than 20 ha of land gets submerged	116	613	1.95	12724	Marginal submergence
Total no. of villages where more than 20 ha of private land gets submerged	77	6087	25.3	2100	Substantial submergence

More details of 77 villages, where substantial submergence was to be caused, are given in table 21.4, below :

Table 21.4. Degree of Submergence in 77 Villages which are Substantially Affected in Madhya Pradesh by SSP

Degree of Agricultural land affected in each village	*No. of villages*	*Private land under submergence in ha*	*Submerged land % age of total Agricultural land*
Less than 20%	31	1431	11.40%
between 20% to 40%	17	1821	28.76%
between 40% to 60%	17	1425	50.00%
between 60% to 80%	7	1023	68.40%
between 80% to 100%	5	387	88.60%
	77	6087	25.3%

It can be seen from table 21.2 that the total area coming under submergence of the Sardar Sarovar dam project will be 34,867 ha extending in three States of Gujarat, Madhya Pradesh and Maharashtra. This submergence would be only 1.82% of the total culturable command area (CCA) of about 1.91 Mha to be benefitted by the project in the three States of Gujarat, Maharashtra and Rajasthan.

The area submerged per unit volume of water stored (ha per M m^3) for SSP is further worked out in table 21.5 to be 6.01, as against the value of 2.15 for Bhakra (H.P.) 4.19 for Nagarjuna Sagar (A.P.), 8.67 for Gandhi Sagar (M.P.) 12.67 for Hirakud (Orissa), 9.86 for Maithon (W.B.), 9.80 for Tungabhadra (Karnataka), and 19.02 for Pochampad *i.e.* Shriram Sagar (A.P.). [Pl. see col. (11) of Appendix Table A-10.

Similarly, the value of submergence per MW of hydro power produced (ha per MW) for SSP is worked out in Table 21.5 to be only 24, as against the figures of 12 for Bhakra (H.P.), 35 for Nagarjuna Sagar (A.P.), 575 for Gandhi Sagar (M.P.), 274 for Hirakud (Orissa), 121 for Maithan, 1050 for Tungabhadra (Karnataka), and 1227 for Pochampad (A.P.).

In the same manner, the value of submergence area as ratio of the irrigated area for SSP comes to about 1.82 as against the value of 2.48 for Bhakra, 3.18 for Nagarjuna Sagar, 13.15 for Gandhi Sagar, 29.42 for Hirakud, 10.16 for Tungabhadra, and 19.14 for Pochampad. These comparisons are shown in col. (10) of the calculation table 21.5.

It can thus, be seen that when compared to all other water resources projects of the country, the SSP is found to be giving better returns, involving lesser submergence.

The strongest objection raised against the project was that the loss of about 11000 ha of valuable forest land would be too high and will cause a great *loss of forest wealth*, besides causing *loss of wild life* and *displacing 1 lakh tribal* population. It was clarified in this connection that virtually all of the 11000 ha (exact figure 10719 ha) of forest land to be submerged by Sardar Sarovar is severely degraded or denuded. It was also argued that the *loss of about 11000 ha of forest* is not a great loss, when compared to the annual loss of about 20,000 ha of forest trees in the country due to illegal felling of trees to quench the commercial, industrial and fuel wood needs of the society. However, while granting clearance to the project, **compensatory afforestation** was made mandatory by the GoI in the States, *to plant new forests in an equivalent area to be submerged, or to plant new trees in an area twice as large if this afforestation is to be done on the existing denuded or degraded forest land*. The points regarding loss of wild life and displacement of tribals is explained below :

(2) Loss of wild life (Flora and Fauna). Regarding loss of wild life, it was clarified that the project will neither cause any loss of flora and fauna of the area, nor any wild life sanctuary will come under the submergence area. On the other hand, the project will help to develop five sanctuaries by making available water throughout the year. These five sanctuaries include the development of one new sanctuary for the Great Indian Bastard in Kachahh, besides helping in the development of 4 existing sanctuaries. One of these 4 existing sanctuaries include the existing wild life sanctuary known as *Dhumkhal Sloth Bear Sanctuary* on the left bank of the Narmada river in an area of about 150 sq km, which is to be extended upto reservoir shore line to provide access to water for the animals. The total area of this extended sanctuary will be about 447 sq km. The sanctuary is renamed as "*Shoolpaneshwar Wildlife Sanctuary*". With provision of fence and other conservation measures, this area will become a rich medium for preservation and protection of flora and fauna. The other existing sanctuaries in the command area of the project, which will be benefitted, include : the *Nal Sarovar Sanctuary*, famous for attracting 120 to 150 species of migratory birds; *Wild Ass Sanctuary* in the little Rann of Kuchchh ; and the *Black Buck Sanctuary* at Velvadar.

Another important benefit of the project, mentioned by the authorities, was in regard to the *development of fisheries*, which is planned for the benefit of the Project Affected Persons (PAP) and persons residing on the reservoir periphery, who are essentially tribals. The work plan includes 10 ha fish farm and hatchery, stocking of major corps and other commercial fish, providing infrastructure fish landing sites, cold storages and transportation, etc.

(3) **Displacement of Tribals Living in Harmony with Forests.** The argument of the anti dam lobby regarding displacement of tribals living in pristine forests in traditional manner as hunter gatherers in harmony with the forest environment, was called by the project formulators as a myth. The satellite imagery supported with field data has infact confirmed that the tribals, at present, are farming barren, stoney and increasingly degraded hill sides, with a small and decreasing part of their income derived from the forest land. Most of them would welcome the opportunity to improve their lot in the more fertile and more sustainable command area of the project. A substantial proportion of tribals, appears to be seeing a little future in the increasingly degraded submergence area. Indeed, it is within this tribal group that there appears to be the least opposition to the dam. However, the greatest opposition to the dam is alleged to be from the wealthy non-tribal farmers and vested interests, who may loose their business of tree felling in the area coming under the submergence zone of the dam.

The statement that about 1 lakh tribals would be affected was also not accepted by the project proponents, as the magnitude of rehabilitation is much smaller, as already explained in tables 21.1 to 21.3. It was also stated that out of the total population to be affected by the dam, hardly 50% can be considered as tribals. Besides the tribals are now influenced by the culture and civilisation of the mainstream of the country's rural life.

The Sardar Sarovar dam with full designed height upto FRL 138.68 m, as a matter of fact, will submerge only 3 villages fully and 16 villages partially in Gujarat State. Thus, the submergence problem in Gujarat is not much. In Madhya Pradesh State also, though the number of villages affected appears to be more, yet it is confirmed that about 7000 families do not loose their agricultural land but only home steads. They are likely to be resettled in the nearby higher and safer areas in Madhya Pradesh. Additional 5724 families will have only marginal submergence, and they would also like to settle in Madhya Pradesh itself. Hardly 2000/3000 families may opt for rehabilitation in Gujarat. In majority of cases, local protection may be sufficient. The improved resources and prosperity after the fruits of the dam become available, will also help in the prosperity of the displaced people. *The rehabilitation work of the displaced persons will however, not be unmanageable, provided the affected States develop a will to implement the project.*

In addition to the above discussed major environmental losses like *submergence of forests and land, and displacement of people and wild life*, other environmental impacts which were studied in the Environmental Impact Statement/Assessment of the project are also briefly discussed below :

(4) **Submergence of the Ancient Monuments.** The survey of the villages likely to be submerged was carried out by the Directorate of Archaeology, and the following findings were given :

(1) *That no protected acrhaeological monument lies in the submergence area of Gujarat*. However, certain local temples and pilgrimage centres, which are likely to be affected in the 12 villages of Gujarat State include :

(*i*) A temple known as *Shoolpaneshwar Mahadev* at the village Supan in Nanded taluka of Bharuch district.

(*ii*) Small temples of surrounding areas, known as *Kamleshwar temple, Rajrajeshwar temple* and *Ranchhodi temple*.

(*iii*) *Hanfeshwar Mahadev temple* at village Harif, taluka Chhotoudepur in Vadodara district.

In the DPR, all such temples were promised to be shifted and relocated to suitable places.

(5) **Water-logging and Salinity of the Land to be Irrigated by the Dam Project.** One of the most adverse impact of a big dam project on its irrigated area, is caused by water-logging of the irrigated land, which reduces the crop yield. Water logging is a consequence of insufficient drainage and is often caused by deep percolation of excess surface irrigation water over the years, leading to rise in water-table near the ground surface. In the extreme, water-logging leads to salinity.

Although the losses of surface irrigation water to the underground are unavoidable to a degree, yet proper field irrigation techniques can reduce them to a minimum. Moreover, such losses can be recovered in large part, through pumping of ground water (*i.e.* by *recycling*), which in any case, will help keep the water-table low, to avoid water-logging.

Conjunctive use of surface and ground water thus proves to be an effective means to increase the overall efficiency of water use, and also to provide sufficient (vertical) sub-surface drainage. Where such conjunctive use is not possible (because of saline ground water), pumping and dumping of groundwater may still be feasible. In other cases, other (horizontal) drainage arrangements can be chosen.

In order to prevent water logging and salinity, the anti water logging and anti salinity measures have been built in the planning, design, implementation and operation of the Sardar Sarovar project. Say for example, the water distribution system (*i.e.* canal network) is designed to achieve the highest efficiency of water use, with a very low Delta (Δ) in the field. This system is designed to supply irrigation water on volumetric basis, ensuring that the farmers get water in proportion to their actual land holdings. Conjuctive use of ground and surface water is also aimed at, with continued monitoring of rise in water-tables with the installation of more than 4000 piezometric or well observations.

The irrigation supplies are basically designed conservatively and conjunctively to provide 53 cm of surface water for one crop in a year (measured at the canal head) and 18-20 cm of ground water for each crop. The pumping of ground water will be regulated by carefully watching the piezometric levels, to ensure that the water-table does not rise close to the ground surface, which causes water logging, and salinity with continued water-logging.

The main canal and branch canals are also proposed to be lined with concrete, and distribution system will be lined with bricks with LDP films wherever necessary. This will help in keeping percolation of water to the minimum.

The Sardar Sarovar Project also emphasised that varying irrigation doses have been fixed in the command area by studying the soil types and their water needs. Detailed soil surveys were hence carried out to find out soils which are not suitable for sustained irrigation, necessitating reduction in irrigation supplies. For this purpose, the entire command is divided into 13 zones.

In the distribution of water also, it is planned to hand over supplies to the **farmer's cooperatives**, which will be distributing water among their members according to well laid out principles, thereby achieving maximum saving in water. The operation and maintenance of the delivery system up to 8 ha blocks will be done by the project authorities ; while the operation and maintenance of the distribution system below 8 ha blocks will be done by the beneficiaries. The system will be operated through a computerised control down to the channels of 8.5 m^3/s. From this limit, down to a Village Service Area (VSA) of about 500 ha, the system will be operated by the project organisation. At VSA, the water will be handed over to the farmer's cooperatives for management at their level.

The water management through group cooperatives is likely to succeed in Gujarat, since Gujarat has the experience of successful working of cooperative societies in milk production, where it has become the world leader. The *water cooperatives* are, thus, likely to succeed like *milk cooperatives*.

Training programmes have also been drawn up for the project operation staff as well as the farmers for efficient water management and improved practices for irrigated agriculture. The State possesses one of the most advanced water and Land Management Institutes (WALM) in the country to help in conducting these trainings.

As stated above, the entire network of main and branch canals are to be lined with cement concrete. Not only that, but for the first time in India, mechanised and qualitatively well controlled concrete lining is adopted on such a large scale. This lining process is similar to the one being followed in advanced countries like U.S.A.

Moreover, the canal banks and their junctions with structures are designed to be constructed on the principles of conventional zoned earth dams, with an impervious core to prevent seepage through the canal banks. Very few canals in the world have been designed and constructed with such meticulous details and care.

Another special feature of this project was stated to be the pari passu development of surface drainage system with the construction of canals and distributaries. The earth required for building canal banks will be borrowed from the drains, new as well as existing, which have been planned after detailed field surveys. The village tanks are proposed to be deepened for the same purpose, which will also help in recharging the ground water.

The Sardar Sarovar Project, thus, claims to have eliminated every source of water-logging, in the project formulation itself. There is, hence, no cause for worry on account of water-logging, although the ground water levels will have to be watched, and preventive measures, such as additional pumping of ground water, taken, if becomes necessary.

(6) Sedimentation Rate Affecting the Life of the Reservoir. One of the adverse view against the construction of big dams is considered to be the high siltation rate, which reduces the life of the reservoir. In the case of Sardar Sarovar also, certain people have alleged that the life of this reservoir will be hardly 60 years. The project authorities, however, clarified in their report that the life of the Sardar Sarovar will be much higher, say between 180 to 230 years, since the project is unique in respect of the availability of space for absorbing the incoming silt as discussed below.

The canal off takes at RL 91.46 m and the low water level from the point of view of power generation is 110.67 m. Thus, a very large dead storage of 3.66 bm^3 (2.97 MAF) is available for accumulation of silt before the silt deposits can have significant encroachment on the live storage capacity of the reservoir. The committee set up by the Govt. of India under the Chairmanship of U.N. expert Mr. M.L. Diwan to study the soil erosion in Narmada catchment area found that the siltation rate is likely to reach the danger level of 5 ham (0.05 bm^3) per 100 sq km. The calculation for the life of the SSP were, infact, made on the basis of silt generation of 5.34 ha. m (0.0534 bm^3) per 100 sq km of catchment, which is more than the likely siltation rate recommended by the Diwan Committee.

The rate of siltation in Sardar Sarovar will infact depend upon the construction of 29 reservoirs in the Narmada basin, as planned in Madhya Pradesh, especially the Narmada Sagar. All the assessments of useful life of the Sardar Sarovar show that, if these upstream reservoirs are built in a reasonable time, the life of the Sardar Sarovar will not be less than 233 years. In the worst case of Narmada Sagar project being inordinately delayed for 25 years after the Sardar Sarovar is completed, the life of the SSP may be reduced to 180 years.

Even with our past experience of siltation of existing dam reservoirs, it has been noticed that the sedimentation rates are not alarming, since the sedimentation rates, even though remain higher for the first 15-20 years, fall off significantly, thereafter. Even some of the reservoirs having completed their planned life are still continuing to serve and provide partial benefits.

The apprehensions about the low life of the reservoir are, thus, unfounded and not correct.

(7) Seismic Effects of the Reservoir. Seismicity and the potential for the reservoir induced seismicity for the Sardar Sarovar and Narmada Sagar dams have been studied by the Geological Survey of India, Roorkee University, and the National Geological Research Institute, Hyderabad. It has been concluded that the formation of water reservoirs in the area may cause an earthquake to occur sooner but not to add to its magnitude. Considering the seismic potential of the region, the dams have been designed to withstand earthquake shocks of magnitude 6.5 (Richter scale), with effective peak ground acceleration of 0.2 g. The proposed construction of the dams, therefore, does not pose any threat of enhanced earthquakes in the region, and the dams are likely to be totally safe in themselves.

Even otherwise, the international study of 425 large dams in different parts of the world has shown that only in case of 1.2% of dams, some cognisable seismic effects were observed, but they too were not reaching any destructive dimensions. World's highest dam, *Nurek*, in USSR (300 m high-earthen dam), has already withstood a seismic shock of intensity 7 on Richter scale, having been situated in a highly seismic region.

Similarly, out of India's 3000 large dams, only *Koyna* and *Bhatsa* dams experienced some earthquakes but not at the time of reservoir filling. They were perhaps the normal earthquakes, just coinciding with the project location, by chance. Seismological observations established at the sites of other major Indian dams like *Bhakra*, *Pong* and *Ramganga*, etc., in the fragile Himalayan terrain, have not registered any increase in seismicity due to impounding of water.

Be that as it may, the risk posed by the triggered earthquakes if any, can be mitigated by careful control of the rate at which the water level in the reservoir is changed. The CWPRS, Pune, has been entrusted with the task of seismic surveillance of the area, for which a network of seismological observatories around the reservoir sites is being set up. There is, however, nothing much to worry on this ground.

While discussing the above impact, during environmental clearance debates, it was also pointed out that although the dam and its abutments have been checked to be safe against earthquakes, yet there was a need for conducting geomorphological studies on the *stability of reservoir rim slopes*, and to suggest and take appropriate engineering measures such as contour bunding, bench terracing, gully plugging etc., whereever required. The project proponents did promise to conduct such studies, which must have been completed by now.

(8) **Health Aspects and Possible Increase in Malaria and other Water Related Diseases.** The construction of a large water reservoir, and the large scale use of the stored water for perennial irrigation is claimed to be a potential factor for increasing diseases like *malaria, schisomiasis*, etc. in the project area of a large dam reservoir project. Considering this, detailed surveys are usually carried out to determine the existing health status along with preparing a future plan to create proper authorities who can take suitable control measures along with doing continuous monitoring of the health of the general public of the area.

While discussing on the above environmental impact, it was clarified that the *National Institute of Communicable Diseases* had carried out detailed studies in the area of Gujarat, and concluded that the future irrigation facilities to be developed over 17 years period, are not likely to create any such risk in the area. Subsequent studies carried out by a team led by the Chief of *Schistosomiasis Division of W.H.O.*, have also concluded that the project area did not have any risk of schistosomiasis entering the area.

It was also stated by the Gujarat Govt. that there was no chance for any increase in the incidence of malaria by the reservoir storage, since the operation of the reservoir itself inhibits the proliferation of the malaria larvae ; since while the reservoir builds up the storage during the monsoon rains, the larvae, which prefer to stay around the periphery, get drowned and thus, destroyed. Similarly, when the water is withdrawn for power generation and irrigation, the larvae are stranded and destroyed.

The various State Govts. also promised that their Directors of Health. Services will be directed to monitor at intervals, the incidence of water borne diseases in their respective project areas, and the project authorities shall remain in touch with the Health authorities to ensure implementation of preventive measures, found necessary.

21.7. Assessment of Environmental Losses and Economic Viability of Sardar Sarovar and Narmada Sagar Projects

While assessing the environmental losses in monetary value, the total land lost in submergence was evaluated by the Deptt. of Environment and Forest, GoI, on a unit rate of ₹ 12.67 bn per ha of fully grown forest having tree density of 1, for 50 years. The density of forest coming under Sardar Sarovar was adopted as 0.4, as against the actual observed value of 0.1 for Sardar Sarovar worked out on actual enumeration of trees of 216 per hectare. The density for Narmada Sagar dam is however evaluated at about 0.48.

On the basis of the above hypothetical rate, the losses for Sardar Sarovar were computed to be ₹ **61 bn**, and for the Narmada Sagar to be ₹ **253 bn**, thereby computing the total loss to be ₹ 61 bn + ₹ 253 bn = ₹ **314 bn.**

As against the above losses, the benefits were worked out as under :

(*i*) Irrigation benefits	₹ 469.1 bn
(*ii*) Power benefits	₹ 113.4 bn
Sub Total	₹ 582.5 bn

Allowing for the additional income in agricultural sector every year on account of additional agricultural employment potential of ₹ 105 bn, the *total benefits* were worked out to be *₹ 687.50 bn*, which would out weigh the environmental losses of ₹ 314 bn. With this estimation, the B.C. ratio was worked out to be 2.19.

Economical appraisal of the project was, infact, further conducted by an independent agency *i.e.* Tata Consultancy Services, in 1981-82. *They have worked out a B.C. ratio of 1.84.*

This project was further examined by the World Bank in 1985, at the time of sanctioning credit. They adopted the method of computing the "*rate of internal return*" for the project, and computed it to be 12%, as against the value of *minimum internal rate of return* recommended to be 9% for water resources projects in general, and 7% for drought prone and chronically flood prone areas. These minimum limiting values of 9 & 7% for minimum internal rate of return, were infact recommended by a committee constituted by the Planning Commission GoI, under the chairmanship of Dr. Nitin Desai. Since the computed rate of internal return was more than the minimum specified value, World Bank termed the project to be economically beneficial and viable.

The B.C. ratio was again worked out by the Planning Commission in 1989, while granting clearance to the project, as 1.999.

—22

Preparing Environmental Impact Assessments of Polluting Developmental Projects in India

22.1. Environmental Impact Assessments (EIA) of Polluting Developmental Projects in India

In order to assess the adverse possible effects of a proposed developmental project on the environment surrounding the proposed project, a detailed analysis is required to be made. This report is known as the *Environmental Impact Assessment (EIA). As a matter of fact, the Impact Assessment attempts to determine the effects of the installation of a development project on the existing relationship between the surroundings and its inhabitants.*

The practice of preparing an Environmental Impact Assessment (EIA) for a project, was infact, first initiated on January 1, 1970, in the United States of America. Since then, more than 100 developing and developed nations, have either passed specific analogous laws or have adopted procedures used elsewhere.

As far as India is concerned, the Ministry of Environment and Forests (MoEF), GoI, for the first time, issued environmental guidelines for setting up specific types of industries in August 1985, relating to areas to be avoided for setting up industries and measures to be taken for site selection, as also the various aspects of environmental protection which should be incorporated during the implementation of the proposed industrial development project.

On 27th January 1994, the MoEF, GoI issued its first important notification, governing the need of submission of *Environmental Impact Assessment (EIA)* with *Environmental Management Plan (EMP)* for specified types of developmental projects, to the MoEF for obtaining its clearance, before the project can be got sanctioned for execution.

The first notification of January 27, 1994 has been superseded by a notification of September 14, 2006. This second notification, however, provides that all the applications submitted till 14.09.2006 under the provisions of the 1994 notification, shall be decided on the basis of the same notification.

The industries/activities which comes within the preview of the 2006 notification are given in article 22.2.

22.2. India's Environmental Impact Assessment Notification, 2006

Under this **notification, dated 15.09.2006**, all projects or activities which need environmental clearance from the regulatory authorities have been divided into two categories; *i.e.* **category A** ; and **category B.** (Pl. see table 22.1)

Projects or activities falling in category A needs environmental clearance from the Central Government on the recommendations of the *Expert Appraisal Committee (EAC). Preparation of EIA is necessary for all such category A projects.*

Projects or activities falling in category B shall, however, be further subdivided into two sub-categories; *i.e. Category B-1* and *Category B-2.* ***Projects falling under category B-2 shall not require any EIA Report; while projects falling under B-1 category shall require EIA reports.*** *In the original notification, only one type of activity given at S. No. 8(b) of Table 22.1* [*i.e. Townships and Area Development Projects*] *has been notified as of category B-1.* Thus, there will be no B-2 category in this type of projects. *For the rest of the activities, the notification has authorised the MoEF to issue appropriate guidelines for dividing B-category projects into B-1 and B-2 categories.*

The B-category projects/activities need environmental clearance from *State level Environmental Impact Assessment Authority* (SEIAA) to be constituted by the Central Govt.

The SEIAA shall grant environmental clearance to B-category projects on the recommendations of the *State Expert Appraisal Committee* (SEAC). In the absence of a duly constituted SEIAA or SEAC, a category B project shall be treated as category A project, needing clearance from MoEF (GoI).

These orders shall be applicable not only to all the New activities/projects of the listed types, but shall also include: (*i*) Expansion and modernization of existing projects

An application for obtaining prior Environmental Clearance (EC) shall be made in ***Form 1*** (given in **Appendix 1** to the original notification), after the project site his been identified, but before commencing any construction activity, or preparation of land at the site by the applicant. The applicant shall furnish, along with his application, copy of the *pre-feasibility project report*. In case of construction projects (item 8 of the Schedule—given in table 22.1), however, no pre-feasibility report is required, but some additional information is required to be submitted in ***From 1A*** (given as **Appendix II** to the original notification) along with a copy of the *conceptual plan.*

22.2.1. Stages in the Environmental Clearance Process for New Projects Under the 2006 Notification. The environmental clearance process for new projects will comprise of a maximum of four stages, all of which may not apply to particular cases as set forth below in this notification. These four stages in sequential order are :

- Stage (1) Screening (Only for Category 'B' projects and activities)
- Stage (2) Scoping
- Stage (3) Public Consultation
- Stage (4) Appraisal

I. Stage (1)—Screening :

In case of Category 'B' projects or activities, this stage will entail the scrutiny of an application seeking prior environmental clearance made in Form 1 by the concerned State Expert Appraisal Committee (SEAC) for determining whether or not the project or activity requires further environmental studies for preparation of an Environmental Impact Assessment (EIA) for its appraisal prior to the grant of environmental clearance, depending upon the nature and location specificity of the project. **The projects requiring an Environmental Impact Assessment report shall be termed Category 'B-1' and remaining projects shall be termed Category 'B-2' and will not require an Environment Impact Assessment report.** *For categorization of projects into B-1 or B-2 except for item 8(b), the Ministry of Environment and Forests, shall issue appropriate guidelines from time to time.*

II. Stage (2)—Scoping :

(*i*) "Scoping": refers to the process by which the Expert Appraisal Committee in the case of Category 'A' projects or activities, and State Expert Appraisal Committee in the case of Category 'B-1' projects or activities, including applications for expansion and/or modernization and/or change in product mix of existing projects or activities, determine detailed and comprehensive Terms Of Reference (TOR) addressing all relevant environmental concerns for the preparation of an Environment Impact Assessment (EIA) Report in respect of the project or activity for which prior environmental clearance is sought. The Expert Appraisal Committee or State Expert Appraisal Committee concerned shall determine the Terms of Reference on the basis of the information furnished in the prescribed application Form 1/Form 1A including Terms of Reference proposed by the applicant, a site visit by a sub-group of Expert Appraisal Committee or State Expert Appraisal Committee concerned (only if considered necessary by the Expert Appraisal Committee or State Expert Appraisal Committee concerned), and other information that may be available with the Expert Appraisal Committee or State Expert Appraisal Committee concerned.

All projects and activities listed as Category 'B' in Item 8 of the Schedule (*i.e.*, Construction/Township/Commercial Complexes Housing) shall not require Scoping and will be appraised on the basis of Form 1/Form 1A and the conceptual plan.

(*ii*) The Terms of Reference (TOR) shall be conveyed to the applicant by the Expert Appraisal Committee or State Expert Appraisal Committee concerned within sixty days of the receipt of Form 1. In the case of Category A Hydroelectric projects [*i.e.* Item 1(*c*) (*i*) of the Schedule], the Terms of Reference shall be conveyed along with the clearance for pre-construction activities. If the Terms of Reference are not finalized and conveyed to the applicant within sixty days of the receipt of Form 1, the Terms of Reference suggested by the applicant in Form 1, shall be deemed as the final Terms of Reference approved for the EIA studies.

III. Stage (3)—Public Consultation

(*i*) All category 'A' and 'B1' projects except the following shall require *public consultation*, prior to their appraisal by the EAC or SEAC, as the case may be :

(*a*) modernization of irrigation projects [item 1(*c*) (*ii*) of the Schedule].

(*b*) all projects or activities located within industrial estates or parks [item 7(*c*) of the Schedule] approved by the concerned authorities, and which are not disallowed in such approvals.

(*c*) expansion of Roads and Highways [item 7(*f*) of the Schedule] which do not involve any further acquisition of land.

(*d*) all Building/Construction projects/Area Development projects and Townships [item 8].

(*e*) all Category 'B2' projects and activities.

(*f*) all projects or activities concerning national defence and security or involving other strategic considerations as determined by the Central Government.

"Public Consultation" refers to the process by which the concerns of local affected persons and others who have plausible stake in the environmental impacts of the project or activity are ascertained with a view to taking into account all the material concerns in the project or activity design as appropriate.

(*ii*) Public Consultation shall ordinarily have two components, comprising of :

(*a*) a ***public hearing*** at the site or in its close proximity—district wise, to be carried out in the manner prescribed in **Appendix IV** to the notification, for ascertaining concerns of local affected persons;

(*b*) obtain responses in writing from other concerned persons having a plausible stake in the environmental aspects of the project or activity.

(*iii*) The public hearing at, or in close proximity to, the site(s) in all cases, shall be conducted by the State Pollution Control Board (SPCB) or the Union Territory Pollution Control Committee (UTPCC) concerned, in the specified manner, who will forward the proceedings to the regulatory authority concerned within 45 days of a request to the effect from the applicant.

(*iv*) In case the State Pollution Control Board or the Union Territory Pollution Control Committee concerned does not undertake and complete the public hearing within the specified period, and/or does not convey the proceedings of the public hearing within the prescribed period directly to the regulatory authority concerned as above, the regulatory authority shall engage another public agency or authority which is not subordinate to the regulatory authority, to complete the process within a further period of forty five days.

(*v*) If the public agency or authority nominated under the sub paragraph (*iii*) above, reports to the regulatory authority concerned that owing to the local situation, it is not possible to conduct the public hearing in a manner which will enable the views of the concerned local persons to be freely expressed, it shall report the facts in detail to the concerned regulatory authority, which may, after due consideration of the report and other reliable information that it may have, decide that the public consultation in the case need not include the public hearing.

(*vi*) For obtaining responses in writing from other concerned persons having a plausible stake in the environmental aspects of the project or activity, the concerned regulatory authority and the State Pollution Control Board (SPCB) or the Union Territory Pollution Control Committee (UTPCC) shall invite responses from such concerned persons by placing on their website the Summary EIA report prepared in the format given in **Appendix III and IIIA** to the notification by the applicant along with a copy of the application in the prescribed form, within seven days of the receipt of a written request for arranging the public hearing. Confidential information including non-disclosable or legally privileged information involving Intellectual Property Right, source specified in the application shall not be placed on the web site. The regulatory authority concerned may also use other appropriate media for ensuring wide publicity about the project or activity. The regulatory authority shall, however, make available on a written request from any concerned person, the Draft EIA report for inspection at a notified place during normal office hours till the date of the public hearing. All the responses received as part of this public consultation process shall be forwarded to the applicant through the quickest available means.

(*vii*) After completion of the public consultation, the applicant shall address all the material environmental concerns expressed during this process, and make appropriate changes in the draft EIA and EMP. The final EIA report, so prepared, shall be submitted by the applicant to the concerned regulatory authority for appraisal. The applicant may alternatively submit a supplementary report to draft EIA and EMP, addressing all the concerns expressed during the public consultation.

Modifications in the Original Notification Dated September 15, 2006 w.r. to Public Hearing

(*i*) **On 24.08.2009,** the Ministry of Environment and Forests (MoEF) makes public hearing mandatory for expansion of any existing project, including the ones cleared before the new EIA notification of 2006.

(*ii*) **On 19.04.2010,** the Ministry of Environment and Forests (MoEF), after a court order, revises the conditions of public hearing, asking project proponents as well as State Pollution Control Boards (SPCBs) to provide adequate time and information about public hearing to project-affected communities.

(*iii*) **On 19.12.2012,** the MoEF issues an office memorandum allowing coal mining projects to skip public-hearings for one time capacity expansion of 25 per cent.

IV. Stage (4)—Appraisal :

(*i*) Appraisal means the detailed scrutiny by the Expert Appraisal Committee or State Expert Appraisal Committee of the application and other documents like the Final EIA report, outcome of the public consultations including public hearing proceedings, submitted by the applicant to the regulatory authority concerned for grant of environmental clearance. This appraisal shall be made by Expert Appraisal Committee or State Expert Appraisal Committee concerned in a transparent manner in a proceeding to which the applicant shall be invited for furnishing necessary clarifications in person or through an authorized representative. On conclusion of this proceeding, the Expert Appraisal Committee or State Expert Appraisal Committee concerned shall make categorical recommendations to the regulatory authority concerned either for grant of prior environmental clearance on stipulated terms and conditions, or rejection of the application for prior environmental clearance, together with reasons for the same.

(*ii*) The appraisal of all projects or activities which are not required to undergo public consultation, or submit an Environment Impact Assessment report, shall be carried out on the basis of the prescribed application Form 1 and Form 1A as applicable, any other relevant validated information available, and the site visit wherever the same is considered necessary by the Expert Appraisal Committee or State Expert Appraisal Committee concerned.

(*iii*) The appraisal of an application shall be completed by the Expert

Appraisal Committee or State Expert Appraisal Committee concerned within sixty days of the receipt of the final Environment Impact Assessment report and other documents or the receipt of Form I and Form 1A, where public consultation is not necessary and the recommendations of the Expert Appraisal Committee or State Expert Appraisal Committee shall be placed before the competent authority for a final decision within the next fifteen days.

The prescribed procedure for Appraisal is given in **Appendix V** to the original notification.

Table 22.1. Schedule Giving List of Projects Requiring Prior Environmental Clearance Under EIA Notification, 2006

S. No.	*Project or Activity*	*Category with threshold limit*		*Conditions if any*
		A	*B*	
(1)	*(2)*	*(3)*	*(4)*	*(5)*
1	**Mining, Extraction of Natural Resources, and Power Generation (for a specified production capacity)**			
1(*a*)	Mining of minerals*	≥ 50 ha. of mining lease area* Asbestos mining irrespective of mining area	< 50 ha and ≥ 5 ha. of mining lease area.*	General Condition shall apply Note Mineral prospecting (not involving drilling) are exempted, provided the concession areas have got previous clearance for physical survey.
1(*b*)	Offshore and onshore oil and gas exploration, development and production	All projects	—	Note Exploration Surveys (not involving drilling) are exempted, provided the concession areas have got previous clearance for physical survey.

*The original notification of Sept. 14, 2006, did include mining of all types of minerals in one category. However, by the ***first amendment dated December 1, 2009***, the MoEF divided mine lease areas into two categories – (*i*) *coal mine areas;* and (*ii*) *non-coal mine areas.* The leases for coal mine areas in less than 150 ha were pulled down to category B projects; while non-coal mining leases between 5 ha to 50 ha remained under category B projects.

The MoEF on **September 9, 2013,** issued the **second amendment to the EIA notification 2006,** recognising ***minor minerals*** (such as sand, coastal sand, gravel, brick-earth, and other stones) for the first time, and brought them under a special category within the previous distinction of coal and non-coal areas. As per this new amendment, leases of all minor minerals in less than 50 ha area (including areas less than 5 ha) have been included in category B projects, needing environmental clearance from State Level Environmental Impact Assessment Authority (SEIAA).

(1)	*(2)*	*(3)*	*(4)*	*(5)*
1(*c*)	River Valley projects	(*i*) ≥ 50 MW hydro-electric power generation ; (*ii*) ≥ 10,000 ha. of culturable command area.	(*i*) < 50 MW and ≥ 25 MW hydro-electric power generation ; (*ii*) < 10,000 ha. of culturable command area.	General Condition shall apply
1(*d*)	Thermal Power Plants	≥ 500 MW (coal/ lignite/naptha and gas based) ; ≥ 50 MW (Peat coke, diesel and all other fuels)	< 500 MW (coal/ lignite/naptha and gas based); < 50 MW & ≥ 5 MW (Peat coke, diesel and all other fuels)	General Condition shall apply.
1(*e*)	Nuclear power projects and processing of nuclear fuel	All projects	—	—
2	**Primary Processing**			
2(*a*)	Coal washeries	≥ 1 million ton/ annum throughput of coal	< 1 million ton/ annum throughput of coal	General Condition shall apply (If located within mining area, the proposal shall be appraised together with the mining proposal)
2(*b*)	Mineral beneficiation	≥ 0.1 million ton/ annum mineral throughput	< 0.1 million ton/ annum mineral throughput	General Condition shall apply (Mining proposal with Mineral beneficiation shall be appraised together for grant of clearance)
3	**Materials Production**			
3(*a*)	Metallurgical industries (ferrous and non ferrous)	(*a*) *Primary metallurgical industry* All projects	—	—
		(*b*) *Sponge iron manufacturing* ≥ 200 TPD	Sponge iron manufacturing < 200 TPD	General Condition shall apply for Sponge manufacturing
		(*c*) *Secondary metallurgical processing industry* All toxic and heavy metal producing units ≥ 20,000 tonnes/ annum	*Secondary metallurgical processing industry* (*i*) All toxic and heavy metal producing units < 20,000 tonnes/ annum (*ii*) All other non-toxic secondary	

Contd...

(1)	*(2)*	*(3)*	*(4)*	*(5)*
			metallurgical processing industries > 5000 tonnes/annum	
3(*b*)	Cement plants	≥ 1.0 million tonnes/ annum of production capacity	< 1.0 million tonnes/ annum of production capacity. All stand alone grinding units	General Condition shall apply
4	**Materials Processing**			
4(*a*)	Petroleum industry	All projects	—	—
4(*b*)	Coke oven plants	≥ 2,50,000 tonnes/annum	< 2,50,000 and ≥ 25,000 tonnes/ annum	
4(*c*)	Asbestos milling and asbestos based products	All projects	—	—
4(*d*)	Chlor-alkali industry	≥ 300 TPD production capacity ; or a unit located outside the notified industrial area/ estate	< 300 TPD production capacity ; or a unit located within a notified industrial area/ estate	Specific Condition shall apply No new Mercury Cell based plants will be permitted and existing units converting to membrane cell technology are exempted from this Notification
4(*e*)	Soda ash Industry	All projects	—	—
4(*f*)	Leather/skin/ hide processing industry	New projects outside the industrial area or expansion of existing units outside the industrial area	All new or expansion of projects located within a notified industrial area/ estate	Specific condition shall apply
5	**Manufacturing/Fabrication**			
5(*a*)	Chemical fertilizers	All projects	—	—
5(*b*)	Pesticides industry and pesticide specific intermediates (excluding formulations)	All units producing technical grade pesticides	—	—
5(*c*)	Petro-chemical complexes (industries based on processing of petroleum fractions and natural gas and/or reforming to aromatics)	All projects	—	—
5(*d*)	Man-made fibres manufacturing	Rayon	Others	General Condition shall apply

Contd...

(1)	*(2)*	*(3)*	*(4)*	*(5)*
5(*e*)	Petrochemical based processing (processes other than cracking and reformation and not covered under the complexes)	Located outside the notified industrial area/estate	Located in a notified industrial area/estate	Specific Condition shall apply
5(*f*)	Synthetic organic chemicals industry (dyes and dye intermediates; bulk drugs and intermediates excluding drug formulations; synthetic rubbers; basic organic chemicals, other synthetic organic chemicals and chemical intermediates)	Located outside the notified industrial area/estate	Located in a notified industrial area/estate	Specific Condition shall apply
5(*g*)	Distilleries	(*i*) All Molasses based distilleries	—	General Condition shall apply
		(*ii*) All Cane juice/non-molasses based distilleries ≥ 30 KLD	All Cane juice/non-molasses based distilleries < 30 KLD	
5(*h*)	Integrated paint industry	—	All projects	General Condition shall apply
5(*i*)	Pulp and paper industry, excluding manufacturing of paper from waste paper and manufacture of paper from ready pulp without bleaching	Pulp manufacturing and Pulp and Paper manufacturing industry	Paper manufacturing industry without pulp manufacturing	General Condition shall apply
5(*j*)	Sugar Industry	—	≥ 5000 tcd cane crushing capacity	General Condition shall apply
5(*k*)	Induction/arc furnaces/cupola furnaces 5 TPH or more	—	All projects	General Condition shall apply
6	**Service Sectors**			
6(*a*)	Oil and gas transportation pipeline (crude and refinery/petrochemical products), passing through national parks/sanctuaries/coral reefs/ecologi-	All projects	—	—

Contd...

(1)	*(2)*	*(3)*	*(4)*	*(5)*
	cally sensitive areas including LNG Terminal.			
6(*b*)	Isolated storage and handling of hazardous chemicals (As per threshold planning quantity indicated in column 3 of schedule 2 and 3 of MSIHC Rules 1989 amended 2000)	—	All projects	General Condition shall apply
7	**Physical Infrastructure including Environmental Services**			
7(*a*)	Air ports	All projects	—	—
7(*b*)	All ship breaking yards including ship breaking units	All projects	—	—
7(*c*)	Industrial estates/ parks/complexes/ areas, export processing Zones (EPZs), Special Economic Zones (SEZs), Biotech Parks, Leather Complexes	If at least one industry in the proposed industrial estate falls under the Category A, entire industrial area shall be treated as Category A, irrespecitve of the area.	Industrial estates housing atleast one Category B industry and area < 500 ha.	Special condition shall apply Note: Industrial Estate of area below 500 ha and not housing any industry of category A or B does not require clearance.
		Industrial estates with area greater than 500 ha. and housing atleast one Category B Industry.	Industrial estates of area > 500 ha. and not housing any industry belonging to Category A or B.	
7(*d*)	Common hazardous waste treatment, storage and disposal facilities (TSDFs)	All integrated facilities having incineration and landfill or incineration alone	All facilities having landfill only	General Condition shall apply
7(*e*)	Ports, Harbours	≥ 5 million TPA of cargo handling capacity (excluding fishing harbours)	< 5 million TPA of cargo handling capacity and/or ports/harbours ≥ 10,000 TPA of fish handling capacity	General Condition shall apply
7(*f*)	Highways	(*i*) New National High-ways; and (*ii*) Expansion of National High-ways greater than 30 km, involving additional right of way greater than 20 m involving	(*i*) New State High-ways; and (*ii*) Expansion of National/State Highways greater than 30 km involving additional right of way greater than	General Condition shall apply

Contd...

(1)	(2)	(3)	(4)	(5)
		land acquisition and passing through more than one State.	20 m involving land acquisition.	
7(*g*)	Aerial ropeways	—	All projects	General Condition shall apply
7(*h*)	Common Effluent Treatment Plants (CETPs)	—	All projects	General Condition shall apply
7(*i*)	Common Municipal Solid Waste Management Facility (CMSWMF)	—	All projects	General Condition shall apply
8	**Building/Construction projects/Area Development projects and Townships**			
8(*a*)	Building and Construction projects	—	≥ 20,000 sq. mtrs. of land area; and ≤ 1,50,000 sqm of built up area.*	# (built up area for covered construction ; in the case of facilities open to the sky, it will be the activity area)*
8(*b*)	Townships and Area Development projects	—	Covering an area ≥ 50 ha and or built up area ≥ 1,50,000 sq. mtrs.*	++*All projects under Item 8(b) shall be appraised as Category B1**

Note I. General Condition (GC) :

Any project or activity specified in Category 'B' will be treated as Category A, if located in whole or in part within 10 km from the boundary of ; (*i*) Protected Areas notified under the Wild Life (Protection) Act, 1972 ; (*ii*) Critically Polluted areas as notified by the Central Pollution Control Board from time to time ; (*iii*) Notified Eco-sensitive area ; (*iv*) inter-State boundaries and international boundaries.

Note II. Special Condition (SC) :

If any Industrial Estate/Complex/Export processing Zones/Special Economic Zones/ Biotech Parks/Leather Complex with homogeneous type of industries such as Items 4(*d*), 4(*f*), 5(*e*), 5(*f*), or those Industrial estates with pre-defined set of activities (not necessarily homogeneous, obtains prior environmental clearance, individual industries including proposed industrial housing within such estates/complexes will not be required to take prior environmental clearance, so long as the Terms and Conditions for the industrial estate/complex are complied with (Such estates/ complexes must have a clearly identified management with the legal responsibility of ensuring adherence to the Terms and Conditions of prior environmental clearance, who may be held responsible for violation of the same throughout the life of the complex/estate).

Note : Certain modification to the above mentioned provisions have also been made by MOEF from time to time. Professional engineers are requested to varify latest orders on various types of projects on the website of the Ministry at http:// www.moef.gov.in and http://www.moef.nic.in.

22.2.1. Draft EIA Notification, 2020. The EIA notification, 2006 described above is now further proposed to be superceded by the MOEF, GoI by new EIA notification, 2020, which has been floated in March 2020 for public feedback. The objections/comments on this proposed draft-notification are to be received till August 11, 2020. After the Govt. receives the public feedback, the same will be considered by the Ministry and final notification will be issued, bringing certain new industrial activities under the control of EIS process, while exempting certain other activities from the EIA process.

In fact, there have occurred periodic amendments in the 2006 notification exempting one category of industries or the other, from scruitiny under the EIA process from time to time. Besides, there are complaints from several quarters that the EIA regime dampens the spirit of liberalisation, leading to red tape and graft seeking.

The new 2020 draft notification proposes to bolster the government's discretionary power while limiting the public engagement in safeguarding the environment, particularly in relation to the projects in border areas and for other strategic requirements.

In the new proposed notification, the government is authorised to grant "strategic" tag not only to the projects concerning national defence and security, which are naturally considered strategic, but also to other projects where the government wants to hasten up execution. This will empower the government to designate any desired project as strategic, even if the project can not be related directly to the country's defence and security. ***All "strategic" projects have infact, been exempted from public scruitiny under the EIA process. The Draft notification 2020, says that no information on such strategic projects shall be placed in the public domain.*** This opens a window for summary clearance for any project deemed strategic, without having to explain why.

Additionally, the new draft notification exempts a long list of projects from public consultation under the EIA process. For example, linear projects such as roads and pipelines in border areas will now not require any public hearing. The border area is defined as "the area falling within 100 km aerial distance from the line of Actual Control with bordering countries of India. **This provision will help the government to sanction projects in border areas, as to help in their quicker execution without under-going the rigors of EIA process.**

Besides, all inland waterways projects and expansion/widening of national highways, will be exempted from prior environmental clearance. These include roads that cut through forests and dredging of major rivers.

This new 2020 draft also exempts most of the building projects of built up area up to 1,50,000 sq. m.

The two most significant changes in the new draft 2020 are the provisions for post-facto project clearance and abandoning the public trust doctrine. Projects operating in violation of the Environment Protection Act, 1986, will now be able to apply for post facto approach. It is a reiteration of a March 2017 notification for projects operating without environmental clearance. All a violator will need are two plans for remediation and resources augmentation corresponding to 1.5—2 times "the ecological damage associated and economic benefit derived due to violation". For such late applications, a developer will have to cough up Rs. 2000-10,000 per day for the period of delay. Such a penalty is of course low in a case such as that of an illegal sand miner who unauthorizedly mines and takes out several truck loads of costly sand every day.

The 2020 draft also spells out how the government will take cognisance of such violations. It has to be reported either by a government authority or the developers themselves. There is no scope for any public complaint. This provision is likely to be objected to strongly in public feedback and the various provisions of the new notification, when once finalised and gazetted, will be available on the web site of the Ministry of Environmental and Forests, Govt. of India, by say October 2020, or so. The details of the final notification shall be incorporated in the next-edition of this book.

22.3. General Details Constituting an Environmental Impact Assessment

In general, an Environmental Impact Assessment report should contain atleast the following details :

(1) *Input information* ;

(2) *Plant operation details* ;

(3) *Details of waste streams* ; and

(4) *Environmental Management Planning, including Disaster Management Plan.*

All these factors are discussed below :

(1) Input Information. The input information w.r.to the location of the industry or the project, shall cover the following factors :

(*i*) *Location and layout of the plant/project* ;

(*ii*) *Meteorology of the area* ;

(*iii*) *Demography and socio-economics of the area* ;

(*iv*) *Important monuments existing in the area* ;

(*v*) *Water-resources of the area* ;

(*vi*) *The nature and type of existing soils in the area* ;

(*vii*) *Land use pattern of the area* ;

(*viii*) *Cropping pattern of the area* ; and

(*ix*) *Forests and species of the area.*

These factors are briefly discussed below :

(*i*) **Location and layout of the plant.** Location and layout of the proposed plant or project is of primary importance, in terms of its geographical location, topography and importance of the area in national context. These factors need be discussed here.

(*ii*) **Meteorology of the area.** Meteorology of the area, where the proposed plant or industry is proposed to be located, plays a vital role in the dissipation of the produced pollutants in air and water, both. Thus, a project or an industry which may prove to be harmless in an arid area may prove to be quite dangerous in tropical or other areas, and *vice versa*. Hence, detailed discussion of meteorology of the area is required in an EIA.

(*iii*) **Demography and socio-economics of the area.** Demography and socio-economic structure of the nearby localities in the area of the proposed industry, plays an important role, in terms of its culture and population. Say for example, installation of a sophisticated plant in a rural area may require migrated urban population to work and man the plant in the rural area, that may often lead to social breakdown or displacement of local population. Such details should therefore be investigated and discussed in an EIA.

(*iv*) **Important monuments existing in the area.** If a polluting industry is allowed to be established near the vicinity of important monuments or structures, then such monuments are likely to get defaced or spoiled by the acidic fumes caused by the polluted air emissions, likely to be released by the proposed plant. The harmful effects caused to the Taj Mahal by the Mathura Refinery is an example of such adverse environmental effects. The monuments or important buildings and structures existing in the area of the proposed industry or the project, should therefore be thoroughly studied, and the possible effects of the project on them, should be thoroughly examined and discussed in the EIA.

(*v*) **Water resources of the area.** The existing water resources like the lakes or the rivers of the area of the proposed industry or the project will have an immediate effect in terms of its relative availability and possible contamination by the proposed project. Moreover, excessive usage of water results in the lowering of the sub-soil water-table or reduced availability of irrigation water, leading to many harmful consequences. All such possibilities should be thoroughly examined and discussed in the E.I.A.

(*vi*) **The nature and types of existing soils in the area.** The types of soils in the area of the proposed project will have an immediate effect interms of percolation of pollutants and possible contamination of underground water. Its capacity to receive toxic pollutants and convert them chemically, is also very important. Hence, these impacts must be thoroughly studied and described in the EIA.

(*vii*) **Land use pattern of the area.** The land use pattern of the area of the proposed project needs to be studied and discussed in an environmental impact assessment. For example, the agricultural land may allow for recycling of various components of the polluted water or air emissions, likely to be released by the plant ; whereas, the urban areas has very little capacity to absorb any additional emissions or depositions.

(*viii*) **Cropping pattern of the area.** The crops and their nature are directly connected to the *green belt* concept surrounding the large plants. Some crops are more sensitive to particular pollutants and are also economically more important than others. Hence, a judicial consideration of the impacts of the proposed plant on the crops of the area has to be examined and discussed in the E.I.A.

(*ix*) **Forest and species of the area.** The existing biological environment of the area, consisting of the flora and fauna, including the *trees, grasses, fishes, birds, and mammals*, etc., must be thoroughly studied, along with the possible impacts of the proposed industry (or the plant) on them. Specific description must be made of the plants and animal species or their habitats, which are likely to be lost by the construction and operation of the proposed project. General biological features, such as the species diversity and over-all eco-system stability must also be considered under this factor.

(2) **Plant Operation Details.** The environmental impact of an industry or a project, ultimately depends on the *processes* to be adopted and used in the plant. It also depends on the raw materials used and the finished products produced. The storage, handling and accidental spillage of raw materials, or sometimes even of the finished products, may cause prominent impact on the environment. Hence, all these factors must be thoroughly studied and described in the Environmental Impact Assessment.

(3) **Waste Streams.** The solid, liquid and air wastes, generated by the proposed project, will evidently affect the environment to a large extent, as summarised below :

(*i*) **Solid wastes.** The impacts of solid wastes can generally be ascertained and checked, in a comparatively easier manner, than those of liquid and air wastes. The solid wastes from the process plant and also from the various domestic and commercial activities of the plant, will need to be studied and analysed. These may require different disposal techniques. Depending on the mode of disposal and chemical composition, the impact of solid waste on soil and land would need to be ascertained.

In certain specific projects, like the project of a thermal power plant, the problem of disposal of solid waste generated as *fly ash*, may pose great problems and environmental hazards, since it may need huge land area for storage/disposal, which may not be available, and may also pose adverse air and land pollution.

(*ii*) **Liquid wastes.** There is every possibility of the water pollution being caused by a project or an industry, either through the discharge of wastewaters of the industrial processes, in the water bodies, like the lakes or rivers ; or of the contamination of the underground water by the seepage of polluted wastewater from the plant. The extent of pollution load likely to be contained in the effluents of the plant (even after recycling, etc.) and the manner of its disposal with reference to the quality and capacity of the receiving water body, will determine the adverse impact on the quality of the water bodies. The possibility of causing *thermal pollution* will also need assessment. These details must be very thoroughly analysed and discussed in the Environmental Impact Assessment.

(iii) **Air Emissions.** The polluted air emissions likely to be released by the plant will need a thorough evaluation and analysis. In case of air emissions, the impact on environment may be local as well as over long distances. These impacts will largely depend on the atmospheric conditions and meteorology of the area. While long range environmental effects may attract a little bit lesser attention, it will be very essential to analyse and assess the local impacts to determine the pollutant levels on the ground level surrounding the plant, where the majority of human population or vegetation may be affected. The dispersion and distribution of air-pollutants can be analysed by Gaussian plume distribution, as discussed in article 18.7. The correlation (Eq. 18.3) gives the max. concentration in the windward direction for a relatively short period of exposure. Correction factors based on wind and atmospheric data can be prescribed for 8 and 24 hours, respectively. These values are quite important, because of relevance of the threshold limits of tolerance.

More long term annual average (geometric mean or arithmetic mean) can be calculated using the climatological data, as given by the *wind rose diagram*. This helps in determining the G.I.C. at all important localities surrounding the plant. Sophistication of this exercise can further be achieved by exact day to day wind data and allowing for the contribution of wind movement in other directions also.

The above exercises lead to the *concentration profiles*. If there are multiple sources within the close vicinity, they can be added together for the cumulative effect. These exercises locate the worst affected areas and also inform the pollution levels in highly sensitive areas. Such exercises were infact carried out in the case of Mathura Refinery to study its effects on the Taj Mahal.

The information on the G.I.C. values can, thus, be utilised to evaluate its impact on the human, vegetation and other flora and fauna in the area. Accordingly, it may become necessary to modify the control mechanisms in the plant to reduce such adverse impacts.

(4) **Environmental Management Planning.** A detailed environmental management planning has to be drawn and submitted with EIA, for formulation, implementation, and monitoring of the environmental protection measures, during and after commissioning of the proposed project.

The *environmental management plan* for commissioning of the plant will ensure that resources are used with maximum efficiency to minimise generation of wastes, by adequately treating the residuals, recovering by-products, and adopting recycling techniques to the maximum possible extent. Such an overall strategy would not only cut down pollution control costs, but also results in savings in the cost of production.

The ***post-project management plan*** has also to be drawn for being executed during the operational phase of the project. It is necessary to monitor both the development and subsequent operational activities, if the investment in EIA is not to be wasted. *Post Project Monitoring (PPM)* will secure an immediate advantage in environmental management. It will indicate weather the conditions imposed on an entrepreneur, such as specific mitigation measures, have been adopted. In the longer term, post project monitoring may help to find and implement remedial measures for unexpected future adverse impacts, and also to relax some stringent conditions imposed, while clearing a project if found to be unnecessary during the operation of the project.

The ***Disaster Management Plan*** **(DMP)** should also be worked out with the Environmental Impact Assessment of each project, since the environmental risks are inherent in the construction and operation of each complex industrial plant, particularly the refineries, or chemical or pharmaceutical units. Any major failure or even minor failures in the system could lead to a disaster, resulting in heavy toll of human life and devastating losses to ecology and property. A comprehensive and scientific approach to risk assessment and management decisions is considered essential, in view of the increase in the number and magnitude of environmental risks. Based on the findings of the risk assessment, the following two disaster management plans, should be prepared :

(*a*) Disaster Management Plan (DMP) for the management of various disasters within the plant site ; and

(*b*) Emergency Preparedness Plan (EPP) for co-ordination with various offsite agencies during emergencies likely to affect the outside environment.

22.4. Methodology for Preparing Environmental Impact Assessment

A nicely prepared Environmental Impact Assessment (EIA) may also establish quantitative values for selected parameters, which indicate the quality of the environment before, during, and after the proposed developmental activity.

EIA essentially involves three steps, *i.e.* (*i*) *identification*, (*ii*) *prediction*, and (*iii*) *evaluation* of impacts, as discussed below :

(*i*) **Impact identification** may be carried out with the help of checklists, matrices or networks. Checklists merely present a list of environmental parameters to be investigated for possible impacts. Matrices are two dimensional checklists in which cause-effect relationships are established by listing possible project activities along one axis and potentially impacted environmental characteristics or conditions along the other. Networks illustrate cause-condition effect linkages as also temporal dimensions, and, therefore, provide the most comprehensive methodology for impact identification.

(*ii*) **Prediction of environmental impacts** requires the greatest degree of scientific application. This step involves projecting the baseline environmental setting into the future, with and without the project, and then performing the necessary calculations for predicting real impacts of the proposed development. Whereas, fairly accurate models are available for prediction with respect to air and water components, predictions of biological, socio-economic and cultural impacts are very often subject to uncertainty. The degree of uncertainty could, however, be ascertained mathematically or indicated in qualitative terms while presenting prediction results.

(*iii*) **The evaluation of impacts** in an EIA calls for conversion of the predicted values for various environmental parameters to a comparable set of units using some system of normalisation. Ideally, the environmental impacts should be expressed in monetary units for easy and objective comparison with other costs and benefits of the project. The major problem, however, lies in assigning monetary values to intangible parameters. Methods involving numerical rating and ranking, or weighting and scaling of environmental impacts are, therefore, commonly used. The latter are preferred as the weighted scores can be added to yield an aggregate score, representing the net environmental impact of a particular alternative. The scores for various alternatives could be compared for determining the environmentally preferred alternative.

The weighting and scaling technique, however, offers only a 'yes' or 'no' choice, whereas consideration of impacts on an individual basis or according to common categories permits more flexibility in the decision-making process. It is, therefore, desirable that numerical scores be accompanied with a summary of interpretation in a decision table, while communicating the evaluation results. Table 22.8 summarizes the various techniques used for impact assessment.

Table 22.8. Impact Evaluation Methodologies

S. No.	*Name*	*Salient features*
1.	*McHarg overlay approach*	Involves mapping an area and the environmental impacts of the project. A shading system indicates the degree of impact. The transparent maps are overlaid to identify visually, the project location giving least impacts. Computerised overlays could also be used.
2.	*Leopold matrix method*	Uses matrix of human action and environmental factors. Significant impacts are shown placing a diagonal line in appropriate cell of the matrix. Magnitude and importance of impacts are evaluated on a 1-10 scale and placed on either side of the diagonal. The values indicated in the matrix cells are not equitable. A written description of significant impacts as well as activities or environmental factors, for which many interactions occur, must be presented to enable overall assessment.
3.	**Battelle environmental evaluation system**	This method is based on checklist of 78 environmental parameters. Predicted value of each parameter is converted to a 0-1 scale of environmental quality (EQ) using value function graphs. Each environmental parameter is assigned a weight out of 1000 parameter importance units (PIU) by ranked pairwise comparison. EQ value is multiplied by PIU to provide environmental score. The final score of the alternative is the difference between the sum of the scores, with and without the project.

		These scores of alternatives are then compared to find the best alternative. It provides a useful technique for comparing the alternatives from environmental angle. However, numerical scores do not allow comparison with project costs and benefits, which are in monetary units.
4.	*Adaptive environmental assessment and management*	A workshop comprising groups of experts is held to consider and define variables, such as objectives, indicators, time and spatial boundaries, alternate actions, etc. The core group of experts then develops an explicit computer model of the system, which is later subjected to validity testing and used to develop resilient policies to avoid adverse impacts. Since model is based on understanding of the process as well as components, it simulates behaviour over time and is responsive to different modes of management.
5.	*Economic valuation input-output modelling*	It stimulates the effects of introducing a project on the economy and the environment, using a two-sector model. One sector is the standard economic input-output model for tracing direct and indirect impacts on an economy of injection of or reduction in expenditure. The other sector is the environmental sector with flows of ecological commodity inputs and outputs as well as those associated with the final demand. Feedback from environment to economic sector is also incorporated.
6.	*Cost-benefit analysis*	As applicable to EIA, the technique involves monetary valuation of environmental components and natural resources through surrogate markets, direct questioning, or voting. These values are used along with the values of commodities consumed by individuals as well as their direct and indirect income to obtain the net change in the aggregate income, based on which, alternatives could be compared.

A methodology developed by taking advantage of the best features of all the available methodologies, uses *cause condition-effect networks* for impact identification, mathematical models for impact prediction, and *Battelle Environmental Evaluation System* (BEES) adopted for industrial projects for evaluation.

22.5. Role of an Environmental Engineering/Consulting Firm in Obtaining Environmental Clearance for a Project

To carry out the many tasks required for environmental assessment and to successfully negotiate in the approval process, an applicant must decide whether to do the job internally or whether to hire an environmental engineering/consulting firm. Consideration of credibility and objectivity are very important. An EIA prepared internally without the help of a credible consultant, may usually be termed as non-objective and self servicing, by the Govt. experts, while examining an EIA for its clearance.

Similarly, if you get into the public hearings or into the court, the defensibility of your environmental data is of prime importance. They will be defensible if they are complete, of high quality, responsive to adverse impacts, responsive to alternatives, and if they can be vouchsafed by some recognised expert witnesses.

Also the matter of availability of appropriate and necessary in- house expertise gives rise to a set of questions :

(*i*) Are all the necessary environmental disciplines available in-house ?

(*ii*) Are there areas of special analysis required ?

(*iii*) Are there special monitoring requirements ?

Generally, the environmental impact process is an infrequent or one-time experience for the applicant. However, prior-experience and familiarity with regulations and agency requirements is necessary for the speedy conclusion of the process. The group undertaking environmental aspects of a project must be aware of all relevant environmental regulations ; they must be experienced in doing EIA studies ; they must have the ability to interpret agency requirements to define the appropriate level of effort for EIA studies ; and they must understand the implications of study efficiencies, scheduling and agency relationships.

An applicant's ability to devote proper personnel to the project for the necessary time interval to carry forward environmental studies is frequently a serious problem. The size of a project and its complexity determines the range of skills, the time duration and the agencies involved. If a project is simple, chances that the applicant make the necessary in-house skills available are pretty good. If the project is complex, it is unlikely that these resources would be available in their entirety or could be devoted to the project for the time needed.

Depending upon the applicant's resources and relevant corporate mandate, the entire environmental study work can be performed in- house, or it can be performed with the assistance of a consultant, or a Consultant may be retained to manage the entire process, making the best use of the applicant's personnel and other resources.

22.6. Role of Regulatory Agencies and Control Boards in Obtaining Environmental Clearance for a Project

Interfacing with regulatory agencies and Pollution Control Boards who have to clear the project, is an evolving and continuing process that should be started as early as possible. The applicant must also recognise that the relationship with agencies will be on different levels at different stages of development.

Basic interaction with the regulatory agencies revolves around approvals. Once the required construction and operation approvals are identified, the applicant must develop an efficient strategy for obtaining the needed approvals.

At early conceptual stages (when feasibility is still a consideration), the applicant may not have a full definition of the project and may be concerned, constrained or apprehensive about how much to reveal to regulatory agencies (or to the public) about plans. However, some information must be revealed to regulatory agencies in order to get useful information in return. For example, at this stage, you will want to know what approvals will be required and you will want to know what attitudes agency personnel will have towards projects similar to yours. You will want to know about environmental problems or constraints of particular concern to each regulatory agency. Some of this information may be readily obtained without having to be too explicit, but the more explicit you can be, the better and more useful will be the responses or information from the regulatory agencies. Experience shows that regulatory agencies are willing to be helpful and are generally respectful of the proprietary nature of your corporate project plans.

Likewise, the more willing you are to talk to regulatory agency personnel about potential plans or options, the more responsive and trustful they will be. This is not to say that they do not have their own viewpoints about control technology etc. They have their mandates and objectives, which may represent views counter to yours.

The objective of discussions at the early stages should be to determine if the proposed project has reasonable potential for complying with the environmental regulations, and whether the project has reasonable chances of obtaining all necessary approvals without incurring undue difficulties or delays (for example, public controversy, expensive control measures, lengthy permitting procedures, etc.). Information obtained at an early stage can be helpful in overall project planning and scheduling, in preparing design schedules, plans for ordering equipment, and preparation of construction schedules, etc.

Contact with regulatory agencies at an early stage of project development is usually one portion of an environmental review. The other major portion of an environmental review is concurred with a rapid survey or screening of potential sites in order to identify site-related environmental problems that would also become potential difficulties in obtaining approvals. The information needed to complete an environmental review is most often available from public sources, such as Central or Research Laboratories and Government regulatory agencies, and can be obtained without any reference to a particular project.

Dealing with regulatory agencies at later stages of the project will require an even greater degree of interaction and may require even more details to describe the project. For example, the next distinct step is the active search for (and evaluation of) potential sites. Discussions at this stage will likely require interaction with agencies not previously contacted. Further, determining compatibility of a project with a specific site will require having more detailed information about the project and revealing more about the project to the concerned agencies.

Selection of a site next lends to detailed studies necessary for impact assessment. At this stage, the applicant should be prepared to discuss the project with the regulatory agencies openly and in detail. These discussions commence with definition of the scope of the environmental monitoring progress, and EIA studies that will be required by the concerned agencies.

One often-voiced concern is that such discussions can lead to exorbitant demands by agencies monitoring the impact studies. Another concern is that revealing areas of potential environmental problems may adversely influence agency attitudes towards the project. These concerns must be referenced to the fact that the EIA report is to be a full disclosure document. In the absence of adequate information about a project, agencies may very well have a tendency to ask for more information rather than less.

It is recommended that the applicant should be fully appraised of the extent of potential problems and defines a study programme adequate to examine these problems. Further, it is recommended that developed work programmes be taken to the agencies for comments and approval, rather than simply asking the personnel to specify a work programme. If the applicant has been open, and has established credibility with the agencies by acknowledging potential problems, it is highly likely that some mutual agreement will be reached upon a work programme, that is not excessive.

It is dangerous not to acknowledge or reveal potential environmental problems early-on, in the in-house discussions with the agencies. Such problems, usually come to light eventually, and it is a far better strategy to identify them, before an outside agency or an NGO or an intervenor does. Early identification underscores the fact that you are aware of the problems and are trying to do something about them. In addition, legal entanglements over EIA are almost always the results of not having described the potential problems and impacts of the project, rather than the result of revealing adverse impacts.

By the time you begin the EIA study programme, your project will most likely be known to the public and will begin to be watched closely by the interested individuals and groups. Thus, throughout the conduct of EIA studies, it is advisable to maintain frequent and close communication with the regulatory Govt. agencies, so that no "surprises" are presented to agencies upon submission of the EIA report, which they have to review and approve.

22.7. Role of the General Public in Environmental Clearance

It is evident that a large degree of latitude exists for interaction with the public at various stages of a project. Recognising that the EIA is a full disclosure document and that it will ultimately be made public, and further recognising that a project which impacts a community will be of public interest, the EIA must be written in simple and effective manner, as to be appreciated by public man.

Perhaps, the greatest concern of Corporations is how to interact with the public in the early stages of project development prior to the preparation of an EIA. There are strategies which may be employed ranging from a very elaborate and highly visible public relations programme to a very low profile. The degree of such interactions and the amount of information given about the project, is largely a matter of Corporate choice.

Moreover, if desired, techniques such as surveys can be employed to sample public attitudes, and thus provide some basis for determining how best to interact with the public. Regardless of which strategy a Corporation chooses to employ, the following general points should be taken into account :

(*i*) Prepare in advance, the type and amount of information you choose to release to the public.

(*ii*) Make sure the information released is consistent with the action of the Corporation that are observable by the public.

(*iii*) Make sure that the information given by various Corporate sources is consistent between the sources.

(*iv*) Be prepared to discuss your project in a manner that is responsive, and which will not be viewed as evasive.

(*v*) Do not oversell the project or avoid discussion of potential problems, but remain prepared to give a positive response about what you are doing about public concerns.

—23

Chapterwise Objective Multichoice Questions

Chapter 1 : Introduction to the Subject

Q. 1. The liquid wastes originating from residential and industrial buildings, are collectively called :

(*a*) domestic sewage (*b*) combined sewage
(*c*) sanitary sewage (*d*) none of these.

Q. 2. The sewer which transports the sewage to the point of treatment, is called :

(*a*) house sewer (*b*) main sewer
(*c*) outfall sewer (*d*) none of these.

Q. 3. A combined sewerage system, collectively carries :

(*a*) domestic sewage and industrial sewage
(*b*) storm sewage and domestic sewage
(*c*) storm sewage and industrial sewage
(*d*) storm sewage and sanitary sewage.

Q. 4. The sewerage system originates from :

(*a*) outfall sewer (*b*) main sewer
(*c*) house sewer (*d*) none of these.

Q. 5. A sewer which receives the discharge from a number of independent houses, is called :

(*a*) house sewer (*b*) intercepting sewer
(*c*) lateral sewer (*d*) none of these.

Q. 6. A sewer pipe carrying sewage from a building to the point of its immediate disposal, is called :

(*a*) intercepting sewer (*b*) lateral sewer
(*c*) house sewer (*d*) none of these.

Q. 7. When rainy season is confined to a few months, like that in India, the preferred sewerage system would be :

(*a*) combined system (*b*) separate system
(*c*) partially separate system (*d*) none of these.

Q. 8. The suitable system of sanitation for an area having uniformly distributed mild rains throughout the year, like that in England, is :

(*a*) separate system (*b*) combined system
(*c*) partially separate system (*d*) partially combined system.

Q. 9. The water carriage system of collection of waste products is preferred to dry conservancy system, because :

(*a*) it is cheaper in initial cost
(*b*) it does not require treatment before disposal
(*c*) it is more hygienic in nature
(*d*) it is easier to maintain.

Q. 10. The wastewater coming from kitchens and bath rooms is popularly known as :

(*a*) domestic sewage discharge (*b*) sludge discharge
(*c*) drainage discharge (*d*) none of these.

Q. 11. In a city drainage system, having drainage to sewage ratio as 20, the peak dry weather flow would amount to :

(*a*) 20% of the design discharge
(*b*) 5% of the design discharge
(*c*) slightly less than 5% of the design discharge
(*d*) slightly more than 5% of the design discharge.

Chapter 2 : Estimating the Design Sewage Discharge

Q. 12. For the design of sewers in India, the percentage of sewage discharge, is assumed as :

(*a*) 25—30% of water supplied from water works
(*b*) 75—80% of water supplied from water works
(*c*) 100% of water supplied from water works.
(*d*) none of these.

Q. 13. The sewer pipes have to be designed and checked for :

(*a*) only maximum flow
(*b*) only minimum flow
(*c*) both maximum and minimum flows
(*d*) none of them.

Q. 14. Sewage treatment works are normally designed for a design period of :

(*a*) 40—50 years (*b*) 30—40 years
(*c*) 15—20 years (*d*) 5—10 years.

Q. 15. Sewage pumping plants are normally designed for a design period of :

(*a*) 40—50 years (*b*) 30—40 years
(*c*) 15—20 years (*d*) 5—10 years.

Q. 16. For Indian cities, like Delhi or Calcutta, the per capita sewage production may be of the order of :

(*a*) 500 litres (*b*) 200 litres
(*c*) 100 litres (*d*) none of these.

Q. 17. The peak design flows, expressed as number of times their average values, will be :

(*a*) greater for smaller lateral sewers, as compared to these for larger trunk sewers (*b*) greater for larger trunk sewers, as compared to these for smaller lateral sewers

(*c*) equal for all sizes and types of sewers.

Q. 18. The ratio of maximum daily sewage flow to the average daily sewage flow for sewer mains of 0.5 to 1 m diameter, is generally taken as :

(*a*) 1.5 (*b*) 2.0

(*c*) 3.0 (*d*) none of these.

Q. 19. The ratio of the maximum daily sewage flow to the average daily sewage flow for trunk sewers, above 1.25 m in diameter, is :

(*a*) 1.5 (*b*) 2.0

(*c*) 3.0 (*d*) 4.0.

Q. 20. The ratio of the max. daily sewage flow to the average daily flow for small sewers and laterals up to say 0.25 m in diameter, is :

(*a*) 1.5 (*b*) 2.0

(*c*) 3.0 (*d*) 4.0.

Q. 21. The ratio of the max. sewage flow to the average flow for branch sewers, up to say 0.5 m in diameter, is

(*a*) 1.5 (*b*) 2.0

(*c*) 3.0 (*d*) 4.0.

Q. 22. The ratio of minimum hourly flow to the average flow of sewage, is :

(*a*) $\frac{1}{3}$ (*b*) $\frac{1}{2}$

(*c*) $\frac{2}{3}$ (*d*) 3.

Q. 23. For moderate sized areas and sewers, say up to 1 m in diameter, the maximum hourly drafts may be considered as equal to :

(*a*) 1.5 times the maximum daily

(*b*) 3 times the maximum daily

(*c*) $\frac{1}{3}$ times the maximum daily

(*d*) none of the above.

Q. 24. The peak flow for the design of sewers, has been connected to the average flow and the population of the city in thousands, by an empirical relation, given by

(*a*) $Q = \dfrac{18 + \sqrt{P}}{4 + \sqrt{P}} Q_{av}$ (*b*) $Q = \dfrac{4 + \sqrt{P}}{18 + \sqrt{P}} Q_{av}$

(*c*) $Q = \dfrac{18 + P}{4 + \sqrt{P}} Q_{av}$ (*d*) none of the above.

Q. 25. Point out the wrong statement, in the following assumptions, generally made in the design of moderate sized sewers :

(*a*) average daily flow = annual average daily flow

(*b*) minimum daily flow = $\frac{2}{3}$ average daily flow

(*c*) minimum hourly flow = $\frac{1}{3}$ average daily flow

(*d*) maximum daily flow = 2 average daily flow

(*e*) maximum hourly flow = 3 average daily flow
(*f*) none of them.

Q. 26. $\frac{1}{4}$ to $\frac{1}{3}$ space is left vacant in the design of sewer pipes at maximum discharge, due to :
(*a*) possible low estimates of maximum and average flows
(*b*) large scale possible infiltration of storm water
(*c*) unexpected increase in population
(*d*) all the above factors.

Q. 27. The minimum and the maximum diameter of sewers, generally adopted in the designs, may be :
(*a*) 15 cm and 100 cm (*b*) 15 cm and 300 cm
(*c*) 25 cm and 450 cm (*d*) 60 cm and 300 cm.

Chapter 3 : Estimating the Peak Drainage Discharge

Q. 28. The Rational formula, for evaluating peak drainage discharge, can be safely applied to catchments, up to :
(*a*) 25 hectares (*b*) 500 hectares
(*c*) 2000 hectares (*d*) up to any limit.

Q. 29. The value of the coefficient of runoff for paved areas is of the order of :
(*a*) zero (*b*) 0.5
(*c*) 0.9 (*d*) greater than 1.

Q. 30. The value of the coefficient of runoff for perfectly impervious areas, tend to :
(*a*) zero (*b*) 0.5
(*c*) 1.0 (*d*) infinity.

Q. 31. Critical rainfall intensity would always be .... the hourly rainfall value, if T_c exceed 1 hr.
(*a*) more than (*b*) less than
(*c*) equal to.

Q. 32. With the increase in its duration, the intensity of rainfall will :
(*a*) decrease (*b*) increase
(*c*) not change.

Q. 33. The Rational formula for peak drainage discharge computations, was evolved by :
(*a*) Kuichling (*b*) Fruhling
(*c*) Lloyd David (*d*) all of them.

Q. 34.* If the overland flow from the critical point to the mouth of a drain is 12 km, and the difference in levels between these points is 14.5 m, then the inlet time would be of the order of :
(*a*) 2 hours (*b*) 4 hours
(*c*) 6 hours (*d*) 8 hours.

*Hints for solving starred questions are given at the end of the chapter.

Q. 35. The drainage discharge of a town of 16 hectares area, consisting of 40% hard paved ($k = 0.80$), 30% unpaved ($k = 0.20$), and remaining as wooded ($k = 0.10$), with a max. rain intensity of 5 cm/hr, would be computed by Rational formula, as equal to :

(*a*) 0.91 cumecs (*b*) 0.091 cumecs
(*c*) 9.1 cumecs (*d*) none of these.

Chapter 4 : Hydraulic Design of Sewers and S.W. Drain Sections

Q. 36. Laying of sewers is usually done with the help of :

(*a*) a theodolite (*b*) a compass
(*c*) a plane table (*d*) sight rails and boning rods.

Q. 37. Pick up the incorrect statement :

(*a*) sewer pipes carry sewage flow by gravity
(*b*) sewer pipe material has to withstand wear and tear due to abrasion
(*c*) sewer pipes should be designed for self-cleansing velocity of at least 0.45 m/sec at minimum discharge, or 0.8 m/sec at full design discharge
(*d*) sewer pipes can be carried up and down the hills and valleys
(*e*) none of the above.

Q. 38.* The diameter of a R.C.C. sewer pipe is 300 mm ; the gradient required to generate self cleansing velocity would be of the order of :

(*a*) 1 in 50 (*b*) 1 in 200
(*c*) 1 in 500 (*d*) 1 in 1000.

Q. 39. The gradient required to generate self cleansing velocity in a circular sewer of 150 mm dia, is of the order of :

(*a*) 1 in 30 (*b*) 1 in 100
(*c*) 1 in 200 (*d*) 1 in 250.

Q. 40. The flow velocity in a circular sewer at minimum design discharge is about x times the full design velocity at maximum design discharge, where x is about :

(*a*) 60% (*b*) 30%
(*c*) 90% (*d*) 100%.

Q. 41. Minimum self cleansing velocity in sewers, calculated as 0.45 m/sec for ensuring non-settlement of 1 mm size sand and 6 mm size organics, should be checked, for a minimum discharge, equal to :

(*a*) max design discharge (*b*) average discharge
(*c*) $\frac{1}{3}$ max design discharge (*d*) $\frac{1}{2}$ average discharge.

Q. 42. Sewer pipes of dia less than 0.4 m, are designed at maximum flow, to be running at :

(*a*) full depth (*b*) $\frac{1}{2}$ full depth
(*c*) $\frac{2}{3}$ full depth (*d*) $\frac{3}{4}$ full depth.

Q. 43. Sewer pipes of dia 0.4 to 0.9 m, are designed at maximum flow, running at :

(*a*) full depth (*b*) $\frac{1}{2}$ full depth

(*c*) $\frac{2}{3}$ full depth (*d*) $\frac{3}{4}$ full depth.

Q. 44. Main trunk sewers, larger than 0.9 m in dia, are designed due to ventilation considerations, at max. discharge, running at :

(*a*) full depth (*b*) $\frac{1}{2}$ full depth

(*c*) $\frac{2}{3}$ full depth (*d*) $\frac{3}{4}$ full depth.

Q. 45. The most suitable section of a sewer in a separate sewerage system is :

(*a*) rectangular (*b*) circular

(*c*) new egg-shape (*d*) parabolic.

Q. 46. The most suitable section of a sewer in a combined sewerage system is :

(*a*) rectangular (*b*) circular

(*c*) new egg-shape (*d*) parabolic.

Q. 47. An egg-shaped sewer, when compared to a circular sewer, is :

(*a*) economical

(*b*) more stable

(*c*) easier to construct

(*d*) provides better self-cleansing velocity at low discharges.

Q. 48. The flow velocity in a sewer does not depend on :

(*a*) its grade (*b*) its length

(*c*) its hydraulic mean depth (*d*) its roughness.

Q. 49. The effect of increasing the diameter of a sewer, on the self- cleansing velocity, is :

(*a*) to increase it (*b*) to decrease it

(*c*) not to cause any effect.

Q. 50. Pick up the correct statement from the following :

With self cleansing velocity in sewers,

(*a*) the silting occurs at the bottom

(*b*) the scoring occurs at the bottom

(*c*) the silting and souring both occur at the bottom

(*d*) neither siting nor scouring occur at the bottom.

Q. 51.* In a circular sewer of dia D, if the depth of flow is $\frac{1}{4}$ D, the wetted perimeter will be equal to :

(*a*) $\frac{\pi D}{4}$ (*b*) $\frac{\pi D}{2}$

(*c*) $\frac{\pi D}{3}$ (*d*) none of these.

Q. 52. Two sewer sections, laid at the same grade, will be hydraulically equivalent, if :

(*a*) their discharging capacity, when running full, are equal
(*b*) their flow velocities, when running full, are equal
(*c*) their flow areas, when running full, are equal
(*d*) all the above factors are equal.

Chapter 5 : Sewers, Their Construction, Maintenance and Required Appurtenances

Q. 53. For trunk and outfall sewers, the type of section used to be preferred in olden days, was :

(*a*) Horse shoe shaped
(*b*) parabolic shaped
(*c*) egg-shaped
(*d*) semi-elliptical shaped.

Q. 54. Boussinesq's equation for computing unit pressure (p_t) at a depth H on the sewers, due to traffic load (P) is given by :

(*a*) $p_t = \frac{3H^3P}{2\pi Z^3}$
(*b*) $\frac{2H^3P}{3\pi Z^3}$
(*c*) $p_t = \frac{3H^3P}{2Z^3}$
(*d*) $\frac{2H^3P}{3Z^3}$

where, Z is the slant height of the considered point from the load P.

Q. 55. Pick up the incorrect statement :

(*a*) The pressure exerted by sewage, when running full in a pipe, is called the internal pressure ; and it may induce tensile stresses in the pipe material
(*b*) The pressure exerted by back fill on an underground sewer, the self weight of sewer, and the superimposed traffic loads, may cause compressive stresses in the pipe material
(*c*) Temperature stresses are usually computed on underground sewers, to avoid the cracking of their joints, by the formula $L \propto T$
(*d*) Flexural stresses, caused by the beam action, are normally ignored, while designing underground sewer pipes.

Q. 56. The most prominent force, acting on the underground sewer pipes, would be :

(*a*) compressive force
(*b*) tensile force
(*c*) bending force
(*d*) all of these.

Q. 57. Pick up the incorrect statement :

Asbestos cement pipes are :

(*a*) light in weight
(*b*) susceptible to corrosion by chemicals in sewage
(*c*) available in different sizes, say 10—90 cm in dia
(*d*) strong enough to withstand traffic and other compressive loads.

Q. 58. Pick up the incorrect statement :

(*a*) R.C.C. pipes are very strong to withstand compressive forces
(*b*) R.C.C. pipes are very strong to withstand tensile forces

(*c*) R.C.C. pipes are quite resistant to erosion and abrasion

(*d*) R.C.C. pipes are not easily corroded by the hydrogen sulphide gas, produced by the flowing sewage.

Q. 59. The longitudinal reinforcement on R.C. sewer pipes is *x*% of the cross-sectional area of concrete, where x is :

(*a*) 0.125 (*b*) 0.25

(*c*) 2.5 (*d*) 25%.

Q. 60. Branch and main sewers are widely made of :

(*a*) R.C.C. (*b*) P.C.C.

(*c*) glazed stone-ware (*d*) cast iron.

Q. 61. House connections and lateral sewers are widely made of :

(*a*) R.C.C. (*b*) P.C.C.

(*c*) glazed stoneware (*d*) cast iron.

Q. 62. Stoneware pipes are generally not used for sewer mains, because they are :

(*a*) weak in compression

(*b*) weak in tension

(*c*) hydraulically less efficient

(*d*) less resistant to organic corrosion.

Q. 63. The best sewer material to resist hydrogen sulphide corrosion, is :

(*a*) R.C.C. (*b*) brick masonry

(*c*) glazed stoneware (*d*) asbestos cement.

Q. 64. Sewers are generally laid, starting from their :

(*a*) off take point (*b*) outfall point

(*c*) mid point (*d*) any point along the alignment.

Q. 65. Sheeting and bracing of excavation trench for laying sewers, will be more a necessity, in :

(*a*) clayey soils (*b*) silty soils

(*c*) sandy soils (*d*) quick sands.

Q. 66. Testing of sewer pipes may involve :

(*a*) water test (*b*) mirror test

(*c*) ball test (*d*) all these tests.

Q. 67. For joining smaller sewer pipes of dia less than 0.6 m or so, the preferable joint is :

(*a*) spigot and socket joint (*b*) collar joint

(*c*) bandage joint (*d*) mechanical joint.

Q. 68. For joining bigger sewer pipes of dia more than, say 0.6 m, the preferred joint is :

(*a*) socket and spigot joint (*b*) collar joint

(*c*) bandage joint (*d*) mechanical joint.

Q. 69. Asbestos cement pipes are normally joined by using :

(*a*) bell and spigot joint (*b*) simplex joint

(*c*) lock joint (*d*) none of these.

Q. 70. The equipment used for checking the levels of the sewer inverts is :

(*a*) dumpy level (*b*) boning rod
(*c*) theodolite (*d*) none of these.

Q. 71. Manhole covers are made circular :

(*a*) for architectural reasons
(*b*) to strengthen the cover
(*c*) to prevent falling of the cover into the manhole
(*d*) to make the entry convenient.

Q. 72. Manholes are generally located

(*a*) at all changes of direction of sewer
(*b*) at all changes of gradient of sewer
(*c*) at all junctions of different sewers
(*d*) all of the above.

Q. 73. A manhole is generally classified as a deep manhole, if its depth is more than :

(*a*) 0.9 m (*b*) 1.2 m
(*c*) 1.5 m (*d*) 2 m.

Q. 74. A manhole is generally classified as a shallow manhole, if its depth is less than :

(*a*) 0.9 m (*b*) 1.2 m
(*c*) 1.5 m (*d*) 2 m.

Q. 75. The maximum spacing of manholes on sewers, having diameter more than 1.5 m, is :

(*a*) 75 m (*b*) 150 m
(*c*) 200 m (*d*) 300 m.

Q. 76. For very small sized sewers, of say upto 0.3 m diameter, the maximum manhole spacing, specified by Indian standards, is :

(*a*) 30 m (*b*) 45 m
(*c*) 75 m (*d*) 100 m.

Q. 77. A drop manhole may be provided along a sewer line :

(*a*) when the sewer drops from a height of more than 0.6 m or so.
(*b*) when a branch sewer outfalls into it from a height of more than 0.6 m or so.
(*c*) to provide inspection chambers in the sewer line
(*d*) for none of these.

Q. 78. Cleanouts help in :

(*a*) admitting sewage into the sewer
(*b*) cleaning the sewer
(*c*) inspecting the sewer
(*d*) none of these.

Q. 79. When a sewer line is dropped below the hydraulic gradient line to pass it through an obstruction, the arrangement is known as :

(*a*) inverted syphon (*b*) sag pipe
(*c*) depressed sewer (*d*) all of these.

Q 80. Of the following, the most effective type of storm relief work to divert excess drainage in combined sewers, is :

(*a*) leaping weir (*b*) overflow weir

(*c*) siphon spillway (*d*) all of these.

Q. 81. Pick up the incorrect statement :

(*a*) Small sewers are cleaned by flushing

(*b*) Medium sewers are cleaned by cane rodding

(*c*) Ventilation columns are provided to create airlocks in the sewers

(*d*) Large sized sewers are cleaned by actually entering into the sewers.

Q. 82. The gas which is generally found present in sewers, is :

(*a*) H_2S (*b*) CO_2

(*c*) CH_4 (*d*) all of these.

Q. 83. When a sheet of paper moistened with lead acetate, held for 5 minutes in a manhole, turns black ; then the sewer certainly contains :

(*a*) hydrogen sulphide gas (*b*) carbon dioxide gas

(*c*) methane gas (*d*) petrol vapours.

Q. 84. Extinguishing of the flame of a miners safety lamp within 5 minutes, when held at the level of sewage in a manhole, indicates in the sewer, the presence of :

(*a*) hydrogen sulphide gas (*b*) carbon dioxide gas

(*c*) methane gas (*d*) petrol vapours.

Q. 85. Insertion of safety lamp in the upper portion of a manhole, causing flames, surely indicates in the sewer, the presence of :

(*a*) hydrogen sulphide gas (*b*) carbon dioxide gas

(*c*) methane gas (*d*) petrol vapours.

Q. 86. In a ventilating column, a cowl is provided :

(*a*) to prevent entry of foul gases

(*b*) to prevent entry of objects

(*c*) to prevent the exit of heat

(*d*) for none of these.

Q. 87. Ventilation columns, in sewers, are provided to :

(*a*) help in escaping of foul gases

(*b*) help in preventing spread of foul gases

(*c*) to provide support to the sewers

(*d*) none of these.

Q. 88. Ventilation columns are placed along a sewer line, at intervals of about :

(*a*) 30—50 m (*b*) 75—100 m

(*c*) 150—300 m (*d*) 500—750 m.

Chapter 6 : Pumps for Lifting Sewage

Q. 89. The pump, which permits the sewage solids to pass out with liquid sewage, without clogging the pump, is a :

(*a*) reciprocating pump (*b*) centrifugal pump

(*c*) pneumatic ejector (*d*) none of these.

Q. 90. A disintegrating pump, which breaks the sewage solids, while passing sewage through it, is a special type of :

(*a*) centrifugal pump (*b*) reciprocating pump

(*c*) air pressure pump (*d*) none of these.

Q. 91. Pick up the incorrect statement :

(*a*) centrifugal pumps are the most widely used type of pumps for lifting sewage

(*b*) reciprocating pumps have become obsolete these days

(*c*) centrifugal pumps are also known as air pressure pumps

(*d*) none of them.

Q. 92.* A cylindrical ejector having a height of 2 m fills after every 10 minutes with a peak sewage discharge of 0.0146 cumecs ; the diameter of the ejector could be chosen as :

(*a*) 2.5 m (*b*) 2 m

(*c*) 1.5 m (*d*) 1 m.

Q. 93.* A $\frac{1}{2}$ cusecs pump, for a lift of 10 m (including losses), should have a horse power of about :

(*a*) 1 (*b*) 3

(*c*) 5 (*d*) 7.

Chapter 7 : Quality and Characteristics of Sewage

Q. 94. The appropriate percentage of water in sewage is :

(*a*) 90% (*b*) 99%

(*c*) 99.9% (*d*) 99.99%

(*e*) 100%. (*GATE, 1990*)

Q. 95. The specific gravity of sewage is :

(*a*) zero (*b*) slightly less than 1

(*c*) equal to 1 (*d*) slightly greater than 1.

Q. 96. Pick up the incorrect statement :

(*a*) aerobic bacteria flourish in the presence of oxygen

(*b*) anaerobic bacteria flourish in the absence of oxygen

(*c*) facultative bacteria flourish in the presence as well as absence of oxygen

(*d*) none of the above.

Q. 97. Well oxidised sewage will contain nitrogen, largely in the form of :

(*a*) nitrites (*b*) nitrates

(*c*) free ammonia (*d*) all of these.

Q. 98. Partially oxidised stale sewage will contain nitrogen mainly in the form of :

(*a*) nitrites (*b*) nitrates
(*c*) free ammonia (*d*) (*a*) and (*c*) both.

Q. 99. Well oxidised sewage will contain sulphur, largely in the form of :

(*a*) sulphites (*b*) sulphates
(*c*) hydrogen sulphide (*d*) all of these.

Q. 100. Gases, which are generally evolved during aerobic decomposition of sewage, are :

(*a*) $CO_2 + NH_3 + H_2S$ (*b*) $CO_2 + NH_3 + H_2S + CH_4$
(*c*) $CO_2 + NH_3 + SO_2$ (*d*) $CO_2 + NH_3 + SO_2 + CH_4$.

Q. 101. Gases, which are generally evolved during anaerobic decomposition of sewage, are :

(*a*) $CO_2 + NH_3 + H_2S$ (*b*) $CO_2 + NH_3 + H_2S + CH_4$
(*c*) $CO_2 + NH_3 + SO_2$ (*d*) $CO_2 + NH_3 + SO_2 + CH_4$.

Q. 102. Temperature variations affect the :

(*a*) biological activity of bacteria in sewage
(*b*) viscosity of sewage
(*c*) solubility of gases in sewage
(*d*) all of the above.

Q. 103. 1000 kg of sewage is estimated to contain, approximately, total solids, equal to :

(*a*) 0.5—1 kg (*b*) 2—5 kg
(*c*) 5—10 kg (*d*) 10—20 kg.

Q. 104. Imhoff cone is used to measure, in sewage :

(*a*) total solids (*b*) total organic solids
(*c*) total inorganic solids (*d*) settleable solids.

Q. 105. The pH of fresh sewage is usually :

(*a*) less than 7 (*b*) more than 7
(*c*) equal to 7 (*d*) equal to zero.

Q. 106. Acidicity of sewage is indicated by a pH value of :

(*a*) less than 7 (*b*) more than 7
(*c*) equal to 7 (*d*) equal to 14.

Q. 107. pH = 3, when compared to pH = 5, will be more acidic by :

(*a*) 2 times (*b*) 20 times
(*c*) 100 times (*d*) none of them.

Q. 108. pH = 4, when compared to pH = 7, will be more acidic by :

(*a*) 3 times (*b*) 300 times
(*c*) 1000 times (*d*) none of them.

Q. 109. Chlorine content in ordinary municipal sewage is :

(*a*) more than the usual tolerable limit
(*b*) less than the usual tolerable limit
(*c*) equals the usual tolerable limit
(*d*) very much high, necessitating its compulsory removal.

Q. 110. Methaemoglobinemia disease is caused in children, by :
(*a*) conversion of nitrites to nitrates
(*b*) conversion of nitrates to nitrites
(*c*) reaction between haemoglobin and carbon dioxide
(*d*) both due to (*a*) and (*c*).

Q. 111. The solubility of oxygen in sewage, when compared to its solubility in distilled water, is :
(*a*) 80% (*b*) 90%
(*c*) 99% (*d*) 99.9%

Q. 112. Pathogenic bacteria, enter wastewaters, primarily from :
(*a*) industrial wastes
(*b*) domestic wastes
(*c*) both industrial as well as domestic wastes
(*d*) infiltration in sewers from the surrounding soils.

Q. 113. Biodegradable organics enter wastewaters, from :
(*a*) industrial wastes
(*b*) domestic wastes
(*c*) both industrial and domestic wastes
(*d*) infiltration from the surrounding soils.

Q. 114. Soluble organics in domestic wastewaters, include :
(*a*) carbohydrates (*b*) proteins
(*c*) lipids (*d*) all of these.

Q. 115. Chemical Oxygen Demand (COD) of sewage is the :
(*a*) oxygen required to oxidise biologically active organic matter
(*b*) oxygen required to oxidise biologically inactive organic matter
(*c*) (*a*) and (*b*) both
(*d*) none of these.

Q. 116. Biochemical Oxygen Demand (BOD) of sewage is the :
(*a*) oxygen required to oxidise biologically active organic matter
(*b*) oxygen required to oxidise biologically inactive organic matter
(*c*) (*a*) and (*b*) both
(*d*) none of these.

Q. 117. Minimum D.O. prescribed for a river stream, to avoid fish kills, is :
(*a*) 2 ppm (*b*) 4 ppm
(*c*) 8 ppm (*d*) 10 ppm.

Q. 118. BOD_5 represents 5 days-biochemical oxygen demand at a temperature of :
(*a*) 0°C (*b*) 20°C
(*c*) 30°C (*d*) none of these.

Q. 119. Standard 5 day BOD at 20°C, when compared to ultimate BOD, is about :
(*a*) 58% (*b*) 68%
(*c*) 98% (*d*) none of these.

Q. 120. If the depletion of oxygen is found to be 2 mg/l after incubating 3 ml of sewage diluted to 300 ml, at 20°C for 5 days, then the BOD_5 of the sewage would be :

(*a*) 200 mg/l (*b*) 300 mg/l
(*c*) 600 mg/l (*d*) none of these.

Q. 121. While testing for COD of sewage, organic matter is oxidised by $K_2Cr_2O_7$ in the presence of :

(*a*) HCl (*b*) H_2SO_4
(*c*) HNO_3 (*d*) none of these.

Q. 122. The microbial utilisation of organics is a :

(*a*) first order reaction (*b*) second order reaction
(*c*) third order reaction. (*d*) none of these.

Q. 123. Between BOD and COD, the greater of the two, is

(*a*) BOD
(*b*) COD
(*c*) both are equal
(*d*) depends on sewage characteristics.

Q. 124. The average BOD_5 of domestic sewage is :

(*a*) 80 kg/day/person (*b*) 8 kg/day/person
(*c*) 0.8 kg/day/person (*d*) 0.08 kg/day/person.

Q. 125. The graphical representation of BOD (*Y*) exerted against time in days (*t*) is given by :

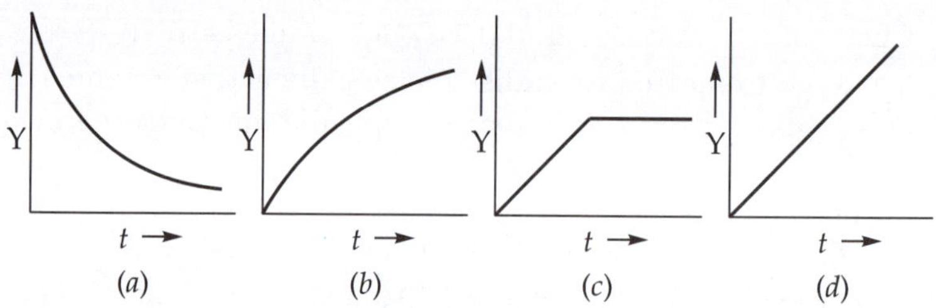

Fig. 23.1

Q. 126. The graph between the amount of organic matter left in sewage, and time elapsed in days, is :

(*a*) linear (*b*) parallel to time axis
(*c*) exponential (*d*) none of these.

Q. 127. If a 2% solution of sewage sample is incubated for 5 days at 20°C, and the dissolved oxygen depletion is 10 mg/l, then the BOD of the sewage would be :

(*a*) 50 mg/l (*b*) 200 mg/l
(*c*) 500 mg/l (*d*) 5000 mg/l.

Q. 128. The relative stability of a sewage sample, whose D.O. equals the total oxygen required to satisfy its BOD, is :

(*a*) zero (*b*) 1%
(*c*) 100% (*d*) infinity.

Chapter 8 : Disposing of the Sewage Effluents

Q. 129. Not withstanding any legal by laws, the sewage can generally be disposed of without any specific treatment, in a water body, if the dilution factor available, is :

(*a*) more than 100
(*b*) more than 300
(*c*) more than 500
(*d*) none of these.

Q. 130. Complete thorough treatment should generally be given to sewage, before its disposal in a stream, if the dilution factor available is :

(*a*) less than 350
(*b*) less than 250
(*c*) less than 150
(*d*) none of these.

Q. 131.* Dissolved exygen in streams is :

(*a*) max. at noon
(*b*) min. at noon
(*c*) max. at midnight
(*d*) same throughout the day.

Q. 132. Algae dies out, though fish life may survive, in a river zone, known as :

(*a*) zone of degradation
(*b*) zone of active decomposition
(*c*) zone of recovery
(*d*) none of these.

Q. 133. D.O. concentration may fall to zero, causing anaerobic conditions in a river reach, called :

(*a*) zone of degradation
(*b*) zone of active decomposition
(*c*) zone of recovery
(*d*) none of these.

Q. 134. If a sewer drain carrying a discharge of 2 cumecs, outfalls into a river carrying a discharge of 10 cumecs, and having DO equal to 8.4 mg/l, the resultant DO of the mix. will be equal to :

(*a*) 5 mg/l
(*b*) 7 mg/l
(*c*) 10.5 mg/l
(*d*) none of these.

Q. 135. As compared to fresh river water, sea water contains :

(*a*) 10% more oxygen
(*b*) 20% more oxygen
(*c*) 10% less oxygen
(*d*) 20% less oxygen.

Q. 136. Disposal of sewage causes formation of sludge banks in :

(*a*) river waters
(*b*) sea waters
(*c*) lake waters
(*d*) none of these.

Q. 137. The most common method of wastewater disposal is :

(*a*) evaporation
(*b*) dilution in surface water
(*c*) rapid infiltration
(*d*) application in irrigation.

Q. 138. The phenomenon by virtue of which a soil is clogged with sewage matter, is called

(*a*) sewage farming
(*b*) sewage sickness
(*c*) sewage bulking
(*d*) none of these.

Q. 139. Most of the bacteria in sewage are :

(*a*) anaerobic
(*b*) parasitic
(*c*) saprophytic
(*d*) pathogenic.

Q. 140. Disposal of sewage for sewage farming will be most favourable, where :
(*a*) rivers run with very low flows during summer
(*b*) climate is wet and rate of evaporation is low
(*c*) area is hilly
(*d*) all of these.

Q. 141. Sewage disposal by dilution is most preferred, where :
(*a*) the sewage arriving at outfall point is fresh and nonseptic
(*b*) there are strong forward currents
(*c*) discharge in water body is not low
(*d*) all of these.

Q. 142. The natural process, under which the flowing river water gets cleaned, is known as :
(*a*) oxidation (*b*) self-purification
(*c*) photosynthesis (*d*) none of these.

Q. 143. When the sewage is distributed over a level area surrounded by dikes, for sewage farming, the method is known as :
(*a*) surface irrigation (*b*) free flooding
(*c*) furrow irrigation (*d*) none of these.

Chapter 9 : Treatment of Sewage
\+
Chapter 10 : Process Design of a Complete Treatment Plant
\+
Chapter 11 : Treatment of Industrial Waste Waters

Q. 144. During preliminary treatment of sewage :
(*a*) oils and greases are removed by skimming tanks
(*b*) floating materials are removed by screening
(*c*) grit and sand are removed by grit chambers
(*d*) all of the above are correct.

Q. 145. Sewage treatment units are generally designed for :
(*a*) maximum flow only
(*b*) minimum flow only
(*c*) average flow only
(*d*) maximum and minimum flow both.

Q. 146. If the sewage contains greases and fatty oils, they are removed in :
(*a*) grit chambers (*b*) detritus tanks
(*c*) sedimentation tanks (*d*) skimming tanks
(*e*) aeration tanks (*f*) none of these.

Q. 147. The detention period adopted for grit chambers is of the order of :
(*a*) 1 minute (*b*) 5 minutes
(*c*) 2—4 hours (*d*) 12 hours.

Q. 148. The detention period adopted for sewage sedimentation tanks is of the order of :

(*a*) 1—2 hours (*b*) 4—8 hours
(*c*) 8—16 hours (*d*) 24—36 hours.

Q. 149. The detention period adopted for oxidation ponds is of the order of :

(*a*) 24—36 hours (*b*) 2—4 days
(*c*) 1—2 weeks (*d*) 2—6 weeks.

Q. 150. The detention period in a septic tank is of the order of :

(*a*) 2—6 hours (*b*) 4—8 hours
(*c*) 12—36 hours (*d*) 2—4 days.

Q. 151. Primary treatment of sewage is meant for :

(*a*) removal of larger suspended matter
(*b*) removal of fine suspended organic matter
(*c*) removal of dissolved organic matter
(*d*) removal of pathogenic bacteria.

Q. 152. The spacing of bars in coarse screens is generally more than :

(*a*) 6 mm or so (*b*) 10 mm or so
(*c*) 25 mm or so (*d*) 50 mm or so.

Q. 153. The settling velocity of a spherical body in still water is given by :

(*a*) Lacey's formula
(*c*) Darcy's law
(*c*) Hazen William's formula
(*d*) Stoke's law.

Q 154. The ratio of 'flowing through period' to 'detention period' in a sedimentation tank, is called :

(*a*) surface loading (*b*) displacement efficiency
(*c*) theoretical efficiency (*d*) settling velocity.

Q. 155. The BOD_5 of the effluents obtained from septic tanks, is of the order of :

(*a*) zero—5 mg/l (*b*) 10—20 mg/l
(*c*) 100—200 mg/l (*d*) 500—2000 mg/l.

Q. 156. The detention period in Imhoff tanks is of the order of :

(*a*) 30 min—1 hour (*b*) 1—2 hours
(*c*) 2—4 hours (*d*) 6—12 hours.

Q. 157. The short-circuiting occurring in a sedimentation tank is represented by :

(*a*) surface loading (*b*) displacement efficiency
(*c*) recirculation ratio (*d*) detention time.

Q. 158. Surface loading for a rectangular sedimentation tank of length L, width B, and height H, is given by :

(*a*) $\frac{Q}{B.H}$ (*b*) $\frac{Q}{B.L}$

(*c*) $\frac{BLH}{Q}$ (*d*) none of them.

(where Q is the design discharge through the tank)

Q. 159. The ratio of the design discharge to the surface area of a sedimentation tank is known as its :

(*a*) surface loading (*b*) overflow velocity
(*c*) overflow rate (*d*) all of these.

Q. 160. Normal values of surface loading for primary sewage sedimentation tanks, ranges between :

(*a*) 10—15 cum/m^2/day (*b*) 25—30 cum/m^2/day
(*c*) 40—50 cum/m^2/day (*d*) none the these.

Q. 161. Normal values of surface loading for secondary sewage sedimentation tanks, ranges between :

(*a*) 10—15 cum/m^2/day (*b*) 25—30 cum/m^2/day
(*c*) 40—50 cum/m^2/day (*d*) none the these.

Q. 162. A grit chamber, having higher detention period (about 3—4 times), is called a :

(*a*) grit basin (*b*) detritus tank
(*c*) vacuator (*d*) grit channel.

Q. 163. The sewage treatment units, which work on anaerobic decomposition of organic matter, are :

(*a*) oxidation ponds (*b*) septic tanks
(*c*) activated sludge plants (*d*) (*b*) and (*c*) both.

Q. 164. The sewage treatment units, which work on aerobic decomposition of organic matter, are :

(*a*) trickling filters (*b*) imhoff tanks
(*c*) sludge digestion tanks (*d*) none of these.

Q. 165. Activated sludge treatment plants are normally preferred for :

(*a*) towns and smaller cities (*b*) medium sized cities
(*c*) large sized cities (*d*) all of them.

Q. 166. Trickling filter plants are preferred for sewage treatment for :

(*a*) towns and smaller cities (*b*) medium sized cities
(*c*) large sized cities (*d*) (*a*) and (*b*) both.

Q. 167. To exclude the floating debris from sewage, screens are used and fixed :

(*a*) parallel to the direction of flow.
(*b*) perpendicular to the direction of flow
(*c*) at an angle 30°—60° to the direction of flow
(*d*) orientation is immaterial.

Q. 168. The grit chambers of a sewage treatment plant, normally need cleaning :

(*a*) every hour (*b*) every day
(*c*) every fortnight (*d*) every year.

Q. 169. For a grit channel, if the recommended flow velocity is 0.25 m/sec, and the detention period is 1 minute, then the length of the tank is :

(*a*) 15 m (*b*) 25 m
(*c*) 240 m (*d*) none of these.

Q. 170. In the design of grit chambers in sewage treatment :

(*a*) baffles are essential
(*b*) temperature control is an important factor
(*c*) the maximum flow velocity should be of the order of 0.3 m/sec
(*d*) the detention period should at least be 10 minutes.

Q. 171. The working conditions in Imhoff tanks are :

(*a*) aerobic only
(*b*) anaerobic only
(*c*) aerobic in lower compartment and anaerobic in upper compartment
(*d*) anaerobic in lower compartment and aerobic in upper compartment.

Q. 172. The secondary treatment of sewage is caused by :

(*a*) bacteria (*b*) algae
(*c*) coagulants (*d*) none of them.

Q. 173. Chlorine is sometimes used in sewage treatment :

(*a*) to aid flocculation
(*b*) to increase biological activity of bacteria
(*c*) to avoid bulking of activated sludge
(*d*) to help in grease separation.

Q. 174. The anaerobic secondary treatment of sewage is essentially a process of :

(*a*) dehydration (*b*) reduction
(*c*) oxidation (*d*) none of these.

Q. 175. The gas, which is evolved in a sludge digestion tank, is mainly composed of :

(*a*) nitrogen (*b*) ammonia
(*c*) hydrogen sulphide (*d*) methane.

Q. 176. The rate of sludge accumulation in a septic tank is generally of the order of :

(*a*) 10 litres/person/yr (*b*) 20 litres/person/day
(*c*) 30 litres/person/day (*d*) 100 litres/person/day.

Q. 177. The flowing through velocity in an Imhoff tank should not exceed :

(*a*) 0.1 m/min. (*b*) 0.3 m/min.
(*c*) 0.7 m/min. (*d*) 1.0 m/min.

Q. 178. Quantity of solids removed by screening depends on the :

(*a*) length of vertical bars (*b*) thickness of vertical bars
(*c*) screen opening size (*d*) none of these.

Q. 179. Oxygen-utilisation rate in biological reactors is a function of :

(*a*) wastewater characteristics only
(*b*) reactor characteristics only
(*c*) both (*a*) and (*b*)
(*d*) atmospheric characteristics.

Q. 180. Dissolved organic solids in wastewater treatment may be removed by :

(*a*) coagulation (*b*) hypochlorination
(*c*) reverse osmosis (*d*) none of them.

Q. 181. A reactor, in which the randomly packed solid medium provides the surface area for biofilm growth, is :

(*a*) an oxidation pond
(*b*) a bio-tower
(*c*) a trickling filter
(*d*) an anaerobic sewage lagoon.

Q. 182. The effluent of a septic tank is :

(*a*) fit for discharge into any open drain
(*b*) highly foul, containing large amount of suspended solids, needing sedimentation
(*c*) as good as that from a complete treatment plant
(*d*) none of the above.

Q. 183. In a shallow waste stabilisation pond, the sewage is treated by :

(*a*) aerobic bacteria only
(*b*) algae only
(*c*) dual action of aerobic bacteria and algae
(*d*) sedimentation.

Q. 184. Which of the following units work on the principle or anaerobic decomposition :

(*a*) sedimentation tanks (*b*) trickling filters
(*c*) sludge digestion tanks (*d*) activated sludge plant.

Q. 185. The maximum efficiency of BOD removal is achieved in :

(*a*) oxidation ditch (*b*) oxidation pond
(*c*) aerated lagoon (*d*) trickling filter.

Q. 186. A septic tank is :

(*a*) a settling tank (*b*) a digestion tank
(*c*) both (*a*) and (*b*) (*d*) none of these.

Q. 187. Composting and lagooning are the methods of :

(*a*) filtration (*b*) sedimentation
(*c*) sludge digestion (*d*) sewage disposal.

Q. 188. For satisfactory working of a sludge digestion unit, the pH range of digested sludge should be maintained in the range of :

(*a*) 3 to 5 (*b*) 6.5 to 8
(*c*) 8.5 to 10 (*d*) any of these.

Q. 189. The gas coming out from a sludge digestion tank is :

(*a*) methane only
(*b*) carbon dioxide only
(*c*) 70% methane and 30% carbon dioxide
(*d*) 30% methane and 70% carbon dioxide.

Q. 190. For the same solids content, if the quantity of sludge with moisture content of 97% is x litres, then the quantity of sludge with moisture content of 94% will be :

(*a*) $\frac{x}{2}$ litres (*b*) $\frac{x}{3}$ litres

(*c*) $2x$ litres (*d*) $3x$ litres.

Q. 191. A given sludge with 98% moisture is x times more bulky than with 95% moisture, where x is :

(*a*) 6 (*b*) 2.5

(*c*) 3 (*d*) none of these.

Q. 192. When there is no recirculation of treated sewage in high rate biological filteration of sewage, then the recirculation factor is :

(*a*) 1 (*b*) 0

(*c*) infinity (*d*) none of the above.

Q. 193.* When the recirculation ratio in a high rate trickling filter is unity, then the recirculation factor is :

(*a*) 1 (*b*) more than 1

(*c*) less than 1 (*d*) zero.

Q. 194. Activated sludge is the :

(*a*) aerated sludge in the aeration unit

(*b*) sludge settled in the humus tank

(*c*) sludge in the secondary tank after aeration and rich in microbial mass

(*d*) sludge in the secondary tank after aeration and rich in nutrients.

(*GATE, 1993*)

Q. 195. In a conventional activated sludge plant, the oxygen demand is highest near the :

(*a*) inlet end of the aeration tank

(*b*) outlet end of the aeration tank

(*c*) inlet end of the clarifier

(*d*) outlet end of the clarifier.

Q. 196. A conventional activated sludge plant involves a mixing regime, which is essentially of :

(*a*) plug flow type (*b*) completely mixed type

(*c*) both (*a*) and (*b*) (*d*) none of the above.

Q. 197. Lower F/M value in a conventional activated treatment plant will mean :

(*a*) lower BOD removal (*b*) higher BOD removal

(*c*) no effect on BOD removal (*d*) none of the above.

Q. 198. The aeration process adopted in an activated sludge plant, which ensures optimal utilisation of air, is known as :

(*a*) conventional process (*b*) step aeration

(*c*) activated aeration (*d*) tapered aeration.

Q. 199. The term sludge age is associated with :
(*a*) sedimentation (*b*) aeration
(*c*) sludge drying (*d*) none of these.

Q. 200. High COD to BOD ratio of an organic pollutant represents :
(*a*) high biodegradability of the pollutant
(*b*) low biodegradability of the pollutant
(*c*) presence of free oxygen for aerobic decomposition
(*d*) presence of toxic material in the pollutant. (*GATE, 1993*)

Chapter 12 : Excreta Disposal in Unsewered and Rural Areas

Q. 201. The minimum horizontal distance for locating a privy pit, from a well or a hand pump, is :
(*a*) 10 m (*b*) 30 m
(*c*) 100 m (*d*) 500 m.

Q. 202. The minimum vertical distance between the bottom of the privy pit and the ground watertable is :
(*a*) 0.3 m (*b*) 1 m
(*c*) 3 m (*d*) none of them.

Q. 203. The most economical and hygienic rural privy is :
(*a*) an aqua-privy (*b*) a privy pit
(*c*) a cess pool (*d*) a soak pit.

Chapter 13 : Sewage Collection from Houses and Buildings

Q. 204. An intercepting trap is provided at the junction of :
(*a*) a house sewer and a municipal sewer
(*b*) an unfoul house drain and a foul house drain
(*c*) any two house drains
(*d*) none of these.

Q. 205. A gully trap is provided at the junction of :
(*a*) a house sewer and a municipal sewer
(*b*) an unfoul roof or room drain and a foul bath or a kitchen drain
(*c*) any two house drains
(*d*) none of these.

Q. 206. A nahni trap is provided :
(*a*) at the head of each house drain
(*b*) at the outfall end of each house drain
(*c*) at the junction of two house drains
(*d*) none of these.

Q. 207. The waste pipe in house plumbing, which carries the discharge from sanitary fittings like bath rooms, etc. ; is called :
(*a*) waste pipe (*b*) soil pipe
(*c*) anti-siphonage pipe (*d*) vent pipe.

Q. 208. A pipe which is installed in house drainage, to preserve the water seal of traps, is called :

(*a*) vent pipe (*b*) waste pipe
(*c*) soil pipe (*d*) antisiphonage pipe.

Q. 209. Two pipe system of providing building drainage consists of :

(*a*) one soil pipe + one waste pipe + one vent pipe + one sullage pipe
(*b*) one soil pipe + one waste pipe + two vent pipes
(*c*) two soil pipes + two waste pipes
(*d*) two soil pipes only.

Q. 210. A pipe installed in house drainage for ventilation purposes is called a :

(*a*) soil pipe (*b*) antisiphonage pipe
(*c*) vent pipe (*d*) none of these.

Q. 211. The pipe in buildings, through which human excreta flows, is called :

(*a*) soil pipe (*b*) waste pipe
(*c*) vent pipe (*d*) none of these.

Q. 212. Ventilation of house drainage is required to :

(*a*) dilute the foul air in the drain
(*b*) reduce the obnoxious effects of foul air
(*c*) relieve the pressure of foul gases
(*d*) all of the above.

Q. 213. In one pipe system of plumbing, wastewater is carried down in soil pipe from :

(*a*) water-closets (*b*) kitchens and bath rooms
(*c*) wash basins and sinks (*d*) all of the above.

Q. 214. In one pipe system of building drainage, we provide :

(*a*) one soil pipe
(*b*) one waste pipe
(*c*) one soil pipe and one vent pipe
(*d*) one waste pipe and one vent pipe.

Q. 215. In single stack drainage system in houses, we provide :

(*a*) one soil pipe only
(*b*) one soil pipe and one vent pipe
(*c*) one sullage pipe only
(*d*) one soil pipe and one sullage pipe.

Q. 216. In house plumbing system, the leakage of different pipes is tested by :

(*a*) smoke test (*b*) air test
(*c*) water test (*d*) all of these.

Q. 217. Point out the incorrect statement, in connection with building drainage :

(*a*) a floor trap prevents the evolution of foul gases
(*b*) an antisiphonage pipe provides the basic water seal of W.C.'s and bath rooms.

(*c*) a vent pipe removes foul gases into the atmosphere above the roof level

(*d*) soil and waste pipes carry down the sewage, produced in the house fixtures.

Chapter 14 : Disposal of Municipal and Industrial Solid Wastes

Q. 218. The anaerobic method of mechanical composting, as practised in India, is called the :

(*a*) Indore method (*b*) Mangalore method

(*c*) Bangalore method (*d*) none of these.

Q. 219. The method of refuse disposal, involving burial in trenches, is called :

(*a*) incineration (*b*) pulverisation

(*c*) composting (*d*) none of these.

Q. 220. The aerobic method of mechanical composting, as practised in India, is called the :

(*a*) Bangalore method (*b*) Nagpur method

(*c*) Bhopal method (*d*) Indore method.

Q. 221. Sanitary landfills may cause troubles during :

(*a*) peak summers (*b*) peak winters

(*c*) peak monsoons (*d*) none of these.

Q. 222. Leachate is a coloured liquid, that comes out of :

(*a*) septic tanks (*b*) sanitary landfills

(*c*) compost plants (*d*) aerated lagoons.

Q. 223. The quantity of refuse produced in an average Indian city or a town is of the order of :

(*a*) $\frac{1}{4} - \frac{1}{5}$ t/c day (*b*) 2 – 5 t/c day

(*c*) 4 – 5 t/c day (*d*) none of these.

Chapter 15 : Disposal of Environmentally Hazardous Biomedical, Radioactive, and E-Wastes

Q. 224. Biomedical wastes consist of :

(*a*) solid wastes (*b*) liquid wastes

(*c*) both (*a*) and (*b*) (*d*) none of these.

Q. 225. The approximate quantity of biomedical waste generated in Indian hospitals per patient per day approximates to :

(*a*) 0.5 kg (*b*) 0.75

(*c*) 1.45 kg (*d*) 4.0 kg.

Q. 226. Biomedical waste (*Management and Handling Rules*) were enacted in India under Environment Protection Act, 1986, in the year :

(*a*) 1986 (*b*) 1989

(*c*) 1998 (*d*) 2000.

Q. 227. Which of the following is not a provision of the Biomedical Waste (*Management and Handling Rules*) :

(a) All the health care institutes are reqd. to make arrangements to ensure that their produced wastes are handled and managed without causing any adverse health effects

(b) It is necessary to segregate the various types of biomedical wastes by collecting the wastes into yellow, red, and black coloured containers.

(c) Sharp wastes should be handled with extra-ordinary care and placed in puncture-resistance containers or should be mutilated at the point of generation by some needle destroyer.

(d) The offenders cannot be punished.

Q. 228. For proper disposal of biomedical wastes, all infected, soiled, pathological, human and sharp wastes should be marked as cytotoxic wastes, by using the symbol :

(a) 

(b) 

(c) 

(d) none of them.

Q. 229. Which of the following methods is generally not adopted for safe disposal of biomedical wastes ?

(a) Shredding after disinfection (b) Hydroclaving

(c) Incineration (d) Landfilling.

Q. 230. Incineration can be adopted for disposal of biomedical wastes of the type :

(a) solid wastes (b) liquid wastes

(c) both (a) and (b) (d) none of them.

Q. 231. Which of the following statements is incorrect in relation to auto-claving for disposal of biomedical wastes ?

(a) Auto-claving is a high heat thermal process

(b) Auto-claving is a low heat thermal process

(c) Auto-claving is meant for disposal of solid wastes only

(d) After auto-claving, the wastes are shredded and disposal off suitably.

Q. 232. Liquid biomedical wastes should be disposed off :

(a) by disinfection and discharge into sewers

(b) by incineration into liquid waste incinerators

(c) both (a) and (b)

(d) none of the above.

Q. 233. Throughout the world, rotary kiln incinerators are being widely used for burning and disposing off biomedical wastes of the type :

(a) solid wastes (b) liquid wastes

(c) glass wastes (d) none of the above.

Q. 234. A conventional rotary kiln incinerator is provided with an induced draft fan to :

(a) maintain a negative pressure in the entire system

(*b*) ensure that all leakages from the kiln occur into the system rather than out of it

(*c*) both (*a*) and (*b*)

(*d*) none of the above.

Q. 235. Which of the following statements in relation to a conventional rotary kiln incinerator is incorrect ?

(*a*) An induced draft fan is provided into the system, and sized to maintain a negative pressure in the entire system

(*b*) The horizontal kiln is mounted on a slight incline of 2 to 4% slope, with higher end at the feed end of the kiln.

(*c*) The kiln designed for incinerating liquid wastes may not have any longitudinal slope or may have a negative slope

(*d*) None of the above.

Q. 236. Which of the following statements is incorrect ?

(*a*) The cell damaging capacity of different types of radioactive radiations is compared by relative biological effectiveness (RBE).

(*b*) The RBE for alpha particles is 20, while electrons have RBE of 1.

(*c*) (*b*) above means that the dense track of ions formed by alpha particles causes about 20 times as much damage to a biological body as is caused by a thin ion trail generated by electrons.

(*d*) None of the above.

Q. 237. The commonly used unit of absorbed radiation is defined as that dose of ionizing radiation that imparts 1 Joule of energy to 1 kg of absorbing material. This unit is known as :

(*a*) rad (*b*) gray (Gr)

(*c*) rem (*d*) silvert.

Q. 238. The biological damage produced in a human body by different types of radiations is obtained by multiplying the radiation dose in rad with RBE of the radiation type. This damage is then measured in units of :

(*a*) rem (*b*) silvert

(*c*) gray (*d*) none of them.

Q. 239. The relation between the old and new units of absorbed radiation is :

(*a*) 1 rad = $\frac{1}{100}$ gray (*b*) 1 rad = 100 gray

(*c*) 1 rad = 1 gray (*d*) none of them.

Q. 240. The relation between the old and new units of biological damage produced in a human body by radiations is given as :

(*a*) 1 rem = $\frac{1}{100}$ silvert (*b*) 1 rem = 100 silvert

(*c*) 1 rem = 1 silvert (*d*) none of these.

Q. 241. Radiation sickness may be produced by radiation dose of about :

(*a*) 100 rem (*b*) 50 rem

(*c*) 20 rem (*d*) 5 rem.

Q. 242. The relation between rem and rad is given as :

(*a*) rem = rad × RBE (*b*) rem = $\frac{\text{rad}}{\text{RBE}}$

(*c*) rem = $\frac{\text{RBE}}{\text{rad}}$ (*d*) none of the above.

Q. 243. An ordinary chest X-ray subjects human body to radiation of 5 rad, and if RBE of X-rays is 1, then the biological damage in radiation terms of rem will be equal to :

(*a*) 5 (*b*) $\frac{1}{5}$

(*c*) $\log_{10} 5$ (*d*) $\log_{10} \frac{1}{5}$.

Q. 244. Exposure of human body to radiation of x rems may cause severe changes in blood cells and hemorrhage, where x may be about

(*a*) 50 (*b*) 100

(*c*) 300 (*d*) 1000.

Q. 245. The no. of rems of radiation exposure of a human being, which may cause hair loss and loss of body's ability to fight infections, usually resulting in death, is about :

(*a*) 50 rem (*b*) 100 rem

(*c*) 300 rem (*d*) 600 rem.

Q. 246. The aquatic animals which are most badly affected by radioactive radiations are :

(*a*) fish (*b*) mollusks

(*c*) algae (*d*) none of them.

Q. 247. The most common impacts of radioactive radiations, which appear late in life are :

(*a*) shortened life span and increased probability of developing cancers and cataracts

(*b*) damage to genes, causing dangerous mutagenic effects leading to birth of abnormal offsprings

(*c*) both (*a*) and (*b*)

(*d*) none of them.

Q. 248. Which of the following radio isotope is capable to contaminate the milk of cows and other animals, due to their exposure to the radiation caused by that isotope ?

(*a*) Strontium-90 (*b*) Cesium-137

(*c*) Iodine-131 (*d*) none of them.

Q. 249. Radioactive radiations may cause mutagenic effects in :

(*a*) humans (*b*) animals

(*c*) both (*a*) and (*b*) (*d*) none of them.

Q. 250. Trees are very badly affected by longer exposure to :

(*a*) α-rays (*b*) β-rays

(*c*) γ-rays (*d*) none of them.

Q. 251. Which of the following is a natural source of radioactivity in the environment ?

(*a*) cosmic rays
(*b*) radioactive rocks and minerals
(*c*) radio nuclides present within the bodies of organisms
(*d*) all of the above.

Q. 252. ...diation sickness may be caused by as low a radiation dose as ...at :

(...) 25 rem (*b*) 50 rem
(*c*) 100 rem (*d*) none of them.

Q. 253. Which among the following is not a man-made source of radioactivity in the environment ?

(*a*) nuclear power plants (*b*) nuclear explosions
(*c*) cosmic rays (*d*) none of the above.

Q. 254. High level radioactive wastes are those which have radio-activities measured in terms of :

(*a*) curies per litre (*b*) millicuries per litre
(*c*) micro-curies per litre (*d*) none of the above.

Q. 255. Low level radioactive wastes are those which have radio-activities measured in terms of :

(*a*) curies per litre (*b*) millicuries per litre
(*c*) micro-curies per litre (*d*) none of the above.

Q. 256. High level radioactive wastes generated in a nuclear power plant in the form of spent fuel remain a great hazard for several thousands of years, because it contains Plutonium-239 which has a very long half life of about :

(*a*) 5,000 years (*b*) 10,000 years
(*c*) 14,000 years (*d*) 24,000 years.

Q. 257. Which of the following is not a stage in disposal of High Level liquid wastes ?

(*a*) Storage for about 5 years in leak proof underground depositories.
(*b*) Solidification by using advanced technological methods.
(*c*) Burial of solidified wastes into permanent deep disposal sites.
(*d*) None of the above.

Q. 258. The quantum of high level wastes, annually generated by the world's 442 nuclear power plants and reactors, is about :

(*a*) 4000 t (*b*) 8000 t
(*c*) 12000 t (*d*) 20000 t.

Q. 259. The country which is not having any permanent site for disposal of high level wastes is :

(*a*) USA (*b*) Russia
(*c*) India (*d*) none of them.

Chapter 16 : Ventilation of Buildings for Controlling Indoor Air Pollution

Q. 260. Air changes for ventilation of occupied rooms and halls in a building is mainly required to :
(*a*) keep carbon dioxide content under control
(*b*) dissipate body heat
(*c*) increase oxygen content
(*d*) reduce body odours.

Q. 261. Fresh air is required to be supplied to an occupied hall at the rate of :
(*a*) 0—10 cum/c/hr (*b*) 15—30 cum/c/hr
(*c*) 50—100 cum/c/hr (*d*) none of these.

Q. 262. Number of air changes required per hour to ensure good ventilation of kitchens, is of the order of :
(*a*) 10—40 (*b*) 6—10
(*c*) 2—5 (*d*) 3.

Q. 263. Number of air changes required per hour to ensure sufficient ventilation of offices and libraries, is of the order of :
(*a*) 10—40 (*b*) 6—10
(*c*) 2—5 (*d*) 3.

Q. 264. Number of air changes required per hour to ensure sufficient ventilation of cinema halls, is of the order of :
(*a*) 10—40 (*b*) 6—10
(*c*) 2—5 (*d*) 3.

Q. 265. The ventilation of kitchens by exhaust fans, as adopted in India, is a method of mechanical ventilation, called :
(*a*) the plenum system (*b*) the vacuum system
(*c*) the A.C. system (*d*) none of these.

Q. 266. The ventilation of public halls, by means of air-coolers and exhaust fans, is known as :
(*a*) the plenum system (*b*) the vacuum system
(*c*) the A.C. system (*d*) none of these.

Q. 267. The most wholesome ventilation of buildings is provided by :
(*a*) natural ventilation (*b*) plenum system of ventilation
(*c*) A.C. system of ventilation (*d*) vacuum system of ventilation.

Q. 268. A cooling room-air-conditioner, modifies the moisture content of the air, before letting it out, by :
(*a*) reducing humidity
(*b*) increasing humidity
(*c*) optimally reducing as well as increasing humidity
(*d*) none of these, as it does not modify humidity.

Q. 269. Point out the incorrect statement :
(*a*) a summer A.C. uses a de-humidifier
(*b*) a winter A.C. uses a humidifier

(*c*) recirculation of room air takes place in a summer A.C. as well as in a winter A.C.

(*d*) all the above are correct.

Chapter 17 : The Environment, Its Pollution, and Present Status

Q. 270. The zone on Earth, where alone life exists, is called :

(*a*) hydrosphere (*b*) lithosphere

(*c*) biosphere (*d*) troposphere.

Q. 271. Point out the incorrect statement :

(*a*) new species are born due to disturbances in the ecological balance of biosphere

(*b*) carnivorous animals are non-veg animals, like lions

(*c*) CO_2 gas produces heating effects on earth

(*d*) Bangalore is the most polluted Indian city, as far as air pollution is concerned.

Q. 272. The polluting gas, which is primarily responsible for causing the green house effect and global warming, is :

(*a*) sulphur dioxide (*b*) hydrogen sulphide

(*c*) carbon dioxide (*d*) none of them.

Q. 273. According to the WHO's reports of 1992, the most polluted city in the world, is :

(*a*) Delhi (*b*) Bangkok

(*c*) Newyork (*d*) Tokyo.

Chapter 18 : Engineering of Air Pollution, Its Control and Monitoring

Q. 274. The average consumption of air by an adult human being is of the order of :

(*a*) 5 kg (*b*) 12 kg

(*c*) 25 kg (*d*) none of them.

Q. 275. Air pollution causes harmful effects on :

(*a*) biotic world (*b*) abiotic world

(*c*) both (*a*) and (*b*) (*d*) none of these.

Q. 276. Air pollution may be caused by :

(*a*) human activities (*b*) natural disastrous events

(*c*) both (*a*) and (*b*) (*d*) none of them.

Q. 277. Which of the following is not an anthropogenic cause of air pollution ?

(*a*) Burning of fossil fuels

(*b*) Burning of firewood

(*c*) Agricultural activities

(*d*) Burning of forests due to lightening.

Q. 278. 'London smog' was primarily caused by :

(*a*) burning of coal (*b*) burning of oil

(*c*) eruption of a volcano (*d*) none of the above.

Q. 279. A major disaster, known as 'London smog', occurred in the British city of London in the year :

(*a*) 1942 (*b*) 1952
(*c*) 1962 (*d*) 1972.

Q. 280. Which of the following statements is incorrect ?

(*a*) Air pollution causes local, regional, as well as global adverse impacts on the environment
(*b*) Air pollution has been in existence at some places even in the ancient times
(*c*) Air pollution does not cause adoptations and environmental imbalances in biological species
(*d*) None of the above.

Q. 281. The factor which is primarily responsible for causing air pollution in modern days is :

(*a*) industries and automobiles (*b*) dust storms
(*c*) forest fires (*d*) none of them.

Q. 282. About 90% of world's air pollution is caused by :

(*a*) primary air pollutants (*b*) secondary air pollutants
(*c*) volcanism (*d*) none of them.

Q. 283. The primary air pollutant which is formed due to incomplete combustion of organic matter is :

(*a*) methane (*b*) sulphur dioxide
(*c*) ozone (*d*) carbon monoxide.

Q. 284. Among the following, the only secondary pollutant is :

(*a*) sulphur dioxide (*b*) hydrocarbons
(*c*) ozone (*d*) sulphur tetraoxide.

Q. 285. Which among the following is not a secondary air pollutant ?

(*a*) sulphuric acid (*b*) formaldehyde
(*c*) peroxyacyl nitrate (*d*) none of the above.

Q. 286. Which among the following is not a primary air pollutant ?

(*a*) oxides of nitrogen
(*b*) volatile organic compounds like hydrocarbons
(*c*) suspended particulate matter
(*d*) PAN.

Q. 287. Which among the following statements is incorrect ?

(*a*) Secondary air pollutants are formed from a simple chemical reaction between two primary pollutants
(*b*) The sun light plays an important role in the formation of secondary pollutants
(*c*) The photochemical reaction is responsible for the formation of secondary pollutants
(*d*) No secondary pollutant was involved in 'London Smog' that occurred in 1952 killing more than 4000 people.

Q. 288. In 1929, several deaths occurred in a fire in a clinic in Cleveland (USA) due to burning of X-ray films, due to evolution of :

(*a*) sulphur dioxide (*b*) sulphur trioxide
(*c*) nitrogen dioxide (*d*) PAN.

Q. 289. More than 4000 people died in December 1952 in 'London smog', which was caused by :

(*a*) fog in the city
(*b*) evolution of smoke containing sulphur dioxide in the city
(*c*) arrival of anticyclone in the city
(*d*) all of the above.

Q. 290. Hairs of human nose can remove all particles of size greater than :

(*a*) 1 mm (*b*) 1 micron
(*c*) 10 micron (*d*) 100 micron.

Q. 291. The chemical compounds which are responsible for production of photochemical smog are :

(*a*) hydrocarbons (*b*) oxides of nitrogen
(*c*) both (*a*) and (*b*) (*d*) none of them.

Q. 292. The poisonous gas responsible for causing the catastrophic Bhopal gas tragedy in 1984 in India was :

(*a*) sulphur dioxide
(*b*) methyl isocyanate
(*c*) carbon monoxide and methane both
(*d*) laughing gas.

Q. 293. Which of the following is not a major constituent of a photochemical smog ?

(*a*) PAN (*b*) CO
(*c*) HC (*d*) O_3.

Q. 294. The major photochemical oxidant is :

(*a*) hydrogen peroxide (*b*) ozone
(*c*) nitrogen oxides (*d*) peroxyl acetyl nitrate (PAN).

Q. 295. The secondary pollutant among the following is :

(*a*) sulphur dioxides (*b*) methane
(*c*) carbon dioxide (*d*) peroxyl acetyl nitrate.

Q. 296. The particles in cigarette smoke are usually smaller than :

(*a*) 1 micron (*b*) 10 micron
(*c*) 25 micron (*d*) 100 micron.

Q. 297. The most hazardous gaseous air pollutant for human health is :

(*a*) carbon dioxide (*b*) oxygen
(*c*) nitrogen (*d*) sulphur dioxide.

Q. 298. The maximum permissible average annual specified standard of SO_2 for residential areas in the New Revised National Ambient Air Quality Standards in India is :

(*a*) 30 $\mu g/m^3$ (*b*) 50 $\mu g/m^3$
(*c*) 60 $\mu g/m^3$ (*d*) 120 $\mu g/m^3$.

Q. 299. During temperature inversion in atmosphere, air pollutants tend to :
(*a*) accumulate above inversion layer
(*b*) accumulate below inversion layer
(*c*) disperse laterally
(*d*) disperse vertically. (*GATE, 1993*)

Q. 300. The primary air pollutant, which is formed due to incomplete combustion of organic matter, is :
(*a*) methane (*b*) sulphur dioxide
(*c*) ozone (*d*) carbon monoxide.

Q. 301. Electrostatic precipitators remove :
(*a*) sulphur dioxide (*b*) particulate matter
(*c*) both (*a*) and (*b*) (*d*) none of them.

Q. 302. Among the following, the only secondary pollutant, is :
(*a*) sulphur dioxide (*b*) ozone
(*c*) hydrocarbons (*d*) sulphur tetroxide.

Q. 303. The percentage of particles eminating from cigarette smoking, which get deposited in lungs, is about :
(*a*) 10% (*b*) 25%
(*c*) 65% (*d*) 100%.

Q. 304. The Respiratory Suspended Particulate Matter (RSPM) concentration includes all particles in an air mass of size up to
(*a*) 10μ (*b*) 25μ
(*c*) 50μ (*d*) none of them.

Q. 305. The RSPM is also designated as
(*a*) PM 2.5 (*b*) PM 10
(*c*) PM 25 (*d*) none of them.

Q. 306. The monitoring of the particles of size lesser than 10 micron (PM 10) or even lesser than 2.5 micron (PM 2.5) is a better indicator of air pollution, because :
(*a*) such smaller particles tend to be deposited in nasopharyngeal region, causing serious health problems
(*b*) such smaller particles are mostly deposited in trachea bronchial and pulmonary region causing bronchitis and other lung diseases
(*c*) both (*a*) and (*b*)
(*d*) none of them.

Q. 307. The air pollutant which causes sever damage to plants, even at much lower concentrations than what may be harmful to human health, is :
(*a*) fluorine (*b*) ozone
(*c*) PAN (*d*) none of them.

Q. 308. The air pollutant which does not cause any major adverse impacts on the cattle is :
(*a*) fluorine (*b*) arsenic
(*c*) lead (*d*) none of them.

Q. 309. The max. permissible average annual standard for NO_2 for residential, industrial and other areas under the new revised NAAQS in India is :

(*a*) 30 μg/m^3 (*b*) 40 μg/m^3
(*c*) 60 μg/m^3 (*d*) 120 μg/m^3.

Q. 310. Sulphur dioxide exposures to quantities higher than permissible in ambient air may cause :

(*a*) breathing and respiratory illnesses
(*b*) hay fever
(*c*) typhoid
(*d*) malaria.

Q. 311. High lead exposures in humans through inhalations or food may cause :

(*a*) mental retardation
(*b*) disorder of central nervous system
(*c*) high blood pressure
(*d*) all of the above.

Q. 312. Higher exposures of humans to nitrogen oxides may cause :

(*a*) bronchitis (*b*) pneumonia
(*c*) allergies and viral attacks (*d*) all of the above.

Q. 313. Carbon monoxide is hazardous to health, because :

(*a*) it causes loss of sense of smell
(*b*) it is carcinogenic in nature
(*c*) it reduces oxygen carrying capacity of blood
(*d*) it may cause conjunctivitis.

Q. 314. Longer exposure to NO_2 even in small concentrations may cause diseases pertaining to :

(*a*) lever (*b*) lung
(*c*) kidneys (*d*) heart.

Q. 315. Heavy loading of pollen grains in air may cause :

(*a*) anaemia (*b*) typhoid
(*c*) hayfever (*d*) influenza.

Q. 316. The permissible TSPM (annual average) standard for residential areas, under the Pre-revised National Ambient Air Quality Standards in India, is :

(*a*) 60 μg/m^3 (*b*) 140 μg/m^3
(*c*) 200 μg/m^3 (*d*) 360 μg/m^3.

Q. 317. The permissible PM 2.5 (annual average) ambient air quality standard for residential, rural and industrial areas in India is :

(*a*) 40 μg/m^3 (*b*) 60 μg/m^3
(*c*) 80 μg/m^3 (*d*) 100 μg/m^3.

Q. 318. The permissible annual average standard of PM 10 (RSPM) ambient air quality standard for residential areas in India is :

(*a*) 60 μg/m^3 (*b*) 80 μg/m^3
(*c*) 90 μg/m^3 (*d*) 100 μg/m^3.

Q. 319. The permissible RSPM (annual average) ambient air quality standard for industrial areas under the Pre-revised NAAQS of India is :

(*a*) 60 μg/m³ (*b*) 120 μg/m³
(*c*) 140 μg/m³ (*d*) 360 μg/m³.

Q. 320. The permissible SO_2 (annual average) ambient air quality standard for residential and industrial areas in India is :

(*a*) 30 μg/m³ (*b*) 50 μg/m³
(*c*) 60 μg/m³ (*d*) 80 μg/m³.

Q. 321. The permissible NO_2 (annual average) ambient air quality standard for residential areas under the old pre-revised NAAQ Standards in India is :

(*a*) 50 μg/m³ (*b*) 60 μg/m³
(*c*) 80 μg/m³ (*d*) 120 μg/m³.

Q. 322. The most significant primary gaseous pollutant, found in vehicular emissions, is :

(*a*) CO (*b*) CO_2
(*c*) SO_2 (*d*) O_3.

Q. 323. The device, which can be used to control gaseous as well as particulate pollutants in the industrial emissions, is known as :

(*a*) cyclone (*b*) spray tower
(*c*) dynamic precipitator (*d*) fabric filter.

Chapter 19 : Global Environmental Issues

Q. 324. The region most affected by acid rains, is :

(*a*) India and China (*b*) Europe and North east America
(*c*) Africa (*d*) South America.

Q. 325. The acid rains are more prominent in :

(*a*) temperate regions (*b*) tropical regions
(*c*) arid regions (*d*) equally prominent in all regions.

Q. 326. A rainfall is generally classified as acidic, if its pH is less than equal to :

(*a*) 5.6 (*b*) 6.5
(*c*) 7 (*d*) 7.5.

Q. 327. An ordinary rain, will generally be :

(*a*) slightly acidic (*b*) slightly alkaline
(*c*) neither acidic nor alkaline.

Q. 328. The important gaseous pollutants, contributing to acid rains, are :

(*a*) SO_2 and NO_x (*b*) CO_2 and H_2S
(*c*) NO_x and O_3 (*d*) none of these.

Q. 329. The most significant environmental issue, threatening human health, is :

(*a*) global warming (*b*) acid rain
(*c*) ozone hole (*d*) none of these.

Q. 330. Man-made emission of chloro fluorocarbons (CFCs), is likely to deplete the ozone layer in the troposphere, leading to increase in :

(*a*) radioactivity (*b*) heat
(*c*) sound (*d*) ultra violet radiations

Q. 331. Two third of acid rains are caused by sulphur dioxide gas ($SO_2 \uparrow$) which is chiefly produced by the burning of :

(*a*) coals (*b*) oils
(*c*) both (*a*) and (*b*) (*d*) none of the above.

Q. 332. The fossil fuels responsible for causing acid rains are used in :

(*a*) industries (*b*) domestic cooking and heating
(*c*) automobiles (*d*) all of the above.

Q. 333. One third of acid rains are caused by nitrogen oxides (NO_x) which are chiefly produced by :

(*a*) auto and industrial emission
(*b*) nitrogenous fertilizers
(*c*) forests and grasslands on fire
(*d*) all of the above.

Q. 334. How many times more acidic a rain having pH = 5.6 will be in comparison to a neutral rain having pH = 7 ?

(*a*) $\frac{7}{5.6} = 1.25$ (*b*) $\frac{7^2}{5.6^2} = 1.56$
(*c*) $\frac{10^7}{10^{5.6}} = 20$ (*d*) none of them.

Q. 335. Which one of the following statements is incorrect ?

(*a*) To neutralise the acidic nature of lake waters, lime is sprayed on lake waters.
(*b*) The damage caused to trees by acidic rains is countered by fertilising the damaged trees with calcium, potassium, magnesium, etc.
(*c*) Electro-static precipitators are installed in thermal power plants to check emission of SO_2 and NO_x gases, to reduce acid rains.
(*d*) None of the above.

Q. 336. Which of the following statements is incorrect in relation to checking and controlling acid rains ?

(*a*) Expensive scrubbers are installed in industries and thermal power stations.
(*b*) Catalytic convertors are installed in cars.
(*c*) Use of petrol cars is promoted by discouraging the use of diesel cars.
(*d*) None of the above.

Q. 337. The ozone depletion in the stratosphere (outer atmosphere) is chiefly caused by the release of :

(*a*) chlorofluorocarbons (CFCs) (*b*) carbon dioxide (CO_2)
(*c*) methane (CH_4) (*d*) none of them.

Q. 338. The chlorofluorocarbons (CFCs) are released into the environment by humans, as they are largely used :

(*a*) in refrigeration and air-conditioning
(*b*) as cleaning solvents in factories
(*c*) in aerosol sprays
(*d*) as blowing agents to create certain kinds of plastic foams
(*e*) all of the above.

Q. 339. Depletion of ozone layer in the outer atmosphere is likely to increase the incidence of :

(*a*) skin cancer (*b*) lung cancer
(*c*) bronchitis (*d*) none of them.

Q. 340. Pollution by depletion of ozone layer in stratosphere, threatening the environment, is caused due to reaction of ozone with :

(*a*) carbon monoxide (*b*) chlorine
(*c*) sulphur dioxide (*d*) nitrous oxide.

Q. 341. The most significant environmental issue, causing serious diseases in humans, is :

(*a*) acid rain (*b*) ozone hole
(*c*) global warming (*d*) none of these.

Q. 342. The chances of development of an ozone hole are more at :

(*a*) arctic and antarctic regions
(*b*) equatorial regions
(*c*) regions of mid latitudes
(*d*) none of the above.

Q. 343. The ozone hole appears in Antarctica during

(*a*) early winter (*b*) peak summer
(*b*) late winter (*c*) autumn.

Q. 344. Which of the following is crucial for the formation of anthropogenic (man-made) ozone hole ?

(*a*) chlorides and bromides in the stratosphere
(*b*) solar radiation and thin ice crystals in the stratosphere
(*c*) (*a*) and (*b*) both
(*d*) none of the above.

Q. 345. Harmful U.V. radiations eminating from the sun are prevented from reaching the Earth by the presence of ozone in the :

(*a*) thermosphere (*b*) mesosphere
(*c*) stratosphere (*d*) troposphere.

Q. 346. The ozone layer thickness is measured in :

(*a*) Decibels (dB) (*b*) Dobson Units (DU)
(*c*) Becquerel (Bq) (*d*) None of them.

Q. 347. One Doboson Unit (DU) is equal to :

(*a*) 1 mm (*b*) 0.1 mm
(*c*) 0.01 mm (*d*) 0.001 mm.

Q. 348. The normal average thickness of stratospheric ozone across the globe is :

(*a*) 150 DU (*b*) 350 DU
(*c*) 500 DU (*d*) none of them.

Q. 349. The possible substitutes of CFCs are :

(*a*) HCFs (*b*) ODSs
(*c*) Halons (*d*) all of the above.

Q. 350. The thickness of stratospheric ozone, particularly above the polar regions, have been noticed to have fallen to as low a value as :

(*a*) 250 DU (*b*) 150 DU
(*c*) 100 DU (*d*) none of them.

Q. 351. The depletion of stratospheric ozone thickness to as low a value as about 150 DU (1.5 mm) as against the usual value of 350 DU (3.5 mm) to 250 DU (2.5 mm), creates ozone hole, the area of which in the southern hemisphere has been observed to be maximum in Sept. 2000 and Sept. 2006, equalling about :

(*a*) 29 million sq. km (*b*) 19 million sq. km
(*c*) 9 million sq. km (*d*) none of them.

Q. 352. In order to control depletion of ozone in the stratosphere, it is necessary to reduce CFC's and other such substances which release chlorine atoms, which, inturn, react with ozone to convert it into ClO and O_2, thereby reducing ozone (O_3). To achieve this objective, an international treaty was signed in 1987, imposing bans on the use of substances which release CFC's. This treaty is known as :

(*a*) Geneva Convention (*b*) London Amendment
(*c*) Kyota Protocol (*d*) Montreal Protocol.

Q. 353. The ozone strongly absorbs ultraviolet light in the region having wave length in the range of :

(*a*) 360–460 nm (*b*) 220–330 nm
(*c*) 135–200 nm (*d*) none of the above.

Q. 354. High altitude aircrafts lead to :

(*a*) destruction of oxygen (*b*) destruction of ozone
(*c*) formation of ozone (*d*) none of the above.

Q. 355. The world ozone day is celebrated on :

(*a*) 16th September (*b*) 31st December
(*c*) 14th January (*d*) 22nd June.

Q. 356. The major green house gas which is responsible for causing about 60% of the green house effect on Earth, among the following, is :

(*a*) carbon monoxide (*b*) methane
(*c*) NO_x (*d*) carbon dioxide.

Q. 357. Increased concentration of carbon dioxide in the Earth's atmosphere :

(*a*) leads to cooling on Earth
(*b*) leads to warming on Earth
(*c*) does not cause any change in Earth's temperature
(*d*) may sometimes cause cooling and sometimes warming, depending upon the climatic conditions.

Q. 358. The green house effect causes :

(*a*) cooling of the Earth
(*b*) warming of the Earth
(*c*) both (*a*) and (*b*) depending upon the season
(*d*) none of the above.

Q. 359. Had there been no or little carbon dioxide in the Earth's atmosphere, the temperature on Earth would have been below freezing as that existing on :

(*a*) Mars (*b*) Venus
(*c*) Moon (*d*) none of them.

Q. 360. Had there been too much quantity of carbon dioxide in the Earth's atmosphere, the temperature on Earth would have been much higher to make life impossible, as that existing on :

(*a*) Mars (*b*) Venus
(*c*) Moon (*d*) none of them.

Q. 361. The concentration of carbon dioxide in the Earth's atmosphere used to be about 316 ppmv (parts per million by volume) in the year 1959, which has now (2008) increased to a value of about :

(*a*) 330 ppmv (*b*) 350 ppmv
(*c*) 380 ppmv (*d*) 480 ppmv.

Q. 362. The average annual rate of increase of carbon dioxide in the Earth's atmosphere, has over the last five decades, been observed to be of the order of :

(*a*) 0.01% (*b*) 0.04%
(*c*) 0.4% (*d*) 4%.

Q. 363. The carbon dioxide in Earth's atmosphere, has during the last about 50 years, increased by about :

(*a*) 1% (*b*) 5%
(*c*) 10% (*d*) 20%.

Q. 364. The warming of the Earth's atmosphere due to increasing concentration of Green house gases (GhGs) is not likely to cause :

(*a*) increased heat conditions leading to warmer weather and long summers on Earth
(*b*) melting of snow of glaciers and of the poles, and rise in sea levels
(*c*) severe climate changes
(*d*) none of the above.

Q. 365. Which one of the following statements is incorrect ?

(*a*) Continued global warming due to increasing concentrations of GhGs at its present rate, is likely to increase the Earth's temperature during the 21st century by about 2°C ; and by about 1°C even if the concentrations of all GhGs and aerosols is kept constant at year 2000 level
(*b*) Continued global warming due to increasing concentration of CO_2 in the ambient air in likely to cause severe climate changes, leading to more floods, droughts, and cyclones.

(*c*) The rising sea levels due to global warming may submerge the coastal habitats and spoil the ground water by the ingress of the salty sea water towards the presently habited land areas

(*d*) None of the above.

Q. 366. Greenhouse effects are produced due to :

(*a*) absorption of infrared radiation in the atmosphere

(*b*) absorption of long range radiation in the atmosphere

(*c*) (*a*) and (*b*) both

(*d*) none of the above.

Q. 367. 80% of carbon dioxide, annually emitted into the atmosphere, comes from :

(*a*) rotting vegetation

(*b*) methane release from cattle sheep and termites

(*c*) landfills

(*d*) fossil fuels.

Q. 368. How much carbon dioxide can a tree remove annually from the air ?

(*a*) 1 kg (*b*) 5 kg

(*c*) 15 kg (*d*) 50 kg.

Q. 369. If all the world's ice caps and glaciers were to melt due to continued global warming, approximately what percentage of land area is likely to be submerged ?

(*a*) 5% (*b*) 10%

(*c*) 20% (*d*) 35%.

Q. 370. The reason(s) for rise of water levels in the world's seas and oceans due to global warming is/are :

(*a*) melting of glaciers

(*b*) thermal expansion of water volume contained in the world ocean

(*c*) both (*a*) and (*b*)

(*d*) none of the above.

Q. 371. The international treaty which aims at reducing global warming is :

(*a*) Montreal Protocol (*b*) Kyoto Protocol

(*c*) Ramsar Convention (*d*) CITES.

Q. 372. Emission Trading and Clean Development Mechanism (CDM) are an inbuilt mechanism of an international treaty for controlling :

(*a*) ozone depletion (*b*) global warming

(*c*) acid rains (*d*) none of the above.

Q. 373. What percentage of the total GhGs emissions into the world's atmosphere is being released by USA ?

(*a*) 10% (*b*) 15%

(*c*) 20% (*d*) 30%.

Q. 374. The country which releases a lot of GhGs into the world's atmosphere, but has refused to sign the international treaty for controlling GhGs emissions to avoid global warming, is :

(*a*) Russian Federation (*b*) USA

(*c*) Japan (*d*) none of them.

Chapter 20 : Noise Pollution

Q. 375. Noise is :
(*a*) loud noise (*b*) constant sound
(*c*) unwanted sound (*d*) sound of high frequency.

Q. 376. Noise is measured in units of :
(*a*) hertz (*b*) decibel
(*c*) doboson (*d*) bacqueral.

Q. 377. Which of the following is not an effect of noise an humans ?
(*a*) general annoyance (*b*) sleeplessness
(*c*) high blood pressure (*d*) hearing loss
(*e*) gastric secretions (*f*) none of the above.

Q. 378. Which of the following is not an effect of noise on wild life ?
(*a*) inactiveness/dullness (*b*) change of habitat
(*c*) environmental imbalances (*d*) none of the above.

Q. 379. 10 decibel (dB) increase in sound level will increase the loudness of sound by :
(*a*) 2 times (*b*) 5 times
(*c*) 10 times (*d*) none of them.

Q. 380. The maximum sound level, beyond which it is certainly regarded as a pollutant, is :
(*a*) 20 dB (*b*) 40 dB
(*c*) 45 dB (*d*) 80 dB.

Q. 381. The unit for measuring the frequency of sound is :
(*a*) decibel (dB) (*b*) hertz (Hz)
(*c*) doboson Unit (DU) (*d*) none of the above.

Q. 382. Although human ear can hear sound of all frequencies varying from 20 to 20,000 Hz, but it is most sensitive to frequencies in the range of :
(*a*) 100 to 500 Hz (*b*) 500 to 5000 Hz
(*c*) 5000 to 10,000 Hz (*d*) 10,000 to 20,000 Hz.

Q. 383. The weighting network that is usually used to filter out certain frequencies of sound in our day to day sound measurement is of :
(*a*) A type (*b*) B type
(*c*) C type (*d*) none of the above.

Q. 384. A-network that is generally fitted in our sound measuring meters of day to day use, severely filters out sound of :
(*a*) low frequency ranging between 10 to 200 Hz
(*b*) medium frequency ranging between 200 to 2000 Hz
(*c*) high frequency ranging between 2000 to 10,000 Hz
(*d*) none of the above.

Q. 385. Fluctuating noise levels from various sources at a place over a period of time can be represented by a constant value over that entire time period, by a value of sound, known as :
(*a*) equivalent noise level (*b*) average noise level
(*c*) both (*a*) and (*b*) (*d*) none of the above.

Q. 386. Acceptable noise level for residential and business urban areas as per IS : 4954-1968 is :

(*a*) 25-35 dB (*b*) 40-50 dB
(*c*) 50-60 dB (*d*) 70-80 dB.

Q. 387. The maximum average permissible noise level during day time hours (6 am to 10 pm) in residential areas, according to specified ambient air quality standards laid down under the Environment Protection Act in India, is :

(*a*) 30 dB (*b*) 45 dB
(*c*) 50 dB (*d*) 55 dB.

Q. 388. Which of the following is not a cause/source of noise ?

(*a*) Traffic
(*b*) Industrial and commercial activities
(*c*) Religious places
(*d*) Schools
(*e*) none of the above.

Q. 389. Which of the following will not help in abating noise pollution ?

(*a*) Raising of thick and high vegetation and trees
(*b*) Specifying and implementing noise standards for industrial activities and motor vehicles
(*c*) Imposing ban an public against blaring of loudspeakers and plying of radiosets
(*d*) Zoning and segregation of residential complexes from the commercial and industrial areas
(*e*) None of the above.

Q. 390. What type of noise can be abated by providing lining on walls and ceiling with sound absorbing material ?

(*a*) Source noise (*b*) Reflection noise
(*c*) Structural noise (*d*) Direct air-borne noise.

Chapter 22 : Preparing Environmental Impact Assessments of Polluting Industrial Projects

Q. 391. The idea of sustainable development was defined for the first time by the World Commission on Environment and Development (*i.e.* the Brundtland Commission) in the year :

(*a*) 1985 (*b*) 1987
(*c*) 1989 (*d*) 1991.

Q. 392. EIA is the abbreviated form for :

(*a*) Energy Impact Assessment (*b*) Ecological Impact Assessment
(*c*) Environmental Impact Assessment
(*d*) Emission Impact Assessment.

Q. 393. Environmental Impact Assessment is prepared for :

(*a*) every specified polluting developmental project
(*b*) every natural disaster to pre-assess its adverse impacts on humanity and environment

(*c*) both (*a*) and (*b*)
(*d*) none of the above.

Q. 394. An Environmental Impact Assessment is intended to identify the environmental, social and economic impacts of a proposed development :
(*a*) prior to the decision to sanction a project is taken
(*b*) during the execution of a project
(*c*) after the execution of a project to assess its beneficiaries
(*d*) none of the above.

Q. 395. An Environmental Impact Statement should include :
(*a*) description of the project with location, design, size, etc.
(*b*) description of its significant effects on environment and society
(*c*) mitigating measures to reduce the possible adverse impacts
(*d*) all the above.

Q. 396. The purpose of 'screening' during the EIA process under Indian EIA notification 2006 is to :
(*a*) decide as to whether an environment impact assessment is required for the project
(*b*) decide as to whether for proposed project falls under category 'A' or 'B'
(*c*) (*a*) and (*b*) both
(*d*) none of the above.

Q. 397. The 'scoping' during EIA process under Indian EIA Notification 2006 is applicable to :
(*a*) 'A' category of projects (*b*) 'B1' category of projects
(*c*) 'B2' category of projects alone
(*d*) (*a*) and (*b*) both
(*e*) (*a*), (*b*) and (*c*) all.

Q. 398. During the EIA process under Indian EIA Notification 2006, no public consultation is required for :
(*a*) All 'A' category of projects (*b*) All 'B1' category of projects
(*c*) All 'B2' category of projects (*d*) none of them.

Q. 399. During the EIA process under India's EIA notification 2006, public consultation is a must for :
(*a*) all 'A' category projects
(*b*) all 'A' category projects except for a few exceptional ones
(*c*) all 'B1' category projects
(*d*) all 'A' and 'B1' category projects except for a few exceptional ones.

Q. 400. The type of project categorised in 'A' category in the Schedule attached with EIA notification 2006 of India needs environmental clearance (before starting any work at site) from :
(*a*) the Central Govt. on the recommendation of the Expert Appraisal Committee
(*b*) the Central Govt. without the recommendation of the Expert Appraisal Committee

(*c*) Govt. of the State in which the given project is to be located on the recommendation of the State Expert Appraisal Committee

(*d*) none of the above.

Q. 401. The type of project categorised in 'B' category in the Schedule attached with EIA notification 2006 of India needs environmental clearance (before staring any work at site) from the :

(*a*) Central Govt. without the recommendation of the Expert Appraisal Committee

(*b*) State Govt. on the recommendation of State Expert Appraisal Committee

(*c*) State Environmental Impact Assessment Authority on the recommendation of State Expert Appraisal Committee

(*d*) None of the above.

Q. 402. Under the Indian EIA notification 2006, preparation of EIA is not at all required for projects falling in category :

(*a*) A (*b*) B-1

(*c*) B-2 (*d*) none of these.

Q. 403. The type of projects, which among the following, will require public consultation under EIA process to be carried out under Indian EIA notification 2006, is :

(*a*) building/construction/area development/township projects

(*b*) expansion of roads and highways which do not involve any further land acquisition

(*c*) projects concerning national defence and security

(*d*) nuclear power projects.

Answers For the Objective Questions

1. (*c*)	**2.** (*c*)	**3.** (*d*)	**4.** (*c*)	**5.** (*c*)
6. (*c*)	**7.** (*b*)	**8.** (*b*)	**9.** (*c*)	**10.** (*b*)
11. (*c*)	**12.** (*b*)	**13.** (*c*)	**14.** (*c*)	**15.** (*d*)
16. (*b*)	**17.** (*a*)	**18.** (*b*)	**19.** (*a*)	**20.** (*d*)
21. (*c*)	**22.** (*a*)	**23.** (*a*)	**24.** (*a*)	**25.** (*f*)
26. (*d*)	**27.** (*b*)	**28.** (*b*)	**29.** (*c*)	**30.** (*c*)
31. (*b*)	**32.** (*a*)	**33.** (*d*)	**34.** (*c*)	**35.** (*a*)
36. (*d*)	**37.** (*d*)	**38.** (*b*)	**39.** (*b*)	**40.** (*a*)
41. (*d*)	**42.** (*b*)	**43.** (*c*)	**44.** (*d*)	**45.** (*b*)
46. (*b*)	**47.** (*d*)	**48.** (*b*)	**49.** (*a*)	**50.** (*d*)
51. (*c*)	**52.** (*a*)	**53.** (*a*)	**54.** (*a*)	**55.** (*c*)
56. (*a*)	**57.** (*d*)	**58.** (*d*)	**59.** (*b*)	**60.** (*a*)
61. (*c*)	**62.** (*b*)	**63.** (*c*)	**64.** (*b*)	**65.** (*d*)
66. (*d*)	**67.** (*a*)	**68.** (*b*)	**69.** (*b*)	**70.** (*b*)
71. (*d*)	**72.** (*d*)	**73.** (*c*)	**74.** (*a*)	**75.** (*d*)
76. (*b*)	**77.** (*b*)	**78.** (*b*)	**79.** (*d*)	**80.** (*c*)
81. (*c*)	**82.** (*d*)	**83.** (*a*)	**84.** (*b*)	**85.** (*c*)
86. (*b*)	**87.** (*a*)	**88.** (*c*)	**89.** (*b*)	**90.** (*a*)

91. (*c*)	**92.** (*a*)	**93.** (*b*)	**94.** (*c*)	**95.** (*d*)
96. (*d*)	**97.** (*b*)	**98.** (*d*)	**99.** (*b*)	**100.** (*a*)
101. (*b*)	**102.** (*d*)	**103.** (*a*)	**104.** (*d*)	**105.** (*b*)
106. (*a*)	**107.** (*c*)	**108.** (*c*)	**109.** (*b*)	**110.** (*b*)
111. (*b*)	**112.** (*b*)	**113.** (*c*)	**114.** (*d*)	**115.** (*c*)
116. (*a*)	**117.** (*b*)	**118.** (*b*)	**119.** (*b*)	**120.** (*a*)
121. (*b*)	**122.** (*a*)	**123.** (*b*)	**124.** (*d*)	**125.** (*a*)
126. (*c*)	**127.** (*c*)	**128.** (*c*)	**129.** (*c*)	**130.** (*c*)
131. (*a*)	**132.** (*a*)	**133.** (*b*)	**134.** (*b*)	**135.** (*d*)
136. (*b*)	**137.** (*b*)	**138.** (*b*)	**139.** (*c*)	**140.** (*a*)
141. (*d*)	**142.** (*b*)	**143.** (*b*)	**144.** (*d*)	**145.** (*c*)
146. (*d*)	**147.** (*a*)	**148.** (*a*)	**149.** (*d*)	**150.** (*c*)
151. (*a*)	**152.** (*d*)	**153.** (*d*)	**154.** (*b*)	**155.** (*c*)
156. (*c*)	**157.** (*b*)	**158.** (*b*)	**159.** (*d*)	**160.** (*c*)
161. (*b*)	**162.** (*b*)	**163.** (*b*)	**164.** (*a*)	**165.** (*c*)
166. (*d*)	**167.** (*c*)	**168.** (*c*)	**169.** (*a*)	**170.** (*c*)
171. (*d*)	**172.** (*a*)	**173.** (*d*)	**174.** (*d*)	**175.** (*d*)
176. (*c*)	**177.** (*b*)	**178.** (*c*)	**179.** (*c*)	**180.** (*c*)
181. (*c*)	**182.** (*d*)	**183.** (*c*)	**184.** (*c*)	**185.** (*a*)
186. (*c*)	**187.** (*d*)	**188.** (*b*)	**189.** (*c*)	**190.** (*a*)
191. (*b*)	**192.** (*a*)	**193.** (*b*)	**194.** (*c*)	**195.** (*a*)
196. (*a*)	**197.** (*b*)	**198.** (*d*)	**199.** (*b*)	**200.** (*b*)
201. (*b*)	**202.** (*b*)	**203.** (*a*)	**204.** (*a*)	**205.** (*b*)
206. (*a*)	**207.** (*a*)	**208.** (*d*)	**209.** (*b*)	**210.** (*c*)
211. (*a*)	**212.** (*d*)	**213.** (*d*)	**214.** (*c*)	**215.** (*a*)
216. (*d*)	**217.** (*b*)	**218.** (*c*)	**219.** (*c*)	**220.** (*d*)
221. (*c*)	**222.** (*b*)	**223.** (*a*)	**224.** (*c*)	**225.** (*c*)
226. (*c*)	**227.** (*d*)	**228.** (*a*)	**229.** (*d*)	**230.** (*c*)
231. (*a*)	**232.** (*c*)	**233.** (*a*)	**234.** (*c*)	**235.** (*d*)
236. (*d*)	**237.** (*a*)	**238.** (*a*)	**239.** (*a*)	**240.** (*a*)
241. (*a*)	**242.** (*a*)	**243.** (*a*)	**244.** (*c*)	**245.** (*d*)
246. (*a*)	**247.** (*c*)	**248.** (*a*)	**249.** (*c*)	**250.** (*c*)
251. (*d*)	**252.** (*c*)	**253.** (*c*)	**254.** (*a*)	**255.** (*c*)
256. (*d*)	**257.** (*d*)	**258.** (*c*)	**259.** (*c*)	**260.** (*b*)
261. (*b*)	**262.** (*a*)	**263.** (*d*)	**264.** (*b*)	**265.** (*b*)
266. (*a*)	**267.** (*c*)	**268.** (*a*)	**269.** (*d*)	**270.** (*c*)
271. (*d*)	**272.** (*c*)	**273.** (*b*)	**274.** (*b*)	**275.** (*c*)
276. (*c*)	**277.** (*d*)	**278.** (*a*)	**279.** (*b*)	**280.** (*c*)
281. (*a*)	**282.** (*a*)	**283.** (*d*)	**284.** (*c*)	**285.** (*d*)
286. (*d*)	**287.** (*d*)	**288.** (*c*)	**289.** (*d*)	**290.** (*c*)
291. (*c*)	**292.** (*b*)	**293.** (*b*)	**294.** (*c*)	**295.** (*d*)
296. (*a*)	**297.** (*d*)	**298.** (*b*)	**299.** (*b*)	**300.** (*d*)
301. (*b*)	**302.** (*b*)	**303.** (*b*)	**304.** (*a*)	**305.** (*b*)

306. (*b*)	**307.** (*a*)	**308.** (*d*)	**309.** (*b*)	**310.** (*a*)
311. (*d*)	**312.** (*d*)	**313.** (*c*)	**314.** (*b*)	**315.** (*c*)
316. (*b*)	**317.** (*a*)	**318.** (*a*)	**319.** (*b*)	**320.** (*b*)
321. (*b*)	**322.** (*a*)	**323.** (*b*)	**324.** (*b*)	**325.** (*a*)
326. (*a*)	**327.** (*a*)	**328.** (*a*)	**329.** (*c*)	**330.** (*d*)
331. (*c*)	**332.** (*d*)	**333.** (*d*)	**334.** (*c*)	**335.** (*c*)
336. (*d*)	**337.** (*a*)	**338.** (*e*)	**339.** (*a*)	**340.** (*b*)
341. (*b*)	**342.** (*a*)	**343.** (*c*)	**344.** (*c*)	**345.** (*c*)
346. (*b*)	**347.** (*c*)	**348.** (*b*)	**349.** (*a*)	**350.** (*b*)
351. (*a*)	**352.** (*d*)	**353.** (*b*)	**354.** (*b*)	**355.** (*a*)
356. (*d*)	**357.** (*b*)	**358.** (*b*)	**359.** (*a*)	**360.** (*b*)
361. (*c*)	**362.** (*c*)	**363.** (*d*)	**364.** (*d*)	**365.** (*d*)
366. (*c*)	**367.** (*d*)	**368.** (*c*)	**369.** (*c*)	**370.** (*c*)
371. (*b*)	**372.** (*b*)	**373.** (*d*)	**374.** (*b*)	**375.** (*c*)
376. (*b*)	**377.** (*f*)	**378.** (*d*)	**379.** (*c*)	**380.** (*d*)
381. (*b*)	**382.** (*b*)	**383.** (*a*)	**384.** (*a*)	**385.** (*a*)
386. (*b*)	**387.** (*d*)	**388.** (*e*)	**389.** (*e*)	**390.** (*b*)
391. (*b*)	**392.** (*c*)	**393.** (*a*)	**394.** (*a*)	**395.** (*d*)
396. (*c*)	**397.** (*d*)	**398.** (*c*)	**399.** (*d*)	**400.** (*a*)
401. (*c*)	**402.** (*c*)	**403.** (*d*)		

Hints For Solving Starred Questions

Q. 34. Use Eqn. (3.2).

Q. 38. Assume 0.8 m/sec as minimum self cleansing velocity at full discharge and $N = 0.015$, and compute S as :

$$0.8 = \frac{1}{0.015} \times \left(\frac{0.3}{4}\right) . S^{1/2}$$

or
$$S = \frac{1}{219} \text{ ; or say 1 in 200. } \textbf{Ans.}$$

Q. 51. Use Eq. (4.12), and find $\theta = 120°$.

Then use Eq. (4.15), as :

$$p = \pi D\left(\frac{120°}{360°}\right) = \frac{\pi D}{3}.$$

Q. 92. Refer example 6.1.

Q. 93.
$$\text{HP} = \frac{wQH}{75} \cdot \frac{1}{\eta}$$

$$= \frac{1000 \times 10}{75 \times 2 \times 354} \times \frac{1}{0.65} = 2.89 \approx 3. \textbf{ Ans.}$$

Q. 131. Plants give oxygen during day time, much more than what they require ; activity being maximum at mid noon, due to sun's effects on photosynthesis.

Q. 193. Use equation 9.33.

—24

Test Paper on Objective Questions

Q. 1*. The correct ratio of the diameter of a circular sewer to the side of a hydraulically equivalent square section of a sewer, is :

(*a*) 1 (*b*) 0.9
(*c*) 0.925 (*d*) 1.095.

Q. 2. A circular sewer section looses most of its merits, when the sewage depth becomes :

(*a*) less than $\frac{3}{4}$ full (*b*) less than $\frac{2}{3}$ full
(*c*) less than $\frac{1}{2}$ full (*d*) none of the above.

Q. 3. The usual mix adopted for cement concrete for R.C.C. pipes is :

(*a*) 1 : 2 : 4 (*b*) 1 : 1.5 : 3
(*c*) 1 : 3 : 6 (*d*) 1 : 4 : 8.

Q. 4. Manhole covers are made circular :

(*a*) to strengthen the cover
(*b*) to make the entry convenient
(*c*) for architectural reasons
(*d*) to prevent falling of the cover into the manhole.

Q. 5*. If in the relative stability test on sewage, the period of incubation at 20°C is 15 days, the relative stability is :

(*a*) 95.95% (*b*) 96.05%
(*c*) 96.86% (*d*) 100%.

Q. 6*. If the period of incubation at 37°C is 15 days in the relative stability test on sewage, the relative stability is :

(*a*) 99% (*b*) 99.9%
(*c*) 99.99% (*d*) 100%.

Q. 7. If 10 ml of raw sewage is diluted to 250 ml, the dilution factor is :

(*a*) 24 (*b*) 25
(*c*) 1/25 (*d*) none of these.

Q. 8*. If 10 ml of raw sewage is diluted to 250 ml, and the dissolved oxygen content of the diluted sample changes from 9 mg/l to 5 mg/l after incubation at 20°C for 5 days, BOD_5 of raw sewage is :

(*a*) 0.16 mg/l (*b*) 4 mg/l
(*c*) 10 mg/l (*d*) 100 mg/l.

*Hints are given for starred questions.

Q. 9*. 2 ml of sewage, having BOD_5 = 500 mg/l, is diluted to 200 ml, and incubated for 5 days at 20°C. If the D.O. content after 5 days is found to be 5 mg/l, what was the initial D.O. content of the diluted sample ?

(*a*) 10 mg/l (*b*) 5 mg/l
(*c*) 1 mg/l (*d*) none of the above.

Q. 10. The oxygen utilisation rate of micro-organisms depends on the characteristics of :

(*a*) the reactor only
(*b*) the wastewater only
(*c*) the wastewater and the reactor both
(*d*) none of the above.

Q. 11. The BOD removal in an oxidation pond may be upto :

(*a*) 100% (*b*) 97%
(*c*) 94% (*d*) 90%.

Q. 12. A reactor, in which the surface area for growth of biofilm is provided by randomly packed solid forms, is called :

(*a*) activated sludge reactor (*b*) trickling filter
(*c*) stabilisation pond (*d*) mixed reactor.

Q. 13. The depth of biofilters varies between :

(*a*) 0.6 to 1.0 m (*b*) 1.2 to 1.5 m
(*c*) 1.5 to 1.8 m (*d*) none of the above.

Q. 14. The flowing through velocity for Imhoff tank, should, generally not exceed :

(*a*) 0.3 m/min (*b*) 3 m/min
(*c*) 30 m/min (*d*) none of the above.

Q. 15. As a result of the stabilisation of sewage effluent, the most appropriate end product produced is :

(*a*) chloride (*b*) plant nutrients
(*c*) hardness (*d*) alkalinity.

Q. 16. When chlorine is added to sewage both at the beginning as well as at the end of the treatment process, the phenomenon is called :

(*a*) post chlorination (*b*) super chlorination
(*c*) split chlorination (*d*) none of the above.

Q. 17. The liquid that has percolated through the solid waste, and has extracted dissolved or suspended materials from it, is called :

(*a*) refuge (*b*) leachate
(*c*) sewage (*d*) particulate
(*e*) none of the above.

Q. 18. The end product formed, after separation and anaerobic bacterial digestion of organic municipal solid wastes, is called :

(*a*) compost (*b*) humus
(*c*) leachate (*d*) ashes.

Q. 19. The process by virtue of which the heat transfer occurs in the troposphere is :

(a) conduction
(b) convection
(c) evaporation-condensation cycle of water
(d) green house effect
(e) all of these.

Q. 20. The transfer of heat from the earth to the atmosphere by direct physical contact between the earth and the air, is called :
(a) convection (b) conduction
(c) green house effect (d) tropospheric heating.

Q. 21. The one among the following, which is not an elemental property of the atmosphere, is :
(a) heat (b) pressure
(c) wind (d) wind speed
(e) moisture.

Q. 22. The study of earth's atmosphere and its changes, is called :
(a) environmental engineering (b) ecology
(c) meteorology (d) philology.

Q. 23. The effect of earth's rotation on wind velocity and direction, is called :
(a) Coriolis effect (b) Vander waal's effect
(c) gravitational effect (d) centrifugal effect.

Q. 24. The permissible TSPM (annual average) standard in industrial areas, under the pre-revised standards, in India, is :
(a) 100 $\mu g/m^3$ (b) 140 $\mu g/m^3$
(c) 360 $\mu g/m^3$ (d) 1000 $\mu g/m^3$.

Q. 25. The process of heating a solid waste and splitting its organic substances by thermal cracking and condensation, is called :
(a) gasification (b) pulverisation
(c) incineration (d) pyrolysis.

Q. 26. Partial combustion of a carbonaceous fuel to generate a combustible fuel gas, rich in carbon monooxide and oxygen, is called :
(a) gasification (b) composting
(c) pyrolysis (d) Bangalore method.

Q. 27. The temperature gradient of ambient air, is called :
(a) adiabatic lapse rate (b) super adiabatic lapse rate
(c) environmental lapse rate (d) dry adiabatic lapse rate.

Q. 28. The taste threshold of sulphur dioxide is about :
(a) 0.1 ppm (b) 0.3 ppm
(c) 0.5 ppm (d) 3 ppm.

Q. 29. The odour threshold of sulphur dioxide, is about :
(a) 0.1 ppm (b) 0.3 ppm
(c) 0.5 ppm (d) 3 ppm.

Q. 30. Heavy loading of pollen grains in air, may cause :
(a) anaemia (b) typhoid
(c) hayfever (d) influenza.

Q. 31. Hairs of our nose can remove all particles of size, greater than :
(*a*) 1 mm (*b*) 1 micron
(*c*) 10 micron (*d*) 100 micron.

Q. 32. A shallow sewage treatment pond, in which dissolved oxygen is present at all depths, may be called :
(*a*) oxidation pond (*b*) aerobic pond
(*c*) unaerated lagoon (*d*) facultative pond.

Q. 33. The chemical compounds, which are essentially responsible for production of smog, are :
(*a*) hydrocarbons (*b*) oxides of nitrogen
(*c*) oxides of sulphur (*d*) (*a*) and (*b*) both
(*e*) all of (*a*), (*b*) and (*c*).

Q. 34. The major photochemical oxidant is :
(*a*) hydrogen peroxide (*b*) ozone
(*c*) nitrogen oxides (*d*) peroxy acetyl nitrate (PAN).

Q. 35. When the ambient lapse rate exceeds the adiabatic lapse rate, the ambient lapse rate, is called the :
(*a*) super adiabatic lapse rate
(*b*) dry adiabatic lapse rate
(*c*) saturated adiabatic lapse rate
(*d*) sub adiabatic lapse rate.

Q. 36. Carbon monoxide is hazardous to health, because :
(*a*) it causes loss of sense of smell
(*b*) it is carcinogenic in nature
(*c*) it reduces oxygen carrying capacity of blood
(*d*) it may cause conjunctivitis.

Q. 37. Particles of around 1 micron (10^{-6} m) size are best removed by :
(*a*) filtration (*b*) plain sedimentation
(*c*) chemical coagulation (*d*) chemical precipitation.

Q. 38*. If a sample of air is analysed at standard temperature and pressure, and is found to contain 0.3 ppm of sulphur dioxide, the equivalent SO_2 concentration in mg/m^3 will be :
(*a*) 8000 (*b*) 800
(*c*) 80 (*d*) 0.8.

Q. 39. Subsidence inversion can be related to a :
(*a*) cyclone (*b*) anticyclone
(*c*) radiation (*d*) tornado.

Q. 40. The minimum specified height of a chimney for evolution of industrial gases other than those in thermal power plants, in the absence of data for the evolved gases, is :
(*a*) 10 m (*b*) 30 m
(*c*) 100 m (*d*) 220 m.

Q. 41. Carbon monoxide concentration of 9 ppm at 20°C and 1 atmospheric pressure, is equivalent to nearly :

(*a*) 1 mg/l (*b*) 10 mg/l
(*c*) 100 mg/l (*d*) 1000 mg/l.

Q. 42. Longer exposure to NO_2 even in small concentrations, may cause diseases pertaining to :
(*a*) lever (*b*) lung
(*c*) kidneys (*d*) heart.

Q. 43. The average NO_2 concentration in tobacco is approximately :
(*a*) 1 ppm (*b*) 5 ppm
(*c*) 10 ppm (*d*) 50 ppm.

Q. 44. The specified ambient air quality standard for NO_2 is about 40 $\mu g/m^3$, which in ppm approximates :
(*a*) 0.01 (*b*) 0.02
(*c*) 0.1 (*d*) 0.5.

Q. 45. The secondary air pollutant among the following is :
(*a*) sulphur dioxide (*b*) methane
(*c*) carbon monoxide (*d*) peroxy acetyl nitrate (PAN).

Q. 46. The most threatening aspect of the pollution of the environment, is :
(*a*) global warming (*b*) acid rain
(*c*) ozone deficit (*d*) none of the above.

Q. 47*. The minimum size of the particulate matter of sp. gravity 2.0 that will get removed by a 100% efficient settling chamber of 10 m length and 1.5 m height, if the viscosity of hot emission is 1.7×10^{-5} kg/m sec and horizontal velocity into the chamber is 0.3 m/sec, will be :
(*a*) 5.2 μm (*b*) 52 μm
(*c*) 520 μm (*d*) 52 mm.

Q. 48*. A decrease in the radius of a cyclone collector will :
(*a*) increase its efficiency (*b*) decrease its efficiency
(*c*) not affect its efficiency.

Q. 49. The filter system which filters out the gaseous emissions, is called a :
(*a*) trickling filter (*b*) moving bed filter
(*c*) baghouse filter (*d*) none of the above.

Q. 50. Electrostatic precipitators are now-a-days installed in thermal power plants, to remove from its gaseous emissions, the :
(*a*) flyash (*b*) SO_2
(*c*) both (*a*) and (*b*) above (*d*) none of the above.

Q. 51. While examining the impact of sewage discharge into a river, the Deoxygenation Rate (K_D) is increased, by keeping all other parameters to be the same. This will :
(*a*) increase the D.O. deficit and displace the critical point upstream
(*b*) increase the D.O. deficit and displace the critical point downstream
(*c*) decrease the D.O. deficit and displace the critical point upstream
(*d*) decrease the D.O. deficit and displace the critical point downstream.

Q. 52. While examining the impact of sewage discharge in to a river, the Reaeration rate (K_R) is increased, by keeping all other parameters to be the same. This will :

(*a*) increase the D.O. deficit and displace the critical point upstream

(*b*) increase the D.O. deficit and displace the critical point downstream

(*c*) decrease the D.O. deficit and displace the critical point upstream

(*d*) decrease the D.O. deficit and displace the critical point downstream.

Q. 53. The deoxygenation rate constant (K_D) for a river differs from the BOD rate constant (K), because there are physical and biological differences between a river and a BOD bottle. The value of K_D is always more than the value of K, because :

(*a*) BOD is exerted more rapidly in river than in the bottle

(*b*) BOD is exerted less rapidly in river than in the bottle

(*c*) BOD is exerted at equal rates in the river and the bottle

(*d*) none of the above.

Q. 54. Bosco has developed the following equation :

$$K = K' + \frac{v}{y} \cdot \eta$$

where v = average vel. of river stream

H = average depth of river

η = bed activity coefficient of the river.

In this equation :

(*a*) K is the Deoxygenation rate and K' is the BOD rate

(*b*) K is the BOD rate and K' is the Deoxygenation rate

(*c*) K is the Deoxygenation rate and K' is the Re-aeration rate

(*d*) K is the Re-aeration rate and K' is the Deoxygenation rate.

Q. 55. The re-aeration rate of a river depends upon the degree of turbulent mixing. It however does not depend upon the :

(*a*) velocity in the river

(*b*) depth of water in the river

(*c*) temperature of water in the river.

(*d*) none of the above.

Q. 56. Ultimate BOD of a given sewage depends upon its :

(*a*) temperature (*b*) initial organic matter

(*c*) flow velocity (*d*) all of the above.

Q. 57. The BOD rate equation is represented as

$$BOD_t = L\left[1 - 10^{-K_D \cdot t}\right]$$

where BOD_t is the BOD of sewage after t days from the start of oxidation of wastes, by consumption of oxygen

K_D is the deoxygenation constant

t is the time in days.

In the above equation :
(*a*) L and K_D both depend upon temperature
(*b*) only L depends upon temperature
(*c*) only K_D depends upon temperature
(*d*) L and K_D both are independent of temperature.

Answers For Test Paper

1. (*d*)	**2.** (*c*)	**3.** (*b*)	**4.** (*b*)	**5.** (*c*)
6. (*b*)	**7.** (*b*)	**8.** (*d*)	**9.** (*a*)	**10.** (*c*)
11. (*d*)	**12.** (*b*)	**13.** (*b*)	**14.** (*a*)	**15.** (*d*)
16. (*c*)	**17.** (*b*)	**18.** (*b*)	**19.** (*e*)	**20.** (*b*)
21. (*d*)	**22.** (*c*)	**23.** (*a*)	**24.** (*c*)	**25.** (*d*)
26. (*a*)	**27.** (*c*)	**28.** (*b*)	**29.** (*c*)	**30.** (*c*)
31. (*c*)	**32.** (*b*)	**33.** (*e*)	**34.** (*c*)	**35.** (*a*)
36. (*c*)	**37.** (*c*)	**38.** (*b*)	**39.** (*b*)	**40.** (*b*)
41. (*b*)	**42.** (*b*)	**43.** (*b*)	**44.** (*b*)	**45.** (*d*)
46. (*c*)	**47.** (*b*)	**48.** (*a*)	**49.** (*c*)	**50.** (*a*)
51. (*a*)	**52.** (*c*)	**53.** (*a*)	**54.** (*a*)	**55.** (*d*)
56. (*b*)	**57.** (*c*)			

Hints For Solving Starred Questions of The Test Paper

Q. 1.

$$Q_1 = Q_2$$

or $$A_1 R_1^{2/3} = A_2 R^{2/3} \qquad (S \text{ and } N \text{ being the same})$$

or $$\frac{\pi}{4} D^2 \left(\frac{D}{4}\right)^{2/3} = b^2 \left(\frac{b}{4}\right)^{2/3}$$

or $$\frac{D}{b} = \left(\frac{4}{\pi}\right)^{3/8} = 1.095$$

Q. 5. Use Eqn. (7.22) as :

$$S = 100\left[1 - (0.794)^{15}\right] = 96.86\%$$

Q. 6. Use Eqn. (7.23) as :

$$S = \left[100 - (0.630)^{15}\right] = 99.9\%$$

Q. 8. Use Eqn. (7.11) and determine :

$$\text{BOD}_5 = \text{DO lost in sample} \times \text{dilution factor}$$
$$= (9 - 5) \text{ mg/l} \times \text{dilution factor}$$
$$= 4 \text{ mg/l} \times \left(\frac{250}{10}\right) = 100 \text{ mg/l}.$$

Q. 9. Use Eqn. (7.11), as :

$$\text{BOD}_5 = \text{D.O. lost in sample} \times \text{dilution factor}$$

or $$500 = (x - 5)\,\frac{200}{2} \quad \text{or} \quad x = 10 \text{ mg/l}.$$

Q. 37. *Plain Sedimentation or Filtration alone is ineffective in removing micro-organisms, like bacteria, virus, etc.* Hence, all such micron sized bacteria and colloids (10^{-5} m to 10^{-8} m) need to be increased in size by *chemical coagulation* for subsequent sedimentation, filtration and removal.

Chemical precipitation is the term, which is generally used to mention precipitation of hardness by lime-soda process, etc., which thus, covers the removal of dissolved micro impurities and is not the case here.

Q. 38. SO_2 in μg/m³ $= \dfrac{SO_2 \text{ in ppm} \times 64000}{24.04}$

$$= \frac{0.3 \times 64000}{24} = 800.$$

Q. 47. Use Eqn. (18.14) as :

$$d = C \,.\, \sqrt{\frac{18\mu \,.\, v_h \,.\, H}{g \,.\, L\rho_p}} \quad \text{where } C = 2$$

$$\therefore \quad d = 2 \times \sqrt{\frac{18 \times (1.7 \times 10^{-5})\ \text{kg/ms} \times 0.3\ \text{m/s} \times 1.5\ \text{m}}{9.81\ \text{m/s}^2 \times 10\ \text{m} \times 2000\ \text{kg/m}^3}}$$

$$= \frac{2}{10^4} \times (0.26)\ \text{m} = \frac{0.52}{10^4}\ \text{m} = 52\ \mu\text{m}.$$

Q. 48. Eqn. (18.15) says that centrifugal force $F_c \propto \dfrac{1}{R}$. Hence, decrease in R will increase F_c, and consequently increase the efficiency of dust collection.

—25

Objective Question of the AMIE (Section B) Exams

Q. 1. Fill in the blanks :

(i) For volatile solids determination, the dried sample is fired at temperature of 550 ± 50°C in a ...... furnace.

(ii) If BOD/COD ratio of a wastewater is more than ...., the wastewater is considered to be amenable to biological treatment.

Q. 2. Match the following sets :

Set A	*Set B*
(*i*) Adiabatic lapse rate	(*a*) Combined sewer
(*ii*) Algal-bacterial symbiosis	(*b*) Automatic exhaust
(*iii*) Aromatic hydrocarbons	(*c*) Atmospheric stability
(*iv*) Leaping weir	(*d*) Oxidation pond
(*v*) Algal bloom	(*e*) Nitrogenous waste-water

Q. 3. Choose the correct statement in the following :

(i) Drop manholes are more frequently provided in the sewerage system, when the ground surface is :

(*a*) more or less flat (*b*) undulating

(*c*) having a steep gradient.

(ii)* Corrosion of concrete sewers are caused by :

(*a*) high flow velocity in sewers

(*b*) anaerobic decomposition of sewage solids

(*c*) high pH value of sewage

(*d*) aerobic oxidation of sewage solids.

(iii)* Iron may be present in significant quantities in

(*a*) septic tank effluent (*b*) irrigation return water

(*c*) rain water

(*d*) fresh water from deep tubewells.

(iv) The grown algae is useful in the operation of :

(*a*) oxidation ponds (*b*) slow sand filters

(*c*) sludge-digestion lagoons (*d*) septic tanks.

(v) Design of an appropriate velocity control device is an essential component in the design of :

*Hints for solving starred questions is given at the end of the chapter

(*a*) activated sludge process (*b*) flow-equalisation tank
(*c*) rapid sand filter (*d*) grit chamber.

Q. 4. Fill in the blanks :

(i) Organic-nitrogen and ammonia-nitrogen, determined together, is usually known as ...... nitrogen.

(ii)* Permissible BOD of the sewage effluents, when discharged into the rivers, is ..... mg/l.

Q. 5. Match the following sets :

Set A	*Set B*
(*i*) Proportional weir	(*a*) Nitrate
(*ii*) Mottled teeth	(*b*) Slow sand filters
(*iii*) Muffle furnace	(*c*) Grit chambers
(*iv*) Methemoglobinemia	(*d*) Volatile solids
(*v*) Schmutzdecke	(*e*) Fluoride.

Q. 6. Indicate whether the following statements are True or False :

(i) Carbon dioxide is relatively transparent to shortwave ultraviolet light from the sun, but restricts the loss of heat by radiation from the ground surface.

(ii) Contact stabilisation is a common variation of 'activated sludge process'.

Q. 7. Choose the correct statement in the following :

(i)* BOD value of medium strength domestic sewage could be above :

(*a*) 12 mg/l (*b*) 60 mg/l
(*c*) 120 mg/l (*d*) 200 mg/l.

(ii) The purpose of the wastewater solid digestion are :

(*a*) to decompose enough of the organic matter to avoid creating a nuisance
(*b*) to separate liquids from solids to facilitate drying
(*c*) to decrease the bulk of the wastewater solids to facilitate handling
(*d*) all of the above.

Q. 8. Choose the correct answer in the following :

(i) Secondary treatment of sewage removes :

(*a*) suspended organic matter
(*b*) dissolved and colloidal organic matter
(*c*) harmful bacteria
(*d*) fats and grease.

Q. 9. Fill in the blanks :

(i) The amount of oxygen consumed by sewage from potassium dichromate is termed as ....

(ii) In stabilisation ponds, the strength of sewage is reduced by the principles of ....

Q. 10. Match the following sets :

Group A	*Group B*
(*i*) Arithmetical increase method	(*a*) Septic tank
(*ii*) Anaerobic treatment	(*b*) Old city
(*iii*) Water bearing strata	(*c*) Pathogenic bacteria
(*iv*) Water borne diseases	(*d*) Particulate matter
(*v*) Electrostatic precipitators	(*e*) Aquifer

Q. 11. Indicate whether the following statements are True or False :

(i) Biochemical oxygen demand of safe drinking water is zero.

(ii) 70 to 90% of suspended solids are removed in plain sedimentation of sewage.

Q. 12. Choose the correct statement in the following :

(i) Ozone layer in the upper atmosphere is getting destroyed owning to its reaction with :

(*a*) carbon dioxide (*b*) hydrogen peroxide
(*c*) oxides of nitrogen (*d*) chlorofluorocarbons.

(ii) The intensification of greenhouse effect is attributed to the increased level of :

(*a*) carbon dioxide (*b*) carbon monoxide
(*c*) chlorofluorocarbons (*d*) sulphur dioxides.

Q. 13. Fill in the blanks :

(i) BOD/COD ratio in a typical domestic wastewater is ....

Q. 14. State whether the following statements are True or false :

(i) Algal growth in the waste stabilization ponds is not desirable.

Q. 15. Match the following sets :

Set A	*Set B*
(*i*) Recorbonation	(*a*) Activated carbon
(*ii*) Chlorination	(*b*) Trihalomethanes
(*iii*) Taste and odour	(*c*) Lime soda process
(*iv*) Temperature inversion	(*d*) Smog

Q. 16. Choose the correct statement in the following :

(i) A sloped floor of a manhole on both sides of and above a channel, on which a man can stand for cleaning the sewers is called :

(*a*) Manhole stand (*b*) Cleaning bench
(*c*) Benching (*d*) Wash bench
(*e*) Sewer stand.

(ii) Corrosion in sewer occurs due to :

(*a*) Methane (*b*) Carbon dioxide
(*c*) Carbon monoxide (*d*) Hydrogen sulphide
(*e*) All of the above.

(iii)* Bacteria which use carbon dioxide as a source of carbon are known as :

(*a*) Micro bacteria (*b*) Autotrophic
(*c*) Heterotrophic (*d*) Aerobic
(*e*) Anaerobic.

(iv) Electrostatic precipitators are used as pollution control device for the separation of :
(*a*) SO_2 (*b*) NO_x
(*c*) Hydrocarbon (*d*) Particulate matter

(v) In the analysis, muffle furnace is used to estimate :
(*a*) Volatile fatty acids (*b*) Kjeldahl nitrogen
(*c*) Volatile solids (*d*) Biodegradable solids
(*e*) Non-biodegradable solids.

(vi) Pumping storm water is usually done by employing :
(*a*) Submergible pump (*b*) Centrifugal pump
(*c*) Axial flow pump (*d*) Turbine pump
(*e*) Reciprocating pump.

(vii) Anaerobic biological treatment of high BOD wastes are suggested to :
(*a*) Obtain end products having a fuel value
(*b*) Yield least amount of sludge
(*c*) Avoid high requirement of compressed air
(*d*) All of the above.

(viii) Recirculation in 'Activated Sludge Process' is done to :
(*a*) dilute the incoming sewage
(*b*) dampen the effect of flow variation
(*c*) operate the plant continuously
(*d*) supply seed materials to the aeration tank.

Q. 17. Choose the correct statement in the following :
The rate of 5 day BOD exerted at any time is :
(*a*) directly proportional to BOD satisfied
(*b*) directly proportional to BOD remaining
(*c*) inversely proportional to BOD remaining
(*d*) inversely proportional of BOD satisfied.

Q. 18. Fill in the blank :
BOD reduction in oxidation pond is based on the principle of ....

Q. 19. State whether the following statements are true or false :
(i) The major contributor to the depletion of ozone layer is chlorofluoro carbon.
(ii) The treatment in sludge digestion tank is based upon aerobic degradation of organic matter.

Q. 20. Match the following sets :

Set A	*Set B*
(*i*) Acidity	(I) Particulate matter
(*ii*) Alum	(II) Oils and greases
(*iii*) Cyclone	(III) Flocculation
(*iv*) Skimming tank	(IV) Carbon dioxide

Q. 21. Choose the correct statement in the following :

(i) Presence of higher algal content in water indicates that water is
(*a*) having high BOD (*b*) free from oxygen
(*c*) saturated with oxygen (*d*) obeying none above.

(ii) In the design of inverted syphon, attempt is made to run all the pipes flowing :
(*a*) $\frac{1}{3}$ rd full (*b*) $\frac{1}{2}$ full
(*c*) $\frac{3}{4}$ th full (*d*) complete full.

(iii) The optimum temperature for sludge digestion is :
(*a*) 10°C (*b*) 25°C
(*c*) 37°C (*d*) 55°C.

(iv) For rural areas, VIP latrines mean
(*a*) Very important persons latrine
(*b*) Ventilated improved pit latrine
(*c*) Vault and inverted syphon pit latrines
(*d*) None as given in (*a*), (*b*) and (*c*).

(v) For rural areas, most suitable solid waste disposal method is :
(*a*) Pyrolysis (*b*) Incineration
(*c*) Land filling (*d*) Composting.

(vi) For conventional activated sludge process, the mixed liquor suspended solid should range between :
(*a*) 10 to 100 mg/l (*b*) 150 to 300 mg/l
(*c*) 1500 to 3000 mg/l (*d*) 5000 to 10000 mg/l.

Q. 22. Fill in the blanks :

(i) Limestone surfaces are damaged by an aggressive air environment caused by the presence of gas ....

(ii) Maximum concentration of oxygen (saturation concentration of dissolved oxygen) in water bodies at about 30°C is ...... mg/l.

Q. 23. Choose the correct answer in the following :

(i) In combined sewer system, storm waters are admitted through :
(*a*) leaping weirs (*b*) storm regulators
(*c*) inlets (*d*) intercepting weirs.

(ii) Ozone layer in the upper atmosphere is getting depleted owing to its reaction with pollutants like
(*a*) hydrogen peroxide (*b*) carbon monoxide
(*c*) chlorofluoro carbons (*d*) volatile fatty acids.

(iii) Corrosions of concrete sewers are generally caused by :
(*a*) high velocity of flow of sewage
(*b*) higher concentration of grits in sewage
(*c*) high pH of the sewage
(*d*) septic condition of the sewage.

(iv) Pumping of storm water is usually done by employing :
(*a*) axial flow pump (*b*) non-clug pump
(*c*) turbine pump (*d*) submersible pump.

(v) Electrostatic precipitators are used as pollution control device for the removal of :

(*a*) SO_2 (*b*) NO_x

(*c*) suspended particulate matter

(*d*) volatile organic substances.

(vi) On mass basis, the largest fraction of the municipal solid wastes from a typical Indian city is made up of :

(*a*) salvageable substances

(*b*) combustible substances

(*c*) compostible substances

(*d*) toxic and hazardous substances.

Q. 24. Fill up the blanks :

(i)* The bacteria which brings about sludge digestion in gobar gas plant are ....

(ii) Radioactive wastes can be disposed of by ....

(iii) Stone-ware sewers are also known as ....

Q. 25. Choose the correct answer :

(i) The colour of fresh domestic sewage is

(*a*) Dark brown (*b*) Red

(*c*) Black (*d*) Grey.

(ii) What is the % of activated sludge mixed with the sewage in primary settling tanks ?

(*a*) 10 to 20 (*b*) 40 to 50

(*c*) 20 to 40 (*d*) 50 to 70.

Q. 26. Match the following :

Column A	*Column B*
(*i*) Aeration fountain	(I) Sullage
(*ii*) Flushing tank	(II) Removes odour and taste by exchange of gases
(*iii*) Finally divided aerosol particles resulting from incomplete combustion	(III) Biochemical reaction by bacteria takes place
(*iv*) Waste water from kitchen	(IV) Smoke
(*v*) Septic tank	(V) Produces self cleaning.

Q. 27. State whether true or false :

Strength of waste water is measured by pH.

Q. 28. Choose the correct answer :

(i) Bacteria that obtain both energy and material from organic sources are called :

(*a*) autotrophs (*b*) heterotrophs

(*c*) phototrophs (*d*) protists.

(ii) Laying of sewers is usually done with the help of :

(*a*) a compass (*b*) a theodolite

(*c*) a plane table (*d*) sight rails and boning rods.

(iii) The specific gravity of sewage is :
(*a*) 1 (*b*) slightly less than 1
(*c*) slightly more than 1 (*d*) 2.65.

(iv) The pipe installed in the house drainage to preserve the water seal is called :
(*a*) waste pipe (*b*) soil pipe
(*c*) vent pipe (*d*) antisiphonage pipe.

(v) The gas from a sludge digestor unit consists of
(*a*) only methane (*b*) only carbon dioxide
(*c*) about 70% methane (*d*) about 30% methane.

(vi) The pH range to be maintained for proper functioning of a sludge digestion tank must be :
(*a*) 3.5 (*b*) 6.5—8.5
(*c*) 8.5—10.5 (*d*) no control needed.

(vii) The sewage in a waste stabilisation pond is treated by
(*a*) aerobic bacteria (*b*) algae
(*c*) aerobic bacteria and algae
(*d*) facultative bacteria.

(viii) Which of the following is not a primary air pollutant ?
(*a*) sulphur dioxide (*b*) nitrogen oxides
(*c*) hydro carbons (*d*) ozone.

Q. 29. Choose the correct answer :

(i) Which prevents the entry of foul gas and allows ventilation ?
(*a*) Intercepting trap (*b*) Fresh air inlet
(*c*) Gully trap (*d*) Flushing cistern.

(ii) The BOD test is an indirect measure of
(*a*) dissolved solids (*b*) volatile acid content
(*c*) organic matter (*d*) none of the above.

(iii)* For self cleansing velocity of 0.90 m/s with a sewer gradient of 1 in 150, the pipe dia is
(*a*) 15 cm (*b*) 30 cm
(*c*) 75 cm (*d*) none of the above.

Q. 30. Match the following :

(*i*) Intercepting trap	(*a*) Discharge into soil pipe through floor trap
(*ii*) Gully trap	(*b*) Cleans suddenly with a large volume of water
(*iii*) Cowl	(*c*) Prevents nesting of birds, etc.
(*iv*) Water closet	(*d*) Prevents gases entering a house from public sewers
(*v*) Flushing cisterns	(*e*) Receives human excreta directly.

Q. 31. Fill in the blanks :

(i) The leakage of sewage from sewers into the surrounding is known as ....

(ii) Composting is a good method of solid waste disposal when waste is ....

(iii) Air pollution can be minimised by ....

(iv) Smog is a mixture of ...... and smoke.

Q. 32. Fill in the blanks :

(i) The main yard stick for measuring strength of sewage is ....

(ii) The stabilisation of sewage in oxidation pond takes place by symbiotic action of ....

Q. 33. State whether the following statements are True or False :

(i) The stable atmosphere helps in the dispersion of air pollutants.

(ii) The Chemical Oxygen Demand (COD) is always lesser than the Biochemical oxygen demand (BOD).

Q. 34. Match the following :

Column-I	*Column-II*
(*i*) Circular ring system	(*a*) Hydrocarbon, carbon monoxide, nitrous oxide
(*ii*) Depletion of oxygen in stream	(*b*) Mottling of teeth
(*iii*) Inverted sewer	(*c*) Carry sewage below road/ pavement
(*iv*) Excess fluorides in water	(*d*) Fish killing
(*v*) Automobile exhaust	(*e*) Suitable layout of pipes for well planned city

Q. 35. Choose the correct answer :

(i)* For controlling the Algae, the most commonly used chemical is :

(*a*) lime (*b*) bleaching powder

(*c*) alum (*d*) copper sulphate.

(ii) Which trap separates the municipal sewer from house drainage system ?

(*a*) Gully trap (*b*) P-trap

(*c*) S-trap (*d*) Intercepting trap.

Q. 36. Choose the correct answer :

(i) Sewage is known to contain a very good amount of water, and solid matter is only :

(*a*) 10% (*b*) 1%

(*c*) 0.1% (*d*) None of the above.

(ii) The normal temperature of sewage than that of water supply is :

(*a*) lower (*b*) higher

(*c*) same (*d*) atmospheric.

(iii)* Septic tank is most suitable unit for :

(*a*) hospitals (*b*) hotels

(*c*) hostels (*d*) All of the above.

(iv) The DO content in self-purification of stream is nearly to its saturation value in :

(*a*) zone of active decomposition

(*b*) zone of recovery

(*c*) zone of degradation
(*d*) None of the above.

(v) Indices of pollution in case of stream are :
(*a*) Colour and turbidity (*b*) DO and BOD
(*c*) Flora and Fauna (*d*) All of the above.

(vi) The maximum damage to the 'Taj Mahal' is because of the gas :
(*a*) CO_2 (*b*) CO
(*c*) SO_2 (*d*) All of the above.

(vii) The popular treatment method of solid waste in India is :
(*a*) Incineration (*b*) Landfill
(*c*) Pyrolysis (*d*) None of the above.

(viii) The mixed liquor suspended solid concentration in the basin of aeration system is 2200 mg/L and sludge volume after settling in a one litre graduated cylinder is 200 ml, the sludge volume index is :
(*a*) 2400 mg/L (*b*) 200 ml/L
(*c*) 91 ml/g (*d*) 91 mg/L.

(ix)* Determine the time of a waste stabilization pond of 1.5 m depth having a flow of 1.5 MLD and 300 mg/L BOD with a loading rate of 150 kg/hectare/day is in between (and how much ?)
(*a*) 10-40 hours (*b*) 1-4 days
(*c*) 10-40 days (*d*) None of the above.

Q. 37. State whether the following statements are True or False :

(i) Increase in temperature with height in the atmosphere is known as lapse rate.

(ii) Anaerobic process of composting is known as Bangalore process, while aerobic process of composting is known as Indore process.

Q. 38. Fill in the blanks :

(i) Sludge gas is produced in digestors due to action of ....

(ii) ...... as a gaseous pollutant combines with haemoglobin in blood to cause toxic effects.

Q. 39. State whether the following statements are True or False :

(i) Tricking filters in waste-water treatment are intended to remove dissolved organics

(ii)* High COD to BOD ratio for waste water indicates its high biodegradability

(iii)* Population equivalent of 2000 m^3/d of waste water with BOD of 200 mg/l at 50 gm/capita/day is 100000.

Q. 40. Match the following :

(*i*) Trickling filter (*a*) Do not require oxygen
(*ii*) Nitrates (*b*) Oxidation pond
(*iii*) Zeolites (*c*) Attached growth process
(*iv*) Anaerobes (*d*) Cause methemoglobinemia
(*v*) Bacterial algal symbiosis (*e*) Remove hardness in water

Q. 41. Select correct answer :

Compared to the activated sludge process, trickling filter

(*a*) withstands shock loads better

(*b*) does not withstand shock loads better

(*c*) effluent is well stabilized

(*d*) is difficult to maintain.

Q. 42. Choose correct answer :

(i) The recommended value of SVI in diffused aeration system should be :

(*a*) 5 to 10 (*b*) 10 to 25

(*c*) 25 to 55 (*d*) 55 to 150.

(ii) Eutrophic conditions in lakes are due to :

(*a*) excess algae (*b*) excess of colour

(*c*) excess of turbidity (*d*) None of the above.

(iii) Electronic precipitators are used to control :

(*a*) particulate matter (*b*) sulphur dioxide

(*c*) lead (*d*) None of the above.

(iv) Methane formers are

(*a*) obligate aerobes (*b*) facultative bacteria

(*c*) obligate anaerobes (*d*) None of the above.

Q. 43. Fill in the blanks :

(i) If the settling velocity of a particle is 1.4 mm/s and overflow rate is 0.056 mm/s, the percentage of solids removed in sedimentation tank is ....

(ii) A BOD_5 test is conducted on 4 ml of river water in 300 ml of BOD bottle. The initial DO is 7 mg/l and final DO is 0 mg/l. The BOD_5 of the waste water is ....

(iii) Pollutants in diesel exhaust are ...... (given in the order as per quantity wise).

(iv) The sanitary fixture provided in water closet to prevent entry of foul gases into houses from septic tank is ....

(v) The type of settling which occurs in grit chamber is ....

Q. 44. State whether the following statements are True or False :

(i) The concentration of BOD_5 of an industrial waste with high concentration of cyanides is zero.

(ii) Trickling filters are an example of suspended growth of aerobic biological treatment.

Q. 45. Match the following sets :

Set A	*Set B*
(*i*) Autotrophic bacteria	(*a*) Dissolved oxygen sag curve
(*ii*) Reflux valve	(*b*) Anaerobic bacteria
(*iii*) Septic tank	(*c*) Suspended particulate matter
(*iv*) Streeter Phelps equation	(*d*) Flow allows one direction only
(*v*) High volume sampler	(*e*) Utilize CO_2 as food carbon source.

Q. 46. Choose the correct answer :

(i) The most commonly used disinfectant of wastewater is :

(*a*) Chlorine dioxide (*b*) Chlorine

(*c*) Ozone (*d*) U.V. radiation.

(ii) Activated sludge process is used to remove :

(*a*) settleable organic matter

(*b*) soluble organic matter

(*c*) pathogens

(*d*) inorganic matter and toxic chemicals.

(iii) Sewage contains a solid content of :

(*a*) 1% (*b*) 0.5%

(*c*) 0.1% (*d*) 5%.

(iv) Major air pollutants from a thermal power station are :

(*a*) Fly ash and PAN (*b*) Fly ash and ozone

(*c*) Fly ash, hydrogen fluoride and H_2S

(*d*) Fly ash and SO_2.

Q. 47. Match the following :

(i) (*a*) Cyclone separators (1) Automobiles

(*b*) Euglena (2) Sulphides

(*c*) Pink sulfur bacteria (3) Algae

(*d*) Benzopyrene (4) Particulate matter

(ii) (*a*) Splash pad (1) Grease removal

(*b*) Ether soluble solids (2) Bulking of sludge

(*c*) SVI (3) Activated sludge process

Q. 48. Choose the correct answer :

(i) Pyrolysis is a process in :

(1) Water treatment (2) Vehicular air pollution

(3) Solid waste management (4) Industrial waste treatment.

(ii) The hydraulic surface loading in a primary sedimentation tank is :

(1) 2.5 m/h (2) 25 m/h

(3) 0.5 m/h (4) 10 m/h.

(iii) The electrostatic precipitators are used in :

(1) waste water treatment (2) thermal power plant

(3) automobiles (4) nuclear power plants.

(iv) Vanadium pentoxide is used as a catalyst to treat :

(1) Sulphur dioxide in exhausts

(2) Carbon monoxide in exhausts

(3) Arsenic in water

(4) Chromium in waste water.

(v) Desulpho vibrio is responsible for :

(1) Denitrification (2) Nitrification

(3) Sulphate reduction (4) None of the above.

(vi) $\overline{Gt}$ value is associated with :

(1) Noise pollution (2) Crank case emission control

(3) Flocculation (4) None of the above.

Q. 49. Choose the correct answers :

(i) A sloped floor of a manhole on both sides of the channel on which a man can stand for cleaning the sewer, is called :

(*a*) manhole stand (*b*) benching

(*c*) wash stand (*d*) cleaning platform.

(ii) Bacteria which use carbon dioxide as a source of carbon, are known as :

(*a*) microbacteria (*b*) heterotrophic

(*c*) aerobic (*d*) anaerobic.

(iii) In ASP unit F/M ratio should be between :

(*a*) 0.2 to 0.4 (*b*) 0.6 to 0.8

(*c*) 0.8 to 1 (*d*) 1.1.

(iv) The recommended value of SVI in diffused aeration system should be :

(*a*) 5 to 10 (*b*) 20 to 30

(*c*) 30 to 40 (*d*) 55 to 150.

(v) For an average Indian city, solid waste generated per capita may be assessed as :

(*a*) 0.1 kg (*b*) 0.4 kg

(*c*) 1.0 kg (*d*) 2.0 kg.

(vi) Silicosis is caused in the :

(*a*) textile industries (*b*) sugar industries

(*c*) storage battery industries

(*d*) stone crushers.

(vii) The sludge does not contain waste water from :

(*a*) both rooms (*b*) wash basins

(*c*) kitchen sinks (*d*) toilets.

(viii) If q is the average sewage flow from a city of population P in thousand, the maximum sewage flow Q is :

(*a*) $Q = \dfrac{4+\sqrt{P}}{18+\sqrt{P}} \cdot q$ (*b*) $Q = \dfrac{18+P}{4+P} \cdot q$

(*c*) $Q = \dfrac{18+\sqrt{P}}{4+\sqrt{P}} \cdot q$ (*d*) $Q = \dfrac{18 . \sqrt{P}}{4\sqrt{P}+1} \cdot q.$

(ix) If D is the diameter of upper circular portion, the overall depth of standard egg-shaped sewer is :

(*a*) D (*b*) $1.25D$

(*c*) $1.50D$ (*d*) $2.0D$.

(x) The first stage of sludge digestion process is :

(*a*) acid fermentation (*b*) acid regression

(*c*) alkaline fermentation (*d*) None of these.

Q. 50. Choose the correct statement/answer(s) :

(i) Sewage sickness of land can be occurred due to :

(*a*) Pores of soil get clogged

(*b*) No air circulation through pores of soil

(*c*) Anaerobic conditions prevailed

(*d*) All of the above.

(ii) Which one of the following solid waste disposal method is ecologically most acceptable ?

(*a*) Sanitary land fill (*b*) Incineration

(*c*) Composting (*d*) Pyrolysis.

Q. 51. Match the pairs in the following sets :

(*i*) Septic tank	(*a*) To prevent entry of foul gases
(*ii*) Water seal	(*b*) Anaerobic treatment unit

Q. 52. Answer briefly :

(i)* What is the role of protozoa in activated sludge process ?

(ii) What is NOD ?

(iii) What is antisiphonage pipe ?

(iv)* An aeration tank has a MLSS concentration of 2000 mg/l. After settling for 30 minutes in one litre of graduated cylinder, the sludge is measured to be 150 ml. Calculate the SVI of the sludge.

(v) What is PM_{10} ?

(vi) What is meant by ozone depletion ?

Answers

1 **(i).** (Muffle) 1 **(ii).** (0.6)

2 (*i*)—(*c*) ; (*ii*)—(*d*) ; (*iii*)—(*b*) ; (*iv*)—(*a*) ; (*v*)—(*e*)

3 **(i).** (*c*) 3 **(ii).** (*b*) 3 **(iii).** (*d*) 3 **(iv).** (*a*) 3 **(v).** (*d*)

4 **(i).** (Kjedahl) 4 **(ii).** (20)

5 (*i*)—(*c*) (*ii*)—(*e*) (*iii*)—(*d*) (*iv*)—(*a*) (*v*)—(*b*)

6 **(i).** (True) 6 **(ii).** (True)

7 **(i).** (*d*) 7 **(ii).** (*d*)

8 **(i).** (*b*)

9 **(i).** COD 9 **(ii).** algal-bacterial symbiosis

10 (*i*)—(*b*) ; (*ii*)—(*a*) ; (*iii*)—(*e*) ; (*iv*)—(*c*) ; (*v*)—(*d*)

11 **(i).** True 11 **(ii).** False.

12 **(i).** (*d*) 12 **(ii).** (*a*)

13 **(i).** 0.6 14 **(i).** False

15 (*i*)—(*c*) (*ii*)—(*b*) (*iii*)—(*a*) (*iv*)—(*d*)

16 **(i)**—(*c*) 16 **(ii)**—(*d*) 16 **(iii)**—(*b*) 16 **(iv)**—(*d*)

16 **(v)**—(*c*) 16 **(vi)**—(*b*) 16 **(vii)**—(*d*) 16 **(viii)**—(*d*)

17 (*b*) 18 Algal symbiosis

19 **(i).** True 19 **(ii).** False

20 (*i*)—IV, (*ii*)—III, (*iii*)—I, (*iv*)—II.

21 (i)—(*d*) **21 (ii)**—(*d*) **21 (iii)**—(*b*) **21(iv)**—(*b*)
21 (v)—(*d*) **21(vi)**—(*c*)
22 (i)—CO_2 **22 (ii)**—7
23 (i)—(*c*) **23 (ii)**—(*c*) **23 (iii)**—(*d*)
23 (iv)— (*a*) **23 (v)**—(*c*) **23 (vi)**—(*c*)
24 (i)—anaerobic in nature
24 (ii)—storage in safe depositories (vaults), to be designed for placement at safe and predetermined sites, either above the ground or below the ground
24 (iii)—vetrified clay sewers or salt glazed sewer pipes.
25 (i)—(*d*) **25 (ii)**—(*c*)
26 (*i*)—(II) ; (*ii*)—(V) ; (*iii*)—(IV) ; (*iv*)—(I) ; (*v*)—(III)
27 False
28 (i)—(*b*) **28 (ii)**—(*d*) **28 (iii)**—(*c*) **28(iv)**—(*d*)
28 (v)—(*c*) **28 (vi)**—(*b*) **28 (vii)**—(*b*) **28(viii)**—(*d*)
29 (i)—(*b*) **29 (ii)**—(*c*) **29 (iii)**—(*d*)
30 (*i*)—(*d*) ; (*ii*)—(*a*) ; (*iii*)—(*c*) ; (*iv*)—(*e*) ; (*v*)—(*b*)
31 (i)—infiltration ; **31 (ii)**—organic
31 (iii)—reducing the emission of pollutants from automobiles and factories by adopting mechanical means, or by using high rise chimneys for better dispersion and dilution of pollutants over a longer range of environment.
31 (iv)—fog.
32 (i)—BOD **32. (ii)**—algae and bacteria/protozoa
33 (i)—False **33. (ii)**—False
34 (*i*)—(*e*) ; (*ii*)—(*d*) ; (*iii*)—(*c*) ; (*iv*)—(*b*) ; (*v*)—(*a*)
35 (i)—(*a*) **35 (ii)**—(*d*).
36 (i)—(*c*) **36 (ii)**—(*b*) **36 (iii)**—(*d*) **36 (iv)**—(*d*)
36 (v)—(*b*) **36 (vi)**—(*c*) **36 (vii)**—(*b*) **36 (viii)**—(*c*)
36 (ix)—(*c*) ; 30 days. **37 (i)**—(False) **37 (ii)**—(True).
38 (i)—methane formers **38 (ii)**—Carbon monoxide
39 (i)—False **39 (ii)**—False **39 (iii)**—False
40 (*i*)—(*c*) ; (*ii*)—(*d*) ; (*iii*)—(*e*) ; (*iv*)—(*a*) ; (*v*)—(*b*)
41 (*a*)
42 (i)—(*d*) **42 (ii)**—(*a*) **42 (iii)**—(*a*) **42 (iv)**—(*c*)
43 (i)—(100%) **43 (ii)**—525 mg/l **43 (iii)**—(NO_x, HC, CO)
43 (iv)—(Trap) **43 (v)**—(Type I settling)
44 (i)—(False) **44 (ii)**—(False)
45 (*i*)—(*e*) (*ii*)—(*d*) (*iii*)—(*b*) (*iv*)—(*a*) (*v*)—(*c*)
46 (i)—(*b*) **46 (ii)**—(*b*) **46 (iii)**—(*c*) **46 (iv)**—(*d*)
47 (i) (*a*)—(4), (*b*)—(3), (*c*)—(2), (*d*)—(1)
47 (ii) (*a*)—(2) (*b*)—(1) (*c*)—(3)
48 (i)—(3) **48 (ii)**—(1) **48 (iii)**—(2) **48 (iv)**—(1)

48 **(v)**—(3) **48 (vi)**—(3)
49 **(i)**—(*b*) **49 (ii)**—(*b*) **49 (iii)**—(*a*) **49 (iv)**—(*d*)
49 **(v)**—(*b*) **49 (vi)**—(*d*) **49 (vii)**—(*d*) **49 (viii)**—(*c*)
49 **(ix)**—(*c*) **49 (x)**—(*a*)
50 **(i)**—(*d*) **50 (ii)**—(*c*)
51 (*i*)—(*b*) (*ii*)—(*a*)
52 **(iv)**—(75 ml/g)

Hints For Solving Starred Questions

Q. 3. (ii) Concrete sewers are corroded by H_2S gas, that may be produced by anaerobic decomposition of sewage, when it becomes stale.

Q. 3. (iii) Large amount of iron is found in soils, and hence the possibility of its presence in deep tubewell water is high.

Q. 4. (*ii*) Pl. refer Col. (3) of Table 8.2.

Q. 7. (*i*) Pl. refer Table 7.3.

Q. 16. (ii) Corrosion of sewers, particularly at the crowns, is caused by the hydrogen sulphide gas, which is usually evolved when sewage becomes stale, particularly in sewers which are constructed on flat gradients in warm climates. The evolved H_2S gas is absorbed in the water condensed on the side walls and crowns of the sewer pipe. Here the ***sulphur bacteria***, able to tolerate pH levels of less than 1.0, oxidise the weak acid H_2S to strong sulphuric acid (H_2SO_4) by using oxygen from air in the sewer. The sulphuric acid may react with concrete sewer, reducing its structural strength. When the sewer pipe gets sufficiently weakened, it may collapse under the heavy overburden loads. Use of corrosion resistant pipe materials such as vitrified clay, or PVC plastic, may offer protection from corrosion. In large sewers, where size and economics dictate the use of concrete sewers, crown corrosion can be reduced by ventilation, to help expel H_2S gas to reduce moisture of condensation. Protection of concrete pipe interior by providing coatings of coal tar, vinyl or epoxy may also sometimes be adopted to reduce corrosion.

Besides H_2S causing corrosion of upper portion of sewer pipe, corrosion of pipe bottom may be caused by the acidic industrial waste waters that may enter the municipal sewers. In such eventualities, the inflow of acidic wastewaters are either limited or excluded. Sometimes, corrosion resistant liners, such as vitrified clay plates, may be placed in the invert of the sewer for protection.

Q. 16. (iii) Depending upon their source of nutrients, bacteria are classified in the following two major groups :

(1) *heterotrophic* ; and

(2) *autotrophic.*

(1) **Heterotrophic bacteria,** also sometimes called as *saprophytes, use organic matter as* both an energy and *a carbon source*, for synthesis. These bacteria are further divided into : aerobic, non-aerobic, and facultative groups.

(2) **Autotrophic bacteria** oxidise inorganic matter for energy and *use carbon dioxide as a carbon source*. The three types of commonly found

autotrophic bacteria, are : (*a*) *Nitrifying bacteria* ; (*b*) *Sulphur bacteria* ; and (*c*) *Iron bacteria*.

The ***nitrifying bacteria*** oxidise Ammonium nitrogen into nitrates.

The ***sulphur bacteria*** cause crown corrosion in concrete sewers as explained in previous question.

The ***iron bacteria*** oxidise soluble inorganic ferrous iron to insoluble ferric iron. They thrive in water pipes, where dissolved iron is available as an energy source, and bicarbonates are available as a carbonate source. These bacteria die and decompose with time, releasing foul tastes and odours.

Q. 24 (i) The Bio-gas is popularly known as **Gobar gas** in India, mainly because cow dung has been the main material for its production. The cattle waste (gobar), piggery waste and poultry droppings are very effectively used in biogas generation. The biogas is produced by anaerobic digestion of wastes which is well discussed in the text, and contains about 65-70% methane and 30% carbon dioxide ; and the rest being the impurities (H_2, H_2S and some N_2). Gobar gas plants are the most effective and useful tools for disposal of cattle and poultry wastes. Standard designs of various types of gobar gas plants are given in author's another book titled "*Ecology & Environmental Studies*", and may be referred to in special needs.

Q. 35 (i) Please read article 8.4.5.

Q. 36 (ix). *Please see Solved Example 9.31 and compute as under :*

$$Q = 1.5 \text{ MLD}$$

$$\text{BOD} = 3000 \text{ mg}/l$$

BOD load of sewage flow/day

$$= (1.5 \times 10^6 \times 3000) \text{ mg/d} = \frac{1.5 \times 10^6 \times 3000}{10^6} \text{ kg/d} = 4500 \text{ kg/d}.$$

Organic loading = 150 kg/ha/d

Surface area of oxidation pond reqd.

$$= \frac{4500 \text{ kg/d}}{150 \text{ kg/ha/d}} = 30 \text{ ha} = 30 \times 10^4 \text{ m}^2$$

If $L = 2B$, then $2B^2 = 30 \times 10^4 \text{ m}^2$

or $$B = \sqrt{15 \times 10^4} = 387.3 \text{ m} ; L = 2 \times 387.3 = 774.6 \text{ m}$$

Effective depth = d = 1.5 m (given)

Capacity = 387.3 × 774.6 × 1.5 = 45000 m³.

$$\text{Detention time in days} = \frac{\text{Capacity in cum}}{\text{Sewage flow in m}^3/\text{d}} = \frac{45000 \text{ m}^3}{\frac{1.5 \times 10^6}{10^3} \text{ m}^3/\text{d}}$$

$$= \frac{45000 \text{ m}^3}{1500 \text{ m}^3} \text{d} = \textbf{30 days.} \quad \textbf{Ans.}$$

Q. 39. (ii) For high biodegradability, BOD should be more ; then eventually $\frac{\text{COD}}{\text{BOD}}$ should be less. Hence, low value $\frac{\text{COD}}{\text{BOD}}$ ratio will represent high biodegradability. The given statement is, thus, false.

Q. 39. (iii) Use Eq. (7.20) :

BOD of industrial waste

= (P.E.) × Standard Domestic BOD per person per day

$$\therefore \quad 2000 \frac{m^3}{d} \times 200 \text{ mg/l} = (\text{P.E}) \times 50 \text{ gm/person/d}$$

$$\text{or} \qquad \text{P.E.} = \frac{2000 \times 10^3 \frac{l}{d} \times \frac{200}{1000} \frac{g}{l}}{50 \text{ gm/person/d}}$$

$$= \frac{400 \times 10^3}{50} \text{ persons} = \mathbf{8000.}$$

Hence, the given answer of 100000 is false.

Q. 52. (i) Protozoa are *motile, micro-scopic single cell* animals, representing the simplest and lowest form of *animal* life. They can be directly counted under a microscope. The majority of protozoa are aerobic heterotrophs, although a few are anaerobic. Protozoa are, generally, inorder of magnitude, larger than bacteria and often consume bacteria as an energy source. *In effect, the protozoa act as polishers of the effluents from biological waste treatment processes by consuming bacteria and particulate organic matter. They thus, play an important role in activated sludge process (aerobic oxidation) by causing oxidation of organic matter, thereby helping in the treatment of waste water.*

Q. 52. (iv) Use Eq. (9.52) as :

$$SVI \text{ in } \frac{\text{mL}}{\text{g}} = \left[\frac{V_{ob} \text{ in } \frac{\text{ml}}{\text{l}}}{X_{ob} \text{ (MLSS) in } \frac{\text{mg}}{\text{l}}} \times 1000 \right]$$

$$= \frac{150}{2000} \times 1000 = \mathbf{75 \text{ ml/g} \quad Ans.}$$

—26

Objective Question of the Combined Engineering Services (CES) Exams

Year 1993

Q. 1.* Match List I with List II and select the correct answer using the codes given below the lists :

List I (*Treatment units*)	List II (*Detention time*)
A. Grit chamber	**1.** Six hours
B. Primary sedimentation	**2.** Two minutes
C. Activated sludge	**3.** Two hours
D. Sludge digestion	**4.** Twenty days

Codes	A	B	C	D
(*a*)	3	1	4	2
(*b*)	2	3	1	4
(*c*)	2	1	3	4
(*d*)	1	2	3	4

Q. 2.* Which one of the following statements is true of tricking filter sludge ?

(*a*) It has a comparatively low sludge volume index.

(*b*) It is more difficult to dewater than activated sludge.

(*c*) It has a comparatively low concentration of sludge solids.

(*d*) It is bulky.

Q. 3. Match List I with List II and select the correct answer using the codes given below the lists :

List I (*Pipe material*)	List II (*Property of material*)
A. Concrete sewer	**1.** Cannot withstand high external load
B. Stone ware sewer	**2.** Corrosion resistance in most natural soils
C. Cast iron sewer	**3.** Resistant to corrosion from most acids
D. Steel sewer	**4.** Unsuitable where soil contains excessive sulphates

*Hints for solving the starred Questions are given at the end of the chapter.

Codes	A	B	C	D
(*a*)	1	2	3	4
(*b*)	4	3	2	1
(*c*)	4	1	2	3
(*d*)	2	1	3	4

Q. 4. BOD is preferred to COD as an index of sewage concentration, because :

(*a*) BOD represents both carbonaceous and nitrogenous organic matter, while COD may indicate carbonaceous matter only.

(*b*) BOD test is easier to perform and gives more reliable results.

(*c*) BOD relates specifically to putrescible organic matter which is the most objectionable sewage constituent.

(*d*) COD relates to the impurities which can only be removed by chemical treatment which is expensive.

Q. 5.* Under Indian conditions, the average per capita contribution of BOD is :

(*a*) 10 to 20 gm/d (*b*) 20 to 35 gm/d

(*c*) 35 to 50 gm/d (*d*) 50 to 70 gm/d.

Q. 6. Match List I with List II and select the correct answer using the codes given below the lists :

List I	List II
A. Waste pipe	**1.** Carries waste water.
B. Soil pipe	**2.** Carries liquid wastes that do not include human excreta.
C. Vent pipe	**3.** Preserves the water seal of traps through access to atmospheric air.
D. Antisiphonage pipe	**4.** Carries liquid wastes including human excreta.
	5. Provides flow of air to or from drainage system in order to prevent vacuum pressures and excessive pressure and provides escape for foul gases.

Codes	A	B	C	D
(*a*)	2	4	5	3
(*b*)	3	5	1	2
(*c*)	4	5	1	2
(*d*)	1	4	5	3

Q. 7. Match List I with List II and select the correct answer using the codes given below the lists :

List I (*Cause*)	List II (*Effect*)
A. Carbon monoxide	**1.** Acid rain
B. Carbon dioxide	**2.** Explosion
C. Methane	**3.** Asphyxiation
D. Sulphur dioxide	**4.** Green house effect

Codes	A	B	C	D
(*a*)	2	3	1	4
(*b*)	3	4	2	1
(*c*)	1	3	4	2
(*d*)	4	2	1	3

Q. 8. Sewage sickness relates to :

(*a*) toxicity of sewage interfering with response to treatment.
(*b*) destruction of aquatics flora and fauna due to gross pollution of receiving bodies of water by sewage.
(*c*) reduction in the waste purifying capacity of the soil.
(*d*) clogging of pores in soil due to excessive application of sewage to land, obstructing aeration and leading to septic conditions.

Q. 9*. Under natural conditions of flow, an unpolluted river would contain

(*a*) more dissolved oxygen in summer than in winter.
(*b*) less dissolved oxygen in summer than in winter.
(*c*) more or less the same amount of dissolved oxygen in winter and summer.
(*d*) the least amount of dissolved oxygen during the floods.

Year 1994

Q. 10*. Consider the data presented in the following table :

Temperature in °C	*BOD reaction rate constant (K)*
20	0.01
30	0.02
10	0.005

In the data presented above, the value of K

(*a*) should have remained constant
(*b*) should have decreased with increase in temperature
(*c*) should have remained the same at 20°C and 30°C
(*d*) has followed the correct trend.

Q. 11. For the combined sewerage system, egg-shaped sewers are preferred because :

(*a*) their construction is economical
(*b*) they are structurally more stable
(*c*) their maintenance is easier
(*d*) they offer good flow velocity during the dry weather flow condition.

Q. 12. Match List I with List II and select the correct answer using the codes given below the lists :

List I (*Terms / Description*)	List II (*Treatment operation / process*)
A. Sludge volume index	**1.** Settling primary sedimentation tank
B. Thickening of sludge	**2.** Settling in secondary sedimentation tank
C. Scum removal	**3.** Filtration in trickling filter
D. Recycling of effluent	**4.** Activated sludge process

Codes	A	B	C	D
(*a*)	2	4	1	3
(*b*)	4	2	3	1
(*c*)	2	4	3	1
(*d*)	4	2	1	3

Q. 13. An aeration basin with a volume of 400 m^3 contains mixed liquor with suspended solid concentration of 1000 mg/l. The amount of mixed liquor suspended solids in the tank is

(*a*) 500 kg (*b*) 250 kg

(*c*) 600 kg (*d*) 400 kg.

Q. 14.* The following reactions take place during anaerobic digestion of organics:

1. Methane production **2.** Alkaline fermentation

3. Acid fermentation **4.** Acid regression

The correct sequence of these reactions is

(*a*) 3, 4, 2, 1 (*b*) 4, 3, 2, 1

(*c*) 3, 4, 1, 2 (*d*) 4, 3, 1, 2.

Q. 15. Traps are used in household drainage systems to :

(*a*) prevent entry of foul gases in the houses

(*b*) restrict the flow of water

(*c*) provide a partial vacuum

(*d*) trap the solid wastes.

Q. 16.* Modern coal-based thermal power stations pollute the atmosphere by adding :

(*a*) NO_x and SO_2 (*b*) NO_x, SO_2 and SPM

(*c*) NO_x, SO_2, SPM and CO (*d*) NO_x, SPM and CO.

Q. 17. Consider the following statements :

Excessive growth of water weeds in a water body is attributed to the:

1. increase in the benthic organisms including bacteria

2. imbalance in aquatic ecosystem

3. excessive inflow of nutrients

Of these statements

(*a*) 1, 2 and 3 are correct (*b*) 1 and 2 are correct

(*c*) 1 and 3 are correct (*d*) 2 and 3 are correct.

Q. 18. Area method of land filling is most suitable when :

(*a*) area is unsuitable for excavation of trenches.

(*b*) adequate depth of cover material is available at the site

(*c*) the water-table is near the surface

(*d*) natural or artificial depressions exist in the vicinity.

Q. 19. Various unit operations exist in a sewage treatment plant. These would include :

1. screening

2. grit removal

3. secondary sedimentation

4. aeration

5. primary sedimentation

The correct sequence of these operations is

(*a*) 1, 2, 3, 4, 5 (*b*) 1, 2, 5, 4, 3

(*c*) 2, 1, 4, 5, 3 (*d*) 2, 1, 4, 3, 5.

Year 1995

Q. 20.* In transition of sewers from smaller diameter sewers to larger diameters sewers, the continuity of sewers is maintained at the :

(*a*) bottom of the concrete bed of sewers

(*b*) inverts of the sewers

(*c*) crowns of the sewers

(*d*) hydraulic gradients of the sewers.

Q. 21. The slope of a 1.0 m diameter concrete sewer laid at a slope of 1 in 1000 develops a velocity of 1 m/s, when flowing full. When it is flowing half full, the velocity of flow through the sewers will be :

(*a*) 0.5 m/s (*b*) 1.0 m/s

(*c*) $\sqrt{2}$ m/s (*d*) 2.0 m/s.

Q. 22.* Match List I with List II and select the correct answer using the codes given below the lists :

List I (*Process*)	List II (*Biological agent*)
A. Oxidation ditch	**1.** Facultative bacteria
B. Waste stabilisation pond	**2.** Anaerobic bacteria
C. Imhoff tank	**3.** Aerobic bacteria (suspended culture)
D. Rotating Biological Contractor (RBC)	**4.** Aerobic bacteria (attached culture)

Codes	A	B	C	D
(*a*)	4	1	2	3
(*b*)	3	1	2	4
(*c*)	1	2	3	4
(*d*)	3	4	1	2

Q. 23.* One litre of sewage, when allowed to settle for 30 minutes gives sludge of 27 cm^3. If the dry weight of sludge is 3.0 gm, then its sludge volume index will be :

(*a*) 9 (*b*) 24

(*c*) 30 (*d*) 81.

Q. 24. An industrial wastewater enters a stream having a BOD concentration of 10 mg/l and flow of 20 m^3/s. If the flow of wastewater is 1.5 m^3/s and its BOD concentration is 250 mg/l, then the BOD concentration in the stream at a point downstream of the point of confluence of wastewater with the stream will be :

(*a*) 2.67 mg/l (*b*) 12.09 mg/l

(*c*) 13.00 mg/l (*d*) 26.74 mg/l.

Q. 25. A polluted stream undergoes self purification in four distinct zones :

1. Zone of clear water
2. Zone of active decomposition
3. Zone of degradation
4. Zone of recovery

The correct sequence of these zones is :

(*a*) 3, 4, 2, 1 (*b*) 2, 3, 4, 1

(*c*) 2, 4, 3, 1 (*d*) 3, 2, 4, 1.

Q. 26. Match List I with List II and select the correct answer using the codes given below the lists :

List I	List II
A. Soil pipe	**1.** Ventilating pipe
B. Intercepting trap	**2.** Wash basin
C. p-trap	**3.** Water closet waste
D. Cowl	**4.** House drainage

Codes	A	B	C	D
(*a*)	3	4	1	2
(*b*)	3	4	2	1
(*c*)	4	3	2	1
(*d*)	4	3	1	2

Q. 27. Match List I with List II and select the correct answer using the codes given below the lists :

List I *(Pollutants)*	List II *(Effect produced)*
A. CO	**1.** Green house effect
B. CO_2	**2.** Acid rains
C. SO_2	**3.** Acute toxicity
D. NO_x	**4.** Ozone liberation at ground level

Codes	A	B	C	D
(*a*)	3	4	1	2
(*b*)	3	4	2	1
(*c*)	4	3	2	1
(*d*)	4	3	1	2

Q. 28.* The atmosphere extends up to a height of 10,000 km. It is divided into the following four thermal layers.

1. Mesosphere
2. Stratosphere
3. Thermosphere
4. Troposphere

The correct sequence of these layers starting from the surface of the earth upwards is :

(*a*) 2, 4, 1, 3 (*b*) 4, 2, 1, 3
(*c*) 4, 2, 3, 1 (*d*) 2, 4, 3, 1.

Q. 29. Which one of the following solid waste disposal methods is ecologically most acceptable ?

(*a*) Sanitary land fill (*b*) Incineration
(*c*) Composting (*d*) Pyrolysis.

Year 1996

Q. 30. The waste stabilisation ponds can be :

(*a*) aerobic (*b*) anaerobic
(*c*) facultative (*d*) any of the above.

Q. 31. Which of the following pairs is not correctly matched ?

(*a*) BOD—Strength of sewage
(*b*) Methane—Product of anaerobic decomposition
(*c*) COD—Biodegradability of wastewater
(*d*) Nitrate—Methemoglobinemia.

Q. 32. The following are the sewage treatment processes :

1. Primary sedimentation
2. Screening
3. Grit removal
4. Secondary sedimentation.

When only preliminary treatment is to be given for sewage, select the required treatment processes including their correct sequence from the codes given below :

Codes :

(*a*) 2, 3 (*b*) 2, 3, 1
(*c*) 1, 2, 3, 4 (*d*) 3, 1, 2, 4.

Q. 33. The formulation for BOD assimilation in a stream should include :

(*a*) BOD rate constant
(*b*) Sedimentation of organic matter
(*c*) BOD rate constant and sedimentation of organic matter
(*d*) Pathogenic bacterial decay coefficient.

Q. 34. When was the Water (Prevention and Control of Pollution) Act enacted by the Indian Parliament ?

(*a*) 1970 (*b*) 1974
(*c*) 1980 (*d*) 1985.

Q. 35. The entry of foul smelling gases into the house coming from the sewers can be prevented by :

(*a*) providing water seals for all the fixtures

(*b*) providing water seals for all the fixtures and a vent pipe in the plumbing system

(*c*) providing efficient vent pipes in the plumbing system

(*d*) exhaust fans.

Q. 36. Ringlemann's scale is used to :

(*a*) measure CO

(*b*) measure CO_2

(*c*) grade density of smoke

(*d*) grade automobile exhaust gas.

Q. 37. Eutrophication of water bodies is caused by the :

(*a*) discharge of toxic substances

(*b*) excessive discharge of nutrients

(*c*) excessive discharge of suspended solids

(*d*) excessive discharge of chlorides.

Q. 38. Non disposal of solid waste may cause the spread of :

(*a*) malaria
(*b*) rodents related plague
(*c*) typhoid
(*d*) dysentery.

Year 1997

Q. 39. Self cleansing velocity is :

(*a*) the minimum velocity of flow required to maintain a certain amount of solids in the flow

(*b*) the maximum velocity of flow required to maintain a certain amount of solids in the flow

(*c*) such flow velocity as would be sufficient to flush out any deposited solids in the sewer

(*d*) such flow velocity as would be sufficient to ensure that sewage does not remain in the sewer.

Q. 40. The following steps are involved in laying a sewer in a trench :

1. Transferring the centre line of the sewer to the bottom of the trench
2. Setting sight rails over the trench
3. Driving pegs to the level of the invert line of the sewer
4. Placing the sewer in the trench.

The correct sequence of these steps is :

(*a*) 1, 2, 3, 4
(*b*) 2, 3, 4, 1
(*c*) 4, 2, 3, 1
(*d*) 2, 3, 1, 4.

Q. 41.* A grit chamber of dimensions 12.0 m × 1.50 m × 0.80 m liquid depth has a flow of 720 m^3/hr. Its surface loading rate and detention time are, respectively :

(*a*) 40,000 $m^3/hr/m^2$ and 1.2 minute

(*b*) 40,000 lph/m^2 and 40 minutes

(*c*) 40 $m^3/hr/m^2$ and 12 minutes

(*d*) 40,000 lph/m^2 and 1.2 minutes.

Q. 42. Match List I (Impurities to be removed from sewage) with List II (Treatment unit used) and select the correct answer using the codes given below the lists :

List I	List II
A. Large floating matter	**1.** Trickling filter
B. Suspended inorganic matter	**2.** Primary clarifier
C. Suspended organic matter	**3.** Grit chamber
D. Dissolved organic matter	**4.** Screens.

Codes	A	B	C	D
(*a*)	3	4	2	1
(*b*)	3	4	1	2
(*c*)	4	3	2	1
(*d*)	4	3	1	2

Q. 43.* Which of the following are claimed as advantages in respect of aerobic sludge digestion as compared to anaerobic sludge digestion ?

1. Lower BOD concentration in supernatant liquor
2. Production of a sludge with excellent dewatering propensity
3. Greater production of methane
4. Lesser operating cost
5. Lesser capital cost.

Select the correct answer using the codes given below :

Codes :

(*a*) 1, 2 and 4 (*b*) 2, 3, 4 and 5

(*c*) 3, 4 and 5 (*d*) 1, 2 and 5.

Q. 44. The following data pertain to a sewage sample :

Initial D.O.	= 10 mg/l
Final D.O.	= 2 mg/l
Dilution	= 1%

The BOD of the given sewage sample is

(*a*) 8 mg/l (*b*) 10 mg/l

(*c*) 100 mg/l (*d*) 800 mg/l.

Q. 45. Match List I with List II and select the correct answer using the codes given below the lists :

List I	List II
A. Sludge disposal	**1.** Seeding
B. Sludge digestion	**2.** Biofilters
C. Aerobic action	**3.** Lagooning
D. Recirculation	**4.** Contact bed

Codes	A	B	C	D
(*a*)	3	1	4	2
(*b*)	3	1	2	4
(*c*)	1	3	2	4
(*d*)	1	3	4	2

Q. 46. Which one of the following would help to prevent the escape of foul sewer gases from a water closet ?

(*a*) Air gap (*b*) Vent pipe

(*c*) Gully trap (*d*) None of the above.

Q. 47. Sewage sickness occurs when :

(*a*) sewage contains pathogenic organisms

(*b*) sewage enters the water supply system

(*c*) sewers get clogged due to accumulation of solids

(*d*) voids of soil get clogged due to continuous application of sewage on a piece of land.

Q. 48. In the design of storm sewers, "time of concentration" is relevant to determine the :

(*a*) rainfall intensity (*b*) velocity in the sewer

(*c*) time of travel (*d*) area served by the sewer.

Q. 49. Self purification of running streams may be due to :

(*a*) sedimentation, oxidation and coagulation

(*b*) dilution, sedimentation and oxidation

(*c*) dilution, sedimentation and coagulation

(*d*) dilution, oxidation and coagulation.

Q. 50. Which of the following waste disposal tasks are achieved by a septic tank with its dispersion trench ?

1. Aerobic sludge digestion
2. Settling and anaerobic sludge digestion
3. Anaerobic sewage stabilisation
4. Bioxidation of effluent.

Codes :

(*a*) 1 and 3 (*b*) 3 and 4

(*c*) 2 and 4 (*d*) 1 and 4.

Q. 51. Which of the following air pollutants is/are responsible for photochemical smog ?

1. Oxides of nitrogen 2. Ozone

3. Unburnt hydrocarbons 4. Carbon monoxide

Codes :

(*a*) 1 alone (*b*) 2, 3 and 4

(*c*) 1, 3 and 4 (*d*) 1 and 3.

Q. 52. Match List I (Equipment) with List II (Pollutants removed) and select the correct answer using the codes given below the lists :

List I	List II
A. Electrostatic precipitators	**1.** Coarse particles
B. Cyclones	**2.** Fine dust
C. Wet scrubbers	**3.** Gas
D. Absorbers	**4.** Sulphur dioxide

Codes	A	B	C	D
(*a*)	1	2	3	4
(*b*)	2	1	3	4
(*c*)	2	1	4	3
(*d*)	1	2	4	3

Q. 53.* Which of the following pairs is/are correctly matched ?

1. Eutrophication — Nutrient accumulation leading to ecosystem change occurring in impounded water
2. Autotrophism — Utilization, rearrangement and decomposition of complex materials predominate
3. Heterotrophism — Predominance of fixation of light energy, use of simple inorganic substances and built up complex substances.

Select the correct answer using the codes given below :

Codes :

(*a*) 1, 2 and 3 (*b*) 1 alone
(*c*) 2 and 3 (*d*) 1 and 3.

Q. 54. The typical density in kg/cu m (insitu) of well compacted municipal solid waste in landfill is in the range of :

(*a*) 100 to 300 (*b*) 310 to 500
(*c*) 550 to 850 (*d*) 900 to 1100.

Q. 55.* Which of the following materials are used as landfill sealants for the control of gas and leachate movements ?

1. Lime 2. Sand
3. Bentonite 4. Fly ash
5. Butyl rubber

Select the correct answer using the codes given below :

(*a*) 1, 2 and 3 (*b*) 4 and 5
(*c*) 3 and 5 (*d*) 1, 2 and 4.

Q. 56.* Two sources generate noise levels of 90 dB and 94 dB respectively. The cumulative effect of these two noise levels on the human ear is :

(*a*) 184 dB (*b*) 95.5 dB
(*c*) 94 dB (*d*) 92 dB.

Year 1998

Q. 57. The maximum flow occurs in an egg shaped sewer when the ratio of depth of flow to vertical diameter is :

(*a*) 0.33 (*b*) 0.50
(*c*) 0.95 (*d*) 1.00.

Q. 58. The following three stages are known to occur in the biological action involved in the process of sludge digestion :

1. Acid fermentation 2. Alkaline fermentation.
3. Acid regression.

The correct sequence of these stages is :

(*a*) 1, 2, 3 (*b*) 2, 3, 1

(*c*) 3, 1, 2 (*d*) 1, 3, 2.

Q. 59. Fresh sludge has moisture content of 99% and, after thickening, its moisture content is reduced to 96%. The reduction in volume of sludge is :

(*a*) 3% (*b*) 5%

(*c*) 75% (*d*) 97.5%.

Q. 60. Match List-I (Nature of the solids) with List-II (Unit operation or process connected with its removal) and select the correct answer using the codes given below the lists :

List-I	List-II
A. Dissolved solids	**1.** Sedimentation
B. Colloidal solids	**2.** Reverse osmosis
C. Volatile solids	**3.** Coagulation
D. Settleable solids	**4.** Digestion

Codes	A	B	C	D
(*a*)	2	3	4	1
(*b*)	3	2	4	1
(*c*)	2	3	1	4
(*d*)	3	2	1	4

Q. 61. In the oxidation ditch, the excess sludge is taken to :

(*a*) anaerobic digester (*b*) aerobic digester

(*c*) drying beds (*d*) incinerator.

Q. 62. The flow sheet of the liquid stream of a sewage treatment scheme consists of :

1. Trickling filter
2. Primary settling tank
3. Grit chamber
4. Screen chamber
5. Secondary settling tank.

The correct sequence of these units in the sewage treatment scheme of a liquid stream is

(*a*) 3, 4, 1, 2, 5 (*b*) 3, 4, 2, 1, 5

(*c*) 4, 3, 2, 1, 5 (*d*) 4, 3, 1, 2, 5.

Q. 63. Which one of the following pairs is NOT correctly matched ?

(*a*) Activated sludge ...... aeration

(*b*) Trickling filters ...... attached growth system

(*c*) Oxidation ditch ...... algae

(*d*) Channel grit chamber...... proportional weir

Q. 64. The two main gases liberated from an anaerobic sludge digestion tank would include :

(*a*) ammonia and carbon dioxide

(*b*) carbon dioxide and methane

(*c*) methane and hydrogen sulphide

(*d*) ammonia and methane.

Q. 65. Which one of the following sewage treatment units has a parshall flume?

(*a*) Trickling filter (*b*) Oxidation ditch

(*c*) Grit chamber (*d*) Aerated lagoon.

Q. 66.* Which one of the following principal types of reactors is related to trickling filter ?

(*a*) Plug flow (*b*) Complete-mix

(*c*) Packed-bed (*d*) Fluidized-bed.

Q. 67. Which one of the following is LEAST important in the activated sludge process ?

(*a*) Proper proportion of the return sludge from the secondary settling tank

(*b*) Adequate aeration in the biological reactor, so as to maintain certain minimum dissolved oxygen

(*c*) Proper food to micro-organisms (F : M) ratio

(*d*) The sludge volume index of the return sludge to be less than 200.

Q. 68. Sludge bulking can be controlled by :

(*a*) chlorination (*b*) coagulation

(*c*) aeration (*d*) denitrification.

Q. 69. The following zones are formed in a polluted river :

1. zone of clear water 2. zone of active decomposition

3. zone of recovery 4. zone of pollution.

The correct sequence in which these zones occur progressively downstream in a polluted river is :

(*a*) 4, 2, 1, 3 (*b*) 4, 2, 3, 1

(*c*) 2, 4, 3, 1 (*d*) 2, 4, 1, 3.

Q. 70. Match List-I (Pollutants) with List-II (Sources) and select the correct answer using the codes given below the lists :

List-I	List-II
A. Acid water	**1.** Volcanoes
B. SO_2	**2.** Automobiles
C. CO	**3.** Thermal power station
D. Fly ash	**4.** Mining

Codes	A	B	C	D
(*a*)	4	1	2	3
(*b*)	4	1	3	2
(*c*)	1	4	3	2
(*d*)	1	4	2	3

Q. 71. Aerosol is :

(*a*) carbon particles of microscopic size

(*b*) dispersion of small solid or liquid particles in gaseous media

(*c*) finely divided particles of ash

(*d*) diffused liquid particles.

Q. 72. Which one of the following methods can be employed for plastic and rubber waste disposal ?

(*a*) Composting (*b*) Incineration

(*c*) Sanitary landfill (*d*) Pyrolysis.

Year 1999

Q. 73. Consider the following statements :

Ventilation of sewer lines is necessary to :

1. avoid building up of sewer gases
2. ensure atmospheric pressure in the waste water surface
3. ensure the safety of sewer maintenance people
4. provide oxidation facility to sewage.

Which of these statements are correct ?

(*a*) 1, 2 and 4 (*b*) 1, 3 and 4

(*c*) 2, 3 and 4 (*d*) 1, 2 and 3.

Q. 74.* For a colony of 10,000 persons having sewage flow rate of 200 l/capita/day, BOD of applied sewage of 300 mg/l and organic loading of 300 kg/day/hectare, the area of an oxidation pond required for treating the sewage of the colony is :

(*a*) 0.2 hectares (*b*) 1 hectare

(*c*) 2 hectares (*d*) 6 hectares.

Q. 75. The second stage BOD as shown in the figure below is due to :

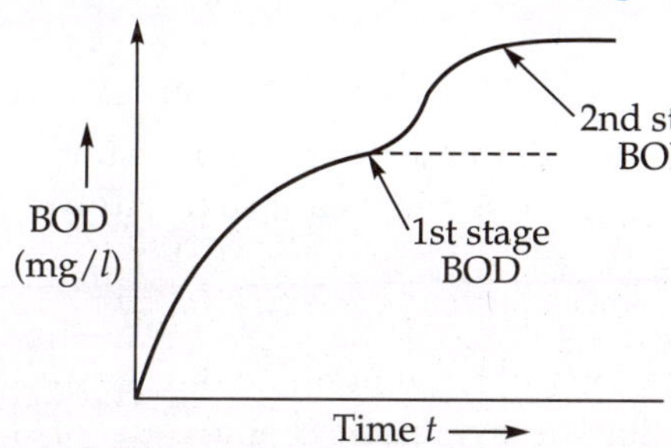

Fig. 26.1

(*a*) experimental error

(*b*) increased activity of bacteria

(*c*) nitrification demand

(*d*) interference by certain chemical reaction.

Q. 76. The purpose of a proportional weir at the effluent end of a channel type grit removal unit is to :

(*a*) provide easy passage of solid particles
(*b*) measure the rate of flow in the channel
(*c*) keep the depth of flow in the channel above a certain value
(*d*) maintain constant mean velocity in the channel.

Q. 77. The correct sequence of the sludge digestion steps is :
(*a*) Acid formation, hydrolysis, methane formation.
(*b*) Methane formation, acid formation, hydrolysis.
(*c*) Hydrolysis, methane formation, acid formation.
(*d*) Hydrolysis, acid formation, methane formation.

Q. 78. For fish habitat in a river, the minimum dissolved oxygen required is :
(*a*) 2 mg/l (*b*) 4 mg/l
(*c*) 8 mg/l (*d*) 10 mg/l.

Q. 79. Sewage sickness is a term used for :
(*a*) persons who become sick after drinking polluted water
(*b*) a treatment plant which does not function properly
(*c*) a stream where the flora and fauna die due to sewage flow
(*d*) the condition of land where sewage is applied continuously for a long period.

Q. 80. The trap used for a water closet is called :
(*a*) gully trap (*b*) *p*-trap
(*c*) intercepting trap (*d*) anti-siphon trap.

Q. 81. In a design of storm sewer, if the time taken by rain water to flow from the farthest point of the watershed to the sewer inlet is t_i, and the time of flow of water from the sewer inlet to the point in the sewer that is under consideration is t_f, then the time of concentration will be :
(*a*) t_i (*b*) t_f
(*c*) $t_i + t_f$ (*d*) t_i or t_f whichever is greater.

Q. 82. A circular sewer of diameter 1 m carries storm water to a depth of 0.75 m. The hydraulic radius is approximately :
(*a*) 0.3 m (*b*) 0.4 m
(*c*) 0.5 m (*d*) 0.6 m.

Q. 83. Consider the following statements about waste stabilization ponds :
1. The pond has a symbiotic behaviour of waste stabilization through algae on one hand and bacteria on the other.
2. The oxygen in ponds is provided by algae through photosynthesis.
3. The detention period for waste stabilization pond is of the order of two to three days.
4. The bacteria which develop in the pond, are aerobic bacteria.

Which of these statements are correct ?

(*a*) 1, 2 and 3 (*b*) 1, 3 and 4

(*c*) 2 and 4 (*d*) 1 and 2.

Q. 84. Match List I with List II and select the correct answer using the codes given below the lists :

List I (*Air Pollution*)	List II (*Environmental effect*)
A. Carbon dioxide	**1.** Respiratory distress for living beings
B. Particulate matter	**2.** Chemical reaction with haemoglobin in blood
C. Nitrogen oxides	**3.** Reduction in visibility and aero allergens
D. Sulphur dioxide	**4.** Photochemical smog in atmosphere

Codes	A	B	C	D
(*a*)	2	3	1	4
(*b*)	3	2	4	1
(*c*)	2	3	4	1
(*d*)	3	2	1	4

Q. 85.* If carbon monoxide is released at the rate of 0.03 m^3/min from a gasoline engine and 50 ppm is the threshold limit for an 8 hour exposure, the quantity of air which dilutes the contaminant to a safe level will be :

(*a*) 60 m^3/min (*b*) 600 m^3/min

(*c*) 600 m^3/s (*d*) 60 m^3/s.

Q. 86. Consider the following statements associated with water pollution parameters :

1. One of the primary indicators of the degree of water pollution is the concentration of organic matter
2. Total organic carbon (TOC), chemical oxygen demand (COD) and biochemical oxygen demand (BOD) are important parameters of water pollution
3. Generally TOC > COD > BOD.

Which of these statements are correct ?

(*a*) 1, 2 and 3 (*b*) 1 and 2

(*c*) 1 and 3 (*d*) 2 and 3.

Q. 87. Match List I (Methods of solid wastes disposal) with List II (Terms pertaining to the methods) and select the correct answer using the codes given below the list :

List I	List II
A. Incineration	**1.** Requires presorting, grinding and turning
B. Sanitary land fill	**2.** Limited to special wastes and selected materials
C. Composting	**3.** High operational and maintenance cost
D. Salvage by sorting	**4.** Tractor
	5. Rat and fly breeding

Codes	A	B	C	D
(*a*)	2	5	4	3
(*b*)	1	4	2	3
(*c*)	3	4	1	2
(*d*)	3	5	4	2

Q. 88.* The sound pressure level for a jet plane on the ground with sound pressure of 2000 m bar should be :

(*a*) 60 decibel (*b*) 100 decibel

(*c*) 140 decibel (*d*) 180 decibel.

Q. 89. Examine the 'Assertion A' and 'Reason R', and decide if the assertion A and the reason R are individually true, and if so, whether the reason is a correct explanation of the Assertion ; then mark your choice using the codes given below :

Codes :

(*a*) Both A and R are true and R is the correct explanation of A

(*b*) Both A and R are true and R is not a correct explanation of A

(*c*) A is true but R is false

(*d*) A is false but R is true.

(i) *Assertion (A) :* Vent should be provided over both the septic tank and the seepage pit.

Reason (R) : The vent pipes remove the gases that are generated due to bacterial action and avoid building up of pressure.

(ii) *Assertion (A) :* Sanitary land fills can no longer be used for the disposal of solid wastes in India.

Reason (R) : The leachates from sanitary landfills may pollute the ground water.

Year 2000

Q. 90. In which one of the following tests is the organic matter in the waste water used as food by micro-organisms ?

(*a*) BOD (*b*) Most probable number

(*c*) COD (*d*) chlorine demand.

Q. 91. The function of algae in an oxidation pond is to :

(*a*) provide a mat over the surface of the oxidation pond so as to prevent evaporation of water

(*b*) provide oxygen for bacteria to degrade organic matter

(*c*) provide a greenish appearance to the pond

(*d*) prevent the odour nuisance.

Q. 92. At an incubation temperature of 20°C, if initial D.O. (dissolved oxygen) and final D.O. values after 5 days incubation period on a 2% sample of sewage, are 8.5 mg/l and 5.5 mg/l, respectively, then the B.O.D. will be :

(*a*) 50 mg/l (*b*) 150 mg/l

(*c*) 250 mg/l (*d*) 350 mg/l.

Q. 93. Consider the following pairs of treatment units and impurities removed:

1. *Grit chamber*	*Sand and silt*
2. *Detritus tank*	*Organic matter*
3. *Primary sedimentation tank*	*Suspended impurities*
4. *Aeration tank of activated sludge process plant*	*Oil and grease*

Which of these pairs are correctly matched ?

(*a*) 1 and 2 (*b*) 1, 2, 3 and 4

(*c*) 2, 3 and 4 (*d*) 1 and 3.

Q. 94. When a sewage is disposed off in a river, the rate of depletion of dissolved oxygen of the river mainly depends on :

(*a*) biochemical oxygen demand of the sewage

(*b*) chemical oxygen demand of the sewage

(*c*) total organic carbon present in the sewage

(*d*) dissolved oxygen present in the sewage.

Q. 95. For a waste water, the 5-day BOD at 20°C is found to be 200 mg/l. For the same waste, 5-day BOD at 30°C will be :

(*a*) less than 200 mg/l

(*b*) more than 200 mg/l

(*c*) 200 mg/l

(*d*) zero, as the bacteria cannot withstand such a high temperature.

Q. 96. The ultimate BOD value of a waste :

(*a*) increase with temperature

(*b*) decreases with temperature

(*c*) remains the same at all temperatures

(*d*) doubles with every 10°C rise in temperature.

Q. 97. For the design of a storm sewer in a drainage area, if the time of concentration is 20 min, then the duration of rainfall will be taken as :

(*a*) 10 min. (*b*) 20 min.

(*c*) 30 min. (*d*) 40 min.

Q. 98. From ecological considerations, the minimum level of dissolved oxygen (D.O.) necessary in the rivers and streams is :

(*a*) 1 mg/l (*b*) 2 mg/l

(*c*) 4 mg/l (*d*) 8 mg/l.

Q. 99. The role of the bed material in a "packed tower" used for removing particulate matter from gaseous emissions is to :

(*a*) act as a filter bed to capture the particulates within the pores

(*b*) provide a large surface area on which the particulate matter can be collected

(*c*) reduce the flow of gas

(*d*) uniformly distribute the spray of water.

Q. 100. Which one of the following plume behaviours occurs when atmospheric inversion begins from the ground level and continues ?

(*a*) Looping (*b*) Fumigation
(*c*) Coning (*d*) Fanning.

Q. 101. Which one of the following pollutants or pairs of pollutants is formed due to photochemical reactions ?

(*a*) CO alone (*b*) O_3 and PAN
(*c*) PAN and NH_3 (*d*) NH_3 and CO.

Q. 102. In the context of basic concept of an ecological system, the most appropriate definition of ecology is that it is a study of the :

(*a*) inter-relationship between organisms and the environment
(*b*) relationship of human species with the industry
(*c*) relationship of human species with natural resources
(*d*) relationship of human species with air.

Q. 103. Organisms that mineralise organic matter in an ecosystem are called :

(*a*) producers (*b*) consumers
(*c*) decomposers (*d*) carnivorous.

Q. 104. Aerobic method of composting practised in India is called :

(*a*) Bangalore method (*b*) Nagpur method
(*c*) Delhi method (*d*) Indore method.

Q. 105. Which of the following statements related to C/N (Carbon/Nitrogen) ratio is not correct ?

(*a*) Lower initial C/N ratio leads to loss of nitrogen and slows down the rate of decomposition
(*b*) Higher initial C/N ratio leads to cell destruction to obtain nutrition
(*c*) Higher initial C/N ratio leads to lower conservation of nitrogen in the finished compost
(*d*) An initial C/N ratio of 30 to 50 is optimal for composting.

Q. 106. Which one of the following sets of processes is a part of self-purification of streams ?

(*a*) Settling, bio-degradation and desalination
(*b*) Settling, bio-degradation and aeration
(*c*) Flotation, ion exchange and desalination
(*d*) Desalination, ion exchange and reverse osmosis.

Q. 107. When sewage enters a flowing river, the rapid depletion of dissolved oxygen is due to :

(*a*) change in temperature of river water
(*b*) the suspended particles in river and waste
(*c*) respiratory activity of aquatic plants
(*d*) microbial activity.

Q. 108. Consider the following advantages :

1. Lower BOD concentration in supernatent liquor
2. Production of sludge with excellent dewatering capacity
3. Recovery of methane gas
4. Lower operation cost

Which of these are the advantages of anaerobic digestion over aerobic digestion ?

(*a*) 1 and 4 (*b*) 1 and 2

(*c*) 2, 3 and 4 (*d*) 1, 2 and 3.

Q. 109.* The MLSS concentration in the aeration tank of extended aeration activated sludge process is 400 mg/l. If one litre of sample settled in 30 minutes and the measuring cylinder showed a sludge volume of 200 ml, then the sludge volume index would be nearly :

(*a*) 200 (*b*) 150

(*c*) 100 (*d*) 50.

Q. 110. Examine Assertion (A) and Reason (R) and select your answer using the codes given below :

(*a*) Both A and R are true and R is the correct explanation of A

(*b*) Both A and R are true but R is NOT the correct explanation of A

(*c*) A is true but R is false

(*d*) A is false but R is true.

(A) *Assertion A :* Sewers are not allowed to flow full.

Reason R : Reserve space in the sewer takes care of fluctuations in the sewage flow.

(B) *Assertion A :* In recent years, there has been a progressive increase in the percentage of solar ultraviolet rays reaching the Earth's surface.

Reason R : In the last few years, there has been a progressive increase in the percentage of carbon dioxide in the Earths' atmosphere.

Year 2001

Q. 111. Match List I (Unit) with List II (Purpose) and select the correct answer using the codes given below the lists :

List I *(Unit)*	List II *(Purpose)*
A. Leaping weir	**1.** To prevent grit, sand, debris etc. from entering the storm sewer
B. Gutter inlet	**2.** To carry the sewer below a stream or railways line
C. Inverted syphon	**3.** To drain rain water from roads to the storm sewer
D. Catch basin	**4.** To separate storm water and the sanitary sewage

Codes	A	B	C	D
(*a*)	4	3	1	2
(*b*)	4	3	2	1
(*c*)	3	4	2	1
(*d*)	3	4	1	2

Q. 112. Match List I (Treatment Units) with List II (Types of processes) and select the correct answer using the codes given below the list :

List I *(Treatment units)*	List II *(Types of processes)*
A. Trickling filter	**1.** Symbiotic
B. Activated sludge process	**2.** Extended aeration
C. Oxidation ditch	**3.** Suspended growth
D. Oxidation pond	**4.** Attached growth

Codes	A	B	C	D
(*a*)	3	4	2	1
(*b*)	4	3	1	2
(*c*)	3	4	1	2
(*d*)	4	3	2	1

Q. 113.* If the moisture content of a sludge is reduced from 98% to 96%, the volume of sludge will decrease by :

(*a*) 2% (*b*) 20%

(*c*) 25% (*d*) 50%.

Q. 114.* In a high rate Trickling filter, the problem of ponding can be solved by :

(*a*) flooding and raking (*b*) chlorination and supply of air

(*c*) raking and chlorination (*d*) flooding and supply of air.

Q. 115. Consider the following statements :

The process of activated sludge can be explained as :

1. a physical action whereby the finer suspended particles of sewage form a sublayer for a bacterial film at the surface.
2. a chemical action whereby the finer suspended particles and colloidal solids are combined into masses of large bulk.
3. a biochemical action whereby the sludge flocs so formed act as vehicle for aerobic bacteria oxidising the organic matter.

Which of these statements are correct ?

(*a*) 1, 2 and 3 (*b*) 1 and 2

(*c*) 2 and 3 (*d*) 1 and 3.

Q. 116. Sewage may be disposed of without treatment into a water body if the available dilution is not less than

(*a*) 150 (*b*) 250

(*c*) 400 (*d*) 500.

Q. 117. A certain waste has a BOD of 162 mg/l and its flow is 1000 cubic metre per day. If the domestic sewage has a BOD of 80 gram per capita, then the population equivalent of the waste would be :

(*a*) 20.25 (*b*) 1296
(*c*) 2025 (*d*) 12960.

Q. 118. In a sanitary plumbing of buildings, a two pipe system signifies :

(*a*) separate soil-pipe and waste pipe without vent pipes
(*b*) a soil cum waste pipe and a ventilating pipe
(*c*) separate soil and waste pipe and a common ventilating pipe
(*d*) separate soil pipe and waste pipe, each with its own vent pipe.

Q. 119. Which of the following are storm water regulators ?

1. Side weir
2. Leaping weir
3. Syphonic spillway
4. Float actuated gates or valves
5. Inverted syphon.

Select the correct answer using the codes given below :

Codes :

(*a*) 1, 2, 3 and 4 (*b*) 1, 3 and 5
(*c*) 2, 3, 4 and 5 (*d*) 1, 2, 4 and 5.

Q. 120. When waste water is disposed of into a running stream, four zones are formed. In which one of the following zones will the minimum level of dissolved oxygen be found ?

(*a*) zone of degradation (*b*) zone of active decomposition
(*c*) zone of recovery (*d*) zone of clear water.

Q. 121. The least expensive and most suitable excreta disposal unit for rural areas would be the :

(*a*) soak pit (*b*) pit privy
(*c*) leaching casspool (*d*) septic tank.

Q. 122. Which of the following pairs are correctly matched ?

1. Ringelmann chart ... To grade density of smoke
2. Pneumoconiosis ... Disease caused due to coal dust
3. PAN ... Secondary air pollutant.

Select the correct answer using the codes given below :

Codes :

(*a*) 2 and 3 (*b*) 1 and 2
(*c*) 1 and 3 (*d*) 1, 2 and 3.

Q. 123.* Which one of the following terms correctly describes 'Biomagnification'?

(*a*) Reproduction of micro-organisms
(*b*) Observation of micro-organisms under a microscope
(*c*) Ability of micro-organisms to form zooleal film
(*d*) Concentration of toxic materials in the food chain.

Q. 124. Which one of the following comprehensive classifications is used for different types of solid waste ?

(*a*) Residential, commercial and treatment plant wastes
(*b*) Food, demolition and construction wastes
(*c*) Municipal, industrial and hazardous wastes
(*d*) Rubbish, special wastes and wastes from open areas.

Q. 125. Which one of the following methods would be the best suited for disposal of plastic and rubber waste ?

(*a*) Composting (*b*) Incineration
(*c*) Pyrolysis (*d*) Sanitary land fill.

Q. 126.* Which of the following pairs are correctly matched ?

1. Reverberation time ... Time required to reduce noise by 60 dB
2. NIPTS ... Responsible for permanent hearing loss
3. Sound foci ... Formed when sound waves are reflected from convex surface
4. TTS ... Responsible for temporary hearing loss.

Codes :

(*a*) 2, 3 and 4 (*b*) 1, 3 and 4
(*c*) 1, 2 and 4 (*d*) 1, 2 and 3.

Year 2002

Q. 127. The correct statement of comparison of ultimate BOD (BOD_u), COD, Theoretical oxygen demand (ThOD) and 5-day BOD (BOD_5) is

(*a*) $BOD_u > COD > ThOD > BOD_5$
(*b*) $COD > ThOD > BOD_u > BOD_5$
(*c*) $ThOD > COD > BOD_u > BOD_5$
(*d*) $COD > BOD_u > BOD_5 > ThOD$.

Q. 128. At a sewage treatment plant for a flow of 3 m^3/s, the cross- sectional area of grit chamber will be about

(*a*) 3 m^2 (*b*) 10 m^2
(*c*) 25 m^2 (*d*) 30 m^2.

Q. 129. Match List I (Different forms of nitrogen in water) with List II (Inferences) and select the correct answer using the codes given below the lists :

List I *(Different forms of nitrogen in water)*	List II *(Inferences)*
A. Nitrate nitrogen	**1.** Unsatisfactory microbial activity
B. Total nitrogen	**2.** Satisfactory microbial activity
C. Nitrite nitrogen	**3.** Eutrophication may result
D. Ammonia nitrogen	**4.** Recent organic pollution

Codes	A	B	C	D
(*a*)	3	2	1	4
(*b*)	1	4	3	2
(*c*)	3	4	1	2
(*d*)	1	2	3	4

Q. 130. Match List I (Standards of sewage effluents for the discharge in surface water sources) with List II (Tolerance limits) and select the correct answer using the codes given below the lists :

List I *(Standards of sewage effluents for the discharge in surface water sources)*	List II *(Tolerance limits)*
A. BOD_5, (mg/l)	12.5
B. COD, (mg/l)	23.0
C. Oil and grease, (mg/l)	32.0
D. Total suspended solids, (mg/l)	41.0

Codes	A	B	C	D
(*a*)	3	4	1	2
(*b*)	2	4	1	3
(*c*)	3	1	4	2
(*d*)	2	1	4	3

Q. 131. In the preliminary treatment of wastewater, skimming tanks are often included in the treatment scheme. Various features of skimming tanks are that these :

1. can remove general floating matter.
2. can remove oily and greasy matter.
3. have detention time of 30 min.
4. employ compressed air blown through diffusers.

Which of these statements are correct ?

(*a*) 1 and 3 (*b*) 2 and 4

(*c*) 2 and 3 (*d*) 1 and 2.

Q. 132.* The following is a well known formula for estimating the plume rise :

$$\Delta h = \frac{V_s d}{u}\left(1.5 + 0.0096\,\frac{Q_h}{V_s d}\right)$$

where the letters have their usual meaning. The estimated plume rise (by the above formula) with a stack gas having heat emission rate 2000 kJ/s, the wind speed 4 m/s, stack gas speed 8 m/s inside a stack diameter of 1 m at the top is

(*a*) 7.8 m (*b*) 8.7 m

(*c*) 3.15 m (*d*) $\frac{8000}{\pi}(1.5 + 0.0024\pi)$.

Q. 133. In sampling of stack gases for measurement of concentration of Suspended Particulate Matter (SPM), the flue gases are sucked in the instrument at

(*a*) any rate of flow from mid diameter of chimney

(*b*) any point of chimney cross-section and at any rate of flow

(*c*) a constant rate of flow but at four equidistant points along the diameter

(*d*) controlled positions and controlled velocities along the chimney diameter to get isokinetic conditions.

Q. 134. Match List I (Air pollutants) with List II (Harmful effects) and select the correct answer using the codes given below the lists :

List I *(Air pollutants)*	List II *(Harmful effects)*
A. SPM	**1.** Blood haemoglobin
B. NO	**2.** Vegetation
C. CO	**3.** Respiratory system
D. SO_2	**4.** Building materials

Codes	A	B	C	D
(*a*)	3	4	1	2
(*b*)	1	2	3	4
(*c*)	3	2	1	4
(*d*)	1	4	3	2

Q. 135. Match List I (Air pollutants) with List II (Emitted mainly by) and select the correct answer using the codes given below the lists :

List I *(Air pollutants)*	List II *(Emitted mainly by)*
A. Hydrocarbons	**1.** Coal burning
B. Particulates and gases	**2.** Gasoline fuel
C. Sulphur dioxide	**3.** Tyres
D. Carbon monoxide	**4.** Carburettor

Codes	A	B	C	D
(*a*)	3	4	2	1
(*b*)	4	3	2	1
(*c*)	3	4	1	2
(*d*)	4	3	1	2

Q. 136. 'Air binding' may occur in :

(*a*) Sewers (*b*) Artesian well

(*c*) Aerator (*d*) Filter.

Q. 137.* The description of solid waste collected is as follows :

Night soil	—	35 t
Rubbish	—	40 t
Debris	—	25 t
Garbage	—	40 t

The organic solids in the above composition is :

(*a*) 35 t (*b*) 60 t

(*c*) 100 t (*d*) 75 t.

Q. 138. In a sanitary landfill, decomposition and chemical changes within organic content of the solid waste goes on. Consequential changes within landfill can be :

1. Temperature changes within landfill.
2. Production of gases like H_2S, CO, CO_2 and CH_4.
3. Destruction of pathogens.
4. Production of other gases like SO_2 and NO_2.

Which of these statements are correct ?

(*a*) 1, 2, 3 and 4 (*b*) 1, 2 and 3

(*c*) 1 and 4 (*d*) 2 and 3.

Q. 139. Direction. The following 4 items [S. No. (*i*) to (*iv*)] consist of two statements, one labelled the 'Assertion A' and the other labelled the 'Reason R'. You are to examine these two statements carefully and decide if the Assertion (A) and the Reason (R) are individually true and if so, whether the Reason is a correct explanation of the Assertion. Select your answers to these items using the codes given below and mark your Answer Sheet accordingly.

Codes :

(*a*) Both A and R are true and R is the correct explanation of A

(*b*) Both A and R are true but R is not a correct explanation of A

(*c*) A is true but R is false

(*d*) A is false but R is true.

(*i*) *Assertion (A) :* U.K. Royal Commission on Sewage Disposal of 1898-1915 classified British rivers on the basis of 65°F, 5-day BOD.

Reason (R) : British rivers do not have flow time to open sea more than 5 days.

(*ii*) *Assertion (A) :* Settled sewage is subjected to biological treatment.

Reason (R) : Any biological treatment is proceeded by a settling tank.

(*iii*) *Assertion (A) :* The surface of sea water is on the rise.

Reason (R) : A thick layer of gases developing the Earth does not allow heat to pass into space from the Earth at a rate which is as much as the rate at which the heat coming from space penetrates it towards the Earth.

(*iv*) *Assertion (A) :* Wet scrubber removes particulates from a gaseous stream.

Reason (R) : In the wet scrubber, water droplets come into contact with the particulates.

Year 2003

Q. 140. A sewer is commonly designed to attain self-cleansing velocity at :

(*a*) Peak hourly rate of flow

(*b*) Average hourly rate of flow

(*c*) Minimum hourly rate of flow

(*d*) Sewer running half full.

Q. 141. Sewage sickness signifies :

(*a*) Diseases caused by sewage

(*b*) Soil pores getting clogged and preventing free circulation of air when sewage is continuously applied on land

(*c*) Raw sewage is applied and used for irrigating vegetables which are eaten raw

(*d*) Disposal of septic sewage on land.

Q. 142.* Consider the following data in the design of grit chamber :

1. Sp. gravity of grit = 2.7
2. Size of grit particle = 0.21 mm
3. Viscosity of water = 1.0×10^{-2} cm^2/s

The settling velocity (cm/s) of the grit particle will be

(*a*) 1 to 2.5 (*b*) 2.6 to 5.0

(*c*) 5.1 to 7.8 (*d*) > 7.8.

Q. 143. Amongst the various sewage treatment methods, for the same discharge, the largest area is needed for :

(*a*) Trickling filter (*b*) Anaerobic pond

(*c*) Oxidation ditch (*d*) Oxidation pond.

Q. 144. Consider the following treatment steps in a conventional wastewater treatment plant :

1. Primary sedimentation
2. Grit removal
3. Disinfection
4. Secondary sedimentation
5. Screening
6. Secondary treatment unit.

The correct answer of these steps is

(*a*) 5–2–1–6–4–3 (*b*) 1–2–4–5–3–6

(*c*) 2–3–4–5–6–1 (*d*) 6–5–4–3–2–1.

Q. 145. In an activated sludge process, the sludge volume index can be controlled by

(*a*) Aeration (*b*) Adding chlorine

(*c*) Reducing recycling

(*d*) Increasing the depth of aeration tank.

Q. 146. A primary sedimentation tank is not required for

(*a*) Activated sludge system (*b*) Extended aeration system

(*c*) Trickling filtration system

(*d*) Tapered activated sludge process using pure oxygen for aeration.

Q. 147. In urban air pollution, the most poisonous gas is supposed to be carbon monoxide. It is hazardous because :

(*a*) It affects our sense of smell

(*b*) It is carcinogenic in nature

(*c*) It combines with haemoglobin

(*d*) It causes blindness.

Q. 148.

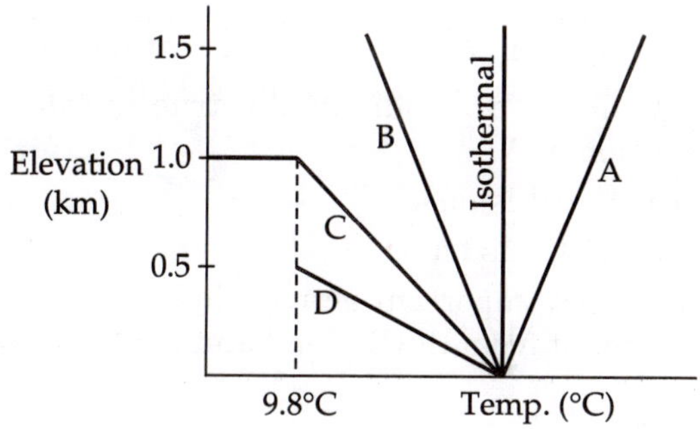

Fig. 26.2

The above graph shows the relationship of ambient lapse rates to the dry adiabatic lapse rate under different conditions of stability. Match stability situations A, B, C and D (as given in the graph) with the classes of stability as follows :

1. Superadiabatic
2. Dry adiabatic
3. Subadiabatic
4. Inversion.

Select the correct answer using the codes given below :

Codes	A	B	C	D
(*a*)	3	4	1	2
(*b*)	4	3	2	1
(*c*)	3	4	2	1
(*d*)	4	3	1	2

Q. 149. The term 'Refuse' generally does not include :

(*a*) Putrescible solid waste

(*b*) Excreta

(*c*) Non-putrescible solid waste

(*d*) Ashes.

Q. 150. Bangalore method and and Indore method of disposing solid wastes are :

(*a*) Identical

(*b*) Different as Bangalore method is an anaerobic method

(*c*) Different as Bangalore method does not contain human excreta

(*d*) Different as Indore method is an incineration method.

Q. 151. Which one of the following statements explains the term pyrolysis ?

(*a*) Solid waste is heated in closed containers in oxygen free atmosphere

(*b*) Solid waste is incinerated in the presence of oxygen

(*c*) Wastewater is treated with oxygen

(*d*) Dissolved solids from water are removed by glass distillation.

Q. 152. Directions : The following item [S. No. (*i*)] consists of two statements, one labelled as the 'Assertion A' and the other labelled as 'Reason (R)'. You are to examine these two statements carefully and select the answers to these items using the codes given below :

Codes :

(*a*) Both A and R are individually true and R is the correct explanation of A

(*b*) Both A and R are individually true but R is not the correct explanation of A

(*c*) A is true but R is false

(*d*) A is false but R is true.

Assertion (A) : Most important activity in sewer line construction is to start constructing it from the tail end and to check levels with a boning rod.

Reason (R) : Construction of sewer line from tail end is recommended because required number of pumping stations may be incorporated in sewer network design.

Year 2004

Q. 153. Consider the following statements :

1. The velocity of flow in the rising main should not be less than 0.8 m/s at any time.
2. Maximum velocity of flow is generally limited to 1.8 m/s and never allowed to exceed 3.0 m/s.

In the design of large sewage pumping stations, which of the above conditions must be satisfied ?

(*a*) 1 only (*b*) 2 only

(*c*) Both 1 and 2 (*d*) Neither 1 nor 2.

Q. 154. Which one of the following statements is correct ? If a sewer X is to be designed to generate equivalent self-cleansing action as in sewer Y, then

(*a*) Velocity in sewer X must be equal to velocity in sewer Y

(*b*) Slope of sewer X must be equal to slope of sewer Y

(*c*) Tractive force intensity generated in sewer X must be same as that in sewer Y

(*d*) The roughness coefficient of X sewer material should be same as that of Y sewer material.

Q. 155. Consider the following statements regarding building manholes :

1. They must be provided at every change of alignment, gradient or diameter.
2. They must be provided at the head of all sewers.
3. They must be provided at every junction of two or more sewers.
4. They must be provided at every 100 m along straight runs of sewers.

Which of the statements given above are correct ?

(*a*) 1, 2, 3 and 4 (*b*) 1, 3 and 4

(*c*) 1, 2 and 3 (*d*) 2 and 4.

Q. 156. Which of the following pairs is correctly matched ?

(*a*) $\frac{BOD}{COD} = 0$ Waste-water is toxic

(*b*) $\frac{BOD}{COD} \le 0.20$ Acclimatization of seed is necessary

(*c*) $\frac{BOD}{COD} \ge 0.6$ Waste-water is non-biodegradable

(*d*) BOD = COD = 0 Waste-water is devoid of organic matter.

Q. 157. In aerobic conditions, the microbial decomposition of organics results in the formation of which one of the following ?

(*a*) Stable and objectionable end products

(*b*) Unstable and objectionable end products

(*c*) Unstable and acceptable end products

(*d*) Stable and unobjectionable end products.

Q. 158. A waste water sample of 2 ml is made upto 300 ml in BOD bottle with distilled water. Initial DO of the sample is 8 mg/l and after 5 days it is 2 mg/l. What is its BOD ?

(*a*) 894 mg/l (*b*) 900 mg/l

(*c*) 300 mg/l (*d*) 1200 mg/l.

Q. 159. An industry manufacturing urea produces waste-water, which largely contains urea and ammonia. The treatment plant consists of the following units for effective control of nitrogen :

1. Nitrification followed by denitrification
2. Hydrolysis unit
3. Ammonia stripping by air
4. Lime dosing and mixing unit.

Which one of the following is the correct sequence of above unit operations in the treatment plant ?

(*a*) 3-2-1-4 (*b*) 2-3-1-4

(*c*) 2-4-3-1 (*d*) 4-2-3-1.

Q. 160.* Which one of the following procedures is used for sampling of flue gas in a chimney for SPM ?

(*a*) Isothermal sampling (*b*) Isokinetic sampling

(*c*) Adiabatic sampling (*d*) Variable rate of sampling.

Q. 161. Which one of the following units is employed for the removal of particulate matter above 50 m in size ?

(*a*) Gravity settling chamber (*b*) Cyclone

(*c*) Fabric filter (*d*) Electrostatic precipitator.

Q. 162. Which one of the following Acts/Rules has a provision for "No Right to appeal" ?

(*a*) Environment (Protection) Act, 1986
(*b*) The Hazardous Waste (Management and Handling) Rules, 1989
(*c*) Manufacture, Storage and Import of Hazardous Chemical Rules, 1989
(*d*) Environment (Protection) Rules, 1992.

Q. 163.* Which one of the following reactors can create a problem of short circuiting due to density currents ?

(*a*) Batch (*b*) Complete mix
(*c*) Plug flow (*d*) Fluidized bed.

Q. 164. Consider the following statements :

The basic difference between water pipes and sewer pipes is :

1. in the material used for pipes
2. in the pressure of the liquid flow
3. in the suspended solids they carry

Which one of the statements given above is/are correct ?

(*a*) 1 and 3 (*b*) 1 only
(*c*) 2 and 3 (*d*) 1, 2 and 3.

Q. 165. Which one of the following statements is correct ?

In dissolved oxygen sag curve, the sag results because :

(*a*) it is a function of a rate of addition of oxygen to the stream
(*b*) it is a function of a rate of addition of oxygen from the stream
(*c*) it is a function of both addition and depletion of oxygen from the stream
(*d*) the rate of addition of oxygen is linear, but the rate of depletion is non-linear.

Q. 166. Which one of the following statements is correct ?

A combined sewer is one, which transports domestic sewage and :

(*a*) storm water (*b*) industrial wastes
(*c*) overhead flow
(*d*) industrial wastes and storm water.

Year 2005

Q. 167. Directions : The following item consist of two statements, one labelled as the 'Assertion (A)' and the other as 'Reason (R)'. You are to examine these two statements carefully and select the answers to these items using the **codes** given below :

(*a*) Both A and R are individually true and R is the correct explanation of A
(*b*) Both A and R are individually true but R is *not* the correct explanation of A
(*c*) A is true but R is false
(*d*) A is false but R is true

Assertion (A) : Composting is the most commonly used process for the decomposition of the organic components of municipal solid wastes.

Reason (R) : Bangalore method is a common anaerobic method used for biological conversion of organic components of municipal solid wastes.

Q. 168. Which type of light energy is effectively absorbed by CO_2 in the lower boundary of the troposphere ?

(*a*) X-rays (*b*) UV-rays

(*c*) Visible light (*d*) Infra-red rays.

Q. 169. Which one of the following is the correct sound-intensity expression with usual notations ?

(*a*) $dB = 10 \log_{10} (I/I_0)^2$ (*b*) $dB = 10 \log_{10} (I/I_0)$

(*c*) $dB = 10 \log_{10} (I - I_0)^2$ (*d*) $dB = 10 \log_{10} (I - I_0)$.

Q. 170. What type of noise can be abated by providing lining on walls and ceiling with sound absorbing material ?

(*a*) Source noise (*b*) Reflection noise

(*c*) Structural noise (*d*) Direct air-borne noise.

Q. 171. In which type of lakes, does a perfect ecological equilibrium among the producers, decomposers, and consumer groups of organisms exist ?

(*a*) Senescent lakes (*b*) Mesotrophic lakes

(*c*) Oligotrophic lakes (*d*) Eutrophic lakes.

Q. 172.* A municipal sewage has BOD_5 of 200 mg/l. It is proposed to treat it and dispose off into a marine environment. For what minimum efficiency should the sewage treatment plant be designed ?

(*a*) 85% (*b*) 60%

(*c*) 50% (*d*) 33.67%.

Q. 173. Which one of the following types of settling phenomenon can be analysed by the classic sedimentation laws of Newton and Stokes ?

(*a*) Discrete setting (*b*) Flocculent settling

(*c*) Hindered settling (*d*) Compression settling.

Q. 174. Match List I (*Treatment Process*) with List II (*Related Terms*) and select the correct answer using the code given below the Lists :

List I	List II
A. Lagoons	**1.** Attached growth system
B. Trickling filter	**2.** Algae-bacteria symbiotic relationship
C. Oxidation ponds	**3.** Extended aeration
D. Activated sludge process	**4.** Low cost treatment method

Codes	A	B	C	D
(*a*)	4	3	2	1
(*b*)	2	1	4	3
(*c*)	4	1	2	3
(*d*)	2	3	4	1

Q. 175. Which of the following pairs is *not* correctly matched ?

Plume Behaviour		*Atmospheric Condition*
(*a*) Looping	:	Stable
(*b*) Fumigation	:	Inversion above and lapse below the stack
(*c*) Fanning	:	Inversion
(*d*) Trapping	:	Inversion above and below the stack with lapse in between

Q. 176. Which is the major pollutant present in photochemical smog ?

(*a*) PAN (*b*) SO_2

(*c*) HC (*d*) NO_2.

Q. 177. Consider the following statement :

In solid waste management :

1. Density separation of solid wastes can be accomplished by air classifiers.
2. Iron recovery from solid wastes can be done by magnetic separators.
3. Aluminium separation from solid wastes can be accomplished by eddy current separators.

Which of the statements given above are correct ?

(*a*) 1 and 2 (*b*) 2 and 3

(*c*) 1 and 3 (*d*) 1, 2 and 3.

Q. 178. Which of the following statement is *not* correct ?

(*a*) Settling and sludge digestion occurs in septic tanks in one compartment

(*b*) Settling and sludge digestion occurs in imhoff tank in different compartments

(*c*) Septic tank is a low-rate anaerobic unit whereas an imhoff tank is a high rate anaerobic unit

(*d*) The rate of sludge accumulation in septic tank is approximately 40–70 litres/capita/year.

Q. 179. Consider the following statements in regard to aerobic and anaerobic treatment processes :

1. Biomass production in the aerobic treatment process is more as compared to the anaerobic treatment process.
2. Start-up period is more in the aerobic treatment process as compared to the anaerobic treatment process.
3. Energy consumption and production is more in the aerobic treatment process as compared to the anaerobic treatment process.

Which of the statements given above is/are correct ?

(*a*) 1 and 2 (*b*) 2 and 3

(*c*) Only 2 (*d*) Only 1.

Q. 180. What is 5 days 20°C BOD equal to ?

(*a*) 3 days 27°C BOD (*b*) 4 days 30°C BOD

(*c*) 6 days 32°C BOD (*d*) 7 days 35°C BOD.

Q. 181. Match List I (*Parameter*) with List II [*General Standard for Discharge into the Inland surface Water in mg/l(max)]* and select the correct answer using the code given below the list :

List I	*List II*
A. BOD (5 day 20°C)	**1.** 250
B. COD	**2.** 100
C. Oil and grease	**3.** 20
D. Suspended Solids	**4.** 10
	5. 30

Codes	A	B	C	D
(*a*)	2	1	4	5
(*b*)	5	4	3	2
(*c*)	2	4	3	5
(*d*)	5	1	4	2

Q. 182.* When the recirculation ratio in a high rate trickling filter is unity, then what is the value of the recirculation factor ?

(*a*) 1 (*b*) > 1

(*c*) < 1 (*d*) Zero.

Q. 183.* Presence of nitrogen in a waste water sample is due to the decomposition of :

(*a*) Carbohydrates (*b*) Proteins

(*c*) Fats (*d*) Vitamins.

Year 2006

Q. 184. Which one of the following is the range of ozonosphere in atmosphere ?

(*a*) Troposphere to Stratosphere (*b*) Tropopause to Stratopause

(*c*) Tropopause to Mesosphere (*d*) Stratosphere to Mesosphere.

Q. 185. Which one of the following pairs is not correctly matched ?

(*a*) Coriolis effect : The effect of Earth's rotation on wind direction and velocity

(*b*) PAN : Found during photochemical smog

(*c*) Cyclone : Employed for particulate matter removal

(*d*) Wind rose : Employed in forecast of pollutant dispersion in ambient air.

Q. 186. In context of water polluted with sewage, what does BOD signify?

(*a*) Biological oxygen demand

(*b*) Bacteriological oxygen demand

(*c*) Biochemical oxygen demand

(*d*) Biology of degradation.

Q. 187.* Match *List-I* with *List-II* and select the correct answer using the codes given below the Lists :

List-I (*Process/Bacteria*)	List-II (*Energy/Material*)
A. Anabolism	**1.** Providing energy for the synthesis of new cells and maintenance of other cell functions
B. Autotrophs	**2.** Obtaining energy and material for growth from organic sources
C. Catabolism	**3.** Providing the material for growth from organic sources
D. Heterotrophs	**4.** Obtaining energy and material for growth from inorganic sources

Codes	A	B	C	D
(*a*)	1	4	3	2
(*b*)	3	2	1	4
(*c*)	1	2	3	4
(*d*)	3	4	1	2

Year 2007

Q. 188. Which among the following brings about the Hazardous Waste Management and Handling Rules in India ?

(*a*) Central Pollution Control Board
(*b*) Ministry of Environment and Forests
(*c*) Ministry of Urban Development
(*d*) Ministry of Rural Development.

Q. 189. In conventional activated sludge process, MLSS is generally kept in which range ?

(*a*) < 100 mg/l
(*b*) 1000-2000 mg/l
(*c*) 2000-3000 mg/l
(*d*) 3000-5000 mg/l.

Q. 190. Which one of the following methods of solid waste management conserves energy most efficiently in the form of gas or oil ?

(*a*) Incineration with heat recovery
(*b*) Composting
(*c*) Fluidized-bed incineration
(*d*) Pyrolysis.

Q. 191. Match *List-I* with *List-II* and select the correct answer using the code given below the Lists :

List-I (*Contaminant*)	List-II (*Environmental significance*)
A. Suspended solids	**1.** May cause eutrophication
B. Nutrients	**2.** Toxic, may interfere with effluent reuse
C. Heavy metals	**3.** May interfere with effluent reuse
D. Dissolved inorganic solids	**4.** Cause sludge deposits

Codes	A	B	C	D
(*a*)	4	1	2	3
(*b*)	2	3	4	1
(*c*)	4	3	2	1
(*d*)	2	1	4	3

Q. 192. Match *List-I* with *List-II* and select the correct answer using the codes given below the Lists :

List-I (*Physical properties of filtering material for trickling filters*)	List-II (*Limiting Value*)
A. Crushing strength, N/mm^2	**1.** 12.0
B. Hardness	**2.** 100.0
C. Per cent wear	**3.** 4.0
D. Specific gravity	**4.** 2.6

Codes	A	B	C	D
(*a*)	3	1	2	4
(*b*)	2	4	3	1
(*c*)	3	4	2	1
(*d*)	2	1	3	4

Q. 193. What is eutrophication of lakes primarily due to ?

(*a*) Multiplication of bacteria

(*b*) Excessive inflow of nutrients

(*c*) Increase in benthic organisms

(*d*) Thermal and density currents.

Q. 194. The daily cover of MSW landfills consists of which one of the following ?

(*a*) Compacted soil (*b*) Geomembrane

(*c*) Geotextile (*d*) Geocomposite.

Q. 195. Match *List-I* with *List-II* and select the correct answer using the codes given below the Lists :

List-I (*Air Pollutant*)	List-II (*Impact on Human Health*)
A. Particulates	**1.** Impairs transport of O_2 in bloodstream
B. Carbon monoxide	**2.** Irritation of mucous membranes of respiratory tract
C. Sulphur oxides	**3.** Causes coughing, shortness of breath, headache, etc.
D. Photochemical oxidants	**4.** Causes respiratory illness

Codes	A	B	C	D
(*a*)	2	3	4	1
(*b*)	4	1	2	3
(*c*)	2	1	4	3
(*d*)	4	3	2	1

Q. 196. Which type of plume may occur during winter nights ?

(*a*) Looping
(*b*) Inversion
(*c*) Coning
(*d*) Lofting.

Q. 197. A machine in a steel plate fabricating industry is found to be producing a sound level of 50 dB. In the expansion plans, one more such machine needs to be added. What will be the combined noise level ?

(*a*) 80–100 dB
(*b*) 101–150 dB
(*c*) 51–70 dB
(*d*) 40–50 dB.

Q. 198. Which one of the following is the correct statement ?

A heterotroph is an organism that obtains

(*a*) its cell carbon from an inorganic source
(*b*) its energy from the oxidation of simple inorganic compounds
(*c*) its cell carbon as well as its energy from organic matter
(*d*) its energy from a natural ecosystem.

Q. 199. The term 'biological magnification' indicates which one of the following?

(*a*) Likelihood of increasing size of animals during evolution
(*b*) Magnification pertaining to microscopy
(*c*) Accumulation of pollutants in soil
(*d*) Accumulation of pollutants in successive consumers.

Year 2008

Q. 200. In the design consideration of sewerage system, the sewers must have which one of the following ?

(*a*) Maximum velocity of flow
(*b*) Only 50 percent of maximum velocity of flow
(*c*) Minimum velocity of not less than cleaning velocity of flow
(*d*) High pressure at all times.

Q. 201. Which one of the following is considered as the thermophilic range of sludge digestion ?

(*a*) 60°C to 70°C
(*b*) 50°C to 57°C
(*c*) 29°C to 40°C
(*d*) 20°C to 30°C.

Q. 202. Consider the following statements :

In context of sludge volume index (SVI)

1. the SVI is expressed as the volume (in cu. cm.) of the activated sludge for one gram of dry weight of the sludge.
2. a higher value of SVI indicates a light and fluffy sludge which is not easily settleable.

Which of the statements given above is/are correct ?

(*a*) 1 only (*b*) 2 only

(*c*) Both 1 and 2 (*d*) Neither 1 nor 2.

Q. 203. Which one of the following can fix atmospheric nitrogen ?

(*a*) Green algae (*b*) Blue green algae

(*c*) Red algae (*d*) Brown algae.

Q. 204.* An industry has sewage plant which produces sludge with a moisture content of 98%. With the solid content remaining the same, the sludge is thickened so that the moisture content now is 96%. If the original quantity of sludge is *P*, what is the quantity of the thickened sludge?

(*a*) 0.98 *P* (*b*) 3 *P*/4

(*c*) 2 *P*/3 (*d*) *P*/2.

Q. 205. In a well operating activated sludge process unit, what is the value of sludge volume index ?

(*a*) < 50 (*b*) 100—150

(*c*) 200—300 (*d*) < 300.

Q. 206. What does the presence of excess nitrates in river water indicates ?

(*a*) Recent pollution of water with sewage

(*b*) Past pollution of water with sewage

(*c*) Intermittent pollution of water with sewage

(*d*) No pollution of water with sewage.

Q. 207. What is the depth of waterseal in the traps ?

(*a*) < 2.5 cm (*b*) 2.5–7.5 cm

(*c*) 7.5–12.5 cm (*d*) Not less than 15 cm.

Q. 208. Bangalore and Indore process of composting are which of the following?

(*a*) Both anaerobic processes

(*b*) Both aerobic processes

(*c*) Anaerobic process and aerobic process, respectively

(*d*) Aerobic process and anaerobic process, respectively.

Q. 209. A solid waste sample has been segregated and one of the components has been subjected to elemental analysis. The result of analysis in percent by mass revealed C(40%), H(6.0%), O(44%), N(0.3%). What is the likely waste component ?

(*a*) Food waste (*b*) Paper and cardboard waste

(*c*) Plastic waste (*d*) Leather waste.

Q. 210. Pneumoconiosis is caused due to inhalation of which one of the following ?

(*a*) Silica (*b*) NO_x

(*c*) Lead (*d*) Cadmium.

Q. 211. Which one of the following conditions of automobile gives maximum unburned hydrocarbons ?

(*a*) Idling (*b*) Cruise

(*c*) Acceleration (*d*) Deceleration.

Q. 212. Electrostatic precipitator is most useful for which one of the following industries ?

Tannery

Hydroelectric power generation

(*c*) Thermal power generation

(*d*) Textile factory.

Q. 213. What will be the resultant decibel level when two sources make noise of equal decibels ?

(*a*) Decibel level will be the same

(*b*) Decibel level will increase by 3 decibels

(*c*) Decibel level will decrease by 3 decibels

(*d*) Decibel level will be equal to the sum of decibels of the two sources.

Q. 214. Match List-I with List-II and select the correct answer using the code given below the lists :

List-I (*Air Pollutant*)	List-II (*Effect*)
A. CO	**1.** Acid rain
B. CO_2	**2.** Acute toxicity
C. SO_2	**3.** Ozone liberation
D. NO_x	**4.** Greenhouse effect

Codes	A	B	C	D
(*a*)	4	3	1	2
(*b*)	4	3	2	1
(*c*)	2	4	1	3
(*d*)	3	4	1	2

Q. 215. Biological magnification of pesticides takes place through which of the following ?

(*a*) Population pyramids (*b*) Hydrologic cycle

(*c*) Food chains (*d*) Air cycle.

Year 2009

Q. 216. Effluent from a wastewater treatment plant (flow rate = 1.2 m^3/s, temperature = 25°C), is discharged to a surface stream (flow rate = 8640 m^3/s, temperature = 15°C). What is the temperature of the stream after mixing ?

(*a*) 10°C (*b*) 15.77°C

(*c*) 20°C (*d*) 24.99°C.

Q. 217. Match List-I with List-II and select the correct answer using the code given below the Lists :

List-I *(Water / Wastewater treatment)*	List-II *(Operating problem)*
A. Trickling filter	**1.** Negative head
B. Activated sludge process	**2.** Fly-breeding
C. Rapid gravity filter	**3.** Sludge bulking
D. Anaerobic sludge digester	**4.** pH reduction

Codes	A	B	C	D
(*a*)	4	3	1	2
(*b*)	2	3	1	4
(*c*)	4	1	3	2
(*d*)	2	1	3	4

Q. 218. Which of the following pollutants are generally **not** removed in a sewage treatment plant ?

(*a*) Inorganic suspended solids (*b*) Dissolved organic solids

(*c*) Oil and grease (*d*) Dissolved inorganic solids.

Q. 219.* What is the theoretical oxygen demand of 300 mg/L glucose solution?

(*a*) 300 mg/L (*b*) 320 mg/L

(*c*) 350 mg/L (*d*) 400 mg/L.

Q. 220. Which one of the following types of samples is relevantly employed for the design of wastewater treatment plant ?

(*a*) Grab sample (*b*) Composite sample

(*c*) Integrated sample (*d*) Any sample.

Q. 221.* A drain carrying sewage of BOD = 200 mg/L and flow rate of 50 m^3/s joins a river whose upstream BOD is 8 mg/L and flow rate is 500 m^3/s. Assume immediate and complete mixing of drain with the river. What is the estimated downstream BOD of the river flow ?

(*a*) 20.4 mg/L (*b*) 25.4 mg/L

(*c*) 104.4 mg/L (*d*) 70.4 mg/L.

Q. 222.* A 12.5 mL sample of treated wastewater requires 187.5 mL of odor-free distilled water to reduce the odor to a level that is just perceptible. What is the **threshold odor number (TON)** for the wastewater sample ?

(*a*) 0.07 (*b*) 1.07

(*c*) 15 (*d*) 16.

Q. 223. Which one of the following parameters is **not** included in the routine characterization of solid waste for its physical composition ?

(*a*) Moisture content (*b*) Density

(*c*) Particle size analysis (*d*) Energy value.

Q. 224. Which one of the following toxic gases has physiological action as asphyxiant ?

(*a*) SO_2 (*b*) NO_2

(*c*) Cl_2 (*d*) CO.

Q. 225. Assuming annual travel for each vehicle to be 20000 km, what is the quantity of NO_x produced from 50000 vehicles with emission rate of 2.0 g/km/vehicle ?

(*a*) 1800 tonnes (*b*) 1900 tonnes

(*c*) 2000 tonnes (*d*) 2100 tonnes.

Q. 226. What are the air pollutants responsible for acid rain within and downwind areas of major industrial emissions ?

(*a*) Hydrogen sulfide and oxides of nitrogen

(*b*) Sulfur dioxide and oxides of nitrogen

(*c*) Carbon dioxide and hydrogen sulfide

(*d*) methane and hydrogen sulfide.

Q. 227. Consider the following air pollutants :

1. NO_x 2. PAN

3. CO_2 4. CO

Which of the above air pollutants is/are present in an auto exhaust gas ?

(*a*) 1 only (*b*) 1 and 2

(*c*) 2 and 3 (*d*) 1, 3 and 4.

Year 2010

Q. 228. ***Directions :*** *Each of the following item(s) consists of two statements, one labelled as 'Assertion (A)' and the other as 'Reason (R)'. You are to examine these two statements carefully and select the answers to these item(s) using the codes given below :*

Codes :

(*a*) *Both A and R are individually true and R is the correct explanation of A*

(*b*) *Both A and R are individually true but R is not the correct explanation of A*

(*c*) *A is true but R is false*

(*d*) *A is false but R is true.*

(i) *Assertion (A) :* Traffic 'smog' is likely to occur in regions where vehicle mileage is considerable and there is low incidence of sunlight.

Reason (R) : Traffic 'smog' is caused by the reaction of oxides of nitrogen and some of the hydrocarbons in the presence of bright sunlight.

Q. 229.* In an aerobic attached-culture system, the biomass of the biofilm-medium surface interface experiences

(*a*) Aerobic and endogenous metabolism

(*b*) Anaerobic and endogenous metabolism

(*c*) Anaerobic and exogenous metabolism

(*d*) Aerobic and exogenous metabolism.

Q. 230. A catchment consists of 30% area with runoff coefficient of 0.40 with remaining 70% area with runoff coefficient of 0.60. The equivalent runoff coefficient will be

(*a*) 0.48 (*b*) 0.54
(*c*) 0.63 (*d*) 0.76.

Q. 231. Secondary clarifier of an activated sludge process must be designed for effluent clarification and solids thickening, both of which relate directly to the

(*a*) surface area
(*b*) transport velocity due to sludge withdrawal
(*c*) gravity settling of solids relative to the water
(*d*) underflow solids concentration.

Q. 232. In the facultative pond system, the aerobic zone may get extended downward due to

(*a*) calm waters along with weak sunlight
(*b*) mixing by wind action along with weak sunlight
(*c*) mixing by wind action along with penetration by sunlight
(*d*) calm waters along with penetration by sunlight.

Q. 233. Noise pollution in a road-side building can be reduced by

(*a*) providing a ditch around the building and filling it with water
(*b*) providing a steel mesh around the building
(*c*) providing a thick bush around the building
(*d*) planting tall trees around the building and fencing them with barbed wires.

Q. 234. Shallow ponds in which dissolved oxygen is present at all depths are called

(*a*) Aerobic lagoons
(*b*) Aerobic ponds
(*c*) Facultative lagoons
(*d*) Facultative ponds.

Q. 235. Consider the following statements :

In a sewage treatment process

1. Detention time required for anaerobic decomposition is more as compared to that for aerobic decomposition.
2. Anaerobic decomposition is more energy-consuming as compared to aerobic decomposition.
3. Anaerobic decomposition is a more sensitive process as compared to aerobic decomposition.

Which of the above statements is/are correct ?

(*a*) 1, 2 and 3 (*b*) 2 only
(*c*) 3 only (*d*) 1 and 3 only.

Q. 236.* The below figure shows BOD curve when the experiment is conducted at 20°C. If the experiment is conducted at 30°C, then the portion AB of the curve

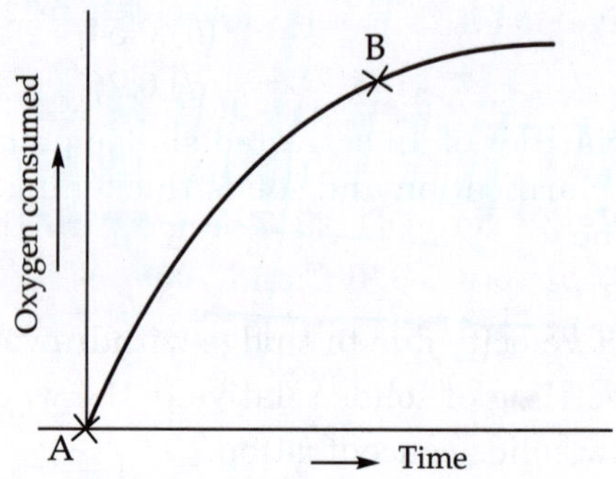

Fig. 26.3

(*a*) shifts to the left (*b*) shifts to the right
(*c*) remains unchanged (*d*) shrinks.

Year 2011

Q. 237. Consider the following statements :

Activated sludge process can be said to comprise :

1. Conversion of dissolved organic matter into biological flocs.

2. Removal of dissolved BOD of the waste water.

3. Digestion of the sludge.

Which of these statements are correct ?

(*a*) 1, 2 and 3 (*b*) 1 and 2 only
(*c*) 2 and 3 only (*d*) 1 and 3 only.

Q. 238 During sewage treatment, effluent from which one of the following treatment units has minimum wt/vol amount of suspended solids ?

(*a*) Detritus channel
(*b*) Primary sedimentation tank
(*c*) Secondary sedimentation tank
(*d*) Activated sludge process aeration tank.

Q. 239.* The group of micro-organisms involved in production of methane from acetic acid (or acetate) in anaerobic waste water treatment process is :

(*a*) Methanothrix and Methanobacterium
(*b*) Methanobacterium and Methanosarcina
(*c*) Methanosarcina and Methanospirillum
(*d*) Methanothrix and Methanosarcina.

Q. 240. When sufficient energy through mechanical mixing is supplied to keep the entire contents, including the sewage solids, mixed and aerated, the reactor is termed :

(*a*) An aerobic lagoon (*b*) An aerobic pond
(*c*) A facultative lagoon (*d*) A facultative pond.

Q. 241. Deep ponds, in which oxygen is absent except, perhaps across a relatively thin surface layer, are called.

(*a*) Aerobic ponds (*b*) Anaerobic ponds

(*c*) Facultative ponds (*d*) Polishing ponds.

Q. 242. The manufacturer of aeration devices reports the oxygen transfer rate of the device obtained through laboratory tests carried under standard conditions. Such standard conditions are :

(*a*) Wastewater at zero DO, 25°C and 760 mm Hg

(*b*) Tap water at zero DO, 0°C and 700 mm Hg

(*c*) Tap water at zero DO, 20°C and 760 mm Hg

(*d*) Wastewater at zero DO, 0°C and 700 mm Hg.

Q. 243. Which of the following are responsible for the formation of photochemical smog ?

1. Light intensity

2. Ratio of hydrocarbons to nitric oxide

3. CO_2

4. Hydrocarbon reactivity

5. SO_2

Which of these statements are correct ?

(*a*) 1, 2, 3, 4 and 5 (*b*) 1, 2 and 4 only

(*c*) 2, 3 and 4 only (*d*) 2, 3 and 5 only.

Directions. *Each of the following items consists of two statements, one labelled as 'Assertion (A)' and the other as 'Reason (R)'. You are to examine these two statements carefully and select the answer to these items using the codes given below :*

Codes :

(*a*) *Both A and R are individually true and R is the correct explanation of A*

(*b*) *Both A and R are individually true but R is not the correct explanation of A*

(*c*) *A is true but R is false*

(*d*) *A is false but R is true.*

Q. 244.* *Assertion (A) :* The BOD gets removed at very fast rate immediately after sewage is discharged into a river.

Reason (R) : A part of BOD in the sewage is due to settleable organic matter therein.

Q. 245. *Assertion (A) :* Laterals of minimum specified diameter in sewage systems have to be laid at slopes designed for self cleansing velocity.

Reason (R) : For the specified minimum lateral diameter at specified slopes, a minimum flow rate is not essential to maintain self-cleansing velocity.

Q. 246. *Assertion (A) :* In the secondary sedimentation tank of sewage treatment plant, the settling particles form a blanket which descends and captures more particles.

Reason (R) : The particles are flocculent and in very high concentration whenever there is an efficient working activated sludge, or trickling filter, process.

Year 2012 (*Engg. Services Objectives*)

Q. 247.* An urban area is located in plains having "average climatic conditions". The impervious area thereof for which drainage must be provided is 3.6 ha and the design rainfall intensity is 2.0 cm/hr. The drains will be designed for a runoff of

(*a*) 0.05 m^3/s (*b*) 0.10 m^3/s

(*c*) 0.20 m^3/s (*d*) 0.40 m^3/s.

Q. 248.* A sewer has a diameter of 300 mm and slope of 1 in 400. While running full it has a mean velocity of 0.7 m/s. If both the diameter and slope are doubled (to respectively be 600 mm and 1 in 200), what will be the changed mean velocity when running half-full ? Use Manning's formula.

(*a*) 1.59 m/s (*b*) 2.80 m/s

(*c*) 0.90 m/s (*d*) 1.00 m/s.

Q. 249.* A sample of sewage is estimated to have a 5 day 20°C BOD of 250 mg/*l*. If the test temperature be 30°C, in how many days will the same value of BOD be obtained ?

(*a*) 1.5 days (*b*) 2.5 days

(*c*) 3.3 days (*d*) 7.5 days.

Q. 250.* Consider the following statements :

The Time of BOD assimilation in a stream can be affected by

1. Ratio of stream depth to flow width.

2. Stream BOD value.

3. BOD rate constant

Which of these statements are correct ?

(*a*) 1, 2 and 3 (*b*) 1 and 2 only

(*c*) 2 and 3 only (*d*) 1 and 3 only

Q. 251. It takes 0.4 hrs to drive from the garage to the head of the route, 0.4 hrs to drive between the route head and disposal site and 0.25 hrs to return from the disposal site. It takes 0.2 hrs to off load a truck at the disposal site. The crew is permitted two 15-minute breaks and a further 30 minutes for miscellaneous delays. If two runs are made to the deposit site each day, how much time is left in an 8-hr nominal duty duration for refuse collection before starting to return to garage from disposal site ? Take loading time as 30 minutes.

(*a*) 4.15 hrs (*b*) 4.25 hrs

(*c*) 4.75 hrs (*d*) 4.85 hrs.

Q. 252. Which of the following factors contribute to formation of photochemical smog ?

1. Stable atmosphere **2.** NO_x

3. Solar insolation **4.** CO

(*a*) 1, 2, 3 and 4 (*b*) 2, 3 and 4 only
(*c*) 1 and 4 only (*d*) 1, 2 and 3 only.

Q. 253. Consider the following statements :

1. Particulates have irregular shapes.

2. Size can be determined by an equivalent aerodynamic diameter by comparing with a perfect sphere.

3. Particulates larger than 10 μ are said to settle relatively quickly since their settling velocity is not less than 10 cm/min.

4. The particles roughly the size of bacteria have aerodynamic diameter of 0.1 μm to 10 μm.

Which of these statements are correct ?

(*a*) 1, 2, 3 and 4 (*b*) 1 and 3 only
(*c*) 1, 2 and 4 only (*d*) 2, 3 and 4 only.

Q. 254.* For noise measurement, formula for sound pressure level (SPL) is 20 log P/P_{ref}. What will be the resultant noise in dB is P is 0.0002 m bar?

(*a*) 0 (*b*) 60
(*c*) 90 (*d*) 100

Q. 255. Environmental impact assessment includes

(*a*) Environmental statement
(*b*) Environmental management plan
(*c*) Risk and hazard assessment and mitigation
(*d*) All of the above

Directions : Each of the next twenty. (20) items consists of two statements, one labelled as the 'Statement (I)' and the other as 'Statement (II)'. You are to examine these two statements carefully and select the answers to these items using the codes given below :

Codes :

(*a*) *Both Statement (I) and Statement (II) are individually true and Statement (II) is the correct explanation of Statement (I)*

(*b*) *Both Statement (I) and Statement (II) are individually true but Statement (II) is not the correct explanation of Statement (I)*

(*c*) *Statement (I) is true but Statement (II) is false*

(*d*) *Statement (I) is false but statement (II) is true*

Q. 256. *Statement (I) :* Air pollutant concentration and time of retention increase due to inversion.

Statement (II) : During winter, the heavy cold layer in the atmosphere retains the hot toxic pollutants for a longer period in the atmosphere.

Q. 257. *Statement (I) :* Virus is living organisms in a natural environment including soil.

Statement (II) : Virus comes to life after entering the host tissue through contamination.

Q. 258. *Statement (I) :* The BOD test in conducted for 5 days at 20°C.

Statement (II) : The amount of oxygen utilized by microorganisms anaerobically is called BOD.

Q. 259. *Statement (I) :* Municipal Solid Waste is disposed off in the Transport Safe Disposal Facility (TSDF) to convert it into organic compost.

Statement (II) : The organic Municipal Solid Waste is converted into compost by worms; and the process is called 'Vermicomposting'.

Q. 260. *Statement (I) :* Chlorophyll-bearing plants take water and carbondioxide to synthesize carbohydrates.

Statement (II) : Wasted food ultimately leads to production of various natural resources like water and sunlight energy.

Year 2013 (*Engg. Services Objectives*)

Q. 261. The self-cleansing velocity in a sewer depends on :

1. BOD (soluble)

2. Slope of the sewer

3. Ratio of depth of flow sewage to sewer diameter ratio

(*a*) 1, 2 and 3

(*b*) 1 and 2 only

(*c*) 2 and 3 only

(*d*) 1 and 3 only.

Q. 262. One sewer has a dia. of 300 mm and another one has a dia. of 600 mm. When both run half-full, what will be the ratio of velocities in the two pipes if the slope of both pipes is the same ?

(*a*) 1 (*b*) $\frac{1}{2}$

(*c*) $\left(\frac{1}{2}\right)^{2/3}$ (*d*) $\left(\frac{1}{2}\right)^{3/2}$

Q. 263. If the slope of sewer A is 1/100 and that of sewer B is 1/400, the velocity of flow in the two sewers will have a ratio of (size of both the sewers being same) :

(*a*) $\frac{1}{2}$ (*b*) 1

(*c*) $(2)^{2/3}$ (*d*) 2

Q. 264. What is the correct sequence of the following treatment units employed in a conventional sewage treatment plant ?

1. Screen chamber

2. Primary settling tank

3. Trickling filter

4. Grit chamber

5. Secondary settling tank

(*a*) 5, 4, 2, 3 and 1 (*b*) 1, 4, 2, 3 and 5

(*c*) 5, 2, 3, 4 and 1 (*d*) 1, 4, 3, 2 and 5

Q. 265. Which of the following parameters are employed in the design of trickling filter ?

1. Hydraulic loading rate
2. Organic loading rate
3. Detention time
4. Weir loading rate

(*a*) 1, 2, 3 and 4 (*b*) 1 and 2 only
(*c*) 2 and 3 only (*d*) 3 and 4 only

Q. 266. In activated sludge process, 1 mld (by volume) of secondary sludge has to be returned to keep the required MLSS concentration in the aeration tank. This sludge has a water content of 99%. If the sludge water content is reduced to 98%, what volume of the sludge will be needed to be recycled?

(*a*) 0.25 mld (*b*) 0.5 mld
(*c*) 0.75 mld (*d*) 1.00 mld

Q. 267. At the same mean velocity of flow, the ratio of head loss per unit length for a sewer pipe running full to that for the same pipe flowing half-full is :

(*a*) 2.0 (*b*) 1.67
(*c*) 1.0 (*d*) 0.67

Q. 268. Consider the following statements :

1. Grease data is a measure of effectiveness of primary settling tanks.
2. Volatile acid data provide information concerning anaerobic degradation of organic matter.
3. Total and volatile solids data are indispensable in the design and operation of sludge digestion, vacuum filter and incineration units.
4. Ammonia and organic nitrogen data are helpful in controlling aerobic treatment costs to a minimum.

Which of these statements are correct ?

(*a*) 1, 2, 3 and 4
(*b*) 1, 2 and 3 only
(*c*) 1, 2 and 4 only
(*d*) 2, 3 and 4 only

Q. 269. Consider the following statements in regard to an electrostatic precipitator :

1. Power requirement is small compared to other devices.
2. Immune to variable particulate loadings and flow rates.
3. 99+ percent efficiency is obtainable. Very small particles can be collected, wet or dry.
4. Can handle both gases and mists at high volume of flow.

Which of these statements are correct ?

(*a*) 1, 2, 3 and 4
(*b*) 1, 3 and 4 only
(*c*) 1, 2 and 3 only
(*d*) 2, 3 and 4 only

Q. 270. Consider the following statements :

1. Noise pollution can be reduced using double-glass window panes.
2. Glass absorbs the noise.
3. The air trapped in the double-glass system acts as an insulator and reduces the noise.
4. The noise totally reflects back due to the two layers of glass.

Which of these statements are correct ?

(*a*) 1, 2, 3 and 4
(*b*) 1, 2 and 3 only
(*c*) 1 and 3 only
(*d*) 2 and 4 only

Q. 271. Consider the following statements :

1. Sound is a form of mechanical energy from a vibrating substance, transmitted by a cyclic series of compressions and rarefactions of the molecules of the material through which it passes.
2. In a pure tone, the wave pattern of the alternating positive and negative sound pressure is of ideal sinusoidal form with fixed wave lentgh, frequency and amplitude.
3. The speed of transmission of sound is a function of the transmitting medium and its temperature.
4. The audible range of 200 of 25000 Hz is considered normal for young adults.

Which of these statements are correct in respect of sound transmission?

(*a*) 1, 2, 3 and 4
(*b*) 1, 2 and 4 only
(*c*) 1, 2 and 3 only
(*d*) 2, 3 and 4 only

Directions : *Each of the next four (4) items consists of two statements, one labelled as the 'Statement (I)' and the other as 'Statement (II)'. You are to examine these two statements carefully and select the answers to these items using the codes given below :*

Codes :

(*a*) *Both Statement (I) and Statement (II) are individually true and Statement (II) is the correct explanation of Statement (I)*
(*b*) *Both Statement (I) and Statement (II) are individually true but Statement (II) is not the correct explanation of Statement (I)*
(*c*) *Statement (I) is true but Statement (II) is false*
(*d*) *Statement (I) is false but Statement (II) is true*

Q. 272. *Statement (I) :* Plastic media of 25 to 100 mm size are employed in super-rate trickling filters for treatment of waste-water having a very high BOD.

Statement (II) : Plastic media have high specific surface area, high void space and low weight. Hydraulic loading rate and organic loading rate are 30 to 90 $m^3/d/m^2$ and 1000 to 2000 $g/d/m^3$ respectively, which are much higher than in stone media trickling filter.

Q. 273. *Statement (I) :* The cross-section of a grit channel is designed for a uniform velocity at all flow rates.

Statement (II) : A U-type of cross-section maintains a uniform velocity at all flow rates.

Q. 274. *Statement (I) :* For aerobic composting of refuse, it is mandatory to maintain the C/N ratio between 50 and 60.

Statement (II) : C/N ratio is one of the important factors in composting. It is an important constituent as loss of nitrogen is to be prevented and maximum proportion of the nitrogen in compost is to be preserved.

Q. 275. *Statement (I) :* 'Environment' includes abiotic and biotic parameters.

Statement (II) : Abiotic parameters include algae, bacteria, animals; and biotic parameters are air, water and soil.

Year 2014 (*Engg. Services Objectives*)

Q. 276. If the average sewage from a city is 95×10^6 *l*/day and the average 5-day BOD is 300 mg/*l* against a standard of 75 mg/day per capita at 20°C, then the population equivalent of the city will be

(*a*) 28,500 (*b*) 285×10^6

(*c*) 3,80,000 (*d*) 380×10^6

Q. 277. As per Manual on Sewerage and Sewage Treatment, Central Public Health and Environmental Engineering Organization, Ministry of Urban Development, Government of India, when a sewer connects with another sewer having a difference of 600 mm in level between the water lines (peak flow levels) of main line and the invert level of branch line, the connection may be provided using a

(*a*) Side entrance manhole (*b*) Junction manhole

(*c*) Flushing manhole (*d*) Drop manhole

Q. 278*. In an aerobic attached-culture system, the biomass present at the biofilm-medium surface interface experiences

(*a*) Aerobic and endogenous metabolism

(*b*) Anaerobic and exogenous metabolism

(*c*) Anaerobic and endogenous metabolism

(*d*) Aerobic and exogenous metabolism

Q. 279. A sewage sludge has a water content of 99%. What will be the concentration of suspended solids in the sludge?

(*a*) 10 mg/*l* (*b*) 100 mg/*l*

(*c*) 1000 mg/*l* (*d*) 10,000 mg/*l*

Q. 280. Consider the following statements :

1. The quality of waste water is determined on the basis of DO.

2. The BOD test is based on DO.

3. Determination of DO helps in controlling corrosion.

Which of the above statements are correct ?

(*a*) 1, 2 and 3 (*b*) 1 and 2 only

(*c*) 1 and 3 only (*d*) 2 and 3 only

Q. 281. The major source of 'Carbon monoxide' in the urban atmosphere is due to

(*a*) Decomposition of organics

(*b*) Chemical reaction between VOC and NO_X

(*c*) Incomplete combustion of fuel

(*d*) Incomplete combustion in the presence of sunlight

Q. 282. Consider the following statements relating zones of atmosphere to altitude :

1. Temperature decreases with altitude in troposhere.

2. Speed of sound decreases with altitude in troposphere.

3. Temperature increases with altitude in stratosphere.

4. Ozone is present in stratosphere which protects people from harmful effects of solar radiation.

Which of the above statements are correct ?

(*a*) 1, 2 and 3 only (*b*) 1, 2 and 4 only

(*c*) 3 and 4 only (*d*) 1, 2, 3 and 4

Q. 283. The principle involved in collection and sampling of particulate matter in which the particles are drawn through a device by deflecting them from their orignal paths is called.

(*a*) Filtration (*b*) Electrostatic precipitation

(*c*) Impaction (*d*) Gravitational settling

Directions : *Each of the next given items consists of two statements, one labelled as the 'Statement (I)' and the other as 'Statement (II)'. You are to examine these two statements carefully and select the answers to these items using the codes given below :*

Codes :

(*a*) *Both Statement (I) and Statement (II) are individually true and Statement (II) is the correct explanation of Statement (I)*

(*b*) *Both Statement (I) and Statement (II) are individually true but Statement (II) is not the correct explanation of Statement (I)*

(*c*) *Statement (I) is true but Statement (II) is false*

(*d*) *Statement (I) is false but Statement (II) is true.*

Q. 284. *Statement (I) :* When waste water is treated at an oxidation ditch, anaerobic sludge digester is not required.

Statement (II) : Oxidation ditch admits an extended aeration process whereby sludge gets mineralized.

Q. 285. *Statement (I) :* Dilution of the wastewater sample with organic-free, oxygen-saturated water is necessary of measure 5-day 20°C BOD values greater than 7 mg/*l*.

Statement (II) : The saturation for oxygen in water at 20°C is approximately 9 mg/*l*.

Q. 286*. *Statement (I) :* Electrostatic precipitators (ESP) can remove particles of sub-micron size.

Statement (II) : Due to high voltage supply, ESPs attract any charged particle.

Year 2015 (*Engg. Services Objectives*)

Q. 287. A sludge had 100 m^3 volume when its moisture content was 95%. What would be its volume of its moisture content changed to 90% ?

(*a*) 200 m^3 (*b*) 50 m^3

(*c*) 94.7 m^3 (*d*) 105.5 m^3

Q. 288. The Sludge Volume Index for mixed liquor having suspended solids concentration of 2000 mg/l and showing a settled volume of 200 ml from a one litre sample would be

(*a*) 0.1 (*b*) 1000

(*c*) 100 (*d*) 10

Q. 289. The different actions that take place in anaerobic decomposition process are

1. Alkaline fermentation **2.** Acid fermentation

3. Acid regression **4.** Methane formation

What is the correct sequence of these actions (from earlier to later) ?

(*a*) 4, 3, 1 and 2 (*b*) 2, 3, 1 and 4

(*c*) 4, 1, 3 and 2 (*d*) 2, 1, 3 and 4

Q. 290. Which of the following help to prevent water pollution due to land-disposal of waste ?

1. Proper consolidation of waste to reduce pore space and permeability

2. Disposal over impervious strata

3. Layer of impervious soil on the top and the sides of the deposited solid waste

(*a*) 1 and 2 only (*b*) 1 and 3 only

(*c*) 1, 2 and 3 (*d*) 2 and 3 only

Q. 291. Consider the following statements in respect of electrostatic precipitators :

1. Power requirement is very small compared to other air pollution control devices and so they are cheaper to perform than other devices :

2. Can handle both gases and mists for high valume flow.

3. Very small particles can be collected, either wet or dry.

Which of the above statements are correct ?

(*a*) 1 and 2 only (*b*) 2 and 3 only

(*c*) 1 and 3 only (*d*) 1, 2 and 3

Q. 292. Consider the following statements :

1. Ammonia nitrogen is a measure of nitrogen present as ammonium hydroxide and ammonium salts. It will progressively decrease as sewage gets treated.

Organic nitrogen is the total nitrogenous matter in sewage excepting that present as ammonia nitrogen, nitrites and nitrates. It becomes ammonia in anaerobic decomposition and nitrites or nitrates in aerobic decomposition.

Which of the above statements is/are correct ?

(*a*) 1 only (*b*) Both 1 and 2

(*c*) 2 only (*d*) Neither 1 nor 2

Q. 293. What is the required plan size of a square sedimentation tank (as the primary sedimentation tank in sewage treatment), given that its effective depth is 3 m, and the flow rate is 40 MLD with admissible surface loading of 100,000 $l/m^2/day$?

(*a*) 23.5 m × 23.5 m (*b*) 30 m × 30 m

(*c*) 20 m × 20 m (*d*) 15 m × 15 m

Q. 294. Consider the following statements related to ozone :

1. Tropospheric ozone is harmful

2. Stratospheric ozone is beneficial

3. During prevalence of photochemical smog, O_3 is formed

Which of the above statements are correct ?

(*a*) 1, 2 and 3 (*b*) 1 and 2 only

(*c*) 1 and 3 only (*d*) 2 and 3 only

Q. 295.* Consider the following statements related to noise :

1. The range of sound power and sound pressures produced is from 0.0002 μ bars to 10000 μ bars.

2. Human ears do not respond linearly to increase in sound pressures.

3. Regular exposure to moderate noise makes the human ear more resistant to occasional exposures of high-intensity noise.

Which of the above statements are correct ?

(*a*) 1 and 2 only (*b*) 1 and 3 only

(*c*) 2 and 3 only (*d*) 1, 2 and 3

Q. 296. Consider the following statements in respect of effect of air pollutants on vegetation :

1. Necrosis refers to killing of tissue

2. Calorosis refers to loss or reduction of green plant pigment

3. Leaf abscission refers to the dropping of leaves

4. Leaf epinasty refers to a downward curvature of a leaf due to a higher rate of growth on the upper surface

Which of the above statements are correct ?

(a) 1, 2 and 3 only (b) 1, 2, 3 and 4

(c) 2, 3 and 4 only (d) 1, 2 and 4 only

Q. 297. Consider the following statements regarding excreta disposal without water carriage system :

1. Pit-Privy is a pit in the ground with the toilet seat located directly over it.
2. Bore-Hole Latrines do not cause nuisance due to flies and odour.
3. Aqua-Privy works on the same principle as a septic tank.
4. In the context of a Bore-Hole Latrine, a pit of about 30 cm to 40 cm diameter is dug to a depth of 4 m to 8 m.

Which of the above statements are correct ?

(a) 1, 2 and 3 only (b) 2 and 4 only

(c) 1, 3 and 4 only (d) 1, 2, 3 and 4

Q. 298.* Consider the following statements regarding ecology :

1. Climax ecosystem is a stage in the evolution of an ecosystem, at which all the species are in dynamic equilibrium among themselves as also with the environment.
2. Ecological niche means all the physical, chemical and biological factors that a species needs in order to live and reproduce exist.
3. Edge effect refers to the presence of rich and unique biological diversity found in an ecotone.

Which of the above statements are correct ?

(a) 1, 2 and 3 (b) 1 and 2 only

(c) 1 and 3 only (d) 2 and 3 only

Q. 299. Consider the following sttements regarding biochemical oxygen demond (BOD) of river water :

1. The BOD rate constant varies with river water temperature.
2. The BOD rate constant does not depend on the BOD of the river water.
3. The BOD rate constant is often different for different river waters.
4. The BOD rate constant cannot be determined in a laboratory.

Which of the above statements are correct ?

(a) 1 and 4 (b) 1 and 3

(c) 2 and 3 (d) 2 and 4

Directions : *Each of the next items consists of two statements, one labelled as the 'Statement (I)' and the other as 'Statement (II)'. Examine these two statements carefully and select the answers to these items using the codes given below :*

Codes :

(a) *Both Statement (I) and Statement (II) are individually true and Statement (II) is the correct explanation of Statement (I)* (*b*) *Both Statement (I) and Statement (II) are individually true but Statement (II) is* ***not*** *the correct explanation of Statement (I)*

(*c*) *Statement (I) is true but Statement (II) is false*

(*d*) *Statement (I) is false but Statement (II) is true.*

Q. 300. *Statement (I) :* The flow in water distribution pipes takes place due to gravity.

Statement (II) : The flow in sewers takes place due to gravity.

Q. 301. *Statement (I) :* Anaerobic digestion of sewage is unsuitable in the vicinity of a crowded locality.

Statement (II) : Aerobic digestion of sewage is costly but is suitable at a crowded locality.

Q. 302. *Statement (I) :* Composting is basically a treatment method for inorganic waste from a community.

Statement (II) : In the incineration method of refuse disposal, the refuse is burnt off and the volume is much reduced.

Q. 303. *Statement (I) :* Gases are normally formless fluids and can be changed to liquid or solid states by change of temperature and pressure.

Statement (II) : Smog refers to the occurrence of a heavy, cloudy, hazy floating layer in the atmosphere formed by a mixture of smoke, dust, fog and mist.

Answers

1. (*b*)	**2.** (*a*)	**3.** (*b*)	**4.** (*c*)	**5.** (*d*)
6. (*a*)	**7.** (*b*)	**8.** (*d*)	**9.** (*b*)	**10.** (*d*)
11. (*d*)	**12.** (*d*)	**13.** (*d*)	**14.** (*a*)	**15.** (*a*)
16. (*b*)	**17.** (*d*)	**18.** (*b*)	**19.** (*b*)	**20.** (*c*)
21. (*b*)	**22.** (*b*)	**23.** (*a*)	**24.** (*d*)	**25.** (*d*)
26. (*b*)	**27.** (*c*)	**28.** (*b*)	**29.** (*c*)	**30.** (*d*)
31. (*c*)	**32.** (*a*)	**33.** (*a*)	**34.** (*b*)	**35.** (*b*)
36. (*c*)	**37.** (*b*)	**38.** (*b*)	**39.** (*c*)	**40.** (*d*)
41. (*d*)	**42.** (*c*)	**43.** (*d*)	**44.** (*d*)	**45.** (*a*)
46. (*d*)	**47.** (*d*)	**48.** (*a*)	**49.** (*b*)	**50.** (*c*)
51. (*d*)	**52.** (*b*)	**53.** (*b*)	**54.** (*b*)	**55.** (*c*)
56. (*b*)	**57.** (*c*)	**58.** (*d*)	**59.** (*c*)	**60.** (*a*)
61. (*c*)	**62.** (*c*)	**63.** (*c*)	**64.** (*b*)	**65.** (*c*)
66. (*c*)	**67.** (*d*)	**68.** (*a*)	**69.** (*b*)	**70.** (*a*)
71. (*b*)	**72.** (*b*)	**73.** (*d*)	**74.** (*c*)	**75.** (*c*)
76. (*d*)	**77.** (*d*)	**78.** (*b*)	**79.** (*d*)	**80.** (*b*)
81. (*c*)	**82.** (*a*)	**83.** (*d*)	**84.** (*c*)	**85.** (*b*)
86. (*b*)	**87.** (*c*)	**88.** (*c*)	**89.** (*i*) (*d*)	**89.** (*ii*) (*d*)
90. (*a*)	**91.** (*b*)	**92.** (*b*)	**93.** (*d*)	**94.** (*a*)
95. (*b*)	**96.** (*c*)	**97.** (*b*)	**98.** (*c*)	**99.** (*d*)
100. (*d*)	**101.** (*b*)	**102.** (*a*)	**103.** (*c*)	**104.** (*d*)
105. (*d*)	**106.** (*b*)	**107.** (*b*)	**108.** (*c*)	**109.** (*d*)
110. A (*c*)	**110.** B (*c*)	**111.** (*b*)	**112.** (*d*)	**113.** (*d*)
114. (*c*)	**115.** (*a*)	**116.** (*d*)	**117.** (*c*)	**118.** (*d*)

119. (*a*)	**120.** (*b*)	**121.** (*b*)	**122.** (*d*)	**123.** (*d*)
124. (*c*)	**125.** (*b*)	**126.** (*a*)	**127.** (*c*)	**128.** (*b*)
129. (*a*)	**130.** (*b*)	**131.** (*b*)	**132.** (*a*)	**133.** (*d*)
134. (*c*)	**135.** (*d*)	**136.** (*d*)	**137.** (*d*)	**138.** (*a*)
139. (*i*) (*b*)	**139. (*ii*)** (*b*)	**139. (*iii*)** (*a*)	**139. (*iv*)** (*a*)	**140.** (*c*)
141. (*b*)	**142.** (*d*)	**143.** (*d*)	**144.** (*a*)	**145.** (*c*)
146. (*b*)	**147.** (*c*)	**148.** (*b*)	**149.** (*b*)	**150.** (*b*)
151. (*a*)	**152.** (*c*)	**153.** (*c*)	**154.** (*a*)	**155.** (*a*)
156. (*d*)	**157.** (*d*)	**158.** (*b*)	**159.** (*d*)	**160.** (*b*)
161. (*a*)	**162.** (*b*)	**163.** (*c*)	**164.** (*d*)	**165.** (*c*)
166. (*d*)	**167.** (*b*)	**168.** (*c*)	**169.** (*b*)	**170.** (*b*)
171. (*c*)	**172.** (*c*)	**173.** (*a*)	**174.** (*b*)	**175.** (*a*)
176. (*a*)	**177.** (*a*)	**178.** (*c*)	**179.** (*d*)	**180.** (*b*)
181. (*d*)	**182.** (*b*)	**183.** (*b*)	**184.** (*b*)	**185.** (*d*)
186. (*c*)	**187.** (*a*)	**188.** (*b*)	**189.** (*c*)	**190.** (*a*)
191. (*a*)	**192.** (*d*)	**193.** (*b*)	**194.** (*a*)	**195.** (*c*)
196. (*b*)	**197.** (*c*)	**198.** (*c*)	**199.** (*d*)	**200.** (*c*)
201. (*b*)	**202.** (*b*)	**203.** (*b*)	**204.** (*d*)	**205.** (*b*)
206. (*b*)	**207.** (*b*)	**208.** (*c*)	**209.** (*a*)	**210.** (*a*)
211. (*a*)	**212.** (*c*)	**213.** (*b*)	**214.** (*c*)	**215.** (*c*)
216. (*b*)	**217.** (*b*)	**218.** (*d*)	**219.** (*b*)	**220.** (*a*)
221. (*b*)	**222.** (*d*)	**223.** (*d*)	**224.** (*b*)	**225.** (*c*)
226. (*b*)	**227.** (*d*)	**228.** (*d*)	**229.** (*a*)	**230.** (*b*)
231. (*b*)	**232.** (*c*)	**233.** (*d*)	**234.** (*b*)	**235.** (*d*)
236. (*a*)	**237.** (*b*)	**238.** (*d*)	**239.** (*d*)	**240.** (*a*)
241. (*b*)	**242.** (*c*)	**243.** (*b*)	**244.** (*b*)	**245.** (*b*)
246. (*c*)	**247.** (*c*)	**248.** (*a*)	**249.** (*c*)	**250.** (*c*)
251. (*a*)	**252.** (*d*)	**253.** (*a*)	**254.** (*a*)	**255.** (*d*)
256. (*a*)	**257.** (*b*)	**258.** (*c*)	**259.** (*a*)	**260.** (*c*)
261. (*c*)	**262.** (*c*)	**263.** (*d*)	**264.** (*b*)	**265.** (*b*)
266. (*b*)	**267.** (*c*)	**268.** (*d*)	**269.** (*d*)	**270.** (*b*)
271. (*a*)	**272.** (*a*)	**273.** (*a*)	**274.** (*d*)	**275.** (*c*)
276. (*d*)	**277.** (*d*)	**278.** (*a*)	**279.** (*d*)	**280.** (*a*)
281. (*c*)	**282.** (*d*)	**283.** (*b*)	**284.** (*c*)	**285.** (*d*)
286. (*a*)	**287.** (*b*)	**288.** (*c*)	**289.** (*b*)	**290.** (*c*)
291. (*d*)	**292.** (*c*)	**293.** (*c*)	**294.** (*a*)	**295.** (*d*)
296. (*b*)	**297.** (*d*)	**298.** (*a*)	**299.** (*b*)	**300.** (*d*)
301. (*b*)	**302.** (*d*)	**303.** (*b*)		

Hints For Solving Starred Objective Questions

Q. 1. The usually adopted values of detention times for different treatment units, as given in our Text, are :

(A) grit chambers = 1 min

(B) primary sedimentation = 1—2 hr.

(C) activated sludge = 4—8 hr.

(D) digestion = 1 month.

The nearest choices available in the given choices of 6 hr, 2 min, 2 hr, and 20 days, fits well respectively to units at S. No. (C), (A), (B) and (D) respectively. Hence correct choice code is C-1, A-2, B-3 and D-4 ; *i.e.* at choice (*b*).

Q. 2. Trickling filter sludge contains lesser water than the activated sludge, and is therefore, less bulky. Choices (*b*), (*c*) and (*d*) are all reverse to this fact. The only choice left is (*a*), which, as a matter of fact, is a correct choice because sludge volume index reflects the volume of 1 gm of settled sludge, and hence is less for less bulky sludge.

Q. 5. The average BOD of domestic sewage produced per person has been considered by us to be 0.08 kg/day, (Refer article 7.5). This factor being 80 gm/d, the nearest choice available in the given choices is (*d*) *i.e.* 50 to 70 gm/d, and hence is the correct choice.

Q. 9. An unpolluted river will have its D.O. equal to the saturation D.O., and since the saturation D.O. reduces with the increase in temperature (Refer Appendix Table A-3 at the end of the book), an unpolluted stream would contain less D.O. in summer. Hence, choice (*b*) is correct. **Ans.**

Q. 10. BOD reaction rate constant, K_D, varies with temperature (T°C) by the equation

$$K_{D(T^\circ)} = K_{D(20^\circ)}\ [1.047]^{T-20^\circ}$$

When T is more than 20°C, $(1.407)^{T-20}$ is more than 1.047 ; and hence $K_{D(T)}$ will be more than $K_{D(20)}$. Similarly, when T is less than 20°, $(1.047)^{T-20}$ will be less than 1.047, and hence $K_{D(T)}$ will be less than $K_{D(20^\circ)}$. In other words, K_D values at temperatures more than 20°C will be more than the K_D value at 20°C ; and K_D values at temperatures lesser than 20°C will be lesser than the K_D value at 20°C ; which exactly is the trend reflected in the given values of K_{30° (0.02) and K_{10° (0.005) when compared to K_{20° (0.01). Hence, choice (*d*) is correct. **Ans.**

Q. 14. Pl. refer and study article 9.22 to ascertain the correct sequence, which is given in choice (*a*). **Ans.**

Q. 16. Coal based thermal power plants, as discussed under article 16.1, produce three major air pollutants, namely (*i*) Fly ash (SPM), (*ii*) SO_2 and (*iii*) NO_x. Although, due to incomplete combustion of coal, there may exist a possibility of emission of CO and unburnt carbon ; yet however, the modern thermal power stations are generally provided with *automatic combustion control systems*, which eliminate the formation of these products. Hence, CO is not attributed to the thermal power stations. Hence, choice (*b*) is considered to be correct, and not choice (*c*), which many students may feel to be correct.

Q. 20. All changes in size, shape, alignment, flow volume, free and submerged discharge at the end of sewer lines, passing of flow through measuring or diversion devices and sewer junctions, etc. are known as **sewer transitions**. Manholes must be located at all such transitions.

The transition from larger dia sewer to smaller dia sewer should not be made. However, in other transitions, such as change from small dia sewer to large dia sewer or change of slope, etc., *the crowns* of sewers are always kept continuous. In no case, the hydraulic flow line (grade line) in the larger sewer should be higher than the incoming one.

Q. 21. Refer Table 4.8 ; and find

for $\frac{d}{D} = 0.5 \frac{v}{V} = 1$

Hence $v = V = 1.0$ m/s. **Ans.**

Q. 22. The *oxidation ditch*, as described in the text, works under aerobic bacteria (suspended culture of course). *Other units which* work on similar aerobic (suspended culture) include : activated sludge process, oxidation pond, and sludge digestion tanks. A waste stabilisation pond works under facultative bacteria, and an Imhoff tank works under anaerobic bacteria, as discussed in the text.

The RBC's (*Rotating Biological Contractors*), as described in text, infact work on the *aerobic bacteria* **(attached culture)** *like that of a trickling filter as discussed*.

Based on this knowledge and without even knowing about RBC's, the correct choice can be marked, which indicate A-3, B-1, and C-2 ; *i.e.* the choice given at (*b*).

Q. 23. 3 gm of sludge occupies 27 cm^3 (27 ml) of volume in 1 litre of sewage settled for 30 min. Hence, 1 gm of such sludge will contain 9 cm^3 (*i.e.* 9 ml) of volume, which infact is nothing but SVI, as defined in article 9.36. Therefore, SVI is 9 ml/gm ; and hence the choice at (*a*) is a correct choice. **Ans.**

Q. 28. We have already stated in article 19.2 that the gaseous layers surrounding the earth up to a distance of about 500 km from the surface of earth, composes what is known as the **atmosphere**. Depending upon the distance from the earth, this atmosphere is divided into four zones, as detailed in the Table below :

Table 26.1

S. No.	*Region / Zone*		*Approximate altitude range in km*	*Approximate temperature range in °C*
1.	*Troposphere*		0—11 to 16 km	15° to – 80°C
2.	*Stratosphere*		11 to 16—50 km	– 80° to 0°C
3.	*Mesosphere*	**Iono-sphere**	50—80 km	0° to – 110°C
4.	*Thermosphere*		80—500 km	– 110°C to 1200°C

The above sequence of layers is shown in choice (*b*), which is the correct choice. **Ans.**

Q. 41. $Q = 720$ m^3/hr

Surface area $= 12.5 \text{ m} \times 1.5 \text{ m} = 18 \text{ m}^2$

Surface loading $= \frac{720 \text{ m}^3/\text{hr}}{18 \text{ m}^2}$

$= 40$ m^3/hr/m^2 = **40000 lp.h/m^2.**

The detention time for Grit chambers is usually 40 to 60 secs or so. In the given choices, the nearest available is 1.2 minutes = 72 secs. Other choices are 12 minutes and 40 minutes, which are ignored.

Hence, the choice (*d*), which indicates correct surface loading of 40,000 lph/ m^2 and 1.2 minute of detention time is correct. **Ans.**

Q. 43. Objective Questions can sometimes be solved very easily by using elimination technique, as we can do here :

As we know, methane gas is produced only in anaerobic digestion, and hence it cannot be claimed as advantageous to aerobic sludge digestion. The assertion at 3 should therefore not be there in the correct code. This eliminates choices (*b*) and (*c*). The choices left are only (*a*) and (*d*). Choice (*a*) includes 1, 2 and 4 ; whereas choice (*d*) includes 1, 2 and 5. Thus, 1 and 2 are correctly the characteristics (advantages) of aerobic sludge digestion. We have now to find, whether characteristic 4 or 5 pertain to aerobic digestion. Characteristic 4 is "lesser operating cost", which certainly cannot belong to aerobic digestion, because aerobic digestion, compared to anaerobic one, requires continuous operation to ensure oxidation of sludge, whereas anaerobic digestion is carried out in closed tanks without any supply of oxygen, etc. This eliminates characteristic 4 from the correct choice, thus excluding choice (*a*). The choice, thus, left is (*d*), which can be marked as the correct choice. **Ans.**

Now, coming to the ***aerobic sludge digestion***, which has not been described in the Text, it is stated that *aerobic sludge digestion is nothing but the stabilisation of sludge by long term aeration that biologically destroys the volatile solids. It is, thus, nothing more than a continuation of the activated sludge process.*

Aerobic sludge digestion is accomplished in an open tank resembling the aeration tank of an activated sludge plant. As in an activated plant, this *aerobic digester* must be followed by a *settling tank*, unless the sludge is to be disposed of on land in liquid form. Unlike that in an activated plant, the effluent from the clarifier is recycled back to the head end of the plant, as it is high in BOD (500 mg/L or 80) and S.S. (100–300 mg/L).

Such an aerobic digester is usually operated by continuously feeding raw sludge. The sludge undergoing digestion in the digester is continuously aerated during filling and for the specified design period after the tank is full. Aeration is then discontinued to allow the stabilised solids to settle down by gravity. The supernatant is decanted and returned to the head of the treatment plant, and a portion of the gravity thickened sludge is removed from the tank bottom for disposal.

In practice, aeration and settlement may be a daily cycle with feed applied early in the day, and the clarified supernatant decanted later in the day. Digested solids are withdrawn when the sludge in the tank does not gravity thicken to provide a supernatant of adequate clarity.

Q. 53. Eutrophication, as explained in article 8.4.5, is the natural process of gradual accumulation and cycling of nutrients in lakes (impounded waters), making them shallower and more productive ; phosphorous being the decisive factor in eutrophication. This definition of eutrophication is correctly reflected in the definition given at S. No. 1. Hence, code 1 is certainly correct.

Autotrophism and *Heterotrophism* infact represent the oxidation (aerobic) or degradation (anaerobic) of the **inorganic** and *organic* matter by bacteria (microorganisms) respectively, since the bacteria that oxidize **inorganic matter** are termed as **autotrophic**, and those that oxidize *organic matter* are called *heterotrophic*.

Heterotrophism, thus involves the use of organic matter, while in the definition given at S. No. 3, it is defined otherwise (*i.e.* uses inorganic substances). Hence, definition at S. No. 3 is incorrect. When we eliminate code containing 3 from correct choice, choices (*a*), (*c*) and (*d*) automatically get eliminated, leaving choice (*b*) as the correct choice. **Ans.**

Q. 55. As stated in article 12.3.1, **leachate** has to be prevented from seeping underground, to prevent contamination of ground water. For this purpose, covers made of plastics or rubber may be used, and impermeable clay or earth or bentonite mud layers may be spread over the filled up land to avoid penetration of leachate below. Under-drain pipes are also laid to collect the leachate before it penetrates below to underground.

Hence, only impermeable soils like clay or bentonite, plastic or rubber covers and under-drains can be said to be the sealants used for preventing leachate and gas in sanitary land fills. The choice (*c*), which includes Bentonite and butyl rubber is therefore the correct choice.

Even by using elimination technique, we can exclude fly ash (code 4) and sand (code 2) being permeable soils. The choice (*a*), (*b*) and (*d*), all containing codes 2 or 4, thus automatically get excluded, leaving choice (*c*) only, as the correct choice. **Ans.**

Q. 56. As discussed in article 20.3

$$\text{dB} = 20 \log \left(\frac{p_{\text{rms}}}{20\ \mu\text{Pa}}\right)$$

$$\therefore \quad 90\ \text{dB} = 20 \log \left(\frac{p_{\text{rms}}}{20}\right)$$

$$\text{or} \quad \frac{p_{\text{rms}}}{20} = \text{Antilog}\left(\frac{90}{20}\right)$$

$$\therefore \quad 90\ \text{dB as } p_{\text{rms}} = 20 \times \text{Antilog}\left(\frac{90}{20}\right) = 6{,}32{,}455.53\ \mu\text{Pa}$$

$$\text{Similarly, } 94\ \text{dB} = 20\ \text{Antilog}\left(\frac{94}{20}\right) = 10{,}02{,}374.47\ \mu\text{Pa}$$

$$\begin{aligned} 90\ \text{dB} + 94\ \text{dB} &= \sqrt{(6{,}32{,}455.53)^2 + (10{,}02{,}374.47)^2} \\ &= 11{,}85{,}223.43\ \mu\text{Pa} \\ &= 20 \log_{10}\left(\frac{11{,}85{,}223.43}{20}\right)\ \text{dB} \\ &= 95.46\ \text{dB} \text{ ; say } \mathbf{95.5\ dB} \end{aligned}$$

which is given in choice (*b*), which is the correct choice. **Ans.**

Q. 66. In addition to the Trickling filters and Rotating Biological Contractors (Fig. 9.44), still another *aerobic attached growth process* is a

packed-bed reactor, used for the removal of carbonaceous BOD as well as biological **nitrification*** of wastewaters.

Typically, a **packed bed reactor** (Fig. 26.4) consists of a container (reactor) that is packed with a medium to which microorganisms can get attached. Wastewater is introduced from the bottom of the container through an appropriate under-drain system or inlet chamber. Air or pure oxygen necessary for the process is also introduced with the wastewater.

Such a reactor is thus, related to a trickling filter, which also represents an aerobic attached growth system and a packing medium.

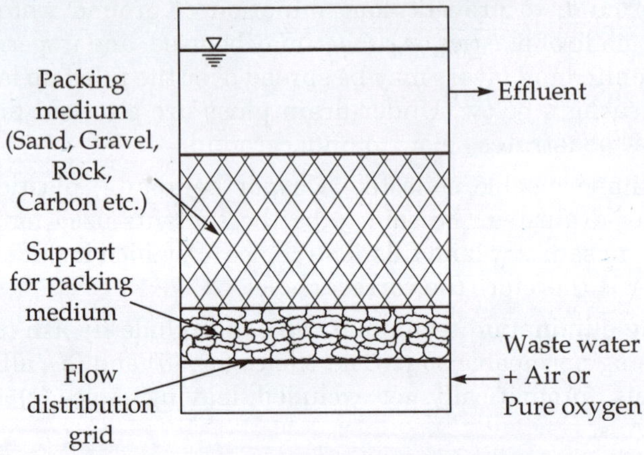

Fig. 26.4. Typical packed bed reactor.

The *plug flow* and *complete mix* reactors relate to the *suspended growth* activated sludge process, which is altogether different from the attached growth systems. Similarly, the *fluidized-bed reactor* is an anaerobic reactor and hence altogether different from an aerobic filter. Thus, the reactor which is related to a trickling filter, is the packed-bed one ; and hence the choice (*c*) is correct. **Ans.**

Q. 74. Sewage produced $= 10{,}000 \times 200$ l/day $= 2 \times 10^6$ l/day

BOD produced $= 2 \times 10^6$ l/d $\times$ 300 mg/l

$= 6 \times 10^8$ mg/d $=$ 600 kg/d

Area of oxidation pond $= \dfrac{600 \text{ kg/d}}{300 \text{ kg/d/ha}} =$ **2 ha. Ans.**

Q. 85. Air required to maintain 50 ppm of CO

$$= 0.03 \frac{\text{m}^3}{\text{min}} \times \left(\frac{10^4}{50}\right)$$

$= \mathbf{600\ m^3/min.}$ **Ans.**

*When wastewater containing ammonia is discharged into a stream, it consumes dissolved oxygen (DO) from the stream to convert ammonia into nitrates, thereby depleting D.O. This depletion of oxygen can be avoided, if the ammonia is oxidised to nitrate before it is discharged. ***Nitrification*** is the process used to accomplish this objective.

Q. 88. The sound pressure level in dB (decibel) is given by Eq. (20.10) as

$$= 20 \log_{10}\left(\frac{p_{rms}}{p_{std.} = 20\ \mu\text{Pa}}\right)$$

where p_{rms} = measured sound pressure
= 2000 μ.bar
= 2000 × 10^5 μ.Pa
= 20 × 10^7 μPa

[Since 1 bar = 100 kPa = 10^5 Pa]

$$\therefore \quad \text{dB} = 20 \log_{10}\left(\frac{20 \times 10^7\ \mu\text{Pa}}{20\ \mu.\text{Pa}}\right)$$

= 20 × 7 = **140 dB. Ans.**

Q. 109. X_t = 4000 mg/l

1 litre of sewage will contain solids = 4000 mg = 4 g

Sludge volume = 200 ml

$$\text{SVI} = \frac{200\ \text{ml}}{4\ \text{g}}$$

= 50 ml/g **Ans.**

Q. 113. $V_2 = V_1 \dfrac{(100-98)}{(100-96)} = \dfrac{V_1}{2}$ = 50% *V*. Hence choice (*d*) is the correct answer.

Q. 114. Field problem of **ponding** in a biological sewage filter is caused by clogging of filter voids, due to fungi and algae. It may be caused by *excessive loading, inappropriate hydraulic loading and inadequate choice and packing of filter media*. The problem can be overcome by raking or forking the filter surface, washing the filter by *applying high pressure, steam of water at the surface, stopping the distributor to allow continuous heavy point by point dosing, or by light chlorination of influent* with a dose not exceeding 5 kg/100 m^2 of the filter area. Ponding is rarely a serious problem even in apparently overloaded plants in hot climates. If overloading is the cause, then the only permanent solution is to provide more filter capacity.

Q. 123. Toxic chemicals particularly the organo metallic chemicals, like *DDT, Dioxan, Minamata, Love canal,* etc. have been found to be entering our food chain through their large scale use in our daily life, or through the consumption of plants and vegetables using pesticides eaten directly by humans, or through the consumption of milk and milk products obtained from animals grazing on such vegetables and grasses. The movement of such toxic chemicals through the environment, ultimately causes their **biotransformation, bioaccumulation** and **biomagnification*** (large scale magnified effects through food chain), creating adverse effects on human, plant and animal health and on the environment in general.

---

*Explained in details in authors' another book titled "Ecology and Environmental Studies".

Q. 126. NIPTS stands for **Noise Induced Permanent Threshold Shift**, which represents permanent hearing loss in human beings due to continued exposure to high noise over a long period. TTS stands for **Temporary Threshold Shift**, which represents temporary hearing loss by exposure to short time high noise. Since human ear has a tendency to recover in due course of time after the subject returns to a quiet environment, such loss in hearing sensitivity is only temporary and may be detected if the hearing loss test is performed immediately after the exposure within a standard interval of 2 minutes. TTS is also called NITTS *i.e.*, ***Noise induced temporary threshold shift.***

Reverberation time is the time required by a sound to die down fully, say up to zero dB, and not only for its decibel reduction by 60 dB. Hence, pair at 1 does not match, while pairs at 2, 3 and 4 do match. Hence, choice (*a*) is the correct answer.

Q. 132. Here V_s = stack gas velocity = 8 m/s

d = inside exit dia. of stack = 1 m

u = wind speed = 4 m/s

Q_h = heat emission rate in kJ/s = 2000 kJ/s.

Substituting values in the given formula, we get

$$\Delta h = \frac{8 \times 1}{4}\left[1.5 + 0.0096 \times \frac{2000}{8 \times 1}\right]$$

= 7.8. Hence choice (*a*) is correct. **Ans.**

Q. 137. The organic solids in the given solid waste shall be caused by Night soil (35 t) and Garbage (40 t) only, since Rubbish (40 t) and Debris (25 t) are non-organic solids. Hence, organic solids amount to 35 t + 40 t = 75 t. Hence, choice (*d*) is correct. **Ans.**

Q. 142. Here $G = 2.7$

$d = 0.21 \text{ mm} = 0.21 \times 10^{-3}$ m

$\nu = 10^{-2}$ cm²/s $= 10^{-6}$ m²/s

Using Eq. (9.5), we have the settling velocity (v_s) as :

$$v_s = \left[\frac{g(G-1)}{13.88}\frac{d^{1.6}}{\nu^{0.6}}\right]^{0.714}$$

$$= \left\{\frac{9.81\,(2.7-1)}{13.88} \times \frac{(0.21 \times 10^{-3})^{1.6}}{(10^{-6})^{0.6}}\right\}^{0.714}$$

$$= \left[1.2015 \times \frac{1.3048 \times 10^{-6}}{2.5119 \times 10^{-4}}\right]^{0.714}$$

$= [6.2412 \times 10^{-3}]^{0.714}$ m/s

$= 0.02267$ m/s = **2.67 cm/s.**

Check

The settling velocity v_s can also be computed from Eq. (9.7) as :

$$v_s = 60.6 \, . \, d \, . \, (G - 1) \left(\frac{3T + 70}{100} \right)$$

The value of viscosity ν becomes 1 × 10^{-6} m^2/sec at around 21°C (Pl. see Appendix Table A-6 at the end of the book)

$$\therefore \qquad v_s = 60.6 \times 0.021 \, (2.7 - 1) \left(\frac{3 \times 21 + 70}{100} \right) \text{ cm/sec}$$

$$= \mathbf{2.87 \text{ cm/sec}}$$

Hence, choice (*b*) is correct. **Ans.**

Q. 160. The sampling procedures for air pollution studies have not been discussed in our text, as it is beyond the syllabus of under-graduate courses, which involves only elementary study of Air Pollution. Hence, to know about sampling procedures and instruments used to collect samples of air/gas, containing pollutants, including their measurements, readers may refer to "Air Pollution" by M.N. Rao and H.V.N. Rao, published by Tata McGraw Hill Publishing Company Ltd.

We may, however point out here that sampling of particulate matter must be carried out under conditions which are as **isokinetic** as possible, inorder to eliminate the sources of errors that may be introduced due to agglomeration or breaking up of particulate matter. Hence, choicc (*b*) is the correct answer. **Ans.**

Q. 163. The tanks or vessels or containers used for carrying out biological or chemical reactions for treatment of sewage, known as **reactors**, may be of the following types :

(*i*) the *batch reactor* ;

(*ii*) the *plug flow reactor*, also known as *tubular-flow reactor* ;

(*iii*) the *continuous flow stirred-tank* reactor ; also known as *complete mix reactor* ;

(*iv*) the *arbitrary-flow reactor* ;

(*v*) the *packed bed reactor* ; and

(*vi*) the *fluidized bed reactor*.

The classification of the first four reactors is based on their hydraulic characteristics. Homogeneous reactions are usually carried out in such reactors. Heterogeneous reactions are however, usually carried out in the latter two types of reactors (S. No. (*v*) and (*vi*)) above.

In a **betch reactor**, the flow neither enters nor leaves the reactor, such as in a BOD test bottle.

The **plug flow reactor**, in practice, is approximated by a very long rectangular tank, such as a *grit channel* or *a large sedimentation tank*. In some cases, these tanks exceed 180 m in length. To use space more effectively, such tanks are often folded in a serpentine fashion.

In a **continuous flow stirred tank** or a **complete mix reactor**, complete mixing occurs, when particles entering the tank are dispersed immediately throughout the tank. Such a mixing can be accomplished in circular or square or rectangular tanks, if the contents of the tank are uniformly and continuously redistributed.

An **arbitrary flow reactor**, is in between both the plug flow and continuous mixing type of reactors, involving *particle mixing*.

A **packed bed reactor** is filled with some type of packing material, such as rock, slag, ceramic, sand etc. With respect to flow, they can be completely filled (*anaerobic filter*) or intermittently dosed (*trickling filter*).

A **fluidized-bed reactor** is similar to the packed bed reactor in many respects, but the packing medium is expanded by the upward movement of the fluid (air or water) through the bed. The porosity of packing can be varied, thereby controlling expanded depth by controlling the flow rate of the fluid.

Plug flow, continuous flow a and *arbitrary flow* reactors are the types of reactors, which are most commonly used in the field of waste water treatment. The most important practical considerations involved in reactor design is how to achieve the ideal conditions postulated in the analysis of their performance. For example, when a continuous flow stirred tank reactor is designed, the important issue would be to decide as to how the flow should be introduced to satisfy the requirement of instantaneous and complete dispersion. Properly designed inlet and outlet arrangements are, hence, required to be provided. In practice, however, there is always some deviation from ideal conditions, and precautions have to be taken to minimize those effects that are important.

A common problem affecting reactor performance is ***short circuiting.*** If the incoming waste water is either colder or warmer than the liquid in the reactor, **short circuiting in plug-flow reactors** can be caused by **density currents.** (See Fig. 26.5*a*)

In warm climates, the performance of large sedimentation tanks can also be affected by density currents. Correction of short circuiting in plug-flow reactors may require the use of submerged deflection baffles located at either the top or bottom of the tank. Alternatively, some form of energy input may be required to equalize the temperature. Both mechanical mixers and diffused air have been used for this purpose.

Inefficient use of the reactor volume and short circuiting caused by inadequate mixing in a *complete mix* or *continuous-flow stirred-tank reactor*, as shown in Fig. 26.5 (*b*), can, similarly, cause problems, especially where rectangular tanks are used.

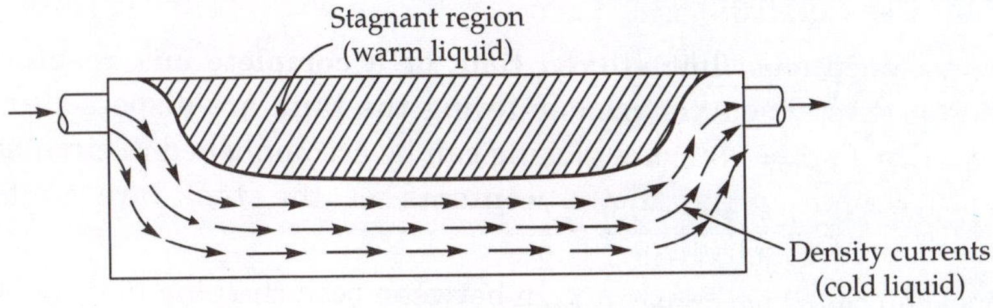

(*a*) Short circuiting caused by density currents in a plug flow reactor

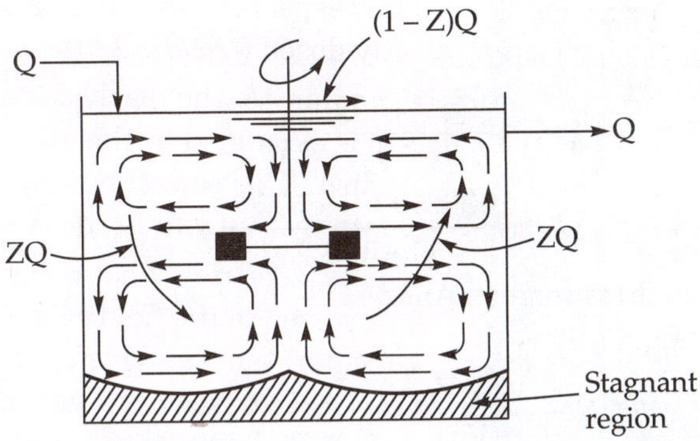

(*b*) Short circuiting caused by inadequate mixing in continuous flow stirred-tank reactor.

Fig. 26.5. Short circuiting reactors.

Q. 172. Max. BOD_5 allowed for discharge of sewage effluents in marine waters = 100 mg/l (Pl. see tables 8.3 and 8.6).

BOD_5 of sewage = 200 mg/l

$$\text{Degree of treatment reqd.} = \frac{100}{200} \times 100\% = 50\%$$

Hence, choice (*c*) is the correct choice. **Ans.**

Q. 182. Pl. see Equation (9.33).

Q. 183. Pl. see Fig. 7.1 (*a*).

Q. 187. ***Autotrophs*** and *heterotrophs* (bacteria) are defined in the Hint given to solve **Q. 53,** as the bacteria that respectively use ***inorganic*** and *organic* matter for their growth.

Anabolism, as you may be knowing, is the process by which the food substances are changed into the tissues of a living animal or plant. This process thus, represents the constructive phase of metabolism.

Catabolism, on the other hand, is the process of breaking down of living tissues into simpler substances or waste matter, thereby producing energy. This process, thus, constitutes the destructive phase of metabolism.

With the above knowledge, you can easily mark the correct choice as (*a*), because B corresponds to 4, D to 2, A to 1, and C to 3.

Q. 204. Use Eq. (9.38) as

$$V = \frac{V_1(100 - p_1)}{100 - p}$$

Here $V_1 = P, p_1 = 98\%, p = 96\%$

$$\therefore \quad V = \frac{P(100-98)}{(100-96)} = \frac{P}{2}$$

Hence, choice (*d*) is correct. **Ans.**

Q. 215. Pl. refer "*Ecology and Environmental Studies*" by the same author to know about magnification of pesticides.

Q. 219. Pl. refer Solved Example 7.1.

Q. 221. Use Eq. (8.1) as

$$C = \frac{C_S Q_S + C_R Q_R}{Q_S + Q_R}$$

or

$$C = \frac{200 \times 50 + 8 \times 500}{50 + 500} = 25.4 \text{ mg/L}$$

Hence, choice (*b*) is correct. **Ans.**

Q. 222. Use Eq. (7.8) as

$$\text{TON} = \frac{V_s + V_D}{V_s}$$

$$= \frac{12.5 + 187.5}{12.5} = 16$$

Hence, choice (*d*) is correct. **Ans.**

Q. 229. Endogenous metabolism means metabolism from inner substances within the cells or tissues without the addition of outside nutrients. ***Exogenous metabolism*** means metabolism helped by outside nutrients.

In an aerobic-attached culture, growth of micro-organisms occur due to consumption of oxygen without the addition of any outside nutrients, and hence it can be considered as aerobic endogenous metabolism.

Q. 236. Since the value of K_D (30°) is more than the value of K_D (20°), evidently the BOD dissipation rate is more at 30°C. Thus, oxygen consumed in a particular given time will be more at 30°C rather than at 20°C. Hence, the curve AB will shift upward, *i.e.* towards the left. Hence, choice (*a*) is correct. **Ans.**

Q. 239. Anaerobic metabolism. As explained in the text, there is no oxygen present in an anaerobic waste water treatment. In the fermentation that occurs, a portion of the energy-rich organic compounds are oxidized, while other organic compounds or CO_2 are used as electron (hydrogen) acceptors. The transfer of electron does release relatively small amount of chemically bound energy, which is used for the growth of anaerobic bacteria. *Since the energy yield of anaerobic fermentation is only about one seventh of the yield for aerobes, the growth of anaerobes is slow.* Because there is no oxygen consumed in the process, there is no direct reduction in COD in an anaerobic process. ***The removal of COD is, however, accomplished by the conversion of***

organics into methane, which is relatively an insoluble gas. Significant production of CO_2 and negligibly small amount of H_2 and other gases also occur.

Anaerobic treatment is realized in the final conversion of metabolic intermediates to methane. Eventually, if the process is stopped midway (short of conversion to methane), the effluent will contain soluble products from intermediate stages of metabolism with the same oxygen demand as the initial material. ***There will, thus, be no reduction in BOD, and treatment will be of no use.***

Anaerobic digestion, infact, involves a complex consortium of micro-organisms. The biochemical processes involved along with their respective microbial species can be divided into three categories. These three stages are :

(*i*) *Hydrolysis* ;

(*ii*) *Acid formation* ; and

(*iii*) *Methanogenesis*.

These stages are briefly explained below :

(*i*) **Hydrolysis.** Large molecules and suspended organic solids can not be directly metabolised by anaerobes. When the concentration of anaerobes become significant, hydrolysis reaction becomes an important first stage of anaerobic metabolism. Hydrolysis is the breakdown of large complex soluble and insoluble molecules into smaller molecules that can be transported into cells and metabolised. Extracellular enzymes associated with the primary *fermentative microorganisms* are used to accomplish the task.

Energy is consumed in hydrolysis and synthesis, and is obtained from the catabolism of the smaller molecules resulting from hydrolysis. The fermenting micro-organisms responsible for hydrolysis do not form methane.

(*ii*) **Acid Formation.** The same micro-organisms that perform hydrolysis reactions carry the fermentation through this stage also. The end products of hydrolysis are further fermented (degraded) by micro-organisms into *organic acids*, and other low molecular mass compounds like H_2 and CO_2. *The primary product of this fermentation is acetic acid. Large quantity of H_2 is also produced.* The bacteria that produce acetic acid and H_2 are called ***acetogenic bacteria.*** Other acid forming bacteria form butyric and propionic acid along with some low molecular mass compounds. These micro-organisms (acetogenic bacteria, etc.) can tolerate wide range of environmental conditions. These micro-organisms, called ***acid formers,*** have an optimal pH range of 5-6; whereas digesters are normally operated at a pH near 7. In spite of this, the *acid formers* can sometimes work more favourably than the *methane formers* responsible for the final conversion of organics into methane. Without proper operation of the process, and particularly pH control, the *acid formers* can create highly unfavourable conditions for the *methane formers. For this very reason, pH in the digester should not be allowed to fall below about 7 (which is the optimal pH range for methane formers), so as not to provide optimal conditions for the acid formers, but to provide optimal pH for the methane formers.*

(*iii*) **Methanogenesis.** The formation of methane, which is the ultimate product of anaerobic digestion, occurs by two major routes. The primary route is the fermentation of the major product of the acid forming phase (*i.e.* acetic acid) to methane and CO_2. *Bacteria that utilize acetic acid* are ***acetoclastic*** (or ***acetophilic***) bacteria. The overall reaction is :

$$\underset{\text{Acetic acid}}{CH_3COOH} \xrightarrow[\text{bacteria}]{\text{Acetoclastic}} \underset{\text{Methane}}{CH_4} + CO_2 \uparrow$$

Based on thermodynamic considerations and from experimental data, Zeikus (1975) proposed the following reaction for conversion of acetic acid (or acetate) to methane :

$$CH_3COOH + 4\,H_2 \longrightarrow \underset{\text{Methane}}{2\,CH_4} + 2\,H_2O$$

The most common acetoclastic methanogens in reactors treating wastes with volatile fatty acid content are from the genera ***Methanosarcina*** and ***Methanosaeta*** (formerly called ***Methanothrix***).

Methanosarcina spp. are coccoid bacteria with doubling times nearly 1.5 days, and Methanosaeta spp. are sheathed rods, sometimes growing as long filaments with doubling times nearly 4 days. These doubling times occur at optimal conditions for the methane formers. Even though methanosaeta (methanothrix) spp grow more slowly, they are most frequently the dominant genus. Hence, the choice of methanothrix and methanosercina, *i.e.* choice at (*d*) is correct. **Ans.**

Q. 244. (*i*) The Assertion (A) that the BOD gets removed at a very fast rate immediately after sewage is discharged into a river is correct because the BOD dissipation rate $\left(i.e., \dfrac{dL_t}{dt}\right)$ *i.e.* rate of deoxygenation is directly proportional to the amount of organic matter present in sewage at that time (*i.e.* L_t). Since the amount of organic matter in disposed sewage will be maximum immediately on its disposal, the BOD removal rate will also be maximum and very fast DO absorption and BOD dissipation will occur.

The given Reason (R) states that part of BOD in sewage is due to settleable organic matter therein. This statement in itself is correct, but is not the correct explanation of the above assertion, since in a flowing stream there is no time for organic settleable organics to settle down due to flow velocity ; and moreover, sewage contains very small amount of settleable solids which do not settle at the points of outflow of sewage to reduce BOD. Hence, although A and R both are correct individually, but R is not the correct explanation of A, and hence correct answer is (*b*).

Q. 247. $Q = \dfrac{1}{36} . K . p_c . A$

For impervious area, $K = 1$

$$\therefore \quad Q = \frac{1}{36} \times 1 \times 2 \times 3.6 = 0.2 \text{ m}^3/\text{s}$$

Hence, choice (*c*) is correct. **Ans.**

Q. 248. From Mannings formula with sewer running full, in the first case, we have :

$$V_1 = \frac{1}{N} . R_1^{2/3} . S_1^{1/2}$$

$$\text{where } R_1 = \frac{D_1}{4} = \frac{0.3\,m}{4} = 0.075 \text{ m}$$

$$\therefore \quad V_1 = \frac{1}{N} \times (0.075)^{2/3}. \frac{1}{\sqrt{400}} = 0.7 \text{ m/s}$$

$$\therefore \quad n = 0.0127$$

For 2nd case, we have

$$V_2 = \frac{1}{n} . R_2^{2/3}. S_2^{1/2}$$

or $$V_2 = \frac{1}{n}.\left(\frac{0.6}{4}\right)^{2/3}.\frac{1}{\sqrt{200}} \left(\text{for sewer running 1/2 full, } R_2 = \frac{D_2}{4} = \frac{0.6m}{4}, \textit{ i.e. } \text{same as for sewer running full} \right)$$

$$n = N = 0.0127$$

$$S_2 = \frac{1}{200}$$

or $$V_2 = \frac{1}{0.0127} \times (0.15)^{2/3}. \frac{1}{\sqrt{200}} = 1.57 \text{ m/sec.}$$

The nearest choice is at (*c*) equal to 1.59 m/sec ; which is the correct answer.

Ans.

Q. 249. $Y_t = L\left[1-(10)^{-K_D.t}\right]$

or Y_5 at 20°C = 250 = $L\left[1-(10)^{-K_{D(20)}\times 5}\right]$ ...(1)

But $K_{D(T)} = K_{D(20)}\,[1.047]^{T-20}$

$\therefore \quad K_{D(30)} = K_{D(20)}\,[1.047]^{30-20} = 1.583\,K_{D(20)}$

$\therefore \quad Y_t$ at 30° = 250 = $L\left[1-(10)^{-K_{D30}\times t}\right]$ ...(*ii*)

Comparing (*i*) and (*ii*), we have

$$1-10^{-K_{D(20)}\times 5} = 1-10^{-K_{D30}\times t}$$

or $$10^{-K_{D(20)}\times 5} = 10^{-K_{D30}\times 5} = (10)^{-1.583\,.\,K_{D(20)}\times t}$$

or $$5 = 1.583t \text{ or } t = 3.16 \text{ days.}$$

The nearest available choice of 3.3 days at (*c*) is hence correct. **Ans.**

Q. 251. (*i*) The time taken in 1st trip from garage to the disposal site and returning to the filling site (*i.e.,* at head of route) = Time taken from garage to head of route (0.4 hr) + Time taken in loading (0.5 hr) + Time taken is driving from head of the route to the disposal site (*i.e.,* 0.4 hr) + Time taken in unloading (0.2 hr) + Time taken is returning from the disposal site (0.25 hr)

$$= 0.4 + 0.5 + 0.4 + 0.2 + 0.25 = 1.75 \text{ h.} \quad ...(i)$$

(*ii*) The time taken in 2nd trip before starting to return from disposal site = Time taken in loading (0.5 hr) + Time taken is driving to disposal site (0.4 hr) + Time taken in unloading (0.2 hr)

$$= 0.5 + 0.4 + 0.2 = 1.1 \text{ hr.} \quad ...(ii)$$

(*iii*) Breaks and Miscellaneous delays

$$= 2 \times 15 \text{ min} + 30 \text{ min} = 60 \text{ min} = 1 \text{ hr} \quad ...(iii)$$

Total time taken = (*i*) + (*ii*) + (*iii*)

$$= 1.75 + 1.1 + 1 = \mathbf{3.85\ hr}$$

Time left in 8 hr nominal duty duration for refuse collection

$$= 8 - 3.85 = \mathbf{4.15\ hr.}$$

Hence, choice (*a*) is correct. **Ans.**

Q. 254. Sound Pressure level (SPL)

$$= 20 \log_{10} \frac{P}{P_{ref}}$$

where $P_{ref} = 20\ \mu.\text{Pa}$ (Eq. 20.10)

Also $P = 0.0002\ \mu$ bar (given)

But 1 bar = 10^5 Pa (Pl. see Table A-11)

$\therefore$ $P = 0.0002 \times 10^5\ \mu.\text{Pa} = 20\ \mu\text{Pa}.$

$$\therefore \quad \text{SPL} = 20 \log_{10} \frac{20\ \mu\text{Pa}}{20\ \mu\text{Pa}} = 20 \log_{10}^{1} = 0$$

Hence, choice (*a*) is correct. **Ans.**

Q. 266. Using Eq. (9.38), we have

$$V = V_1 \left(\frac{100 - p_1}{100 - p} \right)$$

$$\therefore \quad V = V_1 \left(\frac{100 - 99}{100 - 98} \right) = \frac{V_1}{2}$$

$$\therefore \quad V = 0.5 V_1 = 0.5 \times 1 \text{ mld} = 0.5 \text{ mld}$$

Hence, choice (*b*) is correct. **Ans.**

Q. 286. The sub-micron size gaseous particles (*i.e.* particles of size less than one microne), although cannot be removed efficiently in old conventional electrostatic precipitators (ESPs), but in modern days, ESPs have been developed to remove such smaller particles also, since many countries are beginning to limit the emissions of micron and sub-micron particles from incomplete combustion exhaust gas, because such smaller particles can suspend in the atmosphere for a long time, and they are very harmful to human health.

In conventional ESPs, the collection efficiency drops down significantly when the particle size is less than 1 μm (micron), as the collection of particles in an ESP depends upon the electric force. More electrical charges on the particles could be obtained with high electric fields. Generally, a DC-energized ESP is commonly used, but there is a limit on the supplied voltage, which could not be too high due to an occurrence of an electrical breakdown.

A wire-cylinder electrostatic precipitator (ESP) has been developed for control of submicron particles generated in the exhaust gas. However, for such smaller particles, the efficiency is not high, of which the value is in a range of 70-80 percent. Use of a non-thermal plasma (NTP) precharger integrated infront of the ESP, has also been suggested in advances for increasing this efficiency.

Be that as it may, both the statements are correct and statement II is the correct explanation of statement I. Hence the answer is (*a*). **Ans.**

Q. 287. From Eq. (9.38), we have

$$V = V_1 \left[\frac{100 - p_1}{100 - p}\right]$$

Here $V_1 = 100\ m^3$, $p_1 = 95\%$, $p = 90\%$, $V = ?$

$$\therefore \quad V = 100\ m^3 \left[\frac{100 - 95}{100 - 90}\right] = 100 \times \frac{5}{10} = 50\ m^3$$

Hence, choice (*b*) is correct. **Ans.**

Q. 288. From Eq. (9.52), we have

$$SVI = \frac{V_{ab}}{X_{ab}} \times 1000\ ml/g.$$

where V_{ab} = Settled sludge volume of the 1 litre sample of mixed liquor after 30 minutes settling = 200 ml/l

X_{ab} = MLSS, *i.e.* concentration of suspended solids in the above 1 litre sample of mixed liquor collected from the discharge end of the aeration tank = 2000 mg/l

$$\therefore \quad SVI = \text{Sludge Volume Index} = \frac{200\ ml/l}{2000\ mg/l}$$

$$= \frac{1}{10}\ ml/mg = \frac{1}{10} \times 1000\ ml/g$$

$$= 100\ ml/g.$$

Hence, choice (*c*) is correct. **Ans.**

Q. 295. 1 bar = 10^5 Pascal (*i.e.* N/m^2)

The min. sound pressure level (Table 26.1) is

$$20\ \mu Pa = 20 \times 10^{-5}\ \mu\ bar = 0.0002.$$

Similarly, the highest sound level is $10^9\ \mu Pa$

$$= 10^9 \times 10^{-5} = 10^4 = 10{,}000\ \mu\ bar$$

Hence (1) is correct. Since (2) and (3) are also correct, the final choice (*d*) is correct. **Ans.**

Q. 298. Pl. refer to "Ecology and Environmental Studies" by the same anthor.

—27

Conventional Questions of the Civil Services (IAS) Exams for the Past Years

The Union Public Service Commission, Dholpur House, Shahjahan Road, New Delhi-110011, conducts every year two competitions, in which, you as civil engineering graduates (even with exam. results awaited) can participate for securing a direct class I Govt. appointment.

The first competition is named as **Combined Engineering Services (CES) Exam.,** and the other competition is known as **Civil Services Exam.**, popularly called the *IAS competition.*

Press notifications for obtaining and submitting application forms are released by Secretary UPSC, through news papers including "Employment News" sometimes in Feb. every year. The exams are usually held in August.

On the basis of the first competition, you are likely to be appointed as Asstt. Executive Engineer or Asstt. Director or on an equivalent post in the civil engineering side of the various Government Departments like Railways, CPWD, P&T, Central Water Commission, etc. ; whereas, on the basis of the second competition, you are liable to become an IFS, IAS, IPS, or an officer on allied services, such as Asstt. Custom Collector or Asstt. Excise Collector, etc.

Most of the engineering degree students aspire to clear any one of these two competitions, for an excellent career full of promotional avenues, just after doing BE degree. The syllabuses for both these competitions, now-a-days, provide full opportunities to Civil Engineering degree holders, to make through these competitions.

In this chapter, we will take up the questions of the Civil Services Exams of the past years pertaining to the subject of "Waste Water Engineering" including Environmental Engineering and Air Pollution.

Certain Typical Questions from the Old Civil Service Exams from the Years 1980 to 2010.

Q.27.1. *Write an equation for photosynthesis. Why do plants perform this reaction ?* *(Civil Services, 1980)*

Solution. The basic equation governing **photosynthesis** is available in Elementary Books of Biology, and is given as :

$$\underset{}{12H_2O + 6CO_2} \xrightarrow[\text{Chlorophyll in green plants}]{\text{in light}} \underset{\text{Oxygen}}{6O_2} \uparrow + 6H_6O + \underset{\text{Carbohydrates Sugar, etc.}}{C_6H_{12}O_6} \qquad ...(27.1)$$

Green photosynthetic plants absorb moisture (from ground),

light and carbon dioxide gas from the atmosphere, through the process of *photosynthesis,* to release oxygen. Infact, it was Van Niel, who postulated that the moisture (H_2O) absorbed by the plants from the ground is converted into oxygen by a *green pigment*, called **chlorophyll** (present in green photosynthetic plants) *in the presence of light* (particularly sunlight).

The light from sun, or from artificial light sources, contain energy, which is absorbed by such plants, although a part of it is also lost as *heat and fluorescence.* The absorbed light energy triggers the above *photochemical reaction,* that induces splitting of water, release of hydrogen and oxygen ; and the released hydrogen helping in the synthesis of a *reducing agent* which finally converts carbon dioxide (from the surroundings) into carbohydrates, etc. The photochemical reaction is shown in Fig. 27.1.

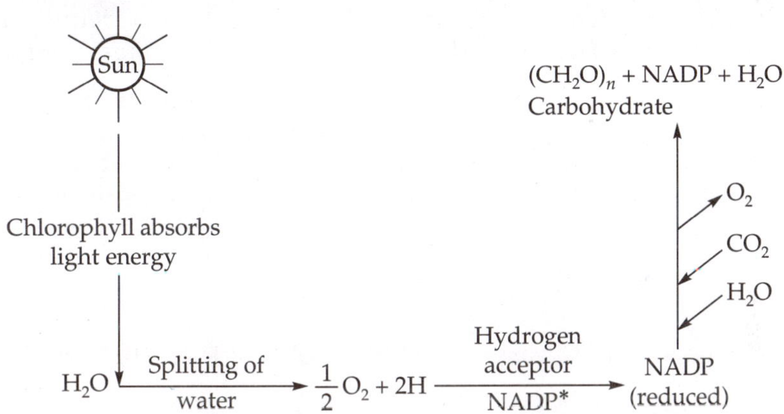

Fig. 27.1. Photosynthesis (reaction).

In other words, the light energy (mainly solar energy) is converted into chemical energy in the form of carbohydrates, proteins, fats, etc., through the process of photosynthesis. **Ans.**

Q. 27.2. *BOD (Biochemical oxygen demand at 20°C after 5 days) of a sewage sample has been found to be 250 mg/l. What is its value on the 15th day at 25°C? Assume $k_{20}' = 0.12$.* *(Civil Services, 1982)*

Solution : $K_{D(20°)} = 0.12$

$Y_5 = 250$ mg/l (at 20°C)

Y_{15} at 25°C = ?

Using $Y_{5 \text{ at } (20°)} = L\left[(1-10)^{-K_{D(20°)} \cdot t}\right]$, we have

$$250 = L\left[1-(10)^{-0.12\times 5}\right] = L\left[1-\frac{1}{(10)^{0.6}}\right]$$

*NADP stands for Nicotinamide adenine dinucleotide phosphate.

$$0.7488\,L = 260$$

or $$L = 333.86 \text{ mg/l}$$

Now, $$K_{D(25°)} = K_{D(20°)}\,[1.047]^{T-(20°)}$$
$$= 0.12\,[1.047]^{25-20}$$
$$= 0.12 \times 1.258 = \mathbf{0.151}$$

Now, $$Y_{15 \text{ at } 25°} = L\left[1-(10)^{-K_{D(25°)}\,.\,t}\right]$$
$$= 336.86\left[1-(10)^{-0.151\times 15}\right]$$
$$= 333.86\left[1-\frac{1}{(10)^{2.265}}\right]$$
$$= \mathbf{332.05 \text{ mg/l.}} \quad \textbf{Ans.}$$

Q. 27.3. *Calculate the discharge of 1.25 m circular sewer laid at a slope of 1 in 360, when it is running half full. Assume n in Mannings formula as 0.011.*

Solution $$D = 1.25 \text{ m}, S = \frac{1}{360}, N = 0.011.$$

At $$\frac{d}{D} = 0.5, \frac{q}{Q} = 0.5$$

where Q is the discharge when pipe runs full.

$$Q = \frac{1}{N}\,AR^{2/3}.\,S^{1/2}$$

$$R = \frac{A}{P} = \frac{\pi D^2}{4\,.\,\pi D} = \frac{D}{4} = \frac{1.25}{4}\,m = 0.3125$$

$$A = \frac{\pi}{4}\,D^2 = \frac{\pi}{4} \times (1.25)^2$$

$$Q = \frac{1}{0.011} \times \frac{\pi}{4} \times (1.25)^2 \times (0.3125)^{2/3}\,.\,\frac{1}{\sqrt{360}}$$
$$= 2.70 \text{ m}^3/\text{sec.}$$

$$\therefore \quad q = \frac{Q}{2} = \frac{2.70}{2} = \mathbf{1.35 \text{ m}^3/\text{sec.}} \quad \textbf{Ans.}$$

Q. 27.4. *A population of 40,000 resides in a town covering an area of 75 hectares. Calculate the discharge for a combined system of sewers. Given :*

(*i*) *Coefficient of run off = 0.70,*

(*ii*) *Rain concentration = 40 minutes,*

(*iii*) *Water consumption = 120 l/d/person ; and*

(*iv*) *70% of wastewater reaches the sewers.* (*Civil Services, 1984*)

Solution $K = 0.70$, $T_c = 40 \text{ minutes} = \frac{40}{60}$ hr. and $A = 75$ hectares

Using rational formula,

$$Q = \frac{1}{36} \cdot K \cdot p_c \cdot A$$

$$= \frac{1}{36} \cdot K \cdot p_0 \left(\frac{2}{1+T_c}\right) \cdot A$$

and assuming design value of p_0 = 4 cm/hr, we have

$$\text{Rain run off} = \frac{1}{36} \times 0.7 \times 4.0 \left(\frac{2}{1+\frac{40}{60}}\right) = 7.00 \text{ m}^3\text{/sec.}$$

$$\text{Av. sewage discharge} = 0.70 \times \frac{120 \times 40{,}000}{1000} \text{ m}^3\text{/day}$$

$$= \frac{0.70 \times 120 \times 40}{24 \times 60 \times 60} \text{ m}^3\text{/sec} = 0.04 \text{ m}^3\text{/s.}$$

Peak rate of sewage = 3 × 0.04 = 0.12 m^3/sec

Hence, total discharge in sewer

= 7.00 + 0.12 = **7.12 m^3/sec. Ans.**

Q. 27.5. *The 5 days 20°C BOD of domestic sewage is 250 mg/l. Estimate the three day BOD at 20°C. The reaction constant k = 0.1.* (*Civil Services, 1999*)

Solution :

$$Y_{5\,(20°)} = 250 \text{ mg/l}$$

$$Y_{3\,(20°C)} = ?$$

$$k = K_D = 0.1\text{/day}$$

Using $Y_t = L\left[1-(10)^{-K_D \cdot t}\right]$, we get

$$250 = L\left[1-(10)^{-0.1\times 5}\right] = L\left[1-\frac{1}{(10)^{0.5}}\right]$$

$$= 0.6837\, L$$

or $L = 365.62$ mg/l

Now, $Y_{3\,(20°)} = L\left[1-(10)^{-0.1\times 3}\right]$

$$= 365.62\left[1-\frac{1}{(10)^{0.3}}\right] = 182.38$$

or $Y_{3\,(20)}$ = **182.38 mg/l. Ans.**

Q. 27.6. *A primary sedimentation sludge has a moisture content of 95% and 65% of the dry solids are volatile. Assuming the specific gravity of the volatile solids as 1.00 and that of fixed solids as 2.2, find (i) the specific gravity of the total dry solids ; and (ii) the specific gravity of the sludge.*

(*Civil Services, 2003*)

Solution: Assume 100 kg of sludge

m.c. of sludge = 95%

$\therefore$ Wt. of total sludge solids = 5 kg

Wt. of water in sludge = 95 kg

5 kg of total solids contain 65% of volatile solids

Wt. of volatile solids = 65% × 5 kg = 3.25 kg

$\therefore$ Wt. of fixed solids = 5 – 3.25 = 1.75 kg

Sp. gr. of volatile solids = 1.00

Sp. gr. of fixed solids = 2.2

$\therefore$ Sp. gr. of total solids

$$= \frac{3.25 \times 1.00 + 1.75 \times 2.2}{5} = \mathbf{1.35.} \quad \textbf{Ans.}$$

Sp. gr. of wet sludge

$$= \frac{95 \text{ kg} \times \text{sp. gr. of water} + 5 \text{ kg} \times \text{sp. gr. total solids}}{100 \text{ kg}}$$

$$= \frac{95 \times 1.00 + 5 \times 1.35}{100} = \mathbf{1.0175.} \quad \textbf{Ans.}$$

Q. 27.7. *Determine the diameter of a storm sewer based on the following data :*

(*a*) *Sewer length = 1 km to be laid at a slope of 0.001.*

(*b*) *Catchment area of the sewer = 50 hectares of which the coefficient of runoff is 0.9, 0.80 and 0.20 for 30% (roof area), 20% (pavements) and 50% (open land) of the area, respectively.*

(*c*) *Velocity of full flow = 1.5 m/sec*

(*d*) *Manning's n = 0.013*

(*e*) *Time of entry = 3 min.*

(*f*) *Intensity of rainfall in cm/hr* $= \dfrac{75}{5 + \text{Time of concentration in minutes}}$.

(*Civil Services, 2008*)

Solution :

$$K = \frac{K_1 A_1 + K_2 A_2 + K_3 A_3}{A_1 + A_2 + A_3}$$

or

$$K = \frac{0.9 \times (30\% \times 50 \text{ ha}) + 0.8(20\% \times 50 \text{ ha}) + 0.20(50\% \times 50 \text{ ha})}{50 \text{ ha}}$$

or $K = 0.27 + 0.16 + 0.10 = \mathbf{0.53}$

Now, $p_c = \dfrac{75}{5 + T_c(\text{min.})}$

where $T_c = T_i + T_f$

where $T_i = 3$ min

$$T_f = \frac{1000 \text{ m}}{1.5 \text{ m/s}} = \frac{1000}{1.5} \text{ s}$$

$$= \frac{1000}{1.5 \times 60} \text{ min}$$

$$= \frac{1000}{90} = 11.11 \text{ min.}$$

$\therefore \quad T_c = 3 + 11.11 = 14.11$ min

$$\therefore \quad p_c = \frac{75}{5 + 14.11} = \frac{75}{19.11} = 3.92 \text{ cm/h}$$

Now, $Q_p = \dfrac{1}{36} \times K.\, p_c.\, A$

where $A = 50$ ha, $p_c = 3.92$ cm/hr, $K = 0.53$

$$= \frac{1}{36} \times 0.53 \times 3.92 \times 50 = \mathbf{2.889\ m^3/s}$$

Now, $Q = \dfrac{1}{n} AR^{2/3}.\sqrt{S}$

$$2.889 = \frac{1}{0.013} \times \left(\frac{\pi}{4}.\, D^2\right)\left(\frac{D}{4}\right)^{2/3}.\sqrt{0.001}$$

or $$\frac{2.889 \times 0.013 \times 4 \times (4)^{2/3}}{\sqrt{0.001}} = D^2.\, D^{2/3} = D^{8/3}$$

or $$11.97 = D^{8/3}$$

or $D = (11.97)^{3/8} = (11.97)^{0.375} = \mathbf{2.54\ m.}$ **Ans.**

Q. 27.8. *Taking W and W_s as the weight of water and dry solids respectively, derive an expression for the specific gravity (S) of sludge if S_s represent the specific gravity of the dry solids present in the sludge, and p represents the percentage of moisture (water) (and W = 1).* *(Civil Services, 2008)*

Solution : If p is the percentage of water in the sludge, then $(100 - p)$ kg of solids will make = 100 kg of sludge.

$\therefore \quad W_s$ kg of solids will make

$$= \frac{100\, W_s}{(100 - p)} \text{ kg of sludge}$$

If S is the sp. gravity of sludge, then its

Unit wt. $\gamma_s = S\, .\, \gamma_w$ (where γ_w = unit wt. of water)

$\therefore$ Vol. of $\dfrac{100\, W_s}{(100 - p)}$ kg of sludge

$$= \frac{100\, W_s \times 1}{(100 - p)\,(S\, .\, \gamma_w)} \quad ...(i)$$

But vol. of sludge = Vol. of dry solids + Vol. of water

$$= \frac{W_s}{\gamma_w\, .\, S_s} + \frac{W}{\gamma_w} = \frac{1}{\gamma_w}\left[\frac{W_s}{S_s} + W\right] \quad ...(ii)$$

Equating (i) and (ii), we get

$$\frac{100\,W_s}{(100-p)\,.\,S\,.\,\gamma_w} = \frac{1}{\gamma_w}\left[\frac{W_s + S_s\,.\,W}{S_s}\right]$$

or
$$\frac{100\,W_s}{(100-p)\,.\,S} = \left[\frac{W_s + S_s\,.\,W}{S_s}\right]$$

or
$$\boxed{S = \frac{100\,W_s}{(100-p)} \times \frac{S_s}{(W_s + S_s W)}}$$

When $W = 1$, then

$$S = \frac{100\,W_s\,.\,S_s}{(100-p)\,(W_s + S_s)}.\quad \textbf{Ans.}$$

Q. 27.9. *Determine using the 'rational formula' with runoff co-efficient equal to 0.80, the peak storm runoff to be carried out by the underground drainage pipeline AB for the drainage basin area of 0.2 ha shown in the given figure (Fig. 27.2). Take the length of pipeline as 100 m and the level difference between A and B as 0.5 m. Assume initially a pipe diameter of 250 mm. Consider Darcy-Weisbach friction factor f = 0.015. The rainfall duration and rainfall intensity is furnished in tabular form below. Consider time of entry at end A as 2 min.*

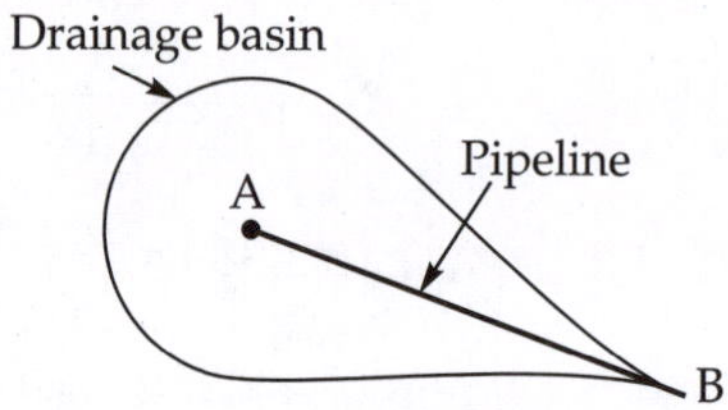

Fig. 27.2

Check whether the pipe size is adequate or not.

Duration (mins)	*Rainfall intensity (mm/hr)*
2.0	122
2.5	108
3.0	97
3.5	89
4.0	82
4.5	77

(Civil Services, 2009)

Solution : $K = 0.80$, $A = 0.2$ ha, $L = 100$ m

$H_L = 0.5$ m, $D = 250$ mm $= 25$ cm $= 0.25$ m

$f = 0.015$

$$H_L = \frac{f.LV^2}{2gD}$$

or
$$0.5\text{ m} = \frac{0.015 \times 100 \times V^2}{2 \times 9.81 \times 0.25}$$

or $V = \mathbf{1.279\ m/s.}$

$$\text{Time of flow} = T_f = \frac{\text{Length of pipe}}{\text{Vel. of flow}} = \frac{100\text{ m}}{1.279\text{ m/s}}$$

= **78.21 sec = 1.3 min.**

$T_c = T_i + T_f$ = 2 min. + 1.3 min. = **3.3 min.**

Rainfall intensity (as per given table) :

For T_1 = 3.0 min. = 97 mm/hr

and for T_2 = 3.5 min. = 89 mm/hr

Interpolating, we get, p_c for T_c of 3.3 min.

$$= \left[97 - \frac{97-89}{0.5} \times 0.3\right] = 92.2 \text{ mm/h} = \mathbf{9.22 \text{ cm/h}}$$

Now, $$Q_p = \frac{1}{36} K p_c \, . \, A$$

$$= \left[\frac{1}{36} \times 0.80 \times 9.22 \times 0.2\right] = 0.041 \text{ m}^3\text{/s}$$

The capacity of the 0.25 m dia pipe

$$= V \, . \, A = 1.279 \times \left(\frac{\pi}{4} \times 0.25^2\right) = \mathbf{0.062 \text{ m}^3\text{/s}}$$

Since, capacity of pipe (0.062 m^3/s) is more than the discharge (0.041 m^3/s), the pipe dia. is adequate. **Ans.**

Year Wise Questions from the Civil Services Exams from the Year 2010 onwards

Year 2010 (*Civil Services*)

Q. 5. (*e*) *Determine the BOD loading of an oxidation pond used for treating the domestic sewage of 20,000 persons when the water supply demand is 100 litres per capita per day. Assume a detention time of 6 days and BOD of sewage as 300 mg/l. The pond is to be 1.2 m deep.* (12)

Solution : Quantity of water supplied = 20000 × 100 litres/day

= 2 MLD.

Assuming sewage generation to be at 80% of water supply, we have

Sewage generated = 2MLD × 0.8 = 1.6 MLD

Vol. of Oxidation Pond Reqd. = 1.6 MLD × Detention Time in days

$$= \left[\frac{1.6 \times 10^6}{10^3} \frac{\text{m}^3}{\text{d}} \times 6\text{d}\right] = 9600 \text{ m}^3.$$

$$\text{Area of oxidation pond reqd.} = \frac{\text{Volume}}{\text{Water depth}} = \frac{9600 \text{ m}^3}{1.2 \text{ m}}$$

= 8000 m^2 = **0.8 ha**

Total BOD Load of sewage

= Sewage volume × BOD of sewage

= 1.6 MLD × 300 Mg/l

$$= \left[\frac{1.6 \times 10^6 \times 300}{10^6}\right] \text{kg/day} = 480 \text{ kg/day}$$

$$\text{BOD loading in kg/day/ha} = \frac{480 \text{ kg/day}}{0.8 \text{ ha}} = \mathbf{600\ kg/ha/d.}$$

Ans.

Q. 5. (*f*) *Stating a use-oriented definition of water pollution, list the various sources of water pollution, along with the major characteristics of each of the stated sources.* (12)

[**Hint:** *Pl see article 9.1 and 9.3 in "Ecology and Environmental Studies" by the same author*]

Q. 6. (*c*) (*ii*) *Estimate the population equivalent for each of*

(*a*) *Combined system for 2000 persons having 100 gm per capita daily BOD on the assumption that the contribution of BOD is 75 gm per capita per day.*

(*b*) *80,000 litres daily contribution of industrial wastes containing 100 mg/litre of suspended solids on the assumption that the contribution of suspended solids is 50 gm per capita per day. State clearly any other assumption needed to be invoked.* (10)

Solution : (*a*) Total BOD of the system = 2000 × 100 g/d = 2 × 10^5 g/d

Population Equivalent

$$= \frac{\text{Total BOD per day}}{\text{Std. BOD of domestic sewage per person per day}}$$

$$= \frac{2 \times 10^5 \text{ g/d}}{75 \text{ g/c/d}} = \frac{2 \times 10^5}{75} = \mathbf{2667\ Ans.}$$

(*b*) Total S.S. of industrial wastes

= 80,000 l/d × 100 mg/l

= 8 × 10^6 mg/d = 8000 g/d

Population Equivalent

$$= \frac{\text{Total S.S. per day in industrial wastes}}{\text{Std. S.S. of domestic sewage per person per day}}$$

$$= \frac{\mathbf{8000\ g/d}}{\mathbf{50\ g/c/d}} = \mathbf{160.\ \ Ans.}$$

Q. 7. (*b*) (*i*) *What will be the diameter of a circular sewer to carry a discharge of 0.624 m³/s, when flowing full at a slope of 10 in 10,000 ? Use n as 0.012 in the Manning's formula.* (10)

Solution : From Manning's formula, we have

$$Q = \frac{1}{N} AR^{2/3}\sqrt{S}$$

For flowing full conditions of a circular sewer, we have

$$A = \frac{\pi}{4} D^2, \text{ where } D \text{ is dia of sewer}$$

$$P = \pi D$$

$$R = \frac{D}{4}$$

$$\therefore \quad 0.624 = \frac{1}{0.012}\left(\frac{\pi}{4} D^2\right)\left(\frac{D}{4}\right)^{2/3} \cdot \frac{1}{\sqrt{1000}}$$

$$\text{or} \quad D^{8/3} = 0.7597$$

$$\text{or} \quad D = (0.7597)^{\frac{3}{8} = 0.375} = \mathbf{0.9\ m.} \quad \textbf{Ans.}$$

Q. 7. (*b*) (*ii*) *In the above sewer, what would be the discharge and velocity if the proportionate depth is held at 0.4. Use the table given herewith*

d/D	*q/Q*	*v/V*
0.4	*0.32*	*0.88*

(10)

Solution : When sewer is running at depth (*d*) equal to 0.4 *D*, then we have

$$\frac{q}{Q} = 0.32 \text{ (given)}$$

$$\therefore \quad q = \text{discharge at 0.4 full depth} = 0.32\,Q$$

$$= 0.32 \times 0.624 = \mathbf{0.2\ m^3/sec.} \quad \textbf{Ans.}$$

Also $\quad \frac{v}{V} = 0.88$

$$\therefore \quad v = \text{Velocity at partial depth} = 0.88\,V$$

(where V is vel. at full depth)

But $\quad V = \frac{Q}{A} = \frac{0.624}{\pi/4 \times (0.9)^2} = 0.98 \text{ m/s}$

$$\therefore \quad v = 0.88 \times 0.98 \text{ m/sec} = \mathbf{0.863\ m/sec.} \quad \textbf{Ans.}$$

Q. 7. (*c*) (*i*) *Write a note on river training in the context of river pollution control at river banks through a typical Indian city.* (10)

Q. 7. (*c*) (*ii*) *Discuss the phenomenon of self-purification of streams. How can it be evaluated and represented ? What are the factors that significantly affect the self-purifying abilities of a stream?* (10)

Q. 8. (*b*) (*ii*) *Explain the meaning of "sustainable development" and its role in project planning. Also discuss the appropriateness of the "separate system of sewerage" in India.* (10)

Year 2011 (*Civil Services*)

Q. 5. (*b*) *Differentiate between Thermophilic and Mesophilic bacteria explaining their roles in anaerobic digestion.* (10)

Q. 5. (*d*) *15 mL of the waste sample has been added into a 300 mL BOD incubation bottle. The initial DO of the diluted sample was 7.8 mg/L and the final DO after 5 days was 1.7 mg/L. The corresponding initial and final DO of* *the seaded dilution water was 8.1 and 6.9, respectively. What is the 5 day BOD of the wastewater sample.* (10)

Solution: While explaining the BOD test under article 7.4.10, we have explained the use of BOD test bottle for determining the BOD_5 of the given waste water, wherein the given waste water sample is mixed with pure distilled water and incubated for 5 days. The DO of the diluted sample before incubation (say D_1) and after incubation (say D_2) are determined, and the BOD_5 of the sewage sample is finally computed by Eq. (7.11) as:

BOD_5 of sewage sample = DO consumed

$$\text{i.e.}\quad (D_1 - D_2) \times \left[\frac{\text{Vol. of diluted sample}}{\text{Vol. of undiluted sewage sample}}\right]$$

$$= (D_1 - D_2) \times \text{Dilution factor} \qquad \text{...(Eq. 7.11)}$$

However, when pure distilled water is not used for dilution of the sewage sample, and the aged domestic water or polluted water containing some organic matter is used, then evidently some of the DO shall be consumed by the diluting water. The BOD_5 of sewage sample can in such a case be calculated by using a **modified equation**, given as.

BOD_5 of sewage sample

$$= \left[(\mathbf{D_1 - D_2}) - (\mathbf{B_1 - B_2})\mathbf{f}\right] \text{. Dilution factor} \qquad \text{...(27.2)}$$

where B_1 = Initial DO of the dilution water

B_2 = DO of the dilution water after incubating for 5 days in a separate BOD bottle.

f = Ratio of dilution water volume in wastewater test to the dilution water volume in BOD test on dilution water.

Using the above equation, and given values, we have

$D_1 = 7.8$ mg/L $\qquad D_2 = 1.7$ mg/L

$B_1 = 8.1$ mg/L $\qquad B_2 = 6.9$ mg/L

$$f = \frac{285\text{ ml}}{300\text{ ml}}$$

$$\text{Dilution factor} = \frac{300\text{ ml}}{15\text{ ml}} = 20$$

$$\therefore \quad BOD_5 \text{ of sewage} = \left[(7.8 - 1.7) - (8.1 - 6.9)\frac{285}{300}\right] 20$$

$$= \left[6.1 - 1.2 \times \frac{285}{300}\right] 20$$

$$= [6.1 - 1.14]20 = \mathbf{99.2\ mg/L.}\quad \textbf{Ans.}$$

Q. 5. (*f*) *Based on type of pollution and climate, differentiate between 'Gray air' and 'Brown air' cities, with examples.* (10)

[Hint : The formation of smog (*i.e.* smoky fog) over a city can be of two types; *i.e.*

(*i*) *Industrial smog* or *sulphurous smog* or '*gray air smog*'; and

(*ii*) *photochemical smog* or *brown air smog.*

The ***gray air smog*** is caused by the large scale burning of coal and other fossil fuels, like oil, etc. in homes, industries, or in power plants. The resulting sulphur dioxide gas ($SO_2\uparrow$), H_2SO_4 vapours, and particulate matter, causes the formation of a thick blanket of gray or blackish coloured smog over the city, if the environment is stable, and the chances of dispersion are minimal. The sulphurious smog is hence termed as gray air smog, and cities affected by such smogs are termed as ***'gray air' cities***.

Formation of such gray air smog was quite common in London and New York cities upto the middle of 20th century, since huge amount of coal was used to be burnt in those cities. The 1952 *London smog,* called the ***Great Smog of 1952***, darkened the streets of London and killed approximately 4000 people in the short time of 4 days, with a further death of about 8000 people due to its after effects in the following weeks and months. In the modern days, however, the problem of such a smog has reduced, particularly in developed cities, since fuel is now-a-days burned with enough pollution control installations and use of high rise chimneys. Still, however, the problem continues in several industrial cities of less developed countries including China, India, Ukraine, and some eastern European countries, etc.

The second type of smog, called the ***photochemical smog***, is primarily caused by the automobile exhausts, which contain large quantities of nitric oxide (NO), nitrogen dioxide (NO_2), unburned hydrocarbons (HC), etc. The photochemical smog is formed due to the chemical reaction of sunlight, nitrogen oxides (NO_x) and volatile organic compounds in the atmosphere, which leaves airborne particles and ground level ozone. The noxious mixture of air pollutants may include : aldehydes, nitrogen oxides, peroxyacyl nitrates, tropo-sphere ozone, and volatile organic compounds (VOCs). All these compounds are usually highly reactive and oxidizing. Photochemical smog is, therefore, considered to be a problem of modern indus-trialization. *It is generally present in all cities but is more common in cities with sunny, warm and dry climates, having a large number of motor vehicles.* ***Thermal inversion*** in urban basin valleys further increases build up of air pollutants at ground level, usually lasting for several days, with harmful and even lethal effects on life.

Photochemical smog is usually of beige or brown colour; the brown colouration is due to the presence of NO_x in this type of smog. For this reason, photochemical smog is also referred to as ***"brown air smog"*** and the cities affected by such smogs are termed as ***'brown air' cities.***

The cities which are most badly affected by brown air smog (photochemical smog) are : *Los Angels* (California, USA), *Denver* (USA), *Mexico city* (Mexico), *Paulo* and *Rio de Janeiro* (Brazil), *Santiago* (Chile), *Beijing and Shenying* (in China), etc. The worst affected cities like Los Angeles, Mexico and Santiago, are infact all located in Valleys, surrounded by high hills and mountains, because the surrounding hills do tend to reduce the flow of air in the valley below, making the environment 'stable', with less chances of winds sweeping away the pollutants to the down wind areas.

The frequency and severity of development of smog blankets in a city or an area, thus, depends not only on the local climate (such as high winds or high precipitation—both reducing pollutants) but also upon the topographical location of the city. Cities located in valleys or having several high rise buildings (which reduce wind speed) tend to have increased air pollution with increased chances of development of gray or brown smogs.]

Q. 7. (*a*) (*i*) *What are the factors that influence the flow of sewage in sewers ? Define time of concentration for storm water drains and explain its significance.* (5)

Q. 7. (*a*) (*ii*) *A vitrified clay sewer is to carry a discharge of 1000 l.p.s, when flowing half full and laid at a gradient of 0.0032. The Mannings coefficient of the sewer is same as that of a triangular notch with unit depth and having a slope of 1 in 2. The velocity in the notch is obtained as 0.804 m/s with a gradient of 1 in 400. Determine the velocity of flow and diameter of sewer.* (15)

Solution : Here the value of Manning's coefficient N is to be determined from the discharge data given for the notch, shown in Fig. 27.3.

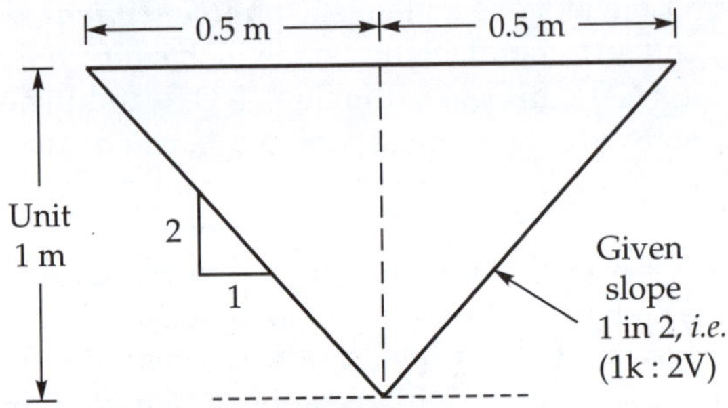

Fig. 27.3

From Fig. 27.3, we have

$$A = \text{Area of cross-section of flow}$$

$$= 2\left[\frac{1}{2} \times 1 \times 0.5\right] \text{ m}^2 = 0.5 \text{ m}^2$$

$$P = \text{Wetted perimeter} = 2\left[\sqrt{1^2 + 0.5^2}\right] \text{ m} = 2.24 \text{ m}$$

$$\therefore \quad R = \frac{A}{P} = \frac{0.5}{2.24} \text{ m} = 0.223 \text{ m}$$

Now using Manning's formula as :

$$V = \frac{1}{N} R^{2/3} . \sqrt{S} \text{ , we have}$$

$$0.804 = \frac{1}{N} . (0.223)^{2/3} . \sqrt{\frac{1}{400}}$$

or $N = \mathbf{0.0229.}$

Now, for the given sewer section running 1/2 full, we have

$Q = \frac{1}{n} .ar^{2/3}\sqrt{s}$ [using smaller letters for half full section]

where $a = \frac{\pi}{8}D^2$ where D is the dia of the sewer

$$p = \frac{\pi D}{2}; \quad r = \frac{D}{4}; \quad n = N = 0.0229$$

$$q = 1000 \text{ lps} = 1 \text{ m}^3/\text{s} ; \; S = 0.0032$$

$$\therefore \quad 1 = \frac{1}{0.0229} \times \left(\frac{\pi}{8}D^2\right)\left(\frac{D}{4}\right)^{2/3} . \sqrt{0.0032}$$

or $1 = 0.385\, D^{8/3}$

or $D^{8/3} = 2.598$

or $D = (2.598)^{3/8 = 0.375} = \mathbf{1.43 \text{ m.}}$ **Ans.**

Velocity of flow (v) in sewer

$$= \frac{q}{a} = \frac{1}{\frac{\pi}{8} \times (1.43)^2} = \mathbf{1.25 \text{ m/sec.}} \quad \textbf{Ans.}$$

Q. 8. (*a*) (*i*) *Differentiate between Garbage and Rubbish.* (6)

(*ii*) *What are the advantages and disadvantages of sanitary land-fills ?* (7)

(*iii*) *Define Municipal solid waste. List out the good practices that should be followed for an efficient management of MSW.* (7)

Q. 8. (*c*) (*i*) *What is radioactive pollution? How is radioactive waste disposed ?* (10)

(*ii*) *Discuss the need of Environmental Impact Assessment study. How do thermal power plants affect the environment ?* (10)

Year 2012 (*Civil Services*)

Q. 7. (*a*) (*i*) *Explain egg shaped sewer section with their relative advantages with circular section.* (5)

Q. 7. (*a*) (*ii*) *Design a sewer for a newly developed town having population of 50,000. The sewer is to be designed for 3 times of dry weather flow when running full. Also find velocity of flow. Available slope for the sewer to be laid is 1 in 700. Assume Manning's coefficient n = 0.01.* (15)

Solution : Pl. follow **Similar Solved Example 4.9.**

Assume daily per capita water supply to the newly developed town at 135 l/c/d of which 80% finds its way into the sewer.

Q. 7. (*b*) (*i*) *Explain significance of* $\frac{BOD}{COD}$ *ratio. How it helps in deciding type* *of treatment to be given to waste water.* (5)

Q. 7. (*b*) (*ii*) *A combined sewer system is serving 20,000 persons having BOD, 70 gm per capita and 50,000 litres industrial effluent per day having BOD 450 mg/l. If avergae standard BOD of domestic sewage is 0.08 kg / day / person, find the population equivalent of sewage.* (15)

Solution : Pl. refer article 7.5 and Solved Examples 7.8 and 7.15; and compute as under :

(*i*) BOD of sewage produced per day

$= 20{,}000 \times 70 \text{ gm} = 1400 \times 10^3 \text{ gm/day}$

$= \mathbf{1400} \text{ kg/day}$ ...(*i*)

(*ii*) BOD of industrial effluent per day

$= (50000\ l \times 450 \text{ mg/l}) \text{ per day} = 22.5 \times 10^6 \text{ mg/day}$

$= 22.5 \text{ kg/d}$ ...(*ii*)

Total BOD entering the sewer = (*i*) + (*ii*)

$= \mathbf{1422.5} \text{ kg/day}$ ...(*iii*)

$$\text{Population equivalent} = \frac{\text{Total BOD}}{\text{Std. BOD of domestic sewage}}$$

$$= \frac{1422.5 \text{ kg / day}}{0.08 \text{ kg / d / p}} = \mathbf{17781\ Ans.}$$

Q. 7. (*c*) (*ii*) *Explain working of skimming tanks for removing oil and grease.* (5)

Q. 7. (*c*) (*ii*) *What are stabilisation ponds? design an oxidation pond for treating sewage from a town having population of 15,000. Sewage contribution per capita per day is 120 l and* BOD_5 *200 mg/l. The town is located in hot climate.* (15)

Hint : Pl. refer **Similar Solved Example 9.31.**

Q. 8. (*b*) (*i*) *Explain the terms garbage and rubbish. How these are different than ashes?* (5)

Q. 8. (*b*) (*ii*) *Discuss composting for disposal of solid wastes. Also explain methods adopted for composting in rural areas in India.* (15)

Q. 8. (*c*) (*i*) *Discuss the status of monitoring and control of air pollution in India.* (10)

Q. 8. (*c*) (*ii*) *What are the sources of particulate matter in the environment? Discuss the control of particulate matters in Industries.* (10)

Year 2013 (*Civil Services*)

Q. 5. (*e*) *Define the following terms :*

(*i*) *Sewage* (*ii*) *Sewers* (*iii*) *Sewerage.*

Explain the Rational method for the design of storm sewers. What are the disadvantages of this method ? (10)

Q. 6. (*c*) *Determine the dimensions of a circular Bio-filter to treat waste water with the following characteristics :*

Daily Flow Rate = 100 m^3/day

Five-day BOD after primary sedimentation = 160 mg/lit

The filter is to be designed in accordance with the following criteria :

Surface loading = 2 $m^3/m^2/day$

Organic loading = 0.15 kg 5-days BOD value per m^3 per day. (15)

Solution : Pl. follow Solved Example 9.9, and compute as under :

Flow rate of sewage to the filter = 100 m^3/d

BOD_5 of sewage (after primary sedimentation) entering the filter = 160 mg/l

∴ Total BOD entering the filter

$$= 100 \frac{m^3}{d} \times 160 \text{ mg/l}$$

$$= 100 \frac{m^3}{d} \times 160 \text{ g/m}^3 = 16000 \text{ g/d} = \mathbf{16\ kg/d.}$$

Vol. of filter medium reqd.

$$= \frac{\text{Total BOD removed}}{\text{Organic loading}} = \frac{16 \text{ kg/d}}{0.15 \text{ kg/m}^3/\text{d}} = \mathbf{106.67\ m^3}$$

Now, assuming a Recirculation Ratio $\left(\frac{R}{I}\right)$ of 1.0,

We have the recirculated sewage volume equal to the original sewage volume of 100 m^3/d.

∴ Total sewage volume in the filter

= Original volume + Recirculated volume

= 100 m^3/d + 100 m^3/d = 200 m^3/d

∴ Filter tank area reqd.

$$= \frac{\text{Total flow volume}}{\text{Surface loading}} = \frac{200 \text{ m}^3/\text{d}}{2 \text{ m}^3/\text{m}^2/\text{d}} = 100 \text{ m}^2$$

Diameter of circular filter tank reqd.

$$= \sqrt{100 \times \frac{4}{\pi}} = 11.28 \text{ m.}$$

Hence, use 11.3 m dia. circular filter tank having area

$$= \frac{\pi}{4} \times (11.3)^2 = 100.28 \text{ m}^2. \quad \textbf{Ans.}$$

∴ Depth of filter media reqd.

$$= \frac{\text{Vol. of filter media}}{\text{Surface area}} = \frac{106.67 \text{ m}^3}{100.28 \text{ m}^2} \approx \mathbf{1.0\ m.}$$

The Efficiency of such a filter (η) can also be computed using the Eq. (9.34) as :

$$\eta = \frac{100}{1 + 0.0044\sqrt{\frac{Y}{VF}}}$$

where Y = Total arginic loading *i.e.* Total BOD applied to filter in kg/d = 16 kg/d

$$V = \text{Vol. of filter in ha.m} = \frac{106.67}{10^4}$$

$$F = \text{Recirculation factor}$$

$$= \frac{1 + R/I}{\left(1 + 0.1\dfrac{R}{I}\right)^2} \quad \text{where } \frac{R}{I} = 1.0$$

$$\therefore \quad F = \frac{1+1}{(1+0.1)^2} = \frac{2}{1.21} = 1.65$$

$$\therefore \quad \eta = \frac{100}{1 + 0.0044\sqrt{\dfrac{16 \times 10^4}{106.67 \times 1.65}}} = 88.3\%. \quad \textbf{Ans.}$$

Q. 7. (*c*) *Find out the effective height of a stack from the following data :*

Physical stack height	*= 200 m*
Inside diameter of the stack	*= 1 m*
Wind velocity	*= 3.5 m/sec*
Air temperature	*= 13°C*
Barometric pressure	*= 1000 millibars*
Stack gas velocity	*= 9 m/sec*
Stack gas temperature	*= 140°C*

Draw a sketch to show the effective height.

[**Hint :** Pl. follow exactly Similar **Solved Example 18.3**, and compute Effective height (H) as = **205.98 m. Ans.**]

Year 2014 (*Civil Services*)

Q. 5. (*b*) *The BOD of a sewage incubated for one day at 30°C has been found to be 110 mg/l. What will be the 5 days 20°C BOD? Assume $K_1 = 0.1$ at 20°C.* (10)

[**Solution :** Pl. refer Solved Example 7.12. **This question was set from this book.**]

Q. 5. (*c*) *Define Air Pollution. Discuss the classification of Air Pollutants and effect of them on Human health.* (10)

[**Hint :** Pl. refer articles 18.1, 18.4 and 18.5.1.]

Q. 8. (*b*) *An Activated Sludge Tank is 30 m long, 9 m wide and has liquid depth of 4 m. The influent sewage flow rate is 3.5 MLD. The primary effluent has BOD of 130 mg/l and suspended load of 150 mg/l. The MLSS concentration in the Aeration Tank is 1800 mg/l. Calculate :*

(*i*) *BOD Loading in kg/Ha-m* (*ii*) *F/M Ratio*
(*iii*) *Aeration Period* (*iv*) *Sludge Age.* (20)

Solution : Pl. refer article 9.35 and compute various parameters as below :

(*i*) **BOD loading** is given by Eq. (9.42) as :

$$\text{BOD loading in kg/Ha-m} = \frac{\text{Mass of BOD applied per day to the aeration tank in kg}}{\text{Volume of aeration tank in Ha-m}}$$

Here BOD of sewage = 130 mg/*l*

Sewage flow rate = 3.5 M*l*/d

$$\therefore \quad \text{BOD load entering per day} = \frac{3.5 \times 10^6\ l}{\text{d}} \times \frac{130\ \text{mg}}{l}$$

$$= \mathbf{455}\ \text{kg/d} \qquad ...(i)$$

$$\text{Volume of Aeration Tank} = 30\ \text{m} \times 9\ \text{m} \times 4\ \text{m} = 1080\ \text{m}^3$$

$$= 1080\ \text{m}^2 \, . \, \text{m} = \frac{1080}{10^4}\ \text{ha.m} \qquad [\because \quad 1\ \text{ha} = 10^4\ \text{m}^2]$$

$$= 0.108\ \text{ha.m}$$

$$\therefore \quad \text{BOD loading} = \frac{455\ \text{kg}}{0.108\ \text{ha.m}} = \mathbf{4213\ kg/ha.m} \quad \textbf{Ans.}$$

(*ii*) **F/M Ratio.** F/M ratio is defined by Eq. (9.43) as:

$$\text{F/M Ratio} = \frac{\text{Daily BOD load applied to the aeration system in kg } (F)}{\text{Total Microbial mass in the system in kg } (M)}$$

Here F = 455 kg [calculated above at Eq. (*i*)]

M, *i.e.* the total microbial mass in the aeration system is computed by multiplying the average concentration of solids in the mixed liquor of the aeration tank (called MLSS - Mixed Liquor Suspended Solids) with the volume of the aeration tank.

Hence, M (Total Microbial Mass)

= MLSS × Vol. of aeration tank

where MLSS = 1800 mg/*l* (given)

Vol. of tank = 1080 m^3

$$\therefore \quad M = 1800\ \text{mg}/l \times (1080 \times 1000)\ l$$

$$= 1800 \times 1080 \times 10^3\ \text{mg}$$

$$= 18 \times 108 \times \frac{10^6}{10^6}\ \text{kg} = 1944\ \text{kg}.$$

$$\therefore \quad \text{F/M Ratio} = \frac{455\ \text{kg}}{1944\ \text{kg}} = \mathbf{0.234} \quad \textbf{Ans.}$$

(*iii*) **Aeration Period.** Aeration Period or Hydraulic Retention Time is given by Eq. (9.41) as :

Aeration Period (t) (in days)

$$= \frac{\text{Volume of the Aeration Tank (m}^3)}{\text{Rate of sewage flow in the Tank} \left(\dfrac{\text{m}^3}{\text{d}}\right)}$$

or $$t = \frac{1080 \text{ m}^3}{(3.5 \times 10^3)\frac{\text{m}^3}{\text{d}}} = 0.3086 \text{ days}$$

$$= 0.3086 \times 24 \text{ h} = \mathbf{7.4 \ hr} \quad \textbf{Ans.}$$

(*iv*) **Sludge Age.** Sludge age (θ_c) is determined from Eq. (9.48) or (9.48 *a*) as:

$$\theta_c = \frac{V \cdot X_T}{Q_W \cdot X_R + (Q - Q_W) X_E} \quad \text{...(Eq. 9.48)}$$

or $$\theta_c = \frac{V \cdot X_T}{Q_W \cdot X_R} \text{ (when } X_E \text{ is small)} \quad \text{...(Eq. 9.48 } a\text{)}$$

where $V = 1080 \text{ m}^3$

$X_T = 1800$ mg/*l*

$Q = 3500 \text{ m}^3\text{/d}$

Q_W = Vol. of wasted sludge = (Not given)

X_R = Conc. of solids in the returned sludge

= (Not given)

X_E = Conc. of solids in the effluent of aeration tank (Not given)

Lets us now assume Q_R/Q = 0.3 (*i.e.* between 0.25 to 0.5 from Table 9.10 for Conventional Activated Plant)

$\therefore \quad Q_R = 0.3 \times 3500 \text{ m}^3\text{/d} = \mathbf{1050 \ m^3}$

Also, the concentration of solids in the influent sewage (X) is given at 125 mg/*l*; while the concentration of solids in the aeration tank (*i.e.* in the mixture of influent sewage + Returned sludge) is equal to 1800 (*i.e.* MLSS) mg/*l*.

$$\therefore \quad \frac{Q \cdot X + Q_R \cdot X_R}{Q + Q_R} = X_{\text{mix}} = \text{MLSS} = 1800 \text{ mg}/l$$

or $$\frac{3500 \times 125 + 1050 X_R}{3500 + 1050} = 1800$$

or X_R = Conc. of solids in the returned sludge

$$= \frac{[1800 \times 4550 - 3500 \times 125]}{1050} = \mathbf{7383 \ mg/l.}$$

Also assuming that the Vol. of wasted sludge (Q_W) is one percent of Q, we have

$$Q_W = \frac{1.0}{100} \times 3500 = 35 \text{ m}^3\text{/d}$$

Also assuming that the conc. of solids in the effluent of the aeration tank (X_E) is small, we can apply Eq. (9.48 *a*) for computing θ_c

$$\therefore \quad \theta_c = \frac{V \cdot X_T}{Q_W \cdot X_R} = \frac{1080 \text{ m}^3\text{/d} \times 1800 \text{ mg}/l}{35 \frac{\text{m}^3}{\text{d}} \times 7383 \text{ mg}/l} = \mathbf{7.5 \ days.} \quad \textbf{Ans.}$$

Year 2015 (*Civil Services*)

Q. 5. (*e*) *Explain the importance of COD/BOD ratio while designing the treatment options for COD/BOD ratio of 1, 3 and 6, and justify the selection of treatment units.* **(10)**

Hint: [**Hint:** Pl. see and read at Page 439 the importance of BOD/COD (or its reverse value *i.e.* COD/BOD) ratio. When $\frac{\text{BOD}}{\text{COD}}$ is more than 0.6, the waste water is biologically treatable without acclimatisation; and if this ratio is less than 0.6 and up to 0.3 then acclimatisation is needed for biological treatment; and if it is less than 0.3 then anaerobic treatment will be reqd.

For the given values, we have **case (*i*)** when $\frac{\text{COD}}{\text{BOD}} = 1$, Then, $\frac{\text{BOD}}{\text{COD}} = 1$, which is more than 0.6. Hence only biological treatment units, *i.e.* aerobic units like cheaper stabilisation ponds, or aerated lagoons, or costlier units like trickling filters or activated sludge plant can be used.

Case (ii): When $\frac{\text{COD}}{\text{BOD}} = 3$, we have $\frac{\text{BOD}}{\text{COD}} = 0.33$, which meeans that the biological treatment units *i.e.* aerobic treatment units can be used with *acclimatisation* (*i.e.* gradual exposure of waste water in increasing concentration to the seed or initial micro-biological population under a controlled condition).

Case (iii): When $\frac{\text{COD}}{\text{BOD}} = 6$, then $\frac{\text{BOD}}{\text{COD}} = 0.167$, which clearly shows that aerobic or biological treatment will not serve any purpose, and given waste water will need to be treated in anaerobic units like anaerobic lagoons and septic or imhoff tanks, etc.

Q. 7. (*c*) *Using the data pertaining to Activated Sludge Process (ASP) given below, determine:*

(*i*) *efficiency;* (*ii*) *volume of aerator;*

(*iii*) *hydraulic retention time (HRT);* (*iv*) *volumetric loading.*

Wastewater flow = 20 MLD

BOD in settled wastewater = 200 mg/l

Effluent BOD required = 15 mg/l

F/M ratio = 0.2

MLSS = 3000 mg/l (20)

Solution: (*i*) Efficiency = Efficiency of BOD removal

$$= \frac{\text{Incoming BOD} - \text{Outgoing BOD}}{\text{Incoming BOD}}$$

$$= \frac{200 \text{ mg/l} - 15 \text{ mg/l}}{200 \text{ mg/l}} = \frac{185}{200} = 0.925 = \mathbf{92.5\%}. \quad \textbf{Ans.}$$

(*ii*) **Volume of aerator:** Use Eq. (9.44) as:

$$\text{F/M ratio} = \frac{Q \cdot Y_0}{V \cdot X_t} \quad \text{...(Eq. 9.44)}$$

where F/M = 0.2 (given)

$$Q = 20 \text{ MLD} = \frac{20 \times 10^6}{10^3 \times 24} \text{ m}^3\text{/day} = 833.33 \text{ m}^3\text{/day}$$

Y_0 = BOD applied to aeration system = 200 mg/l

V = Volume of aeration tank (m^3) = ?

X_T = MLSS in aeration tank = 3000 mg/l

Substituting values, we get

$$0.2 = \frac{833.33 \times 200}{V. \times 3000}$$

$$\text{or} \quad V = \frac{833.33 \times 200}{3000 \times 0.2} = 277.8 \text{ m}^3. \quad \textbf{Ans.}$$

(*iii*) **Hydraulic Retention Time (HRT)**

From Eq. (9.41), we have the detention period (t)

$$\text{or HRT} = \frac{\text{Volume of the tank}}{\text{Rate of sewage flow in the tank}}$$

$$= \frac{V \text{ in m}^3}{Q \text{ in m}^3\text{/day}} = \frac{277.8 \text{ m}^3}{833.33 \text{ m}^3\text{/day}}$$

= 0.333 days = **8 hours. Ans.**

(*iv*) **Volumetric loading.** The volumetric BOD loading is given by Eq. (9.42) as:

Volumetric loading

$$= \frac{\text{Mass of BOD applied per day to the aeration tank through influent sewage in gm}}{\text{Volume of aeration tank}}$$

$$\text{or Volumetric loading} = \frac{Q \cdot Y_0 \text{ (gm)}}{V} \quad \text{...(Eq. 9.42)}$$

where Q = In flow per day = 833.33 m^3

Y_0 = BOD$_5$ in mg/l (or gm/m^3) of the influent sewage = 200 gm/m^3

V = Volume of aeration tank = 277.8 m^3

∴ Volumetric loading

$$= \frac{833.33 \text{ m}^3 \times 200 \text{ gm/m}^3}{277.8 \text{ m}^3}$$

= 600 gm/m^3 = **0.6 kg/m^3. Ans.**

Q. 8. *(a) 5% diluted wastewater was subjected to BOD test and the following observation are recorded:*

DO of water used for dilution = 3.6 mg/l

DO of wastewater sample = 0.8 mg/l

Do in sample after 5 days = 0.7 mg/l

k_{20} *= 0.12/day*

Determine:

(i) BOD of wastewater; *(ii) ultimate BOD.* (15)

Solution : Dilution factor $= \dfrac{100}{\text{Percent of solution}} = \dfrac{100}{5} = 20$

Let the volume of sewage = x ml

Volume of diluted sample = 20 x ml

∴ Vol. of water added = 19 x ml

DO of sewage = 0.8 mg/l

DO of dilution water = 3.6 mg/l

∴ DO of diluted sample $= \dfrac{x \times 0.8 + 19x \times 3.6}{20\,x} = 3.46$ mg/l

Initial DO of diluted sample = 3.46 mg/l

Final DO of sample after incubation for 5 days = 0.7 mg/l

Loss of dissolved oxygen during test

= DO before testing – DO after testing

= 3.46 – 0.7 = 2.76 mg/l

Using Eq. (7.11), we have

BOD of sewage = Loss of oxygen × Dilution factor

= 2.76 mg/l × 20 = **552 mg/l. Ans.**

Ultimate BOD = Y_u = ?

Using Eq. (7.16), we have

$$Y_t = L\left[1 - (10)^{-K_D \cdot t}\right]$$

or $$Y_5 = L\left[1 - (10)^{-K_D \cdot 5}\right]$$

Here Y_5 = 5 day BOD at of 20°C of wastewater

= 552 mg/l (computed above)

$K_D = k_{20}$ = 0.12/day (given)

$L = Y_u$ = ?

∴ $$552 = L\left[1 - (10)^{-0.12 \times 5}\right]$$

or $$552 = L\left[1 - \frac{1}{(10)^{0.6}}\right]$$

or $$L = \frac{552}{0.7488} = 737 \text{ mg/l}$$

∴ Ultimate BOD = $L = Y_u$ = **737 mg/l. Ans.**

Q. 8. (*b*) *Describing the major categories of pollutants generated by thermal power plants, discuss the environmental impacts of thermal power plants.* (15)

Q. 8. (*c*) *With the help of flow diagram, explain various elements involved in municipal solid waste management.* (20)

Year 2016 (*Civil Services*)

Q. 6. (*b*) (*i*) *What do you understand by sludge digestion?* (4)

Q. 6. (*b*) (*ii*) *Briefly explain the distinct stages of sludge digestion and the factors affecting the sludge digestion.* (4)

Q. 6. (*b*) (*iii*) *Design a circular sludge digestion tank with 6 m depth for the following parameters, when 3.5% of the digester volume is filled on daily basis with the fresh sludge, which is mixed with the digested sludge:*

Sludge content per capita per day = 0.68 kg

Moisture of the sludge = 94%

Specific gravity of wet sludge = 1.02

Provide 30% additional capacity for fluctuations. (10)

[**Hint:** Pl. refer Solved Example 9.23 with the same data values. However, the number of people producing sludge has not been given, which should have been given. We will hence have to assume suitable value for the same, as say, 40,000 people, as in Solved Example 9.23. **This question was infact set from this book].**

Q. 7. (*a*) (*i*) *Define primary and secondary air pollutants with appropriate examples.* (3)

Q. 7. (*a*) (*ii*) *Name at least five green-house gases (GHGs). How do green-house gases (GHGs) play role in climate change?* (3)

[**Hint :** Pl. see footnote under article 19.3.]

Q. 7. (*a*) (*iii*) *Illustrate with diagrams the influence of environmental lapse rates on shape and dispersion of plume emitted from stack of a thermal power plant.* (9)

Q. 8. (*b*) (*i*) *What are different unit operations and processes used at a typical waste water treatment plant?* (5)

Q. 8. (*b*) (*ii*) *Draw typical cross-section of a circular trickling filter. Also, briefly describe the merits and demerits of conventional trickling filters.* (5)

Q. 8. (*b*) (*iii*) *Design a high-rate single-stage circular trickling filter to treat wastewater generated by a town having population of 40,000 persons. Necessary data are given below to be used for the design purpose:*

1. *Domestic sewage = 150 lpcd having BOD of 200 mg/L*
2. *Industrial wastewater = 0.25 million litres/day with BOD of 600 mg/L*

3. BOD removal in primary clarifier = 35%

4. Permissible organic loading of the filter = 800 kg/hec-m/day (this does not include recirculated sewage)

5. Recirculation ratio = 1.0

6. Permissible surface loading = 160 million litres/hec/day (this includes recirculated sewage)

Also, find out the efficiency of the filter and BOD of the effluent. (10)

[Hint. Pl. follow exactly similar Solved Example 9.9. **This question was infact set from this book.]**

Year 2017 (*Civil Services*)

Q. 5. (*e*) (*i*) *Explain the impacts of improper disposal of solid wastes on human health.* (5)

Q. 5. (*e*) (*ii*) *Discuss the composition of municipal solid waste (MSW)* (5)

Q. 7. (*a*) *The classification of the surface on which the rain falls in a city is as follows :*

% of total surface area	*Type of surface*	*Coefficient of runoff*
20%	Roof top	0.90
20%	Pavement	0.85
10%	Paved yard	0.80
15%	Macadam road	0.40
35%	Lawns	0.10

Determine the coefficient of runoff of the area. If the total area of the city is 40 ha and the maximum rain intensity is taken as 6 cm/hr, what is the runoff for the city? (15)

Solution. Please refer exactly similar solved example, 3.1, and compute as under :

The total surface area of the city is made of smaller areas A_1, A_2, A_3, A_4 and A_5, having coefficients of runoff as K_1, K_2, K_3, K_4 and K_5. Then the coefficient of runoff K for the entire area may be computed by using.

$$K = \frac{K_1A_1 + K_2A_2 + K_3A_3 + K_4A_4 + K_5A_5}{A_1 + A_2 + A_3 + A_4 + A_5}$$

where

$$K_1A_1 = K_1\left(\frac{20}{100}\right)A = (0.90)\,(0.2A) = 0.18A$$

$$K_2A_2 = K_2\left(\frac{20}{100}\right)A = (0.85)\,(0.2A) = 0.17A$$

$$K_3A_3 = K_3\left(\frac{10}{100}\right)A = (0.80)\,(0.1A) = 0.08A$$

$$K_4A_4 = K_4\left(\frac{15}{100}\right)A = 0.4\,(0.15A) = 0.06A$$

$$K_5A_5 = K_5\left(\frac{35}{100}\right)A = 0.1\,(0.35A) = 0.035A$$

$$\therefore \qquad K = \frac{K_1A_1 + K_2A_2 + K_3A_3 + K_4A_4 + K_5A_5}{A}$$

$$\text{or} \qquad K = \frac{0.18A + 0.17A + 0.08A + 0.06A + 0.035A}{A}$$

$$\text{or} \qquad K = \frac{0.525A}{A} = \mathbf{0.525} \quad \textbf{Ans.}$$

The peak runoff for the city is calculated by Rational formula as :

$$Q = \frac{1}{36} K \cdot p_c A$$

where, Q is incumecs

p_c is in cm/hr

A is in ha

where $K = 0.525$ (computed above)

p_c = critical or max rain intensity = 6 cm/hr

A = 40 ha.

$$\therefore \qquad Q = \frac{1}{36} (0.525)\ (6.0)\ (40)$$

$$\text{or} \qquad Q = 3.5 \text{ cumecs} \qquad \textbf{Ans.}$$

Q. 7. (*b*) (*i*) *Discuss the significance of* $\frac{BOD}{COD}$ *ratio in the treatment of waste water.* (5)

Q. 7. (*b*) (*ii*) *Calculate 2-day 37°C BOD of a sewage sample, whose 5-day 20°C BOD is 150 mg/L. Assume k_D at 20°C as 0.1.* (10)

Solution. Please refer exactly similar solved example 7.11, and compute as under :

Using Eq. (7.16), we have

The BOD at 20°C after t = 5 days, is given as:

$$Y_t = L[1-(10)^{-K_D(20°)\,.\,t}] \text{ [using } y_t = 50 \text{ mg/L (given)]}$$

$$K_D(20°) = 0.1 \text{ (given)}$$

We have $\qquad 150 = L[1-(10)^{-0.1\times 5}]$

$$\text{or} \qquad 150 = L[1-(10)^{-0.5}] = L\left[1-\frac{1}{3.16}\right]$$

$$\text{or} \qquad 150 = L[1-0.316] = 0.684L$$

$$\text{or} \qquad L = \frac{150}{0.684} = \mathbf{219.3\ mg/L}$$

Now, let us work out K_D at 37°C by using eqn. (7.18) as :

$$K_{D(T°)} = K_{D(20°)}\,[1.047]^{T-20}$$

$$\text{or} \qquad K_{D(37°)} = K_{D(20°)}\,[1.047]^{37-20} = 0.1[1.047]^{17}$$

$$\text{or} \qquad K_{D(37°)} = 0.1 \times 2.18 = 0.218$$

Now, we have to workout Y_t for two days *i.e.*, Y_2 at 37°C, using

$$Y_t = L[1-(10)^{-K_D.t}]$$

or $$Y_2 = L[1-(10)^{-K_D \times 2}]$$

or $$Y_2 \text{ at } 37°\text{C} = 219.3\ [1-(10)^{-K_D \text{ at } 37°C \times 2}]$$

or $$Y_2 \text{ at } 37°\text{C} = 219.3\ [1-(10)^{0.218 \times 2}]$$

$$= 219.3 \left[1 - \frac{1}{(10)^{0.436}}\right] = 219.3\left[1-\frac{1}{2.73}\right]$$

or $$Y_2 \text{ at } 37°\text{C} = 138.94$$

Hence, Y_2 at 37°C = 138.94 mg/L.

Q. 7. (*c*) (*i*) *What are the various stages in the sludge digestion process? Also discuss the factors affecting sludge digestion and their control.* (10)

Q. 7. (*c*) (*iii*) *Design an oxidation pond for treating sewage from a hot climatic residential colony with 6000 persons contributing sewage 150 L per capita per day. The 5-day BOD of sewage is 250 mg/L* (10)

[Hint. Please refer Solved Example 9.31. **This question has been set from this book by slightly changing the data values.**]

Q. 8. (*c*) (*i*) *How do mining activities affect the environment? Discuss the environmental impact of mining in detail.* (10)

Q. 8. (*c*) (*ii*) *What are the various pollutents causing pollution of air? Discuss the sources and their effect on human health and environment in general.* (10)

Year 2018 (*Civil Services*)

Q. 5. (*e*) *Design a septic tank for a small colony of 300 persons with an average wastewater flow of 120 litres per head per day.* (*10*)

[Hint : Pl. refer similar solved example 9.34. De-sludging period and L:B ratio may be assumed, when not given].

Q. 6. (*b*) *A rural watershed consists of :*

5 km^2 of forest area with C = 0.1
2 km^2 of agriculutral area with C = 0.25
1 km^2 of unpaved streets with C = 0.30
0.75 km^2 of green cover with C = 0.35
1.25 km^2 of residential area with C = 0.40

where C = Coefficient of runoff estimate the peak runoff rate forrainfall with a return period (T) of 25 years, if the intensity-duration-frequency relationship for the watershed is

$$i = \frac{150T^{1/3}}{(t+12)^{0.5}}$$

where i = *rainfall intensity in mm/hr*
T = *return period in year*
t = *duration of rainfall in minutes*

The watershed has a time of concentration of 50 minutes. (*20*)

[Solution : Let A represents the total area = 5 + 2 + 1 + 0.75 + 1.25 = 10 km^2 and C = coefficient of runoff of the total catchment; and C given by (Eq. 3.20), as

$$C = \frac{C_1A_1 + C_2A_2 + C_3A_3 + ...}{A_1 + A_2 + A_3 + ...}$$

where, $C_1A_1 = 0.1 \times 5 \text{ km}^2$
$C_2A_2 = 0.25 \times 2 \text{ km}^2$
$C_3A_3 = 0.30 \times 1 \text{ km}^2$
$C_4A_4 = 0.35 \times 0.75 \text{ km}^2$
$C_5A_5 = 0.40 \times 1.25 \text{ km}^2$

$$\therefore \quad C = \frac{[0.5 \times 5 + 0.25 \times 2 + 0.3 \times 1 + 0.35 \times 0.75 + 0.4 \times 1.25]}{5 + 2 + 1 + 0.75 + 1.25 = 10}$$

$$\text{or} \quad C = \frac{2.06}{10} = \mathbf{0.2063}$$

Peak runoff rate can be calculated using the *Rational formula* given by Eq. (3.1) as :

$$Q_p = \frac{1}{36} K \cdot p_c \cdot A = \frac{1}{36} \cdot C \cdot p_c \cdot A \qquad \text{(Here } K = C\text{)}$$

where, $C = 0.2063$
p_c = Critical rain intensity in cm/h = ?
A = Total catchment area in hectares = 10 km^2 *i.e.* 1000 ha.

To find p_c, let us use the given equation

$$i = \frac{150T^{1/3}}{(t + 12)^{0.5}}$$

where i = rainfall intensity in mm/hr
T = return period in yr = 25 yrs (given)
t = duration of rainfall in minutes = time of concentration of the catchment = 50 minutes (given)

$$\therefore \quad i = \frac{150(25)^{1/3}}{(50 + 12)^{0.5}} = 55.7 \text{ mm/hr} = 5.57 \text{ cm/h}$$

Hence, p_c = 5.57 cm/h

$$\therefore \quad Q_p = \frac{1}{36} \times (0.2063)\,(5.57)\,1000 \text{ m}^3\text{/s}.$$

$$\text{or} \quad Q_p = 31.92 \text{ m}^3\text{/s} \qquad \textbf{Ans.]}$$

Q.7. (*c*) *Design high-rate and low-rate trickling filters using the following data using empirical formula:*

Waste water flow = 5 MLD
BOD of waste water = 300 mg/L
BOD removal in PST = 25%
Desired BOD of treated wastewater = 50 mg/L
Recirculating ratio = 1.4
BOD of recirculated wastewater may be neglected. (20)

[Solution. ***(i) Design of high rate trickling filters.*** Please follow similar solved example 9.10, and compute as under :

Quantity of sewage flowing into the filter per day
$= 5$ MLD $= 5$ ml/d $= 5000$ m^3/d.

BOD in raw sewage = 300 mg/L

Total daily BOD present in raw sewage
$= 5$ ML × 300 mg/L = 1500 kg ...(*i*)

BOD removed in Primary Settling Tank (PST) = 25%

∴ BOD left in the sewage entering per day in filter unit
$= 1500 \times 0.75 = 1125$ kg ...(*ii*)

BOD concentration desiged in final effluent = 50 mg/L

∴ Total BOD left in the effluent per day = 5 MLD × 50 mg/L
$= (5 \times 10^6)$ L × 50 mg/L $= 250 \times 10^6$ mg
$= 250$ kg ...(*iii*)

∴ BOD removed in the filter (*ii*) – (*iii*)
$= 1125 - 250 = 875$ kg

$$\text{Efficiency of the filter} = \frac{\text{BOD removed}}{\text{Total BOD applied}} \times 100$$

$$= \frac{875 \text{ kg}}{1125 \text{ kg}} \times 100 = 77.77\%$$

Now, using empirical Eq. (9.34), we have

$$\eta = \frac{100}{1 + 0.0044\sqrt{\dfrac{Y}{V.F.}}}$$

where η in % = 77.77
Y = Total BOD in kg = 1125 kg
V = Filter volume in hr.m

$$\text{where,} \quad F = \frac{1 + \dfrac{R}{I}}{\left(1 + 0.1\dfrac{R}{I}\right)^2}$$

where $\dfrac{R}{I} = 1.4$ (given)

$$\text{or} \quad F = \frac{1 + 1.4}{(1 + 0.1 \times 1.4)^2} = 1.68$$

$$\therefore \quad 77.77 = \frac{100}{1 + 0.0044\sqrt{\dfrac{1125}{V \times 1.68}}}$$

$$\text{or} \quad 1 + 0.0044\sqrt{\frac{669.64}{V}} = 1.286$$

$$\text{or} \quad \sqrt{\frac{669.64}{V}} = \frac{0.286}{0.0044} = 65$$

$$\text{or} \quad \sqrt{\frac{669.64}{V}} = 4225$$

or $$V = \frac{669.64}{4225} = 0.158 \text{ hectare . m} = 1580 \text{ m}^3.$$

Assuming the depth of the filter = 1.5 m, we have

The surface area required $= \dfrac{1580 \text{ m}^3}{1.5 \text{ m}} = 1053.3 \text{ m}^3$

Dia of circular filter $= \sqrt{1053.3 \times \dfrac{4}{\pi}} = 36.62$ m Say 37 m

Hence, use a high rate trickling filter with 37 m dia, 1.5 m deep filter media, and with recirculation (single stage) ratio of 1.4. **Ans.]**

(2) ***Design of Low rate (i.e., standard) trickling filter without any recirculation of treated sewage.***

Please follow example 9.7 and compute as under :

Total BOD present in sewage

$$= 5 \times 10^6 \text{ L/d} \times 300 \text{ mg/L} = 1500 \text{ kg}.$$

Volume of the filter media required

$$= \frac{\text{Total BOD}}{\text{Organic loding}}$$

Let us assume organic loading at 1500 kg/ha.m/d (*i.e.*, between 900 to 2200 kg/ha.m/d — Article 9.21.4).

∴ Volume of filter media required

$$= \frac{1500 \text{ kg/d}}{1500 \text{ kg/ha.m/d}} = 1 \text{ ha.m} = 10{,}000 \text{ m}^3.$$

Assuming the effective depth of the filter = 2 m, we have

The surface area of the filter required $= \dfrac{10000}{2} = 5000 \text{ m}^2$

Using a circular trickling filter of dia 40 m, we have the number of units required.

$$= \frac{\text{Total area required}}{\text{Area of one unit}} = \frac{5000 \text{ m}^2}{(\pi/4) \times (40)^2} = 4 \text{ Nos.}$$ **Ans.**

Hence, use 4 units of 40 m dia filter of effective media depth = 2 m **Ans.]**

Q. 8. (*a*) *Develope a relationship between the side of a square section of one sewer and the diameter of a circular section of another sewer when both are hydraulically equivalent.* (15)

[Solution. Please follow solved example 4.17, and compute as under :

Let the circular section of the sewer has a dia of D m and the lengthof the sides of the square sewer section be B.

Then the area of flow of square sewer (A_1) running full = B^2.

The wetted perimeter (P_1) of the square sewer running full = 4 B.

$$\therefore \quad R_1 = \frac{A_1}{P_1} = \frac{B^2}{4B} = 0.25\, B$$

∴ Discharging capacity of the square sewer

$$Q_1 = \frac{1}{N} A_1 R_1^{2/3} . \sqrt{S}$$

or $$Q_1 = \frac{1}{N} B^2 (0.25B)^{2/3} . \sqrt{S}$$

or $$Q_1 = \frac{1}{N} B^2 . 0.4 B^{2/3} \sqrt{S} = \frac{0.4}{N} \sqrt{S} B^{8/3} \quad ...(i)$$

Similarly, for circular sewer section, we have

$$A_2 = \frac{\pi D^2}{4}; \; P_2 = \pi D; \; R_2 = \frac{A_2}{P_2} = \frac{\frac{\pi D^2}{4}}{\pi D} = \frac{D}{4}$$

$$\therefore \quad Q_2 = \frac{1}{N} . \left(\frac{\pi D^2}{4}\right)\left(\frac{D}{4}\right)^{2/3} . \sqrt{S}$$

or $$Q_2 = \frac{1}{N} . \frac{\pi}{4} \frac{D^{8/3}}{2.52} \sqrt{S} = \frac{0.312}{N} \sqrt{S} D^{8/3} \quad ...(ii)$$

For hydraulically equivalent sections, equating (*i*) and (*ii*), we get

$$\frac{0.4}{N} \sqrt{S} B^{8/3} = \frac{0.312}{N} \sqrt{S} . D^{8/3}$$

or $$B^{8/3} = \frac{0.312}{0.4} . D^{8/3} = 0.78 \, D^{8/3}$$

or $$B = (0.78)^{3/8} . D$$

$$B = (0.78)^{0.375} \; D = 0.91 \, D$$

$$B = 0.91 \, D.$$

This is the required relation between B and D **Ans.]**

Q. 8. (*b*) *Write about the various sources of radio-activity in the environment and its effects on the environment* (15)

[Hint. Please summerise article 15.12.]

Q. 8. (*c*) *What are the different types of municipal solid wastes? Explain about each of them.*

Year 2019 (*Civil Services*)

Q. 5. (*d*) *In a BOD analysis, 10.0 mL of industrial wastewater sample was diluted with 290.0 mL of seeded dilution water. The initial DO was 9.2 mg/L and after 5 days of incubation at 20°C, the final DO of the sample with seeded dilution water is 2.0 mg/L. The final DO of the seeded dilution water is 8.0 mg/L. The volume of the BOD bottle was 300.0 mL. Determine the 5-day, 20°C BOD of the sample.* (*10*)

[Solution: Volume of undiluted wastewaster sample = 10 mL

Volume of dilution water = 290 mL

Total volume of diluted sample = 10 + 290 = 300 mL

Initial DO of diluted sample = 9.2 mg/L

Final DO after 5 days incubation = 2.0 mg/L

Since the initial DO of the dilution water is not given, the question cannot be solved as per Q. No. 4(*c*) (*i*) of year 2016 of the Engineering Service Exam. (Page 1083 of the text), *we will assume that distilled water was used for seeding.* Hence, the BOD of the wastewater will be computed as per Eq. (7.11).

BOD_5 of wastewater = [DO consumed in the test by the diluted sample]

$$\times \frac{\text{Vol. of diluted sample}}{\text{Vol. of undiluted sewage sample}} \quad \text{...(Eq. 7.11)}$$

$$\therefore \text{BOD}_5 \text{ of wastewater} = (9.2 - 2)\ \text{mg/L} \times \frac{300\ \text{mL}}{10\ \text{mL}}$$

$$= 7.2 \times 30 = 216\ \text{mg/L} \quad \textbf{Ans.}$$

Q.6.(*a*) *A main sewer is to be laid for a residential cum commercial area of a town. The area has a residential population of 5000 persons and floating population (offices and shops) of another 5000 persons. The residential population consume 150 lpcd of water, while the floating population consume 45 lpcd of water. Assuming that 90% of water consumed emerges as wastewater, and this also includes infiltration and inflows, design the main sewer. Consider the sewer flowing full at peak discharge. Take peak factor of 3 and minimum discharge factor as one-third. The natural slope of the ground is 1 in 400. Assume Manning's n = 0.013. Check the sewer for minimum and maximum velocities, and comment on the suitability of design.* *(20)*

[**Solution.** Residential pupulation = 5000

Daily water used by residential population = 150 lpcd

Floating population = 5000

Daily water used by floating population = 45 lpcd

∴ Total water used per day

$$= [5000 \times 150 + 5000 \times 45]\ \text{l/d}$$

$$= 9{,}75{,}000\ \text{l/d.}$$

Vol. of sewage produced per day = 90% of water consumed

$$= 90\% \times 9{,}75{,}000 = 8{,}77{,}500\ \text{l/d}$$

$$= 877.5\ \text{m}^3\text{/d}$$

Sewer has, thus, to be designed for an average flow of 877.5 m^3/d

$$Q = \frac{877.5}{24 \times 3600}\ \text{m}^3\text{/sec} = 0.010\ \text{m}^3\text{/s}$$

$$\text{Peak sewer discharge} = 3 \times 0.010 = 0.0305\ \text{m}^3\text{/s}$$

$$\text{Minimum sewer discharge} = \frac{1}{3} \times 0.010 = 0.0034\ \text{m}^3\text{/s}$$

Now, sewer is to be desinged running full at peak discharge of 0.0305 m^3/s. Assume a circular sewer section of dia. D (m)

Using Mannings formula, we have

$$Q = \frac{1}{N} A.R^{2/3}.\sqrt{S}$$

where

Q = Discharge through the sewer = 0.0305 m^3/s

N = Mannings roughness constant = 0.013

A = Area of sewer = $\frac{\pi}{4} . D^2$

$$R = \frac{A}{P} = \frac{D}{4}$$

$$\therefore \quad 0.0305 = \frac{1}{0.013} \times \frac{\pi}{4} D^2 . \left(\frac{D}{4}\right)^{2/3} . \frac{1}{\sqrt{400}}$$

$$\text{or} \quad 0.0101 = \frac{D^2 . D^{2/3}}{(4)^{2/3}}$$

$$\text{or} \quad D^{8/3} = 0.02545$$

$$\text{or} \quad D = (0.02545)^{3/8} = 0.375 = 0.25 \text{ m}$$

Hence use 25 cm or 250 mm dia circular sewer **Ans.**

Check for Minimum Flow

Minimum flow discharge (q) = $\frac{1}{3}$ × Mean discharge = $\frac{1}{3} \times 0.010 = 0.0034 \text{ m}^3/\text{s}$

$$\frac{q}{Q} = \frac{0.0034}{0.0305} = 0.11$$

From chart Fig. 4.2, for $\frac{q}{Q} = 0.11$, we have

$$\frac{d}{D} \approx 0.22, \text{ and } \frac{v}{V} \approx 0.64 \text{ (from Table 4.8)}$$

Velocity (V) when sewer runs full is given as :

$$V = \frac{1}{N} . R^{2/3} . \sqrt{S} = \frac{1}{0.013} \times \left(\frac{0.25}{4}\right)^{2/3} , \frac{1}{\sqrt{400}}$$

$$= 0.6 \text{ m/s}$$

$$\therefore \quad v = 0.6 \times 0.64 = 0.38 \text{ m/s}$$

Hence, max velocity is 0.6 m/s and minimum velocity = 0.36 m/s

Both these velocities are lesser than the ideal values of about 0.9 m/s and 0.45 m/s (Page 46 of text) **Ans.**

Year 2020 (*Civil Services*)

Q. 5(*d*) *During BOD test conducted on a 5% dilution of waste, the following observations were taken :*

DO of aerated water used for dilution = 3.6 mg/L

DO of original sample = 0.8 mg/L

DO of diluted sample after 5 days' incubation = 0.7 mg/L

Compute :

(i) 5 days' BOD *(ii) Ultimate BOD*

Assume deoxygenation constant at test temperature as 0.12.

Solution : Pl. follow exactly Similar Solved Example 7.14 and compute

(*i*) 5 days BOD (Y_5) = **55.2 mg/L** **Ans.**

and (*ii*) Ultimate BOD (L) = **737 mg/L** **Ans.**

Q. 7(*c*) *Determine the aeration period, food to microorganism (F/M) ratio (in kg BOD/day/kg MLSS), percentage (%) efficiency of BOD removal, and sludge* *age (in days) for the given operating data for conventional sludge treatment plant:*

Wastewater flow = 45000 m^3/day

Volume of aeration tank = 15000 m^3

Influent BOD = 250 ppm

Effluent BOD = 25 ppm

Mixed liquor suspended solids (MLSS) = 2500 mg/L

Suspended solids in effluent = 50 mg/L

Suspended solids in waste sludge = 11000 mg/L

Quantity of waste sludge = 220 m^3/day

Solution. Pl. follow exactly Similar Solved Example 9.29 and compute as :

(*i*) Aeration period $(t) = \frac{V}{Q} = \frac{15000}{45000} \times 24 = \mathbf{8\ h}$ **Ans.**

(*ii*) F/M ratio $= \frac{QY_o}{V \times X_T} = \frac{45000 \times 250}{15000 \times 2500} = \mathbf{0.3}$ **Ans.**

(*iii*) Efficiency of BOD removal $= \frac{Y_o - Y_E}{Y_o} \times 100 = \frac{250 - 25}{250} \times 100 = \mathbf{90\%}$ **Ans.**

(*iv*) Sludge age $(\theta_c) = \frac{V \cdot X_T}{Q_W \cdot X_R + (Q - Q_W)X_E} = \mathbf{8\ days}$ **Ans.**

—28

Conventional Questions of the Engineering Services Exams for the Past Years

The **revised examination scheme** and syllabus for the Engg. Services Competition is given below :

COMBINED ENGINEERING SERVICES EXAM.(Civil Engg.)

With effect from the year 2017, the examination scheme for this competition has been revised by the UPSC. This examination will now be conducted in the following three stages:

Stage I:	*Preliminary Exam.*	500 marks
Stage II:	*Main Exams.*	600 marks
Stage III:	*Personality Test*	200 marks
	Total	1300 marks

These stages are explained below:

(*a*) **Stage I:** Preliminary Stage I Exam under Stage I will consists of the following two papers:

Paper-I:	*General Studies and Engineering Aptitude Test*	*– 2 hrs – 200 marks*
Paper-II:	*Engineering Discipline Specific Paper*	*– 3 hrs – 300 marks*
	Subtotal	500 marks

Note: (1) Only those candidates qualifying at this stage will be permitted to appear for Stage-II examination.

(2) Minimum qualifying marks for each paper shall be decided at the discretion of the UPSC.

(*b*) **Stage II:** Main Stage II Exam will consists of the following two papers:

Paper-I:	*Engg. Discipline Specific Paper*	*– 3 hrs – 300 marks*
Paper-II:	*Engg. Discipline Specific Paper*	*– 3 hrs – 300 marks*
	Subtotal	600 marks

Subtotal: Stage-I + Stage-II = 1100 marks.

Only those candidates qualifying at (Stage-I + Stage-II) shall be eligible for Stage-III *i.e.*, Personality Test (Interview) of 200 marks.

However, to determine final merit list, marks of Stage-I + Stage-II, and of Stage-III shall be added together, to draw merit list, out of total of 1300 marks.

The revised syllabus for Civil Engineering Stream for both the stages is given below:

BROAD CONTENTS OF THE GENERAL STUDIES AND ENGINEERING APTITUDE PAPER (Stage-I, Paper-I)–Common to all streams.

General Studies and Engineering Aptitude

(Stage I–Paper I, Objective type, Common to all candidates, 2 hours duration, 200 Marks maximum)

The questions from the following topics will be set in Paper-I of Stage-I

1. Current issues of national and international importance relating to social, economic and industrial development.
2. Engineering Aptitude covering logical reasoning and analytical ability.
3. Engineering Mathematics and Numerical Analysis.
4. General Principles of Design, Drawing, Importance of Safety.
5. Standards and Quality practices in production, construction, maintenance and services.
6. Basics of Energy and Environment: Conservation, environmental pollution and degradation, climate change, environmental impact assessment.
7. Basics of Project Management.
8. Basics of Material Science and Engineering.
9. Information and Communication Technologies (ICT) based tools and their applications in Engineering such as networking, e-governance and technology based education.
10. Ethics and values in Engineering profession.

Note: The paper in General Studies and Engineering Aptitude will include knowledge of relevant topics as may be expected from an engineering graduate, without special study.

Questions from all the 10 topics mentioned above shall be set. Marks for each topic may range from 5% to 15% of the total marks in the paper.

Syllabus for both the papers together for Stage-I objective type Paper-II, and separately for Stage-II conventional type Paper-I and Paper-II.

Civil Engineering Paper I (Revised w.e.f. the year 2017)

1. Building Materials:

Stone, Lime, Glass, Plastics, Steel, FRP, Ceramics, Aluminium, Fly Ash, Basic Admixtures, Timber, Bricks and Aggregates: Classification, properties and selection criteria;

Cement: Types, Composition, Properties, Uses, Specifications and various tests; Lime and Cement Mortars and Concrete: Properties and various tests; Design of Concrete Mixes: Proportioning of aggregates and methods of mix design.

2. Solid Mechanics:

Elastic constants, stress, plane stress, strains, plain stress, Mohr's circle of stress and strain, elastic theories of failure, principle stresses, bending, shear and torsion.

3. Structural Analysis:

Basics of strength of materials, types of stresses and strain, bending moments and shear force, concept of bending and shear stresses; Analysis of determinate and indeterminate structures; Trusses, beams, plane frames; Rolling loads, influence lines, unit load method and other methods; Free and forced vibrations of single degree and multi degree freedom system; Suspended cables; Concepts and use of Computer Aided Design.

4. Design of Steel Structures:

Principles of Working Stress methods, Design of tension and compression members, Design of beams and beam column connections, built-up sections, girders, industrial roofs, Principles of Ultimate load design.

5. Design of Concrete and Masonry Structures:

Limit state design for bending, shear, axial compression and combined forces; Design of beams, Slabs, Lintels, Foundation, Retaining walls, Tanks, Staircases; Principles of pre-stressed concrete design including materials and methods; Earthquake resistant design of structures; Design of Masonry Structure.

6. Construction Practice, Planning and Management

Construction–Planning, Equipment, Site investigation and Management including Estimation with latest project management tools and network analysis for different types of works; Analysis of Rates of various types of works; Tendering Process and Contract Management, Quality Control, Productivity, Operation Cost, Land acquisition, Labour safety and welfare.

Syllabus for Civil Engineering Paper-II (Revised wef year 2017)

1. Flow of Fluids, Hydraulic Machines and Hydro Power:

(*a*) Fluid Mechanics, Open Channel Flow, Pipe Flow:

Fluid properties; Dimensional Analysis and Modeling; Fluid dynamics including flow kinematics and measurements; Flow net; Viscosity, Boundary layer and control, Drag, Lift, Principles in open channel flow, Flow controls. Hydraulic jump; Surges, Pipe networks.

(*b*) Hydraulic Machines and Hydro Power

Various pumps, Air vessels, Hydraulic turbines–types, classifications and performance parameters; Power house–classification and layout, storage, pondage, control of supply.

2. Hydrology and Water Resources Engineering:

Hydrological cycle, Ground water hydrology, Well hydrology and related data analysis; Streams and their gauging; River morphology; Flood, drought and their management; Capacity of Reservoirs.

Water Resources Engineering: Multipurpose uses of water, River basins and their potential; Irrigation systems, water demand assessment; Resources–storages and their yields; Water logging, canal and drainage design, Gravity dams, falls, weirs, energy dissipaters, barrage distribution works, cross drainage works and head-works and their design; Concept in canal design, construction and maintenance; River training, measurement and analysis of rainfall.

3. Environmental Engineering:

(*a*) Water Supply Engineering:

Sources, Estimation, quality standards and testing of water and their treatment; Rural, Institutional and industrial water supply; Physical, chemical and biological characteristics and sources of water, pollutants in water and its effects, Estimation of water demand; Drinking water standards, Water Treatment Plants, Water distribution networks.

(b) Waste Water Engineering:

Planning and design of domestic waste water, sewage collection and disposal; Plumbing Systems. Components and layout of sewerage system; Planning and design of Domestic Waste-water disposal system; Sludge management including treatment, disposal and re-use of treated effluents; Industrial waste waters and Effluent Treatment Plants including institutional and industrial sewage management.

(c) Solid Waste Management:

Sources and classification of solid wastes along with planning and design of its management system; Disposal system, Beneficial aspects of wastes and Utilization by Civil Engineers.

(d) Air, Noise pollution and Ecology:

Concepts and general methodology.

4. Geo-technical Engineering and Foundation Engineering:

(a) Geo-technical Engineering: Soil exploration–planning and methods, Properties of soil, classification, various tests and inter-relationships; Permeability and Seepage, Compressibility, consolidation and Shearing resistance, Earth pressure theories and stress distribution in soil; Properties and uses of geo-synthetics.

(b) Foundation Engineering: Types of foundation and selection criteria, bearing capacity, settlement analysis; design and testing of shallow and deep foundations; Slope stability analysis, Earthen embankments, Dams and Earth retaining structures: types, analysis and design, Principles of ground modifications.

5. Surveying and Geology:

(a) Surveying: Classification of surveys, various methodologies, instruments and analysis of measurement of distances, elevation and directions; Field astronomy, Global Positioning System; Map preparation, Photogrammetry; Remote sensing concepts; Survey Layout for culverts, canals, bridges, road/railway alignment and buildings, setting out of curves.

(b) Geology: Basic knowledge of Engineering geology and its application in projects.

6. Transportation Engineering:

Highways: Planning and construction methodology, Alignment and geometric design; Traffic Surveys and Controls; Principles of Flexible and Rigid pavements design.

Tunneling: Alignment, methods of construction, disposal of muck, drainage, lighting and ventilation.

Railway Systems: Terminology, planning, designs and maintenance practices; track modernization.

Harbours: Terminology, layouts and planning.

Airports: Layout, planning and design.

Note: *Evidently, the subject of Hydrology and Water Resources is covered in Paper-II of Main Exam (i.e., Stage-II) for conventional questions. Objective type questions from the subject will be covered in Paper-II of Stage-I (i.e., Preliminary Exam).*

In order to enable you to have a practice for solving questions of this important examination and for your facility and guidance, we are giving below the conventional questions that appeared during the last several years, since 1995, from *Irrigation Engg. portion*, in this competition. The objective type questions of previous years have been given in chapter 26.

Certain Typical Questions from the Old Engineering Services Exams from the Year 1988-2009

Q. 28.1. *Calculate the diameter required for a single-stage trickling filter which is to yield an effluent BOD_5 of 20 mg/l when treating settled domestic sewage with a BOD_5 of 120 mg/l. The wastewater flow is 2200 m^3/day and the recirculation is constant at 4000 m^3/day. The filter depth is 1.5 m.*

(*Engineering Services, 1992*)

Solution : BOD_5 of sewage entering the filter = 120 mg/l

BOD_5 of sewage leaving the filter = 20 mg/l

BOD_5 of sewage removed in filter = 100 mg/l

Total sewage flow being treated

$$= 2200 \text{ m}^3/\text{day} = 2200 \times 1000 \; l/\text{d}$$

$\therefore$ BOD removed from sewage/day

$$= (2200 \times 1000) \times 100 \text{ mg/day} = 220 \text{ kg/day}$$

Assuming on organic loading for this single stage filter (which must be a conventional one) to be 1500 kg/ham/day (between 900—2200 kg of BOD_5 per ham), we have

$$\text{Filter volume required} = \frac{\text{Total BOD removed}}{\text{Organic loading}}$$

$$= \frac{220 \text{ kg / day}}{1500/\text{ham/day}} = \frac{220}{1500} \times 10000 \text{ m}^3$$

$$= 1467 \text{ m}^3.$$

Filter depth = 1.5 m (given)

$$\therefore \quad \text{Area of filter required} = \frac{\text{Volume}}{\text{Depth}} = \frac{1467}{1.5} = \mathbf{978 \; m^2}.$$

If d is the dia of the filter, then

$$\frac{\pi}{4} \,.\, d^2 = 978$$

$$\text{or} \qquad d = \sqrt{\frac{978 \times 4}{\pi}} = 35.2 \text{ m}$$

Hence, use 35 m dia circular filter unit. **Ans.**

Check for hydraulic loading

$$\text{Area of filter provided} = \frac{\pi}{4} \,.\, (35)^2 = \mathbf{962 \; m^2}.$$

If a hydraulic loading of 28 Ml/hr/day is adopted (between 22—44 ; normally 28)*, then filter area required

$$= \frac{2200 \times 1000 \; l/\text{day}}{28 \times 10^6 \; l/\text{ha/day}}$$

$$= \frac{2.2}{28} \times 10^4 \text{ m}^2 = 786 \text{ m}^2.$$

Since the area provided is 962 m^2, which is more than the above requirement of 786 m^2, the adopted area or dia is safe.

Hence, use 35 m dia trickling filter. **Ans.**

[**Note.** The given quantity of recirculation of 4000 m^3/day, will of course affect the efficiency of the filter and design of the distributors but will not affect the filter volume, or the filter area, as discussed in article 9.22.2. Hence, the above value is not required, and is simply given to confuse the students.

Q. 28.2. (*a*) *The treated domestic sewage of a town is to be discharged in a natural stream. Calculate the percentage purification required in the treatment plant with the following data :*

Population = 50,000,

BOD contribution per capita = 0.07 kg / day,

BOD of stream on U/s side = 3 mg/lit,

Permissible maximum BOD of stream on D/s side = 5 mg/lit,

Dry weather flow of sewage = 140 litres per capita per day,

Minimum flow of stream = 0.13 m³/sec,

(*Engineering Services, 1995*)

Solution : Total BOD of raw domestic sewage

$$= 50000 \times 0.07 \text{ kg/day} = 3500 \text{ kg/day}$$

$$Q_S = \text{Sewage discharge} = 140 \times 50{,}000 \ l\text{/day}$$

$$= 7000 \text{ m}^3\text{/d}$$

$$= \frac{7000}{24 \times 60 \times 60} \text{ m}^3\text{/s} = 0.081 \text{ m}^3\text{/s}$$

$$Q_R = \text{River discharge} = 0.13 \text{ m}^3\text{/s}$$

BOD of mix = 5 mg/l

BOD of river = 3 mg/l

BOD of treated discharged sewage = C_S

$$\text{BOD of mix} = \frac{\text{BOD of treated discharged sewage} \times \text{Sewage flow} + \text{BOD of river} \times \text{River flow}}{\text{Sewage flow} + \text{River flow}}$$

or
$$5 = \frac{C_S \times 0.081 + 3 \times 0.13}{0.081 + 0.13}$$

or
$$C_S = \frac{5 \times 0.211 - 0.39}{0.081} = 8.21 \text{ mg/}l$$

Max. permissible BOD of discharged treated sewage

$$= C_S = \mathbf{8.21 \text{ mg/}l}$$

BOD of untreated sewage

$$= \frac{(3500 \times 10^6) \text{ mg/d}}{(7000 \times 1000)\ l\text{/d}} = 500 \text{ mg/}l$$

%age treatment required

$$= \frac{\text{BOD of raw sewage} - \text{Max. permissible BOD of discharged sewage}}{\text{BOD of raw sewage}}$$

$$= \frac{500 \text{ mg}/l - 8.21 \text{ mg}/l}{500 \text{ mg}/l} \times 100$$

$$= \mathbf{98.36\%.} \quad \textbf{Ans.}$$

Q. 28.3. *Consider the case where a noise level of 90 dBA exists for five minutes and is followed by a reduced noise level of 60 dBA for 50 minutes. What is the equivalent continuous equal energy level (L_{eq}) for the 55 minutes period ? Assume a five minute sampling period. Write the concepts of L_{eq}.* (15)

(*Engineering Services, 1995*)

Solution : Using eqn. (20.14), we get

$$L_{eq} = 10 \log \sum (10)^{\frac{L_i}{10}} \times t_i$$

$$= 10 \log \left[(10)^{\frac{90}{10}} \times \frac{5}{55} + (10)^{\frac{60}{10}} \times \frac{50}{55} \right]$$

$$= 10 \log \left[9.0909 \times 10^7 + 9.0909 \times 10^5 \right]$$

$$= 10 \log \left[(10^5)\,(909.09 + 9.0909) \right]$$

Q. 28.4. *Estimate efficiency of a 30 m diameter and 1 m deep single stage high rate trickling filter for the following data :*

(*i*) *Sewage flow = 4.5 Mld (million litres/day)*

(*ii*) *Recirculation ratio = 1.4*

(*iii*) *BOD of raw sewage = 250 mg/l*

(*iv*) *BOD removed in primary clarifier = 25%.* (*Engineering Service, 1997*)

Solution : The efficiency of a high rate single stage trickling filter is given by eqn. (9.34), as :

$$\eta\% = \frac{100}{1 + 0.0044 \sqrt{\dfrac{Y}{V.F}}}$$

where Y = Total organic loading in kg/day

V = Filter vol. in ha.m

F = Recirculation ratio = 1.4 (given)

BOD to be removed in filter

= Total BOD of sewage − BOD removed in primary clarifier

= 250 mg/l − 25% × 250 mg/l

= 250 × 0.75 = 187.5 mg/l

$$Y = \text{Total organic loading}$$
$$= 187.5 \text{ mg}/l \times 4.5 \text{ M.ld} = 843.75 \text{ kg/d}$$

$$\text{Filter vol.} = \left(\frac{\pi}{4}.d^2\right) \times \text{depth} = \frac{\pi}{4} \times (30)^2 \times 1 \text{ m}^3$$
$$= 706.86 \text{ m}^3$$

$$\therefore \quad V = \frac{706.86}{10^4} \text{ ha.m} = 0.070686 \text{ ha.m}$$

$$\therefore \quad \eta = \frac{100}{1 + 0.0044\sqrt{\dfrac{843.75}{0.070686 \times 1.4}}}$$
$$= \mathbf{71.11\%.} \quad \textbf{Ans.}$$

Q. 28.5. *(c) A combined sewer of circular section is to be laid to serve an area of 100 ha with a population of 90,000, supplied with water at 200 litres per day. Assuming an impermeability factor of 0.5 and time of concentration of rainfall 't' as 20 minutes, calculate the size of sewer when it has to run full with a velocity of 0.3 m / sec. Assume suitable coefficients for 'a' and 'b' in the equation for 'R', the intensity of rainfall relating to 't', the time of concentration.*

(Engineering Service, 2003)

Solution : Av. sewage discharge = 80% of average water supply (asumed)

$$= 80\% \ [90{,}000 \times 200 \text{ L/day}]$$
$$= 0.8 \times 90 \times 200 \text{ m}^3\text{/d}$$
$$= \frac{0.8 \times 90 \times 200}{24 \times 3600} \text{ m}^3\text{/s} = 0.1667 \text{ m}^3\text{/s}$$

$$Q_1 = \text{Peak sewage flow} = 3 \times \text{Av. sewage flow}$$
$$= 3 \times 0.1667 = \mathbf{0.5 \text{ cumec}} \quad \ldots(ii)$$

$$Q_2 = \text{Storm water flow} = \frac{1}{36} K \, . \, p_c \, . \, A \quad \text{(Eq. 3.1)}$$

where $K = 0.5$; $A = 100$ ha

p_c is calculated from Eq. (3.6) as

$$p_c \text{ or } R = \frac{a}{T + b}$$

At $T = T_c = 20$ min, assume values of $a = 75$ and $b = 10$

$$\therefore \quad p_c = \frac{75}{20 + 10} = 2.5 \text{ cm/sec.}$$

$$\therefore \quad Q_2 = \frac{1}{36} \times 0.5 \times 2.5 \times 100 = \mathbf{3.47 \text{ m}^3\text{/s}} \quad \ldots(ii)$$

$$Q = Q_1 + Q_2 = 0.5 + 3.47 = \mathbf{3.97 \text{ m}^3\text{/s}}$$

Area of circular sewer

$$= A = \frac{Q}{V} = \frac{3.97 \text{ m}^3\text{/s}}{0.3 \text{ m/s}}$$
$$= 13.23 \text{ m}^2 \quad (\because \ V = 0.3 \text{ m/s, as given})$$

$$\therefore \quad \frac{\pi}{4} . D^2 = 13.23 \text{ m}^2$$

$\therefore$ Dia. of sewer = D = **4.1 m. Ans.**

Q. 28.6. *Compute the effective stack height of a coal burning power plant with physical stack height of 200 m and stack diameter of 8 m, stack gas exit velocity of 18.3 m/sec, temperature of gas 140°C, when the ambient air temperature is 8°C, atmospheric pressure is 1000 millibars and average wind speed is 4.5 m/sec. Also compute the emission rate of sulphur dioxide of the plant assuming burning of 24000 tonnes of coal per day with a sulphur content of 4.2%.*

(*Engineering Services, 2003*)

Solution : We have to use Eq. (18.21) and (18.22) and follow Solved Example 18.3 to compute as under :

Effective *Height* = $H = h + \Delta h$ ...(*i.e.* Eq. 18.21)

where h = actual ht of stack = 200 m

Δh = Plume ht given by Eq. (18.22) as :

$$\Delta h = \frac{v_s . D}{u}\left[1.5 + 2.68 \times 10^{-3}\ P.D.\left(\frac{T_s - T_a}{T_s}\right)\right]$$

where v_s = stack gas exit velocity in m/s = 18.3 m/s

D = Inside exit dia of stack in m = 8 m

u = wind speed in m/s = 4.5 m/sec

P = Atmospheric pressure in millibars = 1000 m.b.

T_s = stack gas temp. in K = 140°C = 140 + 273 = 413K

T_a = Air temperature in K = 8°C = 8 + 273 = 281K

$$\therefore \quad \Delta h = \frac{18.3 \times 8}{4.5}\left[1.5 + 2.68 \times 10^{-3} \times 1000 \times 8\left(\frac{413 - 281}{413}\right)\right]$$

= **271.73 m**

Eff. Ht. = H = 200 + 271.73 = **471.73 m. Ans.**

Emission rate of SO_2 is calculated as below : [Follow Example 18.1] :

Coal burnt per day = 24000 t/d

Sulphur content of coal = 4.2%

Sulphur produced per day

$$= 24000 \times \frac{4.2t}{100} = 1008t$$

The produced sulphur produces SO_2 as :

$$S + O_2 = SO_2$$

Since the molecular mass of both S and O_2 is 32 and they combine on one to one mass basis, 1008*t* of S will combine with 1008*t* of O_2 to produce 2016*t* of SO_2.

Hence, SO_2 produces per day = 2016 t/day

$\therefore$ Emission rate of SO_2 per second

$$= \frac{2016t}{24 \times 3600 \text{ sec}} = \frac{2016 \times 1000 \text{ kg}}{24 \times 3600 \text{ sec}}$$

$= \textbf{23.33 kg/sec.}$ **Ans.**

Q. 28.7. *Find out the volume of an anaerobic digestion tank for 5 MLD of domestic wastewater treatment plant having 60% suspended solids removal efficiency of primary clarifier and 250 mg/L suspended solids in wastewater based on sludge volume reduction in digester. Moisture content of influent sludge is 96% ; initial volatile solids content in sludge is 70% ; volatile solids destroyed is 65% ; digested sludge solid concentration is 8.0% ; specific gravity of primary sludge is 1.03 ; specific gravity of digested sludge is 1.04 ; density of water is 1000 kg/m^3 ; mean cell residence time is 15 days.*

(*Engineering Services, 2006*)

Solution : Study Solved Examples 9.20 and 9.24, and compute as under :

Sewage per day = 5 ML

Total S.S. in sewage = 250 mg/L

$\therefore$ Mass of S.S. in 5 ML of sewage (entering daily)

$= 5 \times 10^6 \text{ L} \times 250 \text{ mg/L}$

$$= \frac{5 \times 10^6 \times 250}{10^6} \text{ kg} = 1250 \text{ kg.}$$

S.S. removed in primary clarifier = 60%

Mass of solids removed daily in primary clarifier

$= 60\% \times 1250 \text{ kg} = 750 \text{ kg}$

M.C. of fresh primary sludge = 96%

$\therefore$ 4 kg of dry solids will make

= 100 kg of fresh primary wet sludge

or 750 kg of dry solids will make

$$= \frac{100}{4} \times 750 \text{ kg}$$

= 18750 kg of fresh primary wet sludge

Specific gravity of fresh (raw) wet sludge = 1.03

$\therefore$ Density of fresh (raw) wet sludge = 1030 kg/m^3

$\therefore$ Volume of fresh influent sludge produced daily (V_1)

$$= \frac{18750 \text{ kg}}{1030 \text{ kg/m}^3} = \mathbf{18.2 \text{ m}^3} \quad ...(i)$$

Now, we have to calculate the Vol. of digested sludge (V_2) as follows :

Mass of solids in raw sludge = 750 kg

Volatile solids = 70%.

∴ Mass of volatile solids = 70% × 750 kg = 525 kg

Mass of non-volatile solids = 750 – 525 = 225 kg

Volatile solids removed/destroyed in sludge digestion = 65%

∴ Mass of volatile solids destroyed

= 65% × 525 kg = 341.25 kg

Mass of volatile solids left

= 525 – 341.25 = 183.75 kg

Mass of non-volatile solids left

= 225 kg (as in original influent sludge)

Total mass of solids in digested sludge

= 183.75 + 225 = **408.75 kg**

But the digested sludge is said to contain 8% solids.

Thus 8 kg of solids make = 100 kg of digested sludge

$$1 \text{ kg of solids make} = \frac{100}{8} \text{ kg of digested sludge}$$

408.75 kg of solids make

$$= \frac{100}{8} \times 408.75 \text{ kg of digested sludge}$$

= 5109.375 kg of digested sludge

Density of digested sludge = 1.04 × 1000 kg/m³ (given)

= 1040 kg/m³

∴ Vol. of digested sludge

$$= \frac{5109.375}{1040} \text{ m}^3$$

$$= \mathbf{4.91 \text{ m}^3} \qquad \ldots(ii)$$

Capacity of digestion tank

$$= \left[V_2 - \frac{2}{3}(V_2 - V_1)\right]t\,; \quad \text{where } t = 15\, d \text{ (given)}$$

$$= \left[18.2 - \frac{2}{3}(18.2 - 4.91)\right]15$$

= **140.1 m³. Ans.**

Q. 28.8. *What is self-cleansing velocity ? A 20 cm diameter sewer with invert slope of 1 in 500 is running full. Calculate the rate of flow in the sewer. Compare the velocity with self-cleansing velocity. Assume Manning's n = 0.012.*

(Engineering Services, 2007)

Solution :

$$V = \frac{1}{n} A R^{2/3} \sqrt{S}$$

$$= \frac{1}{0.012} \times \left(\frac{0.2}{4}\right)^{2/3} \cdot \frac{1}{\sqrt{500}} = \mathbf{0.51 \text{ m/s}}$$

$$Q = A \, . \, V = \frac{\pi}{4} \times (0.2)^2 \times 0.051$$
$$= 1.589 \times 10^{-3} \text{ m}^3\text{/s} = 1.589 \text{ l/s}$$

Self cleansing velocity (V_s) from Eq. (4.10)

$$= \frac{1}{n} \, . \, R^{1/6} . \, \sqrt{kd' \, (G - 1)}$$

Thus, V_s for removing inorganic sand particles of dia 1 mm (d') and specific gravity 2.65 (G) having k = 0.04 (assumed)

$$= \frac{1}{0.012} \left(\frac{0.2}{4} \right)^{1/6} \sqrt{\left(0.04 \times \frac{1}{1000} \right) (2.65 - 1)}$$
$$= 50.58 \times 8.124 \times 10^{-3} = \mathbf{0.41 \text{ m/s}}$$

Hence, the velocity generated in the sewer (0.51 m/s) is self cleansing, since it is more than the reqd. minimum velocity for self cleansing (0.41 m/s).

Q. 28.9. *A river before its entry into a town had a discharge of 100 l/sec, and 20 mg/l as the concentration of a conservative parameter. The town's waste water outfall having 200 mg/l (two hundred mg/l) concentration of the same conservative parameter raised the river concentration to 50 mg/l, after a complete mix with the river water. Determine the dilution ratio resulting from the discharge of the said waste water outfall.* *(Engineering Services, 2007)*

Solution : Q_R = River discharge = 100 l/s

C_R = Concentration of conservative parameter in river flow = 20 mg/l

Q_s = Sewage discharge = ?

C_s = Conc. of parameter in town's sewage outfall = 200 mg/l

C = Conc. of parameter in river after complete mix = 50 mg/l

Using Eq. (8.1), we have

$$C = \frac{C_s \, . \, Q_s + C_R \, . \, Q_R}{Q_s + Q_R}$$

or
$$50 = \frac{200 Q_s + 20 \times 100}{Q_s + 100}$$

$$50 Q_s + 5000 = 200 Q_s + 2000$$
$$150 Q_s = 3000$$

or
$$Q_s = 20 \text{ l/s}$$

Thus, 20 l/s of sewage flow is mixed with 100 l/s of river flow, thereby diluting the sewage flow by $\frac{100 + 20}{20}$ = 6 times.

Hence, dilution ratio = 6. **Ans.**

Yearwise Questions from the Year 2008 onwards

Year 2008 (*Engineering Services*)

Q. 2. (*d*) *Estimate the moisture content of a solid waste sample with the following composition :*

Component	*Per cent by mass*	*Moisture (%)*
Food wastes	*20*	*70*
Paper	*40*	*6*
Cardboard	*10*	*5*
Plastics	*5*	*2*
Garden trimmings	*5*	*60*
Wood	*5*	*20*
Tin cans	*5*	*3*

(10)

Solution : Let the total mass of the waste sample be 100 M. Then Moisture (%) of the sample

$$= \left[\frac{\begin{array}{r}20\text{ M}\times 70\% + 40\text{ M}\times 6\% + 10\text{ M}\times 5\% \\ +5\text{ M}\times 2\% + 5\text{ M}\times 60\% \\ +5\text{ M}\times 20\% + 5\text{ M}\times 3\%\end{array}}{100\text{ M}}\right]\times 100$$

$$= (14 + 2.4 + 0.5 + 0.1 + 3 + 1 + 0.15)\,\frac{\text{M}}{\text{M}}$$

$$= \mathbf{21.15\%.\ Ans.}$$

Q. 4. (*a*) *Design a septic tank for 200 persons with a water supply of 125 litre per capita per day. Assume any other data and mention it.* (10)

[Hint : *Pl. refer exactly similar Solved Question 9.33.*]

Q. 5. (*b*) *Discuss the microbial growth pattern in waste water.* (10)

[Hint : Growth pattern of Micro-organisms

When a small number of a viable bacterial cell (*i.e.* bacteria) contained in sewage are placed in a closed vessel (such as during sewage treatment) containing sufficient food (organic matter or BOD) supply in a suitable environment ; conditions get established in which unrestricted growth of the micro-organisms cell takes place. However, growth of an organism does not go indefinitely, and after a characteristic size is reached, the cell divides (due to hereditary and internal limitations) becoming two. Assuming an adequate food supply, the cells then grow and divide again, like the original cell. Every time a cell splits*, approximately every 20 to 30 minutes, a new generation occurs. This growth is known as the ***exponential*** or *logarithmic* ***growth phase.*** At the exponential growth rate, the largest number of cells are produced in the shortest period of time. In nature, this growth cannot be maintained indefinitely, simply because the optimum environment of growth cannot be maintained. The amount of

*This splitting of bacteria is known as ***binary fission.***

growth is the function of two variables, *i.e.* environment and food. The pattern which actually results is known as the ***bacterial growth rate curve,*** such as the one shown in Fig. 28.1.

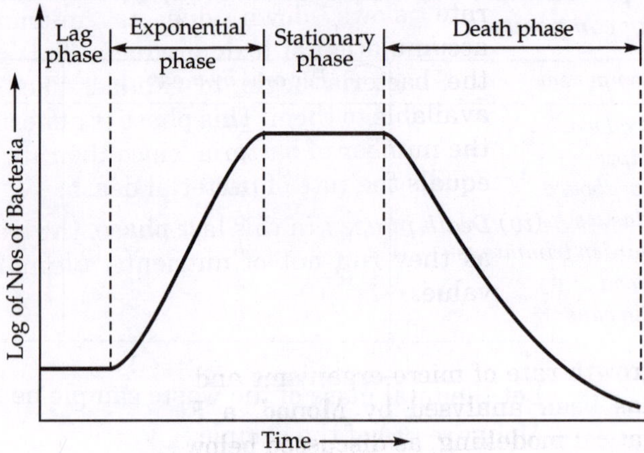

Fig. 28.1. Bacterial growth rate curve.

Such a **bacterial growth pattern curve** may be divided into the following four well defined phases:

(*i*) *Lag phase ;*

(*ii*) *Exponential* or *Logarithmic phase or Log growth phase.*

(*iii*) *Stationary phase ;* and

(*iv*) *Death phase.*

These phases are briefly discussed below :

(*i*) *Lag phase :* During this initial lag phase, the bacteria adapt themselves to growth conditions. It is the period in which the individual bacteria are maturing and not yet able to divide. During this phase of the bacterial growth cycle, synthesis of RNA, enzymes and other molecules occurs.

(*ii*) *Exponential or Log growth phase :* This phase is characterized by *cell doubling*. The number of new bacteria appearing per unit time is proportional to the present population. If growth is not limited, doubling will continue at a constant rate, and, thus, the number of bacterial cells and the rate of population increase, both doubles with each consecutive time period. For this type of economical growth, plotting the natural logarithm of cell numbers against time produces a straight line. The slope of this line represents the ***specific growth rate of the organisms,*** which is a measure of the number of divisions per cell per unit time. The actual rate of this growth ; (*i.e.* the slope of the line in Fig. 28.1) depends upon the growth conditions, which affect the frequency of cell division events and the probability of both daughter cells surviving. Under controlled conditions, cyano bacteria can double their population four times a day. However, exponential growth cannot continue indefinitely, because the median is soon depleted of nutrients and enriched with wastes.

(*iii*) *Stationary phase :* During the stationary phase, the growth rate slows down due to nutrient depletion and accumulation of toxic products. This phase is reached as the bacteria begin to exhaust the resources that are available to them. This phase is almost a constant value of the number of bacteria, since the rate of bacterial growth equals the rate of bacterial death.

(*iv*) *Death phase :* In this last phase, the bacteria starts to dia, as they run out of nutrients, ultimately reaching a nil value.

Monod Analysis

The growth rate of micro-organisms and utilization of substrate (BOD) by them, has been analysed by Monod, a French scientist, on the basis of mathematical modelling, as discussed below :

For the large numbers and mixed cultures of micro-organisms in waste treatment systems, it is convenient to measure biomass rather than the number of micro-organisms. In the log-growth phase, the growth *rate of biomass* $\left(\frac{dX}{dt}\right)$ is given by the expression

$$\frac{dX}{dt} = \mu X \quad \text{...(28.1)}$$

where $\frac{dX}{dt}$ = Rate of microbial growth or Growth rate of biomass (mg/L . t)

μ = Growth rate constant per unit time (*i.e.*, t^{-1})

X = Concentration of the biomass (mg/L)

Monod Equation. Monod has developed an empirical model equation which is given as

$$\boxed{\mu = \frac{\mu_m \cdot Y}{K_s + Y}} \quad \text{...(28.2)}$$

where μ = ***Growth rate constant*** **(t^{-1})**

μ_m and K_s are the empirical coefficients to the Monod equation and defined as :

μ_m = ***Maximum growth rate constant***, t^{-1}. Its usual value is 2.5/d.

K_s = ***Half saturation constant*** (or ***Half velocity constant***) = concentration of limiting food or sub-strate, Y, (BOD_5), when $\mu = 0.5\ \mu_m$. Its usual value is 100 mg of BOD/L

Y = concentration of limiting food in solution, in mg/L *i.e.*, BOD_5 or Soluble BOD.

Substituting the value of μ from Eq. (28.2) in Eq. (28.1), the rate of biomass production (*i.e.*, microbial growth rate)

$$\frac{dX}{dt} = \frac{\mu_m \,.\, Y \,.\, X}{K_s + Y} \quad ...(28.3)$$

Eq. (28.2) of Monod, infact, assumes only growth of micro-organisms, and ignores the natural die-off of organisms. The death or decay of the microbial mass is assumed to be a first order equation in biomass, and hence Eqn. (28.3) can be modified to

$$\frac{dX}{dt} = \left[\frac{\mu_m \,.\, Y \,.\, X}{K_s + Y} - K_e \,.\, X\right] \quad ...(28.4)$$

where K_e = Endogeneous decay rate constant, t^{-1}. Its usual value is 0.06/*d*

Now, if all the food in the system was converted to biomass, the rate of food utilization $\left(\frac{dY}{dt}\right)$ would equal to the rate of biomass production. However, due to inefficiency of the conversion process, the rate of food utilization $\left(\frac{dY}{dt}\right)$ will be greater than the rate of biomass utilization, and is given as

$$-\frac{dY}{dt} = \frac{1}{\alpha_y} \cdot \frac{dX}{dt} \quad ...(28.5)$$

where α_y = decimal fraction of food mass converted to biomass = Yield coefficient in mg/L of biomass/mg/L of food (BOD) utilized.

Usual value of α_y is 1.0 w.r. to TSS, and 0.6 – 0.5 w.r. to VSS (*i.e.*, mg VSS/ mg BOD)

Combining Eq. (28.5) and (28.3), we get

$$-\frac{dY}{dt} = \frac{1}{\alpha_y} \cdot \left[\frac{\mu_m \,.\, YX}{K_s + Y}\right] \quad ...(28.6)$$

Equations (28.4) and (28.6) are a fundamental part of the development of the design equations for a wastewater treatment process.

Year 2009 (*Engineering Services*)

Q. (*c*) *A large stream has a reoxygenation constant of 0.4 per day. At a velocity of 0.85 m/s ; and at the point at which an organic pollutant is discharged, it is saturated with oxygen at 10 mg/L (D_0 = 0). Below the outfall, the ultimate demand for oxygen is found to be 20 mg/L and the deoxygenation constant is 0.2 per day. What is the dissolved oxygen at 48.3 km downstream ?* (10)

Solution : K_R = 0.4/day

K_D = 0.2/day

D_s = Saturation DO = 10 mg/L

D_0 = Initial D.O. deficit = 0

$L = Y_u$ = Ultimate BOD of the mix = 20 mg/L

We have to compute the D.O. Deficit created by the sewage flow at a point 48.3 km downstream. Since the flow velocity of the stream is 0.85 m/s, the sewage discharged will reach this point after time t, given by

$$S = \text{Velocity} \times \text{Time}$$

$$48.3 \text{ km} = 0.85 \text{ m/s} \times \text{Time } (t)$$

or $$\text{Time } (t) = \frac{48.3 \times 1000 \text{ m}}{0.85 \text{ m}} \text{ sec}$$

or $$t = \frac{48.3 \times 1000}{0.85} \times \frac{1}{3600 \times 24} \text{ days} = 0.6577 \text{ days}$$

The D.O. deficit after t day (D_t) is now given by Eq. (8.3) as :

$$D_t = \frac{K_D\, L}{K_R - K_D}\left[(10)^{-K_D.t} - (10)^{-K_R.t}\right] + \left[D_0 \times (10)^{-K_R.t}\right]$$

or $$D_t = \left[\frac{0.2 \times 20}{0.4 - 0.2}\right]\left[(10)^{-0.2 \times 0.6577} - (10)^{-0.4 \times 0.6577}\right] + \left[0 \times (10)^{-0.4 \times 0.6577}\right]$$

or $$D_t = 20\left[\frac{1}{(10)^{0.13154}} - \frac{1}{(10)^{0.26308}}\right]$$

$$= 20\,[0.76 - 0.546] = \mathbf{4.29\ mg/L}$$

The D.O. left = Initial DO – DO deficit

$$= 10 - 4.29 = \mathbf{5.71\ mg/L.}\quad \textbf{Ans.}$$

Q. 3. (c) *Design a rectangular grit chamber for a flow of 40 MLD. Specific gravity = 2.65 and size to be removed is 0.2 mm.*

Find the

(a) settling velocity of 0.2 mm particles,

(b) critical horizontal velocity of flow, and

(c) size of the grit chamber

Assume kinematic viscosity of the liquid = 1.0 × 10^{-2} cm²/s. (10)

[Hint : *Pl. refer Solved Example 9.2 (a).*]

Q. 4. (b) *1200 m long storm sewer collects wastewater from a catchment area of 50 hectare, where 35% area is covered by roof (I = 0.9), 20% area is covered by pavements (I = 0.8) and 45% area is covered by open land (I = 0.13). Determine the average I, and diameter of storm sewer line assuming*

(i) the time of entry = 3 min

(ii) velocity of full flow = 1.50 m/s

(iii) n = 0.013 and slope = 0.001

I = runoff ratio. (10)

Solution : $A = 50$ ha

K = (or I here) varies, and hence

Average K (or I)

$$= \frac{K_1A_1 + K_2A_2 + K_3A_3}{A}$$

$$= \frac{0.9 \times 0.35A + 0.8 \times 0.20\,A + 0.13 \times 0.45A}{A}$$

$$= [0.9 \times 0.35 + 0.8 \times 0.20 + 0.13 \times 0.45]$$

$$= \mathbf{0.5335.\ Ans.}$$

Now, velocity of full flow in sewer = 1.5 m/s.

But $V = \frac{1}{n}.R^{2/3}\sqrt{S}$ (Mannings formula)

$$\therefore \quad 1.5 = \frac{1}{0.013} R^{2/3}.\sqrt{0.001}$$

$$\text{or} \quad R^{2/3} = \frac{1.5 \times 0.013}{\sqrt{0.001}} = 0.6166$$

$$\text{or} \quad \left(\frac{D}{4}\right)^{2/3} = 0.6166$$

$$\text{or} \quad \frac{D}{4} = (0.6166)^{3/2} = 0.484$$

$$\text{or} \quad D = \mathbf{1.94\ m.} \quad \mathbf{Ans.}$$

Q. 7. (*b*) *Draw and discuss the single stack system of plumbing.* (10)

Year 2010 (*Engineering Services*)

Q. 1. (*c*) (*i*) *For a BOD test, raw sewage (3.0 ml) was diluted to 300 ml (capacity of a BOD bottle). The diluted sewage was observed for its dissolved oxygen at the beginning and end of 5 days incubation at 20°C. The respective values were 8.6 mg/L and 4.6 mg/L. Determine the BOD of the raw sewage.* (2)

Solution : Using Eq. (7.11), we have

BOD_5 of raw sewage

$$= \text{DO consumed in the test} \times \frac{\text{Vol. of diluted sample}}{\text{Vol. of undiluted sewage}}$$

$$\therefore \quad BOD_5 = (8.6 - 4.6)\text{ mg/L} \times \frac{300}{3}$$

$$= 4 \times 100 = \mathbf{400\ mg/L.}\quad \textbf{Ans.}$$

Q. 1. (c) (*ii*) *A mixed liquor with 2,000 mg/L of suspended solids has the settled volume of 200 ml from a litre of this mixed liquor. Calculate its sludge volume index. Is it safe ?* (2)

Solution : Pl. refer article 9.36 and equation 9.52, from which we have

$$\text{S.V.I.} = \frac{V_{ob}}{X_{ob}} \times 1000 \text{ mL/g}$$

where SVI = Sludge Vol Index = ?

V_{ob} = settled sludge vol in ml obtained from 30 min settling of 1 litre sample of mixed liquor

= 200 ml (given)

X_{ob} = suspended solids in mixed liquor

= 2000 mg/L (given)

Substituting the given values, we get

$$\text{SVI} = \frac{200}{2000} \times 1000 = \mathbf{100\ mL/g.}\quad \textbf{Ans.}$$

Q. 5. (c) *Using n = 0.015 in Manning's formula, design a sewer running half-full at a flow rate of 650 l/sec and laid at an invert slope of 0.0001.* (8)

Solution : For a sewer, running 1/2 full, we have

$$Q = \frac{\pi}{8} D^2 \text{ (where } D \text{ is the dia. of the sewer)}$$

$$p = \frac{\pi D}{2}$$

$$r = \frac{D}{4}$$

$$Q = 650 \text{ l/s} = 0.65 \text{ m}^3/\text{s}$$

Using Manning's formula as :

$$Q = \frac{1}{n} . a . r^{2/3} . \sqrt{s}$$

Substituting values, we get

$$0.65 = \frac{1}{0.015} \times \frac{\pi}{8} D^2 \left(\frac{D}{4}\right)^{2/3} \times \sqrt{0.0001}$$

$$\text{or} \quad 2.483 = \frac{D^2 . D^{2/3}}{(4)^{2/3}}$$

$$\text{or} \quad D^{8/3} = 6.256$$

$$\text{or} \quad D = (6.256)^{3/8 = 0.375} = \mathbf{1.99\ m;\ say\ 2\ m.}\quad \textbf{Ans.}$$

Q. 6. (*c*) *The BOD rate constant (k) for a river's BOD assimilation was determined to be 2.0 day^{-1} (base e). The BOD of this river after leaving a heavily populated town was determined to be 50 mg/L. Determine the distance after which the river's BOD would become 4 mg/L when the average velocity of the river was 1 m/sec. What would have been the K value if this distance would have been 300 km and state what K manifests ?* (8)

Solution : $K = 2.0$/day (on base e)

$$K_D = 0.434\,K = \frac{K}{2.3}$$

(where K_D is on base 10) from Eq. (7.14)

or K_D = Deoxygenation constant

$$= \frac{2.0}{2.3} = 0.87$$

Now, the initial BOD of river = L = 50 mg/L

BOD of river after say t days = Y_t = 4 mg/L

Now, using Eq. (7.16), we get

$$Y_t = L\left[1-(10)^{-K_D \cdot t}\right]$$

or $$4 = 50\,[1-(10)^{-0.87\times t}]$$

or $$(10)^{-0.87t} = 0.92$$

Taking log, we get

$$-0.87t \times 1 = \log 0.92 = -0.0362$$

or $$t = 0.0416 \text{ day.}$$

Hence, river's BOD will become 4 mg/L after 0.0416 days.

Now, Distance = Velocity of river × Time

∴ Distance after which BOD will become 4 mg/L

$$= 1 \text{ m/sec} \times (0.0416 \times 24 \times 3600 \text{ s})$$

$$= 3596 \text{ m} = \mathbf{3.596\ km.}$$ **Ans.**

If BOD would have been reduced to 4 mg/L after a distance of 300 km, then time t_2 after which this will occur is given by :

Distance = Vel. × Time (t_2)

$$300 \text{ km} = 1\,\frac{\text{m}}{\text{s}} \times (t_2 \text{ in days} \times 24 \times 3600)$$

$$300 \times 1000 \text{ m} = 1 \times t_2 \times 24 \times 3600$$

∴ $$t_2 = 3.472 \text{ days.}$$

Using Eq. (7.16) again, we get the new value of K_D *i.e.* K_{D_2} as:

$$4 = 50\left[1-(10)^{-K_{D_2}\times 3.472}\right]$$

$$(10)^{-K_{D_2}\times 3.472} = 0.92$$

$$-3.472\,K_{D_2} \times 1 = \log 0.92 = -0.0362$$

or $K_{D_2} = 0.01$

$\therefore$ K_2 (on base e) $= K_{D_2} \times 2.3$

$= \mathbf{0.023.}$ **Ans.**

The K_D value represents the rate at which BOD is consumed or dissipated. When K_D value was 0.87, the BOD of river was reduced from 50 mg/L to 4 mg/L only in 0.0416 days at 3.596 km downstream ; while when K_D value was as low as 0.01 (87 times less), the BOD reduction from 50 mg/L to 4 mg/L took 3.472 days and occurred at 300 km downstream. The higher the value of K, the quicker will be the BOD dissipation and restoration of original conditions of the river after its pollution at a heavily populated town].

Q. 7. (*c*) *On the basis of a detention period of 24 hrs, determine the size (assuming length to width ratio of around 2 and depth of waste water about 1 m) of a Septic Tank required for a large house, dwelling 100 persons. The flow into the tank may be assumed at the rate of 70 lpcd. What will be the surface loading and equivalent weir loading of the tank ?* (8)

[**Solution :** Pl refer Solved Example 9.33 and compute as under :

The flow into the tank = 70 × 100 l/d

= 7000 l/d = 7 m³/day

Since detention time is 24 hrs, the quantity of sewage produced during 24 hrs; *i.e.*, the capacity of tank = 7 m³.

Depth of waste water = 1 m (given)

$$\therefore \quad \text{Area of tank reqd} = \frac{7\ \text{m}^3}{1\ \text{m}} = 7\ \text{m}^2.$$

If $L = 2B$

We have $2B \times B = 7\ \text{m}^2$

$$\therefore \quad B = \sqrt{3.5} = 1.87\ \text{m}$$

$$L = 3.74\ \text{m}$$

Hence, use a tank of size 3.74 m × 1.87 m, with depth of 1 m + 0.5 m (sludge zone) + 0.3 m (free board) = 1.8 m.

$$\textit{Surface loading} = \frac{\text{Discharge } (Q)}{\text{Surface area of tank}}$$

$$= \frac{7\ \text{m}^3/\text{d}}{1.87 \times 3.74\ \text{m}^2}$$

$= \mathbf{1\ m/d.}$ **Ans.**

Weir loading

Length of weir along periphery of the rectangular tank

$$= 2\,[3.74 + 1.87] = 11.22 \text{ m}^2$$

$$\therefore \quad \text{Weir loading} = \frac{7 \text{ m}^3/\text{d}}{11.2 \text{ m}^2} = \mathbf{0.62 \text{ m/d.}} \quad \textbf{Ans.}$$]

Q. 8. (*c*) *Discuss the impact of air pollution on*

(*i*) *Monuments*

(*ii*) *Animals*

(*iii*) *Plants and*

(*iv*) Climate. (8)

Year 2011 (*Engineering Services*)

Q. 4. (*a*) *Explain the processing operations that are generally carried out for handling and disposal of solid wastes accumulated at the bottom from the individual apartments of a high rise apartment building with the help of solid waste chute.* (10)

[Hint : Pl. refer article 14.6 on the issue. The operations that are generally carried out for handling and disposal of wastes may be grouped into the following :

1. Segregation of wastes (if no automated segregation/ recycling system is provided)
2. Collection of segregated wastes into different colour-coded community bins (containers)
3. Transfer through trucks or trolleys of the biodegradable wastes to the in-house compost plant within the complex, if provided ; otherwise transfer to recycling centre, Garden wastes also to be treated similarly.
4. Transfer of plastic and paper wastes to the recycling centre ; or its sale to recyclers.
5. The compost production from organic waste to be stored and used as manure or fertilizer, and rejects to be separated for transfer to dumping.
6. The street waste collected with both manual and mechanical sweeping along with the rejects of the compost plant and other non-biodegradable wastes of the apartments to be transported through trucks or trolleys to the city's solid waste Disposal Facility for final dumping/ disposal.

Q. 7. (*d*) (*i*) *Explain the following terms generally adopted in waste water collection system :*

(1) Sewage (2) sewers (3) Sewerage system. Briefly mention the principles involved in 'Combined sewer design'. (5)

Q. 7. (*d*) (*ii*) *A site adjacent to a river is considered for the location of a municipal waste water treatment plant having an average inflow of 2500* $m^3/$ *day. The selected treatment process sequence is as follows : (1) grit channel (2) primary sedimentation (3) activated sludge and (4) secondary sedimentation.*

Design the grit channels assuming maximum hydraulic loading as twice the average inflow rate. Assume other appropriate design parameters, if required.

(5)

Solution : We can design a rectangular grit channel provided with a proportioning flow weir at the effluent end (for controlling velocity of flow) by following the Solved Example 9.2 (*a*), as follows :

Average inflow = 2500 m^3/d

Maximum hydraulic loading on grit channel

$$= 2 \times 2500 \text{ m}^3/\text{d}$$

$$= \frac{5000}{24 \times 3600} \text{ m}^3/\text{s}$$

$$= 0.0579 \text{ m}^3/\text{s}$$

Now, assume that a horizontal flow through velocity (V_h) of 0.3 m/s is maintained in the channel (even at varying discharge) by a proportionate weir :

Then $\quad Q = V_h \times A$

or $\quad 0.0579 = 0.3 \times A$

or $\quad A = \frac{0.0579}{0.3} \text{ m}^2 = 0.193 \text{ m}^2$

Assuming a water depth of 1 m above the crest of the weir, which is kept at 0.3 m above the channel bottom (Fig. 9.5 *a*), we have the width (B) of the channel as

$$1 \times B = 0.193$$

or $\quad$ B = 0.193 m; say 0.2 m.

overall depth of grit channel (D)

= water depth above crest of weir (H)

+ 0.3 m + Free board of 0.45 m

= 1 + 0.3 + 0.45 = **1.75 m**

Now, assume settling velocity (V_s) = 0.02 m/s

$$\text{But Detention Time} = \frac{\text{Water depth in the basin}}{\text{Settling velocity}}$$

$$= \frac{1 \text{ m}}{0.02 \text{ m/s}} = 50 \text{ sec}$$

$\therefore$ Length of the grit channel

$$= V_h \times \text{Detention time}$$

$$= 0.3 \text{ m/s} \times 50 \text{ s} = 15 \text{ m}$$

The length of the grit channel is increased beyond theoretical value of 15 m by a factor of 10–50% to account for non-idealities in the flow and settling of particles. Hence, using about 30% (say) increased length, we get L = 20 m

Hence, use a rectangular channel (tank) with dimensions.

Length (L) = 20 m

Width (B) = 0.2 m

Depth (O) = 1.75 m **Ans.**

The proportional flow weir at the effluent end of the channel can also be designed (Fig. 9.5 *a*) by using Eqs. (9.11) and (9.12). We can thus compute a and b as follows.

For a = 0.025 m (*i.e.* 25 mm), we have

$$b = 1.466\, B.V_h = 1.466 \times 0.2 \times 0.3 = 0.088 \text{ m}$$

$\therefore$ Base width of weir = $2b$ = 0.176 m

Other coordinates (x, y) for the weir can be worked out by using Eq. (9.11) as

$$x = \frac{2\, B.\, V_h}{C_d \cdot \sqrt{2g}\, \pi \,.\, \sqrt{y}} = \frac{2 \times 0.2 \times 0.3}{0.62 \times \sqrt{19.62} \times 3.14\sqrt{y}}$$

or $$x = \frac{0.0139}{\sqrt{y}}$$

For different assumed values of y varying between a to H, *i.e.* 0.025 m to 1.0 m, the corresponding values of x can be worked out to plot the geometry of weir. **Ans.**

Year 2012 (*Engineering Services*)

Q. 6. (*b*) *Explain the process of Anaerobic sludge digestion. Name the experiments to be performed in the laboratory to determine the digestibility of sludge.* (10)

Ans. Although anaerobic sludge digestion and digester have been described in the text, we are again explaining the process in details, giving more details than given in the text.

(1) **Anaerobic Sludge Digestion.** Anaerobic sludge digestion is the most common process adopted for dealing with wastewater sludge containing primary sludge. Primary sludge is the solids which settle out of the wastewater in the sedimentation tanks just after the wastewater passes through the grit chambers. The settled material represents 40% to 60% of the suspended solids that exist in the wastewater. This represents 25% to 35% of the BOD in the wastewater.

BOD is a measure of the biodegradable organic matter in the wastewater. It is determined by the amount of oxygen required to metabolize the organic matter in the water.

Secondary sludge from the clarifier is also sent to the digester. Secondary sludge is generated when the overflow from the settling tanks goes into the aeration chambers and the aerobic bacteria convert the dissolved organics into carbon dioxide, water and solids. The solids settle out in the clarifier.

Anaerobic digestion is preferred to reduce the high organic loading of primary sludge because of the rapid growth of the biomass that would ensue if the sludge were treated aerobically. Anaerobic decomposition creates considerably less biomass than the aerobic process. Anaerobic digestion converts as much of the sludge as possible to end products, such as, liquid and gases while producing as little residual biomass as possible.

The liquids, in the form of supernatant, from the digester are sent back through the plant for further treatment.

Therefore, in a sewage treatment facility, the aerobic and anaerobic processes work together to achieve BOD removal of up to 95%.

An anaerobic sludge digester is designed to encourage the growth of anaerobic bacteria, particularly the ***methane producing bacteria*** that decreases organic solids by reducing them to soluble substances and gases, mostly carbon dioxide and methane.

There are three basic stages in anaerobic digestion. The first stage is production of carbon dioxide and organic acids from fermentation. The second stage is metabolizing of organic acids to hydrogen, carbon dioxide and other organic acids. The third stage utilizes the products of the preceeding stages to produce methane from carbon dioxide, hydrogen and acetic acid.

The sludge that remains is relatively stable and inert. From 50% to 60% of the organics are metabolized with less than 10% converted to biomass.

The anaerobic process is made up of two basic types of bacteria. The *acid formers* and the *methane formers*. The ***acid formers*** are facultative and anaerobic bacteria and include organisms that solubilize organic solids through hydrolysis. Soluble products are then fermented to acids and alcohols of low molecular mass. The ***methane formers*** are strict anaerobics that convert acids and alcohol along with hydrogen and carbon dioxide to methane.

Stability of the anaerobic process is difficult to maintain, because a balance favorable to several microbial populations is necessary. The methane producers are the most sensitive to conditions. *They may be affected by change in the pH of the digesting sludge.* Each species is limited to the use of a few compounds, mostly alcohols and organic acids. The rugged nature of the acid formers and the sensitive nature of the methane formers creates a bio system that may easily get upset.

(2) Description of the Anaerobic Digestion Process. The purpose of the anaerobic process is to convert sludge to end products of liquid and gases while producing as little biomass as possible. The process is much more economical than aerobic digestion.

It was originally thought that anaerobic digestion was accomplished in three stages: (1) hydrolysis of insoluble polymers, (2) fermentation of monomeric breakdown products and (3) generation of methane. Research has found that there is a step between fermentation and methane generation, the fermentation of acetate and hydrogen from volatile fatty acids. These bacteria are necessary to the process and are even more sensitive than the actual methane formers.

The process has now been described by the following four steps :

1. ***Hydrolysis.*** Large polymers are broken down by enzymes.
2. ***Fermentation.*** Acidogenic fermentations are most important, acetate is the main end product. Volatile fatty acids are also produced along with carbon dioxide and hydrogen.
3. ***Acetogenesis.*** Breakdown of volatile acids to acetate and hydrogen.
4. ***Methanogenesis.*** Acetate, formaldehyde, hydrogen and carbon dioxide are converted to methane and water.

The stability of the anaerobic process is very fragile. The balance between several microbial populations must be maintained. The hydrolysis and fermentation phases have the most robust organisms. They have the broadest environment range in which they thrive. They react quickly to increased food availability. Thereby, increasing the amounts of their products. The volatile fatty acid concentration rises very quickly. This is kept in check by the buffering action of the system provided by carbon dioxide in the form of biocarbonate alkalinity. *The pH range is, therefore maintained under normal circumstances.* However, during shock loading, the acid concentration can overcome the buffering action and raise the pH out of the narrow acceptable limits of the acetogens and the methogens. When this happens, methane production stops and the acid levels rise to the tolerance level of the acid formers. *At this point the system fails.*

Temperature is also a critical element. Sudden changes in temperature adversely affect the methane producers.

Several substances are toxic to the system, such as the heavy metals, chlorinated compounds and detergents. Pretreatment would be necessary for a wastewater high in these.

When operating properly, the digester receives sludge, primary and secondary, from the other treatment processes. The sludge is then held in the tank for 10 to 90 days depending on the system. The sludge goes into the digester, methane, carbon dioxide and traces of hydrogen sulfide go out the gas outlet, supernatant, from the water generated by the process and the water in the sludge, in drawn off as necessary and sent back through the plant, and stabilized sludge is pulled off the bottom to go to the drying beds.

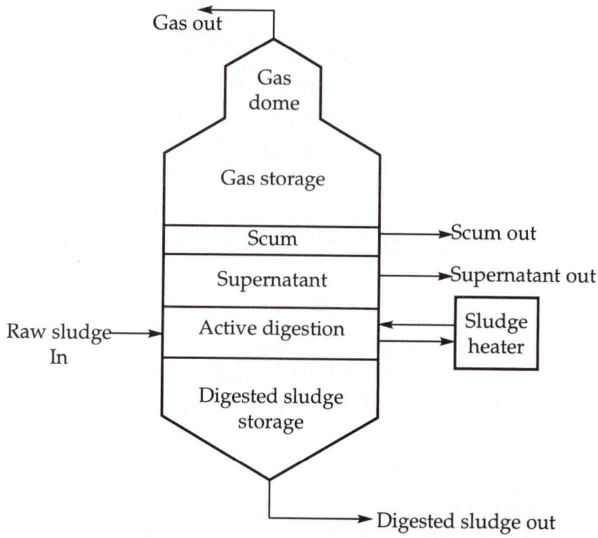

Fig. 28.2. Single Stage, Standard Rate Anaerobic Digester

Fig. 28.2 shows a **standard rate anaerobic digester**. Reactors for anaerobic digesters consist of closed tanks with air tight covers. Treatment plants processing less than 4000 cubic metres/day of wastewater often use standard rate digestion for economic reasons or simplicity of operation. Sludge separates in the reactor as shown, although some mixing occurs in the zone of active digestion and in the supernatant because of withdrawal and return of heated sludge. Sludge is fed to the reactor on an intermittent basis and the supernatant is withdrawn and returned to the secondary treatment unit. The digested sludge accumulates in the bottom to await removal to sludge disposal facilities.

High rate anaerobic digesters (Fig. 28.3) are more efficient

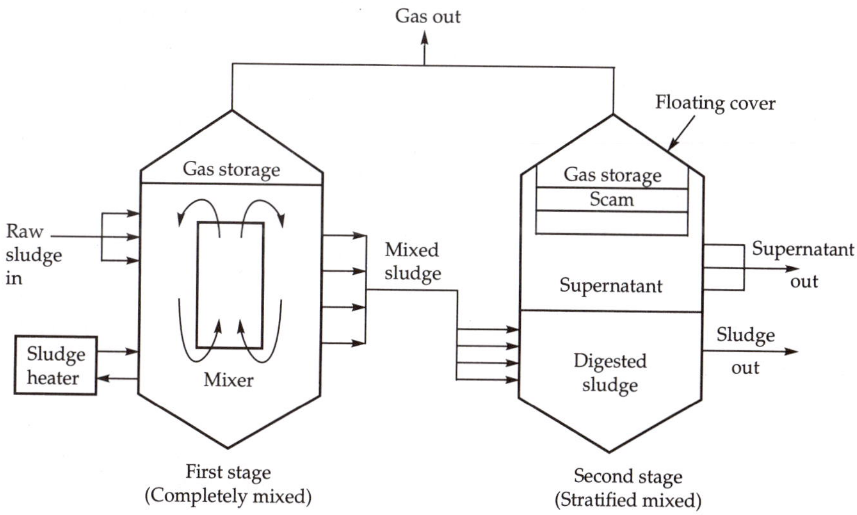

Fig. 28.3. Two Stage, High Rate Anaerobic Digester

and often require less volume than single stage digesters. In the first stage, the sludge is mechanically mixed to ensure better contact between the organics and the bacteria. The unit is heated to increase the metabolic rate of the microorganisms, thus speeding up the digestion process. In the second stage, the sludge is allowed to stratify and separate into layers. Little gas is generated in the second stage. The second stage has a floating cap and is equipped for gas recovery. The second stage is not heated because gas production doesn't occur in this stage. The supernatant, scum and digested sludge are drawn out of this unit.

(3) Operational Limits and Failure Criteria. As *has been previously stated, the methanogens are temperature and pH sensitive.* They operate in a pH range of 6.5 to 7.5. Movement to either side of this range quickly affects their metabolic rates and slows or stops methane production. Also, *the optimum temperature is 35°C.* Methane production drops off either side of this temperature. The methane formers are the bottleneck in the system and must be catered to. Whenever methane production drops, the volatile fatty acids begin to build up quickly due to the robust nature of the acid forming bacteria. Any shift adverse to the methane formers increase acids, which in turn, reduces methane formers.

Pending failure of the anaerobic process is evidenced by decreased gas production, and increase in volatile fatty acids and a drop in pH when acids exceed the buffering capacity of the ammonium biocarbonate in solution.

Failure is caused by:

1. Significant increase in organic loading (shock loading).
2. Sharp decrease in digesting sludge volume (when sludge withdrawn).
3. Sudden increase in operating temperature.
4. Accumulation of a toxic or inhibiting substance.
5. Removal of too much of the supernatant, thus reducing the number of bacteria available to metabolize.

Required Conditions for Sludge Digestion

Temperature	Optimum General range	35°F 30–35°C
pH	Optimum General range Per pound volatile	7.0–7.1 6.7–7.4 8–12 cu. ft.
Gas production	solids added Per pound volatile solids destroyed	16–18 cu. ft.

Gas Composition	Methane Carbon Dioxide Hydrogen Sulfide	65–69% 31–35% Trace
Volatile Acid Concentration as Ascetic Acid	Normal Maximum	200–800 mg/L 2000 mg/L
Alkalinity Concentration as Calcium Carbonate	Normal	2500–3500 mg/L

Loading and Detention Times

	Single Stage Digester	*High Rate Digester (First Stage)*
Loading (kg/m³/d) of VS	0.32–0.80	0.16–0.32
Detention time (days)	30–90	10–15
Capacity of digester (cu. ft./pop. equivalent) (*i*) Primary (*ii*) Primary and Secondary	 2–1 4–6	 0.4–0.6 0.7–1.5
Volatile Solids Reduction	50–70	50

(4) Operational Guidelines and Testing. Anaerobic systems are having a resurgence in use because of a better understanding of process failure and ways to avoid it. The lower operational costs due to the avoidance of aeration and the use of methane as an energy source has helped increase popularity. Also, research has found that anaerobic systems have the ability to tolerate much higher loadings than previously thought possible.

The following guidelines are used to prevent failure :

(*i*) Maintain constant loading rate by using storage capacity of the settling tanks and clarifier to smooth out the raw sludge input.

(*ii*) Make sure retention time is sufficient to accommodate the slowest growing organisms by not drawing off excess amounts of digested sludge.

(*iii*) Maintain temperature at the optimum without allowing it to increase and kill the bacteria. Allowing sludge and supernatant capacity to drop without checking heaters could cause this problem.

(*iv*) At all but the highest organic loading, microbial growth will proceed at submaximal rate. This is difficult with such adverse system.

(*v*) Make sure gas composition is in the proper range.

(*vi*) Maintain pH balance. Use lime if necessary.

The following records should be kept. (Testing)

1. Plot the daily gas production per unit raw sludge feed. This can be done taking the volatile solids of the raw sludge and the metered gas production.
2. Plot the percentage make up of the gases. Methane, carbon dioxide and hydrogen sulfide to ensure that digestion is proceeding properly. This should be done several times a day using a gas analyzer, such as a Rankine Meter.
3. *Plot the pH of the digester sludge* to maintain the proper range. Because of the narrow band (6.7 to 7.4) Litmus paper should be used. Lime may be used to prevent acidic build up.
4. *Plot the percentage of volatile fatty acids in the digester.* Natural buffering will prevent an imbalance from showing up immediately.
5. Plot the *temperature of* the sludge. Use a continuously recording thermometer to indicate trends up or down.
6. Plot the organic make up of digested sludge to make sure that the process is working properly. This can be done by drying the sludge. Weighing a sample and placing it in the oven at 500 degrees, comparing the residue to the original.

Laboratory Testing Required for Determining Digestibility of Sludge

As can be inferred from the above discussions, the following laboratory experiments may be performed to determine the digestibility of sludge :

(*i*) Temperature recording

(*ii*) pH measurement (most important) with continuous measurement during digestion

(*iii*) BOD measurement

(*iv*) Determining total, suspended and volatile solids, and particularly the volatile solids (VS)

(*v*) Determining Alkalinity as $CaCO_3$

(*vi*) Determining volatile acids as Acetic acid during sludge digestion process.

Q. 7. (*b*) *Describe photochemical oxidants. What are the sources of photochemical oxidants and how the detection and analysis for the same is done?* (12)

[Hint. Photochemical oxidants, as indicated in the text, are primarily formed by the action of sunlight on the oxides of nitrogen and hydrocarbons. The prominent of such photochemical oxidants are : **Ozone (O_3),** *Peroxyl Acetyle Nitrate (PAN) and Hydrogen Peroxide* (H_2O_2). But of these well known photochemical oxidants, the main impact on the natural environment is caused by the elevated levels of Ozone (*i.e.* the *Ground level ozone* or *Tropospheric ozone*).

Excessive concentrations of Tropospheric ozone have toxic effects on both plants and human health. It is feared that rising ozone concentrations may lead to depletion of forests and may cause reduced agricultural yields. Hence, ozone is considered to be the symbol of all the photochemical oxidants.

Since Ozone is not emitted into the air from particular activities characteristic of urban or industrial areas, it is referred to as a *secondary pollutant.* The *primary reason for its formation is the reaction of ultra-violet (UV) light from the sun on nitrogen oxides, through a process called* ***photolysis.*** Its production and concentration is hence dependent on the presence of primary pollutants (like nitrogen oxides and hydrocarbons, etc.) as well as ultra-violet light. In the presence of volatile organic compounds (VOCs), high concentrations of ozone are formed.

(1) Sources of Photochemical Oxidants. For the formation of photochemical oxidants (primarily ozone or Tropospheric or Ground level ozone), UV light from the sun is basically required to act upon the primary air pollutants like nitrogen dioxides and hydrocarbons. Eventually, the sources which produce these primary air pollutants, will become the sources of photochemical oxidants.

These sources of photochemical oxidants may be termed as ***oxidant percursors.***

(*i*) ***Man-made sources*** of ***oxidant percursors*** (*i.e.* oxides of nitrogen and hydrocarbons) include :

(*i*) Traffic exhaust of Motor Vehicles ;

(*ii*) Domestic space heating ;

(*iii*) Power plants ;

(*iv*) Industrial combustion ; and

(*v*) Industrial processes.

A balance between oxidizing and reducing agents would, infact, be maintained in the atmosphere, thus avoiding accumulation of ozone and other photochemical oxidants, if it were not for the photochemical degradation of hydrocarbons into peroxy radicals. Peroxy radicals rapidly convert nitric oxide to nitrogen dioxide, thus shifting the equilibrium towards ozone production during daylight. At night, emissions of nitric oxide into the atmosphere serve as a sink for ozone.

(*ii*) ***Indoor Sources,*** such as ultra-violet lamps, electro-static precipitators, photo-copying machines, and odour control equipment, etc. can also lead to increase in indoor concentrations of ozone. *Welding process* and *manufacture of hydrogen peroxide* are the other important indoor sources of ozone in industry, and pose problems in occupational health. However, other indoor activities, such as smoking and cooking with gas stoves tend to produce elevated nitric oxide concentrations that destroy ozone and PANs.

(*iii*) ***Natural Sources.*** In addition to man-made sources of ozone, described above, certain small quantities of ozone may be added to our lower atmosphere by natural sources. These sources are described below :

(*a*) A very small amount of the naturally forming ozone gas in the upper stratosphere in the ozone layer*, may be carried down to the lower troposphere by atmospheric circulation. Such natural intrusions of ozone are associated with cold fronts and atmospheric electrical phenomena.

(*b*) Natural hydrocarbons such as *terpenes* from trees and vegetations are acted upon by sun light to produce ozone, during day light hours. Generally, these processes are difficult to study because of the low concentrations of the hydrocarbons and their short atmospheric residence. However, hazes often associated with certain forest regions may well be explained by photochemical aerosol production from natural airborne hydrocarbons.

(2) Detection and Analysis of Concentrations of photochemical oxidants

Photochemical oxidants are the atmospheric pollutants for which environmental standards have been prescribed. Ozone gas is the main component of these photochemical oxidants (approximately 90%). Ozone is hence ranked as an atmospheric pollutant requiring special caution, around the world. Detection and measurement of ozone concentrations in ambient air is hence required. Detection, measurement, *analysis and monitoring of ozone concentrations become* more important for cities located in valleys, with sunny warm and dry climates, having a large number of motor vehicles.

Colour sensor networks are nowadays used to detect various atmospheric pollutants and gases. *Gas detection tubes* or *papers*, etc. are used as the sensor elements, which change

*Discussed in article 19.2

colour when exposed to the particular airborne pollutant. For detection of ozone, *ozone detection paper* is used, which provides a highly sensitive sensor. The concentration of ozone in the ambient air can be estimated from the degree of fading of the colour of the paper. Before exposure, the ozone detection paper is deep blue in colour. After exposure to ozone, the blue colour fades. Depending upon the extent of its fading, the ozone concentration exposures can be estimated. A linear relationship between the colour strength and its cumulative amount of ozone exposure has been established. When the cumulative ozone exposure exceeds a fixed amount, the ozone detection paper becomes white, and the blue colour is completely lost. The cumulative amount of ozone exposure that turns the paper white depends on the paper's sensitivity. NTT Energy and Environmental Laboratories, Atsugi – shi, Japan uses a highly sensitive paper that turns white when exposed to 100 ppb hr of ozone exposure.

Since the concentration of photochemical oxidants have been increasing year on year in the affected cities, it has become necessary to analyse and compare concentrations measured at various points. For this aim, ***colour sensor networks*** are being developed and installed. The experiments conducted involved a procedure whereby highly sensitive ozone paper was exposed to the ambient air outdoors, images of paper before and after exposure were captured, and image processing was performed to numerically convert the image data into ozone concentrations. By using more and more sensitive detection papers, colour changes could be seen even for relatively low ozone concentrations of 30-60 ppb, making it suitable for use in applications such as raising public awareness.

The experiments conducted with such detection papers by capturing image data and converting the same to the numerical values have been found to compare well with the results obtained from analytical instruments, which are used to test collected samples of air for measuring ozone levels.

(3) Analytical Methods for Measuring Photochemical Oxidants

For analytical measuring purposes, oxidants are generally divided into three categories : (*i*) Ozone ; (*ii*) Total oxidants ; (*iii*) PAN. Ozone and PAN can be measured specifically, while Total oxidants are usually determined as a class of compounds.

The reagent employed to measure the oxidizing property of photochemical oxidants is a solution of **Neutral-buffered potassium iodide (NBKI).** This reagent reacts with ozone, nitrogen dioxide, and PAN. Reducing agents, such as SO_2 have an inhibiting effect on the reagent solution and should be removed from the inlet stream. To this extent, reaction with potassium iodide reagent is a measure of the net oxidising capacity of a sample of ambient air.

The term "oxidant" or "total oxidant" are used to describe the net oxidising capacity of the sampled air, as determined by its reaction with NBKI. The terms "corrected oxidant" or "adjusted oxidant" indicate that the measurements have been corrected for the presence of reducing agents (SO_2) and for other oxidizing agents (NO_2) in the air.

Long term averaging values are usually meaningless for photochemical oxidants, and attention is hence directed to 1-h values. Thus, the use of continuous instruments with automatic data collection systems has an advantage. Manual methods are available for determination of photochemical oxidants, and these may be automated to a certain extent.

(*i*) **Measuring Ozone.** Continuous ozone analysers are usually based on the chemiluminescent reaction of ozone with ethylene. A ***chemiluminescence method*** based on the reaction of ozone with certain dyes has been further developed in the Netherlands. These techniques are not subject to atmospheric interference, and are highly sensitive (2 $\mu g/m^3$ or 0.001 ppm or 1 ppb), and perform well under field conditions. The methods are not absolute and hence some form of calibration involving potassium iodide or U.V. absorption spectroscopy is necessary.

Specific determination of ozone may be accomplished by ***ultraviolet absorption spectroscopy*** in the 200-300 nm wavelength range. The advantages of this method are that it is highly sensitive, does not require cylinders of explosive gases, and that it gives an absolute measurement. Very high concentrations of certain hydrocarbons or mercury vapour may cause interference.

The ***dihydroacridine method*** is a specific, inexpensive, manual method for ozone determination, in which samples are taken every 30 minutes. Interference effects can be eliminated by parallel sampling with an identical system filled with an ozone scrubber. This is the only suitable method for short-term ozone measurements when a chemiluminescent or U.V. adsorption instrument is not available.

(*ii*) **Measuring Total Oxidants.** Total oxidants are generally measured by using acid or neutral buffered solutions of potassium iodide. There are indications that the accuracy of the method depends on the pH of the buffer solution, the concentration of the reagent, and other variables. In addition to these drawbacks, the method is basically unspecific. Any other oxidizing or reducing species can cause interference. Such interference results when sampling air containing sulfur dioxide, chlorine, nitrogen dioxide, or peroxides. Nitrogen dioxide and sulfur dioxide interfere most by giving erroneously high and low values, respectively.

Interference from sulfur dioxide can be eliminated by passing the air sample through glass fibre filters impregnated with a suitable material such as chromium trioxide. Humidity often renders these scrubbers ineffective, because under such conditions, they oxidize nitric oxide to nitrogen dioxide, thus increasing nitrogen dioxide interference. As there is also ozone loss after extended use, these scrubbers are not entirely satisfactory.

Interference from nitrogen dioxide is more difficult to eliminate. Some form of adjustment can be made to the total oxidant reading, if continuous nitrogen dioxide measurements are also available. This correction depends to some extent on experimental conditions, and, in view of the problems previously noted with measurements of oxides of nitrogen, may not be easy to make.

Other methods are available for the determination of oxidants but none of them is commonly used. They include the ferrous ammonium sulphate, alkali potassium iodide, and phenolphthalein methods.

(iii) **Measuring Peroxy-Acetyl-Nitrate (PAN)**

The infrared absorption and electron-capturing properties of peroxyacetylnitrate have been used for its measurement in simulated and real atmospheres. Electron-capture detectors preceded by gas chromatographic separation offer a limit of detection of 0.5 $\mu g/m^3$ (0.0001 ppm or 0.1 ppb) for the automatic determination of ambient concentrations of peroxyacetylnitrate. Water vapours may interfere with the passage of peroxyacetylnitrate along the chromatographic column and trace contaminants in cylinder gases can be a problem because of their slow accumulation on chromatography columns used for continuous measurements.

Peroxyacetylnitrate (PAN) can also be measured by the hydrolysis of peroxyacetylnitrate solution to give nitrite ions that can be determined colorimetrically. However, this method suffers from interference by nitrogen dioxide.]

Year 2013 (*Engineering Services*)

Q. 1. (*f*) *Discuss thermal stratification and its importance in temperature of lakes.* (4)

Q. 2. (*c*) *Why a circular section is most widely adopted for sewer pipes ? Discuss the hydraulic characteristics of circular sewer sections running partially full.* (8)

Q. 5. (*a*) (*i*) *Discuss various methods of landfilling and its operation. Also discuss control of gas movement and leachate movement in landfill sites.* (10)

(*ii*) *What are various secondary air pollutants ? Discuss sources, origin and hazardous effect of sulphur dioxide and carbon monoxide on human body.* (10)

Q. 5. (*c*) *What are different zones of pollution in river stream ? Explain the importance of reoxygenation, deoxygenation and oxygen deficit in problems of stream sanitation.* (10)

Year 2014 (*Engineering Services*)

Q. 1. (*c*) *Determine the area required for a new landfill site with a projected life of 20 years for a population of 150000 generating 25 kg per household per week. Assume that the density of waste is 500 kg/m³. A planning restriction limits the height of the landfill to 10 m.* (8)

Solution : Solid waste produced annually (*i.e.* in 52 weeks) by 1,50,000 people @ 25 kg/c/week,

$$= [1{,}50{,}000 \times 25 \times 52] \text{ kg}$$

$$= \frac{(1{,}50{,}000 \times 25 \times 52) \text{ kg}}{500 \text{ kg/m}^3} = 3{,}90{,}000 \text{ m}^3$$

Solid waste produced in 20 years

$$= 3{,}90{,}000 \text{ m}^3 \times 20 = 78{,}00{,}000 \text{ m}^3$$

Since the height of landfill site will be 10 m, the area required

$$= \frac{78{,}00{,}000 \text{ m}^3}{10 \text{ m}} = 7{,}80{,}000 \text{ m}^2$$

= **78 hectare. Ans.**

Q. 4. (*c*) *Define BOD. Explain how BOD is determined for wastewater. For a BOD test carried out in the laboratory, 5 mL of wastewater having no initial DO was taken in a 300 mL BOD bottle and mixed with dilution water of 9.2 mg/L DO concentration. After incubating the bottle for 5 days at 20°C, the DO of the mixture was found to be 5.0 mg/L. Compute the* BOD_5 *(at 20°C) of wastewater.* (8)

Solution : Please read article 7.4.10, and use Eq. (7.11) as :

$$BOD_5 = \text{D.O. consumed in test by the diluted sample}$$

$$\times \left[\frac{\text{Vol. of diluted sample}}{\text{Vol. of sewage in sample}}\right] \quad ...i.e. \text{ (Eq. 7.11)}$$

Here DO consumed = 9.2 – 5 = 4.2 mg/L

Vol. of diluted sample = 300 ml

Vol. of sewage in sample = 5 ml

$$\therefore \quad BOD_5 = 4.2 \frac{mg}{L} \times \frac{300}{5}$$

$$\text{or} \quad BOD_5 = \mathbf{252\ mg/L.} \quad \textbf{Ans.}$$

Q. 5. (*c*) *A family of four people generates solid waste at the rate of 0.45 kg/cap/day and the bulk density of refuse in a typical garbage can is about 120 kg/m³. If collection is once a week, how many 120 litres garbage cans will be needed for the household?* (8)

Solution : Solid waste (garbage) produced by 4 people @ 0.45 kg/c/d in 7 days.

$$= 0.45 \times (4 \times 7) = 12.6 \text{ kg}$$

Volume of 12.6 kg garbage having density = 120 kg/m³

$$= \frac{\text{Mass}}{\text{Density}} = \frac{12.6 \text{ kg}}{120 \text{ kg/m}^3} = 0.105 \text{ m}^3$$

$$= 0.105 \times 1000 \text{ litres} = 105 \text{ litres}$$

Since garbage produce in one week by the family is 105 litres, only one can of 120 litres capacity will be required. **Ans.**

Q. 6. (*c*) *Given the following data :*

Plant inflow	*= 9000 m³/day*
Recycle flow	*= 6000 m³/day*
Mixed liquid suspended solids (MLSS)	*= 4000 mg/L*
Sludge volume withdrawn	*= 580 m³/day*
Suspended solids	*= 3800 kg/day*

Solid loading rate of 150 kg/m²-day at average flow

Obtain the final settling tank size for an extended aeration plant. (8)

Solution : Pl. follow design of secondary sedimentation tank at the end of Solved Example 9.30, and compute as under :

Total suspended solids = 3800 kg/day (given)

Solids loading rate in settling tank = 150 kg/m²-day (given)

∴ Surface area of tank reqd.

$$= \frac{3800 \frac{kg}{d}}{150 \frac{kg}{m^2.d}} = 25.33 \text{ m}^2$$

Adopting a circular tank, we have

$$\text{dia. of tank reqd.} = \sqrt{\frac{25.33 \times 4}{\pi}} = 5.7 \text{ m}$$

Hence, use 5.7 m dia. settling tank.

Check for weir loading

Weir loading for a circular weir placed along the periphery of the tank of length $\pi D = \pi \times 6.7$ m = 17.9 m will be

$$= \frac{\text{Average flow} = 9000 \text{ m}^3/\text{d}}{17.9 \text{ m}} = 502.59 \text{ m}^3/\text{d/m}$$

which is more than 150.

∴ Use a trough instead of a single weir at the periphery. **Ans.**

Q. 7. (*c*) *A wastewater plant produces 1000 kg of dry solids/day at a moisture content of 95%. The solids are 65% volatile with a specific gravity of 1.05 and the specific gravity of the non-volatile portion is 2.5. Determine the sludge volume—*

(*i*) *as produced ;*

(*ii*) *after digestion reduces the volatile solids by 50% and decreases moisture content by 90% ;*

(*iii*) *after dewatering to 75% moisture ;*

(*iv*) *after drying to 10% moisture ;*

(*v*) *after incineration (only non-volatile solids remain).* (8)

Solution : Pl. follow Solved Example 9.21, and compute as under :

The solids removed in the plant as sludge per day = 1000 kg. Daily sludge produced will, thus, have 1000 kg solids, and the rest will be water. Since the moisture content of the sludge produced is 95% (given), we have 5 kg of solids will produce 100 kg of wet sludge by joining with 95 kg of water.

∴ Water contained in 5 kg of solids = 95 kg.

∴ Water contained in 1000 kg of solids $= \frac{95}{5} \times 1000$ kg

= 19000 kg.

(*i*) Hence, volume of sludge produced

$$= \left[\frac{\text{Mass of solids}}{\text{Density of solids}}\right] + \left[\frac{\text{Mass of water}}{\text{Density of water}}\right]$$

$$= \left[\frac{1000 \text{ kg} \times 65\%}{1.05 \times 1000 \text{ kg/m}^3} + \frac{1000 \text{ kg} \times 35\%}{2.5 \times 1000 \text{ kg/m}^3}\right] + \left[\frac{19000 \text{ kg}}{1000 \text{ kg/m}^3}\right]$$

$$= [0.619 + 0.14] + [19] = 19.759 \text{ m}^3.$$

Hence, the sludge volume produced = 17.759 m³. **Ans.**

(ii) Original mass of volatile solids in produced sludge

= 0.65% × 1000 kg = 650 kg.

During digestion, the mass of volatile solids is reduced by 5%.
Hence, mass of volatile solids left in digested sludge

= 650 × (1 – 0.5) = 650 × 0.5 = 325 kg.

Mass of non-volatile solids remains unchanged

= 0.35 × 1000 kg = 350 kg.

Since the moisture content of digested sludge 90%, we have 10 kg of solids will produce 100 kg of wet digested sludge by joining with 90 kg of water.

∴ Water contained in 10 kg of solids = 90 kg.

∴ Water contained in 325 + 350 = 675 kg of solids

$$= \frac{90}{10} \times 675 = \mathbf{6075\ kg.}$$

Hence, volume of digested sludge

$$= \left[\frac{\text{Mass of solids}}{\text{Density of solids}}\right] + \left[\frac{\text{Mass of water}}{\text{Density of water}}\right]$$

$$= \left[\frac{325 \text{ kg of volatile solids}}{1.05 \times 1000 \text{ kg/m}^3} + \frac{350 \text{ kg of non-volatile solids}}{2.5 \times 1000}\right] + \left[\frac{6075 \text{ kg}}{1000 \text{ kg/m}^3}\right]$$

$$= \left[0.3095 + 0.14 + 6.075 \text{ m}^3\right] = 6.525 \text{ m}^3$$

Hence, vol. of digested sluge = $\mathbf{6.525\ m^3}$. **Ans.**

(iii) *After the digesed sludge is dewatered to contain 75% moisture (as against 90%).*

Here 25 kg of solids will contain 75 kg of water.

∴ Water contained in 25 kg of solids = 75 kg.

∴ Water contained in 675 kg of solids

$$= \frac{75}{25} \times 675 = \mathbf{2025\ kg}$$

Hence, vol. of digested dewatered sludge

$$= \left[\frac{325 \text{ kg of volatile solids}}{1.05 \times 1000 \text{ kg/m}^3} + \frac{350 \text{ kg of non-volatile solids}}{2.5 \times 1000}\right] + \left[\frac{2025 \text{ kg}}{1000 \text{ kg/m}^3}\right]$$

$$= 0.3095 + 0.14 + 2.025 = 2.475 \text{ m}^3$$

Hence, vol. of dewatered digested sludge = $\mathbf{2.475\ m^3}$. **Ans.**

(*iv*) When dewatered digested sludge is dried to contain 10% moisture, we will have

Water contained in 90 kg of solids = 10 kg.

$\therefore$ Water contained in 675 kg of solids = $\frac{10}{90} \times 675 =$ **75 kg.**

Hence, volume of dried dewatered digested sludge

$$= \left[\frac{325 \text{ kg}}{1.05 \times 1000 \text{ kg/m}^3} + \frac{350 \text{ kg}}{2.5 \times 1000 \text{ kg/m}^3}\right] + \left[\frac{75 \text{ kg}}{1000 \text{ kg/m}^3}\right]$$

$= 0.3095 + 0.14 + 0.075 = 0.525 \text{ m}^3.$

Hence, vol. of dried sludge = **0.525 m³. Ans.**

(*v*) After incineration, only non-volatile solids remain and there will be no water. Hence the volume of incinerated sludge

$$= \left[\frac{350 \text{ kg of non-volatile solids}}{2.5 \times 1000 \text{ kg/m}^3}\right] = 0.14 \text{ m}^3.$$

Hence, sludge volume after incineration = **0.14 m³. Ans.**

Q. 7. (*e*) *Traffic noise data are given in the table below. Compute L_{eq}.*

Time (in s)	*10*	*20*	*30*	*40*	*50*	*60*	*70*	*80*	*90*	*100*
Noise (dBA) L(t)	*71*	*75*	*70*	*78*	*80*	*84*	*76*	*74*	*75*	*74*

(8)

Hint: Pl. refer Solved Exampe 20.1.

Year 2015 (*Engineering Services*)

Q. 2. (*c*) *A town on the bank of river Ganga discharges 18000 m³ day⁻¹ of treated wastewater into the river. The treated wastewater has a BOD_5 of 20 mg L^{-1} and a BOD decay constant of 0.12 day^{-1} at 20°C. The river has a flow rate of 0.43 m^3 sec^{-1} and an ultimate BOD of 5.0 mg L^{-1}. The DO of the river is 6.0 mg L^{-1} and the DO of the wastewater is 0.4 mg L^{-1}. Compute the DO and initial ultimate BOD in the river, immediately after mixing.* (8)

Solution : *For sewage wastewater (treated),* we have

$$Y_5 = 20 \text{ mg/L}$$
$$K_{(20°)} = 0.12/\text{d}$$
$$\text{DO} = 0.4 \text{ mg/L}$$

$$\text{Discharge } Q = Q_s = 18000 \text{ m}^3/\text{d} = \frac{18000}{24 \times 3600} \text{ m}^3/\text{s}$$

$$= 0.208 \text{ m}^3/\text{s}.$$

Ultimate BOD = $Y_u = L$ is given by eq. (7.16) as :

$$Y_t = L\left[1 - (10)^{-K_D.t}\right]$$

or $$Y_5 = L\left[1 - (10)^{-K_D.5}\right]$$

$\therefore$ $$Y_5 = 20 \text{ mg/L} = L\left[1 - (10)^{-0.12\times 5}\right]$$

or $$20 \text{ mg/L} = L\left[1 - \frac{1}{(10)^{0.6}}\right]$$

or $$L = 26.71 \text{ mg/L}$$

$\therefore$ *1st stage or initial ultimate BOD of sewage*

$$= L = Y_u = 26.71 \text{ mg/L}$$

For River:

$$Y_u = \text{ultimate BOD} = 5 \text{ mg/L}$$
$$\text{Discharge} = Q_R = 0.43 \text{ m}^3\text{/s}$$
$$\text{DO} = 6 \text{ mg/L}$$

For calculation of DO of mix in river, we have from Eq. (8.1)

$$C = \text{DO of mix} = \frac{C_S . Q_S + C_R\, Q_R}{Q_S + Q_R}$$

$$= \frac{\text{DO of sewage } \times Q_S + \text{DO of river} \times Q_R}{Q_S + Q_R}$$

$$= \frac{0.4 \text{ mg/L} \times 0.208 \text{ m}^3\text{/s} + 6 \text{ mg/L} \times 0.43 \text{ m}^3\text{/s}}{0.208 \text{ m}^3\text{/s} + 0.43 \text{ m}^3\text{/s}}$$

$\therefore$ $$\text{DO of mix in river} = \frac{2.663}{0.638} = 4.17 \text{ mg/L}$$

Hence DO in river immediately after mixing

$$= \mathbf{4.17\ mg/L.}\quad \textbf{Ans.}$$

For calculation of ultimate BOD (1st stage) of mix in river, we have from Eq. (8.1) :

$$C = \text{Ultimate BOD of mix in river}$$

$$= \frac{C_S . Q_S + C_R . Q_R}{Q_S + Q_R}$$

$$= \frac{\text{Ultimate BOD of sewage } \times Q_S + \text{Ultimate BOD of river } \times Q_R}{Q_S + Q_R}$$

$$= \frac{26.71 \text{ mg/L} \times 0.208 \text{ m}^3\text{/s} + 5 \text{ mg/L} \times 0.43 \text{ m}^3\text{/s}}{0.208 \text{ m}^3\text{/s} + 0.43 \text{ m}^3\text{/s}}$$

$$= \frac{77.057}{0.638} = 12.08 \text{ mg/L}$$

Hence, the ultimate BOD of river immediately after mixing

$$= \mathbf{12.08 \text{ mg/L.}} \quad \textbf{Ans.}$$

Q. 5. (c) *What is grit? Why should grit be removed from wastewater? What is the basic principle behind the design of girt chambers ? What is the reason to have constant velocity of flow in grit chamber (conventional horizontal flow) and how is it achieved ?* (8)

Q. 6. (c) *A wastewater treatment plant operating primary settling tank gets an inflow of 12.96 million litres per day with influent suspended solids concentration of 280.0 mg/L. The primary settling tank has a removal efficiency of 59% with sludge concentration of 5%. If the volatile solids concentration in settled sludge is 60% and the specific gravity of volatile solids is 0.990; and the fixed solids concentration is 40% and the specific gravity of fixed solids is 2.65, determine the daily sludge production.* (8)

Solution : Inflow into the settling tank = 12.96 ML/d

Suspended solids in influent = 280 mg/L

∴ Mass of suspended solids (SS) in 12.96 ML of sewage flowing per day into the settling tank

$$= \frac{280 \times (12.96 \times 10^6)}{10^6} \text{ kg} = 3628.8 \text{ kg/d}$$

The S.S. removed in primary settling tank = 59%

∴ Mass of solids removed in the primary settling tank

$$= 59\% \times 3628.8 \text{ kg} = 2141 \text{ kg (say)}$$

Since the sludge collected at the bottom of the primary settling tank has a sludge concentration of 5% (given), it has m.c. of 95%

∴ 2141 kg of dry solids will make

$$= \frac{100}{5} \times 2141 \text{ kg of wet sludge per day}$$

= **42,820 kg** of wet sludge per day.

The wet sludge so obtained, contains 60% of volatile solids having *sp* gravity = 0.99 and 40% of fixed solids having specific gravity = 2.65 ; and, therefore, we have the specific gravity of solids in the resulting sludge

$$= \frac{60\% \times 0.99 + 40\% \times 2.65}{100\%} = \mathbf{1.654}$$

Water contained in 5 kg of solids = 95 kg

∴ Water contained in 2141 kg of solids $= \frac{95}{5} \times 2141$ kg

$= 40{,}679$ kg

$\therefore$ Volume of sludge produced daily

$$= \frac{\text{Mass of solids}}{\text{Density of solids}} + \frac{\text{Mass of water}}{\text{Density of water}}$$

$$= \left[\frac{2141 \text{ kg}}{1.654 \times 1000 \text{ kg/m}^3} + \frac{40679 \text{ kg}}{1000 \text{ kg/m}^3}\right]$$

$$\left[\because \begin{array}{l} \text{Density of solids} \\ = \text{Sp. gr. of solids} \\ \times \text{Density of water} \\ = 1.654 \times 1000 \text{ kg/m}^3 \end{array}\right]$$

$$= \left[\frac{2141}{1.654 \times 1000} + \frac{40679}{1000}\right] \text{m}^3 = 41.97 \text{ m}^3.$$

Hence, the daily sludge production

= 41.97 m^3 or 42820 kg. Ans.

Q. 7. *(c) A town of population of 50,000 generates municipal solid waste of 500 g per person per day. How many trucks would be needed to collect the waste twice weekly? The trucks have a capacity of 4.5 t (metric tonnes) each and operate 5 days per week. Assume that the trucks average two loads per day at 75% capacity. If 20% of the waste generated is recycled, determine the weight of MSW that enters the landfill. If the density of the waste is 280 kg/m^3, what is the volume of MSW?* (8)

Solution : Population of town = 50,000

MSW per capita = 500 g/person

Total daily MSW (Municipal solid waste) generated

$$= \frac{50{,}000 \times 500}{1000} \text{ kg} = 25000 \text{ kg/d} = 25 \text{ Tonne (t)/d}$$

MSW generated per week = 25 × 7 = 175 t/week

MSW carrying capacity of each truck = 75% × 4.5 t/Trip

= 3.375 t

Since each truck takes 2 rounds each day, we have the daily carrying capacity of each truck = 2 × 3.375 t = 6.75 t/d

Since each truck operate for 5 days a week, we have the carrying capacity of each truck per week = 5 × 6.75 t = 33.75 t

$$\therefore \text{ No. of trucks required} = \frac{175 \text{ t}}{33.75 \text{ t}} = 5.185; \textbf{ say 6 Nos. Ans.}$$

Since 20% of MSW is recycled, only 80% of generated waste will reach the landfill site. Hence, MSW reaching the landfill site = 80% × 175 t/week = 140 t. **Ans.**

Volume of MSW reaching landfill

$$= \frac{\text{Mass of MSW}}{\text{Density of MSW}} = \frac{140\,\text{t}}{280\,\text{kg/m}^3}$$

$$= \frac{140 \times 1000\ \text{kg}}{280\,\text{kg/m}^3} = \mathbf{500\ m^3.}\quad \textbf{Ans.}$$

Year 2016 (*Engineering Services*)

Q. 1. (*d*) *Design a circular primary settling tank for a sewage treatment plant. Given data are:*

Flow rate = 27000 m^3/d, *Retention time = 2.5 hr,*
Weir length = 30 m, *Total number of tanks = 2,*
Surface loading rate = 30 $m^3/m^2/d$. (4)

Solution : Pl. follow Solved Example 9.6, and compute as under:

Total quantity of sewage to be treated per day = 27000 m³/d.

∴ Qty. of sewage to be treated per 2.5 hr (retention time)

$$= 27000 \times \frac{2.5}{24}\ \text{m}^3 = 2812.5\ \text{m}^3/\text{d}$$

Since 2 No. Tanks are to be used, capacity of each settling tank

$$= \frac{2812.5\ \text{m}^3}{2} = 1406.25\ \text{m}^3/\text{d}$$

$$\text{Now surface loading} = \frac{Q}{\text{Surface area of tank}} = \frac{Q}{\frac{\pi}{4}d^2}$$

where d = Dia. of settling tank

$$\therefore\quad 30\ \text{m}^3/\text{m}^2/\text{d} = \frac{27000\ \text{m}^3/\text{d}}{\frac{\pi}{4}\times d^2} \qquad \therefore\ d = \sqrt{\frac{27000}{30\times\frac{\pi}{4}}}$$

$$= 33.85 \text{ say } 34 \text{ m.}$$

Net Effective depth of the tank

$$= \frac{\text{Capacity}}{\text{Area of cross-section}} = \frac{1406.25\ \text{m}^3}{\frac{\pi}{4}\times(34)^2} = 1.55\ \text{m}$$

Hence, use two settling tanks, each of 34 m dia and 1.55 m water depth (with extra free board depth of 0. 3 m.) **Ans.**

Q. 1. (*e*) *A 500 mm diameter sewer pipe has been installed with an invert slope of 0.008. Determine the capacity of flow when the sewer pipe is flowing half full. The number of sewer pipes used is 2. Meaning's roughness coefficient is 0.012. Assume that the flow in the sewer pipe is uniform.* (4)

Solution : Pl. refer Solved Example 4.8, and compute as under:

Dia. of sewer = D = 500 mm = 50 cm = 0.5 m

Invert slope = S = 0.008

Manning's roughness coefficient = n = 0.012

Using Manning's formula, we have

$$Q = \frac{1}{n} . A . R^{2/3} . \sqrt{S}$$

or

$$Q = \frac{1}{n} . a . r^{2/3} . \sqrt{S}$$

Here $n = 0.012$

a = area of section in flow (running half full)

$$= \frac{\pi D^2}{8} = \frac{\pi \times (0.5)^2}{8} = 0.0982 \text{ m}^2$$

$$p = \frac{\pi D}{2}, \; r = \frac{a}{p} = \frac{D}{4} = \frac{0.5}{4} \text{ m} = 0.125 \text{ m}$$

$$\therefore \quad Q = \frac{1}{0.012} \times (0.0982)\,(0.125)^{2/3}\,\sqrt{0.008} = 0.183 \text{ m}^3/\text{s}$$

Discharge through 2 sewer pipes (as 2 pipes are used)

$= 2 \times 0.183 = 0.366 \text{ m}^3/\text{s}$ = 366 litres/second

Hence, capacity of installed sewer = **366 *l*/s. Ans.**

Q. 1. (*j*) *Describe briefly the environmental factors considered in designing a land fill.* (4)

Q. 2. (*e*) *Define air pollution as per OECD definition. Differentiate between 'natural' and 'anthropogenic' air pollution sources, with examples. What are the effects of ambient ozone,* PM_{10}, $PM_{2.5}$, *and lead on human health?* (4)

Hints : Pl. see article 18.2 and Table 18.2 for the second part of the question.

As for as definition of air pollution as per OECD is concerned, let us clarify that OECD stands for ***Organisation for Economic Co-operation and Development***, and consists of 35 countries, including EU countries, USA, UK, Canada, Japan, Australia, New Zealand, etc. *As per OECD, air pollution is defined as the presence of contaminants or pollutant substances in the air that do not disperse properly and that interfere with human health or welfare or produce harmful environmental effects.*

Q. 3. (*b*) *Discuss in brief the essentials of rural sanitation. Explain how you dispose off (i) dry refuse, (ii) sullage, and (iii) excretal waste in rural area.* (8)

Q. 4. (*c*) (*i*) *Differentiate between BOD and COD in wastewater.*

(ii) 20 mL of sewage sample was added directly into a 300 mL BOD incubation bottle. The initial DO of the diluted sample was 7.8 mg/L and the final DO after 5d was 1.3 mg/L. The corresponding initial and final DO of the seeded dilution water were 8.1 and 6.9 respectively. What is the BOD_5 of the sewage sample. (8)

Solution : It is clear that the dilution water added to the sewage sample was not distilled water, as is usually used in BOD test bottle testing. Here 20 ml of sewage sample (having BOD = L_0 mg/L say) was mixed with 280 ml of seeded water having BOD_5 = 8.1 – 6.9 = 1.2 mg/L (*i.e.* DO consumed in 5 days). The BOD_5 of the mixture of 300 ml was 7.8 – 1.3 = 6.5 mg/L (*i.e.* DO consumed in 5 days). The value of L_0 can now be computed by Eq. (7.12) as:

BOD of diluted sample (mixture) × Vol. of diluted sample (mixture)

= [BOD of sewage sample × Vol. of sewage sample + BOD of dilution (seeded) water × Vol. of dilution (seeded) water] ...(Eq. 7.12)

or $$6.5 \text{ mg/l} \times 300 \text{ ml} = L_0 \times 20 \text{ ml} + 1.2 \text{ mg/L} \times 280 \text{ ml}$$

$$\therefore \quad L_0 = \frac{300 \times 6.5 - 280 \times 1.2}{20} \text{ mg/L} = \mathbf{80.7 \text{ mg/L}}. \quad \textbf{Ans.}$$

Q. 5. (*d*) *Describe the method of sampling suspended particulate matter by high volume samples. Also explain how the mass concentration of particulates is determined.* (8)

Year 2017 (*Engineering Services*)

Q. 1. (*d*) *A wastewater sample was diluted 10% with tap water for conducting BOD Test. The DO content in wastewater sample was 1.2 mg/l and that of tap water was 6 mg/l. After 5 days incubation at 20°C, the DO of sample was 0.5 mg/l. Calculate*

(i) 5 day BOD of wastewater and

(ii) ultimate BOD. Assume K_{20} = 0.12/day. (12)

Solution : Pl. follow similar Solved Example 7.14 and compute as under:

10% solution means that:

100% content of diluted sewage sample consists of 10% wastewater and 90% of tap water.

Hence it DO = DO of waste water × Its content + DO of dilution water × Its content

$$= 1.2 \text{ mg/l} \times \frac{10}{100} + 6 \text{ mg/l} \times \frac{90}{100} = 5.52 \text{ mg/l}$$

DO of incubated sample after 5 days = 0.5 mg/l.

∴ DO consumed in oxidising organic matter of sewage

$$= 5.52 - 0.5 = 5.02 \text{ mg/l}$$

$\therefore$ BOD of 5 days = DO consumed × Dilution factor

$$= 5.02 \text{ mg/l} \left[\text{Dilution factor} = \frac{100}{\text{Percentage of solution}}\right]$$

$$= 5.02 \text{ mg/l} \times \frac{100}{10} = 50.02 \text{ mg/l} \quad \textbf{Ans.}$$

Ultimate BOD is given by L, in eq. (7.16) as:

$$Y_t = L\left[1 - (10)^{-K_D.t}\right] \text{ where}$$

or $Y_5 = L\left[1 - (-10)^{-K_D.5}\right]$ where K_D at test temp *i.e.*, $K_{D_{20}}$
= 0.12/day (given)

or $$50.02 \text{ mg/l} = L\left[1 - (10)^{-0.12 \times 5}\right] = L\left[1 - \frac{1}{(10)^{0.60}}\right]$$

or $$50.62 \text{ mg/l} = L\left[1 - \frac{1}{3.98}\right] = L\left[1 - 0.251\right] = 0.749 \text{ L}$$

or $$L = 66.8 \text{ mg/l.} \quad \textbf{Ans.}$$

Q. 1. (*e*) *With the help of data given below, estimate the per capita solid waste generated per day in a colony with 20,000 houses.*

Volume of waste / truck load	= *20* m^3
Volume of waste / tractor load	= *2* m^3
Volume of waste/small truck	= *8* m^3
No. of truck loads	= *2/day*
No. of tractor loads	= *1/day*
No. of small truck loads	= *3/day*
Avg. no. of people/house	= *4*
Specific weight of waste in (kg/m^3*)*	= *500 for truck;* *225 for small truck;* *50 for tractor.*

(12)

Solution : No. of houses in the colony = 20,000

Average no. of people/house = 4

$\therefore$ No. of people in the colony = 20,000 × 4 = 80,000

Per day production of solid waste in m³

= 2 Trucks × 20 m^3 + 3 small trucks × 8 m^3 + 1 Tractor × 2 m^3

= 40 + 24 + 2 = 66 m^3.

Per day production of solid waste in kg

= 500 kg/m^3 × 40 m^3 (in Trucks)
+ 225 kg/m^3 × 24 m^3 for small trucks
+ 50 kg/m^3 × 2 m^3 for Tractors

= 20,000 + 5,400 + 100 = 25,500 kg.

Per capita production of solid waste = $\dfrac{25{,}500 \text{ kg}}{80{,}000 \text{ persons}}$

= **0.319 kg/capita Ans.**

Q. 3. (*c*) *Present a flow chart indicating various elements in municipal solid waste management and explain salient aspects of each element.* (20)

Q. 4. (*i*) *Design a high rate trickling filter for treating wastewater with BOD of 160 mg/l. The rate of wastewater generated is 5 MLD. The effluent BOD required is 25 mg/l. Adopt recirculation ratio of 1 : 5.* (10)

Solution. Pl follow similar Solved Example 9.10 and compute as under.

Qty. of sewage flowing into the filter per day = 5 MLD
= 5 Ml/day.

BOD concentration in sewage = 160 mg/l.

∴ Total BOD entering the filter unit per day
= 5 Ml × 160 mg/l = 800 kg ...(*i*)

BOD concentration desired in the effluent = 25 mg/l

∴ Total BOD left in the effluent per day
= 5 Ml × 25 mg/l = 100 kg ...(*ii*)

∴ BOD removed by the filter = (*i*) – (*ii*) = 800 – 100 = **700 kg**

∴ Efficiency of the filter = $\dfrac{\text{BOD removed}}{\text{Total BOD}} \times 100$

$$\eta = \frac{700}{800} = 87.5\%$$

Now, using Eq. (9.34), we have:

$$\eta\ (\%) = \frac{100}{1 + 0.0044\sqrt{\dfrac{Y}{V.F}}}$$

where η = 87.5%

Y = Total BOD applied to filter in kg/day
= 800

$$F = \frac{1 + \dfrac{R}{I}}{\left(1 + 0.1\dfrac{R}{I}\right)^2}\ ; \ (i.e.\ \text{Eq. } 9.33)$$

where $\dfrac{R}{I}$ = 1.5 (given)

$$\because \quad F = \frac{1 + 1.5}{[1 + 0.1 \times 1.5]^2} = 1.89$$

V = Vol. of filter in ha. m = ?

$$\therefore \qquad 87.5 = \frac{100}{1 + 0.0044\sqrt{\frac{800}{V \times 1.89}}}$$

$$\text{or } 1 + 0.0044\sqrt{\frac{423.3}{V}} = \frac{100}{87.5} = 1.43$$

$$\text{or} \qquad \sqrt{\frac{423.3}{V}} = \frac{0.143}{0.0044} = 32.47$$

$$\text{or} \qquad \frac{423.3}{V} = 1054.14$$

or $\qquad V = 0.402$ ha-m $= 4020$ m^3.

Assuming depth of filter as 1.5 m, we have the surface area required

$$= \frac{4020 \text{ m}^3}{1.5 \text{ m}} = 2680 \text{ m}^2$$

Dia. of circular filter required $= \sqrt{2680 \times \frac{4}{\pi}} = 58.4$ m; say 59 m

Hence use a high rate trickling filter with 59 m Dia, 1.5 m deep filter media, and with recirculation (single stage) ratio of 1.5. **Ans.**

Q. 4. (*c*) (*ii*) *Present classification of natural ecosystems and describe the components of ecosystem.* (10)

Year 2018 (*Engineering Services*)

Q. 1. (*d*) (*ii*) *A treatment plant disposes 50 MLD of treated effluent into a river. The river flow rate is 20 m*^{3}*/s and its DO is 8 mg/L before the mixing point. If the BOD of the effluent is 50 mg/L, find the BOD and DO of the river water at the d/s of mixing point. Assume BOD of river water as 0 and DO in WW effluent as 0.* (6)

Solution : Please follow similar Solved Example 8.1, and compute as under for (*i*) BOD, and (*ii*) for DO, separately.

Sewage discharge $= Q_S = 50$ MLD $= \frac{50 \times 10^6}{10^3 \times 24 \times 60 \times 60} = 0.579$ m^3/s.

River discharge $= Q_R = 20$ m^3/s

Now, using concentration values (C) for BOD, we have Eq. 8.1. as C = Concentration of BOD at mixing point, say

$$C_{\text{BOD}} = \frac{C_S Q_S + C_R Q_R}{Q_S + Q_R}$$

where C_S = BOD of sewage effluent = 50 mg/L

C_R = BOD of river = 0.

$$\therefore C_{BOD} = \frac{50 \text{ mg/L} \times 0.579 \text{ m}^3/\text{s} + 0 \times 20 \text{ m}^3/\text{s}}{(0.579 + 20) \text{ m}^3/\text{s}}$$

or C_{BOD} = 1.407 mg/L **Ans.**

Now use C for DO, as C_{DO}

C_{DO} = Concentration of DO at mixing point

$$= \frac{C_S Q_S + C_R Q_R}{Q_S + Q_R}$$

where C_S = DO of effluent = 0

C_R = DO of river = 8 mg/L

$$\therefore \quad C_{DO} = \frac{0 \times 0.579 \text{ m}^3/\text{s} + 8 \text{ mg/L} \times 20 \text{ m}^3/\text{s}}{(0.579 + 20) \text{ m}^3/\text{s}}$$

$$= \frac{160 \text{ mg/L}}{20.579} = 7.79 \text{ mg/L} \quad \textbf{Ans.}$$

Q. 1. (*e*) *What is sludge and how is it classified? With the help of flow chart, explain different processes/operations involved in sludge treatment and indicate the objectives of each process/operation.* (12)

Q. 3. (*c*) *Explain the phenomena responsible for self purification in rivers. Discuss the factors influencing the self purification capacity.* (20)

[**Hint:** Pl. refer article 8.3]

Q. 4. (*c*) (*i*) *Explain in detail how landfills are environmentally safe over open dumps. Explaining landfill operations, state the problems associated with landfills.* (10)

Q. 4. (*c*) (*ii*) *With the help of line sketches, explain the working principles of gravity settlers and electrostatic precipitators used for air pollution control. Also, list their applications and limitations.* (10)

Year 2019 (*Engineering Services*)

Q. 2. (*c*) (*ii*) *Compute the average sound pressure level from the following sound pressure readings:*

(1) 39 dBA (2) 52 dBA (3) 67 dBA (4) 77 dBA

[**Solution.** From article 20.8.2, we have

Average sound pressure level of N readings

$$= \overline{L_p} = 20 \log_{10} \frac{1}{N} \cdot \sum_{n=1}^{n=N} (10)^{L_n/20} \qquad \text{...(Eq. 20.17)}$$

where N = No. of measurement readings

n = 1, 2, 3, ... N

L_n = n^{th} sound pressure level in dB (re . 20 μ Pa)

Substituting the given four readings, we have

$$\overline{L_p} = 20 \log_{10} \frac{1}{N} \cdot \sum_{n=1}^{n=N} (10)^{L_n/20}; \text{ where}$$

$$\sum_{n=1}^{n=N} (10)^{L_n/20} = \left[(10)^{39/20} + (10)^{52/20} + (10)^{67/20} + (10)^{77/20}\right]$$

$$= \left[(10)^{1.95} + (10)^{2.6} + (10)^{3.35} + (10)^{3.85}\right]$$

$$= [89.13 + 398.11 + 2238.72 + 7079.46] = 9805.42$$

$$\therefore \qquad \overline{L_p} = 20 \log_{10} \frac{1}{4} \times 9805.42$$

or $\qquad \overline{L_p} = \mathbf{67.79\ dBA}$ **Ans.**

Q. 3. (*b*) *A wastewater treatment plant consists of primary treatment clarifier followed by an activated sludge treatment unit. The primary and secondary sludge are mixed, thickened in a gravity thickener and sent to further treatment. Wastewater, treatment plant and sludge characteristics are as follows :*

(*a*) *Influent S.S. = 220 mg / L; Primary clarifier diameter = 25 m*

(*b*) *Influent BOD = 250 mg / L; Aerator volume = 3000 m*^{3}

(*c*) *Effluent BOD = 30 mg / L; MLSS in aerator = 3000 mg / L*

(*d*) *Flow = 20000 m*^{3}*/day; Solids in thickener supernatant = negligible*

(*e*) *Primary sludge = 5% solids; secondary sludge = 0.75% solids and thickened sludge = 4% solids*

(*f*) *Efficiency of primary clarifier for S.S. and BOD removal are 58% and 32% respectively.*

(*g*) *Biomass conversion factor in aerator = 0.35.*

Determine—

(*i*) *solids loading in kg / day to the sludge disposal facilities.*

(*ii*) *F / M ratio in aerator*

(*iii*) *percent volume reduction by the thickener.* (20)

[Solution. (*i*) *S.S.* in influent to primary clarifier = 220 mg/L. 58% of these *S.S.* are removed in primary clarifier, while remaining 42% are removed in secondary one (*i.e.*, in aerator).

(*a*) The *S.S.* produced in primary sludge per day

= 220 mg/L × 58% × 20000 m^3/d

= 220 (gm/m^3) × 0.58 × 20000 m^3

= 25,52,000 gm = 2552 kg ...(*i*)

(*b*) The *S.S.* produced in secondary sludge per day

= 220 mg/L (*i.e.*, gm/m^3) × 42% × 20000 m^3/d

= 1848 kg ...(*ii*)

Total solids in primary as well as secondary sludge = (*i*) + (*ii*)

= 2552 + 1848 = **4400 kg**

Since sludge disposal facility receives the primary as well as secondary sludge, the solids loading on sludge disposal facility = **4400 kg/day** **Ans.**

(*ii*) *F*/*M* *Ratio in Aerator*

From Eq. (9.44), we have

$$\frac{F}{M} = \frac{QY_0}{V.X_t} \quad \text{...(Eq. 9.44)}$$

where Q = 20,000 m³/d

Y_0 = BOD applied to aeration system = 250 mg/L

V = Aeration tank volume = 3000 m³

X_t = MLSS = 3000 mg/L

Now, F = *Mass of BOD applied to aeration system*

$$= QY_0 = 20000 \frac{m^3}{d} \times 250 \frac{mg}{L} \text{ (i.e., gm/m}^3\text{)}$$

$$= \frac{20,000 \times 250}{1000} \text{ kg/d} = 5000 \text{ kg/d}$$

$$M = V.X_t = 3000 \text{ m}^3 \times 3000 \text{ mg/L (i.e., gm/m}^3\text{)}$$

$$= \frac{3000 \times 3000}{1000} \text{ kg} = 9000 \text{ kg.}$$

$$\therefore \quad \frac{F}{M} = \frac{5000 \text{ kg/d}}{9000 \text{ kg}} = \mathbf{0.56\ kg} \text{ BOD per day per kg MLSS}$$ **Ans.**

(*iii*) Quantity of solids removed in primary sludge = 2552 kg/d

(*a*) The primary sludge has 5% solids content,

$$\therefore \text{ Wt. of wet primary sludge} = \frac{100 \text{ kg}}{5 \text{ kg}} \times 2552 \text{ kg} = \mathbf{51,040\ kg} \quad \text{...(i)}$$

[∵ 5 kg solids make 100 kg wet sludge, 2552 kg solids will make $= \frac{100 \text{ kg}}{5 \text{ kg}} \times 2552$ kg of wet sludge]

(*b*) S.S. removed in the secondary sludge = 1848 kg

Since the secondary sludge has 0.75% solids contents, the wt. of wet

$$\text{secondary sludge} = \frac{100 \text{ kg}}{0.75 \text{ kg}} \times 1848 \text{ kg} = 2,46,400 \text{ kg} \quad \text{...(ii)}$$

Total mass of primary sludge + secondary sludge = (*i*) + (*ii*)

= 51,040 + 2,46,400 = 2,97,440 kg

$$\text{Vol. of primary and secondary sludge} = \frac{\text{Mass of sludge}}{\text{Density of sludge*}}$$

$$= \frac{2,97,440 \text{ kg}}{1020 \text{ kg/m}^3 *} = \mathbf{2,91.6\ m^3}$$

---

*Assuming specific gravity of unthickened sludge and thickened sludge to be approximately same, as say 1.02.

∴ Density of sludge = sp. gr. × desnity of water = 1.02 × 1000 kg/m³ = 1020 kg/m³

The thickened sludge will have total solids = 4400 kg/day

Since thickened sludge has 4% solids content, the wt. of thickened sludge

$$= \frac{100 \text{ kg}}{4 \text{ kg}} \times 4400 \text{ kg} = 1,10,000 \text{ kg}$$

$$\therefore \text{ Vol. of thickened sludge} = \frac{\text{Mass of slude}}{\text{Density of sludge*}}$$

$$= \frac{1,10,000 \text{ kg}}{1020 \text{ kg/m}^3*} = \mathbf{107.8\ m^3}$$

$$\%\text{age reduction in sludge volume} = \frac{291.6 - 107.8}{291.6} \times 100\% = \mathbf{63\%.} \qquad \textbf{Ans.}$$

Q.4. (*c*) (*i*) *Explain in detail the various process parameters required to control the aerobic composting of solid waste. Discuss the relevance of each parameter also.* *(10)*

Q.4. (*c*) (*ii*) *Discuss the isokinetic sampling process of flue gas stack sampling and explain with the help of diagram, how the results will be affected if the sampling is not done isokinetically.* *(10)*

Year 2020 (*Engineering Services*)

Q. 1.(*a*) (*ii*) *Compute the theoretical oxygen demand of 108.75 mg/l of glucose.* (4)

Solution. Pl. Refer Solved Example 7.1, and compute as under :

The equation of total oxidation of glucose is given as :

$$\underset{180}{C_6\,H_6} + \underset{192}{6O_2} \longrightarrow \underset{264}{6CO_2} + \underset{108}{6H_2O}$$

From the above equation, the theoretical oxygen demand of glucose is found to be $\frac{192}{180} = 1.07$ mg of O_2/mg of glucose.

$\therefore$ For 108.75 mg/L of glucose, the oxygen demand = 1.07 × 108.75 mg/L

= 116.36 mg/L **Ans.**

Q.1. (*e*) (*i*) A town with a population of 3 lakh produces solid waste at a rate of 2.5 kg/capita/day. If the waste is compacted to a density of 1500 kg/m^3, how much volume of land fill site is needed in a year? Assuming that the ratio of solid waste to cover is 4 : 1, what volume of cover soil is needed in a year? What type of soil would you recommend as a cover?

Solution. Population = 3,00,000

Solid waste production = 2.5 kg/c/d

Mass of solid waste produced per year

= (3,00,000 × 2.5 × 365) kg

= (273.75 × 10^6) kg

$$\text{Volume of solid waste} = \frac{\text{Mass of solid waste in kg}}{\text{Density in kg/m}^3}$$

$\therefore$ Volume of solid waste $= \dfrac{273.75 \times 10^6}{1500} \text{ m}^3$

$= 0.1825 \times 10^6 \text{ m}^3 =$ **1,82,500 m³** **Ans.**

Volume of cover soil $= \dfrac{1}{4} \times 0.1825 \times 10^6 \text{ m}^3$

= **45,625 m³** **Ans.**

Type of Soil to be used for cover soil. Good earth to be used for cover should be approximately 50% of clay-silt and 50% of sand. Such a *sandy-loam soil mixture* is porous, compacts well, and is not subject to cracking upon drying. If clay alone is used in cover, then it will crack on drying, and will give access to rodents (rats etc.) and insects **Ans.**

Q.1. (*c*) (*iii*) *The sound power from a voice shouting is 0.002 W. What is the Sound Power Level? What are the Sound Intensity, Sound Intensity Level, the Sound Pressure and, the Sound Pressure Level at a distance of 10 metres from the source? Assume that sound radiates from the source in all directions.* (*8*)

[Hint. Pl. refer Solved Example 20.1]

Q.3. (*a*) (*i*) In a factory, coal is burnt at a rate of 1 kg/second. Analysis of the coal reveals a sulphur content of 3 precent. The sulphur in the ash is 5 percent of the input sulphur. What is the annual rate of emission of sulphur dioxide? (10)

Coal burnt per second = 1 kg/sec

Sulphur content of coal = 3%

Sulphur produced per hour = 1 kg × 3/100 × 3600

= 108 kg/hr

Now $\underset{32}{S} + \underset{32}{O_2} \longrightarrow SO_2$

The molecular mass of both S and O_2 is 32 each, and they combine in one to one mass basis.

$\therefore$ 108 kg of S + 108 kg of O_2 = 216 kg of SO_2 per hour

Emission rate of SO_2 in kg/hr = 216 kg/hr

$\therefore$ Emission rate of SO_2 in a year

= (216 × 24 × 365) kg/year

= 1892.16 × 10³ kg

= 1892.16 tonnes **Ans.**

Q.3. (*a*) (*ii*) Describe various functional elements of a solid waste management system. (10)

Q.3. (*c*) A municipality has directed to upgrade its primary wastewater treatment unit to a secondary unit that can meet an effluent standard of 20 mg/*l* BOD_5 and 20 mg/l total suspended solids. They have selected a completely mixed activated sludge system. BOD_5 of total suspended solids is 63% of TSS concentration. Estimate the required volume of aeration tank. The following data is available from existing primary plant :

Flow = 0.150 m^3/s, BOD_5 = 80 mg/l

Assume the following values for half velocity constant = 95 mg/l of BOD_5; Maximum growth rate constant = 2.5/day; Decay rate of micro-organism = 0.050/day; Yield coefficient = 0.50 mg VSS per mg BOD_5 removed; MLVSS = 2000 mg/l.

[**Hint.** Pl. refer to Solved Example 9.31.]

Q. 4. (*b*) (*i*) *Explain the factors that cause sludge bulking in activated sludge process for wastewater treatment.* (10)

Q. 4. (*b*) (*ii*) *differentiate and compare anaerobic digestion process and composting process used for solid waste treatment.* (10)

Year 2021 (*Engineering Services*)

Q. 1.(*d*) *Define 'noise' and explain as to why and how it should be regarded as an environmental pollutant. Also explain briefly the major factors and actions that may help in noise abatement in a modern society.* (12)

[**Hint :** Pl. study article 20.1 & 20.12]

Q.1(*e*) A wastewater treatment plant discharges 1.0 m^3/sec of effluent having an ultimate BOD of 45.0 mg/L into a river flowing at 10.0 m^3/sec. Just upstream from the discharge point, the river water has an ultimate BOD of 4.0 mg/L. The deoxygenation constant (k_d) is 0.22/day. Answer the following:

(*i*) Assuming complete and instantaneous mixing, what would be the ultimate BOD of the mixture of waste and river just downstream from the outfall?

(*ii*) Assuming a constant cross-sectional area for the river equal to 60 m^2, what ultimate BOD would you expect to find at a point 10.0 km downstream? (12)

[**Solution.** (*i*) Sewage discharge $= Q_S = 1.0\ m^3/s$

Discharge of river $= Q_R = 10\ m^3/s$

Ultimate BOD of sewage $= C_S = 45$ mg/L

Ultimate BOD of river $= C_R = 4$ mg/L

From Eq. (8.1), we have

$$C = \text{Ultimate BOD of diluted mixture} = \frac{C_S Q_S + C_R Q_R}{Q_S + Q_R}$$

$$= \frac{45 \text{ mg/L} \times 1 \text{ m}^3/\text{s} + 4 \text{ mg/L} \times 10 \text{ m}^3/\text{s}}{1 \text{ m}^3/\text{s} + 10 \text{ m}^3/\text{s}}$$

$$= \frac{(45 \times 1 + 4 \times 10)}{11} \text{ mg/L} = \frac{85}{11} = 7.727 \text{ mg/L} \quad \textbf{Ans.}$$

(*ii*) Here, we have to determine ultimate BOD of the river at 10 km downstream. The discharge Q in the stream = 10 m^3/s + 1 m^3/s = 11 m^3/s

Area of stream = A = 60 m^2

Velocity of flow in stream $= \frac{Q}{A} = \frac{11 \text{ m}^3/\text{s}}{60 \text{ m}^2} = 0.183$ m/s

Time taken (t) by water to travel 10 km is given as

$$t = \frac{S}{V} = \frac{\text{Distance}}{\text{Velocity}} = \frac{10 \times 1000 \text{ m}}{0.183 \text{ m/sec}}$$

$$= 54{,}645 \text{ sec} = \mathbf{0.632 \text{ days}}$$

Now, using Eq. (7.15), we have

$$L_t = L\,(10)^{-K_D t} \quad \text{...(Eq. 7.15)}$$

where L = Organic matter or Ultimate BOD at start of BOD reaction

L_t = Organic matter or Ultimate BOD left after t days.

Here, $L = 7.727$ mg/L

$t = 0.632\ d$

$K_D = 0.22/d$ (given)

$\therefore$ $L_t = 7.727[10]^{-0.22 \times 0.632}$

$= 7.727\,(10)^{-0.139}$

$= 7.727 \left[\frac{1}{(10)^{0.139}}\right] = \mathbf{5.61 \text{ mg/L}}$ **Ans.**

Q.3. (*b*) Determine the size and depth of a high rate single stage trickling filter for treating the following domestic and industrial wastewater of a town of 50,000 population. Also determine the efficiency of filter and BOD_5 of the effluent. (20)

- Domestic sewage @150 litres per capita per day having 225 mg/L of BOD_5, and
- Industrial wastewater @2,00,000 litres per day having 650 mg/L of BOD_5.

Assume the following data :

- BOD_5 removal in primary clarifier = 35%
- Permissible organic loading of trickling filter = 900 g/day/m^3 (excluding recirculated sewage)
- Recirculation ratio = 1.0
- Permissible hydraulic loading = 20 m^3/day/m^2 (including recirculated sewage)

[**Solution.** *Please follow exactly similar Solved Example 9.9, and compute as under :*

Quantity of domestic sewage produced per day

$= 150 \times 50{,}000$ L/d

$= 7.5$ ML/day

BOD_5 for domestic sewage = 225 mg/L

Total BOD_5 of domestic sewage per day

$$= 7.5\ \frac{ML}{d} \times 225 \frac{mg}{L} = 1687.5 \text{ kg/d} \quad ...(i)$$

Quantity of industrial sewage produced per day

$$= 2{,}00{,}000 \text{ L/d}$$

BOD_5 of industrial sewage = 650 mg/L

Total BOD of industrial sewage

$$= 2{,}00{,}000 \text{ L/d} \times 650 \text{ mg/L}$$

$$= \frac{2{,}00{,}000 \times 650}{10^6} = 130 \text{ kg/d} \quad ...(ii)$$

Total BOD of domestic and industrial sewage

$$= (i) + (ii) = (1687.5 + 130) \text{ kg/d} = 1817.5 \text{ kg/d}$$

Out of 1817.5 kg/d of BOD_5, 35% is already removed in primary clarifier.

∴ *BOD to be removed by trickling filter*

$$= 0.65 \times 1817.5 \text{ kg/d} = \mathbf{1181.375 \text{ kg/d}}$$

Volume of filter media required

$$= \frac{\text{Total BOD to be removed}}{\text{Organic loading}} = \frac{1181.375 \text{ kg/d}}{900 \text{ g/d/m}^3}$$

$$= \frac{1181.375 \times 1000 \text{ g/d}}{900 \text{ g/d/m}^3} = \mathbf{1312.64 \text{ m}^3}$$

Now, total volume of sewage flowing

$$= 7.5 \text{ ML/d} + 0.2 \text{ ML/d} = 7.7 \text{ ML/d}$$

Recirculation ratio of 1 means that the volume of recirculated sewage (R) = original volume = 7.7 ML/d

Total volume (*i.e.*, original + recirculated)

$$= 2 \times 7.7 \text{ ML/d} = 15.4 \text{ ML/d}$$

Filter area required

$$= \frac{\text{Total flow volume}}{\text{Surface loading or Hydraulic loading}} = \frac{15.4 \text{ ML/d}}{20 \text{ m}^3\text{/d/m}^2}$$

$$= \frac{15.4 \times \frac{10^6}{10^3} \text{ m}^3\text{/d}}{20 \text{ m}^3\text{/d/m}^2} = 770 \text{ m}^2$$

Dia of filter required (d) is given as

$$\frac{\pi}{4} d^2 = \text{Filter Area} = 770 \text{ m}^2$$

$$\therefore \quad d = \sqrt{\frac{770 \times 4}{\pi}} = 31.31 \text{ m}$$

Hence, use say 31.5 m dia tank with area = 779.31 m²

$$\text{Depth of filter media required} = \frac{\text{Volume of filter media}}{\text{Surface area}}$$

$$= \frac{1312.64 \text{ m}^3}{779.31 \text{ m}^2} = 1.68 \text{ m}$$

Size of filter : dia = 31.5 m, depth = 1.68 m **Ans.**

Efficiency of the filter is given by Eq. (9.34) as :

$$\eta = \frac{100}{1 + 0.0044\sqrt{\dfrac{Y}{V.F}}}$$

where Y = total organic load *i.e.*, total BOD applied to filter in kg/d = 1181.375 kg/d

V = volume of filter in ha.m = 1312.64 m^3

$$= \frac{1312.64}{10^4} \text{ ha.m} = 0.131 \text{ ha m}$$

F = Recirculation factor as given by Eq. (9.33) as :

$$= \frac{1 + \dfrac{R}{I}}{\left(1 + 0.1\dfrac{R}{I}\right)^2} \quad \text{where } \frac{R}{I} = 1 \text{ (given)}$$

$$= \frac{1+1}{(1+0.1)^2} = \frac{2}{(1.1)^2} = \frac{2}{1.21} = 1.65$$

$$\therefore \quad \eta = \frac{100}{1 + 0.0044\sqrt{\dfrac{1181.375}{0.131 \times 1.65}}} = \mathbf{75.46\%} \quad \text{Ans.}$$

Now, total BOD of the effluent left

$$= \frac{100 - 75.46}{100} \times 1312.64 \text{ kg} = \mathbf{322.18 \text{ kg}}$$

Total volume of effluent = 7.7 ML/d

BOD concentration in the effluent

$$= \frac{322.15 \times 10^6}{7.7 \times 10^6} \text{ mg/L}$$

= 41.84 mg/L **Ans.**

Q4. (c) (*i*) ***A city has its catchment area of 7500 hectares. If the population density of the city is 200 persons per hectare and the water is supplied at the rate of 175 litres per capita per day, what would be the design flow for a***

combined sewer? Take intensity of rainfall equal to 30 mm/hour, average runoff coefficient equal to 0.50 and only 75% of water supplied contributes to the sewage. Also, peak discharge factor should be taken as 3.0. (10)

[Solution. *Follow Solved Examples 3.2 and compute as under :*

Area of the city = 7500 hectares

Population density = 200 persons/hectare

∴ Population = 7500 × 200 = 15,00,000

= 1.5 million = 1.5 M = 1.5×10^6

Average water supply rate per day = 175 L/c/d

Average quantity of water supplied to the city per day

$$= 175\ L \times (1.5 \times 10^6) = \frac{262.5 \times 10^6}{10^3}\ \text{m}^3/\text{d} = 2{,}62{,}500\ \text{m}^3/\text{d}$$

$$\text{Rate of water supplied} = \frac{2{,}62{,}500}{24 \times 60 \times 60}\ \text{m}^3/\text{s} = 3.038\ \text{m}^3/\text{s}$$

The sewage discharge being 75% of water supplied, we have

Average rate of sewage produced = 0.75 × 3.038 m³/s = 2.28 m³/s

The peak rate of sewage flow = 3 × 2.28 m³/s

= **6.84 m³/s** ...(*i*)

Now, let us calculate storm water discharge that will enter the combined sewer as follows :

Intensity of rainfall = 30 mm/hr

Average runoff coefficient (K) = 0.50

Catchment area = 7500 hectares

Using Rational formula, we have peek rate of runoff from rainfall

$$Q = \frac{1}{36} K \cdot p_c \cdot A \qquad \text{...(Eq. 3.1)}$$

where K = 0.5

A = 7500 hectares

p_c = critical rainfall intensity in cm/hr, during the critical rainfall duration equal to T_c, which is related to one hour rainfall value p_0 with Eq. (3.5).

But since T_c and p_0 are not known we will assume that the given rainfall intensity of 30 mm/hr is the critical rainfall intensity.

Therefore p_c = 30 mm/hr = 3 cm/hr

$$\therefore \qquad Q = \frac{1}{36} \times 0.5 \times 3 \times 7500\ \text{m}^3/\text{s} = 312.5\ \text{m}^3/\text{s}$$

= **312.5 m³/s** ...(*ii*)

Total discharge in sewer = (*i*) + (*ii*)

= (6.84 + 312.5) m³/s

= **319.34 m³/s** **Ans.**

Q4. (c) (ii). *What do you understand by aerobic digestion ? What are its advantages and disadvantages?* (10)

[***Aerobic digestion*** involves aerobic decomposition of organic matter present in solid or even liquid wastes of a society. The municipal solid wastes (refuse) and organic solids of the sludge from a sewage treatment plant, are particularly disposed of by composting.

Composting is a biological method of degradation of solid wastes. The organic sludge solids produced during sewage treatment are also similarly composted. When the decomposition of organic matter is caused by aerobic bacteria or facultative bacteria—operating aerobically, then the process is known as aerobic composting.

The *aerobic bacteria* are those which flourish in the presence of free oxygen and consume organic matter for their food, and thereby oxidising it to stable end products.

The *facultative bacteria* can operate either aerobically or as anaerobically (*i.e.*, in the absence of free oxygen). Hence, they can survive and cause decomposition of organic matter, either through aerobic bacteria or anaerobic bacteria. When the decomposition is caused by aerobic bacteria, it evidently comes under aerobic digestion.

The final end product of composting is a manure, called the *compost* or *humus,* which is in great demand in European countries as fertilizer for farms.

The aerobic composting generally involves piling up of refuse and its regular turning, either manually or by mechanical devices, so as to ensure sufficient supply of air and oxygen during its decomposition by bacteria, fungi and other micro-organisms. Initially the process starts with mesophillic bacteria, which oxidise the organic matter present in the refuse or solid wastes and sludges, to carbon dioxide, and liberate heat. The temperature rises to about 45°C, and at the point, the *thermophillic bacteria* take over and continues decomposition. During this phase, temperature further rises to about 60°C, which has to be maintained for 3 days in order to destroy pathogenic bacteria. *This temperature control is crucial, because optimal decomposition occurs between 56 and 60°C, but if the temperature exceeds 60°C, then decomposition slows down.* In about 4 to 8 weeks, the easily biodegradable fraction gets consumed and the temperature of the compost mass starts falling. Complete stabilization occurs after the compost is allowed to cure for another 2 to 8 weeks. *The entire composting gets completed in about 3-4 months time. The volume reduction of upto 50% of the original organic material is achieved under ideal conditions.*

As far as composting of sewage sludge is concerned, it is done in a reactor, in which oxygen is introduced as fine bubbles of air. The micro-organisms in the sludge convert the organic matter into carbon dioxide and water; and the ammonia and amino species to nitrate.

Advantages and Disadvantages of Aerobic Digestion

(1) Aerobic digestors do not use much system operation and maintenance, as is required by anaerobic digestors. The anaerobic digestors involve system operation and maintenance almost daily, which consumes 30 to 60 minutes of time, to ensure efficient running of the anaerobic system.

(2) Anaerobic digestors provide viable source of renewable energy (biogas) and can serve domestic houses in rural areas with clean and green energy. However, no such benefit is obtained from aerobic digestors.

(3) Aerobic digestores are feasible for smaller farms. As against this anaerobic digestors do need large farms to prove economical.

(4) Low level pollution is not the only disadvantage of the biogas production by an anaerobic plant, but anaerobic plants also present a risk of much more serious environmental damage due to toxic spills. No such hazards are posed by aerobic plants and digestors.

(5) Since aerobic digestion of refuse involves turning and handling of waste mass, and hence is not clean, as is the anaerobic digestion, which does not involve piling on ground and regular turning and handling, but involves dumping the refuse and night soil including gobar (animal wastes) in an underground earthen trench (about 10 m × 1.5 m × 1.5 m). This mass is covered at its top by layer of earth of about 15 cm depth, and is finally left over for decomposition.

(6) Since aerobic digestion of refuse involves piling up of refuse and wastes on the ground with regular handling, the method certainly poses health risks and insanitary conditions.

(7) The stabilized sludge from an aerobic digestor is free of offensive odour, and is an excellent fertilizer.

(8) Supernant BOD concentrations in aerobic sludge digestors are lower than those of anaerobic digestors.

(9) Operation of aerobic sludge digestors is easy.

(10) *The major disadvantage of aerobic digestion is the cost of supplying oxygen to the micro-organisms.* Power cost for supplying oxygen, however, may not be as significant for smaller farms. The end products of aerobic digestion are carbon dioxide, water, ammonia or nitrates. However, the entire cell cannot be biologically transformed to these end products. Approximately 25 percent of the cell is typically non-biodegradable.

(11) Other disadvantages of aerobic digestors are that the stabilized sludge sometimes has poor dewatering characteristics; and process performance depends strongly on temperature. Aerobic digestion is more commonly used at smaller municipal waste water treatment plants with waste water flows less than 20 MLD (million litres per day).

Year 2022 (*Engineering Services*)

Q. 2. (c) A *wastewater treatment plant produces 900 kg of dry solids per day at a moisture content of 96 percent. The solids are 70 percent volatile with a specific gravity of 1.0 and 30 percent non-volatile with a specific gravity of 2.5. What would be the sludge volume under the following conditions:*

(*i*) *After digestion process, which reduces volatile solids content by 50 percent and decreases the moisture content to 91 percent.*

(*ii*) *After dewatering process to 72 percent moisture.*

Solution.

Quantity of dry sludge produced = 900 kg/d

Total mass of sludge = 900 kg/d × $\frac{100}{4}$ = 22500 kg/d

Quantity of volatile solids in the sludge = 900 × 0.7 = 630 kg/d

Quantity of non-volatile solids in the sludge = 900 × 0.3 = 270 kg/d

(*i*)

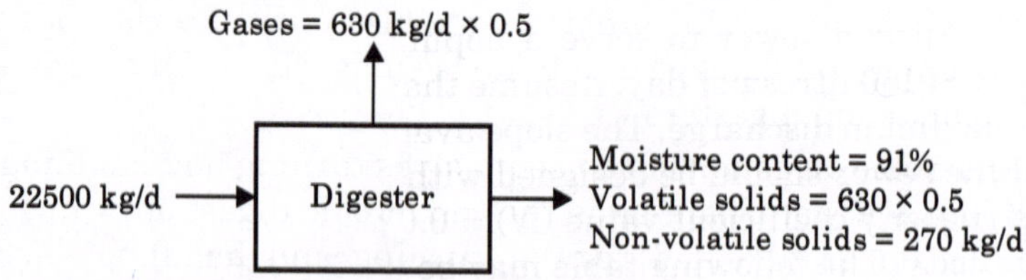

Total solids present in the digested sludge

= (630 × 0.5) + 270 = 585 kg/d

These solids are 9% of total mass

∴ Total mass of sludge = 585 kg/d × $\frac{100}{9}$ = 6500 kg/d

Mass of water in the sludge = 6500 – 585 = 5915 kg/d

Volume of water in the digested sludge = $\frac{5915 \text{ kg/d}}{1000 \text{ kg/m}^3}$ = 5.915 m^3/d

Specific gravity of volatile solids = 1

∴ Volume of volatile solids in the digested sludge $= \frac{(630 \times 0.5) \text{ kg/d}}{1 \times 1000 \text{ kg/m}^3}$

= 0.315 m^3/d

Specific gravity of non-volatile solids = 2.5

∴ Volume of non-volatile solids in the digest sludge = $\frac{270 \text{ kg/d}}{2.5 \times 1000 \text{ kg/m}^3}$

= 0.108 m^3/d

∴ Total volume of digested sludge = $V_{water} + V_{volatile} + V_{non\text{-}volatile}$

= (5.915 + 0.315 + 0.108) m^3/d

= 6.338 m^3/d

(*ii*)

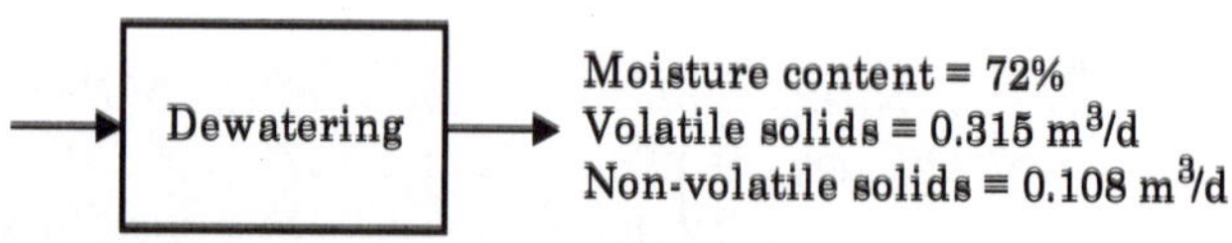

Dewatering is assumed to reduce volume by explusion of water only.

Mass of solids in the dewatered sludge

= 315 + 270 = 585 kg/d

Now, mass of moisture in the dewatered sludge

= 585 kg/d × $\frac{72}{28}$ = 1504.285 kg/d

Volume of moisture in the dewatered sludge

$$= \frac{1504.28 \text{ kg/d}}{1000 \text{ kg/m}^3} = 1.504 \text{ m}^3/\text{d}$$

Thus, volume of dewatered sludge,

$$\text{Volume} = (1.504 + 0.315 + 0.108) \text{ m}^3/\text{d}$$
$$= 1.927 \text{ m}^3/\text{d}$$

Q.3. (*a*) Design a sewer to serve a population of 50,000 with per capita water supply of 150 litres per day. Assume that the sewer should run 0.7 times full at the maximum discharge. The slope available for the sewer to be laid is 1 in 500 and the sewer should be designed with a peaking factor of 3.0. Assume Manning's rugosity coefficient value (N) = 0.012. Also perform check for self-cleaning velocity. The following table may be used if required:

Proportionate depth	*Proportionate velocity*	*Proportionate discharge*
0.20	0.615	0.088
0.30	0.776	0.196

[Solution. Design discharge of the sewer,

$$q = 3 \times (150 \; l/c/d \times 0.8) \times 50000 \text{ c}$$
$$= 18 \times 10^6 \; l/d$$

(A factor of 0.8 is taken for assuming 8070 of water supplied is converted to sewage).

$$\therefore \qquad q = 0.2083 \text{ m}^3/\text{s}$$

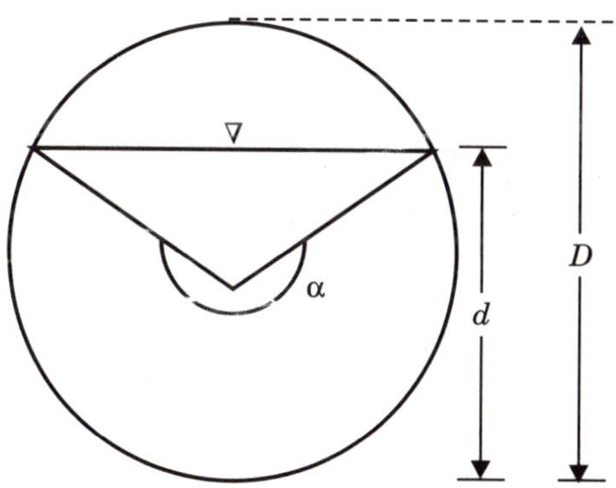

Given $\qquad \dfrac{d}{D} = 0.7$

$$0.7 = \frac{1}{2}\left(1 - \cos\frac{\alpha}{2}\right)$$

$$1.4 = 1 - \cos\frac{\alpha}{2}$$

$$\alpha = 227.156°$$

Now, $\qquad \dfrac{q}{Q} = \left(\dfrac{\alpha}{360°} - \dfrac{\sin\alpha}{2\pi}\right)\left(\dfrac{\dfrac{a}{360°} - \dfrac{\sin\alpha}{2\pi}}{\dfrac{\alpha}{360°}}\right)^{2/3}$

$$= 0.748 \times \frac{(0.748)^{2/3}}{(0.631)^{2/3}} = 0.838$$

For assumed full flow

$$Q = \frac{q}{0.838} = \frac{0.2083 \text{ m}^3/\text{s}}{0.838} = \left(\frac{1}{0.012}\left(\frac{D}{4}\right)^{2/3}\sqrt{\frac{1}{500}}\right)\frac{\pi D^2}{4}$$

$D = 0.561$ m.

Check for self cleansing velocity:

Self cleansing velocity must be generated at minimum flow.

Assuming the minimum flow in the sewer to be 1/3rd the average flow.

$$q_{min} = (150 \; l/c/d \times 0.8 \times 50000c) \times 1/3$$
$$= 2 \times 10^6 \; l/d = 0.0231 \text{ m}^3/\text{s}$$

$$\therefore \quad \frac{q_{\min}}{Q} = \frac{0.0231}{0.2486} = 0.0929$$

$$\therefore \quad Q = \frac{q}{0.838} = 0.2486 \text{ m}^3/\text{s}$$

For $\frac{q_{\min}}{Q} = 0.0929$, from the given table, by interpolation, $\frac{V_{\min}}{V}$ and $\frac{d_{\min}}{D}$ can be found out.

$\frac{d_{\min}}{Q}$	$\frac{V_{\min}}{V}$
0.088	0.615
0.196	0.776
0.0927	X

$$X - 0.615 = \frac{0.776 - 0.615}{0.196 - 0.088}(0.0927 - 0.088)$$

$$X = \frac{V_{\min}}{V} = 0.622$$

$$\therefore V_{min}. = 0.622 \times \frac{1}{n} R^{2/3} S^{1/2} = 0.622 \times \frac{1}{0.012} \times \left(\frac{0.561}{4}\right)^{2/3} \sqrt{\frac{1}{500}}$$
$$= 0.621 \text{ m/s}$$

Usually, minimum velocity greater than 0.6 m/s is satisfactory as per criteria 3.15.1 of GOI manual on sewage and sewage treatement. Hence, OK.

Q.4. (c) *A drinking water treatment plant has a circular sedimentation basin to treat 13 million litres of river water per day. After storms occur upstream, the river often carries 0.010 mm silt particles with an average density of 2300 kg/m³, and the silt must be removed before the water can be used. The sedimentation basin is 3.5 m deep and 21.0 m in diameter. The water is at 15°C. Answer the following:*

(*i*) *What is the hydraulic detention time of the basin?*

(*ii*) *Will the sedimentation basin (clarifier) remove all of the silt particles from the river water? Justify your answer with appropriate calculations. (Take density of water* (ρ) = *999.1 kg/m*^{3} *and viscosity of water* (μ) = *0.00114 kg/m.s at 15°C)*

Solution:

(*i*) Assuming a standard circular tank with side water depth,

$$H = 3.5 \text{ m}$$

and Diameter, $D = 21$ m

Hydraulic detention time,

$$D_t = \frac{\text{Volume of the tank}}{\text{Discharge passing through the tank}}$$

$$D_t = \frac{D^2(0.785H + 0.011D)}{Q}$$

$$= \frac{21^2(0.785 \times 3.5 + 0.011 \times 21)\ \text{m}^3}{13 \times 10^3\ \text{m}^3/d}$$

$$= 0.101039\ \text{m}^3/\text{d} \times 24\ \text{h/d} = 2.424\ \text{h}$$

(*ii*) Surface overflow rate,

$$V_0 = \frac{Q}{\pi \frac{D^2}{4}} = \frac{13 \times 10^3\ \text{m}^3/d}{(\pi \times 21^2 / 4)\ \text{m}^2} = 37.533\ \text{m}^3/\text{m}^2/\text{d}$$

The settling velocity can be found at by Stokes law

$$V_s = \frac{418(G_s - 1)\,(3T + 70)\,d^2}{100} = 0.624\ \text{mm/s}$$

$$= \frac{418(2.3 - 1)\,(3 \times 15 + 70)}{100} \times (0.01)^2$$

$$= 0.0624\ \text{mm/s}$$

$$V_0 = 37.53\ \text{m/d} = 0.4344\ \text{mm/s}$$

$$V_s = \frac{(G - 1)\,\gamma_w d^2}{18\mu} - \frac{(2.3 - 1) \times 9.81 \times 999.1 \times (0.010 \times 10^{-3})^2}{18 \times 0.00114}$$

$$= 0.621\ \text{mm/s}$$

$\therefore$ V_s = minimum {0.0624, 0.0621} = 0.0621 mm/s

$\therefore$ Settling velocity is less than surface overflow rate therefore all silt particles will not be removed.

Year 2023 (*Engineering Services*)

Q. 1. (*d*) *A PST 15 m long, 6 m wide and 3 m deep treats water for a town with a population of 20,000 supplied with 100 lpcd. The raw water sample indicated suspended solids conc. as 60 ppm. The PST worked with efficiency of*

70% SS removal and the average specific gravity of the deposit in PST was 2.6. Compute:

(i) Detention time

(ii) Horizontal velocity

(iii) Rate of dry solids deposited

(iv) Overflow rate

Solution.

Given:

Length of tank, $L = 15$ m

Width of tank, $B = 6$ m

Height of tank, $H = 3$ m

Population of city, $P = 20000$

Per capita demand, $q = 100$ lpcd

Suspended solid concentration, $SS_0 = 60$ mg/L

Efficiency of removal of suspended solid,

$\eta = 70\%$

Specific gravity of deposit, $G = 2.6$

Now, Average discharge, $Q_{avg} = 20000 \times 100$

$= 2 \times 10^6$ Litre/day

Maximum daily discharge, $Q = 1.8 \times 2 \times 10^6$ Litre/day

$= 3.6 \times 10^6$ Litre/day

So, now, Volume of tank, $V = LBH$

$= 15 \times 6 \times 3 = 270 \text{ m}^3$

(*i*) So, Detention time,

$$t_d = \frac{\text{Volume of tank}}{\text{Maximum discharge}}$$

$$= \frac{270 \text{ m}^3}{3600 \text{ m}^3/\text{day}} = 0.075 \text{ day} = 1.8 \text{ hours}$$

(*ii*) Horizontal velocity

$$V_f = \frac{Q}{BH} = \frac{3600 \text{ m}^3/\text{day}}{(6 \times 3) \text{ m}^3} = 200 \text{ m/day}$$

$= 2.32 \times 10^{-3}$ m/s $= 2.32$ mm/sec.

(*iii*) Suspended solids coming in tank $= QSS_0$

$= 3600 \text{ m}^3/\text{day} \times 60 \text{ mg/L} = 216$ kg/day

Deposits in tank $= QSS_0\eta_0 = 216 \times 0.7 = 151.2$ kg/day

(*iv*) Overflow rate,

$$V_0 = \frac{Q}{BL} = \frac{3600 \text{ m}^3/\text{day}}{(15 \times 6) \text{ m}^2} = 40 \text{ m}^3/\text{day/m}^2$$

Q. 2. (c) *Differentiate between Symbiosis and Parasitism relationship. Explaining the working principles of oxidation ponds, discuss the importance of Algal-Bacteria symbiosis relationship in oxidation ponds.*

Solution. The comparison between symbiosis and parasitism relationships are tabulated below :

Paramet	*Symbiosis*	*Parasitism*
Definition	• A close and long term interaction between two species, where at least one species benefits.	• In this type of relationship, one species (parasite) benefits at the expense of another species (host).
Interaction	• Mutualistic, commensalistic or parasitic.	• Predatory or parasitic.
Benefit to host	• Both species can benefit or one species benefits while the other is unaffected.	• The host is harmed or negatively affected.
Dependency	• Both species are dependent on each other for survival or reproduction.	• The parasite is dependent on the host for resources or reproduction. The host is not dependent on the parasite.
Examples	• Lichen (algae and fungi), coral reefs (corals and zooxanthellae), legume plants and nitrogen fixing bacteria.	• Fleas on dogs, ticks on deer, tapeworms in human.
Control	• The interaction is usually well balanced and regulated to prevent over-exploitation or harm to either species.	• The parasite may exert control over the host, often causing harm or disease. The host may evolve defences to minimise the impact of the parasite.

Oxidation ponds, also known as stabilization ponds or waste stabilization ponds are shallow, man-made bodies of water used for wastewater treatment. They rely on the interaction between algae and bacteria to remove pollutants and purify the water.

The key roles of the algae-bacterial symbiosis relationship in oxidation ponds are:

1. Oxygenation: Algae photosynthesize and release oxygen, creating an oxygen rich environment that supports the growth of aerobic bacteria. This oxygenation is crucial for the activity of aerobic bacteria, which are responsible for the breakdown of organic matter and the removal of pollutants.

2. Nutrient cycling: Algae and bacteria work together to cycle and remove nutrients from the wastewater. Algae take up nitrogen and phosphorus during their growth, reducing their concentration in the water. Bacteria then decompose the organic matter provided by the algae, releasing nutrients that can be reused by the algae. This symbiotic relationship helps to maintain a balanced nutrient cycle and prevent excessive nutrient build up in the water.

3. Pollutant removal: The combined activity of algae and bacteria leads to the degradation and removal of organic compounds, including pollutants and pathogens from the wastewater. Algae provide a source of organic matter for bacterial decomposition, while bacteria break down complex organic molecules into simpler, less harmful forms.

4. Algae-Bacteria floc formation: Algae and bacteria contribute to the formation of floc or clumps of organic matter in the oxidation pond. These floes settle to bottom,facilitating the removal of suspended solids and some pollutants through sedimentation. The floc formation is a result of the mutual interactions between algae and bacteria, enhancing the efficiency of solid removal in the pond.

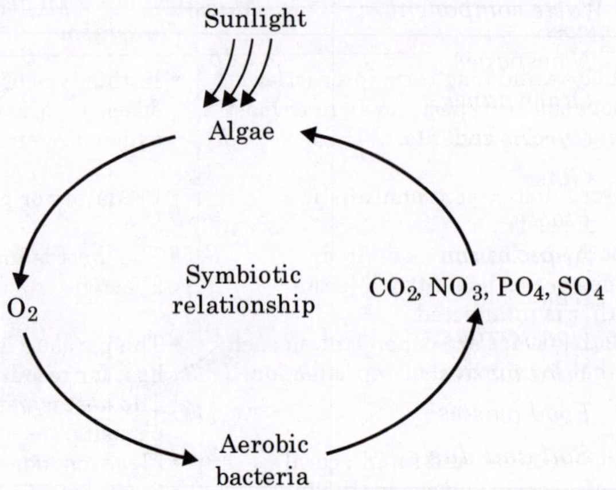

Q.4. (b). ***The MLSS concentration in an aeration tank of ASP system was 3000 mg / L. Sludge volume was 180 mL after 30 minutes of settling in 1000 mL graduated measuring jar. For the above sample, find***

(i) SVI

(ii) SDI

(iii) Return sludge ratio required

(iv) SS conc. in recirculated sludge

Solution. Given, MLSS, $x = 3000$ mg/L

Sludge volume, $v = 180$ mL (Settled after 30 minutes)

(*i*) SVI

In 1 litre jar, 3000 mg of suspended solid has occupied 180 mL in SVI jar.

$$SVI = \frac{v(\text{mL})}{x(g)} = \frac{180 \text{ mL}}{3000 \text{ mg}}$$

$$= \frac{180 \text{ mL}}{3 \text{ g}} = 60 \text{ mL/g}$$

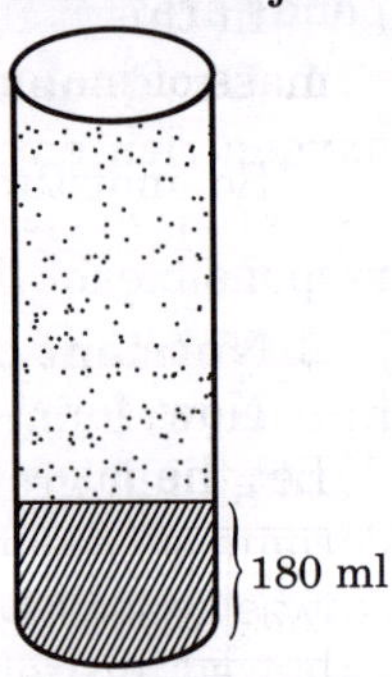

SVI jar (1 lt)

(*ii*) Sludge density index $= \dfrac{100}{SVI} = \dfrac{100}{60} = 1.67$ g/mL

(*iii*) Return sludge ratio, $R = \dfrac{Q_R}{Q_0} = \dfrac{x}{x_u - x}$

$$\text{where } x_u = \frac{10^6}{SVI} = \frac{10^6}{60} = 16.67 \times 10^3 \text{ mg/l}$$

$$\therefore R = \frac{3000}{(16.67 - 3) \times 10^3} = 0.219$$

(*iv*) S.S. concentration in recirculated sludge,

$$x_u = \frac{10^6}{SVI} = \frac{10^6}{60} = 1.67 \times 10^4 \text{ mg/L}$$

Q.4. (c). ***Using the data given below, find the moisture content (Wet and dry basis) of the municipal solid waste. Also estimate the as-discarded density. If the compaction is 3, find the size of collection vehicle required for 1000 kg of MSW.***

S.No.	Waste component	Mass(%)	MC(%)	Density (kg/m^3)
1.	Newspaper	15	6	85
2.	Other paper	24	6	85
3.	Cardboard	33	5	50
4.	Glass	4.2	0.5	195
5.	Plastic	0.49	2	65
6.	Aluminium	0.13	0.5	160
7.	Iron	1.18	0.5	320
8.	Non-ferrous	0.35	0.5	160
9.	Yard wastes	17.97	60	105
10.	Food wastes	1.67	60	290
11.	Soil and dust	2.01	8	480

Solution.

Moisture content of a solid waste is expressed in two ways.

1. Wet mass method

In this method, moisture in a sample is expressed as a percentage of wet mass of material.

So, $\quad \text{moisture content}(\%) = \left(\dfrac{a-b}{a}\right) \times 100$

where, a = Initial mass of sample

b = Mass of sample after drying

2. Dry mass method

In this method, moisture in a sample is expressed as a percentage of dry mass of material.

So, $\text{moisture content}(\%) = \left(\dfrac{a-b}{a}\right) \times 100$

where, a = Initial mass of sample

b = Mass of sample after drying

Now, for the given solid waste sample, calculations are done below in table.

Let the mass of given sample is 100 kg

S. No.	Waste component	Mass (kg)	MC (kg)	Dry mass (kg)	Density (kg/m^3)	Volume (m^3)
1.	Newspaper	15	0.06 × 15 = 0.9	14.1	85	0.1765
2.	Other paper	24	0.06 × 24 = 1.44	22.56	85	0.2824
3.	Cardboard	33	0.05 × 33 = 1.65	31.35	50	0.66
4.	Glass	4.2	0.005 × 4.2 = 0.021	4.179	195	0.0215
5.	Plastic	0.49	0.02 × 0.49 = 0.0098	0.4802	65	0.0075
6.	Aluminium	0.13	0.005 × 0.13 = 0.00065	0.1293	160	0.00081
7.	Iron	1.18	0.005 × 1.18 = 0.0059	1.1741	320	0.00368
8.	Non-ferrous	0.35	0.005 × 0.35 = 0.00175	0.34825	160	0.00218
9.	Yard wastes	17.97	0.6 × 17.97 = 10.782	7.188	105	0.1711
10.	Food wastes	1.67	0.6 × 1.67 = 1.002	0.668	290	0.00576
11.	Soil and dust	2.01	0.08 × 2.01 = 0.1608	1.8492	480	0.00419
				84.02605		1.3356

NOTE: Intable, dry mass is calculated bysubtracting moisture from given mass of individual component and volume is calculated by dividing the given mass of component by its density.

So, $a = 100$ kg

$b = 84.02605$ kg

So, moisture content (on wet basis) $= \left(\dfrac{a-b}{a}\right)\times 100$

$$= \left(\frac{100-84.02605}{100}\right)\times 100 \simeq 15.97\%$$

Moisture content (on dry basis) $= \left(\dfrac{a-b}{a}\right)\times 100$

$$= \left(\frac{100-84.02605}{84.2605}\right)\times 100 = 19.01\%$$

(*ii*) As discarded density

Now, volume of 100 kg sample = 1.3356 m^3

So, Density $= \dfrac{\text{Mass}}{\text{Volume}} = \dfrac{100 \text{ kg}}{1.3356 \text{ m}^3} = 74.87 \text{ kg/m}^3$

(*iii*) Size of collection vehicle,

Compaction ratio $= \dfrac{\text{Density of MSW in vehicle}}{\text{As-discarded density of MSW}}$

$\Rightarrow \quad 3 = \dfrac{\text{Density of MSW in vehicle}}{74.87}$

$\Rightarrow$ Density of MSW in vehicle = 74.87 × 3 = 224.61 kg/m^3

Now, volume of MSW in vehicle

$$= \frac{\text{Mass of MSW}}{\text{Density of MSW}} = \frac{1000 \text{ kg}}{224.61 \text{ kg/m}^3} = 4.45 \text{ m}^3$$

Appendix Table A-1. Density of Dry Air in kg/m³ at Different Temperatures and Different Pressures

	Pressure—kPa										
°C	110.0	100.0	90.0	80.0	70.0	60.0	50.0	40.0	30.0	20.0	10.0
0	1.4029	1.2754	1.1478	1.0203	0.8928	0.7652	0.6377	0.5102	0.3826	0.2551	0.1275
1	1.3978	1.2707	1.1437	1.0166	0.8895	0.7624	0.6254	0.5083	0.3812	0.2541	0.1271
2	1.3927	1.2661	1.1395	1.0129	0.8863	0.7597	0.6331	0.5064	0.3798	0.2532	0.1266
3	1.3877	1.2615	1.1354	1.0092	0.8831	0.7569	0.6308	0.5046	0.3785	0.2523	0.1262
4	1.3827	1.2570	1.1313	1.0056	0.8799	0.7542	0.6285	0.5028	0.3771	0.2514	0.1257
5	1.3777	1.2525	1.1272	1.0020	0.8767	0.7515	0.6262	0.5010	0.3757	0.2505	0.1252
6	1.3728	1.2480	1.1232	0.9984	0.8736	0.7488	0.6240	0.4992	0.3744	0.2496	0.1248
7	1.3679	1.2435	1.1192	0.9948	0.8705	0.7461	0.6218	0.4974	0.3731	0.2487	0.1244
8	1.3630	1.2391	1.1152	0.9913	0.8674	0.7435	0.6195	0.4956	0.3717	0.2478	0.1239
9	1.3582	1.2347	1.1112	0.9878	0.8643	0.7408	0.6174	0.4939	0.3704	0.2469	0.1235
10	1.3534	1.2303	1.1073	0.9843	0.8612	0.7382	0.6152	0.4921	0.3691	0.2461	0.1230
11	1.3486	1.2260	1.1034	0.9808	0.8582	0.7356	0.6130	0.4904	0.3678	0.2452	0.1226
12	1.3439	1.2217	1.0995	0.9774	0.8552	0.7330	0.6109	0.4887	0.3665	0.2443	0.1222
13	1.3392	1.2174	1.0957	0.9740	0.8522	0.7305	0.6087	0.4870	0.3652	0.2435	0.1217
14	1.3345	1.2132	1.0919	0.9706	0.8492	0.7279	0.6056	0.4853	0.3640	0.2426	0.1213
15	1.3299	1.2090	1.0881	0.9672	0.8463	0.7254	0.6045	0.4836	0.3627	0.2418	0.1209
16	1.3253	1.2048	1.0843	0.9638	0.8434	0.7229	0.6024	0.4819	0.3614	0.2410	0.1205
17	1.3207	1.2007	1.0806	0.9605	0.8405	0.7204	0.6003	0.4803	0.3602	0.2401	0.1201
18	1.3162	1.1965	1.0769	0.9572	0.8376	0.7179	0.5983	0.4786	0.3590	0.2393	0.1197
19	1.3117	1.1924	1.0732	0.9540	0.8347	0.7155	0.5962	0.4770	0.3577	0.2385	0.1192
20	1.3072	1.1884	1.0695	0.9507	0.8319	0.7130	0.5942	0.4753	0.3565	0.2377	0.1188
21	1.3028	1.1843	1.0659	0.9475	0.8290	0.7106	0.5922	0.4737	0.3553	0.2369	0.1184
22	1.2984	1.1803	1.0623	0.9443	0.8262	0.7082	0.5902	0.4721	0.3541	0.2361	0.1180
23	1.2940	1.1763	1.0587	0.9411	0.8234	0.7058	0.5882	0.4705	0.3529	0.2353	0.1176
24	1.2896	1.1724	1.0551	0.9379	0.8207	0.7034	0.5862	0.4690	0.3517	0.2345	0.1172

	Pressure—kPa										
°C	110.0	100.0	90.0	80.0	70.0	60.0	50.0	40.0	30.0	20.0	10.0
25	1.2853	1.1684	1.0516	0.9348	0.8179	0.7011	0.5842	0.4674	0.3505	0.2337	0.1168
26	1.2810	1.1645	1.0481	0.9316	0.8152	0.6987	0.5823	0.4658	0.3494	0.2329	0.1165
27	1.2767	1.1607	1.0446	0.9285	0.8125	0.6964	0.5803	0.4643	0.3482	0.2321	0.1161
28	1.2725	1.1568	1.0411	0.9254	0.8098	0.6941	0.5784	0.4627	0.3470	0.2314	0.1157
29	1.2683	1.1530	1.0377	0.9224	0.8071	0.6918	0.5765	0.4612	0.3459	0.2306	0.1153
30	1.2641	1.1492	1.0343	0.9193	0.8044	0.6895	0.5746	0.4597	0.3448	0.2298	0.1149
31	1.2599	1.1459	1.0309	0.9163	0.8018	0.6872	0.5727	0.4582	0.3436	0.2291	0.1145
32	1.2558	1.1416	1.0275	0.9133	0.7991	0.6850	0.5708	0.4567	0.3425	0.2283	0.1142
33	1.2517	1.1379	1.0241	0.9103	0.7965	0.6827	0.5690	0.4542	0.3414	0.2276	0.1138
34	1.2476	1.1342	1.0208	0.9074	0.7939	0.6805	0.5671	0.4537	0.3403	0.2268	0.1134
35	1.2436	1.1305	1.0175	0.9044	0.7914	0.6783	0.5653	0.4522	0.3392	0.2261	0.1131
36	1.2396	1.1269	1.0142	0.9015	0.7888	0.6761	0.5634	0.4507	0.3381	0.2254	0.1127
37	1.2356	1.1232	1.0109	0.8986	0.7863	0.6739	0.5616	0.4493	0.3370	0.2246	0.1123
38	1.2316	1.1196	1.0077	0.8957	0.7837	0.6718	0.5598	0.4479	0.3359	0.2239	0.1120
39	1.2276	1.1160	1.0044	0.8928	0.7812	0.6696	0.5580	0.4464	0.3348	0.2232	0.1116
40	1.2237	1.1125	1.0012	0.8900	0.7787	0.6675	0.5562	0.4450	0.3337	0.2225	0.1112
41	1.2198	1.1089	0.9980	0.8872	0.7763	0.6654	0.5545	0.4436	0.3327	0.2218	0.1109
42	1.2160	1.1054	0.9949	0.8843	0.7738	0.6633	0.5527	0.4422	0.3316	0.2211	0.1105
43	1.2121	1.1019	0.9917	0.8815	0.7713	0.6612	0.5510	0.4408	0.3306	0.2204	0.1102
44	1.2083	1.0984	0.9886	0.8788	0.7689	0.6591	0.5492	0.4394	0.3295	0.2197	0.1098
45	1.2045	1.0950	0.9855	0.8760	0.7665	0.6570	0.5475	0.4380	0.3285	0.2190	0.1095
46	1.2007	1.0916	0.9824	0.8732	0.7641	0.6549	0.5458	0.4366	0.3275	0.2183	0.1092
47	1.1970	1.0882	0.9793	0.8705	0.7617	0.6529	0.5441	0.4353	0.3264	0.2176	0.1088
48	1.1932	1.0848	0.9763	0.8678	0.7593	0.6509	0.5424	0.4339	0.3254	0.2170	0.1085
49	1.1895	1.0814	0.9733	0.8651	0.7570	0.6488	0.5407	0.4326	0.3244	0.2163	0.1081

°C	Pressure—kPa										
	110.0	*100.0*	*90.0*	*80.0*	*70.0*	*60.0*	*50.0*	*40.0*	*30.0*	*20.0*	*10.0*
50	1.1859	1.0780	0.9702	0.8624	0.7546	0.6468	0.5390	0.4312	0.3234	0.2156	0.1078
51	1.1822	1.0747	0.9673	0.8598	0.7523	0.6448	0.5374	0.4299	0.3224	0.2149	0.1075
52	1.1786	1.0714	0.9643	0.8571	0.7500	0.6429	0.5357	0.4286	0.3214	0.2143	0.1071
53	1.1750	1.0681	0.9613	0.8545	0.7477	0.6409	0.5341	0.4273	0.3204	0.2136	0.1068
54	1.1714	1.0649	0.9584	0.8519	0.7454	0.6389	0.5324	0.4259	0.3195	0.2130	0.1065
55	1.1678	1.0616	0.9555	0.8493	0.7431	0.6370	0.5308	0.4247	0.3185	0.2123	0.1062
56	1.1642	1.0584	0.9526	0.8467	0.7409	0.6350	0.5292	0.4234	0.3175	0.2117	0.1058
57	1.1607	1.0552	0.9497	0.8442	0.7386	0.6331	0.5276	0.4221	0.3166	0.2110	0.1055
58	1.1572	1.0520	0.9468	0.8416	0.7364	0.6312	0.5260	0.4208	0.3156	0.2104	0.1052
59	1.1537	1.0488	0.9440	0.8391	0.7342	0.6293	0.5244	0.4195	0.3147	0.2098	0.1049
60	1.1503	1.0457	0.9411	0.8366	0.7320	0.6274	0.5228	0.4183	0.3137	0.2091	0.1046
61	1.1468	1.0426	0.9383	0.8340	0.7298	0.6255	0.5213	0.4170	0.3128	0.2085	0.1043
62	1.1434	1.0394	0.9355	0.8316	0.7276	0.6237	0.5197	0.4158	0.3118	0.2079	0.1039
63	1.1400	1.0364	0.9327	0.8291	0.7255	0.6218	0.5182	0.4145	0.3109	0.2073	0.1036
64	1.1366	1.0333	0.9300	0.8266	0.7233	0.6200	0.5166	0.4133	0.3100	0.2067	0.1033
65	1.1333	1.0302	0.9272	0.8242	0.7212	0.6181	0.5151	0.4121	0.3091	0.2060	0.1030
66	1.1299	1.0272	0.9245	0.8218	0.7190	0.6163	0.5136	0.4109	0.3082	0.2054	0.1027
67	1.1266	1.0242	0.9218	0.8193	0.7169	0.6145	0.5121	0.4097	0.3073	0.2048	0.1024
68	1.1223	1.0212	0.9191	0.8169	0.7148	0.6127	0.5106	0.4085	0.3064	0.2042	0.1021
69	1.1200	1.0182	0.9164	0.8146	0.7127	0.6109	0.5091	0.4073	0.3055	0.2036	0.1018
70	1.1167	1.0152	0.9137	0.8122	0.7107	0.6091	0.5076	0.4061	0.3046	0.2030	0.1015
71	1.1135	1.0123	0.9110	0.8098	0.7086	0.6074	0.5061	0.4049	0.3037	0.2025	0.1012
72	1.1103	1.0093	0.9084	0.8075	0.7065	0.6056	0.5047	0.4037	0.3028	0.2019	0.1009
73	1.1071	1.0064	0.9058	0.8051	0.7045	0.6039	0.5032	0.4026	0.3019	0.2013	0.1006
74	1.1039	1.0035	0.9032	0.8028	0.7025	0.6021	0.5018	0.4014	0.3011	0.2007	0.1004

	Pressure—kPa										
°C	*110.0*	*100.0*	*90.0*	*80.0*	*70.0*	*60.0*	*50.0*	*40.0*	*30.0*	*20.0*	*10.0*
75	1.1007	0.0006	0.9006	0.8005	0.7004	0.6004	0.5003	0.4003	0.3002	0.2001	0.1001
76	1.0976	0.9978	0.8980	0.7982	0.6984	0.5987	0.4989	0.3991	0.2993	0.1996	0.0998
77	1.0944	0.9949	0.8954	0.7959	0.6965	0.5970	0.4975	0.3980	0.2985	0.1990	0.0995
78	1.0913	0.9921	0.8929	0.7937	0.6945	0.5953	0.4960	0.3968	0.2976	0.1984	0.0992
79	1.0882	0.9893	0.8904	0.7914	0.6925	0.5936	0.4946	0.3957	0.2968	0.1979	0.0989
80	1.0851	0.9865	0.8878	0.7892	0.6905	0.5919	0.4932	0.3946	0.2959	0.1973	0.0986
81	1.0821	0.9837	0.8853	0.7870	0.6886	0.5902	0.4918	0.3935	0.2951	0.1967	0.0984
82	1.0790	0.9809	0.8828	0.7847	0.6866	0.5886	0.4905	0.3924	0.2943	0.1962	0.0981
83	1.0760	0.9782	0.8804	0.7825	0.6847	0.5869	0.4891	0.3913	0.2935	0.1956	0.0978
84	1.0730	0.9754	0.8779	0.7803	0.6828	0.5853	0.4877	0.3902	0.2926	0.1951	0.0975
85	1.0700	0.9727	0.8754	0.7782	0.6809	0.5836	0.4864	0.3891	0.2918	0.1945	0.0973
86	1.0670	0.9700	0.8730	0.7760	0.6790	0.5820	0.4850	0.3880	0.2910	0.1940	0.0970
87	1.0640	0.9673	0.8706	0.7738	0.6771	0.5804	0.4836	0.3869	0.2902	0.1935	0.0967
88	1.0611	0.9646	0.8682	0.7717	0.6752	0.5788	0.4823	0.3858	0.2894	0.1929	0.0965
89	1.0582	0.9620	0.8658	0.7696	0.6734	0.5772	0.4810	0.3848	0.2886	0.1924	0.0962
90	1.0552	0.9593	0.8634	0.7674	0.6715	0.5756	0.4797	0.3837	0.2878	0.1919	0.0959
91	1.0523	0.9567	0.8610	0.7653	0.6697	0.5740	0.4783	0.3827	0.2870	0.1913	0.0957
92	1.0495	0.9541	0.8587	0.7632	0.6678	0.5724	0.4770	0.3816	0.2862	0.1908	0.0954
93	1.0466	0.9514	0.8563	0.7612	0.6660	0.5709	0.4757	0.3806	0.2854	0.1903	0.0951
94	1.0437	0.9489	0.8540	0.7591	0.6642	0.5693	0.4744	0.3795	0.2847	0.1898	0.0949
95	1.0409	0.9463	0.8517	0.7570	0.6624	0.5678	0.4731	0.3785	0.2839	0.1893	0.0946
96	1.0381	0.9437	0.8493	0.7550	0.6606	0.5662	0.4719	0.3775	0.2831	0.1887	0.0944
97	1.0353	0.9412	0.8471	0.7529	0.6588	0.5647	0.4706	0.3765	0.2824	0.1882	0.0941
98	1.0325	0.9386	0.8448	0.7509	0.6570	0.5632	0.4693	0.3755	0.2816	0.1877	0.0939
99	1.0297	0.9361	0.8425	0.7489	0.6553	0.5617	0.4681	0.3744	0.2808	0.1872	0.0936
100	1.0270	0.9336	0.8402	0.7469	0.6535	0.5602	0.4648	0.3734	0.2801	0.1867	0.0934

Source : Robert J. List, *Smithsonian Meterological Tables.*

Appendix Table A-2. Viscosity of Dry Air AT 100 kPa (About 1 Atm Pressure)

Temperature (°C)	*Dynamic viscosity* (μ) (*Pa . s*)
0	17.1
5	17.4
10	17.7
15	17.9
20	18.2
25	18.5
30	18.7
35	19.0
40	19.3
45	19.5
50	19.8
55	20.1
60	20.3
65	20.6
70	20.9
75	21.1
80	21.4
85	21.7
90	21.9
95	22.2
100	22.5
150	25.2

$\mu = 17.11 + 0.0536\ T + (P/8280)$ where T is in °C and P is in kPa.

Appendix Table A-3. Saturation Values of Dissolved Oxygen (D.O.) in Fresh Water at Different Temperatures and 1 Atmospheric Pressure (i.e. 101.325 kPa)*

Temperature (°C)	*Dissolved oxygen (mg/l)*	*Temperature (°C)*	*Dissolved oxygen (mg/l)*
0	14.62	20	9.17
1	14.23	21	8.99
2	13.84	22	8.83
3	13.48	23	8.68
4	13.13	24	8.53
5	12.80	25	8.38
6	12.48	26	8.22
7	12.17	27	8.07
8	11.87	28	7.92
9	11.59	29	7.77
10	11.33	30	7.63
11	11.08	31	7.51
12	10.83	32	7.42
13	10.60	33	7.28
14	10.37	34	7.17
15	10.15	35	7.07
16	9.95	36	6.96
17	9.74	37	6.86
18	9.54	38	6.75
19	9.35		

*For other barometric pressures, the solubilities vary approximately in proportion to the ratios of those pressures to the standard pressure.

Appendix Table A-4. Saturation Values of D.O. For Different Types of Waters at Different Temperatures and 1 Atmospheric Pressure

S. No.	*Types of water*	*Temp. in °C*						
		0°	*5°*	*10°*	*15°*	*20°*	*25°*	*30°*
		Saturation D.O. Values in mg/l						
1.	Fresh water *river water* (chlorides = 0 ppm)	14.6	12.8	11.3	10.2	9.2	8.4	7.6
2.	Brackish water *Lake water* (chlorides = 10,000 ppm)	10.0	11.4	10.1	9.1	8.3	7.6	6.9
3.	*Sea water*	11.3	10.1	9.0	8.1	7.4	6.7	6.1

Appendix Table A-5. Properties of Air at Standard Conditions (*i.e.* 1 Atm. pressure and 298° K Temperature)

Molecular weight	M	28.97
Specific gas constant	$R = \frac{R_u}{M_o}$	287 J/kg . K
Specific heat at constant pressure	c_p	1.005 J/kg . K
Specific heat at constant volume	c_v	718 J/kg . K
Density	ρ	1.185 kg/m^3
Dynamic viscosity	μ	1.8515×10^{-5} Pa . s
Kinematic viscosity	ν	1.5624×10^{-5} m^2/s
Thermal conductivity	k	0.0257 W/m . K
Ratio of specific heats, c_p/c_v	k	1.3997
Prandtl number	Pr	0.720

Appendix Table A-6. Important Properties of Water

Temp. °C	*Sp. gravity*	*Unit wt. in kN/m^3 i.e. Sp. gravity × 9.807*	*Vapour pressure in kN/m^2*	*Viscosity* Dynamic (μ) in N-sec/m^2	Kinematic (ν) in m^2/sec $\nu = \frac{\mu}{\rho}$
(1)	(2)	(3)	(4)	(5)	(6)
0	0.99987	9.805	0.611	17.97×10^{-4}	1.78×10^{-6}
5	0.99999	9.807	0.872	15.17×10^{-4}	1.518×10^{-6}
10	0.99975	9.805	1.225	13.06×10^{-4}	1.310×10^{-6}
15	0.99907	9.798	1.703	11.25×10^{-4}	1.124×10^{-6}
20	0.99810	9.788	2.334	10.03×10^{-4}	1.009×10^{-6}
30	0.99574	9.765	4.609	8.02×10^{-4}	0.804×10^{-6}
40	0.99228	9.731	7.600	6.52×10^{-4}	0.654×10^{-6}
60	0.98338	9.644	19.938	4.70×10^{-4}	0.478×10^{-6}
80	0.97196	9.532	46.583	3.56×10^{-4}	0.366×10^{-6}
100	0.95865	9.401	101.502	2.82×10^{-4}	0.295×10^{-6}

Appendix Table A-7. Frequently used Constants

Standard atmospheric pressure	P_{atm}	**101.325 kPa = 101.325 N/m²**
Standard gravitational acceleration	g	9.8067 m/s^2
Universal gas constant	R_u	8,314 J/kilo . mol . K
Electrical permittivity constant	ε_0	8.85×10^{-12} C/V . m
Electron charge	q_e	1.60×10^{-19} C
Boltzmann's constant	k	1.38×10^{-23} J/K
Joule's constant	J_e	4.184 J/cal
Farraday's constant	F	9.6487×10^7 C/kilo mole
Molecular mass of air	M_o	28.964 kg/kilo mole
Specific gas constant dry for air (R)	$= \dfrac{R_u}{M_o}$	$= \dfrac{8314}{28.964}$ = 287 J/kg.K
Specific gas constant for water	$= R_w$	$= \dfrac{R_u}{M_w} = \dfrac{8314}{18}$ = 462 J/kg.K
Mathematical constants	e	= 2.71828
	$\log_e 10 = \ln 10$	= 2.30259
	π	= 3.14159

Appendix Table A-8. Modern Periodic Table of the Elements

Key: 1 ← ATOMIC NUMBER; H; 1.00794 ← ATOMIC MASS

Periods	I A	II A	III B	IV B	V B	VI B	VII B	VIII	VIII	VIII	I B	II B	III A	IV A	V A	VI A	VII A	0
1	1 H 1.00794																	2 He 4.00260
2	3 Li 6.941	4 Be 9.01218											5 B 10.811	6 C 12.011	7 N 14.00674	8 O 15.9994	9 F 18.99840	10 Ne 20.1797
3	11 Na 22.98977	12 Mg 24.3050											13 Al 26.98154	14 Si 28.0855	15 P 30.97376	16 S 32.066	17 Cl 35.4527	18 Ar 39.948
4	19 K 39.0983	20 Ca 40.078	21 Sc 44.95591	22 Ti 47.88	23 V 50.9415	24 Cr 51.9961	25 Mn 54.9380	26 Fe 55.847	27 Co 58.93320	28 Ni 58.69	29 Cu 63.546	30 Zn 65.39	31 Ga 69.723	32 Ge 72.61	33 As 74.92159	34 Se 78.96	35 Br 79.904	36 Kr 83.80
5	37 Rb 85.4678	38 Sr 87.62	39 Y 88.90585	40 Zr 91.224	41 Nb 92.90638	42 Mo 95.94	43 Tc 98.9072	44 Ru 101.07	45 Rh 102.90550	46 Pd 106.42	47 Ag 107.8682	48 Cd 112.411	49 In 114.82	50 Sn 118.710	51 Sb 121.75	52 Te 127.60	53 I 126.90447	54 Xe 131.29
6	55 Cs 132.90543	56 Ba 137.327	57 *La 138.9055	72 Hf 178.49	73 Ta 180.9479	74 W 183.85	75 Re 186.207	76 Os 190.2	77 Ir 192.22	78 Pt 195.08	79 Au 196.96654	80 Hg 200.59	81 Tl 204.3833	82 Pb 207.2	83 Bi 208.98037	84 Po 208.9824	85 At 209.9871	86 Rn 222.0176
7	87 Fr 223.0197	88 Ra 226.0254	89 †Ac 227.0278	104 Rf 261.11	105 Ha 262.114	106 Sg 263.118	107 Ns 262.12	108 Hs	109 Mt									
	Alkali metals	Alkaline earth metals															Halogens	Noble or inert gases

*	58 Ce 140.115	59 Pr 140.90765	60 Nd 144.24	61 Pm 144.9127	62 Sm 150.36	63 Eu 151.965	64 Gd 157.25	65 Tb 158.92534	66 Dy 162.50	67 Ho 164.93032	68 Er 167.26	69 Tm 168.93421	70 Yb 173.04	71 Lu 174.967
†	90 Th 232.0381	91 Pa 231.0359	92 U 238.0289	93 Np 237.0482	94 Pu 244.0642	95 Am 243.0614	96 Cm 247.0703	97 Bk 247.0703	98 Cf 252.0587	99 Es 252.083	100 Fm 257.0951	101 Md 258.10	102 No 259.1009	103 Lr 260.105

Appendix Table A-9. The Basic Units Used in Radiation and Radioactivity

(*i*)	1 *Roentgen* unit of ionizing radiation (*r* or R)	2.58×10^4 calorie per kg of energy.
(*ii*)	1 Grey (Gy)	Radiation with transfer of 1 joule of energy per kg of absorbing medium. Unit of dose to human body.
(*iii*)	1 *rem* (roentgen)	*Dose in rad* × R.B.E. (Relative biological effectiveness for gamma and X radiations = 1 approxi-mately) ≈ 1 rad.
(*iv*)	1 Curie (Ci)	3.7×10^{10} disintegrations per second (dps) of the radioactive material.
(*v*)	1 Pico Curie $= 1 \text{ pCi} = 10^{-12}$ Ci	0.037 dps = 0.037 Bq
(*vi*)	1 Becquerel (Bq)	One disintegration per second (dps) $= \frac{1}{0.037}$ pCi = 27 pCi
(*vii*)	1 AMU or 1 amu (atomic mass unit)	931 MeV $= \frac{1}{12}$ mass of C-12 atom $= \frac{1}{12} \frac{12}{\text{Avogadros No.}} = \frac{1}{12}\left[\frac{12}{6.022 \times 10^{23}}\right]$ **= 1.6606×10^{-21} gm**
(*viii*)	1 MeV	10^6 eV
(*ix*)	1 G. eV	10^9 eV

Table A-10. Statement showing the Height, Live Storage, Submergence and Benefits of Some Major Water Resources Projects in India

Sr. No.	Name of the Project	State	Height (m)	Live storage (Mm^3)	Submergence (ha)	Irrigation (ha)	Power (MW)	Irrigation benefits (ha) per ha of submergence	Percentage of area submerged to area irrigated	Submergence per Mm^3 of storages (ha)	Irrigation per Mm^3 of power (ha)	Submergence per MW generation (ha)
1	2	3	4	5	6	7	8	9	10	11	12	13
1.	Sardar Sarovar	Gujarat	163	5800	34867	1912150	1450	54.85	1.82	6.01	308.97	24
2.	Narmada Sagar (Punasa)	M.P.	125	9750	91300	256000	1000	2.804	35.66	9.36	26.26	91
3.	Sardar Sarovar & Narmada Sagar Combined Complex	Gujarat & M.P.	—	15550	126167	2168150	2450	17.18	5.82	8.11	139.43	51
4.	Tehri	U.P.	261	2615	4200	270000	1000	64.286	1.55	1.61	103.25	4
5.	Kishau	U.P.	253	1230	2325	271000	—	116.559	0.86	1.89	220.32	—
6.	Bhakra	H.P.	226	7800	16800	676000	1354	40.238	2.48	2.15	98.67	12
7.	Idukki	Kerala	169	1459	6000	—	390	—	—	4.11	—	15
8.	Srisailam (H.E. Project)	A.P.	143	4370	59500	—	770	—	—	13.62	—	77
9.	Beas at Pong	H.P.	133	7290	28271	2100000	360	74.281	1.35	3.88	288.06	79
10.	Nagarjun Sagar	A.P.	125	6800	28500	895000	815.6	31.404	3.18	4.19	131.62	35
11.	Koyana (Sivaji Sagar) (H.E. Project)	Maharashtra	103	2680	31363	—	560	—	—	11.70	—	56
12.	Rihand (H.E. Project)	U.P.	93	9000	46620	—	300	—	—	5.18	—	155
13.	Ukai	Gujarat	81	7200	60095	32000	300	5.32	18.78	8.35	44.44	200
14.	Bhadra	Karnataka	72	1785	11251	106000	25	9.421	10.61	6.30	59.38	450
15.	Mahi Bajaj Sagar	Rajasthan	68	1712	13300	80000	37	6.015	16.62	7.77	46.73	359

Contd...

Sr. No.	Name of the Project	State	Height (m)	Live storage (Mm^3)	Submergence (ha)	Irrigation (ha)	Power (MW)	Irrigation benefits (ha) per ha of submergence	Percentage of area submerged to area irrigated	Submergence per Mm^3 of storages (ha)	Irrigation per Mm^3 of power (ha)	Submergence per MW generation (ha)
1	2	3	4	5	6	7	8	9	10	11	12	13
16.	Lower Sileru (H.E. Project)	A.P.	67	378	2787	—	600	—	—	7.37	—	5
17.	Kadana	Gujarat	66	1203	17008	211700	240	12.447	8.03	14.14	175.98	71
18.	Gandhi Sagar	M.P.	64	7630	66166	503200	115	7.605	13.15	8.67	65.95	575
19.	Sharavaty (Linganmakki)	Karnataka	61	4260	33790	—	891	—	—	7.93	—	38
20.	Hidkal (Ghata Prabha)	Karnataka	60	1387	7100	139000	28	19.577	5.11	5.12	100.22	254
21.	Hirakud	Orissa	59	5830	73892	251150	270	3.399	29.42	12.67	43.08	274
22.	Hemawathy	Karnataka	59	927	8500	265000	—	31.176	3.21	9.17	285.87	—
23.	Totladeh (Pench) (H.E. Project)	Maharashtra	61	1088	7750	94000	85	12.129	8.24	7.12	86.40	91
24.	Maithon (H.E. Project)	Bihar	56	736	7257	—	60	—	—	9.86	—	121
25.	Panchet (H.E. Project)	Bihar	49	392	5957	—	40	—	—	15.20	—	149
26.	Almatti (Upper Krishna)	Karnataka	48	843	79000	—	160	—	—	93.71	—	494
27.	Matatila	U.P.	46	770	13085	165750	31	12.667	7.89	16.99	215.26	422
28.	Tilaiya	Bihar	45	216	3160	30351	4	9.047	10.41	14.63	140.51	790
29.	Tungabhadra	Karnataka	43	3860	37814	372000	36	9.838	10.16	9.80	96.37	1050
30.	Pochampad (Shriram Sagar)	A.P.	43	2322	44157	230679	36	5.224	19.14	19.02	99.34	1227
31.	Jaikwadi (Paithan)	Maharashtra	43	2171	35000	278000	12	7.943	12.59	16.12	128.05	2917

M.P.—Madhya Pradesh, U.P.—Uttar Pradesh, H.P.—Himachal Pradesh, A.P.—Andhra Pradesh

Appendix Table A-11. Useful Conversion Factors

Length (m)		
	1 inch (in)	= 0.0254 m
	1 foot	= 0.3048 m
	1 mile	= 1.609 km
	1 micron (m)	= 10^{-6} m
	1 Angstrom (Å)	= 10^{-10} m
Mass (kg)		
	1 ounce	= 28.35 g
	1 ounce	= 0.02835 kg
	1 pound	= 0.4536 kg
	1 ton	= 1000 kg
Area (m^2)		
	1 in^2	= 6.4516 cm^2
	1 ft^2	= 929.03 cm^2
	1 ft^2	= 0.0929 m^2
	1 Acre	= 4046.85 m^2
	1 Acre	= 0.4046 hectare
	1 $mile^2$	= 2.59 km^2
Volume (m^3)		
	1 $inch^3$	= 16.387 cm^3
	1 $inch^3$	= 0.01639 litre (*L*)
	1 ft^3	= 0.0283 m^3
	1 ft^3	= 28.316 *l*
	1 gallon (U.S.)	= 3.785 *l*
	1 gallon (U.S.)	= 0.003785 m^3
	1 litre	= 10^{-3} m^3
	1000 litres	= 1 m^3
Energy (J = N.m)		
	1 joule	= 1 kg-m^2/s^2
	1 erg	= 1 g-cm^2/s^2
	1 joule	= 10^7 ergs
	1 cal	= 4.184 J
	1 ft-lb_f	= 1.3558 J
	1 Btu	= 1.055×10^2 J
	1 hp-hr	= 2.6845×10^6 J
	1 kW-hr	= 3.600×10^6 J
Power (W = J/s)		
	1 hp	= 745.7 W
	1 Btu/s	= 1055 W
	1 ft-lb_f/s	= 1.3558 W
	1 kW	= 1.341 hp

Pressure *(Pa = N/m²)*	
1 kg/m.s²	= 1 N/m²
1 lb_f/ft^2	= 47.88 Pa
1 atm	= 1.0133 × 10^5 Pa
1 mm Hg	= 1.3332 × 10^2 Pa
1 in Hg	= 3.3864 × 10^3 Pa
1 bar	= 10^5 Pa
1 mm H_2O	= 9.806 Pa
1 in H_2O	= 249.09 Pa
Density *(kg/m³)*	
1 g/cc	= 10^3 kg/m³
1 mg/L	= 10^{-3} kg/m³
1 lbm/ft³	= 16 kg/m³
Viscosity *(N s/m²)*	
1 kg/m.s	= 1 N.s/m²
1 cP	= 10^{-2} N.s/m²
1 lbm/ft.s	= 1.488 N.s/m²
1 lbm/ft.hr	= 4.1338 N.s/m²
Thermal conductivity *(W/m K)*	
1 cal/s.cm.K	= 418.4 W/m K (K for Kelvin)
1 kcal/hr.m.K	= 1.1623 W/m K
1 Btu/hr.ft. °F	= 1.7307 W/m K

Appendix Table A-12. List of Top 19 Countries with their Emissions of Green House Gases (as CO_2 equivalents) by 2012 Estimates

S. No.	*Name of Country*	*CO_2 emissions (billion tonne) (bt)*	*Emissions per capita (t)*
(*i*)	Whole World	34.50	4.9
(*ii*)	International Transport	1.06	—
1.	China	9.86	7.1
2.	United States	5.19	16.4
3.	India	1.97	1.6
4.	Russia	1.77	12.4
5.	Japan	1.32	10.4
6.	Germany	0.81	9.7
7.	South Korea	0.64	13.0
8.	Canada	0.56	16.0
9.	United Kingdom	0.49	7.7
10.	Mexico	0.49	4.0
11.	Indonesia	0.49	2.0
12.	Saudi Arabia	0.46	16.2
13.	Brazil	0.46	2.3
14.	Australia	0.43	18.8
15.	Iran	0.41	5.3
16.	Italy	0.39	6.3
17.	France	0.37	5.8
18.	South Africa	0.33	6.3
19.	Poland	0.32	8.4

Bibliography

General Publications

1. *Water and Waste-water Engineering Vol. I & Vol. II* : *G.M. Fair & J.C. Geyer* (John Wiley & Sons, New York)
2. *Water Supply and Waste Disposal* : *W.A. Hardenbergh and E.B. Rodie* (Allied Publishers Pvt. Ltd., New Delhi)
3. *Public Health Engineering* : *P.C.C. Issac* (C.E. & FN Spon, London)
4. *Sewerage and Sewage Treatment* : *H.E. Babit and E.R. Baumann* (John Wiley & Sons, New York)
5. *Water Supply and Sewerage* : *E.W. Steel & Terence J. McGhee* (McGraw Hill Book Co., New York, 1985)
6. *Public Health Engineering Vol. I & Vol. II.* : *E.B. Phelps in Collaboration with C.J. Velj* (John Wiley & Sons, New York)
7. *Waste waterEngineering— Treatment, Disposal and Reuse* : *Metcalf and Eddy* (McGraw Hill Book Co., New York, 1979)
8. *Introduction to Environmental Engineering* : *Mackengzie L. Davis and David A. Cornwell* (McGraw Hill Book Co., New York, 1991)

Special Publications

Chapter 1 & 2 : Introduction to the Subject ; Estimating the Design Sewage Discharge

9. *Elements of Water Supply and Waste-water Disposal* : *G.M. Fair and J.C. Geyer* (John Wiley & Sons, New York)
10. *Sewerage and Sewage Treatment* : *S.R. Kshirsagar* (Roorkee Publishing House, Civil Lines, Roorkee)

Chapter 3 : Estimating the Peak Drainage Discharge

11. *Design and Construction of Sanitary Storm Sewers* : *American Society of Civil Engineers Manual of Practice 37, New York, 1960*
12. *Hydrology and Water Resources Engineering* : *Santosh Kumar Garg* (Khanna Publishers, Delhi)
13. *Handbook of Hydrology* : *Govt. of India's Publication* (Ministry of Agr. & Irr., New Delhi, India) India)

Chapter 4 : Hydraulic Design of Sewers and S.W. Drain Sections

14. *A Treatise on Applied Hydraulics* : *Addison & Harbert* (John Wiley & Sons, New York)

15. *Design of Sewers to Facilitate Flow* : *Comp. T.R.* (Sewage Works, Journal, 18.3, 1946)

Chapter 5 : Sewers, Their Construction, Maintenance and Required Appurtenances

16. *Sewers, Their Theory, Design, Specifications and Construction* : *Bevan and Rees*

17. *Water Supply and Sewerage* : *Terence McGhee* (McGraw Hill Book Co., New York)

18. *A Text-book of Sanitary Engineering* : *R.S. Despande* (United Book Corporation Poona, India)

19. *Manual on Sewerage and Sewage Treatment* : *Expert Committee Constituted by Govt. of India* (Central Public Health & Env. Engg. Organisation, Ministry of Urban Dev. New Delhi, 1993)

20. *IS : 4111 (Part I) : 1986 Code of Practice for Ancillary Structures in Sewerage System Part I : Manholes* : *B.I.S. Code* (Bureau of Indian Standards, New Delhi)

21. *IS : 4111 (Part 3) 1985 ; Part 4-1968 ; Part 5-1993, Code of Practice for Ancillary Structures in Sewerage System Part 3 : Inverted Suphon, Part 4 : Tidal Outfalls, Part 5 : Tidal Outfalls* : *B.I.S. Codes* (Bureau of Indian Standards, New Delhi)

22. *IS : 7740-1985, Code of Practice for Construction & Maintenance of Road Gullies* : *B.I.S. Code* (Bureau of Indian Standards, New Delhi)

23. *IS : 11972-1987, Code of Practice for Safety Precautions to be taken when Entering a Sewerage System* : *B.I.S. Code* (Bureau of Indian Standards, New Delhi)

24. *IS : 13496-1992, General Requirements of Suction Machines for Cleaning Sewers, Man-holes and Ancillary Structures Provided on Sewer Lines and Closed Storm Water Mains* : *B.I.S. Code* (Bureau of Indian Standards, New Delhi)

25. *IS : 11387-1985 Requirements for High Pressure Jetting Machines for Sewer Cleaning* : *B.I.S. Code* (Bureau of Indian Standards, New Delhi)

26. *IS : 10595-1983, Requirements for Power Driven Bucket Type Sewer Cleaning Machines* : *B.I.S. Code* (Bureau of Indian Standards, New Delhi)

27. *IS : 11117-1984, Requirements for Power Driven Rodding Machine for Sewers* : *B.I.S. Code* (Bureau of Indian Standars, NewDelhi)
28. *IS : 11397-1985, Specification for Attachment Tools for Power Driven Rodding Machine for Sewers* : *B.I.S. Code* (Bureau of Indian Standards, New Delhi)
29. *IS : 10552-1983, Specifications for Buckets to be used in Power Driven Bucket Type Sewer Cleaning Machine* : *B.I.S. Code* (Bureau of Indian Standards, New Delhi)
30. *IS : 9110-1979, Specifications for Hand Operated Augers for Cleaning Water Closets, Pipes and Sewer* : *B.I.S. Code* (Bureau of Indian Standards, New Delhi)
31. *IS : 11931-1987, Specifications for Sewer Cleaning Metal Rods* : *B.I.S. Code* (Bureau of Indian Standards, New Delhi)

Chapter 6 : Pumps For Lifting Sewage

32. *Pump Application Engineering* : *Hicks, Tyler G & Theodore W. Edwards* (McGraw Hill Book Co., New York, 1972)
33. *Pump Selection and Application* : *T.G. Karassik* (McGraw Hill Book Co., New York)
34. *Centrifugal and Other Roto-dynamic Pumps* : *Addison and Herbert* (Chapman & Hall, London)

Chapter 7 : Quality Characteristics of Sewage

35. *Laboratory Manual for Chemical & Bacterial Anal. of Water & Sewage* : *F.R. Theoroux* (McGraw Hill Book Co., New York)
36. *Biological Engineering* : *H.P. Schwar* (McGraw Hill Book Co., New York)
37. *Environmental Engineering* : *H.S. Peavy, D.R. Rowe, G. Tchobanoglous* (McGraw Hill International Edition, 1986)
38. *Standard Methods of Water and Sewage Analysis* : *An American Public Health Association's Publication*
39. *Chemistry for Sanitary Engineering* : *C.N. Sawyer and P.L. McCarty* (McGraw Hill Book Co., New York)
40. *Engineering Management and Water Quality* : *P.H. McGauhey* (McGraw Hill Book Co., New York)
41. *Microbiology for Sanitary Engineering* : *R.E. Mckinney* (McGraw Hill Book Co., New York)
42. *A Guide to Study of Fresh Water Biology* : *J.C. Needham and P.R. Needham* (San Francisco, Holden Day 1962)
43. *Hydrology and Quality of Water Resources* : *M.J. Hammer & K.A. Mackichen* (John Wiley & Sons, New York)
44. *Guide to Sanitation and Natural Disasters* : *M. Assar* (WHO Publication, Geneva, 1971)
45. *IS : 7232-1974, Method for Imhoff Cone Test* : *B.I.S. Code* (Bureau of Indian Standards, N. Delhi)

Chapter 8 : Disposing of the Sewage Effluents

46.	*Scientific Stream Pollution Analysis*	: *Nemerow, Nelson Leonard* (McGraw Hill Book Co., New York)
47.	*Stream Sanitation Engineering*	: *E.B. Phelps* (John Wiley & Sons, New York)
48.	*Water Pollution and Disposal of Wastewater on Land*	: *U.N. Mahida* (Tata McGraw Hill Publishing Company Ltd., New Delhi, 1981)
49.	*IS : 4764-1973, Tolerance Limits for Sewage Effluent Discharge into Surface Water Sources*	: *B.I.S. Code* (Bureau of Indian Standards, New Delhi, India)
50.	*IS : 2490-1974, Tolerance Limits for Industrial Effluent Discharge into Inland Surface Waters*	: *B.I.S. Code* (Bureau of Indian Standards, New Delhi, India)
51.	*IS : 3306-1974, Tolerance Limits for Industrial Effluent Discharge into Public Sewers*	: *B.I.S. Code* (Bureau of Indian Standards, New Delhi, India)
52.	*IS : 3307-1965, Tolerance Limits for Various Polluting Constituents of Waste-water Effluents for Their Discharge on Land for Irrigation*	: *B.I.S. Code* (Bureau of Indian Standards, New Delhi, India)

Chapter 9 and 10 : Treatment of Sewage ; Process Design of a Complete Sewage Treatment Plant

53.	*Biological Waste Treatment*	: *W.W. Eckenfelder and D.J.O. Connor* (Pergamon Press, New York)
54.	*The Chemistry of Water and Sewage Treatment*	: *A.M. Buswell* (Chemical Catalog Co., New York)
55.	*Municipal and Rural Sanitation*	: *V.M. Ehlers and E.W. Steel* (McGraw Hill Book Co., New York)
56.	*Operation and Maintenance of Waste-water Collection Systems*	: *WPCF Manual of Practice No. 7* (Water Pollution Control Federation, Washington D.C., 1971)
57.	*Theory and Practice of Water and Waste-water Treatment*	: *Ronald L. Droste* (John Wiley & Sons, New York, 1997)
58.	*Water and Waste-water Technology*	: *Mark J. Hammer* (John Wiley & Sons, New York, 1986)
59.	*Sewage Treatment-Imhoff Tanks*	: *V. Surya Prakash* (Journal of Institution of Engineers India, June 1951)
60.	*Waste-water Treatment for Pollution Control*	: *Soli J. Arceivala* (Tata McGraw Hill Publishing Company Ltd., New Delhi, 1986)
61.	*Waste Treatment Plants*	: *C.A. Sastry, M.A. Hashim and P. Agamuthu* (Narosa Publishing House, N. Delhi, 1995)

62. *IS : 2470 (Part 1)-1985 and Part 2-1985, Code of Practice for Installation of Septic Tanks : Part 1-Design Criteria & Construction Part 2-Secondary Treatment and Disposal of Septic Tank Effuluent* : *B.I.S. Codes* (Bureau of Indian Standards, New Delhi)

63. *IS : 5611-1987, Code of Practice for Waste Stabilisation Ponds (Facultative Type)* : *B.I.S. Code* (Bureau of Indian Standards, New Delhi)

64. *IS : 8403-1977, Code of Practice for Construction of Clarifier Digestor for Treatment of Sewage* : *B.I.S. Code* (Bureau of Indian Standards, New Delhi)

65. *IS : 6279-1971, Equipment for Grit Removal Devices* : *B.I.S. Code* (Bureau of Indian Standards, N. Delhi)

66. *IS : 7208-1992, Guidelines for Flocculator Devices* : *B.I.S. Code* (Bureau of Indian Standards, N. Delhi)

67. *IS : 7090-1985, Guidelines for Rapid Mixing Devices* : *B.I.S. Code* (Bureau of Indian Standards, N. Delhi)

68. *IS : 8413 (Part 1)-1977 and IS : 8413 (Part 2)-1982, Requirements for Biological Treatment Equipments, Part 1 : Trickling Filters, Part 2 : Activated Sludge Process and its Modifications* : *B.I.S. Code* (Bureau of Indian Standards, New Delhi)

69. *IS : 10261-1982, Requirements for Settling Tank (Clarifier Equipment) for Waste-water Treatment* : *B.I.S. Code* (Bureau of Indian Standards, New Delhi)

70. *IS : 10037 (Part 1)-1981, Part 2 and 3-1983, Requirements for Sludge Dewatering Equipment Part 1 : Sludge Drying Beds—Sand Gravel and Under-drains, Part 2 : Vacuum Filtration Equipment, Part 3 : Centrifugal Equipment (Solid Bowl Type)* : *B.I.S. Code* (Bureau of Indian Standards, New Delhi)

71. *IS : 13166-1992, Guidelines for Evaluation and Testing of Mechanical Surface Aerators* : *B.I.S. Code* (Bureau of Indian Standards, New Delhi)

Chapter 11 : Treatment of Industrial Wastewaters

72. *Waste-water Treatment* : *M.N. Rao and A.K. Datta* (Oxford & IBH Publishing Co., Ltd., 1987)

73. *Pollution Control in Process Industries* : *S.P. Mahajan* (Tata McGraw Hill Publishing Co., Ltd., New Delhi, 1987)

74. *Industrial Water Pollution Control* : *W. Wesley Eckenfelder, Jr* McGraw Hill International Edition, 1967)

Chapter 12 : Excreta Disposal in Unsewered and Rural Areas

75. *Water Treatment and Sanitation ; A Hand Book of Simple Methods for Rural Areas in Developing Countries* : *H.T. Mann and D. Williamson* (Intermediate Technology Publications Ltd., London, 1973)

76. *Excreta Disposal for Rural Areas and Small Communities* : *E.G. Wagner and J.N. Lenoix* (WHO Monograph No. 39, 1958)

77. *Sulabh Shauchalaya* (Hand Flush Water Seal Latrine) : *Bindeshwar Pathak, Chairman Sulabh International* (Amola Prakashan, Patna)

78. *Composting* : *H.B. Gotaas* (WHO Monograph No. 31, 1956)

Chapter 13 : Sewage Collection from Houses and Buildings

79. *Water Supply and Sanitary Installations Within Buildings* : *A.C. Panchadhari* (Wiley Eastern Ltd., New Age International Ltd., 1995)

80. *IS : 1742-1983, Code of Practice for Building Drawing* : *B.I.S. Code* (Bureau of Indian Standards, New Delhi)

81. *IS : 1172-1993, Code of Basic Requirements for Water Supply, Drainage and Sanitation (4th revision)* : *B.I.S. Code* (Bureau of Indian Standards, New Delhi)

82. *IS : 12183 (Part I)-1987, Code of Practice for Plumbing in Multi-storeyed Buildings* : *B.I.S. Code* (Bureau of Indian Standards, New Delhi)

83. *IS : 2064-1993, Code of Practice for Selection, Installation and Maintenance of Sanitary Appliances (Second Revision)* : *B.I.S. Code* (Bureau of Indian Standards, New Delhi)

84. *IS : 4111 (Part II), 1985 ; Code of Practice for Ancillary Structures in Sewerage System Part II : Flushing Tanks* : *B.I.S. Code* (Bureau of Indian Standards, New Delhi)

85. *IS : 12251-1987, Code of Practice for Drainage of Building Basements* : *B.I.S. Code* (Bureau of Indian Standards, New Delhi)

86. *IS : 5329-1983, Code of Practice for Sanitary Pipe Work above Ground for Buildings* : *B.I.S. Code* (Bureau of Indian Standards, New Delhi)

87. *IS : 2064-1993, Code of Practice for Selection, Installation and Maintenance of Sanitary Appliances* : *B.I.S. Code* (Bureau of Indian Standards, New Delhi)

88. *IS : 2963-1979, Specifications for Copper alloy Waste fittings for Wash Basins and Sinks* : *B.I.S. Code* (Bureau of Indian Standards, New Delhi)

89. *IS : 773-1998, Specifications for Enamelled Cast Iron Water Closets Railway Coaching Stock Type* : *B.I.S. Code* (Bureau of Indian Standards, New Delhi)

90. *IS : 3489-1985, Specifications for Enamelled Steel Bath Tubes* : *B.I.S. Code* (Bureau of Indian Standards, New Delhi)

91. *IS : 9785-1981, Specifications for Flush Valves and Fittings for Water Closets and Urinals* : *B.I.S. Code* (Bureau of Indian Standards, New Delhi)

92. *IS : 774-1984, Specifications for Flushing Cistern for Water Closets and Urinals (Other then Plastic Cistern)* : *B.I.S. Code* (Bureau of Indian Standards, New Delhi)

93. *IS : 772-1973, Specifications for General Requirements for Enamelled Cast Iron Sanitary Appliances* : *B.I.S. Code* (Bureau of Indian Standards, New Delhi)

94. *IS : 771-(Part 1 to 4 & 6 to 7), Specifications for Glazed Fire Clay Sanitary Appliances* : *B.I.S. Codes* (Bureau of Indian Standards, New Delhi)
Part 1 : General Requirements
Part 2 : Specific Requirements Kitchens and Laboratory Sinks
Part 3 : Specific Requirements of Urinals
Part 4 : Specific Requirements of Postmortom Slabs
Part 6 : Specific Requirements of Bed-pan Sinks
Part 7 : Specific Requirements of Slop Sinks

95. *IS : 13983-1994, Specifications for Stainless Steel Sinks for Domestic Purposes* : *B.I.S. Code* (Bureau of Indian Standards, New Delhi)

Chapter 14 : Disposal of Solid and Industrial Wastes

96. *Solid Wastes : Engineering—Principles and Management Issues* : *Tchobanoglous, G.H. Theisen, and R. Eliassen* (McGraw Hill Book Co., New York, 1977)

97. *Sanitary Land-fill Design Hand Book* : *G. Noble* (Technomic, West Port, Conn, 1976)

98. *Solid Waste Collection Practice* : *American Public Works Association, Institute of Solid Wastes* (American Public Works Association, Chicago, 1975, 4th ed.)

99. *Environmental Pollution Control Engineering* : *C.S. Rao* (Wiley Eastern Ltd., New Delhi, 1992)

100. *Environmental Engineering* : *Arcadio P. Sincero and Gregoria A.Sincero* (Prentice-Hall of India Pvt. Ltd. New Delhi (1999))

101. *IS : 6924-1973, Construction of Refuse Chutes in Multi-Storeyed Buildings, Code of Practice* : *B.I.S. Code* (Bureau of Indian Standards, New Delhi)

102. *IS : 12314-1987, Code of Practice for Sanitation with Leaching Pits for Rural Community* : *B.I.S. Code* (Bureau of India Standards, New Delhi)

Chapter 15 : Ventilation of Buildings for Controlling Indoor Air Pollution

103. *Indoor Air Pollution Control* : *Godish* (John Wiley & Sons, New York)

Chapter 16 : The Environment, Its Pollution and Present Status

104. *Man, Nature and Environmental Laws* : *G.S. Nathawet, Satish Shastri and J.P. Vyas* (RBSA Publishers, Jaipur, 1988)

105. *Global Environmental Monitoring System—Urban Air Pollution* : *WHO Publication, 1988*

Chapter 17 : Disposal of Environmentally Hazardous Biomedical and Radioactive Wastes

106. *Environmental Engineering* : *G.N. Pandey and G.C. Karney* (Tata McGraw Hill Publishing Company Ltd., 1989)

107. *Hazardous Waste Management* : *Charles A. Wentz* (McGraw Hill International Edition, 1989)

108. *Hazardous Waste Incineration* : *Calvin R. Brunner* (McGraw Hill International Edition, 1989)

Chapter 18 : Engineering of Air Pollution, Its Control and Monitoring

109. *Air Pollution* : *M.N. Rao and H.V. Rao* (Tata McGraw Hill Publishing Company Ltd., New Delhi, 1989)

Chapter 19 : Global Environmental Issues

110. *Environmental Science* : *S.H. Anderson, R.E.* (Beiswenger, and P.W. Purdom)

111. *Air Pollution and Climate Change—The Biological Impact* : *A. Wellburn* (Longman Scientific and Technical Essex, U.K., 1994)

Chapter 20 : Noise Pollution

112. *Environmental Noise Control* : *E.B. Magrab* (John Wiley & Sons, New York, 1975)

113. *Fundamentals of Industrial Noise Control* : *L.H. Bell, Trumbull C.N.* (Harmony Publications, 1973)

114. *IS : 4954-1968, Recommendations for Noise Abatement in Town Planning* : *B.I.S. Code* (Bureau of Indian Standards, New Delhi)

115. *IS : 7194-1973, Specifications for Assessment of Noise Exposure during work for Hearing Conservation Purposes* : *B.I.S. Code* (Bureau of Indian Standards, New Delhi)

116. *IS 9779-1981, Specifications for Sound Level Meters* : *B.I.S. Code* (Bureau of Indian Standards, New Delhi)

117. *IS : 9989-1981, Assessment of Noise with Respect to Community Response* : *B.I.S. Code* (Bureau of Indian Standards, New Delhi)

Chapter 21 : Environmental Impacts of Thermal Power Plants, Mining and River Valley Projects

118. *Environmental Concepts and Strategies* : *T.N. Khoshoo* (Ashish Publishing House, New Delhi, 1988)

119. *Environmental Impact Assessment Studies (Case Studies)* : *Publication No 248* (Central Board of Irr. & Power New Delhi, 1995)

120. *Hydropower and River Valley Development—Environmental Management—Case Studies and Policy Issues* : *R.S. Goel and R.N. Srivastava* (Oxford & IBH Publishing Co. Ltd., New Delhi)

121. *Booklet titled "Sardar Sarovar Project ; What it is and what it is not"* : *Dr G.C. Patel Chairman, Sardar Sarovar Narmada Nigam Ltd.* Released at International Conference on "What Future for Large Rivers", Organised by MOE, Govt of France held at Orleans France, Sept. 1991)

Chapter 22 : Preparing Environmental Impact Assessments of Polluting Industrial Projects

122. *Environmental Impact Assessment* : *Jain* (McGraw Hill, Inc, New York)

123. *Environmental Impact Assessment* : *L.W. Kanter* (McGraw Hill, Inc New York)

Index

D

T